Power System
Analysis and Design

For the Students of B.E. / B.Tech., B.Sc. (Engg.), A.M.I.E. and Other Engineering Examinations

Dr. B.R. Gupta

B.Sc. Engineering (Hons.), M.Sc. Engineering,
Ph.D., MIE (India)
Senior Member of IEEE (USA)
Former Professor of Electrical Engineering,
Punjab Engineering College,
Chandigarh

S. CHAND PUBLISHING
empowering minds

S. CHAND & COMPANY PVT. LTD.
(AN ISO 9001: 2008 COMPANY)
RAM NAGAR, NEW DELHI-110 055

S.CHAND
PUBLISHING
empowering minds

S. CHAND & COMPANY PVT. LTD.

(An ISO 9001:2008 Company)

Head Office: 7361, RAM NAGAR, NEW DELHI - 110 055
Phone: 23672080-81-82, 9899107446, 9911310888 Fax: 91-11-23677446
www.schandpublishing.com; e-mail: helpdesk@schandpublishing.com

Branches

Ahmedabad	:	Ph: 27541965, 27542369, ahmedabad@schandpublishing.com
Bengaluru	:	Ph: 22268048, 22354008, bangalore@schandpublishing.com
Bhopal	:	Ph: 4274723, 4209587, bhopal@schandpublishing.com
Chandigarh	:	Ph: 2625356, 2625546, chandigarh@schandpublishing.com
Chennai	:	Ph: 28410027, 28410058, chennai@schandpublishing.com
Coimbatore	:	Ph: 2323620, 4217136, coimbatore@schandpublishing.com (Marketing Office)
Cuttack	:	Ph: 2332580, 2332581, cuttack@schandpublishing.com
Dehradun	:	Ph: 2711101, 2710861, dehradun@schandpublishing.com
Guwahati	:	Ph: 2738811, 2735640, guwahati@schandpublishing.com
Hyderabad	:	Ph: 27550194, 27550195, hyderabad@schandpublishing.com
Jaipur	:	Ph: 2219175, 2219176, jaipur@schandpublishing.com
Jalandhar	:	Ph: 2401630, 5000630, jalandhar@schandpublishing.com
Kochi	:	Ph: 2378740, 2378207-08, cochin@schandpublishing.com
Kolkata	:	Ph: 22367459, 22373914, kolkata@schandpublishing.com
Lucknow	:	Ph: 4026791, 4065646, lucknow@schandpublishing.com
Mumbai	:	Ph: 22690881, 22610885, mumbai@schandpublishing.com
Nagpur	:	Ph: 6451311, 2720523, 2777666, nagpur@schandpublishing.com
Patna	:	Ph: 2300489, 2302100, patna@schandpublishing.com
Pune	:	Ph: 64017298, pune@schandpublishing.com
Raipur	:	Ph: 2443142, raipur@schandpublishing.com (Marketing Office)
Ranchi	:	Ph: 2361178, ranchi@schandpublishing.com
Siliguri	:	Ph: 2520750, siliguri@schandpublishing.com (Marketing Office)
Visakhapatnam	:	Ph: 2782609, visakhapatnam@schandpublishing.com (Marketing Office)

First Edition 1998
Subsequent Editions and Reprints 2001, 2003, 2004 (Twice), 2005, 2006, 2007, 2008, 2009 (Twice) 2010, 2011, 2012, 2013; Seventh Revised Edition 2014; Reprints 2014 (Twice), 2015
Reprint 2016

ISBN: 978-81-219-2238-8 **Code: 1010C 258**

This book is for sale in India & other SAARC countries only.

PRINTED IN INDIA

By Nirja Publishers & Printers Pvt. Ltd., 54/3/2, Jindal Paddy Compound, Kashipur Road, Rudrapur-263153, Uttarakhand and published by S. Chand & Company Pvt. Ltd., 7361, Ram Nagar, New Delhi -110 055.

PREFACE TO THE SEVENTH REVISED EDITION

The book **Power System: Analysis and Design**, now in its seventh edition, continues to provide all essential features of the subject. In this revised edition, six new chapters have been incorporated. They are; Generator, Transformer and Load Models, Economic Operation of Power System and Unit Commitment, Symmetrical Faults, Symmetrical Components, Unsymmetrical Faults, and Load Frequency and MVAR Voltage Control. Besides, now the seventh edition contains twenty-five chapters in all which make this textbook up–to–date.

The author wishes to express his sincere thanks to the Management and Editorial team of S. Chand & Company Pvt. Ltd., New Delhi for their unstinted support.

Although every care has been taken to make this book free of errors, some errors might be crept in. The author shall be highly obliged if the errors are brought to his notice. Any suggestions for the improvement of the book will be gratefully acknowledged.

Author

PREFACE TO THE FIRST EDITION

Power System : Analysis and Design has been written for undergraduate students of Indian and foreign universities to serve as a complete textbook for students undergoing a core or advanced course in transmission, distribution, stability, load flow, surge-phenomena, fault studies, travelling waves and design of transmission systems. The material presented is such that it can be covered in a two-semester course.

New topics like load flow studies, EHV transmission, HVDC transmission, etc., have been incorporated in the text to generate a harmonious blend of the conventional topics as well as the modern topics. It has been primarily written from students' point of view and to accomplish this goal, most of the material of the text has been class tested.

It is known that the ability to use mathematics makes the study of a subject less ambiguous and more meaningful. However, it is author's firm conviction that a qualitative and physical understanding of the problem and its solution is equally necessary. In view of this a balance between mathematical details and qualitative discussion has been kept throughout the book.

A number of solved and unsolved problems (with answers) have been included in each chapter. To help the final year students in their project work, a separate chapter on 'Design of Transmission Lines' has been added. In view of their growing importance objective questions have been given in Appendix C. Coprehensive bibliography has been included for the benefit of those who need further information on this subject.

The author will welcome constructive criticisms of the text and will gratefully acknowledge the comments pertaining to the text. Any suggestions for errata will be most welcome.

Thanks are due to my wife Geeta and daughter Vandana for their constant encouragement and whole-hearted cooperation during the preparation of the manuscript.

B.R. Gupta

CONTENTS

CHAPTER 1

Power System Network

1.1 BASIC STRUCTURE OF POWER SYSTEM

Electrical energy is generated in large hydro-electric, thermal and nuclear power stations. These stations are generally situated far away from the load centres. This necessitates an extensive power supply network between the generating stations and the consumers' loads. Fig. 1.1 shows the layout of a power supply network. This network may be divided into two parts, viz., transmission system and distribution system. The former may be divided into primary and secondary (or sub) transmission and the latter into primary and secondary distribution. Transmission lines transmit bulk electrical power from sending end stations to receiving end stations without supplying any consumers en route: by contrast, a distribution line or distributor supplies consumers directly at short intervals along the line. A distributor is subject to the legal requirement that power must be supplied at a voltage within ±6% of the declared (or rated) voltage whereas a transmission line is not subject to any such restriction and its voltage can vary as much as 10% or even 15% due to variation in loads. Any restriction on transmission voltage is technical and not legal.

The transmission system of an area (or state) is known as a grid. The different grids are interconnected through tie lines to form a regional grid and the different regional grids are further inter-connected to form a national grid. Each grid operates independently. However, power can be transmitted from one grid to another, over the tie lines, under conditions of sudden loss of generation or increase in load.

The maximum generation voltage in advanced countries is 33 kV while that in India is 11 kV. The amount of power that has to be transmitted through transmission lines is very large and if this power is transmitted at 11 kV (or 33 kV) the line current and the power loss would be very large. Therefore, this voltage is stepped up to a higher value by using step up transformers located in sub-stations (known as sending end sub-stations) near the generating stations. The transmission voltages in India are 66 kV, 110 kV, 132 kV, 220/230 kV, 400 kV and 765 kV. The high voltage transmission lines transmit electrical power from the sending end sub-stations to the receiving end sub-stations. At the receiving end sub-stations the voltage is stepped down to a lower value of 66, 33 or 11 kV. The secondary transmission system forms the link between the main receiving end sub-stations and the secondary sub-stations. At the secondary sub-stations the voltage is stepped down to 33, 11 or 3.3 kV and power is fed into the primary distribution system. The 33, 11 or 3.3 kV distribution lines (usually known as feeders) originate from the secondary sub-stations and terminate in distribution sub-stations. The distribution sub-stations consist of step down transformers and protective devices. These are placed at suitable locations in the area in which power is to be supplied. Sometimes these distribution sub-stations consist of pole-mounted transformers located on the road side. In large multi-storeyed buildings, the distribution transformers are located within the building itself. These transformers step down the voltage to 400 V. The 400 V distribution lines (or distributors) are laid along the roads and service connections to the consumers are tapped off from the distributors.

Every power system need not necessarily have all the parts shown in Fig. 1.1. In some cases there is only one level of transmission and the secondary transmission system does not exist. In such cases the feeders directly take off from the main sub-stations. Some large consumers are supplied at 132 kV and 66 kV directly.

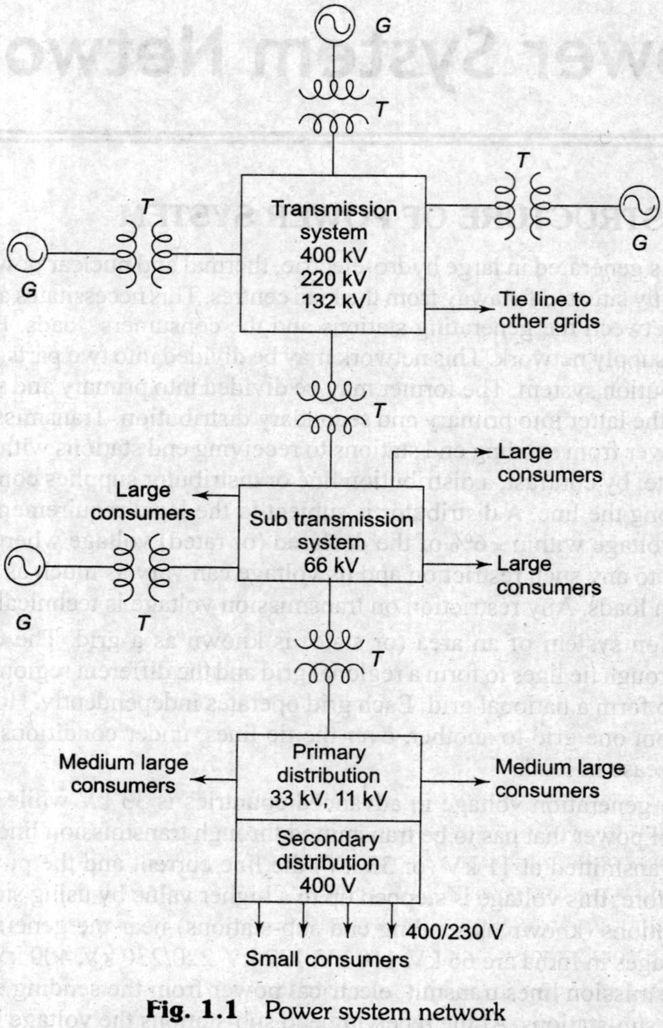

Fig. 1.1 Power system network

A single phase ac circuit requires two conductors. A 3-phase ac circuit, using the same size conductors as the single-phase circuit, can carry three times the power which can be carried by a single-phase circuit and uses three conductors for the three phases and one conductor for the neutral. Thus, a 3-phase circuit is more economical than a single-phase circuit in terms of initial cost as well as the losses. All transmission and distribution systems are, therefore, 3-phase systems. In fact, a balanced 3-phase circuit does not require the neutral conductor, as the instantaneous sum of the three line currents is zero. Therefore, the transmission lines and feeders are 3-phase 3-wire circuits. The distributors are 3-phase 4-wire circuits because a neutral wire is necessary to supply the single-phase loads of domestic and commercial consumers. The standard frequency in India and many other countries is 50 Hz. In connection diagrams and layouts, the three-phase circuits are generally shown by single line diagrams in which only one phase is shown instead of all the three. The layout of Fig. 1.1 is a single line diagram.

1.2 OVERHEAD AND UNDERGROUND SYSTEMS

The three-phase transmission and distribution systems may consist of overhead lines or underground cables or a mixture. The main advantages of an underground system are that it is less prone to natural hazards like rain, wind and lightning and that it does not interfere with other amenities. However, an underground system is more costly as compared to an overhead system. The approximate cost ratios are as under:

System Voltage kV	0.4	11	33	66	132	220	400
Cost ratio cable/OH line	2	3	5	7	9	13	24

Because of the cost consideration, the transmission and sub-transmission system in India are generally overhead. For distribution, the use of underground cables is limited to densely populated areas. The underground cables have lower series impedance and higher shunt capacitance as compared to overhead lines and are usually only a few km long. In overhead lines the power loss is mostly due to the I^2R loss in the conductor. However, the dielectric loss and the loss in the metallic sheath are also significant in the case of cables. The charging current, in cables, also contributes to power loss.

When the distribution system is through overhead lines, tappings can be made at desired points on the distributors to provide connections to the consumers. Thus the cost of providing service mains is also reduced. This cost is much higher when the distribution system uses underground cables.

Voltage regulation is also an important factor in the ac distribution systems. Overhead lines have considerable inductance which causes voltage drop. Cables have small spacing between conductors. Therefore cables have lower inductance and lower voltage drop as compared to overhead lines.

1.3 OVERHEAD LINE CONDUCTORS

The material for the conductor of an overhead line should have a high tensile strength (high breaking load) so that the spans between the towers can be as long as possible and the sag as small as possible, thus reducing the number and height of towers and the number of insulators. The conductor should have low resistivity to reduce the I^2R losses and voltage drop. The cost of its installation and maintenance should be low and it should have a long life. The final choice of material is often a compromise.

Overhead line conductors are invariably stranded to make them more flexible during erection and while in service. A stranded conductor has a central wire and one or more layers of conductors over the central wire. Each layer is twisted in the sense opposite to that of the layer below (the outside layer being always right handed) in order to increase the mechanical strength of the conductor. Fig. 1.2 shows some typical arrangements for stranding.

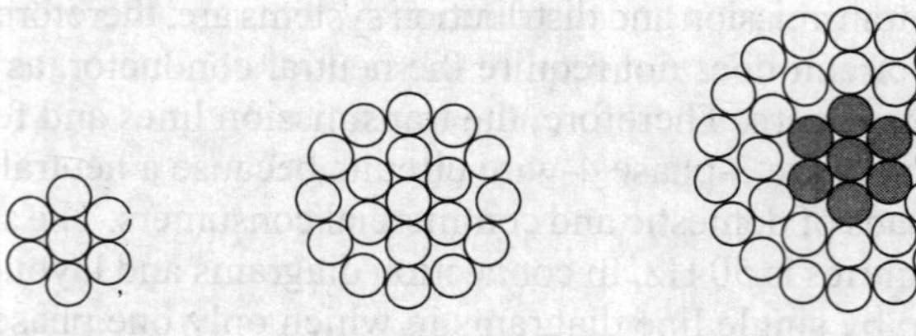

Fig. 1.2 Stranded conductors

The size of a line conductor is determined by its cross-sectional area, generally given in mm^2 or 'circular mils'. One circular mil is defined as the area of a circle of diameter 10^{-3} inch. MCM denotes an area of 1000 circular mils. When considering conductors other than copper, it is a common practice to refer to their size in terms of equivalent copper area. The equivalent copper area of a conductor denotes the cross-sectional area of a solid hard drawn copper conductor having resistance per unit length equal to that of the conductor under consideration. The overall diameter of a stranded conductor is given by $(2n + 1) \times$ diameter of each strand where n is the number of layers outside the central strand.

The two most common conductor materials are hard drawn copper and aluminium. Their properties are given in Table 1.1.

Table 1.1 Properties of copper and aluminium conductors

	Copper	*Aluminium*
Electrical conductivity (Silver = 1.0)	0.975	0.585
Resistivity ($\mu\,\Omega$ – cm)	1.777	2.826
Specific gravity	8.89	2.70
Tensile strength (N/m^2)	384×10^6 to 430×10^6	180×10^6 to 234×10^6
Coefficient of linear expansion per °C	17×10^{-6}	23×10^{-6}
Temperature coefficient of resistance at 20°C	0.00393 per °C	0.004 per °C
Ratio of conductivities for equal area	1	0.6
Ratio of diameters for equal resistance	1	1.29
Ratio of weights for equal resistance	2	1

Hard drawn copper has the advantages of very high conductivity (i.e., very low resistivity), good tensile strength and weather resisting properties. Many years back it was widely used for construction of overhead lines. Due to non-availability and high cost it is generally not used in India. In other countries, too, it is very rarely used.

Aluminium has the advantages of much lower cost and lesser weight as compared to copper. The fact that an aluminium conductor has a larger diameter than a copper conductor of the same resistance is also an extra advantage. A large diameter, for the same voltage, leads to lower voltage gradient at the conductor surface with a tendency of reduced ionization level of air and corona. However, the tensile strength of aluminium is low and, therefore, an all aluminium conductor (AAC) is rarely used except for low voltage distribution lines of short spans.

ACSR (Aluminium Conductor Steel Reinforced) conductor comprises hard drawn aluminium wires stranded around a core of single or multiple strand galvanized steel wire. Fig. 1.2(c) shows an ACSR conductor having 7 strands of steel and 30 strands of aluminium. Aluminium provides the necessary conductivity while steel provides the necessary mechanical strength.

During manufacture, a layer of grease is put between aluminium and steel to reduce electrolytic action (corrosion) between zinc and aluminium (The steel strands are galvanized with zinc). All transmission lines and most of the distribution lines use ACSR conductor. These conductors are manufactured in a wide variety of sizes from about 5 mm to over 40 mm overall diameter.

An ACSR conductor has a larger diameter than an equivalent copper conductor and, therefore, lesser corona loss. In an ACSR conductor the critical voltage is about 30—50% higher than that in an equivalent copper conductor.

The breaking load of an ACSR conductor is considerably higher than that of equivalent copper conductor. Moreover the weight of an ACSR conductor is about 25% lesser than that of equivalent copper conductor. The sag of an overhead line increases with weight but decreases with increase in tension (which is proportional to breaking load). Therefore an overhead line with ACSR conductor has lesser sag than an equivalent copper conductor line. This means that, when ACSR conductor is used, the supporting towers can be shorter or spans can be longer leading to overall economy.

An ACSR conductor is more costly as compared to an all aluminium conductor.

Appendix A gives some characteristics of commonly used ACSR conductors.

1.3.1 New Transmission Line Conductors

Electricity demand is increasing in all countries of the world. The competition in electricity market is also increasing rapidly. Electricity deregulation is throwing new challenges to electric transmission utilities and they have to think of innovative ways to increase transmission capacity. One such innovation is new and better conductors for overhead lines.

During the past about 20 years homogenous all Aluminium Alloy conductor (AAAC) has become very popular. The material used in this conductor is aluminium-magnesium-silicon alloy. It is lighter and has higher conductivity than ACSR conductor. Its higher strength to weight ratio allows the lines to be built with lower sag. These lines can be operated at a higher temperature than lines with ACSR conductor. However, the lower sag means higher conductor tension which results in increased vibrations and early fatigue. This conductor is used when terrain and wind conditions are favourable. Many new lines in national grid of UK use AAAC conductor.

Lot of research has been conducted in Japan to develop conductors which can be operated at high temperatures (about 200°C as compared to 75°C for ACSR conductor). The alloys TAI, ZTAI and XTAI are the basis of these high temperature conductors.

The coefficient of thermal expansion of aluminium is about twice that of steel. As the temperature rises more and more stress is transferred to steel. At a temperature called transition point (or knee point) all the stress is transferred to steel. For ACSR conductor transition point is about 100°C.

TACSR (T aluminium conductor steel reinforced) uses heat resistant material TAI instead of aluminium. TAI is an aluminium alloy which can be used at high temperature. Thus lines built with TACSR conductor can be operated at higher temperatures with no loss in strength and lower sag than lines having ACSR conductor.

Invar is an alloy of iron and nickle. This alloy has lower coefficient of thermal expansion than steel (2.8 micro strain per kelvin upto 100°C and 3.6 over 100°C as compared to 11.5 for steel). T-aluminium conductor invar reinforced (TACIR) can be operated at 150°C while ZTACIR can be operated at 210°C and XTACIR at 230°C. However invar has lower mechanical strength than steel. Therefore ACIR (aluminium conductor invar reinforced) needs more strands of reinforcing material invar.

Another material being used in UK and Japan in GZTACSR (called Gap type ZT-aluminium conductor steel reinforced). In this conductor a small annular gap exists between high strength steel core and first-layer of trapezoidal shaped aluminium strands. This construction has low sag properties above erection temperature and good strength below it. The use of this conductor allows a 30% increase in line rating for the same sag. A drawback of this conductor is its complex installation procedure and, therefore, longer time to build the line. UK has some lines using GZTACSR conductor.

1.4 OPTICAL FIBRE CONDUCTORS

Power line carrier communication system uses phase conductors. Fibre optic communication system has many advantages over PLCC. These include freedom from electromagnetic induction, more transmission capacity, noise immunity and higher speed of transmission. Last few years have seen development of conductor systems incorporating optical fibres. It is more convenient to incorporate the optical fibres in the earthwire of transmission systems than in phase conductors. The optical fibre is embedded inside the earthwire to form the communication channel. However, special jointing techniques are needed for such earthwires. It is necessary to maintain mechanical and electrical integrity of the conductor without prejudice to the optical properties of the core. Moreover, it is also necessary to provide, within the joint design, a suitably protected and accessible cavity within which connection of optical cores may be made effectively and speedily. The technology of using optical fibres in earthwire has become very popular in many countries.

1.5 TRANSMISSION VOLTAGES

The generation and transmission of electrical energy has been increasing at a tremendous rate, throughout the world during the past many decades. In USA the installed capacity has increased from 5×10^4 MW in 1940 to 110×10^4 MW in 2011 and the electrical energy generated annually has increased from about 170×10^9 kWh in 1940 to about 4750×10^9 kWh in 2011. The trend of power generation is very similar in other countries and this trend is likely to continue for many more years. With the development of inter-connections and the growth of loads the transmission network has to be expanded to keep pace with the growth in electrical energy. The transmission facilities tie up the power system so that generating capacity at distant locations in the system can be made available for use. The transmission facilities affect the cost and reliability of energy supplied to the consumers to a great extent.

The increasing need of transmitting greater and greater amounts of power has led to a continuous increase in transmission voltages. In USA the highest transmission voltage was raised to 132 kV in 1916, 220 kV in 1922, 287 kV in 1927, 345 kV in 1953, 500 kV in 1963 and 765 kV in 1969. An experimental 1100 kV line has been built in USA and a commercial 1100 kV line is expected to be installed in near future. In USSR, Japan and Italy also experimental 1000/1100 kV lines are existing and commercial lines of this class of voltage are being planned. Feasibility studies of 1500 kV lines are also been carried out in USA. As per the current terminology, transmission voltages less than 300 kV are categorized as high voltages, those between 300 kV and 765 kV are termed as Extra High Voltages (EHV) and those greater than 765 kV are classified as Ultra High Voltages (UHV).

1.6 BUNDLED CONDUCTORS

To cope with an ever increasing demand, it is necessary to transmit large blocks of power over long distances. An increase in line voltage increases the transmission efficiency. At voltages above 300 kV (referred to as extra high voltages or EHV), corona causes a significant power loss and interference with communication circuits, if the circuit has one conductor per phase. The use of multiple conductors per phase decreases the voltage gradient in the vicinity of the line and, thus, reduces the possibilities of the corona discharge. Such an arrangement is known as bundled conductors. Lines of 400 kV and higher voltages invariably use bundled conductors. Each phase consists of two, three or four (or more) conductors, spaced short distances apart and metally connected by spacers. The three conductor bundle has the conductors at the vertices of an equilateral triangle and the four conductor bundle has the conductors at the corners of square (Fig. 1.3). The intra-bundle spacing 'd' is around 40 cms.

The advantage of using bundled conductors (Fig. 1.3) is to reduce the corona loss and minimize the interference with communication circuits. Further, it leads to reduction in circuit inductance with consequent increase in the stability margin and decrease in voltage drop. This gives rise to an increase in the circuit capacity and boost up of operating voltage.

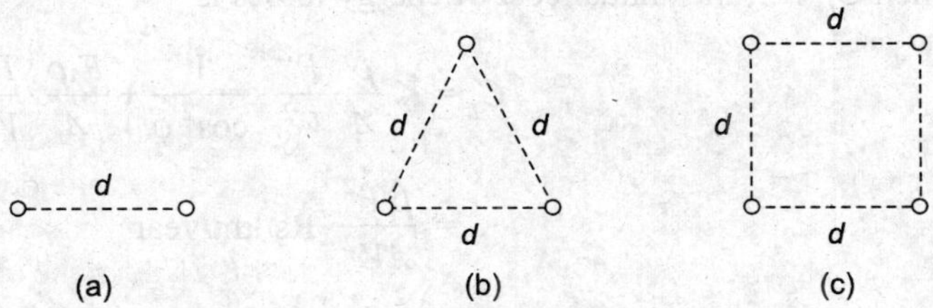

Fig. 1.3 Bundled conductors

1.7 CHOICE OF ECONOMIC VOLTAGE

The choice of the most economical voltage for a transmission line requires a detailed study of many technical and economic factors. An elementary study can be made from the following simplified analysis. A detailed analysis can then be carried out to determine the most economical value.

The capital cost of line depends on the line to line voltage V and conductor cross-section A. A part of the cost is of fixed type and does not depend on voltage and current considerations. Thus, the capital cost can be expressed as

$$CC = [a + bV + cA] \text{ Rs per km} \tag{1.1}$$

where CC is the capital cost, b is the coefficient which determines the influence of voltage on the capital cost, c is the coefficient which determines the influence of conductor cross section A, and a is the fixed cost which is independent of V and A.

The annual fixed cost C_f can be expressed as

$$C_f = \alpha(CC) = \alpha(a + bV + cA) \text{ Rs per km per year} \tag{1.2}$$

where α is a coefficient which accounts for interest rate, depreciation, and maintenance, etc.

If the line current at peak load is I and the active power is P, Then

$$P = \sqrt{3}\ VI \cos\phi \tag{1.3}$$

where $\cos\phi$ is the power factor at peak load. The peak power loss (i.e., power loss in the line when it is delivering peak load) is

$$p = 3\frac{\rho}{A}I^2 = \frac{\rho}{A}\cdot\frac{P^2}{V^2}\cdot\frac{1}{\cos^2\phi} \tag{1.4}$$

where p is the peak power loss per unit length of line and ρ is the resistivity of conductor material.

The energy losses in a certain time interval depend on the line load at different times during the day. A simplified way of calculating the energy losses is to express them as the product of peak power loss p and a suitable factor D (to account for load distribution). Thus,

$$\text{Energy losses} = Dp = \frac{\rho}{A}\cdot\frac{P^2}{V^2}\cdot\frac{D}{\cos^2\phi} \tag{1.5}$$

Let the energy cost be written as

$$\text{Energy cost} = K\,(\text{Peak loss}) + F\,(\text{Energy loss}) \tag{1.6}$$

Then C_L, i.e., the annual cost of energy losses is

$$C_L = K\,\frac{\rho}{A}\cdot\frac{P^2}{V^2}\cdot\frac{1}{\cos^2\phi} + \frac{F.\rho}{A}\cdot\frac{P^2}{V^2}\cdot\frac{D}{\cos^2\phi}$$

$$= \frac{\beta P^2}{AV^2}\ \text{Rs/km/year} \tag{1.7}$$

where $\beta = \dfrac{\rho}{\cos^2\phi}\,(K + FD)$

and the total annual cost C_t per unit length per unit power transfer is given by

$$C_t = \frac{C_f + C_L}{P} = \frac{\alpha}{P}\,(a + bV + cA) + \frac{\beta P}{AV^2}\ \text{Rs/km/year/per unit power} \tag{1.8}$$

The total cost is, thus, mainly a function of conductor cross-sectional area and the operating voltage, i.e.,

$$C_t = f(A, V)$$

and the total incremental cost is given by

$$dC_t = (\partial f/\partial A)\,dA + (\partial f/\partial V)\,dV$$

For the cost to be minimum dC_t must be equal to zero. This leads to $\partial f/\partial A$ and $\partial f/\partial V$ both equal to zero. When $\partial f/\partial A$ is zero we obtain

$$\partial f/\partial A = \frac{\alpha c}{P} - \frac{\beta P}{A^2 V^2} = 0 \quad\text{or}\quad A = A_0 = \left(\frac{\beta}{\alpha c}\right)^{0.5}\frac{P}{V} \tag{1.9}$$

and $\dfrac{\partial f}{\partial V} = 0$ gives $\dfrac{\alpha b}{P} - \dfrac{2\beta P}{AV^3} = 0$ or $V = V_0 = \left(\dfrac{2\beta P^2}{A\alpha b}\right)^{1/3}$ (1.10)

Substituting A_0 for A from Eq. (1.9) in Eq. (1.10) gives

$$V_0 = \left(\frac{2}{b}\right)^{0.5}\left(\frac{\beta c}{\alpha}\right)^{0.25} P^{0.5} \tag{1.11}$$

and substitution of A_0 for A and V_0 for V in Eq. (1.8) gives

$$C_t = C_{t,\,\min} = \frac{\alpha a}{P} + \frac{J}{\sqrt{P}} \tag{1.12}$$

where $J = 2\sqrt{2}\,(\alpha^3 b^2 \beta c)^{0.25}$

From the above analysis, the following conclusions can be drawn:

(i) For a particular value of P, there is one value of conductor cross-section and one value of voltage which give the minimum total cost per unit length per unit power.

(ii) With the increase in power the economic values of conductor cross-section and voltage increase in proportion to $\sqrt{P}$.

The above results are shown graphically in Fig. 1.4. The curve representing the minimum total cost Eq. (1.12) when voltage and conductor cross-section vary as $\sqrt{P}$ (according to Eqs. 1.9 and 1.11) may be approximated by a few curves, each corresponding to a fixed voltage and a fixed cross-section.

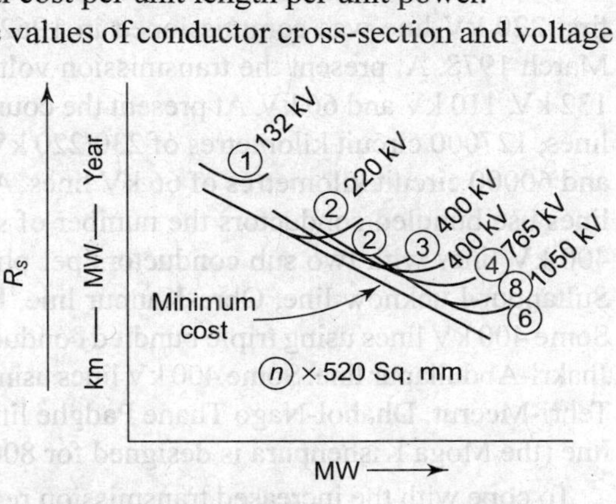

Fig. 1.4 Variation of transmission cost with the power transmitted

The above simplified analysis also confirms that a very limited number of voltage levels with an appropriate value of conductor cross-section may be used in a system without deviating too far from the optimum value.

1.8 TRANSMISSION SYSTEM IN USA

USA has only 6% of world population but accounts for over 30% of the electrical energy consumed in the world. At present the installed capacity is over 1000 GW, annual electricity generated is over 5×10^{12} kWh and per capita consumption is over 14000 kWh per year. There are many electric utilities involved in generation, transmission and distribution. Most of these utilities are invester owned, though those owned by State government, federal government, municipal authorities and cooperatives also exist. The standard frequency is 60 Hz.

The generation is in 11 kV to 33 kV range. This is stepped up to 345 kV, 500 kV and 765 kV and transmitted to load centres. The network of these EHV lines is known as supergrid. This supergrid feeds sub-transmission system operating at 115 kV. The various distribution voltages are 69 kV, 46 kV, 34.5 kV, 23 kV, 14.4 kV, 13.2 kV, 7.2 kV, 4.8 kV and 2.4 kV. The low voltage distribution is at 120 V, single phase.

The transmission voltage has shown an almost continuous increase. In 1892, the highest transmission voltage was 10 kV. The 220 kV system was introduced in 1922 and the 765 kV line was added in the year 1969. No further increase in voltage has occurred after that. At present thousands of circuit kilometres of 765 kV lines exist in USA and Canada. It has been reported that about 20% of transmission capability in USA, is by 765 kV lines.

Extensive studied are being conducted to introduce still higher transmission voltages. The next higher voltage is likely to be 1500 kV.

1.9 TRANSMISSION AND DISTRIBUTION NETWORK IN INDIA

The installed generating capacity in India has increased multifold from a level of 2300 MW in 1950 to about 200000 MW as on 31.3.2011. The present gross generation is about 960000 GWH. The per capita consumption is 813 kWh per year. The growth rate of installed capacity is about 7% per year.

The high voltage transmission network has also developed extensively, during the last 3 decades and particularly during the last 15 years, to keep pace with the growth in the installed generating capacity and to meet the transmission needs arising from the system interconnection.

The transmission system in India entered the high voltage era with the commissioning of a 100 kV line in Maharashtra in 1911. The 132 kV system voltage was introduced in 1931. The first 220 kV line was commissioned in 1962 and the first 400 kV line was commissioned in March 1978. At present the transmission voltages in use are 765 kV, 400 kV, 230 kV, 220 kV, 132 kV, 110 kV and 66 kV. At present the country has about 88000 circuit kilometres of 400 kV lines, 127000 circuit kilometres of 230/220 kV lines, 150000 circuit kilometres of 132 kV lines and 60000 circuit kilometres of 66 kV lines. All these lines are of overhead variety. All 400 kV lines use bundled conductors the number of sub conductors per phase being 2, 3 and 4. Some 400 kV lines with two sub conductors per phase are Dehar-Panipat line, Obra Sultanpur line, Sultanpur-Lucknow line, Obra-Kanpur line, Kanpur-Muradnagar line and Koradi-Kalwa line. Some 400 kV lines using triple bundled conductors are Nathpa Jhakri-Kol Dam line, and Nathpa Jhakri-Abdullapur line. Some 400 kV lines using quad (4) bundle conductor are Dadri-Malerkotla, Tehri-Meerut, Dhabol-Nago Thane Padghe line, Delhi ring main system and Moga Kishenpura line (the Moga Kishenpura is designed for 800 kV but is presently operating at 400 kV).

To cope with the increased transmission requirements, India has added 765 kV transmission system. The first 765 kV line from Seoni to Seepat in MP was commissioned in 2008. This 1500 MVA single circuit line uses Bersimis ACSR conductor having following particulars

Stranding	42/4.57 mm	Aluminium
	7/2.54 mm	Steel
Cross section area	685.9 mm^2	Aluminium
	35.5 mm^2	Steel
Overall diameter	35.05 mm	

A large number of new 765 lines are being constructed and will be commissioned in the next few years.

The country has been divided into five regions: Northern, Western, Eastern, Southern and North-Eastern with the aim of establishing regional grids. Northern Regional Electricity Board (head office New Delhi) covers Haryana, HP, J&K, Punjab, Rajasthan, UP, Delhi, Uttranchal and Chandigarh. Western Regional Electricity Board (head office Bombay) covers Gujarat, MP, Maharashtra, Chhatisgarh, Goa, Daman, Diu, Dadra and Nagar Haveli. Southern Regional Electricity Board (head office Bangalore) covers AP, Karnataka, Tamil Nadu, Kerala, Pondicherry and Lakshdweep. Eastern Regional Electricity Board (head office Calcutta) covers WB, Bihar, Orissa, Sikkim and Jharkhand. North Eastern Regional Electricity Board (head office Shillong) covers Assam, Manipur, Meghalaya, Nagaland, Tripura, Arunachal Pradesh and Mizoram. Regional load despatch centres have been established to coordinate the activities of state load despatch centre so as to ensure optimum utilization of generation and transmission facilities in the regions. These centres have been equipped with computer based telemetry, data processing and load frequency control equipment. Each state has a State Electricity Board (and corporations) who are mainly responsible for generation, transmission and distribution of electricity in their respective States. Many generating plants, transmission lines and sub-stations are under the control of Central Government through central organizations like National Thermal Power Corporation (NTPC), Power Grid Corporation of India Ltd (PGCIL), National Hydroelectric Power Corporation (NHPC) etc. Efforts are being made to set up a national load despatch centre.

Though the first 400 kV line was commissioned only in 1978, it soon became apparent that the capacity of 400 kV lines is limited and a higher voltage is needed to cope with the transmission of power from large pithead superthermal stations to load centres. The next higher voltage of

765/800 kV has been selected. Some 765/800 kV lines being set up are Anapara-Unnao (UP), Vindhayachal-Bina-Nagda (MP), Tehri-Meerut (UP), Moga-Kishenpura.

Indian power system entered the 765 kV era with the commissioning of 765 kV single circuit transmission line between Seoni and Seepat in Madhya Pradesh in 2007-2008. This line uses Bersimis ACSR bundled conductor (Aluminium 42/4.57 and Steel 7/2.54) having 725 mm^2 cross-sectional area and overall diameter 35.05 mm. It uses steel earthwire 7/3.66 size. More 765 kV lines are likely to be commissioned in near future.

High voltage direct current (HVDC) transmission has also been introduced in Indian grid. Rihand Delhi HVDC system and Vindhayachal HVDC back to back system have already been commissioned. ±500 kV has been selected as the voltage for HVDC systems. Some more HVDC systems are being setup.

The country has a fairly extensive distribution system which operates at two levels HT and LT. The HT level operates at 33/22/15/11/6.6/3.3/2.2 kV and has a total length of about 25 lakh circuit kilometres. The LT level operates at 400 V and has a total length of about 40 lakh circuit kilometres. All HT lines are 3 phase 3 wire while LT lines are 3 phase 4 wire.

The supply to each city/town is received from a grid sub-station at 220 kV or 132 kV. At the receiving sub-station, located on the outskirts of each town, the voltage level is stepped down to 33 kV/11 kV for feeding the HT system. The distribution transformers located either in distribution sub-stations or pole mounted stations reduce the voltage level to 400 V and feed the LT system. The transformer capacity in step up (at generating stations) about 160000 MVA, step down (receiving end sub stations) about 622000 MVA and distribution 306000 MVA. The approximate number of consumers are: domestic 130×10^6, commercial 22×10^6, industrial 5×10^6, agricultural 18×10^6 and others 15×10^5. The approximate total connected load in the country is domestic 160000 MW, commercial 45000 MW, industrial 85000 MW, agricultural 100000 MW and others 25000 MW. The country has made significant progress in rural electrification. More than 500000 villages (about 86% of total villages) have been electrified.

1.10 HIGH PHASE ORDER (MULTIPHASE) TRANSMISSION

Polyphase transmission has traditionally meant three-phase transmission although higher phase orders (6 and 12 phase) have been considered for many years. The disadvantages of adding more phases on overhead lines discouraged any serious development of the idea. However, the combination of sophisticated design capabilities and environmental and cost demands have made high phase order transmission one of the most important options for future.

The benefits of high phase order arise from the smaller phase angle between phase voltages e.g. 230.95 kV phase to ground voltage corresponds to 400 kV between phases in a 3 phase system but the same 230.95 kV phase to ground voltage corresponds to 230.95 kV between phases for a 6 phase system. Therefore it is easier to insulate one phase from the other, spacings can be smaller, conductor surface gradients are less giving lower noise levels. The other benefits include narrow sight of way and smaller size leading to attractive appearance. High phase order transmission can be successfully used at all transmission voltages either by converting existing 3 phase double circuit lines into 6 phase lines or by constructing new 6 phase lines.

1.11 SMART GRID

The concept of electrical grid was given by Nicole Tesla about 120 years ago. Since then the generation, transmission and distribution of electrical energy has undergone vast changes. Giant size generators and EHV lines are very common now. The per capita consumption has increased manifold (and is still increasing) and consumers demand reliability and quality in electric supply. The power industry is faced with challenges and opportunity of transforming the electrical grid into a smart grid to manage demand growth, conserve energy, reduced losses, optimise efficiency, improve equipment utilisation, make generation, transmission and distribution ecofriendly and improve power quality and reliability. Smart grid (also called intelligent grid) refers to a combination of hardware, software and communication tools to meet the above objectives.

The main features of a smart grid are:

1. It is self healing (from power disturbances).
2. It ensures power quality and reliability.
3. It accommodates centralized giant size generators as well as distributed generation.
4. It uses conventional energy sources (coal, hydro, nuclear) as well as renewable energy resources (wind, solar, biomass, tidal, fuel cells etc.)
5. It enables active participation by consumers
6. It has energy storage options (storage using MW range batteries)
7. It is flexible
8. It is economical
9. It reduces peak demand by managing consumer demand
10. It supports new entrepreneurial energy markets
11. It optimises grid asset utilisation and efficiency
12. It ensures minimal damage to environment and takes care of climate change issues.
13. It ensures energy security.

A smart grid will use CT, PT, phasor measurement unit, many different transducers, power line carrier, wireless radio, advanced metering facilities, home area network, fibre optic network, wide area monitoring and control, microgrid management tools, optimal power flow and reactive power control techniques, automatic generation control, advanced protection systems, HVDC, facts controllers, distributed generation, distributed energy storage system, reclosers, automatic switches, breakers, on load tap changers, smart integration module.

SUMMARY

1.1 A power system consists of generation system, transmission system, sub-transmission system, primary distribution system and secondary distribution system.

1.2 Underground transmission and distribution systems are more reliable but also more costly as compared to overhead transmission and distribution systems.

1.3 Overhead lines use ACSR conductor. This conductor has hard drawn aluminium wires around single or multiple strand galvanised steel wires.

1.4 Transmission voltages used nowadays are 765 kV, 400 kV, 230 kV, 220 kV and 132 kV. Voltages above 300 kV are called EHV (extra high voltage) while those below 300 kV are called high voltages.

1.5 EHV lines are constructed using bundled conductors.

1.6 Choice of economic voltage involves a number of technical and economic factors.

1.7 Six phase transmission has many advantages over 3 phase transmission.

SELF TEST QUESTIONS

A. State whether the following are true (T) or false (F).

1.1 The permissible voltage drop in a distributor is ± 6%.
1.2 The highest transmission voltage used in India is 230 kV.
1.3 Primary distribution voltages, in India, are 33 kV, 11 kV, 6.6 kV and 3.3 kV.
1.4 At 400 kV, an underground transmission system costs about 5 times more than overhead system.
1.5 In ACSR conductor, the steel core is made of cast iron.
1.6 Distribution lines use copper conductor.
1.7 A bundled conductor line has lesser voltage drop than a line with one conductor per phase.
1.8 The capital cost of a line does not depend on line voltage.

B. Fill in the blanks with suitable words.

1.9 Generation, in India, is mostly at _______ kV.
1.10 Transmission systems are all 3 phase _______ wire systems.
1.11 Underground cable distribution lines have _______ voltage drop than overhead lines.
1.12 For the same current rating ACSR conductor has _______ diameter than copper conductor.
1.13 Optical fibre conductors have better _______ and _______ speed of transmission.
1.14 The use of multiple conductors per phase _______ voltage gradient near the line.
1.15 Six phase transmission has _______ right of way than 3 phase transmission.

ANSWERS

1.1 T	1.2 F	1.3 T	1.4 F	1.5 F	1.6 F
1.7 T	1.8 F	1.9 eleven	1.10 three	1.11 lower	1.12 larger
1.13 noise immunity, higher		1.14 reduces	1.15 lesser		

JOB INTERVIEW QUESTIONS

1.1 Why are EHV lines constructed using bundled conductors?
1.2 Compare underground and overhead transmission.
1.3 What is meant by stranded conductor?
1.4 What are the advantages of 6 phase transmission?
1.5 What voltages are used, in India, for transmission?
1.6 Which is the highest transmission voltage being used?
1.7 What is meant by EHV?

REVIEW QUESTIONS

1.1 Discuss why:
 (a) All ac transmission and distribution systems are 3 phase system.
 (b) The transmission systems are mostly overhead systems.
 (c) All overhead lines use ACSR conductors.
 (d) Overhead line conductors are invariably stranded.

1.2 Give reasons:
 (a) The transmission lines are 3 phase 3 wire circuits while distribution lines are 3 phase 4 wire circuits.
 (b) It is necessary to use high voltages for transmission systems.
 (c) At 400 kV and above the transmission lines have bundled conductors.

(d) The tendency of corona formation is lesser in aluminium conductor lines than in copper conductor lines.

(e) The voltage drop is a very important consideration in distribution lines but not so important in transmission lines.

1.3 Draw a single line diagram showing the essential parts in a modern power system network.

1.4 Bring out the relative advantages and disadvantages of overhead and underground systems.

1.5 Write a short note on 'overhead line conductors' bringing out the reasons for using ACSR conductors.

1.6 What is a bundled conductor? Why is it used? Give a few configurations of such conductors commonly employed.

1.7 Derive expressions for calculating the economic voltage and economic conductor cross-section of a line. Comment on the results.

CHAPTER 2

Line Parameters

2.1 INTRODUCTION

An overhead line consists of a group of conductors running parallel to each other and carried on supports which provide insulation between the different conductors and between each conductor and earth. A transmission line has four parameters—resistance, inductance, capacitance and shunt conductance. The shunt conductance is caused by the leakage current and is so small that it can be neglected. The series resistance causes a power loss in the conductor. The power transmission capacity of the line is mainly governed by the series inductance. The shunt capacitance causes a charging current to flow in the line and assumes importance for medium and long lines. These parameters are uniformly distributed throughout but can be lumped for the purpose of analysis on approximate basis. However, the validity of assumptions for the analysis on lumped basis may fail if the line is very long.

2.2 LINE RESISTANCE

The dc resistance of conductor is given by

$$R_{dc} = \rho\, l/a \tag{2.1}$$

where ρ = resistivity of conductor, Ω—m

l = length in m

a = cross-sectional area, m^2

The dc resistance of a stranded conductor is greater than the value given by Eq. (2.1) because spiralling of the strands makes them longer than the conductor itself. The increase in resistance due to spiralling is around 1% for three strand conductors and around 2% for concentrically stranded conductors.

The distribution of current over the cross-section of the conductor is uniform for dc only. When an alternating current flows through a conductor, the distribution tends to become non-uniform and the degree of non-uniformity increases with increase in frequency. This effect is known as skin effect and is due to a magnetic flux set up by the alternating current inside the conductor. We can imagine the conductor to be composed of a number of annular filaments. The filaments on the surface of the conductor are not linked by the internal flux. The flux linking the filaments decreases progressively from the centre to the surface. As an outcome, the inner filament has higher inductance than the filament on the surface. Therefore, the outer filaments carry more current than the inner filaments. At radio frequencies the skin effect becomes so pronounced that the current resides mostly on the surface of the conductor. Even at 50 Hz, skin effect is significant and causes the ac resistance of a conductor to be higher than the dc resistance. The effective ac resistance is usually referred as the effective resistance of the conductor. The increase in resistance due to skin effect increases with conductor cross-section and frequency. At 50 Hz, the increase in resistance due to skin effect is around 3–5 per cent. For an analytical treatment of skin effect the reader is referred to the bibliography (32). The inductance and, therefore, the current distribution in a conductor is also affected by the presence of other conductors in its vicinity. This effect, known as proximity effect, is negligible for overhead lines but causes an appreciable increase in the resistance of underground cables.

The resistance of metals increases with increase in temperature. Since the value of ρ is given at a specific temperature and the line operates at a higher temperature, the actual resistance is higher than the value found from Eq. (2.1). For small changes in temperature, the resistance increases linearly with temperature and the resistance at a temperature t is given by

$$R_t = R_0 (1 + \alpha_0 t) \tag{2.2}$$

where R_t = resistance at t °C

R_0 = resistance at 0 °C

α_0 = temperature coefficient of resistance at 0 °C

Using Eq. (2.2) the resistance R_2 at a temperature t_2 °C can be found if the resistance R_1 at a temperature t_1 °C is known.

$$\frac{R_2}{R_1} = \frac{1/\alpha_0 + t_2}{1/\alpha_0 + t_1} \tag{2.3}$$

The constant $1/\alpha_0$ equals 228 for aluminium.

2.3 INDUCTANCE—BASIC CONCEPTS

We know from our understanding of electromagnetic field theory that a conductor carrying current has a magnetic field around it. The lines of force of the magnetic field are concentric circles having their centres at the centre of the conductor and are arranged in planes perpendicular to the conductor. When the magnetic flux changes, an emf is induced in the conductor. The induced emf is given by

$$e = \frac{d\lambda}{dt} = \frac{d\lambda}{di} \times \frac{di}{dt} = L\frac{di}{dt} \tag{2.4}$$

where λ is the number of flux linkages in weber-turns. The quantity $\frac{d\lambda}{di}$ is the inductance in henrys. If the flux linkages vary linearly with current

$$L = \frac{\lambda}{i}, H \tag{2.5}$$

If the current is sinusoidal, having r.m.s. value I, the flux linkages are also sinusoidal and of r.m.s. value ψ. Then inductance is given by

$$L = \frac{\Psi}{I}, H \tag{2.6}$$

Ampere's law relates H to I, i.e.,

$$\oint H\text{-}dl = I_{\text{enclosed}} \quad (2.7)$$

The flux density is related to H by the relation

$$B = \mu H \qquad \text{where} \qquad \mu = \mu_0 \mu_r \quad (2.8)$$

$\mu_0 = 4\pi \times 10^{-7}$ H/m is the permeability of free space and μ_r is the relative permeability.

By integrating the normal flux density over any surface bounded by the conductors composing the circuit we find the flux linked with the circuit. Then using Eq. (2.6) we compute the inductance.

Similarly, the mutual inductance between two circuits is defined as the flux linkages of one circuit due to current in the second circuit per ampere of current in the second circuit.

2.4 FLUX LINKAGES WITHIN THE CONDUCTOR PRODUCING THE FLUX

To calculate the inductance of a conductor, it is necessary to consider the flux inside the conductor as well as the external flux. Fig. 2.1 shows the cross-section of a long cylindrical conductor of radius r carrying a sinusoidal current of r.m.s. value I. The return path is assumed to be situated so far away that the return current does not affect the magnetic field of the conductor. Then the magnetic lines of flux are concentric with the conductor.

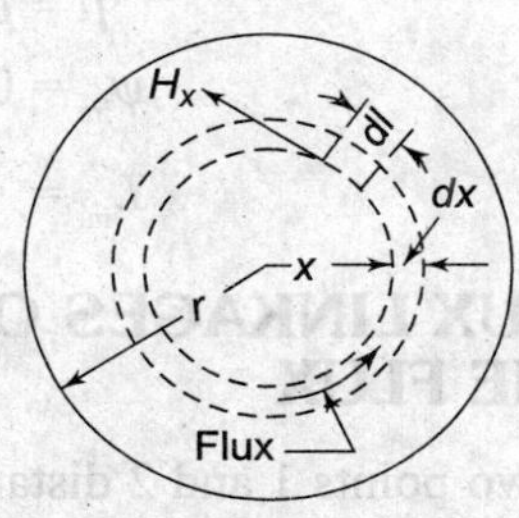

Fig. 2.1 Linkages due to internal flux

Let the field intensity at a distance x metres from the centre of the conductor be H_x. Since the field is symmetrical, H_x is constant for all points equidistant from the centre. If I_x is the current enclosed upto distances x, Eq. (2.7) gives

$$\oint H_x \cdot dl = I_x \quad (2.9)$$

or

$$2\pi x H_x = I_x \quad (2.10)$$

For finding the value of I_x, the current is assumed to be uniformly distributed over the cross-section of the conductor. This assumption is quite valid at a frequency of 50 Hz. Then

$$I_x = \left(\frac{\pi x^2}{\pi r^2}\right) I = \left(\frac{x^2}{r^2}\right) I \quad (2.11)$$

From Eqs. (2.10) and (2.11)

$$H_x = \frac{I x}{2\pi r^2} \quad \text{AT/m} \quad (2.12)$$

The flux density B_x at a distance x from the centre is

$$B_x = \mu H_x = \frac{\mu I x}{2\pi r^2} \quad \text{Wb/m}^2 \quad (2.13)$$

For finding flux linkages, a tubular element of thickness dx may be considered. The cross-sectional area of the element, normal to the flux lines is dx times the axial length. The flux per metre length is

$$d\phi = \frac{\mu I x}{2\pi r^2} dx \quad \text{Wb/m} \quad (2.14)$$

A flux line positioned at x $(0 \le x \le r)$ links with $\frac{\pi x^2}{\pi r^2}$ of the total current. Thus the flux linkage for flux $d\phi$ is given by

$$d\psi = \frac{\pi x^2}{\pi r^2} \cdot d\phi = \frac{\mu I x^3}{2\pi r^4}\, dx \quad \text{Wb-T/m} \tag{2.15}$$

For computing the total internal flux linkages ψ_{int} we integrate Eq. (2.15) from the centre to the surface of the conductor.

$$\psi_{int} = \int_0^r \frac{\mu I x^3}{2\pi r^4}\, dx = \frac{\mu I}{8\pi} \quad \text{Wb-T/m} \tag{2.16}$$

Since the conductor material is non-magnetic μ_r can be taken as unity and

$$\mu = \mu_0 = 4\pi \times 10^{-7} \quad \text{H/m}$$

Then

$$\psi_{int} = 0.5\, I \times 10^{-7} \quad \text{Wb-T/m} \tag{2.17}$$

$$L_{int} = \frac{\Psi_{int}}{I} = 0.5 \times 10^{-7} \quad \text{H/m} \tag{2.18}$$

2.5 FLUX LINKAGES OUTSIDE THE CONDUCTOR PRODUCING THE FLUX

Consider two points 1 and 2 distant D_1 and D_2 from the centre of the conductor (Fig. 2.2). Since the magnetic lines of flux are concentric circles around the conductor, whole of the flux between points 1 and 2 lies within the concentric cylindrical surfaces which pass through these points. The field intensity H_x at a distance x is

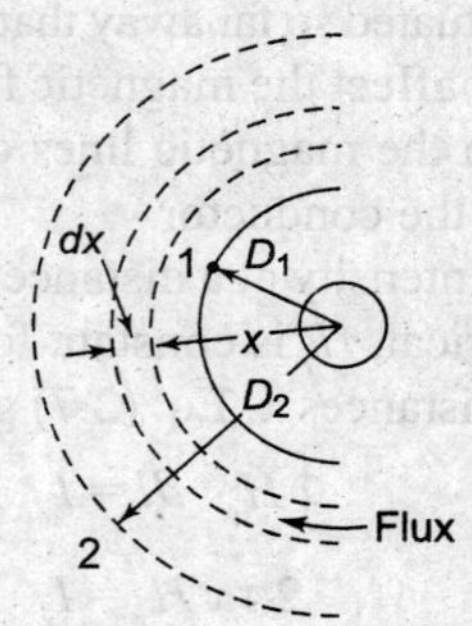

Fig. 2.2 Flux linkages between two external points

$$H_x = \frac{I}{2\pi x} \quad \text{AT/m} \tag{2.19}$$

and

$$B_x = \frac{\mu I}{2\pi x} \quad \text{Wb/m} \tag{2.20}$$

The flux $d\phi$ in a tubular element of thickness dx is

$$d\phi = \frac{\mu I}{2\pi x} \cdot dx \quad \text{Wb/m} \tag{2.21}$$

The flux external to the conductor links the current once. Thus

$$d\psi = \frac{\mu I}{2\pi x}\, dx \quad \text{Wb-T/m} \tag{2.22}$$

For finding total flux linkages between points 1 and 2, we integrate Eq. (2.22) between $x = D_1$ and $x = D_2$. This gives

$$\psi_{12} = \int_{D_1}^{D_2} \frac{\mu I}{2\pi x}\, dx = \frac{\mu I}{2\pi} \ln \frac{D_2}{D_1} \quad \text{Wb-T/m} \tag{2.23}$$

where ln denotes natural logarithm[1]. Taking $\mu_r = 1$

$$\psi_{12} = 2 \times 10^{-7} I \quad \ln \frac{D_2}{D_1} \quad \text{Wb-T/m} \tag{2.24}$$

The inductance of the conductor, contributed by the flux between points 1 and 2 is

$$L_{12} = \frac{\Psi_{12}}{I} = 2 \times 10^{-7} \ln \frac{D_2}{D_1} \quad \text{H/m} \tag{2.25}$$

2.6 INDUCTANCE OF A SINGLE PHASE TWO WIRE LINE

Figure 2.3 shows a single phase line consisting of two solid conductors of radii r_1 and r_2 spaced distance D apart. The conductors carry equal currents but in the opposite directions. The inductance of each conductor is due to internal flux linkages and external flux linkages and regarding external flux linkages the following is noteworthy.

1. A line of flux set up by the current in conductor 1, at a distance equal to or more than $(D + r_2)$ from the centre of conductor 1, links a net zero current (because the current in conductor 2 is equal and opposite to the current in conductor 1).
2. The fraction of the current linked by the above line of flux at a distance equal to or less than $(D - r_2)$ is 1.

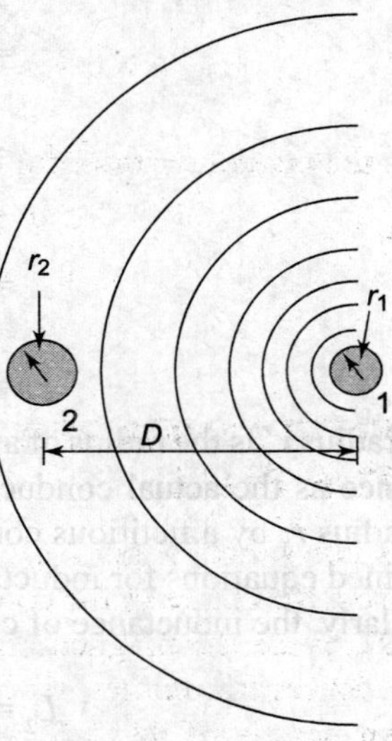

Fig. 2.3 Two-wire line showing magnetic flux lines due to current in conductor 1

3. The fraction of current linked by the above line of flux between distance $(D - r_2)$ and $(D + r_2)$ (i.e., over the surface of conductor 2) varies from 1 to 0.

As a simplifying assumption we can assume that all the external flux set up by current in conductor 1 links all the current upto the centre of conductor 2 and that the flux beyond the centre of conductor 2 does not link any current. This assumption simplifies the calculations and gives quite accurate results especially when D is much greater than r_1 and r_2, as is usually the case in overhead lines. Based on this assumption the inductance of conductor 1 (due to external flux linkages) is determined from Eq. (2.25) by substituting D for D_2 and r_1 for D_1.

For external flux only

$$L_{1,\text{ext}} = 2 \times 10^{-7} \ln \frac{D}{r_1} \quad \text{H/m} \tag{2.26}$$

For internal flux only

$$L_{1,\text{int}} = 0.5 \times 10^{-7} \text{ H/m} \tag{2.27}$$

The total inductance of conductor 1 is

$$L_1 = 10^{-7} \left(0.5 + 2 \ln \frac{D}{r_1} \right) \text{H/m} \tag{2.28}$$

---

1. Throughout this book ln denotes natural logarithm (base ε) and log denotes common logarithm (base 10)

$$= 2 \times 10^{-7} \left(0.25 + \ln \frac{D}{r_1}\right) \text{ H/m}$$

We note the following arithmetic identity:

$$0.25 = \ln \varepsilon^{1/4} \tag{2.29}$$

Therefore, Eq. (2.28) simplifies to

$$L_1 = 2 \times 10^{-7} \left(\ln \varepsilon^{1/4} + \ln \frac{D}{r_1}\right)$$

$$= 2 \times 10^{-7} \ln [D/(r_1\, \varepsilon^{-1/4})]$$

$$= 2 \times 10^{-7} \ln \left(\frac{D}{r_1'}\right) \text{ H/m} \tag{2.30a}$$

$$= 0.4605 \log \left(\frac{D}{r_1'}\right) \text{ mH/km} \tag{2.30b}$$

where $r_1' = r_1 \varepsilon^{-1/4} = 0.7788\, r_1$

The radius r_1' is the radius of a fictitious conductor which has no internal flux but has the same inductance as the actual conductor with radius r_1. The idea of replacing the original conductor of radius r_1 by a fictitious conductor of radius r_1' is quite attractive because we can develop streamlined equations for inductance without bogging down in accounting for the internal flux.

Similarly, the inductance of conductor 2 is

$$L_2 = 2 \times 10^{-7} \ln \left(\frac{D}{r_2'}\right) \text{ H/m} \tag{2.31}$$

The total inductance of the circuit is

$$L = L_1 + L_2 = 4 \times 10^{-7} \ln \left(D / \sqrt{r_1' \, r_2'}\right) \text{ H/m} \tag{2.32}$$

If $r_1 = r_2 = r$, then

$$L = 4 \times 10^{-7} \ln \left(\frac{D}{r'}\right) \text{ H/m} = 0.921 \log \left(\frac{D}{r'}\right) \text{ mH/km} \tag{2.33}$$

Eq. (2.33) gives the inductance of a two-wire line in which one conductor acts as a return conductor for the other. This is known as loop inductance, i.e., inductance per loop metre to distinguish it from the inductance of each conductor. The inductance per conductor is one half of the loop inductance and is given by Eq. (2.30).

Example 2.1

Find the loop inductance and reactance per km of a single phase overhead line consisting of two conductors, each 1.213 cm diameter. The spacing between conductors is 1.25 metres and frequency is 50 Hz.

Solution

$$r = 1.213/2 = 0.6065 \text{ cm}$$

$$r' = 0.7788 \times 0.6065 = 0.4723 \text{ cm}$$

$$L = 4 \times 10^{-7} \ln (125/0.4723)$$
$$= 22.31 \times 10^{-7} \text{ H/m} \quad \text{or} \quad 22.31 \times 10^{-4} \text{ H/km}$$
$$X = 2\pi f L = 2\pi \times 50 \times 22.31 \times 10^{-4} = 0.7 \text{ ohm/km}$$

2.7 FLUX LINKAGES OF ONE CONDUCTOR IN AN ARRAY

Figure 2.4 shows as array of n long round conductors suspended parallel to each other in space and carrying currents $I_1, I_2, \ldots I_n$ such that

$$I_1 + I_2 + I_3 + \cdots + I_n = 0 \tag{2.34}$$

The radii of conductors are $r_1, r_2 \ldots r_n$. The distances of these conductors from a remote point P are $D_{1p}, D_{2p}, \ldots D_{np}$. The distances between the conductors themselves are denoted as shown. We assume that the distances between conductors are large as compared to their radii. The current in each conductor may be credited with setting up a certain flux on its own account. The sum of all these fluxes is the total flux of the system and the total flux linkages of any one conductor is the sum of its linkages with all the individual fluxes set up by the conductors of the system.

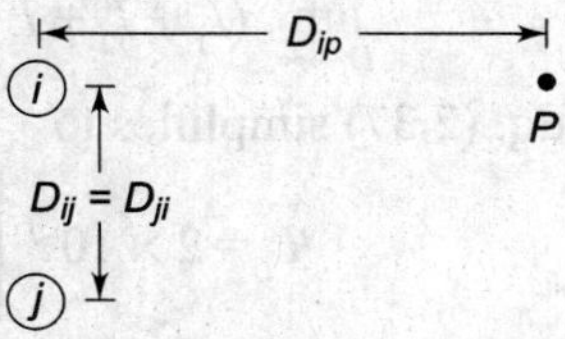

Fig. 2.4 An array of n-current carrying conductors

The flux linkages of ith conductor per unit length, due to current I_1 in conductor 1 up to point P as per Eq. (2.24) is

$$\psi_{i1} = 2 \times 10^{-7} I_1 \ln (D_{1p}/D_{i1}) \text{ Wb-T/m} \tag{2.35a}$$

Similarly the flux linkages of ith conductor per unit length due to currents $I_2, I_3, \ldots, I_n$ up to point P are

$$\psi_{i2} = 2 \times 10^{-7} I_2 \ln \left(\frac{D_2 P}{D_{i2}} \right) \tag{2.35b}$$

$$\psi_{i3} = 2 \times 10^{-7} I_3 \ln \left(\frac{D_3 P}{D_{i3}} \right) \tag{2.35c}$$

....

$$\psi_{ii} = 2 \times 10^{-7} I_i \ln \left(\frac{D_{ip}}{D_{ii}} \right) \tag{2.35i}$$

....

$$\psi_{in} = 2 \times 10^{-7} l_n \ln \left(\frac{D_{np}}{D_{in}} \right) \tag{2.35n}$$

Where $D_{ij} = D_{ji}$ = centre to centre distance between conductors i and j. The term D_{ii} in Eq. (2.35i) stands for the distance of conductor i from itself and equals r_i' or 7788 r_i. The total flux linkages per unit length of the ith conductor is the sum of the above n flux linkages.

$$\Psi_i = 2 \times 10^{-7} \left(I_1 \text{ In} \frac{D_1 P}{D_{i1}} + I_2 \text{ In} \frac{D_2 P}{D_{i2}} + \ldots + I_i \text{ In} \frac{D_{ip}}{D_{ii}} + \ldots I_n \text{ In} \frac{D_{np}}{D_{in}} \right) \text{ Wb-T/m} \tag{2.36}$$

Eq. (2.36) can be re-written as

$$\Psi_i = 2 \times 10^{-7} \left[I_1 \ln \frac{1}{D_{i1}} + I_2 \ln \frac{1}{D_{i2}} + ... + I_i \ln \frac{1}{D_{ii}} + ... \right.$$

$$\left. + I_n \ln \frac{1}{D_{in}} + I_1 \ln D_{1p} + I_2 \ln D_{2p} + ... + I_n \ln D_{np} \right] \text{ Wb-T/m} \quad (2.37)$$

To account for the total flux linkages of conductor *i*, the point *P* must approach infinity. As it does so $D_{1p} \simeq D_{2p} \simeq D_{3p} \simeq D_{np} \simeq D$. Then

$$\lim_{D \to \infty} (I_1 + I_2 + I_3 + ... + I_n) \ln D = 0 \quad (2.38)$$

Eq. (2.37) simplifies to

$$\Psi_i = 2 \times 10^{-7} \left[I_1 \ln \frac{1}{D_{i1}} + I_2 \ln \frac{1}{D_{i2}} + ... + I_i \ln \frac{1}{D_{ii}} + ... I_n \ln \frac{1}{D_{in}} \right] \text{ Wb-T/m} \quad (2.39)$$

2.8 INDUCTANCE OF COMPOSITE CONDUCTOR LINES

Figure 2.5 shows a single phase line with composite conductors, which implies conductors composed of two or more elements or strands electrically in parallel. Conductor *M* consists of *m* similar parallel sub-conductors and conductor *N* consists of *n* similar parallel sub-conductors. We assume all strands/sub-conductors of the conductor to be identical so that they share the current equally. If the line current is *I*, then each strand of conductor *M* carries a current *I*/*m* and each strand of conductor *N* carries a current of – *I*/*n* (the conductor *N* being the return conductor). Using Eq. (2.39) we can write the flux linkages of sub-conductor *a* as

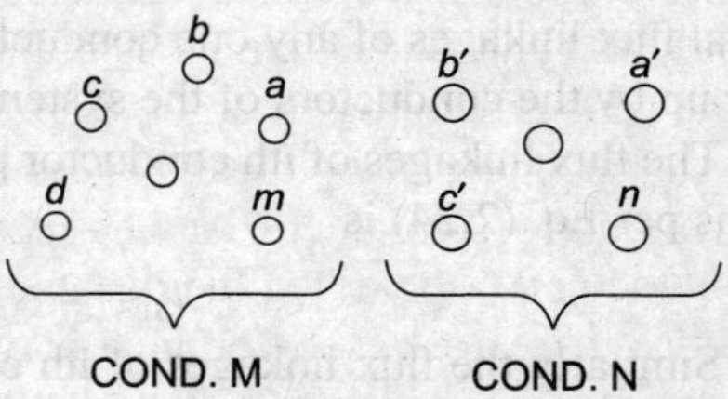

Fig. 2.5 Single phase line having composite conductors

$$\Psi_a = 2 \times 10^{-7} \times \frac{I}{m} \left[\ln \frac{1}{D_{aa}} + \ln \frac{1}{D_{ab}} + ... + \ln \frac{1}{D_{am}} \right]$$

$$- 2 \times 10^{-7} \times \frac{I}{n} \left[\ln \frac{1}{D_{aa'}} + \ln \frac{1}{D_{ab'}} + ... + \ln \frac{1}{D_{am}} \right] \quad (2.40)$$

or

$$\Psi_a = 2 \times 10^{-7} \times I \times \ln \frac{(D_{aa'} D_{ab'} ... D_{an})^{1/n}}{(D_{aa} D_{ab} ... D_{an})^{1/m}} \text{ Wb-T/m} \quad (2.41)$$

The inductance of sub-conductor *a* is

$$L_a = \frac{\Psi_a}{I/m} = 2m \times 10^{-7} \times \ln \frac{(D_{aa'} D_{ab'} ... D_{an})^{1/n}}{(D_{aa} D_{ab} ... D_{an})^{1/m}} \quad (2.42)$$

Similarly, the inductance of sub-conductors *b* is

$$L_b = \frac{\Psi_b}{I/m} = 2m \times 10^{-7} \times \ln \frac{(D_{ba'} D_{bb'} ... D_{bn})^{1/n}}{(D_{ba} D_{bb} ... D_{bm})^{1/m}} \quad (2.43)$$

The average inductance of a sub-conductor of conductor M is

$$L_{av} = (L_a + L_b + \dots\dots\dots + L_m)/m \tag{2.44}$$

Since the conductor M consists of m sub-conductors, its inductance is $1/m$ times the average inductance of its sub-conductors.

$$L_M = L_{av}/m = (L_a + L_b + \dots + L_m)/m^2 \tag{2.45}$$

Substituting the values of L_a, L_b ... L_m in Eq. (2.45) we get

$$L_M = 2 \times 10^{-7} \times \ln \frac{\left[(D_{aa'}\, D_{ab'} \cdots D_{an})(D_{ba'}\, D_{bb'} \cdots D_{bn}) \cdots (D_{ma'}\, D_{mb'} \cdots D_{mn})\right]^{1/mn}}{\left[(D_{aa}\, D_{ab} \cdots D_{am})(D_{ba}\, D_{bb} \cdots D_{bm}) \cdots (D_{ma}\, D_{mb} \cdots D_{mm})\right]^{1/m^2}} \text{ H/m} \tag{2.46}$$

The numerator of the argument of logarithm in Eq. (2.46) is the *mn*th root of product of *mn* terms, these terms being the distances from all the m sub-conductors of conductor M to all the n sub-conductors of conductor N. For each sub-conductors of conductor M there are n distances to the different sub-conductors of conductor N. Since the conductor M has m sub-conductors, the total number of distances amounts to mn. The *mn*th root of the product of mn distances is known as the geometric mean distance between conductor M and conductor N. It is written as D_m or GMD and is also termed as mutual GMD between the two conductors.

The denominator of the argument of the logarithm in Eq. (2.46) is the m^2 root of the product of m^2 terms, these being the distances from each sub-conductor of conductor M to every other sub-conductor of conductor M. The sub-conductor a has m distances to the other sub-conductors of conductor M (including the distance D_{aa} which is 0.7788 times the radius of sub-conductor a). Since there are m sub-conductors of conductor M, the total number of distances is m^2. The denominator is called self GMD and is written as D_s. Self GMD is commonly known as geometric mean radius or GMR.

Eq. (2.46) can be written as

$$L_M = 2 \times 10^{-7} \ln (D_m/D_s) \quad \text{H/m} = 0.4605 \log (D_m/D_s) \text{ mH/km} \tag{2.47}$$

It is to be noted that Eq. (2.47) is similar to Eq. (2.30). As a matter of fact Eq. (2.30) constitutes a special case of Eq. (2.47). When each conductor of a single phase line is a solid conductor, the GMD equals the distance D between the conductors and GMR equals 0.7788 times the radius of the conductor. The term r' in Eq. (2.30) can be termed as the GMR of a solid conductor of radius r.

The inductance L_N of return conductor N can be determined in a similar manner and the loop inductance of the single phase line is the sum of L_M and L_N.

Example 2.2

Figure 2.6 shows a stranded conductor having 7 identical strands each of radius r. Find the geometric mean radius of the conductor and the ratio of GMR to over all conductor radius. Comment on the results.

Solution

The distances from strand 1 to the other strands are

$$D_{12} = D_{16} = D_{17} = 2r,$$

$$D_{14} = 4\, r,$$

$$D_{13} = [(D_{14}^2 - D_{34}^2)]^{0.5}$$

$$= [(4r)^2 - (2r)^2]^{0.5} = 2r\,\sqrt{3}\,,$$

$D_{15} = D_{13}$

The geometric mean radius of the seven-strand conductor is the 49th root of the product of 49 distances.

$$D_s = \sqrt[49]{\left[(r')^7 \left(D_{12}^2 D_{13}^2 D_{14} D_{17}\right)^7 (2r)^6\right]}$$

where r' is the GMR of each strand. It is raised to the seventh power to account for seven strands. The term $D_{12}^2 D_{13}^2 D_{14} D_{17}$ represents the product of distances from one outside strand to every other strand. This term is raised to the sixth power to account for the six outside strands. The term $(2r)^6$ accounts for the product of the distances from the central strand to every outside strand.

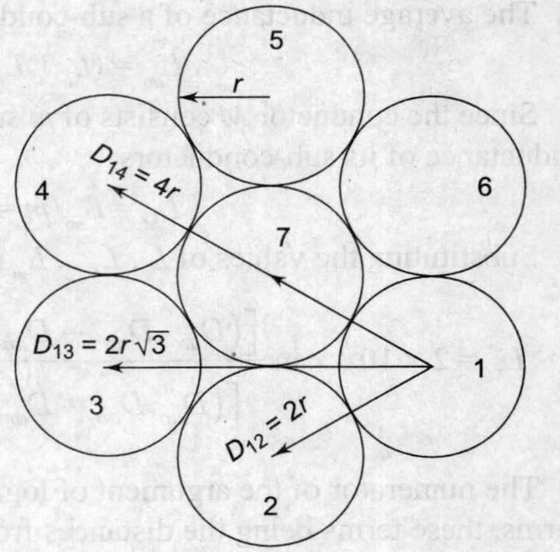

Fig. 2.6 Cross-section of a conductor with seven strands

$$D_s = \sqrt[49]{\left[(0.7788)^7 \left\{(2r)^2 \left(2\sqrt{3}r\right)^2 \cdot 4r \cdot 2r\right\}^6 (2r)^6\right]} = 2.177\, r.$$

The overall conductor radius is $3r$. Therefore, the ratio of GMR to overall conductor radius is $2.177\, r/3r$, i.e., 0.7257. As the number of strands increases, this ratio approaches 0.7788 which is that for a solid conductor. For 3 strands, this ratio is 0.678 and for 169 strands this ratio is 0.777.

2.9 INDUCTANCE OF THREE PHASE LINE WITH EQUILATERAL SPACING

Figure 2.7 shows the conductors of a 3 phase line with equilateral spacing. The conductors are spaced at the corners of an equilateral triangle.

From Eq. (2.39) the flux linkages of conductor a can be written as

$$\Psi_a = 2 \times 10^{-7} \left[I_a \ln (1/D_{aa}) + I_b \ln (1/D_{ab}) + I_c \ln (1/D_{ac})\right]$$

$$= 2 \times 10^{-7} \left[I_a \ln (1/r') + I_b \ln (1/D) + I_c \ln (1/D)\right] \text{ Wb-T/m} \tag{2.48}$$

Assuming balanced currents or in the absence of a neutral wire,

$$I_a + I_b + I_c = 0 \quad \text{or} \quad I_a = -(I_b + I_c)$$

Then $\Psi_a = 2 \times 10^{-7} \left[I_a \ln (1/r') - I_a \ln (1/D)\right]$ Wb-T/m (2.49)

and $L_a = (\Psi_a/I_a) = 2 \times 10^{-7} \ln (D/r')$ H/m (2.50a)

or $L_a = 0.4605 \log (D/r')$ mH/km (2.50b)

Eq. (2.50) is similar to Eq. (2.30) for a single phase line. For stranded conductors D_s replaces r' in the above equation. Because of symmetry the inductances of conductors of phase b and phase c are the same as the inductance of the conductor of phase a. Since each phase has one conductor, Eq. (2.50) gives the inductance per phase of a 3 phase line.

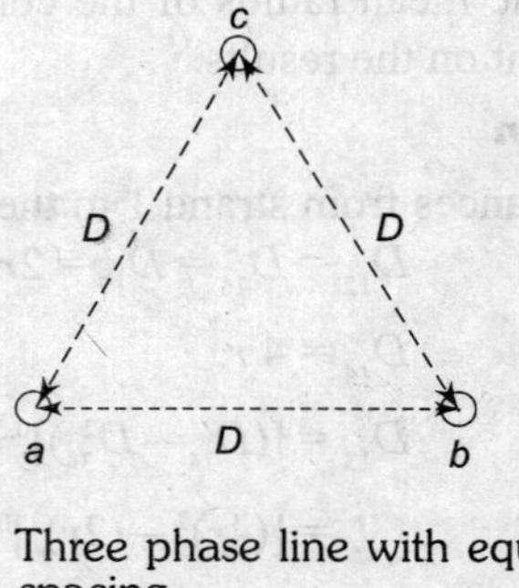

Fig. 2.7 Three phase line with equilateral spacing

2.10 INDUCTANCE OF THREE PHASE LINE WITH UNSYMMET-RICAL SPACING

In actual practice the conductors of a three-phase line are not at the corners of an equilateral triangle. Then the flux linkages and inductance of each phase are not the same. A different inductance in each phase results in an unbalanced circuit. If the line is sufficiently long and the unbalance in the inductance (and capacitance) is quite large, the lines are transposed. In a transposed line, the phase conductors have their positions interchanged at special transposition towers so that the conductor of each phase occupies positions 1, 2 and 3 for about one third of its length as illustrated in Fig. 2.8. Transposition makes it possible for each phase conductor to have the same average inductance. Modern transmission lines are not transposed at regular intervals although an interchange in the positions of the conductors is made at the switching stations to balance the inductance. For all practical purposes the asymmetry between the phases of an untransposed line may be neglected and the inductance of an untransposed line may be taken equal to that of a transposed line.

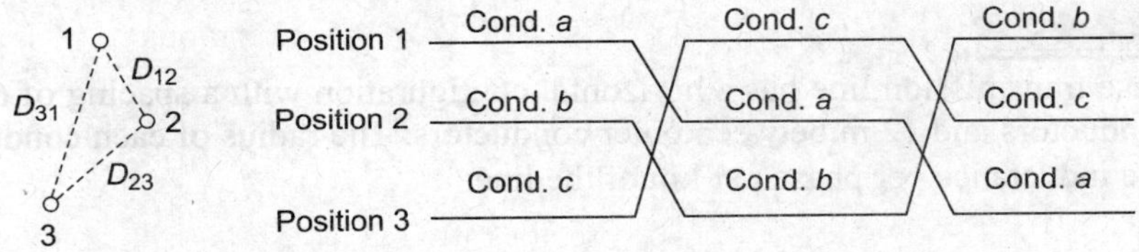

Fig. 2.8 Transposition of conductors

The average inductance of a conductor of a transposed line is found by calculating the flux linkages for each position occupied by the conductor and then finding the average flux linkages.

The flux linkages of conductor *a*, when this conductor is in position 1, conductor *b* is in position 2 and conductor *c* is in position 3, (as shown in Fig. 2.8) are

$$\Psi_{a1} = 2 \times 10^{-7} [I_a \ln (1/r') + I_b \ln (1/D_{12}) + I_c \ln (1/D_{31})] \text{ Wb-T/m} \quad (2.51)$$

with conductor *a* in position 2, *b* in position 3 and *c* in position 1,

$$\Psi_{a2} = 2 \times 10^{-7} [I_a \ln (1/r') + I_b \ln (1/D_{23}) + I_c \ln (1/D_{12})] \text{ Wb-T/m} \quad (2.52)$$

with conductor *a* in position 3, *b* in position 1, and *c* in position 2

$$\Psi_{a3} = 2 \times 10^{-7} [I_a \ln (1/r') + I_b \ln (1/D_{31}) + I_c \ln (1/D_{23})] \text{ Wb-T/m} \quad (2.53)$$

the average flux linkages of conductor *a* are

$$\Psi_a = \frac{\Psi_{a1} + \Psi_{a2} + \Psi_{a3}}{3} = \frac{2\times10^{-7}}{3} [3\, I_a \ln (1/r')$$

$$+ I_b \ln (1/D_{12}\, D_{23}\, D_{31}) + I_c \ln (1/D_{31}\, D_{12}\, D_{23})] \text{ Wb-T/m} \quad (2.54)$$

Since $I_a = -(I_b + I_c)$

$$\Psi_a = \frac{2\times10^{-7}}{3} [3\, I_a \ln (1/r') - I_a \ln (1/D_{12}\, D_{23}\, D_{31})]$$

$$= 2 \times 10^{-7}\, I_a \ln \left(\sqrt[3]{D_{12} D_{23} D_{31}} \,/\, r'\right) \text{ Wb-T/m} \quad (2.55)$$

The inductance of phase *a* is

$$L_a = \frac{\Psi_a}{I_a} = 2 \times 10^{-7} \ln \left(\sqrt[3]{D_{12} D_{23} D_{31}} \,/\, r'\right) \text{ H/m} \quad (2.56a)$$

$$= 2 \times 10^{-7} \ln (D_{eq}/r') \text{ H/m} \tag{2.56b}$$

$$= 0.4605 \log (D_{eq}/r') \text{ mH/km} \tag{2.56c}$$

The quantity $\sqrt[3]{D_{12}D_{23}D_{31}}$ is called the equivalent spacing and is denoted by D_{eq}. It is the geometric mean of the distances of the line.

Example 2.3

A three-phase transmission line has its conductors at the corners of an equilateral triangle with side 3 m. The diameter of each conductor is 1.63 cm. Find the inductance per phase per kilometre of the line.

Solution

$$r' = 0.7788 \times 0.5 \times 1.63 = 0.635 \text{ cm.}$$

$$L = 0.4605 \log (300/0.635) = 1.2315 \text{ mH/km}$$

Example 2.4

A three-phase transmission line has a horizontal configuration with a spacing of 6 m between adjacent conductors and 12 m between outer conductors. The radius of each conductor is 1.81 cm. Find the inductance per phase per km of the line.

Solution

$$r' = 0.7788 \times 1.81 = 1.4096 \text{ cm}$$

$$D_{eq} = \sqrt[3]{(6 \times 6 \times 12)} = 7.5595 \text{ m}$$

$$L = 0.4605 \log (7.5595 \times 100/1.4096) = 1.2569 \text{ mH/km}$$

2.11 INDUCTANCE OF BUNDLED CONDUCTOR LINES

The geometric mean radius of bundled conductors can be found in the same manner as that for stranded conductors.

For a two-conductor (duplex) arrangement (Fig. 1.3a)

$$D_s^b = \sqrt[4]{(D_s \cdot d)^2} = \sqrt{(D_s \cdot d)} \tag{2.57}$$

For a three-conductor (triplex) arrangement (Fig. 1.3b)

$$D_s^b = \sqrt[9]{(D_s \cdot d \cdot d)^3} = \sqrt[3]{(D_s \cdot d^2)} \tag{2.58}$$

For a four-conductor (quadruplex) arrangement (Fig. 1.3c)

$$D_s^b = \sqrt[16]{\left[D_s \cdot d \cdot d\sqrt{2} \times d\right]^4} = 1.09 \times \sqrt[4]{(D_s \times d^3)} \tag{2.59}$$

In Eqs. (2.57 to 2.59) D_s^b denotes the geometric mean radius of bundled conductor, D_s denotes geometric mean radius of each sub-conductor of bundle and d denotes the spacing between the sub-conductors of a bundle.

The geometric mean distance (GMD) of a bundled conductor line can be found by taking the root of the product of distances from each conductor of a bundle to every other conductor of the other bundles. However, it is sufficiently accurate to take the distance from centre of one bundle to the centres of the other bundles as the distance D_{12}, D_{23} and D_{31}.

The use of bundled conductors increases the geometric mean radius. In the formula for calculating inductance, the geometric mean radius is a factor in the denominator. Therefore, the inductance of a bundled conductor line is less than the inductance of the line with one conductor per phase.

Example 2.5

A bundled conductor line has 4 conductors per bundle. The four sub-conductors are placed at the corners of square of side 25 cm. The radius of each sub-conductor is 1.573 cm. Find the geometric mean radius of this configuration.

Solution

For each sub-conductor $D_s = r' = 0.7788 \times 1.573 = 1.225$ cm

Using Eq. (2.59)

$$D_s^b = \sqrt[16]{\left[(1.225\times 25\times 25\times 25\times \sqrt{2})^4\right]} = 12.826 \text{ cm}$$

Example 2.6

A 400 kV 3-phase bundled conductor line with two sub-conductors per phase has a horizontal configuration as shown in Fig. 2.9. The radius of each sub-conductors is 1.6 cm. (a) Find the inductance per phase per km of the line. (b) Compute the inductance of the line with only one conductor per phase having the same cross-sectional area of the conductor of each phase.

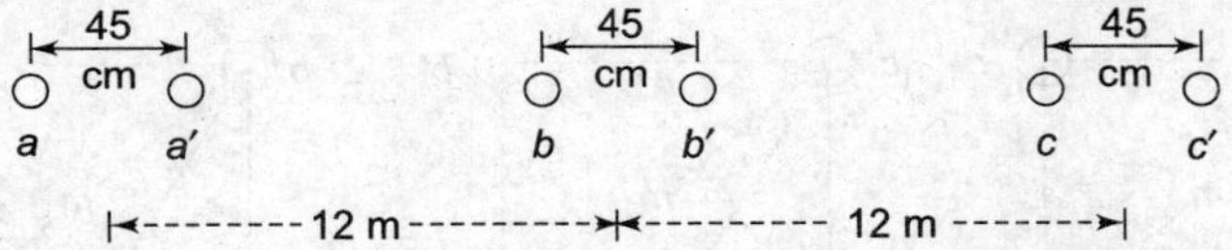

Fig. 2.9 Three-Phase Line with Bundled Conductors

Solution

(a) $r' = 0.7788 \times 1.6 = 1.246$ cm

$$GMR = \sqrt[4]{r'\cdot d\cdot r'\cdot d\cdot} = \sqrt{1.246\times 45} = 7.49 \text{ cm}$$

$$D_{ab} = D_{bc} = \sqrt[4]{12(12+.45)\times 12(12-.45)} = 11.996 \text{ m}$$

$$D_{ca} = \sqrt[4]{24(24+.45)\times 24(24-.45)} = 23.998 \text{ m}$$

$$D_{eq} = \sqrt[3]{11.996\times 11.996\times 23.998} = 15.115 \text{ m}$$

(Taking the centre to centre distances between the bundles as $D_{ab}, D_{bc}, D_{ca}, D_{eq} = \sqrt[3]{12\times 12\times 24}$ $= 15.119$ m which is almost the same as the accurate value of D_{eq})

$$L = 0.4605 \log (D_{eq}/GMR) = 0.4605 \log (15.115 \times 100/7.49) = 1.06 \text{ mH/km}$$

(b) Conductor radius for the same cross-sectional area $= \sqrt{2} \times 1.6 = 2.2624$ cm

$r' = 0.7788 \times 2.2624 = 1.762$ cm

$D_{eq} = 15.119$ m

$L = 0.4605 \log (15.119 \times 100/1.762) = 1.351$ mH/km

It is seen that the inductance of the bundled conductor line is less than that of the line with one conductor per phase. The lower inductance leads to a greater transmission capacity of the bundled conductor line.

2.12 INDUCTANCE OF DOUBLE CIRCUIT THREE-PHASE LINE

A double circuit three-phase line has two parallel conductors for each phase. The use of two three-phase circuits on the same towers leads to a greater reliability and a higher transmission capacity. If the two circuits are very widely separated, the mutual inductance between the circuits can be neglected and the net inductance is half of the inductance of individual circuits. In actual practice the separation is not very wide and the mutual inductance is not negligible. The method of GMD can be used to find the inductance per phase by taking the conductors of one phase to be the strands of one composite conductor. It is desirable to have a configuration which results in minimum inductance (and hence maximum transmission capability). This is possible if GMD is low and GMR is high. Therefore, the individual conductors of a phase should be widely separated (to give high GMR) and the distance between phases should be kept low (to give low GMD).

Figure 2.10 shows the three sections of a double circuit three-phase line, with vertical spacing, over a transposition cycle. The conductors a and a' belong to one phase, b and b' to the second phase and c and c' to the third phase. In each case the conductors of two phases are placed diametrically opposite to each other and those of the third phase are placed horizontally opposite to each other. This configuration gives high value of GMR. To calculate inductance it is necessary to determine D_{eq} or GMD and D_s or GMR.

GMR of conductors of phase a in section 1 is

$$D_{sa} = \sqrt[4]{r' \cdot n \cdot r' \cdot n} = \sqrt{r' \cdot n}$$

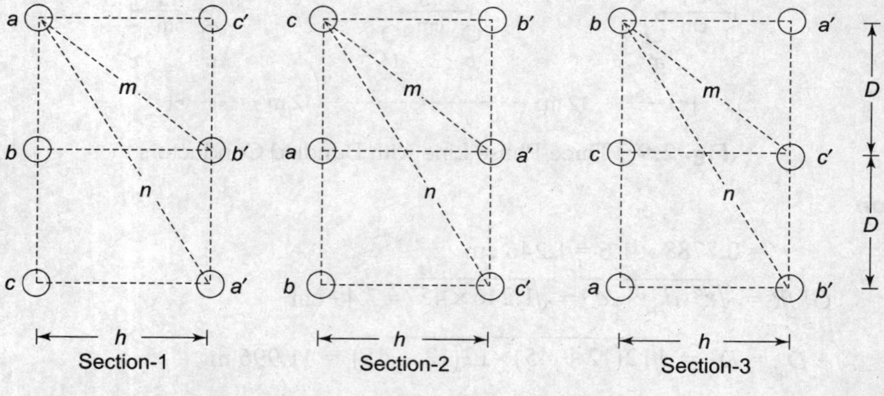

Fig. 2.10 Arrangement of conductors in a double circuit three-phase line

GMR of conductors of phase b and c in section 1 are

$$D_{sb} = \sqrt[4]{r' \cdot h \cdot r' \cdot h} = \sqrt{r' \cdot h}$$

$$D_{sc} = \sqrt[4]{r' \cdot n \cdot r' \cdot n} = \sqrt{r' \cdot n}$$

Equivalent GMR is

$$D_s = \sqrt[3]{D_{sa} \cdot D_{sb} \cdot D_{sc}} = [(r')^{1/2} (n)^{1/3} (h)^{1/6}] \tag{2.60}$$

It can be verified that the value of D_s is the same in all the three sections. GMD is given by

$$D_{eq} = \sqrt[3]{D_{ab} \cdot D_{bc} \cdot D_{ca}} \tag{2.61}$$

where D_{ab} = Geometric mean distance between phases a and b in section 1

$$= \sqrt[4]{D \cdot m \cdot D \cdot m} = \sqrt{D \cdot m}$$

D_{bc} = Geometric mean distance between phases b and c in section 1

$$= \sqrt[4]{D \cdot m \cdot D \cdot m} = \sqrt{D \cdot m}$$

D_{ca} = Geometric mean distance between phases c and a in section 1

$$= \sqrt[4]{(2D)(h)(2D)(h)} = \sqrt{2Dh}$$

$$D_{eq} = \sqrt[6]{(Dm)(Dm)(2Dh)} = 2^{1/6} D^{1/2} m^{1/3} h^{1/6} \quad (2.62)$$

The inductance per phase per m is

$$L = 2 \times 10^{-7} \ln (D_{eq}/D_s)$$

$$= 2 \times 10^{-7} \ln \frac{2^{1/6} D^{1/2} m^{1/3} h^{1/6}}{(r')^{1/2} n^{1/3} h^{1/6}}$$

$$= 2 \times 10^{-7} \ln \left[2^{1/6} \left(\frac{D}{r'}\right)^{1/2} \left(\frac{m}{n}\right)^{1/3} \right] \text{ H/phase/m} \quad (2.63)$$

Example 2.7

A 3-phase double circuit line has vertical configuration shown in Fig. 2.10. Radius of each conductor is 1.1 cm. The horizontal distance h is 5 m and vertical distance D is 3 m. Find the inductance per phase per km of the line.

Solution

$$m = \sqrt{5^2 + 3^2} = 5.83 \text{ m}$$

$$n = \sqrt{5^2 + 6^2} = 7.81 \text{ m}$$

$$r' = (0.7788 \times 1.1/100) = 0.857 \times 10^{-2} \text{ m}$$

$$L = 2 \times 10^{-7} \ln [2^{1/6} (3/.857 \times 10^{-2})^{1/2} (5.83/7.81)^{1/3}]$$

$$= 5.89 \times 10^{-7} \text{ H/phase/m} \quad \text{or} \quad 5.89 \times 10^{-4} \text{ H/phase/km}$$

Example 2.8

A 3-phase double circuit line has the configuration shown in Fig. 2.11. The radius of each conductor is 0.9 cm. Find the inductance per phase per km of line length.

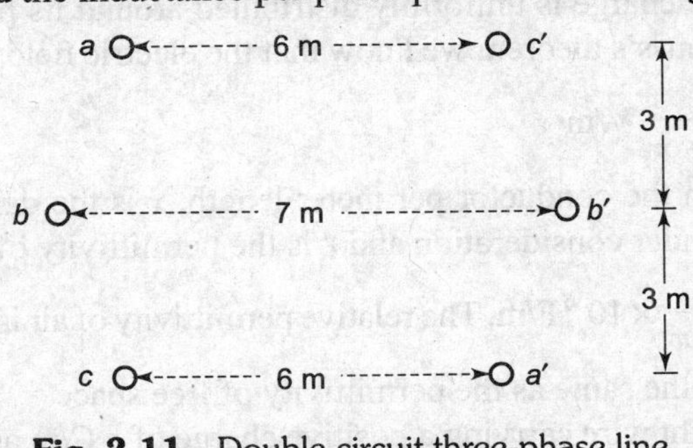

Fig. 2.11 Double circuit three-phase line

Solution

$$r' = 0.7788 \times 0.9 = 0.7 \text{ cm} = 0.7 \times 10^{-2} \text{ m}$$

Distance between a and a' = distance between c and c' = $\sqrt{6^2 + 6^2} = 8.485$ m

GMR of conductors of phase $a = \sqrt{0.7 \times 10^{-2} \times 8.485} = 0.2437$ m

GMR of conductors of phase $b = \sqrt{0.7 \times 10^{-2} \times 7} = 0.2214$ m

GMR of conductors of phase $c = \sqrt{0.7 \times 10^{-2} \times 8.485} = 0.2437$ m

$$D_s = \sqrt[3]{(0.2437)(0.2214)(0.2437)} = 0.236 \text{ m}$$

Distance between conductors a and $b = \sqrt{3^2 + 0.5^2} = 3.041$ m

Distance between conductors a and $b' = \sqrt{3^2 + 6.5^2} = 7.159$ m

Geometric mean distance between phases a and $b = \sqrt[4]{(3.041 \times 7.159)^2} = 4.666$ m

Geometric mean distance between phases b and $c = \sqrt[4]{(3.041 \times 7.159)^2} = 4.666$ m

Geometric mean distance between phases c and $a = \sqrt[4]{(6 \times 6)^2} = 6$ m

$$D_{eq} = \sqrt[3]{4.666 \times 4.666 \times 6} = 5.074 \text{ m}$$

$$L = 2 \times 10^{-7} \ln (5.074/0.236) = 6.136 \times 10^{-7} \text{H/phase/m}$$

$$= 6.136 \times 10^{-4} \text{ H/phase/km}$$

2.13 ELECTRIC FIELD AND POTENTIAL DIFFERENCE

From electromagnetic field theory it is well known that an electric field exists around a conductor carrying current. The lines of electric flux originate on the positive charges on one conductor and terminate on the negative charges on the other conductor. The ratio of charge on the conductors per unit of potential difference between them is the capacitance between the conductors. If a long straight cylindrical conductor has a uniform charge throughout its length and is isolated from other charges so that the charge is uniformly distributed around its periphery, the electric flux lines are radial. From Gauss's theorem we know that the electric field intensity E at any point is

$$E = \frac{q}{2\pi\varepsilon x} \text{ V/m} \tag{2.64}$$

where q is the charge on the conductor per meter length, x is the distance in meters from the conductor to the point under consideration and ε is the permittivity of the medium. The permittivity of free space is $\frac{1}{36\pi} \times 10^{-9}$ F/m. The relative permittivity of air is almost 1 and permittivity of air can be taken to be the same as the permittivity of free space.

Consider a long straight wire carrying a positive charge of q C/m as shown in Fig. 2.12. The electric flux lines are radial. Since the equipotential surface is orthogonal to electric flux lines, the equipotential surfaces are concentric cylinders surrounding the conductor. Points x and y are situated at distance D_x and D_y from the centre of the conductor. The potential difference between two points is numerically equal to the work in joules necessary to move a unit positive charge between the two points. Since the electric field intensity is negative of potential gradient, we can find the potential difference V_{xy} between the two points by integrating Eq. (2.64) over the radial path between the two points. The potential difference between the two points is unique and, thus, independent of the path of integration. We choose $x\,X\,y$ as the path of integration. Since a positive charge exerts a repelling force on a positive charge, there is a drop in potential from point x to X. The points X and y lie on an equipotential surface and are at the same potential. Thus,

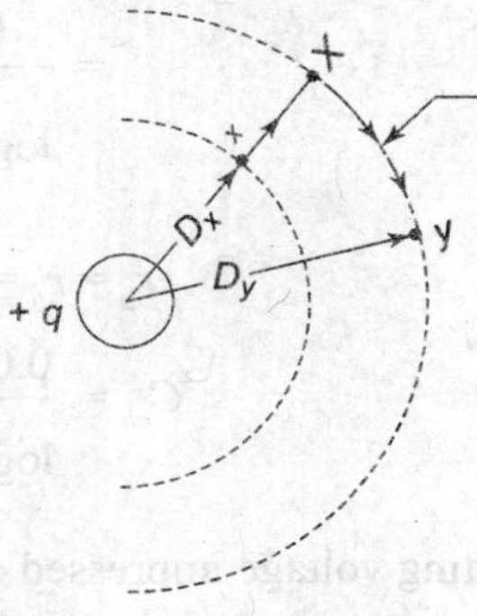

Fig. 2.12 Potential difference between two points in an electric field

$$V_{xy} = \int_{D_x}^{D_y} \frac{q}{2\pi\varepsilon x}\,dx = \frac{q}{2\pi\varepsilon}\,\text{In}\,\frac{D_y}{D_x}\ \text{volts} \tag{2.65}$$

The above concept can be used to find the potential difference between two conductors of an array of parallel conductors. Fig. 2.13 shows an array of m conductors. We assume the distance between the conductors to be large as compared to the radii so that the charge on each conductor is uniformly distributed over its surface. By the principle of superposition, the potential difference between conductors a and b due to all the charges is the sum of potential differences due to each charge acting alone. The potential difference due to each charge is given by Eq. (2.65). Thus,

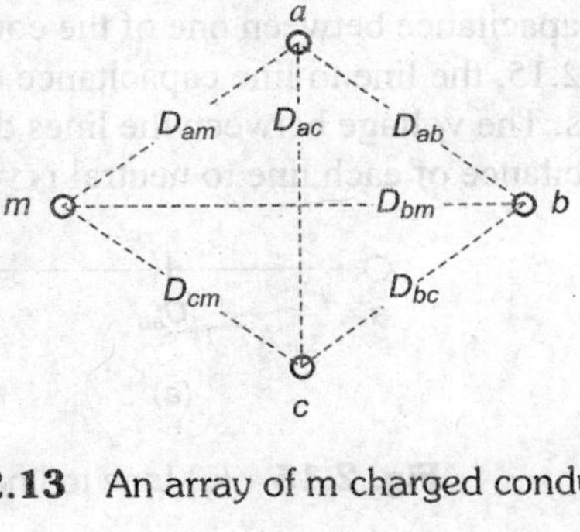

Fig. 2.13 An array of m charged conductors

$$V_{ab} = \frac{1}{2\pi\varepsilon}\left[q_a\,\text{In}\,\frac{D_{ab}}{r_a} + q_b\,\text{In}\,\frac{r_b}{D_{ba}} + q_c\,\text{In}\,\frac{D_{cb}}{D_{ca}} + \cdots + q_m\,\text{In}\,\frac{D_{mb}}{D_{ma}}\right] \tag{2.66}$$

2.14 CAPACITANCE OF TWO WIRE LINE

Figure 2.14 shows the two conductors of a single phase line. If the charge per unit length on conductor a is q, then the charge per unit length on conductor b is $-q$. From Eq. (2.66) the potential difference between conductors a and b is

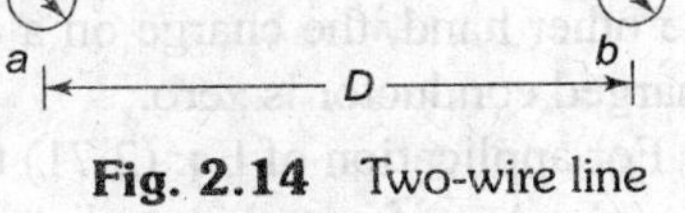

Fig. 2.14 Two-wire line

$$V_{ab} = \frac{1}{2\pi\varepsilon}\left(q\,\text{In}\,\frac{D}{r_a} - q\,\text{In}\,\frac{r_b}{D}\right) = \frac{q}{2\pi\varepsilon}\,\text{In}\,\frac{D^2}{r_a r_b} \tag{2.67}$$

The capacitance per unit length between the conductors, i.e., C_{ab} is the ratio of charge to potential difference per unit length.

$$C_{ab} = \frac{q}{V_{ab}} = \frac{2\pi\varepsilon}{In\,\dfrac{D^2}{r_a\,r_b}} = \frac{\pi\varepsilon}{In\left(\dfrac{D}{\sqrt{r_a\,r_b}}\right)}\ \text{F/m} \tag{2.68}$$

$$= \frac{0.01206}{\log\left(\frac{D}{\sqrt{r_a r_b}}\right)} \ \mu\text{F/km} \tag{2.69a}$$

If $r_a = r_b = r$

$$C_{ab} = \frac{0.01206}{\log\left(\frac{D}{r}\right)} \ \mu\text{F/km} \tag{2.69b}$$

An alternating voltage impressed on a transmission line causes the charge on the conductors at any point to increase and decrease with the increase and decrease of the instantaneous voltage between the conductors at that point. The flow of charge is a current and the current caused by the alternate charging and discharging of the line due to the alternating voltage is called the charging current of the line. Charging current flows in a line even when the line is open circuited and affects the voltage drop, efficiency and power factor of the line.

For the single phase line, the charging current I_c is

$$I_c = j\,\omega\, C_{ab}\, V_{ab} \tag{2.70}$$

Eq. (2.69) gives the capacitance between the conductors. Sometimes it is necessary to find the capacitance between one of the conductors and a neutral point between them. As seen from Fig. 2.15, the line to line capacitance can be considered to be composed of two capacitance in series. The voltage between the lines divides equally and the point n is at ground potential. The capacitance of each line to neutral is twice the line to line capacitance.

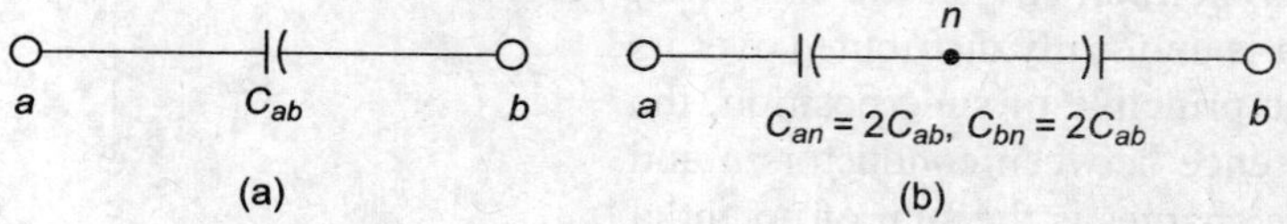

Fig. 2.15 (a) Line to line capacitance, (b) line to neutral capacitances

$$C_n = C_{an} = C_{bn} = 2C_{ab} = \frac{0.02412}{\log(D/r)} \ \mu\text{F/km} \tag{2.71}$$

Eq. (2.71) for capacitance corresponds to Eq. (2.30) for inductance. However, one important difference between the two formulae should be noted carefully. The formula for capacitance uses r, i.e., the actual outside radius of the conductor whereas the formula for inductance uses r', i.e., the geometric mean radius. This essential difference arises from the fact that a magnetic field exists inside the conductor and must be taken into account in calculating inductance. On the other hand, the charge on a conductor resides on its surface and the electric field inside a charged conductor is zero.

For application of Eq. (2.71) to a transmission line, following approximations are involved:

(a) A uniform charge distribution has been assumed on the surface. This is strictly true only if all the other charges are removed to infinity. The error becomes smaller as D/r is increased. For high voltage lines D/r is always more than 300 and the error is less than 0.0003%.

(b) The distance D between the centres of the conductors has been used as one limit of integration rather than $D - r$. This is again justificable when $D >> r$.

(c) In actual practice the conductor is stranded and not smooth. The electric field at the surface of a stranded conductor is not the same as the field at the surface of a smooth round conductor. However, this error is again negligible and the outside radius of a stranded conductor can be used for r in Eq. (2.71).

Example 2.9

Find the capacitance between the conductors of a single phase 10 km long line. The diameter of each conductor is 1.213 cm. The spacing between conductors is 1.25 m. Also find the capacitance of each conductor to neutral.

Solution

$$C_{ab} = \frac{0.01206}{\log(125/1.213\times 0.5)} = 5.21\times 10^{-3}\ \mu\text{F/km}$$

For 10 km length, $C_{ab} = 5.21\times 10^{-3}\times 10 = 0.0521\ \mu\text{F}$

$$C_n = 2C_{ab} = 0.1042\ \mu\text{F}$$

2.15 CAPACITANCE OF THREE-PHASE LINE WITH EQUILATERAL SPACING

In a three-phase line with equilateral spacing (Fig. 2.7) the conductors are spaced at the corners of an equilateral triangle. Let the radius of each conductor be r and the spacing between conductors be D. From Eq (2.66) the expressions for V_{ab} and V_{ac} can be written as

$$V_{ab} = \frac{1}{2\pi\varepsilon}\left[q_a \text{ In}\frac{D}{r} + q_b \text{ In}\frac{r}{D} + q_c \text{ In}\frac{D}{D}\right] \tag{2.72}$$

$$V_{ac} = \frac{1}{2\pi\varepsilon}\left[q_a \text{ In}\frac{D}{r} + q_b \text{ In}\frac{D}{D} + q_c \text{ In}\frac{r}{D}\right] \tag{2.73}$$

Addition of Eqs. (2.72) and (2.73) gives

$$V_{ab} + V_{ac} = \frac{1}{2\pi\varepsilon}\left[2q_a \text{ In}\frac{D}{r} + (q_b + q_c)\text{In}\frac{r}{D}\right] \tag{2.74}$$

If there are no other charges in the vicinity, $q_a + q_b + q_c = 0$ or $q_b + q_c = -q_a$

Then $$V_{ab} + V_{ac} = \frac{3q_a}{2\pi\varepsilon}\text{ In}\frac{D}{r} \tag{2.75}$$

If the three-phase voltages are V_{an}, V_{bn}, and V_{cn}, then

$$V_{an} = V\angle 0,\ V_{bn} = V\angle -120^\circ,\ V_{cn} = V\angle +120^\circ \tag{2.76}$$

$$V_{ab} = V_{an} - V_{bn} = V\angle 0 - V\angle -120^\circ = \sqrt{3}\,V\angle 30^\circ \tag{2.77a}$$

$$V_{ac} = V_{an} - V_{cn} = V\angle 0 - V\angle +120^\circ = \sqrt{3}\,V\angle -30^\circ \tag{2.77b}$$

$$V_{ab} + V_{ac} = \sqrt{3}\,V\angle 30^\circ + \sqrt{3}\,V\angle -30^\circ = 3V = 3V_{an} \tag{2.78}$$

Substituting $3V_{an}$ for $(V_{ab} + V_{bc})$ in Eq. (2.75)

$$V_{an} = \frac{q_a}{2\pi\varepsilon}\text{ In}\frac{D}{r} \tag{2.79}$$

The line to neutral capacitance is

$$C_n = \frac{q_a}{V_{an}} = \frac{2\pi\varepsilon}{\text{In}(D/r)}\ \text{F/m} \tag{2.80a}$$

$$= \frac{0.02412}{\log(D/r)} \; \mu\text{F/km} \tag{2.80b}$$

The charging current per phase is

$$I_c = j\,\omega\, C_n\, V_{an} \tag{2.81}$$

2.16 CAPACITANCE OF THREE-PHASE LINE WITH ASYMMETRICAL SPACING

The capacitance of a line with asymmetrical spacing is found by assuming that the line is transposed. The asymmetry of the untransposed line is small for the usual configurations. In a transposed line the capacitance of all the three phases are equal.

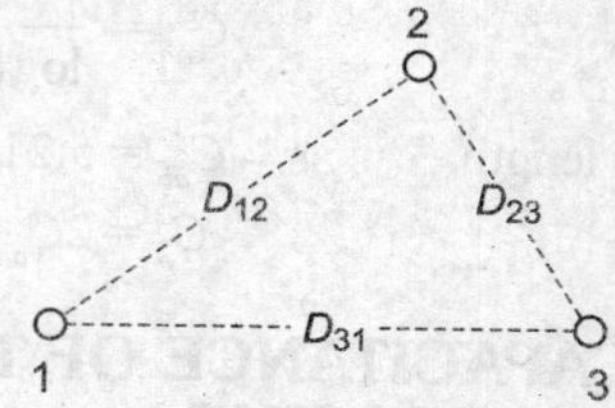

Fig. 2.16 Three-phase line with asymmetrical spacing

For the configuration shown in Fig. 2.16, the three equations for V_{ab} for the three different parts of the transposition cycle can be written as follows. For phase a in position 1, b in position 2 and c in position 3.

$$V_{ab} = \frac{1}{2\pi\varepsilon}\left[q_a \,\text{In}\, \frac{D_{12}}{r} + q_b \,\text{In}\, \frac{r}{D_{12}} + q_c \,\text{In}\, \frac{D_{23}}{D_{31}}\right] \tag{2.82}$$

For phase a in position 2, b in position 3 and c in position 1

$$V_{ab} = \frac{1}{2\pi\varepsilon}\left[q_a \,\text{In}\, \frac{D_{23}}{r} + q_b \,\text{In}\, \frac{r}{D_{23}} + q_c \,\text{In}\, \frac{D_{31}}{D_{12}}\right] \tag{2.83}$$

For phase a in position 3, b in position 1 and c in position 2

$$V_{ab} = \frac{1}{2\pi\varepsilon}\left[q_a \,\text{In}\, \frac{D_{31}}{r} + q_a \,\text{In}\, \frac{r}{D_{31}} + q_c \,\text{In}\, \frac{D_{12}}{D_{23}}\right] \tag{2.84}$$

Truly speaking the charge on each conductor is different when the position of the conductor changes with respect to other conductors. However, for usual spacings and conductors, sufficiently accurate results are obtained by assuming that the charge per unit length of the conductor is the same in every part of the transposition cycle. With this assumption the voltage between a pair of conductors is different for each part of transposition cycle. An average value of voltage is used to calculate capacitance.

From Eqs. (2.82 to 2.84) the average V_{ab} is

$$V_{ab} = \frac{1}{6\pi\varepsilon}\left[q_a \,\text{In}\, \frac{D_{12}D_{23}D_{31}}{r^3} + q_a \,\text{In}\, \frac{r^3}{D_{12}D_{23}D_{31}} + q_c \,\text{In}\, \frac{D_{12}D_{23}D_{31}}{D_{12}D_{23}D_{31}}\right]$$

$$= \frac{1}{2\pi\varepsilon}\left[q_a \,\text{In}\, \frac{D_{eq}}{r} + q_b \,\text{In}\, \frac{r}{D_{eq}}\right] \tag{2.85}$$

where $D_{eq} = \sqrt[3]{D_{12}D_{23}D_{31}}$

Similarly $$V_{ac} = \frac{1}{2\pi\varepsilon}\left[q_a \ln\frac{D_{eq}}{r} + q_c \ln\frac{r}{D_{eq}}\right] \tag{2.86}$$

Adding Eqs. (2.85 and 2.86) and using Eq. (2.78)

$$V_{ab} + V_{ac} = 3\,V_{an}$$

$$= \frac{1}{2\pi\varepsilon}\left[2q_a \ln\frac{D_{eq}}{r} + q_b \ln\frac{r}{D_{eq}} + q_c \ln\frac{r}{D_{eq}}\right] \tag{2.87}$$

But $q_b + q_c = -\,q_a$ in a balanced 3-phase circuit. Then

$$3\,V_{an} = \frac{3}{2\pi\varepsilon} q_a \ln\frac{D_{eq}}{r} \tag{2.88}$$

$$C_n = \frac{q_a}{V_{an}} = \frac{2\pi\varepsilon}{\ln\left(D_{eq}/r\right)} \text{ F/m} \tag{2.89a}$$

$$= \frac{0.02412}{\log\left(D_{eq}/r\right)}\ \mu\text{F/km} \tag{2.89b}$$

Eq. (2.89) is similar to Eq. (2.80) except that when the spacings are asymmetrical, the equivalent spacing D_{eq} is to be used in place of equilateral spacing D. The charging current is given by Eq. (2.81).

Example 2.10

A 220 kV, 50 Hz, 200 km long 3-phase line has its conductors on the corners of a triangle with sides 6 m, 6 m and 12 m. The conductor radius is 1.81 cm. Find the capacitance per phase per km, capacitive reactance per phase, charging current and total charging megavolt-amperes.

[RTM Nagpur Uni. 2011]

Solution

$$D_{eq} = \sqrt[3]{6\times6\times12} = 7.56 \text{ m}$$

$$C_n = \frac{0.02412}{\log(0.756\times100/1.81)} = 9.2\times10^{-3}\ \mu\text{F/phase/km}$$

Total line capacitance $= 9.2\times10^{-3}\times200 = 1.84\ \mu\text{F/phase}$

$$X_c = \frac{1}{2p\times50\times1.84\times10^{-6}} = 1730 \text{ ohms/phase}$$

Charging current $= 2\pi\times50\times1.84\times10^{-6}\times\frac{220}{\sqrt{3}}\times10^{3} = 73.42$ A/phase

Charging megavolt amperes $= \frac{\sqrt{3}\times220\times10^{3}\times73.42}{10^{6}} = 27.97$ Mvar

2.17 CAPACITANCE OF BUNDLED CONDUCTOR LINES

Figure 2.17 shows a 3-phase bundled conductor line with two sub-conductors per phase. The formula for capacitance can be derived in the same way as in the single conductor line except that the charges on all the six sub-conductors are to be considered. Since the two sub-conductors of each phase are in parallel, the charge per bundle can be assumed to be equally divided between the sub-conductors of a bundle. Since $D_{12} >> d$, the distance D_{12} can be used in place of $(D_{12} - d)$ and $(D_{12} + d)$ with similar approximations in the equations of voltage. Then

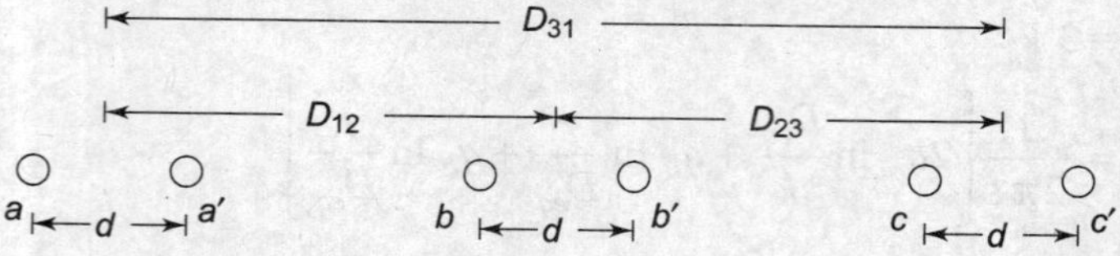

Fig. 2.17 Three-phase bundled conductor line with two sub-conductors per phase

$$V_{ab} = \frac{1}{2\pi\varepsilon}\left[\frac{q_a}{2}\left(\mathrm{In}\frac{D_{12}}{r} + \mathrm{In}\frac{D_{12}}{d}\right) + \frac{q_b}{2}\left(\mathrm{In}\frac{r}{D_{12}} + \mathrm{In}\frac{d}{D_{12}}\right)\right.$$
$$\left. + \frac{q_c}{2}\left(\mathrm{In}\frac{D_{23}}{D_{31}} + \mathrm{In}\frac{D_{23}}{D_{31}}\right)\right] \tag{2.90}$$

or
$$V_{ab} = \frac{1}{2\pi\varepsilon}\left[q_a\ \mathrm{In}\frac{D_{12}}{\sqrt{rd}} + q_b\ \mathrm{In}\frac{\sqrt{rd}}{D_{12}} + q_c\ \mathrm{In}\frac{D_{23}}{D_{31}}\right] \tag{2.91}$$

Eq. (2.91) is the same as Eq. (2.82) except that $\sqrt{rd}$ has replaced r. Therefore, assuming the line to be transposed and proceeding in the same way as in section (2.16) we can find the expression for C_n. The result is

$$C_n = \frac{2\pi\varepsilon}{\mathrm{In}\left(D_{eq}/\sqrt{rd}\right)} \text{ F/m} \tag{2.92a}$$

$$= \frac{0.02412}{\log\left(D_{eq}/\sqrt{rd}\right)} \ \mu\text{F/km} \tag{2.92b}$$

The term $\sqrt{rd}$ is the same as D_s^b for two-conductor bundle (section 2.11) except that outside conductor radius r has replaced D_s. This leads to a very important conclusion that a modified GMD method can be used for capacitance calculations. The modification is that the outside conductor radius is to be used in place of geometric mean radius of the conductor. This change is very logical because in the expressions for capacitance, derived in sections (2.13–2.16) the outside radius r (and not r') appears in the final equations. The physical explanation for this change is the same as given in section (2.14). The capacitance for triplex and quadruplex bundled conductor lines can also be calculated by following this modified GMD method. Thus, for a two-conductor bundle, self GMD of bundle for calculating capacitance, i.e.,

$$D_{sc}^b = \sqrt{rd} \tag{2.93}$$

For a three-conductor bundle, $D_{sc}^b = \sqrt[9]{(r \times d \times d)^3} = \sqrt[3]{rd^2}$ (2.94)

For a four-conductor bundle, $D_{sc}^b = \sqrt[16]{\left(r \times d \times d \times \sqrt{2}d\right)^4} = 1.09\sqrt[4]{rd^3}$ (2.95)

Example 2.11

A 400 kV 3 phase single circuit bundled conductor line with two sub-conductors per phase has a configuration shown in Fig. 2.17. The centre to centre distance between adjacent phases is 12 m and distance between sub-conductors of a phase is 45 cm. The radius of each sub-conductor is 1.6 cm. Find the capacitance per phase per km.

Solution

$$D^b_{sc} = \sqrt{1.6 \times 45} = 8.485 \text{ cm} = 0.0845 \text{ m}$$

$$D_{eq} = \sqrt[3]{12 \times 12 \times 24} = 15.119 \text{ m}$$

$$C_n = \frac{0.02412}{\log(15.119)/0.08485} = 0.0107\ \mu\text{F/km}$$

2.18 DOUBLE CIRCUIT THREE PHASE LINES

Throughout the above analysis, similarity can be found between the expressions for inductance and capacitance. The capacitance of bundled conductor lines is found by modified *GMD* method. It can be shown that the modified *GMD* method holds good for finding the capacitance of transposed double circuit 3-phase lines with equilateral spacing (conductors at the vertices of regular hexagon) and with flat vertical spacing (Fig. 2.10). It is reasonable to assume that the modified *GMD* method can be used to find the capacitance of a line with any configuration intermediate between these two configurations.

Example 2.12

A 3-phase double circuit line has the conductors at the vertices of a hexagon as shown in Fig. 2.18. (a) Find the formula for calculating capacitance per phase per km in terms of side D and conductor radius r. (b) If $D = 3.5$ m and $r = 1.09$ cm find the capacitance per phase per km and capacitance per conductor per km (c) If the line voltage is 132 kV and line length 100 km, find the charging current and charging megavolt amperes.

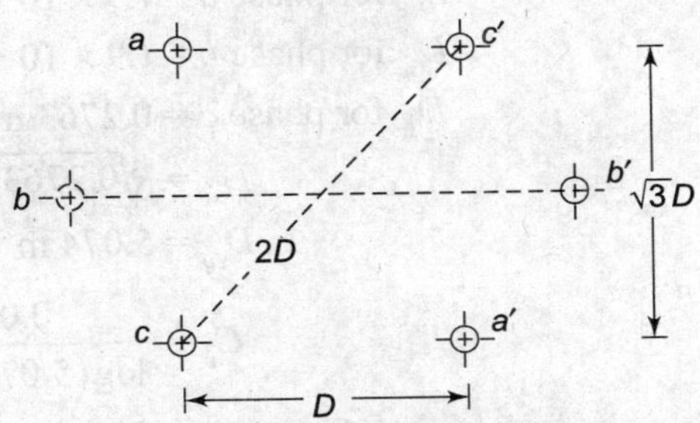

Fig. 2.18

Solution

(a) The distance between different conductors can be found in terms of D and are as shown in Fig. 2.18.

$$D_{eq} = \sqrt[3]{D_{ab} D_{bc} D_{ca}}$$

where D_{ab} = Geometric mean distance between phases a and b

$$= \sqrt[4]{D \times \sqrt{3}D \times D \times \sqrt{3}D} = 3^{1/4} D$$

$$D_{bc} = \sqrt[4]{D \times \sqrt{3}D \times D \times \sqrt{3}D} = 3^{1/4} D$$

$$D_{ca} = \sqrt[4]{D \times \sqrt{3}D \times D \times \sqrt{3}D} = 3^{1/4} D$$

$$D_{eq} = [(3^{1/4} D)(3^{1/4} D)(3^{1/4} D)]^{1/3} = 3^{1/4} D$$

$$D_{sc} = \sqrt{r \times 2D}$$

$$C_n = \frac{2\pi\varepsilon}{\text{In}\left[\left(3^{1/4}\, D\right)/\sqrt{2r\, D}\right]} = \frac{4\pi\varepsilon}{\text{In}\left(\sqrt{3}D/2r\right)} \text{ H/phase/m}$$

$$= (0.04824)/\log\ \frac{\sqrt{3}\, D}{2r}\ \mu\text{F/phase/km}$$

(b) $C_n = (0.04824)/\log\ [\sqrt{3} \times 350/(2 \times 1.09)] = 0.0197\ \mu$F/phase/km

Capacitance per conductor per km = $C_n/2 = 0.00985\ \mu$F/conductor/km.

(c) $I_c = \omega\, C_n\, V_n = 2\pi \times 50 \times .0197 \times 10^{-6} \times \dfrac{132 \times 10^3}{\sqrt{3}} = 0.472$ A/km

Charging current for 100 km line = 0.472 × 100 = 47.2 A

Charging megavolt amperes $= \dfrac{\sqrt{3} \times 132 \times 10^3 \times 47.2}{10^6} = 10.79$ Mvar

Example 2.13

Find the capacitance per phase per km for the double circuit three-phase line of example 2.8.

Solution

D_{sc} for phase $a = (.9 \times 10^{-2} \times 8.485)^{0.5} = 0.2763$ m

D_{sc} for phase $b = (.9 \times 10^{-2} \times 7)^{0.5} = 0.251$ m

D_{sc} for phase $c = 0.2763$ m

$$D_{sc} = \sqrt[3]{0.2763 \times 0.251 \times 0.2763} = 0.2676 \text{ m}$$

$$D_{eq} = 5.074 \text{ m}$$

$$C_n = \frac{0.02412}{\log(5.074/0.2676)} = 0.018875\ \mu\text{F/phase/km}$$

2.19 EFFECT OF GROUND ON LINE CAPACITANCE, METHOD OF IMAGES

The presence of ground alters the electric field of a line and hence affects the line capacitance. Ground can be assumed to be a perfect conductor in the form of a horizontal plane of infinite extent. Evidently the electric field of charged conductors above ground is not the same as it would be if the equipotential surface of ground were not present.

The effect of ground can be modelled by the method of images. Although not exact, experience has shown that this method is based on realistic and practical approximations. This well-known method makes use of the fact that the effect of ground on the charges and potentials of the conductors can be replaced by the effect of images of conductors below the surface of the ground. Any one conductor and its image conductor are located at equal distances, but in opposite directions from the ground surface. Their charges (or charging currents) and potentials are equal but of opposite sign. It is evident that if an image conductor is located for each overhead conductor, the electric flux between the overhead conductors and their images is perpendicular to the plane which replaces the ground (as shown in Fig. 2.19) and this plane forms an equipotential surface. The electric flux above the plane is the same as it is when the ground is present instead of the image conductors.

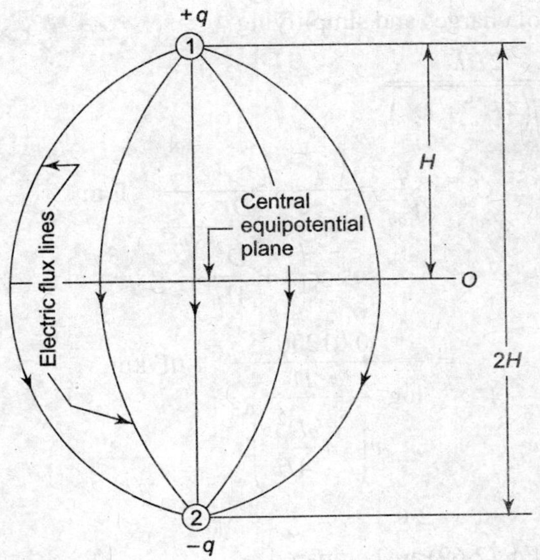

Fig. 2.19 Ground plane and image charge of one line conductor

(a) Single Phase Line

The conductors of a single phase line along with image conductors are shown in Fig. 2.20. The conductor radius is r. Using Eq. (2.66) the equation for voltage drop V_{ab} is

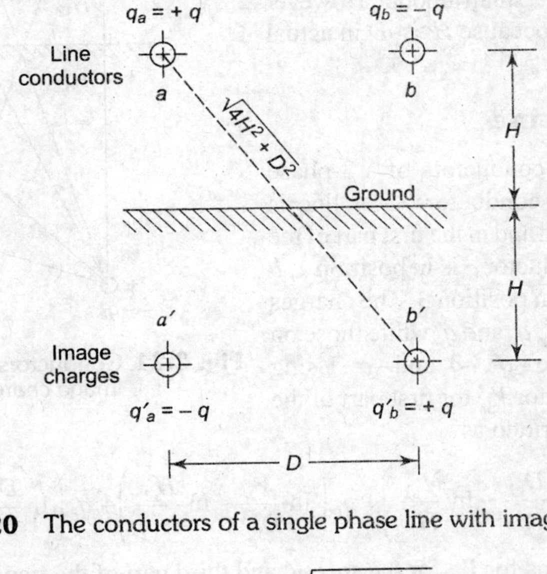

Fig. 2.20 The conductors of a single phase line with image charges

$$V_{ab} = \frac{1}{2\pi\varepsilon}\left[q_a \ln\frac{D}{r} + q_b \ln\frac{r}{D} + q'_a \ln\frac{\sqrt{4H^2+D^2}}{2H} + q'_b \ln\frac{2H}{\sqrt{(4H^2+D^2)}}\right] \quad (2.96)$$

Substituting the values of charges and simplifying

$$V_{ab} = \frac{q}{\pi\varepsilon} \ln \frac{2HD}{r\sqrt{(4H^2 + D^2)}} \tag{2.97}$$

Then

$$C_{ab} = \frac{q}{V_{ab}} = \frac{\pi\varepsilon}{\ln \dfrac{D}{r\left(1+\dfrac{D^2}{4H}\right)^{0.5}}} \quad \text{F/m} \tag{2.98a}$$

$$= \frac{0.01206}{\log \dfrac{D}{r\left(1+\dfrac{D^2}{4H}\right)^{0.5}}} \quad \mu\text{F/km} \tag{2.98b}$$

and $C_n = 2\,C_{ab}$

When Eq. (2.98) and Eq. (2.69) are compared, it is seen that because of the presence of ground, r has been replaced by $r\left(1+\dfrac{D^2}{4H^2}\right)^{0.5}$. When the presence of ground is taken into account, the line capacitance increases by a small amount. However the effect is very small because $H >> D$ in actual practice.

(b) Three Phase Line

Figure 2.21 shows the conductors of a 3-phase line along with image conductors. The line is assumed to be transposed and in the first part of the transposition cycle conductor a is in position 1, b is in position 2 and c is in position 3. The charges on line conductors are q_a, q_b and q_c while those on the image conductors are $-q_a$, $-q_b$ and $-q_c$. Using Eq. (2.66), the equation for V_{ab} for first part of the transposition cycle is written as

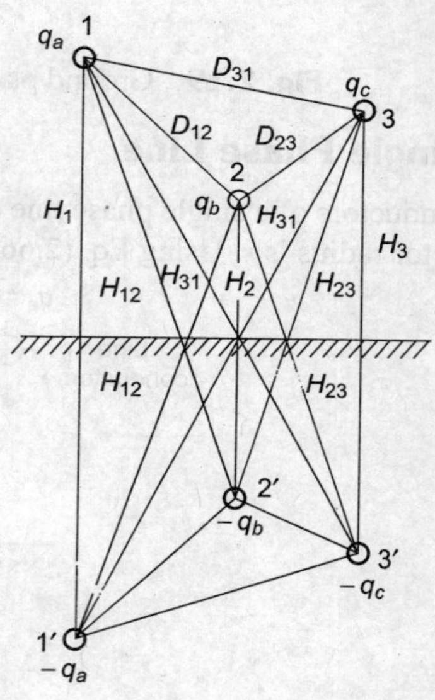

Fig. 2.21 Conductors of a 3-phase line with image charges

$$V_{ab} = \frac{1}{2\pi\varepsilon}\left[q_a\left(\ln\frac{D_{12}}{r} - \ln\frac{H_{12}}{H_1}\right) + q_b\left(\ln\frac{r}{D_{12}} - \ln\frac{H_2}{H_{12}}\right) + q_c\left(\ln\frac{D_{23}}{D_{31}} - \ln\frac{H_{23}}{H_{31}}\right)\right] \tag{2.99}$$

Similarly the equations for V_{ab} for the second and third part of the transposition cycle can be written and the capacitance C_n can be determined (as in section 2.16). The result is

$$C_n = \frac{2\pi\varepsilon}{\ln\dfrac{D_{eq}}{r} - \ln\left(\sqrt[3]{H_{12}H_{23}H_{31}} \,/\, \sqrt[3]{H_1H_2H_3}\right)} \quad \text{F/m} \tag{2.100a}$$

$$= \frac{0.02412}{\log \frac{D_{eq}}{r} - \log\left(\sqrt[3]{H_{12}H_{23}H_{31}} / \sqrt[3]{H_1 H_2 H_3}\right)} \ \mu\text{F/km} \qquad (2.100b)$$

A comparison of the above equation with Eq. (2.89) shows that the presence of ground increases the line capacitance by a small amount. However, the effect is very very small because the heights of conductors are large as compared to the distance between them.

Example 2.14

The conductors of the single phase line of example 2.9 are 5.5 m above ground. Find the capacitance of each conductor to neutral taking the presence of ground into account.

Solution

$$C_n = 2C_{ab} = \frac{2 \times 0.01206}{\log \dfrac{125}{1.213 \times 0.5\left(1 + 1.25^2 / 4 \times 5.5^2\right)^{0.5}}} \ \mu\text{F/km}$$

$$= 0.010436 \ \mu\text{F/km}$$

For 10 km length, capacitance of each conductor

$$= 0.010436 \times 10 = 0.10436 \ \mu\text{F}$$

A comparison with the result of example (2.9) shows that the presence of ground increases the capacitance by about 0.15 per cent.

Example 2.15

Find the resistance per phase per km of otter conductor 6/1/4.22 mm diameter at 20°C and at 50°C. Resistivity of aluminium is 2.826×10^{-8} Ω-m at 20°C. Assume that current is carried only by the aluminium portion of conductor. The increase in length due to stranding is 1.5%. Temperature coefficient of resistance is 0.004 per °C.

Solution

$$\text{Area of cross-section} = \frac{\pi}{4} \times 4.22^2 \times 6$$

$$= 83.92 \text{ mm}^2 = 83.92 \times 10^{-6} \text{ m}^2$$

$$\text{Resistance at 20°C} = \frac{2.826 \times 10^{-8} \times 1000 \times 1.015}{83.92 \times 10^{-6}}$$

$$= 0.3418 \ \Omega/\text{km}$$

$$\text{Resistance at 50°C} = 0.3418\,(1 + 0.004 \times 30)$$

$$= 0.3828 \ \Omega/\text{km}$$

Example 2.16

A 3-phase 132 kV, 50 Hz overhead line has ACSR conductors of equivalent copper area 1.5 cm² and effective diameter 39.8 mm, spaced equilaterally 8 m apart (a) Find line parameters, (b) Find charging current and charging MVA. Resistivity of copper is 1.73×10^{-6} ohm-cm.

Solution

(a)
$$\rho = 1.73 \times 10^{-6}\ \Omega\text{-cm} = 1.73 \times 10^{-8}\ \Omega\text{-m}$$
$$l = 1\text{ km} = 1000\text{ m}$$
$$\text{Copper area} = 1.5\text{ cm}^2 = 1.5 \times 10^{-4}\text{ m}^2$$
$$R = \frac{\rho l}{a} = \frac{1.73\times10^{-8}\times1000}{1.5\times10^{-4}} = 0.1153\ \Omega/\text{km}$$
$$r = 0.5 \times 39.8 \times 10^{-3} = 1.99 \times 10^{-2}\text{ m}$$
$$L = 0.4605 \log\frac{8}{0.7788\times1.99\times10^{-2}} = 1.25\text{ mH/km}$$
$$C = \frac{0.02412}{\log\dfrac{8}{1.99\times10^{-2}}} = 0.00926\ \mu\text{F/km}$$

(b)
$$\text{Charging current} = 2\pi \times 50 \times 0.00926 \times 10^{-6} \times \frac{132}{\sqrt{3}} \times 10^3$$
$$= 0.2217\text{ A/km/phase}$$
$$\text{Charging MVA} = \frac{\sqrt{3}\times132\times10^3\times0.2217}{10^6} = 0.0507\text{ MVAr/km}$$

Example 2.17

A symmetrical 3-phase double circuit line is arranged in horizontal plane as shown in Fig. 2.22. Assume balanced loads equally distributed in two circuits. The radius of each conductor is r. (a) Derive an expression for inductance (b). Find inductance per km per phase if $r = 0.9 \times 10^{-2}$ m and $d = 3.5$ m.

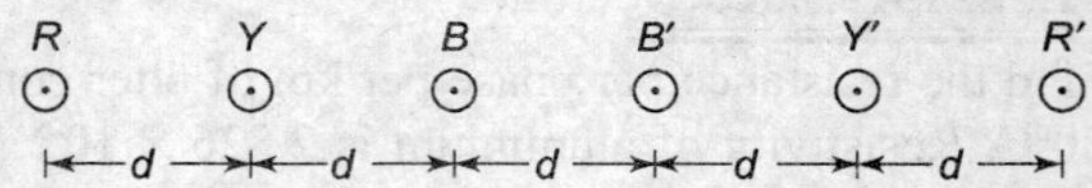

Fig. 2.22

Solution

(a) $r' = 0.7788\, r$

Distance between R and R' $= 5d$

GMR of conductors of phase $R = (r' \times 5d)^{0.5}$

Distance between Y and Y' $= 3d$

GMR of conductors of phase $Y = (r' \times 3d)^{0.5}$

Distance between B and B' $= d$

GMR of conductors of phase $B = (r'd)^{0.5}$

$$D_s = [(5\,r'd)\,(3\,r'd)\,(r'd)]^{1/6} = (15)^{1/6}\,(r'd)^{0.5}$$
$$= (15)^{1/6}\,(0.7788\,rd)^{0.5} = 1.3859\,(rd)^{0.5}$$

Distance between R and Y $= d$

Distance between R and Y' $= 4d$

$$D_{RY} = \sqrt[4]{[(d)(4d)]^2} = 2d$$

Distance between Y and B $= d$

Distance between Y and B' $= 2d$

$$D_{YB} = \sqrt[4]{[(d)(2d)]^2} = 1.414\, d$$

Distance between B and R $= 2d$

Distance between B and R' $= 3d$

$$D_{BR} = \sqrt[4]{[(2d)(3d)]^2} = 2.449\, d$$

$$\text{GMD} = [(2d)\,(1.414d)\,(2.449d)]^{1/3}$$

$$= 1.906\, d$$

$$L = 0.4605 \log \frac{1.906d}{1.3859(rd)^{0.5}}$$

$$= 0.4605 \log [1.375\left(\sqrt{d/r}\right)]\ \text{mH/km}$$

(b)
$$L = 0.4605 \log \left[1.375\left(\frac{3.5}{0.9\times 10^{-2}}\right)^{0.5}\right]$$

$$= 0.66\ \text{mH/km}$$

Example 2.18

A 765 kV single circuit line has four sub-conductors per phase, each sub-conductor having a diameter of 26.88 mm. Spacing between sub-conductors is 45 cm, interphase spacing is 15.25 m. The configuration is horizontal. Find inductance per phase per km.

Solution

$$r = 26.88/2\ \text{mm} = 1.344\ \text{cm}$$

$$r' = 0.7788 \times 1.344 = 1.047\ \text{cm}$$

$$\text{GMR} = 1.09\,(1.047 \times 45^3)^{1/4} = 19.16\ \text{cm}$$

$$\text{GMD} = (15.25 \times 15.25 \times 30.5)^{1/3} = 19.21\ \text{m}$$

$$L = 0.4605 \log (19.21 \times 100/19.16)$$

$$= 0.9215\ \text{mH/km}$$

Example 2.19

A 3-phase 3-wire overhead line consists of 2.5 cm diameter conductors in horizontal configuration. The line is supplying a balanced load (a) Find the inductance of each phase conductor per km length (b) Why are the inductances of the three phases different? What is the significance of imaginary terms in the expressions? Assume that the line is not transposed. Interphase spacing is 3 m.

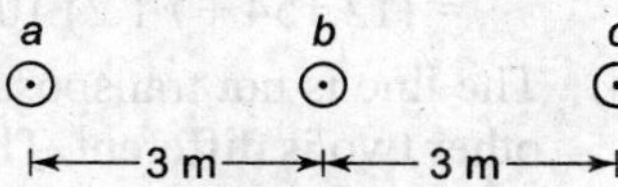

Fig. 2.23

Solution

(a) Since currents are balanced $I_a = I\angle 0$, $I_b = I\angle -120°$, $I_c = I\angle 120°$. The configuration is shown in Fig. 2.23. Using Eq. (2.51)

$$\Psi_a = 2\times10^{-7}\left[I_a \ \text{In}\frac{1}{r'} + I_b \ \text{In}\frac{1}{D_{ab}} + I_c \ \text{In}\frac{1}{D_{ac}}\right]$$

$$r' = 0.7788 \times \frac{2.5}{2} \times 10^{-2} = 0.9735 \times 10^{-2} \text{ m}$$

$$\Psi_a = 2\times10^{-7}\left[I \ \text{In}\frac{1}{0.9735\times10^{-2}} + (I\angle-120)\,\text{In}\frac{1}{3} + (I\angle-120°)\,\text{In}\frac{1}{6}\right]$$

$$= 2\times10^{-7}\, I\,[4.632 + (-1.0986\angle-120°) + (-1.7918\angle120°)]$$

$$L_a = \frac{\Psi_a}{I_a} = \frac{\Psi_a}{I} = 2\times10^{-7}\,[4.632 + .5493 + j\,.9514 + .8959 - j\,1.5517] \text{ H/m}$$

or

$$L_a = (12.154 - j\,1.2)\,10^{-7} \text{ H/m}$$

$$= 1.2154 - j\,0.12 \text{ mH/km}$$

$$\Psi_b = 2\times10^{-7}\left[I_a \ \text{In}\frac{1}{D_{ab}} + I_b \ \text{In}\frac{1}{r'} + I_c \ \text{In}\frac{1}{D_{bc}}\right]$$

$$= 2\times10^{-7}\left[I\angle0\ \text{In}\frac{1}{D_{ab}} + I\angle120°\,.\,\text{In}\frac{1}{r'} + I\angle120°\,.\,\text{In}\frac{1}{D_{bc}}\right]$$

$$= 2\times10^{-7}\left[I\angle0\ \text{In}\frac{1}{3} + I\angle-120°\ \text{In}\frac{1}{0.9735\times10^{-2}} + I\angle120°\ \text{In}\frac{1}{3}\right]$$

$$L_b = \frac{\Psi_b}{I_b} = \frac{\Psi_b}{I\angle-120°} = 2\times10^{-7}\left[1\angle120°\ \text{In}\frac{1}{3} + \text{In}\frac{1}{0.9735\times10^{-2}} + 1\angle240°\ \text{In}\frac{1}{3}\right]$$

$$= 2\times10^{-7}\,[0.5493 - j\,0.9514 + 4.632 + 0.5493 + j\,0.9514]$$

$$= 11.46\times10^{-7} \text{ H/m} = 1.146 \text{ mH/km}$$

$$\Psi_c = 2\times10^{-7}\left[I_a \ \text{In}\frac{1}{D_{ac}} + I_b \ \text{In}\frac{1}{D_{bc}} + I_c \ \text{In}\frac{1}{r'}\right]$$

$$= 2\times10^{-7}\left[I\angle0\ \text{In}\frac{1}{6} + I\angle-120°\ \text{In}\frac{1}{3} + I\angle120°\ \text{In}\frac{1}{0.9735\times10^{-2}}\right]$$

$$L_c = \frac{\Psi_c}{I_c} = \frac{\Psi_c}{I\angle120°} = 2\times10^{-7}\,[0.8959 + j\,1.5517 + .5493 - j\,0.9514 + 4.632]$$

$$= (12.154 + j\,1.2)\,10^{-7} \text{ H/m} = 1.2154 + j\,0.12 \text{ mH/km}$$

(b) The line is not transposed. The relative position of each conductor with respect to the other two is different. The flux linkages of each conductor are due to currents in all the three conductors. Therefore flux linkages of each conductor are different. Inductance is flux linkages per ampere. Therefore, inductances of the three lines are different.

The inductive voltage drop is equal to $j\,\omega\,\psi$. Since the flux linkages are different, the inductive voltage drop in each line is different. Moreover, the inductive voltage drop in each phase does not lead the current in that phase by 90° but by an angle which is either slightly less or more than 90°. This leads to imaginary terms in the expression for inductances and the inductive voltage drop. These imaginary terms represent the transfer of power between phases due to mutual effect and they add upto zero.

Example 2.20

Find the capacitance per phase per km of the line of example 2.18.

Solution

$$r = 1.344 \text{ cm}$$

GMR for calculating capacitance of a bundled conductor line with 4 sub-conductors per phase is given by Eq. (2.95)

$$\text{GMR} = 1.09\,(1.344 \times 45^3)^{1/4} = 20.39 \text{ cm}$$

$$\text{GMD} = 19.21 \text{ m}$$

$$C = \frac{0.02412}{\log(19.21 \times 100/20.39)} = 0.0122\ \mu\text{F/km}$$

Example 2.21

Calculate the inductance per km per phase and inductive reactance per phase per km for a 3 phase, 50 Hz, bundled conductor line shown in Fig. 2.24. Each subconductor has a diameter of 25 mm and subconductor spacing is 0.3 m. Assume that each phase group shares total current and charge equally and the line is completely transposed. *(AMIE 2001)*

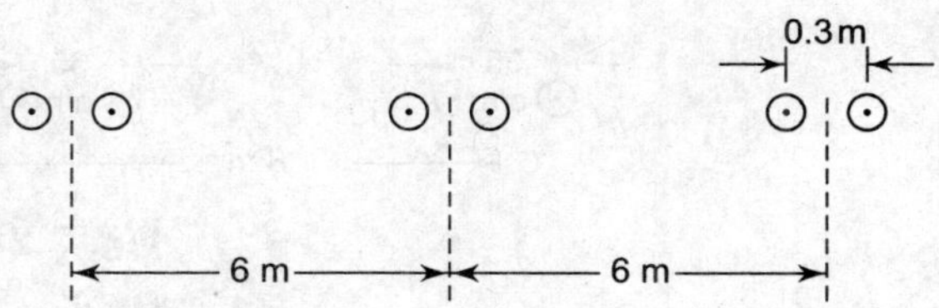

Fig. 2.24

Solution

$$r' = 0.7788 \times \frac{2.5}{2} = 0.9735 \text{ cm}$$

$$\text{GMR} = (r'.d.r'.d)^{0.25} = (r'.d)^{0.5} = (0.9735 \times 30)^{0.5}$$

$$= 5.404 \text{ cm}$$

$$D_{ab} = D_{bc} = [6(6 + 0.30)\ 6(6 - 0.30)^{0.25} = 5.996 \text{ m}$$

$$D_{ac} = [12\ (12 + 0.3)12\ (12 - 0.3)]^{0.25} = 11.998 \text{ m}$$

$$D_{eq} = [(5.996)\ (5.996)\ (11.998)]^{1/3} = 7.556 \text{ m}$$

$$L = 0.4605 \log\,(D_{eq}/\text{GMR})$$

$$= 0.4605 \log\,(7.556 \times 100/5.404)$$

$$= 0.988 \text{ mH/phase/km}$$

$$X_L = 2\pi f L = 2\pi\,(50)\,(0.988) \times 10^{-3}$$

$$= 0.31 \text{ ohm/phase/km}$$

Example 2.22

A three phase 50 Hz OH line has regularly transposed conductors equilaterally placed at 4m spacing. The capacitance in 0.01 μF/km. Recalculate the capacitance per km to neutral when conductors are in the same horizontal plane with successive spacing of 4m and are regularly transposed. *[RTM Nagpur University 2010]*

Solution

$$C = \frac{0.02412}{\log\left(D_{eq}/r\right)}$$

$$0.01 = \frac{0.02412}{\log(4\times100/r)} \quad \text{or} \quad r = 1.549 \text{ cm}$$

In new configuration, $D_{ab} = D_{bc} = 4\text{m}$, $D_{ca} = 8$ m

$$D_{eq} = (4 \times 4 \times 8)^{1/3} = 5.0397 \text{ m}$$

$$C = \frac{0.02412}{\log(5.0397\times100/1.549)} = 0.0096\ \mu\text{F/km}$$

Example 2.23

A 400 kV 3 phase bundled conductor line has configuration as shown in Fig. 2.25. The radius of each such conductor is 1.4 cm. Find inductance per phase per km of line.

[RTM Nagpur University 2011]

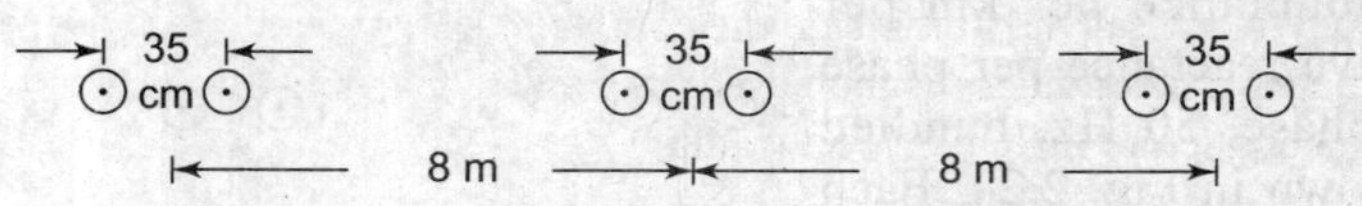

Fig. 2.25

Solution

$$r' = 0.7788 \times 1.4 = 1.09 \text{ cm}$$

$$\text{GMR} = (r'd\,r'd)^{0.25} = (r'd)^{0.5}$$

$$= (1.09 \times 35)^{0.5} = 6.176 \text{ cm}$$

$$D_{eq} = (8 \times 8 \times 16)^{1/3} = 10.079 \text{ m}$$

$$L = 0.4605 \log(D_{eq}/\text{GMR})$$

$$= 0.4605 \log(10.079 \times 100/\ 6.176)$$

$$= 1.0189 \text{ mH/km}$$

SUMMARY

2.1. The resistance of a transmission/distribution line conductor is higher than dc resistance due to skin effect and temperature rise.

2.2. Inductance is flux linkage per ampere of current.

2.3. The flux linkages of a line conductor can be divided into two parts i.e due to flux inside the conductor and flux outside the conductor.

2.4. Inductance of a line conductor $= 0.4605 \log \dfrac{\text{GMD}}{\text{GMR}}$ mH/km.

2.5. For a 3 phase line with equilateral spacing

GMD = line spacing

2.6. GMR of a solid conductor = 0.7788 (radius)

2.7. For a 3 phase line with spacings D_{12}, D_{23} and D_{31},

$$GMD = (D_{12}\, D_{23}\, D_{31})^{1/3}$$

2.8. Inductance of a bundled conductor line is less than that of a similar line with one conductor per phase.

2.9. For a 3 phase line with equilateral spacing

$$\text{Capacitance per phase} = \frac{0.02412}{\log(D/r)}\ \mu\text{F/km}$$

2.10. The presence of ground increases the line capacitance slightly.

SELF TEST QUESTIONS

A. State whether the following are true (T) or false (F).

2.1 At 50 Hz the increase in resistance due to skin effect is about 20%.
2.2 Skin effect is more pronounced at high frequencies.
2.3 Magnetic field lines are concentric with conductor.
2.4 As frequency increases the flux inside the conductor decreases.
2.5 If the diameter of a conductor is increased, its inductance is increased.
2.6 For a solid conductor GMR = 0.7788 (radius).
2.7 Self GMD is same as GMR.
2.8 For a line with interphase spacings D_{12}, D_{23}, D_{31}

$$GMD = (D_{12}\, D_{23}\, D_{31})^{1/3}$$

2.9 Loop inductance of a single phase overhead line is about 1 mH/km.
2.10 Concept of GMD and GMR can be used to calculate capacitance of line.

B. Fill in the blanks with suitable words.

2.11 If r is radius and d is bundle spacing, GMR of a 3 sub-conductor bundle = ______
2.12 The constant $1/\alpha_0$ for aluminium = ______
2.13 Power transmission capacity of a line depends mainly on line ______.
2.14 For a solid conductor of radius r, GMR = ______
2.15 For a line with horizontal spacing D between adjacent conductors GMD = ______
2.16 The use of bundled conductor ______ the GMR.
2.17 Presence of ground ______ the line capacitance slightly.
2.18 For a 3 phase line with conductor radius r and equilateral spacing D, capacitance per phase = ______.
2.19 Capacitance of a 3 phase line is about ______ μF/km/phase.
2.20 The increase in length of conductor due to stranding is about ______%.

ANSWERS

2.1 F	2.2 T	2.3 T	2.4 T	2.5 F	2.6 T
2.7 T	2.8 T	2.9 F	2.10 T	2.11 $\sqrt[3]{(0.7788r)(d)^2}$	
2.12 228	2.13 inductance		2.14 $0.7788\,r$	2.15 $(2)^{1/3}D$	2.16 increases
2.17 increases	2.18 $\dfrac{0.02412}{\log(D/r)}$		2.19 0.01	2.20 1.5	

JOB INTERVIEW QUESTIONS

2.1 What is the effect of using bundled conductors on line inductance?
2.2 What is meant by terms GMD and GMR?
2.3 How is GMD calculated?
2.4 What is the procedure to find inductance of a double circuit line?
2.5 How is line capacitance calculated?
2.6 What is charging current? What is its effect?
2.7 What is skin effect?

REVIEW QUESTIONS

2.1 Give reasons:
   (a) The ac resistance of a conductor is more than the dc resistance.
   (b) The resistance of a conductor is more for high frequency currents than for low frequency currents.
   (c) The resistance of a stranded conductor is higher than the resistance of a solid conductor of the same length and diameter.
   (d) The inductance of a conductor is lesser for high frequency currents than for low frequency currents.

2.2 Discuss with reasons:
   (a) Bundled conductor lines have lower inductance than single conductor lines.
   (b) In double circuit lines the configurations of conductors is so selected that the distance between the conductors of the same phase is maximum.
   (c) The inductances of the three phases of a transmission line are generally not equal.
   (d) The geometric mean radius of a stranded conductor is lower than that of a solid conductor of the same overall diameter.

2.3 Explain with reasons:
   (a) It is necessary to use geometric means radius in the formula for the inductance of a single conductor whereas the formula for the capacitance uses the actual outside radius.
   (b) The presence of ground on the capacitance can be taken into account by the method of images.
   (c) The effect of ground on the capacitance of a line is negligible.
   (d) A charging current exists in a line even when it is open circuited at the load end.
   (e) The effect of line capacitance is usually small for lines with short lengths but appreciable for long lines.

2.4 What factors must be taken into account while calculating the resistance of overhead line conductors? How are these factors accounted for?

2.5 Derive the expression for calculating the internal and external flux linkages for a conductor carrying current. Use these expressions to derive the equation for the inductance of a single phase line.

2.6 Discuss the concept of geometric mean distance. How is this concept used to find the inductance of composite conductor line?

2.7 How can the inductance of a bundled conductor line be calculated? Derive expressions for geometric mean radii of duplex, triplex and quadruplex arrangement.

2.8 What is equivalent spacing of a 3-phase line? What is its significance?

2.9 Derive an expression for the inductance of a double circuit line whose conductors are placed at the vertices of a regular hexagon.

2.10 Derive an expression for the capacitance between conductors of a single phase line. Deduce the expression for line to neutral capacitance. Discuss the approximation involved in deriving the above expressions.

2.11 What is modified *GMD* method? How is it used to find the capacitance of (a) 3-phase single circuit bundled conductor line (b) 3-phase double circuit line?

2.12 What is charging current? How can it be calculated? What is its effect?

2.13 What is the method of images? How can it be used to take into account the presence of ground in calculating the capacitance of a single phase line?

2.14 What is skin effect? How does it influence the resistance, inductance and capacitance of a conductor?

PROBLEMS

2.1 Calculate the loop inductance per km of a single phase overhead line consisting of two conductors each 1 cm dia and 1.2 m apart. [2.292 mH]

2.2 A 3-phase transposed line has conductors of diameter 2 cm and spaced at distance of 3.65, 5.5 and 8.2 m between the centres. Calculate the inductance per phase per km of line length. [1.31 mH]

2.3 A 3-phase single circuit 132 kV overhead line has conductors of diameter 1.8 cm each. The spacing between conductors is 4 m, 6 m and 9 m. Find the inductance per phase per km. [1.35 mH]

2.4 A 3-phase 50 km long single circuit 66 kV transposed overhead line has horizontal spacing with 3 m between adjacent conductors and 6 m between outer conductors. The conductor diameter is 2 cm. Find the Inductance per phase. [61.85 mH]

2.5 A 3-phase single circuit bundled conductor line with two sub-conductors per phase has horizontal spacing with 6.1 m between the centre lines of adjacent phases. The distance between the sub-conductors of each phase is 30.5 cm and each sub-conductor has a diameter of 2.54 cm. Find the inductance per phase per km. [0.988 mH]

2.6 Repeat problem 2.5 if each phase has 3 sub-conductors. The diameter of each sub-conductor and spacing remain the same. [0.874 mH]

2.7 A 3-phase double circuit line has its conductors at the vertices of a regular hexagon with side 3.5 m. The diameter of each conductor is 2 cm. Find the inductance per phase per km. [0.596 mH]

2.8 A double circuit 3-phase overhead line has a configuration as shown in Fig. 2.11. The distance between conductors *a* and *c*′ is 7.5 m, between *b* and *b*′ is 10 m and between *a* and *c* is 8 m. The radius of each conductor is 1.284 cm. Find the inductance per phase per km. [0.607 mH]

2.9 A 3-phase line has an equilateral spacing of 6 m. It is desired to rebuild the line with same D_{eq} and horizontal configuration so that the central conductor is midway between the outers. Find the spacing between the outer conductor and central conductor. [4.762 m]

2.10 Find the capacitance of each conductor to neutral for the single phase line of problem 2.1. [.01 μF/km]

2.11 A 3-phase 132 kV, 100 km, 50 Hz, single circuit line has horizontal spacing with 3.5 m between adjacent conductors. The conductor diameter is 1.2 cm. Find the line capacitance per phase and charging current per phase. [0.8415 μF, 20.15 A]

2.12 Find the capacitance per phase per km of the line of problem 2.8. [0.0191 μF]

2.13 A long straight cylindrical wire of radius *r* meter, in a medium of permittivity ε is parallel to a horizontal plane conducting sheet. The axis of the wire is *h* metres above the sheet. (a) Derive an expression of the capacitance per unit length between the wire and the sheet. (b) If $r = 0.3 \times 10^{-2}$ m, $h = 0.12$ m find the capacitance per metre length. (c) If the potential difference between the wire and sheet is 5 kV, find the magnitude and direction of electric stress in the medium at the upper surface of the sheet at a distance 20 cm from the axis of the wire.

$$\text{Take } \varepsilon = \frac{1}{36\pi} \times 10^{-9} \text{ F/m}$$

[(a) $C = \dfrac{2\pi\varepsilon}{\ln \dfrac{2h-r}{r}}$ F/m (b) 0.0127×10^{-9} F/m (c) 6.85 kV/m acting vertically downward]

2.14 A double circuit 3-phase line has its conductors at the vertices of a regular hexagon of side 2 m. The conductor radius is 1.25 cm. Find the capacitance per phase per km. [0.0225 μF/km]

2.15 Find the capacitance per phase per km of the line of problem 2.5. [.0115 μF]

2.16 Find the capacitance per phase per km of the line of problem 2.6. [.01296 μF]

2.17 The 400 kV single circuit Obra-Sultanpur transmission line has the following data:

Number of sub-conductors per phase	2
Sub-conductor spacing	0.45 m
Conductor configuration	horizontal
Interphase spacing from the centre of one bundle to the centre of adjacent bundle	11.3 m
sub-conductor size	54/7/3.53 mm

Assuming the line to be fully transposed, find the inductance and capacitance per phase per km of the line. Neglect the effect of ground. [1.05 mH/km, 0.0108 μF/km]

2.18 Find the resistance per phase per km of Moose conductor at 20 °C and 50 °C. The conductor size is 54/7/3.53 mm. Resistivity of aluminium at 20 °C is 2.826×10^{-8} Ω-m. Assume that increase in length due to stranding is 1.5% and increase in resistance due to skin effect is 3.5%. Temperature coefficient of resistance is 0.004 per °C. [0.0562 Ω/km, 0.0629 Ω/km]

2.19 Find the resistance per phase per km of Markulla conductor at 20 °C and 50 °C. The conductor size is 42 strands of aluminium of 4.13 mm diameter each. Resistivity at 20 °C is 2.826×10^{-8} Ω-m. Assume that increase in length due to spiralling is 1.5% and skin effect increases the resistance by 4%. [0.053 Ω/km, 0.0594 Ω/km]

2.20 A double circuit 3-phase line has the configuration shown in Fig. 2.11. The distance between conductors a and c' is 7.5 m, between b and b' is 9 m and between a and c is 8 m. Find the inductance per km per phase. The radius of each conductor is 1.25 cm. [0.61 mH/km]

2.21 Find the capacitance per phase per km of the line of problem 2.20. [0.019×10^{-6} F/km]

2.22 A conductor 6 mm diameter is suspended at 10 m above sea water which forms the return conductor. Find the inductance of the system per km length. [1.81 mH]

2.23 A 765 kV single circuit line has 3 sub-conductors per phase. Each sub-conductor has a diameter of 31.68 mm and sub-conductor spacing is 45 cm. The interphase spacing is 15.25 m. The configuration is horizontal. Find inductance per phase per km. [0.99 mH/km]

2.24 Find capacitance per phase per km of the line of problem 2.23. [0.0114 μF/km]

CHAPTER 3

Generator, Transformer and Load Models

3.1 INTRODUCTION

A power system consists of electrical power plants, transformers, transmission lines and loads. In addition of these main equipments, many other equipments for protection and safety are also essential. We will discuss the transmission lines in chapter 4. The main features of generators, transformers and loads and their modelling are discussed in this chapter.

GENERATORS

3.2 GENERATORS FOR HYDRO PLANTS

Alternators for use in hydro-electric plants have the following main features.

(1) Speed is low but the inherent danger of runaway speeds has to be guarded against. For vertical machines the speeds range from around 50 to 500 rpm. For horizontal machines the speed range is 100 to 1000 rpm.

(2) The machines may have horizontal or vertical configuration. The alternators used in conjunction with impulse turbines have usually horizontal configuration while those used alongwith Francis and Kaplan turbines have vertical configuration.

(3) The machines have large diameters and small lengths.

(4) The machines have usually salient pole construction. The number of poles varies from 6 to 120.

(5) They are built in a variety of sizes from 0.5 MW onwards. The largest size, used in India at present, is 165 MW at Dehar Power Plant of Beas Project. The largest size operating in U.S.A. has 615 MW rating at Grand Coulee Dam Power Plant. It has been reported that very large size water wheel generators (rating more than 1000 MW) have been installed in Russia.

(6) The rotor consists of laminated poles projecting radially from a steel core carried on spider. The core consists of mild steel stampings about 2 mm thick. The poles may be

integral with rotor core or bolted or keyed to the rotor core. The stampings are clamped axially under heavy pressure between thick steel end plates by rivets or bolts. The field winding is usually of copper strip, the turns lying edge wise relative to the pole body. The turns are insulated with asbestos paper impregnated with varnish. To damp out the oscillations, damper windings are housed in pole shoes. The damper windings consist of low resistance copper bars with their projecting ends connected to short circuiting end rings.

(7) The stator is built from low carbon high resistivity silicon steel stampings, around 0.5 mm thick, insulated with varnish. The laminations are separated, at intervals, by spacers to form radial ventilating ducts and clamped between end plates. The stator core is assembled on axial ribs on the inner periphery of welded steel frame. The stator winding is usually double layer housed in open slots. The number of coils is the same as the number of slots, one side of each coil lying at the top of a slot, its other side being at the bottom of another slot. Each coil may consist of single turn or more turns wound continuously in series according to the voltage and current rating of the machine. Generators of large ratings may have two or more circuits in parallel even with single turn coils. If conductor size is large, the conductors are laminated to reduce eddy current loss in conductors. Roebel bar construction is generally used. The stator windings are connected in star.

(8) The short circuit ratio is around unity.

(9) The machine reactances have following ranges :

Direct axis reactance	0.6	to	1.5 p.u.
Quad. axis reactance	0.4	to	0.8 p.u.
Sub transient reactance	0.15	to	0.4 p.u.
Transient reactance	0.2	to	0.5 p.u.
Negative sequence reactance	0.15	to	0.7 p.u.
Zero sequence reactance	0.05	to	0.2 p.u.

(10) The machine specifications include MVA rating, frequency (50 Hz in India), number of phases (always three); voltage rating, power factor, current rating, speed, type of exciter, excitation voltage, method of cooling, different reactances, short circuit ratio, position of shaft, type of voltage regulator and its response time, efficiency, field current at full load and rated power factor, regulation at full load, mechanical details (runaway speed, direction of rotation, overall size, flywheel effect, number and location of bearings etc.) temperature rise and details of accessories.

(11) The machines are generally air cooled. In open ventilation system, fresh air from the atmosphere enters from one or both ends, flows through ducts between windings and between core sections and discharges at the periphery. It is necessary to clean incoming air. Passages are provided so that air can be taken from out of doors and discharged out of doors. The air intake is preferably located near the tail race but sufficiently above the highest water level in it. The requirement is around 3 cubic metres per minute per kW loss and air velocity around 400 metres per minute. Large machines mostly use closed air circuit. The same air is continually recirculated after being cooled in water cooled heat exchanger. An advantage of closed circuit is freedom from contamination, permitting the machine to run for much longer periods without the necessity of cleaning. The exclusion of moisture is also an advantage. The limited amount of air available provides a natural protection against extensive damage by fire, in the event of electrical breakdown. The ventilating system also includes fire fighting equipment using carbon dioxide.

3.3 TURBO-GENERATORS

The generators used in steam plants have the following main features :

(1) They are all high speed machines, the usual value of speed being 3000 or 1500 rpm for 50 Hz systems.

(2) The machines have horizontal configuration.

(3) The machines have cylindrical rotors wound usually for 2 poles or 4 poles. The outside diameter of rotor is limited by maximum peripheral speed which should not exceed 175 m/sec. Thus the rotor of a turbo-alternator is of small diameter and is very long (length about 5 times the diameter) as compared to the rotor of a water wheel generator. The maximum diameter for 2 pole machines is around 1 metre. The length of rotor between the bearings is limited by the considerations of critical speed. Modern turbo alternators are designed such that first critical speed lies between 1250 and 1400 rpm and the second critical speed is not less than 3700 rpm.

(4) The machines are built in variety of sizes from 10 MW to over 1500 MW. In India large size turbo-alternators have been successively introduced during the past two decades. During 3rd plan a number of turbo alternators of about 60 MW rating were commissioned. During the fourth and fifth plan a number of 100 – 120 MW sets were commissioned. Many 200 MW units have been commissioned at Obra, Badarpur Ropar and Koradi and other power plants and a number of other 200 MW plants are being installed at present. 500 MW units have been installed at Trombay, Singrauli, Anpara, Chandarpur, Ramagundem, Farakka and Korba. More 500 MW units will be commissioned during the next decade.

(5) The rotor is made from solid steel forgings of high grade nickel chrome molybdenum steel. It is necessary to provide passages in the rotor body for the flow of cooling medium. These passages are provided either in the teeth or below the rotor slots. The circumference of rotor has large number of radial deep slots machined parallel to the axis to accommodate the field winding which has to carry currents of the order of many kiloamperes. The rotor winding is of copper strip insulated with mica. Around two-third of rotor circumference is taken up by the slots. A wedge is driven into the mouth of each slot to keep the winding in position. The damper windings are not needed in turbo-alternators.

(6) The stator is built from high permeability sheet steel laminations. One side of each lamination is treated with insulating material (varnish or enamel or paper) to prevent the flow of eddy currents. The air gap in turbo alternators is quite large (around 5 cm or so). The large air gap results in high reactance and makes the machine more stable. To reduce the skin effect and to obtain a uniform distribution of current in the conductors, the stator windings are divided into a number of thin conductors each wrapped with insulating material. The conductors are transposed in slots. The number of slots per pole per phase, pitch of winding and disposition of conductors are so selected as to make the terminal voltage as near sinusoidal as possible. The stator windings are connected in star. 500 MW units generally use hollow conductors.

(7) The short circuit ratio is 0.4 to 0.6.

(8) The machine reactances have following ranges :

Synchronous reactance	1.5	to	2.5 p.u.
Transient reactance	0.15	to	0.35 p.u.
Sub-transient reactance	0.1	to	0.15 p.u.
Negative sequence reactance	0.1	to	0.15 p.u.
Zero sequence reactance	0.02	to	0.15 p.u.

(9) The small machines are air cooled. However hydrogen cooling is invariably used for both stator and rotor of medium and large size turbo-alternators. Hydrogen cooling offers a number of advantages : the windage loss is reduced to about one tenth thus improving the efficiency of machines by 0.5 to 1 per cent, for the same physical size a hydrogen cooled turbo-alternator can develop about 25 per cent more power as compared to an air cooled turbo-alternator, life of insulation is increased due to the absence of oxygen, hydrogen has higher heat transfer coefficient than air and thus cooling is more efficient and faster, since hydrogen does not support combustion fire hazard is reduced. The stator frame, of a hydrogen cooled alternator, must be gas tight to avoid the possibility of air mixing with hydrogen and causing an explosion. The hydrogen pressure is kept higher than outside atmospheric pressure to eliminate the leakage of air into the machine. The stator is made sufficiently strong not only to withstand the gas pressure but also to withstand the pressure due to explosion. In 500 MW units the stator windings are cooled by water.

3.4 PHASOR DIAGRAM OF CYLINDRICAL ROTOR ALTERNATOR

The phasor diagram of an alternator can be developed by first considering the operation at no load. In the next step the effect of armature reaction is included. Then the effect of armature resistance and armature leakage reactance are included.

3.4.1 No Load Condition

When the generator is an no load (i.e. no current in armature winding) the field current sets up mmf F and flux ϕ. Figure 3.1 shows a 2 pole generator with 2 stator coils per pole per phase. The magnetic flux due to field current is also shown.

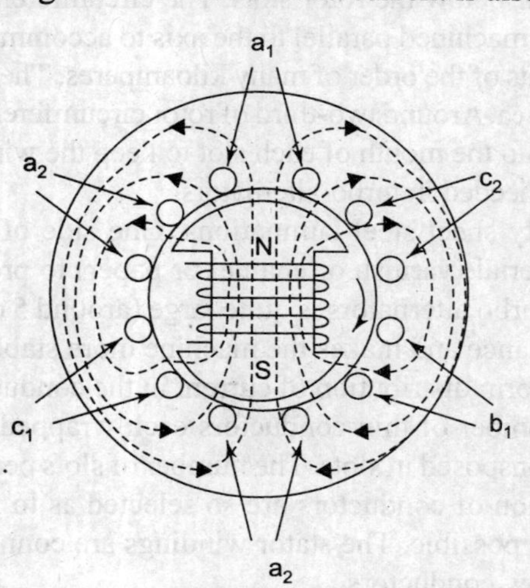

Fig. 3.1 Magnetic field in an alternator on no-load

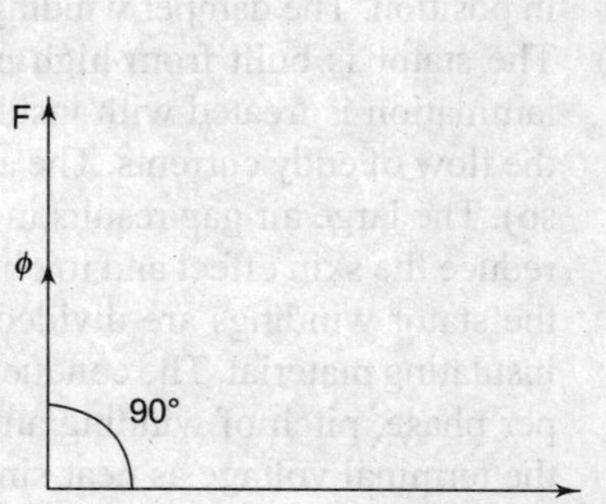

Fig. 3.2 Phasor diagram of alternator on no-load

Since the rotor is rotating clockwise, the emf induced in phase a is maximum at the instant shown and is into the paper in coil a_1 and out of the paper in coil a_2.

Figure 3.2 shows the phasor diagram under no-load condition. The induced emf E lags the the mmf and flux by 90° as shown. E is the open circuit voltage per phase. Since this voltage is completely due to excitation of the generator, E is known as excitation voltage or generated voltage.

3.4.2 Effect of Armature Reaction

When a balanced 3-phase load is connected at the alternator terminals, an rms current I flows through each phase of armature winding. The three phase currents flowing in the 3-phase winding produce a rotating magnetic field. The armature mmf rotates at synchronous speed with respect to stator winding. Since the rotor is also moving at synchronous speed with respect to stator, the rotor and stator mmfs are stationary with respect to each other. Both these mmfs act upon magnetic circuits which have equal reluctance in all axes and may be combined as phasors.

Figure 3.3 shows the distribution of flux due to armature currents at the instant the current in phase a is at maximum value and is at unity power factor. By right hand rule, the armature flux is clockwise due to currents in top conductors and anticlockwise due to currents in the bottom conductors.

The resultant flux due to super imposition of rotor and stator fluxed is shown in Fig. 3.4. A comparison of Figs 3.1 and 3.3 shows that the two fluxes are in opposition over the leading half of each pole face and are in the same direction over the trailing half of each pole face. Therefore, the effect is to distort the magnetic fluxes as shown in Fig. 3.4. It is seen that the direction of most of the lines of flux in the air gap has been skewed and lengthened. This causes a backward pull on the rotor. To overcome the tangential component of this pull, the prime mover driving the generator has to exert a larger torque than that required at no-load. As seen in Fig. 3.4 the armature mmf has a cross-magnetising effect on field poles.

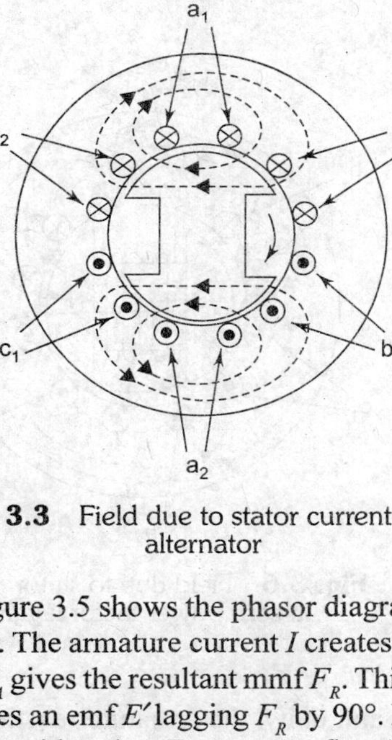

Fig. 3.3 Field due to stator current in an alternator

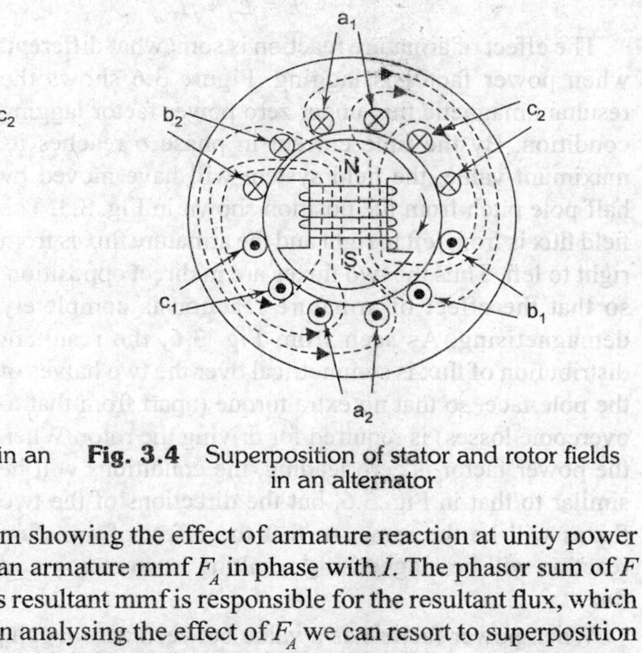

Fig. 3.4 Superposition of stator and rotor fields in an alternator

Figure 3.5 shows the phasor diagram showing the effect of armature reaction at unity power factor. The armature current I creates an armature mmf F_A in phase with I. The phasor sum of F and F_A gives the resultant mmf F_R. This resultant mmf is responsible for the resultant flux, which induces an emf E' lagging F_R by 90°. In analysing the effect of F_A we can resort to superposition and consider, that F_A sets up a flux ϕ_R which is synchronously rotating and cuts the armature winding, there by inducing an armature reaction emf E_{ar} per phase. This emf lags behind mmf F_A

and flux ϕ_R, by 90°. The resultant sum of E and E_{ar} gives emf E'. The armature reaction voltage can be considered as a voltage drop across a fictitious inductive reactance x_a, such that

$$E_{ar} = -jIx_a \qquad ...(3.1)$$

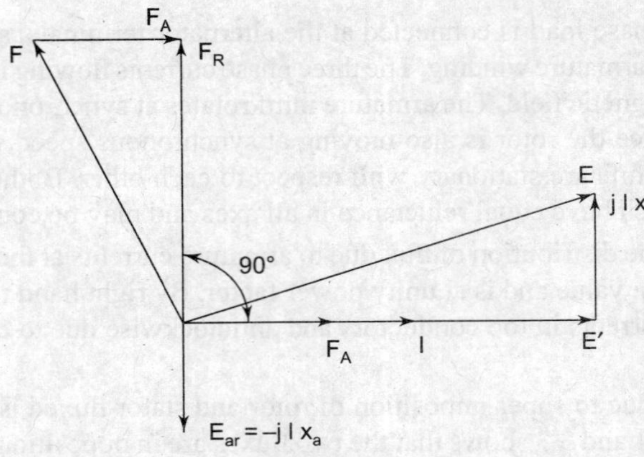

Fig. 3.5 Phasor diagram showing effect of armature reaction at unity power factor

The minus sign in Eq. (3.1) has been used for defining E_{ar}, so that an equal and opposite quantity can be handled with a plus sign (i.e. plus sign is used for a voltage drop and minus sign for a voltage rise). It should always be remembered that x_a is a fictitious reactance equivalent in its effect to the armature reaction. Evidently

$$E = E' + jIx_a \qquad ...(3.2)$$

The effect of armature reaction is somewhat different, when power factor is lagging. Figure 3.6 shows the resultant magnetic flux under zero power factor lagging condition. By the time current in phase *a* reaches its maximum value, the field system will have moved by half pole pitch from the position shown in Fig. 3.3. The field flux is from left to right and the armature flux is from right to left. Thus the two fluxes are in direct opposition, so that the effect of armature reaction is completely demagnetising. As seen from Fig. 3.6, the resultant distribution of flux is symmetrical over the two halves of the pole face, so that no extra torque (apart from that to overcome losses) is required for driving the rotor. When the power factor is zero leading, the conditions will be similar to that in Fig. 3.6, but the directions of the two fluxes will be the same, so that the effect of armature reaction will be magnetising resulting in an increase in flux.

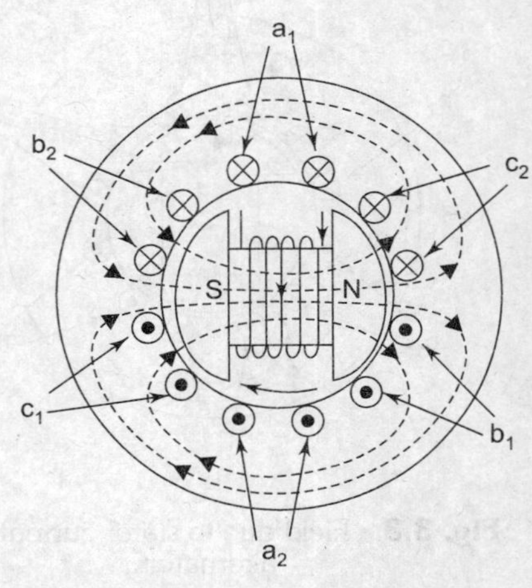

Fig. 3.6 Field due to stator current at zero power factor lagging

When power factor has a value between zero lagging and unity, the armature reaction will be partly demagnetising and partly cross-magnetising. Figure 3.7 shows the phasor diagram. The armature mmf F_A is in phase with I and the phasor sum of field mmf F and armature mmf F_A gives the resultant mmf F_R. The effect of F_A can be replaced by a fictitious reactance x_a. The phasor sum of E' and jIx_a gives the excitations emf E.

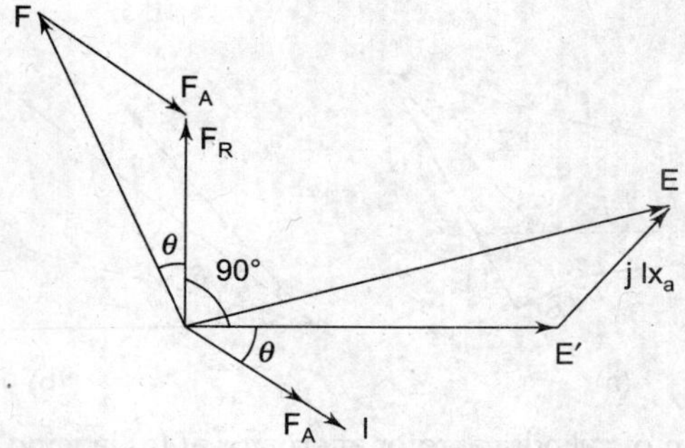

Fig. 3.7 Phasor diagram showing the effect of armature reaction at a lagging power factor

3.4.3 Effect of Armature Resistance and Leakage Reactance

When the alternator is loaded, a voltage drop takes place in the stator winding, due to its resistance R_a and leakage reactance x_l. When these voltage drops are subtracted from E', the result is V, i.e. the terminal voltage per phase. In other words, the voltage drops IR_a and jIx_l, when added to V give the emf E'.

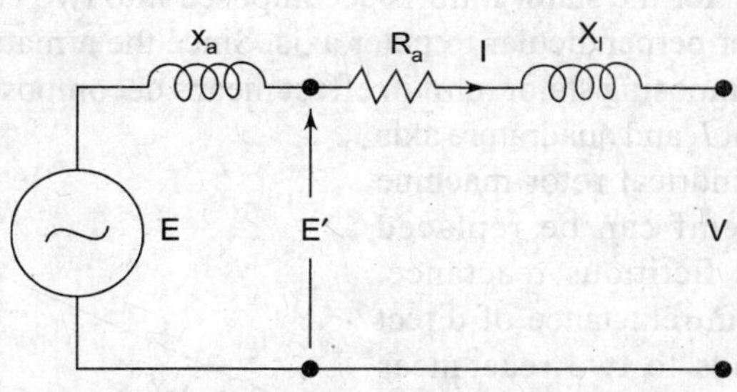

Fig. 3.8 Equivalent circuit of alternation

Figure 3.8 shows the equivalent circuit of the alternator. The current flow is shown in the conventionally positive direction, i.e. emerging from the positive terminal. The voltage drops IR_a, jIx_l and jIx_a when added to terminal voltage V give the open circuit emf E. Thus

$$E = V + IR_a + jIx_l + jIx_a \qquad ...(3.3)$$

or
$$E = V + IR_a + jI(x_l + x_a)$$

or
$$E = V + IR_a + jIx_s \qquad ...(3.4)$$

or
$$E = V + IZ_s$$

The quantity x_s is the sum of leakage reactance x_l and fictitious armature reactance x_a and is known as synchronous reactance. The quantity Z_s is known as synchronous impedance.

Figures 3.9 (*a* and *b*) show the phasor diagram under lagging and leading power factor conditions, respectively. The phasor sum of V, IR_a, jIx_l and jIx_a gives the emf E.

E and V have a phase difference of δ, which is known as power angle of the machine.

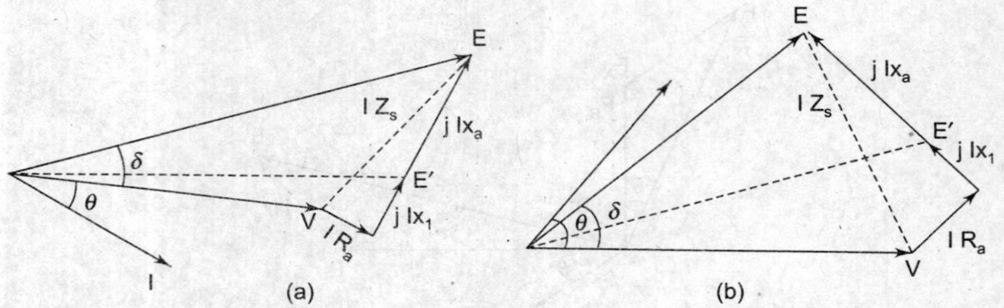

Fig. 3.9 Phasor diagram of cylindrical rotor alternator at (a) lagging power factor, (b) leading power factor

3.5 Phasor Diagram of Salient Pole Alternator

In a cylindrical rotor alternator the reluctance of magnetic path is same in all directions and field mmf and armature mmf can be combined directly. However in a salient pole machine the reluctance of the magnetic current along polar (i.e. direct) axis is much less than that along the interpolar (i.e. quadrature axis).

In a salient pole alternator the stator mmf is decomposed into two components, one parallel to polar axis and the other perpendicular to polar axis. Since the armature mmf is proportional to armature current, decomposing stator mmf in effect means decomposing the armature current into direct axis component I_d and quadrature axis component I_q. As in cylindrical rotor machine the armature reaction emf can be replaced by voltage drop across fictitious reactance. However the difference in reluctance of direct and quadrature axis leads to two reactances i.e. direct axis synchronous reactance and quadrature axis synchronous reactance. In view of above discussion the phasor diagram of a salient pole generator is shown in Fig. 3.10. θ is the power factor angle. The armature current is resolved into two components i.e. I_d which is perpendicular to E and I_q which is in phase with E. The phasor sum of terminal voltage V, IR_a, $jI_d x_d$ and $j I_q x_q$ gives the excitation emf E.

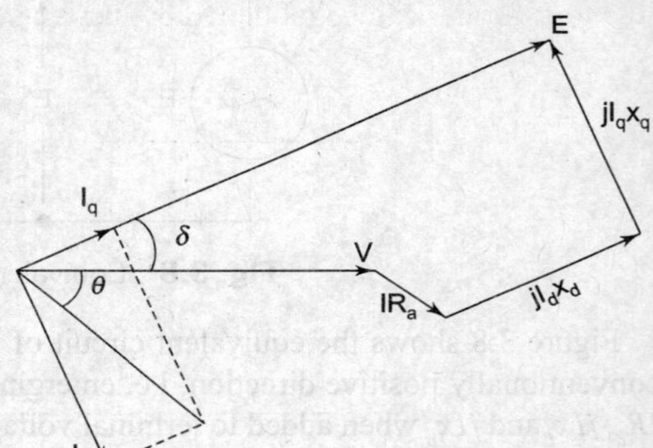

Fig. 3.10 Phasor diagram of salient pole generator as per two reaction theory

3.6 SINGLE PHASE TWO WINDING TRANSFORMER

Figure 3.11 shows the exact equivalent circuit of a single phase two winding transformer. The equivalent circuit is drawn by assuming the transformer to be ideal Resistance R_0 represents the effect of core losses and reactance X_0 accounts for magnetising current I_m. Winding resistances and leakage reactances are represented separately. (It must be remembered that these resistances and reactances are actually present in the windings. But they are shown separately for ease of representation and calculations).

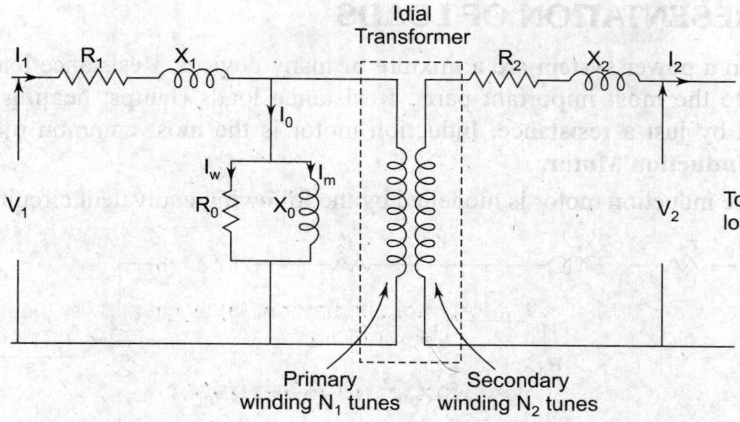

Fig. 3.11 Equivalent circuit of single phase two winding transformer

The no load current I_0 is generally very small (about 5% of primary current I_1). Therefore the shunt branch consisting of R_0 and X_0 can be shifted to the extreme left as shown in Fig. 3.12. Moreover the secondary winding resistance R_2 and leakage reactance X_2 can be transferred to the primary side to give the simplified approximate equivalent circuit shown in Fig. 3.12.

Moreover the shunt resistance R_0 and reactance X_0 can, most of the time, be neglected so that the transformer can be modelled by just a series resistance and reactance.

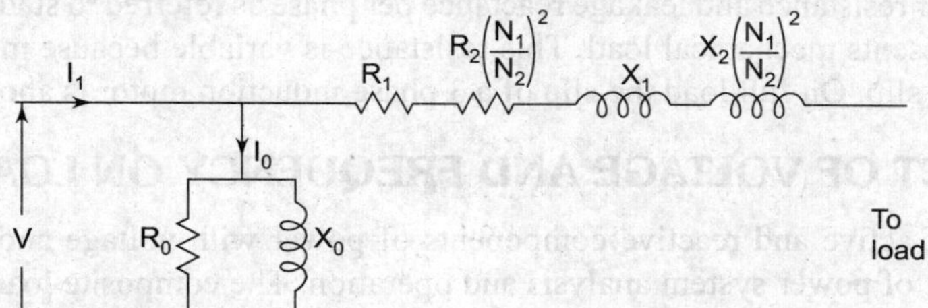

Fig. 3.12 Approximate equivalent circuit of two winding transformer

3.7 THREE PHASE TRANSFORMERS

A three phase transformer generally consists of a single unit having a 3 limbed core, one limb for each phase winding. On each limb the low voltage winding is placed over the core and the high voltage winding is placed over the low voltage winding with suitable insulation between core and low voltage winding as well as between low voltage and high voltage windings. The three windings may be connected in star-star, delta-delta, star-delta or delta-star. Generally star-delta and delta-star connections are used with high voltage winding connected in star. However whenever neutral connection is needed that winding is connected in star.

For steady state analysis a three phase transformer is analysed on per phase basis. As a matter of fact all computations for steady state are done on per phase basis because all phases are similar. However unsymmetrical fault analysis cannot be done on per phase basis. Moreover the method of neutral grounding affects unsymmetrical fault analysis.

3.8 REPRESENTATION OF LOADS

The loads on a power system are a mixture of many devices. Resistance loads and motor loads constitute the most important parts. Resistance loads (lamps, heating devices) can be represented by just a resistance. Induction motor is the most common of motor loads.
Modelling of Induction Motor.

A three phase induction motor is modelled by the following equivalent circuit, (Fig. 3.13).

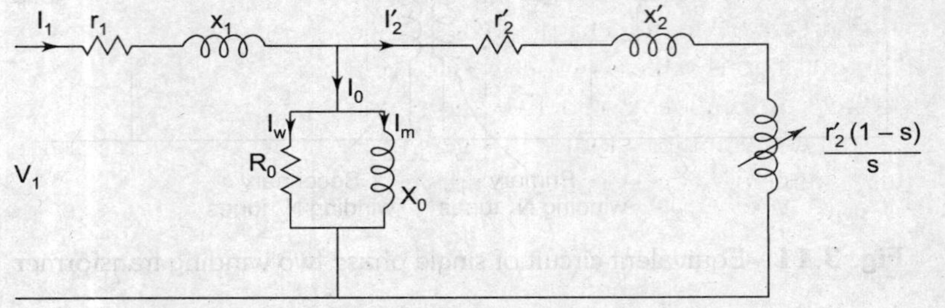

Fig. 3.13 Equivalent circuit of induction motor

V_1 is per phase input voltage and I_1 is input (stator) current per phase, r_1 and x_1 are stator resistance and leakage reactance per phase. R_0 is the resistance which represents stator core loss and reactance X_0 is magnetising reactance. I_0 is the no load current and is phasor sum of core loss component I_w and magnetising current I_m. I_2' is rotor current per phase as referred to stator. r_2' and x_2' are rotor resistance and leakage reactance per phase as referred to stator. The resistance r_2' $(1 - s)/s$ represents mechanical load. This resistance is variable because mechanical load is variable. s is the slip. On full load the slip of a 3 phase induction motor is about 3-5%.

3.9 EFFECT OF VOLTAGE AND FREQUENCY ON LOADS

The variation of active and reactive components of power with voltage and frequency is an important aspect of power system analysis and operation. The composite load at a sub-station usually consists of : Induction motors 50-70%, Lighting and heating 20-25%, Synchronous motors 10%, Transmission losses etc. 10-20%.

3.9.1 Effect of Voltage

(*a*) *Lighting load.* Lighting loads do not consume any reactive power. Since the resistance of lighting devices is not constant, the active power does not vary as $(\text{voltage})^2$ but approximately as $(\text{voltage})^{1.6}$.

(*b*) *Heating.* The resistance is almost constant and active power varies as $(\text{voltage})^2$.

(*c*) *Induction motors.* The performance of an induction motor can be analysed from its equivalent circuit. The active and reactive power vary with voltage in a rather complex manner. The effect also depends on whether the shaft torque is constant or varies with speed and also whether the motor is running at full load or less than full load. The decrease of voltage causes an increase in current so that the effect on power is small. If the shaft torque is constant, a 10% drop in voltage at full load condition may cause about 0.5% decrease in power. However if the voltage decreases appreciably (to about 0.8 pu) the power increases with further reduction in voltage.

(*d*) *Synchronous motors.* The active power does not depend on voltage and remains, more or less, constant. A reduction in voltage causes the vars to increase in the leading direction.

In most of the analysis the composite sub-station loads are represented by constant impedances so that both P and Q are assumed to vary as (voltage)^2. However, it is only an approximate representation.

3.9.2 Effect of Frequency

The power consumed by a resistive load does not depend on frequency. However most of the loads are inductive in nature. The impedance of an inductive load depends on the frequency and therefore the power drawn by an inductive load depends on frequency. The analysis of the effect of frequency on motor load is rather complex. Since only very small changes in frequency are permissible, the effect of frequency on active and reactive power is neglected in calculations. However in load frequency control studies it is generally assumed that a one per cent decrease in frequency causes a one per cent decrease in active power.

Example 3.1

A load can be modelled by a constant impedance $Z = R + jX$. By what percent will the real power decrease if voltage decreases by 1%.

Solution

$$S = P + jQ = VI^* = \frac{VV^*}{Z^*} = \frac{|V|^2}{R - jX} = |V|^2 \frac{R + jX}{(R - jX)(R + jX)}$$

$$= |V|^2 \left[\frac{R + jX}{R^2 + X^2}\right]$$

$$P = |V|^2 \frac{R}{R^2 + X^2}$$

For small change in voltage

$$\frac{\Delta P}{\Delta |V|} = \frac{\partial P}{\partial |V|} = 2|V| \frac{R}{R^2 + X^2} = \frac{2P}{|V|}$$

or
$$\frac{\Delta P}{P} = \frac{2\Delta |V|}{|V|} \qquad \text{...}(i)$$

Eq. (*i*) shows that if $\frac{\Delta |V|}{|V|}$ is 1% $\frac{\Delta P}{P}$ will be 2% or 1% change in voltage causes 2% change in real power i.e. 1% decrease in voltage causes 2% decrease in real power.

Example 3.2

A passive load can be modelled as $Z = R + jX$. The power factor is 0.8. How would 1% drop in frequency affect real power.

Solution

As in example 3.1

$$P = |V|^2 \frac{R}{R^2 + X^2} = |V|^2 \frac{R}{R^2 + (2\pi f L)^2} = |V|^2 \frac{R}{R^2 + 4\pi^2 L^2 f^2}$$

$$\frac{\Delta P}{\Delta f} = \frac{\partial P}{\partial f} = -|V|^2 \left[\frac{R 8\pi^2 L^2 f}{\left(R^2 + 4\pi^2 L^2 f^2\right)^2} \right] = -|V|^2 \frac{2RX^2}{f\left(R^2 + X^2\right)^2}$$

or
$$\frac{\Delta P}{\Delta f} = -P \frac{2X^2}{f\left(R^2 + X^2\right)}$$

or
$$\frac{\Delta P}{P} = \frac{-2X^2}{R^2 + X^2}\left(\frac{\Delta f}{f}\right)$$

When $\cos\theta = 0.8$, $\dfrac{X^2}{R^2 + X^2} = \sin^2\theta = (0.6)^2 = 0.36$

$$\therefore \quad \frac{\Delta P}{P} = -2(0.36)\frac{\Delta f}{f} = -0.72\frac{\Delta f}{f}$$

∴ 1% drop in frequency means 0.72% increase in load.

Example 3.3

A 400V 3 phase induction motor drives a compressor whose torque is speed independent. The initial slip is 0.03. How will the motor load change if the voltage drops by 1%.The parameters of the motor are as under. Neglect Z_m . $R_1 = 0.290$ ohm/phase, $R_2 = 0.15$ ohm/phase, $X_1 + X_2 =$ 0.7 ohm/phase.

All the parameters are referred to stator.

Solution

Figure 3.14 shows the equivalent circuit.

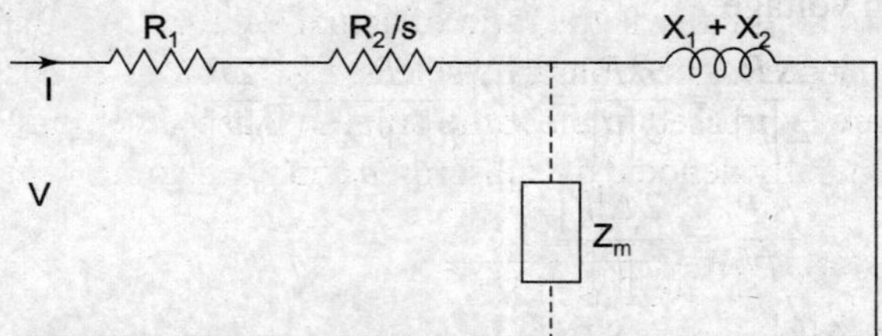

Fig. 3.14

From induction motor theory we know that

$$T \propto |V|^2 s$$

Assuming that motor is operating in linear region of its torque-slip curve and since torque is speed independent.

$$\frac{T_1}{T_2} = 1 = \frac{|V_1|^2}{|V_2|^2}\frac{s_1}{s_2}$$

Taking $\quad V_1 = 1.0$, $V_2 = 0.99$. Also $s_1 = 0.03$

$$\therefore \quad 1 = \frac{1}{(0.99)^2}\left(\frac{0.03}{s_2}\right) \text{ or } s_2 = 0.0306$$

From Fig. 3.14

$$I = \frac{V}{\left[\left(R_1 + \frac{R_2}{s}\right)^2 + (X_1 + X_2)^2\right]^{0.5}}$$

Substituting the values

New Current $\qquad I_2 = 1.0084\, I_1$

Thus current increases by 0.84%

$$P = \left(R_1 + \frac{R_2}{s}\right) I^2$$

or
$$\frac{P_1}{P_2} = \frac{0.29 + 0.15/0.03}{0.29 + 0.15/0.0306}\left[\frac{1}{1.0084}\right]^2$$

or
$$P_2 = 0.998 P_1.$$

Thus, 1% drop in voltage increases current by 0.84% and decreases power input by 0.2%.

3.10 PER UNIT METHOD

Power system calculations can be done by taking either the actual values of different quantities or in per unit values. Taking the different quantities in per unit simplifies the calculations to a great extent. Moreover the system data is generally available in per unit. Therefore it is always convenient to use per unit method of calculations.

The per unit value of a quantity is the ratio of actual value of that quantity and an arbitrarily selected value of that quantity. The arbitrarily selected value is called base value. Both the actual value and base value are expressed in the same units so that the per unit value is dimensionless. The base values are generally denoted by subscript *b* and the per unit value by subscript p.u. Thus

$$\text{Per unit kVA} = \frac{\text{Actual kVA}}{\text{Base kVA}} \qquad \text{...(3.5)}$$

$$\text{Per unit impedance} = \frac{\text{Actual impedance (ohms)}}{\text{Base impedance (ohms)}}$$

Out of the 4 quantities i.e. kVA, kV, current and impedance only two are independent. It is convenient to select base values of kVA and kV and find base values of current and impedance from the chosen base kVA and base kV. In 3 phase systems 3 phase kVA and line to line kV are chosen as base. Then

$$\text{Base current} = \frac{\text{Base kVA}}{\sqrt{3}\,(\text{Base kV})} \qquad \text{...(3.6)}$$

$$\text{Base impedance} = \frac{\text{Line to neutral value of base voltage}}{\text{Base current}}$$

$$= \frac{(\text{Base kV})^2 \times 1000}{\text{Base kVA}} = \frac{(\text{Base kV})^2}{\text{Base MVA}} \text{ ohms} \tag{3.7}$$

For circuits containing transformers, it is convenient to select the same kVA base for both the sides. However, the ratio of base voltage on the two sides is kept the same as the transformation ration. Such a selection gives the same per unit impedance on either side of the transformer.

It is usually necessary to transform the per unit impedance from one set of base values to a new set of base values. This can be done by using the equation:

Per unit impedance referred to new base

$$= \begin{bmatrix} \text{Per unit impedance} \\ \text{referred to old base} \end{bmatrix} \left[\frac{\text{Base kV}_{\text{old}}}{\text{Base kV}_{\text{new}}} \right]^2 \left[\frac{\text{Base kVA}_{\text{new}}}{\text{Base kVA}_{\text{old}}} \right] \tag{3.8}$$

The three-phase circuits are always solved in terms of per phase values. Since the base values are 3-phase kVA and line to line kV, doubt may arise regarding the relation between per unit value of line to line voltage and per unit value of line to neutral voltage. It is sufficient to mention that the p.u. value of line to neutral voltage on line to neutral voltage base is equal to p.u. value of line to line voltage on line to line voltage base. Moreover the p.u. value of total 3-phase kVA on 3-phase kVA base is equal to p.u. value of kVA per phase on kVA per phase base. The following example illustrates the above observation.

Let 3-phae base kVA = 60,000 kVA

Line to line base kV = 220 kV

Then 1 phase base kVA = 60,000/3 = 20,000 kVA

Line to neutral base kV = $220/\sqrt{3}$ = 127.02 kV

If actual 3-phase kVA is 30000 kVA (so that per phase kVA is 10000) and line to line kV is 176 kV (so that actual line to neutral kV is $176/\sqrt{3}$ = 101.617) then

$$\text{p.u. kVA} = \frac{30000}{60000} = \frac{10000}{20000} = 0.5$$

$$\text{p.u. kV} = \frac{176}{220} = \frac{101.617}{127.02} = 0.8$$

3.11 ADVANTAGES OF PER UNIT SYSTEM

1. Manufacturers usually specify the impedance values of equipments in per unit of the equipment's rating. If any data is not available, it is easier to assume its per unit value than its numerical value.
2. When expressed in per unit, system parameters tend to fall in relatively narrow numerical ranges. Therefore, any erroneous data can be easily identified.
3. Per unit data representation yields important information about relative magnitudes.
4. Power systems contain a large number of transformers. The ohmic value of impedance as referred to secondary is different from the value as referred to primary. However, if base values are selected properly, the per unit impedance is the same on the two sides of the transformer.
5. The transformer connections in 3-phase circuits do not affect the per unit value of impedance although the base voltages on the two sides do depend on the connections.

Since the data of different equipments is available in per unit of its rating, it is necessary to convert all the data to the common base kVA and kV. After the analysis it is usually necessary convert the per unit values of result to actual values.

The calculations can also be done by using per cent values. However, when quantities expressed in per cent are multiplied, it is necessary to divide the product by some power of 100 to obtain correct result in per cent. The confusion caused by introduction of power of 100 is avoided in p.u. system.

3.12 PER UNIT IMPEDANCES OF TRANSFORMERS

(a) Two Winding Transformers

In circuits containing transformers the base volt-amp (or kVA) is the same on the two sides. However, the voltage base on the two sides must have the same ratio as the turn ratio. This selection makes the p.u. impedance equal on the two sides.

Consider a single phase transformer having V_1 and V_2 as the primary and secondary voltages and I_1 and I_2 as primary and secondary currents. Let impedance as referred to primary be Z_1. Then

$$\frac{V_1}{V_2} = \frac{I_2}{I_1}$$

$$\text{Base volt ampere} = V_1 I_1 = V_2 I_2$$

$$\text{Base voltage} = V_1 \text{ for primary and } V_2 \text{ for secondary}$$

$$\text{Base impedance for primary} = \frac{V_1}{I_1}$$

$$\text{P.u. impedance as referred to primary} = \frac{Z_1}{V_1/I_1} = I_1 Z_1/V_1$$

$$\text{Actual impedance as referred to secondary} = Z_1 (V_2/V_1)^2$$

$$\text{Base impedance for secondary} = V_2/I_2$$

$$\text{P.u. impedance as referred to secondary} = \frac{Z_1(V_2/V_1)^2}{V_2/I_2} = \frac{Z_1 V_2 I_2}{(V_1)^2} = \frac{Z_1 V_1 I_1}{(V_1)^2}$$

$$= I_1 Z_1/V_1 = \text{p.u. impedance as referred to primary.}$$

For circuits containing 3-phase transformers the base values are selected as under:

1. kVA (or MVA) base is the same on the two sides.
2. The line to line kV on the primary side is selected as the base kV for primary. The line to line kV on the secondary side is selected as base kV for secondary.
3. For single phase transformers connected as 3-phase units, the 3-phase rating of the transformer is the sum of the single phase rating of each transformer. Impedance in per unit for the 3-phase unit is the same as that for each single phase transformer.
4. The per unit impedance on one base can be converted to another base by using Eq. (3.8).

(b) Three-Winding Transformer

The equivalent circuit of a 3-winding transformer is shown in Fig. 3.15. The impedances Z_1, Z_2 and Z_3 are the impedances of primary, secondary and tertiary winding. These may be available in ohm or in per unit of the respective ratings of the windings. The kVA and kV ratings of the three windings are different.

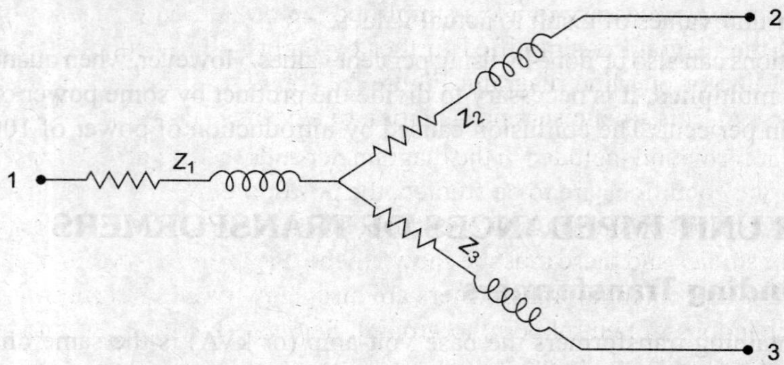

Fig. 3.15 Equivalent circuit of 3-winding transformer

To express these impedances in per unit, the base values are selected as under:

1. kVA base is the same for all the three windings.
2. The rated line to line kV for each of the three windings is selected as base kV for that winding.

3.13 SINGLE LINE DIAGRAM

A power system consists of a number of generators, transformers, lines, etc., connected together. In drawing a diagram of such a system, the equipments are shown by standard symbols. For 3.16 shows some symbols of commonly used equipments.

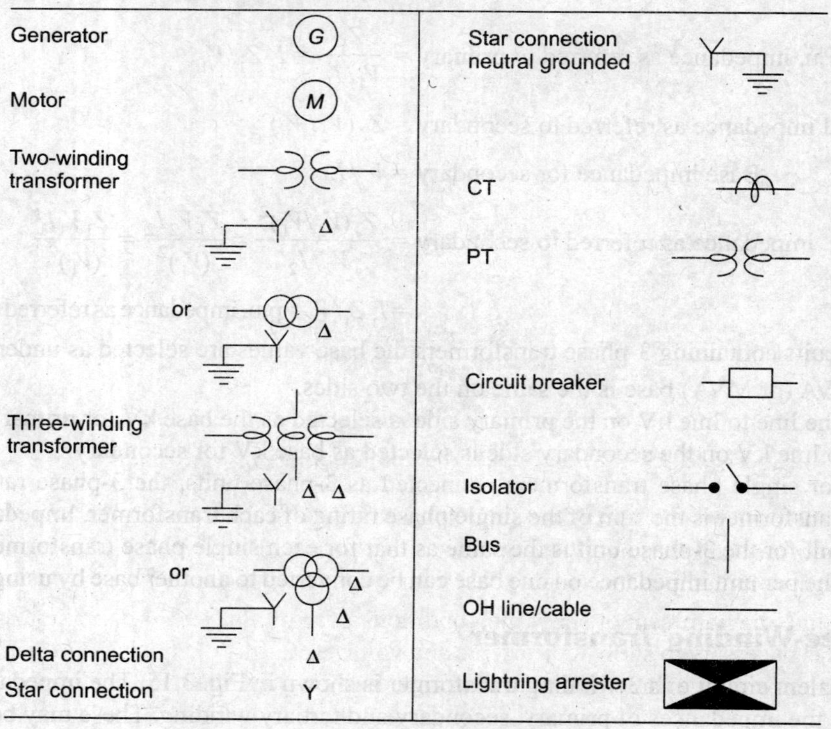

Fig. 3.16 Symbols of Electrical Equipments

All power systems are 3-phase systems. If all the equipments of the three phases are shown, the diagram would be rather complicated and difficult to understand. Since all the three phases are exactly similar, it is necessary to use a simplified representation using only one phase and neutral. Usually the diagram is simplified further by omitting the neutral. Such a diagram is known as single line or one-line diagram. Figure 3.17(a) shows the single line diagram of a simple power system. It is a concise representation of system.

The extent of information included in the diagram depends on the purpose for which it is to be used. If steady state conditions are to be studied, the position of relays, circuit breakers, etc., is unimportant and need not be shown. However the position of relays and breakers is important in fault and stability studies and these must be shown in the diagrams for fault and stability studies. Sometimes current and potential transformers are also shown. In many studies it is important to know the connections of neutral point to ground. Sometimes the ratings and impedances of equipments are also shown on the diagram.

As seen in Fig. 3.17(a) a single line diagram consists of generators, transformers, lines, buses, etc. A bus is a node at which one or more lines, or loads or generators are connected. One of the nodes in every system is taken as a reference node. The other independent nodes are the buses. In a single line diagram a bus is shown by a small vertical line.

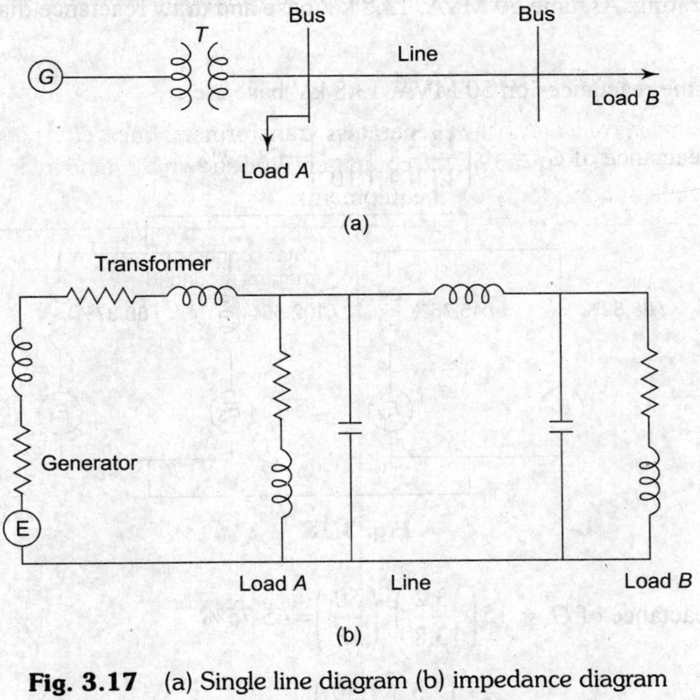

Fig. 3.17 (a) Single line diagram (b) impedance diagram

As mentioned above one single line diagram is not sufficient for all the types of studies.

The transformer connections (delta or star) affect the steady state as well as transient state and must be shown on the diagram.

A single line diagram cannot show the conditions during unbalanced operation of a power system. If it is necessary to show the currents and voltages during an unbalanced operation, it is necessary to show all the three phases and a single line diagram is not sufficient.

3.14 IMPEDANCE AND REACTANCE DIAGRAM

In order to analyze a system it is necessary to draw its equivalent circuit (also known as impedance diagram). A generator is represented by an e.m.f in series with an impedance. The two- and three-winding tranformers are represented as discussed above. The shunt admittance representing the magnetising current of a transformer is usually omitted in power system studies. A short line is represented by its series impedance. Medium and long lines are represented by nominal π circuits. Very long lines are represented by equivalent π circuits. A static load is represented by a resistance and reactance in series. Motors are represented by their equivalent circuits. Fig. 3.17(b) shows an impedance diagram. If the resistance components of the impedances are neglected, an impedance diagram reduces to a reactance diagram. Sometimes the values are expressed in percent (i.e., p.u. value × 100). However, it is preferable to use per unit values.

Example 3.4

Two generators rated at 10 MVA, 13.2 kV and 15 MVA, 13.2 kV, respectively are connected in parallel to a bus. The bus feeds two motors rated at 8 MVA and 12 MVA respectively. The rated voltage of motors is 12.5 kV. The reactance of each generator is 15% and that of each motor is 20% on its own rating. Assume 50 MVA, 13.8 kV base and draw reactance diagram.

Solution

Using Eq. (3.8), the reactances on 50 MVA, 13.8 kV base are

$$\text{Reactance of } G_1 = 15\left(\frac{13.2}{13.8}\right)^2\left(\frac{50}{10}\right) = 68.62\%$$

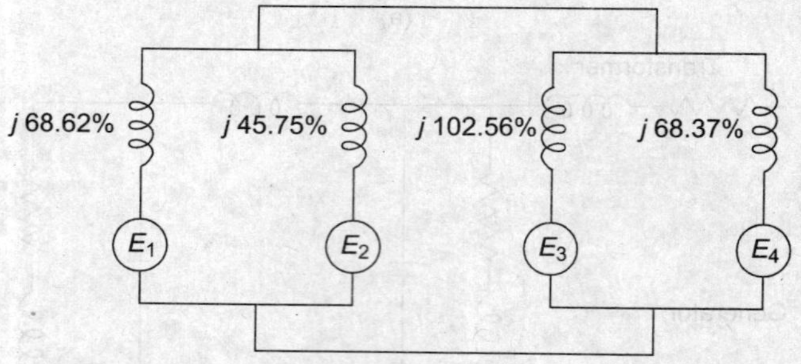

Fig. 3.18

$$\text{Reactance of } G_2 = 15\left(\frac{13.2}{13.8}\right)^2\left(\frac{50}{15}\right) = 45.75\%$$

$$\text{Reactance of } M_1 = 20\left(\frac{12.5}{13.8}\right)^2\left(\frac{50}{8}\right) = 102.56\%$$

$$\text{Reactance of } M_2 = 20\left(\frac{12.5}{13.8}\right)^2\left(\frac{50}{12}\right) = 68.37\%$$

Fig. 3.18 shows the reactance diagram. The values indicate reactances in per cent. E_1, E_2 are e.m.fs of generators while E_3, E_4 are e.m.fs of motors.

Example 3.5

A 90 MVA 11 kV 3-phase generator has a reactance of 25%. The generator supplies two motors through transformers and transmission line as shown in Fig. 3.19. The transformer T_1 is a 3-phase transformer, 100 MVA, 10/132 kV, 6% reactance. The transformer T_2 is composed of 3 single phase units each rated at 30 MVA, 66/10 kV with 5% reactance. The connections of T_1, T_2 are as shown. The motors are rated at 50 MVA and 40 MVA both 10 kV and 20% reactance. Taking the generator rating as base, draw reactance diagram and indicate the reactances in per unit. The reactance of line is 100 ohms.

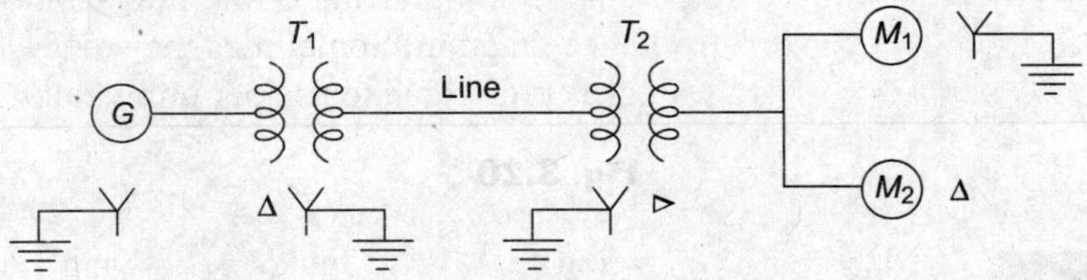

Fig. 3.19

Solution

Base MVA 90

Base voltage for generators $= 11$ kV

Base voltage for line $= 11 \times \dfrac{132}{10} = 145.2$ kV

Line to line voltage ratio of T_2 $= \sqrt{3} \times 66 : 10 = 114.31 : 10$ kV

Base voltage for motors $= \dfrac{145.2 \times 10}{114.31} = 12.7$ kV

For the above base values, the reactances are

Reactances of generator $= 25\% = 0.25$ p.u.

Reactances of T_1 $= 6 \times \left(\dfrac{10}{11}\right)^2 \left(\dfrac{90}{100}\right) = 4.46\% = 0.0446$

The reactance of T_2 is 5% on $\sqrt{3} \times 66 : 10$ kV, i. e., 114.31: 10 kV and 3×30, i.e., 90 MVA base. For 90 MVA, 145.2: 12.7 kV base,

$$\text{Reactance of } T_2 = 5 \times \left(\frac{10}{12.7}\right)^2 \left(\frac{90}{90}\right) = 3.1\% \text{ or } 0.031 \text{ p.u}$$

$$\text{Base reactance of line} = \frac{(\text{Base kV})^2}{\text{Base MVA}} = \frac{(145.2)^2}{90} = 234.26 \text{ ohm}$$

$$\text{Reactance of line} = \frac{100}{234.26} = 0.427 \text{ p.u.}$$

$$\text{Reactance of Motor } M_1 = 20\left(\frac{10}{12.7}\right)^2 \left(\frac{90}{50}\right) = 22.3\% \text{ or } 0.223 \text{ p.u.}$$

$$\text{Reactance of Motor } M_2 = 20\left(\frac{10}{12.7}\right)^2\left(\frac{90}{40}\right) = 22.9\% \text{ or } 0.279 \text{ p.u.}$$

Fig. 3.20 shows the reactance diagram. E_g is the e.m.f. of generator. E_1 and E_2 are e.m.fs of motors.

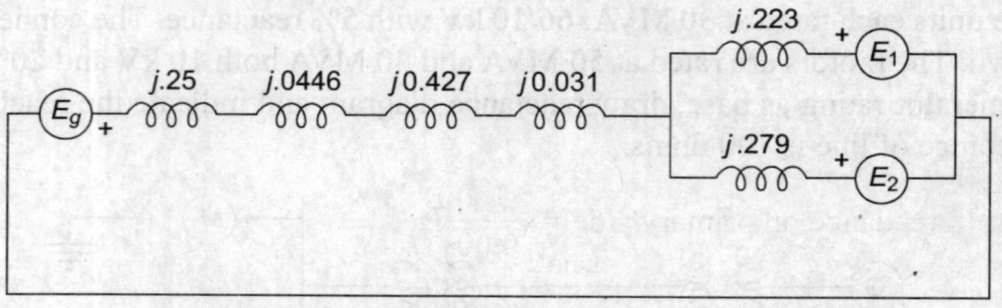

Fig. 3.20

Example 3.6

A 3-phase 3-winding Y–Y–Δ transformer is rated as under:

Primary	132 kV,	30 MVA
Secondary	11 kV,	20 MVA
Tertiary	6.6 kV,	10 MVA

The results of short circuit tests are under:

Secondary short circuited, tertiary open circuited, primary excited

$$X = 7\% \text{ on } 132 \text{ kV, } 30 \text{ MVA base}$$

Secondary open circuited, tertiary short circuited, primary excited

$$X = 9\% \text{ on } 132 \text{ kV, } 30 \text{ MVA base.}$$

Tertiary short circuited, primary open circuited, secondary excited

$$X = 4\% \text{ on } 11 \text{ kV, } 20 \text{ MVA base}$$

Find X_1, X_2, X_3 in p.u. Neglect resistances.

Solution

Base MVA 30

Base voltage for primary = 132 kV, base voltage for secondary = 11 kV,

Base voltage for tertiary = 6.6 kV.

The first two results are on the above base. The reactance of 4% on above base is equal to

$$4 \times 30/20 = 6\% \quad \text{or} \quad 0.06 \text{ p.u.}$$

It is seen from the equivalent circuit of Fig. 3.15 that

$$X_1 + X_2 = 0.07$$
$$X_1 + X_3 = 0.09$$
$$X_2 + X_3 = 0.06$$
$$X_1 = 0.5\,(0.07 + 0.09 - 0.06) = 0.05 \text{ p.u.}$$
$$X_2 = 0.5\,(0.07 + 0.06 - 0.09) = 0.02 \text{ p.u.}$$
$$X_3 = 0.5\,(0.09 + 0.06 - 0.07) = 0.04 \text{ p.u.}$$

Example 3.7

A single phase 11 kV/415 V, 100 kVA transformer has primary impedance $(3 + j5)$ ohm and secondary impedance $(0.01 + j0.04)$ ohm. Find p.u. impedance of primary and secondary. [*JNTU 2010*]

Solution

$$\text{Primary current} = \frac{100}{11} = 9.091 \text{ A}$$

$$\text{Base impedance on primary side} = \frac{11000}{9.091}$$

$$= 1210 \text{ ohm}$$

$$\text{P.u. impedance of primary} = \frac{3 + j5}{1210}$$

$$= 0.002479 + j0.004132$$

$$\text{Secondary current} = \frac{100000}{415} = 240.96 \text{ A}$$

$$\text{Base impedance on secondary side} = \frac{415}{240.96}$$

$$= 1.722 \text{ ohm}$$

$$\text{P.u. impedance of secondary} = \frac{0.01 + j0.04}{1.722}$$

$$= 0.0058 + j0.0232$$

Example 3.8

Figure 3.21 shows a single line diagram of a power system. Draw impedance diagram. Choose a base of 100 MVA, 220 kV in 50 ohm line. Ratings of generators etc. are as under:

Generator 40 MVA, 25 kV, $X'' = 20\%$

Synch. motor 50 MVA, 11 kV, $X'' = 30\%$

Y–Y transformer 40 MVA, 33/220 kV, $X = 15\%$

Y–Δ transformer 30 MVA, 11/220 kV, $X = 15\%$ [VTU 2011]

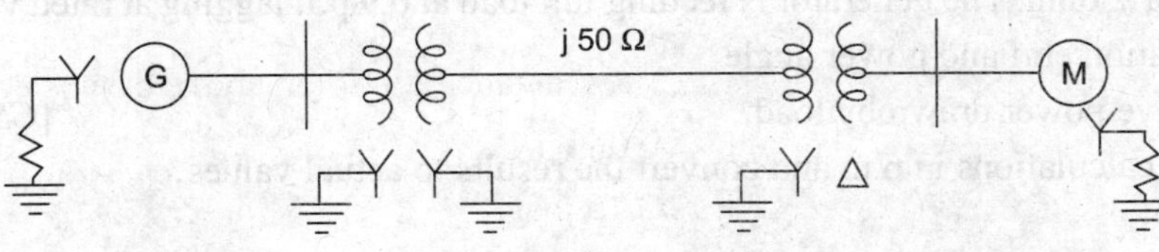

Fig. 3.21

Solution

$$\text{Base impedance of line} = \frac{(220)^2}{100} = 484\ \Omega$$

$$\text{P.u. impedance of line} = \frac{j50}{484} = j0.1033 \text{ p.u.}$$

Base values for Y–Y transformer 100 MVA, 33/220 kV

P.u. impedance of Y–Y-transformer on new base

$$= 0.15\left(\frac{100}{40}\right) = 0.375 \text{ p.u.}$$

Base values for generator 100 MVA, 33 kV

P.u. impedance of generator on new base

$$= j0.2\left(\frac{100}{40}\right)\left(\frac{25}{33}\right)^2 = j0.287 \text{ p.u.}$$

Base values for Y–Δ transformer 100 MVA, 220/11 kV

$$\text{P.u. impedance of } Y\text{–}\Delta \text{ transformer} = j0.15\left(\frac{100}{30}\right)$$

$$= j0.5 \text{ p.u.}$$

Base values for motor 100 MVA, 11 kV

$$\text{P.u. impedance of motor} = j0.3\left(\frac{100}{50}\right) = j0.6 \text{ p.u.}$$

Figure 3.22 shows the impedance diagram.

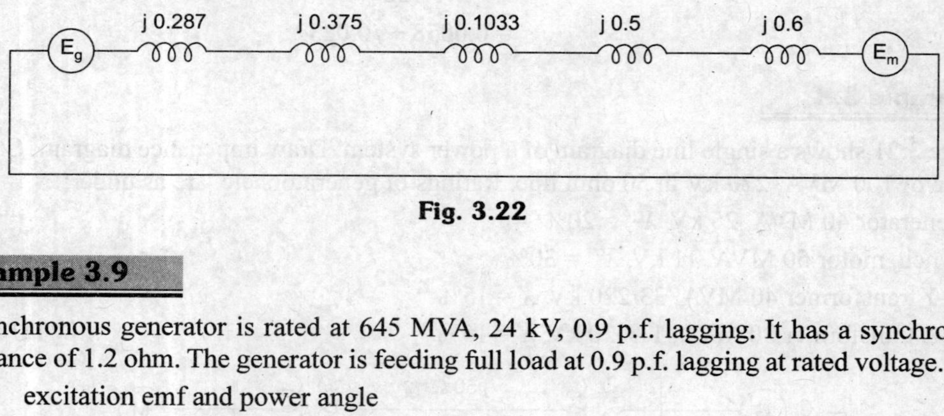

Fig. 3.22

Example 3.9

A synchronous generator is rated at 645 MVA, 24 kV, 0.9 p.f. lagging. It has a synchronous reactance of 1.2 ohm. The generator is feeding full load at 0.9 p.f. lagging at rated voltage. Find

(*i*) excitation emf and power angle

(*ii*) reactive power drawn by load. [*CSVTU 2010*]

Carry out calculations in p.u. and convert the results to actual values.

Solution

(*i*) Base MVA 645

Base kV 24

$$\text{Base impedance} = \frac{(\text{Base kV})^2}{\text{Base MVA}} = \frac{24^2}{645} = 0.893 \text{ p.u.}$$

Load current 1 p.u.

Let $V_r = 1 \angle 0$ p.u.

$$I_r = 1\,(0.9 - j0.436) \text{ p.u.}$$

$$X = \frac{1.2}{0.893} = 1.348 \text{ p.u.}$$

$$E = V_r + jIX$$

$$= 1 \angle 0 + (0.9 - j0.436)\,(j1.348)$$

$$= 1 + 0.5877 + j1.213$$

$$= 1.998 \angle 37.38° \text{ p.u.}$$

Excitation emf = 1.998 × 24 = 47.952 kV

Power angle = 37.38°

(*ii*) Reactive power of load = $Q = 1 \sin \phi = 0.436$ p.u.

= 281.22 Mvar

Example 3.10

For the system shown in Fig. 3.23 find generator voltage. [*CSVTU, 2011*]

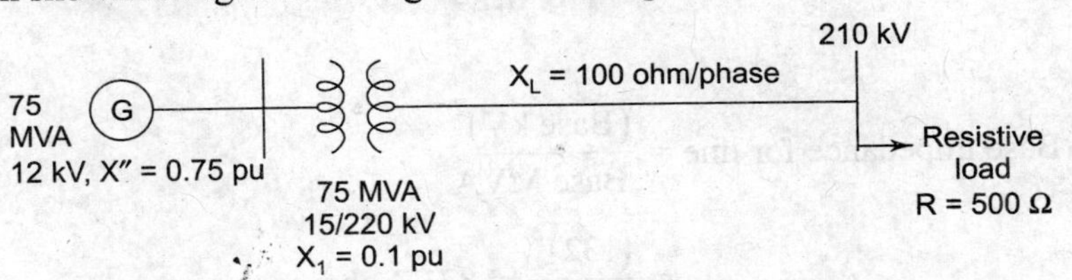

Fig. 3.23

Solution

Let base MVA = 75 MVA

Base kV for generator = 15 kV

Base kV for line = 220 kV

$$V_r = \frac{210}{220} = 0.9545 \text{ p.u.}$$

$$\text{Base impedance for line and load} = \frac{(220)^2}{75}$$

$$= 645.33 \text{ ohm}$$

$$\text{Load impedance} = \frac{500}{645.33} = 0.7748 \text{ p.u.}$$

$$\text{Load current} = \frac{0.9545}{0.7748} = 1.232 \text{ p.u.}$$

$$X_L = \frac{100}{645.33} = 0.155 \text{ p.u.}$$

$$\text{Total } X = 0.155 + 0.1 + 0.75 = 1.005 \text{ p.u.}$$

$$\text{Generator voltage} = 0.9545 + (1.232)(j1.005)$$

$$= 0.9545 + j1.238$$

$$= 1.563 \angle 52.37° \text{ p.u.}$$

$$= 1.563 \times 15 = 23.445 \text{ kV}$$

Example 3.11

Draw p.u. impedance diagram for the system in Fig. 3.24. Choose base voltage of 132 kV and base MVA 20 for transmission line. [*RTM Nagpur Uni. 2010*]

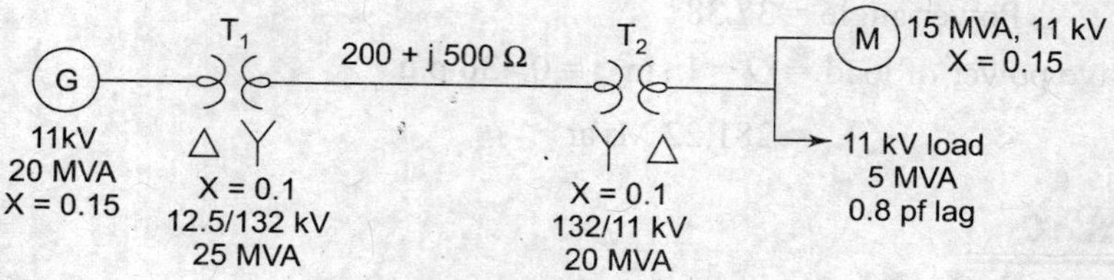

Fig. 3.24

Solution

$$\text{Base impedance for line} = \frac{(\text{Base kV})^2}{\text{Base MVA}}$$

$$= \frac{(132)^2}{20} = 871.2 \text{ ohm}$$

$$\text{P.u. impedance of line} = \frac{200 + j500}{871.2} = 0.2296 + j0.574 \text{ p.u.}$$

$$\text{P.u. impedance of } T_1 = \frac{j0.1 \times 20}{25} = j0.08 \text{ p.u.}$$

Base voltage for generator = 12.5 kV

Base MVA for generator = 20 MVA

P.u. impedance of generator on new base

$$= j0.15 \left(\frac{11}{12.5}\right)^2 = j0.116 \text{ p.u.}$$

$$\text{P.u. impedance of motor} = j0.15 \times \frac{20}{15} = j0.2 \text{ p.u.}$$

$$\text{P.u. impedance of } T_2 = j0.1 \text{ P.u.}$$

$$\text{Current drawn by load} = \frac{5000}{\sqrt{3} \times 11} = \angle -\cos^{-1} 0.8$$

$$= 262.44 \angle -36.87° \text{ A}$$

$$\text{Impedance of load} = \frac{11000}{\sqrt{3}\times 262.44} = 24.2\ \angle 36.87°\ \Omega$$

$$\text{Base impedance for load} = \frac{(11)^2}{20} = 6.05 \text{ ohm}$$

$$\text{P.u. impedance of load} = \frac{24.2\angle 36.87°}{6.05}$$

$$= 4\ \angle 36.87° \text{ p.u.}$$

Figure 3.25 shows the p.u. impedance diagram.

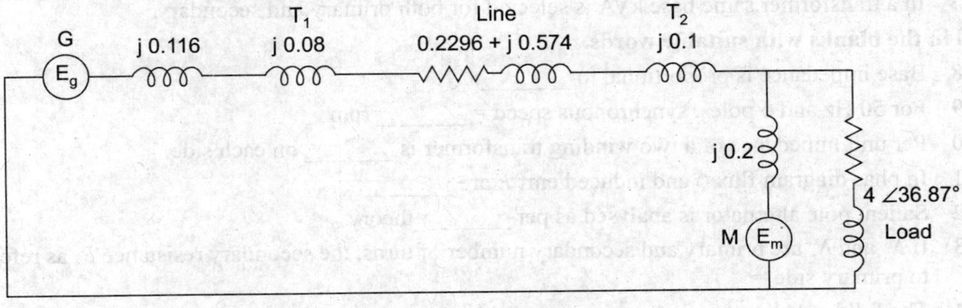

Fig. 3.25

SUMMARY

3.1 Generators used in hydroplants have salient pole construction and are wound for large number of poles and are slow speed machines.

3.2 Generators used in steam (coal fired and nuclear) plants are high speed cylindrical rotor machines. They have 2 or 4 poles and 3000 or 1500 rpm speed for 50 Hz frequency.

3.3 The equivalent circuit of cylindrical rotor generator has generated emf E in series with armature winding resistance, leakage reactance and fictitious reactance to account for armature reaction.

3.4 For cylindrical rotor generator

$$E = V + IR_a + jIx_s$$

3.5 A salient pole generator is analysed as per Blondel's two reaction theory.

$$E = V + IR_a + jI_d x_d + jI_q x_q$$

3.6 In a single phase two winding transformer all resistances and reactances can be transferred to the primary side (as also to secondary side) for ease of calculations. Moreover the shunt resistance R_0 and shunt reactance X_0 can sometimes be neglected.

3.7 A three phase transformer is also analysed on per phase basis.

3.8 Lighting and heating loads can be represented by a resistance.

3.9 Induction motor is analysed by drawing its equivalent circuit.

3.10 Approximately both P and Q vary as per $(\text{voltage})^2$

3.11 Power system calculations are usually carried out on per unit basis.

3.12 Per unit impedance of a two winding transformer is same on both sides.

3.13 A power system can be represented by a single line diagram.

3.14 An impedance diagram shows per phase emf and all series and shunt impedances per phase. If resistances are neglected, an impedance diagram becomes a reactance diagram.

SELF TEST QUESTIONS

A. State whether the following are true (T) or false (F)

3.1 Generators used in hydroplants are slow speed machines.

3.2 Per unit calculations are easy and useful.

3.3 At lagging power factor the armature reaction in generators is demagnetising.

3.4 Armature reaction can be replaced by a fictitious resistance.

3.5 Salient pole generator has higher reluctance in polar axis as compared to interpolar axis.

3.6 In per unit calculations on 3 phase systems, 3 phase kVA and line to line voltage are usually chosen as base values.

3.7 In a transformer same base kVA is selected for both primary and secondary.

B. Fill in the blanks with suitable words.

3.8 Base impedance is proportional to ________.

3.9 For 50 Hz and 6 poles, synchronous speed = ________ rpm

3.10 Per unit impedance of a two winding transformer is ________ on each side.

3.11 In phas diagram flux ϕ and induced emf E are ________.

3.12 Salient pole alternator is analysed as per ________ theory.

3.13 If N_1 and N_2 are primary and secondary number of turns, the secondary resistance R_2 as referred to primary side = ________.

3.14 The full load slip of an induction motor is about ________.

3.15 For a resistive load the effect of frequency on power consumed is ________.

ANSWERS

3.1 T	3.2 T	3.3 T	3.4 F	3.5 F
3.6 T	3.7 T	3.8 $(V)^2$	3.9 1000	3.10 same
3.11 perpendicular		3.12 two reaction		3.13 $R_2(N_1/N_2)^2$
3.14 5%	3.15 nil			

JOB INTERVIEW QUESTIONS

3.1 What is per unit system?

3.2 What is a salient pole generator?

3.3 What is leakage reactance?

3.4 What is a single line diagram?

3.5 What is equivalent circuit?

3.6 What is per phase analysis?

REVIEW QUESTIONS

3.1 Discuss the main features of cylindrical rotor alternators.

3.2 Bring out the constructional differences between salient pole and cylindrical rotor alternators.

3.3 What are the advantages of per unit system?

3.4 Draw and explain equivalent circuit of two winding transformer.

3.5 Draw and explain phasor diagram of cylindrical rotor alternator.

3.6 Discuss the effect of voltage and frequency on active and reactive power.

3.7 What is an impedance diagram?

3.8 Draw and explain equivalent circuit of induction motor.

PROBLEMS

3.1 Draw reactance diagram for the system shown in Fig. 3.26. Select suitable base.

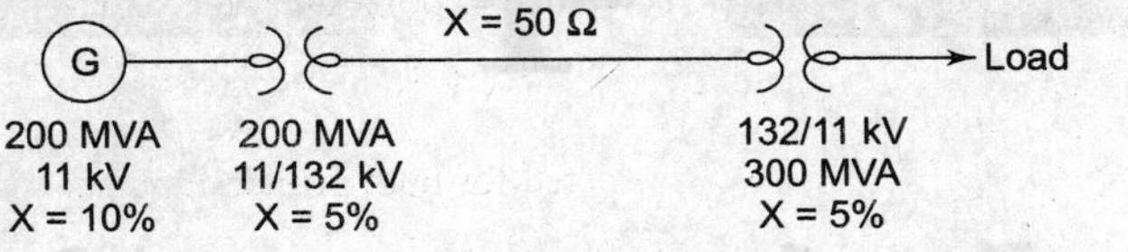

Fig. 3.26

3.2 Select a base of 500 MVA, 11 kV for generator in Fig. 3.26 and draw reactance diagram.

3.3 A generator is rated at 500 MVA, 11 kV and 0.2 p.u. reactance. Find reactance on 200 MVA, 33 kV base. [0.008888]

3.4 A line has a reactance of 35 ohms. Find per unit reactance on 10 MVA, 220 kV base. [0.00723]

3.5 A 40 kVA, 6600/250 V, single phase transformer has R_1 = 8 ohm, R_2 = 0.02 ohm, X_1 = 15 ohm and X_2 = 0.05 ohm. Find p.u. resistance and reactance as referred to secondary side. [0.0315 ohm, 0.0715 ohm]

3.6 A 5 kVA, 200/400 V single phase transformer has a total impedance of 0.12 + j0.32 ohm as referred to low voltage side. Show that p.u. impedance is the same on both sides. Find the p.u. impedance. [0.015 + j0.04 p.u.]

3.7 A single phase transformer 50 kVA, 2000/200 V has impedance drop of 10% and resistance drop of 4%. Find p.u. resistance and reactance. [0.04 + j0.0917 p.u.]

3.8 For the system in Fig. 3.27 draw p.u. impedance diagram. Select suitable base values.

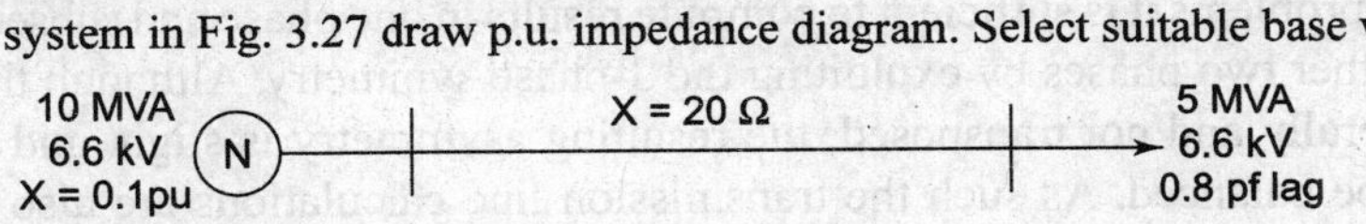

Fig. 3.27

3.9 For the system in Fig. 3.27 select a base of 20 MVA and 6.6 kV and draw p.u. impedance diagram.

CHAPTER 4

Performance of Transmission Lines

4.1 REPRESENTATION OF LINES

The performance of a power system under normal balanced steady state conditions is of primary importance in power system engineering. The transmission line is the main energy corridor in a power system. The performance of a power system is mainly dependent on the performance of the transmission lines in the system. It is necessary to calculate the voltage, current and power at any point on a transmission line provided the values at one point are known. We are aware that in 3-phase circuit problems it is sufficient to compute results in one phase and subsequently predict results in the other two phases by exploiting the 3-phase symmetry. Although the lines are not spaced equilaterally and not transposed, the resulting asymmetry is slight and the phases are considered to be balanced. As such the transmission line calculations are also carried out on per phase basis.

The transmission line performance is governed by its four parameters—series resistance and inductance, shunt capacitance and conductance. All these parameters are distributed over the length of the line. The insulation of a line is seldom perfect and leakage currents flow over the surface of insulators especially during bad weather. This leakage is simulated by shunt conductance. The shunt conductance is in parallel with the system capacitance. Generally the leakage currents are small and the shunt conductance is ignored in calculations.

Transmission lines may be classified as short, medium and long.[1] When the length of the line is less than about 80 km, the effect of shunt capacitance can be ignored and the line is designated as a short line.

When the length is between 80 and 250 km, the shunt capacitance can be considered as lumped and the line is termed as a medium length line. Lines more than 250 km long require calculations

1. The classification is mainly based on the ratio of length to wavelength. The wavelength of a 50 Hz line is 6×10^6 m or 6000 km. For 100 km line, the ratio equals 1/60 and for 300 km it is 1/20. Lines of about 500 km length have a ratio 1/12.

in terms of distributed parameters. Such lines are known as long lines. This classification on the basis of length is more or less arbitrary and the real criterion is the degree of accuracy required.

While working with balanced 3-phase circuits it is usual to assume that all the given voltages are line to line values, that all currents are line currents, that all volt amperes, power and reactive power are 3-phase and that all power factors are phase power factors (cosine of angle between phase voltage and corresponding phase current). However, all calculations are done in per phase values. The nomenclature used in this and the following chapters is as under:

V_s and I_s	Sending end per phase voltage and current
V_r and I_r	Receiving end per phase voltage and current
l	Line length
r	Line resistance per unit length per phase
L	Line inductance per unit length per phase
C	Line capacitance per unit length per phase to neutral
z	Line series impedance per unit length per phase
y	Shunt admittance per unit length per phase to neutral
Z	Total series impedance per phase $= zl = (r + j\,\omega L)\,l = R + j\,X$
Y	Total shunt admittance per phase to neutral $= yl = j\,\omega\,Cl$
δ	Power angle (i.e., angle between V_s and V_r)
ϕ_s	Phase angle between sending end voltage and current
ϕ_r	Phase angle between receiving end voltage and current

Generally the calculations are done in per unit.

4.2 SHORT TRANSMISSION LINE

Fig 4.1 (a) shows the equivalent circuit of a short transmission line. The resistance and inductance are actually distributed along the length of line but are taken as concentrated for purpose of analysis. In short lines the effect of shunt capacitance is neglected. It is apparent that

$$I_s = I_r \tag{4.1}$$

$$V_s = V_r + I_r Z \tag{4.2}$$

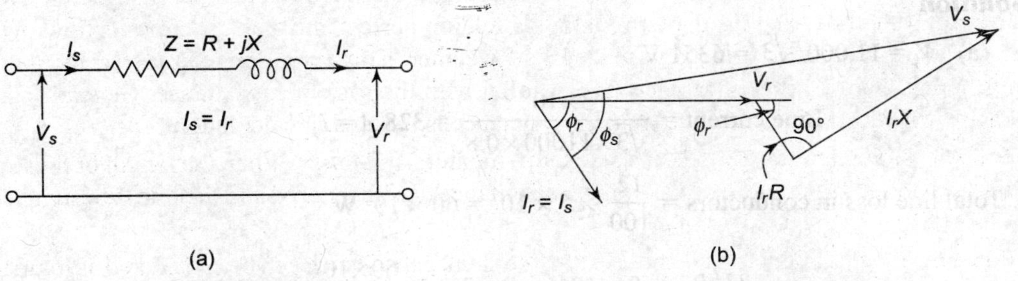

Fig 4.1 Short Transmission line (a) Equivalent circuit (b) Phasor diagram

Eqs (4.1) and (4.2) can be written in matrix form as

$$\begin{bmatrix} V_s \\ I_s \end{bmatrix} = \begin{bmatrix} 1 & Z \\ 0 & 1 \end{bmatrix} \begin{bmatrix} V_r \\ I_r \end{bmatrix} \tag{4.3}$$

From the phasor diagram in Fig 4.1 (b) we get

$$V_s = (V_r + I_r R \cos \phi_r + I_r X \sin \phi_r)^2 + (I_r X \cos \phi_r - I_r R \sin \phi_r)^2 \quad (4.4)$$

The quantities $I_r R$ and $I_r X$ are very small as compared to V_r. Therefore an approximate expression for V_s is

$$V_s = V_r + I_r R \cos \phi_r + I_r X \sin \phi_r \quad (4.5)$$

Voltage regulation of a line is defined as the rise in receiving end voltage when full load at a specified power factor is removed when sending end voltage is kept constant. It is expressed as a fraction or percentage of full load rated receiving end voltage. For short line the receiving end voltage at no load is V_s.

Hence regulation of short line is

$$\frac{|V_s| - |V_r|}{|V_r|} = \frac{I_r R \cos\phi_r + I_r X \sin\phi_r}{V_r} \; p.u. \quad (4.6)$$

The phasor diagram of Fig 4.1 (b) and Eq (4.6) are for lagging power factor. For leading power factor voltage regulation is

$$\text{Voltage regulation} = \frac{I_r R \cos\phi_r - I_r X \sin\phi_r}{V_r} \; p.u. \quad (4.7)$$

Equations (4.6) and (4.7) depict that power factor has a great influence on the voltage regulation. For lagging power factor the regulation is positive (i.e., V_r is less than V_s). For leading power factors regulation may even be negative. Regulation is zero when p.f. is leading and

$$I_r R \cos \phi_r = I_r X \sin \phi_r \quad \text{or} \quad \tan \phi_r = \frac{R}{X} \quad (4.8)$$

A certain amount of power is dissipated as I^2R loss in the line conductors. The ratio of power at the receiving end to the power at the sending end is known as efficiency of line.

Example 4.1

A15 km long 3-phase overhead line delivers 5 MW at 11 kV at a power factor of 0.8 lagging. Line loss is 12% of the power delivered. Line inductance is 1.1 mH per km per phase. Calculate (a) sending end voltage and regulation (b) power factor of the load to make regulation zero, (c) The value of capacitor to be connected at the receiving end to reduce regulation to zero.

Solution

(a) $V_r = 11{,}000/\sqrt{3} = 6351$ V

$$\text{Line current} = \frac{3\times10^6}{\sqrt{3}\times11000\times0.8} = 328\,A = I_r$$

$$\text{Total line loss in conductors} = \frac{12}{100} \times 5 \times 10^6 = 60 \times 10^4 \text{ W}$$

$$3I_r^2 R = 60 \times 10^4 \quad \text{or} \quad R = \frac{60\times10^4}{3\times328\times328} = 1.86 \text{ ohm}$$

$$X = 2\pi \times 50 \times 1.1 \times 10^{-3} \times 15 = 5.184 \text{ ohm}$$

$$V_s \approx V_r + I_r R \cos \phi_r + I_r X \sin \phi_r$$

$$= 6351 + 328 \times 1.86 \times 0.8 + 328 \times 5.184 \times 0.6 = 7859.27 \text{ V}$$

$$\text{Sending end line voltage} = \frac{\sqrt{3}\times 7859.27}{1000} = 13.61 \text{ kV}$$

$$\text{Regulation} = \frac{7859.27-6351}{6351} = 0.2375 \text{ or } 23.75\%$$

(b) For 0 regulation, $\tan\phi_r = \frac{1.86}{5.184} = 0.359, \cos\phi_r = 0.941$

and $\phi_r = 19.75°$ leading

(c) When a capacitor has been added across each phase at the receiving end, the phasor diagram is as shown in Fig 4.2. The current drawn by the capacitor is I_c. The phasor sum of I_r and I_c equals I, the new line current. From the phasor diagram it is seen that

$$I_r \cos 36.87° = I \cos 19.75°$$

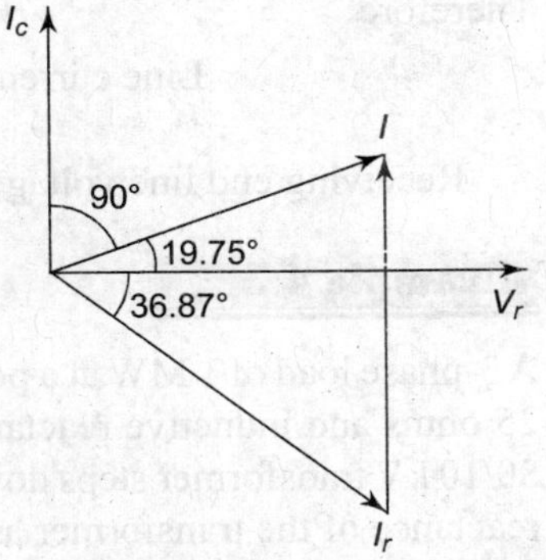

Fig. 4.2 Phasor diagram for example 4.4(c)

or
$$I = \frac{328 \cos 36.87°}{\cos 19.75°} = 278.85 \text{ A}$$

$$I_c = I_r \sin 36.87° + I \sin 19.75$$
$$= 328 \times 0.6 + 278.85 \times 0.338 = 291.05 \text{ A}$$

$$X_c = \frac{V_r}{I_c} = \frac{6351}{291.05} = 21.82 \text{ ohm}$$

$$\frac{1}{2\pi \times 50 \times C} = 21.82 \quad \text{or} \quad C = 145.88 \times 10^{-6} \text{ F}$$

Three capacitors each of 145.88 μF, connected in star at the receiving end will improve the receiving end power factor to 0.941 leading so as to result in zero regulation.

Example 4.2

A 15 km long 3-phase line has a resistance of 5.31 ohms per phase and inductive reactance of 5.54 ohms per phase. The sending end voltage is 11 kV. The receiving end load is 1200 kW at a power factor of 0.8 lagging. Find the receiving end voltage and line current.

Solution

It is a problem with mixed end conditions. The sending end voltage and the receiving end power are given. It is not necessary to derive formulae for solving all such cases. It is always better to tackle the problem from basic fundamentals and obtain a solution through mathematical manipulations.

$$V_s = 11000/\sqrt{3} = 6351 \text{ V}$$

$$I_r = \frac{1200\times 10^3}{3\times V_r \times 0.8} = 500 \times 10^3/V_r$$

$$V_s = V_r + I_r R \cos\phi_r + I_r X \sin\phi_r$$

$$6351 = V_r + \frac{500\times 10^3 \times 5.31\times 0.8}{V_r} + \frac{500\times 10^3 \times 5.54\times 0.6}{V_r}$$

or $$6351\,V_r = V_r^2 + 500 \times 10^3 \times 7.572$$

or $$V_r^2 - 6351\,V_r + 500 \times 10^3 \times 7.572 = 0$$

Hence $$V_r = 5685 \text{ V} \quad \text{or} \quad 665.96 \text{ V}$$

Evidently the value of V_r cannot be 665.96 (why ?).

Therefore $$V_r = 5685 \text{ V}$$

$$\text{Line current} = I_r = 500 \times 10^3/5685 = 87.95 \text{ A}$$

$$\text{Receiving end line voltage} = \frac{\sqrt{3} \times 5685}{1000} = 9.846 \text{ kV}$$

Example 4.3

A 3-phase load of 1 MW at a power factor of 0.8 lagging is supplied by a 30 kV line of resistance 25 ohms and inductive reactance 12 ohms per phase. The voltage across the load is 10 kV. A 30/10 kV transformer steps down the voltage at the receiving end. The equivalent resistance and reactance of the transformer as referred to 10 kV side are 0.8 ohms and 2.5 ohms respectively. Find the sending end voltage and regulation.

Solution

It is convenient to do all the calculations on the HV side of the transformer.

$$\text{Transformer resistance as referred to 30 kV side} = 0.8\left(\frac{30}{10}\right)^2 = 7.2 \text{ ohms}$$

$$\text{Transformer reactance as referred to 30 kV side} = 2.5\left(\frac{30}{10}\right)^2 = 22.5 \text{ ohms}$$

Total resistance (line and transformer) i.e., $R = 25 + 7.2 = 32.2$ ohms

Total reactance (line and transformer) i.e., $X = 12 + 22.5 = 34.5$ ohms

$$I_r = \frac{1 \times 10^6}{\sqrt{3} \times 30 \times 10^3 \times 0.8} = 24.057 \text{ A}$$

$$V_r = \frac{30000}{\sqrt{3}} = 17321 \text{V}$$

$$V_s = 17321 + 24.057 \times 32.2 \times 0.8 + 24.057 \times 34.5 \times 0.6 = 18438.7 \text{ V}$$

$$\text{Sending end line voltage} = \frac{\sqrt{3} \times 18438.7}{1000} = 31.94 \text{ kV}$$

$$\text{Regulation} = \frac{18438.7 - 17321}{17321} = 0.0645 \text{ or } 6.45\ \%$$

Example 4.4

A short 3-phase transmission line connected to 33 kV generating station at the sending end is to supply a load of 10 MW at 0.8 power factor lagging at 30 kV at the receiving end. If the minimum transmission efficiency is to be 0.96, determine the per phase values of line resistance and reactance.

Solution

$$|V_s| = \frac{33000}{\sqrt{3}} = 19052.6 \text{ volts}$$

$$|V_r| = \frac{30000}{\sqrt{3}} = 17320.5 \text{ volts}$$

$$|I_r| = \frac{10\times10^6}{\sqrt{3}\times30000\times0.8} = 240.56 \text{ A}$$

$$\text{Sending end power} = \frac{10\times10^6}{0.96} = 10.4167\times10^6 \text{ W}$$

$$\text{Line losses} = 10.4167\times10^6 - 10\times10^6 = 0.4167\times10^6 \text{ W}$$

$$3I_r^2R = 0.4167\times10^6$$

$$R = \frac{0.4167\times10^6}{3\times240.56\times240.56} = 2.4 \text{ ohm per phase}$$

$$|V_s - V_r| = I_rR\cos\phi_r + I_rX\sin\phi_r$$

$$X = \frac{|V_s - V_r| - I_r R\cos\phi_r}{I_r\sin\phi_r}$$

$$= \frac{19052.6 - 17320.5 - 240.56\times2.4\times0.8}{240.56\times0.6} = 8.8 \text{ ohm per phase.}$$

4.3 MEDIUM LENGTH TRANSMISSION LINE

The effect of shunt capacitance becomes more and more pronounced with the increase in the length of the line. For lines between 80 to 250 km length, the shunt capacitance can be considered as lumped.

In the nominal T circuit, the total capacitance of each conductor is concentrated at the centre of the line. In the nominal π circuit, one half of the total capacitance of each conductor is placed at each end. The qualification nominal infers that these are arbitrary but convenient circuits giving an accuracy which is sufficient for most practical purposes.

4.3.1 Nominal *T* Circuit

Figure. 4.3 shows the nominal T circuit of a line. The shunt admittance is assumed to be concentrated in the middle while the series impedance is split into two equal parts. We can write the following equations for this circuit

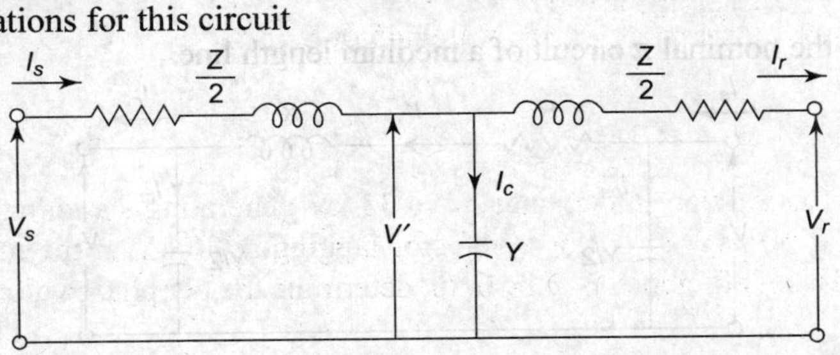

Fig. 4.3 Nominal *T* circuit of a medium length transmission line

$$V' = V_r + I_r \frac{Z}{2} \tag{4.9}$$

$$I_c = V'Y = \left(V_r + I_r \frac{Z}{2}\right)Y \tag{4.10}$$

$$I_s = I_r + I_c = I_r + V_r Y + I_r \frac{ZY}{2} = V_r Y + I_r\left(1 + \frac{ZY}{2}\right) \tag{4.11}$$

$$V_s = V' + I_s \frac{Z}{2} = V_r + I_r \frac{Z}{2} + \left(V_r Y + I_r + I_r \frac{ZY}{2}\right)\frac{Z}{2}$$

$$= V_r\left(1 + \frac{ZY}{2}\right) + I_r Z\left(1 + \frac{ZY}{4}\right) \tag{4.12}$$

Figure 4.4 shows the phasor diagram for this circuit. AC represents the voltage drop $I_r \frac{Z}{2}$ which when added to V_r gives the voltage V'. The phasor OF represents I_c and leads the phasor V' by 90°. The phasor sum of OG (i.e., I_r) and OF gives the current I_s represented by the phasor OH. The phasor CD represents $I_s \frac{Z}{2}$ which when added to V′ gives the sending end voltage V_s.

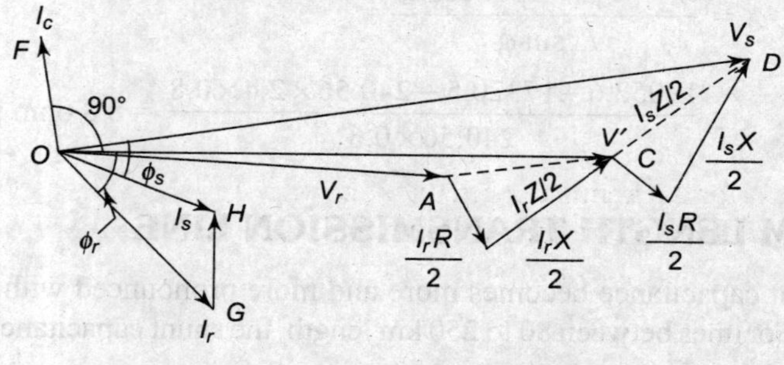

Fig. 4.4 Phasor diagram for nominal *T* circuit

Eqs. (4.12) and (4.11) can be written in the matrix form as

$$\begin{bmatrix} V_s \\ I_s \end{bmatrix} = \begin{bmatrix} 1 + \frac{ZY}{2} & Z\left(1 + \frac{ZY}{4}\right) \\ Y & 1 + \frac{ZY}{2} \end{bmatrix} \begin{bmatrix} V_r \\ I_r \end{bmatrix} \tag{4.13}$$

4.3.2 Nominal π Circuit

Figure 4.5 shows the nominal π circuit of a medium length line.

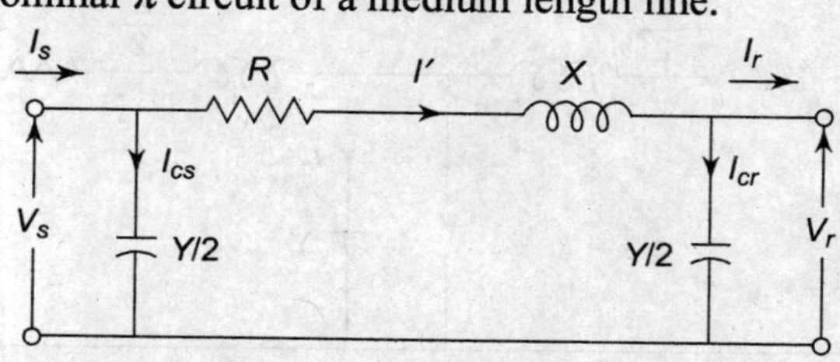

Fig. 4.5 Nominal π circuit of a medium length line

We can write the following equations for this circuit.

$$I' = I_r + I_{cr} = I_r + V_r \frac{Y}{2} \tag{4.14}$$

$$V_s = V_r + I'Z = V_r + \left(I_r + V_r \frac{Y}{2}\right)Z$$

$$= V_r\left(1 + \frac{ZY}{2}\right) + I_r Z \tag{4.15}$$

$$I_s = I' + I_{cs} = I' + V_s \frac{Y}{2}$$

$$= I_r + V_r \frac{Y}{2} + \left[V_r\left(1 + \frac{ZY}{2}\right) + I_r Z\right]\frac{Y}{2}$$

$$= V_r Y\left(1 + \frac{ZY}{4}\right) + I_r\left(1 + \frac{ZY}{2}\right) \tag{4.16}$$

Eqs. (4.15 and 4.16) can be written as:

$$\begin{bmatrix} V_s \\ I_s \end{bmatrix} = \begin{bmatrix} 1 + \dfrac{ZY}{2} & Z \\ Y\left(1 + \dfrac{ZY}{4}\right) & 1 + \dfrac{ZY}{2} \end{bmatrix} \begin{bmatrix} V_r \\ I_r \end{bmatrix} \tag{4.17}$$

The phasor diagram for the nominal π circuit is shown in Fig. 4.6. The receiving end voltage, V_r, is represented by OA and OB is the receiving end current I_r. The phasor BE represents the current I_{cr} and leads the phasor OA by 90°. The phasor sum of I_r and I_{cr} gives I' (shown by phasor OE). AD represents $I'Z$ which when added to V_r gives V_s. EF represents current I_{cs} and leads the phasor OD by 90°. The phasor sum of OE and EF gives OF which represents the sending end current I_s.

It may be noted that the nominal T and π circuits are not equivalent representations. These are different approximations for a real line.

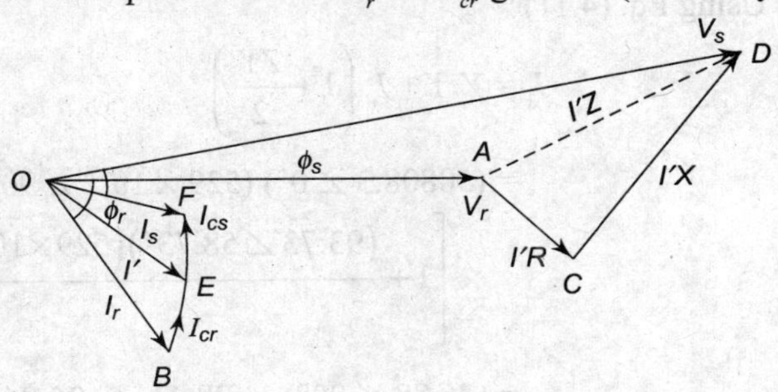

Fig. 4.6 Phasor diagram for nomial π circuit

Example 4.5

A 200 km long 3-phase overhead line has a resistance of 48.7 ohms per phase, inductive reactance of 80.20 hms per phase and capacitance (line to neutral) 8.42 nF per km. It supplies a load of 13.5 MW at a voltage of 88 kV and power factor 0.9 lagging. Using nominal T circuit, find the sending end voltage, current, regulation and power angle.

Solution

$$Z = 48.7 + j\,80.2 \text{ ohms} = 93.83 \angle 58.73° \text{ ohms}$$

$$Y = j\,\omega\,Cl = 2\pi \times 50 \times 8.42 \times 10^{-9} \times 200 \angle 90°$$

$$= 529 \times 10^{-6} \angle 90° \text{ siemens}$$

$$V_r = \frac{88 \times 1000}{\sqrt{3}} = 50808.3 \angle 0 \text{ V}$$

$$I_r = \frac{13.5 \times 10^6}{\sqrt{3} \times 88 \times 1000 \times 0.9} \angle (-\cos^{-1} 0.9)$$

$$= 98.41 \angle -25.84^\circ \text{A}$$

Using Eq. (4.12)

$$V_s = V_r \left(1 + \frac{ZY}{2}\right) + I_r Z \left(1 + \frac{ZY}{4}\right)$$

$$= 50808.3 \left[1 + \frac{(93.83 \angle 58.73^\circ)(529 \times 10^{-6} \angle 90^\circ)}{2}\right] + (98.41 \angle -25.84^\circ)$$

$$\times (93.83 \angle 58.73^\circ) \left[1 + \frac{(93.83 \angle 58.73^\circ)(529 \times 10^{-6} \angle 90^\circ)}{4}\right]$$

$$= 50808.3\,(1 + 0.024818 \angle 148.73^\circ) + (9233.81 \angle 32.89^\circ)$$

$$\times (1 + 0.012409 \angle 148.73^\circ)$$

$$= 50808.3\,(0.9788 + j\,0.0129) + (9233.81 \angle 32.89^\circ) \times (0.9894 + j\,0.00645)$$

$$= 57369.73 + j\,5648.77 = 57647.2 \angle 5.62^\circ$$

Sending end line voltage = $\sqrt{3} \times 57647.2$ V = 99.85 kV

Power angle = 5.62°

Using Eq. (4.11)

$$I_s = V_r Y + I_r \left(1 + \frac{ZY}{2}\right)$$

$$= (50808.3 \angle 0^\circ)(529 \times 10^{-6} \angle 90^\circ) + (98.41) \angle -25.84^\circ)$$

$$\times \left[1 + \frac{(93.73 \angle 58.73^\circ)(529 \times 10^{-6} \angle 90^\circ)}{2}\right]$$

$$= (26.88 \angle 90^\circ) + (98.41 \angle -25.84^\circ)(0.9788 + j\,0.0129)$$

$$= j\,26.88 + 87.25 - j\,40.83 = 87.25 - j\,13.95 = 88.36 \angle -9.08^\circ \text{ A}$$

Sending end line current = 88.36 A

From Eq. (4.18) the receiving end voltage at no load is

$$|V_r| \text{ at no load} = \frac{|V_s|}{\left|1 + \frac{ZY}{2}\right|} = \frac{57647.2}{|0.9788 + j\,0.0129|} = 58890.98$$

$$\text{Regulation} = \frac{58890.98 - 50808.3}{50808.3} = 0.159 \text{ or } 15.9\%$$

Example 4.6

Solve example 4.5 using nominal π circuit.

Solution

Using Eq. (4.15)

$$V_s = V_r\left(1+\frac{ZY}{2}\right)+I_r Z$$

$$= (50808.3 \angle 0°\left[1+\frac{(93.83 \angle 58.73°)(529\times 10^{-6} \angle 90°)}{2}\right]$$

$$+ (98.41 \angle -25.84°)\ (93.83 \angle 58.73°)$$

$$= 50808.3\ (0.9788 + j\ .0.0129) + 9233.81 \angle 32.89°$$

$$= 49731.16 + j\ 655.43 + 7753.76 + j\ 5014.21$$

$$= 57484.92 + j\ 5669.64 = 57763 \angle 5.63°\ \text{V}$$

Sending end line voltage

$$= \sqrt{3} \times 57763\ \text{V} = 100.05\ \text{kV}$$

Power Angle = 5.63°

Using Eq. (4.16)

$$I_s = V_r Y\left(1+\frac{ZY}{4}\right)+I_r\left(1+\frac{ZY}{2}\right)$$

$$= (50808.3 \angle 0°)\ (529\times 10^{-6} \angle 90°)\left[1+\frac{(93.83 \angle 58.73°)(529\times 10^{-6} \angle 90°)}{4}\right]$$

$$+ (98.41 \angle -25.84°)\left[1+\frac{(93.83 \angle 58.73°)(529\times 10^{-6} \angle 90°)}{2}\right]$$

$$= (26.88 \angle 90°)\ (0.9894 + j\ 0.00645) + (98.41 \angle -25.84°)\times(0.9788 + j.0129)$$

$$= j\ 26.6 - 0.17 + 87.2 - j\ 40.84 = 87.03 - j\ 14.24 = 88.19 \angle -9.29°\ \text{A}$$

$$V_r \text{ at no load} = \frac{|V_s|}{\left|1+\frac{ZY}{2}\right|} = \frac{57763}{[0.9788 + j0.0129]}$$

$$= 59009.4\ \text{V}$$

$$\text{Regulation} = \frac{59009.4 - 50808.3}{50808.3} = \frac{8201.1}{50808.3} = 0.1614 \text{ or } 16.140\%$$

Example 4.7

A 3-phase 50 Hz 100 km long overhead line has the following line constants: resistance per phase per km = 0.153 ohm, inductance per phase per km = 1.21 mH, capacitance per phase per km = 0.00958 μF. The line supplies a load of 20 MW at 0.9 power factor lagging at a line voltage of 110 kV at the receiving end. Using nominal π representation calculate sending end voltage, current, power factor, regulation and efficiency.

Solution

$$V_r = \frac{110000}{\sqrt{3}} = 63508.3 \angle 0$$

$$I_r = \frac{20 \times 10^6}{\sqrt{3} \times 110000 \times .9} \angle -\cos^{-1} 0.9 = 116.64 \angle -25.84° \text{ A}$$

$$Z = (0.153 + j\,2\,\pi \times 50 \times 1.21 \times 10^{-3})\,100 = 15.3 + j\,38.0 = 40.96 \angle 68.07° \;\Omega$$

$$Y = (j\,2\,\pi \times 50 \times 0.00958 \times 10^{-6})\,100 = 300.96 \times 10^{-6} \angle 90° \text{ Siemens}$$

Using Eq. (4.15)

$$V_s = V_r\left(1 + \frac{ZY}{2}\right) + I_r Z$$

$$= (63508.5 \angle 0° \left[1 + \frac{(40.96 \angle 68.07°)(300.96 \times 10^{-6} \angle 90°)}{2}\right]$$

$$+ (116.64 \angle -25.84°)\,(40.96 \angle 68.07°)$$

$$= (63508.5 \angle 0°)\,(1 + 6163.66 \times 10^{-6} \angle 158.07)$$

$$+ (116.64 \angle -25.84°)\,(40.96 \angle 68.07°),$$

$$= (63508.5 \angle 0°)\,(0.994282 + j\,0.0023) + 4777.57 \angle 42.23°$$

$$= 63145.4 + j\,146.07 + 3537.57 + j\,3211.04$$

$$= 66682.97 + j\,3357.11 = 66767.42 \angle 2.88° \text{ V}$$

Using Equation (4.16)

$$I_s = V_r Y\left(1 + \frac{ZY}{4}\right) + I_r\left(1 + \frac{ZY}{2}\right)$$

$$= (63508.5 \angle 0°)\,(300.96 \times 10^{-6} \angle 90°)$$

$$\times \left[1 + \frac{(40.96 \angle 68.07°)(300.96 \times 10^{-6} \angle 90°)}{4}\right]$$

$$+ (116.84 \angle -25.84°) \times \left[1 + \frac{(40.96 \angle 68.07°)(300.96 \times 10^{-6} \angle 90°)}{2}\right]$$

$$= (19.11 \angle 90°)\,(1 + 3081.83 \times 10^{-6} \angle 158.07°)$$

$$+ (116.84 \angle -25.84°)\,(1 + 6163.66 \times 10^{-6} \angle 158.07°)$$

$$= (j19.11)\,(0.997141 + j\,0.00115) + (116.84 \angle -25.84°) \times (0.994284 \angle 0.13°)$$

$$= (-0.022 + j\,19.055) + (116.17 \angle -25.71°)$$

$$= -0.022 + j\,19.055 + 104.67 - j\,50.4$$

$= 104.648 - j\,31.345$

$= 109.24 \angle -16.67° \text{ A}$

Sending end line voltage $= \sqrt{3} \times 66767.42 \text{ V} = 115.64 \text{ kV}$

Sending end line current $= 109.24 \text{ A}$

Sending end power factor $= \cos (2.88° + 16.67°)$

$= \cos (19.55°) = 0.942$ lagging

$$V_r \text{ at no load} = \frac{|V_s|}{\left|1 + \dfrac{ZY}{2}\right|} = \frac{66767.42}{0.994282} = 67151.39 \text{ V}$$

$$\text{Regulation} = \frac{67151.39 - 63508.5}{63508.5} \times 100 = 5.74\%$$

$$\text{Efficiency} = \frac{20 \times 10^6}{3\, V_s I_s \cos \phi_s} \times 100$$

$$= \frac{20 \times 10^6}{3 \times 66767.42 \times 109.24 \times 0.942}$$

$$= 97\%$$

4.4 LONG TRANSMISSION LINE

As stated earlier, the line parameters are distributed uniformly over the entire length. The analysis for transmission systems, when the length of the line has a magnitude comparable to wavelength $\left(\dfrac{l}{\lambda} >> \dfrac{l}{20}\right)$ becomes more difficult in contrast to short or medium length lines. The validity of lumped circuit analysis fails in this case and the lumped circuit is to be considered for an elemental section of length only. A differential equation is obtained which is to be solved for boundary conditions.

Figure 4.7 shows a distributed parameter line with series impedance z per unit length and shunt admittance y (siemens) per unit length. x is the distance from the receiving end. $V(x)$ and $I(x)$ denote the phasor voltage and phasor current at location x on the line. For an element Δx, the series impedance is $z\,\Delta x$ and the shunt admittance is $y\,\Delta x$.

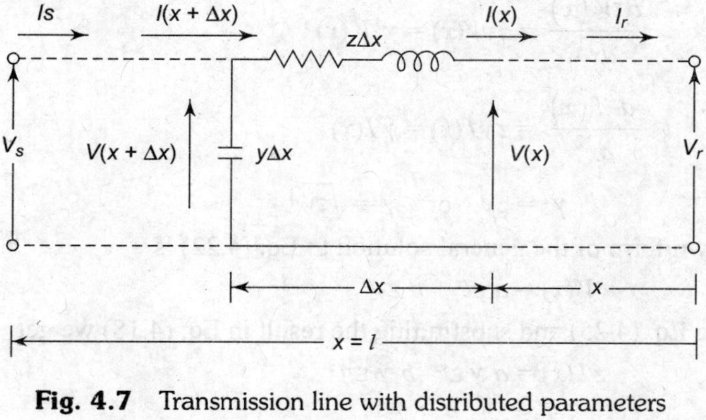

Fig. 4.7 Transmission line with distributed parameters

From Kirchhoffs voltage law:

$$V(x+\Delta x) = V(x) + z\,\Delta x \cdot I(x)$$

Rearranging
$$\frac{V(x+\Delta x)-V(x)}{\Delta x} = z \cdot I(x)$$

Taking the limit as $\Delta x \to 0$

$$\lim_{\Delta x\to 0}\frac{V(x+\Delta x)-V(x)}{\Delta x} = z \cdot I(x)$$

or
$$\frac{dV(x)}{dx} = z \cdot I(x) \tag{4.18}$$

By Kirchhoffs current law:

$$I(x+\Delta x) = I(x) + y\,\Delta x \cdot V(x+\Delta x)$$

Rearranging
$$\frac{I(x+\Delta x)-I(x)}{\Delta x} = y\,V(x+\Delta x)$$

Taking the limit as $\Delta x \to 0$

$$\lim_{\Delta x\to 0}\frac{I(x+\Delta x)-I(x)}{\Delta x} = \lim_{\Delta x\to 0} yV(x+\Delta x)$$

or
$$\frac{dI(x)}{dx} = y\,V(x) \tag{4.19}$$

Differentiate Eqs. (4.18) and (4.19) with respect to x

$$\frac{d^2V(x)}{dx^2} = z\frac{dI(x)}{dx} \tag{4.20}$$

$$\frac{d^2I(x)}{dx^2} = y\frac{dV(x)}{dx} \tag{4.21}$$

Substitute the values of $\frac{dI(x)}{dx}$ and $\frac{dV(x)}{dx}$ from Eqs. (4.19) and (4.18) into Eqs. (4.20) and (4.21) thus giving

$$\frac{d^2V(x)}{dx^2} = zyV(x) = \gamma^2 V(x) \tag{4.22}$$

$$\frac{d^2I(x)}{dx^2} = zyI(x) = \gamma^2 I(x) \tag{4.23}$$

where
$$\gamma^2 = zy \quad \text{or} \quad \gamma = \sqrt{zy} \tag{4.24}$$

One well-known form of the general solution to Eq. (4.22) is

$$V(x) = a\,\varepsilon^{\gamma x} + b\,\varepsilon^{-\gamma x} \tag{4.25}$$

Differentiating Eq. (4.25) and substituting the result in Eq. (4.18) we get

$$zI(x) = a\,\gamma\,\varepsilon^{\gamma x} - b\,\gamma\,\varepsilon^{-\gamma x}$$

or $$I(x) = \frac{a\gamma}{z}\varepsilon^{\gamma x} - \frac{b\gamma}{z}\varepsilon^{-\gamma x}$$

or $$I(x) = \frac{a\varepsilon^{\gamma x}}{\sqrt{z/y}} - \frac{b\varepsilon^{-\gamma x}}{\sqrt{z/y}} \tag{4.26}$$

It is convenient at this state to define the characteristic impedance Z_c

$$Z_c = \sqrt{z/y} \tag{4.27}$$

What are the unit of Z_c? ohms (note that per unit lengths cancel). Since z and y are complex, Z_c is also, in general, complex. The constants a and b can be evaluated from terminal conditions, viz., voltage at receiving end (i.e., at $x = 0$) is V_r and current at receiving end is I_r. Substituting these values in Eqs. (4.25) and (4.26) we get

$$V(0) = V_r = a + b \tag{4.28}$$

$$I(0) = I_r = \frac{1}{\sqrt{z/y}}(a - b) = \frac{1}{Z_c}(a - b) \tag{4.29}$$

From Eqs. (4.28) and (4.29)

$$a = 0.5(V_r + I_r Z_c)$$
$$b = 0.5(V_r - I_r Z_c) \tag{4.30}$$

Substituting the values of a and b from Eq. (4.30) in Eqs. (4.25) and (4.26) we get

$$V(x) = \frac{V_r + I_r Z_c}{2}\varepsilon^{\gamma x} + \frac{V_r - I_r Z_c}{2}\varepsilon^{-\gamma x} \tag{4.31}$$

$$Z_c I(x) = \frac{V_r + I_r Z_c}{2}\varepsilon^{\gamma x} - \frac{V_r - I_r Z_c}{2}\varepsilon^{-\gamma x} \tag{4.32}$$

Rearranging, $$V(x) = \left(\frac{\varepsilon^{\gamma x} + \varepsilon^{-\gamma x}}{2}\right)V_r + \left(\frac{\varepsilon^{\gamma x} - \varepsilon^{-\gamma x}}{2}\right)Z_c I_r \tag{4.33}$$

$$Z_c I(x) = \left(\frac{\varepsilon^{\gamma x} - \varepsilon^{-\gamma x}}{2}\right)V_r + \left(\frac{\varepsilon^{\gamma x} + \varepsilon^{-\gamma x}}{2}\right)Z_c I_r \tag{4.34}$$

Introducing hyperbolic functions for the terms within brackets in Eqs. (4.33) and (4.34):

$$V(x) = (\cosh \gamma x)\, V_r + Z_c (\sinh \gamma x) I_r \tag{4.35}$$

$$I(x) = \frac{1}{Z_c}(\sinh \gamma x)\, V_r + (\cosh \gamma x)\, I_r \tag{4.36}$$

Eqs. (4.35 and 4.36) or (4.33 and 4.34) are completely adequate to find the effective values and phase angles of voltage and current at any point on the line provided V_r, I_r and line parameters are known. To obtain sending end values we set x equal to l. The result, written in matrix form, is:

$$\begin{bmatrix} V_s \\ I_s \end{bmatrix} = \begin{bmatrix} \cosh \gamma l & Z_c \sinh \gamma l \\ \frac{1}{Z_c}\sinh \gamma l & \cosh \gamma l \end{bmatrix} \begin{bmatrix} V_r \\ I_r \end{bmatrix} \tag{4.37}$$

Eq. (4.37) can be written in general form as:

$$\begin{bmatrix} V_s \\ I_s \end{bmatrix} = \begin{bmatrix} A & B \\ C & D \end{bmatrix} \begin{bmatrix} V_r \\ I_r \end{bmatrix} \tag{4.38}$$

where

$$A = D = \cosh \gamma l$$
$$B = Z_c \sinh \gamma l \tag{4.39}$$
$$C = \frac{1}{Z_c} \sinh \gamma l$$

The *ABCD* constants are known as generalized circuit constants and are, in general, complex numbers. *A* and *D* are dimension less and equal if the line is the same as viewed from either end.

The dimensions of *B* and *C* are ohms and siemens respectively. For all two port linear passive bilateral networks

$$AD - BC = 1 \tag{4.40}$$

Eq. (4.38) can be inverted to give the receiving end conditions in terms of sending end conditions.

$$\begin{bmatrix} V_r \\ I_r \end{bmatrix} = \begin{bmatrix} D & -B \\ -C & A \end{bmatrix} \begin{bmatrix} V_s \\ I_s \end{bmatrix} \tag{4.41}$$

4.5 EVALUATION OF *ABCD* PARAMETERS

The propagation constant γ is a complex number and can be written as

$$\gamma = \alpha + j\beta \tag{4.42}$$

α, the real part of γ is known as attenuation constant and is measured in nepers per unit length, β, the imaginary part of γ, is called the phase constant and is measured in radians per unit length. The general line constants are hyperbolic functions of the complex variable γ. The hyperbolic sines and cosines of complex variables can be expanded in terms of circular hyperbolic functions of real variables:

$$\sinh \gamma l = \sinh(\alpha l + j\beta l) = \sinh \alpha l \cos \beta l + j \cosh \alpha l \sin \beta l \tag{4.43}$$
$$\cosh \gamma l = \cosh(\alpha l + j\beta l) = \cosh \alpha l \cos \beta l + j \sinh \alpha l \sin \beta l \tag{4.44}$$

An alternative method of evaluation is the expansion by Maclaurin' series

$$\sinh \gamma l = \gamma l + \frac{\gamma^3 l^3}{3!} + \frac{\gamma^5 l^5}{5!} + \ldots\ldots$$

$$= \sqrt{YZ}\left[1 + \frac{YZ}{3!} + \frac{(YZ)^2}{5!} + \ldots\ldots\right] \tag{4.45}$$

$$\cosh \gamma l = 1 + \frac{\gamma^2 l^2}{2!} + \frac{\gamma^4 l^4}{4!} + \ldots\ldots$$

$$= 1 + \frac{YZ}{2!} + \frac{(YZ)^2}{4!} + \ldots\ldots \tag{4.46}$$

These series converge rapidly for usual values of γl. Sufficient accuracy can be obtained by using the first two or three terms.

A third method of evaluation uses the following equations:

$$\sinh \gamma l = \sinh (\alpha l + j\beta l) = \frac{\varepsilon^{\alpha l}\varepsilon^{j\beta l} - \varepsilon^{-\alpha l}\varepsilon^{-j\beta l}}{2}$$

$$= 0.5\,(\varepsilon^{\alpha l}\angle \beta l - \varepsilon^{-\alpha l}\angle -\beta l) \tag{4.47}$$

$$\cosh \gamma l = \cosh (\alpha l + j\beta l) = \frac{\varepsilon^{\alpha l}\varepsilon^{j\beta l} - \varepsilon^{-\alpha l}\varepsilon^{-j\beta l}}{2}$$

$$= 0.5\,(\varepsilon^{\alpha l}\angle \beta l + \varepsilon^{-\alpha l}\angle -\beta l) \tag{4.48}$$

The infinite series formulae (Eqs. 4.45 and 4.46) show clearly the errors involved in using short and nominal representations. Table 4.1 lists the *ABCD* constants for short, medium and long lines.

Example 4.8

A 3-phase line is 500 km long. The line constants are $z = 0.105 + j\,0.3768$ ohms/km. $y = 0 + j\,2.822 \times 10^{-6}$ siemens per km. The line delivers 40 MVA at 0.9 pf lagging at 220 kV. Find the sending end voltage, current, power factor, MVA and power angle.

Solution

$$Z = (0.105 + j\,0.3768)\,500 = 52.5 + j\,188.4 \text{ ohms} = 195.578 \angle 74.429°$$

$$Y = (0 + j\,2.822 \times 10^{-6})\,500 = 0.001411 \angle 90° \text{ siemens}$$

$$YZ = 0.27596 \angle 164.429°$$

$$A = 1 + \frac{YZ}{2} + \frac{(YZ)^2}{24}$$

$$= 1 + \frac{0.27596 \angle 164.429°}{2} + \frac{(0.27596 \angle 164.429°)^2}{24}$$

$$= 0.8698 + j\,0.035397 = 0.8705 \angle 2.33° = D$$

$$B = Z\left[1 + \frac{YZ}{6} + \frac{(YZ)^2}{120}\right]$$

$$= (195.578 \angle 74.429°)\left[1 + \frac{0.27596 \angle 164.429°}{6} + \frac{(0.27596 \angle 164.429)^2}{120}\right]$$

$$= 47.937 + j\,180.786 = 187.03 \angle 75.15°$$

$$C = Y\left[1 + \frac{YZ}{6} + \frac{(YZ)^2}{120}\right]$$

$$= -\,0.000017 + j\,0.001349 = 0.0013494 \angle 90.722°$$

$$V_r = \frac{220000}{\sqrt{3}} = 127021 \angle 0 \text{ V}$$

$$I_r = \frac{40 \times 10^6}{\sqrt{3} \times 200 \times 1000} \angle -25.84° = 104.976 \angle -25.84° \text{A}$$

Table 4.1 Transmission line *ABCD* Constants

S. No.	*Description*	*Parameters*
1.	Short line $A = 1$, $C = 0$	$B = Z$, $D = A = 1$
2.	Medium line (nominal T representation) $A = \left(1+\frac{ZY}{2}\right)$, $C = Y$	$B = Z\left(1+\frac{ZY}{4}\right)$, $D = A$
3.	Medium line (nominal π representation) $A = 1+\frac{ZY}{2}$, $C = Y\left(1+\frac{ZY}{4}\right)$	$B = Z$, $D = A$
4.	Long line $A = \cosh \gamma l = \cosh \sqrt{YZ} = 1+\frac{YZ}{2!}+\frac{(YZ)^2}{4!}+\ldots$, $C = \frac{1}{Z_c}\sinh \gamma l = y\frac{\sinh \sqrt{YZ}}{\sqrt{YZ}} = Y\left(1+\frac{YZ}{3!}+\frac{(YZ)^2}{5!}+\ldots\right)$	$B = Z_c \sinh \gamma l = \frac{\sinh \sqrt{YZ}}{\sqrt{YZ}} = Z\left[1+\frac{YZ}{3!}+\frac{(YZ)^2}{5!}+\ldots\right]$, $D = A$

$$\begin{aligned}
V_s &= AV_r + BI_r \\
&= (0.8705 \angle 2.33°)(127021 \angle 0) \\
&\quad + (187.03 \angle 75.15)(104.976 \angle -25.84°) \\
&= 110571.8 \angle 2.33° + 19633.66 \angle 49.31° \\
&= 123280.86 + j\,19382.48 = 124795.23 \angle 8.935° \text{ V} \\
I_s &= CV_r + D\,I_r \\
&= (.0013494 \angle 90.722°)(127021 \angle 0) + (0.8705 \angle 2.33°) \\
&\quad \times (104.976 \angle -25.84°) = 81.64 + j\,134.939 = 157.71 \angle 58.8° \text{ A}
\end{aligned}$$

$$\text{Sending end line voltage} = \frac{\sqrt{3}\times 124795.23}{1000} = 216.15 \text{ kV}$$

Sending end line current = 157.71 A

Sending end power factor = cos (58.8° – 8.935°) = 0.645 leading

$$\text{Sending end MVA} = \frac{\sqrt{3}\times 216.15\times 157.71}{1000} = 59.042 \text{ MVA}$$

Power angle = 8.935°

The first three terms of the infinite series have been used in evaluating *A, B* and C. However, it is sufficiently accurate to use only the first two terms.

Example 4.9

Find the sending end voltage, current and power factor for the transmission line of Example 4.8, when the line is open circuited at the receiving end. The voltage at receiving end is 220 kV.

Solution

$$V_r = 127021 \angle 0, I_r = 0$$

$$V_s = AV_r + BI_r = (0.8705 \angle 2.33°)\,(127021 \angle 0)$$

$$= 110571.8 \angle 2.33°$$

$$I_s = CV_r + DI_r = (0.0013494 \angle 90.722°)\,(127021 \angle 0)$$

$$= 171.4 \angle 90.722° \text{ A}$$

$$\text{Sending end line voltage} = \frac{110571.8 \times \sqrt{3}}{1000} = 191.51 \text{ kV}$$

Sending end line current = 171.4 A

Sending end power factor = cos (90.722° – 2.33°) = 0.0281 leading

Example 4.10

A 3-phase 200 km long high voltage line has $Z = 14.1 + j\,51.48$ ohms and $y = 0 + j\,1.194 \times 10^{-3}$ siemens. Find the characteristic impedance, propagation constant and constants *A, B* and *C* for the line.

Solution

$$z = \frac{14.1 + j\,51.48}{200} = 0.0705 + j\,0.2574 \text{ ohms per km}$$

$$y = \frac{0 + j\,1.194 \times 10^{-3}}{200} = 0 + j\,5.97 \times 10^{-6} \text{ siemens per km}$$

$$Z_c = \sqrt{z/y} = \frac{\sqrt{0.0705 + j0.2574}}{\sqrt{0 + j\,5.97 \times 10^{-6}}} = 212 \angle -7.68° \text{ ohms}$$

$$\gamma = \sqrt{z\,y} = \sqrt{(0.0705 + j0.2574)\,(0 + j\,5.97 \times 10^{-6})}$$

$$= (0.16826 + j\,1.251) \times 10^{-3}$$

$$\alpha l = 0.16826 \times 10^{-3} \times 200 = 0.0337$$

$$\beta l = 1.251 \times 10^{-3} \times 200 = 0.25 \text{ radian}$$

$$A = \cosh \gamma l = \cosh\,(.0337 + j\,0.25)$$

$$= \cosh\,(0.0337)\cos\,(0.25) + j\,\sinh\,(0.0337)\sin\,(0.25)$$

$$= (1.0006)\,(0.969) + j\,(0.0337)\,(0.247)$$

$$= 0.9696 + j\,.0083 = 0.9696 \angle 0.49°$$

$$B = Z_c \sinh \gamma l = (212 \angle -7.68°) [\sinh (0.0337) \cos (0.25) + j \cosh (0.0337) \sin (0.25)]$$
$$= (212 \angle -7.68°) (0.0337 \times 0.969 + j\, 1.0006 \times 0.247)$$
$$= 52.84 \angle 74.79° \text{ ohms}$$
$$C = \frac{1}{Z_c} \sinh \gamma l$$
$$= \frac{1}{212 \angle -7.68°} (0.0337 \times 0.969 + j\, 1.0006 \times 0.247)$$
$$= 0.001176 \angle 90.15° \text{ siemens}$$

Example 4.11

A 3-phase 300 km long line has a total series impedance of 200 ∠ 80° ohms per phase and a total shunt admittance of 0.0013 ∠ 90° siemens per phase to neutral. The sending end voltage is 220 kV. The line supplies 60 MW at 0.8 power factor lagging. Find the receiving end voltage and current.

Solution

$$Z = 200 \angle 80°, \quad Y = .0013 \angle 90°$$
$$A = 1 + \frac{YZ}{2} + \frac{(YZ)^2}{24}$$
$$= 1 + \frac{(.0013 \angle 90°)(200 \angle 80°)}{2} + \frac{[(.0013 \angle 90°)(200 \angle 80°)]^2}{24}$$
$$= 0.875 \angle 1.41°$$
$$B = Z\left[1 + \frac{YZ}{6} + \frac{(YZ)^2}{120}\right]$$
$$= 200 \angle 80° \left[1 + \frac{(.0013 \angle 90°)(200 \angle 80°)}{6} + \frac{\{(.0013 \angle 90°)(200 \angle 80°)\}^2}{120}\right]$$
$$= 191.56 \angle 80.44°$$

Let receiving end phase voltage $= V_r \angle 0\ V$

$$I_r = \frac{60 \times 10^6}{3\, V_r \times 0.8} \angle -36.9° \text{ A} = \frac{25 \times 10^6}{V_r} \angle -36.9° \text{ A}$$
$$V_s = A\, V_r + B\, I_r = \frac{220000}{\sqrt{3}} \angle \delta = 127000 \angle \delta$$
$$127000 \angle \delta = (0.875 \angle 1.41°)(V_r \angle 0) + (191.56 \angle 80.44) \times \left(\frac{25 \times 10^6}{V_r} \angle -36.9°\right)$$
$$127000 \angle \delta = 0.875\, V_r + j\, 0.02153\, V_r + \frac{348 \times 10^7}{V_r} + j\frac{330 \times 10^7}{V_r}$$

This equation cannot be solved directly because value of δ is not known. However, solution can be obtained by comparing the magnitude of the two sides.

$$127000^2 = \left(0.875\,V_r + \frac{348\times 10^7}{V_r}\right)^2 + \left(0.02153\,V_r + \frac{330\times 10^7}{V_r}\right)^2$$

or $$0.7656\,V_r^4 - 9.9\times 10^9\,V_r^2 + 2.294\times 10^{19} = 0$$

which gives $$V_r = 99500 \quad \text{or} \quad 55000 \text{ V}$$

The value of 55000 V for V_r represents very poor regulation and efficiency. Hence the proper value of V_r is 99500 V.

$$\text{Receiving end line voltage} = \frac{\sqrt{3}\times 99500}{1000} = 172.33 \text{ kV}$$

$$\text{Receiving end line current} = \frac{60\times 10^6}{3\times 995000\times 0.8} = 251.26 \text{ A}$$

4.6 EQUIVALENT π AND T CIRCUITS

A long line can be replaced by an equivalent π circuit shown in Fig. 4.8. The equivalence is valid for only one frequency and only for the terminal conditions.

For the equivalent π circuit of Fig. 4.8 the sending end quantities can be written in terms of the receiving end quantities by using Eq. (4.17)

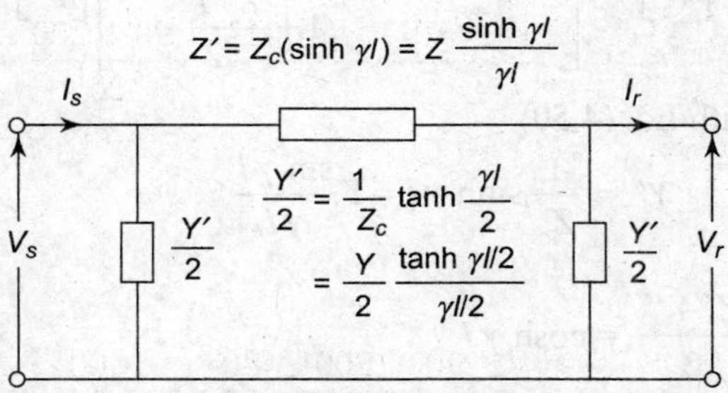

Fig. 4.8 Equivalent π circuit of a line

$$\begin{bmatrix} V_s \\ I_s \end{bmatrix} = \begin{bmatrix} 1+\dfrac{Z'Y'}{2} & Z' \\ Y'\left(1+\dfrac{Z'Y'}{4}\right) & 1+\dfrac{Z'Y'}{2} \end{bmatrix} \begin{bmatrix} V_r \\ I_r \end{bmatrix} \tag{4.49}$$

The exact solution as given by Eq. (4.37) is

$$\begin{bmatrix} V_s \\ I_s \end{bmatrix} = \begin{bmatrix} \cosh\gamma l & Z_c \sinh\gamma l \\ \dfrac{1}{Z_c}\sinh\gamma l & \cosh\gamma l \end{bmatrix} \begin{bmatrix} V_r \\ I_r \end{bmatrix} \tag{4.50}$$

Comparing the above two equations

$$Z' = Z_c \sinh \gamma l = \sqrt{\frac{z}{y}}\ \sinh\gamma l = Z\frac{\sinh\gamma l}{\gamma l} \tag{4.51}$$

$$1+\frac{Z'Y'}{2} = \cosh \gamma l \text{ or } 1+\frac{1}{2}Y' Z_c \sinh \gamma l - \cosh \gamma l = 0$$

or
$$\frac{Y'}{2} = \frac{1}{Z_c}\left(\frac{\cosh \gamma l - 1}{\sinh \gamma l}\right) = \frac{1}{Z_c} \tanh \frac{\gamma l}{2} = \frac{Y}{2} \frac{\tanh \frac{\gamma l}{2}}{\gamma l/2} \tag{4.52}$$

The long line can also be replaced by an equivalent T circuit shown in Fig. 4.9. For this circuit the sending end quantities can be written in terms of the receiving end quantities by using Eq. (4.13).

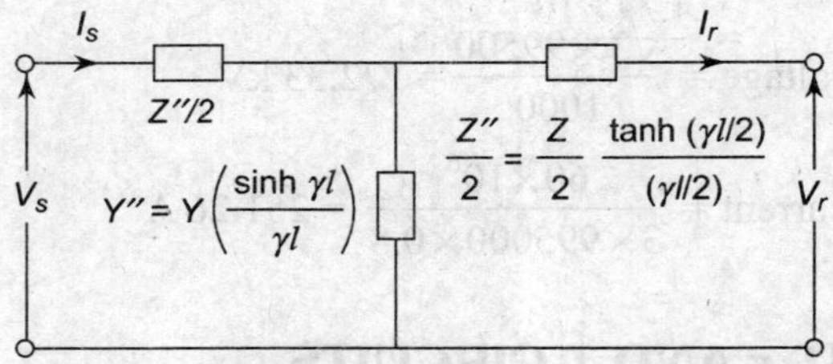

Fig. 4.9 Equivalent T circuit of a line

$$\begin{bmatrix} V_s \\ I_s \end{bmatrix} = \begin{bmatrix} 1+\frac{Z''Y''}{2} & Z''\left(1+\frac{Z''Y''}{4}\right) \\ Y'' & 1+\frac{Z''Y''}{2} \end{bmatrix} \begin{bmatrix} V_r \\ I_r \end{bmatrix} \tag{4.53}$$

Comparing Eq. (4.53) with Eq. (4.50)

$$Y'' = \frac{1}{Z_c} \sinh \gamma l = Y \frac{\sin \gamma l}{\gamma l} \tag{4.54}$$

$$1+\frac{Z''Y''}{2} = \cosh \gamma l$$

$$\frac{Z''}{2} = Z_c\left(\frac{\cosh \gamma l - 1}{\sinh \gamma l}\right) = \frac{Z}{2} \frac{\tanh (\gamma l/2)}{(\gamma l/2)} \tag{4.55}$$

The equivalent π circuit is very suitable for load flow studies and is therefore, more commonly used. As the lines become shorter, $\sinh \gamma l \to \gamma l$ and $\tanh \frac{\gamma l}{2} \to \frac{\gamma l}{2}$ and the equivalent π and T circuit degenerate to nominal π and T circuits.

Example 4.12

A 350 km long line has total series impedance of $180 \angle 75°$ ohms per phase, a total shunt admittance to neutral perphase of $1 \times 10^{-3} \angle 90°$ Siemens, (a) Find the constants of the equivalent π circuit (b) Compare these constants with the constants of nominal π circuit and comment on the result.

Solution

(a)
$$z = \frac{180}{350}\angle 75° = \frac{18}{35}\angle 75° \text{ ohms per km}$$

$$y = \frac{1\times10^{-3}}{350}\angle 90° = \frac{10^{-4}}{35}\angle 90° \text{ Siemens per km}$$

$$\gamma = \sqrt{zy} = \sqrt{\frac{18}{35}\times\frac{10^{-4}}{35}}\angle (90° + 75°)/2$$

$$= \frac{4.24\times10^{-2}}{35}\angle 82.5°$$

$$\gamma l = \left(4.24\times\frac{10^{-2}}{35}\angle 82.5°\right)(350) = 0.424\angle 82.5° = 0.0553 + j\,0.42$$

$$Z_c = \sqrt{\frac{z}{y}} = \left[\left(\frac{18}{35}\angle 75°\right)\left(\frac{35}{10^{-4}}\angle -90°\right)\right]^{0.5} = 424\angle -7.5°$$

$$\sinh \gamma l = \sinh(0.0553 + j\,0.42)$$
$$= \sinh(0.0553)\cos(0.42) + j\cosh(0.0553)\sin(0.42)$$
$$= 0.05533\times0.913 + j\,1.0014\times0.40776$$
$$= 0.0505 + j\,0.4083$$

$$\cosh \gamma l = \cosh(0.0553 + j\,0.42)$$
$$= (\cosh 0.0553)(\cos 0.42) + j(\sinh 0.0553)\sin(0.42)$$
$$= 1.0014\times0.913 + j\,0.05533\times0.40776 = 0.914 + j\,0.0226$$

$$Z' = Z_c \sinh \gamma l = (424\angle -7.5°)(0.0505 + j\,0.4083)$$
$$= 174.44\angle 75.45°\ \Omega$$

$$\frac{Y'}{2} = \frac{1}{Z_c} = \left[\frac{\cosh\gamma l - 1}{\sinh \gamma l}\right] = \frac{0.914 + j\,0.0226 - 1}{(424\angle -7.5°)(0.0505 + j\,0.4083)}$$
$$= 0.5096\times10^{-3}\angle 89.8° \text{ siemens}$$

(b) For the nominal π circuit the constants are Z and $\frac{Y}{2}$, i.e., $180\angle 75°$ and $0.5\times10^{-3}\angle 90°$.

It is seen that the difference between the constants of nominal π and equivalent π circuits is very small. This means that even for this 350 km long line, sufficiently accurate results can be obtained by using nominal π circuit.

Example 4.13

A 300 km 132 kV 3-phase overhead line has a total series impedance of $52 + j\,200$ ohms per phase and a total shunt admittance of $j\,1.5\times10^{-3}$ siemens per phase to neutral. The line is supplying 40 MVA at 0.8 p.f. lagging at 132 kV. Find sending end voltage, current, power factor and power. Use (a) nominal π circuit (b) long line equations.

Solution

$$V_r = 132000/\sqrt{3} \qquad = 76210 \angle 0°$$

$$I_r = \frac{40\times10^6}{\sqrt{3}\times132\times10^3} \qquad = 174.96 \angle -\cos^{-1}(0.8)\text{ A}$$

$$= 174.96 \angle -36.87° \text{ A}$$

$$Z = 52 + j\,200 \qquad = 206.6 \angle 75.43° \text{ ohms}$$

$$Y = 1.5\times10^{-3} \angle 90° \text{ siemens}$$

(a) For the nominal π circuit (Table 4.1)

$$A = D = 1 + \frac{YZ}{2} = 1 + \frac{(1.5\times10^{-3} \angle 90°)(206.6 \angle 75.43°)}{2}$$

$$= 0.85 \angle 2.63°$$

$$B = Z = 206.6 \angle 75.43° \text{ ohms}$$

$$C = Y\left(1 + \frac{ZY}{4}\right)$$

$$= (1.5\times10^{-3} \angle 90°)\times\left(1 + \frac{1.5\times10^{-3} \angle 90° \times 206.6 \angle 165.43°}{4}\right)$$

$$= 1.3875\times10^{-3} \angle 91.2° \text{ siemens}$$

$$V_s = A\,V_r + B\,I_r$$

$$= (0.85 \angle 2.63°)(76210 \angle 0) + (206.6 \angle 75.43°)(174.96 \angle -36.87°)$$

$$= 92975 + j\,25504 = 96410 \angle 15.34° \text{ V}$$

$$I_s = C\,V_r + D\,I_r$$

$$= (1.3875\times10^{-3} \angle 91.2°)(76210 \angle 0) + (0.85 \angle 2.63°)(174.96 \angle -36.87°)$$

$$= 120.73 + j\,22.0 = 122.73 \angle 10.35° \text{ A}$$

$$\text{Sending end line voltage} = \frac{\sqrt{3}\times96410}{1000} = 166.99 \text{ kV}$$

Sending end current = 122.73 A

Sending end power factor = cos (15.34° – 10.35°) = 0.996 lagging

$$\text{Sending end power} = \frac{\sqrt{3}\times166.99\times122.73\times0.996}{1000} = 35.35 \text{ MW}$$

(b) Long line equations

$$\gamma l = \sqrt{YZ} = \sqrt{(1.5\times10^{-3} \angle 90°)(206.6 \angle 75.43°)}$$

$$= 0.5567 \angle 82.715° = 0.0706 + j\,0.5522$$

$$\beta\,l = 0.5522 \text{ radians and } \alpha\,l = 0.0706$$

$$\cosh(\alpha l + j\,\beta l) = \cosh(0.0706)\cos(0.5522) + j\sinh(0.0706)\times\sin(0.5522)$$

$$= (1.0025\times0.85137) + j\,(0.07\times0.52456) = 0.8535 + j\,0.03672$$

$$= 0.854 \angle 2.46° = A = D$$

$$\sinh \gamma l = \sinh(\alpha l + j\beta l) = \sinh \alpha l \cos \beta l + j \cosh \alpha l \sin \beta l$$
$$= 0.07 \times 0.85137 + j\,1.0025 \times 0.52456$$
$$= 0.06 + j\,0.526 = 0.53 \angle 83.5^\circ$$

$$Z_c = \sqrt{\frac{Z}{Y}} = \sqrt{\left(\frac{206.6 \angle 75.43^\circ}{0.0015 \angle 90^\circ}\right)} = 371.1 \angle -7.28^\circ$$

$$B = Z_c \sinh \gamma l = (371.1 \angle -7.28^\circ)(0.53 \angle 83.5^\circ)$$
$$= 196.68 \angle 76.22^\circ \text{ ohms}$$

$$C = \frac{1}{Z_c} \sinh \gamma l = \frac{0.53 \angle 83.5^\circ}{371.1 \angle 7.28^\circ}$$
$$= 1.428 \times 10^{-3} \angle 90.78 \text{ siemens}$$

$$V_s = A\,V_r + B\,I_r$$
$$= (0.854 \angle 2.46^\circ)(76210 \angle 0)^\circ + (196.68 \angle 76.22^\circ)(174.96 \angle -36.87^\circ)$$
$$= 65085 \angle 2.46^\circ + 34411 \angle 39.55^\circ$$
$$= 91632.6 + j\,24612.0 = 94880 \angle 15.03^\circ \text{ V}$$

$$I_s = C\,V_r + D\,I_r$$
$$= (1.428 \times 10^{-3} \angle 90.78^\circ)(76210 \angle 0^\circ) + (0.854 \angle 2.46)(174.96 \angle -36.87^\circ)$$
$$= 108.83 \angle 90.78^\circ + 149.42 \angle -34.41^\circ$$
$$= 121.79 + j\,24.38 = 124.2 \angle 11.32^\circ \text{ A}$$

$$\text{Sending end line voltage} = \frac{\sqrt{3} \times 94882}{1000} = 164.3 \text{ kV}$$

Sending end line current = 124.2 A

Sending end power factor = cos (15.03° – 11.32°) = 0.998 lagging

$$\text{Sending end power} = \frac{\sqrt{3} \times 164.3 \times 124.2 \times 0.998}{1000} = 35.27 \text{ MW}$$

4.7 APPLICATION OF MATRIX METHODS

The equations of the 4 terminal network are well suited to manipulations by matrix methods. The matrix consisting of *ABCD* parameters (Eq. 4.38) is known as transfer matrix of the network. The transfer matrices of some special networks can be derived as under:

(a) *Series impedance* Z: Short line or synchronous alternator (Fig. 4.10a)

$$V_s = V_r + Z I_r \qquad (4.56)$$
$$I_s = I_r$$

The transfer matrix is $\begin{bmatrix} 1 & Z \\ 0 & 1 \end{bmatrix}$

(b) *Shunt admittance Y:* Transformer on no load (Fig. 4.10b)

$$V_s = V_r$$
$$I_s = Y V_r + I_r \qquad (4.57)$$

The transfer matrix is $\begin{bmatrix} 1 & 0 \\ Y & 1 \end{bmatrix}$

(c) *Asymmetrical π network:* (Fig. 4.10c)

This network is a cascade connection of a shunt admittance, followed by a series impedance followed by a shunt admittance. The overall transfer matrix can be obtained by multiplying together the transfer matrices of the three component parts.

$$\begin{bmatrix} 1 & 0 \\ Y_1 & 1 \end{bmatrix} \begin{bmatrix} 1 & Z \\ 0 & 1 \end{bmatrix} \begin{bmatrix} 1 & 0 \\ Y_2 & 1 \end{bmatrix}$$

$$= \begin{bmatrix} 1 + Y_2 Z & Z \\ Y_1 + Y_2 + Y_1 Y_2 Z & 1 + Y_1 Z \end{bmatrix} \tag{4.58}$$

If the π network is symmetrical and $Y_1 = Y_2 = \frac{Y}{2}$, where Y is the total shunt admittance, the transfer matrix reduces to that for nominal π representation of a transmission line. (Table 4.1)

(d) *Asymmetrical T network:* (Fig. 4.10d)

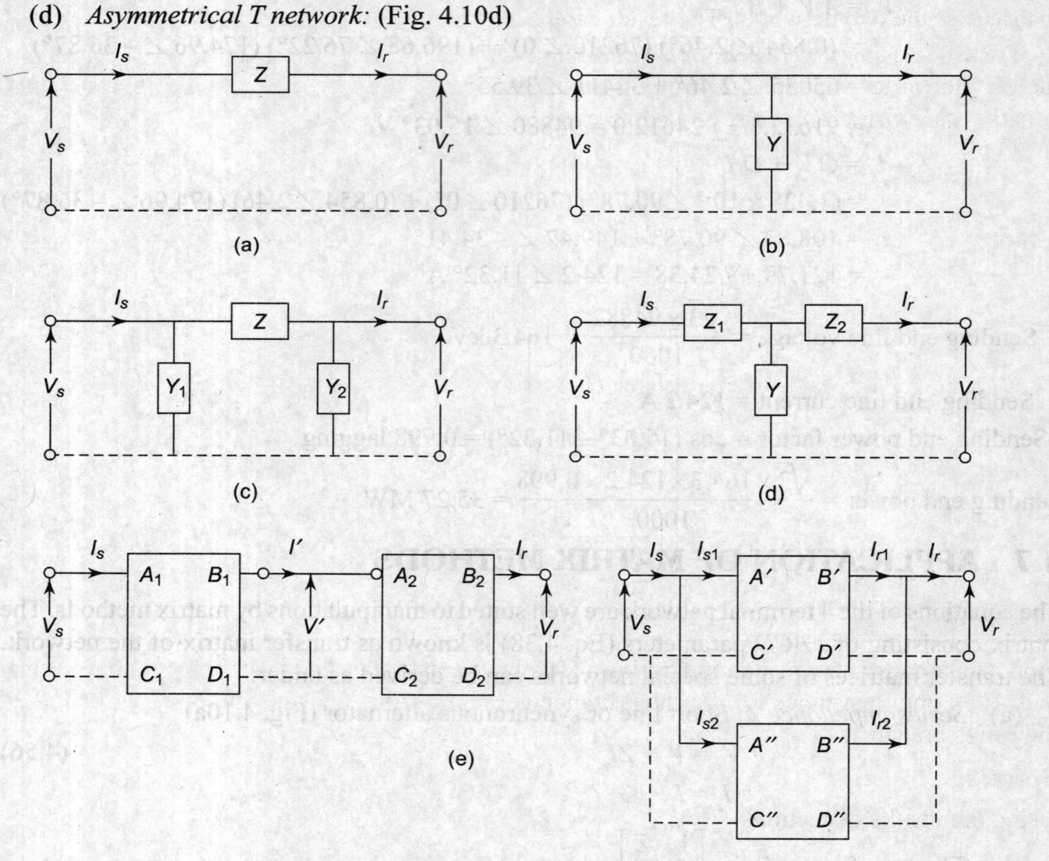

Fig. 4.10 Four terminal networks

This network is a cascade connection of a series impedance Z_1 followed by a shunt admittance Y followed by a series impedance Z_2. The overall transfer matrix can be obtained by multiplying together the transfer matrices of three component parts.

$$\begin{bmatrix} 1 & Z_1 \\ 0 & 1 \end{bmatrix} \begin{bmatrix} 1 & 0 \\ Y & 1 \end{bmatrix} \begin{bmatrix} 1 & Z_2 \\ 0 & 1 \end{bmatrix} = \begin{bmatrix} 1 + YZ_1 & Z_1 + Z_2 + YZ_1 Z_2 \\ Y & 1 + YZ_2 \end{bmatrix} \tag{4.59}$$

If the T network is symmetrical and $Z_1 = Z_2 = Z/2$ where Z is the total series impedance, the transfer matrix reduces to that for a nominal T representation of a transmission line (Table 4.1).

(e) *Two 4 terminal networks in cascade:* (Fig. 3.16e)

When two transmission lines are connected such that the output of the first line serves as input to the second line and the output of the second line is fed to the load, the two lines behave as two 4 terminal network in cascade. The overall transfer matrix can be obtained by expressing V_s and I_s in terms of V' and I' and then expressing V' and I' in terms of V_r and I_r. Thus

$$\begin{bmatrix} V_s \\ I_s \end{bmatrix} = \begin{bmatrix} A_1 & B_1 \\ C_1 & D_1 \end{bmatrix} \begin{bmatrix} V' \\ I' \end{bmatrix} = \lambda \begin{bmatrix} A_1 & B_1 \\ C_1 & D_1 \end{bmatrix} \begin{bmatrix} A_2 & B_2 \\ C_2 & D_2 \end{bmatrix} \begin{bmatrix} V_r \\ I_r \end{bmatrix}$$

$$= \begin{bmatrix} (A_1\,A_2 + B_1\,C_2) & (A_1\,B_2 + B_1\,D_2) \\ (C_1\,A_2 + D_1\,C_2) & (C_1\,B_2 + D_1\,D_2) \end{bmatrix} \begin{bmatrix} V_r \\ I_r \end{bmatrix} \tag{4.60}$$

Thus the resultant transfer matrix for the two cascaded networks is the product of transfer matrices of the two networks. The result can be extended to any number of networks.

(f) *Two 4 terminal networks in parallel.* Fig. 4.10 (f) shows this arrangement. The inputs to the two networks as well as the outputs are in parallel. It is easier to obtain the solution if the equations are written in terms of y parameters.

$$\begin{bmatrix} I_{s1} \\ I_{r1} \end{bmatrix} = \begin{bmatrix} y'_{11} & y'_{12} \\ y'_{21} & y'_{22} \end{bmatrix} \begin{bmatrix} V_s \\ V_r \end{bmatrix}$$

$$\begin{bmatrix} I_{s2} \\ I_{r2} \end{bmatrix} = \begin{bmatrix} y''_{11} & y''_{12} \\ y''_{21} & y''_{22} \end{bmatrix} \begin{bmatrix} V_s \\ V_r \end{bmatrix}$$

$$\begin{bmatrix} I_s \\ I_r \end{bmatrix} = \begin{bmatrix} I_{s1} + I_{s2} \\ I_{r1} + I_{r2} \end{bmatrix} = \begin{bmatrix} (y'_{11} + y''_{11}) & (y'_{12} + y''_{12}) \\ (y'_{21} + y''_{21}) & (y'_{22} + y''_{22}) \end{bmatrix} \begin{bmatrix} V_s \\ V_r \end{bmatrix}$$

$$= \begin{bmatrix} \left(\dfrac{D'}{B'} + \dfrac{D''}{B''}\right) & -\left(\dfrac{1}{B'} + \dfrac{1}{B''}\right) \\ \left(\dfrac{1}{B'} + \dfrac{1}{B''}\right) & -\left(\dfrac{A'}{B'} + \dfrac{A''}{B''}\right) \end{bmatrix} \begin{bmatrix} V_s \\ V_r \end{bmatrix} \tag{4.61}$$

Example 4.14

Find $ABCD$ constants of a π circuit having a 500 ohm resistance for the shunt branch at the sending end, 1000 ohm resistance for the shunt branch at the receiving end and 100 ohm resistance for the series branch.

Solution

Using the results of section 4.7(c)

$$Y_1 = \frac{1}{500} = 2 \times 10^{-3} \angle 0° \text{ siemens}$$

$$Y_2 = \frac{1}{1000} = 1 \times 10^{-3} \angle 0° \text{ siemens}$$

$$Z = 100 \angle 0° \text{ ohms}$$

$$A = 1 + Y_2 Z = 1 + (1 \times 10^{-3} \angle 0°)\,(100 \angle 0°) = 1.1 \angle 0°$$

$$B = Z = 100 \angle 0° \text{ ohms}$$

$$C = Y_1 + Y_2 + Y_1 Y_2 Z$$
$$= 2 \times 10^{-3} \angle 0° + 1 \times 10^{-3} \angle 0° + (2 \times 10^{-3} \angle 0°) \times (1 \times 10^{-3} \angle 0°)\,(100 \angle 0°)$$
$$= 3.2 \times 10^{-3} \angle 0 \text{ siemens}$$
$$D = 1 + Y_1 Z = 1 + (2 \times 10^{-3} \angle 0°)\,(100 \angle 0°) = 1.2 \angle 0°$$

Example 4.15

Two three-phase lines have the following constants:

$$A_1 = D_1 = 0.98 \angle 2°,\ B_1 = 28 \angle 69° \text{ ohms},\ C_1 = 0.0002 \angle 80° \text{ siemens}$$
$$A_2 = D_2 = 0.95 \angle 3°,\ B_2 = 40 \angle 85° \text{ ohms},\ C_2 = 0.0004 \angle 90° \text{ siemens}$$

The two lines are connected is cascade. Find (a) *ABCD* constants for the composite system (b) Sending end voltage, current and power factor if the composite system delivers 200 A at 110 kV and 0.95 p.f. lagging.

Solution

(a) Using the results of section 4.7(e)

$$A = A_1 A_2 + B_1 C_2$$
$$= (0.98 \angle 2°)\,(0.95 \angle 3°) + (28 \angle 69°)\,(0.0004 \angle 90°)$$
$$= 0.931 \angle 5° + 0.0112 \angle 159° = 0.917 + j\,0.085 = 0.921 \angle 5.3°$$
$$B = A_1 B_2 + B_1 D_2$$
$$= (0.98 \angle 2°)\,(40 \angle 85°) + (28 \angle 69°)\,(0.95 \angle 3°)$$
$$= 39.2 \angle 87° + 26.6 \angle 72° = 10.27 + j\,64.44 = 65.26 \angle 80.95°$$
$$C = C_1 A_2 + D_1 C_2$$
$$= (0.0002 \angle 80°)\,(0.95 \angle 3°) + (0.98 \angle 2°)\,(0.0004 \angle 90°)$$
$$= 0.00019 \angle 83° + 0.000392 \angle 92°$$
$$= 9.47 \times 10^{-6} + j\,5.8 \times 10^{-4} = 5.8 \times 10^{-4} \angle 89.1°$$
$$D = C_1 B_2 + D_1 D_2$$
$$= (0.0002 \angle 80°)\,(40 \angle 85°) + (0.98 \angle 2°)\,(0.95 \angle 3°)$$
$$= 0.008 \angle 165° + 0.931 \angle 5° = 0.92 + j\,0.083 = 0.924 \angle 5.2°$$

(b)
$$V_r = \frac{110000}{\sqrt{3}} = 63508.5 \angle 0° \text{ V}$$
$$I_r = 200 \angle (-\cos^{-1} 0.95) = 200 \angle -18.2° \text{ A}$$
$$V_s = A\,V_r + B\,I_r$$
$$= (0.921 \angle 5.3°)\,(63508.5 \angle 0°) + (65.26 \angle 80.95°)\,(200 \angle -18.2°)$$
$$= 58491.3 \angle 5.3° + 13052 \angle 62.75°$$
$$= 64217.27 + j\,17006.45 = 66431 \angle 14.83° \text{ V}$$
$$I_s = C\,V_r + D\,I_r = (5.8 \times 10^{-4} \angle 89.1°)\,(63508.5 \angle 0°)$$
$$+ (0.924 \angle 5.2°)\,(200 \angle -18.2°)$$
$$= 36.8 \angle 89.1° + 184.8 \angle -13° = 180.7 \angle -1.51° \text{ A}$$

Sending end line voltage $= \dfrac{\sqrt{3} \times 66431}{1000} = 115.06$ kV

Sending end line current = 180.7 A

Sending end power factor = cos (14.83° + 1.51°) = 0.96 lagging

4.8 FERRANTI EFFECT

A long transmission line draws a substantial quantity of charging current. If such a line is open circuited or very lightly loaded at the receiving end, the voltage at the receiving end may become higher than the voltage at the sending end. This is known as Ferranti effect and is due to the voltage drop across the line inductance (due to the charging current) being in phase with the sending end voltage. Thus both capacitance and inductance are necessary to produce this phenomenon. The capacitance (and charging current) is negligible in short line but significant in medium length lines and appreciable in long lines. Therefore, this phenomenon occurs only in medium and long lines.

The phasor diagram of Fig. 4.11 illustrates the Ferranti effect. The line may be represented by a nominal π circuit so that half the total line capacitance is assumed to be concentrated at the receiving end. *OM* represents the receiving end voltage. *OC* represents the current drawn by capacitance assumed to be concentrated at the receiving end. *MN* is the resistance drop and *NP* is the inductive reactance drop. *OP* is the sending end voltage under no load condition and is less than the receiving end voltage. The line resistance is usually small as compared to the line inductive reactance. Moreover, the resistance drop *MN* is in quadrature with *OM* and *NP*. Therefore, the resistance can be neglected in calculating the Ferranti effect.

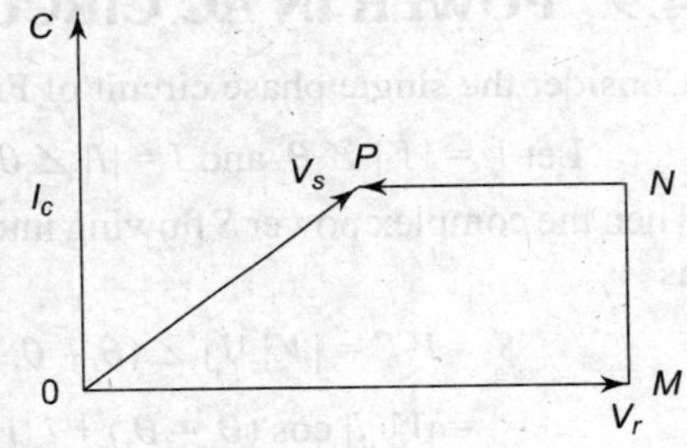

Fig. 4.11 Ferranti effect

Using the constants of nominal π representation (Table 4.1).

$$V_s = V_r\left(1 + \frac{ZY}{2}\right)$$

$$V_s - V_r = V_r\frac{ZY}{2} = V_r\frac{(r + j\omega L)\, l\, (j\omega Cl)}{2} \tag{4.62}$$

Neglecting the line resistance

$$V_s - V_r = \frac{-V_r\,\omega^2 l^2 LC}{2} \tag{4.63}$$

The quantity $\dfrac{1}{\sqrt{(LC)}}$ is constant for all overhead lines and is the velocity of propagation of electromagnetic waves on the line. It is nearly equal to the speed of light, i.e., 3×10^5 km per second. Then

$$V_s - V_r = -V_r\,\omega^2 l^2 \frac{1}{2(3\times10^5)^2} = \frac{-V_r \cdot \omega^2 l^2 \times 10^{-10}}{18} \text{ volts} \tag{4.64}$$

The negative sign in Eq. (4.64) indicates that V_r is more than V_s.

Example 4.16

Find the no load sending end voltage and the voltage rise from sending end to receiving end for a 50 Hz, 300 km long line if the receiving end voltage is 220 kV.

Solution

$$V_r = 220000/\sqrt{3} = 127017 \text{ V}$$

$$V_s - V_r = -\frac{127017\times(2\pi\times 50)^2\times 300\times 300\times 10^{-10}}{18} = -6268 \text{ V}$$

$$V_s = 127017 - 6268 = 120749 \text{ V/phase}$$

$$\text{Sending end line voltage} = \frac{\sqrt{3}\times 120749}{1000} = 209.14 \text{ kV.}$$

$$\text{Voltage rise} = 6268 \text{ volts/phase}$$

4.9 POWER IN AC CIRCUITS

Consider the single phase circuit of Fig. 4.12.

Let $V = |V| \angle \theta_1$ and $I = |I| \angle \theta_2$

Then the complex power S flowing into the network is defined as

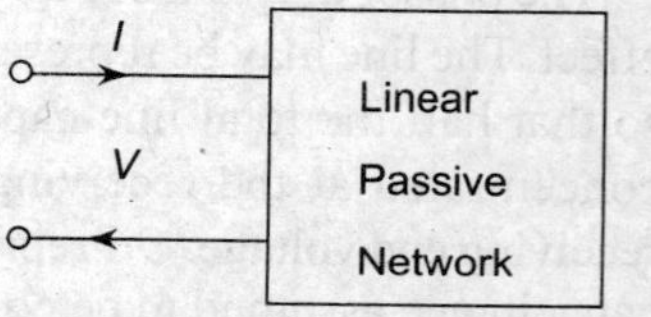

Fig. 4.12 Single phase circuit

$$S = VI^* = |V|\,|I| \angle (\theta_1 - \theta_2)$$

$$= |V|\,|I| \cos(\theta_1 - \theta_2) + j\,|V|\,|I| \sin(\theta_1 - \theta_2) \tag{4.65}$$

The complex power S has two components. The first component $|V|\,|I| \cos(\theta_1 - \theta_2)$ is known as real (active) power P and is measured in watts (or kW or MW). The second component $|V|\,|I| \sin(\theta_1 - \theta_2)$ is known as reactive power Q and is measured in vars (or kvars or Mvars). The angle $\angle \theta_1 - \theta_2$ is the phase angle between voltage and current.

Real power is positive as long as the angle $(\theta_1 - \theta_2)$ is not larger than 90° in either direction $(-90 < \theta_1 - \theta_2 < 90°)$. The sign of P is determined by the positive direction of voltage and current. The usual selection of these directions is such that P is positive when flowing into a load or out of a generator. For a transmission line the real power is positive at the sending end if it is flowing into the line, but at the receiving end the real power is considered positive if it is flowing out of the line.

The sign of Q depends on the sign of $\sin(\theta_1 - \theta_2)$. When $(\theta_1 - \theta_2)$ is positive, Q is positive and when $(\theta_1 - \theta_2)$ is negative, Q is negative. $(\theta_1 - \theta_2)$ is positive when the current phasor lags the voltage phasor which is true for an inductive load. Therefore, inductive load absorbs positive reactive power. On the other hand the leading vars, absorbed by a capacitive load are taken to be negative. Thus a capacitor is a sink of leading (negative) vars. The action of a capacitor connected across the load (for power factor improvement) can be explained by saying that the capacitor generates lagging vars which it supplies to the load, thereby reducing the demand for lagging vars from the supply.

In a transmission line the series inductance is a sink of lagging vars equal to $I^2 \times X_L$ while the shunt capacitance is a sink of leading vars (source of lagging vars) equal to V^2 / X_c.

In a 3-phase circuit the total active power is 3 times the active power per phase and the total reactive power is 3 times the reactive power per phase.

Example 4.17

Two three-phase machines have generated emfs of $11\angle 0$ kV and $11 \angle 40°$ kV. They are connected through a 3-phase line having an impedance of $0 + j15\ \Omega$ per phase. Find (a) whether each machine is acting as generator or motor and the real power generated or consumed by it (b) whether each machine is delivering or consuming reactive power and the amount of reactive power (c) the real and reactive power absorbed by line impedance.

Solution

Fig. 4.13 shows the equivalent circuit (impedance diagram) of the system. Let E_1 and E_2 be the phase emfs of the two machines.

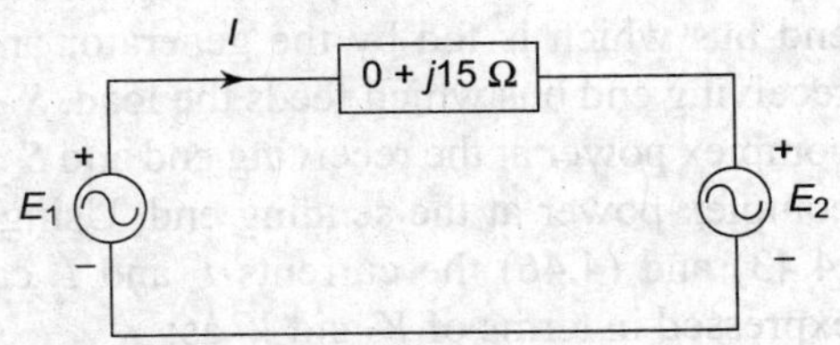

Fig. 4.13

$$E_1 = \frac{11000}{\sqrt{3}} \angle 0 = 6351.04° \text{ V}$$

$$E_2 = \frac{11000}{\sqrt{3}} \angle 40° = 6351.04 \angle 40° \text{ V}$$

$$I = \frac{E_1 - E_2}{Z} = \frac{6351.04 - 6351.04 \angle 40°}{0 + j15} = -272.16 - j\,99.06 = 289.63 \angle -160° \text{A}$$

The current I is being delivered by machine 1 (i.e., it is flowing out of the terminals of machine 1). At the same time current I is being absorbed by machine 2.

Complex power S_1 being delivered by machine 1 is

$$\begin{aligned} S_1 &= 3\,E_1 I^* = 3(6351.04)\,(289.63 \angle -160)^* \\ &= 3\,(6351.04)\,(289.63 \angle 160°) \text{ VA} \\ &= -5185557.6 + j\,1887388.6 \text{ VA} \\ &= -5.185 + j\,1.887 \text{ MVA} \end{aligned}$$

Complex power S_2 being absorbed by machine 2 is

$$\begin{aligned} S_2 &= 3\,E_2 I^* = 3\,(6351.04 \angle 40°)\,(289.63 \angle -160°)^* \\ &= 3\,(6351.04 \angle 40)\,(289.63 \angle 160°) \text{ VA} \\ &= 5518355.15 \angle 200° \text{ VA} \\ &= -5185557.6 - j\,1887388.6 \text{ VA} \\ &= -5.185 - j\,1.887 \text{ MVA} \end{aligned}$$

(a) The real power delivered by machine 1 is negative. Thus this machine is absorbing real power, i.e., machine 1 is acting as motor and consuming 5.185 MW.
The real power absorbed by machine 2 is negative. Thus machine 2 is delivering real power, i.e., machine 2 is acting as generator and supplying 5.185 MW.

(b) Machine 1 is supplying positive reactive power of 1.887 Mvars. Machine 2 is absorbing negative reactive power of 1.887 Mvars. In other words it is supplying reactive power of 1.887 Mvars.

(c) Since the line does not have resistance the real power consumed by line is zero. Reactive power absorbed by line inductance

$$\begin{aligned} &= 3\,I^2 X = 3\,(289.63)^2\,(15) \text{ Var} \\ &= 3.774 \text{ Mvars} \end{aligned}$$

It is seen that real and reative power balance is existing in the system. The real power delivered by machine 2 equals the real power absorbed by machine 1. The total reactive power supplied by the two machines equals the reactive power absorbed by the line inductance.

4.10 POWER FORMULAE FOR TRANSMISSION LINES

Figure. 4.14 shows the single line diagram of a three-phase transmission line. The ends of a transmission line can terminate at generating stations, grid switching stations or grid supply points. The ends of transmission lines are designated as busses. The concept of a bus in a one line diagram is essentially the same as that of a node in a circuit diagram. The system in Fig. 4.14 is a two bus system having the sending end bus which is fed by the generator and the receiving end bus which feeds the load. S_r is the complex power at the receiving end and S_s is the complex power at the sending end. Using Eqs. (4.43) and (4.46) the currents I_r and I_s can be expressed in terms of V_r and V_s as:

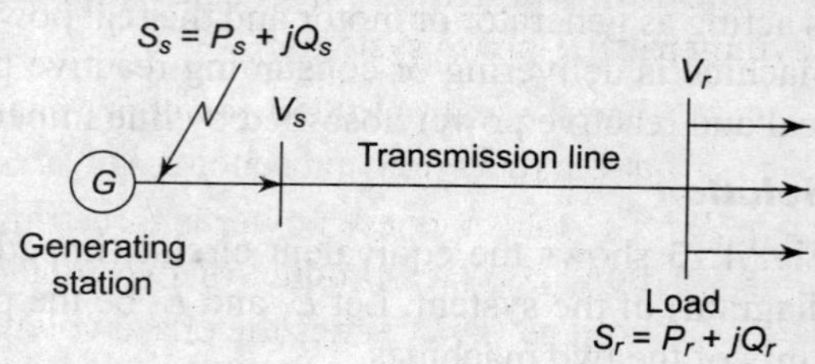

Fig. 4.14 A two bus power system

$$I_r = \frac{1}{B}V_s - \frac{A}{B}V_r \tag{4.66}$$

$$I_s = \frac{D}{B}V_s - \frac{1}{B}V_r = \frac{A}{B}V_s - \frac{1}{B}V_r \tag{4.67}$$

Let $\quad V_r = |V_r| \angle 0,\ V_s = |V_s| \angle \delta,\ D = A = |A| \angle \alpha,\ B = |B| \angle \beta$

Then
$$I_r = \frac{|V_s|}{B}\angle(\delta-\beta) - \frac{|A||V_r|}{|B|}\angle(\alpha-\beta) \tag{4.68}$$

$$I_s = \frac{|A||V_s|}{|B|}\angle(\alpha+\delta-\beta) - \frac{|V_r|}{|B|}\angle -\beta \tag{4.69}$$

The conjugates of I_r and I_s are

$$I_r^* = \frac{|V_s|}{|B|}\angle(\beta-\delta) - \frac{|A||V_r|}{|B|}\angle(\beta-\alpha) \tag{4.70}$$

$$I_s^* = \frac{|A||V_s|}{|B|}\angle(\beta-\alpha-\delta) - \frac{|V_r|}{|B|}\angle\beta \tag{4.71}$$

The complex power per phase at the receiving end and sending end are

$$S_r = P_r + jQ_r = V_r I_r^*$$

$$= |V_r| \angle 0 \left[\frac{|V_s|}{|B|}\angle(\beta-\delta) - \frac{|A||V_r|}{|B|}\angle(\beta-\alpha)\right]$$

$$= \frac{|V_s||V_r|}{|B|}\angle(\beta-\delta) - \frac{|A||V_r|^2}{|B|}\angle(\beta-\alpha) \tag{4.72}$$

$$S_s = P_s + jQ_s = V_s I_s^*$$

$$= |V_s| \angle \delta \left[\frac{|A||V_s|}{|B|}\angle(\beta-\alpha-\delta) - \frac{|V_r|}{|B|}\angle\beta\right]$$

$$= \frac{|A||V_s|^2}{|B|} \angle(\beta - \alpha) - \frac{|V_r||V_s|}{|B|} \angle(\beta + \delta) \tag{4.73}$$

The following important observations about Eqs. (4.72) and (4.73) are applicable to symmetrical 3-phase systems:

1. These equations yield power per phase if the phase voltage (i.e., line to neutral voltage) at the receiving and sending end are used for V_r and V_s.
2. The total 3-phase power is three times the power per phase. Each term in these equations contains product of two voltages or the square of a voltage. Since line to line voltage is $\sqrt{3}$ times the phase voltage, these equations yield 3-phase power directly if V_r and V_s represent line to line voltage at the receiving end and sending end.
3. If the voltage is expressed in volts, the power is expressed in watts or vars. If the voltages are considered in kilovolts, the power is expressed in mega watts or megavars.

From Eq. (4.72) the real and reactive power at receiving end are:

$$P_r = \frac{|V_s||V_r|}{|B|} \cos(\beta - \delta) - \frac{|A||V_r|^2}{|B|} \cos(\beta - \alpha) \tag{4.74 a}$$

$$Q_r = \frac{|V_s||V_r|}{|B|} \sin(\beta - \delta) - \frac{|A||V_r|^2}{|B|} \sin(\beta - \alpha) \tag{4.74b}$$

The real and reactive power at sending end are

$$P_s = \frac{|A||V_s|^2}{|B|} \cos(\beta - \alpha) - \frac{|V_r||V_s|}{|B|} \cos(\beta + \delta) \tag{4.75a}$$

$$Q_s = \frac{|A||V_s|^2}{|B|} \sin(\beta - \alpha) - \frac{|V_r||V_s|}{|B|} \sin(\beta + \delta) \tag{4.75b}$$

For fixed values of V_s and V_r the receiving end real power is maximum when $\delta = \beta$. Thus

$$P_{r\max} = \frac{|V_s||V_r|}{|B|} - \frac{|A||V_r|^2}{|B|} \cos(\beta - \alpha) \tag{4.76}$$

The corresponding value of Q_r at this power limit is

$$Q_r = -\frac{|A||V_r|^2}{|B|} \sin(\beta - \alpha) \tag{4.77}$$

The load must draw the leading vars given by Eq. (4.77) to achieve the condition of maximum real power at the receiving end. Very important conclusions can be drawn by considering the power transfer over a short line which has $A = D = 1 \angle 0$, $B = Z = |Z| \angle \theta$. Substituting these values in Eqs. (4.74) and (4.75) we have

$$P_r = \frac{|V_s||V_r|}{|Z|} \cos(\theta - \delta) - \frac{|V_r|^2}{\|Z\|} \cos\theta \tag{4.78a}$$

$$Q_r = \frac{|V_s||V_r|}{|Z|} \sin(\theta - \delta) - \frac{|V_r|^2}{|Z|} \sin\theta \tag{4.78b}$$

$$P_s = \frac{|V_s|^2}{|Z|}\cos\theta - \frac{|V_s||V_r|}{|Z|}\cos(\theta+\delta) \tag{4.79a}$$

$$Q_s = \frac{|V_s|^2}{|Z|}\sin\theta - \frac{|V_s||V_r|}{|Z|}\sin(\theta+\delta) \tag{4.79b}$$

The resistance of a transmission line is usually very small as compared to inductive reactance. Thus $|Z| \approx X$ and $\theta = 90°$. Substituting these values in the above equations

$$P_r = \frac{|V_s||V_r|}{X}\sin\delta \tag{4.80}$$

$$Q_r = \frac{|V_s||V_r|}{X}\cos\delta - \frac{|V_r|^2}{X} \tag{4.81}$$

The power or load angle δ is generally small from considerations of system stability. Then $\cos\delta \simeq 1$ and Eq. (4.81) reduces to

$$Q_r = \frac{|V_s||V_r|}{X} - \frac{|V_r|^2}{X} = \frac{|V_r|}{X}[|V_s| - |V_r|] \tag{4.82}$$

Eqs. (4.78 to 4.82) are for short lines. However a long line can be represented by an equivalent π circuit (Fig. 4.8). The shunt admittance at the receiving end can be combined with the load and the shunt admittance at the sending end can be combined with the generator network. Thus the above equations can also be applied to long lines. A study of Eqs. (4.80 to 4.82) leads to following important conclusions:

1. For fixed values of V_s, V_r and X the real power depends on angle δ which is the phase angle between V_s and V_r. This angle δ is, therefore, known as power angle. P is maximum when $\delta = 90°$. However, δ has to be kept to values well below 90° from considerations of system stability.
2. Power can be transferred over line even when $|V_s| = |V_r|$. The phase difference between V_r and V_s causes the flow of current in the line. Modern power systems are operated with almost the same voltage magnitudes (i.e., 1 p.u.) at important busses by using methods of voltage control.
3. The maximum real power transferred over a line increases with increase in V_s and V_r. An increase of 10% in V_r and V_s increases the power transfer by 21%. This explains the reason for adopting high and extra high transmission voltages.
4. The maximum real power depends on the reactance X which is directly proportional to line inductance. A decrease in inductance increases the line capacity. The line inductance can be decreased by using double circuit lines and by the use of bundled conductors. Another method of reducing line inductance is by inserting series capacitance in the line. This method is known as series compensation. The series capacitors are usually installed at the mid point of the line.
5. The reactive power transferred over a line is directly proportional to $(|V_s| - |V_r|)$, i.e., voltage drop along the line and is independent of power angle. This means that the major cause of voltage drop on the line is the transfer of reactive power over the line. To maintain a good voltage profile, the control of reactive power is necessary.

4.11 LINE VOLTAGE REGULATION AND COMPENSATION

The voltage, relative to nominal voltage, at the different busses of the power system is known as voltage profile. Most of the equipments, especially at 400 kV level and above are operated close to the present limits of design. In view of this the modern power systems are operated with a voltage profile of ± 5%.

It follows that the transmission lines have to transfer power with a voltage drop of less than 10%. The major cause of voltage drop in the line is the flow of reactive power. Moreover reactive currents cause I^2R losses in the system but produce no revenue.

Most loads absorb positive (i.e., lagging) vars to supply the magnetizing current of transformers, induction motors, etc. At any moment the maximum vars which can be transferred over the lines are fixed by the voltage profile. It follows that at times of peak loads the vars demanded by the loads greatly exceed the vars which can be transmitted over the lines from the consideration of voltage drop. Therefore, additional equipment is necessary to generate lagging vars to meet the reactive power requirement of the consumers. If this is not done, the system voltage at some of the busses is likely to become lower than the nominal voltage.

The shunt capacitance of the line absorbs leading vars (i.e., generates lagging vars). At the time of light loads the lagging vars produced by the lines are much larger than that required by consumers loads. These surplus lagging vars must be absorbed by additional equipment to keep the voltage profile within limits. If this is not done the system voltage at some of the busses is likely to become higher than the nominal voltage.

From the above discussion it follows, that it is necessary to provide additional equipment, called reactive power compensating equipment to generate or absorb vars. A shunt connected inductance absorbs lagging vars while a shunt capacitor generates lagging vars (by absorbing leading vars). Specially designed synchronous machines called synchronous compensators (or phase modifiers) can generate or absorb vars.

Applications of compensation equipment in a system has the following effects:

1. Reduction in reactive component of circuit current.
2. Maintenance of voltage profile within limits.
3. Reduction of copper losses in the system due to reduction of current.
4. Reduction in investment in the system per kW of load supplied.
5. Decrease in kVA loading of generators and circuits. This decrease in kVA loading may relieve an overload condition or release capacity for additional load growth.
6. Improvement in power factor of generators.
7. Reduction in kVA demand charges for large consumers.

4.11.1 Static Compensation

The static compensating equipments consist of shunt capacitors, shunt reactors and series capacitors.

Shunt capacitors are used in ratings from 15 kvar to over 100 Mvar. Small banks of capacitors, upto a few hundred kvar rating are used on individual distribution circuits of consumers. Capacitor banks of 500-3000 kvar are used in small distribution sub-stations and those with still larger ratings are used in large sub-stations. They are arranged in 3-phase banks and are connected in star or delta. Capacitor banks may be permanently connected. However for voltage regulation they have to be switched on and off as per the system demands. They have to be switched on at times of peak loads and switched off at times of light loads. Switching may be done manually or automatically controlled either by time clocks or in response to voltage or reactive power requirements.

Shunt connected inductors are used to absorb the lagging vars generated by the line during light load conditions. The inductor may be a coreless type or a gagged core type. The coreless type is linear and is used for the large ratings and voltages (viz., 100 Mvar at 400 kV). The gagged core type is designed to be linear upto a voltage of about 1.3 per unit. It is also possible to use a tapped saturable reactor which saturates at selected voltages within a range of ± 10% of the system nominal voltage. When saturated, the vars absorbed by the reactor change rapidly for small changes in voltage.

The static compensation equipment is often connected to the tertiary delta winding of the sub-station transformer as shown in Fig. 4.15. Some capacitors are permanently connected. The remaining capacitors and inductor are provided with on load tap changing gear for switching as and when required.

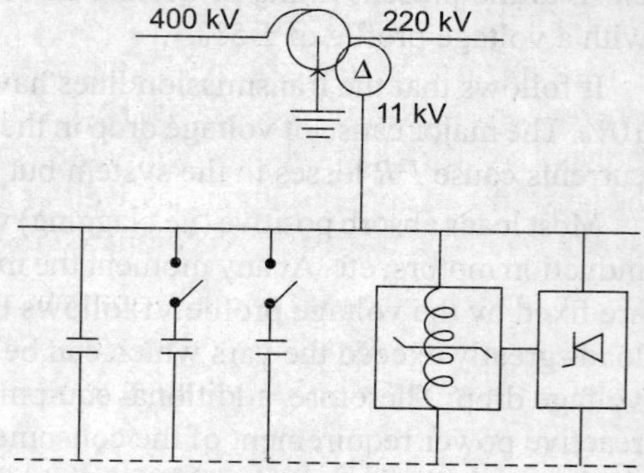

Fig. 4.15 Static compensation

If the capacitors are connected in star,

$$I_c = \frac{|V|}{\sqrt{3}X_c} \text{ kA}$$

$$Q_c = \frac{3|V||V|}{\sqrt{3}X_c\sqrt{3}} = \frac{|V|^2}{X_c} \text{ Mvar} \tag{4.83}$$

where V is tertiary line voltage in kV, X_c is the capacitive reactance per phase and Q_c is the lagging 3-phase vars generated by capacitors.

For star connected inductors:

$$Q_L = \frac{-|V|^2}{X_L} \text{ Mvar} \tag{4.84}$$

where Q_L is the leading (i.e., negative) 3-phase vars generated by the inductors and X_L is the inductive reactance per phase.

The use of series capacitors which are connected in series with the line (usually at the mid point) is more recent than the use of shunt capacitors. They generate vars proportional to the square of current in them. Their primary use is to reduce the series inductance of the line in order to raise the power limit and to improve system stability. The ability of the series capacitor to supply the reactive power of the loads is negligible whereas the contribution of the shunt capacitor in this respect is very major. This difference results from the fact that the shunt capacitor actually supplies the reactive power requirement of the loads while the series capacitor merely provides compensation for a part of the reactive power requirement of the line.

4.11.2 Synchronous Compensators

A synchronous compensator (or synchronous phase modifier) is essentially a synchronous motor with no mechanical output. It takes just enough real power to supply its losses. They are usually of salient pole design (6 or 8 poles) with ratings upto 60 MVA, 11 kV and connected to high voltage systems through transformers. The excitation of the machine is varied to provide the necessary amount of compensation.

A synchronous compensator delivers lagging (or absorbs leading) vars when its excitation is large and delivers leading (or absorbs lagging) vars when its excitation is small. Thus it is a very flexible source of reactive power because its reactive power can be continuously varied in magnitude and direction (from lagging to leading) with utmost ease simply by varying its excitation.

A transmission line requires lagging vars at the receiving end during peak load conditions to prevent a voltage drop. During off-peak conditions it requires leading vars at the receiving end to prevent a voltage rise. Thus synchronous compensator is excellently suited for voltage regulation because it can supply lagging as well as leading vars. The advantages and disadvantages of synchronous compensator as compared to static compensation equipment are as under:

1. A synchronous compensator can supply lagging vars, equal to its rating, to the system and can usually absorb lagging vars up to 50% of its rating. Thus a synchronous compensator of a certain rating is equal to a static capacitor of that rating and a shunt reactor of half its rating.
2. The control of synchronous compensator is fast and continuous.
3. A synchronous compensator can be overloaded for short periods but a static capacitor cannot be overloaded.
4. A drop in voltage causes the synchronous compensator to supply greater vars (because difference between generated e.m.f. and terminal voltage increases) which helps in arresting the voltage dip. In the case of static capacitor a drop in voltage causes a reduction in vars delivered to the system.
5. The inertia of synchronous compensator improves system stability and reduces the effect of sudden changes in load.
6. Static capacitor installations can be distributed in the system. However small size synchronous compensators are very uneconomical. Thus static capacitors can be located near the loads and are more effective.
7. Synchronous compensators add to short circuit currents in the system and increase the circuit ratings.
8. A failure of one unit of a static capacitor bank affects that unit only and the remaining units continue to operate. However, the failure of a synchronous compensator means loss of complete unit.
9. For small reactive power requirements static capacitors are economical and preferable. For requirements, above 10 Mvar or so, a synchronous compensator is more economical and preferable.

4.12 NATURAL LOAD

The maximum amount of power which can be transmitted over a line is dependent on many engineering and economic variables and cannot be expressed by a single formula. Many a time it is convenient to express the power transmitted as fraction of natural power.

A line is said to be operating at its natural load when it is terminated by a resistance equal to its characteristic impedance Z_c. The power transmitted under this condition is called natural power (Characteristic impedance is sometimes called surge impedance. However, the term surge impedance usually refers to the value of characteristic impedance of a loss-less line as discussed in chapter 12).

For a receiving end line voltage of V_r kV, the natural power P_N is

$$P_N = \frac{V_r^2}{Z_c} \text{ MW} \tag{4.85}$$

The characteristic impedance of all single circuit high voltage lines is always about 400 ohms.

For this value of Z_c the natural power is $\frac{V_r^2}{400}$ MW or 2.5 V_r^2 kW.

It is known that the load carrying capability of a line decreases with increase in line length. It has been established by extensive studies and experience that very long lines (more than 400 km or so) can be loaded (within reasonable technical and economic limitations of voltage drop and cost of reactive power compensation equipment) only upto about natural load. A major problem with very long lines is the system stability. However, the short lines can be loaded to several times natural load. The most obvious way of increasing natural power is to increase the line voltage, since natural power is proportional to the square of voltage. This explains the historical increase in transmission voltage as longer and longer transmission line projects are taken up.

4.13 POWER CIRCLE DIAGRAMS

The complex power at the receiving and sending ends of a line are given by Eqs. (4.72 and 4.73) which can be re-written as

$$S_r = P_r + j\,Q_r$$

$$= \frac{-|A|\,|V_r|^2}{|B|} \angle(\beta-\alpha) + \frac{|V_s|\,|V_r|}{|B|} \angle(\beta-\delta) \tag{4.86}$$

$$S_s = P_s + j\,Q_s = \frac{|A|\,|V_s|^2}{|B|} \angle(\beta-\alpha) - \frac{|V_r|\,|V_s|}{|B|} \angle(\beta+\delta) \tag{4.87}$$

Each of the powers is the sum of two phasors. Since the real parts of these phasors represent real power P and the imaginary part represents reactive power Q, it is possible to plot S_r and S_s in the $x-y$ plane whose horizontal and vertical coordinates represent the real and reactive powers respectively. The transmission lines are usually operated with constant sending and receiving end voltages. Then one component of each power is a constant phasor (i.e., phasor of constant magnitude and phase) while the second component is a phasor of constant magnitude but variable angle. The loci of S_r and S_s are, therefore, circles drawn from the tip of constant phasors as centres.

The centre of the receiving end circle is located at the tip of the phasor $\frac{-|A|\,|V_r|^2}{|B|} \angle(\beta-\alpha)$.

Thus the x-coordinate of the centre is $\frac{-|A|\,|V_r|^2}{|B|} \cos(\beta-\alpha)$. MW and the y-coordinate is $\frac{-|A|\,|V_r|^2}{|B|} \sin(\beta-\alpha)$ Mvar. The radius of the receiving end circle is $\frac{|V_s|\,|V_r|}{|B|}$ MVA.

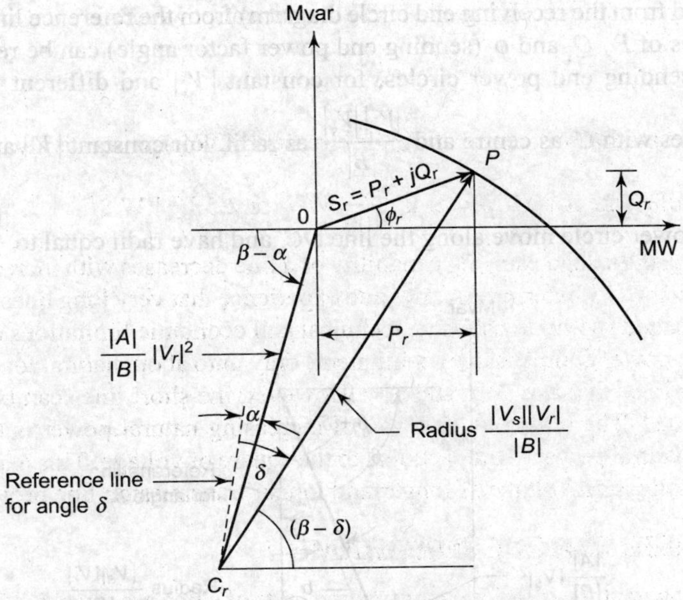

Fig. 4.16 Receiving end power circle diagram

The receiving end circle is drawn in Fig. 4.16. The centre C_r is located by drawing OC_r equal to $\frac{|A||V_r|^2}{|B|}$ inclined at $\angle (\beta - \alpha)$ in the positive (anticlockwise) direction from the negative x-axis. From centre C_r the receiving end circle is drawn with radius $\frac{|V_s||V_r|}{|B|}$. The operating point P on the circle is located by the amount of real power delivered to the load, i.e., P_r. The corresponding value of Q_r can be read from the diagram. The power angle is the angle between the reference line shown and phasor $C_r P$. Many other useful information, e.g., capacity of compensation equipment, maximum receiving end power, etc., can also be obtained from the power circle diagram. This is illustrated in examples (4.18 and 4.19).

The receiving end power circles for constant V_r but varying V_s are concentric circles with C_r as centre and $\frac{|V_r||V_s|}{|B|}$ as radii.

It is seen from Eq. (4.86) that the centre of the sending end circle is located at the tip of the phasor $\frac{|A||V_s|^2}{|B|} \angle (\beta - \alpha)$ and the radius of this circle is $\frac{|V_r||V_s|}{|B|}$. The sending end circle is drawn in Fig. 4.17. The centre C_s is located by drawing OC_S equal to $\frac{|A||V_s|^2}{|B|}$ inclined at an angle $\angle (\beta - \alpha)$ to the positive x-axis. With C_s as centre, the sending end circle is drawn with $\frac{|V_r||V_s|}{|B|}$ radius. The operating point M is located by drawing the line $C_s M$ inclined at an angle

δ (as determined from the receiving end circle diagram) from the reference line as shown in Fig. 4.21. The values of P_s, Q_s and ϕ_s (sending end power factor angle) can be read from the circle diagram. The sending end power circles, for constant $|V_s|$ and different values of $|V_r|$ are concentric circles with C_s as centre and $\frac{|V_r||V_s|}{|B|}$ as radii. For constant $|V_r|$ and varying $|V_s|$ the centres of the power circle move along the line OC_s and have radii equal to $\frac{|V_r||V_s|}{|B|}$.

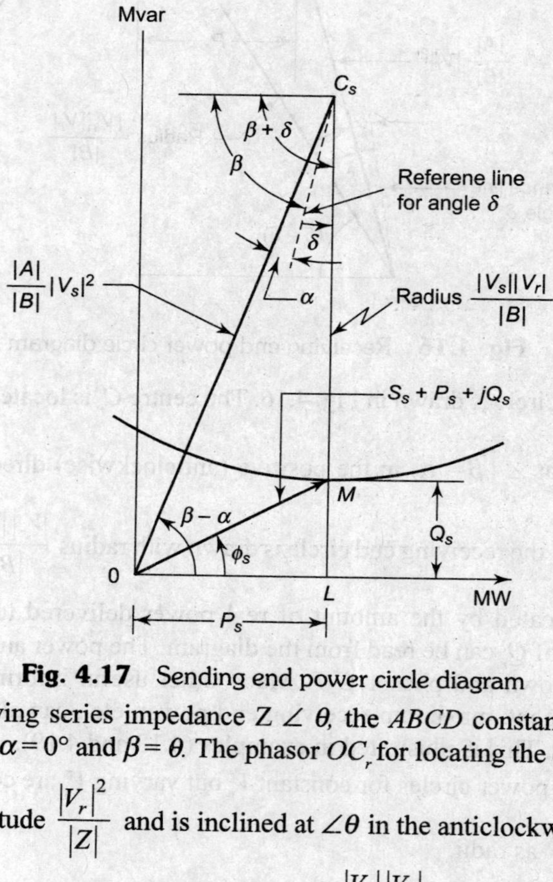

Fig. 4.17 Sending end power circle diagram

For a short line having series impedance $Z \angle \theta$, the *ABCD* constants assume the values: $|A| = |D| = 1$, $|B| = |Z|$, $\alpha = 0°$ and $\beta = \theta$. The phasor OC_r for locating the centre of the receiving end circle, has a magnitude $\frac{|V_r|^2}{|Z|}$ and is inclined at $\angle\theta$ in the anticlockwise direction from the negative x axis. The radius of receiving end circle is $\frac{|V_s||V_r|}{|Z|}$. Similarly the phasor OC_S in the sending end circle diagram, has a magnitude $\frac{|V_s|^2}{|Z|}$ and is inclined at $\angle\theta$ to the positive x-axis and the radius of the sending end circle is $\frac{|V_s||V_r|}{|Z|}$.

Example 4.18

A 3-phase overhead line has a series impedance of 10 + *j* 30 ohms per phase. For receiving and sending end voltages of 132 kV and 140 kV respectively draw the receiving end power circle and determine (a) the maximum real power which the line can supply and the load power factor for drawing this maximum power (b) the capacity of shunt compensation equipment for supplying a load of 150 MVA at 0.8 p.f lagging and power angle for this load condition (c) the capacity of shunt compensation equipment needed to maintain the above voltages under no load condition (d) the unity p.f. load which the line can supply with voltages at above values.

Solution

$$Z = 10 + j\,30 = 31.62 \angle 71.6° \text{ ohms}$$

$$\frac{|V_r|^2}{|Z|} = \frac{132\times132}{31.62} = 551 \text{ MVA}$$

To ensure accuracy, the circle diagram should be reasonably big. Choosing a scale of 1 cm = 100 MVA, line OC_r is drawn equal to 5.51 cm at an angle of 71.6° in the anticlockwise direction from the negative *x*-axis.

The radius of the receiving end circle is $\frac{132\times140}{31.62}$ i.e., 584.4 MVA or 5.84 cm. The circle diagram is shown in Fig. 4.18.

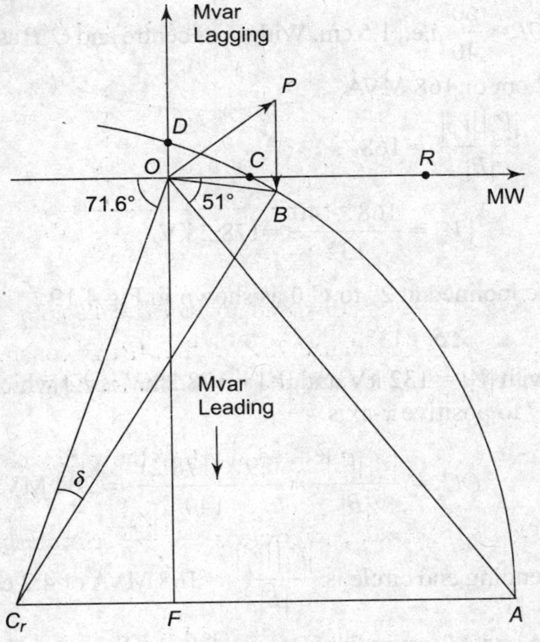

Fig. 4.18

(a) The maximum real power is *FA* which is 4.15 cm or 415 MW. The load power factor angle is $\angle ROA$, i.e., 51°. Power factor is cos (51°) = 0.63 leading.

(b) Line *OP* inclined at $\cos^{-1}$ 0.8 is the 0.8 lagging power factor line. *OP* is 150 MVA or 1.5 cm. From *P* draw *PB* parallel to *y*-axis. For the specified voltage, load should be *OB*. The capacity of the shunt compensation equipment is *PB* which is 1.15 cm or 115 MVA leading. The power angle δ is 12° approximately.

(c) The operating point for zero load is D. OD represents the amount of shunt compensation to maintain the specified voltages under no load condition. OD is 0.35 cm or 35 Mvar lagging. Therefore, the rating of the shunt reactor is 35 Mvar.

(d) C represents the unity power factor operating point with the specified sending end and receiving end voltages. OC is the unity p.f. Load and is 0.85 cm or 85 MW.

Example 4.19

The constants of a 3-phase line are $A = 0.9 \angle 2°$ and $B = 140 \angle 70°$ ohms per phase. The line delivers 60 MVA at 132 kV and 0.8 p.f. lagging. Draw circle diagrams and find (a) sending end voltage and power angle (b) the maximum power which the line can deliver with the above values of sending and receiving end voltages (c) the sending end power and power factor (d) line losses.

Solution

$$OC_r = \frac{|A||V_r|^2}{|B|} = \frac{0.9\,(132)^2}{140} = 112 \text{ MVA}$$

$$\angle(\beta - \alpha) = 68°$$

Select a scale of 1 cm = 40 MVA and draw OC_r equal to $\frac{112}{40}$ i.e., 2.8 cm inclined at 68° to the negative real axis. (a) OP is the 0.8 lagging p.f. line inclined at $\cos^{-1} 0.8$, i.e., 36.87° to positive x-axis. Take $OP = \frac{60}{40}$ i.e., 1.5 cm. With C_r as centre and C_rP as radius draw the receiving end circle. $C_r P = 4.2$ cm or 168 MVA

$$\frac{|V_s||V_r|}{|B|} = 168$$

or

$$|V_s| = \frac{168 \times 140}{132} = 178.2 \text{ kV}$$

Draw reference line inclined at 2° to C_r0 as shown in Fig 4.19

$$\angle\delta = 13°.$$

(b) Maximum power with $|V_r| = 132$ kV and $|V_s| = 178.2$ kV is FA which is 3.25 cm or 130 MW.

(c) Draw OC_s at $\angle$ 68° to positive x-axis.

$$OC_s = \frac{|A||V_s|^2}{|B|} = \frac{0.9 \times (178.2)^2}{140} = 204 \text{ MVA}$$

or 5.1 cm. Radius of sending end circle is $\frac{|V_s||V_r|}{|B|} = 168$ MVA or 4.2 cm. Draw the sending end circle with C_s as centre and 4.2 cm radius. Draw reference line at $\angle$ 2° to the line C_sO as shown and then C_sM *at* $\angle\,\delta = 13°$ to the reference line. M is sending end operating point. OM is 1.5 cm or 60 MVA and $\angle\,\phi_s = 20°$

Sending end p.f. = cos (20°) = 0.94 lagging. S_s, i.e., sending end MVA = 60 MVA

P_s, i.e., sending end power = 60 × 0.94 = 58.4 MW

(d) Line losses = $P_s - P_r = 58.4 - 60 \times 0.8 = 10.4$ MW

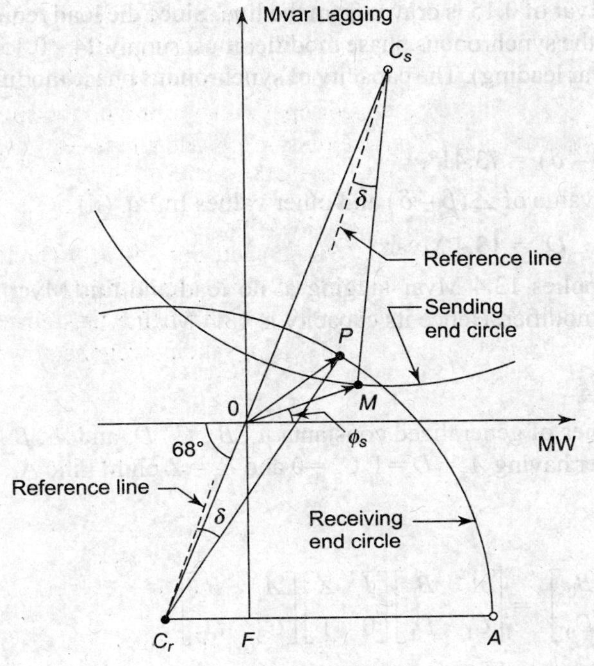

Fig. 4.19

Example 4.20

A 3-phase 132 kV line delivers 40 MVA at 0.8 p.f. lagging. The line constants are $A = D = 0.98 \angle 3°$, $B = 110 \angle 75°$, $C = 5 \times 10^{-4} \angle 88°$. Find the capacity of phasemodifier at (a) full load, (b) no load, $V_s = 140$ kV and $V_r = 132$ kV.

Solution

(a) $$P_r = 40 \times 0.8 = 32 \text{ MW}$$

$$P_r = \frac{|V_s||V_r|}{|B|}\cos(\beta - \delta) - \frac{|A||V_r|^2}{|B|}\cos(\beta - \alpha) \quad \text{(i)}$$

or $$32 = \frac{(140)(132)}{110}\cos(\beta - \delta) - \frac{0.98(132)^2}{110}\cos(75° - 3°)$$

or $$32 = 168\cos(\beta - \delta) - 155.23\cos 72°$$

or $$\angle(\beta - \delta) = 61.58°$$

$$Q_r = \frac{|V_s||V_r|}{|B|}\sin(\beta - \delta) - \frac{|A||V_r|^2}{|B|}\sin(\beta - \alpha) \quad \text{(ii)}$$

$$= \frac{(140)(132)}{110}\sin(61.58°) - \frac{0.98(132)^2}{110}\cos 72°$$

$$= 147.75 - 147.6 = 0.15 \text{ Mvar}$$

Thus a lagging Mvar of 0.15 is drawn from the line. Since the load requires a lagging Mvar of $40 \times 0.6 = 24$ Mvar, the synchronous phase modifier must supply $24 - 0.15 = 23.85$ Mvar lagging (or absorb 23.85 Mvar leading). The capacity of synchronous phase modifier is thus 23.85 Mvar.

(b) $P_r = 0$

Using Eq (i), $\angle(\beta - \delta) = 73.41°$.

Substituting this value of $\angle(\beta - \delta)$ and other values In Eq. (ii)

$$Q_r = 13.4 \text{ Mvar}$$

Thus the line supplies 13.4 Mvar lagging at no load and this Mvar must be absorbed by synchronous phase modifier. Hence its capacity is 13.4 Mvar.

Example 4.21

Two transmission lines of generalized constants A_1, B_1, C_1, D_1 and A_2, B_2, C_2, D_2 are connected through a transformer having $A_3 = D_3 = 1$, $C_3 = 0$ and $B_3 = Z$ ohm. Find A_0, B_0, C_0, D_0 parameters for the total system.

Solution

$$\begin{bmatrix} A_0 & B_0 \\ C_0 & D_0 \end{bmatrix} = \begin{bmatrix} A_1 & B_1 \\ C_1 & D_1 \end{bmatrix} \begin{bmatrix} I & Z \\ 0 & 1 \end{bmatrix} \begin{bmatrix} A_2 & B_2 \\ C_2 & D_2 \end{bmatrix}$$

$$= \begin{bmatrix} A_1A_2 + A_1Z\,C_2 + B_1C_2 & A_1B_2 + A_1Z\,D_2 + B_1D_2 \\ A_2C_1 + C_1C_2\,Z + D_1C_2 & C_1B_2 + C_1Z\,D_2 + D_1D_2 \end{bmatrix}$$

Example 4.22

A 3-phase overhead line has $A = D = 0.9\,\angle 1°$.and $B = 140\,\angle 84°$ ohm. It is operating with V_s = 240 kV and V_r = 220 kV (line to line). Find (a) maximum power which can be received at the receiving end (b) rating of synchronous phase modifier at the receiving end if load at the receiving end is 80 MW at 0.8 p.f. lagging (c) angle δ under above conditions.

Solution

$$P_r = \frac{|V_s||V_r|}{|B|}\cos(\beta - \delta) - \frac{|A||V_r|^2}{|B|}\cos(\beta - \alpha)$$

(a) For maximum P_r, $\dfrac{d\,P_r}{d\delta} = 0$

$$\frac{d\,P_r}{d\delta} = -\frac{|V_s||V_r|}{|B|}\sin(\beta - \delta) = 0$$

or $\quad \sin(\beta - \delta) = 0$ or $\beta = \delta$

Hence $\quad \delta = 84°$

$$P_r = \frac{(240)\,(220)}{140} - \frac{0.9(220)^2}{140}\cos(84° - 1°)$$

$$= 377\text{–}14 - 37.92 = 339.22 \text{ MW}$$

(b) $\quad P_r = 80$ MW, $S_r = 100$ MVA

$$80 = \frac{(240)(220)}{140}\cos(\beta - \delta) - \frac{0.9(220)^2}{140}\cos(84° - 1°)$$

or $$80 = 377.14 \cos(\beta - \delta) - 37.92$$

or $$\cos(\beta - \delta) = 0.313 \quad \text{and} \quad \angle(\beta - \delta) = 71.76°$$

$$Q_r = \frac{|V_s||V_r|}{|B|}\sin(\beta - \delta) - \frac{|A||V_r|^2}{|B|}\sin(\beta - \alpha)$$

$$= \frac{(240)(220)}{140}\sin(71.76°) - \frac{0.9(220)^2}{140}\sin(84° - 1°)$$

$$= 358.19 - 308.82 = 49.37 \text{ Mvar}$$

Since the load requires $(100^2 - 80^2)^{0.5}$, i.e., 60 Mvar lagging, the difference, i.e., 60 – 49.37 = 10.73 Mvar must be supplied by phase modifier. Thus its capacity is 10.73 Mvar.

(c) $\angle \beta - \delta = 71.76°$ or $\angle 5 = 84° - 71.76° = 12.24°$.

Example 4.23

A line has $A = D = 0.93 \angle 3°$ and $B = 150 \angle 70°$ ohm. At the receiving end a step down transformer having a shunt admittance of $0.00025 \angle -75°$ siemens and series impedance of $100 \angle 70°$ ohms is connected. Find the constants A_0, B_0, C_0, D_0 to represent the line with transformer. Derive the expression used.

Solution

The circuit is shown in Fig. 4.20.

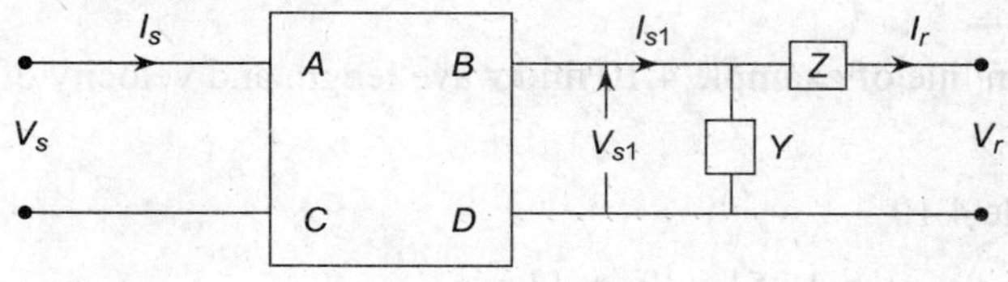

Fig. 4.20

$$V_{s1} = V_r + I_r Z$$
$$I_{s1} = V_{s1} Y + I_r = (V_r + I_r Z) Y + I_r$$
$$= V_r Y + I_r (1 + ZY)$$
$$V_s = AV_{s1} + B I_{s1}$$
$$= A(V_r + I_r Z) + B[V_r Y + I_r (1 + ZY)]$$
$$= (A + BY)V_r + (AZ + B + B YZ) I_r$$
$$I_s = CV_{s1} + DI_{s1}$$
$$= C(V_r + I_r Z) + D[V_r Y + I_r (1 + ZY)]$$
$$= (C + DY) V_r + [CZ + D + DYZ] I_r$$

Therefore $$\begin{bmatrix} A_0 & B_0 \\ C_0 & D_0 \end{bmatrix} = \begin{bmatrix} A + BY & AZ + B + BYZ \\ C + DY & CZ + D + DYZ \end{bmatrix}$$

Alternatively $\begin{bmatrix} A_0 & B_0 \\ C_0 & D_0 \end{bmatrix} = \begin{bmatrix} A & B \\ C & D \end{bmatrix} \begin{bmatrix} 1 & 0 \\ Y & 1 \end{bmatrix} \begin{bmatrix} 1 & Z \\ 0 & 1 \end{bmatrix} = \begin{bmatrix} A+BY & AZ+B+BYZ \\ C+DY & CZ+D+DYZ \end{bmatrix}$

$$A = D = 0.93 \angle 3°, B = 150 \angle 70°, Z = 100 \angle 70°, Y = .00025 \angle -75°$$

$$AD - BC = 1 \quad \text{or} \quad C = \frac{AD-1}{B} = \frac{(0.93 \angle 3°)^2 - 1}{150 \angle 70°}$$

or
$$C = 11.09 \times 10^{-4} \angle 77.15°$$

$$A_0 = A + BY = 0.93 \angle 3° + (150 \angle 70°)(.00025 \angle -75°)$$
$$= 0.967 \angle 2.7°$$

$$B_0 = AZ + B + BZY = (.93 \angle 3°)(100 \angle 70°) + 150 \angle 70°$$
$$+ (150 \angle 70°)(100 \angle 70°)(.00025 \angle -75°)$$
$$= 246.65 \angle 71.05° \text{ ohm}$$

$$C_0 = C + DY = (11.09 \times 10^{-4} \angle 77.15°) + (.93 \angle 3°)(.00025 \angle 75°)$$
$$= 0.00092 \angle 70° \text{ siemens}$$

$$D_0 = CZ + D + DZY$$
$$= (11.09 \times 10^{-4} \angle 77.15°)(100 \angle 70°) + .93 \angle 3°$$
$$+ (.93 \angle 3°)(100 \angle 70°)(.00025 \angle -75°) = 0.865 \angle 7.7°$$

It is seen that A_0 is not equal to D_0. This is because of the reason that the circuit (Fig. 4.30) is not bilateral.

Example 4.24

For the transmission line of example 4.10 find wave length and velocity of propagation.

Solution

As found is example 4.10

$$\beta = 1.251 \times 10^{-3} \text{rad/km}$$

Wavelength
$$\lambda = \frac{2\pi}{\beta} = 5.0225 \times 10^3 \text{ km}$$

Velocity of propagation
$$= f\lambda = 50 \times 5.0225 \times 10^3$$
$$= 2.51 \times 10^5 \text{ km/sec.}$$

Example 4.25

Solve example 4.8 using per unit method

Solution

$$\text{Base voltage} = 220 \text{ kV line to line}$$
$$= \frac{220}{\sqrt{3}} \text{ kV line to neutral}$$

Base MVA = 40 MVA.

$$\text{Base current} = \frac{40 \times 10^6}{\sqrt{3} \times 220 \times 1000} = 104.976 \text{ A}$$

$$\text{Base impedance} = \frac{(220)^2}{40} = 1210\ \Omega$$

$$\text{Base admittance} = \frac{1}{1210}\ \text{siemens}$$

$$Z = \frac{195.578 \angle 74.429}{1210}\ \text{p.u.}$$

$$Y = (0.001411 \angle 90°)\ (1210)\ \text{p.u.}$$

$$YZ = 0.27596 \angle 164.429°\ \text{p.u.}$$

$$A = 1 + \frac{YZ}{2} + \frac{(YZ)^2}{24}$$

$$= 0.8705 \angle 2.33° = D$$

$$B = Z\left[1 + \frac{YZ}{6} + \frac{(YZ)^2}{120}\right] = 0.15457 \angle 75.15°\ \text{p.u.}$$

$$C = Y\left[1 + \frac{YZ}{6} + \frac{(YZ)^2}{120}\right] = 1.6328 \angle 90.722°\ \text{p.u.}$$

$$V_r = 1 \angle 0\ \text{p.u.}$$

$$I_r = 1 \angle -\cos^{-1} 0.9 = 1 \angle -25.84°\ \text{p.u.}$$

$$V_s = A\,V_r + B\,I_r$$

$$= (0.8705 \angle 2.33°)\ (1 \angle 0) + (0.15457 \angle 75.15°)\ (1 \angle -25.84)$$

$$= (0.8705 \angle 2.33°) + (0.15457 \angle 49.31°)$$

$$= 0.982467 \angle 8.935°\ \text{p.u.}$$

$$I_s = CV_r + DI_r$$

$$= (1.6328 \angle 90.722)\ (1 \angle 0) + (0.8705 \angle 2.33°)\ (1 \angle -25.84°)$$

$$= 1.6328 \angle 90.722 + 0.8705 \angle -23.51°$$

$$= 0.777\,665 + j\,1.28542 = 1.5023 \angle 58.8°\ \text{p.u.}$$

Sending end line voltage $= 0.982467 \times 220 = 216.14$ kV

Sending end line current $= 1.5023 \times 104.976 = 157.71$ A

$$\text{Sending end MVA} = \frac{\sqrt{3} \times 216.14 \times 157.71}{1000} = 59.04\ \text{MVA}$$

Sending end power factor $= \cos(58.8 - 8.935°) = 0.645$ leading

Example 4.26

Fig. 4.25 (a) shows a single phase system having three parts *A, B, C.* The ratings of transformers are

A–B : 10 MVA, 13.8/138 kV, $X = 10\%$

B–C : 10 MVA, 138/69 kV, $X = 8\%$

If the base in circuit *B* is chosen as 10 MVA, 138 kV, find the per unit impedance of 300 ohm resistance load in circuit *C* referred to circuits *C*, *B* and *A*. Find voltage on side *A* if voltage at load is 66 kV.

Solution

Base MVA 10

Base Voltage 138 kV in circuit *B*,

$$138\ (13.8/138) = 13.8 \text{ kV in circuit } A$$

and $$138\ (69/138) = 69 \text{ kV in circuit } C$$

Base impedance in circuit $C = 69^2/10 = 476.1\ \Omega$

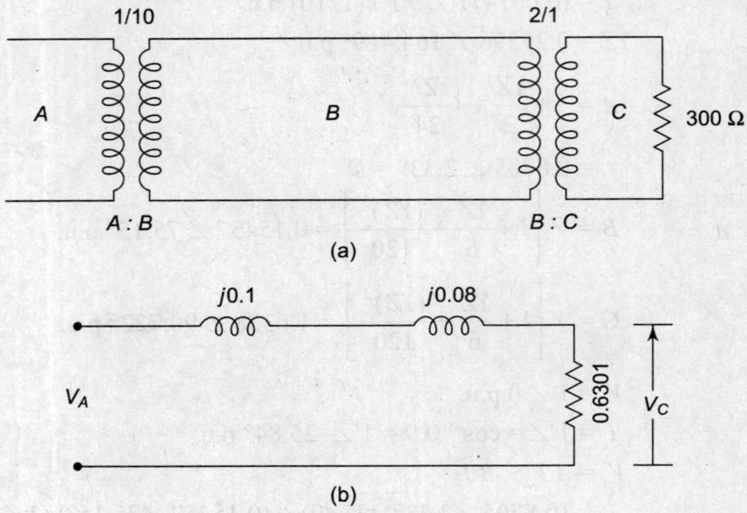

Fig. 4.21

Load resistance = 300/476.1 = 0.6301 pu. When referred to circuits *B* and *A*, the p.u., load resistance remains the same.

$$V_C = 66/69 = 0.9565\ \angle 0 \text{ p.u.}$$

$$I = 0.9565/0.6301 = 1.518\ \angle 0 \text{ p.u.}$$

$$V_A = 0.9565\ \angle 0 + (1.518\ \angle 0)\ (j\,0.18)$$

$$= 0.9948\ \angle 15.94 \text{ p.u.}$$

$$\text{Voltage in circuit } A = 0.9948 \times 13.8$$

$$= 13.728 \text{ kV}$$

Fig. 4.21 (b) shows the equivalent circuit with p.u. values.

Example 4.27

Fig 4.22(a) shows a power system. Redraw the circuit with p.u. impedance represented on 5000 VA base and common system base voltage of 250 V for G_1. Also draw impedance diagram.

Solution

Base VA = 5000 VA

Base voltage 250 V for G_1 and G_2, 250/800 V for T_1 800/400 V for T_2 and 800 V for line.

Reactance of $$G_1 = (j\,0.2)\left(\frac{5000}{2000}\right) = j\,0.5 \text{ p.u.}$$

$$\text{Reactance of } G_2 = (j\,0.3)\left(\frac{5000}{2000}\right) = j\,0.75 \text{ p.u.}$$

$$\text{Reactance of } T_1 = (j\,0.2)\left(\frac{5000}{2000}\right) = j\,0.25 \text{ p.u.}$$

$$\text{Reactance of } T_2 = (j\,0.06)\left(\frac{5000}{2000}\right)\left(\frac{1000}{800}\right)^2 = j\,0.0586 \text{ p.u.}$$

$$\text{Base impedance of line} = \frac{(800/1000)^2 \times 1000}{5000/1000} = 128\ \Omega$$

$$\text{Impedance of line} = \frac{40 + j\,150}{128} = 0.3125 + j\,1.1719 \text{ p.u.}$$

The single line diagram and impedance diagram with p.u. values are shown in Fig. 3.32 (b and c)

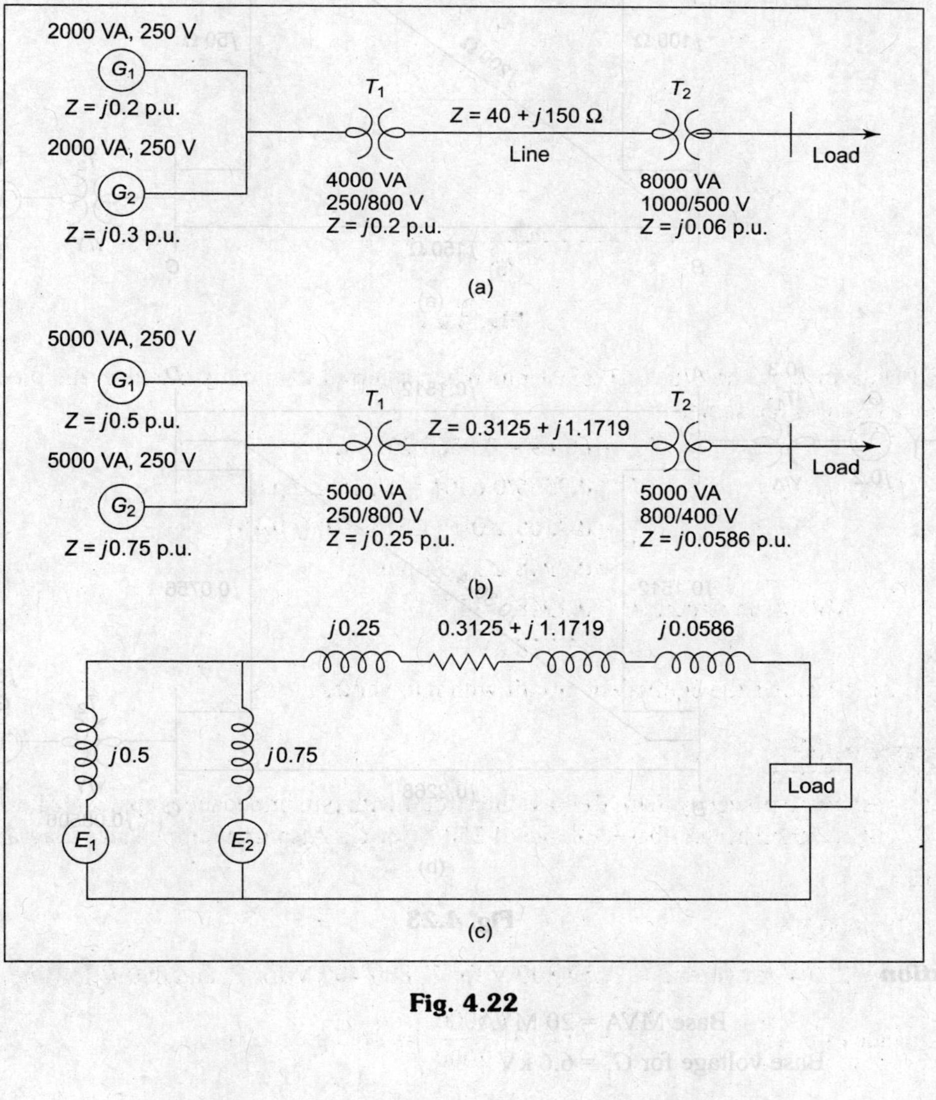

Fig. 4.22

Example 4.28

Fig. 4.23(a) shows a system. Draw a single line diagram and insert the reactances in p.u. on 20 MVA, 6.6 kV base in the circuit of generator G_1. The data is

Generator G_1 : 10 MVA, 6.6 kV, j 0.1 p.u.

Generator G_2 : 20 MVA, 11.5 kV, j 0.1 p.u.

Transformer T_1 : 10 MVA, 3-phase, 6.6/115 kV, j 0.15 p.u.

Transformer T_2 : Three single phase units, each 10 MVA, 75/7.5 kV, j 0.1 p.u.

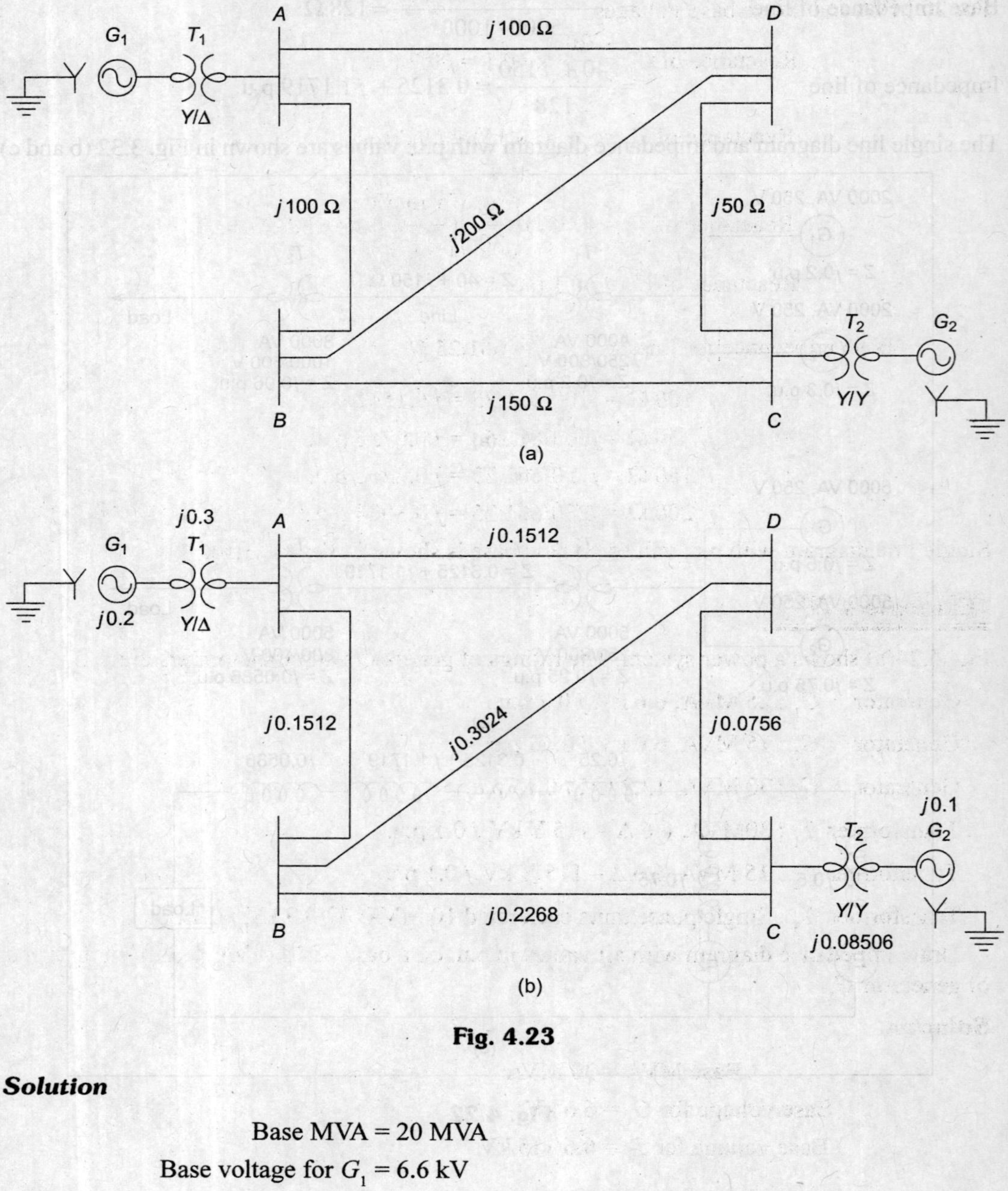

Fig. 4.23

Solution

$$\text{Base MVA} = 20 \text{ MVA}$$

$$\text{Base voltage for } G_1 = 6.6 \text{ kV}$$

$$\text{Base voltage for } T_1 = 6.6/115 \text{ kV}$$
$$\text{Base voltage for lines} = 115 \text{ kV}$$
$$\text{Line to line voltage ratio of } T_2 = \sqrt{3} \times 75 / \sqrt{3} \times 7.5$$
$$= 129.9/12.99 \text{ kV}$$
$$\text{Base voltage for } T_2 = 115/11.5 \text{ kV}$$
$$\text{Rating of } T_2 = 3 \times 10 = 30 \text{ MVA}$$
$$\text{Base voltage for } G_2 = 11.5 \text{ kV}$$

For 20 MVA and above base voltages

$$\text{Reactance of } G_1 = \frac{20}{10} \times j\,0.1 = j\,0.2 \text{ p.u.}$$
$$\text{Reactance of } T_1 = \frac{20}{10}\,(j\,0.15) = j\,0.3 \text{ p.u.}$$
$$\text{Reactance of } T_2 = (j\,0.1)\left(\frac{20}{30}\right)\left[\frac{129.9}{115}\right]^2 = j\,0.08506 \text{ p.u.}$$
$$\text{Reactance of } G_2 = j\,0.1 \text{ p.u.}$$
$$\text{Base impedance for lines} = \frac{115^2}{20} = 661.25\ \Omega$$
$$j\,100\ \Omega = j100/661.25 = j\,0.1512 \text{ p.u.}$$
$$j\,50\ \Omega = j50/661.25 = j\,0.0756 \text{ p.u.}$$
$$j\,150\ \Omega = j150/661.25 = j\,0.2268 \text{ p.u.}$$
$$j\,200\ \Omega = j200/661.25 = j\,0.3024 \text{ p.u.}$$

Single line diagram with p.u. values on new base is shown in Fig. 4.23(b).

Example 4.29

Fig. 4.24(a) shows a power system. The ratings of generators and transformers are:

Generator G_1 : 25 MVA, 6.6 kV, $j\,0.2$ p.u.

Generator G_2 : 15 MVA, 6.6 kV $j\,0.15$ p.u.

Generator G_3 : 30 MVA, 13.2 kV, $j\,0.15$ p.u.

Transformer T_1 : 30MVA, 6.6 Δ – 115 Y kV, $j\,0.1$ p.u.

Transformer T_2 : 15 MVA, 6.6 Δ – 115 Y kV, $j\,0.1$ p.u.

Transformer T_3 : Single phase units each rated 10 MVA, 69/6.9 kV, $j\,0.1$ p.u.

Draw impedance diagram with all values in p.u. on a base of 30 MVA, 6.6 kV in the circuit of generator G_1.

Solution

$$\text{Base MVA} = 30 \text{ MVA}$$
$$\text{Base voltage for } G_1 = 6.6 \text{ kV}$$
$$\text{Base voltage for } T_1 = 6.6/115 \text{ kV}$$

Base voltage for T_2 = 115/6.6 kV

Base voltage for G_2 = 6.6 kV

MVA rating of $T_3 = 3 \times 10 = 30$ MVA

Voltage rating of $T_3 = \sqrt{3} \times 69 / \sqrt{3} \times 6.9$ kV

= 119.51/11.951 kV

Base voltage for T_3 = 115/11.5 kV

Base voltage for G_3 = 11.5 kV

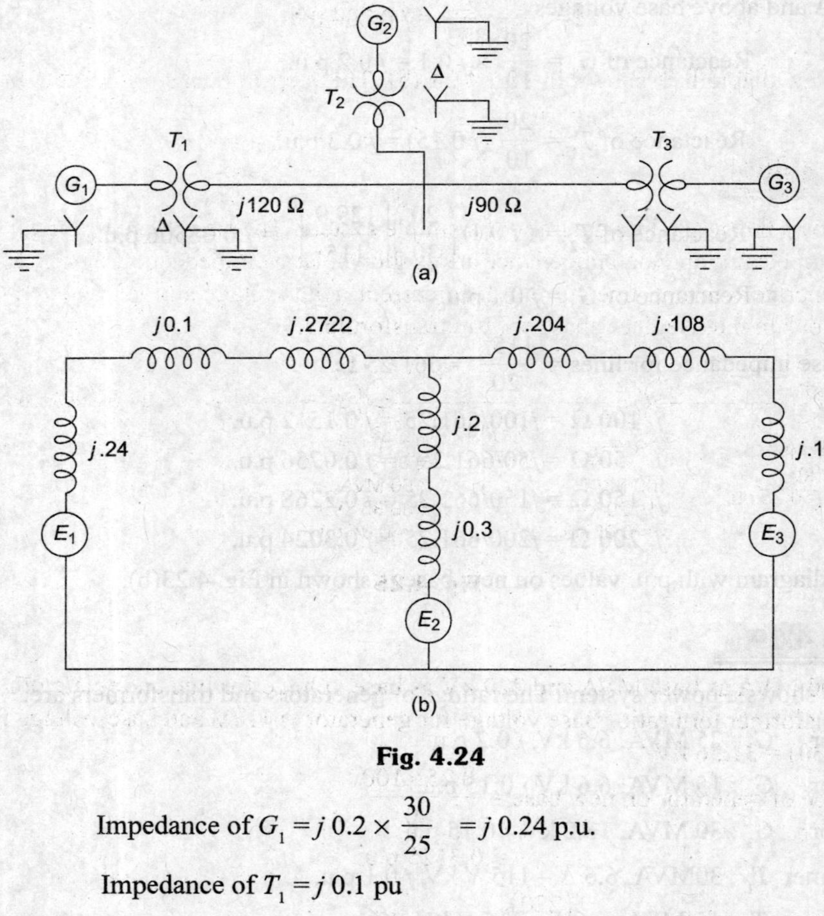

Fig. 4.24

$$\text{Impedance of } G_1 = j\,0.2 \times \frac{30}{25} = j\,0.24 \text{ p.u.}$$

$$\text{Impedance of } T_1 = j\,0.1 \text{ pu}$$

$$\text{Impedance of } G_2 = j\,0.15 \times \frac{30}{15} = j\,0.3 \text{ p.u.}$$

$$\text{Impedance of } T_2 = j\,0.1 \times \frac{30}{15} = j\,0.2 \text{ p.u.}$$

$$\text{Impedance of } T_3 = \left(\frac{119.51}{115}\right)^2 (j\,.1) = j\,0.108 \text{ p.u.}$$

$$\text{Impedance of } G_3 = j\,0.15\left(\frac{13.2}{11.5}\right)^2 = j\,0.1976 \text{ p.u.}$$

$$\text{Base impedance for lines} = \frac{115^2}{30} = 440.833\ \Omega$$

$$j120\ \Omega = \frac{j\,120}{440.833} = j0.2722 \text{ p.u.}$$

$$j90\ \Omega = \frac{j\,90}{440.833} = j\,0.204 \text{ p.u.}$$

The impedance diagram is shown in Fig. 4.24 (b). The p.u., impedances on new base are as shown.

Example 4.30

Fig. 4.25 shows the single line diagram of a simple system. The transformer ratings, generator rating, line impedance and load impedance are as shown. Draw impedance diagram. Find actual values of generator current, line current, load current, load voltage and load power. Carry out the calculations in p.u. Neglect phase shift in transformers.

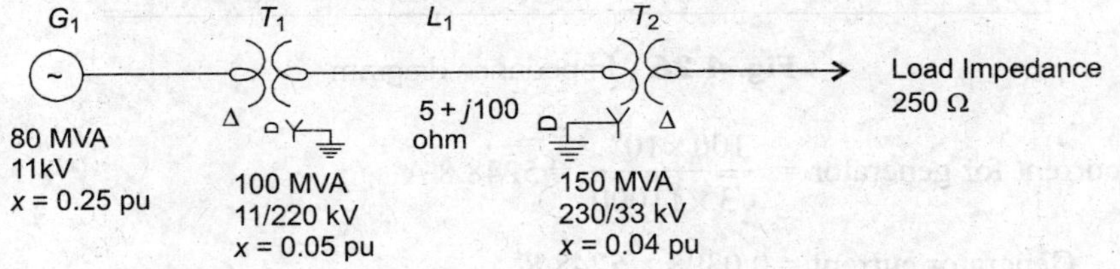

Fig. 4.25

Solution

Select 100 MVA as base MVA and 220 kV as base voltage for line and relate other voltage basis by transformer turn ratio. Base voltage for generator is 11kV and base voltage for load is = 220 (33/230) = 31.56 kV

$$\text{Impedance of generator on new base} = \frac{0.25\times100}{80} = 0.3125 \text{ p.u.}$$

$$\text{Base impedance for line} = \frac{(220)^2}{100} = 484 \text{ ohm}$$

$$\text{Per unit impedance of line} = \frac{5 + j100}{484} = 0.01 + j0.2067 \text{ p.u.}$$

Per unit impedance of T_1 = 0.05 p.u

$$\text{Per unit impedance of } T_2 = 0.04\left(\frac{230}{220}\right)^2\left(\frac{100}{150}\right) = 0.029 \text{ p.u.}$$

$$\text{Base impedance for load} = \left(\frac{31.56}{100}\right)^2 = 9.96 \text{ ohm}$$

$$\text{Per unit impedance of load} = \frac{250}{9.96} = 25.1 \text{ ohm}$$

Impedance diagram is shown in Fig. 4.26

Assume generator voltage as $1\angle 0$ p.u.

$$\text{Per unit current} = \frac{1\angle 0}{j0.3125 + j0.05 + .01 + j.2067 + j.029 + 25.1}$$

$$= 0.0398 \text{ p.u.}$$

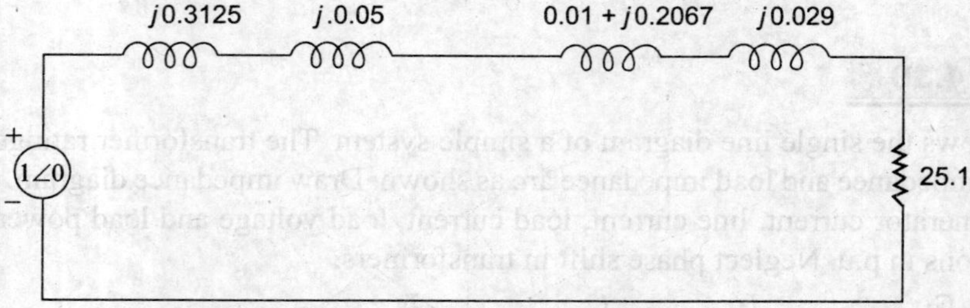

Fig. 4.26 Impedance diagram

$$\text{Base current for generator} = \frac{100\times10^6}{\sqrt{3}\times11000} = 5248.8 \text{ A}$$

$$\text{Generator current} = 0.0398 \times 5248.8$$

$$= 208.9 \text{ A}$$

$$\text{Base current for line} = \frac{100\times10^6}{\sqrt{3}\times220000} = 262.44 \text{ A}$$

$$\text{Line current} = 0.0398 \times 262.44$$

$$= 10.445 \text{ A}$$

$$\text{Base current for load} = \frac{100\times10^6}{\sqrt{3}\times31.56\times1000} = 1829.5 \text{ A}$$

$$\text{Load current} = 0.0398 \times 1829.4$$

$$= 72.81 \text{ A}$$

$$\text{Load voltage} = \frac{72.81\times250}{1000} = 18.2 \text{ kV per phase}$$

$$\text{Line value of load voltage} = \sqrt{3}\times18.2 = 31.52 \text{ kV}$$

$$\text{Load power} = \frac{3(72.81)^2(250)}{10^6} = 3.976 \text{ MW}$$

Example 4.31

A 3-phase 132 kV line delivers 45 MW at 0.8 pf lagging. Using power circle diagram find (a) sending end voltage (b) capacity of synchronous condenser required if sending end voltage is increased to 180 kV. State whether synchronous condenser is over excited or under excited, (c) capacity of synchronous condenser needed at no load if both sending and receiving end voltages are 132 kV. The line constants are $A = 0.9\ \angle 2.5°$, $B = 100\ \angle 70°$

Solution

(a) $$OC_r = \frac{|A||V_r|^2}{|B|} = \frac{.9(132)^2}{100} = 156.82 \text{ MVA}$$

$$\angle \beta - \alpha = 70\text{–}2.5 = 67.5°$$

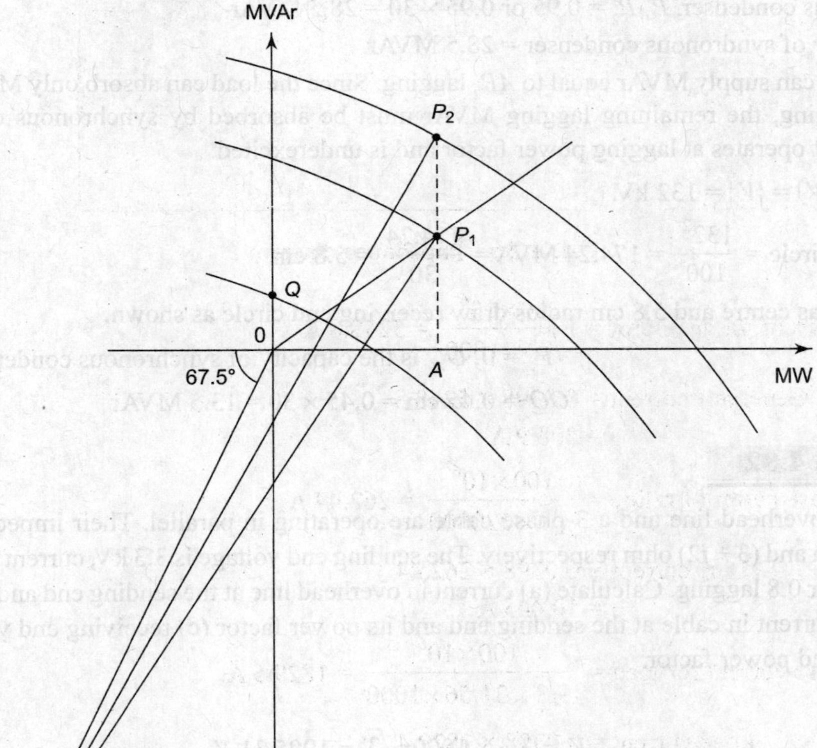

Fig. 4.27

Select a scale of 1 cm = 30 MVA and draw $OC_r = 156.82/30 = 5.23$ cm inclined at 67.5° to negative real axis as shown in Fig. 4.27. OP_1 is the 0.8 p.f. line inclined at $\cos^{-1} 0.8 = 36.87°$ to positive x-axis

$$OP_1 = \frac{45}{0.8} = 56.25 \text{ MVA or } 1.875 \text{ cm.}$$

$$C_r P_1 = 7.1 \text{ cm} = 7.1 \times 30 = 213 \text{ MVA}$$

$$\frac{|V_s||132|}{100} = 213 \quad \text{or} \quad V_s = 161.36 \text{ kV}$$

The circle with $C_r P_1$ as radius is the receiving end circle,

(b) $$\frac{|V_s||V_r|}{|B|} = \frac{180 \times 132}{100} = 237.6 \text{ MVA}$$

$$C_r P_2 = \frac{237.6}{300} = 7.92 \text{ cm}$$

With $C_r P_2$ as radius, draw receiving end circle as shown in Fig. 4.31. $P_1 P_2$ is the capacity of synchronous condenser. $P_1 P_2 = 0.95$ or $0.95 \times 30 = 28.5$ MVAr

Capacity of syndronous condenser = 28.5 MVAr.

The line can supply MVAr equal to AP_2 lagging. Since the load can absorb only MVAr equal to AP_1 lagging, the remaining lagging MVAr must be absorbed by synchronous condenser. Therefore it operates at lagging power factor and is underexcited.

(c) If $|V_r| = |V_s| = 132$ kV.

$$\text{Radius of circle} = \frac{132^2}{100} = 174.24 \text{ MVA} = \frac{174.24}{30} = 5.8 \text{ cm}$$

With C_r as centre and 5.8 cm radius draw receiving end circle as shown.

Since $P_r = 0$, OQ is the capacity of synchronous condenser

$$OQ = 0.45 \text{ cm} = 0.45 \times 30 = 13.5 \text{ MVAr}$$

Example 4.32

A 3-phase overhead line and a 3-phase cable are operating in parallel. Their impedances are $(4 + j6)$ ohm and $(3 + j2)$ ohm respectively. The sending end voltage is 3.3 kV, current 250 A and power factor 0.8 lagging. Calculate (a) current in overhead line at the sending end and its power factor (b) current in cable at the sending end and its power factor (c) receiving end voltage (d) receiving end power factor.

Solution

$$V_s = 3.3 \times 1000/\sqrt{3} = 1905.31 \; V$$

$$I = 250\,(0.8 - j\,0.6) = 200 - j\,150A$$

Let I_1 and I_2 be the currents in overhead line and cable respectively.

(a) $$I_1 = \frac{(200 - j\,150)\,(3 + j\,2)}{4 + j\,6 + 3 + j2} = 84.83 \angle -52° \text{ A}$$

$$\text{Power factor} = \cos(-52°) = 0.616 \text{ lagging}$$

(b) $$I_2 = I - I_1 = (200 - j\,150) - (84.83 \angle -52°)$$

$$= 169.57 \angle -29.37°$$

$$\text{Power factor} = \cos\,(-29.37°) = 0.87 \text{ lagging}$$

(c) Fig 4.28 shows the phasor diagram.

$$\text{Voltage drop} = (84.83 \angle -52°)(4 + j\,6) = 611.72 \angle 4.3° \text{ V} = 610 + j\,4.63 \text{ V}$$

$$V_s = V_r \angle -\delta + 610 + j\,4.63$$

Since $\angle\delta$ is very small, we can write

$$V_s^2 = (V_r + 610)^2 + (4.63)^2$$

or $$1905.31^2 = (V_r + 610)^2 + (4.63)^2$$

or $$V_r = 1295.3\text{V}$$

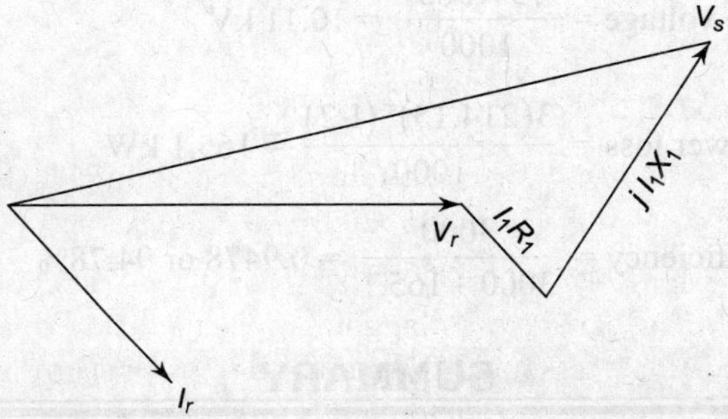

Fig 4.28

$$\text{Receiving end line voltage} = \frac{\sqrt{3}(1295.3)}{1000} = 2.243 \text{ kV}$$

(d) $$\tan\delta = \frac{4.63}{1295.3 + 610} = 0.00243$$

$$\delta = 0.14°$$

Phase angle between receiving end voltage and current = $(\cos^{-1} 0.8) - 0.14 = 36.73°$

Receiving end power facter = cos (36.73°) = 0.8 lagging

Example 4.33

A 3 phase 3 km long line delivers 3000 kW at a power facter of 0.8 lagging to a load. If voltage at sending end is 11 kV, find receiving end voltage and efficiency of transmission line resistance and reactance are 0.4 ohm and 0.8 ohm per km per conductor respectively. (AMIE 2003)

Solution

$$V_s = \frac{11000}{\sqrt{3}} = 6351 \text{ V}, \cos\phi_r = 0.8$$

$$I_r = \frac{3000 \times 10^3}{3\,V_r \times 0.8} = \frac{1250 \times 10^3}{V_r}$$

$$R = 0.4 \times 3 = 1.2 \text{ ohm}, X = 2.4 \text{ ohm}$$

$$V_s = V_r + I_r R \cos\phi_r + I_r \times \sin\phi_r$$

or $$6351 = V_r + \frac{1250 \times 10^3 \times 1.2 \times 0.8}{V_r} + \frac{1250 \times 10^3 \times 2.4 \times 0.6}{V_r}$$

or $$6351\ V_r = V_r^2 + 3000000$$

or $$V_r^2 - 6351\ V_r + 3000000 = 0$$

Hence $$V_r = 5837 \text{ or } 514\text{V}$$

V_r can not so low as 514 V.

Therefore $$V_r = 5837 \text{ V}$$

$$I_r = \frac{1250\times10^3}{5837} = 214.15 \text{ A}$$

$$\text{Receiving end line voltage} = \frac{\sqrt{3}\times5837}{1000} = 10.11 \text{ kV}$$

$$\text{Power loss} = \frac{3(214.15)^2(1.2)}{1000} = 165.1 \text{ kW}$$

$$\text{Efficiency} = \frac{3000}{3000+165.1} = 0.9478 \text{ or } 94.78\%$$

SUMMARY

4.1 The performance of a transmission line depends on line parameters.

4.2 The line parameters are series resistance, series inductance, shunt conductance and capacitance. All these parameters are distributed over the length of line. Shunt conductance is neglected in all calculations.

4.3 The line performance is calculated on single phase basis. It is assumed that 3 phase system is balanced.

4.4 If the length of line is less than 80 km, it is termed as short line. Lines having length between 80 km and 250 km are called medium length lines and those longer then 250 km are called long lines.

4.5 Calculations in per unit have many advantages. As such most transmission system calculations are done in per unit.

4.6 A 3 phase system in balanced steady state can be represented by a single line diagram.

4.7 Voltage regulation $= \dfrac{|V_s|-|V_r|}{|V_r|}$

4.8 A medium length line can be represented by a nominal T circuit or nominal π circuit.

4.9 In a long line the line resistance, inductance and capacitance are taken as distributed over the line length.

4.10 A long line can be analysed using A, B, C, D parameters. These A, B, C, D parameters are functions of R, L, C and line length.

4.11 In a short line the effect of shunt capacitance is negligible. As such series resistance and inductance can be treated as lumped.

4.12 Characteristic inpedance of line $= \sqrt{z/y}$

4.13 Propagation constant $\gamma = \alpha + j\beta$

4.14 Equivalent T and π circuit can be used to represent a long line.

4.15 When a long line is under no load condition, the receiving end voltage may be more than sending end voltage. This is called Ferranti effect.

4.16 $S = VI^*$

4.17 A line needs shunt compensation to maintain a good voltage profile and to reduce losses.
4.18 Shunt compensation may use static capacitors or synchronous condenser.
4.19 Tap changing can be used to improve voltage profile.
4.20 A power circle diagram is a graphical method to predict line performance and calculate the shunt compensation requirement.

SELF TEST QUESTIONS

A. State whether the following are true (T) or false (F)

4.1 In per unit system for 3 phase systems, 3 phase kVA and line to line voltage are usually chosen as base quantities.
4.2 Lines of length more than 250 km are called long lines.
4.3 In medium length lines, the effect of shunt capacitance is neglected.
4.4 The voltage regulation of a short line supplying an inductive load may be negative.
4.5 If a suitable capacitor is connected at the end of a short line, voltage regulation can be made zero.
4.6 Calculations using nominal *T* circuit are exact and accurate.
4.7 For a two port linear passive bilateral network AD – BC = 1
4.8 The units of parameter *B* are ohms.
4.9 Shunt compensation causes a reduction in system investment per kW of load supplied.
4.10 The reactive power transferred over a line is proportional to $|V_s - V_r|$.

B. Fill in the blanks with suitable words

4.11 A synchronous compensator is a ______.
4.12 Static compensation equipment is generally connected to ______ winding of substation transformer.
4.13 If receiving end voltage is V_r and characteristic impedance is Z_c, natural load = ______
4.14 For fixed values of V_s and V_r the real power depends on ______.
4.15 The shunt capacitance of line absorbs ______ vars.
4.16 In a 3 phase system base current = ______.
4.17 In a short line voltage regulation can be zero or negative only if the power factor is ______.
4.18 The units of constant *C* are ______.
4.19 In terms of γ constant A = ______.
4.20 If δ is power angle, real power $\propto$ ______.

ANSWERS

4.1 T 4.2 T 4.3 F 4.4 F 4.5 T 4.6 F
4.7. T 4.8 T 4.9 T 4.10 T 4.11 synchronous motor
4.12 tertiary 4.13 V_r^2/Z_c 4.14 power angle 4.15 leading
4.16 $\dfrac{\text{Base kVA}}{\sqrt{3}\,(\text{Base kV})}$ 4.17 leading 4.18 siemens 4.19 $\cosh(\gamma l)$ 4.20 $\sin\delta$

JOB INTERVIEW QUESTIONS

4.1 Differentiate between nominal T and equivalent T circuits.

4.2 Differentiate between series and shunt compensation.

4.3 What is the significance of power angle?

4.4 Name the advantages of series compensation.

4.5 What is meant by per unit system?

4.6 What is a synchronous compensator?

4.7 What is Ferranti effect?

4.8 What is the significance of natural load?

4.9 What are *ABCD* constants?

4.10 What is meant by propagation constant?

REVIEW QUESTIONS

4.1 Discuss why:

(a) The representation of a line by lumped parameters gives inaccurate results for long lines.

(b) Equivalent π circuit of a long line is preferred to the equivalent T circuit.

(c) The receiving end voltage of an unloaded long line may be more than the sending end voltage.

(d) Long lines usually need reactive power compensation equipment for proper operation.

(e) Complex power in an ac circuit is given by VI^*.

4.2 Give reasons:

(a) Eqs. (4.77) and (4.78) can be used to directly give single phase power as well as 3-phase power.

(b) Power can be transferred over ac lines even when the magnitudes of sending end and receiving end voltages are equal.

(c) The real power transferred over a line depends on the power angle and not on the difference between sending end and receiving end voltage.

(d) The reactive power transferred over a line does not depend on power angle but on the difference between the sending end and receiving end voltage.

(e) In order to maintain a good voltage profile on the line, the control of reactive power flow is necessary.

4.3 Discuss why:

(a) The nature of reactive power compensation is different for peak load and off peak load conditions.

(b) Synchronous compensators can supply leading as well as lagging vars.

(c) A drop in system voltage causes the synchronous compensator to supply more vars.

(d) Series compensation increases the power transmission capacity of a line.

(e) Ferranti effect is significant only in medium and long lines.

4.4 What is per unit system? How are the base quantities selected ?

4.5 Draw the phasor diagram of a short line and derive an expression for voltage regulation.

4.6 What is a nominal circuit? Find *ABCD* constants for nominal π and nominal T circuits of a transmission line

4.7 Draw phasor diagrams for a nominal π circuit and a nominal T circuit of a transmission line. Derive expressions for sending end voltage and current in each case.

4.8 Starting from first principles deduce expressions for $ABCD$ constants of a long line in terms of its parameters. Define propagation constant and characteristic impedance.

4.9 What is an equivalent π and equivalent T circuit of a long line ? Derive expressions for parameters of these circuits in terms of line parameters.

4.10 What is Ferranti effect? Deduce a simple expression for the voltage rise of an unloaded line.

4.11 Deduce expressions for the sending end and receiving end power of a line in terms of voltages and $ABCD$ constants. Show that the real power transferred is dependent on the power angle and the reactive power transferred is dependent on the voltage drop in the line.

4.12 What is line compensation ? Why is it necessary? Compare static capacitors and synchronous compensators.

4.13 What is a receiving end power circle-diagram? How can it be drawn ? What information does it provide ?

PROBLEMS

4.1 A 3-phase transmission line having a series impedance of 20 + j 30 ohms delivers 7 MW at 33 kV and 0.85 lagging power factor. Find sending end voltage, regulation and power angle. Neglect capacitance. [41.4 kV, 25.45%, 5.18°]

4.2 A 3-phase transmission line has a series reactance of 24 ohms per phase. The voltages at the sending end and receiving end are 140 kV and 132 kV. The sending end current is 500 A. Find the active and reactive power delivered at the receiving end. Neglect resistance and capacitance. [105.5 MW, 44 Mvar lagging]

4.3 A 3-phase overhead line has a series impedance of 20 + j 50 ohms per phase. Find the maximum active power which can be delivered at a receiving end voltage of 31.2 kV if the sending end voltage is 33 kV. [12.3 MW]

4.4 A 10 km long 3-phase line delivers 1 MW at 0.8 p. f. lagging. The series impedance of the line is 0.5 + j 0.56 ohm per km per phase. The sending end voltage is 11 kV. Find the line current, receiving end voltage and transmission efficiency. [71.5 A, 10.1 kV. 0.93]

4.5 A 3-phase line delivers 1 MW at 11 kV and 0.8 lagging power factor. The series impedance of the line is 4 + j 6 ohms per phase. Calculate the sending end voltage if the load voltage is to be maintained and addition of load at the receiving end increases the line current by 50% and raises the power factor to 0.9 lagging. Also calculate the equivalent series impedance of the total load and voltage regulation. [12.06 kV, 58.08 + j 28.14 ohms, 9.6%]

4.6 A 3-phase 66 kV 50 Hz line has a resistance of 9.6 ohm, inductance of 0.097 H and capacitance of 0.765 μF per phase respectively. It delivers 24 MVA at 66 kV at 0.8 power factor lagging. Find voltage regulation and transmission efficiency. Use nominal T circuit. [14.7 %, 93.6%]

4.7 A 3-phase line has a series impedance of 300 $\angle$ 78° ohms per phase and a shunt admittance of 24 × 10^{-4} $\angle$ 90° siemens per phase. The voltage at the receiving end is 220 kV but there is no load at receiving end. A load of 100 MW at unity p.f. is connected at the mid point of the line. Using nominal π circuit find sending end voltage. [316.5 kV]

4.8 A 3-phase overhead line delivers 75 MVA at 132 kV and 0.8 lagging power factor. The series impedance of the line is 28 + j 63 ohms per phase and the shunt admittance is 4 × 10^{-4} $\angle$ 90° siemens per phase. Find sending end voltage, current and power factor and voltage regulation and efficiency. Use nominal π circuit. [165.7 kV, 306.9 A, 0.779 lagging, 27.1%, 87.48 %]

4.9 A 300 km 3-phase overhead line has a series impedance of $11 + j\,106$ ohms per phase and a shunt admittance of $1.13 \times 10^{-3} \angle 90°$ siemens per phase. The line delivers 250 MW at unity p.f. and at 275 kV. Find (a) constants A, B, C, D (b) sending end voltage (c) power angle

[(a) $A = D = 0.941 \angle 0.4°$, $B = 104 \angle 84.3°$ ohms, $C = 1.11 \times 10^{-3} \angle 90.2°$ siemens (b) 285 kV (c) 19.6°]

4.10 A 275 kV 3-phase line has parameters $A = 0.95 \angle 3°$ and $B = 80 \angle 75°$ ohms per phase. For receiving end voltage of 275 kV (a) calculate sending end voltage for a receiving end load of 250 MW at 0.85 lagging power factor (b) calculate the additional reactive power which the compensation plant at the receiving end must supply for a load of 400 MVA at 0.9 lagging power factor if the sending end voltage is to be maintained at 300 kV.

[(a) 331 kV (b) 222 Mvar lagging.]

4.11 A 3-phase long line has a series impedance of $200 \angle 80°$ ohms/phase and a shunt admittance of $0.0013 \angle 90°$ siemens per phase to neutral. The line supplies 90 MW at a voltage of 200 kV at 0.9 lagging power factor. Find (a) sending end voltage, sending end current, power angle, efficiency of transmission (b) the receiving end voltage at no load with the sending end voltage at the value found in (a).

[(a) 245 kV, 232 A, 19.7°, 92.6% (b) 280 kV]

4.12 A 3-phase long line has constants $A = 0.8698 \angle 2.36°$ and $B = 186.9 \angle 75.15°$ ohms. Both the sending end and the receiving end line voltages are to be maintained at 154 kV. Find the reactive vars which the compensation equipment must supply to maintain the voltage profile (a) when the line supplies a load of 50 MVA at 0.85 p.f. lagging (b) when the line is open circuited at the receiving end.

[(a) 29.63 Mvar lagging (b) 17.1 Mvar leading]

4.13 A 132 kV 3-phase line has constants $A = 0.98 \angle 3°$ and $B = 110 \angle 75°$ ohms per phase. The line is to be operated with both the sending end and receiving end voltages at 132 kV. Draw receiving end power circle diagram and determine (a) the reactive vars which the shunt compensation equipment must supply if load is 50 MVA, 0.8 lagging p.f. (b) the maximum power which the line can deliver.

[(a) 45.6 Mvar lagging (b) 112 MW]

4.14 Solve problem 4.12 by drawing the receiving end circle diagram.

[(a) 30 Mvar lagging (b) 17 Mvar leading]

4.15 A 3-phase line delivers 600 MW to a load at p.f. 0.98 lagging. The line reactance is 21.83 ohms/phase. Draw the receiving end power circle and determine the reactive power supply at the receiving end to maintain both the sending and receiving end voltages at 275 kV. Neglect line resistance and capacitance.

[180 Mvar lagging]

4.16 A 3-phase overhead line has constants $A = 0.97 \angle 0.49°$ and $B = 52.88 \angle 74.79°$ ohms. The receiving end voltage is 132 kV. Plot the variations of sending end voltage and line losses with the received power as the received power varies from 0 to 150 MW at 0.85 p.f. lagging. Use power circle diagram method.

4.17 The line of problem 4.16 is to be operated with both sending end and receiving end voltages at 132 kV. Draw receiving end circle diagram and plot the variation of reactive power compensation needed to maintain the voltage profile as the received power varies from 0 to 150 MW. The load power factor is constant at 0.85 lagging.

4.18 Fig. 4.33 shows a power system having three generators, four transformers and one transmission line. Take base MVA = 30 and rated voltages as base voltage. Draw reactance diagram. Indicate the reactances in p.u. on the diagram.

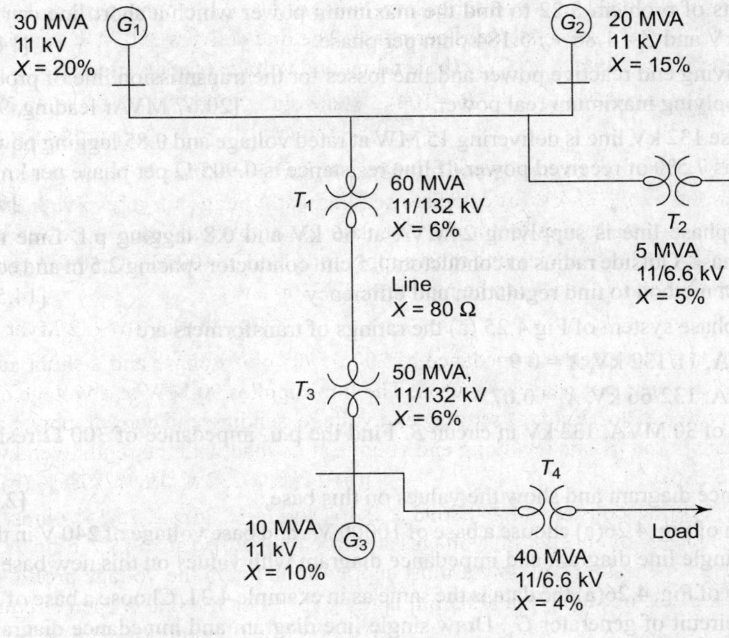

Fig. 4.33

4.19 Fig. 4.34 shows a generator feeding two motors through transformers and line. The ratings and reactances are as under.

Generator : 100 MVA 11 kV, 3 phase, $X = 20\%$

Transformer T_1: 3 phase, 100 MVA, 11/132 kV, $X = 5\%$

Transformer T_2: Bank of 3 single phase transformers, each rated at 35 MVA, 66/11 kV, $X = 4\%$

Motor M_1: 40 MVA, 3 phase, 10 kV, $X = 20\%$

Motor M_2: 60 MVA, 3 phase, 11 kV, $X = 15\%$ The line reactance is 80 ohms.

Select suitable base values. Draw reactance diagram. Indicate p.u. reactances on the diagram.

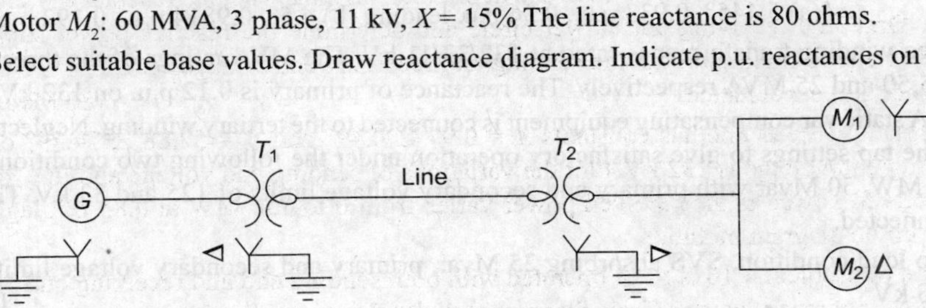

Fig. 4.34

4.20 A generator is rated at 200 MVA, 11 kV. Its reactance is 0.3 p.u. Find the p.u. reactance on 100 MVA, 33 kV base. [0.01666 p.u.]

4.21 A transmission line has a reactance of 37.7 Ω. Find the p.u. reactance if base MVA = 10 and base voltage = 132kV [0.216 p.u.]

4.22 A short transmission line has a series impedance $Z = R + jX$ ohms per phase. The voltages at sending and receiving ends are V_s and V_r per phase respectively. Find a formula for the maximum real power which the line can supply. $\left[3V_r^2\left(\frac{V_{s/}}{ZV_r} - \frac{R}{Z^2}\right)\right]$

4.23 Use the results of problem 3.22 to find the maximum power which a short line can deliver if $V_s = V_r = 11$ kV and $Z = 1.86 + j\,5.184$ ohm per phase. [14.55 MW]

4.24 Find the receiving end reactive power and line losses for the transmission line of problem 4.23 when it is supplying maximum real power. [20.67 MVAr leading, 9.82 MW]

4.25 A short 3-phase 132 kV line is delivering 15 MW at rated voltage and 0.85 lagging power factor. The line loss is 7.5% of received power. If line resistance is 0.905 Ω per phase per km, find the length of line. [69.55 km]

4.26 An 80 km 3-phase line is supplying 24MVA at 66 kV and 0.8 lagging p.f. Line resistance 0.12 Ω/km/ phase. Outside radius of conductors 1.5 cm. conductor spacing 2.5 m and equilateral. Use nominal π method to find regulation and efficiency [14.5%, 94%]

4.27 In the single phase system of Fig 4.25 (a) the ratings of transformers are

$A - B$: 20 MVA, 11/132 kV, $X = 0.9$

$B - C$: 18 MVA, 132/66 kV, $X = 0.07$

Choose a base of 30 MVA, 132 kV in circuit B. Find the p.u. impedance of 300 Ω resistance in circuit C.

Draw impedance diagram and show the values on this base. [2.066 p.u.]

4.28 For the system of Fig. 4.26(a) choose a base of 10000 VA and base voltage of 240 V in the circuit of G_1. Draw single line diagram and impedance diagram with values on this new base.

4.29 For the system of Fig. 4.26(a) the data is the same as in example 4.31. Choose a base of 30 MVA, 11 kV in the circuit of generator G_1. Draw single line diagram and impedance diagram. Show the values on above base in the diagrams.

4.30 For the system of Fig. 4.28(a) the data is the same as in example 4.32 choose a base of 40 MVA, 6.9 kV in the circuit of generator G_1. Draw single line diagram and impedance diagram and insert the values on this new base.

4.31 A 3-phase interconnector connecting two stations has an impedance of $(3 + j\,9)$%. Find the phase angle between the two-station voltages and p.u. real and reactive power received when the voltages at the two stations are (a) 1 p.u. each (b) 1.05 and 1 p.u. The current in interconnector is rated full load current.

[(a) 5.445°, 0.933 p.u., 0.36 p.u. leading (b) 4.5°, 0.9814 p.u., 0.192 p.u. lagging]

4.32 A three-winding transformer is rated at 132/33/11 kV. The MVA ratings of the three windings are 75,50 and 25 MVA respectively. The reactance of primary is 0.12 p.u. on 132 kV, 75 MVA base. A static var compensating equipment is connected to the tertiary winding. Neglecting losses find the tap settings to give satisfactory operation under the following two conditions (a) load of 60 MW, 50 Mvar with primary and secondary voltage limits of 125 and 33 kV. The SVS is disconnected.

(b) No load condition, SVS absorbing 25 Mvar, primary and secondary voltage limits 140 kV and 33 kV [± 8.96%]

4.33 A 10 MW, 0.8 lagging p.f. load is supplied by two 3-phase overhead lines A and B operating in parallel. Line A supplies 100 A. Line B supplies 5.5 MW. If the impedance of line B is $(8 + j\,12)$ ohm per phase find the impedance of line A per phase. The receiving and voltage is 30 kV. [$14.6 + j\,14.4$ ohm]

CHAPTER 5

Overhead Line Insulators

5.1 INSULATOR MATERIALS

The overhead line conductors are bare and not covered with any insulation coating. To provide necessary clearances between conductors and metal work, the conductors are attached to insulators mounted on cross-arms. The insulator must provide the necessary clearance against the highest voltage and worst atmospheric conditions to which the line is likely to be subjected. It also provides the necessary mechanical support for the conductor.

The two most suitable materials for insulators are toughened glass and porcelain. In toughened glass, the outer surface is in compression and the interior is in tension. This state is achieved by rapidly cooling the glass insulator after shaping and allowing the interior to cool slowly. Toughened glass can withstand greater tension than annealed glass. Glass is cheaper than porcelain in the simpler shapes. Owing to its transparency, flaws in the material can be detected easily by visual examination. However, moisture readily condenses on the surface giving rise to high surface leakage. Another disadvantage is that in high voltage insulators the larger mass of the material combined with the irregular shape may result in internal strain after cooling thus reducing reliability. In view of this glass insulators are used very rarely and that too for lines below 25 kV.

Porcelain is made from a fine homogeneous mix of wet plastic clay which is shaped, covered with glaze and then fired in a kiln.

Glazing is necessary to provide a surface which can be kept relatively free from dirt and moisture. Any sealed in impurities decrease the dielectric strength. It is, therefore, essential that porcelain must be completely air free and impervious to the entry of gases and liquids. It is difficult to manufacture porcelain in the thickness required for some types of insulators and it is then necessary to adopt a two or more piece construction, the various pieces being fired and glazed separately and then cemented together. Porcelain is mechanically stronger than glass, gives less trouble from leakage and is less affected by changes of temperature. Porcelain is the most commonly used material for manufacture of insulators.

5.2 TYPES OF INSULATORS

Neglecting the shackle type insulators used to support 400/230V lines between houses, insulators can be classified as pin type, suspension type, post type and strain insulators.

5.2.1 Pin Type Insulators

A pin type insulator is small, simple in construction and cheap. It is used on lines upto and including 33 kV lines. The conductor is bound into a groove on the top of the insulator which is cemented on to a galvanized steel pin attached to the crossarm on the pole or tower. To avoid a direct contact between the porcelain and the metal pin, a soft metal (generally lead) thimble is used. An adequate length of leakage path is obtained by providing the insulator with two or three petticoats or rain sheds. These are so designed that even when the outer surface of the insulator is wet due to rain, sufficient leakage resistance is still given by the inner dry surfaces. Fig. 5.1 shows two pin type insulators. In its electrical behaviour, a pin type insulator may be compared to a complicated series of condensers with resistances in series and shunt. The petticoats with the intervening air spaces form the condenser system and the leakage paths over the surface and through the body of the material are represented by the resistances. In modern pin type insulators the surfaces of the petticoats or rain sheds, conform to the equipotential surfaces of the electrostatic field between the pin and the conductor while the body of the insulator conforms to the lines of electrostatic field. Pin type insulators are used only upto about 33kV because for higher voltages they tend to be very heavy and more costly than suspension type insulators.

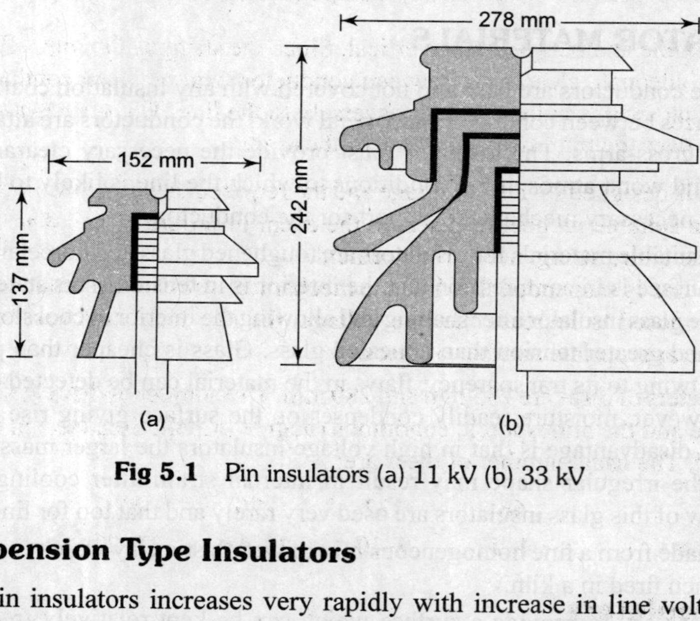

Fig 5.1 Pin insulators (a) 11 kV (b) 33 kV

5.2.2 Suspension Type Insulators

The cost of a pin insulators increases very rapidly with increase in line voltage. Therefore, suspension insulators are used for lines above 33 kV. They are also known as disc insulators or string insulators.

A suspension insulator consists of porcelain disc units mounted one above the other. Each disc consists of a single shed of porcelain grooved on the under surface to increase the creepage distance. The upper surface of each disc is inclined at a suitable angle to the horizontal in order to ensure free drainage of water. Each disc is provided with a metal cap at the top and a metal pin underneath. The cap is recessed so as to take the pin of another unit and thus a string of any required number of units can be built up. The most commonly used disc is the cemented cap type shown in Fig. 5.2. The conductor is suspended below the point of support by means of the insulator string (and hence the name suspension insulator).

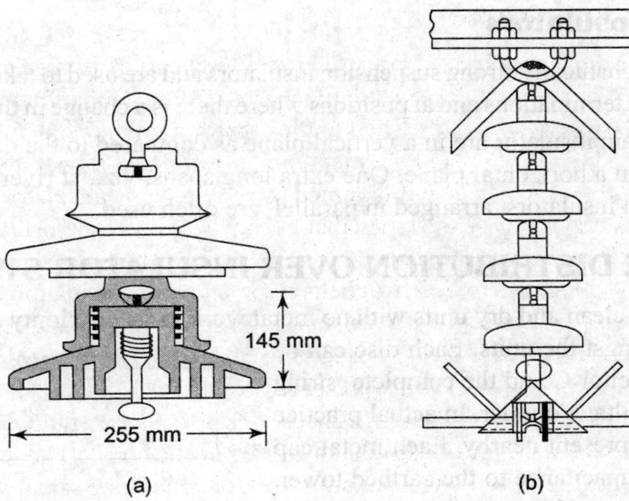

Fig. 5.2 (a) Suspension insulator unit (b) A suspension insulator string with arching horns

The string is flexible and hangs almost vertical. Since the string is flexible, allowances must be made to ensure adequate clearances between conductors and between conductor and metal work when the string swings in the wind, transversely to the line. The suspension insulator has many advantages, over the pin insulator. Some of these are:

(a) Each disc is designed for a low voltage and the required degree of insulation is achieved by using a suitable number of discs. In the event of failure of one disc, only that disc, and not the entire string, needs replacement.
(b) Since the line is suspended flexibly, the mechanical stresses are reduced.
(c) If it is proposed to increase the line voltage, the line insulation can be increased by adding the appropriate number of discs.

The usual diameters of discs are 255 mm and 280 mm. The number of discs in a string depends on the line voltage and the atmospheric conditions (degree of pollution, etc.) in the area where the line is situated. The usual number of discs are:

Voltage kV	66	132	220	400
Number of discs	4–5	9–10	15–16	22–23

5.2.3 Post Insulators

These are used for supporting the bus bars, and disconnecting switches in sub-stations. A post insulator is similar to a pin type insulator but has a metal base and frequently a metal cap so that more than one unit can be mounted in series.

In extra high voltage sub-stations (400 kV and above) polycon post insulators are used. Also known as multicone insulator, this insulator is a puncture proof, solid core insulator for outdoor use. In this insulator the porcelain elements are in the form of cones smugly fitting one inside the other and bonded by special cement. The puncture path is through many layers of porcelain cones and the voltage required to puncture this path is many times the external flash over voltage so that the insulator is almost puncture proof.

5.2.4 Strain Insulators

These are special mechanically strong suspension insulators and are used to take the tension of the conductors at the line terminations and at positions where there is a change in the direction of line.

The discs of a strain insulator are in a vertical plane as compared to the discs of suspension insulator which are in a horizontal plane. One extra long spans, viz., at river crossings, two or three strings of strain insulators, arranged in parallel, are often used.

5.3 VOLTAGE DISTRIBUTION OVER INSULATOR STRING

A string of identical, clean and dry units with no metal work in its proximity shares the applied voltage equally amongst the units. Each disc can be considered as a capacitor C and the complete string acts as a capacitor voltage divider. In actual practice the earthed tower is present nearby. Each metal cap and pin joint has a capacitance to the earthed tower. These stray capacitances have an important effect on the voltage distributions as a result of which the voltage across a unit nearest to the line is more than the voltage across a unit situated near the tower end. The string efficiency of an insulator of n units is defined as:

String efficiency

$$= \frac{\text{flashover voltage of the string}}{n \times \text{flashover voltage of one unit}}$$

$$= \frac{\text{Voltage across string}}{n \times \text{Voltage across lowermost disc}} \quad (5.1)$$

The string efficiency decreases with an increase in the number of units. Fig. 5.3 illustrates the procedure for calculating the voltage distribution. The capacitance of each joint to the earthed tower can be assumed to be kC where C is the capacitance of each unit and k lies between 0.1 and 0.2.

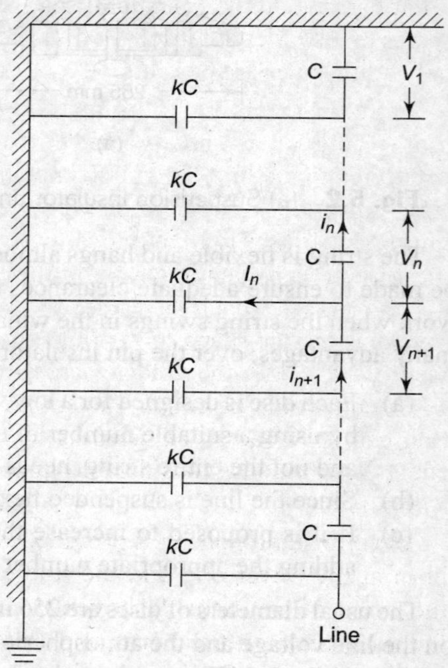

Fig. 5.3 Voltage distribution across an insulator string

From Fig. 5.3 $\quad i_{n+1} = i_n + I_n \quad (5.2)$

In general $\quad i = V \cdot j\omega C \quad (5.3)$

Then $\quad V_{n+1} j\omega C = (V_1 + V_2 + \ldots + V_n) j\omega k C + V_n \, j\omega C \quad (5.4)$

or $\quad V_{n+1} = (V_1 + V_2 + V_3 + \ldots V_{n-1}) k + V_n \cdot (1 + k) \quad (5.5)$

Using Eq. (5.5) the voltage across each unit can be written in terms of V_n. The value of V_n can be found by equating the sum of voltages across all the units to the line to neutral voltage. Then the voltages across different units can be found.

5.4 IMPROVEMENT OF STRING EFFICIENCY

The voltage distribution across an insulator string is not uniform. The units nearest to the line end are stressed to their maximum allowable value while those near the tower end are considerably understressed resulting in a waste of insulating material. The string efficiency indicates the extent of this wastage.

Though string efficiency can never be made 100 per cent, an improvement in its value is necessary to minimize the wastage. Some methods to improve string efficiency are:

1. Reduction in the shunt capacitance relative to the capacitance of each unit. This can be done only by increasing the length of crossarm. Due to the limitations of mechanical strength and cost of towers the value of k cannot be reduced to less than 0.1.

2. Capacitance grading: Capacitance grading means an increase in the capacitance of each unit from the tower end towards the line end. Theoretically the voltages across the different units can be made exactly equal by correct capacitance grading. However, this is seldom done because it implies that all the insulator discs are different from one another. This involves maintaining spares of all variety of insulator discs which is contrary to the tendency of standardization since only a few types are possible.

3. Static shielding: This method uses a large metal ring surrounding the bottom unit and connected to the line. This ring, known as a grading ring, introduces capacitances between different joints and line. This has the effect of increasing the effective capacitance of bottom units. From Fig. 5.4 it is seen that if the value of S_n can be adjusted (by adjusting the shape of the ring) such that $I_n = I'_n$, then $i_n = i_n + 1$ or the voltage across the nth and (n +1)th unit would be equal. Though the shape of the ring can never be so adjusted as to give perfectly equal distributions, considerable improvements are still possible.

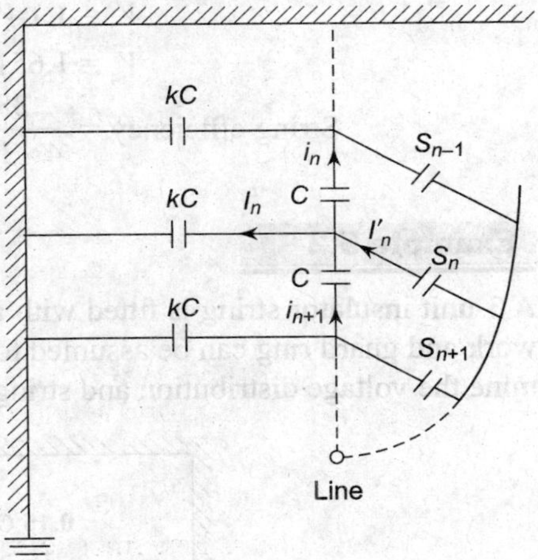

Fig. 5.4 Static shielding

Every insulator string is provided with an arcing horn at the tower end and the line end so that in the event of insulator flashover, the arc may take the path between the horns and stay clear of the insulator string. Static shielding is achieved by changing the arcing horn at the line end to a grading ring.

Example 5.1

An insulator string for 66 kV line has 4 discs. The shunt capacitance between each joint and metal work is 10% of the capacitance of each disc. Find the voltage across the different discs and string efficiency.

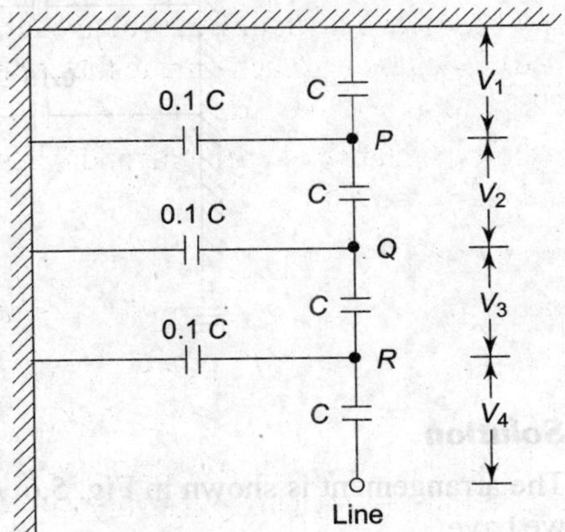

Fig. 5.5

Solution

The arrangement is shown in Fig. 5.5. If the capacitance of each unit is, C, then each shunt capacitance equals 0.1C. Applying Kirchhoff's current law to different junctions we have

At junction P, $\omega CV_2 = \omega(CV_1 + 0.1\ CV_1)$ or $V_2 = 1.1\ V_1$

At junction Q, $\omega CV_3 = \omega[CV_2 + 0.1C(V_1 + V_2)]$

or $V_3 = 1.1\ V_2 + 0.1\ V_1 = 1.1 \times 1.1\ V_1 + 0.1\ V_1 = 1.31\ V_1$

At junction R, $\omega CV_4 = \omega[CV_3 + 0.1\ C(V_1 + V_2 + V_3)]$

or
$$V_4 = 1.1\ V_3 + 0.1\ V_2 + 0.1\ V_1$$
$$= 1.1 \times 1.31\ V_1 + 0.1 \times 1.1\ V_1 + 0.1\ V_1$$
$$= 1.441\ V_1 + 0.11\ V_1 + 0.1\ V_1 = 1.651\ V_1$$

$$V_1 + V_2 + V_3 + V_4 = \frac{66}{\sqrt{3}}\ \text{kV} = 38.1\ \text{kV}$$

or $V_1(1 + 1.1 + 1.31 + 1.651) = 38.1$

or
$$V_1 = \frac{38.1}{5.061} = 7.53\ \text{kV}$$
$$V_2 = 1.1\ V_1 = 8.28\ \text{kV},\ V_3 = 1.31\ V_1 = 9.86\ \text{kV},$$
$$V_4 = 1.651\ V_1 = 12.43\ \text{kV}$$
$$\text{String efficiency} = \frac{38.1}{4 \times 12.43} = 0.766 \text{ or } 76.6\%$$

Example 5.2

A 3-unit insulator string is fitted with a guard ring. The capacitances of the link pins to metal work and guard ring can be assumed to be 15% and 5% of the capacitance of each unit. Determine the voltage distribution and string efficiency. *[RTM Nagpur Uni. 2011]*

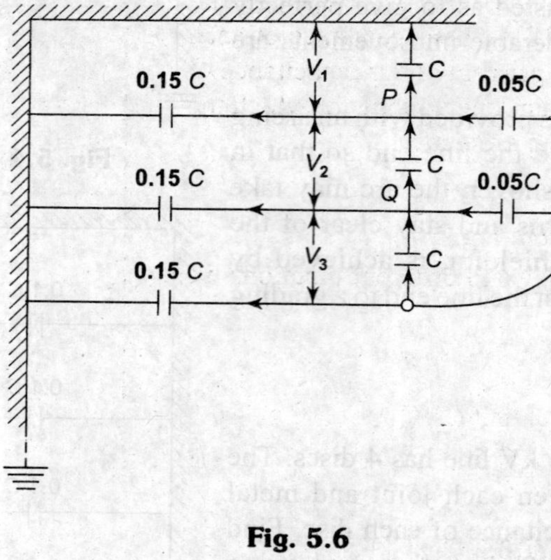

Fig. 5.6

Solution

The arrangement is shown in Fig. 5.6. Applying Kirchhoff's current law to different junctions we have

At junction P,

$$\omega CV_2 = \omega CV_1 + 0.15\ \omega CV_1 - 0.05\ \omega C\ (V_2 + V_3)$$

or
$$V_2 = V_1 + 0.15\ V_1 - 0.05\ V_2 - 0.05\ V_3$$

or
$$1.05\ V_2 = 1.15\ V_1 - 0.05\ V_3 \tag{i}$$

At junction Q, $\omega CV_3 = \omega CV_2 + 0.15\,\omega C(V_1 + V_2) - 0.05\,\omega CV_3$

or $1.05\,V_3 = 1.15\,V_2 + 0.15\,V_1$ (ii)

Substituting the value of V_3 from (ii) into (i)

$$1.05\,V_2 = 1.15\,V_1 - 0.05\left(\frac{1.15V_2}{1.05} + \frac{0.15V_1}{1.05}\right)$$

or $V_2 = 1.0345\,V_1$ (iii)

Substituting the value of V_2 from (iii) into (ii)

$$1.05\,V_3 = 1.15 \times 1.0345\,V_1 + 0.15\,V_1$$

or $V_3 = 1.276\,V_1$ (iv)

$$V_1 + V_2 + V_3 = 100$$

or $V_1 + 1.0345\,V_1 + 1.276\,V_1 = 100$

or $V_1 = 30.2\%$

$$V_2 = 1.0345\,V_1 = 31.24\%$$

$$V_3 = 1.276\,V_1 = 38.54\%$$

$$\text{String efficiency} = \frac{100}{3 \times 38.54} = 0.865 \text{ or } 86.5\%$$

Example 5.3

A suspension string has 3 units. Each unit can withstand a maximum voltage of 11 kV. The capacitance of each joint and metal work is 20 per cent of the capacitance of each disc. Find (a) maximum line voltage for which the string can be used (b) string efficiency.

Solution

The arrangement is shown in Fig. 5.7.

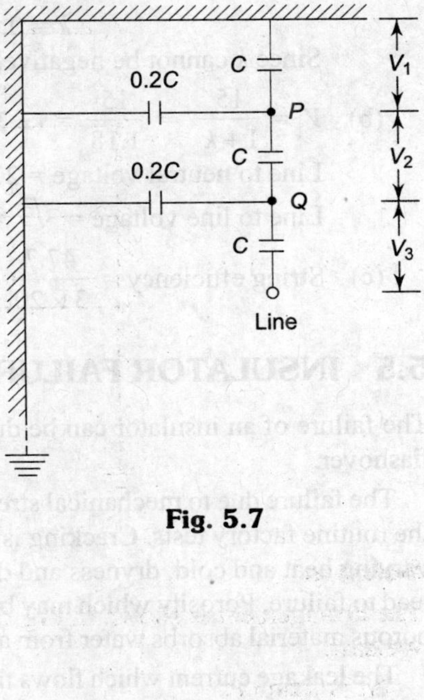

Fig. 5.7

Applying Kirchhoff's laws to different junctions we have

At junction P,

$$\omega CV_2 = \omega CV_1 + \omega(0.2\,C)\,V_1$$

or $V_2 = 1.2\,V_1$ (i)

At junction Q,

$$\omega CV_3 = \omega CV_2 + \omega(0.2\,C)\,(V_1 + V_2)$$

or $V_3 = 1.2\,V_2 + 0.2\,V_1$ (ii)

From Eqs. (i) and (ii)

$$V_3 = 1.64\,V_1$$

Since $V_3 = 11\text{kV}$

$$V_1 = \frac{11}{1.64} = 6.7\text{ kV}$$

$$V_2 = 1.2 \times 6.7 = 8.04\text{ kV}$$

Maximum line to neutral voltage = $V_1 + V_2 + V_3 = 25.74$ kV

Maximum line to line voltage = $\sqrt{3} \times 25.74 = 44.58$ kV

$$\text{String efficiency} = \frac{25.74}{3 \times 11} \times 100 = 78\%$$

Example 5.4

A 3-phase line is supported by suspension string having three units. The voltage across the unit nearest to the line is 20 kV and that across the adjacent unit is 15 kV. Find (a) ratio of the capacitance of joint to capacitance of disc (b) system line voltage (c) string efficiency.

[RTM Nagpur Uni. 2010]

Solution

The arrangement is similar to that shown in Fig. 5.7. Let the capacitance between each joint and metal work be kC.

(a) $V_3 = 20$ kV, $V_2 = 15$ kV, $V_1 = ?$

$V_3 = V_1 k + V_2 (1 + k)$

$V_2 = (1 + k) V_1$

Substituting the values of V_3 and V_2 we have

$$20 = V_1 k + 15 (1 + k) \quad \text{(i)}$$

$$15 = V_1 (1 + k) \quad \text{(ii)}$$

From Eqs. (i) and (ii)

$$k = 0.18 \text{ or } -1.847$$

Since k cannot be negative, $k = 0.18$

(b) $V_1 = \dfrac{15}{1+k} = \dfrac{15}{1.18} = 12.7$ kV

Line to neutral voltage = 20 + 15 +12.7 = 47.7 kV

Line to line voltage = $\sqrt{3} \times 7.7 = 82.6$ kV

(c) $$\text{String efficiency} = \frac{47.7}{3 \times 20} \times 100 = 79.5\%$$

5.5 INSULATOR FAILURE

The failure of an insulator can be due to mechanical stresses, cracking, porosity, puncture and flashover.

The failure due to mechanical stresses is very rare because defective pieces are weeded out in the routine factory tests. Cracking is due to continued action of alternate and sometimes rapidly varying heat and cold, dryness and dampness which produce high stresses on the porcelain and lead to failure. Porosity which may be due to under firing or other causes leads to failure because porous material absorbs water from atmosphere or cement and its electrical resistance decreases.

The leakage current which flows through the porcelain tends to heat up a localized portion to a high temperature. This local heat may then cause mechanical failure followed by the passage of a power arc through the insulator shell. A porosity test is a necessary routine factory test for insulators.

The failure of an insulator by puncture is quite rare. Insulators are so designed that they never puncture. Nevertheless puncture does occur occasionally due to improper vitrification in the material. The puncture voltage of an insulator is the voltage which will cause a breakdown in the porcelain between the conductor and the insulator base, causing the insulator to be destroyed. The breakdown through puncture is affected by the electrical characteristics of the material and the care taken in the manufacture. The puncture causes a permanent damage to the insulator.

The most common cause of the insulator damage is the flashover of the insulator. The flashover voltage is the voltage which will cause an arc through the air surrounding the insulator. In actual practice insulators are often subjected to power frequency and impulse over voltages due to switching and lightning. If the over voltage exceeds the flashover voltage of the insulator, an arc occurs between the line conductor and insulator base. As and when a flashover occurs the surrounding air is so ionized that there is virtually a follow through power arc due to the system voltage, after the transient voltage has subsided. After a flashover an insulator continues to function properly unless it is shattered by the heat of the arc. For this reason insulators are often fitted with arcing horns to keep this power arc away from the insulator.

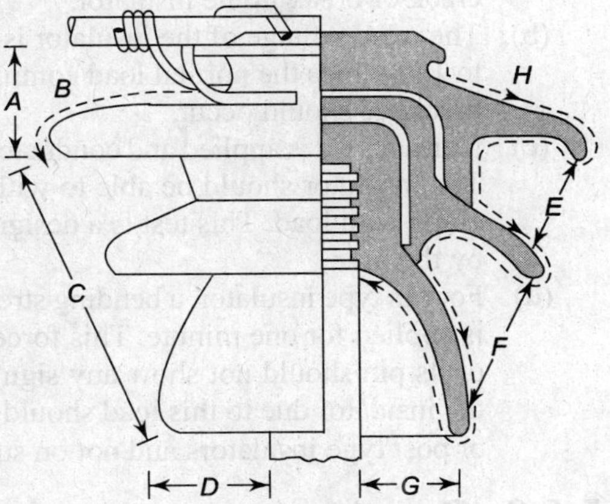

Fig. 5.8 Flash over distances of a pin insulator

In service the upper surface of an insulator can become damp and polluted by industrial dirt, salt laden coastal spray, frost and fog to such an extent that these upper surfaces offer greatly reduced resistance to surface leakage current due to the system voltage. This has the effect of decreasing the voltage necessary to cause insulator flashover. The wet flashover voltage is, therefore, lower than the dry flashover voltage. The flashover distance is the minimum distance, measured by a taut tape, from the conductor to the metal base. Fig. 5.8 shows the flashover distance as usually measured on a pin insulator. The distance $A + C + D$ of $B + C + D$ (depending on whether A and B is the shorter distance) gives the length of path for a dry flashover. When the exposed upper surfaces of the insulator are wet due to rain they become conducting and the distance $E + F + G$ gives the wet flashover distance. The line H shows the leakage path.

5.6 TESTING OF INSULATORS

An insulator should have good mechanical strength to withstand the load conditions. It should not have any pores or voids which may lead to breakdown. Its electrical strength should be large enough to withstand the normal voltage and usual overvoltages but the insulator should flashover at such voltages which are likely to cause damage to other equipment. The following tests are performed on the insulators.

5.6.1 Mechanical Strength

The insulator is mounted on the crossarm as in the normal working conditions. A conductor of length 1.5 times the length of the insulator as in use (for string insulator, total length of string is counted as insulator length) or 1 metre, whichever is less, is fitted in the normal direction. The conductor should not be less than one cm. diameter and for pin type the binding wire should be of at least 0.3 cm. diameter. Then the following tests for mechanical strength are performed:

(a) A mechanical load equal to 1.5 times the load of the conductor, expected in a single span is applied, in compression for pin type insulator and in tension for suspension insulator. For strain insulator the load applied should be 1.5 times the tension that would come on the insulator while in use. The load, maintained for one minute, should not cause any crack or break in the insulator.

(b) The rated voltage of the insulator is applied to the attached conductor and a load equal to 1.25 times the normal load coming on insulator is applied for 1 minute. No crack or breakage should occur.

(c) Normal load is applied and conductor is supplied with 75% of the dry-flashover voltage. The insulator should be able to withstand power frequency voltage of this magnitude along with load. This test is a design test and may not be performed as acceptance test by the user.

(d) For pin type insulator, a bending stress equal to three times the normal load on insulator is applied for one minute. This force is in the direction of the conductor. The insulator or its pin should not show any sign of damage after the test and the permanent set to the insulator due to this load should not exceed 1%. This test is made only on pin type or post type insulators and not on strings.

5.6.2 Porosity

This is a destructive test and is performed on very few selected pieces, usually one or two out of a production batch. The insulator to be tested is broken into pieces and the pieces are immersed in a 1% solution of fuchsin dye in alcohol under pressure of 15×10^6 N/m^2. After sufficient time (the product of test duration in hours and pressure in N/m^2 should not be less than 18×10^7) the pieces are removed from the container and examined. A deep penetration of dye indicates porosity in insulator.

5.6.3 Power Frequency Dry Flashover Voltage

The conductor of a size not less than that used for mechanical strength measurement is used. For pin insulators the binding wire of not less than 3 mm diameter has to be used. The insulator should be mounted or suspended from a crossarm extending at least 3 ft. (0.9m) on either side and there should be no other conducting object in the vicinity of the conductor. A steadily increasing power frequency voltage is applied and raised till the flashover occurs. The test is repeated 5 or 10 times. For 5 readings, the mean is taken as the correct flashover voltage. The other more commonly used method is to take the median of 10 values of flashover voltage. The median can be obtained by arranging the ten values in ascending order and taking the average of 5th and 6th reading. The voltage so obtained is called the 50% dry flashover voltage.

The 50% power frequency dry flashover test can also be performed by applying the same voltage to 10 insulators together. The spacing between the insulators should be at least 4 times the length of each. All the insulators may not flashover simultaneously. The voltage till which 5 insulators have flashed over is called the 50 per cent power frequency flashover voltage. The flashover voltage obtained by the mean of 5 or median of 10 readings should not be less than that given in IS: 731 (1971).

The voltage may be applied constantly and the voltage which causes flashover after one minute is called one minute dry flashover voltage.

5.6.4 Power Frequency Wet Flashover Voltage

In this test water is sprayed on the insulator at an inclination of 45∞ and equivalent to a precipitation of 3 mm per minute. Then the test is continued in the same way as in dry flashover test, the spray being continued till the actual flashover occurs. The resistivity of water in this test should be 10,000 Ω-cm. and if the resistivity of pure water be more, it can be reduced by adding small quantities of a mineral salt or an inorganic acid. This voltage should not be less than the values given in IS: 731 (1971) for the smallest value noted for 10 tests.

5.6.5 Impulse Voltage Flashover Test

Voltage impulses of increasing magnitudes are applied to the conductor in succession. The peak value of impulse voltage giving flashover is called the impulse flashover voltage. The impulse ratio is given by impulse flashover voltage/power frequency dry flashover voltage. Minimum acceptable flashover voltages are given in standards. If a surge generator for high voltage, of the order required, be not available, a high frequency (usually 200 kHz) damped voltage may also be applied.

5.6.6 Puncture Voltage Test

When used as design test, this test is a destructive test and very few pieces out of a batch are selected for this test. The insulator with the pin in proper position and conductor at the other end is immersed in an insulating oil and a steadily increasing power frequency voltage is applied. The voltage at which conduction starts is called the puncture voltage. In air, flashover occurs at a voltage lower than the puncture voltage, so the test is conducted in oil at room temperature. The puncture voltage for strings should be at least 30% higher than the power frequency wet flashover voltage and for pin insulators it should not be less than that given in IS: 731 (1971). As an acceptance test the voltage can be rapidly increased to the value given in standards for pin insulators and 1.3 times the wet flashover voltage for string insulators. No conduction should take place. The acceptance test as such, is a non destructive test and the insulator can be reused if the puncture has not actually taken place.

Example 5.5

Each line of a 3-phase system is suspended by a string of 3 identical insulators of self capacitance C Farads. The shunt capacitance of metal work of each insulator is 0.26 C to earth and 0.15 C to line. Find string efficiency if a guard ring increases the shunt capacitance to line of the metal work of the lowest insulator to 0.35C.

Solution

The arrangement is shown in Fig. 5.9. Applying Kirchhoff's current law to different junctions we have At Junction P,

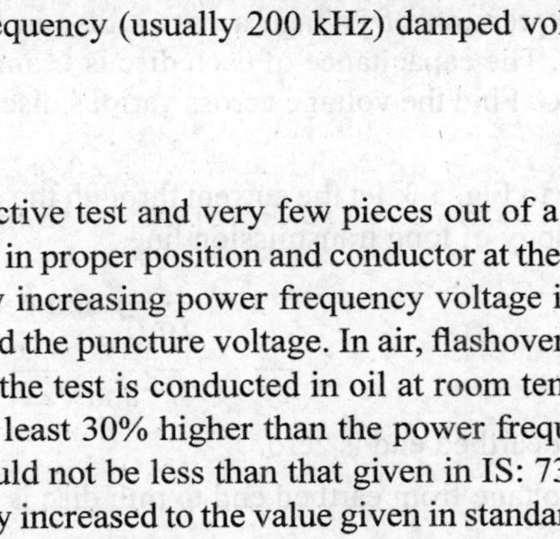

Fig. 5.9

$$\omega CV_2 = \omega CV_1 + .26\,\omega CV_1 - .15\,\omega C(V_2 + V_3)$$

or
$$V_2 = 1.26\,V_1 - .15\,V_2 - .15\,V_3$$

or
$$1.15\,V_2 = 1.26\,V_1 - .15\,V_3 \qquad \text{(i)}$$

At junction Q,

$$\omega CV_3 = \omega CV_2 + .26\,\omega C(V_1 + V_2) - .35\,\omega C V_3$$

or $$1.35\,V_3 = 1.26\,V_2 + .26\,V_1 \qquad \text{(ii)}$$

Eliminating V_3 from (i) and (ii), we get

$$V_2 = 0.954\,V_1 \text{ and } V_3 = 1.086\,V_1$$

$$\text{String efficiency} = \frac{V_1 + 0.954\,V_1 + 1.086\,V_1}{3\,V_3} = \frac{3.04 V_1}{3 \times 1.086 V_1} = 0.933$$

Example 5.6

Derive expression of voltage across a disc in a string of n insulators in terms of hyperbolic functions. The capacitance of each disc is C and the capacitance between each joint and metal work is kC. Find the voltage across various discs if $n = 4$, $k = 0.1$ and voltage across string is V.

Solution

Referring to Fig. 5.3, let the current through the disc 1 be I_1. The voltage equation can be written from analogy of long transmission line.

$$z = \frac{1}{j\omega C} \text{ and } y = j\,\omega\,kC$$

$$\sqrt{zy} = \sqrt{k} \text{ and } \sqrt{ZY} = n\sqrt{k}$$

Voltage at earthed end is zero.

V_m, i.e., voltage from earthed end to mth disc is

$$V_m = I_1\,Z_c \sinh m\sqrt{k}$$

V_{m-1} i.e., voltage from earthed end to $(m-1)$th disc is

$$V_{m-1} = I_1\,Z_c \sinh(m-1)\sqrt{k}$$

Voltage across mth disc is

$$v_m = V_m - V_{m-1} = I_1\,Z_c\,[\sinh m\sqrt{k} - \sinh(m-1)\sqrt{k}\,]$$

V i.e., voltage across entire string is

$$V = I_1\,Z_c \sinh n\sqrt{k}$$

$$\frac{v_m}{V} = \frac{\sinh m\sqrt{k} - \sinh(m-1)\sqrt{k}}{\sinh n\sqrt{k}}$$

$$= \frac{2\cosh[m-0.5)\sqrt{k}\,]\sinh[0.5\sqrt{k}\,]}{\sinh n\sqrt{k}}$$

Substituting $n = 4$ and $k = 0.1$ we get

$$\frac{v_m}{V} = \frac{2\cosh[(m-0.5)\,0.316]\sinh[0.5 \times 0.316]}{\sinh(4 \times 0.316)}$$

For $m = 1, 2, 3, 4$ we get

$$v_1 = 0.1975\,V,\; v_2 = 0.217\,V$$

$$v_3 = 0.259\,V,\; v_4 = 0.343\,V$$

Example 5.7

A string of 5 insulators is fitted with a gradingring. All discs are similar and capacitance of each pin to earth is C. Find the values of line to pin capacitances so that voltage distribution is uniform.

Solution

The configuration is shown in Fig. 5.10. The voltage across each disc is V. This can be possible only if $i_1 = i_2$, $i_3 = i_4$; $i_5 = i_6$ and $i_7 = i_8$.

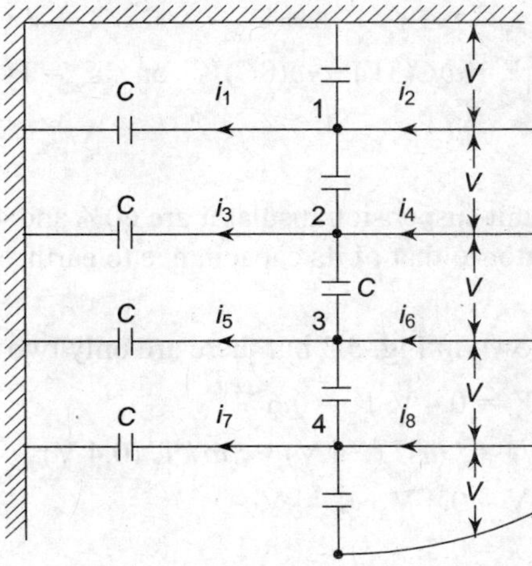

Fig. 5.10

At junction 1, $\quad i_1 = i_2$

or $\quad \omega CV = \omega C_1 (4V) \quad$ or $\quad C_1 = C/4$

At junction 2, $\quad i_3 = i_4$

or $\quad \omega C(2V) = \omega C_2(3V) \quad$ or $\quad C_2 = \boxed{\frac{2}{3}} C$

At junction 3, $\quad i_5 = i_6$

or $\quad \omega C(3V) = \omega C_3(2V) \quad$ or $\quad C_3 = 1.5\,C$

At junction 4, $\quad i_7 = i_8$

or $\quad \omega C(4V) = \omega C_4(V) \quad$ or $\quad C_4 = 4C$

Example 5.8

A graded string of 4 insulators has a uniform voltage across each disc. The capacitance of each pin to earth is C. The mutual capacitance of the top disc is 5 C. Find the mutual capaci-tance of the remaining discs.

Solution

The configuration is shown in Fig. 5.11. Equal voltage distribution is possible only if the mutual capacitances of different discs are unequal. Let the mutual capacitances of second, third and fourth discs be C_2, C_3 and C_4 respectively. Then

$$I_2 = I_1 + i_1$$

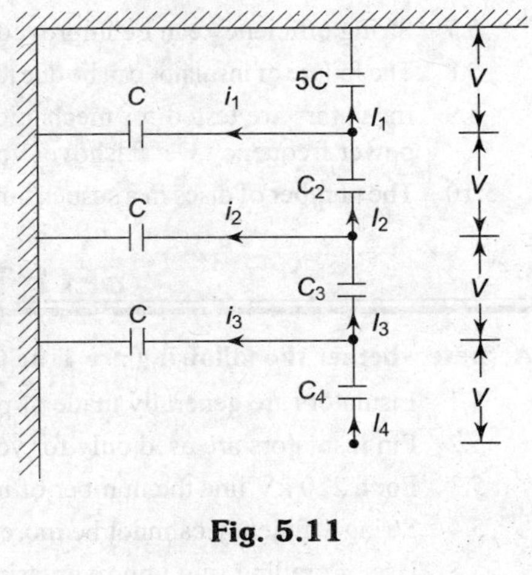

Fig. 5.11

or $$\omega C_2 V = \omega CV + \omega(5C)V \quad \text{or} \quad C_2 = 6C$$

$$I_3 = I_2 + i_2$$

$$\omega C_3 V = \omega C(2V) + \omega(6C)V \quad \text{or} \quad C_3 = 8C$$

$$I_4 = I_3 + i_3$$

or $$\omega C_4 V = \omega C(3V) + \omega(8C)V \quad \text{or} \quad C_4 = 11C$$

Example 5.9

If the voltage across the units in a 2-unit suspension insulator are 60% and 40% of line voltage, find the ratio of capacitance of insulator to that of its capacitance to earth.

Solution

The arrangement is similar to that shown in Fig. 5.3 but there are only two discs.

$$V_1 = 0.4 \text{ V}, V_2 = 0.6 \text{ V}$$

$$j\omega C\,(0.6 \text{ V}) = j\,\omega C\,(0.4 \text{ V}) + j\,\omega k\,C\,(0.4 \text{ V})$$

or $$0.6 \text{ V} = 0.4 \text{ V} + 0.4 \text{ kV}$$

or $$k = 0.5$$

$$\frac{\text{Capacitance of insulator}}{\text{Capacitance to earth}} = \frac{C}{0.5C} = 2$$

SUMMARY

5.1 Insulators provide the necessary clearance between conductors and metal work on overhead lines.

5.2 Porcelain the most commonly used material for insulators.

5.3 For lines upto 33kV pin insulators are used. For lines above 33 kV suspension insulators are used.

5.4 Post insulators are used for supporting the bus bars, disconnecting switches etc. in substations.

5.5 Because of stray capacitances, the voltage distribution across the discs of a suspension insulator is not uniform.

5.6 String efficiency $= \dfrac{\text{voltage across string}}{n \times \text{voltage across the lower most disc}}$

5.7 String efficiency can be improved by the use of grading ring.

5.8 The failure of insulator can be due to mechanical stresses, cracking, porosity, puncture and flash-over.

5.9 Insulators are tested for mechanical strength, porosity, power frequency dry flashover voltage, power frequency wet flashover voltage, impulse voltage and puncture voltage

5.10 The number of discs in a suspension insulator depends on line voltage and atmospheric conditions.

SELF TEST QUESTIONS

A. State whether the following are True (T) or False (F)

5.1 Insulators are generally made of porcelain.

5.2 Pin insulators are used only for voltages upto 33 kV.

5.3 For a 220 kV line the number of insulator discs is about 10.

5.4 String efficiency cannot be more than 1.

5.5 Use of grading ring improves string efficiency.

5.6 Wet flashover strength of an insulator is less than dry flashover strength.

5.7 Porosity test on insulators is a destructive test.

5.8 Impulse flashover voltage may be higher or lower than power frequency flashover voltage.

B. Fill in the blanks with suitable words

5.9 The most common cause of insulator failure is _______.

5.10 To keep the insulator surface free from dirt and moisture it is necessary to _______ the insulator surface during manufacture.

5.11 As the line voltage increases the number of insulator discs _______.

5.12 In a suspension insulator the conductor is suspended _______ the point of support.

5.13 The discs of strain insulator are in _______ plane.

5.14 For a 400 kV line the number of discs is about _______.

5.15 In wet flashover test, water is sprayed on insulator surface of an angle of _______.

ANSWERS

5.1 T	5.2 T	5.3 F	5.4 T	5.5 T	5.6 T
5.7 T	5.8 F	5.9 Flashover	5.10 Glaze	5.11 Increases	5.12 Below
5.13 Vertical	5.14 Twenty Two	5.15 45°			

JOB INTERVIEW QUESTIONS

5.1 Differentiate between pin and suspension insulators.

5.2 What is string efficiency?

5.3 What is meant by 'glazing'?

5.4 Name the tests performed on an insulator.

5.5 What are the causes of insulator failure?

5.6 How does grading ring improve string efficiency?

5.7 What is meant by insulator flashover?

5.8 Why is wet flashover voltage always less than dry flashover voltage?

REVIEW QUESTIONS

5.1 Discuss why:

(a) Porcelain is the most commonly used material for insulators.

(b) Pin insulators are used for low voltages while suspension insulators are used for high voltages.

(c) The voltage distribution across the units of a string insulator is not uniform.

(d) Provision of a grading ring improves the string efficiency.

(e) Insulators are often fitted with arcing horns.

(f) Transmission lines in heavily polluted area are provided with a greater number of insulator discs.

5.2 Name the different types of insulators used in power systems. Give the application of each insulator. Describe the main constructional features of pin and string insulators.

5.3 What is string efficiency? Why is it necessary to have a high string efficiency? How can it be achieved?

5.4 Write a short note on insulator failure.

5.5 What are the important factory tests conducted on insulators? Give a brief procedure for conducting each of these tests.

PROBLEMS

5.1 An insulator for 66 kV line is provided with 5 discs. The capacitance between each joint and tower is one sixth of the self capacitance of each disc. Find the voltage across each disc and string efficiency. [4.25 kV, 5.0 kV, 6.5 kV, 9.11 kV, 13.3 kV, 57.3%]

5.2 Astring insulator for 66 kV line has 4 discs. The capacitance from each joint to tower is 25% of the self capacitance of each unit. Find the voltage distribution across the units and string efficiency. [5.53, 6.91, 10, 15.6, kV, 61%]

5.3 A string insulator has 3 units. The capacitance from each joint to tower is 12.5% of the capacitance of each unit. It is desired that the voltage across any unit should not exceed 11 kV. Find the maximum voltage for which the string can be used. [27.83 kV]

5.4 A string has 6 insulator units. The capacitance between each joint and earth is 10% of self capacitance of each unit. Find the voltage across the lowest unit as a percentage of total voltage. Also find string efficiency. [28.5%, 58.5%]

5.5 In a string of 3 units each unit has a capacitance *C*. The capacitance between each joint and tower is 0.2 C. A grading ring is also provided. The capacitance between the grading ring and the lowest joint is 0.4 C. The capacitance between grading ring and second lowest joint is 0.1C. Find string efficiency. [100%]

5.6 In a 3-unit string insulator, the joint to tower capacitance is 20% of the capacitance of each unit. By how much should the capacitance of the lowest unit be increased to get a string efficiency of 90%. The remaining two units are left unchanged. [26.7%]

5.7 The string insulator disc in an overhead line is composed of (a) 3 (b) 9 similar units. The flashover voltage of one disc is 100kV. The capacitance from each link to earth is 10% of the capacitance of each disc. Find the flashover voltage in each case. [(a) 260 kV (b) 366 kV]

5.8 A string of eight suspension discs is fitted with a grading ring. Each pin to earth capacitance is *C*. If the voltage distribution is uniform find the values of line to pin capacitances.

$$\left[\frac{C}{7}, \frac{C}{3}, \frac{3C}{5}, C, \frac{5C}{3}, 3C, 7C\right]$$

5.9 A suspension insulator with 6 discs has a uniform voltage distribution. Capacitance grading is used to give the uniform voltage distribution. Each pin to earth capacitance is *C*. Capacitance of the topmost disc is 10 *C*. Find mutual capacitances of the remaining discs. [From top to bottom disc 10 *C*, 11 *C*, 13 *C*, 16 *C*, 20 *C*, 25 *C*]

5.10 A string has 5 suspension discs. The capacitance between each unit and earth is one-fifth of the mutual capacitance; (a) Find the voltages across different discs as percent of total string voltage (b) Also find string efficiency. [(a) 10.1, 12.1, 16.56, 24.3, 36.92% (b) 54.17%]

5.11 A string has two discs. The capacitance of pin to earth is half the capacitance of disc. Find the ratio of voltages across the discs and string efficiency [1.5, 83.33%]

5.12 The string insulator of problem 5.2 is fitted with a grading ring. The capacitance between grading ring and each pin is one-eighth of self capacitance of each disc. Find the voltage across each disc and string efficiency. [8.839 kV, 7.353 kV, 8.687 kV, 13.22 kV, 72%]

CHAPTER 6

Mechanical Design of Overhead Lines

6.1 GENERAL CONSIDERATIONS

The design of a transmission line has to be satisfactory from electrical as well as mechanical considerations. The line should have sufficient current carrying capacity so that the required power transfer can take place without excessive voltage drop or overheating. The line losses should be small and the insulation of the line should be adequate to cope with the system voltage. As far as mechanical aspects are concerned, the line conductors, supports and cross-arms should have sufficient mechanical strength to cope with the worst probable (but not worst possible) weather conditions. The cost of a line which would withstand a worst possible weather condition (e.g., very severe hurricane) would be extremely high. From the considerations of economy it is better to run the risk of failure under such extreme conditions. The line conductor, supports and cross-arms must be strong enough to give satisfactory service over a long period of time without the necessity of too much maintenance. The tension in the conductor should be well below the breaking load and reasonable factor of safety should be used. To achieve this, an appreciable amount of sag has to be allowed. Adequate clearance between the lowest point on the line and ground must be maintained.

6.2 LINE SUPPORTS

The supports for an overhead line must be capable of carrying the load due to the conductors and insulators (including the ice and wind loads on the conductors) together with the wind load on the support itself. The supports generally used are wooden poles, RCC poles, steel tubular poles and steel towers.

The design of a support depends to a great extent on whether the support is rigid or has a certain amount of flexibility in the direction of the line. Wooden poles and some steel structures fall in the latter category and in designing these it is usual to consider only the transverse wind pressure which occurs on the conductors and support. The longitudinal pull of conductors is usually balanced on either side of the support. In the event of a break in one or more conductors, an unbalanced load occurs due to the excess of the transverse wind pressure. With flexible supports this unbalanced load is absorbed by the line. A few supports on each side of the broken conductor bend towards the adjoining spans and this movement of the support tops causes a reduction in the tension in the wires of the adjoining span owing to the increased sag of these wires. A certain degree of rigidity in the longitudinal direction is provided by using rigid anchoring towers at intervals of about 2 kms. These rigid towers are designed to withstand the failure of one of the conductors on one side as well as the transverse load. In the case of rigid supports such as lattice steel tower it is usual to provide equal strength in both longitudinal and transverse directions and to design each tower to withstand the breaking of one conductor on one side.

(a) *Wooden Poles*: Poles made of chemically treated wood are used for distribution lines especially in areas where ample supplies of good quality wood are available. They are very economical but susceptible to decay. To reduce decay the wood poles are protected by an aluminium or zinc cap at the top and bitumen coating over the portion of the pole in the ground. For low voltage lines only one pole is used but for 33 kV lines two poles in *A* or *H* formation are used. Fig. 6.1 shows some wooden pole supports.

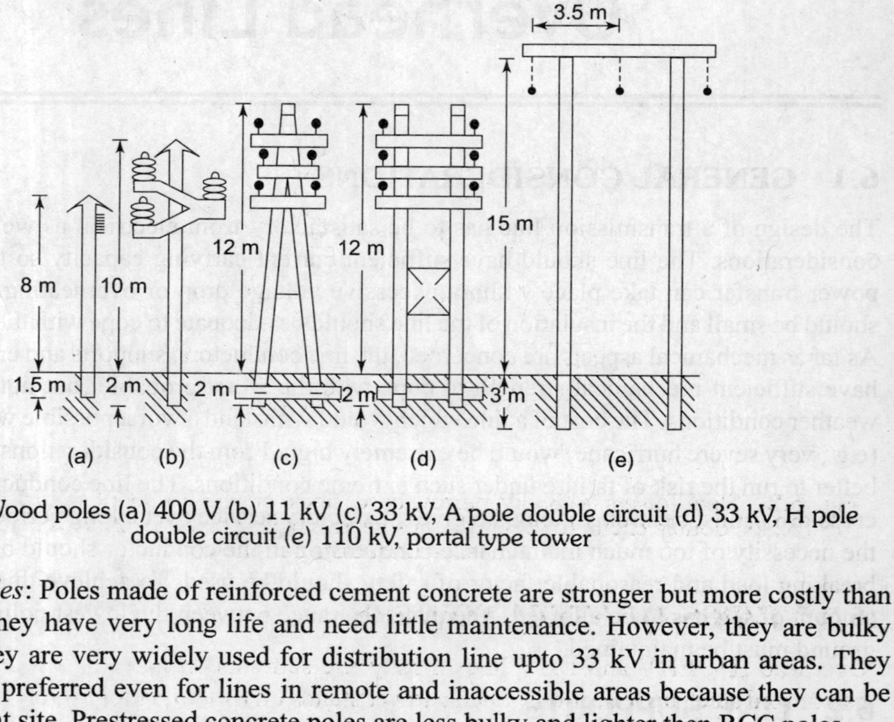

Fig. 6.1 Wood poles (a) 400 V (b) 11 kV (c) 33 kV, A pole double circuit (d) 33 kV, H pole double circuit (e) 110 kV, portal type tower

(b) *RCC poles*: Poles made of reinforced cement concrete are stronger but more costly than wood poles. They have very long life and need little maintenance. However, they are bulky and heavy. They are very widely used for distribution line upto 33 kV in urban areas. They are sometimes preferred even for lines in remote and inaccessible areas because they can be manufactured at site. Prestressed concrete poles are less bulky and lighter then RCC poles.

(c) *Steel Tubular Poles*: These may be stepped or swaged type. A stepped pole is manufactured from a single tube, the diameter being reduced in parallel steps by passing the tube through a series of dies. A swaged pole is manufactured from poles of different diameters which are swaged together when hot. They are more costly than RCC and wood poles but possess the distinct advantages of light weight, high strength to weight ratio and long life. The use of a cap at the top, concrete muff in the ground and regular painting prolongs their life. They are very widely used for lines upto 33 kV.

(d) *Steel Towers*: Lines of 66 kV and above are invariably supported on steel towers. They are fabricated from painted or galvanized angle sections which can be transported separately and the erection done on site. Steel towers have the advantage of a very long life and a high degree of reliability. They can withstand very severe weather conditions. The risk of service interruptions, due to damage in insulators, is greatly reduced since large spans can be used. They are very suitable for double circuit lines. Towers of many shapes and sizes are used for different types of lines. Fig. 6.2 shows a few typical steel towers. The height of the tower depends on the line voltage and span length. The legs of the towers are set in special concrete foundations. The forces which have to be taken into account in the tower design include vertical loads of conductor, insulators, fittings and tower itself, wind pressure on conductor and wind pressure on tower itself. Fig. 6.3 shows a 765 kV tower.

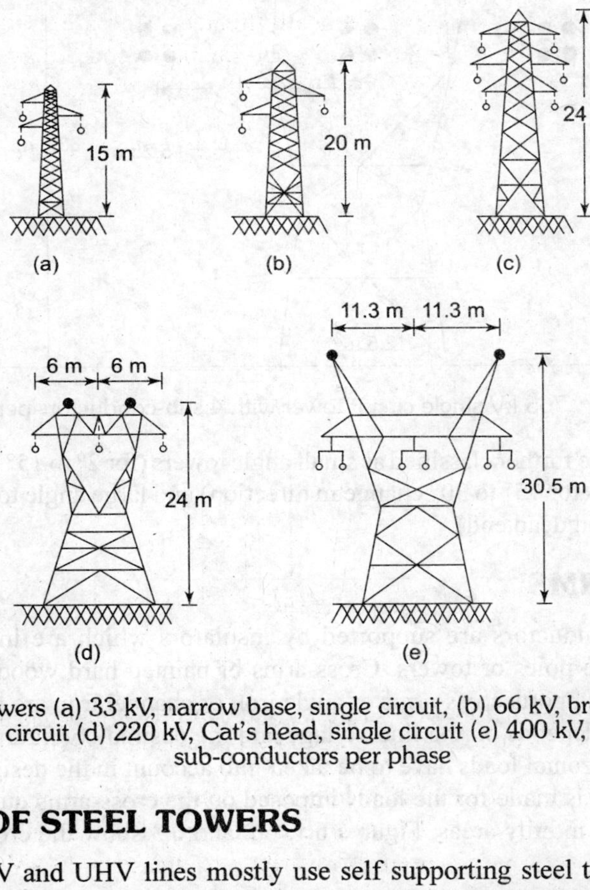

Fig. 6.2 Steel Towers (a) 33 kV, narrow base, single circuit, (b) 66 kV, broad base, single circuit (c) 132 kV, double circuit (d) 220 kV, Cat's head, single circuit (e) 400 kV, single circuit with two sub-conductors per phase

6.3 TYPES OF STEEL TOWERS

Overhead HV, EHV and UHV lines mostly use self supporting steel towers fabricated from galvanized steel angle sections. Such a tower stands on its foundation rigidly and vertically and does not need any other support. (On the other hand a Guyed tower is supported by guy wires from different sides.) A wide variety of configurations are used. They can be broadly classified as (a) tangent towers (b) deviation towers.

(a) *Tangent Towers*: They are used for straight runs of the line and upto 2° line deviation from the straight run. The line is straight or along the tangent to the line route. In these towers the stress is due to the weight of the line, ice and wind loads. In addition extra forces due to a break in the line on one side of the tower has also to be considered in tower design. The base of the tower may be square or rectangular. Suspension type insulators are used with these towers.

(b) *Deviation Towers*: They are used at points where the transmission line changes direction. In addition to forces to which the tangent towers are subjected, these towers have also to withstand the resultant forces due to change in direction. They have broader base, stronger members and are costlier as compared to tangent towers. Strain insulators are used with these towers.

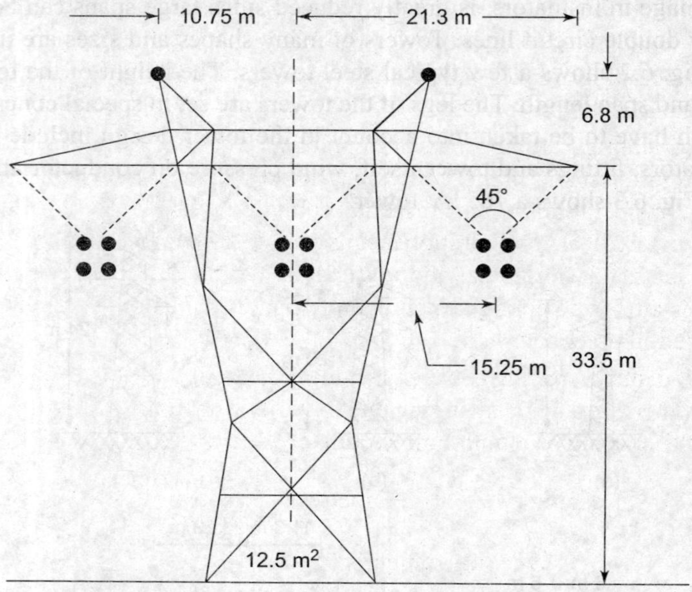

Fig. 6.3 765 kV single circuit tower with 4 sub-conductors per phase

Deviation towers are further classified as small angle towers (for 2° to 15° change in direction), medium angle towers (for 15° to 30° change in direction) and large angle towers (for 30° to 60° change in direction and dead end).

6.4 CROSS-ARMS

Transmission line conductors are supported by insulators which are in turn supported by cross-arms attached to poles or towers. Cross-arms of painted hard wood are used for wood pole lines. For other supports cross-arms of angle iron or channel iron are used. The shape and length of cross-arms depends on the line voltage and the desired configuration of conductors. Both vertical and horizontal loads have to be taken into account in the design of cross-arms. In addition an allowance is made for the loads imposed on the cross-arms due to a change in the line profile especially in hilly areas. Figures 6.1, 6.2 and 6.3 show the cross-arms along with the towers and poles.

6.5 SPAN, CONDUCTOR CONFIGURATION, SPACINGS AND CLEARANCES

A longer span means a smaller number of towers but the towers are taller and more costly. The higher the operating voltage, the greater should be the span to reduce the high cost of insulators. Moreover the insulators constitute the weakest part of a transmission line and a reduction in the number of towers per km (by using longer span) increases the reliability of the line. For every proposed line there is a definite length of span which will give the minimum cost of the line. From mechanical consideration there is a maximum value of span for each conductor size. Many a time it happens that the conductor size, as determined from electrical calculations is very small and it is possible to reduce the cost of line by using thicker and stronger conductor so that a longer span may be employed. Sometimes it is not feasible to determine the tower height and span length on the basis of line cost alone because lightning hazards increase greatly as the height of conductors above ground is increased. Modern high voltage lines, have spans between 200 and 400 m. For river and ravine crossings exceptionally long spans upto 800 m or so have been satisfactory employed.

Many conductor configurations are used in practice. There is no special advantage in using symmetrical configuration and in most cases flat horizontal or vertical configurations are used from mechanical considerations. A flat horizontal configuration means a lesser tower height but a wider right of way. A vertical configuration means a taller tower and increased lightning hazards. In spite of these facts, flat horizontal and vertical configurations are used in many cases. For single circuit lines an *L* type configuration is quite popular. Figs. 6.1, 6.2 and 6.3 depict some configurations used in practice.

There must be adequate spacing between conductors so that they do not come within sparking distance of each other even while swinging due to wind. An empirical formula commonly used for determining the spacing of aluminium conductor lines is

$$\text{Spacing} = \sqrt{S} + \frac{V}{150} \text{ metres} \tag{6.1}$$

where S = Sag in metres

V = Line voltage in kV

Some typical values of spacings are:

Line voltage (kV)	0.4	11	33	66	132	220	400	765
Spacing (m)	0.2	1.2	2.0	2.5	3.5	6.0	11.5	14

The Indian Electricity Rules specify the minimum clearance between the ground and the conductor. These values are:

kV	*0.4*	*11*	*33*	*66*	*132*	*220*	*400*
Clearance to ground							
(a) Across Street (m)	5.8	5.8	6.1	6.1	6.1	7.0	8.4
(b) Along Street (m)	5.5	5.5	5.8	6.1	6.1	7.0	8.4
(c) Other Areas (m)	4.6	4.6	5.2	5.5	6.1	7.0	8.4

These rules also specify the minimum clearances for power lines from buildings, railway tracks, and telecommunication lines, etc.

6.6 SAG AND TENSION CALCULATIONS

6.6.1 Supports at Same Level

Fig 6.4 shows an overhead line suspended between two supports *A* and *B* at the same level. The line is assumed to be flexible and sags below the level *AB* due to its weight. The exact shape of the line is that of a catenary. Except for lines with very very long span and large sag, it is sufficiently accurate to assume that the shape of the line is that of a parabola $y = ax^2$ where a is constant for a given line and O is the origin.

Let l = length of span, i.e., horizontal distance between supports, in metres

S = sag at mid span, in metres

T = conductor tension (assumed constant over the whole span) in Newtons

w = conductor weight, N/m

$y = S$ when $x = \dfrac{l}{2}$. It follows that $S = a\,(l/2)^2$

or $$a = 4S/l^2 \tag{6.2a}$$

Then $$y = 4S\left(\frac{x}{l}\right)^2 \tag{6.2b}$$

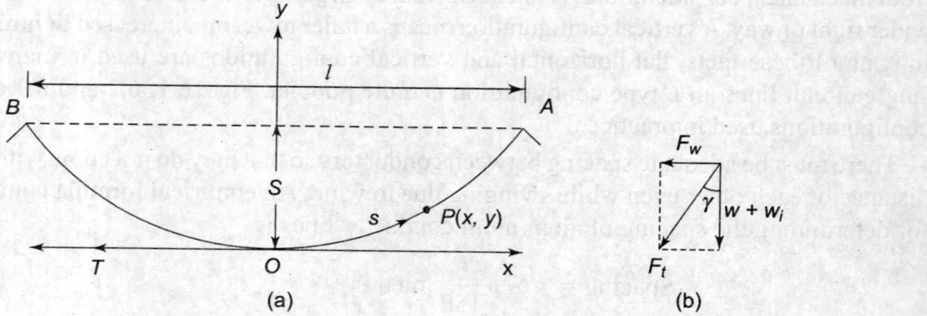

Fig. 6.4 (a) Parabolic form (b) Effect of wind force F_w

Consider the equilibrium of half line *OA*. Assuming that the conductor is almost horizontal and taking moments about *A*

$$TS = \left(\frac{wl}{2}\right)\left(\frac{l}{4}\right) \text{ or } S = \frac{wl^2}{8T} \tag{6.3}$$

Substituting this value of *S* in Eqs. (6.2 a and b)

$$a = \frac{w}{2T} \tag{6.4 a}$$

and $$y = \frac{wx^2}{2T} \tag{6.4b}$$

6.6.2 Effect of Ice and Wind

In addition to its own weight, a transmission line conductor is also subject to wind pressure. A coating of ice may also be formed on the conductors of the lines in hilly areas during severe winter season. These two factors must be taken into account to calculate the sag.

Consider 1 m length of conductor.

d = diameter of conductor, metres

t = radial thickness of ice, metres

Then the overall diameter of ice covered conductor as seen from Fig. 6.5 is $D = d + 2t$

Volume of ice per metre length of conductor $= \dfrac{\pi}{4}(D^2 - d^2)$ m³.

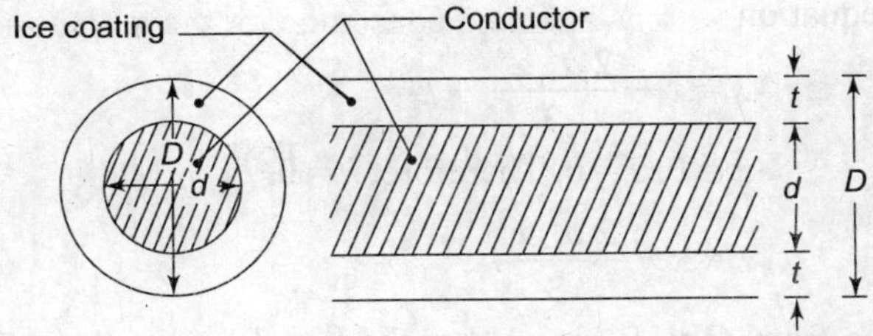

Fig. 6.5 Ice coated conductor

The weight of ice is approximately 8920 N/m³. Then weight of ice per metre length of conductor, i.e.,

$$w_i = \frac{8920\,\pi}{4}(D^2 - d^2) = 2230\,\pi(d^2 + 4\,t^2 + 4\,dt - d^2)\text{ N/m}$$

or
$$w_i = 2.8 \times 10^4\, t(d + t)\text{ N/m} \tag{6.5}$$

The wind pressure is assumed to act horizontally on the projected area of the ice covered conductor. From Fig. 6.5 it is seen that this projected area is D square metres per metre length of conductor. For a wind pressure of p Newton per sq. m of projected area, wind load F_w is

$$F_w = p\,D\text{ N/m} \tag{6.6}$$

The effective area of a conductor (for calculating wind load) is two-third of the projected area for cylindrical surfaces like poles and conductors. Now-a-days it is usual to specify the wind pressure as N/m² of projected are and Eq. (6.6) can be directly used to calculate the wind load.

If the wind velocity is specified, the wind pressure p is taken as 0.059 V^2 N/m where V is wind velocity in km/hr. The total force F_t acting on the conductor per metre length is given by:

$$F_t = [(w + w_i)^2 + F_w^{\,2}]^{0.5}\text{ N/m}$$

The force F_t lies in the new plane of conductor and is inclined to the vertical at an $\angle\gamma$ (Fig. 6.4b) which is given by:

$$\tan\gamma = \frac{F_w}{w + w_i}$$

Eqs. (6.2a) to (6.4b) are still valid if all the measurements are made in the new plane of conductor and w is replaced by F_t. If T is the limiting tension and F_t is the total force per metre on the conductor under worst conditions, then using Eq. (6.3), the sag in the new plane is

$$S = \frac{F_t\, l^2}{8T} \tag{6.7}$$

The vertical sag is $S\cos\gamma$.

6.6.3 Total Length of Conductor

Consider a small right angled triangle on the parabola $y = a\,x^2$ in Fig. 6.4. This gives;

$$(ds)^2 = (dx)^2 + (dy)^2$$

or
$$\left(\frac{ds}{dx}\right)^2 = 1 + \left(\frac{dy}{dx}\right)^2 = 1 + 4a^2\,x^2$$

or
$$ds = (1 + 4\,a^2\,x^2)^{0.5}\,(dx) \approx (1 + 2a^2\,x^2)\,dx$$

integrating the above equation

$$s = x + \frac{2\,a^2\,x^3}{3} + k$$

when $x = 0$, $s = 0$. Therefore, $k = 0$.

Thus,
$$s = x + \frac{2\,a^2\,x^3}{3} \tag{6.8}$$

s is the curved distance from O to P $(x,\ y)$. Let the length of line be z. If we put $x = l/2$ in Eq. (6.8) and substitute the value of a from Eq. (6.2a) we get the length of half line, i.e., $z/2$

Thus,
$$\frac{z}{2} = \frac{l}{2} + \frac{2\times 16S^2}{l^4}\times\frac{l^3}{8\times 3} = \frac{l}{2} + \frac{4}{3}\frac{S^2}{l}$$

or
$$z = l + \frac{8S^2}{3l} \tag{6.9}$$

Substituting the value of S from Eq. (6.7) into Eq. (6.9)

$$z = l\left(1 + \frac{F_t^2\,l^2}{24T^2}\right) \tag{6.10}$$

6.6.4 Equivalent Span

A section of an overhead line may consist of a number of spans of different lengths due to the changing profile and the desire to locate the towers in such positions as to cause minimum interference with the normal use of land. In order to ensure that the insulators hang vertically, the horizontal components of the tension at the support points of adjacent sections should be equal. If each section is strung according to the formulae derived above, the resulting tensions would not be necessarily equal.

The tensions are made equal by defining an equivalent span l_e. and basing all calculations on the equivalent span l_e. The two conditions which have to be satisfied are that the sum total of horizontal spans should be equal and that the sum total of curved lengths should be equal for the actual line and the equivalent line.

If there are n sections, then

$$n\,l_e = \Sigma\, l \tag{6.11}$$

Using Eq. (6.10)

$$n\left(l_e + \frac{F_t^2\,l_e^3}{24T^2}\right) = (\Sigma\, l) + \frac{F_t^2\,(\Sigma l^3)}{24T^2} \tag{6.12}$$

Substituting Eq. (6.11) into Eq. (6.12) we get $l_e^2 = \dfrac{\Sigma l^3}{nl_e} = \dfrac{\Sigma l^3}{\Sigma l}$

or
$$l_e = \left[\frac{(l_a^3 + l_b^3 + l_c^3 + \cdots\cdots)}{l_a + l_b + l_c}\right]^{0.5} \tag{6.13}$$

where a, b, c, etc., are the sections.

Knowing l_e the tension T to which all the sections would be erected can be calculated.

6.6.5 Supports at Different Levels

Many a time (especially in hilly areas) the two supports of a span may be at different levels. Fig. 5.6 shows such a section suspended between two supports *B* and *C* which are at different levels. The curve *BOCA* is the complete parabola with *A* and *B* at the same level.

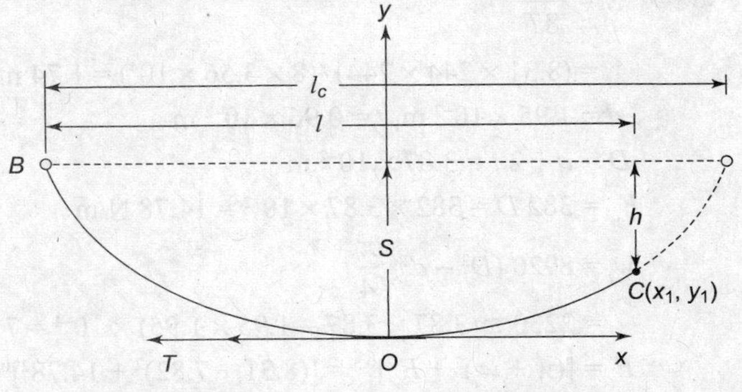

Fig. 6.6 Supports at different levels

The actual line *BOC* is a part of this complete parabola.

Let l be the actual span (horizontal distance between *B* and *C*), l_c be the span of the complete parabola and x_1, y_1 be the coordinates of point *C*.

Then $$x_1 = l - \frac{l_c}{2} \tag{6.14}$$

The equation of parabola $y = ax^2$ is valid for the curve *BOCA* as well as *BOC*.

From Eq. (6.3), $$S = \frac{w l_c^2}{8T} \tag{6.15}$$

Substituting the coordinates of point *C* in Eq. (6.2b)

$$y_1 = S - h = \frac{4 S x_1^2}{l_c^2} \quad \text{or} \quad \frac{x_1^2}{S-h} = \frac{l_c^2}{4S} \tag{6.16}$$

Substituting the values of *S* and x_1 from Eqs. (6.15) and (6.14) into Eq. (6.16) and re-arranging we have

$$l_c = l + \frac{2Th}{wl} \tag{6.17}$$

Since *h*, *T*, *l*, *w* are known, l_c can be found out. Then all the results derived in the previous section can be applied. The above theory is valid even when the two supports *B* and *C* fall on the same side of origin *O* (i.e., span l is less than $\frac{l_c}{2}$).

Example 6.1

An overhead line has *ACSR* conductor of 1.95 cm diameter and a span of 244 m. The allowable tension is 3.56×10^4 N. Find (a) sag in still air condition with no ice covering (b) vertical sag when there is an ice covering of 0.96 cm thickness and a horizontal wind pressure of 382 Newton per square metre of projected area. Ice weighs 8920 N/m³ (c). The line is carried by insulator strings 1.43 m long. What should be the height of lowest cross-arm to give a minimum ground clearance of 7.62 m under bad weather conditions? The conductor weight is 0.847 kg/m.

(AMIE 2004)

Solution

$$w = 0.847 \times 9.81 = 8.31 \text{ N/m}, \ l = 244 \text{ m}$$

$$T = 3.56 \times 10^4 \text{ N}$$

(a)
$$S = \frac{wl^2}{8T}$$

$$= (8.31 \times 244 \times 244)/(8 \times 3.56 \times 10^4) = 1.74 \text{ m}$$

(b)
$$d = 1.95 \times 10^{-2} \text{ m}, \ t = 0.96 \times 10^{-2} \text{ m}$$

$$D = d + 2\,t = 3.87 \times 10^{-2} \text{ m}$$

$$F_w = 382\,D = 382 \times 3.87 \times 10^{-2} = 14.78 \text{ N/m}$$

$$w_i = 8920\,(D^2 - d^2)\,\frac{\pi}{4}$$

$$= 2230\,\pi\,(3.87 \times 3.87 - 1.95 \times 1.95) \times 10^{-4} = 7.82 \text{ N/m}$$

$$F_t = [(w + w_i)^2 + F_w^2]^{0.5} = [(8.31 + 7.82)^2 + 14.78^2]^{0.5} = 21.88 \text{ N/m}$$

$$s = \frac{F_t l^2}{8T} = (21.88 \times 244 \times 244)/(8 \times 3.56 \times 10^4) = 4.57 \text{ m}$$

$$\gamma = \tan^{-1}[F_w/(w + w_i)] = \tan^{-1}\frac{14.78}{8.31 + 7.82} = 42.5°$$

Vertical sag = $S \cos \gamma = 3.37$ m

(c) Height of lowest cross-arm

= Minimum ground clearance + vertical sag + length of insulator string

= 7.62 + 3.37 + 1.43 = 12.42 m

Example 6.2

An overhead line has a span of 336 m. The line is supported, at a water crossing, from two towers whose heights are 33.6 m and 29 m above water level. The weight of conductor is 8.33 N/m and tension in the conductor is not to exceed 3.34×10^4 N. Find (a) clearance between the lowest point on the conductor and water (b) horizontal distance of this point from the lower support.

Solution

Referring to Fig. 6.6

$$l_c = l + \frac{2Th}{wl}$$

$$= 336 + 2 \times 3.34 \times 10^4\,(33.6 - 29)/(8.33 \times 336)$$

$$= 445.79 \text{ m}$$

(a)
$$S = \frac{wl_c^2}{8T}$$

$$= (8.33 \times 445.79 \times 445.79)/(8 \times 3.34 \times 10^4)$$

$$= 6.195 \text{ m}$$

The lowest point O is 6.195 m below support B which is 33.6 m above water level. Therefore, the clearance between point O and water is 33.6 – 6.195 = 27.405 m.

(b) Distance of point O from higher support B

$$= 0.5 \times 445.79 = 222.89 \text{ m}$$

Distance of point O from lower support = 336 – 222.89 = 113.11 m

Example 6.3

An overhead line, over a river crossing, is supported by two towers 50 m and 80 m above water level. The horizontal span is 300 m. If the weight of conductor is 8.28 N/m and the tension in the conductor is 19620 N, find the height of mid point of the line above water level.

Solution

$$l_c = l + \frac{2Th}{wl}$$

$$= 300 + 2 \times 19620 \times (80 - 50)/(8.28 \times 300) = 773.9 \text{ m}$$

l_c is more than twice the horizontal span of 300 m. This means that both the supports are on the same side of point O (Fig. 6.7).

Horizontal distance between O and C

$$= \frac{773.9}{2} - 300 = 86.95 \text{ m}$$

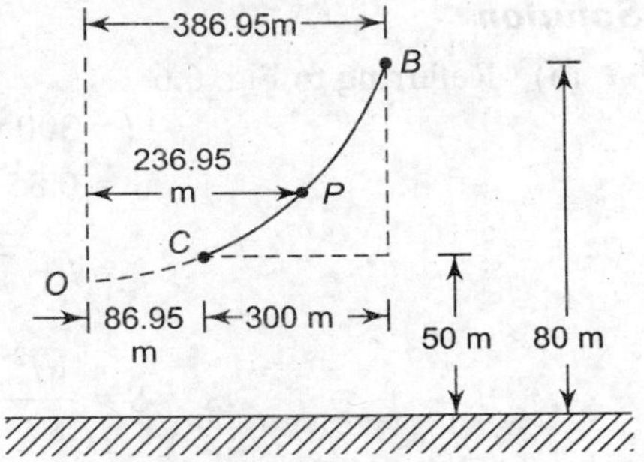

Fig. 6.7

From Eq. 6.4(b)

Height of point C above $O = \dfrac{wx^2}{2T}$

$$= \frac{8.28 \times 86.95 \times 86.95}{2 \times 19620} = 1.595 \text{ m}$$

Horizontal distance between O and P = 86.95 + 150 = 236.95 m

Height of point P above O

$$\frac{8.28 \times 236.95 \times 236.95}{2 \times 19620} = 11.847 \text{ m}$$

Height of mid point P above point C = 11.847 – 1.595 = 10.252 m

Height of mid point P above water level = 10.252 + 50 = 60.252 m

Example 6.4

An overhead line has a span of 152 m and is supported on level supports. The conductor has an effective diameter of 2.068 cm, cross-sectional area of 3.065 sq. cm and weighs 2.292 kg/m. The line is subjected to a wind pressure of 39.063 kg/sq m of projected area. Assuming a maximum stress of 1054.63 kg/sq cm, find sag under the given conditions. Also find vertical component of sag.

Solution

$$w = 2.292 \times 9.81 = 22.484 \text{ N/m}$$

$$T = 1054.63 \times 3.065 = 3232.44 \text{ kg} = 3.171 \times 10^4 \text{ N}$$

$$F_w = 39.063 \times 9.81 \times 2.068 \times 10^{-2} = 7.92 \text{ N/m}$$

$$F_t = \sqrt{(w^2 + F_w^2)} = (22.484^2 + 7.92^2)^{0.5} = 23.838 \text{ N/m}$$

$$S = \frac{F_t l^2}{8T} = \frac{23.838 \times 152 \times 152}{8 \times 3.171 \times 10^4} = 2.17 \text{ m}$$

Vertical component of sag $= 2.17 \times \dfrac{22.484}{23.838} = 2.047$ m

Example 6.5

A transmission line at a river crossing is supported by two towers 50 m and 55 m above water level. The horizontal distance between towers is 300 m. The tension in the conductor is 2000 kg and weight of conductor is 0.85 kg/m (a) Find the minimum clearance between conductor and water (b) Determine the position of minimum clearance.

Solution

(a) Referring to Fig. 6.6

$$l = 300 \text{ m}, h = 5 \text{ m}, T = 2000 \times 9.81 = 19.62 \times 10^3 \text{ N}$$

$$w = 0.85 \times 9.81 = 8.34 \text{ N/m}$$

$$l_c = l + \frac{2Th}{wl} = 300 + \frac{2 \times 19.62 \times 10^3 \times 5}{8.34 \times 300} = 378.4 \text{ m}$$

$$S = \frac{wl_c^2}{8T} = \frac{8.34 \times 378.4 \times 378.4}{8 \times 19.62 \times 10^3} = 7.61 \text{ m}$$

The lowest point on the conductor is 7.61 m below the higher support *B* which is 55 m above water level. Therefore, minimum clearance between conductor and water is 55 – 7.61 = 47.39 m

(b) Distance of point *O* (position of minimum clearance) from support *B*. i.e., higher support = 0.5 × 378.4 = 189.2 m
Distance of point *O* from support *A* i.e., lower support
= 300 – 189.2 = 110.8 m

6.7 ERECTION CONDITIONS

The regulations require that sag must be calculated for worst probable conditions and the minimum ground clearance should be maintained for these conditions. Moreover, the maximum tension in the conductor should not exceed half the breaking load (so as to allow a factor of safety of two). In practice the ambient conditions, when the line is erected, are very different from the worst probable conditions. The erection tension should be such that when loading conditions subsequently occur, there will be no infringement on the factor of safety.

Let the different quantities under one set of physical conditions (e.g., the worst probable) be denoted by subscript 1 and those under the second set of physical conditions (e.g., erection conditions) be denoted by subscript 2. Thus, z_1, T_1, θ_1, F_{t1} denote the total conductor length, tension, temperature and total load per metre under first set of physical conditions and z_2, T_2, θ_2, F_{t2} denote the same quantities under the second set of physical conditions. A relationship between them can be derived as under.

An increase in tension from T_1 to T_2 elongates the conductor by $z_1 (T_2 - T_1)/AE$ where A is cross-sectional area and E is Young's modulus of elasticity. The increase in temperature from θ_1 to θ_2 increases the conductor length by $\alpha z_1 (\theta_2 - \theta_1)$ where α is coefficient of linear expansion. Thus

$$z_2 = z_1 + z_1 \frac{(T_2 - T_1)}{AE} + \alpha z_1 (\theta_2 - \theta_1) \tag{6.18}$$

Substituting for z_1 and z_2 from Eq. (6.10), using the *binomial* expansion and neglecting second order terms, Eq. (6.18) reduces to

$$\left(\frac{l^2}{24}\right)\left(\frac{F_{t_2}^2}{T_2^2} - \frac{F_{t_1}^2}{T_1^2}\right) = \frac{(T_2 - T_1)}{AE} + \alpha(\theta_2 - \theta_1) \tag{6.19}$$

After rearranging the terms, Eq. (6.19) can be written as

$$T_2^2\left[T_2 - \left\{T_1 - \alpha AE(\theta_2 - \theta_1) - \frac{AEF_{t_1}^2 l^2}{24T_1^2}\right\}\right] = \frac{AEF_{t_2}^2 l^2}{24} \tag{6.20}$$

This cubic equation can be solved graphically or analytically. Using this equation we can find erection tension T_1 such that the tension T_2 (i.e., tension under bad weather conditions) will not exceed the safe limit of tension. Using this value of tension and Eq. (6.7) the sag for erection conditions can be calculated. The sag and height of the line above ground at the time of erection must be such that the ground clearance will not fall below the statutory minimum at any future time.

Using Eqs. (6.20) and (6.7) the graph of tension vs. temperature and sag vs. temperature can be plotted. This graph is plotted for a fixed span and is known as stringing chart and is shown in Fig. 6.8. From this chart the values of sag and tension can be found for any temperature. A stringing chart is very useful in the field for stringing the conductors and adjusting them to the proper sag and tension.

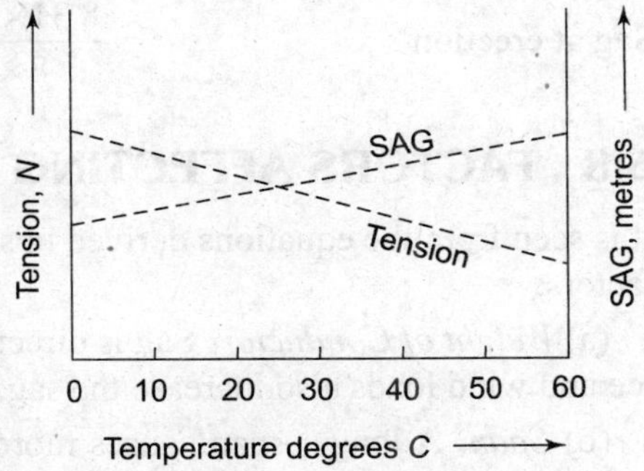

Fig. 6.8 Stringing chart

Example 6.6

An overhead line having a span of 250 m is to be erected at a temperature of 40°C in still air conditions. It is desired that a factor of safety of 2 should be maintained under bad weather conditions when the temperature is 10°C and wind load is 378 N/m² of projected area. The data for the ACSR conductor used for the line is as under: diameter 1.95 cm, area 2.25 sq. cm., weight = 8.31 N/m, breaking load 77900 N, coefficient of linear expansion 18.44×10^{-6} per degree C, Young's modulus 91.4×10^3 N/mm². Find the sag and tension under erection conditions.

Solution

$$d = 1.95 \text{ cm} = 1.95 \times 10^{-2} \text{ m}$$

$$A = 2.25 \text{ sq. cm} = 2.25 \times 10^{-4} \text{ sq.m}$$

$$E = 91.4 \times 10^3 \text{ N/mm}^2 = 91.4 \times 10^3 \times 10^6 = 91.4 \times 10^9 \text{ N/m}^2$$

For bad weather conditions (subscript 1)

$$F_w = 378 \times 1.95 \times 10^{-2} = 7.37 \text{ N/m}$$

$$F_{t1} = (7.37^2 + 8.31^2)^{0.5} = 11.11 \text{ N/m}$$

$$T_1 = 0.5 \times 77900 = 38950 \text{ N}$$

$$\theta_1 = 10\ °\text{C}$$

For erection conditions (subscript 2)

$$F_{t2} = 8.31 \text{ N/m}, T_2 = ? \ \theta_2 = 40 \text{ °C}$$

$$T_2^2 \left[T_2 - \left\{ T_1 - \alpha AE(\theta_2 - \theta_1) - \frac{AEF_{t_1}^2 l^2}{24 T_1^2} \right\} \right] = \frac{AEF_{t_2}^2 l^2}{24}$$

$$\alpha AE(\theta_2 - \theta_1) = 18.44 \times 10^{-6} \times 2.25 \times 10^{-4} \times 91.4 \times 10^9 \times 30 = 11376.6$$

$$\frac{AEF_{t1}^2 l^2}{24 T_1^2} = \frac{2.25 \times 10^{-4} \times 91.4 \times 10^9 \ (11.11)^2 \ (250)^2}{24 \ (38950)^2} = 4357.2$$

$$AEF_{t2}^2 l^2 / 24 = 2.25 \times 10^{-4} \times 91.4 \times 10^9 \times 8.31^2 \times 250^2/(24) = 36.99 \times 10^{11}$$

$$T_2^2 \ [T_2 - 38950 + 11376.6 + 4357.2] = 36.99 \times 10^{11}$$

or $\quad T_2^2 \ (T_2 - 23216.2) = 36.99 \times 10^{11}$

or $\quad T_2^3 - 23216.2 \ T_2^2 - 36.99 \times 10^{11} = 0$

By hit and trial, $\quad T_2 = 23800$ N

Sag at erection $\quad = \dfrac{8.31 \times 250 \times 250}{8 \times 23800} = 2.73\text{m}$

6.8 FACTORS AFFECTING SAG

It is seen from the equations derived in section 6.6 and 6.7 that sag is affected by the following factors.

(a) *Weight of Conductor:* Sag is directly proportional to weight per unit length of conductor. Ice and wind loads also increase the sag.

(b) *Span:* A longer span causes more sag. Since sag is proportional to square of span, an increase of 25% in span increases the sag by 56.25%.

(c) *Conductor Tension:* Sag is inversely proportional to conductor tension. However, an increase in conductor tension causes more stresses in the conductor and more load on insulators and towers.

(d) *Ground Clearance:* Electricity rules specify minimum clearance between the ground and conductor (Section 6.5). To maintain this clearance, it may be necessary to increase the height of the towers if higher value of sag is desired (so as to keep conductor tension within safe limit).

(e) *Temperature:* A decrease in temperature reduces the sag. However when a decrease in temperature is accompanied by snow and wind, the sag may increase. Generally heavy snow and high wind pressure create worst conditions for the line.

6.9 SAG TEMPLATE

A sag template (Fig. 6.9) is a very convenient way of locating the towers in the field. This template is prepared on celluloid or tracing cloth and the curves are drawn to the same horizontal and vertical scales as the profile of the line. The upper curve represents the conductor. The middle curve is below the upper curve by a uniform vertical distance equal to the desired minimum vertical clearance to ground. The lower curve is below the upper one by a uniform vertical distance equal to the height of a standard tower measured to the point of support of the conductor. If the template is made of celluloid, it may be kept properly oriented by keeping its base against a *T*-square. If the left tower in Fig. 6.9 has been located, then the position of the right hand tower is found by adjusting the sag template so that the conductor line passes through the point of support on the left hand tower and the clearance line is tangent to ground (at one or more points) and never dips below. The tower footing line indicates the actual position of tower on ground. Sometimes it may happen that one or more intermediate points between the supports on the profile are cut by the clearance line on the template. If this happens, it is obvious that the base of the support or

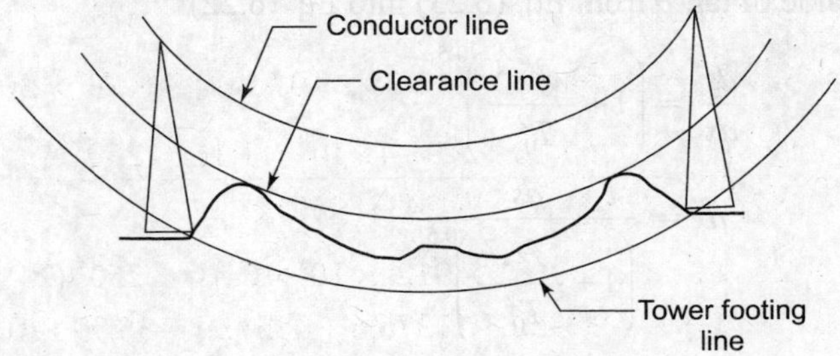

Fig. 6.9 Sag template

supports require extension in order to ensure the necessary ground clearance. In such cases a fourth line may be drawn on the template to indicate the position of the extension footing.

6.10 CATENARY

As mentioned earlier the exact shape of a transmission line is that of a catenary. For extra high voltage lines, when sag is greater than 10% of the span, formulae based on catenary should be used.

Fig. 6.10 (a) shows a line with supports A and B at the same level. T_0 is the tension at the lowest point O. T_p is the tension at point $P(x, y)$ and s is the length of conductor from O to P. The portion OP is in equilibrium under the action of three forces T_0, T_p and the weight ws of the conductor portion OP. The horizontal and vertical components of forces must balance for equilibrium.

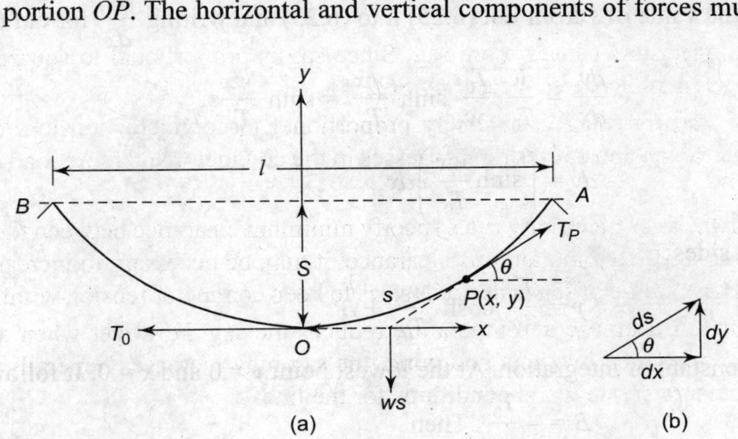

Fig. 6.10 Catenary

Thus $$T_p \sin\theta = ws \tag{6.21a}$$

And $$T_p \cos\theta = T_0 \tag{6.21b}$$

Consider a small right angled triangle on the catenary (Fig. 6.10 b)

$$(ds)^2 = (dx)^2 + (dy)^2$$

or $$\frac{ds}{dx} = \sqrt{1+\left(\frac{dy}{dx}\right)^2} = \sqrt{(1+\tan^2\theta)} \tag{6.22}$$

From Eqs. (6.21a) and (6.21b)

$$\tan\theta = \frac{ws}{T_0} \tag{6.23}$$

Substituting the value of tan θ from Eq. (6.23) into Eq. (6.22)

$$\frac{ds}{dx} = \left(1 + \frac{w^2 s^2}{T_0^2}\right)^{0.5}$$

or

$$dx = \frac{ds}{\left(1 + \dfrac{w^2 s^2}{T_0^2}\right)^{0.5}}$$

Integrating both sides

$$x = \left(\frac{T_0}{w}\right) \sinh^{-1} \frac{ws}{T_0} + A \tag{6.24}$$

where A is the constant of integration. At the lowest point on the curve, $x = 0$ and $s = 0$. It follows that $A = 0$.

Then

$$x = \left(\frac{T_0}{w}\right) \sinh^{-1} \frac{ws}{T_0}$$

or

$$s = \left(\frac{T_0}{w}\right) \sinh \frac{wx}{T_0} \tag{6.25}$$

Substituting the value of s from Eq. (6.25) into (6.23) and writing $\dfrac{dy}{dx}$ for tan θ we have

$$\frac{dy}{dx} = \frac{w}{T_0} \cdot \frac{T_0}{w} \sinh \frac{wx}{T_0} = \sinh \frac{wx}{T_0}$$

or

$$dy = \left(\sinh \frac{wx}{T_0}\right) dx$$

Integrating both sides

$$y = \frac{T_0}{w} \cosh \frac{wx}{T_0} + B$$

where B is the constant of integration. At the lowest point $y = 0$ and $x = 0$. It follows that

$B = -\dfrac{T_0}{w}$. Then

$$y = \left(\frac{T_0}{w}\right)\left[\cosh \frac{wx}{T_0} - 1\right] \tag{6.26}$$

From Eqs. (6.21 a) and (6.21 b)

$$T_p^2 \sin^2\theta + T_p^2 \cos^2\theta = T_p^2 = T_0^2 + w^2 s^2 \tag{6.27}$$

Substituting the value of s from Eq. (6.25) into Eqs. (6.27)

$$T_p^2 = T_0^2 + w^2\left[\frac{T_0^2}{w^2} \sinh^2\left(\frac{wx}{T_0}\right)\right] = T_0^2 \cosh^2\left(\frac{wx}{T_0}\right)$$

or $$T_p = T_0 \cosh\left(\frac{wx}{T_0}\right) \qquad (6.28)$$

As x increases T_p increases for a given value of T_0. The minimum tension in the line is T_0. The sag S can be found by putting $x = l/2$ in Eq. (6.26).

$$S = \frac{T_0}{w}\left[\cosh\left(\frac{wl}{2T_0}\right) - 1\right]$$

or $$S = \frac{wl^2}{8T_0}\left[1 + \left(\frac{l^2}{48}\right)\left(\frac{w}{T_0}\right)^2 + \cdots\cdots\right] \qquad (6.29)$$

A comparison of Eq. (6.29) with Eq. (6.3) gives the extent of error involved in using the approximation of treating the line as a parabola. The sag for the worst loading conditions can be found from Eq. (6.29) by substituting F_t for w and safe tension T_1 for T_0. Eq. (6.25) gives the curved distance from O to P. If we put $x = l/2$ we get the length of half line, i.e., $z/2$.

$$\frac{z}{2} = \left(\frac{T_0}{w}\right)\sinh\left(\frac{wl}{2T_0}\right)$$

or $$z = \left(\frac{2T_0}{w}\right)\sinh\left(\frac{wl}{2T_0}\right)$$

$$= l\left\{1 + \left(\frac{l^2}{24}\right)\left(\frac{w}{T_0}\right)^2 + \left(\frac{l^4}{1920}\right)\left(\frac{w}{T_0}\right)^4 + \cdots\cdots\right\} \qquad (6.30)$$

The tension at support point A is

$$T_A = T_0 \cosh\left(\frac{wl}{2T_0}\right) = T_0 + w\,S \qquad (6.31)$$

If wind and ice loads exist, F_t will replace w in all above equations.

Example 6.7

A 600 m span over a river is to have its two supports at the same level. The conductor weighs 12 N/m and ice load is 14 N/m. The tension in the conductor under the above conditions should not exceed 50000 N. Find the sag by (a) representing the line as a parabola (b) representing the line as catenary and compare the results.

Solution

(a) $$S = \frac{F_t l^2}{8T_0}$$

Total vertical load = 12 + 14 = 26 N/m

Horizontal load = 0

$$S = 26 \times 600 \times 600/(8 \times 50000) = 23.4 \text{ m}$$

(b) $$S = \frac{F_t l^2}{8T_0}\left[1 + \frac{l^2}{48}\left(\frac{F_t}{T_0}\right)^2\right] = \frac{26\times600\times600}{8\times50000}\left[1 + \frac{600\times600}{48}\left(\frac{26}{50000}\right)^2\right] = 23.447 \text{ m}$$

It is seen that even for this long span the difference between the results of exact and approximate formulae is only 0.2%. Therefore, the exact formulae are seldom used in practice.

6.11 CONDUCTOR VIBRATION

In addition to normal swinging in wind, overhead line conductors may be subject to vibrations which can be broadly classified as resonant vibrations (also known as Aeolian vibrations) and dancing or galloping.

6.11.1 Resonant Vibration

These vibrations have low amplitudes (upto 5 mm) and high frequencies (50 – 100 Hz). They are caused by vortex phenomenon in light winds (6 – 20 km/hr). The frequency f is empirically given by 50 (v/d) where v is wind velocity (km/hr) and d is conductor diameter (mm). The length of a loop (half wave length) depends on tension T(N) and conductor weight w (kg f/m) and is given by $\frac{1}{2f}\sqrt{\frac{T}{w}}$. Thus a 30 mm conductor weighing 1.6 kgf/m and strung to a tension of 50 kN will vibrate at about 50 Hz with a loop length of 1.8 m when the wind velocity is 30 km/hr. These vibrations are common to all conductors and are more or less always present. The ACSR conductor has low weight to diameter ratio and is subject to fatigue by these vibrations.

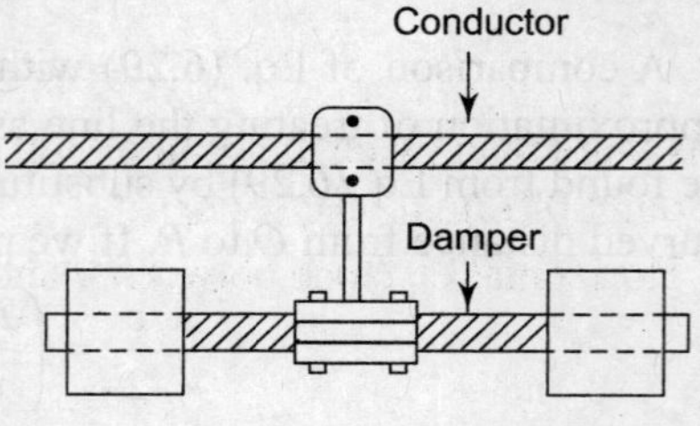

Fig. 6.11 Stockbridge damper

To minimize these vibrations, a stock bridge damper (Fig. 6.11) is used. It comprises two masses at the end of a short length of stranded steel cable suspended from the conductor about midway between two nodes of the vibrations, i.e., where the amplitude of the vibration would be a maximum. The vibration causes movement of the damper and energy is absorbed by the interstrand friction in the steel cable. A typical damper for use on 132 kV lines is about 60 cm long and weighs about 5 kg.

6.11.2 Galloping or Dancing Conductors

These vibrations have high amplitude (upto 3 m) and low frequency (1 –2 Hz) and are generally caused by asymmetrical layers of ice formation. When the ice coated conductor is acted upon by a light drift wind particularly where the ground slopes at right angles to the line, the wind travels up the slope and appears to get underneath the conductor and initiate vibrations. The stranding of conductors also con-tributes to these vibrations. The travel of the conductor follows the path of an ellipse. There is no method to prevent these vibrations but the adoption of horizontal conductor configuration reduces the danger due to these vibrations.

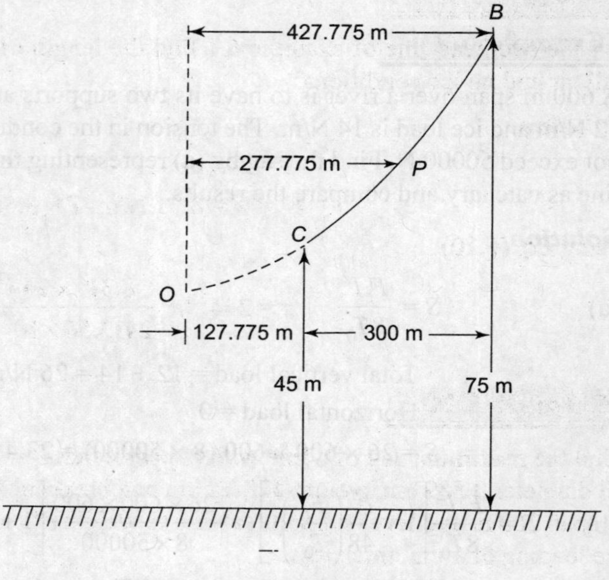

Fig. 6.12

Example 6.8

An overhead line spanning over a river is supported at two ends by towars of 45 m and 75 m. Find the clearance between the conductor and water at a point midway between towers. The tension is to be limited to 2500 kg. The weight of the conductor is 0.9 kg/m and the distance between the towers is 300 m.

Solution

Using the notation of section 6.6.5

$$l_c = l + \frac{2Th}{wl} = 300 + \frac{2\times 2500\times 9.81(75-45)}{0.9\times 9.81\times 300}$$

$$= 855.55 \text{ m}$$

l_c is more than twice the span of 300 m. This means that both the supports are on the same side of point 0 (Fig. 6.12).

Horizontal distance between 0 and C

$$= \frac{855.55}{2} - 300 = 127.775 \text{ m}$$

$$\text{Height of } C \text{ above } O = \frac{wx^2}{2T} = \frac{0.9\times 9.81\times 127.775^2}{2\times 2500\times 9.81} = 2.94 \text{ m}$$

Horizontal distance between O and P = 127.775 + 150 = 277.775 m

$$\text{Height of } P \text{ above } O = \frac{0.9\times 9.81\times (277.775)^2}{2\times 2500\times 9.81} = 13.89 \text{ m}$$

Height of midpoint P above C = 13.89 – 2.94 = 10.95 m

Height of midpoint P above water level = 10.95 + 45 = 55.95 m

Example 6.9

For the overhead line of example 6.1 find the length of conductor between two towers under still air and no ice conditions.

Solution

$$l = 244 \text{ m}$$

$$F_t = w = 8.31 \text{ N/m}, T = 3.56\times 10^4 \text{ N}$$

Using Eq. (6.10)

$$z = 244\left[1 + \frac{8.31^2\times 244^2}{24(3.56\times 10^4)^2}\right] = 244.033 \text{ m}$$

Example 6.10

Find the maximum sag of a line with copper conductor 7/0.295 cm size, area 0.484 cm², over all diameter 0.889 cm, weight 428 kg/km and breaking strength 1973 kg. Use a safety factor of 2, span 200 m and level supports (a) due to weight of conductor (b) due to additional weight of ice loading of 1 cm thickness. *(AMIE 2002)*

Solution

$$w = \frac{428}{1000} \times 9.81 = 4.199 \text{ N/m}$$

$$l = 200 \text{ m}$$

$$T = \frac{1973 \times 9.81}{2} = 9677.6 \text{ N}$$

(a) $$S = \frac{wl^2}{8T} = \frac{4.199(200)^2}{8(9677.6)} = 2.169 \text{ m}$$

(b) $$d = 0.889 \times 10^{-2} \text{ m}$$

$$t = 1 \times 10^{-2} \text{ m}$$

$$D = d + 2t = 2.889 \times 10^{-2} \text{ m}$$

$$w_i = 8920\,(D^2 - d^2)\,\frac{\pi}{4}$$

$$= 8920\left[\left(2.889 \times 10^{-2}\right)^2 - \left(0.889 \times 10^{-2}\right)^2\right]\frac{\pi}{4}$$

$$= 8920\left[8.346 \times 10^{-4} - 0.79 \times 10^{-4}\right]\frac{\pi}{4}$$

$$= 5.294 \text{ N/m}$$

$$w + w_i = 4.199 + 5.294 = 9.493 \text{ N/m}$$

$$S = \frac{(9.493)(200)^2}{8(9677.6)} = 4.9 \text{ m}$$

SUMMARY

6.1 Distribution lines are supported on poles. Transmission lines are supported on towers.

6.2 Poles are made of wood or RCC or steel tubes.

6.3 Towers are fabricated from galvanized iron angle sections.

6.4 Towers are classified as tangent towers and deviation towers.

6.5 Indian electricity rules specify clearances of line conductors.

6.6 Weight of conductor, ice and wind cause sag of overhead line conductors.

6.7 Because of sag, the length of conductor is more than span.

6.8 Tension in the conductor has to be kept within permissible limits. Lower the value of tension higher is the sag.

6.9 Stringing chart helps us to adjust line sag so that tension will be within safe limits even under bad weather conditions.

6.10 Sag template is a convenient tool to locate towers.

6.11 When a line sags between the supports, the exact shape is that of catenary. However, in most calculations the shape is assumed to be parabola.

6.12 Conductor vibrations can be classified as resonant vibrations and galloping.

SELF TEST QUESTIONS

A. State whether the following are true (T) or false (F)

6.1 RCC poles are used for overhead lines upto 33 kV.

6.2 Tangent towers are used at locations where a transmission line changes direction.

6.3 Crossarms are generally made of iron.

6.4 A flat horizontal configuration means wider right of way.

6.5 Ice coating increases the sag of conductor.

6.6 Increase in conductor tension increases the sag of conductor.

6.7 As temperature increases the sag in the conductor increases.

6.8 Sag is not affected by temperature.

6.9 Sag template can be used in the field to locate towers.

6.10 Resonant vibrations have low amplitudes.

B. Fill in the blanks with suitable words

6.11 Galloping vibrations have amplitudes about ______

6.12 The frequency of resonant vibrations is about ______

6.13 Stock bridge damper is used to reduce ______ vibrations.

6.14 A decrease in temperature ______ sag.

6.15 Sag is proportional to ______ of span.

6.16 The maximum tension in the conductor should be about ______ the breaking load.

6.17 An increase in tension ______ the length of conductor.

6.18 If V is wind velocity in km/hour, wind pressure is taken as ______

6.19 If d is diameter of conductor and t is thickness of ice, the overall diameter of ice coated conductor is ______.

6.20 The height of a 400 kV single circuit line with 2 sub conductors per phase and two earth wires is about ______.

ANSWERS

6.1 T	6.2 F	6.3 T	6.4 T	6.5 T	6.6 F
6.7 F	6.8 F	6.9 T	6.10 T	6.11 3 m	6.12 50 – 100 Hz
6.13 resonant	6.14 reduces	6.15 square	6.16 half	6.17 decreases	6.18 $0.059\ V^2$
6.19 $(d + 2t)$	6.20 30 m				

JOB INTERVIEW QUESTIONS

6.1 What are the factors which affect sag?

6.2 What is stock bridge damper?

6.3 How are conductor vibrations classified?

6.4 What are the effects of line configuration on tower height and right of way?

6.5 What is a deviation tower?

6.6 How is conductor spacing related to line voltage?

6.7 What is the effect of wind on sag?

REVIEW QUESTIONS

6.1 Discuss why?

(a) Modern high voltage lines have long spans.

(b) The lines are generally designed for worst probable but not worst possible weather conditions.

(c) Though the exact shape of a line conductor is that of a catenary, the sag calculations are usually done by assuming it to be a parabola.

(d) In some cases it becomes necessary to extend the base of supports of overhead lines.

(e) The sag calculations are usually based on equivalent span.

(f) High voltage lines are supported on steel towers and not on poles.

6.2 Discuss the considerations which govern the selection of span and conductor configuration of a high voltage line.

6.3 Assuming that the shape of an overhead line can be approximated by a parabola, deduce expressions for calculating sag and conductor length. How can the effect of wind and ice loadings be taken into account?

6.4 What is a stringing chart? How can it be prepared? What is its use?

6.5 Derive an equation to calculate the conductor tension under erection conditions if the conductor tension and loading under bad weather conditions are known.

6.6 (a) What is sag template? What is its use?

(b) Write a short note on conductor vibrations.

PROBLEMS

6.1 An overhead stranded galvanized steel conductor has a 183 m span. The conductor has 37 strands each of 0.259 cm. diameter. The weight of conductor is 7.15 N/m and the breaking strength is 67700 N. The factor of safety should be 2.5. Calculate the sag under ice and wind condition if the radial thickness of ice is 0.9 cm and the wind load is 382 N/m^2 of projected area (coated with ice). The weight of ice is 8920 N/m^2. [3.16 m]

6.2 An overhead line has a span of 122 m, the diameter of conductor is 1.15 cm. Calculate the sag at mid span when the conductors have an ice coating 0.96 cm thick and the wind pressure is 382 N/m^2, of projected area. The weight of conductor is 5.83 N/m, ice weighs 8920 N/m^3 and the permissible tension is 3.56×10^4 N. [0.865 m]

6.3 An overhead line has a span of 300 m. The conductor diameter is 1.953 cm and conductor weight is 0.844 kg/m. Calculate the vertical sag when a wind pressure of 736 N/m^2 of projected area acts on conductor. The breaking strength of conductor is 77990 N and the conductor tension should not exceed half the breaking strength. [2.39 m]

6.4 An overhead line with ACSR conductor and a span of 275 m is to be erected in still air conditions at a temperature of 50 °C. The conductor diameter is 1.953 cm, area is 226 mm^2, the breaking strength is 77990 N and conductor weight is 0.844 kg/m. It is desired that a factor of safety of 2 must be maintained under bad weather conditions when temperature is likely to be – 55 °C, wind pressure of 382 N/m^2 of projected are and an ice coating of 0.95 cm. thickness around the conductor. Find the tension and sag under erection conditions. Weight of ice 913 kg/m^3, coefficient of linear expansion 18.44 $\angle \times 10^{-6}$ per degree C, Young's modulus 91.43×10^3 N/mm^2.

[Tension 15650 N. sag 5 m]

6.5 An overhead line is erected on a gradient of 1 to 45. The height of supports is 7.62 m and horizontal span is 137 m. The conductor area is 0.893 cm^2 and maximum allowable stress 6900 N/

cm^2. Calculate the maximum sag relative to the lower support. Assume that conductor along with ice weights 5.1 N/m and wind pressure is 382 N/m^2 of projected area. [1.2 m]

6.6 Estimate the vertical sag for an overhead line with following data: ACSR conductor, diameter 2.226 cm, weight 1.097 kg/m, permissible tension 3.5×10^4 N, wind pressure 380 N/m^2 of projected area, ice coating 0.96 cm thick, weight of ice 8900 N/m^3, span 325 m. [7.29 m]

6.7 Estimate the sag for the line of problem 5.6 under still air conditions with no ice and wind and a permissible tension of 2×10^4 N. [7.10 m]

6.8 An overhead line at a river crossing is supported from two towers 20 m and 14 m above water level. The distance between towers is 250 m and conductor diameter is 1.036 cm. Find the minimum chlorines between conductor and water surface and the distance of the lowest point of conductor from lower level tower. Weight of conductor 590 kg/km, breaking load 2600 kg and factor of safety 2.5. [12.06 m, 82.71 m]

6.9 A high voltage line has a span of 350 m and is supported by towers having a level difference of 35 m. The ultimate strength of conductor is 8500 kg and factor of safety is 3.5. Find the clearance between conductor and ground at a point midway between the towers. The heights of towers are 50 m and 85 m above ground and weight of conductor is 0.82 kg per metre length. [62.33 m]

6.10 Find the total length of conductor between towers for the line of Problem 5.7. [325.414 m]

CHAPTER 7

Corona

7.1 THE PHENOMENON OF CORONA

Air is not a perfect insulator and contains a minute number of electrons and ions as a result of various effects such as ultra violet radiations from sun, cosmic rays, radio activity of soil, etc. When an electric gradient is set up in air, between two large parallel conducting planes, the electrons and ions are set in motion by the electric field and they maintain a small current by convection between the conducting planes. When the electric field intensity is less than 30 kV/cm, this current is very small and hence negligible.

The phenomenon undergoes a radical change when the electric field intensity reaches the critical value of 30 kV/cm. At this intensity, the ions attain a sufficiently high velocity and on striking another neutral molecule dislodge one (or more) electrons from this neutral molecule. This produces a new electron and a positive ion which in turn are accelerated until they collide with other molecules and produce more ions. An ion avalanche (Town-send effect) results. If the field is uniform everywhere (as in the case of parallel plate electrodes) the conditions necessary to produce such avalanches are reached simultaneously everywhere in the gap. Therefore, a complete electric breakdown occurs and an arc is established between the two electrodes.

The electric field in the vicinity of a cylindrical conductor is not uniform. The intensity is maximum at the surface of conductor and then decreases in inverse proportion to the distance from the centre of the conductor. As the voltage applied to the conductor is increased, a layer adjacent to the conductor gets ionized as soon as the electric field intensity at the surface of the conductor becomes more than 30 kV/cm. The field intensity must reach this value in a layer, around the conductor, at least as thick as an average free path before ionization can start. The ionization will be restricted only to this layer because outside this layer, the field intensity being less than 30 kV/cm, the ions will not have sufficient energy to cause further ionization. Consequently, in contrast to the case of uniform field, no complete breakdown of the gap will take place but a region of sustained ionization will persist around the conductor. This ionization is accompanied by a luminous glow around the conductor. If the conductors are rough or dirty, usually the brightest parts of the luminous envelope are near the rough or dirty spots. At the same time a hissing noise can be heard and the characteristic smell of ozone can be detected. The glow increases in size and brightness and the intensity of hissing noise increases with increase in voltage. The term corona is used to designate in particular, this visible luminous envelope and in general all the phenomena accompanying its formation. Corona occurs in power lines of 100 kV and above. The ions produced by the corona result in space charges which are being moved around by the ac field. The energy required for this movement is taken from the conductor and this energy loss is known as corona loss. The energy is dissipated in the form of light, heat, sound and chemical action. The rate of ionization in the corona discharge is not uniform but has certain fluctuations and sudden changes which produce similar other changes of the electric field. These changes can disturb radio reception and thus, bring, about the radio interference (RI) aspect of corona.

If a high direct voltage is applied between two conductors there is a difference in the appearance of the positive and negative wires. The positive wire has a uniform glow about it while the glow around the negative wire is more spotty and even sometimes in the form of short streamers from the rough or sharp spots on the wire. With alternating voltage applied to the conductors, the appearance of the two conductors is the same since the conductors are alternately positive and negative.

As the voltage applied to conductors is increased the corona envelopes grow larger and larger. The corona extends until the increase in the effective diameter of the conductor is sufficient to bring the potential gradient at the edge of corona down to the critical gradient and corona can then spread no further at this particular voltage under consideration. If the voltage is increased to a very high value the envelope grows so large that finally there is a spark over between the conductors. If the spacing between conductors is less than three times the conductor radius, the spark over may occur even before any corona is noticed.

7.2 DISRUPTIVE CRITICAL VOLTAGE

The potential difference between conductors, at which the electric field intensity at the surface of the conductor exceeds the critical value and generates corona is known as disruptive critical voltage. Corona begins when the peak value of critical field intensity equals 30 kV/cm or 3×10^6 V/m in case of smooth conductors of large diameter in air at normal temperature and pressure.

The critical intensity is proportional to density of air and, therefore, proportional to the barometric pressure and is inversely proportional to the absolute temperature. Taking the relative density at θ_0°C and 76 cm of mercury as 1.0, the relative density at a barometric pressure of p cm and a temperature of θ °C is

$$\delta = \frac{p}{76}\left(\frac{273+\theta_0}{273+\theta}\right) \tag{7.1a}$$

Taking $\theta_0 = 20$ °C, $\delta = \dfrac{3.86\,p}{273+\theta}$ (7.1b)

The surface of the conductor also affects the critical intensity. A roughness of conductor surface causes a distortion in electric field and gives rise to local regions of high potential gradient resulting in a lowering of the critical intensity. Stranding and the condition of surface are taken into account by the use of a surface factor (also known as roughness factor) m_0. Thus, after taking into account the effect of pressure, temperature and conductor surface, the r.m.s. value of critical intensity can be written as

$$E_0 = \left(3\times\frac{10^6}{\sqrt{2}}\right)\cdot\delta\cdot m_0 \ \text{V/m} \tag{7.2}$$

where $m_0 = 1.0$ for smooth conductors and 0.93 to 0.98 for rough conductors exposed to atmosphere.

From Eq. (2.64) the electric intensity E_r at the conductor surface of radius r metres is

$$E_r = q/(2\pi\,\varepsilon_0 r) \quad \text{volts/m} \tag{7.3}$$

where q is the charge on the conductor per metre length and ε_0 the permittivity of free space.

From Eq. (2.67) the voltage V between the two conductors of a single phase line, each of radius r metres and spaced a D metres, is

$$V = \frac{q}{\pi\,\varepsilon_0} \ln \frac{D}{r} \tag{7.4}$$

Combining Eqs. (7.3) and (7.4)

$$V = 2\,r\,E_r \ln (D/r) \tag{7.5}$$

When E_r reach the value E_0, the voltage between the conductors is the disruptive critical voltage V_d. Substituting the value of E_0 from Eq. (7.2), for E_r in Eq. (7.5) the equation of disruptive critical voltage for a single phase line is given by

$$V_d = \frac{6\times 10^6}{\sqrt{2}}\, r\,\delta\, m_0 \ln \frac{D}{r} \text{ volts} \tag{7.6}$$

For a 3-phase line with phase (line to neutral) voltage V and equivalent spacing D_{eq}, Eq. (2.88) gives

$$V = \frac{q}{2\pi\,\varepsilon_0} \ln \frac{D_{eq}}{r} \tag{7.7}$$

Combining Eqs. (7.3) and (7.7)

$$V = r\,E_r \ln \frac{D_{eq}}{r} \tag{7.8}$$

When E_r becomes equal to E_0, voltage V is the disruptive critical voltage V_d. Thus, for a 3-phase line with equivalent spacing D_{eq}, the phase value of critical disruptive voltage is

$$V_d = \frac{3\times 10^6}{\sqrt{2}}\, r\,\delta\, m_0 \ln \left(\frac{D_{eq}}{r}\right) \text{ volts} \tag{7.9}$$

7.3 VISUAL CRITICAL VOLTAGE

Experimental investigations have revealed that visual corona does not occur when the electric intensity becomes equal to critical value E_0 but starts at a higher value of intensity E_v. This is because the dielectric breakdown of air requires a finite volume of over stressed air. In other words, a finite amount of dielectric energy is necessary to cause visual corona. The empirical relation for calculating E_v is

$$E_v = 3\times \frac{10^6}{\sqrt{2}}\,\delta\, m_v \left(1 + \frac{0.03}{\sqrt{\delta r}}\right) \text{ V/m} \tag{7.10}$$

Where r is the conductor radius in metres and m_v is the surface factor for calculating visual critical voltage. m_v is generally called irregularity factor.

m_v = 1.0 for smooth conductors, 0.93 to 0.98 for rough conductors exposed to air, 0.72 for local corona on stranded conductors, and 0.82 for decided corona on stranded conductors.

Substituting E_v from Eq. (7.10) in place of E_r in Eqs. (7.5 and 7.8) the visual critical voltage V_v for single phase and 3-phase lines, can be obtained as;

$$V_v = \frac{6\times10^6}{\sqrt{2}} r\,\delta\,m_v \left(1+\frac{0.03}{\sqrt{\delta r}}\right)\ln\frac{D}{r} \quad \text{volts (for single phase lines)} \tag{7.11}$$

and

$$V_v = \frac{3\times10^6}{\sqrt{2}} r\,\delta\,m_v \left(1+\frac{0.03}{\sqrt{\delta r}}\right)\ln\frac{D_{eq}}{r} \quad \text{volts (for 3-phase lines)} \tag{7.12}$$

Note that r and D are to be taken in metres in the above equations.

7.4 CORONA LOSS

Formation of corona is associated with loss of power which affects the efficiency of line. However, the corona loss does not have any appreciable effect on voltage regulation of line. Peek determined the following empirical formula for calculating corona loss under fair weather conditions.

$$P_c = 243.5\,\frac{(f+25)}{\delta}\sqrt{\frac{r}{D}}\,(V-V_d)^2\times10^{-5}\ \text{kW/km/phase} \tag{7.13}$$

where
V = Phase voltage in kV (rms value)
V_d = Disruptive critical voltage in kV (rms value)
r = radius of conductor in metres
D = Spacing (or equivalent spacing) between conductors in metres
f = system frequency

The corona loss under stormy weather conditions is calculated by taking V_d as 80% of its fair weather value.

Peek's formula for corona loss is valid when corona losses are predominant and the ratio (V/V_d) is greater than 1.8. When this value is less than 1.8 the following formula, known as Peterson's formula, gives better results.

$$P_c = \frac{21\times10^{-6} f V^2}{\left(\log_{10}\dfrac{D}{r}\right)^2}\times F \quad \text{kW/phase/km} \tag{7.14}$$

where V is the phase voltage in kV and F is a factor which varies with the ratio (V/V_d) as under

(V/V_d)	0.6	0.8	1.0	1.2	1.4	1.6	1.8	2.0	2.2
F	.012	.018	.05	.08	0.3	1.0	3.5	6.0	8.0

It has been found that the results obtained by the above two methods (Eqs. 7.13 and 7.14) may be very different. The insulator losses can approach the same order of magnitude as the conductor corona loss. Insulator losses are not included in Peterson's equation. Corona losses under bad weather conditions, may be many times more than that under fair weather conditions.

Example 7.1

Find the disruptive critical voltage and visual corona voltage (local corona as well as general corona) for a 3-phase 220 kV line consisting of 22.26 mm diameter conductors spaced in a 6 m delta configuration. The following data can be assumed.

Temperature 25°C, pressure 73 cm of mercury, surface factor 0.84, irregularity factor for local corona 0.72, irregularity factor for general (decided) corona 0.82.

Solution

$$r = 0.5 \times 22.26 = 11.13 \text{ mm} \quad \text{or } 1.113 \times 10^{-2} \text{ m.}$$

$$\delta = \frac{3.86 \times 73}{273 + 25} = 0.9456$$

$$m_0 = 0.84, \quad D = 6 \text{ m}$$

$$V_d = \frac{3 \times 10^6}{\sqrt{2}} r\, \delta\, m_0 \ln \frac{D}{r} \text{ volts}$$

$$V_d = \frac{3 \times 10^6}{\sqrt{2}} \times 1.113 \times 10^{-2} \times 0.9456 \times 0.84 \ln \frac{6.0}{1.113 \times 10^{-2}}$$

$$w = 11.8 \times 10^4 \text{V} \quad \text{or} \quad 118 \text{ kV per phase}$$

For local visual corona, $m_v = 0.72$ and

$$V_v = \frac{3 \times 10^6}{\sqrt{2}} \times 1.113 \times 10^{-2} \times 0.9456 \times 0.72 \times \left(1 + \frac{0.03}{\sqrt{(0.9456 \times 1.113 \times 10^{-2})}}\right) \ln \frac{6.0}{1.113 \times 10^{-2}}$$

$$W = 13.07 \times 10^4 \text{ V} \quad \text{or} \quad 130.7 \text{ kV per phase}$$

For general visual corona $m_v = 0.82$ and

$$V_v = \frac{3 \times 10^6}{\sqrt{2}} \times 1.113 \times 10^{-2} \times 0.9456 \times 0.82 \times \left(1 + \frac{0.03}{\sqrt{(0.9456 \times 1.113 \times 10^{-2})}}\right) \ln \frac{6.0}{1.113 \times 10^{-2}}$$

$$V = 14.88 \times 10^4 \text{ V} \quad \text{or} \quad 148.8 \text{ kV/phase.}$$

Example 7.2

The 3-phase 220 kv, 50 Hz line of example 7.1 is 250 km long. Find the total loss in fair weather and bad weather using Peek's formula.

Solution

For fair weather conditions, $V = 220/\sqrt{3} = 127$ kV per phase and $V_d = 118$ kV per phase (as found in example 7.1)

$$P_c = 243.5 \left(\frac{f + 25}{\delta}\right) \sqrt{\frac{r}{D}}\, (V - V_d)^2 \times 10^{-5} \text{ kW/km/phase}$$

$$= 243.5 \left(\frac{50 + 25}{0.9456}\right) \sqrt{\left(\frac{1.113 \times 10^{-2}}{6}\right)}\, (127\text{–}118)^2 \times 10^{-5} \text{ kW/km/Phase}$$

$$= 0.6738 \text{ kW/ phase/km}$$

Corona loss per phase $= 0.6738 \times 250 = 168.45$ kW/phase

Total corona loss $= 168.45 \times 3 = 505.35$ kW

For bad weather, $V_d = 0.8 \times 118 = 94.4$ kV/phase

$$P_c = 243.5\left(\frac{50+25}{0.9456}\right)\sqrt{\frac{1.113\times10^{-2}}{6}}(127-94.4)^2 \times 10^{-5} \text{ kW/km/phase}$$

$$= 8.84 \text{ kW/phase/km}$$

Corona loss per phase $= 8.84 \times 250 = 2210$ kW/phase

Total corona loss $= 3 \times 2210 = 6630$ kW

Example 7.3

A 3-phase overhead line consists of 3 conductors in equilateral formation with 2.44 m spacing. The conductor diameter is 1.04 cm and surface factor (m) is 0.85. The air temperature and pressure are 21.1°C and 74 cm of mercury. Find visual critical corona voltage

Solution

$$r = 0.5 \times 1.04 = 0.52 \text{ cm} = 0.52 \times 10^{-2} \text{ m}$$

$$\delta = \frac{3.86\times74}{273+21.1} = 0.971$$

using Eq. (7.12)

$$V_v = \frac{3\times10^6}{\sqrt{2}} \times 0.52\times10^{-2} \times 0.971 \times 0.85$$

$$\times\left(1+\frac{0.03}{\sqrt{0.971\times.52\times10^{-2}}}\right)\ln\frac{2.44}{.52\times10^{-2}} = 79.65 \text{ kV/phase}$$

7.5 FACTORS AND CONDITIONS AFFECTING CORONA LOSS

Corona loss depends on a number of factors, viz., system frequency system voltage, conductivity of air, density of air, presence of rain, dust, radius of conductor, surface of conductor, and heating of conductor by load current.

Experimental investigations have revealed that the effect of these factors is as under:

1. *Effect of Frequency*: From Eq. (7.13) it is quite evident that corona loss is directly proportional to system frequency.

2. *Effect of System Voltage*: The potential difference between the conductors governs the electric field in the space around the conductors. Greater the potential difference, greater is the electric field and greater is the corona loss. In the region near the disruptive critical voltage, the rate of increase of corona loss with increase in system voltage is small. As seen from Eqs. (7.13) and (7.14) when V is large as compared with V_d, corona loss increases at a very fast rate with increase in voltage.

3. *Effect of Conductivity of Air*: The conductivity of air depends on the number of ions per unit volume of air, the size and charge per ion. Both these factors vary with altitude and atmospheric conditions. At sea level the number of ions per c.c of air is about 650 (positive) and 550 (negative). At 2000 m above MSL (mean sea level) this number is about 1050 (positive) and 850 (negative) while 5000 m above MSL this number is about 2350 (positive) and 2000 (negative) Thunder storms increase the ion content and make the atmosphere more conducting. Higher conductivity leads to greater corona loss.

4. *Effect of Density of Air*: Corona loss increases with decrease in density of air. The corona loss of a line passing through a hilly area is higher than that of a similar line in plains due to the decreased value of δ at high altitudes.

5. *Effect of Rain, Dust, etc.*: Rain and dust increase the corona loss since rain drops and dust particles settling on the conductor surface establish protruding points which become the regions of local corona discharges. The effect is more pronounced if the conductor diameter is large because on small conductors the radius of a rain drop may not be very much smaller than the radius of the conductor itself.

6. *Effect of Conductor Radius*: A higher conductor radius means lesser surface field intensity and, hence, reduced corona loss. An ACSR conductor, for the same current capacity, has a larger radius than copper conductor. Therefore, ACSR conductor lines have lower corona loss than copper conductor lines. In bundled conductor lines the effective radius is high and, therefore these lines depict lower corona loss than the lines with one conductor per phase.

7. *Effect of Conductor Surface*: The potential gradient at the surface of stranded conductor is greater than that for the equivalent solid conductor. As a result of this the breakdown voltage is low and corona loss is high for stranded conductors. Any roughness on the conductor surface causes a field distortion and gives rise to local regions of high potential gradient and higher corona loss as compared to smooth surface. Stranding and roughness of surface are taken into account by the use of surface factor m_0. When a new line is put into operation, a gradual reduction in corona, due to weathering effect, takes place for the first six months or so. The ionization of air produces ozone. The oxidation of conductor due to ozone is more pronounced at roughnesses protruding from the conductor surface. Consequently, the surface irregularities get blunted and corona loss gets reduced. This reduction in corona loss takes place during the first six months or so. After this time the loss remains almost constant.

8. *Effect of Conductor Heating by Load Current*: The load current raises the conductor temperature leading to an indirect reduction in corona loss. If the conductor temperature is low, tiny dew drops collect on the conductor surface during fog and humid weather and cause larger corona loss. The heating of the conductor due to load current prevents such condensation and reduces corona loss.

7.6 RADIO INTERFERENCE DUE TO CORONA

The radio interference due to corona refers to the adverse effect produced by corona on wireless signals. It has been found that the current pulses generated by corona discharges, as they propagate along the conductor, cause interference with the wireless signals. Amplitude modulated broadcasting, power line carrier, aviation, marine, ship to shore SOS calls and other similar services are affected due to this interference. The interfering field, at any point, is dominated by the aggregation of the effect of all corona discharges spread over many kilometers of line on both sides of the point and not by the direct effect of local discharges at that point. The increase in radio interference field is gradual upto a voltage slightly below the minimum voltage at which measurable corona loss occurs. Above this voltage the increase in radio interference is very rapid. The rate of increase in radio interference is affected by conductor surface and its diameter, which is higher in case of smooth and large diameter conductors. Above a certain voltage the magnitude of radio interference field tends to level off. However for practical conductors the levelling off value is too high to be acceptable and the lines have to be designed to operate below the voltage at which radio interference starts rapidly during fair weather.

A transmission line with clean conductors, with good hardware and having the fair weather gradient less than corona inception gradient, should not present radio interference problems during fair weather. However, experience has shown that line conductors are always not clean and radio interference may occur. It is known that mechanical defects like scratches raised from dents and other surface discontinuities exist. The weather processes attack these projections and corona activity hastens the attack. After these initial conditions the conductor surface intercepts or attracts a varying amount of airborne material. Some material adheres to the conductor surface and forms a transient source of radio interference generation. Later on, the remains of airborne particles cease as corona sources and the remains become a layer of semi-carboniferrous material producing a darkish coat on the conductor surface. This layer can be beneficial since it may cover some of the initial mechanical imperfections. Thus, during the life of a line, a gradual surface stabilization occurs with respect to corona sources but the degree of this surface stabilization is different for different lines.

In determining the RI (Radio Interference) performance of a proposed line the magnitude of RI factors for the entire frequency range of communication sources has to be evaluated. An evaluation of these factors in terms of their effect on the various communication services must take into account many factors like available signal intensities along the line, satisfactory signal to noise ratios, effect of weather on RI factors and the importance of the communication services, etc.

The radio interference field generated from a transmission line varies mostly as the inverse of the radio frequency. Thus, services in the higher frequency band (television, frequency modulated broadcasting, radar, micro wave relay, etc.) are less likely to be affected. The directional antenna which are generally used at these frequencies increase the signal to noise radio.

The degree of acceptable radio noise is not entirely technical decision and depends on factors like density of population, radio listening habits of people, powers of radio transmitters, other noise levels present in the area, and effective coverage, etc.

The radio interference due to corona assumes importance for lines above 200 kV. For EHV lines the line design is affected to a large extent by this phenomenon. A method to assess radio interference in *EHV* lines is discussed in chapter 14.

7.7 PRACTICAL CONSIDERATIONS

Corona has many disadvantages. The corona loss reduces the transmission efficiency though this reduction is quite small in well designed lines. The loss is quite high during bad weather but the overall effect is of little importance. Another disadvantage is that the leakage current due to corona takes place only during peak periods of voltage waves. This non-sinusoidal current causes a non-sinusoidal voltage and large third harmonics are produced. Harmonics are particularly troublesome in causing inductive interference with parallel communications circuits. Ozone, produced by corona, may cause deterioration in any organic material near the line.

An advantage of corona is to cause attenuation of travelling waves due to lightning. The charges induced on the line by lightning (and other causes) are partially dissipated as corona loss. In this way corona acts as a safety valve.

It is considered satisfactory to design a line for a critical voltage which is about 10% higher than operating voltage. In such a line some corona loss will take place during bad weather but the effect will be negligible. In a region where there are only a few bad weather periods it might be better to allow considerable corona loss during these bad weather periods so as to enjoy the continuous benefit of higher voltage without increasing the cost of line construction appreciably. The critical voltage of a line can be increased by increasing the spacing or the size of the conductor. The increase in spacing is not desirable because it increases the tower costs and inductive reactance drop. The use of higher size conductor is sometimes considered desirable with a view to increase the critical voltage. However, many designers consider it more economical to tolerate the corona losses to some extent, than to eliminate them absolutely by using heavy conductors. The radio interference aspect of corona is a very important consideration (more important than even corona loss) in evaluating line design.

7.8 CORONA IN BUNDLED CONDUCTOR LINES

The maximum surface electric field is less for bundled conductor lines than for single conductor lines. It is, therefore, possible, by replacing a single conductor by two or more conductors with the same total cross-section to obtain an increase in corona starting voltage. This leads to reduction in corona loss and radio interference. Alternatively a higher working voltage can be used.

7.8.1 Comparison of Maximum Surface Electric Field for Bundled Conductor and Single Conductor Lines

(a) Bundled Conductor Line: Figure 7.1 shows a single circuit line with two sub-conductors per phase. The radius of each sub-conductor is r. The spacing between the sub-conductors of a phase is d and the spacing between adjacent phases is D. A simplified analysis is possible by assuming that the sub-conductors of one phase have the same total charge on them at any instant and that these charges are distributed uniformly over the surface.

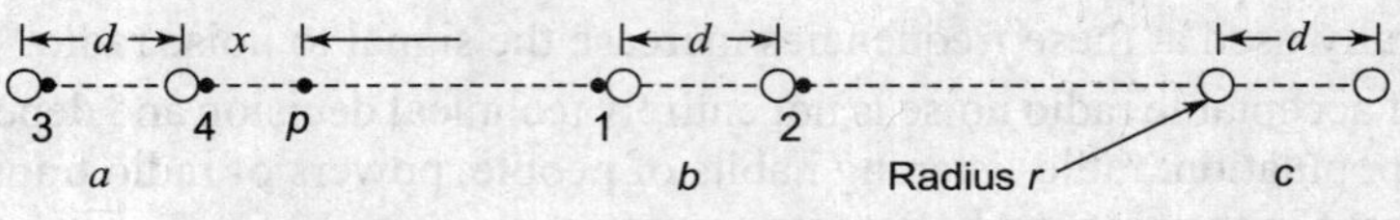

Fig. 7.1 Three-phase single circuit line with two sub-conductors per phase and horizontal configuration

Maximum stress will occur on the surface of a conductor on the line joining the conductor centres at the instant when the voltage on that phase is a maximum with respect to the neutral. The expressions for maximum surface fields at points 1, 2, 3 and 4 can be written from principles of electrostatics and it will be found that $E_4 > E_1\ (=E_2) > E_3$.

When the potential of the sub-conductors of phase a has its maximum value $\hat{V}_{an}$ (with respect to neutral), the potential difference between phases a and b is 0.866 times its maximum value, i.e., $v_{ab} = 0.866\ \hat{V}_{ab}$. If each sub-conductor of phase a has charge $\hat{q}$ at this instant, then the charge on each sub-conductor of phases b and c is $-\hat{q}/2$. Using the Eq. (7.3) and the principle of superposition, the electric intensity at surface of sub-conductor 4, at this instant, can be written as:

$$E_4 = \frac{\hat{q}}{2\pi\varepsilon_0}\left[\frac{1}{r} + \frac{1}{r+d} + \frac{1}{2(D-d-r)} + \frac{1}{2(D-r)} + \frac{1}{2(2D-d-r)} + \frac{1}{2(2D-r)}\right] \tag{7.15}$$

The electric intensity E_x at point p situated at a distance x from the centre of sub-conductor 4 is

$$E_x = \frac{\hat{q}}{2\pi\varepsilon_0}\left[\frac{1}{x} + \frac{1}{x+d} + \frac{1}{2(D-d-x)} + \frac{1}{2(D-x)} + \frac{1}{2(2D-d-x)} + \frac{1}{2(2D-x)}\right] \tag{7.16}$$

At this instant the potential difference between phases a and b is;

$$v_{ab} = 0.866\ \hat{V}_{ab} = \int_{r}^{D-d-r} E_x\, dx \tag{7.17}$$

Substituting the expression for E_x from Eq. (7.16) into Eq. (7.17) and integrating we have:

$$0.866\ \hat{V}_{ab} = \frac{\hat{q}}{4\pi\varepsilon_0}\left[3\ln\frac{D-d-r}{r} + 3\ln\frac{D-r}{d+r} - \ln\frac{D+r}{2D-d-r} - \ln\frac{D+d+r}{2D-r}\right] \tag{7.18}$$

Substituting the value of $\hat{q}$ from Eq (7.18) into Eq. (7.15)

$$E_4 = \left[\frac{\sqrt{3}\hat{V}_{ab}}{3\ln\dfrac{D-d-r}{r} + 3\ln\dfrac{D-r}{d+r} - \ln\dfrac{D+r}{2D-d-r} - \ln\dfrac{D+d+r}{2D-r}}\right]$$
$$\times\left[\frac{1}{r} + \frac{1}{r+d} + \frac{1}{2(D-d-r)} + \frac{1}{2(D-r)} + \frac{1}{2(2D-d-r)} + \frac{1}{2(2D-r)}\right] \tag{7.19}$$

(b) Single Conductor Line: Figure 7.2 shows a 3-phase line with one conductor per phase. If the total cross-sectional area per phase for this line to be the same as for the bundled conductor line of Fig. 7.1, then the radius of each conductor of this line must be $\sqrt{2}r$.

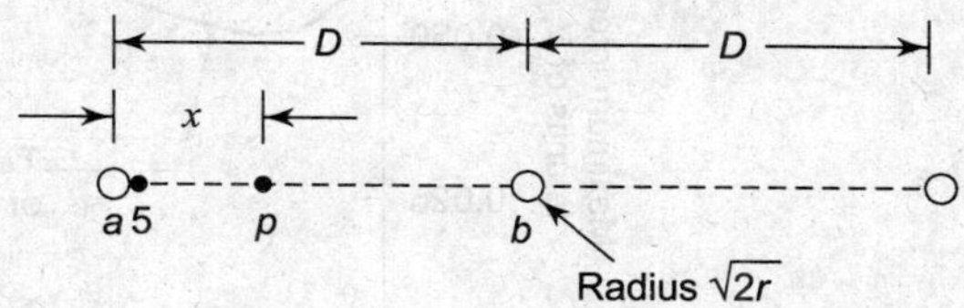

Fig. 7.2 Three-phase single circuit line with one conductor per phase and horizontal configuration

The maximum electric intensity at point 5 occurs at instant when the potential of phase a (with respect to neutral) has its maximum value $\hat{V}_{an}$. At this instant $v_{ab} = 0.866\ \hat{V}_{ab}$. If the charge on conductor a is $\hat{q}_1$, the charge on conductors b and c is $\frac{-\hat{q}_1}{2}$. The maximum electric intensity at point 5 is:

$$E_5 = \frac{\hat{q}_1}{2\pi\varepsilon_0}\left[\frac{1}{\sqrt{2}r} + \frac{1}{2(D-\sqrt{2}r)} + \frac{1}{2(2D-\sqrt{2}r)}\right] \tag{7.20}$$

The electric intensity at point p situated at a distance of x from the centre of conductor a is:

$$E_x = \frac{\hat{q}_1}{2\pi\varepsilon_0}\left[\frac{1}{x} + \frac{1}{2(D-x)} + \frac{1}{2(2D-x)}\right] \tag{7.21}$$

At this instant the potential difference between phases a and b is:

$$v_{ab} = 0.866\ \hat{V}_{ab} = \int_{\sqrt{2}r}^{D-\sqrt{2}r} E_x\ d_x \tag{7.22}$$

Substituting the expression for E_x from Eq. (7.21) into (7.22) and integrating

$$0.866\ \hat{V}_{ab} = \frac{\hat{q}_1}{4\pi\varepsilon_0}\left[3\ln\frac{D-\sqrt{2}r}{\sqrt{2}r} + \ln\frac{2D-\sqrt{2}r}{D+\sqrt{2}r}\right] \tag{7.23}$$

Substituting the value of $\hat{q}_1$ from Eq. (7.23) into Eq. (7.20)

$$E_5 = \frac{\sqrt{3}\hat{V}_{ab}}{3\ln\dfrac{D-\sqrt{2}r}{\sqrt{2}r} + \ln\dfrac{2D-\sqrt{2}r}{D+\sqrt{2}r}} \times \left[\frac{1}{\sqrt{2}r} + \frac{1}{2(D-\sqrt{2}r)} + \frac{1}{2(2D-\sqrt{2}r)}\right] \quad (6.24)$$

By using Eqs. (7.19) and (7.24) the maximum electric intensities for a 3-phase bundled conductor line (with two sub-conductors per phase) and 3-phase line with one conductor per phase can be compared. Figure 7.3 shows this comparison for $\frac{D}{r} = 700$ which is a typical value of $\frac{D}{r}$ for EHV lines. As seen from Fig. 7.3, there is a range of value of $\frac{d}{r}$ for which the intensity in bundled conductor line is less than that for a line with one conductor per phase. Calculations reveal that a reduction upto 11% can take place in this case.

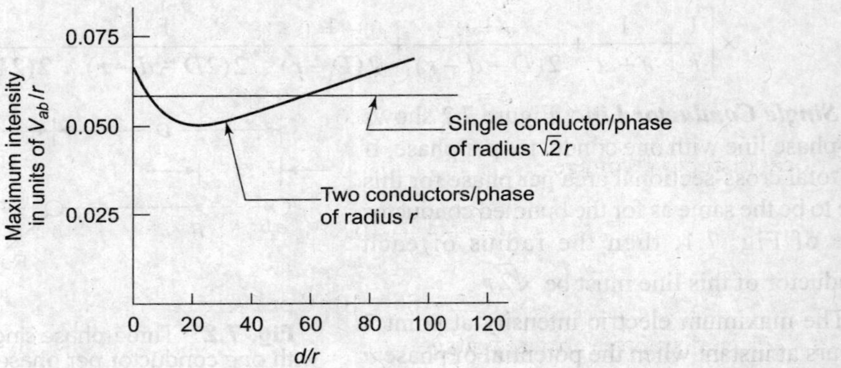

Fig. 7.3 Variation of maximum electric intensity with the ratio $\frac{d}{r}$ for a line with $\frac{D}{r} = 700$

Example 7.4

A 3 phase over head line has conductors of 30 mm diameter and arranged in the form of equilateral triangle. Assume fair weather conditions, air density factor of 0.95 and irregularity factor of 0.95. Find minimum spacing between conductors if disruptive critical voltage is not to exceed 230 kV between lines. Breakdown strength of air is 30 kV/cm (peak).

Solution

$$V_d = \frac{3\times10^6}{\sqrt{2}} r\delta m_0 \ln\left(\frac{D}{r}\right) \text{ volts}$$

$$\frac{230\times10^3}{\sqrt{3}} = \frac{3\times10^6}{\sqrt{2}}\left(\frac{30}{2}\times10^{-3}\right)(0.95)(0.95)\ln\left(\frac{D}{r}\right)$$

or
$$\ln\left(\frac{D}{r}\right) = 4.62$$

$$\frac{D}{r} = 101.49$$

$$D = (101.49)\left(\frac{30}{2}\times 10^{-3}\right) = 1.52 \text{ m}$$

Example 7.5

Find (i) critical disruptive voltage (ii) visual critical disruptive voltage (iii) corona loss under foul weather condition for 3 phase line 160 km long conductor diameter 1.036 cm, 2.44 m delta spacing, air temperature 26.6°C corresponding to approximate barometric pressure of 73.15 cm of mercury, operating voltage 110 kV at 50 Hz, surface irregularity factor 0.85, Assume $m_v = 0.72$

[Mumbai Uni. 2010]

Solution

(i) Critical disruptive voltage

$$= \frac{3\times 10^6}{\sqrt{2}}\; r\,\delta\, m_0\, ln\left(D_{eq}/r\right)$$

$$D_{eq} = 2.44$$

$$r = 0.5 \times 1.036 = 0.518 \text{ cm},\ m_0 = 0.85$$

$$\delta = \frac{3.86(73.15)}{273 + 26.6} = 0.9424$$

Critical disruptive voltage

$$= \frac{3\times 10^6}{\sqrt{2}}\left(0.518\times 10^{-2}\right)\left(0.9424\times 0.85\right)\left[ln\frac{2.44}{0.518\times 10^{-2}}\right]$$

$$= 5.4185 \times 10^4 \text{ V or } 54.185 \text{ kV/phase}$$

(ii) Visual disruptive critical voltage

$$= \frac{3\times 10^6}{\sqrt{2}}\left(0.518\times 10^{-2}\right)\left(0.9424\times 0.72\right)$$

$$\times\left[1+\frac{0.03}{\sqrt{0.9424\times 0.518\times 10^{-2}}}\right]\times ln\left[\frac{2.44}{0.518\times 10^{-2}}\right]$$

$$= 65.6 \text{ kV}$$

(iii) Assume that under bad weather

$$V_d = 0.8 \times 54.185 = 43.348 \text{ kV}$$

$$\text{System voltage} = 110/\sqrt{3} = 63.51 \text{ kV}$$

$$\text{Corona loss} = 243.5\left(\frac{f+25}{\delta}\right)\sqrt{\frac{r}{D}}\left(V - V_d\right)^2 \times 10^{-5} \text{ kW/km/phase}$$

$$= 243.5\left(\frac{50+25}{0.9424}\right)\left(\frac{0.518\times 10^{-2}}{2.44}\right)^{0.5}(63.51 - 43.348)^2 \times 10^{-5}$$

$$= 3.62 \text{ kW/km/phase}$$

$$\text{Total corona loss} = 3.62 \times 3 \times 160 = 1737.6 \text{ kW}$$

SUMMARY

7.1 When the electric field intensity near a conductor surface becomes more than 30 kV/cm, the air in the vicinity of conductor breaks down. This phenomenon is called corona.

7.2 Disruptive critical voltage depends on conductor radius, condition of conductor surface and relative density of air.

7.3 Visual critical voltage is higher than disruptive critical voltage.

7.4 Corona loss can be calculated by Peak's formula or Peterson's formula.

7.5 Corona loss increases with system frequency.

7.6 Rain and dust increase corona loss.

7.7 Corona causes radio interference.

7.8 Corona loss in bad weather is more than that in fair weather.

7.9 Thunderstorms increase corona loss.

7.10 Use of bundled conductors reduces corona loss.

SELF TEST QUESTIONS

A. State whether the following are true (T) or false (F).

7.1 Corona loss increases with increase in density of air.

7.2 A bigger conductor radius means lesser corona loss.

7.3 Corona loss is directly proportional to system frequency.

7.4 Visual critical voltage is less than disruptive critical voltage.

7.5 Relative air density increases with the pressure of air.

7.6 The electric field in the vicinity of conductor is uniform.

7.7 As voltage applied to conductor is increased the corona envelope grows bigger.

7.8 Corona in bundled conductor lines is lesser than in lines with one conductor per phase.

B. Fill in the blanks with suitable words.

7.9 Corona causes _________ of travelling waves.

7.10 Increase in conductor temperature _________ corona loss.

7.11 Rain _________ corona loss.

7.12 As conductor radius is increased, corona loss _________.

7.13 Thunderstorms _________ corona loss.

7.14 A suitable value of conductor roughness factor is about _________.

7.15 Visual critical voltage is _________ than disruptive critical voltage.

ANSWERS

7.1 F	7.2 T	7.3 T	7.4 F	7.5 T
7.6 F	7.7 T	7.8 T	7.9 attenuation	7.10 reduces
7.11 increases	7.12 decreases	7.13 increase	7.14 0.8	7.15 more

JOB INTERVIEW QUESTIONS

7.1 What is meant by corona? Why does it occur?
7.2 What are the factors which affect corona loss?
7.3 What is the difference between visual critical voltage and disruptive critical voltage?
7.4 What is meant by radio interference?
7.5 How can corona loss be calculated?
7.6 What is meant by roughness factor?
7.7 What are the disadvantages of corona?

REVIEW QUESTIONS

7.1 Give reasons for the following:

(a) When the voltage applied between two parallel plate electrodes is gradually increased to a high value, corona does not occur but a complete breakdown of air occurs and an arc is established between the plates.

(b) There is usually a difference in the appearance of positive and negative conductors of a high voltage dc line.

(c) The disruptive critical voltage is less than visual critical voltage.

(d) The disruptive critical voltage for a rough conductor is less than that for smooth conductor of the same radius.

(e) Corona loss is more in stormy weather than during fair weather.

7.2 Discuss why:

(a) Lines with aluminium conductor depict less corona than similar lines with copper conductor.

(b) Corona loss in a line in a hilly area is more than that in a similar line in plains.

(c) When a new line is put into operation, the corona loss in the initial few months is more than that later on.

(d) Bundled conductor lines have less corona loss than the lines with one conductor per phase.

(e) Corona introduces third harmonics in the system.

(f) The heating of conductor due to load current reduces corona loss.

7.3 Describe the phenomenon of corona. Discuss the factors which affect corona loss.

7.4 Define disruptive critical voltage and visual critical voltage. On what factor do they depend? Write equations for calculating these voltages.

7.5 What is corona loss? Why is it different in different weather conditions? How can it be estimated?

7.6 Derive an equation for calculating the maximum electric intensity on the conductor surface of a 3-phase single circuit horizontal configuration line with two sub-conductors per phase.

7.7 What are the disadvantages of corona? Explain how the corona considerations affect the design of a line.

7.8 Write a short note on radio interference due to corona.

PROBLEMS

7.1 Find the disruptive critical voltage and visual critical voltage (for local and general corona) for a 3-phase line consisting of 21 mm diameter conductors spaced in a 6 metre delta configuration. Take temperature 25 °C, pressure 73 cm of mercury, surface factor 0.85, irregularity factor for local visual corona 0.72 and irregularity factor for general (decided) visual corona 0.82.

[113.67 kV/phase, 125.27 kV/phase, 142.67 kV/phase]

7.2 Using Peterson's formula calculate the total corona loss for a 3-phase 220 kV, 200 km long line having D/r ratio of 650. Take factor F as 0.05. [64.2 kW]

7.3 Find the disruptive critical voltage and visual critical voltage (for local and general corona) for 3-phase line having 1 cm diameter conductors spaced in 3 m delta arrangement. Assume temperature 26 °C, pressure 74 cm of mercury, surface factor 0.85 and irregularity factors for local visual corona 0.72 and for general visual corona 0.82.

[55.6 kV/phase, 67.5 kV/phase, 76.9 kV/phase]

7.4 Using Eq. (7.19) plot the variation of maximum electric field intensity with d/r ratio for a 400 kV bundled conductor line having the following data: radius of each sub-conductor 1.573 cm, number of subcontractors phase 2, inter phase spacing 12 m.

CHAPTER

8

Interference Between Power and Communication Lines

8.1 INTRODUCTION

In many situations the power and communication lines may exist in close proximity. Sometimes they exist on the same supports. Under certain circumstances the power line may produce interference with the communication line. The interference is mainly due to electromagnetic and electrostatic induction. The electromagnetic induction may induce currents in the communication line. These currents are superimposed on the true speech currents in the communication line and cause distortion. The electrostatic induction raises the potential of the communication line thereby endangering the equipment and the user. Both these effects depend on the distance between the power and communication lines and the route length over which they run in parallel. It is necessary to ensure that the distortion and maximum voltage rise are within limits. If these limits are exceeded, remedial measures have to be taken by power supply authorities.

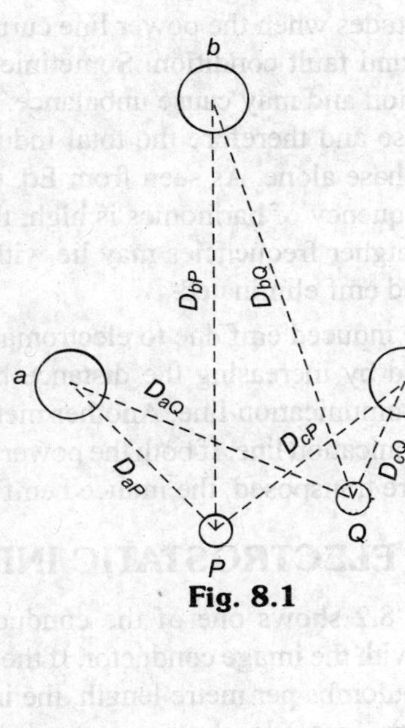

Fig. 8.1

8.2 ELECTROMAGNETIC INDUCTION

Figure 8.1 shows the three-phase conductors a, b, c, of a power line and two conductors P and Q of a nearby communication (say telephone) line. The currents in the conductors of the power line are I_a, I_b and I_c. Using Eq. (2.39), the flux linkages of conductors P and Q due to currents in the power line, can be written as:

$$\Psi_p = 2 \times 10^{-7} \left(I_a \ln \frac{1}{D_{aP}} + I_b \ln \frac{1}{D_{bP}} + I_c \ln \frac{1}{D_{cP}} \right) \text{ Wb-T/m} \tag{8.1a}$$

$$\Psi_Q = 2 \times 10^{-7} \left(I_a \ln \frac{1}{D_{aQ}} + I_b \ln \frac{1}{D_{bQ}} + I_c \ln \frac{1}{D_{cQ}} \right) \text{ Wb-T/m} \tag{8.1b}$$

The net flux linkages of the communication line are:

$$\Psi = \Psi_p - \Psi_Q = 2 \times 10^{-7} \left(I_a \ln \frac{D_{aQ}}{D_{aP}} + I_b \ln \frac{D_{bQ}}{D_{bP}} + I_c \ln \frac{D_{cQ}}{D_{cP}} \right) \text{ Wb-T/m} \tag{8.2}$$

The emf induced the communication line loop is:

$$E = 2\pi f \Psi \text{ V/m} \tag{8.3}$$

The magnitude of the flux linkages and the induced emf greatly depend on balanced or unbalanced nature of currents I_a, I_b and I_c. The following cases need to be examined.

(a) *Balanced currents:* I_a, I_b and I_c. When the currents I_a, I_b and I_c are balanced, the net flux linkages and induced emf is very small. It can be shown that if the power line is transposed and the currents are balanced, the net flux linkage are zero. When the power line in not transposed (as is usually the case), some emf is induced in the communication line even when the currents are balanced. However this emf is very small and negligible.

(b) *Unbalanced currents:* I_a, I_b and I_c. The flux linkages and induced emf can assume high magnitudes when the power line currents are heavily unbalanced as in the case of single phase to ground fault condition. Sometimes third harmonic currents may be present during healthy condition and may cause unbalance. These harmonic currents in the three-line conductors are in phase and therefore the total induced emf, for these harmonics, is three times that due to each phase alone. As seen from Eq. (8.3) the induced emf is proportional to frequency. Since the frequency of harmonics is high, the induced emf becomes all the more greater. Moreover, these higher frequencies may lie within the audible range. Therefore, it is essential to get the induced emf eliminated.

The induced emf due to electromagnetic induction can be reduced by increasing the distance between the power and the communication line. Another method is to transpose the communication line. If both the power and the communication lines are transposed, the induced emf becomes almost zero.

8.3 ELECTROSTATIC INDUCTION

Figure 8.2 shows one of the conductors of a 3-phase line along with the image conductor. If the conductor has a charge of q coulombs per metre length, the image conductor $q-$ has a charge $-q$ coulombs per metre length. The electric field intensity at a point situated at a distance x from the centre of conductor a is give by:

$$E_x = \frac{q}{2\pi\varepsilon_0 x} + \frac{q}{2\pi\varepsilon_0 (2h - x)} \tag{8.4}$$

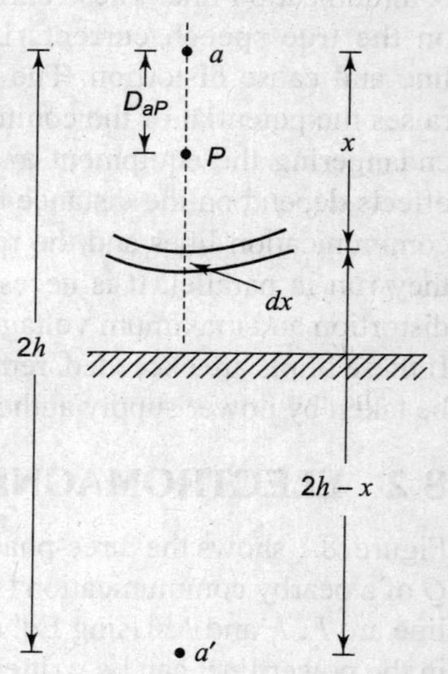

Fig. 8.2

V_a, i.e., the potential of conductor a with respect to ground is:

$$V_a = \int_r^h E_x \, dx = \frac{q}{2\pi\varepsilon_0}\int_r^h \left(\frac{1}{x} + \frac{1}{2h-x}\right) dx$$

or $$V_a = \frac{q}{2\pi\varepsilon_0} \ln \frac{2h-r}{r} \text{ volts} \tag{8.5}$$

where r is radius of conductor a.

Let another conductor P be situated vertically below conductor a at a distance D_{aP} from it. The potential of point P (V_{Pa}) with respect to ground, because of charge on conductor a and its image conductor a' is:

$$V_{Pa} = \int_{D_{aP}}^h E_x \, dx = \frac{q}{2\pi\varepsilon_0}\int_{D_{aP}}^h \left(\frac{1}{x} + \frac{1}{2h-x}\right) dx$$

$$= \frac{q}{2\pi\varepsilon_0} \ln \frac{2h - D_{aP}}{D_{aP}} \text{ volts} \tag{8.6}$$

Substituting the value of q from Eq. (8.5) into Eq. (8.6) we have:

$$V_{Pa} = V_a \frac{\ln[(2h - D_{aP})/D_{aP}]}{\ln[(2h-r)/r]} \text{ volts} \tag{8.7}$$

Similarly the potential of point P (with respect to ground) due to conductors b and c can be found. Then the total potential of conductor P is the phasor sum of the potentials due to conductors a, b and c. In a similar way the potential of conductor Q (the second conductor of the communication circuit) can be found.

Example 8.1

A 3-phase 11 kV 50 Hz feeder has horizontal configuration with 1.2 m between adjacent conductors. The height of the conductor above ground is 10 m. A telephone line is run on the same supports. The horizontal distance between the telephone wires is 0.6 m. The vertical distance between the power line and the telephone line is 1m. Find the magnitude of the voltage induced in the telephone circuit due to electromagnetic induction if the power line carries a balanced current of 300 amperes.

Solution

The configuration is shown in Fig. 8.3. The distances are:

$$D_{aP} = D_{cQ} = 1.345 \text{ m}$$

$$D_{bP} = D_{bQ} = 1.044 \text{ m}$$

$$D_{aQ} = D_{cP} = 1.803 \text{ m}$$

$$I_a = 300 \angle 0°,\ I_b = 300 \angle{-120°},\ I_c = 300 \angle 120° \text{A}$$

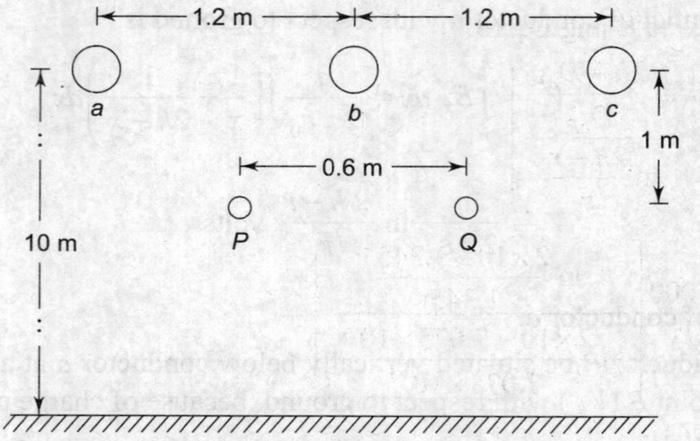

Fig. 8.3

Using Eq. (8.2)

$$\Psi = 2 \times 10^{-7}\left[(300\angle 0)\ln\frac{1.803}{1.345} + (300\angle -120°)\times\ln\frac{1.044}{1.044} + (300\angle 120°)\ln\frac{1.345}{1.803}\right] \text{ Wb-T/m}$$

$$= 2 \times 10^{-7}\,(87.9\angle 0 - 87.9\angle 120°) = 304.5 \times 10^{-7}\angle -30°$$

The voltage induced in the telephone line

$$= 2\pi f\Psi = 2\pi \times 50 \times 304.5 \times 10^{-7} = 9.57\times 10^{-3} \text{ V/m}$$

Example 8.2

Calculate the voltage induced due to electromagnetic induction in the telephone line of example 8.1 under the condition of a single line to earth fault on phase *a* if the fault current is 2000 amperes.

Solution

Using Eq. (8.2)

$$\Psi = 2 \times 10^{-7}\left[(2000\angle 0)\ln\frac{D_{aQ}}{D_{aP}}\right]$$

$$= 4 \times 10^{-4}\ln\frac{1.803}{1.345} = 1.17 \times 10^{-4} \text{ Wb-T/m}$$

Voltage induced $= 2\pi f\Psi = 2\pi \times 50 \times 1.17 \times 10^{-4} = 0.368$ V/m

Example 8.3

Find the potential of conductor *P* above earth due to electrostatic effect in the configuration of example 8.1. The power line conductors have a diameter of 14.15 mm each.

Solution

$$r = \frac{14.15}{2} \text{ mm} = 7.075 \times 10^{-3} \text{ m}$$

$$V_a = \frac{11000}{\sqrt{3}}\angle 0°,\; V_b = \frac{11000}{\sqrt{3}}\angle -120°,\; V_c = \frac{11000}{\sqrt{3}}\angle 120° \text{ V}$$

Referring to Fig. 8.3 and Using Eq. (8.7)

$$V_{Pa} = V_a \left[\frac{\ln \dfrac{2h - D_{aP}}{D_{aP}}}{\ln \dfrac{2h - r}{r}} \right]$$

$$= \frac{11000}{\sqrt{3}} \left[\frac{\ln \dfrac{2\times10 - 1.345}{1.345}}{\ln \dfrac{2\times10 - 7.075\times10^{-3}}{7.075\times10^{-3}}} \right] = 2100 \angle 0 \text{ volts}$$

$$V_{pb} = \left[\frac{11000}{\sqrt{3}} \angle{-120°} \right] \left[\frac{\ln \dfrac{2\times10 - 1.044}{1.044}}{\ln \dfrac{2\times10 - 7.075\times10^{-3}}{7.075\times10^{-3}}} \right] = 2317 \angle{-120°} \text{ volts}$$

$$V_{pc} = \left[\frac{11000}{\sqrt{3}} \angle 120° \right] \left[\frac{\ln \left(\dfrac{2\times10 - 1.803}{1.803} \right)}{\ln \left(\dfrac{2\times10 - 7.075\times10^{-3}}{7.075\times10^{-3}} \right)} \right] = 1847 \angle 120° \text{ volts}$$

The total potential of conductor P (with respect to earth) due to the three-power conductors a, b, c is the phasor sum of V_{pa}, V_{pb}, and V_{pc}.

Total potential of conductor P with respect to earth

$$= 2100 \angle 0 + 2317\angle{-120°} + 1847 \angle 120° = 407 \text{ volts}$$

Example 8.4

A 132 kV, 3-phase, 50 Hz line has its conductors in a symmetrical delta configuration with an inter phase spacing of 3.5 m. The diameter of each conductor is 9 mm. The height of the lowest conductors is 16 m above ground. A communication line is also supported on the same towers as shown in Fig. 8.4. Calculate the voltage induced in the communication circuit due to electromagnetic effect of the power line if the power line carries a balanced current of 200 A.

Solution

From Fig. 8.4

$$D_{ao} = 3.03 \text{ m}, D_{ap} = 7.03 \text{ m}, D_{aQ} = 7.53 \text{ m},$$

$$D_{bp} = 4.37 \text{ m} = D_{cp}, D_{bQ} = 4.83 \text{ m} = D_{cQ}$$

$$I_a = 200 \angle 0°, I_b = 200 \angle{-120°}, I_c = 200 \angle 120$$

$$\Psi = 2 \times 10^{-7} \left[(200\angle 0) \ln \frac{7.53}{7.03} + (200\angle{-120}) \ln \frac{4.83}{4.37} + (200\angle 120) \ln \frac{4.83}{4.37} \right]$$

$$= 12.52 \times 10^{-7} \text{ Wb-T/m}$$

Voltage induced in the telephone circuit

$$= 2\,\pi f \Psi = .393 \times 10^{-3} \text{ V/ m}$$

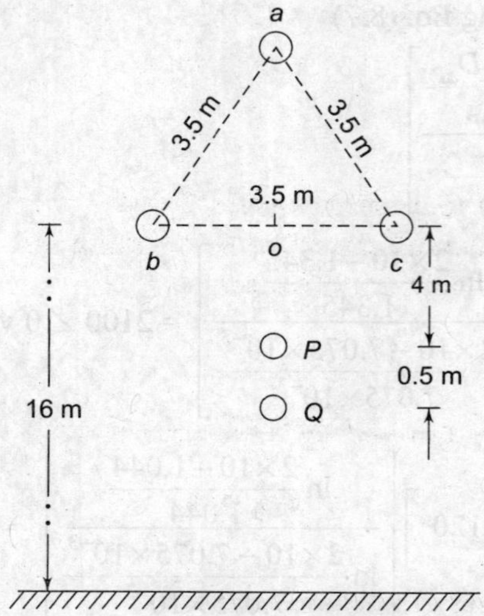

Fig. 8.4

Example 8.5

For the configuration shown in Fig. 8.4 find the voltage induced in conductor *P* due to electrostatic effect.

Solution

The distances are shown in Fig. 8.4

$$V_a = \frac{132000}{\sqrt{3}} \angle 0,\ V_b = \frac{132000}{\sqrt{3}} \angle -120°,\ V_c = \frac{132000}{\sqrt{3}} \angle +120°$$

$$r = \frac{9}{2} \times 10^{-3} = 4.5 \times 10^{-3}\,\text{m}$$

Using Eq. (8.7)

$$V_{pa} = \frac{132000}{\sqrt{3}} \left[\frac{\ln \dfrac{2 \times 19.03 - 7.03}{7.03}}{\ln \dfrac{2 \times 19.03 - 4.5 \times 10^{-3}}{4.5 \times 10^{-3}}} \right] = 12513 \angle 0\ \text{V}$$

$$V_{pb} = \frac{132000}{\sqrt{3}} \angle -120° \left[\frac{\ln \dfrac{2 \times 16 - 4.37}{4.37}}{\ln \dfrac{2 \times 16 - 4.5 \times 10^{-3}}{4.5 \times 10^{-3}}} \right] = 15845 \angle -120°$$

$$V_{pc} = \frac{132000 \angle -120°}{\sqrt{3}} \left[\frac{\ln \dfrac{2 \times 16 - 4.37}{4.37}}{\ln \dfrac{2 \times 16 - 4.5 \times 10^{-3}}{4.5 \times 10^{-3}}} \right] = 15845 \angle 120°$$

Total potential of conductor P above earth $= 12513 \angle 0 + 15845 \angle -120° + 15845 \angle 120°$

$$= 3332 \text{ volts.}$$

Example 8.6

A telephone line runs parallel to an untransposed 3 phase transmission line as shown in Fig. 8.5. The power line carries balanced current of 400 A/phase. Find mutual inductance between circuits and voltage induced in telephone line per km [*RTM Nagpur Uni. 2010*]

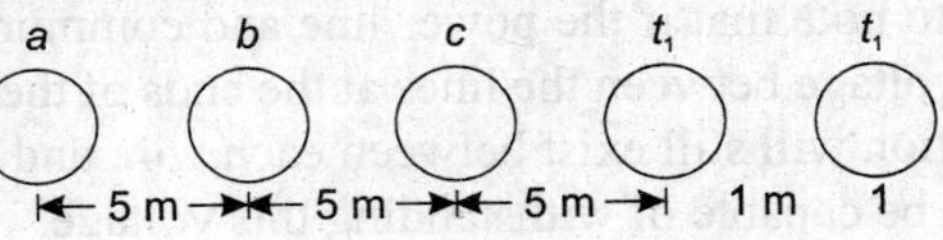

Fig. 8.5

Solution

Using Eq. (8.2) flux linkages of telephone line are

$$\Psi = 2 \times 10^{-7} \left[I_a \ln \frac{D_{at_2}}{D_{at_1}} + I_b \ln \frac{D_{bt_2}}{D_{bt_1}} + I_c \ln \frac{D_{ct_2}}{D_{ct_1}} \right] \text{Web turns/m}$$

$$I_a = 400\angle 0\ A,\ I_b = 400\ \angle -120°\ A,\ I_c = 400\ \angle 120° A$$

$$D_{at_2} = 16 \text{ m},\ D_{at_1}, = 15 \text{ m},\ \Delta_{bt_2} = 11 \text{ m},\ \Delta_{bt_1} = 10 \text{ m}$$

$$D_{ct_2} = 6 \text{ m},\ D_{ct_1}, = 5 \text{ m}$$

$$\Psi = 2 \times 10^{-7} \left[400 \ln \frac{16}{15} + 400 \angle -120° \ln \frac{11}{10} + 400 \angle 120° \ln \frac{6}{5} \right]$$

$$= 2 \times 10^{-7} [400 \times 0.0645 + 400 \angle -120° \times 0.0953 + 400 \angle 120° \times 0.182]$$

$$= 2 \times 10^{-7} [25.8 + 38.12 \angle -120° + 72.8 \angle 120°]$$

$$= 84.36 \times 10^{-7}\ Wb\text{–}T/_m = 84.36 \times 10^{-4}\ Wb\text{–}T/\text{km}$$

$$\text{Mutual inductance } M = \frac{\Psi}{I} = \frac{84.36 \times 10^{-4}}{400}$$

$$= 21.09 \times 10^{-6} \text{ H}$$

Voltage induced in telephone line

$$= \omega MI = 2\pi \times 50 \times 21.09 \times 10^{-6} \times 400$$

$$= 2.65 \text{ V per km}$$

8.4 REDUCTION OF INTERFERENCE

Sometimes the electromagnetic and electrostatic interference between a power line and a communication line may be considerable. Of course the interference can be reduced by ensuring that the power and the communication circuits are far away from each other.

When the power and communication circuits run parallel to each other, interference can be reduced by transposing the conductors of the power line and the communication line. The transposition of power line neutralizes the unbalances in the capacitances of the lines so that the electrically induced voltages are balanced out in a complete set of transpositions. Transposition also results in the equalization of the magnetic fluxes due to the positive and negative sequence currents so that their sum in all the three phases is zero. This neutralizes the magnetically induced voltages in the communication line in one complete set of transposition. The currents due to zero sequence currents do not add up to zero by transposition because these currents are in phase. The induced voltage due to the zero sequence currents can be eliminated by transposition of the conductors of communication line. The length along one complete set of transpositions is known as barrel. It is important to note that if the power line and communication line are transposed there will be no induced voltage between the lines at the ends of the barrel. However, a voltage due to electrostatic induction will still exist between each wire and earth. The insulators of the communication line must be capable of withstanding this voltage.

SUMMARY

8.1 When power and communication lines exist in close vicinity electromagnetic and electrostatic interference may occur.

8.2 When power line carries balanced currents in the three phases electromagnetic interference is negligible.

8.3 When power line carries unbalanced currents, the electromagnetic interference may be high especially during single phase to ground fault condition.

8.4 If both power and communication lines are transposed the induced emf due to electromagnetic induction is almost zero.

8.5 Total potential at any conductor of communication line is the phasor sum of potentials caused by the three power line conductors due to electrostatic induction.

8.6 Interference can be reduced by increasing the distance between power and communication lines and also by transposition.

SELF TEST QUESTIONS

A. State whether the following are true (T) or false (F)

8.1 Induced emf due to electromagnetic induction increases with frequency.

8.2 When power line carries balanced 3 phase currents, the electromagnetic induction is negligible.

8.3 There is no electromagnetic induction during single phase to ground fault condition.

8.4 If the power and communication lines are transposed electromagnetic induction becomes negligible.

8.5 Electrostatic induction increases with increase in line voltage.

B. Fill in the blanks with suitable words

8.6 Transposition of power line conductors neutralizes the unbalanced ________ of the line.

8.7 The induced voltage due to zero sequence currents in power line can be eliminated by transposing the conductors of ________ line.

8.8 The length along one complete set of transposition is called a ________.

8.9 The induced emf, due to electromagnetic induction, on a communication line ________ with frequency.

8.10 ________ harmonic currents are in phase.

ANSWERS

8.1	T	8.2	T	8.3	F	8.4	T	8.5	T
8.6	Capacitances	8.7	communication	8.8	barrel	8.9	increases	8.10	third.

JOB INTERVIEW QUESTIONS

8.1 What is the cause of electromagnetic interference between power and communication line?

8.2 Why is electromagnetic interference high during single phase to ground fault conduction?

8.3 How can interference between power and communication lines be reduced?

8.4 What is meant by electrostatic induction?

REVIEW QUESTIONS

8.1 What is meant by electromagnetic interference between power and communication lines? What is its effect?

8.2 Derive a formula for voltage induced by electromagnetic induction in a communication line due to a nearby power line.

8.3 What is meant by electrostatic interference between a power and communication line? What is its effect?

8.4 Derive a formula for calculating the voltage induced by electrostatic induction in a communication line due to a power line.

8.5 The electromagnetic interference between power and communication lines is small during normal operation of power line but may be high during single phase to ground fault? Why?

8.6 Why is it necessary to avoid interference between communication and power circuits? What steps can be taken for this purpose?

PROBLEMS

8.1 The two conductors of a single phase 50 Hz power line are placed horizontally with a spacing of 4 meters. A telephone line is also supported on the same tower, 2 m below the power line. The conductors of the telephone line are also horizontally placed with a spacing of 0.6 m. Find the voltage induced due to electromagnetic induction in the telephone circuit when the power line carries a current of 200 A. [3.76 V]

8.2 A 3-phase power line and a telephone line are supported on the same tower and have a configuration similar to Fig. 8.4 but the dimensions are as under: interphase spacing between power line conductors = 4 m, Height of conductor *c* above ground 18 m, Distance between point *O* and conductor *P* = 5 m, Distance between conductors *P* and *Q* = 1.5 m, Radius of each power line conductor = 0.5 cm, Find the voltage per m length induced in the telephone line due to electromagnetic effect when the power line carries a balanced current of 250 A. [0.361 V]

8.3 In problem 8.2 find the potential of conductor *P* above earth due to electrostatic effect. [2645 V]

8.4 In problem 8.2 find the voltage induced per m in the telephone line due to electromagnetic induction under the condition of single line to ground fault in conductor of *a* if the fault current is 1000 A. [3.23 V]

CHAPTER

9

Underground Cables

9.1 INTRODUCTION

Large scale use of cables has been made for transmission and distribution of electrical energy in densely populated areas. A cable consists of a central conductor with an insulation to isolate the conductors from each other and their surroundings. External protection is provided for protecting against mechanical damage, chemical reactions and moisture, etc. The materials employed for conductors generally are annealed copper and aluminium. The use of copper is limited to cables used for control circuits, signalling and communications. Power cables use stranded or sector shaped aluminium conductors. Since aluminium has a tendency to oxidize when exposed to air and the oxide film forms a high resistance, special jointing techniques are necessary. Several jointing techniques such as soldering, welding and compression jointing are employed.

9.2 INSULATION

The conductors of a cable have to be covered by an insulation to isolate the conductors from each other and from their surroundings. The insulating material should have high dielectric strength, high insulation resistance, good mechanical strength and should be able to withstand temperatures from about –30°C to over 100°C. Many insulating materials have been developed and are used in cable manufacture. Table 9.1 lists typical properties of some cable insulations.

Table 9.1 Typical properties of cable insulations

Material	*Dielectric strength kV/mm*	ε_r	*Power factor*	*Temperature rating °C*		
				Max. contin-uous	*Short time overload*	*Short circuit*
XLPE	18	2.5	.008	90	130	250
Polyethylene thermoplastic	21	2.3	.0004	75	95	150
Ethylene propylene (rubber)	15	2.8	.009	90	130	250
Butyl rubber	13	3.2	0.015	85	105	200
PVC	17	5	0.045	75	95	150
Oil impregnated paper	21	3.6	0.008	80	110	200

9.2.1 VIR Cables

Vulcanized India rubber insulation was developed in 1870 and has a dielectric strength of about 10–20 kV/mm and a dielectric constant of about 2.5. Sulphur is mainly used as the vulcanizing agent. This insulation is limited to wiring cables for lighting and minor power installations.

9.2.2 Elastomer Insulated Cables

Elastomeric insulation includes–rubber, butyl rubber, silicone rubber and ethylene propylene rubber.

Elastomers have rubber like characteristics which is achieved by compounding the basic-polymer with selective additives. Since natural rubber can be used only up to an operating temperature of 60°C, considerable effort has been devoted to the development of synthetic elastomeric compounds with superior properties.

Polychloroprene (p.c.p) or Neoprene is obtained by polymerization of chloroprene. The raw polymer contains about 35 per cent chlorine as a result of which p.c.p. is self extinguishing if ignited externally. P.c.p. compounds can be used upto a conductor temperature of 60°C and are more resistant to outdoor weathering and to deterioration by oil than natural rubber.

Butyl rubber, a co-polymer of isobutylene and a small quantity of isoprene, is more resistant to oxidation and ozone than natural rubber and can be used upto a conductor temperature of 85°C.

Ethylene propylene rubber (EPR). Ethylene propylene rubber is a saturated copolymer of ethylene and propylene in approximately equimolecular preparations. It has good electrical properties, and is resistant to ozone, heat and chemicals. Cables with this insulation can be buried directly and exposed to weather and contaminated atmospheres and are, therefore, suitable for service in industry. They can be used upto a conductor temperature of 85°C.

9.2.3 PVC Cables

Polyvinyl chloride is a synthetic material obtained from acetylene and can be produced in a number of grades, depending on the polymerization process used. It is in the form of a white powder which is odourless, tasteless, chemically inert, non-inflammable and insoluble at ordinary temperatures in all liquids. In pure form it is rigid and brittle at low temperature. When used for cable insulation it must be combined with a plasticiser whose function is to form a gel and make the material plastic over the desired range of temperature. Its mechanical properties are not as good as those of rubber and the insulation resistance is also lower than that of rubber. However, it is inert to oxygen and almost inert to oils, many acids and alkalies. Therefore, it is preferable to rubber in many environmental conditions.

9.2.4 Polythene Insulated Cables

Polythene is a straight chain polymer derived from ethylene. Its electrical properties are very good, e.g., very low power factor at all frequencies, low dielectric constant and high resistivity. Its main use is in the manufacture of high frequency cables. It has a limited application as a power cable insulant. However, a 3-core 11 kV self supporting aerial cable which can be used for rural distribution has been developed. Due to its low weight it can be suspended from the existing low voltage poles.

9.2.5 XLPE Cables (Cross Linked Polythene Cables)

Low density polythene, when vulcanized under controlled conditions (roughly in the same way as in vulcanizing rubber), results in cross linking of carbon atoms and produces cross-linked polythene. This new material does not melt but carbonizes at 250 –300°C and has become very popular as an insulating material. XLPE cables possess the advantages of light weight, small overall cable dimensions, low dielectric constant and good mechanical strength. These cables permit conductor temperatures of 90°C and 250°C under normal and short circuit conditions and can be buried directly in soil as this insulation has low water absorption. These cables have been found to be very suitable for all voltages up to 33 kV.

9.2.6 Paper Insulated Cables

Paper is the most important insulating material used in the manufacture of power cables upto 500 kV rating. The paper insulation has the advantages of superior heat conductivity, ability to withstand high temperature, durability, high dielectric strength, low cost and low electrostatic capacitance.

For providing insulation over the conductor, the paper tape is lapped on to the conductor until the required thickness is obtained, it is then dried and impregnated with insulating compound. The insulating compounds are required to have, besides high electrical insulation properties, a high viscosity at working temperatures (50–80°C), a low viscosity at impregnation temperatures (105 –120°C), a smooth change in viscosity with temperature and very low coefficient of thermal expansion. The insulating compounds are all based on a paraffinic or on a naphthenic mineral oil, resin being often added to improve the impregnating quality since it is important that no air voids should be left in the finished cable. The impregnation of paper insulation increases its dielectric strength from about 3–5 to 10–80 kV/mm. The dielectric constant of impregnated paper insulation is about 3.6. In a belted cable a part of the insulation is applied to each core individually and the remainder is in the form of an overall belt (Fig. 9.2).

The thickness of insulation depends on the mechanical and electrical requirements which it has to withstand in service. The low voltage cables (less than 1 kV rating) should have sufficient insulation thickness to meet the mechanical treatment which the cable may undergo during manufacture, installation and service. At higher voltages the insulation thickness is determined by electrical requirements.

9.3 SHEATH, ARMOUR AND COVERING

A metal sheath is provided over the cable insulation so that moisture may not be able to affect the insulation and may also protect the insulation from mechanical damage. A metal is essential for sheath because no organic material is sufficiently impervious to moisture. Many years ago lead was established as a suitable material for sheathing. The main advantages of lead sheaths are the comparative ease with which they are made in lead presses, their flexibility and high corrosion resistance. Some of the disadvantages of lead are its large mass (the specific gravity of lead is 11.37), low mechanical strength, fluidity and small resistance to vibrations. The hardness, mechanical strength and resistance to vibrations of lead sheaths can be considerably increased by adding alloying mixtures. Many alloys of lead having about one per cent of copper, tin, bismuth, etc., can be used for sheathing.

Aluminium is now being increasingly used as a sheathing material. Aluminium sheaths have a much greater mechanical strength than lead sheaths, low weight (specific gravity of aluminium is 2.71) and low fluidity. The aluminium sheaths cost about 3.5 times less than the lead sheaths.

Steel tape or wire armour is necessary to protect the sheath against mechanical damage. Steel tape armour consists of two steel tapes coated with preservative compound, applied helically in the same direction over a fibrous bedding, the outer layer covering the spaces between the turns of the inner layer. This provides a good protection against mechanical damage but steel wire armour is recommended where additional longitudinal stresses may occur during installation or in service. Steel wire armour consists of galvanized steel wires applied over a compounded bedding. Double wire armours are used for high degree of mechanical protection. Single core cables for ac systems are usually not provided with armour because of its effect in increasing the losses. In these cables a plastic or PVC oversheath is used for mechanical protection. Aluminium sheathed cables, owing to the comparative hardness of sheath, are not required to be armoured in the same way as the lead sheathed cables but for special conditions armouring is provided for these cables too. As an anticorrosion measure, power cables are usually protected by bitumen impregnated fibrous wrapping (or covering or serving). Generally two layers of such covering are provided, one layer over the sheath (i.e., on the inner side of armour) and the second layer over the armour.

9.4 CLASSIFICATION OF CABLES

Paper insulated metal sheathed power cables can be classified as solid type and pressurized cables.

(a) *Solid Type Cables:* In these cables the pressure within the cables does not go above atmospheric pressure and may even fall below it locally, e.g., in voids. This can lead to cable breakdown when the electrostatic stress is high. The cables for voltages upto 33 kV belong to this category.

(b) *Pressurized Cables:* In these cables pressure is maintained above atmospheric either by oil in oil filled cables or by gas in gas-pressure cables. Gas-pressure cables are used upto 275 kV while oil filled ones are used upto 500 kV or so.

Either of these may be single core or multi-core. A solid single core cable has a central conductor, with an insulation of impregnated paper, a metallic sheath and a plastic oversheath. Fig. 9.1 shows such a cable.

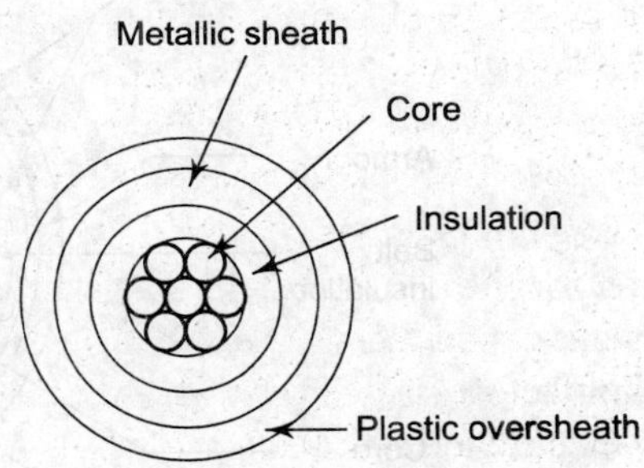

Fig. 9.1 Single core cable

A multi-core cable may have 3 cores (for 3-phase 3-wire systems) or 4 cores (for 3-phase 4-wire systems) and may be a belted type cable or *H* type cable.

A belted cable has a part of the insulation applied to each core individually and the remainder in the form of an overall belt as shown in Figs. 9.2 (*a* and *b*). The cable is provided with a metallic sheath, an anti corrosion serving, an armour and a second layer of anti corrosion serving. Belted cables are used upto 11 kV. *H* type cables (also known as screened cables) were introduced by Hochstadter in 1914. In these cables each core has an insulation layer but no belt insulation is used. Instead a metal conducting surface is provided over the insulation of each core. This screen is a perforated aluminium foil or copper tape with a narrow gap between the turns. The cores are then laid up with jute or paper fillers (as in the belted cable) taking care to see that the metallic screens of the three cores are in good contact with each other. Then a cotton tape, with strands of copper wire interwoven within it, is wrapped around making sure that the copper wire is in good contact with the metallic screens of the cores. Then the lead sheath and armour are applied. The perforations in the screens assist in complete impregnation of the cable which in fact, becomes a homogeneous mass in which there is little likelihood of air or gas pockets. All the four screens and the lead sheath are at earth potential with the result that the electrostatic stresses are purely radial. This reduces the possibility of breakdown. The dielectric losses are also reduced considerably due to radial electrostatic stresses. *H* type cables have been used upto 66 kV but are mostly used only upto 33 kV because pressurized cables have better performance than *H* type cables for voltages above 33 KV.

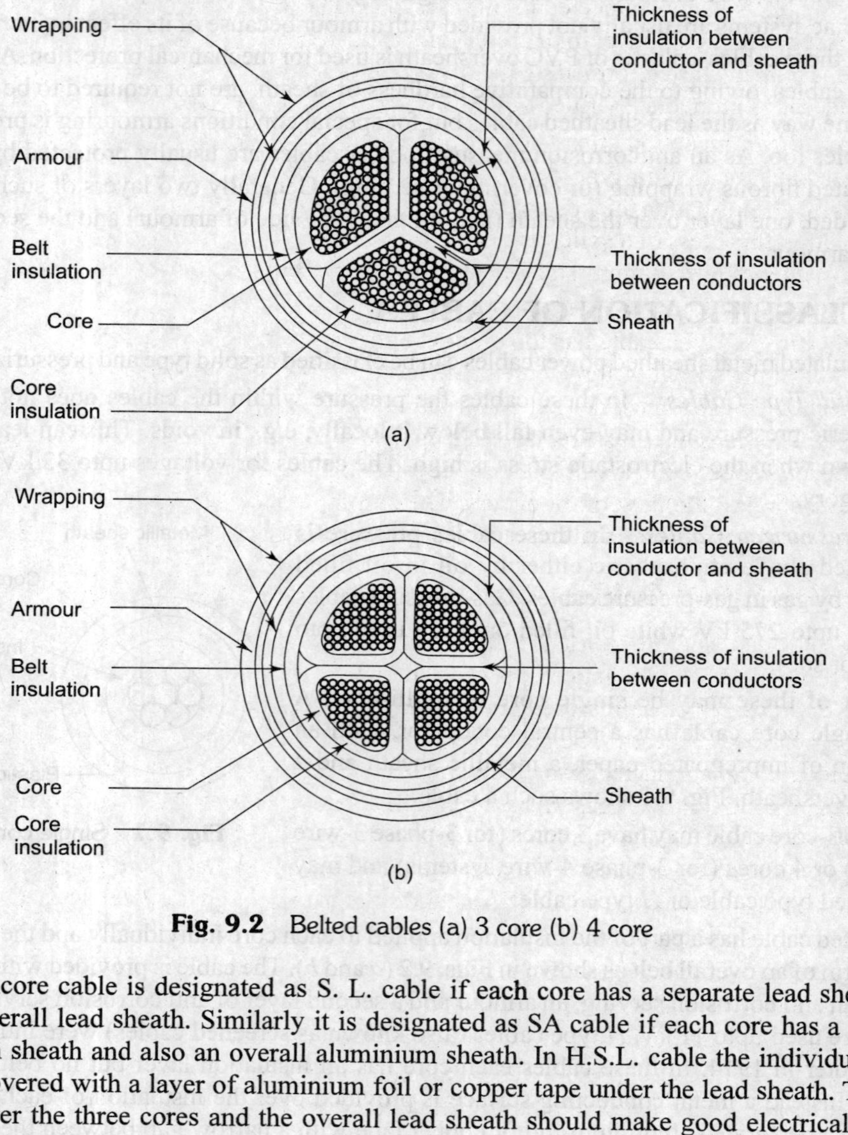

Fig. 9.2 Belted cables (a) 3 core (b) 4 core

A multicore cable is designated as S. L. cable if each core has a separate lead sheath and also an overall lead sheath. Similarly it is designated as SA cable if each core has a separate aluminium sheath and also an overall aluminium sheath. In H.S.L. cable the individual cores are also covered with a layer of aluminium foil or copper tape under the lead sheath. The lead sheaths over the three cores and the overall lead sheath should make good electrical contact with each other.

9.5 PRESSURIZED CABLES

Solid type cables are suitable only upto 33 kV. At higher voltages the chances of breakdown in cables due to ionization in voids become quite high. In pressurized cables voids are eliminated by maintaining a higher pressure than atmosphere. Pressurized cables are either oil filled cables or gas-pressure cables.

9.5.1 Oil Filled Cables

Single core and 3 core oil filled cables are suitable for 33 kV as well as higher voltages upto 500 kV. An oil filled cable is kept constantly supplied with low viscosity mineral oil through ducts in the cable, the ducts being connected to oil reservoirs. The expansions and contractions merely produce changes in the oil level in the reservoirs and, therefore, do not result in voids. Another advantage is that any space formed by the distortion of lead sheath or by relative movement of paper dielectric is immediately filled with oil.

The breakdown inception stress, for oil filled cables, is 300 to 400 kV/cm which is much higher than that for a solid type cable. Therefore, the normal operating stresses are raised considerably. The insulation thickness is mainly governed by requirement of impulse breakdown. (The insulation thickness of a solid type cable is governed by requirement of withstand voltage at power frequency). Maximum stresses upto 130 kV/cm are acceptable for 275 kV and 400 kV oil filled cables.

A single core oil filled cable has the oil duct either within the conductor (i.e., the conductor is hollow) or in the sheath. A three core cable has three oil ducts, one in each filler space. The pressure within the cable must not fall below atmospheric and a lower limit of about 21.7 kN/m^2 gauge is set. The transient pressure, developed when the load is suddenly increased, should not increase beyond about 8600 kN/m^2 gauge. The joints in the cable have to be properly made such that no restriction to the flow of oil occurs. The current carrying capacity of oil-filled cable is determined by allowing a normal maximum conductor temperature of 85°C.

9.5.2 Gas-Filled Cables

In these cables the paper dielectric is impregnated with petroleum jelly so that it may not have any free compound. The interstices between the layers of the paper are then filled with gas, generally dry nitrogen, at a pressure upto 1380 kN/m^2 forming a composite dielectric. Pressure is retained by the lead sheath. The gas flows throughout the cable via the filler spaces, butgaps in the dielectric and conductor strands. In single core cables a small clearance to sheath is left for allowing the axial flow of gas. In case of multi core cables such a clearance is not necessary because the filler spaces and strands provide a sufficiently low resistance path for the flow of gas. The advantage of gas filled cable over the oil filled able is that oil pressure reservoirs and other extraneous apparatus are not necessary and still the ionization is avoided because the voids are filled with gas at high pressure.

Another cable used for high voltages is external pressure cable. In these cables the conductor cores, after being insulated, impregnated and covered with sheath, are placed in a steel pipe which is filled with gas, generally, nitrogen, at a pressure of about 12–15 atmospheres.

Because of external gas pressure, the voltage required to set up ionization inside the voids is increased. Moreover the external pressure tends to close the voids. In comparison with a solid type cable, this construction gives twice the working voltage and about one and a half times the working current. The increased cost is more than offset by the increased operating voltage and current. In addition the steel pipe provides an ideal mechanical protection for the cable.

9.6 EFFECTIVE CONDUCTOR RESISTANCE

The effective resistance R_{eff} of each conductor of a cable has to be calculated by taking into account the dc resistance, skin effect, proximity effect, sheath loss and armour loss.

(a) *DC Resistance:* In calculating the dc resistance the increase in resistance due to temperature, effects of stranding and laying should be taken into account. The conductor temperature must not exceed 85°C for pressurized cables. For other cables the specified temperatures vary from 50°C to 75°C. The dc resistance at the maximum temperature may be about 20 per cent more than that at 20°C. The resistance is multiplied by 1.02 to account for stranding and in case of multi core cables again multiplied by 1.02 to account for additional length of conductor resulting from the lay of strands and of core in manufacture.

(b) *Skin Effect:* The increase in resistance due to skin effect depends mainly on the cross-sectional area. For 400 mm^2 and 600 mm^2 cross-sectional areas, the increase in resistance at 50 Hz is about 3.5 per cent and 6.5 per cent respectively.

(c) *Proximity Effect:* The increase in resistace due to non-uniformity of current density, over the cross-section of the conductor, caused by the magnetic field of currents in other phase conductors is greater in single core cables than in three core belted cables for a given size. For three core belted cables the increase in resistance due to proximity effect is about the same magnitude as that due to skin effect. The dc resistance at the operating temperature as modified for stranding, laying, skin effect and proximity effect is termed as ac resistance, i.e., R_{ac}.

(d) *Sheath Losses:* When alternating current flows in the core of a cable, the core and the sheath act as the primary and secondary of an air core transformer. An emf is induced in the sheath leading to flow of eddy currents and longitudinal circulating currents. These currents give rise to sheath losses. The losses due to eddy current in the sheath are usually negligible except in very large cables.

For 3-phase cables circulating currents are important only when 3 separate single core cables or a three core cable with each core having separate sheath (SL or SA cable) are used. In 3-core belted cable or 3-core screened cable with only a single sheath covering all the three cores, the magnetic fields of the three cores are partially mutually compensated and the circulating currents are reduced. Sheath losses are greater in aluminium sheathed cables than in lead sheathed ones.

When the sheaths of the three single core cables are not bonded or bonded at one end only, circulating currents cannot flow but induced voltages appear between one sheath and another. The magnitude of these voltages may be 50–200 volts per km (and much more at short circuits). In order to avoid the risk of damage or danger arising from the presence of these sheath voltages especially adjacent to terminals, sheaths have to be bonded at both ends. If this is done, circulating currents flow and sheath losses occur. As the spacing between the cores increases the loss also increases. Methods of cross-bonding at joints have been developed by which circulating current sheath losses can be virtually eliminated and the intersheath voltages kept small. Cross-bonding employs a transposition method involving systematic interruption and cross-connection of sheaths at each interruption (i.e., the sheath of conductor 1 connected to that of conductor 2, sheath of conductor 2 connected to that of conductor 3 and sheath of conductor 3 connected to that of conductor 1) at suitable intervals. Since the sheath voltages are equal and displaced by 120° (for balanced loads), the circulating currents are eliminated. For unbalanced loads as well, the circulating currents are very small. Inter-sheath voltage is limited to that of each section. Fig. 9.3 shows crossbonding.

(e) *Armour Losses:* These losses are partly due to the eddy currents in the armour and partly due to hysteresis. The losses due to eddy currents are of greater importance. It is not a standard practice to provide single core cables with armour of magnetic materials because of large armour losses and larger inductive line reactance. In multi core cables the armour losses are generally negligible for conductor sections less than 200 mm^2 but may be as high as 20 per cent of conductor resistance loss for some 3-core cables.

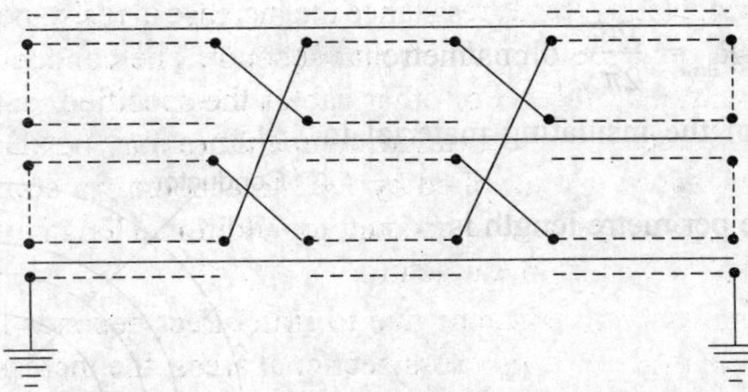

Fig. 9.3 Cross bonding of cable sheaths

If sheath loss = $\lambda_1 \times$ conductor loss and armour loss is $\lambda_2 \times$ conductor loss, then

$$R_{eff} = R_{ac}(1+\lambda_1+\lambda_2).$$

9.7 CONDUCTOR INDUCTIVE REACTANCE

Equations (2.30) and (2.56) of Chapter 2 can be used to calculate the inductive reactance of each conductor of single core and 3-core cables respectively. Equation (2.56) can also be used if the three-phase circuit consists of 3 single core cables. The results have to be modified to take into account the following effects:

(a) Reduction due to mutual coupling with the sheath. This is generally small except in some cases such as large single core cables with aluminium sheath.

(b) Reduction due to the lower effective spacing between cables with conductors of cross-section other than circular.

(c) Increase due to mutual coupling with the armour of 3-core cables. This may be to an extent of 10 per cent.

Because of the need of modification due to reasons given above, it is better to determine the inductance experimentally.

Example 9.1

Calculate the inductance per conductor per km of a 3-core belted cable with 37/0.238 cm conductor and core insulation 0.5 cm thick. Neglect the effect of mutual coupling with sheath and armour.

Solution

A 37-strand conductor has a central strand surrounded by 3 layers containing 6, 12 and 18 strands respectively. The overall conductor radius is 0.238 × 3.5 = 0.834 cm.

Geometric mean radius $r' = 0.7788 \times 0.834 = 0.6495$ cm.

The distance between conductor centres is uniform and, therefore, D_{eq} equals

$$2\,(0.5 + 0.834) = 2.668 \text{ cm.}$$

Using Eq. (2.56 c)

$$L = 0.4605 \log\left(\frac{D_{eq}}{r'}\right) = 0.4605 \log\,(2.668/0.6495) = 0.2825 \text{ mH/km}$$

9.8 PARAMETERS OF SINGLE CORE CABLES

9.8.1 Insulation Resistance

Figure 9.4 shows a *single core cable of conductor radius r*. The cable has a sheath of inside radius R. The insulation resistance $d\,R_{ins}$ of an annulus of thickness dx at radius x is

$$dR_{ins} = \frac{\rho dx}{2\pi x} \text{ ohms/metre} \tag{9.1}$$

where ρ is the resistivity of the insulating material in ohm-metres.

The insulation resistance per metre length is

$$R_{ins} = \int_r^R \frac{\rho dx}{2\pi x}$$

$$= \frac{\rho}{2\pi} \ln \frac{R}{r} \text{ ohms/metre} \tag{9.2}$$

If the cable has a length of l metres

Insulation resistance

$$= \frac{\rho}{2\pi l} \ln \frac{R}{r} \text{ ohms} \tag{9.3}$$

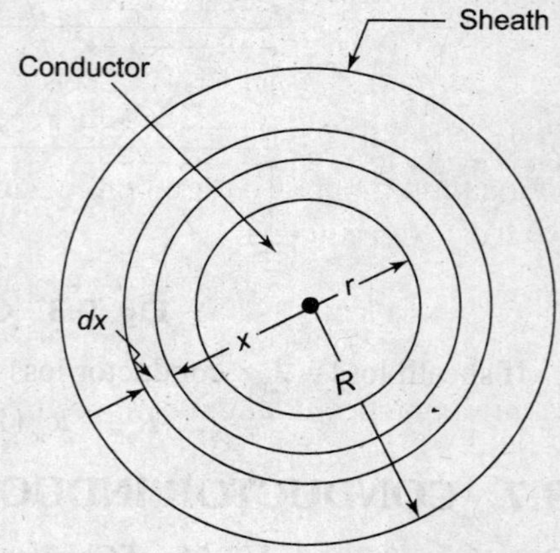

Fig. 9.4 Cross-section of a single core cable

Average value of ρ for impregnated paper varies from 5×10^{12} to 8×10^{12} ohm-metres at 15°C.

The change in resistivity of insulating materials with temperature is described by the equation.

$$\rho_t = \rho_0 \varepsilon^{-at} \tag{9.4}$$

where ρ_t is resistivity at t° C. ρ_0 is resistivity at 0°C and α is a constant.

9.8.2 Capacitance

Since the single core cable (Fig. 9.4) has an earthed metallic sheath, there is an electric field between the conductor and sheath. Let the charge on the surface of the conductor be q coulomb per metre length of cable. The electric flux density D_x at a radius x is

$$D_x = \frac{q}{2\pi x} \text{ C/m}^2$$

The electric field intensity E_x at radius x is

$$E_x = \frac{D_x}{\varepsilon_r \varepsilon_0} = \frac{q}{2\pi \varepsilon_r \varepsilon_0 x} \text{ V/m} \tag{9.5}$$

where ε_r is the relative permittivity (dielectric constant) of the cable insulation and $\varepsilon_0 = 8.85 \times 10^{-12}$ F/m.

The potential difference between the core and the sheath is

$$V = \int_r^R E_x dx = \frac{q}{2\pi \varepsilon_0 \varepsilon_r} \ln \frac{R}{r} \text{ volts} \tag{9.6}$$

Capacitance between core and sheath is

$$C = \frac{q}{V} = \frac{2\pi \varepsilon_0 \varepsilon_r}{\ln \frac{R}{r}} \text{ F/m} \tag{9.7}$$

Eq. (9.7) is for a single core cable but it is also applicable to most of the 3-core cables used at voltages above 11 kV because these cables have an earth metallic screen around each core and/or have separate sheaths for each core. If the conductor is not circular, the radius r may be taken as the $\frac{1}{2\pi} \times$ actual periphery.

9.8.3 Electric Stresses

Substituting the value of q from Eq. (9.6) into Eq. (9.5), the potential gradient at a radius x is

$$E_x = \frac{V}{x \ln \frac{R}{r}} \tag{9.8}$$

The maximum potential gradient occurs at $x = r$, i.e., at the surface of the conductor and is given by:

$$E_{max} = E_r = \frac{V}{r \ln \frac{R}{r}} \tag{9.9}$$

The minimum potential gradient occurs at $x = R$ and is given by:

$$E_{min} = E_R = \frac{V}{R \ln \frac{R}{r}} \tag{9.10}$$

From Eqs. (9.9) and (9.10) the ratio of maximum and minimum potential gradient is

$$\frac{E_{max}}{E_{min}} = \frac{R}{r} \tag{9.11}$$

The above analysis assumes that the conductor is perfectly circular. Stranding of the conductor increases E_{max} by about 20 per cent. For a given value of V and R, there is a certain value of r which gives a minimum value of E_{max}.

Equation (9.9) shows that E_{max} is minimum when $r \ln (R/r)$ is maximum. This happens when

$$\frac{d}{dr}\left[r \ln \frac{R}{r}\right] = 0, \text{ i.e., } R = 2.718\, r \tag{9.12}$$

The practical application of this result is limited but it indicates that in high voltage cables, for given values of V and R there is a minimum value of the conductor radius which must be used to keep E_{max} within limit. The design of high voltage cables, based on Eq. (9.12) gives a high value of r, much too large from the point of view of current carrying capacity. Methods like using stranded conductors around a central core of hemp and using a central lead tube instead of hemp, etc., have been suggested to increase the value of r so that E_{max} is low. However, it may be noted that the variation of E_{max} with r is not large. When r changes from 0.5 R to 0.25 R, E_{max} changes by about 6 per cent.

9.8.4 Dielectric Loss

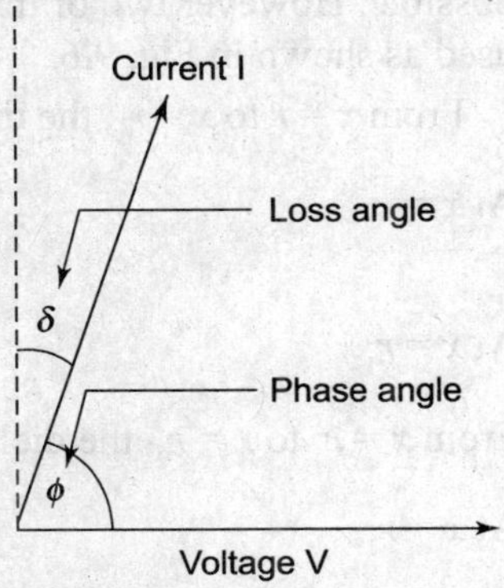

Fig. 9.5 Phasor diagram of imperfect capacitor

A cable has a capacitance between the core and the sheath. When a voltage is applied to an unloaded cable a capacitive current (or charging current) flows. Since the resistivity of the insulation is not infinite, leakage current flows and a power loss occurs. With ac voltages the phenomenon of dielectric absorption also contributes to the power loss. Thus, a cable behaves as an imperfect capacitor and the total current, under unloaded conditions, leads the voltage not by 90° but by an angle (90 – δ) as shown in Fig. 9.5. The angle δ is termed as loss angle of dielectric. The dielectric loss P_d is

$$P_d = VI\cos\phi = VI\sin\delta$$
$$= \omega CV^2 \sin\delta \qquad (9.13)$$

where C is the capacitance of the cable and V is the line to neutral voltage.

9.9 GRADING OF CABLES

The value of E_{max} has to be kept within limits which depend on the margin of safety and the permissible degree of dielectric heating. Since the insulation away from the core is under stressed, there is an avoidable waste of insulation. It seems attractive to use methods which reduce the amount of insulation by redistribution of stress so as to increase the stress in outer layers of insulation without increasing it at the conductor surface. Two methods for such a grading of insulation have been suggested. These are:

(a) Capacitance grading
(b) Intersheath grading

9.9.1 Capacitance Grading

This method involves the use of two or more layers of dielectrics having different permittivities, those with higher permittivities being nearer to the conductor.

The electric field intensity E_x at any radius x is given by Eq (9.5). If it were possible to vary permittivity with radius x such that

$$\varepsilon_r \propto \frac{1}{x} = \frac{m}{x}$$

Then
$$E_x = \frac{q}{2\pi\varepsilon_0\left(\frac{m}{x}\right)(x)}$$

$$= \frac{q}{2\pi\varepsilon_0 m} \qquad (9.14)$$

Thus E_x is constant throughout the thickness of insulation. Such a gradation is evidently, not possible. However two or three dielectrics with different values of relative permittivity can be used as shown in Fig. 9.6.

From $x = r$ to $x = r_1$, the dielectric with relative permittivity ε_1 is used.

At $x = r$,
$$E = \frac{q}{2\pi\varepsilon_0\,\varepsilon_1\, r} \qquad (9.15)$$

At $x = r_1$,
$$E = \frac{q}{2\pi\varepsilon_0\,\varepsilon_1\, r_1} \qquad (9.16)$$

From $x = r_1$ to $x = r_2$, the dielectric with relative permittivity ε_2 is used

At $x = r_1$,
$$E = \frac{q}{2\pi\varepsilon_0\,\varepsilon_2\, r_1} \qquad (9.17)$$

At $x = r_2$,
$$E = \frac{q}{2\pi\varepsilon_0\,\varepsilon_2\, r_2} \qquad (9.18)$$

From $x = r_2$ to $x = R$, the dielectric with relative permittivity ε_3 is used

At $x = r_2$,
$$E = \frac{q}{2\pi\varepsilon_0\,\varepsilon_3\, r_2} \qquad (9.19)$$

At $x = R$,

$$E = \frac{q}{2\pi\varepsilon_0\,\varepsilon_3\,R} \tag{9.20}$$

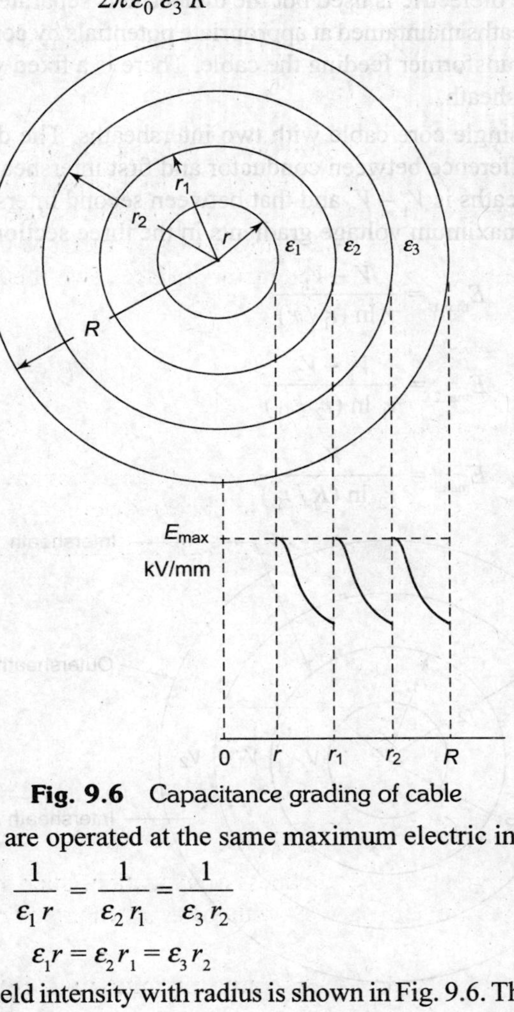

Fig. 9.6 Capacitance grading of cable

If all the three dielectrics are operated at the same maximum electric intensity, then

$$\frac{1}{\varepsilon_1 r} = \frac{1}{\varepsilon_2 r_1} = \frac{1}{\varepsilon_3 r_2}$$

or

$$\varepsilon_1 r = \varepsilon_2 r_1 = \varepsilon_3 r_2 \tag{9.21}$$

The variation of electric field intensity with radius is shown in Fig. 9.6. The operating voltage V is

$$V = \int_r^{r_1} E_x\,dx + \int_{r_1}^{r_2} E_x\,dx + \int_{r_2}^{R} E_x\,dx$$

$$= \int_r^{r_1} \frac{q}{2\pi\varepsilon_0\varepsilon_1 x}\,dx + \int_{r_1}^{r_2} \frac{q}{2\pi\varepsilon_0\varepsilon_2 x}\,dx + \int_{r_2}^{R} \frac{q}{2\pi\varepsilon_0\varepsilon_3 x}\,dx$$

$$= \frac{q}{2\pi\varepsilon_0\varepsilon_1}\ln\frac{r_1}{r} + \frac{q}{2\pi\varepsilon_0\varepsilon_2}\ln\frac{r_2}{r_1} + \frac{q}{2\pi\varepsilon_0\varepsilon_3}\ln\frac{R}{r_2}$$

$$V = \frac{q}{2\pi\varepsilon_0\varepsilon_1 r}\,r\ln\frac{r_1}{r} + \frac{q}{2\pi\varepsilon_0\varepsilon_2 r_1}\,r_1\ln\frac{r_2}{r_1} + \frac{q}{2\pi\varepsilon_0\varepsilon_3 r_2}\,r_2\ln\frac{R}{r_2}$$

$$= E_{max}\left[r\ln\frac{r_1}{r} + r_1\ln\frac{r_2}{r_1} + r_2\ln\frac{R}{r_2}\right] \tag{9.22}$$

9.9.2 Intersheath Grading

In this method only one dielectric is used but the dielectric is separated into two or more layers by thin metallic intersheaths maintained at appropriate potentials by connecting them to tappings on the winding of the transformer feeding the cable. There is a fixed voltage between the inner and outer radii of each sheath.

Figure 9.7 shows a single core cable with two intersheaths. The different radii are r, r_1, r_2, and R. The potential difference between conductor and first intersheath is $V - V_1$, that between first and second intersheaths is $V_1 - V_2$ and that between second intersheath and outer sheath is V_2. Using Eq (9.9) the maximum voltage gradients in the three sections are

$$E_{\max 1} = \frac{V - V_1}{r \ln (r_1 / r)} \tag{9.23a}$$

$$E_{\max 2} = \frac{V_1 - V_2}{r_1 \ln (r_2 / r_1)} \tag{9.23b}$$

$$E_{\max 3} = \frac{V_2}{r_2 \ln (R / r_2)} \tag{9.23c}$$

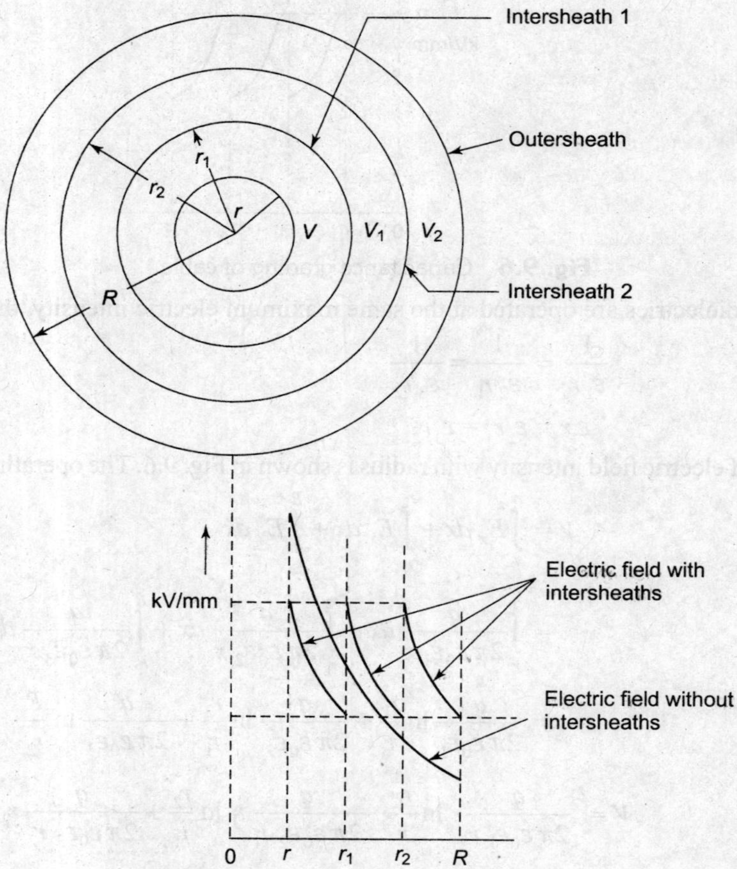

Fig. 9.7 Intersheath grading of cable

If the values of maximum and minimum potential gradients in the three sections are kept the same, we get

$$\frac{r_1}{r} = \frac{r_2}{r_1} = \frac{R}{r_2} = \alpha \tag{9.24}$$

Then

$$\frac{V - V_1}{r \ln \alpha} = \frac{V_1 - V_2}{r_1 \ln \alpha} = \frac{V_2}{r_2 \ln \alpha} \tag{9.25}$$

If the cable does not have any intersheath, the maximum voltage gradient is

$$E_{max} = \frac{V}{r \ln (R / r)} \tag{9.26}$$

Using Eqs. (9.23) and (9.26) the ratio of maximum stress with and without intersheath can be calculated. The radii of intersheaths can be found from Eq (9.24). The voltages V_1 and V_2 can be found from Eq (9.25).

Both the methods for grading of cables involve practical difficulties. With capacitance grading the difficulty exists in obtaining different insulating materials having required values of permittivities. The use of rubber ($\varepsilon_r = 4 - 6$) and impregnated paper ($\varepsilon_r = 3 - 4$) have been suggested. However, rubber is very expensive and the cost of dielectric becomes prohibitive. Moreover, the possible changes of permittivity with time may alter the stress distribution and lead to insulation breakdown even at the working voltage. With intersheath grading it may be difficult to arrange for proper voltages of intersheaths. The jointing of cable having intersheaths also poses problems. The possibility of damage to the thin intersheaths during cable laying also exists. Moreover, the charging current may cause overheating of intersheaths especially in long cables. In view of these practical difficulties neither of these methods of grading is used in practice. Wherever necessary oil filled and gas filled cables are used.

Example 9.2

Find the most economical conductor diameter for single core cables to be used on 33 kV, 3-phase system if the r.m.s. value of maximum stress is not to exceed 35 kV/cm. Also find the radial thickness of insulation.

Solution

For most economical condition

$$R = 2.718\, r \text{ and } \ln (R/r) = 1$$

Using Eq (9.9)

$$35 = \frac{33}{\sqrt{3}} \Big/ [r \ln (R/r)] = \frac{33}{\sqrt{3}} \Big/ r$$

or $$r = 0.544 \text{ cm}$$

$$R = 2.718\, r = 1.479 \text{ cm}$$

Thickness of insulation = 1.479 – 0.544 = 0.935 cm.

Example 9.3

Find the most economical conductor diameter of a single core cable to be used on 132 kV, 3-phase system. Also find the overall diameter of insulation if the permissible stress is not to exceed 60 kV/cm (peak value).

Solution

$$R = 2.718\,r \text{ and } \ln (R/r) = 1$$

Peak value of operating voltage per phase

$$= \frac{132}{\sqrt{3}}(\sqrt{2}) = 107.76 \text{ kV}$$

$$60 = 107.76/[r \ln (R/r)] = 107.76/r$$

$$r = 107.76/60 = 1.796 \text{ cm}$$

$$R = 2.718\,r = 4.88 \text{ cm}$$

$$d = 2r = 3.592 \text{ cm}$$

$$D = 2R = 9.76 \text{ cm}$$

Example 9.4

A single core cable for 66 kV 3-phase system has a conductor of 2 cm diameter and sheath of inside diameter 5.3 cm. It is required to have two intersheaths so that stress varies between the same maximum and minimum values in the three layers of dielectric. Find the positions of intersheaths, maximum and minimum stress and voltages on the intersheaths. Also find the maximum and the minimum stress if the intersheaths are not used.

Solution

Referring to Fig. 9.7 peak value of per phase voltage = V

$$V = \frac{66}{\sqrt{3}}(\sqrt{2}) = 53.8 \text{ kV}$$

$$r = 1 \text{ cm}, R = 2.65 \text{ cm}$$

Since maximum and minimum stresses are the same,

$$\frac{r_1}{r} = \frac{r_2}{r_1} = \frac{R}{r_2} = \alpha$$

or

$$\frac{R}{r} = \alpha^3$$

or

$$\alpha = (2.65/1)^{1/3} = 1.384$$

The radii of intersheaths are

$$r_1 = \alpha r = 1.384 \text{ cm}$$

$$r_2 = \alpha r_1 = 1.915 \text{ cm}$$

Using Eq. (9.25)

$$\frac{V - V_1}{(1) \ln (1.384)} = \frac{V_1 - V_2}{(1.384) \ln (1.384)} = \frac{V}{(1.915) \ln (1.384)}$$

Since $V = 53.8$ kV

We get $V_1 = 41.3$ kV (peak), $V_2 = 23.94$ kV (peak)

$$E_{\max} = \frac{V - V_1}{r \ln \alpha} = \frac{53.8 - 41.3}{\ln 1.384} = 38.46 \text{ kV/cm}$$

$$E_{\min} = \frac{V - V_1}{r_1 \ln 1.384} = \frac{53.8 - 41.3}{1.384 \ln 1.384} = 27.79 \text{ kV/cm}$$

If intersheath are not used

$$\text{Maximum stress} = \frac{53.8}{r \ln (R/r)} = 55.2 \text{ kV/cm}$$

$$\text{Minimum stress} = \frac{53.8}{R \ln (R/r)} = 20.83 \text{ kV/cm}$$

Example 9.5

A single core cable has a conductor radius r and inside sheath radius R. The operating voltage is V kV (r.m.s.). It is provided with an intersheath at radius r_1 such that maximum and minimum electric stresses in the two protions of dielectric are the same. (a) Find radius r_1 and voltage V_1 of the intersheath (b) Find the ratio of maximum electric stress with and without intersheaths (c) If $r = 1$ cm, $R = 3$ cm, $V = 60$ kV (r.m.s.). Find the values of r_1, V_1 and the ratio of maximum electric stress with and without intersheath.

Solution

(a) For same maximum and minimum values of electric stresses in the two regions of dielectric

$$\frac{r_1}{r} = \frac{R}{r_1} = \alpha$$

or

$$\frac{R}{r} = \alpha^2$$

or

$$\alpha = (R/r)^{0.5}$$

Radius of intersheath = $r\alpha$

$$\frac{V - V_1}{r \ln \alpha} = \frac{V_1}{r_1 \ln \alpha}$$

or

$$\frac{V - V_1}{r} = \frac{V_1}{r_1} \text{ or } V - V_1 = \frac{V_1}{r_1 / r} = \frac{V_1}{\alpha}$$

or

$$V_1 = V\left(\frac{\alpha}{\alpha + 1}\right)$$

(b) When intersheath is used

$$\text{Max. electric stress} = \frac{V_1}{r_1 \ln \alpha} = \frac{V[\alpha / (\alpha + 1)]}{r_1 \ln \alpha} = \frac{V[1/(\alpha + 1)]}{r \ln \alpha}$$

When intersheath is not used

$$\text{Max. electric stress} = \frac{V}{r \ln (R/r)} = \frac{V}{r \ln (\alpha^2)} = \frac{V}{2r \ln (\alpha)}$$

$$\frac{\text{Max. electric stress with intersheath}}{\text{Max. electric stress without intersheath}}$$

$$= \frac{V[1/(\alpha+1)]}{r \ln (\alpha)} \times \frac{2r \ln (\alpha)}{V} = \frac{2}{1+\alpha}$$

(c)

$$\alpha^2 = \frac{R}{r} = 3 \text{ or } \alpha = 1.732$$

$$r_1 = \alpha r = 1.732 \text{ cm}$$

$$V_1 = V\left(\frac{\alpha}{\alpha+1}\right) = 60\left[\frac{1.732}{2.732}\right] = 38.04 \text{ kV}$$

$$\frac{\text{Max. electric stress with intersheath}}{\text{Max. electric stress without intersheath}} = \frac{2}{2.732} = 0.732$$

Example 9.6

Find the maximum working voltage of a single core cable having two insulating materials A and B and the following data. Conductor radius 0.5 cm, inside radius 2.5 cm, maximum working potential gradient of A 60 kV/cm, maximum working potential gradient of B 50 kV/cm, relative permittivities of A and B, 4 and 2.5 respectively.

Solution

$$r = 0.5 \text{ cm}, R = 2.5 \text{ cm}$$

Let the first dielectric layer A be upto radius r_1. Between radii r_1 and R is the layer of dielectric B. Then,

$$60 = \frac{q}{2\pi\varepsilon_0\varepsilon_A r} = \frac{q}{2\pi\varepsilon_0 \times 4 \times 0.5}$$

or $$\frac{q}{2\pi\varepsilon_0} = 120$$

$$50 = \frac{q}{2\pi\varepsilon_0\varepsilon_B r_1} = \frac{120}{2.5 r_1}$$

or $$r_1 = 0.96 \text{ cm}$$

$$V = \frac{q}{2\pi\varepsilon_0\varepsilon_A}\ln\frac{r_1}{r} + \frac{q}{2\pi\varepsilon_0\varepsilon_B}\ln\frac{R}{r_1}$$

$$= \frac{120}{4}\ln\frac{0.96}{0.5} + \frac{120}{2.5}\ln\frac{2.5}{0.96}$$

$$= 65.51 \text{ kV}$$

Example 9.7

A 33 kV 3-phase underground feeder, 3.4 km long, uses three single core cable. Each cable has a conductor diameter of 2.5 cm and the radial thickness of insulation is 0.6 cm. The relative permittivity of dielectric is 3.1. Find (a) capacitance of the cable per phase (b) charging current per phase (c) total charging kVar (d) dielectric loss per phase if the power factor of unloaded cable is 0.03 (e) maximum stress in the cable.

Solution

(a) $r = 1.25$ cm, $R = 1.25 + .6 = 1.85$ cm.

Using Eq. (9.7)

$$C = \frac{2\pi \times 8.85 \times 10^{-12} \times 3.1}{\ln\frac{1.85}{1.25}} = 439.7 \times 10^{-12} \text{ F/m}$$

for 3.4 km length, $$C = 439.7 \times 10^{-12} \times 3.4 \times 1000$$

$$= 1.495 \times 10^{-6} \text{ F/ phase}$$

(b) Charging current $= 33 \times \frac{10^3}{\sqrt{3}} \times 2\pi \times 50 \times 1.495 \times 10^{-6}$

$= 8.95$ A per phase

(c) Total charging kVar for 3 phases

$= 3 \times (33000/\sqrt{3}) \times 8.95 \times 10^{-3} = 511.5$ kVar

(d) $\cos \phi = .03$ $\phi = 88.281°$, $\delta = 90 - \phi = 1.719°$

Dielectric loss per phase

$= 2\pi \times 50 \times 1.495 \times 10^{-6} \times (33000/\sqrt{3})^2 \sin(1.719°)$

$= 5114.6$ watts

(e) Using Eq. (9.9)

$$E_{max} = (33/\sqrt{3})/[1.25 \ln(1.85/1.25)] = 38.88 \text{ kV/cm (rms)}$$

9.10 CAPACITANCE OF THREE CORE BELTED CABLE

Three core belted cables are used only upto 11 kV. In this cable a potential difference exists between any two pairs of conductors and also between each conductor and sheath. Thus, there is an electric field between any two pairs of conductors and also between each conductor and sheath. Consequently, there is capacitance, C_c between any two pairs of conductors and capacitance C_s between each conductor and sheath as shown in Fig. 9.8. The overall field pattern is very complicated and can be studied only in an electrolytic tank. The capacitances can also be more easily obtained by measurement.

The three delta connected capacitance C_c in Fig. 9.8 can be replaced by three star connected capacitances each of value 3 C_c using delta star transformation as shown in Fig. 9.9. The capacitance to sheath can be assumed to be in parallel with the star connected capacitance of each core to earthed neutral, i.e., $C = 3\ C_c + C_s$.

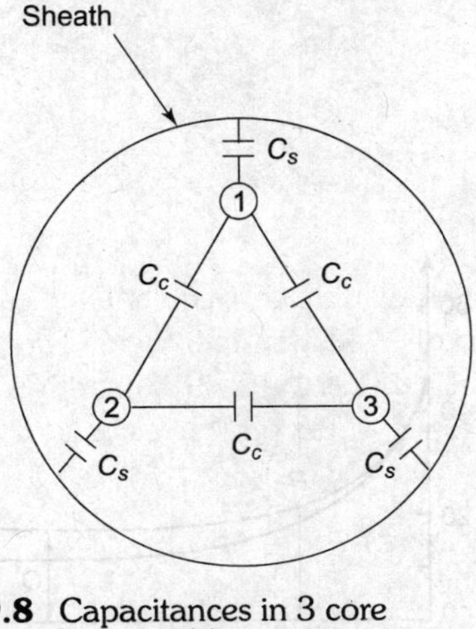

Fig. 9.8 Capacitances in 3 core belted cable

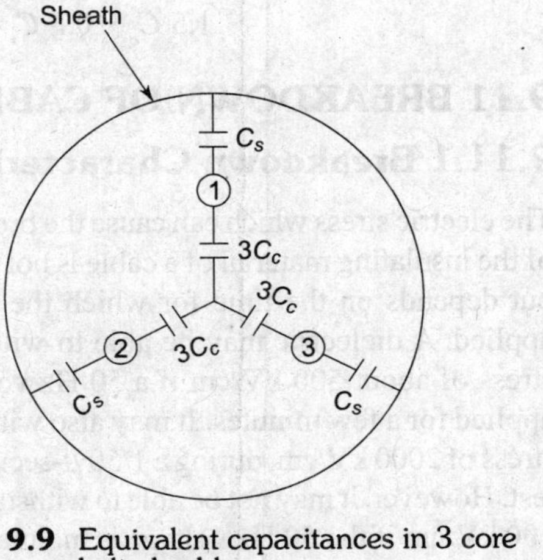

Fig 9.9 Equivalent capacitances in 3 core belted cable

The capacitances C_c and C_s are obtained by the following measurements:

(a) Any two conductors are connected to the sheath and the capacitance C_a between this combination and the third conductor is measured. A little observation will show that

$$C_a = 2\ C_c + C_s \tag{9.27}$$

(b) All the three conductors are joined together and the capacitance C_b between this combination and sheath is measured. This measured value is:

$$C_b = 3\, C_s \tag{9.28}$$

From Eqs. (9.27) and (9.28)

$$C_s = C_b/3 \text{ and } C_c = (C_a/2) - (C_b/6)$$

The effective capacitance between each core and earthed neutral is

$$C = 3\, C_c + C_s = 1.5\, C_a - 0.167\, C_b \tag{9.29}$$

Example 9.8

A one kilometer long 11 kV, 3-phase 3-core, metal sheathed cable gave the following results in a test for capacitance.

(i) Capacitance between two conductors joined to sheath and the third conductor 0.65 μF.
(ii) Capacitance between all the three conductors joined together and the sheath 0.75 μF.

Find (a) effective capacitance of each core to the neutral (b) charging current per phase (c) capacitance between any two conductors.

Solution

(a)
$$C_s = 0.75/3 = 0.25\ \mu\text{F}.$$
$$C_s = 0.65/2 - (0.75/6) = 0.2\ \mu\text{F}$$
$$C = C_s + 3\, C_c = 0.85\ \mu\text{F}$$

(b)
$$\text{Charging current} = (11 \times 10^3/\sqrt{3}) \times 2 \times \pi \times 50 \times 0.85 \times 10^{-6} = 1.696 \text{ A}$$

(c) From Fig. 9.9, capacitance between any two conductors is

$$1.5\, C_c + 0.5\, C_s = 0.425\ \mu\text{F}$$

9.11 BREAKDOWN OF CABLES

9.11.1 Breakdown Characteristics

The electric stress which can cause the breakdown of the insulating material of a cable is not constant but depends on the time for which the stress is applied. A dielectric may be able to withstand a stress of about 500 kV/cm if a 50 Hz voltage, is applied for a few minutes. It may also withstand a stress of 2000 kV/cm, during a 1/50 μ-sec. impulse test. However, it may not be able to withstand even 100 kV/cm if the 50 Hz voltage is maintained for many days.

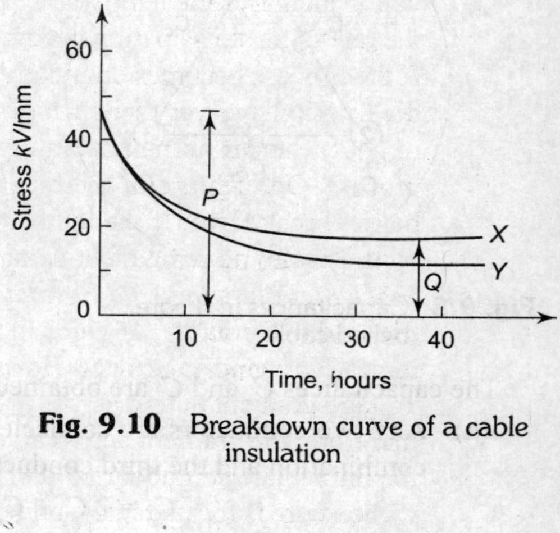

Fig. 9.10 Breakdown curve of a cable insulation

Figure 9.10 shows the electric stress versus time to breakdown curve for two dielectrics. The curve X is for a well impregnated homogeneous paper dielectric. P is the instantaneous breakdown strength for this dielectric and Q is the long term breakdown strength. The long term breakdown strength is reached in about 10 hours. The curve Y is for a poorly impregnated paper. In this case, the long term breakdown strength is reached after a long time. Moisture and airfilled voids decrease the breakdown strength and lead to cable breakdown.

The useful life of a cable depends on the electric stress and the degree of homogeneity of the dielectric. Lower the electric stress greater is the life. Greater the degree of homogeneity, greater is the life. The temperature variations displace the stress time curve downwards, the degree of displacement depending on the maximum temperature and the number of heat cycles per unit time.

9.11.2 Mechanism of Void Formation

The formation of air spaces, called voids, in the dielectric may be due to faulty material, imperfect impregnation and successive expansions and contractions due to the load cycles. As the temperature is increased, the insulating compound expands and the sheath becomes slightly distorted. When the temperature decreases, the insulating compound contracts but the distortion in the sheath remains and leads to void formation. The temperature variations across the cross-section of the cable lead to variations in the degree of viscosity. After a number of heating and cooling cycles, voids are almost inevitable.

9.11.3 Types of Breakdown

Cable breakdown may be due to puncture, thermal instability and tracking following void formation.

(a) *Puncture or Disruptive Breakdown:* In this breakdown a discharge between the conductor and the sheath at a weak point in the dielectric and a hole is formed in the dielectric. A defect in the dielectric which may lead to puncture is indicated in the initial laboratory test on a specimen of the insulation when the applied voltage is raised quickly. However, this breakdown is rare in service.

(b) *Thermal Instability:* The charging current of a cable leads the voltage by an angle $(90° - \delta)$ where δ is the loss angle of the dielectric (as discussed in section 9.8.4). The dielectric loss and the ohmic loss in the conductor heat the insulation and raise its temperature. After a certain temperature is reached, a further increase in temperature increases the loss angle which causes an increased power dissipation in the dielectric which increases the temperature even further. The process becomes cumulative unless the rate of heat dissipation becomes equal to or greater than the rate of heat generation. If the process becomes cumulative the dielectric breakdown may occur. Modern cable dielectrics have very small loss angles (less than 0.5°) and thermal instability does not occur under normal operating conditions except perhaps after the start of tracking process. One feature of thermal instability is that no permanent damage occurs until before breakdown. If the temperature rise is halted (by a decrease in the load carried by the cable) no permanent damage will be caused to the cable.

(c) *Tracking Following Void Formation:* The repeated heating and cooling of the dielectric lead to formation of voids in the dielectric. If the body of the dielectric is uniform, ionization cannot take place. However, the presence of voids can result in ionization. The voltage gradient acting across a void is inversely proportional to the permittivity and, therefore, the gradient across a void may be three to four times that in the dielectric in the immediate neighbourhood of a void. As a result of ionization in the voids, a minute carbon core is formed in the oil between the fibres of the paper at the point of maximum stress near the conductor surface. This leads to tracking, i.e., formation of carbonized track which gradually extends outwards both radially and longitudinally with tree burning pattern on the surface of paper tape till the dielectric is bridged and breaks down. The carbonized track may take a long times to build up but once carbon core has been formed, the breakdown cannot be prevented. Most of the cable breakdowns belong to this category.

9.12 CABLE INSTALLATION

Underground cables are laid either direct in ground or in ducts. Laying the cables directly in the ground is the cheapest, simplest and the most commonly used method. A trench is dug in the ground and the cable is laid on a bedding of a sand. The cable is covered with soil to a depth of about 10 cms and then a trench cover, usually of reinforced concrete, is placed. Then the remaining dug out soil is filled again in the trench. The depth of burial is about 0.45 m for cables below 1.1 kV rating 0.75 m for cables between 3.3 to 11 kV rating and 1 m for cables above 11 kV rating. There should be a sand layer, about 8 cm, below the cable. For cables laid adjacent to each other, there should be about 0.3 m horizontal spacing, filled with sand or riddled soil, between the cables. This spacing is necessary to reduce the effect of mutual heating and to ensure that fault in one cable does not affect the other. The advantages of this method are: low initial cost, ease of installation and higher current carrying capacity of cables (due to better heat dissipation). The disadvantages are: difficulty in maintenance, repair and replacement, danger of mechanical damage and deterioration due to chemical and electrolytic corrosion. It is necessary to preserve a detailed drawing showing location, type and size of cable, details of junctions and joints for future references.

Laying of cables in ducts made of stone ware or reinforced concrete enables easy maintenance, repair and replacement. However the initial cost of installation is high and the current carrying capacity of cables is reduced due to poor heat dissipation. This method is used only for short installations where ease of maintenance and replacement is of primary importance, e.g., road crossings, railways crossings, etc.

9.13 CURRENT RATING OF CABLES

9.13.1 Analogy between Current Flow and Heat Flow

The problem of determining the current rating of cables involves the computations of heat flow. An analogy between current flow and heat flow exists because the electrical and thermal fields are similar, the isothermals and heat flow lines in the thermal case corresponding to the equipotentials and flux lines in the electrical case. The current corresponds to heat flow (watts) and potential difference corresponds to temperature difference (degrees centigrade). The thermal form of Ohm's law gives the amount of heat transferred across a layer whose opposite faces have a temperature difference and is given by:

$$\text{Heat flow in watts} = \frac{\text{Temperature difference (°C)}}{\text{Thermal resistance in thermal ohms}} \tag{9.30}$$

A thermal ohm is that thermal resistance which needs a temperature difference of 1 °C across it to cause a heat flow of 1 watt. Thus the thermal ohm has the units of °C/watt. The thermal resistivity (symbol g) is analogous to electrical resistivity and has the units of °C – m/watt or (°C – cm/watt).

9.13.2 Thermal Field

The heat generated in the cable due to various losses raises the temperature of the cable. This heat is dissipated to the soil through the dielectric, sheath, armour and serving. The dielectric, serving and soil have thermal resistances through which this heat flows. Metals can be regarded as having zero thermal resistance so that metal sheath and armour can be left out of the calculations except in so far as they are sources of heat. The maximum permissible temperature rise in the cable depends on type of cable (belted, screened, sheathed, etc.), number of cores, sheath metal (lead or aluminium), method of installation (directly buried or laid in ducts or in air) and presence of armouring. The specified maximum temperature rise is known for a variety of above factors. The heat generated can be expressed as a function of the current in the cable. If the total thermal resistance of the heat path can be calculated, the maximum current rating can be determined. So the problem of calculating the current rating reduces to the problem of determining the thermal resistances of different components of the heat path. These components are cable insulation, protective covering and soil.

9.13.3 Thermal Resistances

(a) *Insulation of Single Core Cable*

The thermal resistance of the insulation of a single core cable can be determined exactly in the same way as the insulation resistance. Referring to Fig. 9.4 the thermal resistance dG_i of an annulus of thickness dx at radius x is

$$dG_i = \frac{g_1\, dx}{2\pi x} \text{ thermal ohms/metre}$$

where g_1 is the thermal resistivity of the dielectric and is about 5.5 °C–m for most of high voltage cables. The thermal resistance per metre length is

$$G_i = \int_r^R \frac{g_1\, dx}{2\pi x} = \frac{g_1}{2\pi} \ln \frac{R}{r} \text{ thermal ohms/metre} \tag{9.31}$$

(b) *Insulation of Multi Core Cable*

The calculation of thermal resistance of insulation of a multi core cable is quite difficult because the thermal field is quite complex.

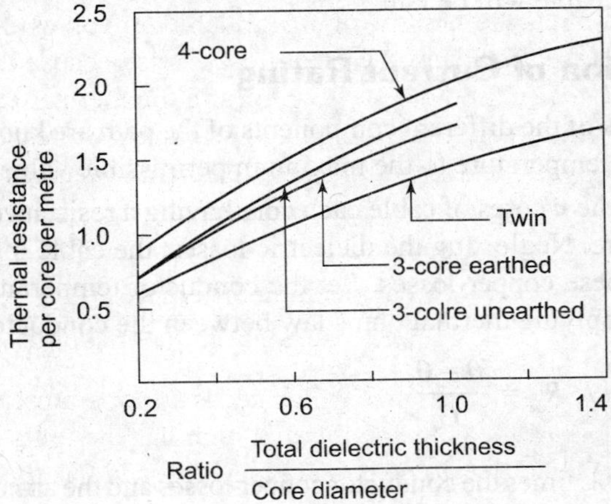

Fig. 9.11 Thermal resistance of belted cable for a thermal resistivity of 5.5°C–m/W

Figure 9.11 shows a set of curves, based on experimental results, for determining the thermal resistance of belted cable for a thermal resistivity of 5.5 °C – m/watt. In these curves the abcissa is the ratio of the total dielectric thickness (i.e., thickness of core insulation + belt insulation) to the core diameter. The ordinate is thermal resistance per core per metre length. The value of thermal resistance, as obtained from these curves must be divided by the number of cores before using it in further calculations.

(c) *Protective Covering*

Since the protective covering (bedding and serving) is in the form of cylinder, the expression for its thermal resistance is similar to Eq. (9.31). The armour thickness is generally very small and can be neglected. Thus, the expression for thermal resistance of protective covering is:

$$G_p = \frac{g_2}{2\pi} \ln \frac{R_2}{R_1} \text{ thermal ohms/metre} \tag{9.32}$$

Where g_2 = thermal resistivity of bedding and serving (about 5°C–m/watt)

R_1 = radius over the sheath, m

R_2 = external radius of the cable, m

(d) *Soil*

The heat transmitted through the cable passes into the surrounding soil. In the ideal case of a single cable of external radius *R*, buried *h* meter below the ground surface in a soil of constant thermal resistivity g_3, the thermal field is exactly analogous to the electric field between a single overhead line conductor and the ground plane. The expression for thermal resistance G_s is, therefore, derived in a manner similar to the calculation of capacitance of an overhead conductor with respect to ground. The expression is

$$G_s = \frac{g_3}{2\pi} \ln \frac{2h}{R_2} \text{ thermal ohms/metre}$$

where g_3 is the thermal resistivity of soil in °C – m/watt. The thermal resistivity of soil depends on the type of soil and the amount of moisture present in the soil and varies from about 0.3 °C – m/watt (for water logged soil) to about 3.0 °C – m/watt for sandy soils with small moisture content. Since the thermal resistivity of soil varies between very wide limits and also varies with season, it is often desirable to survey the cable route and determine the actual value for the soil in which the cable will be laid.

9.13.4 Calculation of Current Rating

If the thermal resistances of the different components of the path are known, the current *I* which will raise the conductor temperature to the maximum permissible value can be calculated.

The copper losses in the *n* cores of cable each core having a resistance of R_{ac} ohms per metre is $n I^2 R_{ac}$ watts per metre. Neglecting the dielectric losses, the cable insulation has to pass the heat generated due to these copper losses. Let the conductor temperature be θ °C and sheath temperature be θ_s °C. Applying thermal ohms law between the conductor and sheath we have

$$n I^2 R_{ac} = \frac{\theta - \theta_s}{G_i} \tag{9.34}$$

If the sheath losses are λ_1 times the conductor copper losses and the armour losses are neglected, the total losses become $(1 + \lambda_1)\, n I^2 R_{ac}$ watts. The protective covering and the soil has to pass the heat generated due to these total losses. Let ambient temperature be θ_a °C. Applying thermal ohm's law between sheath and ground surface we have

$$(1 + \lambda_1)\, nI^2 R_{ac} = \frac{\theta_s - \theta_a}{G_p + G_s} \tag{9.35}$$

From Eqs. (9.34) and (9.35)

$$nI^2 R_{ac}\, G_i + (1 + \lambda_1)\, n\, I^2 R_{ac}\,(G_p + G_s) = \theta - \theta_a$$

or

$$I = \left\{ \frac{\theta - \theta_a}{n R_{ac}\, [G_i + (1+\lambda_1)\,(G_p + G_s)]} \right\}^{0.5} \tag{9.36}$$

9.13.5 Effect of Dielectric Losses

It is necessary to consider the dielectric losses for cables of 66 kV and above. These losses are independent of load current except in so far as the current causes an increase in the temperature of the dielectric. The method is to calculate the dielectric loss per metre length at the maximum permissible temperature. Assume that this loss is concentrated at the core, calculate the temperature rise θ_d which it would produce if it were the only source of heat and use $\theta - \theta_d$ in place of θ in Eq. (9.36). Thus, Eq. (9.36) becomes

$$I = \left\{ \frac{\theta - \theta_d - \theta_a}{n R_{ac}\, [G_i + (1+\lambda_1)\,(G_p + G_s)]} \right\}^{0.5} \tag{9.37}$$

Example 9.9

A 3-core 11 kV aluminium conductor paper insulated lead sheathed belted cable is laid 0.75 m below ground. Find the current rating of the cable using the following data:

radius of each core	13 mm
nominal area of cross-section	400 mm^2
thickness of core insulation	2.7 mm
thickness of belt insulation	1.2 mm
outside radius of cable	40 mm
radius of cable over sheath	35.2 mm
increase in resistance due to skin effect	3.5 per cent
increase in resistance due to proximity effect	3.5 per cent
resistivity of aluminium at 20°C	0.02826 $\mu\Omega$ – m
temperature coefficient of resistance	0.004 per °C
increase in resistance due to stranding and laying	2 per cent each
sheath losses	10% of total conductor loss
thermal resistivity of dielectric	5.5°C– m/watt
thermal resistivity of protective covering	5°C– m/watt
thermal resistivity of soil	1.5°C – m/watt
ambient temperature	20°C
maximum permissible conductor temperature	65°C

neglect dielectric losses.

Solution

$$\text{Conductor resistance} = \frac{0.02826 \times 10^{-6} \times 1000 \times 1000}{400} = 7.065 \times 10^{-5} \text{ ohm/m}$$

Allowing for the increase in resistance due to stranding and laying

$$R_{dc} \text{ at } 20\text{ °C} = 1.02 \times 1.02 \times 7.065 \times 10^{-5} = 7.35 \times 10^{-5} \text{ ohm/m}$$

$$R_{dc} \text{ at } 65\text{ °C} = 7.35 \times 10^{-5} (1 + 0.004 \times 45) = 8.673 \times 10^{-5} \text{ ohm/m}$$

Allowing for the increase in resistance due to skin effect and proximity effect

$$R_{ac} = 1.035 \times 1.035 \times 8.673 \times 10^{-5} = 9.29 \times 10^{-5} \text{ ohm/m}$$

total insulation thickness = 3.9 mm

$$\text{ratio of total insulation thickness to core diameter} = \frac{3.9}{13 \times 2} = 0.15$$

Using Fig. 9.11 (and extrapolating) thermal resistance per core per metre = 0.65 thermal ohms per metre:

Since it is a three-core cable

$$G_i = 0.65/3 = 0.217 \text{ thermal ohms per metre}$$

$$R_2 = 40 \text{ mm} = 40 \times 10^{-3} \text{ m}$$

$$R_1 = 35.2 \text{ mm} = 35.2 \times 10^{-3} \text{ m}$$

$$G_p = \frac{5}{2\pi} \ln \frac{40 \times 10^{-3}}{35.2 \times 10^{-3}}$$

$$= 0.102 \text{ thermal ohms per metre}$$

$$G_s = \frac{1.5}{2\pi} \ln \frac{2 \times 0.75}{40 \times 10^{-3}} = 0.865 \text{ thermal ohms per metre}$$

$$\lambda_1 = 0.1$$

Using Eq. (9.36)

$$I = \left\{ \frac{65 - 20}{3 \times 9.29 \times 10^{-5} \, [0.217 + 1.1\,(.102 + .865)]} \right\}^{0.5}$$

$$= 355\text{A}$$

9.14 AERIAL BUNCHED CABLES

Introduction

Aerial Bunched Cables (ABC) is a very novel concept for Over Head Power distribution. When compared to the conventional bare conductor over head distribution system. ABC provides higher safety and reliability, lower power losses and ultimate system economy by reducing installation, maintenance and operative cost. This system is ideal for rural distribution and specially attractive for installation in difficult terrains such as hilly areas, forest areas, coastal areas etc.

Aerial Bunched Cables is also considered to be the best choice for power distribution in congested urban areas with narrow lanes and by-lanes. In developing urban complex, Aerial Bunched Cables is the better choice because of flexibility for rerouting as demanded by changes in urban development plan.

9.14.1 Construction of ABC

XLPE/HDPE insulated power conductors of Aluminium (Neutral conductor and street lighting conductor if and when necessary) are laid together (twisted) around a high tensile standard and galvanized steel (Aluminium Alloy may be used) insulated or bare messenger wire to form the Aerial Bunched Cable. This assembly is directly strung on to distribution pole/towers by means of standard hardware available in the market but care shall be taken to render the messenger wire completely insulated from Earthing at any point of distribution in case of HT ABC.

The XLPE (Cross-linked Polyethylene) insulation is black in colour and is stabilized against deterioration caused by exposure to direct sunlight and ultraviolet radiation. XLPE is cross-linkable low-density polyethylene, which is made thermo set by special formulation from base polymer of thermoplastic low-density polyethylene. XLPE combines the best electrical properties of LDPE and superior thermo mechanical properties.

9.14.2 Materials

- Aluminium conductors conform to IS:8130 (Class-II)
- Standard high tensile galvanized steel messenger wire conforms to IS:398 (Part-2). Alternatively Aluminium Alloy messenger wire conforms to IS:398 (Part-4)
- XLPE and HDPE insulation of power conductors specify IS:7098 (Part-I and II) and IS:6474 respectively.

Since, the tension from the current carrying conductor is totally removed by introduction of messenger wire the operating temperature of the conductor is 90°C as against 75°C of the bare conductor of the same size.

9.14.3 Stringing

No difficulty is envisaged during stringing of Aerial Bunched Cables in the conventional method but care shall be taken that insulated conductors do not get damaged during installation. Dragging the ABC on the ground is to be avoided. Tension to be applied during stringing shall be 25% of the breaking load of the messenger wires. This will allow line to have sag within specified limit of 1.5% of the span at the lowest ambient temperature.

9.14.4 Jointing

While mid-span jointing permissible for LT ABC system by conventional technique, the recommendation is to draw the line in such a way as to bring the joints at the supports. Mid-span joining is not at all recommended in the case of HT lines the recommendation is for outdoor type HV terminations only. Under unavoidable circumstances, line tapping at the support points may be allowed through suitably designed clamp connectors/PG clamps. The semi conducting screen continuity shall be maintained at all joints as far as possible to avoid fluctuations during system disturbances. The 3-phase screens may be shorted and earthed through suitable non-linear surge arrestor. Attempting to make a tap off from power conductors in the region where catenary is under tension in not recommended.

9.14.5 Application

Aerial Bunched Cables can be conveniently used:

- As replacement of bare lines in Rural Areas, in woods and in other localities and narrow streets where the space limited.
- As replacement of bare lines where reliability of supply is of prime importance.
- As replacement of bare lines where high degree of stability of supply voltage is of importance.
- In hilly terrains where cost of erection of overhead lines or under ground cable becomes very high.
- As reinforcement of existing system without increasing voltage.
- For temporary supplies.

9.14.6 Advantages

In comparison to bare overhead power distribution lines, ABC has very high reliability in maintaining services because power and neutral conductors are insulated with the best dielectric medium, resulting in the following advantages:

- Less fault rate on account of good protection against line and ground fault by high winds or falling trees or birds especially in hilly areas and forests as encountered in rural distribution networks.
- High insulation resistance to earth in all seasons and polluted atmospheres. Negligible leakage currents and low losses.
- Multiple circuits of Power and Telephone cables could be strung in the same set of poles or any other supports like wall etc.
- Better adaptability to run concurrently with existing overhead bare conductor system without any interference.
- High capacitance and low inductance leading to low impedance of lines.
- Lower voltage drop, higher current capacities. vis-a-vis Better Voltage Regulation.
- Longer spans and longer distance lines are possible with better system stability.
- ABC cables are much safer than bare conductors.
- It can be over hung in dense vegetation and forests.
- Additional connections can be easily and quickly made with hot-line connectors.
- Total line costs are reduced.
- Maintenance is very easy.

9.15 SYSTEM OPERATING PROBLEMS WITH UNDERGOUND CABLES

(a) *Charging current:* A cable has high capacitance which results in charging current and reactive power. If V is line to line voltage, the charging current I_c is

$$I_c = 2\,\pi f C\, V/\sqrt{3} \tag{9.38}$$

The 3-phase reactive power is $\sqrt{3}\ V I_c$. Using Eq. (9.7) we get

$$\begin{aligned}\text{Reactive power} &= \sqrt{3}\ V I_c = \sqrt{3}\ V\,(2\,\pi f V/\sqrt{3})\,[2\,\pi\,\varepsilon_0\,\varepsilon_r/\ln{(R/r)}] \\ &= 4\,\pi^2 f V^2\,\varepsilon_0\,\varepsilon_r/[\ln{(R/r)}] \text{ vars/metre}\end{aligned} \tag{9.39}$$

A 33 kV underground feeder using 3 single core cables each having a conductor diameter of 2.5 cm and 0.6 cm as radial thickness of insulation (with $\varepsilon_r = 3.1$) has a kvar requirement of 148 kvar/km and charging current 2.6 A/km.

The flow of charging current causes heating of cable. Therefore, the load current capability of cable is decreased (i.e., derating occurs) if the temperature rise is not to be exceeded. A further reduction in current carrying capacity occurs due to dielectric losses. These factors limit the lengths of cables, used in actual practice, to less than 50 km.

(b) *Overvoltage due to switching:* The interruption of capacitance currents gives rise to overvoltage (the phenomenon of capacitance switching) as discussed in Chapter 12.

(c) *Ferranti effect:* If a significant portion of transmission and distribution system has underground cables, a rise in voltage at load point may occur under no load conditions due to Ferranti effect (section 3.13). Thus, the voltage at load points may vary considerably from full load to no load conditions. It may be necessary to use shunt inductive reactors in the system to avoid increase in voltage due to Ferranti effect.

(d) *Generator stability* The ability of large generators to generate leading reactive power is limited by stability considerations. Large size generators have high synchronous reactance (about 1.5 p.u.) and their leading Mvar capability is low.

9.16 HVDC CABLES

When cables are used in direct current transmission, the main operational features are:

(a) There is no charging current.

(b) Only $I^2 R$ losses are present.

(c) The electric stress distribution in ac cables is governed by dielectric capacitance. In dc cables the stress distribution is determined by resistance of dielectric.

(d) The resistance of dielectric decreases appreciably with increase in temperature. The resistance of dielectric at 60 °C may be about 1% of the resistance at 20 °C. Therefore, the thermal considerations influence the electric stress distribution in the dielectric considerably.

(e) In ac cables the electric stress in the dielectric decreases from conductor to the sheath. In dc cables the electric stress increases and can be large at the sheath. This phenomenon known as stress inversion may lead to troubles at terminations and joints where longitudinal stresses are created.

Because of above features dc cables can have very high current and voltage ratings. These cables have lesser insulation thickness and are more compact. DC cables are always single core and construction is similar to Fig. 9.1. The conductor is always stranded. A cotton tape covers the plastic oversheath and galvanized steel armour is put over the cotton tape for mechanical protection. As an anti-corrosion measure impregnated fibrous wrapping (or serving) covers the armour.

DC cables may have plastic or impregnated paper insulation. Oil filled and gas filled dc cables are also used. Cables having length more than 100 km have been used in many countries.

Example 9.10

A single core lead sheathed cable is graded by using three dielectrics of relative permittivities 5, 4 and 3. The conductor diameter is 2 cm. If the three dielectrics are worked at the same maximum stress of 40 kV/cm find safe working voltage of cable. What will be the value of safe working voltage if grading is not used. Assume same conductor overall diameter and maximum dielectric stress. [*RTM Nagpur Univ. 2010*]

Solution

$$r = 1 \text{ cm}, \varepsilon_1 = 5, \varepsilon_2 = 4, \varepsilon_3 = 3$$

Using Eq. (9.21)

$$\varepsilon_1 r = \varepsilon_2 r_1 = \varepsilon_3 r_2$$

or
$$5 \times 1 = 4r_1 = 3r_2$$

$$r_1 = \frac{5}{4} \text{ cm}, r_2 = \frac{5}{3} \text{ cm}$$

Assuming same minimum stress

$$\varepsilon_2 r_2 = \varepsilon_3 R \quad \text{or} \quad R = \frac{4}{3} \times \frac{5}{3} = \frac{20}{9}$$

Using Eq. (9.22)

$$\text{Safe voltage } V = 40 \left[1 \ln 1.25 + 1.25 \ln \frac{5}{3 \times 1.25} + \frac{5}{3} \ln \frac{20}{9 \times 5/3} \right]$$

$$= 40\ (0.223 + 0.3596 + 0.4795)$$

$$= 42.484 \text{ kV}$$

If grading is not used

$$V = 40 \left(r \ln \frac{R}{r} \right) = 40 \ln \left(\frac{20}{9} \right)$$

$$= 31.94 \text{ kV}$$

SUMMARY

9.1 The insulation of cables may be PVC, oil impregnated paper, rubber, XLPE etc.

9.2 Underground cables are provided with sheath, armour and covering.

9.3 In pressurised cables either oil or gas pressure above atmosphere is maintained.

9.4 A belted cable has a part of insulation applied to each core individually and the remaining insulation in the form of an overall belt.

9.5 In cables sheath losses and armour losses are also important.

9.6 Insulation resistance = $\dfrac{\rho}{2\pi l}$ ln (R/r).

9.7 In a cable the ratio of maximum and minimum potential gradient is equal to (R/r)

9.8 E_{max} is minimum when $R = 2.718\, r$.

9.9 Capacitance grading and intersheath grading can help in reducing the insulation required. However these methods are rarely used in practice.

9.10 Cables may breakdown due to puncture, thermal instability and voids.

9.11 Cables may be installed directly in ground or in ducts.

9.12 Ampacity of cables can be determined by anology between heat flow and current flow.

9.13 Operating problems in system having cables include charging MVAr, switching overvoltages, ferranti effect and generator instability.

SELF TEST QUESTIONS

A. State whether the following are true (T) or false (F)

9.1 Dielectric strength of oil impregnated cable is about 20 kV/mm.
9.2 PVC is inert to oxygen and almost inert to oil, many acids and alkalies.
9.3 Gas pressure cables are used upto 100 kV only.
9.4 Aluminium sheath has lesser mechanical strengths than lead sheath.
9.5 Armour in necessary so that moisture may not affect the insulation.
9.6 Inductance of a 3 core belted cable may be about 0.3 mH/km.
9.7 Maximum potential gradient in a cable occurs at surface of conductor.
9.8 Dielectric losses in a cable are proportional to $(\text{voltage})^2$.
9.9 Grading in cables is used very commonly.
9.10 Cable breakdown is generally due to void formation.

B. Fill in the blanks units suitable words

9.11 Thermal resistivity of sandy soil is about ________.
9.12 Thermal resistivity has the units of ________
9.13 The reactive power due to capacitance of cable is proportional to square of ________.
9.14 In VIR cables the vulcanizing agent is generally ________.
9.15 For PVC cables the maximum continuous temperature can be upto about ________.
9.16 The cost of aluminium sheath is ________ than that of lead sheaths.
9.17 Sheath losses and intersheath losses can be eliminated by ________ of cable sheaths.
9.18 Maximum potential gradient in a cable insulation is minimum when $R =$ ________.
9.19 Insulation resistance of cable is ________ to length.
9.20 Loss angle of a good dielectric is about ________.

ANSWERS

9.1 T	9.2 T	9.3 F	9.4 F	9.5 F	9.6 T
9.7 T	9.8 T	9.9 F	9.10 T	9.11 3 °C-m/watt	9.12 °C- m/watt
9.13 voltage	9.14 sulphur	9.15 75 °C	9.16 more	9.17 cross bonding	9.18 2.718 r
9.19 inversely proportional	9.20 2				

JOB INTERVIEW QUESTIONS

9.1 How are cables installed?
9.2 Name the various types of cables.
9.3 What is a belted cable ?
9.4 What is insulation resistance ?
9.5 Which is the most commonly used insulating material for cables and why?
9.6 Why is sheath used in cables?

9.7 What materials are used for sheaths?

9.8 What is the necessity of armour in cables?

9.9 On what factors does ampacity of cables depend?

9.10 What is a gas pressure cable?

REVIEW QUESTIONS

9.1 Give reasons:
 (a) Impregnated paper is the most commonly used dielectric material for power cables.
 (b) Power cables for 66 kV and above are mostly pressurized cables.
 (c) The effective resistance of a cable is higher than its dc resistance.
 (d) Cross bonding of sheaths is very commonly used for long high voltage cables.
 (e) Single core cables are usually not provided with steel armour.
 (f) Grading of insulation is not used in actual practice.

9.2 Discuss why:
 (a) The current rating of a cable buried in soil is less than that of a similar cable in air.
 (b) Power cables are generally laid directly in soil and very rarely laid in duct.
 (c) Power cables are always provided with a metal sheath.
 (d) XLPE cables are becoming very popular for underground distribution systems.
 (e) The capacitance of multi core cable is generally obtained experimentally.
 (f) Loss angle of a cable should be very small.

9.3 What materials are used for the insulation of cables? Discuss the advantages of each.

9.4 Draw the cross-section of a 3-core belted cable. Discuss the function of each part.

9.5 How are cables classified? Give the application of each type of cable.

9.6 What are oil filled cables? Describe the construction of an oil filled cable.

9.7 What is the function of sheath in cables? How are sheath losses reduced in modern multicore cables.

9.8 Derive a formula for (a) insulation resistance (b) capacitance of a single core cable.

9.9 Dervie a formula for the electric stress in a single core cable. Where is the stress maximum? Where is it minimum?

9.10 Derive a relation between the conductor radius and inside sheath radius of a single core cable so that the electric stress of the conductor surface may be minimum.

9.11 Describe an experiment to determine the capacitance of a belted cable.

9.12 What are the causes of cable breakdown? What are voids? How are they formed? Why do voids lead to cable failure? Explain the terms 'thermal instability' and 'tracking'.

9.13 Discuss the analogy between heat flow and current flow. Prepare a table giving the corresponding analogous quantities.

9.14 How can the current rating of a cable be determined? What factors affect this rating?

9.15 Derive a formula for calculating the current rating of a cable.

9.16 Discuss the methods for grading of cables. Why are they not used generally?

PROBLEMS

9.1 A single core cable, composed of a conductor of 2 mm diameter, has an insulation thickness of 1 mm and an insulation resistance of 450×10^6 ohms per km. What thickness of a similar material would be needed for a 3 mm diameter conductor cable in order to have an insulation resistance of 900×10^6 ohms per km. [4.5 mm]

9.2 A single core 2 km long cable has a conductor radius of 13 mm and an insulation thickness of 4.6 mm. If the resistivity of dielectric is 7×10^{12} ohm-m, find the insulation resistance of the cable. [168.75×10^6 ohms]

9.3 A single core cable has a conductor radius of 14.5 mm and an insulation thickness of 4.6 mm. Find the capacitance per metre length of cable if the dielectric has a relative permittivity of 3.2. [645.79×10^{-12} F/m]

9.4 A single core cable, 2.1 km long, has a conductor radius of 13 mm and an insulation thickness of 5.8 mm. The dielectric has a relative permittivity of 2.8. Find the capacitance per metre length of cable. [0.866 μF]

9.5 3 single core cables of problem 8.3 are used for an 11 kV, 3-phase 2 km long feeder. Find (a) charging current (b) total charging kvar. [2.577 A, 49.1 kvar]

9.6 An 11 kV single core cable has a conductor radius of 8.8 mm and an insulation thickness of 4.6 mm. Find the maximum and minimum electric stress. [29.73 kV/cm, 19.5 kV/cm]

9.7 A 33 kV, 3-phase, 2.5 km long feeder uses single core cables having a conductor radius of 13 mm and an insulation thickness of 8.8 mm. The dielectric has a relative permittivity of 3.05 and the power factor of the unloaded cable is 0.03. Find the capacitance per phase, charging current per phase, total charging kvar, dielectric loss per phase and maximum electric stress in the cable. [0.82×10^{-6} F, 4.91 A, 280.6 k Var, 2.805 kW, 28.35 kV/cm]

9.8 A 2.2 km long 11 kV, 3-phase, 3-core, belted cable gave the following results in a test for capacitance:

(i) Capacitance between two conductor joined to sheath and the third conductor 1.5 μF.

(ii) Capacitance between all the three conductors joined and sheath 1.8 μF.

Find (a) effective capacitance of each core to neutral (b) capacitance between any two cores. [1.95 μF. 0.975 μF]

9.9 A single core cable laid in ground has the following data: conductor diameter 2.5 cm, insulation thickness 4 cm, thermal resistivity of insulation 5 °C– m/watt, R_{ac} 100 $\times$ 10^{-6} ohms per metre, current rating 500A. Find the thermal resistance between the sheath and ground surface for a 50 °C temperature difference between conductor and ground surface. [0.86 thermal ohms per metre length]

9.10 Using the data of example 9.9 plot a curve between the current rating of cable and the thermal resistivity of soil. Assume that the thermal resistivity of soil varies from 0.5°C– m/watt to 3.0°Cm/watt.

9.11 Find the maximum working voltage of a single core cable having two insulating material A and B and following data: conductor radius 0.5 cm, inside sheath radius 3 cm, material A: relative permittivity 5, maximum electric field strength 58 kV/cm, material B, relative permittivity 3.5, maximum electric field strength 46 kV/cm. [66.92 kV]

9.12 A single core cable has a conductor radius 2 cm and inside sheath radius 4 cm. It is provided with one intersheath so that limits of maximum and minimum electric stresses is the same in the two layers of dielectric. The system voltage is 66kV 3-phase (a) Find the radius of intersheath and its voltage (b) Find the ratio of maximum electric stresses with and without intersheath. [(a) 1.414 cm, 22.32 kV (b) 0.8285]

9.13 A single core cable for 132 kV 3-phase system has conductor radius of 0.9 cm and inside sheath radius of 6 cm. It has two intersheaths. The stress varies between the same maximum and minimum limits in the three layers of dielectric. Find radii of the intersheaths and their voltages. [1.694 cm, 3.188 cm, 64.35 kV, 42 kV]

9.14 A single core cable has two intersheaths. Its operating voltage is V, conductor radius r and inside sheath radius R. The stress varies in the same ratio in the three layers of dilelectric. Find expressions for radii and voltages of intersheaths.

$$\left[r_1 = \alpha r, \quad V_1 = V\left[\frac{1+1/\alpha}{1+1/\alpha+1/\alpha^2}\right] \quad r_2 = \alpha^2 r, \quad V_2 = \frac{V}{1+1/\alpha+1/\alpha^2} \quad \text{where } \alpha = (R/r)^{1/3} \right]$$

CHAPTER 10

Load Flow Studies

10.1 INTRODUCTION

The load flow study in a power system constitutes a study of paramount importance. The study reveals the electrical performance and power flows (real and reactive) for specified conditions when the system is operating under steady state. The load flow study also provides information about the line and transformer loads (as well as losses) throughout the system and voltages at different points in the system for evaluation and regulation of the performance of the power system under conditions known *a priori*. Further alternative plans for future expansion to meet new load demands can be analyzed and a complete information is made available through this study.

A bus is a node at which one or many lines, one or many loads and generators are connected. It is not necessary that all of them be connected at every bus. The buses are classified as P–Q (Load Bus), P–V (Generator or Voltage Controlled Bus) and V–Q (Slack Bus). The bus is indicated by a vertical line at which several components are connected.

10.2 BUS ADMITTANCE MATRIX

Load flow calculations can be made by using bus admittance matrix or bus impedance matrix. However, it is simpler to use the bus admittance matrix in these studies since it can be generated by inspection.

A power system consists of a large number of buses interconnected through transmission lines. Some of these buses have generators connected to them while loads are connected to the remaining buses. Some buses may have both generators and loads while some others may have static capacitors (or synchronous condensers) for reactive power compensation or voltage control.

Consider the simple 4-bus system of Fig. 10.1(a). S_{Gi} denotes the 3-phase complex generator power flowing into the ith bus and S_{Li} denotes the 3-phase complex load power flowing out of the ith bus.

$$S_{Gi} = P_{Gi} + jQ_{Gi} \tag{10.1a}$$

$$S_{Li} = P_{Li} + jQ_{Li} \tag{10.1b}$$

where P and Q denote the real and reactive power. At each bus the generator and load power can be combined so that the net 3-phase complex power flowing into ith bus can be written as:

$$S_i = S_{Gi} - S_{Li} = (P_{Gi} - P_{Li}) + j(Q_{Gi} - Q_{Li}) = P_i + jQ_i \tag{10.2}$$

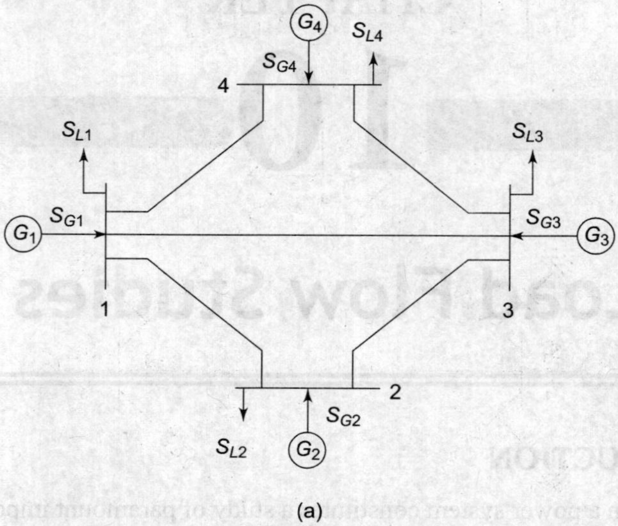

Fig. 10.1 Four bus system (a) One line diagram

Figure 10.1 (b) shows the equivalent circuit of the network of Fig. 10.1 (a). S_1, S_2, S_3, S_4, denote the net 3-phase complex power flowing into the buses and I_1, I_2, I_3, I_4 denote the currents flowing into the buses. Each transmission line is represented by a π circuit.

The equivalent circuit of Fig. 10.1 (b) has been redrawn in Fig. 10.1 (c) where all the sources have been shown connected to a common reference at ground potential and the shunt admittances at the buses have been lumped. Since the system has 4 buses, the number of nodes (m) is 5. In general y_{io} denotes the admittance of ith node with respect to reference, y_{ip} denotes the admittance between ith and pth nodes ($y_{ip} = y_{ip}$) and V_i denotes the voltage of the ith node. We assume that there is no mutual coupling between the lines. Application of Kirchhoff's current law to the four nodes of Fig. 10.1 (c) gives us the following equations:

$$\begin{aligned}
I_1 &= V_1 y_{10} + (V_1 - V_2)y_{12} + (V_1 - V_3)y_{13} + (V_1 - V_4)y_{14} \\
I_2 &= V_2 y_{20} + (V_2 - V_1)y_{12} + (V_2 - V_3)y_{23} \\
I_3 &= V_3 y_{30} + (V_3 - V_1)y_{13} + (V_3 - V_2)y_{23} + (V_3 - V_4)y_{34} \\
I_4 &= V_4 y_{40} + (V_4 - V_1)y_{14} + (V_4 - V_3)y_{34}
\end{aligned} \tag{10.3}$$

Equation (10.3) can be re-arranged and written in matrix form as:

$$\begin{bmatrix} I_1 \\ I_2 \\ I_3 \\ I_4 \end{bmatrix} = \begin{bmatrix} (y_{10}+y_{12}+y_{13}+y_{14}) & -y_{12} & -y_{13} & -y_{14} \\ -y_{12} & (y_{20}+y_{12}+y_{23}) & -y_{23} & 0 \\ -y_{13} & -y_{23} & (y_{30}+y_{13}+y_{23}+y_{34}) & -y_{34} \\ -y_{14} & 0 & -y_{34} & (y_{40}+y_{14}+y_{34}) \end{bmatrix} \begin{bmatrix} V_1 \\ V_2 \\ V_3 \\ V_4 \end{bmatrix} \tag{10.4}$$

Equation (10.4) can be written as:

$$\begin{bmatrix} I_1 \\ I_2 \\ I_3 \\ I_4 \end{bmatrix} = \begin{bmatrix} Y_{11} & Y_{12} & Y_{13} & Y_{14} \\ Y_{21} & Y_{22} & Y_{23} & Y_{24} \\ Y_{31} & Y_{32} & Y_{33} & Y_{34} \\ Y_{41} & Y_{42} & Y_{43} & Y_{44} \end{bmatrix} \begin{bmatrix} V_1 \\ V_2 \\ V_3 \\ V_4 \end{bmatrix} \tag{10.5a}$$

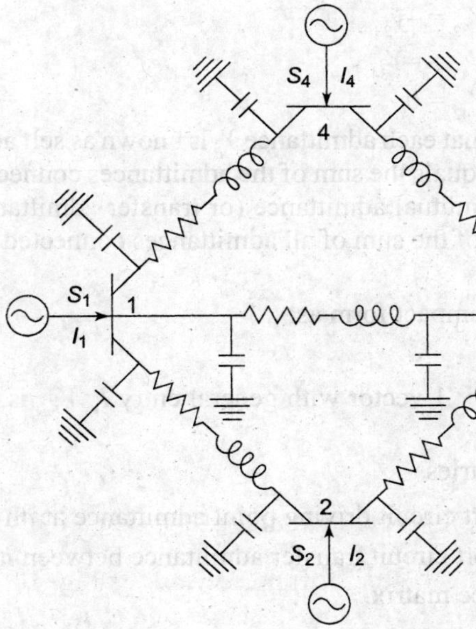

(b)

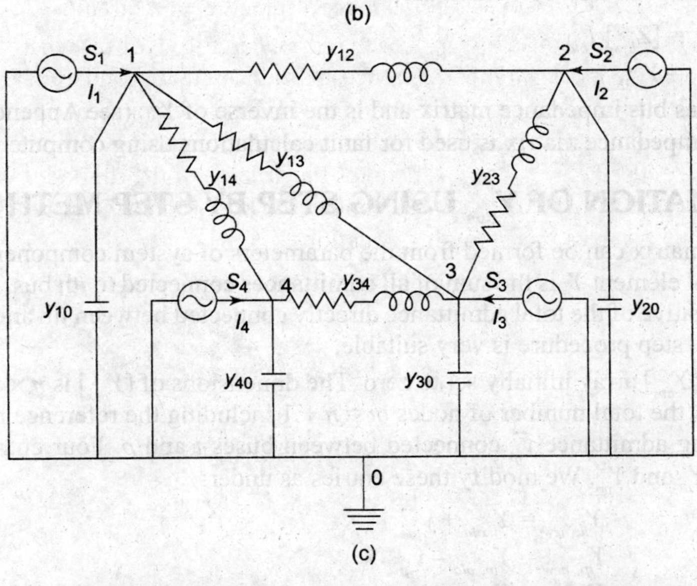

(c)

Fig. 10.2 (b) Equivalent circuit (c) Modified equivalent circuit

where $$Y_{11} = y_{10} + y_{12} + y_{13} + y_{14}$$

$$Y_{22} = y_{20} + y_{12} + y_{23}$$

$$Y_{33} = y_{30} + y_{13} + y_{23} + y_{34}$$
$$Y_{44} = y_{40} + y_{14} + y_{34}$$
$$Y_{12} = Y_{21} = -y_{12}$$
$$Y_{13} = Y_{31} = -y_{13}$$
$$Y_{14} = Y_{41} = -y_{14}$$
$$Y_{23} = Y_{32} = -y_{23}$$
$$Y_{34} = Y_{43} = -Y_{34}$$
$$Y_{24} = Y_{42} = 0$$

From network theory we know that each admittance Y_{ii} is known as self admittance (or driving point admittance) of *i*th node and equals the sum of the admittances connected to *i*th node. Each off diagonal term y_{ip} is known as mutual admittance (or transfer admittance) between *i*th and *p*th nodes and equals the negative of the sum of all admittances connected directly between the *i*th and *p*th nodes.

Eq. (10.5.a) can be written in compact form as:

$$I_{bus} = [Y_{bus}]V_{bus} \tag{10.5b}$$

For an *n* bus system I_{bus} is an $n \times 1$ vector with general entry I_i, V_{bus} is an $n \times 1$ vector with general entry V_i

$[V_{bus}]$ is an $n \times n$ matrix with entries

$Y_{ii} = |Y_{ii}| \angle \gamma_{ii}$ = short circuit driving point admittance at *i*th node

$Y_{ip} = |Y_{ip}| \angle \gamma_{ip}$ = short circuit transfer admittance between *i*th and *p*th nodes.

$[Y_{bus}]$ is known as bus admittance matrix.

Equation (10.5b) can also be written as:

$$V_{bus} = [Z_{bus}]\, I_{bus} \tag{10.6}$$

where $\quad Z_{bus} = Y^{-1}_{bus}$

Z_{bus} is known as bus impedance matrix and is the inverse of Y_{bus} (see Appendix *B* for matrix inversion). Bus impedance matrix is used for fault calculations using computer.

10.3 FORMATION OF Y_{bus} USING STEP BY STEP METHOD

The admittance matrix can be formed from the parameters of system components as discussed above. A diagonal element Y_{ii} is the sum of all admittances connected to *i*th bus. An off diagonal element Y_{ip} is negative of the total admittance directly connected between *i*th and *p*th buses. The following step by step procedure is very suitable.

We start with $[Y_{bus}]$ array initially set to zero. The dimensions of $[Y_{bus}]$ is $n \times n$ where *n* is the number of buses (the total number of nodes $m = n + 1$ including the reference node). Consider an element having admittance y_{ip} connected between buses *i* and *p*. Four entries in $[Y_{bus}]$ are affected: Y_{ii}, Y_{ip}, Y_{pi} and Y_{pp}. We modify these entries as under:

$$Y_{ii,\,new} = Y_{ii,old} + y_{ip} \tag{10.7a}$$
$$Y_{ip,\,new} = Y_{ip,\,old} - y_{ip} \tag{10.7b}$$
$$Y_{pi,\,new} = Y_{pi,\,old} - y_{ip} \tag{10.7c}$$
$$Y_{pp,\,new} = Y_{pp,\,old} + y_{ip} \tag{10.7d}$$

We add the elements one by one and modify the entries of $[Y_{bus}]$ as per Eq. (10.7). If an element is connected from *i*th bus to reference, only entry Y_{ii} is affected.

Example 10.1

Formulate $[Y_{bus}]$ for the network shown in Fig. 10.2. The values shown are admittances.

Solution

Since the network has two buses, $[Y_{bus}]$ is a 2×2 matrix.

$$Y_{11} = -j\,0.5 - j\,0.4 = -j\,0.9$$

$$Y_{22} = -j\,0.5 - j\,0.3 = -j\,0.8$$

$$Y_{12} = Y_{21} = -(-j\,0.5) = j\,0.5$$

$$[Y_{bus}] = \begin{bmatrix} -j0.9 & +j0.5 \\ +j0.5 & -j0.8 \end{bmatrix}$$

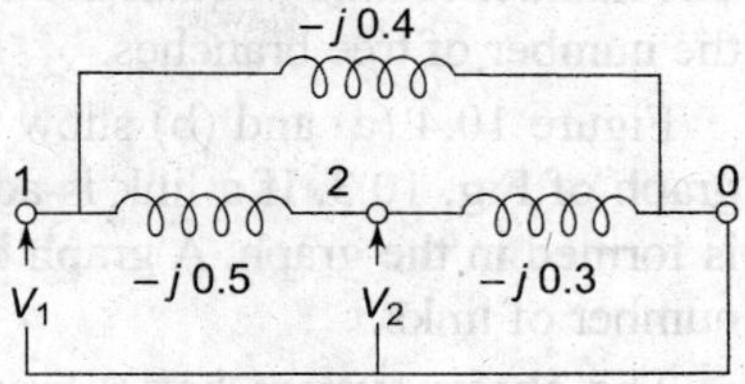

Fig. 10.2

As an alternative we can use the step by step approach (Eq. 10.7). We start with

$$[Y_{bus}] = \begin{bmatrix} 0 & 0 \\ 0 & 0 \end{bmatrix}$$

Add $-j\,0.5$ element

$$[Y_{bus}] = \begin{bmatrix} -j0.5 & +j0.5 \\ +j0.5 & -j0.5 \end{bmatrix}$$

Add $-j\,0.3$ element

$$[Y_{bus}] = \begin{bmatrix} -j0.5 & +j0.5 \\ +j0.5 & -j0.8 \end{bmatrix}$$

Add $-j\,0.4$ element

$$[Y_{bus}] = \begin{bmatrix} -j0.9 & +j0.5 \\ +j0.5 & -j0.8 \end{bmatrix}$$

10.4 FORMATION OF Y_{bus} AND Z_{bus} USING SINGULAR TRANSFORMATION

Another method for formation of Y_{bus} is by singular transformation. To understand this method we first briefly review the graph theory.

10.4.1 Graph, Tree and Co-tree

If each element of a network is represented just by a line, the resulting diagram of the network is known as a graph of the network. Each source and the shunt admittance across it are taken as a single element. If each element of the graph is assigned a direction, it is known as oriented graph. The direction is, generally, so assigned as to coincide with the assumed positive direction of current in the element. Fig. 10.3 shows the oriented graph of the network of Fig. 10.1.

A tree of a graph is that part of the graph which has sufficient number of branches to connect all the nodes without forming a closed path. The tree branches form a sub-set of the elements of the connected graph. The number of branches b required to form a tree equals the number of buses in the network (the total number of nodes, including the reference node, is one more than the number of buses).

Those elements of the graph which are not included in the tree are known as links (or link branches) and form a sub-graph, not necessarily connected, known as co-tree. The number of links l equals the number of elements minus the number of tree branches.

Figure 10.4 (a) and (b) show a tree and co-tree of the graph of Fig. 10.3. If a link is added to the tree one loop is formed in the graph. A graph has as many loops as the number of links.

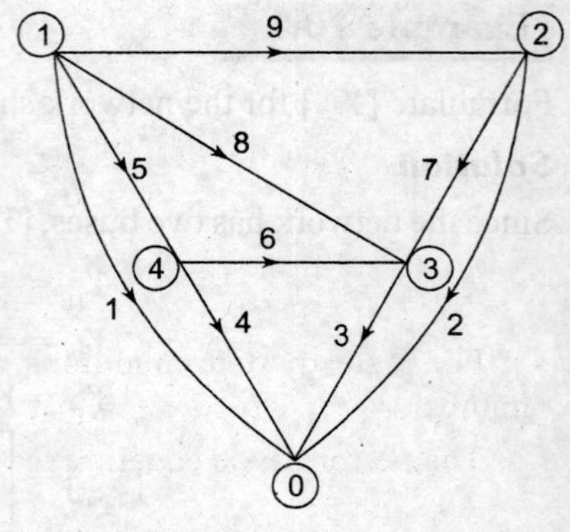

Fig. 10.3 Oriented graph of the network of Fig. 10.1

The above system has 9 branches. Therefore, it has 18 variables (9 branch voltages and 9 branch currents). However, it is easy to see that all these 18 variables are not independent. The number of independent variables is found from the concept of tree. The number of tree branches gives the number of independent voltages. For any system the number of tree branches equals the number of buses n. The number of links gives the number of independent current variables.

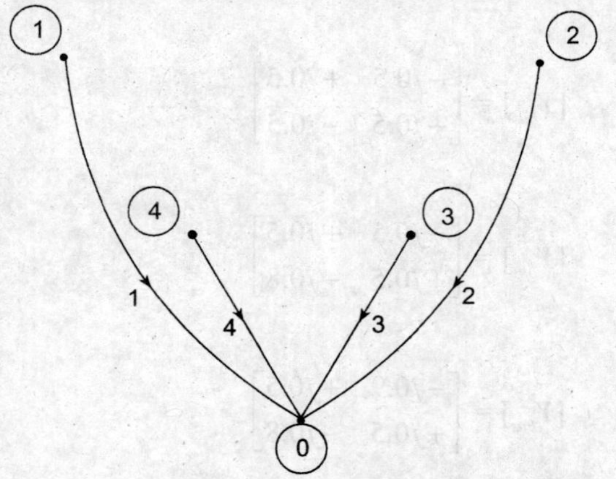

(a)

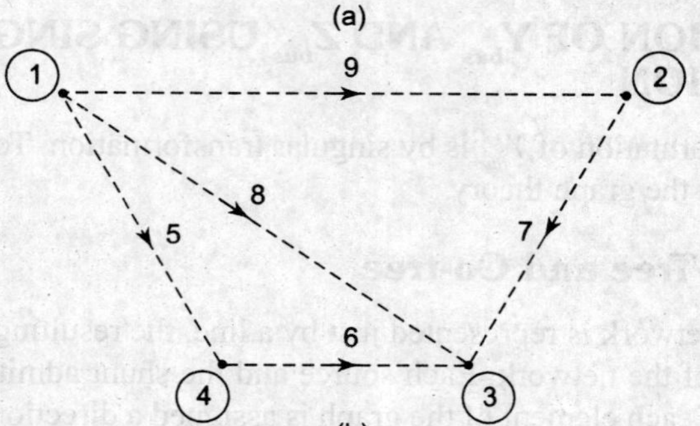

(b)

Fig. 10.4 (a) Tree (b) Co-tree of the graph of Fig. 10.3

10.4.2 Primitive Network

A network consists of many branches. Each branch consists of passive and/or active elements. Fig. 10.5 (a) and (b) show a network branch, containing both active and passive elements in impedance and admittance representation. Let

z_{ip} = self impedance of element $i-p$

y_{ip} = self admittance of element $i-p$

V_{ip} = voltage across element $i-p$

E_{ip} = voltage source in series with element $i-p$

I_{ip} = current through element $i-p$

J_{ip} = current source in parallel with element $i-p$

For ac steady state conditions, all voltages and currents are phasors and z_{ip} and y_{ip} are complex numbers.

The performance equation for impedance representation (Fig. 10.5a) is

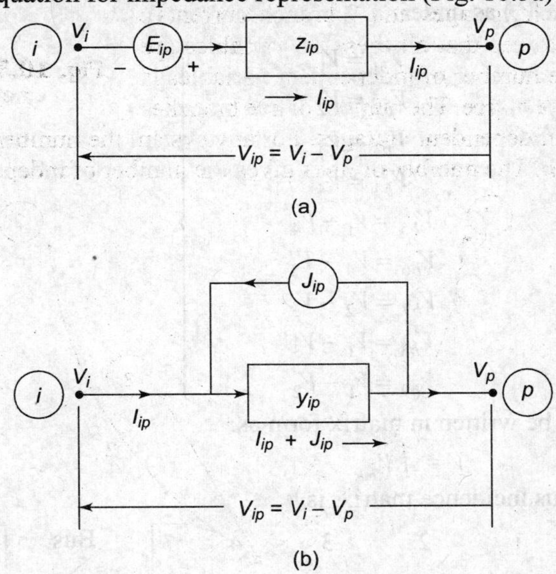

Fig. 10.5 Representation of network element (a) Impedance form, (b) Admittance form

$$V_{ip} + E_{ip} = z_{ip} I_{ip} \tag{10.8a}$$

and for admittance representation (Fig. 10.5b)

$$I_{ip} + J_{ip} = y_{ip} V_{ip} \tag{10.8b}$$

The impedance and admittance representations are related by

$$J_{ip} = -y_{ip} E_{ip} \tag{10.9a}$$

$$y_{ip} = \frac{1}{z_{ip}} \tag{10.9b}$$

A set of unconnected elements is known as primitive network. The performance equations in admittance (or impedance) form can be written for all the branches. The set of these equations in impedance form is

$$V + E = zI \tag{10.10a}$$

and in admittance form

$$I + J = yV \tag{10.10b}$$

where V and E are branch voltage and source voltage matrices, I and J are branch current and source current matrices, z is primitive impedance matrix (i.e., a matrix whose elements are branch self impedances) and y is primitive admittance matrix (i.e., matrix whose elements are branch self admittances). Since no mutual couplings have been assumed between lines, z and y are diagonal matrices.

10.4.3 Bus Incidence Matrix

For the graph of Fig. 10.3 the 9-branch voltages (V_{b1}, V_{b2} .... V_{b9}) can be expressed in terms of 4-bus voltages (V_1 ..... V_4) as under:

$$\left.\begin{aligned} V_{b1} &= V_1 \\ V_{b2} &= V_2 \\ V_{b3} &= V_3 \\ V_{b4} &= V_4 \\ V_{b5} &= V_1 - V_4 \\ V_{b6} &= V_4 - V_3 \\ V_{b7} &= V_2 - V_3 \\ V_{b8} &= V_1 - V_3 \\ V_{b9} &= V_1 - V_2 \end{aligned}\right\} \tag{10.11}$$

Equation (10.11) can be written in matrix form as:

$$V = A\,V_{bus} \tag{10.12}$$

where A, known as bus incidence matrix, is

$$A = \begin{array}{c} \text{Bus} \backslash e \\ 1 \\ 2 \\ 3 \\ 4 \\ 5 \\ 6 \\ 7 \\ 8 \\ 9 \end{array} \left[\begin{array}{cccc} 1 & 2 & 3 & 4 \\ \hline 1 & 0 & 0 & 0 \\ 0 & 1 & 0 & 0 \\ 0 & 0 & 1 & 0 \\ 0 & 0 & 0 & 1 \\ 1 & 0 & 0 & -1 \\ 0 & 0 & -1 & 1 \\ 0 & 1 & -1 & 0 \\ 1 & 0 & -1 & 0 \\ 1 & -1 & 0 & 0 \end{array}\right] \quad \begin{array}{c} \text{Bus} \backslash e \\ \\ \textit{Tree branches} \\ = \\ \\ \textit{Link branches} \\ \\ \end{array} \left[\begin{array}{c} \text{Buses} \\ \hline A_b \\ \hline A_l \end{array}\right] \tag{10.13}$$

The bus incidence matrix depicts the incidence of elements (branches) to the buses in the graph of a network. The element a_{ip} of the bus incidence matrix is written according to the rules given as under:

$a_{ip} = 1$ if the ith element is incident to and oriented away from pth bus.

$a_{ip} = -1$ if the ith element is incident to and oriented towards pth bus.

$a_{ip} = 0$ if the ith element is not incident to the pth bus.

Thus bus incidence matrix is rectangular and, therefore, singular. Its dimension is $e \times n$ where e is the number of elements (branches) and n is the number of buses (excluding the reference). If the rows of the bus incidence matrix are arranged according to a particular tree, the matrix can be partitioned into two sub-matrices A_b of dimension $b \times n$ and A_l of dimension $l \times n$ where b is the number of tree branches and l is the number of link branches.

10.4.4 Formulation of Y_{bus} and Z_{bus}

Substituting Eq. (10.12) into Eq. (10.10b)

$$I + J = y\,A\,V_{bus} \tag{10.14}$$

Premultiplying Eq. (10.14) by A^t (i.e., transpose of the bus incidence matrix)

$$A^t I + A^t J = A^t yA\,V_{bus} \tag{10.15}$$

Since the matrix A shows the incidence of elements to buses, $A^t I$ is a phasor in which each element is the algebraic sum of currents through the network elements terminating at a bus. As per Kirchhoff's current law, the algebraic sum of currents at a bus is zero. Thus

$$A^t I = 0 \tag{10.16a}$$

Similarly, $A^t J$ gives the algebraic sum of source currents at each bus. We have designated this as I_{bus} in writing Eq. 10.5 (b). Thus

$$A^t J = I_{bus} \tag{10.16b}$$

Substituting Eq. 10.16 (a) and (b) into Eq. (10.15)

$$I_{bus} = A^t\,y\,A\,V_{bus} \tag{10.17}$$

Comparing Eq. (10.17) and Eq. 10.5 (b)

$$[Y_{bus}] = A^t\,y\,A \tag{10.18}$$

Eq. (10.18) suggests the formulation of Y_{bus}. Since matrix A is singular, $A^t\,y\,A$ is a singular transformation of y. The bus incidence matrix can be obtained through a computer programme. Standard matrix multiplication and matrix transpose sub-routines can be used to compute Y_{bus} using Eq. (10.18). As mentioned earlier Z_{bus} is the inverse of Y_{bus}.

Example 10.2

Formulate Y_{bus} for the network of Fig. 10.1 (c) using singular transformation.

Solution

$$y = \begin{bmatrix} y_{10} & & & & & & & 0 \\ & y_{20} & & & & & & \\ & & y_{30} & & & & & \\ & & & y_{40} & & & & \\ & & & & y_{14} & & & \\ & & & & & y_{34} & & \\ & & & & & & y_{23} & \\ & & & & & & & y_{13} & \\ 0 & & & & & & & & y_{12} \end{bmatrix}$$

Using A from Eq. (10.13) we have

$$yA = \begin{bmatrix} y_{10} & 0 & 0 & 0 \\ 0 & y_{20} & 0 & 0 \\ 0 & 0 & y_{30} & 0 \\ 0 & 0 & 0 & y_{40} \\ y_{14} & 0 & 0 & -y_{14} \\ 0 & 0 & -y_{34} & y_{34} \\ 0 & y_{23} & -y_{23} & 0 \\ y_{13} & 0 & -y_{13} & 0 \\ y_{12} & -y_{12} & 0 & 0 \end{bmatrix}$$

$$A^t = \begin{bmatrix} 1 & 0 & 0 & 0 & 1 & 0 & 0 & 1 & 1 \\ 0 & 1 & 0 & 0 & 0 & 0 & 1 & 0 & -1 \\ 0 & 0 & 1 & 0 & 0 & -1 & -1 & -1 & 0 \\ 0 & 0 & 0 & 1 & -1 & 1 & 0 & 0 & 0 \end{bmatrix}$$

$$[Y_{bus}] = A^t yA$$

$$= \begin{bmatrix} (y_{10}+y_{14}+y_{13}+y_{12}) & -y_{12} & -y_{13} & -y_{14} \\ -y_{12} & (y_{20}+y_{23}+y_{12}) & -y_{23} & 0 \\ -y_{13} & -y_{23} & (y_{30}+y_{34}+y_{23}+y_{13}) & -y_{34} \\ -y_{14} & 0 & -y_{34} & (y_{40}+y_{14}+y_{34}) \end{bmatrix}$$

The elements of $[Y_{bus}]$ as obtained above are (as it should be) identical with the elements of $[Y_{bus}]$ given in Eq. (10.4).

Example 10.3

Figure 10.6 shows a 4-bus system. The shunt admittances at the buses are negligible. The line impedances are as under:

Line (bus to bus)	1–2	2–3	3–4	1–4
$R(pu)$	0.025	0.02	0.05	0.04
$X(pu)$	0.10	0.08	0.20	0.16

(a) Assume that the line shown dotted (from bus 1 to bus 3) is not present. Formulate Y_{bus}.

(b) Which elements of the Y_{bus} obtained above are affected when the line from bus 1 to bus 3 is added? (The new line has no mutual coupling with the other lines). If the pu impedance of this line is $0.1 + j0.4$, find the new Y_{bus}.

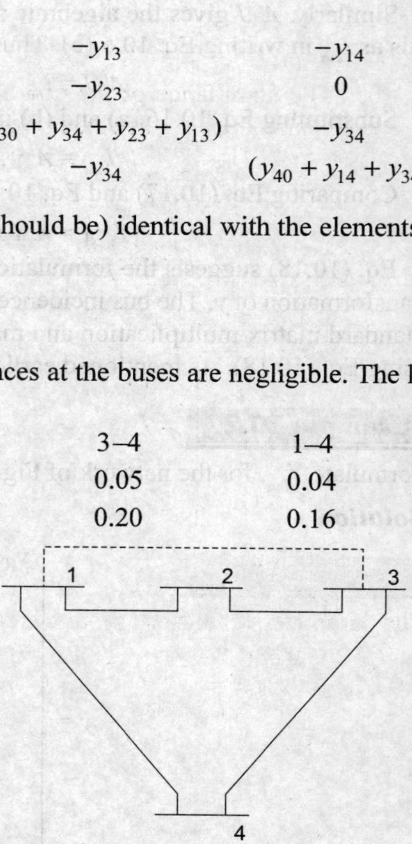

Fig. 10.6 4 Bus system for Example 10.3

Solution

(a) From elementary ac circuit theory we know that $G = \dfrac{R}{R^2 + X^2}$ and $B = -\dfrac{X}{R^2 + X^2}$.

For the given values of line impedances the values of G and B are:

Line (bus to bus)	1–2	2–3	3–4	1–4
G	2.35	2.94	1.176	1.47
B	–9.41	–11.76	–4.706	–5.88

Recalling that the element Y_{ii} of the $[Y_{bus}]$ is the sum of all admittances connected to the ith bus and the element Y_{ip} equals the negative of sum of all admittances connected directly between the ith and pth buses, $[Y_{bus}]$ can be written as:

$$[Y_{bus}] = \begin{bmatrix} y_{12}+y_{14} & -y_{12} & 0 & -y_{14} \\ -y_{12} & y_{12}+y_{23} & -y_{23} & 0 \\ 0 & -y_{23} & y_{23}+y_{34} & -y_{34} \\ -y_{14} & 0 & -y_{34} & y_{34}+y_{14} \end{bmatrix}$$

$$= \begin{bmatrix} 3.82-j15.29 & -2.35+j9.41 & 0 & -1.47+j5.88 \\ -2.35+j9.41 & 5.29-j21.17 & -2.94+j11.76 & 0 \\ 0 & -2.94+j11.76 & 4.116-j16.466 & -1.176+j4.706 \\ -1.47+j5.88 & 0 & -1.176+j4.706 & 2.646-j10.586 \end{bmatrix}$$

(b) The admittance of the new line between buses 1 and 3 is $0.588 - j\,2.352$ *pu*. The elements of Y_{bus} affected by the addition of this line are Y_{11}, Y_{33}, Y_{13} and Y_{31}.

$$Y_{11,\,new} = Y_{11,\,old} + y_{13} = 3.82 - j\,15.29 + 0.588 - j\,2.352$$
$$= 4.408 - j\,17.642$$
$$Y_{33,\,new} = Y_{33,\,old} + y_{13} = 4.116 - j\,16.466 + 0.588 - j\,2.352$$
$$= 4.704 - j\,18.818$$
$$Y_{13,\,new} = Y_{13,\,old} - y_{13} = -0.588 + j\,2.352 = Y_{31,\,new}$$

The new Y_{bus} is

$$Y_{bus} = \begin{bmatrix} 4.408-j17.642 & -2.35+j9.41 & -0.588+j2.352 & -1.47+j5.88 \\ -2.35+j9.41 & 5.29-j21.17 & -2.94+j11.76 & 0 \\ -0.588+j2.352 & -2.94+j11.76 & 4.704-j18.818 & -1.176+j4.706 \\ -1.47+j5.88 & 0 & -1.176+j4.706 & 2.646-j10.586 \end{bmatrix}$$

Example 10.4

The parameters of a 4-bus system are as under:

Bus Code	Line impedance (*pu*)	Charging Admittance (*pu*) *ypq*/2
1–2	$0.2 + j\,0.8$	$j\,0.02$
2–3	$0.3 + j\,0.9$	$j\,0.03$
2–4	$0.25 + j\,1$	$j\,0.04$
3–4	$0.2 + j\,0.8$	$j\,0.02$
1–3	$0.1 + j\,0.4$	$j\,0.01$

Draw the network and find bus admittance matrix

Solution

The network is shown in Fig. 10.7. The line series admittances are:

$$y_{12} = \frac{1}{z_{12}} = \frac{1}{0.2 + j0.8}$$

$$= 0.294 - j\,1.176\ pu$$

$$y_{13} = \frac{1}{0.1 + j0.4} = 0.588 - j\,2.352\ pu$$

$$y_{23} = \frac{1}{0.3 + j0.9} = 0.333 - j1\ pu$$

$$y_{24} = \frac{1}{0.25 + j1} = 0.235 - j0.94\ pu$$

$$y_{34} = \frac{1}{0.25 + j0.8} = 0.294 - j1.176\ pu$$

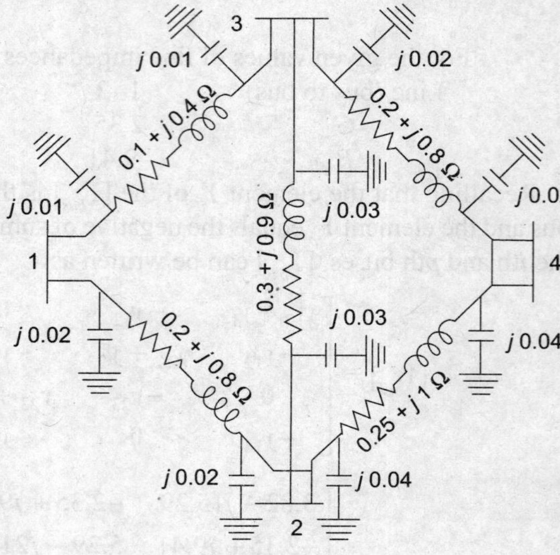

The shunt admittances at different buses are

Fig. 10.7

$$y_{10} = j\,0.03\ pu,\ y_{20} = j\,0.09\ pu$$

$$y_{30} = j\,0.06\ pu,\ y_{40} = j\,0.06\ pu$$

$$Y_{BUS} = \begin{bmatrix} y_{12}+y_{13}+y_{10} & -y_{12} & -y_{13} & 0 \\ -y_{12} & y_{12}+y_{23}+y_{24}+y_{20} & -y_{23} & -y_{24} \\ -y_{13} & -y_{23} & y_{13}+y_{23}+y_{34}+y_{30} & -y_{34} \\ 0 & -y_{24} & -y_{34} & y_{24}+y_{34}+y_{40} \end{bmatrix}$$

$$= \begin{bmatrix} .882 - j3.498 & -.294 + j1.176 & -.588 + j2.352 & 0 \\ -.294 + j1.176 & .862 - j3.026 & -.333 + j1 & -.235 + j.94 \\ -.588 + j2.352 & -.333 + j1 & 1.215 - j4.468 & -.294 + j1.176 \\ 0 & -.235 + j.94 & -.294 + j1.176 & .529 - j2.056 \end{bmatrix}$$

Example 10.5

An incomplete nodal admittance matrix for a 4-bus system with negligible charging admittance is given below. Find the missing terms.

$$\begin{bmatrix} .7 - j3 & -.2 + j1 & -.5 + j2 & Y_{14} \\ Y_{21} & Y_{22} & -.3 + j2 & -.5 + j3 \\ Y_{31} & Y_{32} & Y_{33} & -1 + j4 \\ Y_{41} & Y_{42} & Y_{43} & Y_{44} \end{bmatrix}$$

Solution

$$.7 - j3 = -(-.2 + j1 - .5 + j2 + Y_{14}) \text{ or } Y_{14} = 0$$

$$Y_{21} = Y_{12} = -.2 + j1$$

$$Y_{22} = -(Y_{21} + Y_{23} + Y_{24})$$
$$= -(-.2 + j1 - .3 + j2 - .5 + j3) = 1 - j6$$
$$Y_{31} = Y_{13} = -.5 + j2$$
$$Y_{32} = Y_{23} = -.3 + j2$$
$$Y_{33} = -(Y_{31} + Y_{32} + Y_{34})$$
$$= -(-.5 + j2 - .3 + j2 - 1 + j4) = 1.8 - j8$$
$$Y_{41} = Y_{14} = 0$$
$$Y_{42} = Y_{24} = -.5 + j3$$
$$Y_{43} = Y_{34} = -1 + j4$$
$$Y_{44} = -(Y_{41} + Y_{42} + Y_{43})$$
$$= -(-.5 + j3 - 1 + j4) = 1.5 - j7$$

Example 10.6

Fig. 10.8 (*a*) shows the single line representation of a single system having an off nominal turns ratio transformer. Find bus admittance matrix. Use this matrix to draw equivalent circuit for the system.

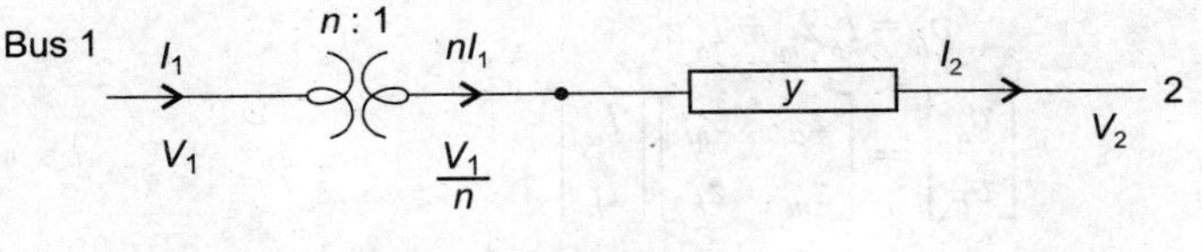

Fig. 10.8 (a)

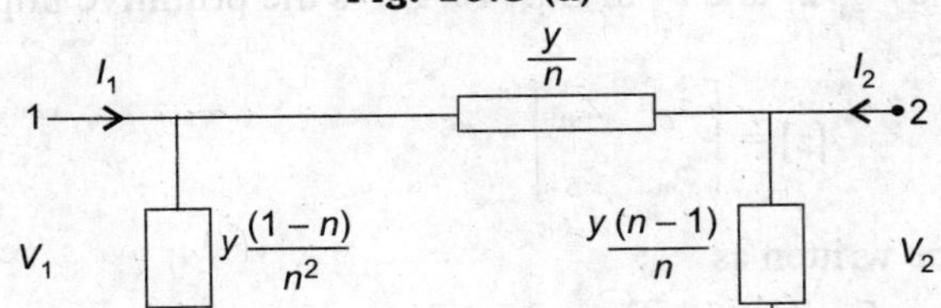

Fig. 10.8 (b) Equivalent circuit for off-nominal ratio transformer

Solution

The currents $n\,I_1$, I_2 and voltages $\frac{V_1}{n}$ and V_2 are related by bus admittance matrix which can be written from the discussion in section 10.3

$$\begin{bmatrix} nI_1 \\ I_2 \end{bmatrix} = \begin{bmatrix} y & -y \\ -y & y \end{bmatrix} \begin{bmatrix} V_1/n \\ V_2 \end{bmatrix} \qquad (i)$$

By elementary matrix operations, the above matrix can be rewriteen as

$$\begin{bmatrix} I_1 \\ I_2 \end{bmatrix} = \begin{bmatrix} \frac{y}{n^2} & \frac{-y}{n} \\ \frac{-y}{n} & y \end{bmatrix} \begin{bmatrix} V_1 \\ V_2 \end{bmatrix} \qquad \text{(ii)}$$

The matrix in Eq. (*ii*) is the bus admittance matrix for the system of Fig. 10.8. (*a*)

From Eq. (*ii*) we can draw the equivalent circuit by following the basic principles of network theory. The equivalent circuit is shown in Fig. 10.8 (*b*)

10.5 INCLUSION OF MUTUALLY COUPLED BRANCHES IN Y_{BUS}

Transmission lines run in different directions from various buses in a power system. In general they do not have mutual coupling with each other. Nevertheless one or two lines in the total system may have some mutual coupling with another one or two lines. It is necessary to include the effect of such mutual coupling for correct evaluation of system performance.

The easiest approach of the formation of Y_{bus} in a system having mutually coupled branches is the use of singular transformation. Eq (10.18), i.e., $(Y_{bus}) = A^t\, y\, A$ is valid in this case also. However, the primitive admittance matrix should be properly formulated by including the mutual coupling.

Fig. 10.9 shows two branches having self impedances z_a and z_b, mutually coupled through impedance z_m. KVL equations are

$$v_a = I_a z_a + I_b z_m$$

$$v_b = I_a z_m + I_b z_b$$

or

$$\begin{bmatrix} v_a \\ v_b \end{bmatrix} = \begin{bmatrix} z_a & z_m \\ z_m & z_b \end{bmatrix} \begin{bmatrix} I_a \\ I_b \end{bmatrix} \qquad (10.19)$$

The matrix formed by z_a, z_b and z_m in Eq. (10.19) is the primitive impedance matrix for the system of Fig. 10.9.

Thus

$$[z] = \begin{bmatrix} z_a & z_m \\ z_m & z_b \end{bmatrix}$$

Equation (10.19) can be written as

$$\begin{bmatrix} I_a \\ I_b \end{bmatrix} = \begin{bmatrix} y_a & y_m \\ y_m & y_b \end{bmatrix} \begin{bmatrix} v_a \\ v_b \end{bmatrix}$$

where the matrix formed by y_a, y_m and y_b in the primitive admittance matrix

Thus

$$[y] = \begin{bmatrix} y_a & y_m \\ y_m & y_b \end{bmatrix}$$

The primitive admittance matrix [*y*] is the inverse of primitive impedance matrix [*z*]. Thus

$$[y] = \begin{bmatrix} z_a & z_m \\ z_m & z_b \end{bmatrix}^{-1} = \frac{1}{z_a z_b - z_m^2} \begin{bmatrix} z_b & -z_m \\ -z_m & z_a \end{bmatrix}$$

It is seen that the primitive impedance matrix [*z*] can be written from Fig. 10.9 by inspection.

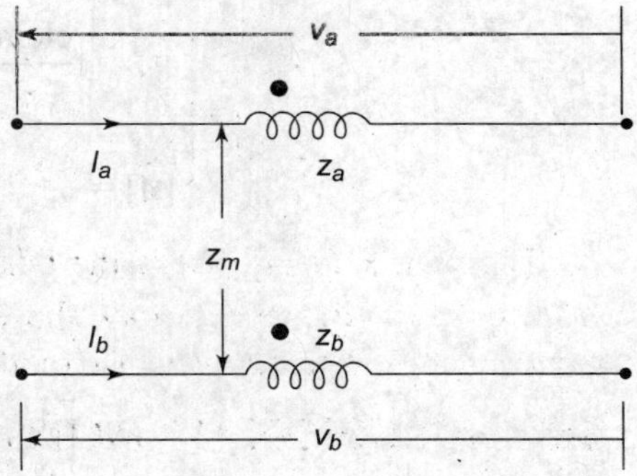

Fig. 10.9

Proper care has to be exercised in assigning the proper + or – sign to the z_m terms. The z_m terms are positive when currents in both branches enter or leave the dot marked terminals. Otherwise the z_m terms are negative. But the primitive [y] matrix cannot be written by inspection. The elements of [z] have direct relation with impedances z_a, z_b and z_m. However, the elements of [y] do not have any such direct relation. The primitive admittance matrix can only be obtained by taking inverse of primitive impedance matrix.

The matrixes [z] and [y] are diagonal if there is no mutual coupling. In this case the primitive self admittances are equal to reciprocals of the corresponding primitive self impedances. The off-diagonal terms in [z] and [y] represent the mutual impedances. In a large network having some mutual coupled branches, the primitive impedance matrix can be straightaway written by inspection. The primitive admittance matrix [y] is the inverse of primitive impedance matrix. Once the matrix [y] has been formed Eq. (10.18) can be used to give $[Y_{bus}]$.

Example 10.7

Figure 10.10 (a) shows a system with one source and four lines. The reactances of source and lines are as shown. Lines *c* and *b* are mutually coupled through a reactance of *j* 0.2 p.u. Formulate bus admittance matrix.

Solution

The primitive impedance matrix of the system is

$$[z] = \begin{array}{c|ccccc} e/e & a & b & c & d & e \\ \hline a & j1 & 0 & 0 & 0 & 0 \\ b & 0 & j0.4 & j0.2 & 0 & 0 \\ c & 0 & j0.2 & j0.5 & 0 & 0 \\ d & 0 & 0 & 0 & j0.2 & 0 \\ e & 0 & 0 & 0 & 0 & j0.25 \end{array}$$

$$[y] = [z]^{-1} \begin{array}{c|ccccc} e/e & a & b & c & d & e \\ \hline a & -j1 & 0 & 0 & 0 & 0 \\ b & 0 & -j3.125 & j1.25 & 0 & 0 \\ c & 0 & j1.25 & -j2.5 & 0 & 0 \\ d & 0 & 0 & 0 & -j5 & 0 \\ e & 0 & 0 & 0 & 0 & -j4 \end{array}$$

The graph is shown in Fig. 10.10 (b). Elements *a*, *b*, *c*, *d* are tree branches while element *e* is link. The bus incidence matrix *A* is

$$[A] = \begin{array}{c|cccc} \text{element/bus} & 1 & 2 & 3 & 4 \\ \hline a & 1 & 0 & 0 & 0 \\ b & -1 & 1 & 0 & 0 \\ c & -1 & 0 & 1 & 0 \\ d & 0 & 0 & 1 & -1 \\ e & 0 & -1 & 0 & 1 \end{array}$$

and
$$[A^t] = \begin{bmatrix} 1 & -1 & -1 & 0 & 0 \\ 0 & 1 & 0 & 0 & -1 \\ 0 & 0 & 1 & 1 & 0 \\ 0 & 0 & 0 & -1 & 1 \end{bmatrix}$$

$$yA = \begin{bmatrix} -j1 & 0 & 0 & 0 \\ j1.875 & -j3.125 & j1.25 & 0 \\ j1.25 & j1.25 & -j2.5 & 0 \\ 0 & 0 & -j5 & j5 \\ 0 & j4 & 0 & -j4 \end{bmatrix}$$

$$[Y_{bus}] = A^t yA = \begin{bmatrix} -j4.125 & j1.875 & j1.25 & 0 \\ j1.875 & -j7.125 & j1.25 & j4 \\ j1.25 & j1.25 & -j7.5 & j5 \\ 0 & j4 & j5 & -j9 \end{bmatrix}$$

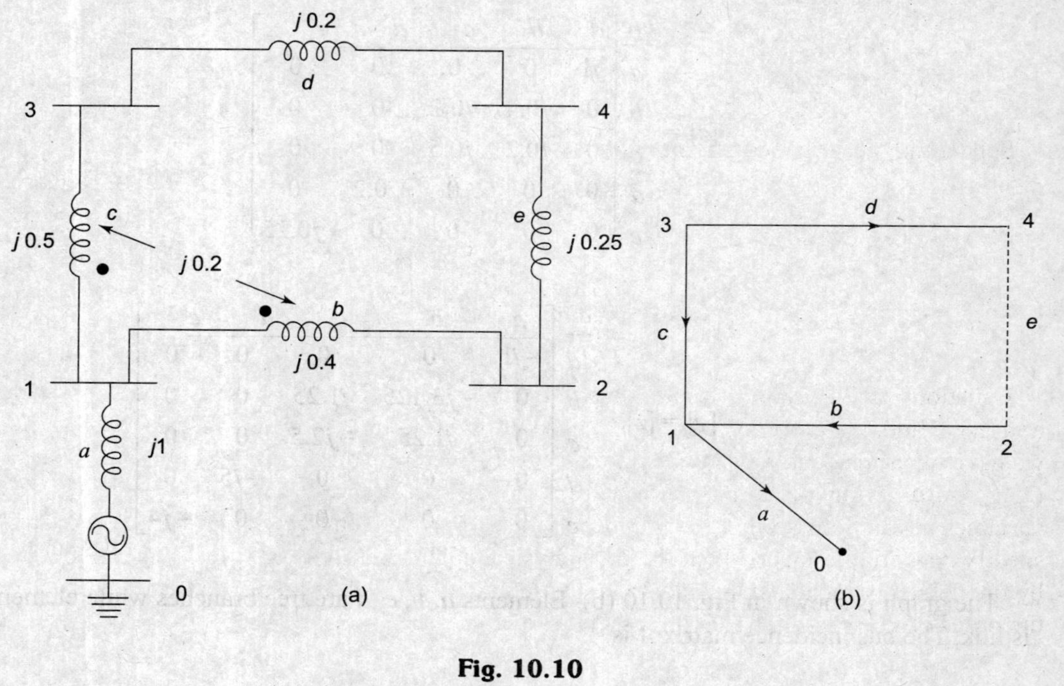

Fig. 10.10

10.6 LOAD FLOW EQUATIONS AND METHODS OF SOLUTION

Recalling that $I_{bus} = [Y_{bus}]\, V_{bus}$, (Eq. 10.5b), the current entering the ith bus of an n bus system is given by:

$$I_i = Y_{i1} V_1 + Y_{i2} V_2 + \dots\dots + Y_{ii} V_i + \dots\dots + Y_{in} V_n \qquad (10.20a)$$

$$= \sum_{p=1}^{n} Y_{ip} V_p \qquad (10.20b)$$

$$= \sum_{p=1}^{n} |Y_{ip}||V_p| \angle(\delta_p + \gamma_{ip}) \qquad (10.20c)$$

where $\quad V_p = |V_p| \angle \delta_p$ and $Y_{ip} = |Y_{ip}| \angle \gamma_{ip}$

The complex power injected into ith bus is

$$S_i = P_i + j\,Q_i = V_i\, I_i^* \qquad (10.21)$$

The load flow computations can be done more easily if I_i (rather than I_i^*) is used.

Taking the complex conjugate of Eq. (10.21)

$$S_i^* = P_i - j\,Q_i = V_i^* I_i \qquad (10.22a)$$

Substituting the value of I_i from Eq. (10.20b) into Eq. (10.22a)

$$S_i^* = P_i - j\,Q_i = V_i^* \sum_{p=1}^{n} Y_{ip} V_p \qquad (10.22b)$$

Since $\quad V_i = |V_i| \angle \delta_i$, $V_i^* = |V_i| \angle -\delta_i$ and $Y_{ip} = Y_{pi} = |Y_{ip}| \angle \gamma_{ip}$ we have

$$P_i - jQ_i = |V_i| \sum_{p=1}^{n} |Y_{ip}||V_p| \angle(\delta_p + \gamma_{ip} - \delta_i)$$

$$= |V_i| \sum_{p=1}^{n} |Y_{ip}||V_p| \angle -(\delta_i - \delta_p - \gamma_{ip}) \qquad (10.23)$$

Separating the real and imaginary parts

$$P_i = |V_i| \sum_{p=1}^{n} |Y_{ip}||V_p| \cos(\delta_i - \gamma_{ip} - \delta_p) \qquad (10.24a)$$

$$Q_i = |V_i| \sum_{p=1}^{n} |Y_{ip}||V_p| \sin(\delta_i - \gamma_{ip} - \delta_p) \qquad (10.24b)$$

Equations 10.24 (a) and (b) are known as static load flow equations. These equations are non-linear and, therefore, only a numerical solution is possible. For each of the n system buses we have two such equations giving a total of $2n$ equations. At each bus we have 4 variables P_i, Q_i, V_i and δ_i giving a total of $4n$ variables. To obtain a solution it is necessary to specify two variables at each bus so that the number of unknowns is reduced to $2n$. Evidently we should specify the variables over which we have physical control. The choice is influenced some what by the devices which are connected to a particular bus. Depending on the quantities specified, the buses can be classified into three types shown in Table 10.1.

Table 10.1 Bus types for load flow studies

Bus type	*Specified quantities*	*Unknown quantities*	*Approximate number*
Load bus or *PQ* bus	P_i, Q_i	V_i, δ_i	85%
Generator bus or voltage controlled bus or *PV* bus	P_i, V_i	Q_i, δ_i	15%
Reference bus or slack bus or swing bus	V_i $\delta_i = 0$	P_i, Q_i	1

The load buses are the most common in the power supply network. At these buses P_i and Q_i are known because P_{Li}, Q_{Li} are known from the load forecast data and P_{Gi}, Q_{Gi} are either zero (no generation at these buses) or specified. At the generator buses, the governor control and excitation control allow us to specify P_{Gi} and V_i. Since P_{Li} is known (from load forecast data), P_i is known.

The need to designate one of the buses as swing or slack bus is evident from the fact that the system power losses are not known initially. Therefore, the net power flow into the system cannot be fixed in advance. The swing bus is a generator bus and the generators at this bus supply the difference between the specified real power injected into the system at other buses and the total system output plus losses. Since the voltages throughout the system must be close to 1 *pu*, we generally specify the voltage at the swing bus equal to 1 *pu*. From elementary ac theory we know that any one phasor can be selected as reference and we select the voltage of the swing bus as reference making its angle δ_i zero. Generally, the bus of the largest generating station is selected as swing bus and numbered as bus 1.

The solution to the load flow problem consists in assuming a certain initial bus load configuration, specifying the $2n$ known variables listed in Table 10.1 and using some numerical method to find the remaining $2n$ variables for the system with known $[Y_{bus}]$. The final solution must satisfy some constraints. These constraints are (1) voltage magnitude at various buses must be within limits (2) active and reactive generator power at different buses must be within the minimum and maximum limits (3) total generation must equal total load plus losses. (4) In addition the system stability considerations impose a limit on the maximum value with δ_i can have. The two most commonly used methods for load flow studies are Gauss-Seidel method and Newton-Raphson method. An approximate but easy solution is possible if some simplifying assumptions are made.

10.7 APPROXIMATE LOAD FLOW STUDY

A simple but approximate solution is possible if the following assumptions are made in load flow analysis.

1. Line resistances are neglected which means that the active power loss in the lines is zero. This reduces the complexity of equations because the total active power generation becomes equal to the total active power demand. The effect of this assumption in Eqs. 9.24 (a and b) is that $\gamma_{ip} = 90°$ and $\gamma_{ii} = -90°$.
2. The angle δ_i is small so that $\sin \delta_i = \delta_i$. This approximation converts the non-linear load flow equations into linear ones so that analytical solution is possible.
3. All buses except the swing bus are voltage controlled buses. This means that the voltages of all buses are specified.

With the above assumptions, Eq. 10.24 (a and b) get modified to the following equations:

$$P_i = |V_i| \sum_{p=1}^{n} |Y_{ip}||V_p| \angle (\delta_i - \delta_p) \tag{10.25a}$$

$$Q_i = -|V_i| \sum_{\substack{p=1 \\ p \neq i}}^{n} |Y_{ip}| \, V_p| \cos(\delta_i - \delta_p) + |V_i|^2 \, |Y_{ii}| \tag{10.25b}$$

Since we have specified V_i, Eq. (10.25a) is a linear equation. The only unknowns are angles δ. For bus 1 (slack bus) $\delta_i = 0$. Thus, we have $n - 1$ linear equations from which the values of δ at all buses can be found. These values of δ when substituted in the set of Eqs. (10.25b) give the values of Q_i at all buses.

Example 10.8.

Figure 10.11 shows a 4-bus system. Line resistances may be neglected and *pu* line reactances are as shown. The voltages at all buses should be 1 *pu*. The *pu* values of loads and active power generated at the different buses are also shown. Find (a) reactive power generation at buses (b) active and reactive line losses, (c) real and reactive power transfer over the lines.

Solution

(a) Let bus 1 be the swing bus. The active and reactive power generation at this bus can be controlled. At buses 2 and 4 there is no active power generation but these buses have reactive power compensation equipment. Bus 3 has a generator whose active power output is 3.5 *pu* but reactive power output can be controlled. The generation and loads at the buses can be combined so that the net power injected into the buses are:

$$S_1 = S_{G1} - S_{L1} = (P_{G1} - 1.5) + j\,(Q_{G1} - 1)$$
$$S_2 = S_{G2} - S_{L2} = -2 + j(Q_{G2} - 0.5)$$
$$S_3 = S_{G3} - S_{L3} = 3.5 + j\,Q_{G3}$$
$$S_4 = S_{G4} - S_{L4} = -2 + j\,(Q_{G4} - 0.8)$$

Since there is no active power loss, the net total active power generated must be equal to the active power of all loads. Thus, P_{G1} can be straightaway found.

$$P_{G1} = 2 + 2 + 1.5 - 3.5 = 2 \; pu$$

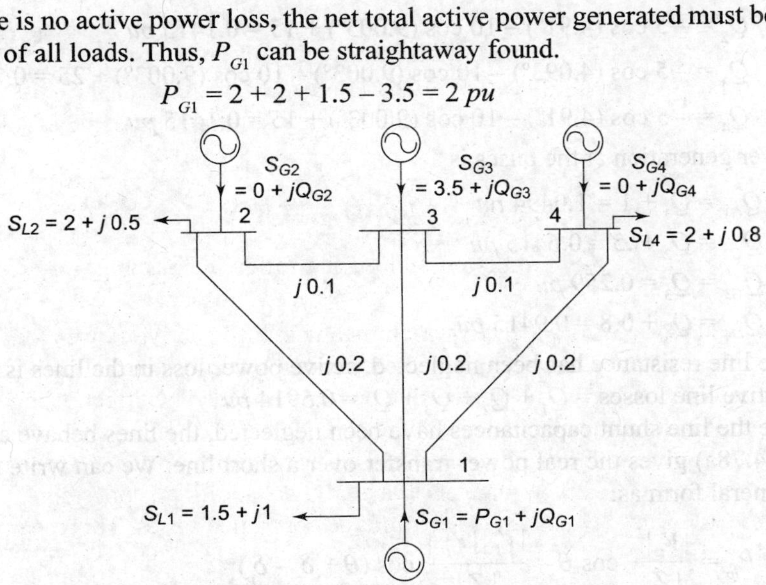

Fig. 10.11 4 Bus system for example 10.8

Thus, the net active power injected into the buses are

$P_1 = 0.5, P_2 = -2, P_3 = 3.5, P_4 = -2$. The admittances between the different buses are

$$y_{12} = y_{13} = y_{14} = -j5.0\ pu$$
$$y_{23} = y_{34} = -j\ 10.0\ pu$$

Using the method of section 10.3, Y_{bus} can be formulated. The result is

$$Y_{bus} = \begin{bmatrix} -j15.0 & j5.0 & j5.0 & j5.0 \\ j5.0 & -j15.0 & j10.0 & 0 \\ j5.0 & j10.0 & -j25.0 & j10.0 \\ j5.0 & 0 & j10.0 & -j15.0 \end{bmatrix}$$

Voltages at all buses is 1 *pu*. $\delta_1 = 0$. Using Eq. (10.25a) the net active powers injected into the buses can be written as:

$$P_1 = 0.5 = -5\delta_2 - 5\delta_3 - 5\delta_4$$
$$P_2 = -2 = 5\delta_2 + 10(\delta_2 - \delta_3) = 15\delta_2 - 10\delta_3$$
$$P_3 = 3.5 = 5\delta_3 + 10(\delta_3 - \delta_2) + 10\ (\delta_3 - \delta_4)$$
$$= -10\delta_2 + 25\delta_3 - 10\delta_4$$
$$P_4 = -2 = 5\delta_4 + 10(\delta_4 - \delta_3) = -10\delta_3 + 15\delta_4$$

Only three of the above four equations are independent. If we add the equations for P_2, P_3, and P_4 we get the equation for P_1. Solving the equations for P_2, P_3 and P_4 we get

$$\delta_2 = \delta_4 = -.0857 \text{ rad.} = -4.91^\circ$$
$$\delta_3 = 0.071428 \text{ rad.} = 4.093^\circ$$

Substituting these values in Eq. (10.25b) we get

$$Q_1 = -5\cos(4.91^\circ) - 5\cos(4.093^\circ) - 5\cos(4.91^\circ) + 15 = 0.0494\ pu$$
$$Q_2 = -5\cos(4.91^\circ) - 10\cos(9.003^\circ) + 15 = 0.1415\ pu$$
$$Q_3 = -5\cos(4.093^\circ) - 10\cos(9.003^\circ) - 10\cos(9.003^\circ) + 25 = 0.259\ pu$$
$$Q_4 = -5\cos(4.91^\circ) - 10\cos(9.003^\circ) + 15 = 0.1415\ pu$$

Reactive power generation at the buses is

$$Q_{G1} = Q_1 + 1 = 1.0494\ pu$$
$$Q_{G2} = Q_2 + .5 = 0.6415\ pu$$
$$Q_{G3} = Q_3 = 0.259\ pu$$
$$Q_{G4} = Q_4 + 0.8 = 0.9415\ pu$$

(b) Since line resistance has been neglected, active power loss in the lines is zero. Reactive line losses = $Q_1 + Q_2 + Q_3 + Q_4 = 0.5914\ pu$

(c) Since the line shunt capacitances have been neglected, the lines behave as short lines. Eq. (4.78a) gives the real power transfer over a short line. We can write this equation in general form as:

$$P_{ip} = \frac{|V_i|^2}{|Z|}\cos\theta - \frac{|V_i||V_p|}{|Z|}\cos(\theta + \delta_i - \delta_p)$$

Substituting $|Z| = |X_{ip}|$ and $\theta = 90°$

$$P_{ip} = \frac{|V_i|\,|V_p|}{|X_{ip}|} \sin(\delta_i - \delta_p)$$

where P_{ip} is the real power transferred from ith bus to pth bus, X_{ip} is the total reactance of line connecting the ith and pth buses and δ_i, δ_p are the bus voltage angles with respect to a common reference (swing bus voltage).

Substituting the values of X and δ

$$P_{12} = -P_{21} = \frac{1}{0.2} \sin(\delta_1 - \delta_2) = 5 \sin(4.91°) = 0.428\ pu$$

$$P_{14} = -P_{41} = \frac{1}{0.2} \sin(\delta_1 - \delta_4) = 5 \sin(4.91°) = 0.428\ pu$$

$$P_{13} = -P_{31} = \frac{1}{0.2} \sin(\delta_1 - \delta_3) = 5 \sin(-4.093°) = -0.357\ pu$$

$$P_{23} = -P_{32} = \frac{1}{0.1} \sin(\delta_2 - \delta_3) = 10 \sin(-9.003°) = -1.565\ pu$$

$$P_{34} = -P_{43} = \frac{1}{0.1} \sin(\delta_3 - \delta_4) = 10 \sin(9.003°) = 1.565\ pu$$

The reactive power transfer is given by Eq. (4.78b). We can write this equation in general form as:

$$Q_{ip} = \frac{|V_i|^2}{|Z|} \sin\theta - \frac{|V_i|\,|V_p|}{|Z|} \sin(\theta + \delta_i - \delta_p)$$

Substituting $\theta = 90°$ and $|Z| = |X_{ip}|$

$$Q_{ip} = \frac{|V_i|^2}{|X_{ip}|} - \frac{|V_i|\,|V_p|}{|X_{ip}|} \cos(\delta_i - \delta_p)$$

where Q_{ip} is the reactive power flow from ith bus to pth bus. It is evident that $Q_{ip} = Q_{pi}$. Substituting the values of X and δ

$$Q_{12} = Q_{21} = \frac{1}{.2} - \frac{1}{.2} \cos(\delta_1 - \delta_2) = 5 - 5 \cos(4.91°) = 0.0183\ pu$$

$$Q_{13} = Q_{31} = \frac{1}{.2} - \frac{1}{.2} \cos(\delta_1 - \delta_3) = 5 - 5 \cos(-4.093°) = 0.01275\ pu$$

$$Q_{14} = Q_{41} = \frac{1}{.2} - \frac{1}{.2} \cos(\delta_1 - \delta_4) = 5 - 5 \cos(4.91°) = 0.0183\ pu$$

$$Q_{23} = Q_{32} = \frac{1}{.1} - \frac{1}{.1} \cos(\delta_2 - \delta_3) = 10 - 10 \cos(-9.003°) = 0.123\ pu$$

$$Q_{34} = Q_{43} = \frac{1}{.1} - \frac{1}{.1} \cos(\delta_3 - \delta_4) = 10 - 10 \cos(9.003°) = 0.123\ pu$$

$P_{12} = -P_{21} = 0.428$ means that bus 1 exports 0.428 *pu* real power to bus 2 or bus 2 imports 0.428 *pu* real power from bus 1. $Q_{12} = Q_{21} = 0.0183$ *pu* means that bus 1 exports 0.0183 *pu* reactive power (lagging) to bus 2. Also bus 2 exports 0.0183 *pu* reactive power (lagging) to bus 1 or imports – 0.0183 *pu* reactive power (leading) from bus 1. Therefore, the complex power export from bus 1 to bus 2 has been shown as 0.428 + j0.0183 *pu* and the complex power import by bus 2 from bus 1 has been shown as 0.428 – *j*0.0183 *pu*. Following a similar reasoning for other lines, the real and reactive generation, loads and line flows can be shown as in Fig. 10.12.

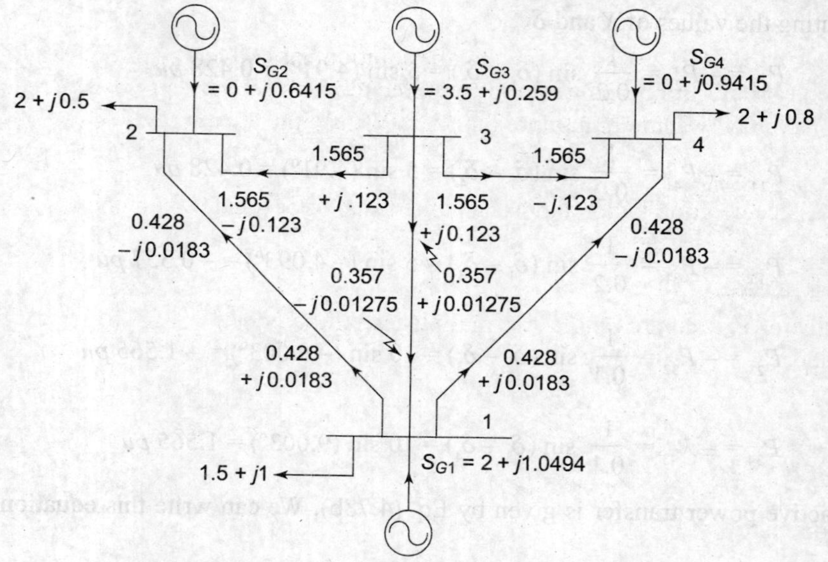

Fig. 10.12 Generation, load and line flows for example 10.7

10.8 GAUSS-SEIDEL METHOD–SOLUTION OF ALGEBRAIC EQUATIONS

A set of simultaneous non-linear algebraic equations cannot be solved by the usual methods. For the solution of these equations iterative techniques are necessary. In an iterative technique we progressively compute more accurate estimates of the unknown until results are obtained to a desired degree of accuracy in a finite number of iterations. When this happens the solution is said to converge. The iterative methods are applicable to those systems of equations in which the diagonal elements of the coefficient matrix are large in comparison with the off-diagonal elements.

Consider a system of *n* non-linear equation having $x_1, x_2 \ldots x_n$ as *n* unknowns. After some manipulations these equations can be put in the form.

$$x_i = f_i(x_1, x_2, \ldots, x_i, \ldots x_n) \qquad (10.26)$$

where $1 \le i \le n$

We start with initial known (or assumed) approximate values of *x*'s written as $x_1^0, x_2^0, \ldots x_n^0$. Our first approximation for x_1 is

$$x_1^1 = f_1(x_1^0, x_2^0, x_3^0 \ldots \ldots x_n^0) \qquad (10.27a)$$

the first approximation for x_2 is

$$x_2^1 = f_2(x_1^1, x_2^0, x_3^0 \ldots \ldots x_n^0) \qquad (10.27b)$$

and the first approximation for x_i is

$$x_i^1 = f_i(x_1^1, x_2^1, \dots \dots x_{i-1}^1, x_i^0, x_{i+1}^0 \dots \dots x_n^0) \tag{10.27c}$$

In general the kth approximation for x_i is computed as:

$$x_i^k = f_i(x_1^k, x_2^k \dots x_{i-1}^k, x_i^{k-1}, x_{i+1}^{k-1} \dots x_n^{k-1}) \tag{10.27d}$$

We examine the change in each variable

$$\Delta x_i = x_i^k - x_i^{k-1} \tag{10.28}$$

When $\Delta x_i < \varepsilon$ for $i = 1, 2, \dots n$, we say that the solution has converged. From experience with practical problems the number of iterations required to reach convergence can be reduced if we update the old value by something more than Δx_i. We use the equation

$$x_i^k = x_i^{k-1} + \alpha \Delta x_i \tag{10.29}$$

α is known as acceleration factor and has a value equal to or greater than unity.

Example 10.9

Find the value of x, accurate upto 5 decimal places, in the equation $2x - \log x = 7$ by Gauss-Seidel method.

Solution

$$2x - \log x = 7$$

or

$$x = 0.5\,(7 + \log x)$$

We assume that $x^0 = 1$

$$x^1 = 0.5(7 + \log 1) = 3.5$$
$$x^2 = 0.5\,(7 + \log 3.5) = 3.772034022$$
$$x^3 = 0.5\,(7 + \log 3.772034022) = 3.7882878$$
$$x^4 = 0.5\,(7 + \log 3.7882878) = 3.789221482$$
$$x^5 = 0.5\,(7 + \log 3.789221482) = 3.789274995$$
$$x^6 = 0.5\,(7 + \log 3.789274995) = 3.789278062$$

Since the values of x^5 and x^6 are the same upto 5 decimal places, the iteration need not be continued further and 3.789278 can be taken as the value of x.

Example 10.10.

Obtain the values of x and y after 4 G–S iterations for the equations

$$x = 0.7 \sin x + 0.2 \cos y$$
$$y = 0.7 \cos x - 0.2 \sin y$$

Assume $x^0 = y^0 = 0.5$

Solution

$$x^0 = 0.5, y^0 = 0.5$$
$$x^1 = 0.7 \sin(0.5) + 0.2 \cos(0.5) = 0.5111143893$$
$$y_1^1 = 0.7 \cos(0.5111143893) - 0.2 \sin(0.5) = 0.5146548547$$
$$x^2 = 0.7 \sin(0.5111143893) + 0.2 \cos(0.5146548547) = 0.5164971976$$

$$y^2 = 0.7 \cos(0.5164971976) - 0.2 \sin(0.5146548547) = 0.5102411391$$
$$x^3 = 0.7 \sin(0.5164971976) + 0.2 \cos(0.5102411391) = 0.5202114603$$
$$y^3 = 0.7 \cos(0.5202114603) - 0.2 \cos(0.5102411391) = 0.5097223258$$
$$x^4 = 0.7 \sin(0.5202114603) + 0.2 \sin(0.5097223258) = 0.5225205507$$
$$y^4 = 0.7 \cos(.5225205507) - 0.2 \sin(0.5097223258) = 0.509007831$$

Thus, after 4 *G–S* iterations

$$x = 0.5225205507, y = 0.509007831$$

10.9 GAUSS-SEIDEL METHOD FOR LOAD FLOW STUDIES

Gauss-Seidel method was one of the most common methods used in load flow studies. The advantages of the method are (1) The simplicity of the technique (2) Small computer memory requirement (3) Less computational time per iteration. The disadvantages of this method are (1) Slow rate of convergence and, therefore, large number of iterations (2) increase of number of iterations directly with the increase in the number of buses. (3) Effect on convergence due to choice of slack bus. In view of these disadvantages *G–S* method is used only for systems having small number of buses.

10.9.1 G–S Method When PV Buses are Absent

We start the discussion of *G–S* algorithm for a system in which voltage controlled buses are absent. This implies that out of the n buses, one bus is slack bus and the remaining $n - 1$ buses are $P - Q$ buses. Initially we assume the magnitudes and angles at these $n - 1$ buses and update these voltages at every step of iteration.

From Eq. (10.20),

$$I_i = Y_{ii} V_i + \sum_{\substack{p=1 \\ p \neq i}}^{n} Y_{ip} V_p$$

or

$$V_i = \frac{1}{Y_{ii}} \left[I_i - \sum_{\substack{p=1 \\ p \neq i}}^{n} Y_{ip} V_p \right] \tag{10.30}$$

From Eq. (10.22a)

$$I_i = \frac{P_i - jQ_i}{V_i^*} \tag{10.31}$$

Substituting for I_i from Eq. (10.31) into Eq. (10.30) we have

$$V_i = \frac{1}{Y_{ii}} \left[\frac{P_i - jQ_i}{V_i^*} - \sum_{\substack{p=1 \\ p \neq i}}^{n} Y_{ip} V_p \right] \tag{10.32}$$

Since bus 1 is the slack bus, Eq. (10.32) represents a set of $n - 1$ equations, for $i = 2, 3, \dots n$, which are to be solved simultaneously for $V_2, V_3 \dots V_n$. The voltages and admittances in these equations are complex quantities. In most of the computers the complex number operations are available.

In order to save computer time it is more appropriate to write Eq. (10.32) in the following form:

$$V_i = \frac{K_i}{V_i^*} - \sum_{\substack{p=1 \\ p \neq i}}^{n} L_{ip} V_p \tag{10.33}$$

where $$K_i = \frac{P_i - jQ_i}{Y_{ii}} \quad \text{for } i = 2, 3 \ldots n$$

$$L_{ip} = \frac{Y_{ip}}{Y_{ii}} \quad \text{for } i = 2, 3 \ldots n$$

$$p = 1, 2 \ldots n$$
$$p \varpi i$$

The values of K_i and L_{ip} are computed once in the beginning and then used in every iteration.

It is necessary to assume initial values of $V_2, V_3 \ldots V_n$. We know that the voltages at different buses in a power system are close to 1 *pu*. Therefore, we assume $V_2, V_3 \ldots V_n$ to be $1 + j0$ initially. At every step of iteration we use the most updated values of the bus voltages to compute the new values of the bus voltages. When $(k + 1)$th iteration is being performed to calculate V_i, the most updated values of the bus voltages are the values obtained in $(k + 1)$th iteration for all buses previous to ith bus (i.e., for $p = 1, 2 \ldots i - 1$) and the values obtained in kth iteration for ith bus and the buses after the ith bus (i.e., for $p = i, i + 1 \ldots n$). In view of this observation Eq. (10.33) can be written as under for the $(k + 1)$ iteration

$$V_i^{k+1} = \frac{K_i}{(V_i^k)} - \sum_{p=1}^{i-1} L_{ip} V_p^{k+1} - \sum_{p=i+1}^{n} L_{ip} V_p^k \tag{10.34}$$

for $i = 2, 3 \ldots n$

The iterative process is continued till the changes in the bus voltages in the successive iterations become less than a certain pre-specified tolerance level for all buses, i.e.,

$$|\Delta V_i^{k+1}| = |V_i^{k+1} - V_i^k| < \varepsilon \text{ for } i = 2, 3 \ldots n \tag{10.35}$$

To reduce the number of iterations, an acceleration factor α can be used. $\alpha = 1.6$ is considered to be a good value for load flow studies.

When all bus voltages have been determined S_i^* is obtained from Eq. (10.22b) and then conjugate of S_i^* gives S_i.

The final step in load flow study is the calculation of line flows. Fig. 10.13 shows a line connecting ith and pth buses. The line is represented by a π circuit. The current I_{ip} and complex power S_{ip} can be found as

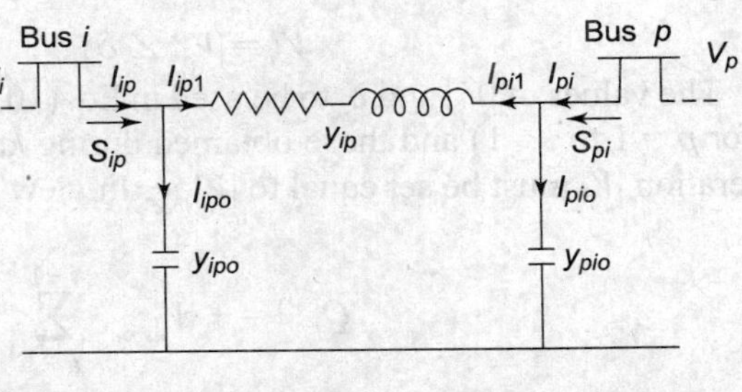

Fig. 10.13 Representation of a line between two buses

$$I_{ip} = I_{ip1} + I_{ipo} = (V_i - V_p)\, y_{ip} + V_i y_{ipo} \tag{10.36}$$

$$S_{ip} = P_{ip} + jQ_{ip} = V_i I_{ip}^* = V_i(V_i^* - V_p^*)\, y_{ip}^* + V_i V_i^* y_{ipo}^* \tag{10.37a}$$

Similarly, the power fed by pth bus into the line connecting pth and ith buses is

$$S_{pi} = V_p (V_p^* - V_i^*)\, y_{ip}^* + V_p V_p^* y_{pio}^* \tag{10.37b}$$

Thus, power flow over all the lines can be computed.

The power loss in the line connecting ith and pth buses is given by $S_{ip} + S_{pi}$ and the total transmission loss is the sum of losses over all the lines.

10.9.2 Modification of G–S Method when PV Buses are Present

Some of the buses in a power system are voltage controlled buses where P and V are specified and Q and δ are unknowns. It is necessary to modify the above computational procedure. Let the buses be numbered as

$$i = 1 \text{ slack bus}$$

$$i = 2, 3, \ldots m \ \ PV \text{ or voltage controlled buses}$$

$$i = m + 1, m + 2, \ldots n \ \ PQ \text{ buses}$$

For the voltage controlled buses, the bus voltage must be equal to the specified voltage $|V_i|_{spec}$. Moreover the maximum and minimum reactive powers at these buses are also specified and the value of Q_i (for $i = 2, 3 \ldots m$) must lie between these limits. Thus, the conditions to be met with are

(1) $|V_i| = |V_i|_{spec}$ for $i = 2, 3 \ldots m$

(2) $Q_{i,\,min} < Q_i < Q_{i,\,max}$ for $i = 2, 3 \ldots m$

The second requirement may be violated if the specified bus voltage $|V_i|_{spec}$ is either too high or too low. It is to be noted that it is possible to control $|V_i|$ only by controlling Q_i. If during any iteration we find that Q_i is outside the limits specified, Q_i is fixed at the minimum or the maximum value as the case may be and then that bus is treated as a $P\,Q$ bus. In view of these observations the procedure of calculation for $P\,V$ buses is as under:

(a) Calculate Q_i using Eq. (10.24b) which is reproduced below

$$Q_i = |V_i| \sum_{p=1}^{n} |Y_{ip}|\,|V_p| \sin(\delta_i - \gamma_{ip} - \delta_p) \tag{10.38}$$

At the start of $(k + 1)$th iteration, $|V_i|^k$, i.e., the magnitude of V_i obtained during kth iteration may not necessarily be equal to $|V_i|_{spec}$. We save this value of V_i^k.

$$V_i^k = |V_i|^k \angle \delta_i^k \tag{10.39}$$

The values of $|V_p|$ and δ_p to be used in Eq. (10.39) are those obtained during $(k + 1)$th iteration (for $p = 1$ to $i - 1$) and those obtained during kth iteration (for $p = i$ to n). Moreover for every iteration $|V_i|$ must be set equal to $|V_i|_{spec}$. In view of this Eq. (10.38) can be written as:

$$Q_i^{k+1} = +|V_i|_{spec} \sum_{p=1}^{i-1} |Y_{ip}|\,|V_p^{k+1}| \sin(\delta_i^k - \gamma_{ip} - \delta_p^{k+1})$$

$$+ |V_i|_{spec} \sum_{p=i}^{n} |Y_{ip}|\,|V_p^k| \sin(\delta_i^k - \gamma_{ip} - \delta_p^k) \tag{10.40}$$

(b) Check Q_i^{k+1} to see if it is within limits, i.e., $Q_{i,min} < Q_i^{k+1} < Q_{i,\,max}$. If so calculate new value of K_i and then use Eq. (10.34) to compute V_i^{k+1} using $|V_i|_{spec}$ and δ_i^k for the magnitude and phase angle of V_i^{k+1}. Reset $|V_i^{k+1}|$ to $|V_i|_{spec}$., but retain the phase angle δ_i^{k+1} and continue to the next bus.

If $Q_i^{k+1} > Q_{i,max}$, set $Q_i^{k+1} = Q_{i,max}$, find new value of K_i and treat the ith bus as PQ bus and continue the computations as in section (10.9.1). Similarly, if $Q_i^{k+1} < Q_{i,min}$, set $Q_i^{k+1} = Q_{i,min}$, find new value of K_i and treat the ith bus as PQ bus and continue the computations as in section (10.9.1).

The flowchart for a load flow study using *G–S* iteration is drawn in Fig. 10.14. The slack bus is numbered as bus 1, the *PV* buses are numbered from $i = 2$ to m and the *PQ* buses are numbered from $i = m + 1$ to n.

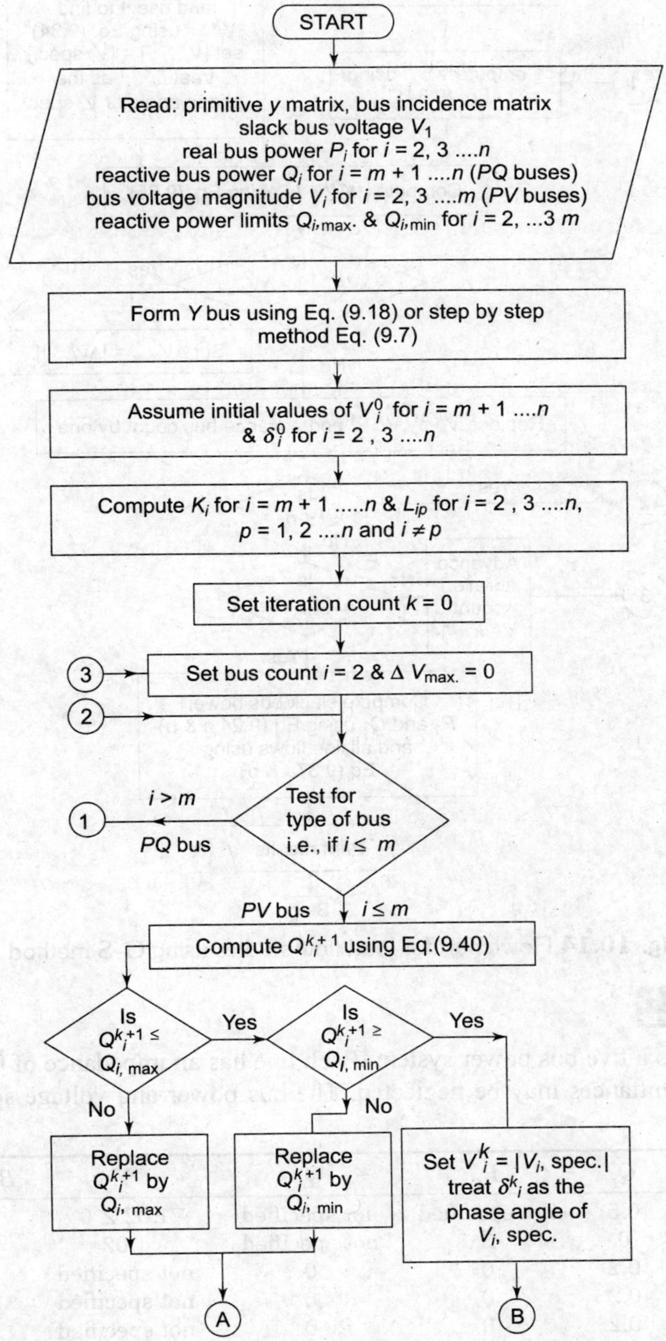

Fig. 10.14 (Continued on page 264)

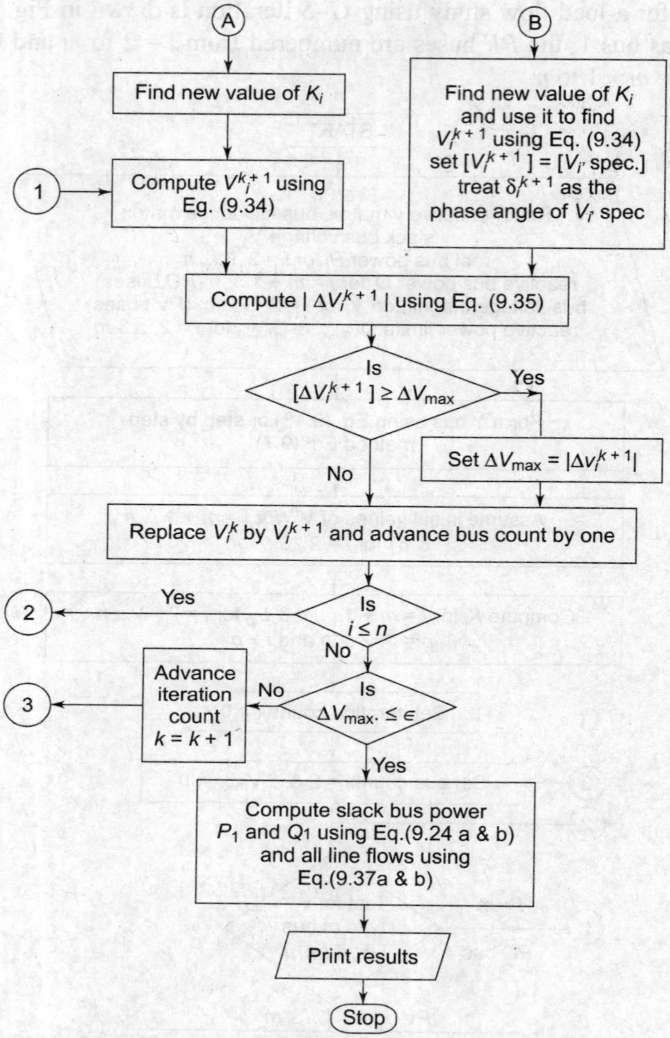

Fig. 10.14 Flow chart for load flow studies using G–S method

Example 10.11

Figure 10.15 shows a five bus power system. Each line has an impedance of $0.05 + j\,0.15\ pu$. The line shunt admittances may be neglected. The bus power and voltage specifications are given below:

Bus	P_L	Q_L	P_G	Q_G	V	*Bus specification*
1	1.0	0.5	not specified	not specified	$1.02 \angle 0$	Slack bus
2	0	0	2	not specified	1.02	*PV* bus
3	0.5	0.2	0	0	not specified	*PQ* bus
4	0.5	0.2	0	0	not specified	*PQ* bus
5	0.5	0.2	0	0	not specified	*PQ* bus

(a) From Y_{bus} (b) Find Q_2, δ_2, V_3, V_4 and V_5 after the first iteration using Gauss-Seidel method. Assume $Q_{2,\min} = 0.2\ pu$ and $Q_{2,\max} = 0.6\ pu$.

Solution

(a) For each line $y = \dfrac{1}{0.05 + j0.15} = (2 - j6)$ siemens

$$Y_{11} = y_{12} + y_{15} = 4 - j12 \text{ siemens}$$
$$Y_{12} = -y_{12} = -2 + j6 \text{ siemens}$$
$$Y_{15} = -y_{15} = -2 + j6 \text{ siemens}$$
$$Y_{13} = Y_{14} = 0$$

Similarly, the other elements of Y_{bus} can be calculated. The result is

$$Y_{bus} = \begin{bmatrix} 4-j12 & -2+j6 & 0 & 0 & -2+j6 \\ -2+j6 & 6-j18 & -2+j6 & 0 & -2+j6 \\ 0 & -2+j6 & 4-j12 & -2+j6 & 0 \\ 0 & 0 & -2+j6 & 4-j12 & -2+j6 \\ -2+j6 & -2+j6 & 0 & -2+j6 & 6-j18 \end{bmatrix}$$

(b) The power injected into the network by each bus is

$$S_1 = P_1 + jQ_1$$
$$S_2 = P_2 + jQ_2 = 2 + jQ_2$$
$$S_3 = P_3 + jQ_3 = -0.5 - j0.2$$
$$S_4 = P_4 + jQ_4 = -0.5 - j0.2$$
$$S_5 = P_5 + jQ_5 = -0.5 - j0.2$$

Also $-2 + j6 = 6.32455 \angle 108.43°$

$6 - j18 = 18.97366 \angle -71.57°$

Assume $V_3^{\circ} = V_4^{\circ} = V_5^{\circ} = 1 \angle 0$

and $\delta_2^{\circ} = \delta_3^{\circ} = \delta_4^{\circ} = \delta_5^{\circ} = 0$

Then $V_2^{\circ} = 1.02 + j0,\ |V_2|_{spec.} = 1.02$

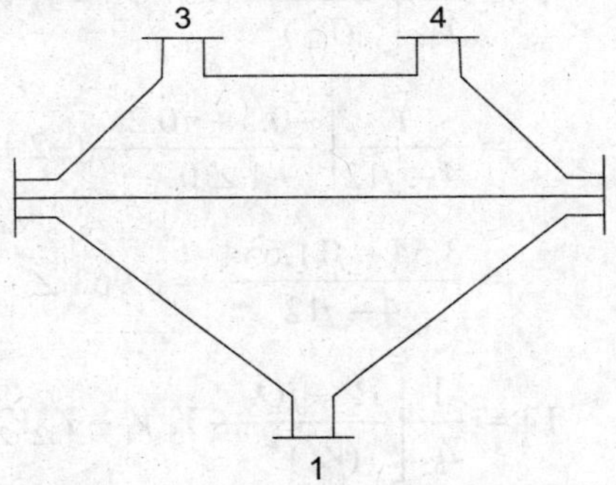

Fig. 10.15 Five bus system for example 10.11

$$Y_{21} = Y_{23} = Y_{25} = -2 + j6 = 6.32455 \angle 108.43°$$
$$Y_{22} = 6 - j18 = 18.97366 \angle -71.57°$$

Using Eq. (10.40)

$$\begin{aligned} Q_2^1 &= -|V_2|_{spec.} [|Y_{21}|V_1| \sin(\delta_1^{\circ} + \gamma_{12} - \delta_2^{\circ}) + |Y_{22}|\,|V_2^{\circ}| \sin(\delta_2^{\circ} + \gamma_{22} - \delta_2^{\circ}) \\ &\quad + |Y_{23}|V_3^{\circ}| \sin(\delta_3^{\circ} + \gamma_{23} - \delta_2^{\circ}) + |Y_{25}|\,|V_5^{\circ}| \sin(\delta_5^{\circ} + \gamma_{25} - \delta_2^{\circ})] \\ &= -1.02\,[(6.32455)(1.02)\sin(108.43°) + (18.97366)(1.02)\sin(-71.57°) \\ &\quad + (6.32455)(1)\sin(108.43°) + (6.32455)(1)\sin(108.43°)] \\ &= 0.2448\ pu \end{aligned}$$

The value of Q_2^1 is within the limits imposed by $Q_{2,min}$ and $Q_{2,max}$. Using the above value of $Q^1{}_2$ we find V_2^1 from Eq. (10.32). Putting $i = 2$ and $k = 0$, this equation can be written as:

$$V_2^1 = \frac{1}{Y_{22}}\left[\frac{P_2 - jQ_2^1}{(V_2^{\circ})^*} - Y_{21}V_1 - Y_{23}V_3^{\circ} - Y_{25}V_5^{\circ}\right]$$

$$= \frac{1}{6 - j18}\left[\frac{2 - j0.2448}{1.02 - j0} - (-2 + j6)(1.02 \angle 0) - (-2 + j6) \times (1 + j0)\right.$$
$$\left. - (-2 + j6)(1 + j0)\right]$$

$$= \frac{8 - j18.36}{6 - j18} = 1.0555 \angle 5.11°. \text{ Therefore, } \delta_2^1 = 5.11° \text{ or } 0.0892 \text{ rad.}$$

As discussed in section 10.9.2 we set $|V_2^1| = |V_2|_{spec}$. and retain the phase angle δ_2^1. Thus, the revised value of V_2^1 for subsequent iteration is

$$V_2^1 = 1.02 \angle 5.11°$$

$$V_3^1 = \frac{1}{Y_{33}}\left[\frac{P_3 - jQ_3}{(V_3^{\circ})^*} - Y_{32}V_2^1 - Y_{34}V_4^{\circ}\right]$$

$$= \frac{1}{4 - j12}\left[\frac{-0.5 + j0.2}{1\angle 0} - (-2 + j6)(1.02 \angle 5.11°) - (-2 + j6)(1 \angle 0°)\right]$$

$$= \frac{4.0765 - j11.714}{4 - j12} = 0.98 \angle 0.76°$$

$$V_4^1 = \frac{1}{Y_{44}}\left[\frac{P_4 - jQ_4}{(V_4^{\circ})^*} - Y_{43}V_3^1 - Y_{45}V_5^{\circ}\right]$$

$$= \frac{1}{4 - j12}\left[\frac{-0.5 + j0.2}{1\angle 0} - (-2 + j6)(0.98 \angle 0.76°) - (-2 + j6)(1 \angle 0°)\right]$$

$$= \frac{3.54 - j11.654}{4 - j12} = 0.963 \angle -1.53°$$

$$V_5^1 = \frac{1}{Y_{55}}\left[\frac{P_5 - jQ_5}{(V_5^{\circ})^*} - Y_{51}V_1 - Y_{52}V_2^1 - Y_{54}V_4^1\right]$$

$$= \frac{1}{6 - j18}\left[\frac{-0.5 + j0.2}{1\angle 0} - (-2 + j6)(1.02 \angle 0)\right.$$

$$\left. - (-2 + j6) \times (1.02 \angle 5.11°) - (-2 + j6)(0.963 \angle -1.53°)\right]$$

$$= \frac{5.887 - j17.71}{6 - j18} = 0.9836 \angle -0.04°$$

Example 10.12.

If the value of $Q_{2,\,min}$ is 0.3 *pu* in example 10.11, find the values of V_2, V_3, V_4 and V_5 after first iteration using Gauss-Seidel method. Assume a flat voltage start at buses.

Solution

The value of Q_2^1 found in example 10.10 is 0.2448 *pu*. Since the minimum permissible value of Q_2 is 0.3 *pu*, we fix $Q_2 = 0.3$ *pu* and treat bus 2 as *P–Q* bus with $V_2^{\circ} = 1 \angle 0$.

$$V_2^1 = \frac{1}{Y_{22}}\left[\frac{P_2 - jQ_2}{(V_2^{\circ})^*} - Y_{21}V_1 - V_{23}V_3^{\circ} - Y_{25}V_5^{\circ}\right]$$

$$= \frac{1}{6-j18}\left[\frac{2-j0.3}{1\angle 0} - (-2+j6)(1.02\angle 0) - (-2+j6)(1\angle 0) - (-2+j6)(1\angle 0^\circ)\right]$$

$$= \frac{8.04-j18.42}{6-j18} = 1.059\angle 5.15^\circ$$

$$V_3^1 = \frac{1}{Y_{33}}\left[\frac{P_3 - jQ_3}{(V_3^\circ)^*} - Y_{32}V_2^1 - Y_{34}V_4^\circ\right]$$

$$= \frac{1}{4-j12}\left[\frac{-0.5+j0.2}{1\angle 0} - (-2+j6)(1.059\angle 5.15^\circ) - (-2+j6)(1\angle 0^\circ)\right]$$

$$= \frac{4.18-j11.94}{4-j12} = 1\angle 0.864^\circ$$

$$V_4^1 = \frac{1}{Y_{44}}\left[\frac{P_4 - jQ_4}{(V_4^\circ)^*} - Y_{43}V_3^1 - Y_{45}V_5^\circ\right]$$

$$= \frac{1}{4-j12}\left[\frac{-0.5+j0.2}{1\angle 0} - (-2+j6)(1\angle 0.864^\circ) - (-2+j6)(1\angle 0)\right]$$

$$= \frac{3.5897-j11.769}{4-j12} = 0.9727\angle -1.47^\circ$$

$$V_5^1 = \frac{1}{Y_{55}}\left[\frac{P_5 - jQ_5}{(V_5^\circ)^*} - Y_{51}V_1 - Y_{52}V_2^1 - Y_{54}V_4^1\right]$$

$$= \frac{1}{6-j18}\left[\frac{-0.5+j0.2}{1\angle 0} - (-2+j6)(1.02\angle 0)\right.$$

$$\left. - (-2+j6)(1.059\angle 5.15^\circ) - (-2+j6)(0.9727\angle -1.47^\circ)\right]$$

$$= \frac{6.015-j17.945}{6-j18} = 0.9975\angle 0.1^\circ$$

10.10 NEWTON-RAPHSON METHOD–SOLUTION OF ALGEBRAIC EQUATIONS

Consider the equation

$$f(x) = 0 \tag{10.41}$$

Assume that x° is the initial estimate of the unknown x and Δx° is the correction which when added to x° gives the correct solution. In other words, $x^\circ + \Delta x^\circ$ is the solution of Eq. (10.41).

Therefore,

$$f(x^\circ + \Delta x^\circ) = 0 \tag{10.42}$$

Equation (10.42) can be expanded into Taylor's series. The result is

$$f(x^o + \Delta x^o) = f(x^o) + \Delta x^o \left(\frac{df}{dx}\right)^0 + \frac{1}{2!}(\Delta x^o)^2 \left(\frac{d^2 f}{dx^2}\right)^0 + \cdots \tag{10.43}$$

The symbol $\left(\frac{df}{dx}\right)^0$ stands for the value of $\frac{df(x)}{dx}$ at $x = x^o$.

A first estimate of the correction is made by truncating the series after the first two terms. Then

$$f(x^o) + \Delta x^o \left(\frac{df}{dx}\right)^0 \cong 0$$

or

$$\Delta x^o \cong \frac{-f(x^o)}{\left(\frac{df}{dx}\right)^o} \tag{10.44}$$

The correction Δx^o should take us from the initial guess to the exact solution. Since we have made an approximation in calculating Δx^o it is necessary to make repeated use of Eq. (10.44) till the solution converges.

The first estimate of the root is

$$x^1 = x^o + \Delta x^o = x^o + \left[\frac{-f(x^o)}{\left(\frac{df}{dx}\right)^o}\right] \tag{10.45}$$

The $(k + 1)$th estimate of the root is

$$x^{k+1} = x^k + \Delta x^k = x^k - \left[\frac{f(x^k)}{\left(\frac{df}{dx}\right)^k}\right] \tag{10.46}$$

The iterations are continued till

$$x^{k+1} - x^k \leq \varepsilon$$

where ε is the specified tolerance.

Now consider a set of two equations

$$f_1(x_1, x_2) = 0 \text{ and } f_2(x_1, x_2) = 0 \tag{10.47}$$

Let x_1^o an x_2^o be the initial estimates of unknowns x_1 and x_2. $\Delta x^o{}_1$ and Δx_2^o are the corrections which when added to x_1^o and $x^o{}_2$ give the correct solution. Then

$$f_1(x_1^o + \Delta x^o{}_1, x_2^o + \Delta x^o{}_2) = 0$$

$$f_2(x_1^o + \Delta x^o{}_1, x_2^o + \Delta x^o{}_2) = 0 \tag{10.48}$$

Expanding Eq. (10.48) into Taylor's series

$$f_1(x_1^o, x^o{}_2) + \Delta x_1^o \left(\frac{\partial f_1}{\partial x_1}\right)^o + \Delta x_2^o \left(\frac{\partial f_1}{\partial x_2}\right)^o + \cdots = 0$$

$$f_2(x_1^o, x_2^o) + \Delta x_1^o \left(\frac{\partial f_2}{\partial x_1}\right)^o + \Delta x_2^o \left(\frac{\partial f_2}{\partial x_2}\right)^o + \cdots = 0 \tag{10.49}$$

where $\left(\frac{\partial f_1}{\partial x_1}\right)^{\text{o}}$ denotes the partial derivative $\frac{\partial f_1(x_1, x_2)}{\partial x_1}$ at $x_1 = x_1^{\text{o}}$.

The other partial derivatives also have similar meanings. Since we have more than one variable, we use partial derivatives in this case. Truncating the series after the first two terms, Eq. (10.49) can be written as

$$\begin{bmatrix} f_1(x_1^{\text{o}}, x_2^{\text{o}}) \\ f_2(x_1^{\text{o}}, x_2^{\text{o}}) \end{bmatrix} + \begin{bmatrix} \left(\frac{\partial f_1}{\partial x_1}\right)^{\text{o}} & \left(\frac{\partial f_1}{\partial x_2}\right)^{\text{o}} \\ \left(\frac{\partial f_2}{\partial x_1}\right)^{\text{o}} & \left(\frac{\partial f_2}{\partial x_2}\right)^{\text{o}} \end{bmatrix} \begin{bmatrix} \Delta x_1^{\text{o}} \\ \Delta x_2^{\text{o}} \end{bmatrix} \simeq \begin{bmatrix} 0 \\ 0 \end{bmatrix} \tag{10.50}$$

The matrix containing partial derivatives is called Jacobian J^{o} so that Eq. (10.50) can be written as

$$\begin{bmatrix} f_1(x_1^{\text{o}}, x_2^{\text{o}}) \\ f_2(x_1^{\text{o}}, x_2^{\text{o}}) \end{bmatrix} \simeq -[J^{\text{o}}] \begin{bmatrix} \Delta x_1^{\text{o}} \\ \Delta x_2^{\text{o}} \end{bmatrix} \tag{10.51}$$

The values of Δx_1^{o} and Δx_2^{o} can be found by rewriting Eq. (10.51) as:

$$\begin{bmatrix} \Delta x_1^{\text{o}} \\ \Delta x_2^{\text{o}} \end{bmatrix} \simeq -[J^{\text{o}}]^{-1} \begin{bmatrix} f_1(x_1^{\text{o}}, x_2^{\text{o}}) \\ f_2(x_1^{\text{o}}, x_2^{\text{o}}) \end{bmatrix} \tag{10.52}$$

Since we have used approximation of finding Δx_1^{o} and Δx_2^{o}, these values are only initial estimates of correction. The first estimates of the roots are

$$\begin{aligned} x_1^1 &= x_1^{\text{o}} + \Delta x_1^{\text{o}} \\ x_2^1 &= x_2^{\text{o}} + \Delta x_2^{\text{o}} \end{aligned} \tag{10.53}$$

The $(k + 1)$th estimates of the roots are

$$\begin{aligned} x_1^{k+1} &= x_1^k + \Delta x_1^k \\ x_2^{k+1} &= x_2^k + \Delta x_2^k \end{aligned} \tag{10.54}$$

where $\begin{bmatrix} \Delta x_1^k \\ \Delta x_2^k \end{bmatrix} = -[J^k]^{-1} \begin{bmatrix} f_1(x_1^k, x_2^k) \\ f_2(x_1^k, x_2^k) \end{bmatrix}$

The iterative process is continued till convergence is obtained.

Now consider a system of n equations

$$\begin{aligned} f_1(x_1, x_2 \dots x_n) &= 0 \\ f_2(x_1, x_2 \dots x_n) &= 0 \\ &\dots\dots\dots\dots \\ f_n(x_1, x_2 \dots x_n) &= 0 \end{aligned} \tag{10.55}$$

Let $x_1^{\text{o}}, x_2^{\text{o}} \dots x_n^{\text{o}}$ be the initial estimates of unknowns $x_1, x_2 \dots x_n$. $\Delta x_1^{\text{o}}, \Delta x_2^{\text{o}} \dots \Delta x_n^{\text{o}}$ are the corrections. Proceeding in the same manner we get an equation similar to Eq. (10.50), i.e.,

$$\begin{bmatrix} f_1(x_1^o,\ x_2^o \ldots.x_n^o) \\ f_2(x_1^o,\ x_2^o \ldots.x_n^o) \\ \text{------------} \\ f_n(x_1^o,\ x_2^o \ldots.x_n^o) \end{bmatrix} + \begin{bmatrix} \left(\frac{\partial f_1}{\partial x_1}\right)^o & \left(\frac{\partial f_1}{\partial x_2}\right)^o & \ldots & \left(\frac{\partial f_1}{\partial x_n}\right)^o \\ \left(\frac{\partial f_2}{\partial x_1}\right)^o & \left(\frac{\partial f_2}{\partial x_2}\right)^o & \ldots & \left(\frac{\partial f_2}{\partial x_n}\right)^o \\ \text{-----} & \text{-----} & & \text{-----} \\ \left(\frac{\partial f_n}{\partial x_1}\right)^o & \left(\frac{\partial f_n}{\partial x_2}\right)^o & \ldots & \left(\frac{\partial f_n}{\partial x_n}\right)^o \end{bmatrix} \times \begin{bmatrix} \Delta x_1^o \\ \Delta x_2^o \\ \vdots \\ \Delta x_n^o \end{bmatrix} \simeq \begin{bmatrix} 0 \\ 0 \\ \vdots \\ 0 \end{bmatrix} \tag{10.56a}$$

or

$$f^o + J^o \Delta x^o \simeq 0 \tag{10.56b}$$

The first estimate of correction is

$$\Delta x^o = -[J^o]^{-1} f^o \tag{10.57}$$

The first estimate of roots is

$$x^1 = x^o + \Delta x^o \tag{10.58}$$

The $(k+1)$th estimate of roots is

$$x^{k+1} = x^k + \Delta x^k \tag{10.59}$$

As before the iterations are continued till

$$x^{k+1} - x^k \le \varepsilon$$

Example 10.13

Find the value of x in the equation $2x - \log x = 7$. Use *NR* method.

Solution

Changing the logarithm to the base e, the equation is

$$f(x) = 2x - 0.43429 \ln x - 7 = 0$$

$$\frac{df(x)}{dx} = 2 - \frac{0.43429}{x}$$

Let

$$x^o = 1$$

$$f(x)^o = -5$$

$$\left(\frac{\partial f}{\partial x}\right)^o = 2 - 0.43429 = 1.56571$$

$$x^1 = 1 - \frac{-5}{1.56571} = 4.1934394$$

$$f(x^1) = 2(4.1934394) - 0.43429 \ln (4.1934394) - 7 = 0.7643149$$

$$\left(\frac{df}{dx}\right)^1 = 2 - \frac{0.43429}{4.1934394} = 1.896435847$$

$$x^2 = 4.1934394 - \frac{0.7643149}{1.896435847} = 3.7904124$$

$$f(x^2) = 2(3.7904124) - 0.43429 \ln(3.7904124) - 7 = 0.002144308$$

$$\left(\frac{df}{dx}\right)^2 = 2 - \frac{0.43429}{9.7904124} = 1.885424$$

$$x^3 = 3.7904124 - \frac{0.002144308}{1.885424} = 3.789275$$

It is seen that we have reached the correct estimate only in three iterations as compared to 6 iterations necessary in *G–S* method (example 10.8). However, the calculations in each iteration are lengthy because calculations of and division by the derivative is required.

Example 10.14

Solve the equations using *NR* method

$$f(x, y) = x + 3 \log x - y^2 = 0$$

$$g(x, y) = 2x^2 - xy - 5x + 1 = 0$$

Assume $x^o = 3.4$ and $y^o = 2.2$

Solution

$$f(x, y) = x + 3 \log x - y^2$$

$$= x + 3 \times 0.43429 \ln x - y^2 = x + 1.30287 \ln x - y^2$$

$$\frac{\partial f}{\partial x} = 1 + (1.30287)/x$$

$$\frac{\partial f}{\partial y} = -2y$$

$$g(x, y) = 2x^2 - xy - 5x + 1$$

$$\frac{\partial g}{\partial x} = 4x - y - 5$$

$$\frac{\partial g}{\partial y} = -x$$

$$J = \begin{bmatrix} \frac{\partial f}{\partial x} & \frac{\partial f}{\partial y} \\ \frac{\partial g}{\partial x} & \frac{\partial g}{\partial y} \end{bmatrix} = \begin{bmatrix} 1 + \frac{1.30287}{x} & -2y \\ 4x - y - 5 & -x \end{bmatrix}$$

Taking $x^o = 3.4$ and $y^o = 2.2$

$$[J^o] = \begin{bmatrix} 1.383 & -4.4 \\ 6.4 & -3.4 \end{bmatrix}$$

$$\begin{bmatrix} f(3.4, 2.2) \\ g(3.4, 2.2) \end{bmatrix} = \begin{bmatrix} 0.1545 \\ -0.72 \end{bmatrix}$$

$$\begin{bmatrix} \Delta x^o \\ \Delta y^o \end{bmatrix} = -\begin{bmatrix} 1.383 & -4.4 \\ 6.4 & -3.4 \end{bmatrix}^{-1} \begin{bmatrix} 0.1545 \\ -0.72 \end{bmatrix} = \begin{bmatrix} 0.157 \\ 0.085 \end{bmatrix}$$

$$x^1 = x^o + \Delta x^o = 3.4 + 0.157 = 3.557$$

$$y^1 = y^o + \Delta y^o = 2.2 + 0.085 = 2.285$$

$$[J^1] = \begin{bmatrix} 1.367 & -4.57 \\ 6.943 & -3.557 \end{bmatrix}$$

$$\begin{bmatrix} f(3.557), 2.285 \\ g(3.557), 2.285 \end{bmatrix} = \begin{bmatrix} -0.01097 \\ 0.39175 \end{bmatrix}$$

$$\begin{bmatrix} \Delta x^1 \\ \Delta y^1 \end{bmatrix} = -\begin{bmatrix} 1.367 & -4.57 \\ 6.943 & -3.557 \end{bmatrix}^{-1} \begin{bmatrix} -0.01097 \\ 0.39175 \end{bmatrix} = \begin{bmatrix} -0.0685 \\ -0.0229 \end{bmatrix}$$

$$x^2 = x^1 + \Delta x^1 = 3.557 - 0.0685 = 3.4885$$

$$y^2 = y^1 + \Delta y^1 = 2.285 - 0.0229 = 2.2621$$

The values of Δx^1 and Δy^1 are small. The values of Δx^2 and Δy^2 (if calculated) will be found to be still smaller. Hence, the iterative process can be stopped here. The solution is $x = 3.4885$, $y = 2.2621$.

Example 10.15

Use *NR* method to find the value of x correct upto 4 decimal places in the equation

$$3x - \cos x - 1 = 0.$$

Solution

$$f(x) = 3x - \cos x - 1$$

$$\frac{df(x)}{dx} = 3 + \sin x$$

Let

$$x^o = 0$$

$$f(x^o) = -2$$

$$\left(\frac{\partial f}{\partial x}\right)^o = 3$$

$$x^1 = x^o - \frac{f(x^o)}{(df/dx)^o} = 0 - \frac{-2}{3} = \frac{2}{3}$$

$$f(x^1) = 3\left(\frac{2}{3}\right) - \cos\left(\frac{2}{3}\right) - 1 = 0.214127$$

$$\left(\frac{df}{dx}\right)^1 = 3 + \sin\left(\frac{2}{3}\right) = 3.6183698$$

$$x^2 = x^1 - \frac{f(x^1)}{(df/dx)^1} = \frac{2}{3} - \frac{0.214127}{3.6183698} = 0.6074889$$

$$f(x^2) = 3(0.6074889) - \cos(0.6074889) - 1 = 1.38274 \times 10^{-3}$$

$$\left(\frac{df}{dx}\right)^2 = 3 + \sin(0.6074889) = 3.570807$$

$$x^3 = x^2 - \frac{(fx^2)}{\left(\frac{df}{dx}\right)^2} = 0.6074889 - \frac{1.38274\times10^{-3}}{3.570807} = 0.6071017$$

$$f(x^3) = 3(0.6071017) - \cos(0.6071017) - 1 = 1.853 \times 10^{-7}$$

$$\left(\frac{df}{dx}\right)^3 = 3 + \sin(0.6071017) = 3.570489$$

$$x^4 = x^3 - \frac{f(x)^3}{\left(\frac{df}{dx}\right)^3} = 0.6071017 - \frac{1.853\times10^{-7}}{3.576489} = 0.60710175$$

The value of x correct to 4 decimal places is 0.6071.

10.11 NEWTON-RAPHSON METHOD FOR LOAD FLOW STUDIES

Newton-Raphson (*NR*) method is very suitable for load flow studies on large systems. The advantages of this method are (1) More accuracy and surety of convergence (2) Only about 3 iterations are required as compared to more than 25 or so as required by *G–S* method (3) The number of iterations is almost independent of the system size (4) This method is insensitive to factors like slack bus selection, regulating transformers, etc. The disadvantages of this method are (1) The solution technique is difficult (2) The calculations in each iteration are more involved and, thus, computer time per iteration is large (3) The computer memory requirement is large.

NR method can be applied to load flow problems in a number of ways, the most common being those using (a) rectangular co-ordinates (b) polar co-ordinates.

10.11.1 NR Method using Rectangular Co-ordinates

In this formulation the quantities are expressed in rectangular form,

Let $$V_p = |V_p| \angle \delta_p = e_p + jf_p$$

The active and reactive powers at each bus are functions of e and f. Thus,

$$P_i = u_1(e, f) \tag{10.60}$$

$$Q_i = u_2(e, f)$$

For a system of n buses and bus 1 designated as slack bus, the difference equations which relate the changes in active and reactive power to changes in e and f take the form

$$\Delta P_i = \sum_{p=2}^{n} \frac{\partial P_i}{\partial e_p} \Delta e_p + \sum_{p=2}^{n} \frac{\partial P_i}{\partial f_p} \Delta f_p \tag{10.61a}$$

$$\Delta Q_i = \sum_{p=2}^{n} \frac{\partial Q_i}{\partial e_p} \Delta e_p + \sum_{p=2}^{n} \frac{\partial Q_i}{\partial f_p} \Delta f_p \tag{10.61b}$$

ΔP_i and ΔQ_i represent the differences between the specified and the calculated values of P_i and Q_i. There are two equations similar to Eq. (10.61) for each of the $(n - 1)$ buses. In matrix form these equations can be written as

$$\left[\frac{\Delta P}{\Delta Q}\right] = \left[\frac{J_1 \mid J_2}{J_3 \mid J_4}\right]\left[\frac{\Delta e}{\Delta f}\right] \tag{10.62}$$

Where J_1, J_2, J_3, J_4, are the elements of the Jacobian and are partial derivatives similar to those appearing in Eq. (10.61). Eq. (10.62) can be used to determine the components of bus voltages from an arbitrary set of values through an iterative procedure.

For a PV bus the reactive power Q_i is not specified but the voltage magnitude $|V_i|$ is specified. Thus,

$$|V_i|^2 = e_i^2 + f_i^2 \tag{10.63}$$

Therefore, the following equation replaces Eq. (10.61b) for each PV bus

$$\Delta |V_i|^2 = \frac{\partial |V_i|^2}{\partial e_i} \Delta e_i + \frac{\partial |V_i|^2}{\partial f_i} \Delta f_i \tag{9.64}$$

The total number of equations is $2\ (n - 1)$. As will be seen in the next section, the use of polar form results in a smaller number of equations. Therefore, the NR method in rectangular co-ordinates is rarely used. In actual practice only polar form is used.

10.11.2 NR Method using Polar Co-ordinates

The Newton Raphson method can be applied to the load flow studies when the bus voltages are expressed in polar form. The active and reactive power at each bus are functions of magnitudes and phase angles of bus voltages. Thus

$$P_i = f_1\,(\delta, |V|) \tag{10.65a}$$

$$Q_i = f_2\,(\delta, |V|) \tag{10.65b}$$

For a system of n buses and bus 1 designated as slack bus, the equations which relate the changes in active and reactive power to changes in bus voltage magnitude and angles take the form

$$\Delta P_i = \sum_{p=2}^{n} \frac{\partial P_i}{\partial \delta_p} \Delta \delta_p + \sum_{p=2}^{n} \frac{\partial P_i}{\partial |V_p|} \Delta |V_p| \tag{10.66a}$$

$$\Delta Q_i = \sum_{p=2}^{n} \frac{\partial Q_i}{\partial \delta_p} \Delta \delta_p + \sum_{p=2}^{n} \frac{\partial Q_i}{\partial |V_p|} \Delta |V_p| \tag{10.66b}$$

ΔP_i and ΔQ_i represent the differences between the specified and the calculated values of P_i and Q_i. Equation (10.66) can be used to determine the bus voltage magnitudes and angles from an arbitrary set of values through an iterative procedure. For each load bus both P and Q are specified and, therefore, Eqs. (10.66 a and b) exist. However, for a PV bus, Q is not specified and, therefore, there is no equation corresponding to Eq. (10.66b) for PV bus. For an n bus system having 1 slack bus and g voltage controlled (PV) buses, the total number of equations is $(2n - 2 - g)$. Thus, the use of the polar form results in lesser number of equations and a smaller size of Jacobian as compared to the rectangular form. This is a definite advantage of polar form over the rectangular form and, therefore, only polar form is used.

It has been found convenient to replace $\Delta |V|$ by $\dfrac{\Delta |V|}{|V|}$ in Eq. (10.66) as a result of which Eqs. (10.66) become

$$\Delta P_i = \sum_{p=2}^{n} \frac{\partial P_i}{\partial \delta_p} \Delta \delta_p + \sum_{p=2}^{n} \frac{\partial P_i}{\partial |V_p|} |V_p| \frac{\Delta |V_p|}{|V_p|}$$

$$\Delta Q_i = \sum_{p=2}^{n} \frac{\partial Q_i}{\partial \delta_p} \Delta \delta_p + \sum_{p=2}^{n} \frac{\partial Q_i}{\partial |V_p|} |V_p| \frac{\Delta |V_p|}{|V_p|} \tag{10.67}$$

The set of Eq. (10.67) for all the $n - 1$ buses can be written in the following matrix form

$$\begin{bmatrix} \Delta P \\ \Delta Q \end{bmatrix} = \begin{bmatrix} H & N \\ J & L \end{bmatrix} \begin{bmatrix} \Delta \delta \\ \dfrac{\Delta |V|}{|V|} \end{bmatrix} \tag{10.68}$$

where $\quad H_{ip} = \dfrac{\partial P_i}{\partial \delta_p}, \quad N_{ip} = \dfrac{\partial P_i}{\partial |V_p|} |V_p|,$

$$J_{ip} = \frac{\partial Q_i}{\partial \delta_p}, \quad L_{ip} = \frac{\partial Q_i}{\partial |V_p|} |V_p|$$

The partial derivatives H, N, J, L are real functions of admittance matrix and the bus voltages. It has been found that computer computations are faster if rectangular complex arithmetic is used. Therefore, the partial derivatives are computed by rectangular complex arithmetic.

From Eq. (10.22b)

$$P_i - jQ_i = V_i^* \sum_{p=1}^{n} Y_{ip} V_p \tag{10.69}$$

Since $\quad V_i = |V_i| \varepsilon^{j\delta i}, V_i^* = |V_i| \varepsilon^{-j\delta 1}$

also $\quad V_p = |V_p| \varepsilon^{j\delta p}$ and $Y_{ip} = |Y_{ip}| \varepsilon^{j\gamma ip}$

Substituting these values, Eq. (10.69) takes the form

$$P_i - j Q_i = |V_i| \varepsilon^{-j\delta i} \sum_{p=1}^{n} (|Y_{ip}| \varepsilon^{j\gamma ip}) (|V_p| \varepsilon^{j\delta p}) \tag{10.70}$$

Differentiating partially with respect to δ_p, $p \varpi i$, we have

$$\frac{\partial P_i}{\partial \delta_p} - j \frac{\partial Q_i}{\partial \delta_p} = j (|V_i| \varepsilon^{-j\delta i}) (|Y_{ip}| \varepsilon^{j\gamma ip}) (|V_p| \varepsilon^{j\delta p}) \tag{10.71}$$

Let $\quad |V_i| \varepsilon^{-j\delta i} = e_i - jf_i$

$$|Y_{ip}| \varepsilon^{j\gamma ip} = G_{ip} + jB_{ip}$$

$$|V_p| \varepsilon^{j\delta p} = e_p + jf_p$$

Then Eq. (10.71) takes the form

$$\frac{\partial P_i}{\partial \delta_p} - j \frac{\partial Q_i}{\partial \delta_p} = j (e_i - jf_i) (G_{ip} + jB_{ip}) (e_p + jf_p)$$

$$= j\,(e_i - jf_i)\,(a_p + jb_p) \tag{10.72}$$

where $a_p + jb_p = (G_{ip} + jB_{ip})\,(e_p + jf_p)$

From Eq. (10.72) the values of H_{ip} and J_{ip}, when $i \varpi p$, are

$$H_{ip} = \frac{\partial P_i}{\partial \delta_p} = a_p f_i - b_p e_i \tag{10.73}$$

$$J_{ip} = \frac{\partial Q_i}{\partial \delta_p} = -(a_p e_i + b_p f_i) \tag{10.74}$$

For the case of $i = p$ we have

$$\frac{\partial P_i}{\partial \delta_i} - j\frac{\partial Q_i}{\partial \delta_i} = -j|V_i|\varepsilon^{-j\delta i}\left[\sum_{p=1}^{n}(|Y_{ip}|\varepsilon^{j\gamma_{ip}})\times(|V_p|e^{j\delta_p})\right] + j|V_i|\varepsilon^{-j\delta_i}[(|Y_{ii}|\varepsilon^{j\gamma_{ii}}(|V_i|\varepsilon^{j\delta_i})$$

$$= -j\,(P_i - jQ_i) + j\,|V_i|^2\,(G_{ii} + j\,B_{ii}) \tag{10.75}$$

From Eq. (10.75)

$$H_{ii} = \frac{\partial P_i}{\partial \delta_i} = -Q_i - |V_i|^2 B_{ii} \tag{10.76}$$

$$J_{ii} = \frac{\partial Q_i}{\partial \delta_i} = P_i - |V_i|^2 G_{ii} \tag{10.77}$$

Differentiating Eq. (10.70) with respect to $|V_p|$, $p \varpi i$, we have

$$\frac{\partial P_i}{\partial |V_p|} - j\frac{\partial Q_i}{\partial |V_p|} = |V_i|\varepsilon^{-j\delta i}\,(|Y_{ip}| \times \varepsilon^{j\gamma ip}\,\varepsilon^{j\delta p}) \tag{10.78}$$

Multiplying both sides by $|V_p|$

$$\frac{\partial P_i}{\partial |V_p|}\,|V_p| - j\,\frac{\partial Q_i}{\partial |V_p|}\,|V_p| = |V_i|\,\varepsilon^{-j\delta i}\,(|Y_{ip}|\,\varepsilon^{j\gamma ip})\,(|V_p|\,\varepsilon^{j\partial p})$$

$$= (e_i - jf_i)\,(a_p + jb_p) \tag{10.79}$$

From Eq. (10.79) the values of N_{ip} and L_{ip}, for $p \varpi i$, are

$$N_{ip} = \frac{\partial P_i}{\partial |V_p|}\,|V_p| = a_p e_i + b_p f_i = -J_{ip} \tag{10.80}$$

$$L_{ip} = \frac{\partial Q_i}{\partial |V_p|}\,|V_p| = a_p f_i - b_p e_i = H_{ip} \tag{10.81}$$

For the case of $i = p$, we have

$$\frac{\partial P_i}{\partial |V_i|} - j\frac{\partial Q_i}{\partial |V_i|} = \varepsilon^{-j\delta i}\sum_{p=1}^{n}(|Y_{ip}|\,\varepsilon^{j\gamma ip}\,|V_p|\,\varepsilon^{j\delta p}) + (|V_i|\,\varepsilon^{-j\delta i})\,(|Y_{ii}|\,\varepsilon^{j\gamma ii}\,\varepsilon^{j\delta i}) \tag{10.82}$$

Multiplying both sides by $|V_i|$

$$\frac{\partial P_i}{\partial |V_i|}\,|V_i| - j\,\frac{\partial Q_i}{\partial |V_i|}\,|V_i| = |V_i|\varepsilon^{-j\delta i}\sum_{p=1}^{n}(|Y_{ip}|\,\varepsilon^{j\gamma ip})\,(|V_p|\,\varepsilon^{j\delta p}) + |V_i|^2\,(|Y_{ii}|\,\varepsilon^{j\gamma ii})$$

$$= P_i - j\,Q_i + |V_i|^2\,(G_{ii} + j\,B_{ii}) \tag{10.83}$$

From Eq. (10.83), we have

$$N_{ii} = \frac{\partial P_i}{\partial |V_i|}\,|V_i| = P_i + G_{ii}\,|V_i|^2 \tag{10.84}$$

$$L_{ii} = \frac{\partial Q_i}{\partial |V_i|} |V_i| = Q_i - B_{ii} |V_i|^2 \tag{10.85}$$

The equations for calculating the elements of Jacobian are summarized below:

when $p \varpi i$

$$H_{ip} = L_{ip} = a_p f_i - b_p e_i$$
$$N_{ip} = -J_{ip} = a_p e_i + b_p f_i \tag{10.86}$$

where
$$Y_{ip} = G_{ip} + jB_{ip}$$
$$V_i = e_i + jf_i$$
$$(a_p + jb_p) = (G_{ip} + jB_{ip})(e_p + jf_p)$$

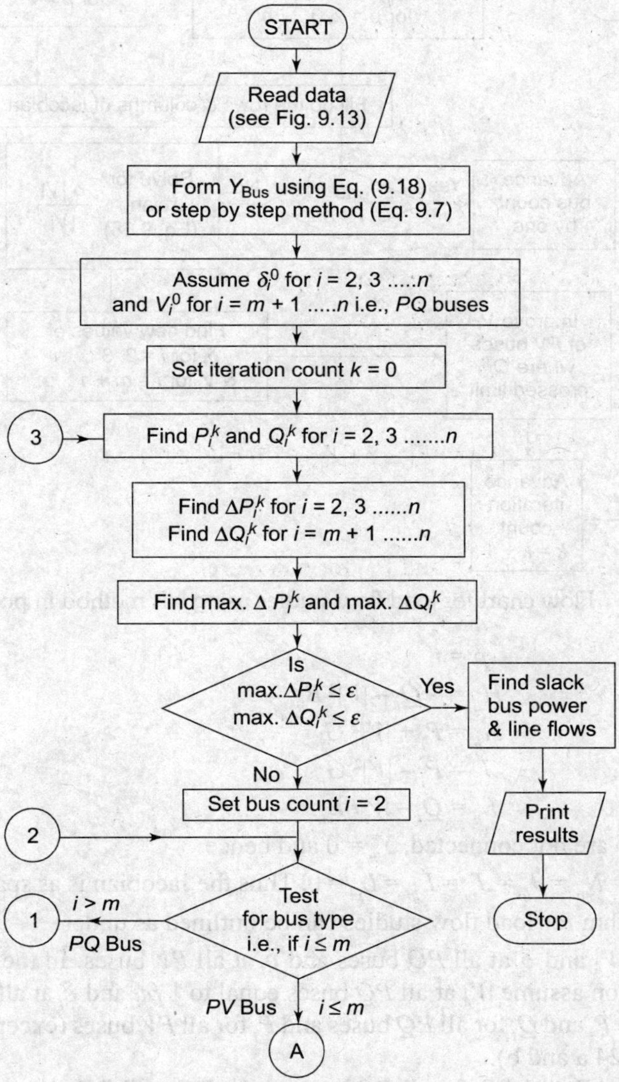

Fig. 10.16 Continued on page 278

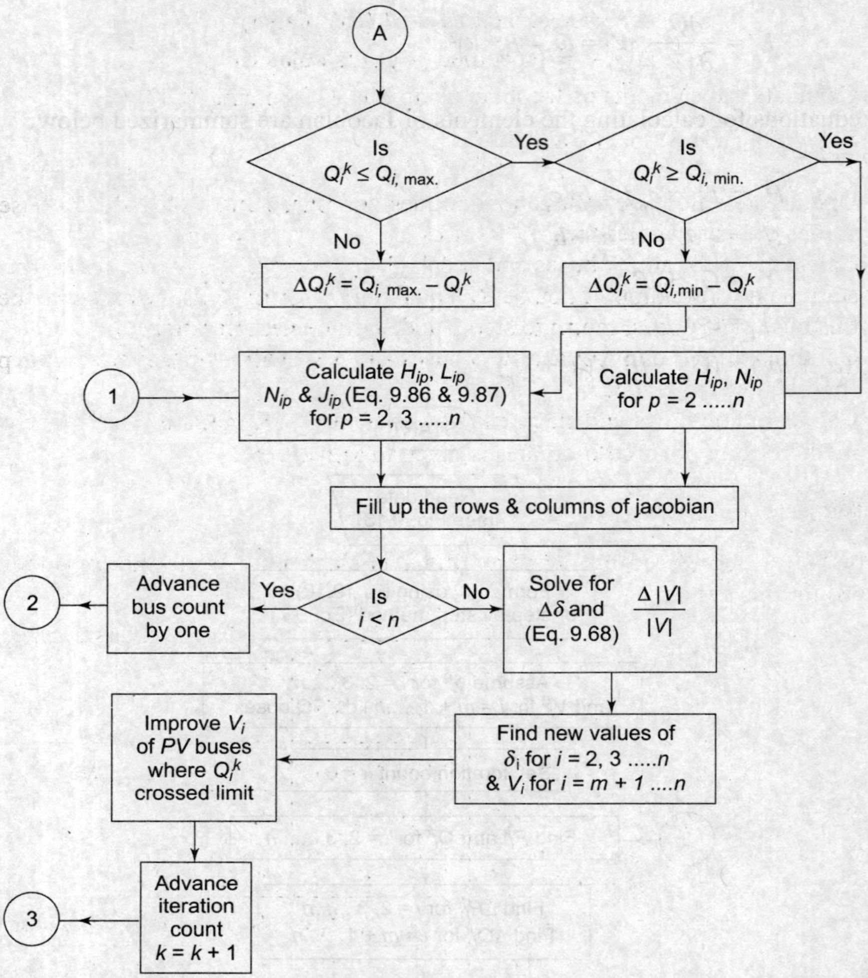

Fig. 10.16 Flow chart for load flow studies using NR method in polar co-ordinates

when

$$
\begin{aligned}
p &= i \\
H_{ii} &= -Q_i - |V_i|^2 B_{ii} \\
N_{ii} &= P_i + |V_i|^2 G_{ii} \\
J_{ii} &= P_i - |V_i|^2 G_{ii} \\
L_{ii} &= Q_i - |V_i|^2 B_{ii}
\end{aligned}
\tag{10.87}
$$

If buses i and p are not connected, $Y_{ip} = 0$ and hence

$H_{ip} = H_{pi} = N_{ip} = N_{pi} = J_{ip} = J_{pi} = L_{ip} = L_{pi} = 0$. Thus the Jacobian is as sparse as the Y_{bus} matrix.

The NR algorithm for load flow studies can be outlined as under:

1. Assume $|V_i|$ and δ_i at all PQ buses and δ_i at all PV buses. In the absence of any other information assume $|V_i|$ at all PQ buses equal to 1 pu and δ_i at all buses equal to zero.
2. Calculate P_i and Q_i for all PQ buses and P_i for all PV buses (except the slack bus) using Eqs. (10.24 a and b).
3. Calculate ΔP_i and ΔQ_i for all PQ buses and ΔP_i for all PV buses (except the slack-bus) by using the equations

$$\Delta P_i = P_i \text{ (specified)} - P_i \text{ (calculated)}$$
$$\Delta Q_i = Q_i \text{ (specified)} - Q_i \text{ (calculated)}$$

4. Calculate the elements of Jacobian using Eqs. (10.86 and 10.87)
5. Solve Eq. (10.68) for $\Delta\,\delta$ and $\dfrac{\Delta |V|}{|V|}$.
6. Update the values of voltage magnitudes and phase angles for all *PQ* buses and the values of phase angles for all *PV* buses.
7. Calculate Q_i for all *PV* buses and check $Q_{i,\,min} < Q_i < Q_{i,\,max}$. If yes, return to step 2 and start the next iteration. If not, set Q_i equal to $Q_{i,min}$ or $Q_{i,max}$ as the case may be and treat this bus as a *PQ* bus, return to step 2 and start the next iteration.
8. Continue till ΔP_i and ΔQ_i at all *PQ* buses and ΔP_i at all *PV* buses are within prescribed (or assumed) tolerance.
9. Calculate line flows and slack bus power as in *G – S*. method. The flow chart for *NR* method using polar co-ordinates is given in Fig. 10.16.

Example 10.16

Figure 10.17 (a) shows a 6-but system. Bus 1 is slack bus. Write the expanded form of Eq. (10.68) for this system.

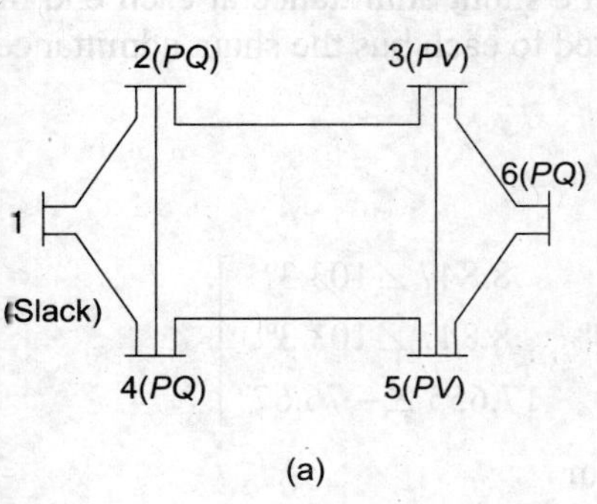

(a)

$$\begin{array}{c|c|}
 & \\
2 & \Delta P_2 \\
 & \Delta Q_2 \\
3 & \Delta P_3 \\
 & \Delta P_4 \\
4 & \Delta Q_4 \\
5 & \Delta P_5 \\
 & \Delta P_6 \\
6 & \Delta Q_6
\end{array}
=
\begin{array}{cc|c|cc|c|cc}
\multicolumn{2}{c}{2} & 3 & \multicolumn{2}{c}{4} & 5 & \multicolumn{2}{c}{6} \\
H_{22} & N_{22} & H_{23} & H_{24} & N_{24} & & & \\
J_{22} & L_{22} & J_{23} & J_{24} & L_{24} & & & \\
H_{32} & N_{32} & H_{33} & & & H_{35} & H_{36} & N_{36} \\
H_{42} & N_{42} & & H_{44} & N_{44} & H_{45} & & \\
J_{42} & L_{42} & & J_{44} & L_{44} & J_{45} & & \\
 & & H_{53} & H_{54} & N_{54} & H_{55} & H_{56} & N_{56} \\
 & & H_{63} & & & H_{65} & H_{66} & N_{66} \\
 & & J_{63} & & & J_{65} & J_{66} & L_{66}
\end{array}
\begin{array}{|c|}
\\
\Delta\delta_2 \\
\dfrac{\Delta V_2}{|V_2|} \\
\Delta\delta_3 \\
\Delta\delta_4 \\
\dfrac{\Delta V_4}{|V_4|} \\
\Delta\delta_5 \\
\Delta\delta_6 \\
\dfrac{\Delta V_6}{|V_6|}
\end{array}$$

(b)

Fig. 10.17

Solution

The complete set of equations is shown in Fig. 10.17 (b). Since bus 1 is the slack bus, it does not find a place in these equations. However, its effect enters the equations through the H_{ii}, N_{ii}, J_{ii} and L_{ii} values for the buses connected to bus 1. Buses 2, 4 and 6 are *PQ* buses and there are two equations for each of these buses. Buses 3 and 5 are *PV* buses and there is only one equation for each of these buses. Since the buses 2 and 5, 3 and 4, 2 and 6, 4 and 6 are not connected, the corresponding elements in the Jacobian are zero.

Example 10.17

Figure 10.18 shows a 3-bus system. The series impedance and shunt admittance of each line are $0.026 + j0.11$ *pu* and $j\,0.04$ *pu* respectively. The bus specification and power input, etc., at the buses is as under:

Bus	P_G	Q_G	P_L	Q_L	*Bus voltage*
1	unspecified	unspecified	1.0	0.5	$1.03 + j\,0$ (Slack bus)
2	1.5	unspecified	0	0	$V = 1.03$ (*PV* bus)
3	0	0	1.2	0.5	unspecified (*PQ* bus)

For bus 2 the minimum and maximum reactive power limits are 0 and 0.8 *pu*.

(a) Form $[Y_{bus}]$

(b) Find $P_2^o, Q_2^o, P_3^o, Q_3^o$

(c) Find $[J^o]$

(d) Form the general equation for calculating the changes in variables by *NR* method. Write the equation for first iteration.

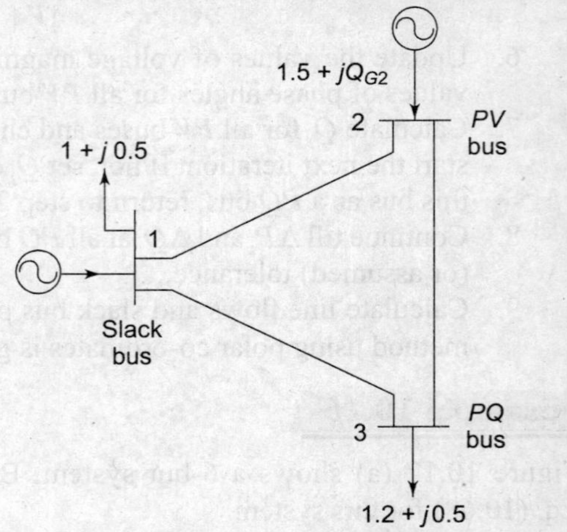

Fig. 10.18 Three-bus system for example 10.17

Solution

(a) The series admittance of each line is $\dfrac{1}{0.026 + j0.11} = 2.035 - j\,8.61$

$$= 8.847 \angle -76.7^o$$

$$Y_{12} = Y_{21} = Y_{13} = Y_{31} = Y_{23} = Y_{32}$$
$$= -(2.035 - j\,8.61) = -2.035 + j\,8.61 = 8.847 \angle 103.3^o$$

Each line can be represented by a nominal π circuit. The shunt admittance at each end of line is $0.5\,(j\,0.04) = j\,0.02\ pu$. Since two lines are connected to each bus the shunt admittance between each bus and ground is $j\,0.04\ pu$.

$$Y_{11} = Y_{22} = Y_{33} = 2(2.035 - j8.61) + j\,0.04\ pu$$
$$= 4.07 - j\,17.18 = 17.655 \angle -76.67^o$$

$$[Y_{bus}] = \begin{bmatrix} 17.655 \angle -76.67^o & 8.847 \angle 103.3^o & 8.847 \angle 103.3^o \\ 8.847 \angle 103.3^o & 17.655 \angle -76.67^o & 8.847 \angle 103.3^o \\ 8.847 \angle 103.3^o & 8.847 \angle 103.3^o & 17.655 \angle -76.67^o \end{bmatrix}$$

(b) For $n = 3$ and $i = 2$ and 3 Eq. (10.70) takes the form

$$P_2 - jQ_2 = |V_2|\varepsilon^{-j\delta_2}\,[|Y_{21}|\varepsilon^{j\gamma_{21}}\,|V_1|\varepsilon^{j\delta_1} + |Y_{22}|\varepsilon^{j\gamma_{22}}\,|V_2|\varepsilon^{j\delta_2} + |Y_{23}|\varepsilon^{j\gamma_{23}}\,|V_3|\varepsilon^{j\delta_3}|]$$
$$P_3 - jQ_3 = |V_3|\varepsilon^{-j\delta_3}\,[|Y_{31}|\varepsilon^{j\gamma_{31}}\,|V_1|\varepsilon^{j\delta_1} + |Y_{32}|\varepsilon^{j\gamma_{32}}\,|V_2|\varepsilon^{j\delta_2} + |Y_{33}|\varepsilon^{j\gamma_{33}}\,|V_3|\varepsilon^{j\delta}$$
$$V_1 = 1.03 \angle 0,\ \text{i.e.,}\ |V_1| = 1.03,\ \delta_1 = 0,\ |V_2| = 1.03$$

Assume $|V_3^o| = 1.0,\ \delta_2{}^o = 0$ and $\delta_3{}^o = 0$

Substituting the values of voltages, admittances and angles in the above equations and separating the real and imaginary terms we get the following values

$$P_2{}^o = 0.063\ pu,\ Q_2{}^o = 0.224\ pu,\ P_3{}^o = -0.122\ pu,\ Q_3{}^o = -0.557\ pu$$

(c) The elements of $[J^o]$ can be found by substituting the values in Eqs. (10.86) and (10.87), e.g.,

$$H_{23}^o = a_3^o f_2^o - b_3^o\, e_2{}^o$$
$$e_2^o = 1.03, f_2{}^o = 0$$
$$a_3^o + jb_3^o = (G_{23} + j\,B_{23})\,(e_3{}^o + jf_3^o)$$

$$= (-2.035 + j\,8.61)(1 + j0) = -2.035 + j\,8.61$$

Thus, $a_3^{\circ} = -2.035,\ b_3^{\circ} = 8.61$

$$H_{23}^{\circ} = -8.61 \angle 1.03 = -8.868$$

$$H_{22}^{\circ} = -Q_2^{\circ} - |V_2|^2 B_{22} = -0.224 - (1.03)^2(-17.18) = 18.002$$

Similarly, the values of other elements can be found. The complete result is

$$[J^{\circ}] = \begin{bmatrix} H_{22}^{\circ} & H_{23}^{\circ} & N_{23}^{\circ} \\ H_{32}^{\circ} & H_{33}^{\circ} & N_{33}^{\circ} \\ J_{32}^{\circ} & J_{33}^{\circ} & L_{33}^{\circ} \end{bmatrix} = \begin{bmatrix} 18.002 & -8.868 & -2.035 \\ -8.868 & 17.736 & 3.948 \\ 2.096 & -4.192 & 16.623 \end{bmatrix}$$

(d) The specified active and reactive powers are

$$P_2 = 1.5,\ P_3 = -1.2,\ Q_3 = -0.5.$$

The residuals are

$$\Delta P_2^{\circ} = P_{2(\text{specified})} - P_2^{\circ} = 1.5 - 0.063 = 1.437\ pu$$

$$\Delta P_3^{\circ} = P_{3(\text{specified})} - P_3^{\circ} = -1.2 - (0.122) = -1.078\ pu$$

$$\Delta Q_3^{\circ} = Q_{3(\text{specified})} - Q_3^{\circ} = -0.5\ (-0.557) = 0.057\ pu$$

The general equation for finding the changes in variables is

$$\begin{bmatrix} \Delta P_2^k \\ \Delta P_3^k \\ \Delta Q_3^k \end{bmatrix} = \begin{bmatrix} H_{22}^k & H_{23}^k & N_{23}^k \\ H_{32}^k & H_{33}^k & N_{33}^k \\ J_{32}^k & J_{33}^k & L_{33}^k \end{bmatrix} = \begin{bmatrix} \Delta\delta_2^{k+1} \\ \Delta\delta_3^{k+1} \\ \dfrac{\Delta |V_3|^{k+1}}{|V_3|} \end{bmatrix}$$

The equation for the first iteration (i.e., $k = 0$) is

$$\begin{bmatrix} 1.437 \\ -1.078 \\ 0.057 \end{bmatrix} = \begin{bmatrix} 18.002 & -8.868 & -2.035 \\ -8.868 & 17.736 & 3.948 \\ 2.096 & -4.192 & 16.623 \end{bmatrix} \begin{bmatrix} \Delta\delta_2^1 \\ \Delta\delta_3^1 \\ \dfrac{\Delta |V_3|^1}{|V_3|} \end{bmatrix}$$

The changes in variables can be calculated from the equation

$$\begin{bmatrix} \Delta\delta_2^1 \\ \Delta\delta_3^1 \\ \dfrac{\Delta |V_3|^1}{|V_3|} \end{bmatrix} = \begin{bmatrix} 18.002 & -8.868 & -2.035 \\ -8.868 & 17.736 & 3.948 \\ 2.096 & -4.192 & 16.623 \end{bmatrix}^{-1} \begin{bmatrix} 1.437 \\ -1.078 \\ 0.057 \end{bmatrix}$$

10.12 FAST DECOUPLED LOAD FLOW

Load flow studies can be made faster and more efficient when the advantage of the physical properties of the system is taken into consideration. One such property is the network sparsity. Since each bus is connected to only a small number (usually two or three) of other buses, Y_{bus} of a large network is very sparse (i.e., it has a large number of zero elements). The sparsity feature of Y_{bus} minimizes the computer memory requirement (because only non-zero terms need to be stored) and results in faster computations. Another property is the loose physical interaction between MW and Mvar flows in a power system. Therefore, the MW – δ and Mvar – V calculations can be decoupled. This decoupling results in a very simple, fast and reliable algorithm. The accuracy is comparable to that of *NR* method.

The first step in applying the decoupling principle is to neglect the sub-matrices *N* and *J* in Eq. (10.68) resulting in two separate equations

$$[\Delta P] = [H]\,[\Delta\,\delta] \tag{10.88a}$$

$$[\Delta\,Q] = [L]\left[\frac{\Delta|V|}{|V|}\right] \tag{10.88b}$$

The next step in deriving the algorithm is to make suitable assumptions in deriving the expressions for H and L.

Since $\quad |Y_{ip}|\,\varepsilon^{j\gamma ip} = G_{ip} + J\,B_{ip}$, Eq. (10.71) can be re-written as

$$\frac{\partial P_i}{\partial \delta_p} - j\frac{\partial Q_i}{\partial \delta_p} = j\,|V_i|\,|V_p|,\ \varepsilon^{j(\delta p - \delta i)}\,.(G_{ip} + jB_{ip}) \tag{10.89}$$

Since the angle $(\delta_p - \delta_i)$ is small, $\varepsilon^{j(\delta p - \delta i)} \cong 1$. Then

$$\frac{\partial P_i}{\partial \delta_p} - j\frac{\partial Q_i}{\partial \delta_p} = j|V_i|\,|V_p|\,(G_{ip} + jB_{ip}) \tag{10.90}$$

Separating the real and imaginary terms, when $i \varpi p$

$$H_{ip} = \frac{\partial P_i}{\partial \delta_p} = -\,|V_i|\,|V_p|\,|B_{ip}| \tag{10.91}$$

As seen in Eq. (10.86) $L_{ip} = H_{ip}$. Therefore,

$$L_{ip} = H_{ip} = -\,|V_i|\,|V_p|\,|B_{ip}| \tag{10.92}$$

In the expressions for L_{ii} and H_{ii} (Eq. 10.87) Q_i is generally very small as compared to $|V_i|^2$ B_{ii}. Therefore,

$$H_{ii} = L_{ii} = -\,|V_i|^2\,B_{ii} \tag{10.93}$$

In view of Eqs. (10.92) and (10.93), the equation for ΔP and ΔQ (Eq. 10.88a and b) can be written as

$$[\Delta P] = [|V|\,|B'|\,|V|]\,[\Delta\delta] \tag{10.94a}$$

$$[\Delta Q] = [|V|\,|B''|\,|V|]\,[\Delta\,V/V] \tag{10.94b}$$

The elements of matrices B' and B'' are the elements of $[-\,B_{ip}]$ matrix (See Example 10.17). The final algorithm is obtained by making the following approximations:

(i) Omit from $[B']$ the representation of those network elements which predominantly affect Mvar flow only and do not affect MW flow significantly, i.e., shunt rectances and off-nominal in phase transformer taps.

(ii) Omit from $[B'']$ the angle shifting effects of phase shifters.

(iii) Take the left hand V terms in Eq. (10.94a and b) to the left hand side of equations and set the other V term in Eq. (10.94a) to 1 *pu*.

(iv) Neglect the series resistances in calculating the elements of $[B]$. With these modifications Eqs. (10.94a and b) becomes

$$\left[\frac{\Delta P}{|V|}\right] = [B']\,[\Delta\delta] \tag{10.95a}$$

$$\left[\frac{\Delta Q}{|V|}\right] = [B'']\,[\Delta V] \tag{10.95b}$$

In the above equations $[B']$ and $[B'']$ are real and sparse. Since $[B']$ and $[B'']$ contain the network admittance terms only, they are constant and need to be evaluated only once in the beginning. Eqs. (10.95a and b) can be solved through an iterative procedure independently and alternately. Each cycle of iteration consists of one solution for $[\Delta\delta]$ to update δ and one solution for $[|\Delta V|]$ to update $[|V|]$. The iterations are continued till ΔP and ΔQ at all PQ buses and ΔP

at all PV buses are **within prescribed** (or assumed) tolerances. Fig. 10.19 shows a flowchart for load flow studies using the above technique.

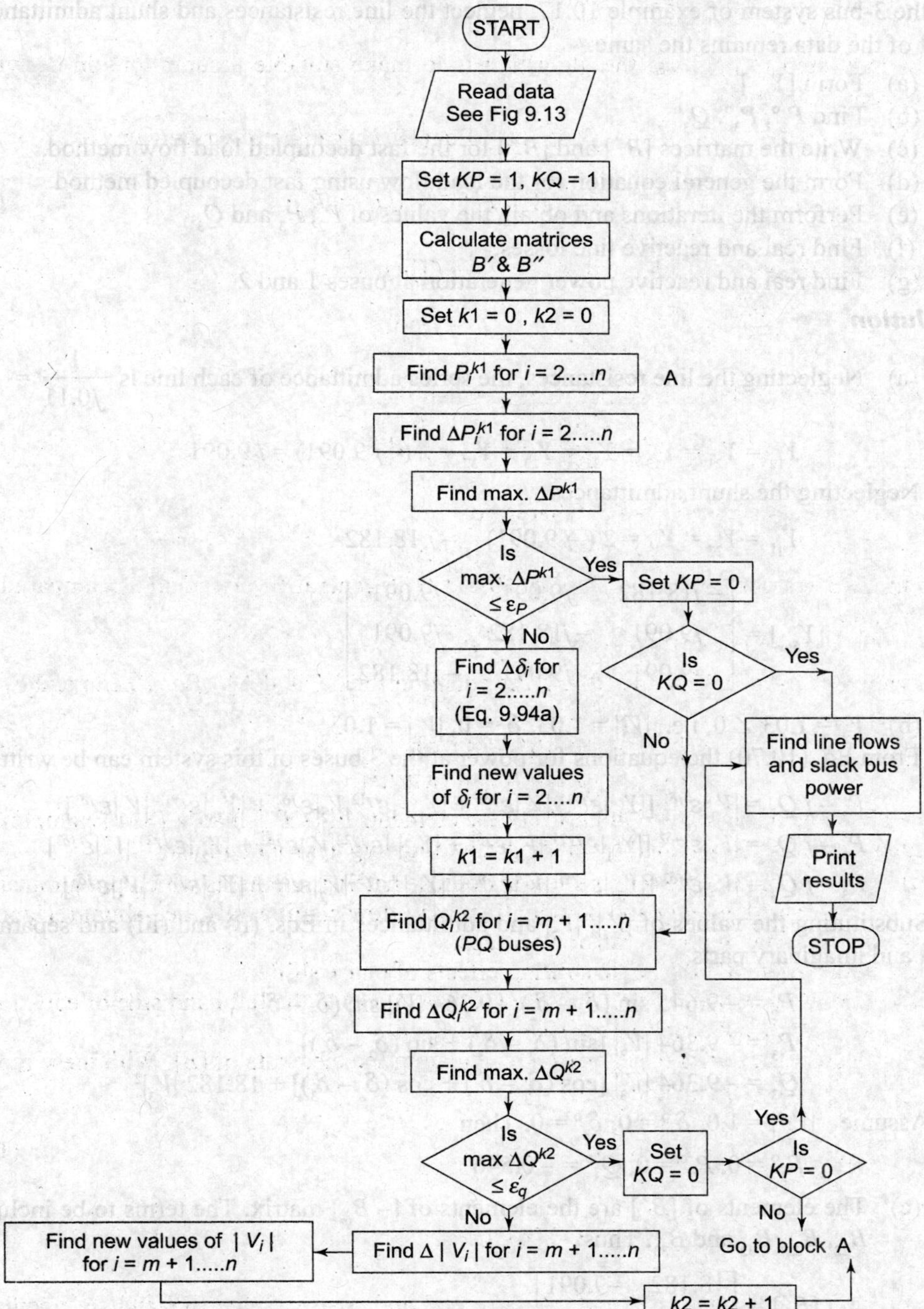

Fig. 10.19 Flow chart for load flow studies using fast decoupled method. KP = 1 means that improvement of δ is needed. KP = 0 means that improvement of δ is not needed. KQ = 1 means that improvement of V is needed. KQ = 0 means that improvement of V is not needed. k1 is iteration count for improvement of δ. k2 is iteration count for improvement of V.

Example 10.18

In the 3-bus system of example 10.17, neglect the line resistances and shunt admittances. The rest of the data remains the same.

(a) Form $[Y_{bus}]$
(b) Find P_2^o, P_3^o, Q_3^o
(c) Write the matrices $[B']$ and $[B'']$ for the fast decoupled load flow method.
(d) Form the general equation for the load flow using fast decoupled method.
(e) Perform the iterations and obtain the values of P_2, P_3 and Q_3.
(f) Find real and reactive line losses.
(g) Find real and reactive power generation at buses 1 and 2.

Solution

(a) Neglecting the line resistances, the series admittance of each line is $\dfrac{1}{j0.11} = -j\,9.091$

$$Y_{12} = Y_{21} = Y_{13} = Y_{31} = Y_{23} = Y_{32} = -(-j\,9.091) = j\,9.091$$

Neglecting the shunt admittances

$$Y_{11} = Y_{22} = Y_{33} = 2\,(-j\,9.091) = -j\,18.182$$

$$[Y_{bus}] = \begin{bmatrix} -j18.182 & j9.091 & j9.091 \\ j9.091 & -j18.182 & j9.091 \\ j9.091 & j9.091 & -j18.182 \end{bmatrix}$$

(b) $V_1 = 1.03 \angle 0$, i.e., $|V_1| = 1.03$, $\delta_1 = 0$, $|V_2| = 1.03$

From Eq. (10.70) the equations for power at the 3 buses of this system can be written as

$$P_1 - j\,Q_1 = |V_1|\varepsilon^{-j\delta 1}\,[|Y_{11}|\varepsilon^{j\gamma 11}\,|V_1|\varepsilon^{j\delta 1} + |Y_{12}|\,|\varepsilon^{j\gamma 12}|V_2|\varepsilon^{j\delta 2} + |Y_{13}|\varepsilon^{j\gamma 13}\,|V_3|\varepsilon^{j\delta 3}] \quad \text{(i)}$$

$$P_2 - j\,Q_2 = |V_2|\varepsilon^{-j\delta 2}\,[|Y_{21}|\varepsilon^{j\gamma 21}\,|V_1|\varepsilon^{j\delta 1} + |Y_{22}|\,|\varepsilon^{j\gamma 22}|V_2|\varepsilon^{j\delta 2} + |Y_{23}|\varepsilon^{j\gamma 23}\,|V_3|\varepsilon^{j\delta 3}] \quad \text{(ii)}$$

$$P_3 - j\,Q_3 = |V_3|\varepsilon^{-j\delta 3}\,[|Y_{31}|\varepsilon^{j\gamma 31}\,|V_1|\varepsilon^{j\delta 1} + |Y_{32}|\,|\varepsilon^{j\gamma 32}|V_2|\varepsilon^{j\delta 2} + |Y_{33}|\varepsilon^{j\gamma 33}\,|V_3|\varepsilon^{j\delta 3}] \quad \text{(iii)}$$

Substituting the values of $|V_1|$, $|V_2|$ and admittances in Eqs. (ii) and (iii) and separating the real and imaginary parts

$$P_2 = -\,9.645 \sin(\delta_1 - \delta_2) - 9.364\,|V_3| \sin(\delta_3 - \delta_2) \quad \text{(iv)}$$

$$P_3 = -\,9.364\,|V_3|\,[\sin(\delta_1 - \delta_3) + \sin(\delta_2 - \delta_3)] \quad \text{(v)}$$

$$Q_3 = -\,9.364\,|V_3|\,[\cos(\delta_1 - \delta_3) + \cos(\delta_2 - \delta_3)] + 18.182\,|V_3|^2 \quad \text{(vi)}$$

Assume $|V_3^o| = 1.0$, $\delta_2^o = 0$, $\delta_3^o = 0$. Then

$$P_2^o = 0,\ P_3^o = 0,\ Q_3^o = -\,0.546$$

(c) The elements of $[B']$ are the elements of $[-B_{ip}]$ matrix. The terms to be included are B_{22}, B_{23}, B_{32} and B_{33}. Thus

$$[B'] = \begin{bmatrix} 18.182 & -9.091 \\ -9.091 & 18.182 \end{bmatrix}$$

the matrix $[B'']$ contains only one term, i.e., $-B_{33}$

$$[B''] = [18.182]$$

(d) The general equations for load flow using fast decoupled method are

$$\begin{bmatrix} \dfrac{\Delta P_2}{|V_2|} \\ \dfrac{\Delta P_3}{|V_3|} \end{bmatrix} = \begin{bmatrix} 18.182 & -9.091 \\ -9.091 & 18.182 \end{bmatrix} \begin{bmatrix} \Delta\delta_2 \\ \Delta\delta_3 \end{bmatrix}$$

or $$\begin{bmatrix} \Delta\delta_2 \\ \Delta\delta_3 \end{bmatrix} = \begin{bmatrix} 18.182 & -9.091 \\ -9.091 & 18.182 \end{bmatrix}^{-1} \begin{bmatrix} \dfrac{\Delta P_2}{1.03} \\ \dfrac{\Delta P_3}{|V_3|} \end{bmatrix}$$

$$= \begin{bmatrix} 0.0733 & 0.03665 \\ 0.03665 & 0.0733 \end{bmatrix} \begin{bmatrix} \dfrac{\Delta P_2}{1.03} \\ \dfrac{\Delta P_3}{|V_3|} \end{bmatrix} \quad \text{(vii)}$$

$$\begin{bmatrix} \dfrac{\Delta Q_3}{|V_3|} \end{bmatrix} = [18.182]\,[\Delta|V_3|]$$

or $$\Delta|V_3| = 0.055 \begin{bmatrix} \dfrac{\Delta Q_3}{|V_3|} \end{bmatrix} \quad \text{(viii)}$$

(e) First iteration

$\Delta P_2^{o} = P_{2(\text{specified})} - P_2^{o} = 1.5 - 0 = 1.5\ pu$

$\Delta P_3^{o} = P_{3(\text{specified})} - P_3^{o} = -1.2 - 0 = -1.2\ pu$

$\Delta Q_3^{o} = Q_{3(\text{specified})} - Q_3^{o} = -(0.5) - (-0.56) = 0.046\ pu$

$|V_3^{o}| = 1.0$

Substituting the values in Eqs. (vii and viii)

$$\begin{bmatrix} \Delta\delta_2^1 \\ \Delta\delta_3^1 \end{bmatrix} = \begin{bmatrix} 0.0733 & 0.03665 \\ 0.03665 & 0.0733 \end{bmatrix} \begin{bmatrix} \dfrac{1.5}{1.03} \\ -1.2 \end{bmatrix}$$

whence $\Delta\,\delta_2^1 = 0.06276$ radians

$\Delta\,\delta_3^1 = -0.03459$ radians

$\Delta\,|V_3^1| = 0.055\ (0.046) = 0.00253\ pu$

Second Iteration

The updated values of δ_2, δ_3 and $|V_3|$ are

$\delta_2^1 = \delta_2^0 + \Delta\,\delta_2^1 = 0.06276$ radians

$\delta_3^1 = \delta_3^0 + \Delta\,\delta_3^1 = -0.03459$ radians

$|V_3^1| = V_3^0 + \Delta\,|V_3^1| = 1.00253\ pu$

Substituting the above values in Eqs. (iv, v, vi)

$P_2^1 = 1.51737,\ P_3^1 = -1.234,\ Q_3^1 = -0.404$

$\Delta P_2^1 = 1.5 - 1.51737 = -0.01737$

$$\Delta P_3^1 = -1.2 - (-1.234) = 0.034$$
$$\Delta Q_3^1 = -0.5 - (-0.404) = -0.096$$

Substituting the values of Eqs. (vii and viii)

$$\begin{bmatrix} \Delta\delta_2^2 \\ \Delta\delta_3^2 \end{bmatrix} = \begin{bmatrix} 0.0733 & 0.03665 \\ 0.03665 & 0.0733 \end{bmatrix} \begin{bmatrix} \dfrac{-0.01737}{1.03} \\ \dfrac{0.034}{1.00253} \end{bmatrix}$$

whence $\Delta\,\delta_2^2 = 6.82 \times 10^{-6}$ radians

$$\Delta\,\delta_3^2 = 1.868 \times 10^{-3} \text{ radians}$$
$$\Delta\,|V_3^2| = 0.055 \times (-0.096) = -0.00528\ pu$$

The updated values of δ_2, δ_3 and $|V_3|$ are

$$\delta_2^2 = \delta_2^1 + \Delta\delta_2^2 = 0.0627668 \text{ radians}$$
$$\delta_3^2 = \delta_3^1 + \Delta\delta_3^2 = -0.032722 \text{ radians}$$
$$|V_3^2| = 0.99725\ pu$$

Substituting the values in Eqs. (iv, v, vi)

$$P_2^2 = 1.499,\ P_3^2 = -1.199,\ Q_3^2 = -0.497$$

Since the values of P_2^2, P_3^2 and Q_3^2 are very close to the specified values, the iterative process need not be continued further.

Hence, $\delta_2 = 0.0627668$ radians $= 3.598°$

$\delta_3 = -0.032722$ radians $= -1.876°$

$|V_3| = 0.99725\ pu$

(f) Substituting the values in Eqs. (i and ii)

$$P_1 = -0.3,\ Q_1 = 0.33,\ Q_2 = 0.37$$

Real line losses $= P_1 + P_2 + P_3 = 0$ (This has to be so since line resistances have been neglected)

Reactive line losses $= Q_1 + Q_2 + Q_3 = 0.2\ pu$

(g) Generation at the buses is

$$P_{G1} = P_1 + P_{L1} = 0.7\ pu$$
$$Q_{G1} = Q_1 + Q_{L1} = 0.83\ pu$$
$$P_{G2} = 1.5\ pu$$
$$Q_{G2} = Q_2 + Q_{L2} = 0.37\ pu$$

10.13 LOAD FLOW COMPUTATIONS IN LARGE SYSTEMS

(a) *Sparsity:* In large power systems each bus is connected to only a small number of other buses (may be two or three). Therefore, bus admittance matrix of a large power system is very sparse (i.e., it has a large number of zero elements). The sparsity feature of Y_{bus} matrix extends to Jacobian matrix also. The sparsity of an $n \times n$ matrix is defined as

$$\text{Sparsity} = \frac{\text{Total number of zero elements}}{n^2} \times 100$$

In a large power system sparsity may be as high as 97%. It may be noted that though Y_{bus} is sparse, Z_{bus} is full.

Modern load flow studies take advantage of sparsity to reduce computer memory requirement and save computer time.

(b) *Compact storage scheme:* In a compact storage scheme only non-zero elements are stored and it is ensured that full position identification can take place. Consider a sparse matrix.

$$Y_{bus} = \begin{bmatrix} 20 & -12 & 0 & -8 & 0 & 0 \\ -12 & 12 & 0 & 0 & 0 & 0 \\ 0 & 0 & 17 & -10 & -7 & 0 \\ -8 & 0 & -10 & 18 & 0 & 0 \\ 0 & 0 & -7 & 0 & 16 & -9 \\ 0 & 0 & 0 & 0 & -9 & 9 \end{bmatrix}$$

We define two vectors

$$\text{YDIAG} = [20, 12, 17, 18, 16, 9]$$
$$\text{YOFFD} = [-12, -8, -12, -10, -7, -8, -10, -7, -9, -9]$$

The vector YDIAG contains the diagonal elements while vector YOFFD contains the off diagonal elements of Y_{bus}. To position the elements in YOFFD, we define two more vectors

$$\text{JROW} = [1, 3, 4, 6, 8, 10]$$
$$\text{JCOL} = [2, 4, 1, 4, 5, 1, 3, 3, 6, 5]$$

The vector JROW identifies the start of different rows in the vector YOFFD, e.g., first row starts with off diagonal element – 12 which is first element in YOFFD, second row starts with – 12 which is third element in YOFFD, fourth row starts with – 8 which is sixth element in YOFFD and so on.

The vector JCOL identifies the column position of each element in the vector YOFFD, e.g., the element – 12 is in second column, – 8 is in the fourth column, – 12 is in first column, – 10 is in fourth column and so on.

(c) *Gauss elimination, Triangular factorization:* The load flow solution requires obtaining inverse of a matrix. When a matrix is sparse, the matrix inversion is a very inefficient approach. Instead, it is better to use Gauss elimination which uses triangular factorization and back substitution. In triangularization the off diagonal elements in each row below the main diagonal are made zero and the diagonal element of each row is normalized as soon as the processing of that row is completed. After triangularization it is easy to obtain the solution by back substitution. We illustrate the procedure by the solution of equation obtained in the first iteration of example 10.16. This equation is

$$\begin{bmatrix} 18.002 & -8.868 & -2.035 \\ -8.868 & 17.736 & 3.948 \\ 2.096 & -4.192 & 16.623 \end{bmatrix} \begin{bmatrix} \Delta\delta_2^1 \\ \Delta\delta_3^1 \\ \dfrac{\Delta|V_3|}{|V_3|} \end{bmatrix} = \begin{bmatrix} 1.437 \\ -1.078 \\ 0.057 \end{bmatrix}$$

Step 1: Divide the elements of first row by a_{11} (i.e., 18.002). The result is

$$\begin{bmatrix} 1 & -0.4926 & -.11304 \\ -8.868 & 17.736 & 3.948 \\ 2.096 & -4.192 & 16.623 \end{bmatrix} \begin{bmatrix} \Delta\delta_2^1 \\ \Delta\delta_3^1 \\ \dfrac{\Delta|V_3|^1}{|V_3|} \end{bmatrix} = \begin{bmatrix} .07982 \\ -1.078 \\ 0.057 \end{bmatrix}$$

Step 2: Divide the elements of second row by a_{21} (i.e., – 8.868) and subtract it from first row. This makes a_{21} zero. The result is

$$\begin{bmatrix} 1 & -0.4926 & -0.11304 \\ 0 & 1.5074 & .33216 \\ 2.096 & -4.192 & 16.623 \end{bmatrix} \begin{bmatrix} \Delta\delta_2^1 \\ \Delta\delta_3^1 \\ \dfrac{\Delta|V_3|^1}{|V_3|} \end{bmatrix} = \begin{bmatrix} .07982 \\ -.04174 \\ 0.057 \end{bmatrix}$$

Step 3: Divide the elements of third row by a_{31}, (i.e., 2.096) and subtract it from first row. This makes a_{31} zero. The result is

$$\begin{bmatrix} 1 & -0.4926 & -0.11304 \\ 0 & 1.5074 & 0.33216 \\ 0 & 1.5074 & -8.04386 \end{bmatrix} \begin{bmatrix} \Delta\delta_2^1 \\ \Delta\delta_3^1 \\ \dfrac{\Delta|V_3|^1}{|V_3|} \end{bmatrix} = \begin{bmatrix} 0.07982 \\ -0.04174 \\ 0.05263 \end{bmatrix}$$

Step 4: Divide the elements of second row by the new value of a_{22} (i.e., 1.5074). The result is

$$\begin{bmatrix} 1 & -0.4926 & -0.11304 \\ 0 & 1 & 0.22035 \\ 0 & 1.5074 & -8.04386 \end{bmatrix} \begin{bmatrix} \Delta\delta_2^1 \\ \Delta\delta_3^1 \\ \dfrac{\Delta|V_3|^1}{|V_3|} \end{bmatrix} = \begin{bmatrix} 0.07982 \\ -0.02769 \\ 0.05263 \end{bmatrix}$$

Step 5: Divide the elements of third row by new value of a_{32} (i.e., 1.5074) and subtract it from second row. This makes a_{32} zero. The result is

$$\begin{bmatrix} 1 & -0.4926 & -0.11304 \\ 0 & 1 & 0.22035 \\ 0 & 0 & 5.5566 \end{bmatrix} \begin{bmatrix} \Delta\delta_2^1 \\ \Delta\delta_3^1 \\ \dfrac{\Delta|V_3|^1}{|V_3|} \end{bmatrix} = \begin{bmatrix} 0.07982 \\ -0.02769 \\ -.0626 \end{bmatrix}$$

Step 6: The triangularization is now complete. From the last equation

$$5.5566\ [\Delta|V_3|^1/|V_3|] = -.0626$$

Since $|V_3| = 1.0,\ \Delta|V_3|^1 = -0.011266$

On back substitution of this value in second equation we get, $\Delta\delta_3^1 = -0.0252$

On back substitution of above values in first equation we get, $\Delta\delta_2^1 = 0.06613$

In the above example we have proceeded by rows for elimination. It is also possible to process by columns. However, the processing by rows leads to lesser computer memory requirement and more efficient computations.

The use of Gauss elimination for solution is advantageous only if Jacobian is sparse (as it generally is) and the rows are ordered properly.

(d) *Optimal ordering:* To obtain full advantage of Gauss elimination, the rows of matrix should be numbered properly. Optimal ordering leads to saving in computer memory and time. Some schemes for optimal ordering are:

1. Number the rows starting with that having the least non-zero terms and ending with one having the most non-zero terms.
2. Number the rows at each step of elimination such that the row to be eliminated is the one having the fewest non-zero terms.
3. Number the rows at each step so that the next row to be eliminated introduces fewest new non-zero terms.

The benefit of scheme 3 has not been fully established as yet. Scheme 1 is used where a straight solution is possible and iteration is not needed. Scheme 2 is the most commonly used one.

Example 10.19

The line admittances of a 4-bus system are as under:

Bus code	Admittance
1–2	$2 - j8$
1–3	$1 - j4$
2–3	$0.666 - j2.664$
2–4	$1 - j4$
3–4	$2 - j8$

The schedule of active and reactive powers is

Bus code	P	Q	V	Bus specification
1	–	–	$1.06 \angle 0$	slack
2	0.5	0.2	not specified	PQ
3	0.4	0.3	not specified	PQ
4	0.3	0.1	not specified	PQ

Form Y_{bus}, compute the voltages at buses 2, 3 and 4 at the end of first iteration using G–S method. Table $\alpha = 1.6$

Solution

The system is shown in Fig. 10.20. The line admittances and power infected into various buses are shown. The elements of bus admittance matrix are:

$$Y_{12} = Y_{21} = -2 + j8$$
$$Y_{13} = Y_{31} = -1 + j4$$
$$Y_{11} = 2 - j8 + 1 - j4 = 3 - j12$$
$$Y_{14} = Y_{41} = 0$$
$$Y_{23} = Y_{32} = -.666 + j2.664$$
$$Y_{24} = Y_{42} = -1 + j4$$
$$Y_{22} = 3.666 - j\,14.664$$
$$Y_{34} = Y_{43} = -2 + j8$$
$$Y_{33} = 3.666 - j\,14.664$$
$$Y_{44} = 3 - j12$$

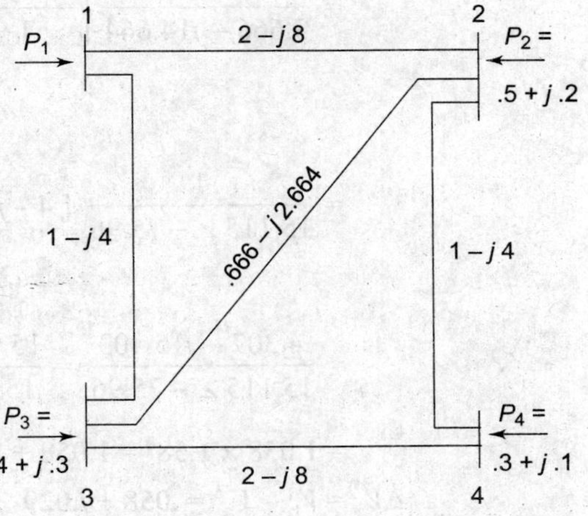

Fig. 10.20

The bus admittance matrix is

$$[Y_{bus}] = \begin{bmatrix} 3-j12 & -2+j8 & -1+j4 & 0 \\ -2+j8 & 3.666-j14.664 & -.666+j2.664 & -1+j4 \\ -1+j4 & -.666+j2.664 & 3.666-j14.664 & -2+j8 \\ 0 & -1+j4 & -2+j8 & 3-j12 \end{bmatrix}$$

Assume $V_2^{o} = V_3^{o} = V_4^{o} = 1 \angle 0$

Using Eq. (10.32) with $i = 2$

$$V_2^1 = \frac{1}{Y_{22}}\left[\frac{P_2 - jQ_2}{V_2^*} - Y_{21}V_1 - Y_{23}V_3^{o} - Y_{24}V_4^{o}\right]$$

$$= \frac{1}{3.666 - j14.664}\left[\frac{.5 - j.2}{1} - (-2 + j8)(1.06) - (-.666 + J2.664)(1) - (-1 + j4)(1)\right]$$

$$= \frac{1}{15.115 \angle -75.96^{o}}\ [.5 - j\,.2 + 2.12 - j8.48 + .666 - j2.664 + 1 - j4]$$

$$= \frac{1}{15.115 \angle -75.96}\ [4.286 - j15.344]$$

or $$V_2^1 = \frac{15.931 \angle -74.39^{o}}{15.115 \angle -75.96^{o}} = 1.054 \angle 1.57^{o}$$

$$\Delta V_2 = V_2^1 - V_2^{o} = .054 + j0.029$$

$$\alpha(\Delta V_2) = 1.6(.054 + j\,.029) = .0864 + j\,.0464$$

For calculation of V_3 we have

$$V_2^1 + V_2^{o} + \alpha\,(\Delta V_2) = 1.0864 + j\,.0464$$

$$V_3^1 = \frac{1}{Y_{33}}\left[\frac{P_3 - jQ_3}{V_3^*} - Y_{31}V_1 - Y_{32}V_2^1 - Y_{34}V_4^{o}\right]$$

$$= \frac{1}{3.666 - j14.664}\left[\frac{.4 - j\,.3}{1} - (-1 + j4)(1.06) - (-.666 + j2.664) \times (1.0864 + j\,.0464) - (-2 + j8)(1)\right]$$

$$= \frac{1}{15.115 \angle -75.96}\ [.4 - j\,.3 + 1.06 - j4.24 + .7235 + .1236 + j\,.031 - j2.894 + 2 - j8]$$

$$= \frac{4.307 - j15.403}{15.115 \angle -75.96} = \frac{15.994 \angle 74.38^{o}}{15.115 \angle 75.96}$$

$$= 1.058 \times 1.58^{o} = 1.058 + j\,.029$$

$$\Delta V_3 = V_3^1 - V_3^{o} = .058 + j.029$$

$$\alpha(\Delta V_3) = 1.6\,(.058 + j\,.029) = .0928 + j\,.0464$$

$$V_3^1 = 1 + \alpha(\Delta V_3) = 1.0928 + j\,.0464$$

$$V_4^1 = \frac{1}{Y_{44}}\left[\frac{P_4 - jQ_4}{V_4^*} - Y_{42}V_2^1 - Y_{43}V_3^1\right]$$

$$= \frac{1}{3 - j12}\left[\frac{.3 - j.1}{1} - (-1 + j4)(1.0864 + j.0464)\right.$$

$$\left. -(-2 + j8)(1.0928) + j.0464\right]$$

or

$$V_4^1 = \frac{1}{12.37\angle -75.96^\circ}\ [.3 - j\,.1 + 1.0864 + .1856 - j4.3456 + j.0464$$

$$+ 2.1856 + .3712 + j\,.0928 - j8.7424]$$

$$= \frac{4.1288 - j13.0488}{12.37\angle -75.96} = \frac{13.686\angle -72.44^\circ}{12.37\angle -75.96^\circ}$$

$$= 1.106\angle 3.52^\circ$$

Example 10.20.

Find bus admittance matrix for the system shown in Fig. 10.21.

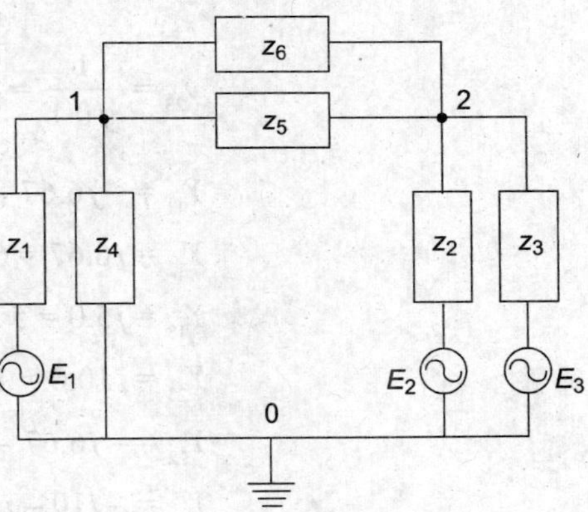

Fig. 10.21

Solution

Let

$$y_1 = \frac{1}{z_1}, y_2 = \frac{1}{z_2}, y_3 = \frac{1}{z_3}$$

$$y_4 = \frac{1}{z_4}, y_5 = \frac{1}{z_5} \text{ and } y_6 = \frac{1}{z_6}$$

Then

$$Y_{11} = y_1 + y_4 + y_5 + y_6$$

$$Y_{22} = y_2 + y_3 + y_5 + y_6$$

$$Y_{12} = Y_{21} = -y_5 - y_6$$

$$Y_{bus} = \begin{bmatrix} y_1 + y_4 + y_5 + y_6 & -y_5 - y_6 \\ -y_5 - y_6 & y_2 + y_3 + y_5 + y_6 \end{bmatrix}$$

Example 10.21.

Figure 10.22 shows a power system. Determine load flows after first iteration using fast decoupled load flow. Limit for generator at bus 2 is $0 < Q_{G2} < 5$ p.u.

Bus 1	Slack bus	$V_{spec} = 1.04\angle 0$ p.u.
Bus 2	*PV* bus	$V_{spec} = 1$ p.u. $P_{G2} = 3$ p.u.
Bus 3	*PQ* bus	$P_{D3} = 4.0$ p.u., $Q_{D3} = 2.0$ p u.

[*JNTU 2010*]

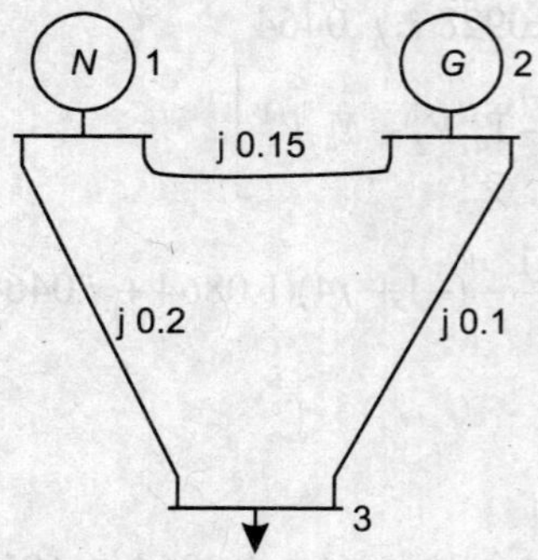

Fig. 10.22

Solution

$$y_{12} = \frac{1}{j0.15} = -j6.67 \text{ p.u.}$$

$$y_{13} = \frac{1}{j0.2} = -j50 \text{ p.u.}$$

$$v_{23} = \frac{1}{j0.1} = -j10.0 \text{ p.u.}$$

$$Y_{11} = -j6.67 - j50 = -j11.67 = 11.7 \angle -90°$$

$$Y_{12} = j6.67 = 6.67 \angle 90°$$

$$Y_{13} = j5.0 = 5 \angle 90°$$

$$Y_{23} = j10.0 = 10 \angle 90°$$

$$Y_{22} = -j6.67 - j10 = -j16.67 \text{ p.u.} = 16.67 \angle -90°$$

$$Y_{33} = -j10 - j5 = -j15.0 \text{ p.u.} = 15 \angle -90°$$

$$[Y_{bus}] = \begin{bmatrix} 11.67\angle -90° & 6.67\angle -90° & 5\angle -90° \\ 6.67\angle -90° & 16.67\angle -90° & 10\angle -90° \\ 5\angle -90° & 10\angle -90° & 15\angle -90° \end{bmatrix}$$

$$P_1 - jQ_1 = |V_1| \varepsilon^{-j\delta_1} [|Y_{11}| \varepsilon^{jr_{11}} |V_1| \varepsilon^{j\delta_1} + |Y_{12}| \varepsilon^{jr_{12}} |V_2| \varepsilon^{j\delta_2} + |Y_{13}| \varepsilon^{jr_{13}} |V_3| \varepsilon^{j\delta_3}]$$

$$P_2 - jQ_2 = |V_2| \varepsilon^{-j\delta_2} [|Y_{21}| \varepsilon^{jr_{21}} |V_1| \varepsilon^{j\delta_1} + |Y_{22}| \varepsilon^{jr_{22}} |V_2| \varepsilon^{j\delta_2} + |Y_{23}| \varepsilon^{jr_{23}} |V_3| \varepsilon^{j\delta_3}]$$

$$P_3 - jQ_3 = |V_3| \varepsilon^{-j\delta_3} [|Y_{31}| \varepsilon^{jr_{31}} |V_1| \varepsilon^{j\delta_1} + |Y_{32}| \varepsilon^{jr_{32}} |V_2| \varepsilon^{j\delta_2} + |Y_{33}| \varepsilon^{jr_{33}} |V_3| \varepsilon^{j\delta_3}]$$

$$V_1 = 1.04 \angle 0 \qquad V_2 = 1.0$$

Substituting the values of V_1, V_2 and admittances in above equations we get

$$P_2 = -6.97 \sin(\delta_1 - \delta_2) - 10.0 |V_3| \sin(\delta_3 - \delta_2)$$

$$Q_2 = -6.97 \cos(\delta_1 - \delta_2) + 16.67 - 10 V_3 \cos(\delta_3 - \delta_2)$$

$$P_3 = -5.2 |V_3| \sin(\delta_1 - \delta_3) - 10.0 |V_3| \sin(\delta_2 - \delta_3)$$

$$Q_3 = -5.2\cos(\delta_1 - \delta_3) - 10\,|V_3|\cos(\delta_2 - \delta_3) + 15\,|V_3|^2$$

Assume $|V_3^0| = 1.0,\ \delta_2^0 = \delta_3^0 = 0$. Then

$$P_2^0 = 0,\ Q_2^0 = -0.3 \text{ set it at } 0$$

$$P_3^0 = 0,\ Q_3^0 = -0.2$$

$$[B'] = \begin{bmatrix} 16.67 & -10 \\ -10 & 15 \end{bmatrix}$$

$$[B''] = [15]$$

$$\begin{bmatrix} \dfrac{\Delta P_2}{|V_2|} \\ \dfrac{\Delta p_3}{|V_3|} \end{bmatrix} = \begin{bmatrix} 16.67 & -10 \\ -10 & 15 \end{bmatrix} \begin{bmatrix} \Delta\delta_2 \\ \Delta\delta_3 \end{bmatrix}$$

or

$$\begin{bmatrix} \Delta\delta_2 \\ \Delta\delta_3 \end{bmatrix} = \begin{bmatrix} 16.67 & -10 \\ -10 & 15 \end{bmatrix}^{-1} \begin{bmatrix} \dfrac{\Delta P_2}{1} \\ \dfrac{\Delta P_3}{|V_3|} \end{bmatrix}$$

$$= \begin{bmatrix} 0.1 & 0.0667 \\ 0.0667 & 0.111 \end{bmatrix} \begin{bmatrix} \Delta P_2 \\ \dfrac{\Delta P_3}{|V_3|} \end{bmatrix}$$

$$\begin{bmatrix} \dfrac{\Delta Q_3}{|V_3|} \end{bmatrix} = [15]\,[\Delta\,|\mathrm{V}_3|\,]$$

or

$$\Delta|V_3| = [0.0667] \begin{bmatrix} \dfrac{\Delta Q_3}{|V_3|} \end{bmatrix}$$

First iteration

$$\Delta P_2^0 = 3 - 0 = 3 \text{ p.u.}$$

$$\Delta P_3^0 = -4 \text{ p.u.}$$

$$\Delta Q_3^0 = -2 - (-0.2) = -1.8 \text{ p.u.}$$

$$\begin{bmatrix} \Delta\delta_2^1 \\ \Delta\delta_3^1 \end{bmatrix} = \begin{bmatrix} 0.1 & 0.0667 \\ 0.0667 & 0.111 \end{bmatrix} \begin{bmatrix} 3 \\ -4 \end{bmatrix}$$

Solution is

$$\Delta\delta_2^1 = 0.0332 \text{ rad.}$$

$$\Delta\delta_3^1 = -0.244 \text{ rad.}$$

$$\Delta|V_{13}| = -0.12 \text{ p.u.}$$

Updated values are

$$\delta_2^1 = 0.0332 \text{ rad},\ \delta_3^1 = -0.244 \text{ rad}$$

$$V_3^1 = 1 - 0.12 = 0.88 \text{ p.u.}$$

Example 10.22

Bus admittance matrix for a system is

$$Y_{bus} = \begin{bmatrix} 3-j9 & -2+j6 & -1+j3 & 0 \\ -2+j6 & 3.666-g11 & -0.666+j2 & -1+j6 \\ -1+j3 & -0.666+j2 & 3.666-j11 & -2+j6 \\ 0 & -1+j3 & -2+j6 & 3-j9 \end{bmatrix} p11$$

$P_2 = 0.5$ p.u., $Q_2 = -0.$ p.u., $P_3 = -1$ p.u., $Q_3 = 0.5$ p.u.

$P_4 = 0.3$ p.u., $Q_4 = -0.1$ p.u. find V_2 after first iteration of *G–S* method. Assume $V_1 = 1.04\angle 0$ p.u.

[*JNTU 2011*]

Solution

$$V_i = \frac{1}{Y_{ii}}\left[\frac{P_i - jQ_i}{V_i^0} \sum_{\substack{p=i \\ p \neq i}}^{n} Y_{ip}V_p\right]$$

Assuming flat start i.e. $V_2^0 = 0$, $V_3^0 = 0$, $V_4^0 = 0$ and substituting $i = 2$ and $p = 1, 3, 4$ we get

$$V_2 = \frac{1}{Y_{22}}\left[\frac{p_2 - jQ_2}{V_2^0} - Y_{21}V_1 - Y_{23}V_3 - Y_{24}V_4\right]$$

$$= \frac{1}{3.666 - j11}\left[\frac{0.5 + 0.2}{1} - (-2 + j6)(1.04) - (-0.666 + j2)(1) - (-1 + j3)(1)\right]$$

$$= \frac{4.246 - j10.24}{3.666 - j11} = \frac{11.0854\angle -67°}{11.5948\angle -71.56\%}$$

$$= 0.9561\angle 4.568° \text{ p.u.}$$

Example 10.23.

Figure 10.23 shows a 4 bus system. Treat bus 4 as reference bus. Find bus admittance matrix. p.u. reactances are as shown in Fig. 10.23.

[*VTU 2011*]

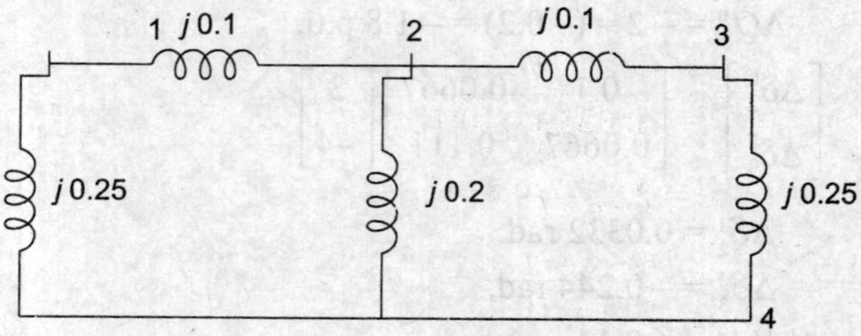

Fig. 10.23

Solution

$$y_{12} = \frac{1}{j0.1} = -j10 = y_{23} = y_{21} = y_{32}$$

$$y_{13} = y_{31} = 0, y_{14} = \frac{1}{j\,0.25} = -j_4$$

$$y_{24} = \frac{1}{j0.2} = -j1, y_{34} = \frac{1}{j\,0.25} = -j4$$

$$Y_{11} = -j10 - j4 = -j14$$

$$Y_{12} = Y_{21} = -y_{21} = j10$$

$$Y_{22} = -j10 - j10 - j5 = -j25$$

$$Y_{13} = Y_{31} = 0$$

$$Y_{23} = Y_{32} = -y_{23} = j10$$

$$Y_{33} = -j10 - j4 = j14$$

$$Y_{\text{bus}} = \begin{bmatrix} -j14 & j10 & 0 \\ j10 & -j25 & j10 \\ 0 & j10 & -j14 \end{bmatrix}$$

Example 10.24.

Figure 10.24 shows a 4 bus system. The per unit impedances are as shown. Form Y bus-admittance matrix. Neglect shunt admittances. *[CSVTU, 2010]*

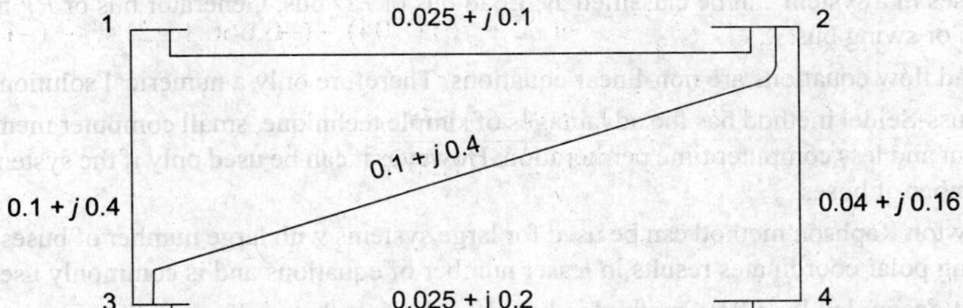

Fig. 10.24

Solution

$$y_{12} = \frac{1}{0.025 + j10.1} = 9.7 \angle -75.96$$

$$= 2.35 - j9.41 = y_{21}$$

$$y_{13} = \frac{1}{0.1 + j0.4} = 2.425 \angle -75.96$$

$$= 0.588 - j2.35 = y_{31}$$

$$y_{24} = \frac{1}{0.04 + j0.1} = 1.47 - j5.88 = y_{42}$$

$$y_{23} = y_{13} = 0.588 - j2.35 = y_{32}$$

$$y_{34} = \frac{1}{0.05 + j0.2} = 1.176 - j4.71 = y_{43}$$

$$\begin{bmatrix} 2.938 - j11.76 & -2.35 + j9.41 & -0.588 + j2.35 & 0 \\ -2.35 + j9.41 & 4.408 - j17.64 & -j0.588 + j2.35 & -1.47 + j5.88 \\ -0.588 + j2.35 & -0.588 + j2.35 & 4.408 - j9.41 & -1.176 + j4.71 \\ 0 & -1.47 + j5.85 & -1.176 + j4.71 & 2.646 - j10.59 \end{bmatrix}$$

SUMMARY

10.1. The aim of load flow study is to find power flows (active and reactive) line and transformer loads and losses and voltages at different buses for specified conditions under steady state.

10.2. Bus admittance matrix is used in load flow studies.

10.3. Bus admittance matrix can be formed by step by step method. Another method uses singular transformation.

10.4. Singular transformation uses the concept of graph theory.

10.5. $[Y_{bus}] = A^t\, y\, A$ where A is incidence matrix, y is primitive admittance matrix and A^t is transpose of A.

10.6. A few lines in a power system may have mutual coupling with some other lines. It is necessary to include the effect of mutual coupling for accurate analysis.

10.7. Buses in a system can be classified as: Load bus or *PQ* bus, Generator bus or *PV* bus and slack bus or swing bus.

10.8. Load flow equations are non-linear equations. Therefore only a numerical solution is possible.

10.9. Gauss-Seidel method has the advantages of simple technique, small computer memory requirement and less computer time per iteration. However it can be used only if the system has a small number of buses.

10.10. Newton Raphson method can be used for large systems with large number of buses. *NR* method using polar coordinates results in lesser number of equations and is commonly used.

10.11. Fast decoupled load flow method is based on the fact that real power flows mainly depend on power angles and reactive power flows mainly depend on voltage magnitudes.

10.12. Power system are sparse i.e. each bus is connected to only a small number of other buses. Sparsity helps in reducing computer memory requirement and computer time.

SELF TEST QUESTIONS

A. State whether the following are true (T) or false (F)

10.1 Load flow equations are non-linear.

10.2 Majority of buses in a power system are load buses.

10.3 At a generator bus P and Q are specified.

10.4 Slack bus is a load bus.

10.5 *NR* method in polar coordinates results in lesser number of equations.

10.6 *NR* method is based on Taylor series.

10.7 Acceleration factor is less than 1.

10.8 $V_{bus} = [Z_{bus}]I_{bus}$

10.9 $[Y_{bus}] = A^t yA$

10.10 Bus incidence matrix is singular.

B. Fill in the blanks with suitable words

10.11 In terms of $[Y_{bus}]$ and V_{bus}, I_{bus} =

10.12 In a network a set of unconnected elements is called a ________.

10.13 If a new line is added between buses 1 and 3 in a system, the elements of Y_{bus} affected by addition of this line are ________.

10.14 The most common buses in a power system are ________ buses.

10.15 The angle δ at a slack bus is assumed to be ________.

10.16 A generator bus is a ________ bus.

10.17 Total generation is equal to total load plus ________.

10.18 In ________ method of load flow, convergence is dependent on choice of slack bus.

10.19 *NR* method requires ________ number of iterations than *GS* method of load flow.

10.20 At *PQ* buses we assume ________ and ________.

ANSWERS

10.1 T	10.2 T	10.3 F	10.4 F	10.5 T	10.6 T
10.7 F	10.8 T	10.9 T	10.10 T	10.11 $[Y_{bus}]\ V_{bus}$	
10.12 primitive network		10.13 $Y_{11}, Y_{33}, Y_{13}, Y_{31}$		10.14 load	10.15 zero
10.16 *PV*	10.17 losses	10.18 *G–S*	10.19 lesser	10.20 $\lvert V\rvert$ and δ	

JOB INTERVIEW QUESTIONS

10.1 Why is load flow study necessary?

10.2 How are buses classified?

10.3 What is meant by sparsity?

10.4 How is sparsity useful in load flow analysis?

10.5 How is $[Y_{bus}]$ formed?

10.6 What is meant by singular transformation?

10.7 What is meant by 'compact storage scheme' in load flow analysis?

10.8 What is decoupled load flow method?

10.9 What is meant by optimal ordering?

10.10 What are the advantages of *NR* method over *GS* method?

10.11 What is the difference between Z_{bus} and Y_{bus}?

10.12 What is meant by term 'Jacobian.'

REVIEW QUESTIONS

10.1 Give reasons:

(a) A direct solution of load flow problem is not possible.

(b) *NR* method is preferred to *G–S* method for load flow studies in power systems.

(c) A majority of the buses in power systems are load buses.

(d) An acceleration factor is commonly used in load flow studies using *G–S* method.
(e) The order of Jacobian in *NR* method using polar coordinates is smaller than that in *NR* method using rectangular coordinates.
(f) Bus admittance matrix is a sparse matrix.
(g) One of the buses is taken as slack bus in load flow studies.

10.2 Starting from first principles show that a diagonal element of Y_{bus} equals the sum of admittances connected to that bus and an off diagonal element equals the negative of the sum of admittances directly connected between the buses.

10.3 Discuss the step by step method of formation of Y_{bus}.

10.4 Explain the terms bus incidence matrix, primitive network.

10.5 Explain in brief the procedure for formulation of Y_{bus} using singular transformation. Derive the necessary equations.

10.6 Starting from first principles develop the equations for real and reactive bus powers.

10.7 Explain the *G–S* method for solution of non-linear algebraic equations.

10.8 Give a flow chart for a load flow study on a power system having only *P–Q* buses using *G–S* method. How does the flow chart get modified to account for *PV* buses?

10.9 Discuss the *NR* method for solving non-linear algebraic equations.

10.10 Derive equations for elements of Jacobian using (a) rectangular coordinates (b) polar coordinates.

10.11 Give a flow chart for a load flow study using *NR* method. How does the method get modified when *PV* buses are also present?

PROBLEMS

10.1 Use *G–S* method to solve the equation

$$1.5x + 2\cos x = 2.225 \qquad [x = 1.848]$$

10.2 Use (a) *G–S* method (b) *NR* method to solve the equation

$$f(x) = x^3 + 2x^2 + 10x - 20 \qquad [1.368808108]$$

10.3 Use *NR* method to solve the equations

$$x = x^2 + y^2$$
$$y = x^2 - y^2 \qquad [x = .772, y = .42]$$

Given that the solution lies near the point (0.8, 0.4)

10.4 Use *NR* method to solve the equations

$$x = \sin x \cosh y$$
$$y = \cos x \sinh y$$

Assume $x^o = 7, y^o = 3$ [$x = 7.4977, y = 2.7687$]

10.5 Assuming an initial value of $x = -1.9$ find the values of x after first and second *NR* iterations in the equation $x^2 + 4\sin x = 0$ [$x^1 = -1.93, x^2 = -1.9388$]

10.6 Use *NR* method to solve the equations

$$\sin x = y - 1.32$$
$$\cos y = x - .85 \qquad [x = .567325, y = 1.857378]$$

10.7 Form Y_{bus} for the 4-bus system of Fig. 10.10 if the line series impedances are as under:

line (bus to bus)	*Impedance*
1–2	$0.15 + j\,0.6$ *pu*
1–3	$0.1 + j\,0.4$ *pu*
1–4	$0.15 + j\,0.6$ *pu*
2–3	$0.05 + j\,0.2$ *pu*
3–4	$0.05 + j\,0.2$ *pu*

Neglect the shunt capacitance of line

$$\begin{bmatrix} 1.372-j5.491 & -.392+j1.569 & -.588+j2.353 & -.392+j1.569 \\ -.392+j1.569 & 1.568-j6.275 & -1.176+j4.706 & 0 \\ -.588+j2.353 & -1.176+j4.706 & 2.94-j11.765 & -1.176+j4.706 \\ -.392+j1.569 & 0 & -1.176+j4.706 & 1.568-j6.275 \end{bmatrix}$$

10.8 In problem 10.7 the lines from bus 1–2, 1–3, 1–4 have each a total shunt admittance of j .08 *pu*. The shunt admittances of the remaining lines are negligible. The series impedances are the same as in problem 10.7. Form Y_{bus}.

$$\begin{bmatrix} Y_{11}=1.372-j5.371, Y_{22}=1.568-j6.235, \\ Y_{33}=2.94-j11.725, Y_{44}=1.568-j6.235 \\ \text{The rest of the entries remain the same.} \end{bmatrix}$$

10.9 In the system of problem 10.7, a new line having a series impedance of $0.15 + j\,0.6$ *pu* is added to connect buses 2 and 4. The new line has no mutual coupling with the existing lines. The remaining data is the same. Form Y_{bus}.

$$\begin{bmatrix} Y_{22}=1.96-j7.844, Y_{44}=1.96-j7.844 \\ Y_{24}=Y_{42}=-.392+j1.569 \\ \text{The rest of the entries remain the same.} \end{bmatrix}$$

10.10 In the system of problem 10.7, the bus powers and voltages are as under:

Bus	P_L	Q_L	P_G	Q_G	V	*Bus specification*
1	2	0.5	not specified	not specified	$1 \angle 0$	slack bus
2	0	0	1	not specified	1	*PV* bus
3	0.7	0.3	0	0	not specified	*PQ* bus
4	0.7	0.3	0	0	not specified	*PQ* bus

$Q_{2,\,min} = 0$ and $Q_{2,\,max} = 0.6$

Find Q_2, δ_2, V_3, V_4 after first iteration using *G–S* method

[0, 8.23°, 0.958 $\angle$ 0.43°, 0.902 $\angle$ – 5.96°]

10.11 Write a computer program and solve problem 9.10 using *G–S* method. Obtain final values of Q_2, δ_2, V_3, V_4, P_{G1}, Q_{G1}.

10.12 Write a computer program for complete solution of problem 10.10 by *G–S* method using an acceleration factor of 1.6. Obtain the final values of Q_2, δ_2, V_3, V_4, P_{G1} and Q_{G1}.

10.13 Write a computer program for complete solution of problem 10.10 using *NR* method in polar coordinates. Obtain final values of Q_2, δ_2, V_3, V_4, P_{G1}, and Q_{G1}.

10.14 In the system shown in Fig. 10.17 each line has a series reactance of j 0.25 ohms. Line resistance and shunt admittance can be neglected. The bus power and voltage data is

Bus	P_G	Q_G	P_L	Q_L	*Bus voltage*
1	not specified	not specified	1.5	0.5	$1 \angle 0$ slack bus
2	0	not specified	1.5	0.5	1 (*PV* bus)
3	2	not specified	1.0	0.5	1 (*PV* bus)

Carry out an approximate load flow and find (a) reactive power generation at the buses (b) reactive line losses. [(a) 0.559, .642, .59 (b) 0.291]

10.15 A power system has the impedances between various buses

bus 1 to reference $j\,2\,\Omega$

bus 2 to reference $j\,2\,\Omega$

bus 3 to reference $j\,2\,\Omega$
bus 1 to bus 3 $j\,0.2\,\Omega$
bus 2 to bus 3 $j\,0.4\,\Omega$
bus 1 to bus 4 $j\,0.2\,\Omega$
bus 2 to bus 4 $j\,0.2\,\Omega$
bus 3 to bus 4 $j\,0.1\,\Omega$

Draw a configuration of the system and find bus admittance matrix

$$\begin{bmatrix} -j10.5 & 0 & j5.0 & j5.0 \\ 0 & -j8.0 & j2.5 & j5.0 \\ j5.0 & j2.5 & -j18.0 & j10.0 \\ j5.0 & j5.0 & j10.0 & -j20.0 \end{bmatrix}$$

10.16 Find bus impedance matrix for the system of problem 10.15. Use matrix inversion.

$$\begin{bmatrix} j0.724 & j0.62 & j0.656 & j0.644 \\ j0.62 & j0.738 & j0.642 & j0.66 \\ j0.656 & j0.642 & j0.702 & j0.676 \\ j0.644 & j0.66 & j0.676 & j0.719 \end{bmatrix}$$

10.17 The impedances between various buses in a system are

Each bus to reference	$j\,2.0\,\Omega$
bus 1 to bus 2	$j\,0.8\,\Omega$
bus 2 to bus 3	$j\,0.5\,\Omega$
bus 2 to bus 3	$j\,1.0\,\Omega$

Find bus admittance matrix

$$\begin{bmatrix} -j3.75 & j1.25 & j2 \\ j1.25 & -j2.75 & j1 \\ j2 & j1 & -j3.5 \end{bmatrix}$$

10.18 A four bus system has the following line reactances (the resistances are negligible)

Line code	Reactance
Bus 1–2	$j\,0.5\,pu$
Bus 2–3	$j\,1.0\,pu$
Bus 3–4	$j\,1.0\,pu$
Bus 1–4	$j\,0.5\,pu$

The bus specifications and values of injected powers are as under

Bus	*P*	*Q*	*V*	*Bus specification*
1	–	–	1.05 *pu*	slack
2	– 0.1 *pu*	Zero	–	load bus
3	0.5 *pu*	0.02 *pu*	–	generator bus
4	– 0.4 *pu*	– 0.05 *pu*	–	load bus

Use *G–S* method and find values of V_2, V_3 and V_4 after one iteration.

$$\begin{bmatrix} V_2 = 1.03387\angle -1.847^\circ\ pu \\ V_3 = 1.14078\angle 11.8^\circ\ pu \\ V_4 = 1.05556\angle .15^\circ\ pu \end{bmatrix}$$

CHAPTER 11

Economic Operation of Power Systems and Unit Commitment

11.1 INTRODUCTION

A modern power system is invariably fed from a number of power plants. Research and development has led to efficient power plant equipment. A generating unit added to the system today is likely to be more efficient than the one added some time back. With a very large number of generating units at hand, it is the job of the operating engineers to allocate the loads between the units such that the operating costs are the minimum.

11.2 METHODS OF LOADING TURBO-GENERATORS

A number of methods of deciding the amount of load to be carried by different turbo-units have been proposed and are used.

(1) Base load to capacity. The turbo units are loaded to full capacity, one after the other, in the order of their efficiencies. As demand increases it is supplied by the most efficient unit till this unit is loaded to its rated capacity. The further increase in demand is met from the next most efficient unit till this unit also gets loaded to full capacity and so on.

(2) Base load to most efficient load. The turbo units are loaded to the point of maximum efficiency in the order of their efficiencies. As demand increases, the most efficient unit is loaded till its point of maximum efficiency is reached. The further increase in demand is met by loading the next most efficient unit to the point of its maximum efficiency and so on. When all the units are operating at maximum efficiencies, further demand is met by loading the units to their rated capacities in the same order.

(3) Loading proportional to capacity. The total load is divided among the units in the ratio of their rated capacities.

(4) Loading proportional to most efficient load. This is similar to base loading to the most efficient load. However, after each unit has been loaded to the point of maximum efficiency, further load is divided among the units in proportion to the difference between the rated capacity and the most efficient load of each unit.

(5) Equal incremental cost operation. The total system load is so divided among the units that all the units operate at equal incremental costs. This is the most scientific and rational method. Operation of units at equal incremental costs results in the most economic operation.

11.3 THERMAL PLANT COST MODELLING, INPUT-OUTPUT CURVE, HEAT RATE, INCREMENTAL COST

In a thermal plant, fuel is burnt and heat energy generated in the boiler. The turbine converts a part of this energy into mechanical energy. The generator converts this mechanical energy into electrical energy. The input is the fuel and output is the electrical power. The output-input relation is basically an efficiency type model. The overall efficiency depends on boiler efficiency, turbine efficiency and generator efficiency.

The boiler efficiency depends on losses in the boiler which can be broadly classified as :

1. Stack heat losses including the losses in flue gas, moisture in gas, incomplete combustion of carbon-monoxide etc.
2. Heat loss in ash (unburnt fuel).
3. Heat in steam for soot blowing.
4. Heat in rejects of pulverising mill.
5. Auxiliary power for pulverising mill, fans, ash disposal, pumps etc.

The above losses are determined by variables related to the fuel, air and feed water. Some variables are uncontrollable while others are controllable. Air and water temperature, moisture in air and coal and calorific value of coal are uncontrollable variables. The air quantity, air pressure, fuel preparation etc. are controllable variables. The condition of boiler, which depends on maintenance, is also an important factor. The efficiencies of turbine and generator can be found out more precisely and are not subject to random changes. The turbine efficiency depends on temperature and pressure of steam at the turbine stop valve, high pressure, low pressure, cylinder exhaust, condenser and low pressure and high pressure heaters etc. The condition of turbine is also an important consideration. The deposition of solids on the blades and blade erosion tend to reduce turbine efficiency. The generator efficiency depends on various losses (*i.e.*, copper, iron, mechanical) in the generator.

Since a number of losses are variable, a thermal plant cost model is liable to change. Therefore periodic updating of the model is necessary.

To compute economic load division between units, it is necessary to express the operating costs in terms of output. For thermal plants, fuel cost is the main operating cost and accounts for 60-70 per cent of the total operating cost. Hence, the fuel cost forms the basis of the load division. It is understood that other costs, which are also function of output, can be represented as a percentage of fuel cost. Cost of nuclear fuel can also be expressed as a function of output.

The input-output curve (Fig. 11.1) expresses fuel input in terms of k-cal per hour against power output in MW. The fuel input can be converted in Rs. per hour by multiplying k-cal per hour by the cost of fuel in Rs. per k-cal. The actual input-output curve is a non-linear curve but can be considered as a linear one as shown in Fig. 11.1.

The heat rate is the ratio of input to the corresponding output and the heat rate curve is shown in Fig. 11.2*.

* Is heat rate related to efficiency ? How ?

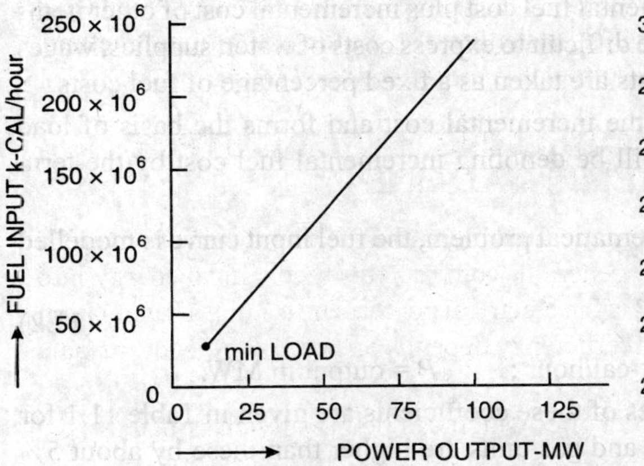

Fig. 11.1. Input-output curve of a generating unit.

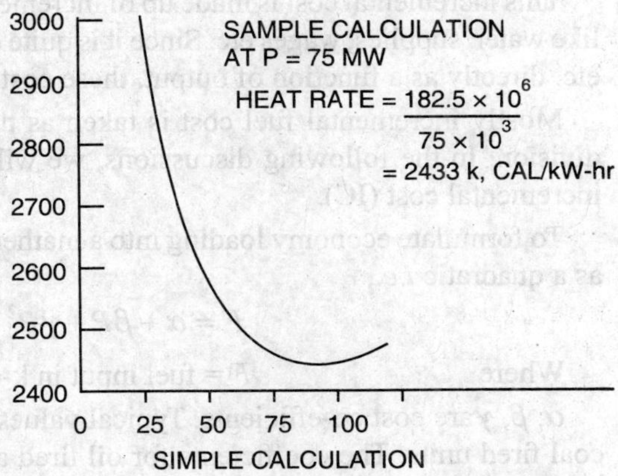

Fig. 11.2. Heat rate curve for a generating unit whose input-output curve is shown in Fig. 11.1

The incremental fuel cost is the ratio of small change in input to a small change in output.

$$\text{Incremental fuel cost} = \frac{\Delta \text{ input}}{\Delta \text{ output}} \qquad \text{... (11.1}a\text{)}$$

As the changes in input and output become progressively smaller :

$$\text{Incremental fuel cost} = \frac{d\,(\text{input})}{d(\text{output})} \qquad \text{... (11.1}b\text{)}$$

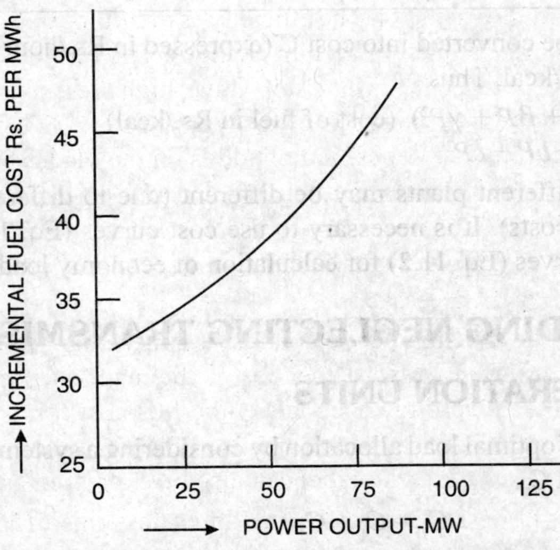

Fig. 11.3. Incremental fuel cost curve for a generating unit whose input-output curve is shown in Fig. 11.1

The incremental fuel cost is given by the slope of the input-output curve*. The unit of incremental fuel cost is Rs. per MW-hour (Fig. 11.3).

* In the input-output characteristics if input is represented in k-cal/hr., and output in MW, the slope of the curve gives incremental fuel rate in k-cal/MWh. If the input is in Rs./hour, the slope gives incremental fuel cost in Rs./MWh.

This incremental cost is made up of incremental fuel cost plus incremental cost of other items like water, supplies, wages etc. Since it is quite difficult to express costs of water, supplies, wages etc. directly as a function of output, these costs are taken as a fixed percentage of fuel costs.

Mostly incremental fuel cost is taken as the incremental cost and forms the basis of load division. In the following discussions, we will be denoting incremental fuel cost by the term incremental cost (IC).

To formulate economy loading into a mathematical problem, the fuel input curve is modelled as a quadratic *i.e.*,

$$F = \alpha + \beta P + \gamma P^2 \qquad ...(11.2)$$

Where F = fuel input in k-cal/hour ; P = output in MW.

α, β, γ are cost co-efficients. Typical values of these coefficients are given in Table 11.1 for coal fired units. The coefficients for oil fired and gas units are higher than these by about 5% and 7% respectively.

Table 11.1. Typical values of cost co-efficients for modern coal fired units.

Unit size *MW*	α	β	γ
50	12×10^6	2.5×10^6	2.5×10^3
100	19×10^6	2.25×10^6	1.1×10^3
200	41.5×10^6	2×10^6	0.55×10^3
500	100×10^6	1.85×10^6	0.28×10^3

The fuel input F can be converted into cost C (expressed in Rs./hour) by multiplying α, β, γ by the cost of fuel in Rs./kcal. Thus

$$C = (\alpha + \beta P + \gamma P^2) \text{ (cost of fuel in Rs./kcal)}$$

or

$$C = a + bP + cP^2 \qquad ... (11.3)$$

The cost of fuel at different plants may be different (due to different quality of coal and different transportation costs). It is necessary to use cost curves (Eq. 11.3) for different units and not the fuel input curves (Eq. 11.2) for calculation of economy loading.

ECONOMY LOADING NEGLECTING TRANSMISSION LOSSES

11.4 TWO GENERATION UNITS

We start the discussion of optimal load allocation by considering a system having two generating units having costs C_1 and C_2.

Total cost $\qquad C_T = C_1 + C_2 \qquad ... (11.4)$

Total output $\qquad P_T = P_1 + P_2 \qquad ... (11.5)$

C_1 and C_2 are the fuel costs of the two units in Rs. per hour. Total output P_T is equal to active power demand and is constant. It is desired to find P_1 and P_2 so that C_T is minimum.

$$\frac{dC_T}{dP_1} = \frac{dC_1}{dP_1} + \frac{dC_2}{dP_1} \qquad ... (11.6)$$

From Eq. (11.5), $\qquad dP_2 = -\,dP_1 \qquad ... (11.7)$

For minimum total cost C_T

$$\frac{dC_T}{dP_1} = 0 \qquad \text{... (11.8)}$$

Combining Eqs. (11.6), (11.7) and (11.8), we get

$$\frac{dC_1}{dP_1} = \frac{dC_2}{dP_2} \qquad \text{... (11.9)}$$

Thus we reach the conclusion that the loads should be so allocated that the two units operate at equal incremental costs.

We can also reach the same conclusion through a qualitative analysis. Let the two units operate at different incremental costs. Transfer some load from the unit operating at higher incremental cost to the unit operating at lower incremental cost. Reduction of load on the unit with higher incremental cost will mean a greater reduction in cost than the increase in cost due to the increase in load on the unit with lower incremental cost. In the limiting case, when the loads on the two units are such that both operate at the same incremental cost, maximum economy will result.

11.5 LARGE NUMBER OF UNITS

The above qualitative discussion of the two unit system can be extended to a system with any number of units. The loads should be so allocated among the different units that every unit operates at the same incremental cost. This criteria can be developed mathematically as under :

For a plant with k units

$$C_T = C_1 + C_2 + ... + C_k = \sum_{n=1}^{k} C_n \qquad \text{...(11.10)}$$

$$P_T = P_1 + P_2 + ... + P_k = \sum_{n=1}^{k} P_n \qquad \text{...(11.11)}$$

Since total load P_T is constant

$$dP_T = 0 = dP_1 + dP_2 + ... + dP_k$$

$$\therefore \quad dP_1 = - dP_2 - dP_3 - ... - dP_k \qquad \text{... (11.12)}$$

Our aim is to find $P_1, P_2, ... P_k$ such that C_T is a minimum. The input or cost of a unit is dependent only on the output of that unit.

$$dC_T = 0 = \sum_{n=1}^{k} \frac{dC_n}{dP_n}.dP_n \qquad \text{... (11.13)}$$

or $$\frac{dC_1}{dP_1}.dP_1 + \sum_{n=2}^{k} \frac{dC_n}{dP_n}.dP_n = 0 \qquad \text{... (11.14)}$$

or $$\frac{dC_1}{dP_1}(- dP_2 - dP_3 ... - dP_k) + \sum_{n=2}^{k} \frac{dC_n}{dP_n}.dP_n = 0 \qquad \text{... (11.15)}$$

or $$\sum_{n=2}^{k} \left(\frac{dC_n}{dP_n} - \frac{dC_1}{dP_1} \right) dP_n = 0 \qquad \text{... (11.16)}$$

This equation is satisfied only if each term is zero.

$\therefore$ For $n = 2, 3, ... k$

$$\frac{dC_1}{dP_1} = \frac{dC_n}{dP_n} \quad \text{... (11.17)}$$

or

$$\frac{dC_1}{dP_1} = \frac{dC_2}{dP_2} = ... = \frac{dC_k}{dP_k} = \lambda \quad \text{... (11.18)}$$

Thus optimum economy is achieved if every unit operates at the same incremental cost. A graphical representation of this result is given in Fig. 11.4.

The load allocation between a large number of units can be found by preparing an operating schedule. This can be done by assuming different values of λ, finding the individual loads of different units and adding different loads to give total load on the system.

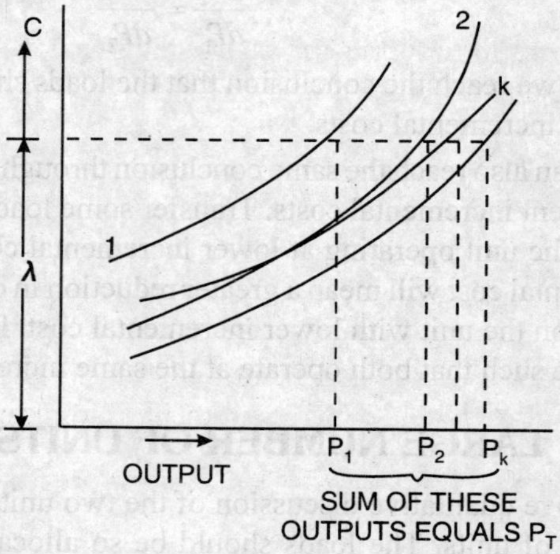

Fig. 11.4 Graphical representation of equal incremental cost criterion.

As the system load increases, the individual loads will also increase. One or several units will reach their rated capacities earlier than other units. After this happens, the remaining units should operate at equal incremental costs. If the minimum and maximum loads of some or all the units are specified, some units will not be able to operate at the same incremental cost and still meet the constraint of the minimum and maximum outputs.

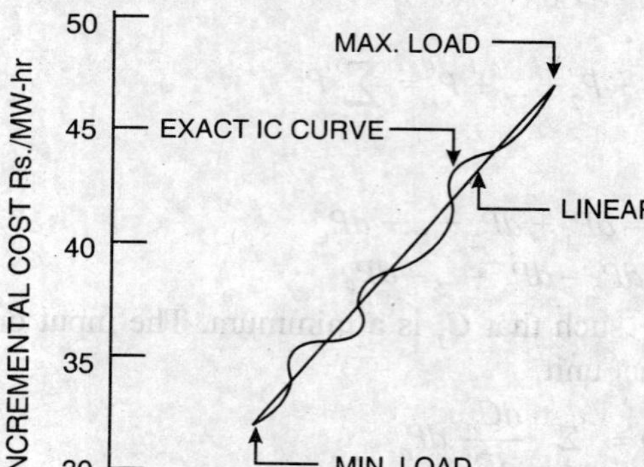

Fig. 11.5 A non-linear incremental cost characteristics and approximate linear characteristics.

The incremental costs are generally non-linear functions of output. Most of the time it is possible to convert these non-linear functions into linear functions without much loss of accuracy and the operating schedule can be obtained analytically. However, if this cannot be done, it becomes necessary to use a digital technique and device for computations. Fig. 11.5 shows a non-linear incremental cost characteristics and the approximate linear characteristics. In each approximation, the area under the curve from the minimum to the maximum output is kept the same and the difference between the ordinates of the exact curve and the approximate curve is kept the minimum.

11.6 LARGE NUMBER OF UNITS-ALTERNATIVE PROOF (METHOD OF LAGRANGIAN MULTIPLIER)

The operating criteria for a large number of units can also be developed by the method of lagrangian multiplier.

$$C_T = C_1 + C_2 + ... + C_k = \sum_{n=1}^{k} C_n \quad ...(11.19)$$

$$P_T = P_1 + P_2 + ... + P_k = \sum_{n=1}^{k} P_n \quad ...(11.20)$$

$$dP_T = 0 = dP_1 + dP_2 + ... + dP_k \quad ...(11.21)$$

Cost of a unit is dependent only on the output of that unit. The total cost depends on the output of each unit. The requirement of minimum C_T means that total differential dC_T is zero.

$$\therefore \quad dC_T = \frac{\partial C_T}{\partial P_1} . dP_1 + \frac{\partial C_T}{\partial P_2} dP_2 + ... + \frac{\partial C_T}{\partial P_k} . dP_k = 0 \quad ...(11.22)$$

Multiply Eq. (11.21) by λ, subtract the resulting equation from Eq. (11.22) and re-arrange the terms.

$$\left(\frac{\partial C_T}{\partial P_1} - \lambda\right) dP_1 + \left(\frac{\partial C_T}{\partial P_2} - \lambda\right) dP_2 + ... + \left(\frac{\partial C_T}{\partial P_k} - \lambda\right) dP_k = 0 \quad ...(11.23)$$

This equation is satisfied if each term is equal to zero. Since the cost of a unit depends on the output of that unit only, the partial derivative becomes full derivative *i.e.*, $\frac{\partial C_T}{\partial P_k}$ becomes $\frac{dC_k}{dP_k}$.

Eq. (11.23) is satisfied if

$$\frac{dC_1}{dP_1} = \lambda, \quad \frac{dC_2}{dP_2} = \lambda, ... \frac{\partial C_k}{\partial P_k} = \lambda \quad ...(11.24)$$

The constant λ is referred to as Lagrangian multiplier and in this case, it is the incremental cost in Rs. per MW-hour. Thus optimum economy is achieved if every unit operates at the same incremental cost.

Example 11.1

The incremental costs in Rs. per MW-hour of two 250 MW units are as under:

$$\frac{dC_1}{dP_1} = 0.2\ P_1 + 30$$

$$\frac{dC_2}{dP_2} = 0.15\ P_2 + 40$$

The minimum load of each is 20 MW. Find the load division between the two units as the total load varies from 40 MW to 500 MW. (*P.U. 1988*)

Solution

When the total load is 40 MW, it will be shared equally by the two units. The IC of unit 1 will be Rs. 34 per MWh and that of unit 2 will be Rs. 43 per MWh. As load increases more load will be added to unit 1 till IC of unit 1 also becomes Rs. 43 per MWh. This will happen when load on unit 1 is 65 MW and total load 85 MW. After this the load division is calculated by assuming successively higher values of IC and finding P_1, P_2 and the total load. The results are given in Table 11.2.

When IC reaches a value of 77.5, unit 2 will supply its maximum load of 250 MW. At this point unit 1 will supply 237.5 MW. After this unit 2 will continue to supply 250 MW and the unit 1 will meet the remaining requirement. The load division is plotted in Fig. 11.6.

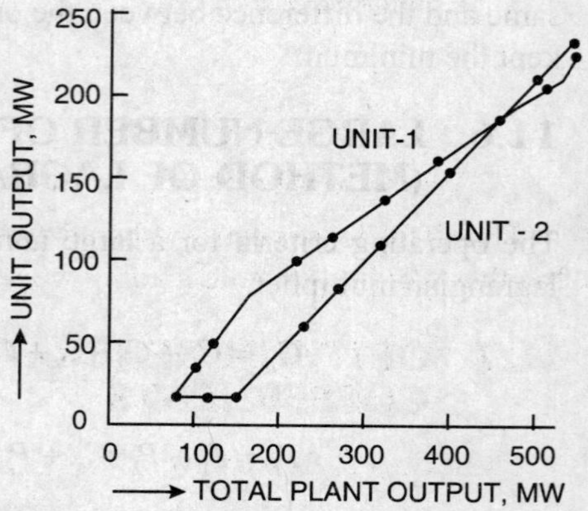

Fig. 11.6 Load division between the two generating units of Example 11.1.

Table 11.2 Load division for Example 11.1

Plant IC *Rs./MWh*	P_1 *MW*	P_2 *MW*	$P_1 + P_2$ *MW*
	50.0	20.0	70.0
43	65.0	20.0	85.0
50	100.0	66.66	166.66
55	125.0	100.0	225.0
60	150.0	133.33	283.33
65	175.0	166.66	341.66
70	200.0	200.0	400.0
75	225.0	233.33	458.33
77.5	237.5	250.0	487.5
80	250.0	250.0	500.0

Example 11.2

Compute the saving per year for economical load allocation compared to equal load sharing for a total system load of 225 MW for the units of Example 11.1. Assume that the system load remains constant throughout the year.

Solution

For equal incremental costs, unit 1 should supply 125 MW and unit 2 should supply 100 MW. Equal sharing means that each unit will supply 112.5 MW

$$\text{Change in cost of unit 1} = \int_{125}^{112.5} (.2P_1 + 30)dP_1$$

$$= \left| .10\, P_1^2 + 30\, P_1 \right|_{125}^{112.5} = -\text{Rs. } 671.88. \text{ per hour}$$

$$\text{Change in cost of unit 2} = \int_{100}^{112.5} (.15\,P_2 + 40)\,dP_2$$

$$= \left| .075\,P_2^2 + 40P_2 \right|_{100}^{112.5} = \text{Rs. } 699.22 \text{ per hour}$$

$$\text{Yearly extra cost} = (699.22 - 671.88)\,8760 = \text{Rs. } 239498 \text{ per year}$$

Thus if the load is equally shared by the two units, an extra cost of Rs. 239498 will be incurred. In other words economic loading would result in saving of Rs. 239498 per year.

11.7 OPTIMUM GENERATION AS FUNCTION OF λ AND COST COEFFICIENTS

From Eq. (11.3) we have

$$\begin{aligned} C_1 &= a_1 + b_1P_1 + c_1P_1^2 \\ C_2 &= a_2 + b_2P_2 + c_2P_2^2 \\ C_k &= a_k + b_kP_k + c_kP_k^2 \end{aligned} \qquad \text{... (11.25)}$$

or
$$\begin{aligned} \frac{dC_1}{dP_1} &= \lambda = b_1 + 2c_1P_1 \\ \frac{dC_2}{dP_2} &= \lambda = b_2 + 2c_2P_2 \\ \frac{dC_k}{dP_k} &= \lambda = b_k + 2c_kP_k \end{aligned} \qquad \text{... (11.26)}$$

or
$$\lambda\left(\frac{1}{c_1} + \frac{1}{c_2} + \ldots \frac{1}{c_k}\right) = \left(\frac{b_1}{c_1} + \frac{b_2}{c_2} + \ldots + \frac{b_k}{c_k}\right) + 2(P_1 + P_2 + \ldots + P_k) \qquad \text{... (11.27)}$$

or
$$\lambda\left(\sum_{n=1}^{k} \frac{1}{c_n}\right) = \sum_{n=1}^{k} \frac{b_n}{c_n} + 2P_T \qquad \text{... (11.28)}$$

or
$$\lambda = \frac{\displaystyle\sum_{n=1}^{k} \frac{b_n}{2c_n} + P_T}{\displaystyle\sum_{n=1}^{k} \frac{1}{2c_n}} \qquad \text{... (11.29)}$$

and
$$P_n = \frac{\lambda - b_n}{2c_n} \qquad \text{... (11.30)}$$

11.8 SINGLE EQUIVALENT MACHINE REPRESENTATION

Eqs (11.29 and 11.30) are simplified results because in this derivation we have neglected the transmission losses. These results can be used to find the initial values of P_1, P_2 etc. so that these values can be used for the final solution when transmission losses are considered. The later solution requires iterative methods.

When a number of thermal units are on the same bus, they can be replaced by a single equivalent machine. Let k such units exist on the same bus. For equivalent single machine

$$C_T = a_e + b_eP_T + c_eP_T^2 \qquad \text{... (11.31)}$$

The addition of set of Eq (11.25) is

$$C_1 + C_2 + ... + C_k = C_T = \sum_{n=1}^{k} a_n + \sum_{n=1}^{k} b_n P_n + \sum_{n=1}^{k} c_n P_n^2 \quad ... (11.32)$$

The conditions of optimum generation (*i.e.*, Eqs 11.29 and 11.30) can be incorporated into Eq. (11.32) and the terms re-arranged. The resulting expression when compared with Eq. (11.31) gives the following values of equivalent parameters.

$$\frac{1}{c_e} = \sum_{n=1}^{k} \frac{1}{c_n}$$

$$\frac{b_e}{c_e} = \sum_{n=1}^{k} \frac{b_n}{c_n} \quad ... (11.33)$$

$$a_e = \sum_{n=1}^{k} \left[a_n - \frac{b_n^2}{4c_n} \right] + \frac{b_e^2}{4c_e} \quad ... (11.34)$$

When the k units are similar having coefficients a, b, c we get

$$c_e = \frac{c}{k} \quad ... (11.35)$$

$$b_e = b \quad ... (11.36)$$

$$a_e = ka \quad ... (11.37)$$

11.9 SOLUTION OF COORDINATION EQUATIONS

Equation (11.30) gives the optimal generation of a generation unit. This equation is

$$P_n = \frac{\lambda - b_n}{2c_n} \quad ... (11.38)$$

The set of equations (11.38) are called coordination equations (These equations donot involve transmission losses. Coordinations equations including transmission losses will be solved in next sections).

Eq. (11.38) involves λ which is given by the Eq. (11.29).

Three methods of solution are available :

(*a*) analytical method

(*b*) graphical method

(*c*) iterative technique using the gradient method

(*a*) **Analytical method :** In this method the value of λ is found from Eq. (11.29). Using this value of λ, the optimum generations P_1, P_2........ are found from Eq. (11.38).

(*b*) **Graphical method :** In this method the incremental cost of each generator is plotted on the same graph as shown in Fig. 11.7. Various values of λ are tried till ΣP_n = total load to be supplied P_D. If ΣP_n is less than P_D we increase λ. If ΣP_n is more than P_D we reduce λ. Thus the horizontal dotted line in Fig. 11.8 is moved up and down till a point is reached where the value of λ is such that $\Sigma P_n = P_D$. Generally MATLAB is used to plot the graph.

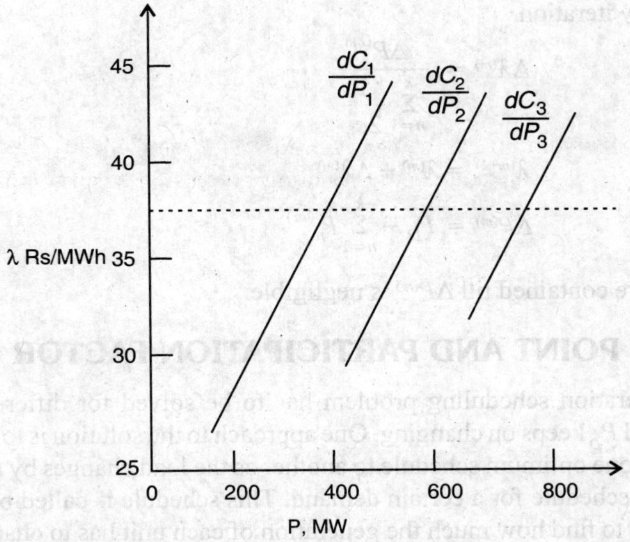

Fig. 11.7. Graphical Method to calculate optimum Generation

(*c*) Iterative Technique using Gradient Method

We can write Eq. (11.38) as

$$P_1 + P_2 + \ldots\ldots + P_k = \sum_{n=1}^{k} \frac{\lambda - b_n}{2C_n} = P_D \quad \ldots(11.39)$$

or

$$f(\lambda) = P_D \quad \ldots(11.40)$$

Assume that λ^0 is the initial estimate of the unknown λ and $\Delta\lambda^0$ is the correction which when applied to λ^0 gives the correct solution. In other words $\lambda^0 + \Delta\lambda^0$ is the solution of Eq (11.40). Expanding Eq. (11.40) into Taylor's series and neglecting the higher order terms we get

$$f(\lambda^0) + \Delta(\lambda^0)\left[\frac{df(\lambda)}{d\lambda}\right]^0 = P_D$$

or

$$\Delta(\lambda^0) = \frac{\Delta P^0}{\left[\frac{df(\lambda)}{d\lambda}\right]^0} = \frac{\Delta P^0}{\Sigma\left[\frac{dP_n}{d\lambda}\right]^0}$$

or

$$\Delta(\lambda^0) = \frac{\Delta P^0}{\sum_{n=1}^{k} \frac{1}{2c_n}} \quad \ldots(11.41)$$

where $\left[\frac{df(\lambda)}{d\lambda}\right]^0$ is the value of $\frac{df(\lambda)}{d\lambda}$ when $\lambda = \lambda_0$

The correction $\Delta(\lambda^0)$ should take us from an initial guess λ^0 to the exact value of λ. Since we have made an approximation in calculating $\Delta(\lambda^0)$, by neglecting higher order terms, we make repeated use of Eq. (11.41) till the solution converges.

In general in any iteration

$$\Delta\lambda^{(m)} = \frac{\Delta P^{(m)}}{\sum_{n=1}^{k} \frac{1}{2c_n}} \quad ...(11.42)$$

$$\lambda^{(m+1)} = \lambda^{(m)} + \Delta\lambda^{(m)} \quad ...(11.43)$$

and

$$\Delta P^{(m)} = P_D - \sum_{n=1}^{k} P_n^{(m)} \quad ...(11.44)$$

The iterations are contained till $\Delta P^{(m)}$ is negligible.

11.10 BASE POINT AND PARTICIPATION FACTOR

The optimum generation scheduling problem has to be solved for different load conditions because the demand P_D keeps on changing. One approach to the solution is to move the operation of generators from one optimum schedule to another as the load changes by a small amount. We start with a known schedule for a certain demand. This schedule is called base point. Then the calculation is made to find how much the generation of each unit has to change to meet a given increase in demand. This means that the additional amount each generator has to contribute or participate to meet the increase in demand has to be found Fig. 11.8, shows this contribution. P_n denotes generation of nth unit and P_n^0 is the generation of nth unit at base point. The incremental cost corresponding to generation P_n^0 is λ^0. As the generation is increased by ΔP_n, the incremental cost increased by $\Delta\lambda$.

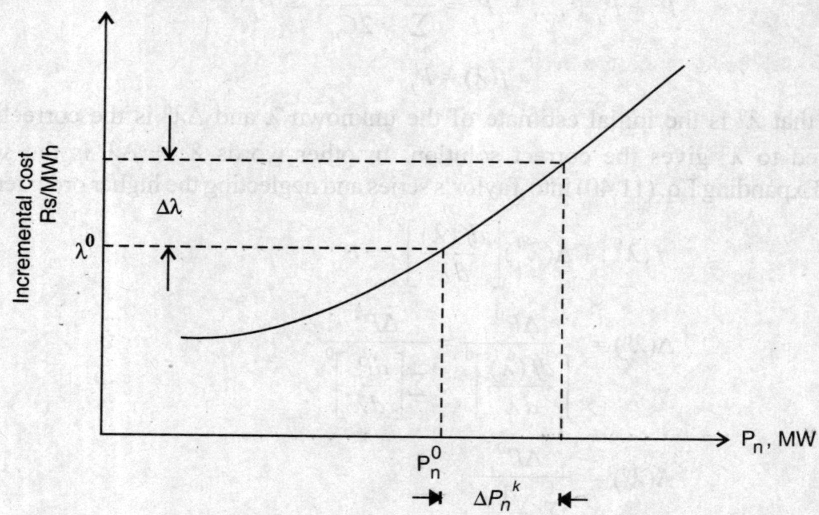

Fig. 11.8 Change in incremental cost due to change in generation.

For a small change in generation

$$\Delta\lambda = (2c_n)\,\Delta P_n \quad (11.45)$$

Eq. (11.45) is valid for each of k units. Thus

$$\Delta P_1 = \frac{\Delta\lambda}{2c_1}$$

$$\Delta P_2 = \frac{\Delta\lambda}{2c_2} \tag{11.46}$$

$$\Delta P_k = \frac{\Delta\lambda}{2c_k}$$

The total change in generation must equal the total change in demand. Thus

$$\Delta P_D = \Delta P_1 + \Delta P_2 + ... + \Delta P_k$$

or
$$\Delta P_D = \Delta\lambda\left[\frac{1}{2c_1} + \frac{1}{2c_2} + ... \frac{1}{2c_k}\right] \tag{11.47a}$$

or
$$\Delta P_D = \Delta\lambda \sum_{n=1}^{k}\left[\frac{1}{2c_n}\right] \tag{11.47 b}$$

Combining Eqs. (11.46) and (11.47*b*) we get

$$\frac{\Delta P_n}{\Delta P_D} = \frac{\left[\dfrac{1}{2c_n}\right]}{\sum\limits_{n=1}^{k}\left[\dfrac{1}{2c_n}\right]} \tag{11.48}$$

The ratio $\frac{\Delta P_n}{\Delta P_D}$ as given by Eq. (11.48) is called Participation Factor because it indicates the extent to which each generator participates to meet the increase in demand.

Example 11.3

The generators in example 11.1 are operating as per equal incremental cost criterion and sharing the total load of 400 MW. Use the participation factors to find increase in generation of each unit for an increase in demand by 50 MW.

Solution

Using Eq. (11.48)

$$\frac{\Delta P_1}{\Delta P_D} = \frac{\dfrac{1}{0.2}}{\dfrac{1}{0.2} + \dfrac{1}{0.15}} = \frac{5}{11.667} = 0.42857$$

$$\frac{\Delta P_2}{\Delta P_D} = \frac{\dfrac{1}{0.15}}{\dfrac{1}{0.2} + \dfrac{1}{0.15}} = \frac{6.6667}{11.6667} = 0.57143$$

Hence
$$\Delta P_1 = 0.42857 \times 50 = 21.42 \text{ MW}$$

$$\Delta P_2 = 0.57143 \times 50 = 28.58 \text{ MW}$$

The total load on the two units would be 221.42 MW and 228.58 MW respectively. It is easy to check that incremental cost will be same for the two units at these loading.

$$\frac{dC_1}{dP_1} = 0.2 \times 221.42 + 30 = 74.28 \text{ Rs/MWh}$$

$$\frac{dC_2}{dP_2} = 0.15 \times 228.58 + 40 = 74.28 \text{ Rs/MWh.}$$

11.11 LOSS OF ECONOMY DUE TO INACCURATE MEASUREMENT OF COST CHARACTERISTICS

It is possible that the cost characteristics of one or more unit have been determined or measured inaccurately, Evidently such an error in representation of cost characteristics will lead to erroneous loading of units and loss of economy.

General equations for calculation of loss of economy in such cases can be derived. However it is always more convenient to calculate the loss of economy by finding the generator loadings in the two cases and hence the difference in costs from basic considerations as illustrated in example 11.14.

11.12 FLOW CHART FOR ECONOMIC DISPATCH NEGLECTING TRANSMISSION LOSSES USING λ ITERATION METHOD

When the number of generators is large, the solution is obtained by a computer program. Fig. 11.9 shows a flow chart for finding economic loading of generators when transmission losses are neglected.

Block 1 : The system data is read. This data includes cost coefficients a_n, b_n, c_n, total system demand P_D, tolerance $\in$, number of generators k, maximum power limit $P_{n, \max}$ and minimum power limit $P_{n, \min}$ for all generators (n = 1 to k.)

Blocks 2 and 3 : After setting iteration count, the program starts by assuming initial value of λ.

Block 4 : Generator loadings P_n are found for the above assumed value of λ for all generators (*i.e.* $n = 1$ to k) using Eq. (11.30).

Blocks 5 and 7 : It is checked whether P_n is less than $P_{n, \max}$. If not, P_n is set equal to $P_{n, \max}$

Blocks 6 and 8 : It is checked whether P_n is more than $P_{n, \min}$. If not, P_n is set equal $P_{n, \min}$.

Block 9 : It is checked whether total generation minus system demand is less, than tolerance $\in$.

Blocks 10, 11 : If balance between total generation and system load is not there, the program computes $\Delta\lambda$ using Eq. (11.42) and updates the value of λ.

Block 12 : The iteration count is increased and program goes back to black 4 for finding new values of generator loadings.

Blocks 13, 14 : The generation cost is calculated using Eq. (11.3) for all generators and results are printed.

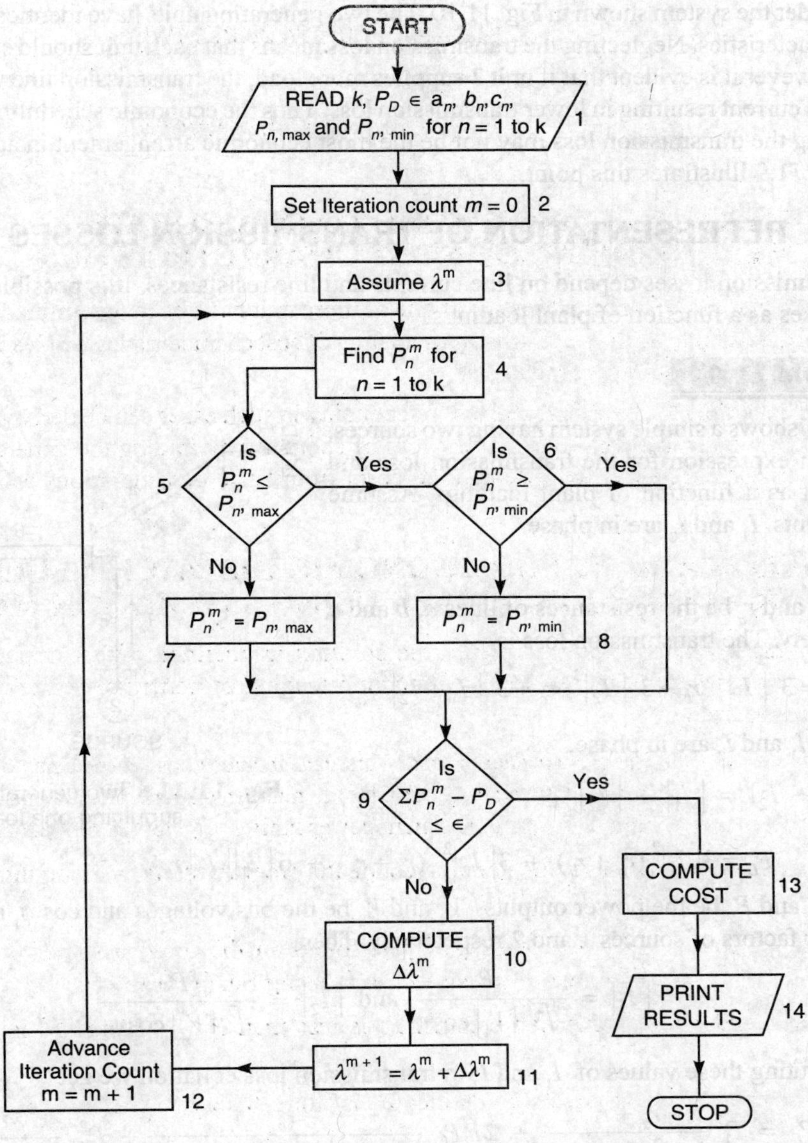

Fig. 11.9 Flow Chart for Economic Dispatch Neglecting Transmission Losses.

ECONOMY LOADING INCLUDING TRANSMISSION LOSSES

11.13 EFFECT OF TRANSMISSION LOSSES

The optimum scheduling for different power stations situated close to each other can be done without considering the losses in the transmission network. However the power stations are situated far from each other, the economic scheduling must take transmission losses into account.

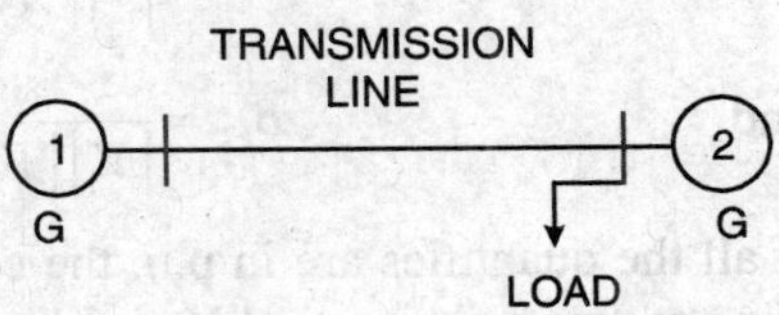

Fig. 11.10 Two generating units connected through transmission line.

Consider the system shown in Fig. 11.10. The two generating units have identical incremental cost characteristics. Neglecting the transmission loss means that each unit should supply half the load. However it is evident that if unit 2 supplies more load, the transmission line would have to carry less current resulting in lower transmission loss. Thus the economic scheduling obtained by neglecting the transmission loss may not be the most economic arrangement in actual practice. Example 11.9 illustrates this point.

11.14 REPRESENTATION OF TRANSMISSION LOSSES

The transmission losses depend on line currents and line resistances. It is possible to represent these losses as a function of plant loadings.

Example 11.4

Fig. 11.11 shows a simple system having two sources. Derive an expression for the transmission loss and express it as a function of plant loadings. Assume that currents I_1 and I_2 are in phase.

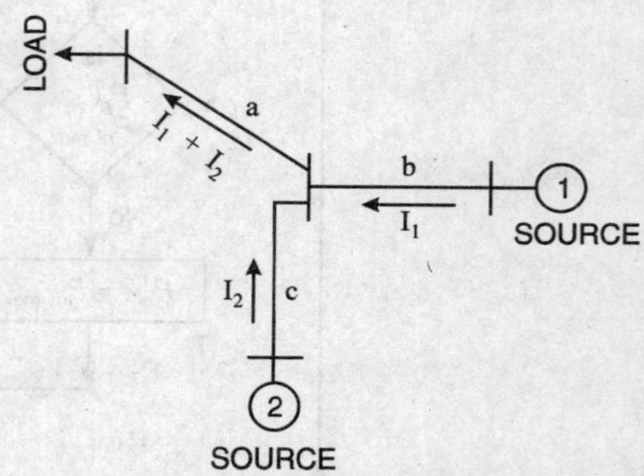

Fig. 11.11 Two generation sources supplying one load

Solution

Let r_a, r_b and r_c be the resistances of lines a, b and c respectively. The transmission loss is

$$P_L = 3\,|I_1|^2\,r_b + 3\,|I_2|^2\,r_c + 3\,|I_1 + I_2|^2\,r_a$$

Since I_1 and I_2 are in phase,

$$|I_1 + I_2| = |I_1| + |I_2|$$

$$\therefore \qquad P_L = 3|I_1|^2(r_a + r_b) + 3|I_2|^2(r_a + r_c) + 6|I_1||I_2|r_a$$

Let P_1 and P_2 be the power outputs, V_1 and V_2 be the bus voltages and cos ϕ_1 and cos ϕ_2 be the power factors of sources 1 and 2 respectively. Then

$$|I_1| = \frac{P_1}{\sqrt{3}\,|V_1|\cos\phi_1} \text{ and } |I_2| = \frac{P_2}{\sqrt{3}\,|V_2|\cos\phi_2}$$

Substituting these values of I_1 and I_2 in transmission loss equation we get

$$P_L = P_1^2\frac{r_a + r_b}{|V_1|^2(\cos\phi_1)^2} + 2P_1P_2\frac{r_a}{|V_1||V_2|(\cos\phi_1)(\cos\phi_2)} + P_2^2\frac{r_a + r_c}{|V_2|^2(\cos\phi_2)^2}$$

or
$$P_L = P_1^2B_{11} + 2P_1P_2B_{12} + P_2^2B_{22}$$

where
$$B_{11} = \frac{r_a + r_b}{|V_1|^2(\cos\phi_1)^2},\quad B_{22} = \frac{r_a + r_c}{|V_2|^2(\cos\phi_2)^2}$$

and
$$B_{12} = \frac{r_a}{|V_1||V_2|(\cos\phi_1)(\cos\phi_2)}$$

If all the quantities are in p.u. the coefficients will be in per unit. However if the voltages are line to line voltages in kV and line resistances are in ohms, the B coefficients will be in reciprocal megawatts.

It is seen that B coefficients depend on source voltages and power factors. The source voltages and power factors depend on and vary with system operating conditions. Nevertheless it has been found that it is sufficiently accurate to find B coefficients for some average operating conditions and use this one set of B coefficients for optimum allocation and scheduling for all the load variations. However extremely wide shifts of loads or major system changes require the calculation of new set of loss coefficients.

Example 11.5

For the system shown in Fig. 11.11, $I_1 = 0.8 \angle 0$ p.u. and $I_2 = 1.0 \angle 0$ p.u. Line impendances are $0.04 + j\,0.12$ p.u. $0.03 + j\,0.1$ p.u. and $0.03 + j\,0.12$ p.u. for the sections a, b and c respectively. The voltage at load bus is $1 \angle 0$ p.u. Find the loss coefficients and the transmission loss.

Solution

$$I_1 + I_2 = 1.8 \angle 0 \text{ p.u.}$$

$$V_1 = 1 \angle 0 + 1.8\,(0.04 + j\,0.12) + 0.8\,(0.03 + j\,0.1)$$

$$= 1.096 + j\,0.296 \text{ p.u.}$$

$$V_2 = 1 \angle 0 + 1.8\,(0.04 + j\,0.12) + 1\,(0.03 + j\,0.12)$$

$$= 1.102 + j\,0.336 \text{ p.u.}$$

$$P_1 = \text{Re}\,[(0.8 + j\,0)\,(1.096 + j\,0.296)] = 0.8768 \text{ p.u.}$$

$$P_2 = \text{Re}\,[(1 + j\,0)\,(1.102 + j\,0.336)] = 1.102 \text{ p.u.}$$

$$\cos \phi_1 = \frac{1.096}{(1.096^2 + 0.296^2)^{0.5}} = 0.965$$

$$\cos \phi_2 = \frac{1.102}{(1.102^2 + 0.336^2)^{0.5}} = 0.957$$

$$B_{11} = \frac{0.04 + 0.03}{\left|(1.096 + j0.296)^2\right|(0.965)^2} = 0.05832 \text{ p.u.}$$

$$B_{22} = \frac{0.04 + 0.03}{\left|(1.102 + j0.336)^2\right|(0.957)^2} = 0.0576 \text{ p.u.}$$

$$B_{12} = \frac{0.04}{|1.096 + j0.296|\,|1.102 + j0.336|(0.965 \times 0.957)} = 0.0331 \text{ p.u.}$$

$$P_L = P_1^2 B_{11} + P_2^2 B_{22} + 2\,P_1 P_2 B_{12}$$

$$= (0.8768)^2(0.05832) + (1.102)^2(0.0576) + 2(0.8768)(0.102)(0.0331)$$

$$= 0.178745 \text{ p.u.}$$

11.15 ECONOMIC SCHEDULING CONSIDERING TRANSMISSION LOSSES

To include the effect of transmission losses in deciding the load allocation, it is necessary to represent the loss as a function of the plant loading. The general form of loss equation is

$$P_L = \sum_m \sum_n P_m B_{mn} \; P_n \qquad \text{... (11.49)}$$

where P_L = Transmission losses in p.u.

P = Plant loading in p.u.

B = Loss coefficients in p.u.

and $\sum_m$ and $\sum_n$ indicate independent summation to include all sources.

For a two generation source system

$$P_L = P_1^2 \, B_{11} + 2P_1 \, P_2 \, B_{12} + P_2^2 \, B_{22} \qquad \text{... (11.50 a)}$$

For a three generation source system

$$P_L = P_1^2 B_{11} + P_2^2 \, B_{22} + P_3^2 \, B_{33} + 2P_1 \, P_2 \, B_{12} + 2P_2 \, P_3 \, B_{23} + 2P_3 \, P_1 \, B_{13} \qquad \text{... (11.50 b)}$$

The matrix representation of the above loss equation is

$$P_L = P^T BP \qquad \text{... (11.51)}$$

where for a total of k sources

$$P = \begin{bmatrix} P_1 \\ P_2 \\ \cdot \\ \cdot \\ \cdot \\ P_k \end{bmatrix} \quad \text{and} \quad B = \begin{bmatrix} B_{11} & B_{12} & \ldots B_{1k} \\ B_{21} & B_{22} & \ldots B_{2k} \\ \cdot & \cdot & \cdot \\ \cdot & \cdot & \cdot \\ \cdot & \cdot & \cdot \\ B_{k1} & B_{k2} & B_{kk} \end{bmatrix}$$

and P^T is the transpose of P.

The B coefficients depend on the system network parameters, configuration, plant power factors and voltages etc. It is generally necessary to make a number of assumptions to simplify the otherwise cumbersome equations. Determination of B coefficients is illustrated through a few simple, yet illustrative, cases in Examples 11.4 to 11.7. For a full treatment of the topic, the reader is referred to the bibliography.

11.16 CO-ORDINATION EQUATIONS

$$C_T = C_1 + C_2 + \ldots + C_k \qquad \text{... (11.10)}$$

$$P_T = P_1 + P_2 + \ldots + P_k \qquad \text{... (11.11)}$$

Assuming that the load P_D at the receiving end is constant

$$P_T - P_L = \text{constant} = P_D \qquad \text{... (11.52)}$$

$$dP_T - dP_L = 0 \qquad \text{... (11.53)}$$

$$dP_T = dP_1 + dP_2 + \ldots + d\,P_k \qquad \text{... (11.54)}$$

Transmission loss P_L is a function of plant outputs

$$\therefore \quad d\,P_L = \frac{\partial P_L}{\partial P_1} . d\,P_1 + \frac{\partial P_L}{\partial P_2} . dP_2 + \ldots + \frac{\partial P_L}{\partial P_k} . d\,P_k \qquad \text{... (11.55)}$$

From Eqs. (11.53), (11.54) and (11.55) we get

$$d\,P_T - d\,P_L = 0$$

$$= \left(1-\frac{\partial P_L}{\partial P_1}\right) dP_1 + \left(1-\frac{\partial P_L}{\partial P_2}\right) d P_2 + \ldots + \left(1-\frac{\partial P_L}{\partial P_k}\right) d P_k \quad \ldots (11.56)$$

or
$$- d P_1 = \frac{1-\frac{\partial P_L}{\partial P_2}}{1-\frac{\partial P_L}{\partial P_1}} . dP_2 + \frac{1-\frac{\partial P_L}{\partial P_3}}{1-\frac{\partial P_L}{\partial P_1}} .dP_3 + \ldots + \frac{1-\frac{\partial P_L}{\partial P_k}}{1-\frac{\partial P_L}{\partial P_1}} .dP_k \quad \ldots (11.57)$$

For C_T to be minimum

$$dC_T = 0 = \frac{\partial C_T}{\partial P_1} .dP_1 + \frac{\partial C_T}{\partial P_2} .dP_2 + \ldots + \frac{\partial C_T}{\partial P_k} .dP_k \quad \ldots (11.58)$$

Substitute the value of $d P_1$ from Eq. (11.57) into Eq. (11.58)

$$\left[\frac{\partial C_T}{\partial P_2} - \frac{1-\frac{\partial P_L}{\partial P_2}}{1-\frac{\partial P_L}{\partial P_1}} \cdot \frac{\partial C_T}{\partial P_1}\right] d P_2 + \left[\frac{\partial C_T}{\partial P_3} - \frac{1-\frac{\partial P_L}{\partial P_3}}{1-\frac{\partial P_L}{\partial P_1}} \cdot \frac{\partial C_T}{\partial P_1}\right] dP_3 +$$

$$\ldots + \left[\frac{\partial C_T}{\partial P_k} - \frac{1-\frac{\partial P_L}{\partial P_k}}{1-\frac{\partial P_L}{\partial P_1}} \cdot \frac{\partial C_T}{\partial P_1}\right] dP_k = 0 \quad \ldots (11.59)$$

This equation is satisfied only if each term is zero. Since the cost of a unit depends on the output of that unit only, the partial derivatives become full derivates *i.e.*, $\frac{\partial C_T}{\partial P_k}$ becomes $\frac{dC_k}{dP_k}$.

$$\therefore \quad \frac{dC_1}{dP_1}\left(\frac{1}{1-\frac{\partial P_L}{\partial P_1}}\right) = \frac{dC_2}{dP_2}\left(\frac{1}{1-\frac{\partial P_L}{\partial P_2}}\right) = \ldots = \frac{dC_k}{dP_k}\left(\frac{1}{1-\frac{\partial P_L}{\partial P_k}}\right) = \lambda \quad \ldots (11.60)$$

i.e.,
$$\frac{dC_k}{dP_k} = \lambda\left(1-\frac{\partial P_L}{\partial P_k}\right)$$

or
$$\frac{dC_k}{dP_k} + \lambda.\frac{\partial P_L}{\partial P_k} = \lambda \quad \ldots (11.61)$$

Also
$$\Sigma P_k - P_D - P_L = 0 \quad \ldots (11.62)$$

The minimum fuel cost for a given total load is obtained by the solution of simultaneous equations represented by Eq. (11.61). λ represents the incremental cost of received power. The k optimum loading equations together with the power balance Eq. (11.62) are sufficient to determine the $k + 1$ unknowns $P_1 \ldots P_k$ and λ. The partial derivative $\frac{\partial P_L}{\partial P_k}$ is referred to as incremental transmission loss $(\text{ITL})_k$ associated with kth generating unit.

For a system with two sources

$$\frac{\partial P_L}{\partial P_1} = 2P_1 B_{11} + 2B_2 B_{12} \quad \ldots (11.63)$$

For a three source system

$$\frac{\partial P_L}{\partial P_2} = 2P_2 B_{22} + 2P_1 B_{12} + 2P_3 B_{23} \quad \text{... (11.64)}$$

For a system with k sources, the transmission loss at nth plant is given by Eq. (11.49).

$$\therefore \quad \frac{\partial P_L}{\partial P_n} = \frac{\partial}{\partial P_n} \sum_{m=1}^{k} \sum_{n=1}^{k} P_m B_{mn} P_n$$

$$= 2 \sum_{m=1}^{k} P_m B_{mn} \quad \text{...(11.65)}$$

The solution of the coordination equations can be obtained by assuming a value of λ. For the assumed value of λ, economy loading of different sources can be obtained through an iterative procedure. By assuming different value of λ, data can be obtained for plotting generation at each source against total generation. Then the transmission loss for each value of λ can be calculated and the generation at different sources can be plotted against the total load or received power.

The iterative procedure for solution of coordination equations becomes very tedious as the number of sources increases. As such the use of digital methods for solution of problems of economic dispatch is preferred.

The transmission losses in systems are generally small. Even when the losses are less than 5%*, the load division between units is considerably affected by the consideration of losses. The allocation of loads between different units, after considering transmission losses, may result in considerable savings as illustrated by example (11.9). Thus the importance of considering transmission losses in computing economic load division cannot be over-stressed.

11.17 PENALTY FACTORS

The coordination equation obtained in section (11.16) is

$$\frac{dC_k}{dP_k}\left(\frac{1}{1-\dfrac{\partial P_L}{\partial P_k}}\right) = \lambda \quad \text{... (11.66)}$$

The factors $\dfrac{1}{1-\dfrac{\partial P_L}{\partial P_1}}, \dfrac{1}{1-\dfrac{\partial P_L}{\partial P_2}}, \ldots \dfrac{1}{1-\dfrac{\partial P_L}{\partial P_k}}$

are called penalty factors of units 1, 2...k and denoted by $L_1, L_2, \ldots L_k$. Then

$$\frac{dC_1}{dP_1} . L_1 = \frac{dC_2}{dP_2} L_2 = \ldots = \frac{dC_k}{dP_k} L_k = \lambda \quad \text{... (11.67)}$$

The penalty factor of kth unit is the ratio of a small change in power at unit k to the small change in the received power when only unit k supplies this small change in the received power.

The solution of Eq. (11.67) leads to results which are identical to the results obtained by solving Eq. (11.61).

* In India the transmission losses are higher than this value.

11.18 ITERATIVE PROCEDURE TO SOLVE COORDINATION EQUATIONS

Since the incremental transmission loss depends on power outputs from all the units, the coordination Eq. (11.61) cannot be solved directly. Instead an iterative procedure involving a method of successive approximations is necessary.

The coordination equation for *n*th unit is

$$\frac{dC_n}{dP_n} + \lambda \frac{\partial P_L}{\partial P_n} = \lambda \quad \text{... (11.68)}$$

Since

$$\frac{dC_n}{dP_n} = 2c_n P_n + b_n \quad \text{... (11.69)}$$

$$\therefore \quad 2c_n P_n + b_n + \lambda \sum_m 2P_m B_{mn} = \lambda \quad \text{... (11.70)}$$

$$\therefore \quad P_n = \frac{1 - \frac{b_n}{\lambda} - \sum\limits_{m \neq n} 2P_m B_{mn}}{\frac{2c_n}{\lambda} + 2B_{mn}} \quad \text{... (11.71)}$$

For a system having two units, the power outputs P_1 and P_2 are given by :

$$P_1 = \frac{1 - \frac{b_1}{\lambda} - 2P_2 B_{12}}{\frac{2c_1}{\lambda} + 2B_{11}} \quad \text{... (11.72)}$$

$$P_2 = \frac{1 - \frac{b_2}{\lambda} - 2P_1 B_{12}}{\frac{2c_2}{\lambda} + 2B_{22}} \quad \text{... (11.73)}$$

The iterative procedure to find P_1 and P_2 for a particular value of λ will run as under

(*i*) Assume $P_2 = 0$

(*ii*) Find P_1 from Eq. (11.72) taking $P_2 = 0$

(*iii*) Find P_2 from Eq. (11.73) using the value of P_1 found in previous step

(*iv*) Find P_1 from Eq. (11.71) using the value of P_2 found in the previous step

(*v*) Repeat steps (*iii*) and (*iv*) one after the other till the values of P_1 and P_2 found in any iteration match the values of P_1 and P_2 found in the previous iteration.

The iterative procedure is further illustrated through example (11.9). The procedure can be extended to cover any number of unit.

Example 11.6

Fig. 11.12 shows a power system consisting of two sources, two loads and a network of 5 branches. Assume that $\frac{\text{resistance}}{\text{reactance}}$ ratio for each of the 5 branches is the same and both the load currents have the same phase angle. Formulate equations for loss coefficients.

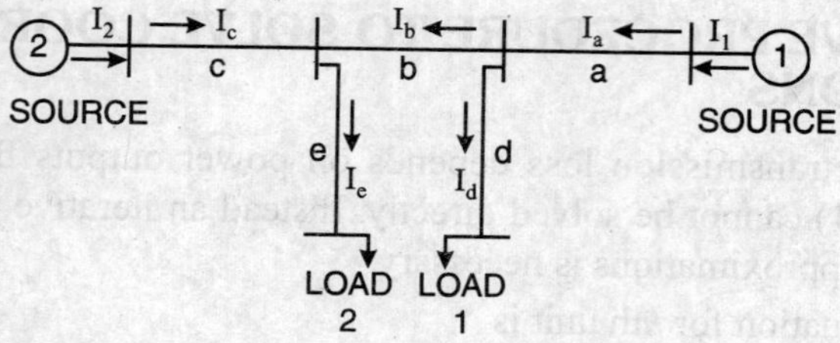

Fig. 11.12 Two generation sources and two loads.

Solution

The currents in the various branches due to sources 1 and 2 can be found by applying the principle of superposition. This is explained in Fig. 11.13.

In Fig. 11.13 (*a*) only source 1 is supplying the total load current I_L. The current in *k*th branch is I_{k1}

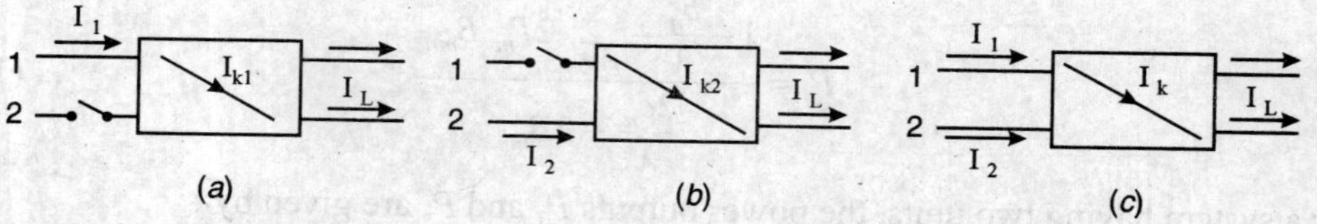

Fig. 11.13 Application of superposition principle to determine total current in *k*th branch of the network.

Let $$N_{k1} = \frac{I_{k1}}{I_L}$$

Similarly in Fig. 11.13 (b) source 2 is supplying the total load current I_L.

Let $$N_{k2} = \frac{I_{k_2}}{I_L}$$

The total current in branch *k*, when both sources are feeding the system, is I_k.

$$I_k = N_{k1}\, I_1 + N_{k2}\, I_2$$

Since $\frac{R}{X}$ ratio for all the branches is the same and both load currents have the same phase angle, the current distribution factors N_{k1} and N_{K2} are real.

Let $$I_1 = |I_1| \cos\theta_1 + j|I_1| \sin\theta_1$$

and $$I_2 = |I_2| \cos\theta_2 + j|I_2| \sin\theta_2$$

θ_1 are θ_1, the phase angles of I_1 and I_2 with respect to a common reference.

$$|I_k|^2 = (N_{k1}\,|I_1|\cos\theta_1 + N_{k2}\,|I_2|\cos\theta_2)^2 + (N_{k1}\,|I_1|\sin\theta_1 + N_{k2}\,|I_2|\sin\theta_2)^2$$

$$= N_{k1}^2\,|I_1|^2 + N_{k2}^2\,|I_2|^2 + 2N_{k1}\,N_{k2}\,|I_1|\,|I_2|\cos(\theta_1 - \theta_2)$$

Let P_1 and P_2 be power outputs, V_1 and V_2 be voltages and $\cos\varphi_1$ and $\cos\varphi_2$ be power factors of sources 1 and 2.

$$|I_1| = \frac{P_1}{\sqrt{3}\,|V_1|\,\cos\varphi_1} \text{ and } |I_2| = \frac{P_2}{\sqrt{3}\,|V_2|\,\cos\varphi_2}$$

The transmission loss P_L is given by :

$$P_L = \sum_k 3 |I_k|^2 r_k$$

$$= \frac{P_1^2}{|V_1|^2 (\cos\varphi_1)^2} \sum_k N_{k1}^2 r_k + \frac{P_2^2}{|V_2|^2 (\cos\varphi_2)^2} \sum_k N_{k2}^2 r_k$$

$$+ \frac{2P_1 P_2 \cos(\theta_1 - \theta_2)}{|V_1|^2 (\cos\varphi_1)(\cos\varphi_2)} \sum_k N_{k1} N_{k2} r_k$$

where Σ indicates summation to include all the branches of the net-work and r_k is the resistance of kth branch.

The loss coefficients are given by

$$B_{11} = \frac{1}{|V_1|^2 (\cos\varphi_1)^2} \sum_k N_{k1}^2 r_k$$

$$B_{22} = \frac{1}{|V_2|^2 (\cos\varphi_2)^2} \sum_k N_{k2}^2 r_k$$

$$B_{12} = \frac{\cos(\theta_1 - \theta_2)}{|V_1||V_2| (\cos\varphi_1)(\cos\varphi_2)} \sum_k N_{k1} N_{k2} r_k$$

Example 11.7

The impedances of the various branches and the currents flowing in the network shown in Fig. 11.12 are as under :

$Z_a = 0.03 + j\,0.09$ pu, $\quad I_a = 1.5 - j\,0.4$ pu

$Z_b = 0.10 + j\,0.30$ pu, $\quad I_b = 0.5 - j\,0.2$ pu

$Z_c = 0.03 + j\,0.09$ pu, $\quad I_c = 1 - j\,0.1$ pu

$Z_d = 0.04 + j\,0.12$ pu, $\quad I_d = 1 - j\,0.2$ pu

$Z_e = 0.04 + j\,0.12$ pu, $\quad I_e = 1.5 - j\,0.3$ pu

The voltage at the bus which is at the junction of branches a, b and d is $1 \angle 0$ p.u. Calculate the loss formula coefficients in p.u. and in reciprocal mega watts if the base is 100 MVA.

Solution

The currents of loads 1 and 2 are 0.4 and 0.6 p.u. of the total load current. Therefore, the current distribution factors are

$N_{a1} = 1,\ N_{b1} = 0.6 \qquad N_{c1} = 0,\ N_{d1} = 0.4 \qquad N_{e1} = 0.6$

$N_{a2} = 0,\ N_{b2} = -0.4 \qquad N_{c2} = 1,\ N_{d2} = 0.4 \qquad N_{e2} = 0.6$

The bus voltages at the two sources are

$$V_1 = 1.0 + (1.5 - j\,0.4)\,(.03 + j.09)$$

$$= 1.081 + j.123 = 1.088 \angle 6.49°$$

$$V_2 = 1.0 - (0.5 - j.2)(.1 + j.\,3) + (1 - j.1)\,(.03 + j.\,09)$$

$$= 0.929 - j\,0.043 = 0.93 \angle -2.65°$$

The phase angles of the source currents are

$$\theta_1 = \tan^{-1}\frac{-0.4}{1.5} = -14.93°$$

$$\theta_2 = \tan^{-1}\frac{-0.1}{1.0} = -5.73°$$

$$\cos(\theta_1 - \theta_2) = \cos(9.2°) = 0.987$$

Power factors at sources 1 and 2 are

$$\cos\varphi_1 = \cos(6.49 + 14.93°) = 0.931$$

$$\cos\varphi_2 = \cos(5.73 - 2.65°) = 0.998$$

Loss coefficients are

$$B_{11} = \frac{1^2 \times 0.3 + .6^2 \times .1 + .4^2 \times 0.4 + .6^2 \times .04}{(1.088)^2(.931)} = 0.07876 \text{ p.u.}$$

$$B_{22} = \frac{.4^2 \times .1 + 1^2 \times .03 + .4^2 \times .04 + .6^2 \times .04}{(.93)^2(.998)} = 0.0774 \text{ p.u.}$$

$$B_{21} = \frac{-.4 \times .6 \times .1 + .4 \times .4 \times .04 + .6 \times .6. \times .03}{1.088 \times .93 \times .931 \times .998} \times (0.987) = -0.00713 \text{ p.u.}$$

For converting the above per unit values, which are to a base of 100 MVA, to actual values the per unit values have to be divided by 100.

$$B_{11} = \frac{0.07876}{100} = 7.876 \times 10^{-4} \text{ MW}^{-1}$$

$$B_{22} = \frac{0.0774}{100} = 7.74 \times 10^{-4} \text{ MW}^{-1}$$

$$B_{12} = \frac{-0.0713}{100} = -0.713 \times 10^{-4} \text{ MW}^{-1}$$

Example 11.8

A system consists of two plants connected by a transmission line as shown in Fig. 11.10. The load is at plant 2. The transmission line loss calculations reveal that a transfer of 100 MW from plant 1 to plant 2 means a loss of 15 MW. Find the required generation at each plant for $\lambda = 60$. Assume that the incremental costs of the two plants are given by :

$$\frac{dC_1}{dP_1} = 0.2\,P_1 + 22 \text{ Rs./MWh}$$

$$\frac{dC_2}{dP_2} = 0.15 P_2 + 30 \text{ Rs./MWh}$$

Solution

The load is at the bus of plant 2. Evidently transmission loss is not affected by variation of P_2.

$\therefore$ $\quad B_{22} = B_{12} = 0$

when $\quad P_1 = 100$ MW, $P_L = 15$ MW

$\therefore$ $\quad 15 = B_{11}(100)^2$

or $\quad B_{11} = 0.0015$ MW^{-1}

$$\mathbf{P_L = 0.0015\,P_1^2}$$

$$\frac{\partial P_L}{\partial P_1} = 0.003\, P_1$$

$$\frac{\partial P_L}{\partial P_2} = 0$$

$$L_1 \text{ i.e., Penalty factor of plant } 1 = \frac{1}{1-\frac{\partial P_L}{\partial P_1}} = \frac{1}{1-0.003\, P_1}$$

L_2 *i.e.*, Penalty factor of plant 2 = 1

The generation at each plant is required to be calculated for $\lambda = 60$

$$\frac{dC_1}{dP_1} L_1 = \lambda$$

$$\frac{dC_2}{dP_2} L_2 = \lambda$$

$$\therefore \quad \frac{0.2\, P_1 + 22}{1-0.003\, P_1} = 60 \text{ or } P_1 = 100 \text{ MW}$$

and $\quad 0.15\, P_2 + 30 = 60 \quad \text{or} \quad P_2 = 200 \text{ MW}$

$$\text{Total load} = P_1 + P_2 - B_{11}\, P_1^2 = 285 \text{ MW.}$$

Example 11.9

For the system of example 11.8 find the saving in Rs. per hour by scheduling the generation by considering the transmission loss rather than neglecting the transmission loss in determining the outputs of the two generators. Assume a load of 285 MW connected at bus of plant 2.

Solution

The optimum generation found in example 2.8 (transmission loss included) is

$$P_1 = 100 \text{ MW } P_2 = 200 \text{ MW}$$

If transmission loss is neglected in determining scheduling

$$0.2P_1 + 22 = 0.15\, P_2 + 30$$

and $\quad P_1 + P_2 - 0.0015\, P_1^2 = 285$

Solution of these two equations for P_1 and P_2 gives

$$P_1 = 161.80 \text{ MW} \quad P_2 = 162.50 \text{ MW}$$

The load on plant 1 is increased from 100 MW to 161.80 MW. Additional cost is

$$\int_{100}^{161.80} (0.2\, P_1 + 22)\, dP_1 = \left|0.1\, P_1^2 + 22P_1\right|_{100}^{161.80} = \text{Rs. } 2977.50$$

The load on plant 2 is decreased from 200 MW to 162.50 MW. The change in cost is

$$\int_{200}^{162.50} (0.15\, P_2 + 30)\, dP_2 = \left|0.075\, P_2^2 + 30P_2\right|_{200}^{162.5} = \text{Rs.} - 2144.50$$

Net change in cost = Rs. 833 per hour

Thus scheduling the generation by taking transmission losses into account would mean a saving of Rs. 833 per hour in fuel cost.

Example 11.10

For a simple two unit system, the loss coefficients are

$$B_{11} = 0.001 \text{ MW}^{-1}, B_{12} = -0.0005 \text{ MW}^{-1}$$

$$B_{22} = 0.0024 \text{ MW}^{-1}$$

The incremental production costs of the two units are

$$\frac{dC_1}{dP_1} = 0.08\, P_1 + 16 \text{ Rs./MW-hour}$$

$$\frac{dC_2}{dP_2} = 0.08\, P_2 + 12 \text{ Rs./MW-hour}$$

Find the generations P_1 and P_2 for $\lambda = 20$

Also compute the transmission loss and received power.

Solution

Substituting the different values in Eqs. (11.72) and (11.73)

$$P_1 = \frac{1 - 0.8 + 0.001\, P_2}{0.004 + 0.002} = \frac{0.2 + 0.001\, P_2}{0.006}$$

$$P_2 = \frac{1 - 0.6 + 0.001\, P_1}{0.004 + 0.0048} = \frac{0.4 + 0.001\, P_1}{0.0088}$$

The values of P_1 and P_2 can be found from the above equations. When the number of equations is large, an iterative procedure is preferable.

As explained in section (11.18), iteration is started by setting $P_2 = 0$. The successive results are shown in the following table.

Iteration Number	*Results*
1	$P_1 = \frac{0.2 + 0}{0.006} = 33.3$
	$P_2 = \frac{0.4 + .0333}{0.0088} = 49.2$
2	$P_1 = \frac{0.2 + 0.0492}{0.006} = 41.5$
	$P_2 = \frac{0.4 + 0.0415}{0.0088} = 50.2$
3	$P_1 = \frac{0.2 + 0.0502}{0.006} = 41.7$
	$P_2 = \frac{0.4 + 0.0417}{0.0088} = 50.2$
4	$P_1 = \frac{0.2 + 0.0502}{0.006} = 41.7$

$$P_2 = \frac{0.4 + 0.0417}{0.0088} = 50.2$$

Thus $\quad P_1 = 41.7$ MW, $P_2 = 50.2$ MW

$$P_L = B_{11}\,P_1^2 + 2B_{12}P_1\,P_2 + B_{22}\,P_2^2$$

$$= 0.001\,(41.7)^2 + 2(-0.0005)(41.7)(50.2) + 0.0024(50.2)^2$$

$$= 5.7 \text{ MW}$$

$$\text{Power received} = 41.7 + 50.2 - 5.7 = 86.2 \text{ MW}$$

Example 11.11

Two generating units of a thermal station have cost characteristics as under:

$$C_1 = 561 + 7.92\,P_1 + 0.001562\,P_1^2 \text{ Rs./hour}$$

$$C_2 = 310 + 7.85\,P_2 + 0.00194\,P_2^2 \text{ Rs./hour}$$

Obtain the cost characteristics of the composite unit for a total demand P_T.

Solution

Using Eqs. (11.33 to 11.34)

$$\frac{1}{c_e} = \frac{1}{c_1} + \frac{1}{c_2} = \frac{1}{0.001562} + \frac{1}{0.00194}$$

or $\quad c_e = 8.653 \times 10^{-4}$

$$\frac{b_e}{c_e} = \frac{b_1}{c_1} + \frac{b_2}{c_2} = \frac{7.92}{0.001562} + \frac{7.85}{0.00194} = 9116.814$$

or $\quad b_e = 9116.814 \times 8.653 \times 10^{-4}$ or $b_e = 7.8888$

$$a_e = a_1 - \frac{b_1^2}{4c_1} + a_2 - \frac{b_2^2}{4c_2} + \frac{b_e^2}{4c_e}$$

$$= 561 - \frac{7.92^2}{4 \times 0.001562} + 310 - \frac{7.85^2}{4 \times 0.00194} + \frac{7.8888^2}{4 \times 8.653 \times 10^{-4}}$$

$$= 561 - 10039.437 + 310 - 7941.044 + 17980.23$$

or $\quad a_e = 870.753$

Therefore the cost characteristics of the composite unit for a demand P_T is

$$C_T = 870.753 + 7.8888\,P_T + 8.653 \times 10^{-4}\,P_T^2$$

Example 11.12

The fuel cost characteristics of two thermal plants are as under

$$C_1 = 7700 + 52.8\,P_1 \times 5.5 \times 10^{-3}\,P_1^2 \text{ Rs./hour}$$

$$C_2 = 2500 + 15\,P_2 + 0.05\,P_2^2 \text{ Rs./hour}$$

The limits of generation for the two units are $200 \leq P \leq 800$ MW. The load curve is shown in Fig. 11.14. Find the daily operating schedule to minimize the operating costs. The cost of taking a unit off and then putting it on is Rs. 1000.00.

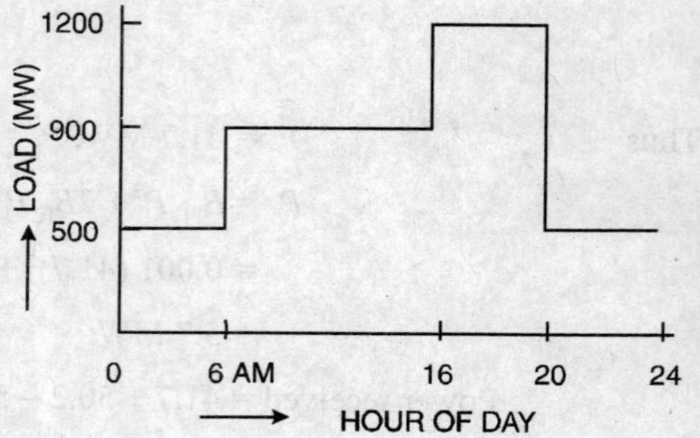

Fig. 11.14.

Solution

For 1200 MW load period it is necessary to run both units

$$\frac{dC_1}{dP_1} = 52.8 + 11 \times 10^{-3} P_1 = \lambda$$

$$\frac{dC_2}{dP_2} = 15 + 0.1 P_2 = \lambda$$

Let the loads be P_1 and $1200 - P_1$. For economic loading incremental costs should be equal

$$52.8 + 11 \times 10^{-3} P_1 = 15 + 0.1 (1200 - P_1)$$

The solution is

$$P_1 = 740.54 \text{ MW}, \quad P_2 = 459.46 \text{ MW}$$

For 900 MW load also it is necessary to run both units. Let loads be P_1 and $(900 - P_1)$.

Then $\quad 52.8 + 11 \times 10^{-3} P_1 = 15 + 0.1 (900 - P_1)$

The solution is

$$P_1 = 470.27 \text{ MW and } P_2 = 429.73 \text{ MW}$$

For the 500 MW load, let both units be run. Then loads are P_1 and $(500 - P_1)$. We get

$$52.8 + 11 \times 10^{-3} P_1 = 15 + 0.1(500 - P_1)$$

The result is

$$P_1 = 109.9 \text{ MW}, \quad P_2 = 390.1 \text{ MW}$$

However the minimum load on any unit is 200 MW. Therefore

$$P_1 = 200 \text{ MW}, \quad P_2 = 300 \text{ MW}$$

The operating cost for the 500 MW load period is

$$C_1 = 7700 + 52.8 \times 200 + 5.5 \times 10^{-3} \times 200^2 = \text{Rs. } 18480 \text{ per hour}$$

$$C_2 = 2500 + 15 \times 300 + 0.05 (300)^2 = \text{Rs. } 11500 \text{ per hour}$$

Total operating cost for the 10 hour period of 500 MW is $10(C_1 + C_2)$ and equals Rs.299800.

The other option is to run only one unit during 500 MW load period.

It is easy to see that it is cheaper to run unit 2. Then the cost for 10 hour period is

$$\text{Rs. } 10 [2500 + 15(500) + (0.05) (500)^2] = \text{Rs. } 225000$$

Add the cost of shutting down and starting unit 1 *i.e.* Rs. 1000. Therefore, total cost for 500 MW load duration is Rs. 225000 + Rs. 1000 or Rs. 226000. As compared to this, the cost of operation if both units are operating is Rs. 299800. Hence it is economical to run only unit 2 during 500 MW load. Hence the complete operating schedule is

1200 MW load : $P_1 = 740.54$ MW, $P_2 = 459.46$ MW
900 MW load : $P_1 = 470.27$ MW, $P_2 = 429.73$ MW
500 MW load : Shut down unit 1, $P_2 = 500$ MW

Example 11.13

Two areas are interconnected as shown in Fig. 11.15. Total generating capacity in each is 600 MW. The cost characteristics are

$$C_1 = 2000 + 20\,P_1 + 0.05\,P_1^2 \quad \text{Rs./hour}$$
$$C_2 = 2750 + 26\,P_2 + 0.03091\,P_2^2 \quad \text{Rs./hour}$$

Find the cost of generation if (*a*) each area supplies its own load (*b*) units are loaded as per economic scheduling. Assume that line has unlimited capacity and transmission losses are negligible. Also find the saving and tie line loading.

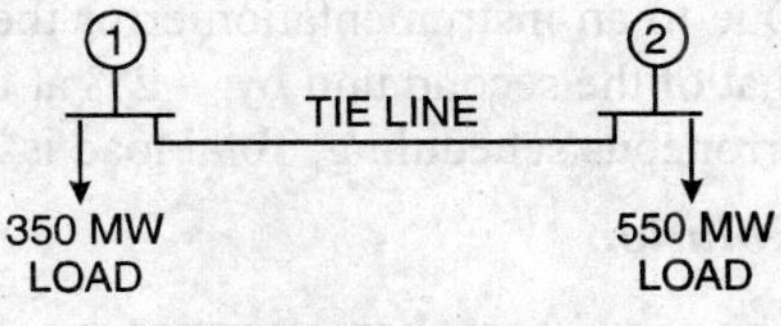

Fig. 11.15.

Solution

(*a*)
$$P_1 = 350 \text{ MW}, \quad P_2 = 550 \text{ MW}$$
$$C_1 = 2000 + 20(350) + 0.05(350)^2 = \text{Rs. } 15125 \text{ per hour}$$
$$C_2 = 2750 + 26\,(550) + 0.03091\,(550)^2 = \text{Rs. } 26400.275 \text{ per hour}$$
$$\text{Total cost} = C_1 + C_2 = \text{Rs. } 41525.275 \text{ per hour}$$

(*b*)
$$\frac{dC_1}{dP_1} = 20 + 0.1\,P_1 \qquad \frac{dC_2}{dP_2} = 26 + 0.06182\,P_2$$

For economic scheduling

$$\frac{dC_1}{dP_1} = \frac{dC_2}{dP_2}$$

Total load is 350 + 550 = 900 MW

Let loads be P_1 and $900 - P_1$. Then

$$20 + 0.1\,P_1 = 26 + 0.06182\,(900 - P_1)$$

or
$$P_1 = 380.9 \text{ MW}, \quad P_2 = 519.1 \text{ MW}$$
$$C_1 = 2000 + 20(380.9) + 0.05\,(380.9)^2$$
$$= \text{Rs. } 16872.24 \text{ per hour}$$
$$C_2 = 2750 + 26(519.1) + 0.03091\,(519.1)^2$$
$$= \text{Rs. } 24575.76 \text{ per hour}$$
$$\text{Total cost} = C_1 + C_2 = \text{Rs. } 41448 \text{ per hour}$$
$$\text{Saving} = \text{Rs. } 41525.275 - 41448$$
$$= \text{Rs. } 77.275 \text{ per hour}$$

$$\text{Tie line load} = 380.9 - 350$$
$$= 30.9 \text{ MW from Plant 1 to Plant 2}$$

Example 11.14

Two units of a thermal station have each the following cost characteristics

$$C = 5000 + 450\,P + 0.5\,P^2 \text{ Rs./hour}$$

Due to an instrumentation error the cost characteristics of first unit is in error by + 2% and that of the second unit by – 2% at the time of scheduling. Find the extra operating cost due to erroneous scheduling. Total load is 200 MW.

Solution

The actual cost characteristics are

$$C_1 = (5000 + 450\,P_1 + 0.5\,P_1^2)\,0.98$$

or
$$C_1 = 4900 + 441\,P_1 + 0.49\,P_1^2$$

$$C_2 = (5000 + 450\,P_2 + 0.5\,P_2^2)\,1.02$$
$$= 5100 + 459\,P_2 + 0.51\,P_2^2$$

$$P_1 + P_2 = 200 \text{ or } P_2 = 200 - P_1$$

$$\frac{dC_1}{dP_1} = \frac{dC_2}{dP_2}$$

or
$$441 + 0.98\,P_1 = 459 + 1.02\,P_2 = 459 + 1.02\,(200 - P_1)$$

The solution is $P_1 = 111$ MW, $P_2 = 89$ MW

If there were no instrumentation error the total operating cost would have been.

$$C = C_1 + C_2 = 4900 + 44\,(111) + 0.49\,(111)^2 + 5100 + 459\,(89) + 0.51\,(89)^2$$
$$= \text{Rs. } 109879 \text{ per hour}$$

Due to instrumentation error, the cost characteristics have been taken as similar. Therefore erroneous scheduling is 100 MW load on each unit. For this loading the operating cost is

$$C = C_1 + C_2 = 4900 + 441\,(100) + 0.49\,(100)^2 + 5100 + 459\,(100) + 0.51\,(100)^2$$
$$= \text{Rs. } 110000 \text{ per hour}$$

Extra operating cost due to erroneous scheduling = Rs. 110000 – Rs. 109879

$$= \text{Rs. } 121 \text{ per hour}$$

Example 11.15

A power plant has 3 units with the following input output curves

$$Q_1 = 0.002\,P_1^2 + 0.86\,P_1 + 20 \text{ tons/hour}$$
$$Q_2 = 0.004\,P_2^2 + 1.08\,P_2 + 20 \text{ tons/hour}$$
$$Q_3 = 0.0028\,P_3^2 + 0.64\,P_3 + 36 \text{ tons/hour}$$

Fuel cost is Rs. 500 per ton. Maximum and minimum generation level for each unit is 120 MW and 36 MW. Find optimum scheduling for a total load of 200 MW.

Solution

The cost curves are

$$C_1 = 500\ Q_1 = P_1^2 + 430\ P_1 + 10000 \text{ Rs./hour}$$

$$C_2 = 500\ Q_2 = 2\ P_2^2 + 540\ P_2 + 10000 \text{ Rs./hour}$$

$$C_3 = 500\ Q_3 = 1.4\ P_3^{\ 2} + 320\ P_3 + 18000 \text{ Rs./hour}$$

The incremental costs are

$$\frac{dC_1}{dP_1} = 2P_1 + 430 \text{ Rs./MWh}$$

$$\frac{dC_2}{dP_2} = 4P_2 + 540 \text{ Rs./MWh}$$

$$\frac{dC_3}{dP_3} = 2.8P_3 + 320 \text{ Rs./MWh}$$

For minimum overall cost

$$\frac{dC_1}{dP_1} = \frac{dC_2}{dP_2} = \frac{dC_3}{dP_3}$$

i.e., $2P_1 + 430 = 4P_2 + 540 = 2.8\ P_3 + 320$ ... (*i*)

and $P_1 + P_2 + P_3 = 200$ ... (*ii*)

Solving (*i*) and (*ii*) we get

$$P_1 = 85 \text{ MW}, \quad P_2 = 15 \text{ MW}, \quad P_3 = 100 \text{ MW}$$

The load on generator 2 is less than the minimum level of 36 MW. Therefore

$P_2 = 36$ MW ...(*iii*)

$P_1 + P_3 = 200 - 36 = 164$ MW ...(*iv*)

$$2P_1 + 430 = 2.8\ P_3 + 320$$

Solving (*iii*) and (*iv*) we get

$$P_1 = 72.75 \text{ MW}, P_3 = 91.25 \text{ MW}$$

Example 11.16

The input output curve of a 30 MW power station is as under.

$$I = (32 + 32\ L + 1.68\ L^2)\ 10^5$$

where I is heat input in *k* J/h and L is load in MW. (*a*) Find heat input per day if the plant works at full load for 18 hours and half load for the remaining period (*b*) find the savings in heat input per kWh of energy produced if the plant works at uniform load for all the 24 hours generating the same amount of energy.

Solution

(*a*) For full load condition for 18 hours

$$\text{Input } I_1 = [32 + 32 \times 30 + 1.68\ (30)^2]\ 10^5 \times 18$$
$$= 45072 \times 10^5 \text{ kJ}$$

For half load condition for 6 hours

$$\text{Input } I_2 = [32 + 32 \times 15 + 1.68(15)^2]10^5 \times 6 = 5340 \times 10^5 \text{ kJ}$$

$$\text{Total heat input} = I_1 + I_2 = (45072 + 5340)\ 10^5 = 50412 \times 10^5 \text{ kJ.}$$

(*b*) Total energy produced in 24 hours

$$= 30 \times 18 + 15 \times 6 = 630 \text{ MWh}$$

$$\text{Uniform load} = \frac{630}{24} = 26.25 \text{ MW}$$

$$\text{Heat input} = [32 + 32 \times 26.25 + 1.68\ (26.25)^2]\ 10^5 \times 24 \text{ kJ}$$

$$= 48711 \times 10^5 \text{ kJ}$$

$$\text{Saving in heat energy} = (50412 - 48711)\ 10^5 \text{ kJ} = 1701 \times 10^5 \text{ kJ}$$

Saving in heat energy per kWh

$$= \frac{1701 \times 10^5}{630 \times 10^3} = 270 \text{ kJ/kWh}$$

Example 11.17

(*a*) How do the equations for economic operation of thermal plant get modified if the minimum and maximum loadings of generators are specified (*b*) A system has 3 generators whose cost curves can be described by the following equations.

$$C_1 = 450 + 6.5\ P_1 + 0.0013\ P_1^2 \qquad \text{Rs/hour}$$
$$C_2 = 300 + 7.8\ P_2 + 0.0019\ P_2^2 \qquad \text{Rs/hour}$$
$$C_3 = 80 + 8.1\ P_2 + 0.005\ P_3^2 \qquad \text{Rs/hour}$$

where C_1, C_2, C_3 are the costs and P_1, P_2, P_3 are outputs in MW of units 1,2,3 respectively.

If total load of 800 MW is to be shared by the three units, find optimum scheduling (*c*) For the above three generator system, the maximum and minimum outputs are specified as under:

Unit 1	Max. output	600 MW
	Min. output	100 MW
Unit 2	Max. output	400 MW
	Min. output	50 MW
Unit 3	Max. output	200 MW
	Min. output	50 MW

Find optimum scheduling.

Solution

(*a*) When the maximum and minimum loadings, of different generating units are specified, the equations for economic operation get modified as under

$$\frac{dC_i}{dP_i} = \lambda \qquad \text{for } P_{i,\,min} < P_i < P_{i,\,max}$$

$$\frac{dC_i}{dP_i} \leq \lambda \qquad \text{for } P_i = P_{i,\,max}$$

$$\frac{dC_i}{dP_i} \geq \lambda \qquad \text{for } P_i = P_{i, min}$$

(*b*)
$$\frac{dC_1}{dP_1} = 6.5 + 0.0026\ P_1 = \lambda \qquad \text{Rs/MWh} \qquad ...(i)$$

$$\frac{dC_2}{dP_2} = 7.8 + 0.0038\ P_2 = \lambda \qquad \text{Rs/MWh} \qquad ...(ii)$$

$$\frac{dC_3}{dP_3} = 8.1 + 0.01\ P_3 = \lambda \qquad \text{Rs/MWh} \qquad ...(iii)$$

$$P_1 + P_2 + P_3 = 800 \qquad ...(iv)$$

Solution of Eqns (*i*) to (*iv*) gives

$$P_1 = 669.73 \text{ MW}$$

$$P_2 = 116.13 \text{ MW}$$

$$P_3 = 14.14 \text{ MW}$$

(*c*) The results of part (*b*) indicate that output P_2 is within the maximum and minimum loading limits of unit 2. However P_1, P_3 are outside the limits. Therefore set the output of unit equal to $P_{1, max}$ and output of unit 3 equal to $P_{3, min}$. Then the dispatch becomes

$$P_1 = 600 \text{ MW},\ P_2 = 150 \text{ MW},\ P_3 = 50 \text{ MW}$$

λ must be equal to incremental cost of unit 2 since it is not at its maximum and minimum limit.

$$\lambda = \frac{dC_2}{dP_2} = 7.8 + 0.0038 \times 150 = 8.37 \text{ Rs./MWh}$$

Then
$$\frac{dC_1}{dP_1} = 6.5 + .0026 \times 600 = 8.06 \text{ Rs/MWh}$$

and
$$\frac{dC_3}{dP_3} = 8.1 + 0.01 \times 50 = 8.6 \text{ Rs/MWh}$$

It is seen that $\frac{dC_1}{dP_1}$ is less than λ and $\frac{dC_3}{dP_3}$ is more than λ. Thus all conditions mentioned in part

(*a*) are satisifed. Hence $P_1 = 600$ MW, $P_2 = 150$ MW, $P_3 = 50$ MW.

11.19 GENERAL TRANSMISSION LOSS FORMULA

Eq. (11.49) expresses transmission loss P_L in terms of plant loadings and *B* coefficients. *A* more general formula containing a linear term and a constant term is as under. It is known as Kron's loss formula.

$$P_L = \sum_m \sum_n P_m\, B_{mn}\, P_n + \sum_n B_{no}\, P_n + B_{oo} \qquad ...(11.74\ a)$$

The first term in Eq. (11.74) is the same as in Eq. (11.49). The second term is the linear term and the third term is the constant term.

Eq. (11.74) can be written in matrix form as under

$$P_L = \begin{bmatrix} P_1 \; P_2 \; .... \; P_k \end{bmatrix} \begin{bmatrix} B_{11} & B_{12}.....B_{1k} \\ B_{21} & B_{22}.....B_{2k} \\ \vdots & \vdots & \vdots \\ B_{k1} & B_{k2}.....B_{kk} \end{bmatrix} \begin{bmatrix} P_1 \\ P_2 \\ \vdots \\ P_k \end{bmatrix} + \begin{bmatrix} P_1 \; P_2 \; ..... \; P_k \end{bmatrix} \begin{bmatrix} B_{10} \\ B_{20} \\ \vdots \\ B_{k0} \end{bmatrix} + B_{00} \quad ... (11.74b)$$

Example 11.18

In a system having 3 plants, the loss coefficients are as under:

$$B_{mn} = \begin{bmatrix} 0.0676 & 0.00953 & -0.00507 \\ 0.00953 & 0.0521 & 0.00901 \\ -0.00507 & 0.00901 & 0.0294 \end{bmatrix}$$

$$B_{no} = \begin{bmatrix} -0.0766 \\ -0.00342 \\ 0.0189 \end{bmatrix} \text{ and } B_{oo} = 0.040357$$

If $P_1 = 107.9$ MW, $P_2 = 50.0$ MW and $P_3 = 60.0$ MW find transmission loss.

Solution

Substituting the value in Eq. (11.74 *b*) we get

$$P_L = [107.9 \quad 50 \quad 60] \begin{bmatrix} +0.0676 & 0.00953 & -0.00507 \\ 0.00953 & 0.0521 & 0.00901 \\ -0.00507 & 0.00901 & 0.00294 \end{bmatrix}$$

$$+ [107.9 \quad 50 \quad 60] \begin{bmatrix} -0.0766 \\ -0.00342 \\ 0.0189 \end{bmatrix} + 0.040357$$

Carrying out the above matrix multiplication and adding the terms we get $P_L = 7.878$ MW.

It is important to note that if linear term and constant term are neglected and P_L is calculated using Eq. (11.49) the inaccuracy is only 0.1%. Thus P_L can be calculated sufficiently accurately by using Eq. (11.49).

11.20 FLOW CHART FOR ECONOMIC DISPATCH INCLUDING TRANSMISSION LOSSES USING λ ITERATION

Fig. 11.16 shows a flow chart for finding economic loading of generators when transmission losses are considered.

Block 1 : The system data is read. This data includes, number of generators *k*, total system demand P_D, tolerance $\in$, cost coefficients a_n, b_n, c_n, maximum power limit $P_{n,\,max}$ and minimum power limit $p_{n,\,min}$ for all generators ($n = 1$ to k) and $[B]$.

Block 2 : Compute P_1, P_2 ... P_k by neglecting transmission losses. This step is necessary since transmission losses depend on plant loading, P_1, P_2 ... P_k. An initial guess of plant loadings is made by neglecting losses.

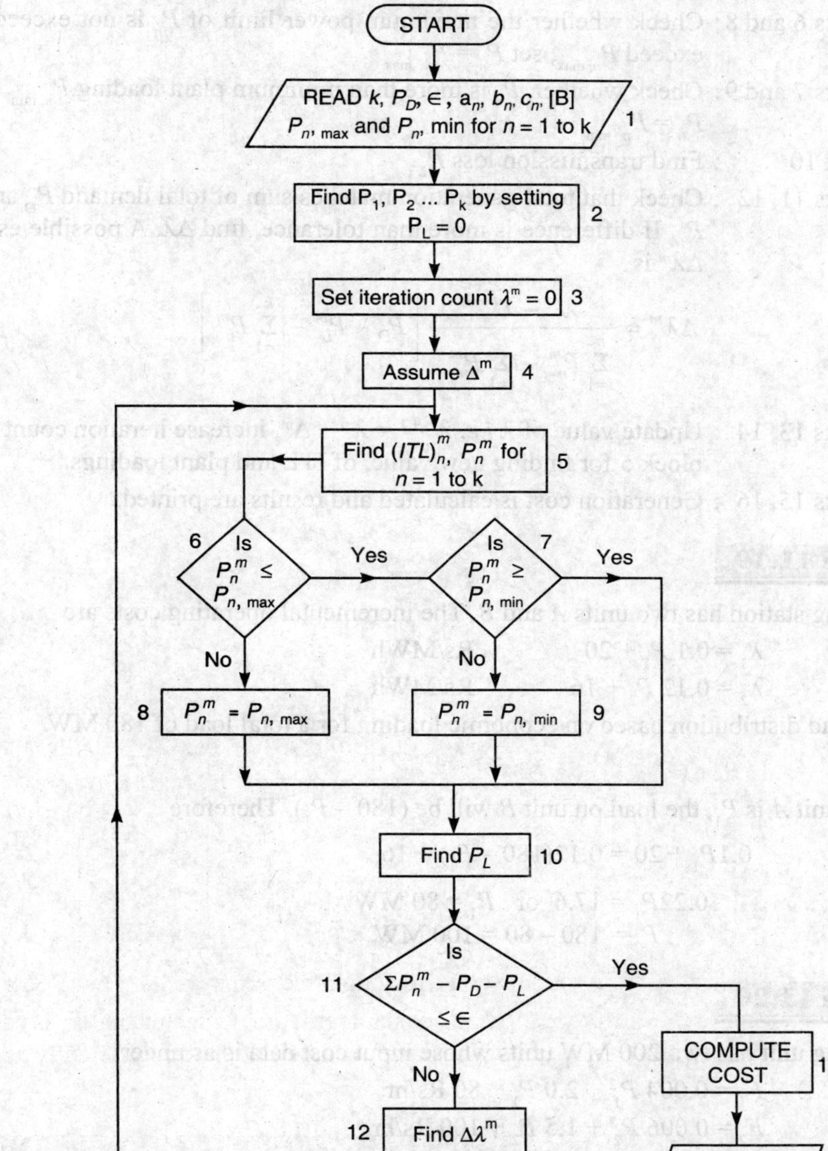

Fig. 11.16 Flow Chart for economic dispatch including transmission losses.

Blocks 3 and 4 : After setting iteration count, the program starts by assuming initial value of λ. A simple way is to make the initial value of λ the same as was obtained in step 2 (for finding P_1, P_2 ... P_k after neglecting losses).

Block 5 : Find incremental transmission loss $(ITL)_n$ and plant loadings P_1, P_2 ... P_k using Eq. (11.60).

Blocks 6 and 8: Check whether the maximum power limit of P_n is not exceeded. If P_n exceed $P_{n,\,max}$, set $P_n = P_{n,\,max}$.

Blocks 7 and 9: Check whether, P_n is more than minimum plant loading $P_{n,\,min}$. If not set $P_n = P_{n,\,min}$.

Block 10 : Find transmission loss P_L.

Blocks 11, 12 : Check that total generation matches sum of total demand P_D and losses, P_L. If difference is more than tolerance, find $\Delta\lambda$. A possible estimate of $\Delta\lambda^m$ is

$$\Delta\lambda^m = \frac{\lambda^m - \lambda^{m-1}}{\sum_{i=1}^{n} P_n^m - \sum_{i=1}^{n} P_n^{m-1}}\left[P_D - P_L^m - \sum_{i=1}^{k} P_n^m\right]$$

Blocks 13, 14 : Update value of λ *i.e.* $\lambda^{m+1} = \lambda^m + \Delta^m$, increase iteration count and goto block 5 for finding new value, of ITL and plant loadings.

Blocks 15, 16 : Generation cost is calculated and results are printed.

Example 11.19

A generating station has two units A and B. The incremental operating costs are

Unit A	$\lambda_1 = 0.1\,P_1 + 20$	Rs/MWh
Unit B	$\lambda_2 = 0.12\,P_2 + 16$	Rs/MWh

Find the load distribution based on economic loading for a total load of 180 MW.

Solution

If load on unit A is P_1, the load on unit B will be $(180 - P_1)$. Therefore

$$0.1P_1 + 20 = 0.12(180 - P_1) + 16$$

or

$$0.22P_1 = 17.6 \text{ or } P_1 = 80 \text{ MW}$$

$$P_2 = 180 - 80 = 100 \text{ MW.}$$

Example 11.20

A generating unit has two 200 MW units whose input cost data is as under:

$$F_1 = 0.004\,P_1^2 + 2.0\,P_1 + 80 \text{ Rs/hr}$$

$$F_2 = 0.006\,P_2^2 + 1.5\,P_2 + 100 \text{ Rs/hr}$$

For a total load of 250 MW find the load division between the two units for economic operation.

Solution

$$\frac{dF_1}{dP_1} = 0.008P_1 + 2$$

$$\frac{dF_2}{dP_2} = 0.012\,P_2 + 1.5$$

$$P_1 + P_2 = 250$$

For economic operation $\dfrac{dF_1}{dP_1} = \dfrac{dF_2}{dP_2}$

or $\quad 0.008P_1 + 2 = 0.012(250 - P_1) + 1.5$

or $\quad P_1 = 125$ MW

$P_2 = 250 - 125 = 125$ MW.

Example 11.21

The Fuel inputs per hour for unit I and unit II are given by

$$F_1 = (8P_1 + 0.024\,P_1^2 + 80)\,10^6 \text{ Btu/hour}$$

$$F_2 = (6P_2 + 0.04\,P_2^2 + 120)\,10^6 \text{ Btu/hour}$$

The maximum and minimum loads on the units are 100 MW and 10 MW. Determine the minimum total cost of generation per day when the following loads is supplied. The cost of fuel is Rs. 2.5 per million Btu. The load during the day is shown in Fig. 11.17. Also find operating cost per unit of energy.

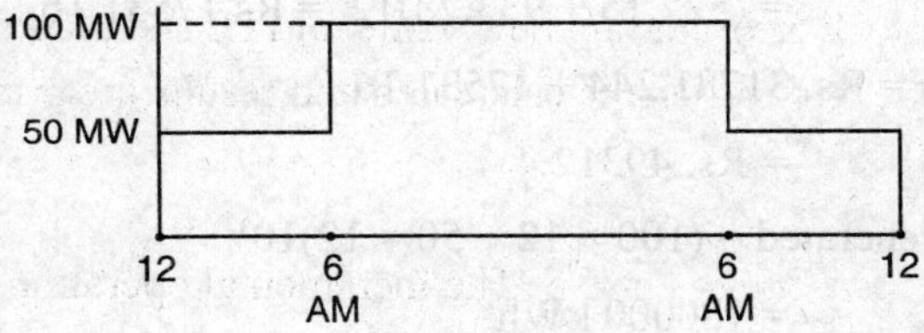

Fig. 11.17.

Solution

The operating costs are

$$C_1 = F_1\left(\frac{2.5}{10^6}\right) = 20\,P_1 + 0.06\,P_1^2 + 200 \text{ Rs/ hour}$$

$$C_2 = F_2\left(\frac{2.5}{10^6}\right) = 15\,P_2 + 0.1\,P_2^2 + 300 \text{ Rs/ hour}$$

$$\frac{dC_1}{dP_1} = 20 + 0.12P_1$$

$$\frac{dC_2}{dP_2} = 15 + 0.2\,P_2$$

For minimum cost $\dfrac{dC_1}{dP_1} = \dfrac{dC_2}{dP_2}$

or $\quad 20 + 0.12P_1 = 15 + 0.2\,(100 - P_1)$

or $\quad P_1 = 46.875$ MW, $P_2 = 100 - P_1 = 53.125$ MW

For 50 MW load

$$20 + 0.12P_1 = 15 + 0.2(50 - P_1)$$

or $\quad P_1 = 15.625$ MW, $P_2 = 50 - P_1 = 34.375$ MW

The constraints of maximum and minimum loads are also satisfied. Hence the operating schedules are proper.

For 12 hours of 100 MW load

$$C_1 = 20(46 - 875) + 0.6\,(46.875)^2 + 200 \text{ Rs/hour}$$

$$C_2 = 15(53.125) + 0.1\,(53.125)^2 + 300 \text{ Rs/hour}$$

or $$C_1 = \text{Rs } 1269.34 \text{ Rs/hour},\ C_2 = 1379.1 \text{ Rs/hour}$$

Total cost for 12 hour period

$$= (1269.34 + 1379.1)12 = \text{Rs. } 31781.244$$

For 12 hours of 50 MW load

$$C_1 = 20(15.625) + .06(15.625)^2 + 200 = 527.15 \text{ Rs/hr.}$$

$$C_2 = 15(34.375) + 0.1(34.375)^2 + 300 = 933.78 \text{ Rs/hr.}$$

Total cost for 12 hour period

$$= (527.15 + 933.78)12 = \text{Rs } 17531.16$$

Total cost for 24 hours = Rs. 31781.244 + 17531.16

$$= \text{Rs. } 49312.4$$

Total units of energy generated = $(100 \times 12 + 50 \times 12)10^3$

$$= 180000 \text{ kWh}$$

$$\text{Operating cost per unit of energy} = \frac{49312.4}{180000} = \text{Rs. } 2.74 \text{ per kWh.}$$

Example 11.22

A power plant has three units with following cost characteristics

$$F_1 = 0.05P_1^2 + 21.5P_1 + 800 \text{ Rs/hr.}$$
$$F_2 = 0.1P_2^2 + 27P_2 + 500 \text{ Rs/hr.}$$
$$F_3 = 0.07P_3^2 = 16P_3 + 900 \text{ Rs/hr.}$$

Find optimum scheduling and total cost per hour for a total load of 200 MW. P_{max} = 120 MW, P_{min} = 39 MW.

Solution

Using Eq. (11.29)

$$\lambda = \frac{\sum_{n=1}^{3} \frac{b_n}{2c_n} + P_T}{\sum_{n=1}^{3} \frac{1}{2c_n}} = \frac{\frac{1}{2}\left[\frac{2.15}{.05} + \frac{27}{.1} + \frac{16}{.07}\right] + 200}{\frac{1}{2}\left[\frac{1}{.05} + \frac{1}{.1} + \frac{1}{.07}\right]} = 30$$

$$\lambda = \frac{dF_1}{dP_1} = \frac{dF_2}{dP_2} = \frac{dF_3}{dP_3}$$

or $$30 = 21.5 + 0.1P_1 = 27 + 0.2\,P_2 = 16 + 0.14P_3$$

or $$P_1 = 85\text{MW},\ P_2 = 15 \text{ MW},\ P_3 = 100 \text{ MW}$$

Since load on P_2 is less than P_{min}, set P_2 = 39 MW.

Then $$P_1 + P_2 = 200 - 39 = 161 \text{ MW} \qquad ...(i)$$

$$\frac{dF_1}{dP_1} = \frac{dF_3}{dP_3}$$

or $$21.5 + 0.1P_1 = 16 + 0.14P_3 \qquad ...(ii)$$

Solving Eqs. (*i*) and (*ii*)

$$P_1 = 71 \text{ MW}, P_3 = 90 \text{ MW}$$

Therefore $$P_1 = 71 \text{ MW}, P_2 = 39 \text{ MW}, P_3 = 90 \text{ MW}$$

For the above schedule

$$F_1 = 0.05(71)^2 + 21.5(71) + 800 = 2578.55$$

$$F_2 = 0.1(39)^2 + 27(39) + 500 = 1705.1$$

$$F_3 = 0.07(90)^2 + 16(90) + 900 = 2907.0$$

$$\text{Toal cost} = F_1 + F_2 + F_3 = \text{Rs } 7190.65 \text{ per hr.}$$

Example 11.23

Fig. 11.18 shows a two bus system. If a load of 125 MW is transferred from plant 1 to load, a power loss of 15.625 MW occurs. Find generation schedule and load demand if cost of received power is Rs. 24 per MWh. The incremental production cost are

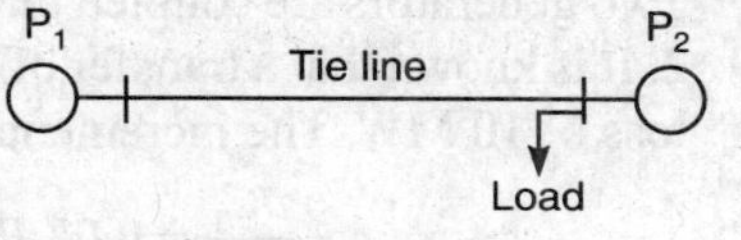

Fig. 11.18.

$$\frac{dF_1}{dP_1} = 0.025\,P_1 + 15$$

$$\frac{dF_2}{dP_1} = 0.05\,P_2 + 20$$

Solution

Since load at bus is P_2, transmission loss is not affected by P_2. Hence $B_{22} = B_{12} = 0$.

If $$P_1 = 125, P_L = 15.625$$

or $$15.625 = B_{11}(125)^2 \text{ or } B_{11} = 0.001$$

$$P_L = 0.001\,P_1^{\,2} \text{ or } \frac{\partial P_L}{\partial P_1} = 0.002P_1$$

and $$\frac{\partial P_L}{\partial P_2} = 0$$

Penalty factor plant $$1 = \frac{1}{1 - \frac{\partial P_L}{\partial P_1}} = \frac{1}{1 - 0.002P_1}$$

Penalty factor of Plant $$2 = 1$$

$$\lambda = 24$$

$$\frac{dF_1}{dP_1} L_1 = \lambda$$

$$\frac{dF_2}{dP_2} L_2 = \lambda$$

or $$0.05\,P_2 + 20 = 24 \quad \text{or} \quad P_2 = 80 \text{ MW}$$

$$\frac{0.025\,P_1 + 15}{1 - 0.002P_1} = 24$$

or $$0.025P_1 + 15 = 24 - 0.048P_1$$

or $$0.073P_1 = 9 \quad \text{or} \quad P_1 = 123.29 \text{ MW}$$

$$\text{Load Demand} = 123.29 - P_L + 80$$
$$= 123.29 - 0.001(123.29)^2 + 80$$
$$= 123.29 - 15.2 + 80 = 188.09 \text{ MW.}$$

Example 11.24

Two generators are coupled through a tie line as shown in Fig. 11.18. Load is at bus of generator 2. It is known that a transfer of 100 MW from generator 1 over the tie line means a transmission loss of 10 MW. The incremental costs are

$$\frac{dC_1}{dP_1} = 0.02\,P_1 + 16$$

$$\frac{dC_2}{dP_2} = 0.04\,P_2 + 20$$

Find the optimum schedule, total generation and total load if $\lambda = 25$ Rs/MWh.

Solution

Since load is at bus of generator 2, the transmission loss is not affected by P_2.

Hence, $$B_{22} = B_{12} = 0$$

If $$P_1 = 100, P_L = 10 \text{ MW}$$

or $$10 = B_{11}\,(100)^2 \quad \text{or} \quad B_{11} = 0.001$$

$$P_L = 0.001P_1^{\,2} \quad \text{or} \quad \frac{\partial P_L}{\partial P_1} = 0.002\,P_1$$

and $$\frac{\partial P_L}{\partial P_2} = 0$$

Penalty factor for plant 1 = $\dfrac{1}{1 - \dfrac{\partial P_L}{\partial P_1}} = \dfrac{1}{1 - 0.002P_1}$

Penalty factor for Plant 2 = 1

$$\lambda = 25$$

$$0.04P_2 + 20 = 25 \quad \text{or} \quad P_2 = 125 \text{ MW}$$

$$\frac{0.02P_1 + 16}{1 - 0.002P_1} = 25 \quad \text{or} \quad 0.02P_1 + 16 = 25 - 0.05P_1$$

or $$P_1 = 128.57 \text{ MW}$$

$$\text{Total generation} = P_1 + P_2 = 128.57 + 125 = 253.57 \text{ MW}$$

$$P_L = \text{Transmission Loss} = 0.001(128.57)^2 = 16.53 \text{ MW}$$

$$\text{Total load} = P_1 + P_2 - P_L = 253.57 - 16.53 = 237.04 \text{ MW.}$$

Example 11.25

Two generators are operating in parallel. The incremental fuel costs are

$$\frac{dC_1}{dP_1} = 0.2P_1 + 40 \qquad 30 \text{ MW} \le P_1 \le 175 \text{ MW}$$

$$\frac{dC_2}{dP_2} = 0.4P_2 + 30 \qquad 20 \text{ MW} \le P_2 \le 125 \text{ MW}$$

How will the load be shared when load varies from minimum to maximum? What is the corresponding incremental fuel cost. [*Mumbai Uni. 2010*]

Solution

The minimum load on the system can be 20 MW and it will be supplied by generator 2. Incremental fuel cost of generator 2 will be 0.4 × 20 + 30 i.e. Rs. 38.

When total load is 100 MW

$$0.2\,P_1 + 40 = 0.4\,(100 - P_1) + 30$$

or $$P_1 = 50 \text{ MW}, \quad P_2 = 50 \text{ MW}$$

$$\text{Incremental fuel cost} = 0.2 \times 50 + 40 = \text{Rs. } 50/\text{MWh}$$

When total load is 150 MW

$$0.2\,P_1 + 40 = 0.4\,(150 - P_1) + 30$$

or $$P_1 = 83.3 \text{ MW}, \quad P_2 = 150 - P_1 = 66.7 \text{ MW}$$

$$\text{Incremental fuel cost} = 0.2 \times 83.3 + 40$$

$$= \text{Rs. } 56.66 \text{ per MWh}$$

When total load is 200 MW

$$0.2\,P_1 + 40 = 0.4\,(200 - P_1) + 30$$

or $$P_1 = 116.67 \text{ MW}, \quad P_2 = 83.33 \text{ MW}$$

$$\text{Incremental fuel cost} = \text{Rs. } 63.33 \text{ per MWh}$$

When total load is 250 MW

$$0.2\,P_1 + 40 = 0.4\,(250 - P_1) + 30$$

or $$P_1 = 150 \text{ MW}, \quad P_2 = 100 \text{ MW}$$

$$\text{Incremental fuel cost} = \text{Rs. } 70 \text{ per MWh}$$

When total load is 300 MW

$$0.2\,P_1 + 40 = 0.4\,(300 - P_1) + 30$$

or $\quad P_1 = 183.33$ but P_1 can be loaded only upto 175 MW.

Hence $\quad P_1 = 175$ MW, $\quad P_2 = 125$ MW

Incremental fuel cost of generator 1 = Rs. 75 per MWh

Incremental fuel cost of generator 2 = Rs. 80 per MWh

Example 11.26

A power system has two generating units and power dispatch is economical with $P_1 = 150$ MW and $P_2 = 275$ MW. The loss coefficients are

$$B_{11} = 0.1 \times 10^{-2}\ \text{MW}^{-1}, \quad B_{12} = -0.01 \times 10^{-2}$$

$$B_{22} = 0.13 \times 10^{-2}\ \text{MW}^{-1}$$

To raise total load on the system by 1 MW will cost an additional Rs. 200 per hour. Find

(*a*) Penalty factor of plant 1

(*b*) Additional cost per hour to increase output of plant 1 by 1 MW [*CSVTU 2010*]

Solution

$$P_L = B_{11}P_1^2 + 2B_{12}\,P_1P_2 + B_{22}P_2^2$$

$$\frac{\partial P_L}{\partial P_1} = 2\,B_{11}P_1 + 2B_{12}\,P_2$$

or

$$\frac{\partial P_L}{\partial P_1} = 2 \times 0.1 \times 10^{-2} \times 150 - 2 \times 0.01 \times 10^{-2} \times 275$$

$$= 0.3 - .055 = 0.245$$

$$\text{Penalty factor for plant 1} = \frac{1}{1 - \dfrac{\partial P_L}{\partial P_1}}$$

$$= \frac{1}{1 - 0.245} = 1.3245$$

(*b*) Additional cost = 200 × 1.3245 = Rs. 264.9 per hour.

Example 11.27

The incremental fuel costs for 2 units of a plant are

$$\frac{dF_1}{dP_1} = 0.01\,P_1 + 11$$

$$\frac{dF_2}{dP_2} = 0.012P_2 + 8$$

Find savings in Rs./hr. for economical distribution of load compared with equal load sharing when total output is 750 MW [*CSVTU 2011*]

Solution

Let load on plant 1 be P_1. Then

$$P_2 = 750 - P_1$$

$$0.01\,P_1 + 11 = 0.012\,(750 - P_1) + 8$$

$$0.01\,P_1 + 0.012\,P_1 = 8 - 11 + 9 = 6$$

or $$P_1 = 272.73$$

$$P_2 = 750 - P_1 = 477.27$$

Equal load sharing $P_1 = P_2 = \dfrac{750}{2} = 375$ MW

$$\text{Change in cost of unit 1} = \int_{272.73}^{375} (0.01P_1 + 11)\,dP_1$$

$$= \left|0.005\,P_1^2 + 11P_1\right|_{272.73}^{375}$$

$$= \text{Rs. } 1456.185$$

$$\text{Change in cost of unit 2} = \int_{477.27}^{375} (.012\,P_2 + 8)\,dP_2$$

$$= \left|.006\,P_2^2 + 8P_2\right|_{477.27}^{375}$$

$$= \text{Rs. } -1341.13$$

Savings = 1456.185 – 1341.13 = Rs. 115.055 per hour.

UNIT COMMITMENT

11.21 DEFINITION

The total load in any power system varies from instant to instant. During certain time of the day the load is at peak value, during certain other times it is very low and during certain other times it is at some intermediate value. The load division between the generating units in operation is determined from the principle of equal incremental costs. The topic of unit commitment deals with specifying the units which should be operated for a given load *i.e.* which units should be committed to supply a given load. It is possible that it may be economical to decommit certain units when load is low. To commit a unit means to bring the boiler to the required temperature, bring the turbine and generator to synchronous speed and synchronise the generator to the system.

11.22 CONSTRAINTS IN UNIT COMMITMENT

The problem of deciding which units to commit for a certain load is beset with many constraints. Some of these are as under:

1. *Spinning Reserve:* Some system capacity has to be kept as spinning reserve to meet an unexpected increase in demand and to ensure power supply in the event of any generating unit suffering a forced outage. It is important to mention that the generating unit may have to be shut down because of fault in generator or any of its auxiliaries.
2. *Minimum up time:* When a thermal generating unit is operating, it cannot be turned off immediately.

3. *Minimum downtime:* When a thermal generating unit is decommitted, it cannot be turned on immediately.
4. *Crew constraints:* A plant always have two or more generating units. It may not be possible to turn on more than one generating unit at the same time due to non-availability of operating personnel.
5. Transition cost. Restoring a unit to the system involves extra expenditure. During shut down of a unit it is expendient to keep the turbine shaft rotating at slow speed to prevent unequal temperature distribution in turbine casing. This also involves energy consumption. All these costs are termed as transition cost.
6. *Hydro constraints:* It is rare that a system has all thermal generating units only. Most systems have some hydroelectric units also. The operation of these units depends on availability of water. Moreover hydro projects are multipurpose projects. Irrigation requirements also determine the operation of hydro plants.
7. *Nuclear constraints:* If a nuclear power plant is a part of the system, another constraint is added. A nuclear plant has to be operated as a base load plant only.
8. *Must run units:* One or two units may be must run units. It may be necessary to run them from consideration of voltage support and system stability.
9. *Fuel Supply Constraint:* Some plants may not be operated due to deficient fuel supply. Some plant may require some special fuel which may not be available.
10. *Auxiliary Power Consumption:* Addition of a unit results in increased auxiliary power consumption. This is especially true for thermal generating units. This aspect generally results in extending the load at which it becomes economical to add a new unit to the system.

11.23 METHODS AVAILABLE FOR UNIT COMMITMENT

The methods available for unit commitment are:

1. Priority List Method.
2. Dynamic Programming.

11.24 PRIORITY LIST METHOD FOR UNIT COMMITMENT

As the name suggests this method means that we prepare a priority list of the order in which the generating units should be added to the system. This list would also include the order of shutting down the units when the system demand decreases.

When the load is very low, the unit with the lowest heat rate supplies the entire load. As the load increases the units having heat rates in ascending order are successively added to the system.

We consider a system fed by 4 generating units *A*, *B*, *C* and *D* each of 150 MW. The unit *B* has the lowest heat rate, *D* has next higher heat rate, *A* has still higher and C has the highest heat rate.

It is necessary to plot the following heat rate curves :

1. Heat rate curve of unit *B*.
2. Combined heat rate curve of units *B* and *D*.
3. Combined heat rate curve of units *B*, *D* and *A*.
4. Combined heat rate curve of units *B*, *D*, *A* and *C*.

The heat rate curve of unit *B* shows the input in k-cal per MW hour for different values of outputs in MW for the unit *B*. The combined heat rate curve of units *B* and *D* shows the total input in *k*-cal per MW hour for different values of outputs in MW if the two units *B* and *D* are in operation and share the load as per the principle of equal incremental costs. The combined heat rate of units *B*, *D* and *A* shows the total input for all these three units in operation at the same incremental cost and sharing the load and so on. The loads at which the units are to be added are given by the points of intersection as shown in Fig. 11.19.

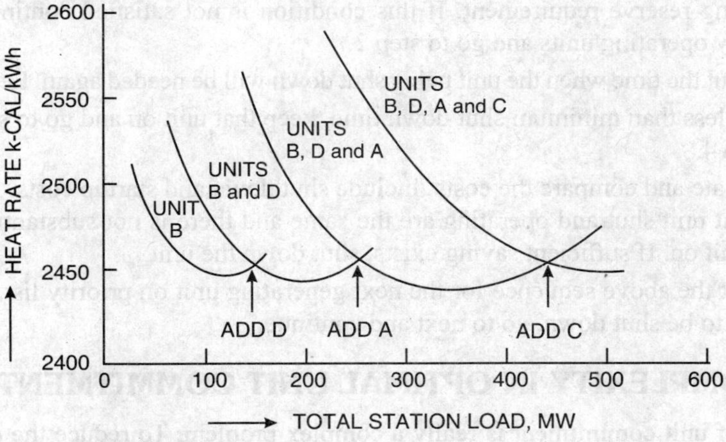

Fig. 11.19 Sequence of adding units.

Let us consider a load of 600 MW and 4 generating units each of 210 MW capacity. If all the four units are to be operated, the operating cost will be minimum when the loads are allocated as per the equal incremental cost criterion. However it is possible that only if units 1, 2 and 3 are operated and load allocated to these three units as per equal incremental cost criterion, the operating cost is lesser than that when all the four are operated.

One possible method for determining optimal unit commitment is to try all possible combinations of units for supplying the load. In the above case of 4 units each of 210 MW to meet 600 MW load demand, the possibilities are :

1. all units 1 to 4 operated
2. units 1, 2, 3 operated and unit 4 shut down
3. units 1, 3, 4 operated and unit 2 shut down
4. units 1, 2, 4 operated and unit 3 shut down
5. units 2, 3, 4 operated and unit 1 shut down

If operating costs is calculated for all the above 5 combinations, the scheme giving the minimum operating cost can be implemented. When the number of units is very large, this simple method becomes very lengthy and a better technique is necessary.

We have discussed this problem above and have seen that by plotting the heat rates of different combinations, the problem of unit commitment can be solved. However, this method also becomes very lengthy when the number of units is large.

For a system having a large number of units, it is preferable to use a method which can be used on a digital computer. Dynamic programming is one such technique.

As discussed in section 11.22, the problem of unit commitment is also governed by the consideration of auxiliary power consumption, reserve requirements, starting losses, stability considerations and voltage instability. Many of these considerations cannot be expressed in economic terms. The technical aspects of the problem of unit commitment are as important as the economic aspects. As such, the decision on plant capacity to be operated and unit commitment must be based on both technical and economic aspects.

The algorithm for unit commitment as per priority list method can be summarised as under:

1. When the load is decreasing, examine whether shutting down a generating unit would still leave enough generation to supply the load plus auxiliary power consumption plus spinning reserve requirement. If this condition is not satisfied continue running the already operating units and go to step 2.
2. Find out the time when the unit being shut down will be needed again. Let this time be T.
3. It T is less than minimum shut down time, keep that unit on and go to step 5. If not go to step 4.
4. Calculate and compare the costs. Include shut down and startup costs. If the total cost for that unit shut and operating are the same and there is not substantial saving keep that unit on. If sufficient saving exists, shut down the unit.
5. Repeat the above sequence for the next generating unit on priority list. If this unit requires to be shut down, go to next and continue.

11.25 COMPLEXITY IN OPTIMAL UNIT COMMITMENT

The problem of unit commitment is really a complex problem. To reduce the complexity the transmission losses are neglected in solving unit commitment problem. Thus the total generation equals the load. Moreover at least one unit has to be kept on, all the time, because minimum load is never zero.

We can divide the 24 hour period into time intervals as shown in Fig. 11.20 where it is shown that day can be divided into 5 intervals with loads as shown. The problem of unit commitment aims to find the feasible combination of generating units to supply the load at minimum operating cost during each time interval.

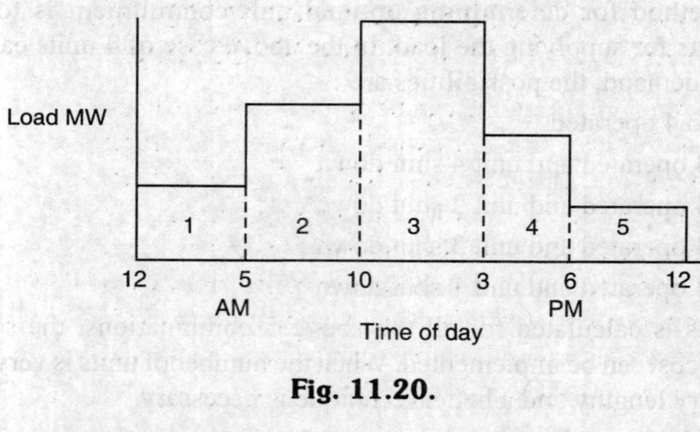

Fig. 11.20.

Let us assume that the system has 4 generating units. One of these must be kept ON to supply the minimum load. For the remaining 3 units there are a total of $(2^3 - 1) = 7$ combinations of units to supply the load. These are shown in table 11.2. Where the load ON denotes that the unit is on and OFF denotes that the unit is off.

Table 11.2

Unit	x_1	x_2	x_3	x_4	x_5	x_6	x_7
1	ON	ON	ON	OFF	ON	OFF	OFF
2	ON	ON	OFF	ON	OFF	ON	OFF
3	ON	OFF	ON	ON	OFF	OFF	ON

It has to be remembered that all the above combinations may not be feasible for a certain time interval. This is, evidently, due to the simple fact that a combination of total capacity less than system load is not feasible and can be disregarded for that time interval. Let

$$x_n(i) = \text{combination } x_n \text{ for } i\text{th interval}$$

Then $x_m(i+1)$ represents combination x_m for $(i+1)$th interval. If $i = 2$ and $x = 4$, it means that units 2 and 3 are ON during this interval and unit 1 is off.

The cost $C_n(i)$ denotes minimum operating cost of nth combination for ith interval. Then $C_m(i + 1)$ will denote minimum operating cost of $x_m(i+1)$ and so on.

The operating cost is not the only cost which has to be considered in optimising unit commitment. Some cost is incurred in changing the state. This cost is usually referred to as start up or shut down cost. Another terminology is transition cost. Let this cost be denoted by symbol S. Then

$$S_{nm,\,i} = \text{cost of transition from combination } x_n(i) \text{ to combination } x_m\ (i+1)$$

Since the number of generating units in any system is pretty large (may be 20 or even more), the combinations of units can be $(2^{20} - 1)$ which is a very large number. Combining it with the number of time intervals (say N) in which the day is divided makes the solution all the more complex. Therefore the unit commitment problem is solved for each time interval separately by a subprogram and then the results for all these N intervals are combined.

The total cost $t_{nm}(i)$ for any time interval can be written as

$$t_{nm}(i) = C_n(i) + S_{nm}(i)$$

Where $C_n(i)$ is the minimum operating cost of nth combination for ith interval and $S_{nm}(i)$ is the cost of transition from combination $x_m(i-1)$ to combination $x_n(i)$.

11.26 SOLUTION OF UNIT COMMITMENT PROBLEM BY FORWARD DYNAMIC PROGRAMMING

The method of dynamic programming can be applied to solve unit commitment problem in two ways *i.e.* Forward and Backward. In forward dynamic programming (FDP) we start with the first duration obtain the unit commitment schedule for this duration, go to second duration till we reach the last duration. In backward dynamic programming we start with the last duration obtain the unit commitment schedule for this duration and proceed backward till we reach the first duration. FDP has some distinct advantages over backward dynamic programming. In using FDP, initial conditions are easily specified, calculations can be carried out for any desired length of time and the previous history of each unit can be calculated at every stage.

The minimum cost in duration i for nth combination can be written as

$$T_{\text{cost}}(i, n) = \min$$

[Least total cost to reach state (i, n) + operating cost for state (i, n) + transition cost from state $(i-1, m)$ to state (i, n)]. ... (11.74)

The operating cost for any stage needs the method of economic dispatch. This is, evidently, due to the fact that for any given combination of units, the operating cost is minimum if all the units in this combination are operating at equal incremental cost.

Two other variables enter the strategy for unit commitment by forward dynamic programming. This is because a number of possible combinations exist for every state. Let this be denoted by K. Another variable is the number of paths or strategies to save at every step. Let this variable be denoted by L.

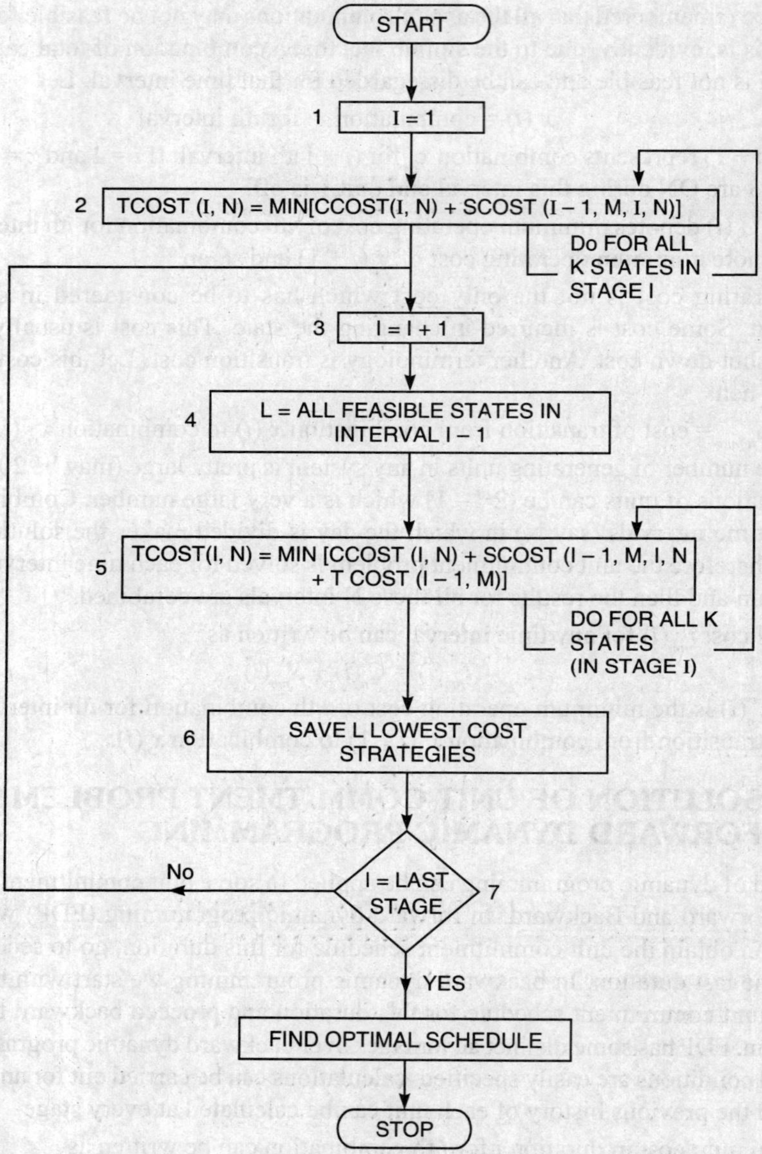

Fig. 11.21 Flow chart for unit commitment using Forward Dynamic Programming

Fig. 11.21 shows a simple flow chart for unit commitment through forward dynamic programming.

Block 1: The stage in load cycle is specified as i. We start with $i = 1$ *i.e.* fist stage.

Blcok 2: The operating cost for this stage is computed. This has to be repeated for all possible combinations K.

Block 3: We now go to $(i + 1)$th stage.

Block 4: The number of feasible paths in duration $(i - 1)$ is found and stored.

Block 5: The minimum total cost as mentioned in Eq. (11.74) is calculated. This has to be repeated for all states in ith interval.

Block 6: The lowest cost paths (number L) are saved in computer memory.

Block 7: Check whether the program has reached the last stage in load cycle. If yes optimal schedule is printed. Otherwise i is updated to $(i + 1)$ and program is repeated.

SUMMARY

11.1 Methods of loading thermal power units are : base load to capacity, base load to the most efficient load, loading proportional to capacity and equal incremental cost operation. The operation on equal incremental costs leads to maximum economy.

11.2 The input-output curve shows the variation of fuel input (k cal/hour) with power output (MW).

11.3 Heat rate is the ratio of heat input to the corresponding output. Its units are k-cal per kWh.

11.4 The slope of input-output curve is the incremental fuel cost. Its units are Rs. per MWh.

11.5 The cost C (Rs/hour) can be written as $C = a + bP + cP^2$ where P is power output and a, b, c are constants which depend on the unit size.

11.6 Equal incremental cost criterion can be written as $\frac{dC_1}{dP_1} = \frac{dC_2}{dP_2} = \ldots = \frac{dC_k}{dP_k} = \lambda$ where $\frac{dC_k}{dP_k}$ is the incremental cost of kth generating unit. λ is known as Langrangian multiplier.

11.7 Equal incremental cost criterion is applicable to the thermal units in operation. This principle cannot specify the units which should be operated for a given load demand. The problem of deciding the units to be operated can be solved in a number of ways. Use of dynamic programming is the best approach.

11.8 When a system consists of a number of power plants interconnected by transmission lines, the transmission losses must be considered to find the economy loading.

11.9 The total plant capacity to be operated also depends on auxiliary power consumption, reserve requirements, starting losses, stability and voltage considerations.

SELF TEST QUESTIONS

A. State whether the following are True or False:

11.1. The incremental cost is the ratio of small change in input to a small change in output.

11.2. The heat rate is the ratio of output to input.

11.3. As the output increases, incremental fuel cost decreases.

11.4. The fuel cost is directly proportional to power output only.

11.5. The cost characteristics of thermal units is non-linear.

11.6. λ is called Langrangian multiplier.

11.7. Equal incremental cost criterion can be used to specify the units to be operated for a given load.

11.8. The total plant capacity to be operated depends on reserve requirements also.

11.9. Restoring a thermal unit to service involves losses.

11.10 An error in representation of cost characteristics of units does not affect economy.

11.11 For a system having two units with outputs P_1 and P_2, the transmission loss P_L can be written as $P_L = B_{11}P_1^2 + B_{22}P_2^2$.

11.12 B coefficients in a system depend on system operating conditions.

B. Fill in the blanks with suitable words:

11.13 Penalty factor is equal to ___________.

11.14 Economic scheduling in a large interconnected system should take ________ in to account.

11.15 Fuel input curve of a thermal unit is generally modelled as a ________.

11.16 Incremental cost characteristics of a thermal generating unit is generally ________.

11.17 B coefficients depend on ________ and power factors.

11.18 Cost of fuel at different plants are always ________.

11.19 Cost coefficients for gas generating units are generally ________ than those for coal fined units.

11.20 In economic scheduling cost of fuel is generally taken in ________.

ANSWERS

11.1 T	11.2 F	11.3 F	11.4 F	11.5 T	11.6 T
11.7 F	11.8 T	11.9 T	11.10 F	11.11 F	11.12 T

11.13 $\dfrac{1}{1-\dfrac{\partial P_L}{\partial P_k}}$ 11.14 transmission losses 11.15 quadratic

11.16 non linear 11.17 source voltages 11.18 different

11.19 higher 11.20 Rs. per kcal.

JOB INTERVIEW QUESTIONS

11.1 What is incremental cost criterion?

11.2 What are the units of incremental cost?

11.3 What is Lagrangian multiplier?

11.4 What is base point?

11.5 What is heat rate?

11.6 What is meant by unit commitment?

11.7 What is meant by economic dispatch.

11.8 Write equation for transmission loss.

11.9 What is penalty factor?

11.10 Define up time and down time.

11.11 Name methods for unit commitment.

11.12 What are thermal unit constraints?

11.13 What is meant by participation factor?

11.14 How is incremental cost calculated?

11.15 What is meant by cascading outages?

REVIEW QUESTIONS AND PROBLEMS

11.1 Discuss why :

(*a*) Generally the fuel cost is taken as the operating cost in economic scheduling.

(*b*) It is necessary to consider the transmission loss in economic scheduling of steam plants.

(*c*) The penalty factor of a plant is always more than 1.

(*d*) The penalty factor of a plant situated far away from load centre is higher than that of a plant situated near the load centre.

(e) Power systems always have some reserve capacity.

(f) The principle of equal incremental costs is applicable only for load allocation between units in operation and not for deciding the number of units to be operated.

11.2. Discuss the importance of proper load allocation in power plants. What information must be available for optimum load allocation?

11.3 How is the sequence of adding units in a thermal plant decided?

11.4 Discuss the factors which govern the plant capacity to be operated in a power system.

11.5 Define input-output characteristics, heat rate and incremental cost.

11.6 Why is incremental fuel cost generally taken as total incremental cost for purpose of load allocation in thermal plant?

11.7 Explain the equal incremental costs criteria for optimum scheduling in power plants.

11.8 Discuss the methods commonly used for deciding the load allocation between the units of a power plant.

11.9 Why is it necessary to consider transmission loss in optimum scheduling?

11.10 How can the effect of transmission loss be included in optimum scheduling in power plants?

11.11 The fuel inputs F_1 and F_2 in kcal/hour for units 1 and 2 are given by :

$$F_1 = (2P_1 + 0.005P_1^2 + 20)\ 10^4$$
$$F_2 = (1.5P_2 + 0.01P_2^2 + 30)\ 10^4$$

where P_1 and P_2 are outputs in MW.

(a) Plot the variation of input with output for the two units as the load varies from 20 MW to 100 MW in each case.

(b) Find heat rate in k cal/MW-hour for each unit and plot it against output.

(c) If the fuel cost is Rs. 1.0 per 10,000 kcal, calculate incremental fuel cost in Rs. per MW-hr. for each unit and plot it against output.

11.12 The fuel inputs F_1 and F_2 in kcal/hour for two units having output P_1 and P_2 MW are as under:

$$F_1 = P_1 + 0.02P_1^2 + 20$$
$$F_2 = 1.2P_2 + 0.015P_2^2 + 10$$

The maximum and minimum loads on the two units are 250 MW and 25 MW respectively. Find and plot the variation of load on each unit as a function of the total load which varies from 50 MW to 500 MW. Also plot incremental cost against total load.

11.13 The fuel input in k cal/hour for a generating unit with output P (MW) is given by :

$$F = (0.00005\ P^3 + 0.004\ P_2 + P + 20)10^6$$

(a) If the fuel cost is Rs. 90 per million kcal, find the expression for incremental fuel cost.

(b) Find a good linear approximation of incremental fuel cost as a function of output.

(c) For an output of 50 MW, what is the average fuel cost per MW-hour ?

(d) Find the additional fuel cost per hour if load increases from 50 to 51 MW.

[(a) $1.35 \times 10^{-2}\ P^2 + 0.72\ P + 90$ Rs./MWh
(c) Rs.155.25 (d) Rs.159.75]

11.14 The incremental costs of two units are

$$\frac{dF_1}{dP_1} = 0.3\ P_1 + 15$$

$$\frac{dF_2}{dP_2} = 0.2\ P_2 + 40$$

where P_1 and P_2 are outputs in MW.

The minimum and maximum loads on the two units are 25 MW and 150 MW respectively. Plot individual loads versus total load as the total load varies from 50 MW to 300 MW. Also plot incremental cost against total load. [P.U. 1987]

11.15 Find the saving per hour for economical load allocation compared to equal load sharing for a total load of 200 MW for the units of Prob. 11.14. [Rs. 225 per hour.]

11.16 The incremental costs of two units are

$$\frac{dF_1}{dP_1} = 0.25\,P_1 + 50 \text{ Rs./MW-hour}$$

$$\frac{dF_2}{dP_2} = 0.15P_2 + 60 \text{ Rs./MW-hour}$$

where P_1 and P_2 are outputs is MW.

The system load varies as under :

12 Midnight to 6 am	60 MW
6 am to 9 pm	200 MW
9 pm to 12 midnight	60 MW

The capacity of each unit is 150 MW

(*a*) Would it be more economical to keep both units operating during the whole period of 24 hours or shut down one unit during low load period. A cost of Rs. 200 is incurred in taking a unit off and returning it to service.

(*b*) What is the economic loading during the high load period ?

(*c*) Find the total fuel cost for the 24 hour period if loads are allocated as per economic scheduling.

[(*a*) keep both running (*b*) $P_1 = P_2 = 100$ MW (*c*) Rs. 225768]

11.17 For the system shown in Fig. 11.11, $I_1 = 1\angle 0°$ p.u., $I_2 = 0.8\angle 0°$ p.u. The voltage at junction of lines *a*, *b* and *c* is $1\angle 0°$ p.u. Line impedances for the lines *b*, *c* and *a* are $0.04 + j\,0.16$ p.u., $0.03 + j\,0.12$ p.u. and $0.02 + j\,0.08$ p.u. respectively. Find the loss coefficients.

[$B_{11} = 0.0554$ p.u., $B_{21} = 0.0188$ p.u., $B_{22} = 0.0477$ p.u.]

11.18 For the system shown in Fig. 11.12, the impedances of lines and currents are as under:

$I_a = 1.2 - j\,0.4$ p.u.	$Z_a = 0.02 + j\,0.08$ p.u.
$I_b = 0.4 - j\,0.2$ p.u.	$Z_b = 0.08 + j\,0.32$ p.u.
$I_c = 0.8 - j\,0.1$ p.u.	$Z_c = 0.02 + j\,0.08$ p.u.
$I_d = 0.8 - j\,0.2$ p.u.	$Z_d = 0.03 + j\,0.12$ p.u.
$I_e = 1.2 - j\,0.3$ p.u.	$Z_e = 0.03 + j\,0.12$ p.u.

The voltage at the bus and junction of lines *a*, *b* and *c* is $1\angle 0°$. Find the loss formula coefficients in p.u. and in reciprocal megawatts if base is 50 MVA.

$$\begin{bmatrix} B_{11} = 0.0678 \text{ p.u. and } 0.1365 \times 10^{-2}\ \text{MW}^{-1} \\ B_{12} = -0.004 \text{ p.u. and } -0.008 \times 10^{-2}\ \text{MW}^{-1} \\ B_{22} = 0.0562 \text{ p.u. and } 0.1124 \times 10^{-2}\ \text{MW}^{-1} \end{bmatrix}$$

11.19 For the system shown in Fig. 11.10, the incremental costs for the two plants are

$$\frac{dC_1}{dP_1} = 0.08\,P_1 + 15 \text{ Rs./MW-hour}$$

$$\frac{dC_2}{dP_2} = 0.1\,P_2 + 13 \text{ Rs./MW-hour}$$

The transmission line loss formula indicates that for a transfer of 150 MW, a loss of 15 MW occurs. If $\lambda = 25$ Rs./MWh. find (*a*) B_{11}, (*b*) P_1 and P_2 for minimum fuel cost (*c*) Power supplied to the load (*d*) Loss in transmission line

$$\begin{bmatrix} (a)\ B_{11} = 0.00067\ \text{MW}^{-1} \quad (b)\ P_1 = 88.5\ \text{MW}, \\ P_2 = 120\ \text{MW},\ (c)\ 203.3\ \text{MW}\ (d)\ 5.2\ \text{MW} \end{bmatrix}$$

11.20 For the system shown in Fig. 11.10, the incremental costs are as under :

$$\frac{dC_1}{dP_1} = 0.056\, P_1 + 32.8 \text{ Rs./MW-hour}$$

$$\frac{dC_2}{dP_2} = 0.112\, P_2 + 36.8 \text{ Rs./MW-hour}$$

When 100 MW power is transmitted over the line, a loss of 10 MW takes place. Find the economic generation of each plant for λ = Rs. 48 per MW-hour and the loss in line.

(*P.U.1989*) [P_1 = 100 MW, P_2 = 100 MW, P_L = 10 MW]

11.21 For the problem .20 find the savings in Rs. per hour obtained by coordinating rather than neglecting the transmission loss in computing the generation of each plant if the load is 190 MW.

[Rs. 522.88 per hour]

11.22 In a system consisting of two generating units connected through a transmission line, the incremental costs are

$$\frac{dC_1}{dP_1} = 0.16 P_1 + 32 \text{ Rs./MW-hour}$$

$$\frac{dC_2}{dP_2} = 0.24\, P_2 + 36 \text{ Rs./MW-hour}$$

The system is operating on economic dispatch with $P_1 = P_2 = 100$ MW and $\dfrac{\partial P_L}{\partial P_2} = 0.2$. Find the penalty factor of plant 1.

[1.5625]

11.23 A system having two generating units connected through a line has following loss coefficients:

$$B_{11} = 0.1 \times 10^{-2}\ \text{MW}^{-1}$$
$$B_{12} = -0.01 \times 10^{-2}\ \text{MW}^{-1}$$
$$B_{22} = 0.13 \times 10^{-2}\ \text{MW}^{-1}$$

Power is being dispatched economically with P_1 = 120 MW and P_2 = 200 MW. To raise the system load by 1 MW, it costs additional Rs. 40 per hour. Find (*a*) penalty factor for plant 1 (*b*) The additional cost per hour to increase the output of plant 1 by 1 MW.

[(*a*) 1.25 (*b*) Rs. 32]

11.24 The cost curves of two generating units can be approximated by second order polynomials as under :

$$C_1 = 0.08\, P_1^2 + 16 P_1 + 400 \text{ Rs./hour}$$

$$C_2 = 0.04 P_1^2 + 20 P_1 + 600 \text{ Rs./hour}$$

(*a*) Find the economic generation of each generator for a total load of 150 MW. The minimum and maximum loads of the units are 20 and 100 MW.

(*b*) Find the cost per hour if the generators are operated as per above schedule.

$$\begin{bmatrix} (a)\ P_1 = 66.7\ \text{MW},\ P_2 = 83.3\ \text{MW} \\ \text{(b) Rs. 4369 per hour} \end{bmatrix}$$

11.25 The loss coefficients for a two unit system are : $B_{11} = 0.0008\ MW^{-1}$, $B_{22} = 0.001\ MW^{-1}$, $B_{12} = -0.0004\ MW^{-1}$.

The incremental costs are given by

$$\frac{dC_1}{dP_1} = 0.3\,P_1 + 15 \quad \text{Rs./MW-hour}$$

$$\frac{dC_2}{dP_2} = 0.2P_2 + 40 \quad \text{Rs./MW-hour}$$

The capacities of each of the units are 250 MW each. By assuming different values of λ and following an iterative procedure, plot the individual loadings versus total load as the total load varies from 50 MW to 400 MW. Also plot the transmission loss versus the total generation.

11.26 The loss coefficients for a two unit system are :

$B_{11} = 0.0006\ MW^{-1}$, $B_{22} = 0.0005\ MW^{-1}$ and $B_{12} = -0.0003\ MW^{-1}$.

The incremental costs of the two units are

$$\frac{dC_1}{dP_1} = 0.15\,P_1 + 40 \quad \text{Rs./MW-hour}$$

$$\frac{dC_2}{dP_2} = 0.2\,P_2 + 35 \quad \text{Rs./MW-hour}$$

The unit sizes are 200 MW each. Assume different values of λ and plot the loads of individual units versus the total load as the total load varies from 150 MW to 350 MW.

CHAPTER

12

Power System Transients

12.1 INTRODUCTION

A power system operates in steady state for a major part of time. However, transients are also initiated off and on. A change in circuit conditions (*e.g.*, opening or closing a switch) or occurrence of a fault or a lightning discharge may result in currents and voltages which are higher in magnitude as compared to those in steady state conditions. A clear appreciation of the events taking place during transient periods helps in understanding the effects of these transients on power system equipment and methods of protection against them.

12.2 CIRCUIT CLOSING TRANSIENT

Fig. 12.1 shows a simple circuit in which a load, represented by a series combination of resistance and inductance, is fed from a sinusoidal a.c source. When the switch is closed, the circuit equation is

$$Ri + L\frac{di}{dt} = V_m \sin(\omega t + \alpha) \tag{12.1}$$

where i is instantaneous value of the current. The inclusion of the arbitrary angle α permits closing of the switch at any instant in the voltage cycle. Eq. (12.1) can be easily solved. The auxiliary equation is

$$(R + pL)\, i = 0 \tag{12.2}$$

The complementary function is

$$i = A\, e^{-Rt/L} \tag{12.3}$$

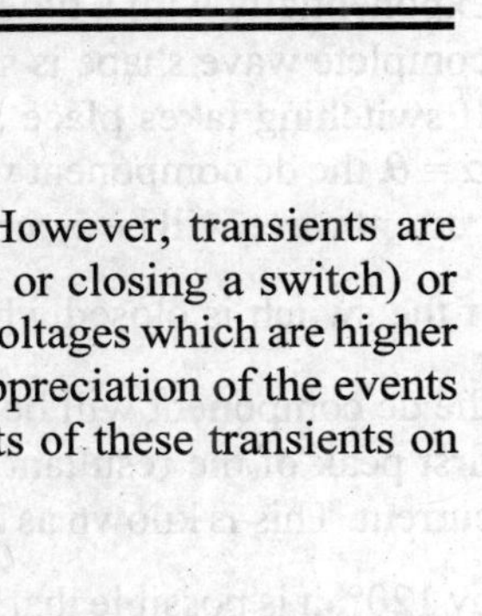

Fig. 12.1 Series RL circuit

The particular solution is

$$i = \frac{V_m}{\sqrt{(R^2 + \omega^2 L^2)}} \sin(\omega t + \alpha - \theta) \tag{12.4}$$

where $\theta = \tan^{-1}(\omega L/R)$

The complete solution is

$$i = A \in^{-Rt/L} + \frac{V_m}{\sqrt{(R^2 + \omega^2 L^2)}} \sin(\omega t + \alpha - \theta) \tag{12.5}$$

The constant A can be found from the initial condition that at $t = 0$, $i = 0$. The value of A is

$$A = \frac{-V_m}{\sqrt{(R^2 + \omega^2 L^2)}} \sin(\alpha - \theta) \tag{12.6}$$

Thus, the final solution is

$$i = \frac{V_m}{\sqrt{(R^2 + \omega^2 L^2)}} [\sin(\omega t + \alpha - \theta) - \sin(\alpha - \theta)\, \varepsilon^{-Rt/L}] \tag{12.7}$$

In Eq. (12.7) the current i has two components. The first component is the ac sinusoidal component having an amplitude $V_m / \sqrt{R^2 + \omega^2 L^2}$ and lagging the voltage phasor by an angle θ. The second is the dc (or transient) component which decays exponentially with a time constant L/R. The complete wave shape is shown in Fig. 12.2. If switching takes place at an instant when $\alpha = \theta$, the dc component will be zero and the current wave will be symmetrical. However, if the switch is closed when $\alpha - \theta = \pm \frac{\pi}{2}$, the dc component will be maximum and the first peak of the resultant current i will approach twice the peak value of the final steady state current. This is known as doubling effect. Since the voltages in a 3-phase circuit are out of phase by 120°, it is possible that the worst switching condition $\left(i.e., \alpha - \theta = \pm \frac{\pi}{2}\right)$ may be met on some phase at every closing operation. When switching a heavy current, *e.g.*, closing a circuit breaker to a fault, this possible doubling of current may cause considerable mechnical forces which must be taken care of in the design of circuit breakers.

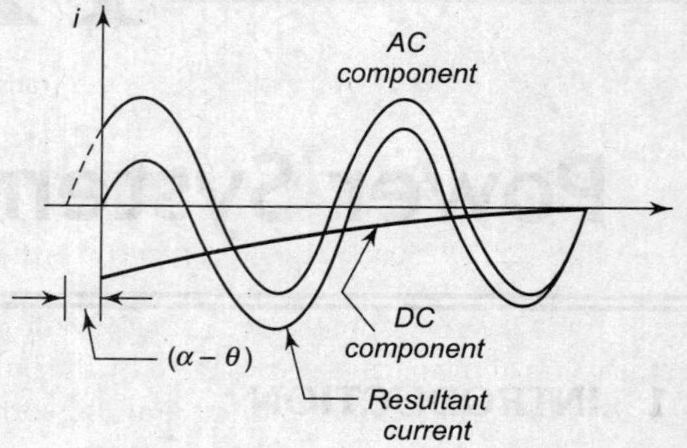

Fig. 12.2 Asymmetrical alternating current in a series RL circuit

12.3 SUDDEN SYMMETRICAL SHORT CIRCUIT OF ALTERNATOR

Consider a 3-phase alternator running at synchronous speed and with the field excited by a constant dc voltage. Let a 3-phase short circuit be suddenly applied to the stator terminals which were previously open. The short circuit current in each phase will consist of an ac component and a dc component. Since the voltages of the three phases are 120° out of phase from each other, the short circuit occurs at different points on the voltage wave of each phase. Therefore, the dc component will be different in each phase. If the dc component is subtracted, the oscillogram of the armature current has a typical wave shape shown in Fig. 12.3. The wave may be divided into three distinct time periods: (1) subtransient period lasting for only about 2 cycles during which the current decrement is very rapid; (2) transient period lasting for about 30 cycles or so; (3) steady state period.

The rms value of initial current (*i.e.*, at the instant of short circuit) is known as sub-transient current I″. The corresponding reactance of the winding is called the direct axis sub-transient reactance. The extrapolation of the transient envelope cuts the y-axis at point b. The rms value of current, represented by the intercept ob, *i.e.*, 0.707 × *ob* in amperes, is known as transient

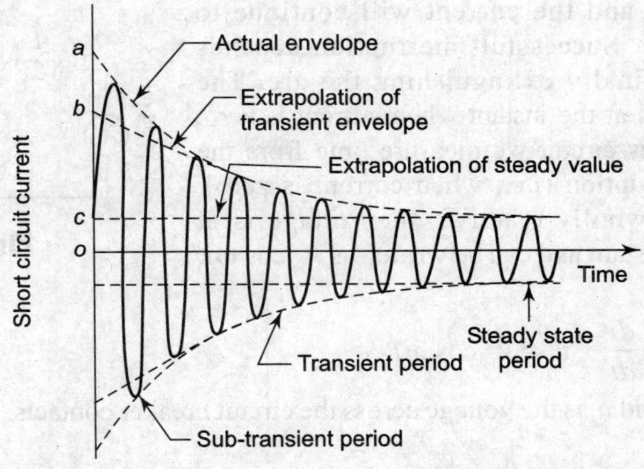

Fig. 12.3 Symmetrical short circuit armature current in a synchronous machine

current I' and the corresponding reactance is known as direct axis transient reactance. Similarly, the rms value of the current represented by the intercept oc, *i.e.*, $0.707 \times oc$ in amperes is known as steady state short circuit current and the corresponding reactance is known as direct axis synchronous reactance. Thus,

$$|I| = \frac{oc}{\sqrt{2}} = \frac{|E_0|}{X_d} \tag{12.8a}$$

$$|I'| = \frac{ob}{\sqrt{2}} = \frac{|E_0|}{X'_d} \tag{12.8b}$$

$$|I''| = \frac{oa}{\sqrt{2}} = \frac{|E_0|}{X''_d} \tag{12.8c}$$

where $|E_0|$ = rms value of open circuit phase voltage
$|I|$ = rms value of steady state short circuit current
$|I'|$ = rms value of transient current
$|I''|$ = rms value of sub-transient current
X_d = direct axis synchronous reactance
X'_d = direct axis transient reactance
X''_d = direct axis sub-transient reactance.

12.4 RECOVERY TRANSIENT DUE TO REMOVAL OF SHORT CIRCUIT

The opening of a circuit under fault conditions results in a transient known as recovery voltage transient. This transient has an important effect on the behaviour and rating of circuit interrupting and protective devices. A simple circuit which illustrates this phenomenon is shown in Fig. 12.4. The load being fed through the circuit breaker is suddenly isolated by the occurrence of a fault. During the time of fault a heavy fault current flows in the circuit. This fault current is limited only by the inductance. The condenser *C* represents the natural capacitance of the circuit adjacent to the circuit breaker. The resistance of the circuit has been neglected.

The fault is cleared by the opening of the circuit breaker contacts. The parting of the circuit breaker contacts does not in itself interrupt the current because an arc will be established between the parting contacts and the current will continue to flow through the arc. Successful interruption depends on controlling and finally extinguishing the arc. The interruption is affected at the instant when current is zero. For the analysis of this circuit we measure time from the instant of fault interruption (*i.e.*, when current is zero). Since the circuit is wholly reactive, the voltage is at its peak value at this instant and can be written as $V_m \cos \omega t$. The circuit equation is

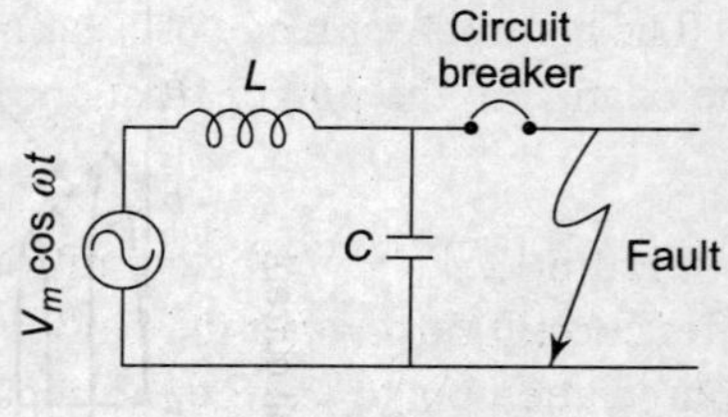

Fig. 12.4

$$L\frac{di}{dt} + v_c = V_m \cos \omega t \tag{12.9}$$

where i is the current and v_c is the voltage across the circuit breaker contacts. i and v_c are related by the equation

$$i = C'\frac{dv_c}{dt} \tag{12.10}$$

Combining Eqs. (12.9 and 12.10)

$$\frac{d^2v_c}{dt^2} + \frac{v_c}{LC} = \frac{V_m}{LC}\cos \omega t \tag{12.11}$$

Taking Laplace transform of Eq. (12.11) and substituting ω_0^2 for $1/LC$ we have

$$p^2 v_c(p) - p\,v_c(0) - v_c'(0) + \omega_0^2 v_c(p) = \omega_0^2 V_m \frac{p}{p^2+\omega^2} \tag{12.12}$$

where $v_c(p)$ is the Laplace transform of v_c, $v_c(0)$ is the value of v_c at $t = 0$ and $v_c'(0)$ is the value of $\frac{dv_c}{dt}$ at $t = 0$.

If we neglect arc voltage, $v_c(0)$ is zero. Also from Eq. (12.10)

$$v_c'(0) = \frac{i(o)}{C} = 0$$

Setting $v_c(0)$ and $v_c'(0)$ equal to zero. Eq. (12.12) can be written as

$$v_c(p) = \omega_0^2 V_m \frac{p}{(p^2+\omega^2)(p^2+\omega_0^2)}$$

$$= \frac{\omega_0^2 V_m}{\omega_0^2 - \omega^2}\left[\frac{p}{p^2+\omega^2} - \frac{p}{p^2+\omega_0^2}\right] \tag{12.13}$$

Taking inverse form of Eq. (12.13)

$$v_c = \frac{\omega_0^2}{\omega_0^2-\omega^2} V_m (\cos \omega t - \cos \omega_0 t) \tag{12.14}$$

Generally, $\omega_0 >> \omega$ so that $\frac{\omega_0^2}{\omega_0^2-\omega^2} \simeq 1$. Then

$$v_c = V_m (\cos \omega t - \cos \omega_0 t) \tag{12.15}$$

The natural frequency oscillations persist for only a small interval of time. During this interval the change in the power frequency term is very little and, hence, negligible. Then,

$$v_c = V_m (1 - \cos \omega_0 t) \qquad (12.16)$$

Figure 12.5 shows the conditions before and after current zero. After the fault is removed, the source attempts to charge the capacitance to its own potential. The inertial effect of the circuit inductance causes an over-swing of the voltage. The oscillations are damped by the system losses. The voltage v_c is known as transient recovery voltage. If the natural frequency ω_0 is high (*i.e.*, if L or C or both are small), the voltage across the circuit breaker contacts will rise very quickly. If this rate of rise of recovery voltage (r.r.r.v) exceeds the rate of rise of build up of the dielectric strength of the space between the contacts, the arc will restrike and the circuit interruption will be delayed for a further half cycle. Because of the possibility of restriking of the arc, the transient recovery voltage is also known as restriking voltage.

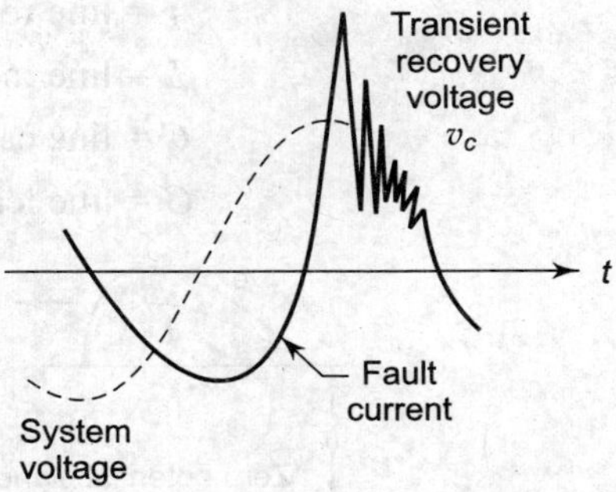

Fig. 12.5 Transient recovery voltage across the circuit breaker in Fig. 12.4

12.5 TRAVELLING WAVES ON TRANSMISSION LINES

A transmission line is a distributed parameter circuit. A distinguishing feature of such a circuit is its ability to support travelling waves of voltage and current. A circuit with distributed parameters has a finite velocity of electromagnetic field propagation. In such a circuit the changes in voltage and current, due to switching and lightning, do not occur simultaneously in all parts of the circuit but spread out in the form of travelling waves or surges. When a line is connected to a source of electrical energy, a travelling wave of voltage passes along the line at a finite velocity (which depends on the inductance and capacitance per unit length) establishing an electric field between the conductors as it does so. The voltage wave is accompanied by a current wave which in turn sets up a magnetic field in the surrounding space. At junctions and terminations these surges undergo reflections and refractions. In an extensive network with many lines and junctions, the number of travelling waves initiated by a single incident wave will mushroom at a considerable rate as the waves split and multiple reflections occur. It is true that the total energy of the resultant waves cannot exceed the energy of the incident wave. However, it is possible for the voltage to build up at certain junctions due to the reinforcing action of several waves. For a complete study of the phenomenon the use of Bewley lattice diagram or digital computer is necessary. In the following sections we will see some of the elementary aspects of wave propagation, reflections and refractions.

12.6 WAVE EQUATIONS

In Chapter 3 we discussed the behaviour of long lines under steady state conditions. In the equations for steady state, time does not enter explicitly but sinusoidal time variation is implied by the use of complex numbers. In the equations for transient conditions, both time and position enter as independent variables. Therefore, it is necessary to use partial differential equations in developing the theory of travelling waves.

Fig. 12.6 shows a pair of voltage and current travelling waves *e* and *i* on a single circuit of two conductors. One of the conductors may be the ground image of the overhead conductor. The line constants to neutral (*i.e.*, zero potential plane) are

r = line resistance per unit length, ohms/metre

L = line inductance per unit length, H/m

C = line capacitance per unit length, F/m

G = line leakage conductance per unit length, siemens/metre

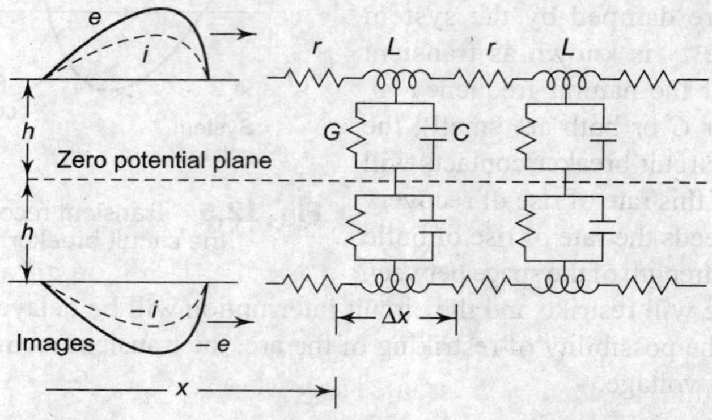

Fig. 12.6 Voltage and current travelling waves on a line

For an element Δx of the line the parameters are $r\,\Delta x$, $L\,\Delta x$, $C\,\Delta x$ and $G\,\Delta x$. The voltage drop in the element Δx in the positive direction of x is

$$-\Delta e = r\,\Delta x i + L\,\Delta x\,\frac{\partial i}{\partial t} \tag{12.17}$$

which in the limit as the element shrinks in length to infinitesmal dx, can be written as

$$\frac{\partial e}{\partial x} = -ri - L\frac{\partial i}{\partial t} \tag{12.18}$$

The change in current in the element Δx is

$$-\Delta i = e\,G\,\Delta x + C\,\Delta x\frac{\partial e}{\partial t} \tag{12.19}$$

which in the limit becomes

$$\frac{\partial i}{\partial x} = -\,Ge - C\,\frac{\partial e}{\partial t} \tag{12.20}$$

Introducing Laplace transform and substituting the operator p for $\dfrac{\partial}{\partial t}$, Eqs. (12.18) and (12.20) become

$$\frac{\partial e}{\partial x} = -\,(r + Lp)\,i = -\,Zi \tag{12.21}$$

$$\frac{\partial i}{\partial x} = -\,(G + C\,p)\,e = -\,Ye \tag{12.22}$$

where $Z = r + Lp$ is the impedance operator and $Y = G + Cp$ is the admittance operator. The negative signs in the above equations arise from the convention being used (both e and i decrease with increasing x).

Differentiate Eq. (12.21) with respect to x and substitute for $\frac{\partial i}{\partial x}$ from Eq. (12.22)

$$\frac{\partial^2 e}{\partial x^2} = -Z\frac{\partial i}{\partial x} = (-Z)(-Ye) = ZYe = \gamma^2 e \tag{12.23}$$

where γ is propagation constant and equals $\sqrt{ZY}$

Similarly,
$$\frac{\partial^2 i}{\partial x^2} = -Y\frac{\partial e}{\partial x} = ZYi = \gamma^2 i \tag{12.24}$$

The solution of Eq. (12.23) is of the form

$$e = \varepsilon^{x\gamma} f_1(t) + \varepsilon^{-x\gamma} f_2(t) \tag{12.25}$$

where $f_1(t)$ and $f_2(t)$ are integration constants with respect to x only and are, therefore, arbitrary functions of t. From Eqs. (12.22) and (12.25)

$$i = -Y\int e\,dx = -\sqrt{\frac{Y}{Z}}\,[\varepsilon^{x\gamma} f_1(t) - \varepsilon^{-x\gamma} f_2(t)] \tag{12.26}$$

To simplify the analysis, the line is assumed as lossless, *i.e.*, $r = 0$ and $G = 0$. Then

$$\gamma = \sqrt{(r + Lp)(G + Cp)} = p\sqrt{LC} = \frac{p}{v} \tag{12.27}$$

where
$$v = 1/\sqrt{LC}$$

Then,
$$e = \varepsilon^{xp/v} f_1(t) + e^{-xp/v} f_2(t) \tag{12.28}$$

$$i = -\sqrt{\frac{C}{L}}\,[\varepsilon^{xp/v} f_1(t) - \varepsilon^{-xp/v} f_2(t)] \tag{12.29}$$

By Taylor's theorem

$$f(t + a) = f(t) + af'(t) + \frac{a^2}{2!} f''(t) + \ldots$$

$$= \left[1 + ap + \frac{a^2}{2!}p^2 + \ldots\right] f(t) = \varepsilon^{ap} f(t) \tag{12.30}$$

By comparison with Eq. (12.30) we can write Eqs. (12.28) and (12.29) as

$$e = f_1\left(t + \frac{x}{v}\right) + f_2\left(t - \frac{x}{v}\right) \tag{12.31}$$

$$i = -\sqrt{\frac{C}{L}}\left[f_1\left(t + \frac{x}{v}\right) - f_2\left(t - \frac{x}{v}\right)\right] \tag{12.32}$$

A function of the form $f\left(t \pm \frac{x}{v}\right)$ or $f(x \pm vt)$ represents a travelling wave because for any value of t, a corresponding value of x can be found such that $(x \pm vt)$ has a constant value and, therefore, defines a fixed point on $f(x \pm vt)$. Corresponding values of x and t which define the same points on a wave are given by

$t - x/v = \lambda_1$ for waves travelling in the direction of $+x$ (forward waves) (12.33)

$t + x/v = \lambda_2$ for waves travelling in the direction of $-x$ (backward waves)

The velocity of propagation can be found by differentiating Eq. (12.33) keeping in view that λ_1 and λ_2 are constants.

$$\left.\begin{aligned} \frac{dx}{dt} &= v = 1/\sqrt{LC} && \text{for forward wave} \\ \frac{dx}{dt} &= -v = -1/\sqrt{LC} && \text{for backward wave} \end{aligned}\right] \tag{12.34}$$

From Eqs. (12.31) and (12.32)

$$\left.\begin{aligned} \frac{e}{i} &= \sqrt{L/C} && \text{for forward wave} \\ \frac{e}{i} &= -\sqrt{L/C} && \text{for backward wave} \end{aligned}\right] \tag{12.35}$$

The ratio $\sqrt{L/C}$ is known as surge impendace (or characteristic impedance) and is written as Z_c. The reciprocal of surge impedance is surge admittance Y_c. Surge impedance has the characteristics of a resistance and has the dimensions of ohm.

The voltage and current waves are replicas of each other. The forward waves are written with a subscript f and the backward waves are written with a subscript b. Fig. 12.7 shows the schematic representation of polarities and directions of propagation for forward and backward waves. The forward voltage and current waves (Fig. 12.7a and c) propagate in the positive direction of x and have the same polarity because the ratio e/i, for a forward wave is positive as given by Eq. (12.35). The backward voltage and current waves (Fig. 12.7b and d) propagate in the negative x direction and have different polarities because the ratio e/i for a backward wave is negative. Thus, if both forward and backward voltage waves are positive the backward current wave differs from the forward one in two respects, *viz.*, in its direction of flow in the conductor as well as in the direction in which the current wave propagates along the line.

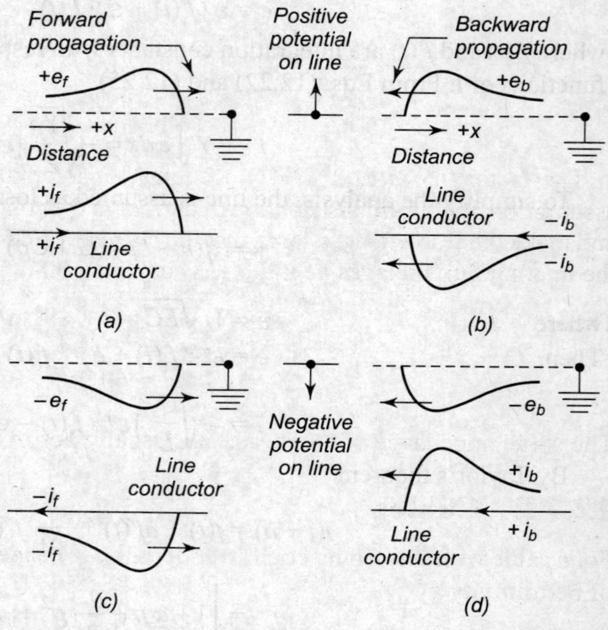

Fig. 12.7 Schematic representation of polarities and directions of voltage and current waves

Each of the Eqs. (12.31) and (12.32) contains one backward and one forward wave. In a particular case, however, only one of them may be present. Also the complete wave may consist of more than one backward and forward waves.

An alternative method for finding the expression for velocity of propagation is as under:

The charge ΔQ on the element Δx is

$$\Delta Q = Ce\Delta x \tag{12.36}$$

$$i = \frac{\Delta Q}{\Delta t} = Ce\frac{\Delta x}{\Delta t} = Cev \tag{12.37}$$

The flux $\Delta\phi$ on element Δx is

$$\Delta\phi = L\,\Delta x\, i = L\,\Delta x\, Cev \tag{12.38}$$

$$\text{Emf } e = \frac{\Delta\phi}{\Delta t} = L\,Cev\,\frac{\Delta x}{\Delta t} = L\,Cev^2 \tag{12.39}$$

$\therefore$
$$e = LC\,e\,v^2$$

or
$$v = 1/\sqrt{LC} \tag{12.40}$$

12.7 SURGE IMPEDANCE AND WAVE VELOCITY

12.7.1 Overhead Lines

For a single conductor of radius r metres and situated at a height h metres above a perfect ground, the line constants are

$$\mathrm{L} = \left(0.5 + 2\ln\frac{2h}{r}\right)10^{-7} \text{ H/m} \tag{12.41}$$

$$\mathrm{C} = 10^{-9}/18 \ln\frac{2h}{r} \text{ F/m} \tag{12.42}$$

The factor 0.5 in Eq. (12.41) accounts for internal flux linkages in the conductor on the assumption of uniform distribution of current in the conductor. The surges have high frequencies and make the skin effect so large that the current resides on the surface of the conductor making the internal flux linkages negligible. Thus, the term 0.5 in Eq. (12.41) can be neglected.

Then
$$v = 1/\sqrt{LC} = 3 \times 10^8 \text{ m/s or } 300 \text{ m/}\mu\text{s} \tag{12.43}$$

$$Z_c = \sqrt{L/C} = 60 \ln\frac{2h}{r} \text{ ohms} \tag{12.44}$$

The surge impedance of overhead lines usually lies in the range 400 to 600 ohms.

12.7.2 Cables

For a cable with solid inner conductor of radius r, inside a sheath of inner radius R and insulation of permittivity k

$$L \cong [2 \ln (R/r)] \times 10^{-7} \text{ H/m} \tag{12.45}$$

$$C = (k \times 10^{-9})/[18 \ln (R/r)] \text{ F/m} \tag{12.46}$$

$$v = 1/\sqrt{LC} = 3 \times 10^8/\sqrt{k} \text{ m/s} \tag{12.47}$$

$$Z_c = \sqrt{L/C} = (60/\sqrt{k})\,[\ln (R/r)] \text{ ohms} \tag{12.48}$$

In most of the commercial cables k varies from 2.5 to 4. Thus, velocity of propagation on cables is about one half to two-third of that on overhead lines. The surge impedance of a cable is around 50 ohms.

12.8 SPECIFICATIONS OF TRAVELLING WAVES

A travelling wave is characterized by four specifications as shown in Fig. 12.8.

1. Crest is the maximum amplitude and is generally expressed in kilovolts or kiloamperes.
2. Front is the portion of the wave before crest and is expressed in time from beginning of the wave to its crest value in μ-sec or millisec. However, for waves having a slow

initial rate of rise, *i.e.*, a long toe, it is better to consider virtual front as determined by straight lines between 30% and 90% points (Fig. 12.8b). The virtual front is $1.67 \times t_1$. The extension of this straight line to *x*-axis gives O_1, *i.e.*, the virtual zero.

3. Tail of the wave is the portion beyond the crest. It is designated by the time in, μ-sec from beginning of the wave to the point where the wave has decreased to half of its value at crest. For waves having a slow initial rate of rise, *i.e.*, long toe, the tail time is measured from O_1 to 50% value on tail (Fig. 12.8b).
4. Polarity is the polarity of crest voltage or current.

A positive wave of 1000 kV crest, 1μ-sec front and 20 μ-secs tail is generally written as + 1000/1.0/20.0.

A travelling wave can be represented mathematically in a number of ways. The simplest and most commonly used representation is the infinite rectangular or step wave shown in Fig. 12.9. This wave jumps suddenly from zero to full value and is maintained at that value thereafter. Such a wave is most dangerous to apparatus since its abrupt front causes maximum gradients and sustained tail produces maximum oscillations in machine windings. Hence, the analysis based on it is liable to err on the side of safety.

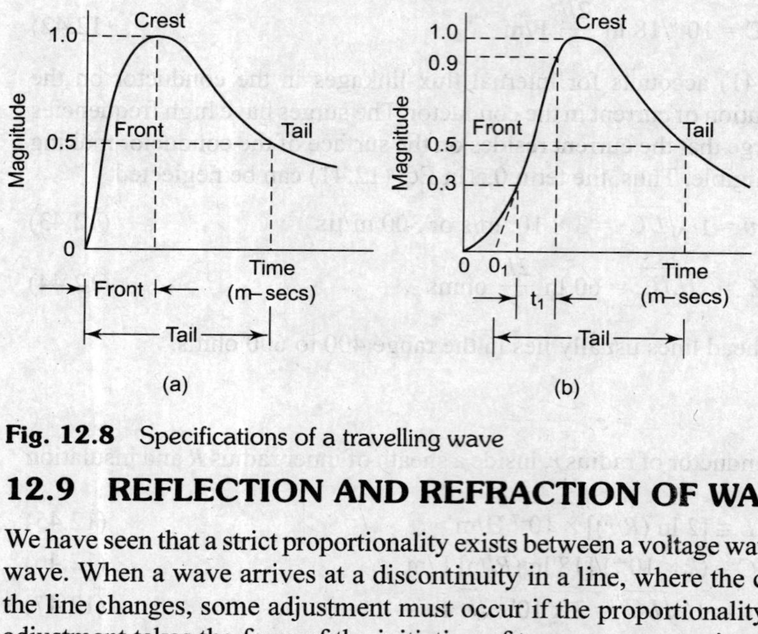

Fig. 12.8 Specifications of a travelling wave

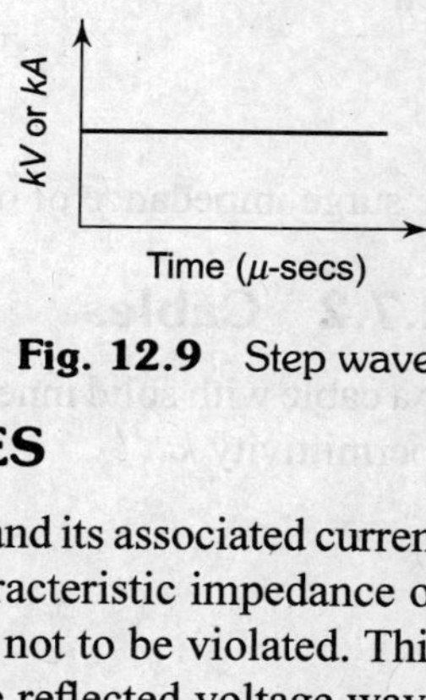

Fig. 12.9 Step wave

12.9 REFLECTION AND REFRACTION OF WAVES

We have seen that a strict proportionality exists between a voltage wave and its associated current wave. When a wave arrives at a discontinuity in a line, where the characteristic impedance of the line changes, some adjustment must occur if the proportionality is not to be violated. This adjustment takes the form of the initiation of two new wave pairs. The reflected voltage wave and its associated current wave travel back down the line, superimposed on the initial or forward or incident wave. The refracted or transmitted wave penetrates beyond the discontinuity. The amplitudes of reflected and refracted waves are such that the voltage to current proportionality is preserved for each as demanded by the surge impedances of the lines on which they are travelling, current and voltages at the discontinuity are themselves continuous and energy is conserved.

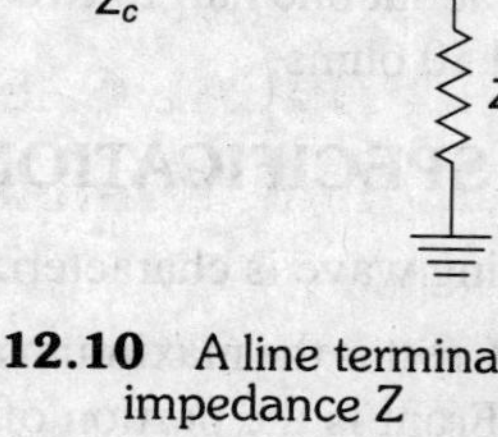

Fig. 12.10 A line terminated by impedance Z

To develop general formulae for reflection and refraction coefficients we consider a line with surge impedance Z_c terminated by an impedance Z (Fig. 12.10). We denote incident or forward voltage and current waves by e_f and i_f, reflected voltage and current waves by e_r and i_r and refracted (or transmitted) voltages and current waves by e_t and i_t. Then

$$\frac{e_f}{i_f} = Z_c \tag{12.49a}$$

$$\frac{e_r}{i_r} = -Z_c \tag{12.49b}$$

$$\frac{e_t}{i_t} = Z \tag{12.49c}$$

The negative sign in Eq. (12.49 b) is because of the reason that e_r and i_r are travelling in the negative direction of x or backwards on the same line.

Since voltage and current waves are continuous at the junction

$$e_t = e_f + e_r \tag{12.50}$$

$$i_t = i_f + i_r \tag{12.51}$$

Substitute the values of i_f, i_r and i_t from Eqs. (12.49 a, b, c) into Eq. (12.51)

$$\frac{e_t}{Z} = \frac{e_f}{Z_c} - \frac{e_r}{Z_c} \tag{12.52}$$

From Eqs. (12.50) and (12.52)

$$e_t\left(1+\frac{Z_c}{Z}\right) = 2e_f$$

$$e_t = e_f \frac{2Z}{Z+Z_c} \tag{12.53}$$

$$e_r = e_t - e_f = e_f \frac{Z-Z_c}{Z+Z_c} \tag{12.54}$$

$$i_t = \frac{e_t}{Z} = e_f \frac{2}{Z+Z_c} \tag{12.55}$$

$$i_r = -\frac{e_r}{Z_c} = -\frac{e_f}{Z_c}.\frac{Z-Z_c}{Z+Z_c} = i_f \frac{Z_c - Z}{Z_c + Z} \tag{12.56}$$

The coefficients $\frac{2Z}{Z+Z_c}$ and $\frac{Z-Z_c}{Z+Z_c}$ are called refraction and reflection coefficients respectively.

12.10 TYPICAL CASES OF LINE TERMINATIONS

12.10.1 Line Terminated by a Resistance Equal to Surge Impedance

If we substitute Z_c for Z in Eqs. (12.53-12.56) we find that $e_t = e_f$, $i_t = i_f$, $e_r = 0$ and $i_r = 0$. Thus, the incident wave continues as it is and there is no reflection.

12.10.2 Open Circuited Line

If we substitute $Z = \infty$ in Eqs. (12.53-12.56) we find that voltage at open end $e_t = 2e_f$.

$$\begin{aligned} i_t &= 0 \\ e_r &= e_f \\ i_r &= -i_f \end{aligned} \tag{12.57}$$

An open circuit at the end of a line demands that the current at that point be zero at all times. Thus, when an incident current wave i_f arrives at the open circuit, a reflected wave equal to $-i_f$ is at once initiated to satisfy this boundary condition. The resultant voltage at the open end is twice the incident voltage. The waves are illustrated in Fig. 12.11.

12.10.3 Short Circuited Line

If we substitute $Z = 0$ in Eqs. (12.53-12.56)

we get

$$\begin{aligned} e_t &= 0 \\ i_t &= 2i_f \\ e_r &= -e_f \\ i_r &= i_f \end{aligned} \tag{12.58}$$

The unique characteristic of the short circuit is that the voltage across it is zero. When an incident voltage wave e_f reaches a short circuit, the reflected voltage wave must be $-e_f$ to satisfy the condition that the voltage across the short circuit is zero. The waves are shown in Fig. 12.12.

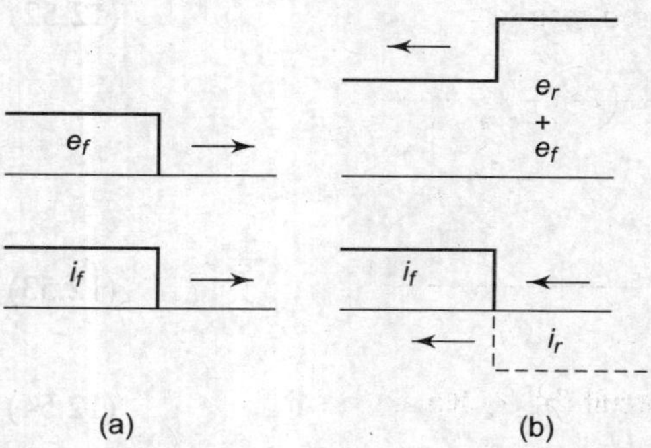

Fig. 12.11 Reflection of waves at open circuit

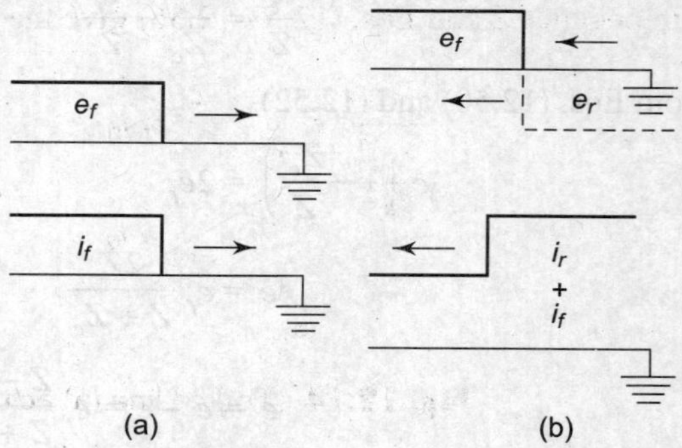

Fig. 12.12 Reflection of waves at short circuit

12.11 EQUIVALENT CIRCUIT FOR TRAVELLING WAVE STUDIES

An equivalent circuit for travelling wave studies can be easily developed through the application of Thevenin's theorem. As per this theorem any network, as seen from two terminals, can be replaced by open circuit voltage across these terminals, connected in series with the impedance of the network as seen from these terminals.

Figure 12.13 (a) shows a line with a surge impedance Z_c. A surge wave e_f travels on the line towards the terminals, XY. The open circuit voltage across the terminals XY is $2e_f$ (section 12.10.2.) The impedance as seen from terminals XY, with the sending end of the line short circuited, is Z_c. Therefore, the circuit of Fig. 12.13 (a) can be replaced by the equivalent circuit of Fig. 12.13 (b). From this equivalent circuit

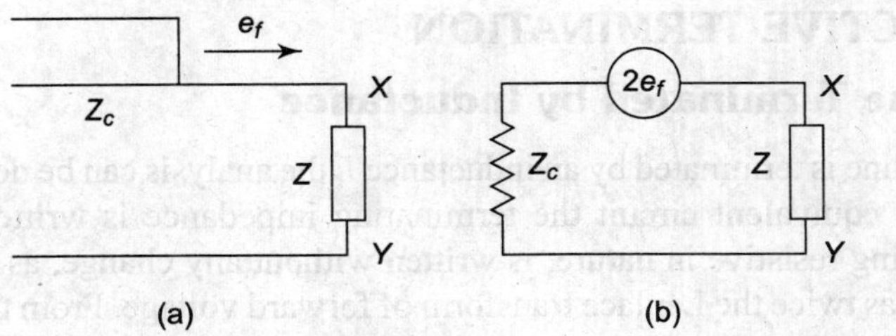

Fig. 12.13 Equivalent circuit on travelling waves (a) actual circuit (b) equivalent circuit

Current through Z, *i.e.*, $i_t = 2e_f/(Z + Z_c)$

Voltage across Z, *i.e.*, $e_t = 2e_f Z/(Z + Z_c)$

The above results, as expected, are the same as Eqs. (12.55) and (12.53), e_r and i_r can be found out from above values of e_t and i_t by using Eqs. (12.50) and (12.51). The equivalent circuit is very useful for travelling wave studies and can be used for resistive as well as reactive terminations.

12.12 FORKED LINE

When a surge voltage travels on a line of surge impedance Z_c and reaches the junction of the line with another line of surge impedance Z, the analysis is exactly similar to that of a line terminated by impedance Z and Eqs. (12.53-12.56) give the desired results.

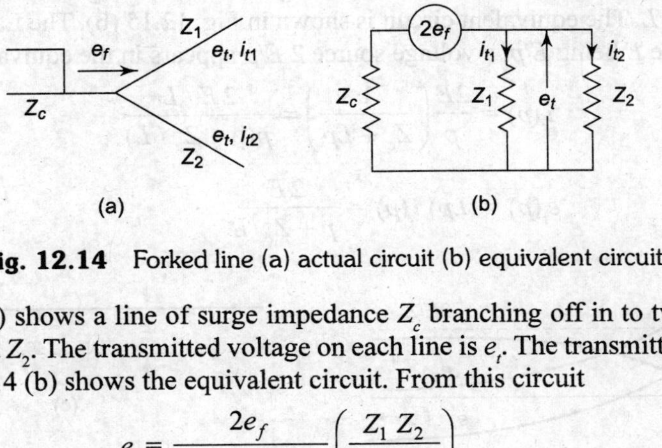

Fig. 12.14 Forked line (a) actual circuit (b) equivalent circuit

Figure 12.14 (a) shows a line of surge impedance Z_c branching off in to two lines of surge impedances Z_1 and Z_2. The transmitted voltage on each line is e_t. The transmitted currents are i_{t1} and i_{t2}. Figure 12.14 (b) shows the equivalent circuit. From this circuit

$$e_t = \frac{2e_f}{Z_c + \dfrac{Z_1 Z_2}{Z_1 + Z_2}} \left(\frac{Z_1 Z_2}{Z_1 + Z_2} \right) \quad (12.59)$$

$$i_{t1} = \frac{e_t}{Z_1} \quad (12.60)$$

$$i_{t2} = \frac{e_t}{Z_2} \quad (12.61)$$

$$e_r = e_t - e_f$$
$$i_r = i_{t1} + i_{t2} - i_f \quad (12.62)$$

The above procedure can be extended to analyze the case of a line branching off into any number of lines.

12.13 REACTIVE TERMINATION

12.13.1 Line Terminated by Inductance

When a loss-free line is terminated by an inductance L the analysis can be done by using Laplace transform. In the equivalent circuit the terminating impedance is written as pL. The surge impedance Z_c, being resistive in nature, is written without any change, as Z_c. The open circuit voltage is written as twice the Laplace transform of forward voltage. From the equivalent circuit the values of $e_t(p)$ and $i_t(p)$, *i.e.*, the transmitted voltage and current in the p domain can be directly written. e_t and i_t are found by taking the inverse transforms of $e_t(p)$ and $i_t(p)$. Standard Laplace transform tables can be used for analysis.

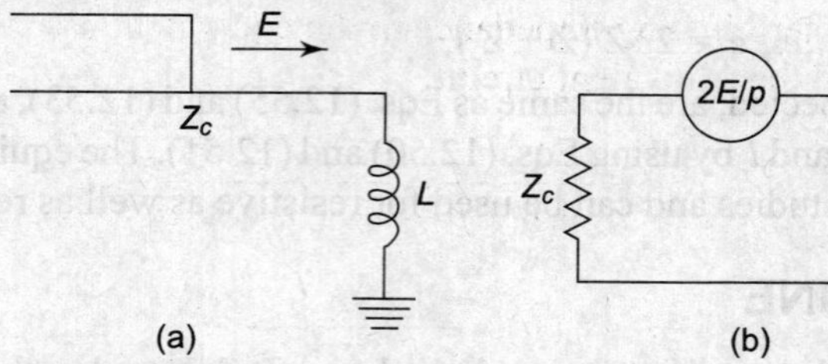

Fig. 12.15 Line terminated by inductance (a) actual circuit (b) equivalent circuit

Figure 12.15 (a) shows a step wave E travelling on a line of surge impedance Z_c terminated by an inductance L. The equivalent circuit is shown in Fig. 12.15 (b). The Laplace transform of a constant voltage E being E/p, a voltage source 2 E/p appears in the equivalent circuit.

$$i_t(p) = \frac{2E}{p}\left(\frac{1}{Z_c + Lp}\right) = \frac{2E/L}{p(\rho + Z_c/L)} \tag{12.63}$$

$$e_t(p) = (Lp)\, i_t(p) = \frac{2E}{p + Z_c/L} \tag{12.64}$$

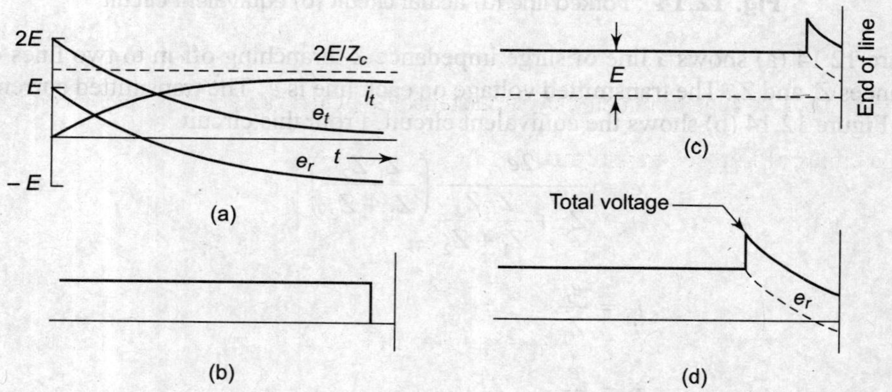

Fig. 12.16 Travelling waves on a line with inductive termination (a) transmitted and reflected waves as function of time (b, c, d) Disposition of waves at different instants.

From the tables of inverse transforms

$$i_t = \frac{2E}{Z_c}\left(1 - \varepsilon^{-z_c t/L}\right) \tag{12.65}$$

$$e_t = 2E\ \varepsilon^{-z_c t/L} \tag{12.66}$$

$$e_r = e_t - e_f = 2E\ \varepsilon^{-z_c t/L} - E = E\,(2\varepsilon^{-z_c t/L} - 1) \tag{12.67}$$

$$i_r = i_t - i_f = \frac{2E}{Z_c}(1 - \varepsilon^{-z_c t/L}) - \frac{E}{Z_c} = \frac{E}{Z_c}(1 - 2\varepsilon^{-z_c t/L}) \tag{12.68}$$

e_t, i_t and e_r as functions of time are shown in Fig. 12.16 (a). The disposition of voltage waves at different instants is shown in Fig. 12.16 (b, c and d).

12.13.2 Line Terminated by Capacitance

The analysis of a line terminated by capacitance is similar to that of a line terminated by inductance. In this case the terminating impedance in the equivalent circuit is $1/Cp$. Fig. 12.17 shows the actual circuit and the equivalent circuit.

$$i_t(p) = \frac{2E}{P}\,\frac{1}{Z_c + 1/Cp} = \frac{2E}{Z_c}\left(\frac{1}{p + 1/Z_c C}\right) \tag{12.69}$$

$$e_t(p) = \left(\frac{1}{Cp}\right) i_t(p) = \frac{1}{Cp}\cdot\frac{2E}{Z_c}\left(\frac{1}{p + 1/Z_c C}\right)$$

$$= \frac{2E}{CZ_c}\cdot\frac{1}{p\left(p + \dfrac{1}{Z_c C}\right)} \tag{12.70}$$

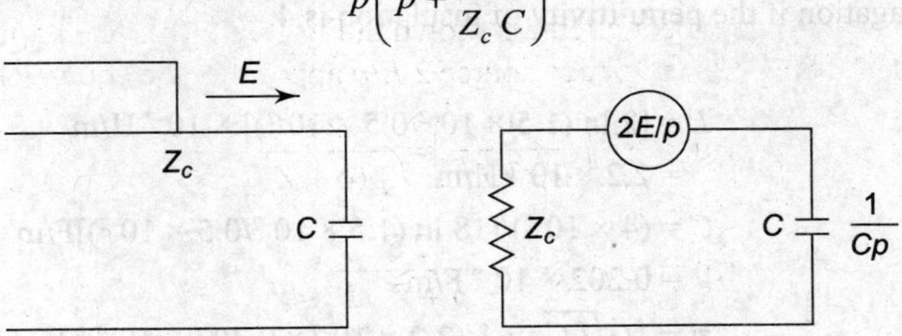

Fig. 12.17 Line terminated by capacitance (a) actual circuit (b) equivalent circuit

From the tables of Laplace transforms

$$i_t = \frac{2E}{Z_c}\,\varepsilon^{-t/CZ_c} \tag{12.71}$$

$$e_t = 2E\,(1 - \varepsilon^{-t/CZ_c}) \tag{12.72}$$

$$e_r = e_t - e_f = e_t - E = E\,(1 - 2\varepsilon^{-t/CZ_c}) \tag{12.73}$$

$$i_r = i_t - i_f = \frac{2E}{Z_c}\,\varepsilon^{-t/CZ_c} - \frac{E}{Z_c} = \frac{E}{Z_c}(2\varepsilon^{-t/CZ_c} - 1) \tag{12.74}$$

e_t, i_t and e_r, as functions of time are shown in Fig. 12.18 (a). The disposition of voltage wave at different instants is shown in Fig. 12.18 (b, c and d).

In travelling wave studies, time t is usually expressed in μ-secs. Therefore, L and C should be taken in, μH and, μF.

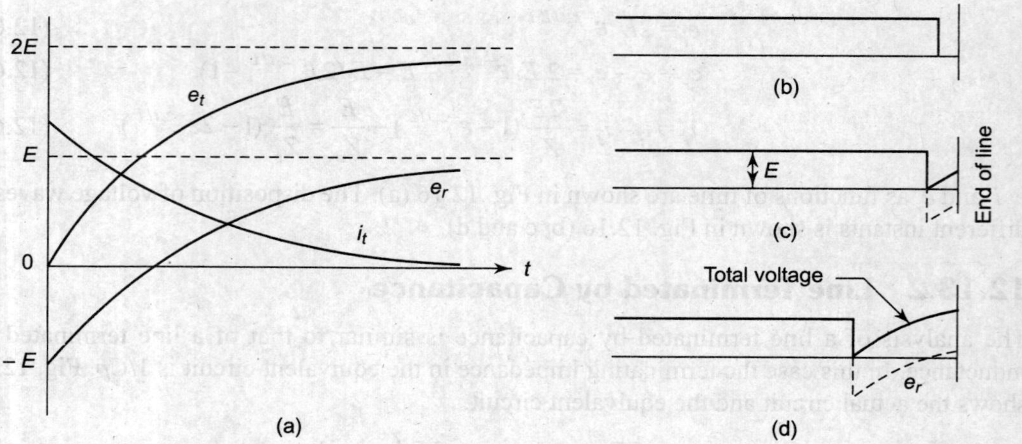

Fig. 12.18 Travelling waves on a line with capacitive termination (a) transmitted and reflected waves as function of time (b, c, d) Disposition of waves at different instants

Example 12.1

A cable has an inner conductor of radius 0.5×10^{-2} m inside a sheath of inner radius 1.5×10^{-2} m. Find the inductance per metre length, capacitance per metre length, surge impedance and velocity of propagation if the permittivity of insulation is 4.

Solution

$$L = [2 \ln (1.5 \times 10^{-2}/0.5 \times 10^{-2})] \times 10^{-7} \text{ H/m}$$
$$= 2.2 \times 10^{-7} \text{ H/m}$$
$$C = (4 \times 10^{-9})/[18 \ln (1.5 \times 10^{-2}/0.5 \times 10^{-2})]\text{F/m}$$
$$= 0.202 \times 10^{-9} \text{ F/m}$$
$$v = 1/\sqrt{LC} = 1/(2.2 \times 10^{-7} \times 0.202 \times 10^{-9})^{0.5}$$
$$= 1.5 \times 10^{8} \text{ m/s}$$
$$Z_c = \sqrt{L/C} = (2.2 \times 10^{-7}/0.202 \times 10^{-9})^{0.5} = 33 \text{ ohm}$$

Example 12.2

A surge of 100 kV travels along an overhead line towards its junction with a cable. The surge impedances for the overhead line and cable are 400 ohms and 50 ohms respectively. Find the magnitude of the surge transmitted on the cable. Comment on the results.

Solution

$$e_t = e_f \frac{2Z}{Z + Z_c}$$

$$e_f = 100 \text{ kV}, Z_c = 400\ \Omega, Z = 50\ \Omega$$
$$e_t = 100 \frac{2 \times 50}{50 + 400} = 22.22 \text{ kV}$$

It is seen that the magnitude of the surge transmitted on the cable is much less than that of the surge on the overhead line. The transformers in indoor sub-stations are connected to an incoming overhead line through a short length of cable. Since the surge transmitted on the cable is less than that on the line, this method is very effective in reducing the amplitude of surge waves. The transformers connected at the end of the cable are subjected to a much lesser surge voltage than the original surge voltage.

Example 12.3

A surge of 200 kV travelling on a line of surge impedance 400 ohms reaches a junction of the line with two branch lines of surge impedances 500 ohms and 300 ohms respectively. Find the surge voltage and current transmitted into each branch line. Also find the reflected surge voltage and current.

Solution

With reference to Fig. 12.14

$$e_t = \frac{2e_f}{Z_c + \dfrac{Z_1 \times Z_2}{Z_1 + Z_2}} \times \frac{Z_1 \times Z_2}{Z_1 + Z_2} = \frac{2 \times 200}{400 + \dfrac{500 \times 300}{500 + 300}} \left[\frac{500 \times 300}{500 + 300}\right]$$

$$= 127.66 \text{ kV}$$

$$i_{t1} = \frac{e_t}{Z_1} = \frac{127.66}{500} = 0.255 \text{ kA}$$

$$i_{t2} = \frac{e_t}{Z_2} = \frac{127.66}{300} = 0.425 \text{ kA}$$

$$e_r = e_t - e_f = 127.66 - 200 = -72.34 \text{ kV}$$

$$i_r = i_{t1} + i_{t2} - i_f = 0.255 + 0.425 - \frac{200}{400} = 0.18 \text{ kA}$$

Example 12.4

A step wave of 100 kV travels on a line having a surge impedance of 400 ohms. The line is terminated by an inductance of 4000 μH. Find the voltage across the inductance and the reflected voltage wave.

Solution

As given by Eqs. (12.66) and (12.67)

$$e_t = 2\,E\,\varepsilon^{-z_c t/L}$$

$$e_r = E\,(2\,\varepsilon^{-z_c t/L} - 1)$$

$$E = 100 \text{ kV},\ Z_c = 400\ \Omega,\ L = 4000\ \mu\text{H}$$

$$e_t = 200\ \varepsilon^{-400t/4000} = 200\ \varepsilon^{-0.1t} \text{ kV}$$

$$e_r = 100\ (2\varepsilon^{-0.1t} - 1) \text{ kV}$$

12.14 SUCCESSIVE REFLECTIONS, BEWLEY LATTICE DIAGRAM

In an extensive network with many junctions and terminations, the number of transmitted and reflected waves, initiated by a single incident wave, increases as the wave meets various junctions.

It is true that we cannot have more energy than that possessed by the initial surge and this energy will be shared by many waves that the incident wave gives rise to. Nevertheless the voltage at some points may build up due to reinforcing action of the waves. A systematic way of keeping track of all the transmitted and reflected waves is through the use of Bewley lattice diagram.

Fig. 12.19 shows the method of drawing this diagram. A line with three sections having surge importances Z_1, Z_2 and Z_3 is shown. These three sections have different surge impedances and different velocities of wave propagation. At the top of the diagram we write the reflection and refraction coefficients for waves approaching the junction from right and left. These coefficients are functions of surge impedance. Thus

a = reflection coefficient for waves approaching from left

a' = reflection coefficient for waves approaching from right

b = refraction coefficient for waves approaching from left

b' = refraction coefficient for waves approaching from right

In drawing the lattice, the horizontal distances between junctions are laid proportional to the time passage of wave on these sections (and not proportional to lengths of sections). The advantage of doing this is that all diagonals have the same slope and the time scale is applicable to each branch. The vertical time scale can be suitably selected depending on the time for which computations are to be made.

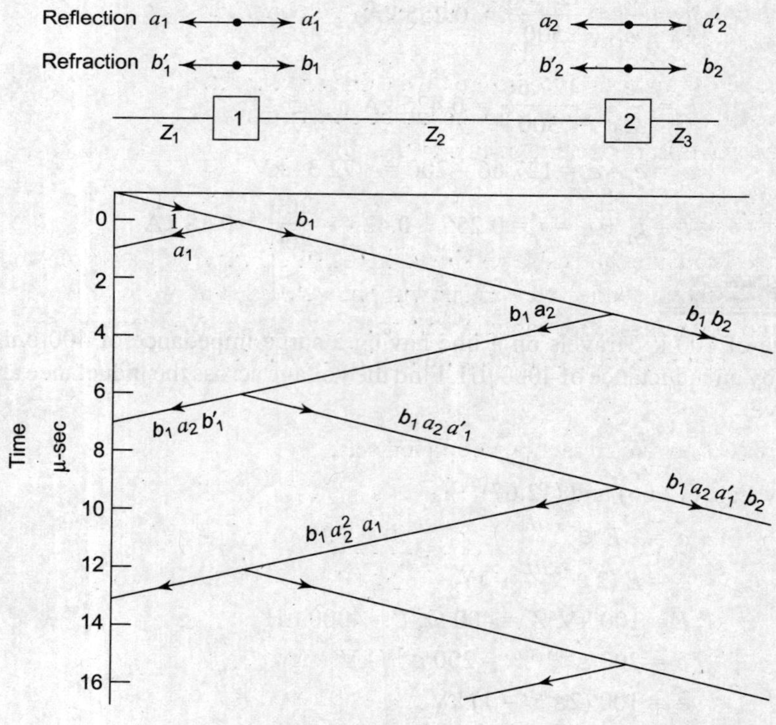

Fig. 12.19 Bewley lattice diagram

A unit incident wave approaching junction 1 from left gives rise to a refracted wave b_1 and

reflected wave a_1. The wave b_1 reaches junction 2 and gives rise to refracted wave $b_1\ b_2$ and reflected wave $b_1\ a_2$. The wave $b_1\ a_2$ approaches junction 1 from right and gives rise to refracted wave $b_1\ a_2\ b'_1$ and reflected wave $b_1\ a_2\ a'_1$ and so on. Thus starting with the initial incident wave, the various reflected and refracted waves are determined and indicated on the diagram.

From the diagram the previous history of any wave can be traced easily. The total potential at any point, at any time, is the superposition of all the waves which have arrived at that point, till the instant of time under consideration, displaced in position from each other by intervals equal to differences in their time of arrival. If necessary the amount by which a wave is attenuated in travelling between junctions can also be incorporated in the diagram.

Example 12.5

A line of surge impedance 400 ohms is charged from a battery of constant voltage 135 V. The line is 300 m long and is terminated in a resistance of 200 ohms. Plot reflection lattice and the voltage across the terminating resistance.

Solution

At the terminating point the reflection coefficient a is given by

$$a = \frac{200 - 400}{200 + 400} = -\frac{1}{3}$$

At the battery end the reflection coefficient is – 1 since the battery voltage is constant and any wave reaching the battery must be nullified by an equal and opposite wave. The reflection lattice is shown in Fig. 12.20 (b).

The voltage across the resistance is zero till the first wave reaches the termination. The voltages across the resistance are also shown on the lattice and are plotted in Fig. 12.20 (c). T is the time of travel of the wave from one end of line to the other.

Example 12.6

A system of overhead lines and cables is shown in Fig. 12.21 (a). The values of surge impedances and lengths of different sections are as shown. The velocities of propagation are 300 m/μ-sec for OH line and 200 m/μ-sec for cable. A unit step wave approaches the system from left. Draw reflection lattice and plot voltage v_4 till a time of 8 μ-seconds.

Solution

The various reflection and refraction coefficients are

$$a_1 = -a_2 = a_3 = -a_4 = a'_4 = -a'_3 = a'_2 = -a'_1$$

$$= \frac{Z_2 - Z_1}{Z_2 + Z_1} = \frac{50 - 500}{50 + 500} = -0.818$$

$$b_1 = b_3 = b'_4 = b'_2 = \frac{2Z_1}{Z_1 + Z_2} = 0.182$$

$$b_2 = b_4 = b'_3 = b'_1 = \frac{2Z_1}{Z_1 + Z_2} = 1.82$$

The time of travel of the wave on each section of cable is 1 μ-sec. For the 150 m overhead line the time of travel of the wave is 0.5 μ-sec.

The reflection lattice is shown in Fig. 12.21 (b). Voltage v_4 is plotted in Fig. 12.21 (c).

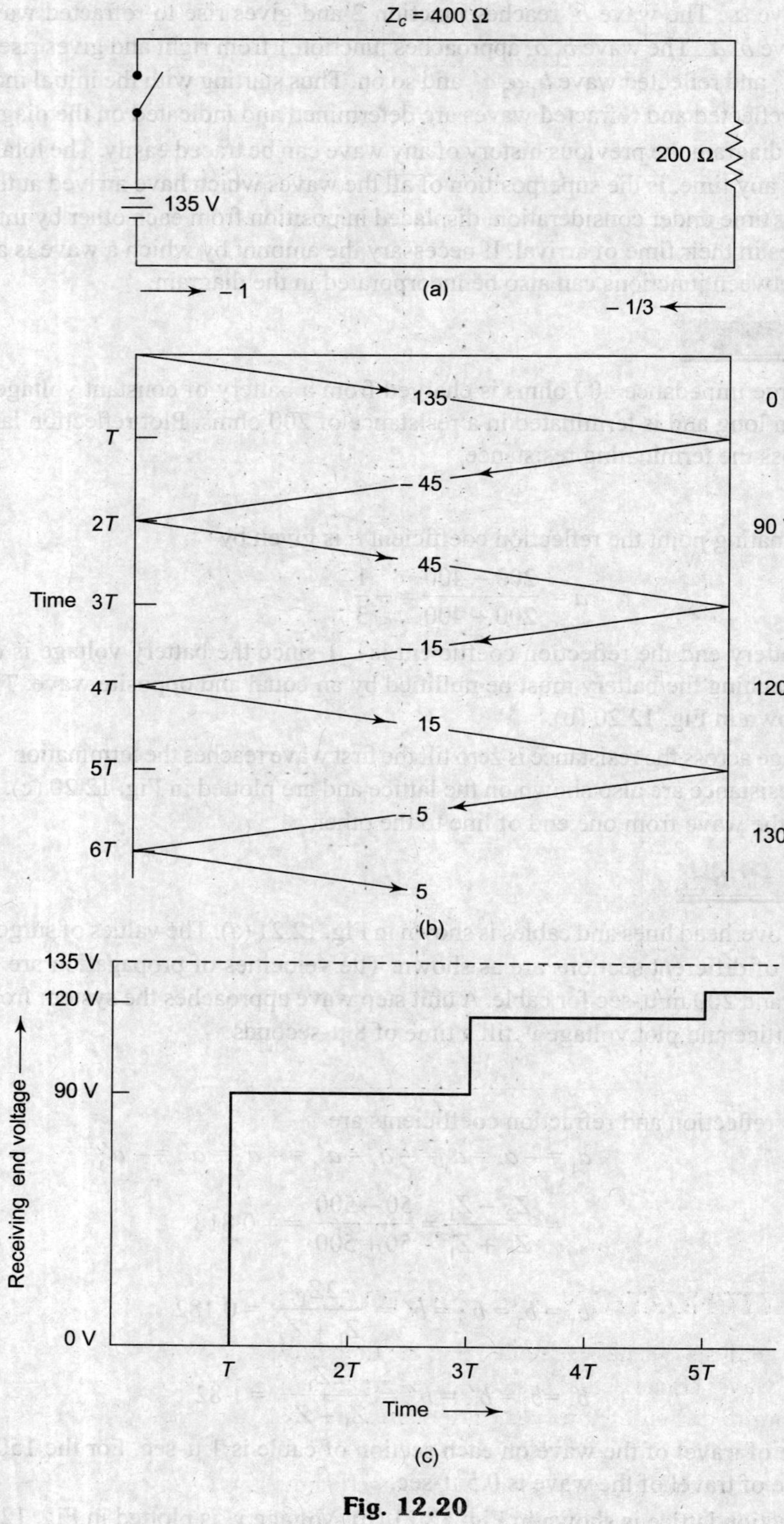
Z_c = 400 Ω
200 Ω
135 V
→ −1
(a)
−1/3 ←
135
0
T
−45
2T
90 V
45
Time 3T
−15
4T
120 V
15
5T
−5
6T
130 V
5
(b)
135 V
120 V
90 V
0 V
Receiving end voltage →
T
2T
3T
4T
5T
Time →
(c)

Fig. 12.20

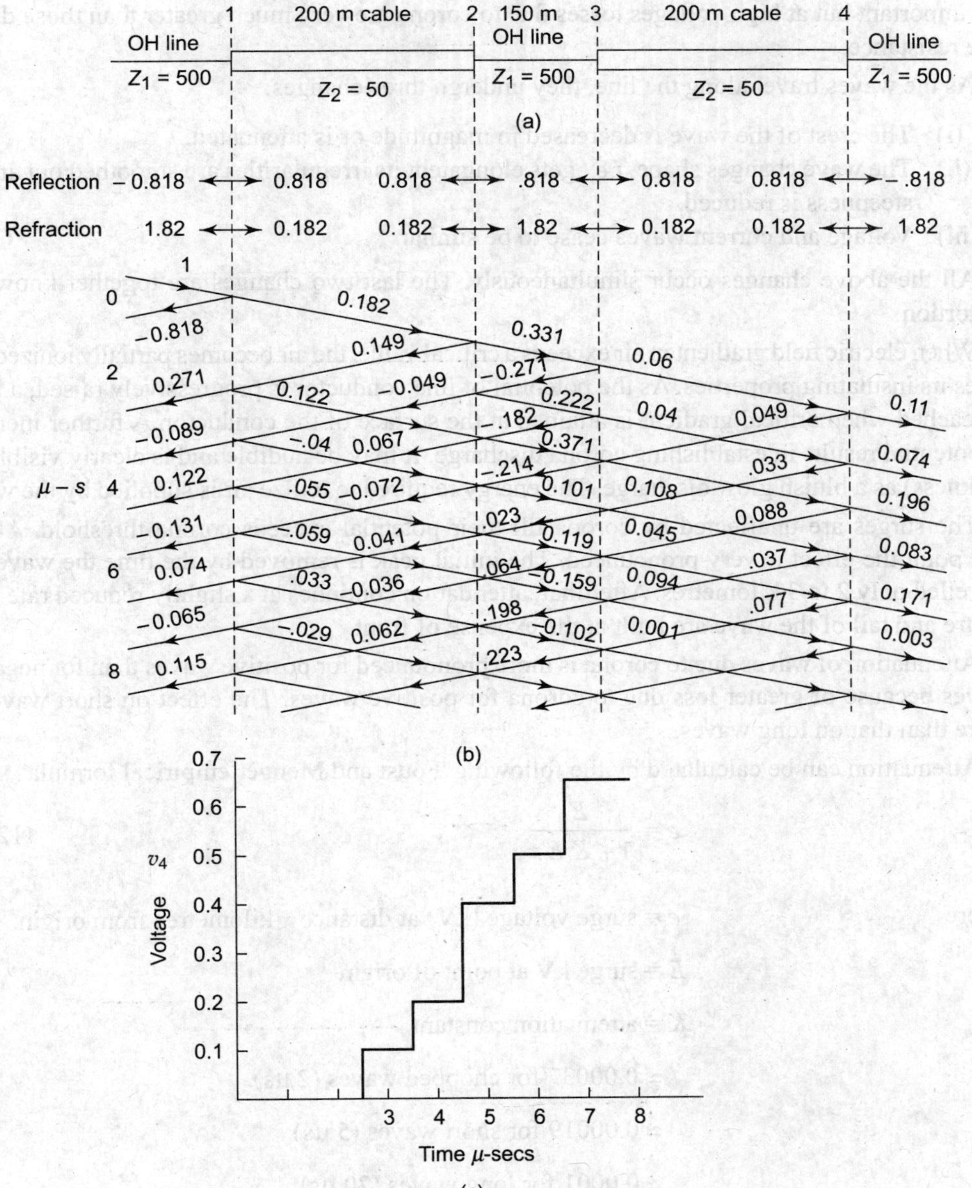

Fig. 12.21

12.15 ATTENUATION AND DISTORTION

In the above analysis, we have assumed that the lines are lossless and there is no attenuation or distortion of waves. In an actual system losses are always there. The presence of losses makes the analysis more difficult. However, these losses are very much desirable because the energy of waves is dissipated through these losses.

These losses are due to line resistance, leakage conductance and corona. The losses due to leakage conductance are usually very small. At low voltages the losses due to line resistance are important but at high voltages losses due to corona are very much greater than those due to line resistance.

As the waves travel along the line, they undergo three changes:

(i) The crest of the wave is decreased in magnitude or is attenuated.
(ii) The wave changes shape, *i.e.*, gets elongated, its irregularities are smoothed out and its steepness is reduced.
(iii) Voltage and current waves cease to be similar.

All the above changes occur simultaneously. The last two changes are together known as distortion.

When electric field gradient in air exceeds a critical value, the air becomes partially ionized and loses its insulating properties. As the potential of line conductor is progressively raised, a level is reached when critical gradient is attained at the surface of the conductor. A further increase in potential results in establishing corona discharge. It may be audible and is clearly visible (in darkness) as a bluish glow discharge. The energy required to ionize air is supplied by the wave.

The surges are unaffected by corona till their potential exceeds corona threshold. Above this point the effect is very pronounced. The initial peak is removed by the time the wave has travelled only 2 to 3 kilometres. After that, attenuation continues at a slightly reduced rate. The centre and tail of the wave are built at the expense of front.

Attenuation of waves due to corona is more pronounced for positive waves than for negative waves because of greater loss due to corona for positive waves. The effect on short waves is more than that on long waves.

Attenuation can be calculated by the following 'Foust and Menger empirical formula'.

$$e = \frac{E}{1 + K E x} \tag{12.75}$$

where e = surge voltage (kV) at distance x kilometres from origin.

E = surge kV at point of origin

K = attenuation constant

$\simeq 0.00037$ for chopped waves (2 μs)

$\simeq 0.00019$ for short waves (5 μs)

$\simeq 0.0001$ for long waves (20 μs)

x = kilometres of travel

12.16 ARCING GROUNDS

It has been found that circuit changes which involve arc paths may produce very high transient voltages. In an isolated neutral system intermittent arcing may take place during a single line to ground fault. This phenomenon is known as arcing ground.

Fig. 12.22 (a) shows a 3-phase isolated neutral system. Assuming that the lines are perfectly transposed, each line conductor will have the same capacitance to ground. When the system is healthy, the three capacitance currents will be equal and balanced as shown in Fig. 12.22 (b).

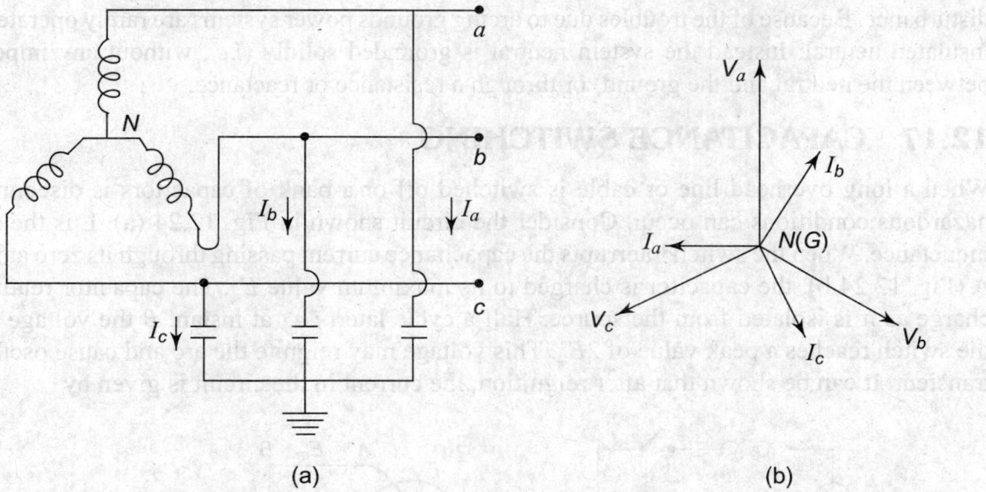

Fig. 12.22 3-phase isolated neutral system under healthy conditions (a) circuit (b) phasor diagram showing capacitance currents

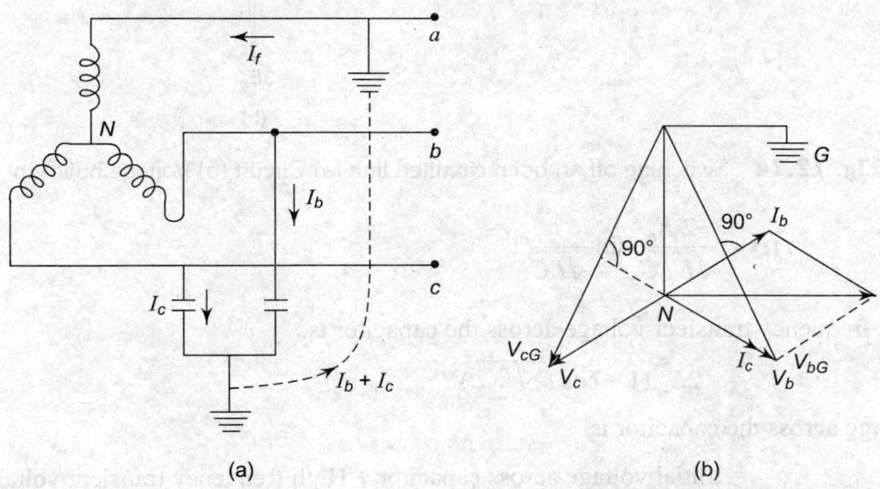

Fig. 12.23 3-phase isolated neutral system (a) circuit showing fault current (b) phasor diagram

A single line to ground fault disturbs the above balanced condition. The capacitance of the faulted phase (say phase *a*) becomes short circuited (Fig. 12.23 a) and the potential of this phase becomes zero. The potentials of the other two phases rise to line to line values. The currents flowing from these healthy phases to ground become $\sqrt{3}$ times the previous values and have a different phase relationship as shown in Fig. 12.23(b). These currents return through the fault and may be sufficient to support an arc. Thus, an arcing ground is established. It may take the form of a flashover of a support insulator. Trouble arises when the arc acts as its own circuit breaker. If the arc is momentarily interrupted at a current zero, the system will attempt to come to its former balanced condition. At the moment of arc extinction C_a is discharged whereas C_b and C_c have considerable voltages across them. The charges associated with these voltages will remain trapped so that the voltage will appear superimposed on the power frequency voltage variation. Following the arc extinction the change in the voltage across the capacitance is of oscillatory nature as the circuit has inherent inductance. As a consequence the arc may restrike and cause further transient disturbance. Because of the troubles due to arcing grounds power systems are rarely operated with insulated neutral. Instead the system neutral is grounded solidly (*i.e.*, without any impedance between the neutral and the ground) or through a resistance or reactance.

12.17 CAPACITANCE SWITCHING

When a long overhead line or cable is switched off or a bank of capacitors is disconnected, hazardous conditions can occur. Consider the circuit shown in Fig. 12.24 (a). L is the source inductance. When the switch interrupts the capacitance current passing through its zero at instant A (Fig. 12.24 b), the capacitor is charged to its maximum value E_m. The capacitor retains this charge as it is isolated from the source. Half a cycle later, *i.e.*, at instant *B* the voltage across the switch reaches a peak value of $2E_m$. This voltage may reignite the arc and cause oscillatory transient. It can be shown that after reignition, the current in the circuit is given by

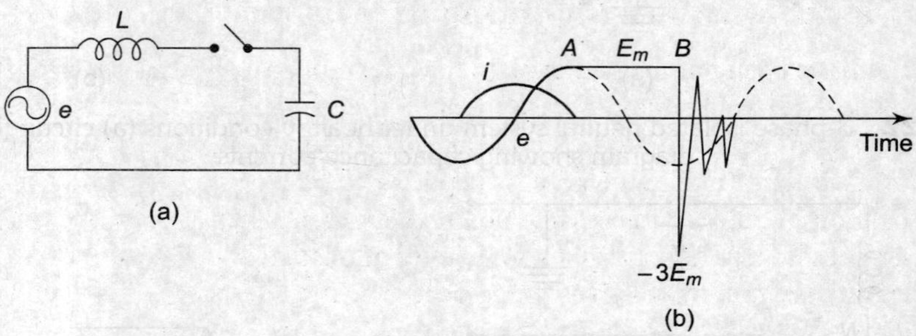

Fig. 12.24 Switching off an open circuited line (a) Circuit (b) Voltage build-up

$$i\,(t) = \frac{-2\,E_m}{\sqrt{L/C}}\,\sin\frac{t}{\sqrt{LC}} \tag{12.76}$$

The high frequency transient voltage across the capacitor is

$$= -\,2E_m\,(1 - \cos t/\sqrt{LC}\,) \tag{12.77}$$

The voltage across the capacitor is

= Initial voltage across capacitor + High frequency transient voltage

$$= E_m - 2E_m\,(1 - \cos t/\sqrt{LC}\,) \tag{12.78}$$

This can reach a maximum value of $-\,3E_m$ (Fig. 12.24b). Interruption of circuit and a second restrike may cause still higher voltage.

12.18 CURRENT CHOPPING

When a small current is interrupted by a circuit breaker, the current may be brought to zero abruptly before the natural current zero. This is known as current chopping and may cause very high voltages. If resistance of the circuit is neglected, the equivalent circuit is shown in Fig. 12.25. If the current at the instant of chopping is i_o, the voltage produced is

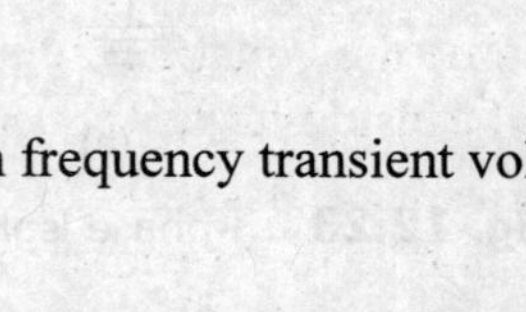

Fig. 12.25 Equivalent circuit for switching off small currents

$$e(p) = \frac{i_0}{p} \frac{L/C}{pL + \frac{1}{Cp}} = \frac{i_0}{C\left(p^2 + \frac{1}{LC}\right)} \tag{12.79}$$

Taking inverse transform.

$$e = i_0 \sqrt{\frac{L}{C}} \sin \frac{t}{\sqrt{LC}} \tag{12.80}$$

This phenomenon is often observed when no load current of a transformer is switched off by an air blast circuit breaker.

12.19 LIGHTNING PHENOMENON

12.19.1 Mechanism of Lightning Stroke

Many theories have been put forth to explain the phenomena whereby a thundercloud becomes electrically charged. Simpson and Scrase theory explains that electrification process is due to breaking of rain drops because of the action of air currents on them. When rain drops break up, they acquire a positive charge and in the process negative charge is imparted to the air. Observations made with balloons, equipped to measure potential gradients, have shown that the great mass of the cloud becomes negatively charged with a layer of positive charge at the top. Occasionally a small concentrated region of positive charge occurs near the bottom of the cloud (Fig. 12.26).

The critical breakdown gradient of dry air at sea level is 30 kV/cm. The presence of moisture reduces it to 10 kV/cm. Whenever a local charge concentration near a cloud causes the gradient to exceed the critical breakdown value, an electric streamer starts from cloud towards ground. This streamer is known as stepped leader. The leader tip moves for about 50 metres before coming to a halt and emitting a bright spot at its head. The streamer halts because the voltage gradient at its head is not capable of causing further breakdown.

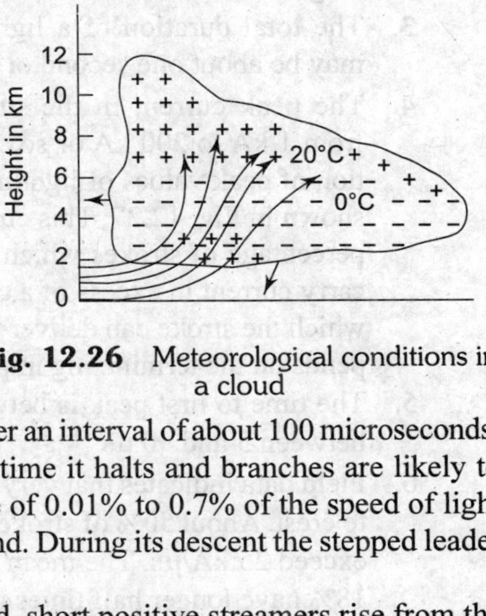

Fig. 12.26 Meteorological conditions in a cloud

But the streamer starts again and repeats performance after an interval of about 100 microseconds. The direction of the streamer is likely to change each time it halts and branches are likely to form. The velocity of the leader stroke lies in the range of 0.01% to 0.7% of the speed of light and it may take about 20 milliseconds to reach the ground. During its descent the stepped leader deposits a negative charge along its path.

When the stepped leader comes very near the ground, short positive streamers rise from the ground. When the negative stepped leader encounters the positive streamer, an intensely luminous discharge starts from ground to cloud. This stroke is known as return stroke and travels at a velocity varying from 10% to 50% of that of light. The negative charge laid down by the leader stroke along its path is rapidly neutralized by the upward moving positive charge previously induced in the ground. The current at the point of contact can be considered either as a negative current flowing into the ground or a positive current flowing out of it. The current in the return stroke may be in the range of a few kilo-amperes to about 250 kA. The return stroke is also known as main stroke and its duration varies from a few μs to about 100 μs.

After the first return stroke there is usually enough charge in the higher region of the cloud to initiate another leader. This leader follows the path of the previous stroke. Owing to the remnant ionization of the path this stroke, known as dart leader, darts to earth directly in about 1 m-sec. The velocity of the dart leader lies in the range of 0.13% to 10% of the speed of light. When the dart leader strikes ground, a second return stroke travels back to cloud. This process may repeat itself many times. Most of the lightning discharges have two or three return strokes.

12.19.2 Characteristics of Lightning Strokes

A large number of field investigations have confirmed the following characteristics of lightning strokes:

1. The lightning strokes are predominantly of negative polarity. However, the surges on transmission lines can be of negative as well as positive polarity depending on whether they are caused by the direct lightning stroke or by induction.
2. The percentage of positive to total surges reported are: strokes to transmission lines 7 to 18%, distribution lightning arrester surges 37%, station lightning arrester surges 12%.
3. The total duration of a lightning stroke may be about one second or so.
4. The peak current in the stroke may be from 1 kA to 200 kA or so. The distribution of peak values of lightning current is shown in Fig. 12.27. This curve gives the percentage of strokes which are likely to carry current in excess of a certain value. This curve is to be interpreted as the current which the stroke can deliver to zero resistance ground. The actual current delivered depends on the terminating impedance and the surge impedance of the lightning channel.

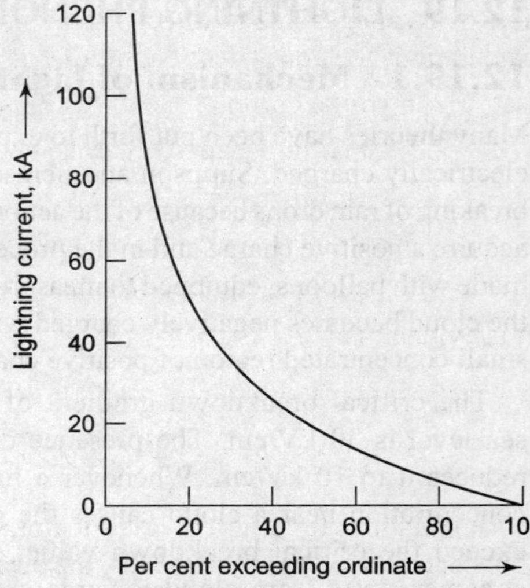

Fig. 12.27

5. The time to first peak is between zero and 10 μs. The time to half value on the tail is between 5 and 90 μs.
6. Field data indicates that very high stroke currents do not coincide with very short times to crest. About 50% of stroke currents have a rate of rise exceeding 7.5 kA/μs and 10% exceed 25 kA/μs. The mean duration of stroke current, above half value, is 30 μs and 18% have longer half times than 50 μs.

12.20 OVERVOLTAGES DUE TO LIGHTNING

The overvoltages due to lightning are generally confined to overhead transmission and distribution systems. These overvoltages may occur due to indirect strokes and direct strokes.

12.20.1 Indirect Strokes

The negative charge on lower part of the cloud and the leader stroke induce positive charges on the ground surface, earth conductor and other grounded objects. Some of these positive charges flow to the phase conductors by leakage over insulators and through neutral points of transformers.

A lightning discharge in the vicinity of the line will cause the field to collapse and the positive charge on the conductors to be released and initiate indirect lightning transients. The charges on the earth wire travel down to earth through the towers and those on the phase conductors travel along the conductors and are dissipated through corona and resistance loss. The induced surges are equal on all the three phases, have usually positive polarity, and their wave front is typically 10 μs. The crest values depend on the stroke current, distance from the stroke, height of phase conductors and the presence of shield wires and rarely exceed 200 kV. Thus, the indirect strokes, while capable of producing flashover on 11 kV lines, have little practical significance for lines of 33 kV and higher operating voltages.

12.20.2. Direct Strokes

The overvoltages due to direct strokes can be very high and may cause flashovers. For simple calculations the equivalent circuit of Fig. 12.28 is very useful.

Let I_0 = stroke current to zero resistance earth
Z_s = surge impedance of stroke channel
Z = terminating impedance
I = current flowing into the terminating impedance
V = voltage across the terminating impedance.

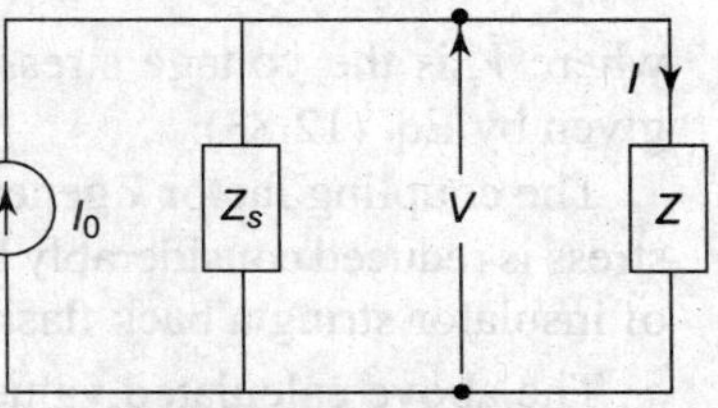

Fig. 12.28

It is evident that

$$V = IZ = I_0 \frac{Z_s}{Z + Z_s} Z = I_0 Z \Big/ \left(1 + \frac{Z}{Z_s}\right) \tag{12.81}$$

Calculations by different investigators show that Z_s lies between 1000 and 3000 ohms. Since Z is generally small relative to Z_s, the current I differs only very little from the stroke current to zero resistance earth (*i.e.*, I_0). The value of the terminating impedance depends on whether the direct stroke hits the phase conductor or the tower or the earth wire.

(a) Direct stroke to phase conductor

When a lightning stroke strikes a phase conductor, surges travel away from the point of strike in each direction. The terminating impedance is equal to half the surge impedance of the phase conductor. The conductor potential can be calculated by substituting $Z_c/2$ for Z in Eq. (12.81). Since $Z_c/2$ is small as compared to Z_s,

$$V \cong \frac{1}{2} I_0 Z_c \tag{12.82}$$

A stroke current as low as 10 kA, which according to Fig. 12.27 has more than 65% probability of being exceeded, will cause a potential of about 2000 kV (because $Z_c \cong 400$ ohms). This is a very high voltage and only the lines insulated for ultra high voltages can withstand this stress. Thus, a direct stroke to phase conductor will invariably lead to flashover. It is an accepted practice to seek to prevent direct strokes to phase conductors by using one or two ground wires.

(b) Direct stroke to tower

For a direct stroke to tower the terminating impedance is the resultant of the tower surge impedance Z_t in parallel with half the ground wire surge impedance Z_g (assuming a ground wire in each direction). If the system has two ground wires in each direction, Z_g should be taken as the equivalent surge impedance. The tower top voltage V_t can be calculated from Eq. (12.81) by substituting the following value of Z.

$$Z = Z_t \Big/ \left(1 + \frac{2Z_t}{Z_g}\right) \tag{12.83}$$

When a lightning surge travels on a ground wire, the associated current and voltage waves induce in any parallel phase conductor a surge of the same polarity and of k times the ground wire voltage where k is the coupling factor between the phase conductor and the ground wire. The voltage stressing the tower insulation is the difference between the tower top potential and the voltage induced in the phase conductor. Thus,

$$V_i = (1 - k)\, V_t = (1 - k)\, I_0 Z \Big/ \left(1 + \frac{Z}{Z_s}\right) \tag{12.84}$$

where V_i is the voltage stressing the tower insulation and Z is the terminating impedance as given by Eq. (12.83).

The coupling factor k generally lies between 0.15 and 0.3. Thus, it is seen that the insulation stress is reduced considerably by the coupling effect. If the voltage V_i exceeds the impulse strength of insulator string a back flashover from tower to the phase conductor may occur.

The above calculated values of V_t and V_i are subsequently modified by the reflections from the tower base and eventually by reflections from the adjacent towers.

(c) *Direct stroke to ground wire at mid-span*

A lightning stroke within a quarter span of a tower is considered equivalent to a stroke at the tower itself whereas a stroke within a quarter span of the mid-span is considered equivalent to a stroke to mid-span.

When a lightning stroke hits the ground wire at mid-span, the voltage V_m on the ground wire can be found from Eq. (12.81) by setting the terminating impedance equal to half the self surge impedance of the ground wire, *i.e.*, $Z = Z_g/2$. Thus,

$$V_m = \frac{1}{2}\, I_0\, Z_g/(1 + Z_g/2\, Z_s) \tag{12.85}$$

The voltage on the phase conductor is k V_m. The voltage $(1 - k)V_m$, which the air gap must withstand is considerably higher than the voltage across the tower insulator strings caused by a stroke of equal intensity striking the tower. This is because of the fact that the terminating impedance in this case (*i.e.*, stroke to ground wire) is higher than that in the case of stroke to tower. For this reason it is a usual practice to increase the mid-span clearance by sagging the ground wire less than the phase conductors. Assuming that there is no flashover at mid-span, the voltage waves V_m and kV_m travel on the ground wire and phase conductor respectively and may cause flashover at the tower. Successive reflections from adjacent towers tend to decrease the voltage stressing the insulation.

12.21 LINE DESIGN BASED ON DIRECT STROKES

The insulation level of a transmission line is governed by the surge voltages due to switching and lightning. Surge voltages are higher in magnitude as compared to the power frequency voltages and if the line is properly insulated to withstand surges, it can easily withstand the highest expected power frequency voltages.

The induced strokes due to lightning assume importance for 11 kV lines only. For high voltage lines (*i.e.*, lines up to 220 kV) the surges due to direct lightning stroke determine the line insulation design. If a high voltage line is capable of withstanding the direct lightning strokes, it can easily withstand the switching surges and the induced lightning strokes. However, for lines in the extra high voltage class (*i.e.*, 400 kV and above), the switching surges assume greater importance as compared to lightning surges. The insulation design of extra high voltage lines is discussed in Chapter 14.

The basic principles of the design of a transmission line based on direct stroke theory are:

(a) The line must be provided with ground wires, of sufficient mechanical strength, to shield the phase conductors from direct lightning strokes.
(b) There must be adequate clearance from the phase conductor to the tower (or ground) so that full effectiveness of the insulating structure can be obtained.
(c) There must be adequate clearance between the phase conductor and ground wire, especially at mid-span, so that flashovers between phase conductors and ground wire are eliminated.
(d) The tower footing resistance should be as low as economically justified. A low tower footing resistance helps in maintaining the tower top voltage (under lightning discharge conditions) to a safe value.

Considerable work has been done to determine the insulation design of high voltage lines. Standard curves are available to fix the insulation level for a desired immunity against failures due to lightning. For a detailed study of this topic the reader is referred to [10, 31] of the bibliography.

12.22 PROTECTION OF SYSTEMS AGAINST SURGES

The surge voltages, especially those due to lightning, have very high magnitudes. Therefore, power systems require adequate protection against direct lightning strokes. Ground wires are provided on almost all transmission lines to shield the phase conductors from direct lightning strokes. Surge diverters are provided in sub-stations to divert the surges to ground.

12.22.1 Ground Wires

Every transmission line is provided with one (sometimes two) ground wires to shield the phase conductors. They are placed above the phase conductors and are grounded at every tower. The ground wire shields the phase conductor by attracting to itself the lightning strokes which, in the absence of ground wire, would strike the phase conductors.

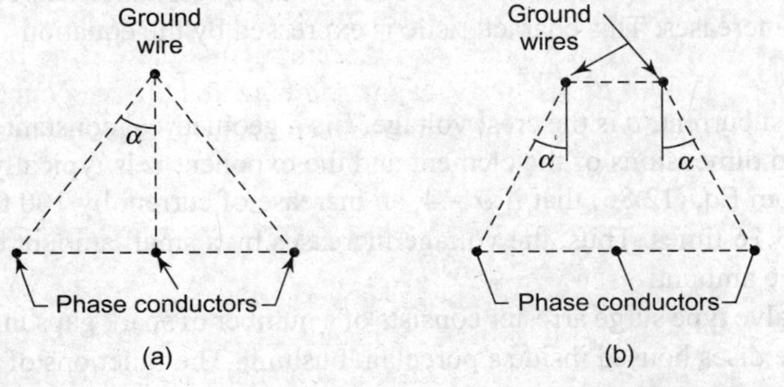

Fig. 12.29 Shielding angle of ground wire (a) Single ground wire (b) Double ground wire

The selection of size, number and arrangement of ground wires is of great importance in line design. The selection of size is based on the consideration of mechanical strength. The number and arrangement depends on the degree of lightning protection needed for the line.

The protective angle or shielding angle of a ground wire is defined as the angle between a vertical plane through the ground wire and slanting plane connecting the ground wire with the outermost phase conductor (Fig. 12.29). A shielding angle of about 30° gives adequate lightning protection. A double ground wire provides better lightning protection as seen from Fig. 12.29 and is preferable in zones prone to severe lightning strokes.

12.23 SURGE ARRESTERS

A surge arrester (also known as surge diverter or lightning arrester) is a device connected between line and earth, *i.e.*, in parallel with the equipment to be protected at the substation. It is a safety valve which limits the magnitude of lightning and switching over voltages at the substation and provides a low resistance path for the surge current to flow to the ground. It is installed adjacent to the equipment to be protected.

An ideal surge arrester should have the following characteristics:

(i) It should not take any current under normal conditions, *i.e.*, its sparkover voltage must be higher than the normal or abnormal power frequency voltage that may occur in the system.

(ii) Any abnormal transient voltage above the breakdown value must cause it to break down as quickly as possible in order to provide an alternative path to earth.

(iii) Breakdown having taken place it must be able to carry the resultant discharge current without causing damage to itself and without the voltage across it exceeding the breakdown value.

(iv) Immediately after the surge voltage falls below the breakdown value, it must interrupt the following power frequency current instantaneously so that the circuit outage is avoided.

The simple type of surge arresters like rod gap, horn gap, etc., are very cheap devices but fulfil only the first three of the above mentioned requirements. The fourth requirement is, however, not fulfilled and every gap operation results in an outage.

12.24 THYRITE SURGE ARRESTER

All modern surge arresters use non-linear resistance elements made of silicon carbide (also known as thyrite) in conjunction with series gaps. It is also known as valve type arrester.

A non-linear resistance element has the property that its resistance decreases sharply as the voltage across it increases. This characteristic is expressed by the equation

$$i = kv^{\alpha} \tag{12.86}$$

where i is the crest current, v is the crest voltage, k is a geometrical constant which depends on design, rating and dimensions of the element and the exponent α is typically in the range of 4 to 6. It is seen from Eq. (12.86) that if $\alpha = 4$, an increase of current by 100 times increases the voltage only by 3.16 times. Thus, the voltage increases by a small amount even if the current changes by a large amount.

The modern valve type surge arrester consists of a number of spark gaps in series with silicon carbide resistance discs housed inside a porcelain bushing. The functions of spark gaps are:

1. To withstand normal operating voltage without sparkover under all operating conditions including those in which the arrester surface may be contaminated.

2. To spark over at a predetermined voltage to provide a path for the flow of discharge current through the valve element to ground.
3. To regain its insulating property after discharging the surge current.

An arrester with a number of gaps can withstand high rate of rise of recovery voltage (RRRV). However, the voltage distribution across the gaps tends to be non-uniform. To proportion the surge voltage equally across the series gaps, they are shunted by non-linear (high resistance) voltage grading resistors. Grading capacitors are also used in more advanced designs. These capacitors supplement the inherent shunt capacity of the arrester. A single phase 3-unit arrester is shown in Fig. 12.30.

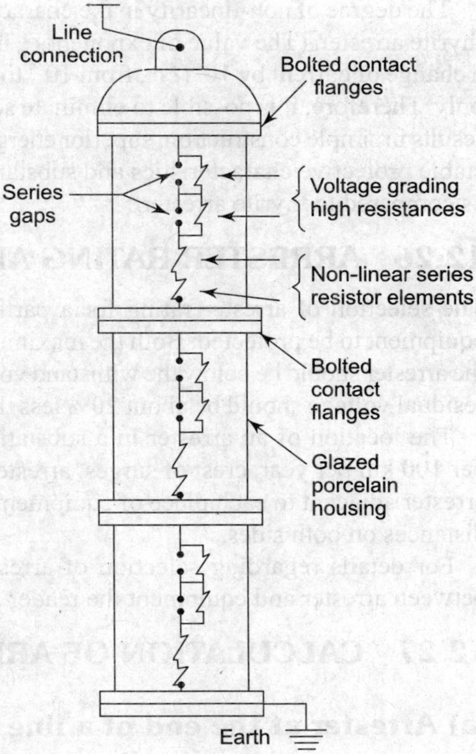

Fig. 12.30 A single phase unit valve type arrester

There are three general classes of arresters: (1) distribution type with voltage rating of 3 to 30 kV (rms); (2) intermediate (or line) type having voltage rating of 3 to 120 kV and (3) station type having voltage rating 3 to 800 kV. The station type arresters have more advanced design features, better discharge characteristics and greater reliability.

The station and intermediate class arresters are all valve type. Some distribution class arresters are expulsion type. Table 12.1 gives an approximate range of characteristics of these arresters.

Table 12.1 Surge arrester protective characteristics

Arrester class	*Rating available kV rms*	*Front of wave sparkover voltage*	*1.2 × 50 sparkover voltage*	*Switching surge sparkover voltage*	*Discharge voltage at 10 kA 8 × 20 wave*
		pu of arrester Crest rating			
Distribution	3 to 30	2 to 6	—	—	2.5 to 3.25
Intermediate	3 to 120	1.92 to 3.66	1.5 to 2.6	1.9 to 2.5	1.82 to 2.55
Station	3 to 800	1.82 to 2.83	1.56 to 1.92	1.45 to 2	1.54 to 2.4

12.25 METAL OXIDE GAPLESS ARRESTER

A new non-linear arrester element with zinc oxide as a major constituent was developed in Japan in 1968. This non-linear resistance offers excellent characteristics and economy as compared to thyrite.

The characteristic valve element of a metal oxide gapless arrester is composed of disc unit having metallic film electrodes at top and bottom. The valve element is a dense ceramic body made up of zinc oxide (more than 90%) and some other metallic oxides.

The degree of non-linearity in the characteristics of this arrester is much more than that in the thyrite arrester. (The value of exponent α, in Eq. 12.86, for this material is more than 25.) Thus, a change of current by 10^5 (*i.e.*, from 10^{-3} to 10^2) changes the voltage across it by a factor of 1.56 only. Therefore, it is possible to eliminate series gaps in arrester design using this material. This results in simple construction, superior energy absorption capability, better surge protection, more stable protective characteristics and substantial reduction in over-voltage across the equipment as compared to thyrite arrester.

12.26 ARRESTER RATING AND LOCATION

The selection of arrester rating for a particular application depends on the impulse level of equipment to be protected. Both the maximum sparkover voltage and the residual voltage across the arrester should be below the withstand voltage of the equipment to be protected. The maximum residual voltage should be about 20% less than the impulse withstand voltage of the equipment.

The location of an arrester in a substation depends on probability of flashover on the line per 100 km per year, crest of surges, arrester characteristics, etc. It is not necessary to apply an arrester adjacent to each piece of equipment. An arrester has the ability to protect up to suitable distances on both sides.

For details regarding selection of arrester rating and calculation of permissible distance between arrester and equipment the reader is referred to [27] of bibliography.

12.27 CALCULATION OF ARRESTER VOLTAGE AND CURRENT

(a) Arrester at the end of a line

Figure 12.31 (a) shows an arrester connected at the end of a line of surge impedance Z_c and a wave e_f travelling on the line. Fig. 12.31 (b) shows the equivalent circuit. The circuit equations are

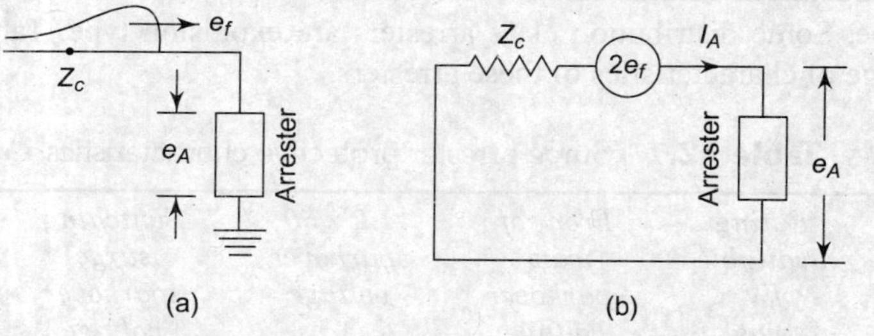

Fig. 12.31 Surge arrester at the end of a line (a) actual circuit (b) equivalent circuit

$$2e_f = e_A + Z_c I_A \tag{12.87}$$

$$I_A = k\,(e_A)^{\alpha} \tag{12.88}$$

e_f, k, α and Z_c being known, e_A and I_A can be calculated. Since Eq. (12.87) is non-linear, a graphical or iterative solution is necessary.

(b) Arrester at junction of two lines

Figure 12.32 (a) shows an arrester at the junction of two lines having surge impedances Z_1 and Z_2. Figure 12.32 (b) shows the equivalent circuit. The circuit equations are

$$2e_f = e_A + Z_1 I = e_A + Z_1\left(I_A + \frac{e_A}{Z_2}\right)$$

or
$$2e_f = e_A\left(1+\frac{Z_1}{Z_2}\right) + Z_1 I_A \qquad (12.89)$$

and
$$I_A = k\,(e_A)^{\alpha} \qquad (12.90)$$

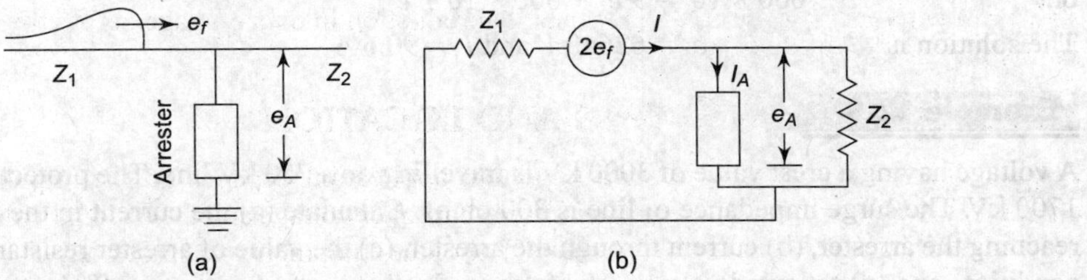

Fig. 12.32 Surge arrester at the junction of two lines (a) actual circuit (b) equivalent circuit

It is seen from the above equations that surge impedance plays a very important role in limiting the voltage across the arrester. It is evident from Eqs. (12.87) and (12.89) that if Z_c or $Z_1 = 0$, *i.e.*, line is absent, the voltage e_A is equal to $2e_f$. An arrester as a separate entity does not limit voltage. It is only through the combination of arrester characteristics and line surge impedance that we can limit the voltage across the equipment.

Example 12.7

A surge arrester installed at the end of a line has the following characteristics:

$$I_A = (1.5 \times 10^{-27})\, e_A^6$$

The surge impedance of the line is 400 ohms. Determine the arrester voltage and current if a 300 kV crest step voltage travels on the line and approaches the arrester.

Solution

Using Eqs. (12.87) and (12.88)

$$2 \times 300 \times 10^3 = e_A + 400\, I_A$$

$$I_A = (1.5 \times 10^{-27})\, e_A^6$$

or
$$600 \times 10^3 = e_A + (1.5 \times 10^{-27})\, 400\, e_A^6$$

or
$$600 \times 10^3 = e_A + 600 \times 10^{-27}\, e_A^6$$

We use an iterative method to solve the above equation. The result is

$$e_A = 97100\ V \text{ and } I_A = 1257\ \text{A}$$

Example 12.8

A surge arrester is installed at the junction of an overhead line and a cable having surge impedances of 400 and 50 ohms respectively. The characteristics of arrester is

$$I_A = (1.5 \times 10^{-27})\, e_A^6$$

A 300 kV step wave travels on the overhead line towards the junction. Find arrester voltage and current.

Solution

Using Eqs. (12.89) and (12.90)

$$2 \times 300 \times 10^3 = e_A\left(1+\frac{400}{50}\right) + 400\, I_A$$

$$I_A = (1.5 \times 10^{-27})\, e_A^6$$

or $$600 \times 10^3 = 9\, e_A + 600 \times 10^{-27}\, e_A^6$$

The solution is $$e_A = 62640\ V \text{ and } I_A = 90.6\ \text{A}$$

Example 12.9

A voltage having a crest value of 3000 kV is travelling on a 750 kV line. The protective level is 1700 kV. The surge impedance of line is 300 ohms. Calculate (a) the current in the line before reaching the arrester, (b) current through the arrester, (c) the value of arrester resistance for this condition, and (d) reflected voltage. Verify the reflection and refraction coefficients.

Solution

$$Z_c = 300,\ e_f = 3000\ \text{kV},\ e_A = 1700\ \text{kV}$$

(a) $i_f = e_f/Z_c = 3000 \times 10^3/300 = 10^4\ A$ or 10 kA

(b) The voltage equation is

$$2e_f = Z_c I_A + e_A$$

or $$2 \times 3000 \times 10^3 = 300\, I_A + 1700 \times 10^3$$

or $$I_A = 14333\ A \text{ or } 14.333\ \text{kA}$$

(c) R = Resistance of arrester

$$= \frac{e_A}{I_A} = \frac{1700 \times 10^3}{14.333 \times 10^3} = 118.61\ \text{ohm}$$

(d) $1700 = 3000 + e_r$

or $$e_r = -1300\ \text{kV}$$

$$\text{Reflection coefficient} = \frac{-1300}{3000} = -0.433$$

$$\text{Refraction coefficient} = \frac{1700}{3000} = 0.567$$

or Using the results of section 12.9

$$\text{Reflection coefficient} = \frac{R - Z_c}{R + Z_c} = \frac{118.61 - 300}{118.61 + 300} = -0.433$$

$$\text{Refraction coefficient} = \frac{2R}{R + Z_c} = \frac{2 \times 118.61}{118.61 + 300} = 0.567$$

12.28 EFFECT OF SHORT LENGTH OF CABLE

Indoor transformers are connected to an incoming overhead line through a short length of cable. When a surge voltage e_f, travelling on an overhead line of surge impedance Z_1, reaches its junction with a cable having surge impedance Z_2, the voltage e_t transmitted on the cable is:

$$e_t = e_f \frac{2Z_2}{Z_1 + Z_2} \tag{12.91}$$

Typical values of Z_1 and Z_2 are 400 ohms and 50 ohms, respectively. Thus, the initial value of e_t is only about 22 per cent of e_f. However, for long waves after a few reflections, the voltage rises to almost full value of incident wave but the front is elongated (Fig. 12.33a). If the wave is short as compared to length of cable, *i.e.*, if it is chopped, a reduction in amplitude also occurs (Fig. 12.33b). Thus, this method is very effective in reducing the amplitude of short waves. For long waves, though the amplitude is not reduced, yet the wave from is elongated and this also represents a real advantage as far as stresses to the windings are concerned.

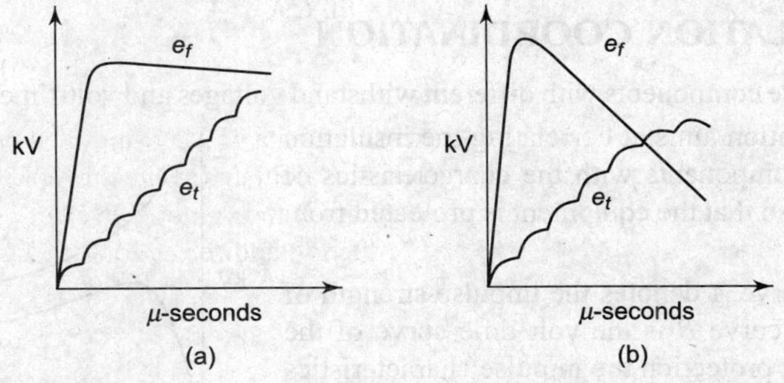

Fig. 12.33 Line terminated by a cable (a) effect on long wave (b) effect on short wave

12.29 SURGE CAPACITOR, SURGE REACTOR AND SURGE ABSORBER

A capacitor connected in parallel with an equipment provides some degree of protection against surges because before the surge can impress a high voltage on the equipment, it must charge the capacitor. This reduces the steepness of the wave front since the voltage across the equipment can rise only as quickly as the capacitor can be charged through the line. Capacitors are particularly effective against short duration surges. The energy of these surges is such that it can be absorbed by the capacitors. However, for high energy surges, the size and cost of capacitors becomes prohibitive.

The combination of a capacitor and an arrester can be very useful because they complement each other. The capacitor reduces the steepness of the wave.

A reactor connected in series with the equipment also serves an identical purpose. When a surge reaches the reactor, it initially appears only across the reactor because the reactor offers a high impedance to the high frequency currents. The surge energy is initially absorbed in the magnetic field of current in the reactor. Its ability to store energy is limited but it can take care of low energy surges.

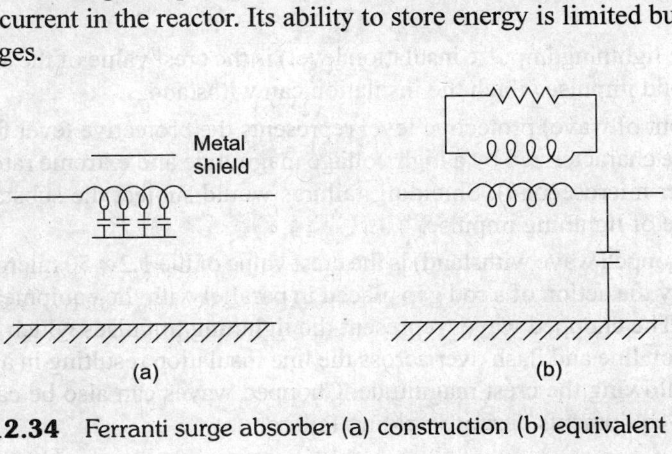

Fig. 12.34 Ferranti surge absorber (a) construction (b) equivalent circuit

In a Ferranti surge absorber (Fig. 12.34) an inductor is enclosed in an earthed metal shield. The damping is due to the eddy currents induced in the shield. The charging of the capacitance between the coil and earthed shield also reduces the energy of the wave. The absorber is effective not only in reducing the steepness of the wave but has some effect on amplitude also as some energy is dissipated by the resistance of the metallic shield.

12.30 INSULATION COORDINATION

Power Systems have components with different withstand voltages and volt-time characteristics. Insulation coordination aims at correlating the insulation of these various components with the characteristics of protective devices so that the equipment is protected from overvoltages.

In Fig. 12.35 curve *A* denotes the impulse strength of an equipment and curve *B* is the volt time curve of the arrester. For proper protection the impulse characteristics of the equipment must be higher than that of arrester over the whole range of volt-time characteristics. A margin of 20% is considered desirable and adequate.

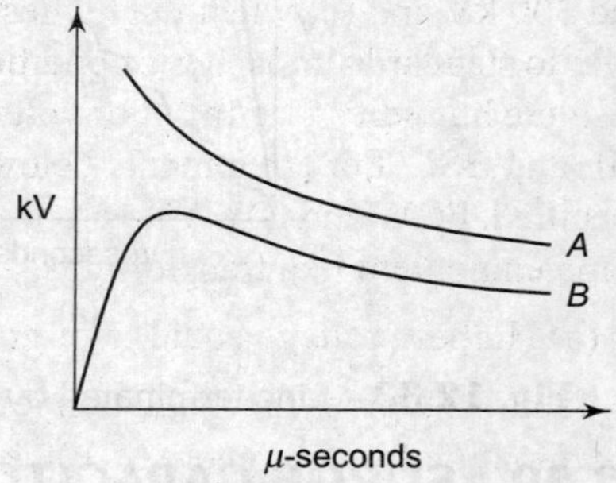

Fig. 12.35 Insulation levels of equipment and arrester

Insulation coordination involves three basic steps:

1. Selection of standard insulation level.
2. Making sure that every equipment has a breakdown strength equal to or higher than insulation level.
3. Application of protective devices which will provide as good protection as can be justified economically.

12.30.1 Basic Insulation Level

Insulation levels for power system equipments have been standardized by American National Standards Institute and published under the title 'Standard Voltage Values for Preferred Transient Insulation Levels ANSI C92.1–1972'. These values reflect the transient overvoltages that the stations are exposed to. The insulation levels are specified in terms of BIL, FOW, CWW and BSL.

1. BIL (basic lightning impulse insulation level) is the crest value of the standard 1.2 × 50 microsecond impulse which the insulation can withstand.
2. FOW (Front of wave) protective level represents the protective level for a direct lightning stroke characterized by a high voltage magnitude and extreme rate of rise (around 100 kV per microsecond). Shielding failures would subject the substation equipment to this type of lightning impulse.
3. CWW (Chopped wave withstand) is the crest value of the 1.2 × 50 microsecond impulse chopped by the action of a rod gap placed in parallel with the equipment, after a specified time. The chopped waves represent the lightning impulses which originate on the transmission line and flash over across the line insulation resulting in a rapid change in voltage following the crest magnitude. Chopped waves can also be caused across the substation equipments during shielding failure.

4. BSL (basic switching impulse insulation level) is the crest value of a standard switching impulse which the insulation can withstand. BSL is more of a concern in EHV and UHV system. For high voltage systems (below 300 kV) BIL generally fulfils the requirements of BSL.

The ratio of withstand voltage of the equipment insulation to arrester protective level is known as protective ratio. As per American practice this ratio should be at least 1.2 for lightning impulses (BIL, FOW and CWW) and 1.15 for switching impulses.

Indian standards classify the insulation level in three categories: (a) when the highest system voltage is equal to or more than 300 kV; (b) when the highest system voltage is between 100 kV and 300 kV and (c) when the highest system voltage is less than 100 kV. The insulation levels refer to standard atmospheric conditions (*i.e.*, temperature 20°C, pressure 760 mm, mercury and absolute humidity 11 g/m^3). For voltage class of 300 kV and above, the Indian standards specify BIL and BSL. For equipments below 300 kV, BIL and power frequency withstand voltage are specified. FOW and CWW levels are not specified. However, chopped wave test is specified for some equipment like transformers.

(a) Highest voltage equal to or greater than 300 kV

Table 12.2 gives the insulation levels for this class as per BIS: 2165 (1973) titled 'Insulation coordination for equipment of 100 kV and above'.

Table 12.2 Standard insulation levels for V_m equal to or greater than 300 kV (BIS: 2165 Part II-1973)

Highest voltage V_m *(kV rms)*	*Base for pu. value* $\frac{\sqrt{2}V_m}{\sqrt{3}}$	*Switching impulse withstand level*		*Lightning impulse withstand level, kV(Peak)*
		pu	*kV (peak)*	
420	343	2.76	950	1050 and 1175
		3.06	1050	1175, 1300, 1425
525	429	2.45	1050	1175, 1300, 1425
		2.74	1175	1300, 1425, 1550
765	625	2.08	1300	1425, 1550, 1800
		2.28	1425	1550, 1800, 2100
		2.48	1550	1800, 2100, 2400

In Table 12.2 it is seen that more than one value of lightning impulse withstand level is specified for each voltage class. The lower values are recommended for equipments protected by surge arresters and the highest value is recommended for equipments not protected by surge arresters.

(b) Highest voltage between 100 and 300 kV.

Table 12.3 gives the insulation levels for this class as per BIS: 2165 (Part II) - 1973. For this class the switching impulse levels are not specified. Instead the one minute power frequency test (and power frequency withstand voltage) are specified. In this class also more than one value of lightning impulse withstand voltage is specified for each voltage class. It is recommended that in isolated neutral systems or resonant earthed systems or in positions in a system where the neutral is not effectively earthed, the highest values be used. The choice of lower value of insulation supposes that the equipment is adequately protected against surges.

Table 12.3 Standard insulation levels for V_m between 100 kV and 300 kV (BIS: 2165 Part II-1973)

Highest voltage for equipment V_m *(kV, rms)*	*Base for pu. value* $\frac{\sqrt{2}V_m}{\sqrt{3}}$	*Lightning impulse withstand voltage kV (peak)*	*Power frequency withstand voltage kV (rms)*
100	81.5	450	185
		380	150
123	100	550	230
		450	185
145	118	650	275
		550	230
		450	185
170	139	750	325
		650	275
		550	230
245	200	1050	460
		950	395
		850	360
		750	325

(c) Highest voltage less than 100 kV

Table 12.4 gives the insulation levels for this class as per BIS : 2165 (Part-III) - 1973 titled "Insulation coordination for highest voltage for equipments above 1 kV and below 100 kV". Two values of lightning impulse voltages (list 1 and list 2) are given. The choice between these two lists depends on degree of exposure to surge, type of neutral earthing, type of overvoltage protective devices, etc.

Table 12.4 Standard insulation levels for V_m less than 100 kV (BIS: 2165-Part III-1973)

Highest voltage for equipments V_m*, (kV, rms) (kV, rms)*	*Power frequency withstand voltage*	*Lightning impulse withstand voltage*	
		List 1 (kV peak)	*List 2 (kV peak)*
3.6	10	20	40
7.2	20	40	60
12	28	60	75
17.5	38	75	95
24	50	95	125
36	70	145	170
52	95	250	250
72.5	140	325	325

12.30.2 Insulation Coordination Between Lines and Substations

Either the line or the substation is overinsulated with normal insulation for the other.

(a) Lines normally insulated, substation overinsulated.

This scheme is used for earthed overhead lines. It ensures that flashover occurs at or near the position of strike. Precautions are taken to prevent damage to line insulation. If adequate shielding against direct strokes is provided at substation, the substation will be subjected only to chopped waves (received through the lines entering the substation). These chopped waves are limited to safe values.

(b) Lines over insulated, substation normally insulated.

This scheme, applicable to wooden pole lines possibly with wooden cross-arms and large conductor spacings, aims at reducing to a minimum, the risk of flashover and damage to the overhead line. The substation is, then, likely to be subjected to high surges and adequate protection at substations is necessary. This is provided through selection of proper rating and location of surge arresters.

12.30.3 Insulation Coordination within a Substation

The coordination of insulation within a substation is very important. The following points must noted in this regard:

1. The provision of surge protective devices must be based on the weakest component in the substation equipment under impulse conditions. The weakest component is not necessarily the costliest component, *i.e.*, transformer. If a current transformer breaks down, the subsequent power current may damage a large part of the stations.
2. The dispersion of volt-time curves of individual components must be taken into account. Thus if a solid piece of insulation is to be protected, its withstand voltage must be higher than the highest sparkover voltage of the arrester (over the whole range of volt-time characteristics) by a suitable margin.
3. The flashover voltage of an insulator under bad weather conditions may be about 20% less than that under fair weather conditions. However, the impulse flashover voltage of an air gap varies very little under different weather conditions. Therefore, for outdoor equipments the insulation coordination should be based on worst probable atmospheric conditions.
4. The protective devices must safeguard the equipment over the whole range of volt-time characteristics and for both positive and negative waves. Solid insulation has a flat volt-time characteristic (curve T of Fig. 12.36) and is also independent of polarity. The characteristics of surge arrestor (curve L of Fig. 12.36) is similar to that of solid insulations. However, for the air gaps the characteristics have the shape shown by R_1 and R_2 in Fig. 12.36. Moreover, in case of air gap, the withstand voltage for positive waves is lower.
5. Provision of rod gaps may increase the number of supply interruptions. Thus if the equipment with insulation T (Fig. 12.36) is to be protected by a rod gap, the setting of the gap would have to be reduced till the curve reaches position R_2. Waves with amplitudes between A and B will then cause unnecessary supply interruptions. Such voltages, though cause flashover of rod gap, are safe for insulation T.

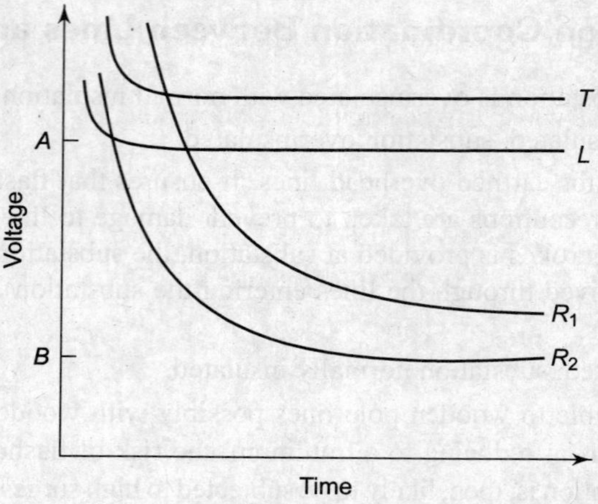

Fig. 12.36

Example 12.10

A 10 kV surge travels on a line having a surge impedance of 400 ohms. (a) Find the magnitude of incident current wave. (b) The line is terminated by a resistance of 1000 Ω. Find the rate of energy dissipation, the reflected voltage and current waves and rate of energy reflection. State the form of reflection. (c) What should be the value of terminating resistance so that energy of wave is completely dissipated? (d) The line is terminated by a long cable having a surge impedance of 50 ohms. Find the transmitted voltage and current waves, reflected voltage and current waves, transmitted power and reflected power. State the form of reflection.

Solution

(a)
$$e_f = 10 \text{ kV}, Z_c = 400\ \Omega$$
$$i_f = \frac{10 \times 1000}{400} = 25 \text{ A}$$

(b)
$$e_t = e_f \left[\frac{2R}{R + Z_c}\right] = 10\left[\frac{2000}{1400}\right]$$
$$= 14.286 \text{ kV}$$
$$i_t = \frac{e_t}{R} = \frac{14.286 \times 1000}{1000} = 14.286 \text{ A}$$
$$e_r = e_t - e_f = 4.286 \text{ kV}$$
$$i_r = \frac{-e_r}{Z_c} = \frac{-4.286 \times 1000}{400}$$
$$= -10.715 \text{ A}$$

Rate of energy dissipation $= e_t\, i_t$
$$= 14.286 \times 1000 \times 14.286$$
$$= 204090 \text{ W} = 204.09 \text{ kW}$$

Rate of energy reflection $= 4.286 \times 10.715 \times 10^3$
$$= 45925 \text{ W} = 45.925 \text{ kW}$$

The reflected voltage wave is a positive wave while the reflected current wave is a negative wave.

(c) If the energy of the wave is to be completely dissipated $R = Z_c$. Hence resistance should be 400 ohms.

(d)
$$e_t = 10 \times \frac{2 \times 50}{400 + 50} = 2.222 \text{ kV}$$

$$i_t = \frac{2.222 \times 1000}{50} = 44.44 \text{ A}$$

$$e_r = e_t - e_f = -7.778 \text{ kV}$$

$$i_r = -\frac{e_r}{Z_c} = 19.445 \text{ A}$$

Rate of transmitted energy $= e_t i_t = 2.222 \times 10^3 \times 44.44$

$= 98750 \text{ W} = 98.75 \text{ kW}$

Rate of reflected energy $= 7.778 \times 19.445 \times 10^3$

$= 151.24 \times 10^3 \text{ W} = 151.24 \text{ kW}$

The reflected voltage wave is negative while the reflected current wave is positive.

Example 12.11

An overhead line has a surge impedance of 400 ohms. A step wave of 20 kV travels on the line. At a certain point the line terminates into a junction and is continued by 2 cables in parallel. The surge impedances of cables are 150 Ω and 200 Ω. Find the voltage and current transmitted into each cable. Also find reflected voltage and current.

Solution

$$\frac{150 \times 200}{150 + 200} = 85.71\ \Omega$$

$$e_t = \frac{2 \times 20 \times 85.71}{400 + 85.71} = 7.0585 \text{ kV}$$

$$i_{t1} = \frac{7.0585 \times 1000}{150} = 47.057 \text{ A}$$

$$i_{t2} = \frac{7.0585 \times 1000}{200} = 35.29 \text{ A}$$

$$e_r = e_t - e_f = 7.0585 - 20$$

$$= -12.9415 \text{ kV}$$

$$i_r = \frac{-e_r}{400} = 32.35 \text{ A}$$

Example 12.12

A 100 kV surge travels on an overhead line of surge impedance 400 ohms. The line is connected to a cable of surge impedance 50 ohms through a resistance R. Find (a) the value of R so that it absorbs maximum energy and the amount of this energy; (b) voltage and current waves transmitted into cable; (c) voltage and current waves reflected; (d) power in incident wave, reflected wave, and transmitted into cable and (e) show the different voltage and current waves on a diagram.

Solution

Figure 12.37 (a and b) show the circuit and equivalent circuit

(a) $$i_t = \frac{2 \times 100}{400 + R + 50} \text{ kA}$$

Energy absorbed by resistance

$$= \left(\frac{200}{400 + R + 50}\right)^2 R \times \text{time}$$

Energy will be maximum when

$$\frac{d}{dR}\left[\frac{R}{(450 + R)^2}\right] = 0$$

The result is $R = 450$ ohms

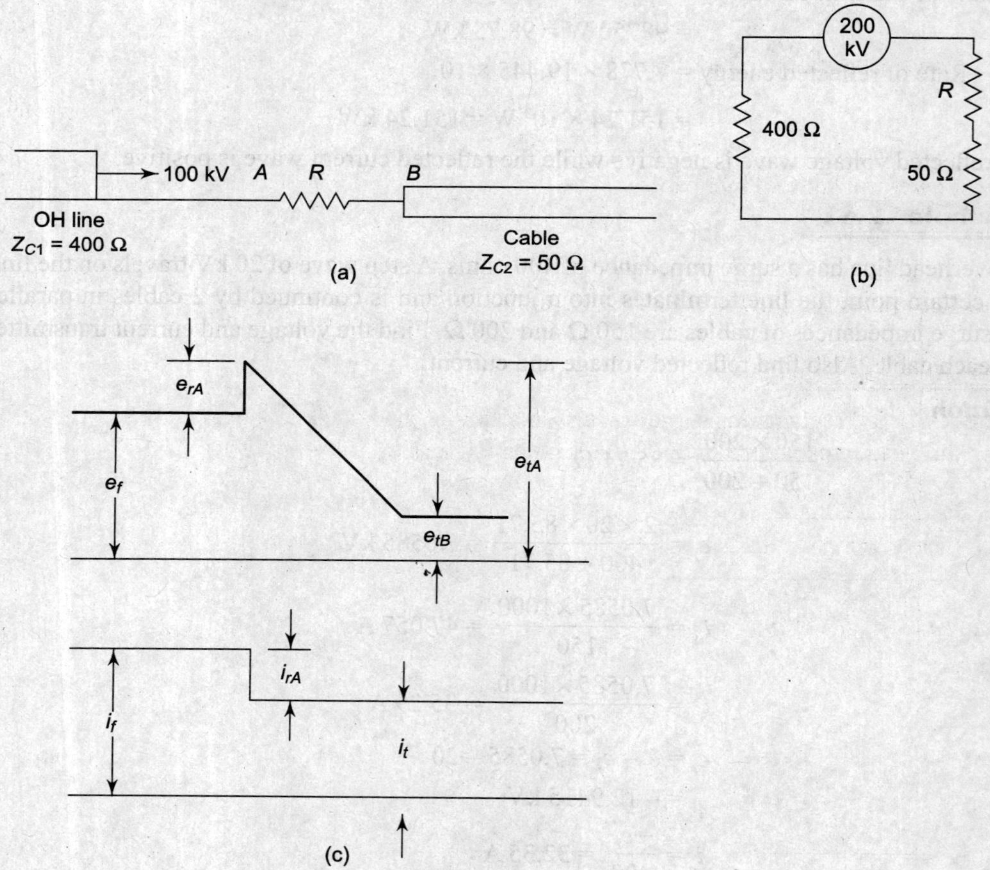

Fig. 12.37

(b) $$e_{tB} = \frac{2 \times 100 \times 50}{400 + 450 + 50} = 11.111 \text{ kV}$$

$$i_t = \frac{11.111}{50} \times 1000 = 222.22 \text{ A}$$

(c) $$e_{tA} = \frac{2\times 100\,(450+50)}{400+50+450} = 111.11 \text{ kV}$$

$$e_{rA} = e_{tA} - e_f = 11.11 \text{ kV}$$

$$i_{rA} = \frac{-e_r \times 1000}{400} = -27.775 \text{ A}$$

(d) $$i_f = \frac{100\times 1000}{400} = 250 \text{ A}$$

Power in incident wave = 100 × 250 × 10^3 W

= 25000 kW

Power in reflected wave = 11.11 × 10^3 × 27.775

= 308.58 × 10^3 W

= 308.58 kW

Power dissipated in resistance R

= $(222.22)^2$ × 450 W

= 22221.78 kW

Power transmitted into cable

= 11.111 × 222.22 × 10^3 W

= 2469.09 kW

(e) The waves are shown in Fig. 12.37 (c)

Example 12.13

A 200 kV, 3 μs, rectangular surge travels on a line of surge impedance of 400 Ω. The line is terminated in a capacitance of 3000 pF. Find an expression for voltage across the capacitance.

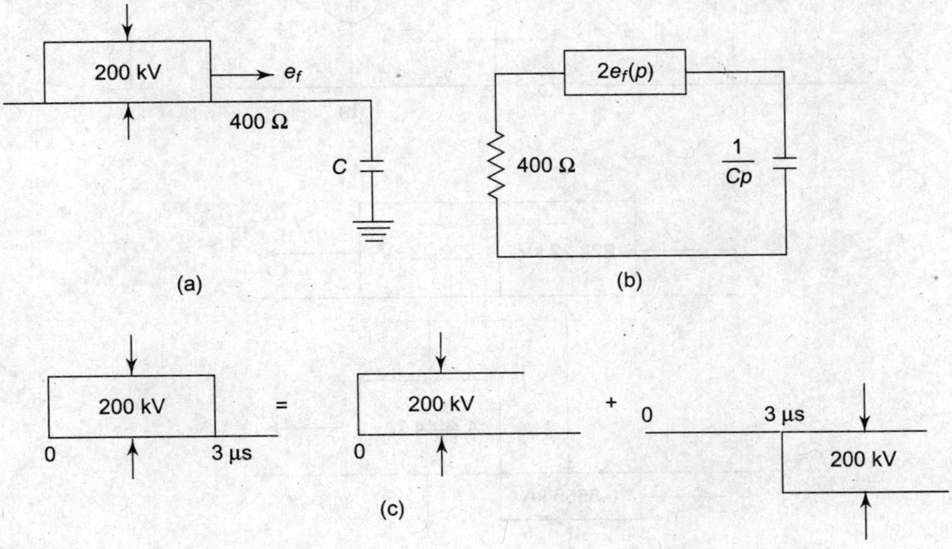

Fig. 12.38

Solution

Figure 12.38 (a) shows the system. The equivalent circuit is shown in Fig. 12.38(b). $e_f(p)$ denotes the Laplace transform of e_f

Since time is in microseconds, C should be in microfarads. The rectangular pulse can be expressed as sum of two waves as shown in Fig. 12.38 (c). Therefore

$$e_f = 200\,[u(t) - u(t-3)]$$

Using the results of section 12.13.2

$$C Z_c = 3000 \times 10^{-12} \times 400 = 1.2 \times 10^{-6}\text{ s} = 1.2\ \mu\text{s}$$

for $0 < t < 3\ \mu\text{s}$

$$e_t = 2 \times 200\,(1 - \varepsilon^{-t/1.2})$$
$$= 400\,(1 - \varepsilon^{-t/1.2})\text{ kV}$$

for $3 < t < \infty$

$$e_t = 400\,(1 - \varepsilon^{-t/1.2}) - 400\,[1 - \varepsilon^{-(t-3)/1.2}]$$

Example 12.14

Two lines having surge impedances Z_1 and Z_2 meet at a junction. A fault leading to fault current I occurs at the junction. Examine the voltage and current surges in the two lines due to clearance of fault. Find the magnitudes of voltage and current surges if $I = 5$ kA, $Z_t = 400$ ohms (OH line) and $Z_2 = 50$ ohms (cable). Show the voltage and current surges on a diagram.

Solution

Fig. 12.39 (a) shows the system. The clearance of fault can be simulated by injecting an equal and opposite current at point of fault. The equivalent circuit is shown in Fig. 12.39(b).

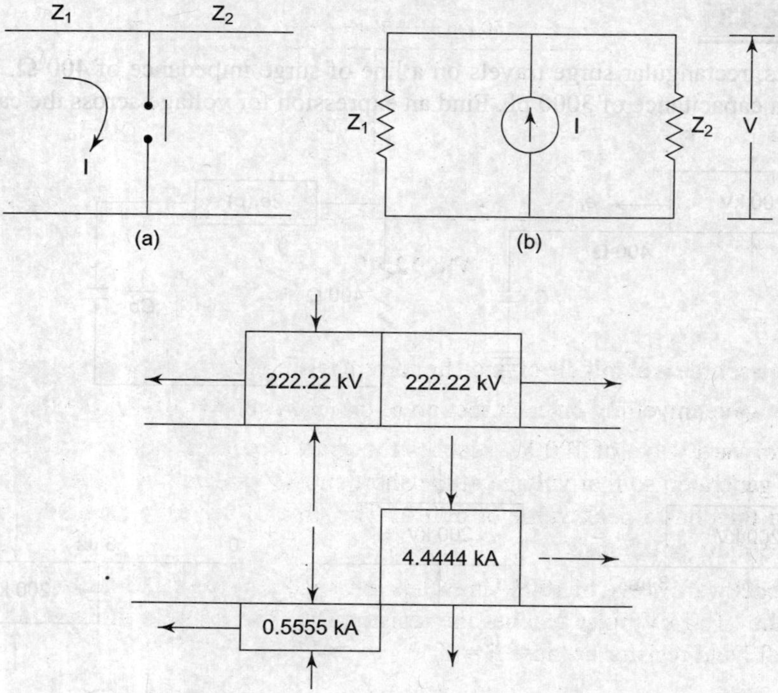

Fig. 12.39

$$\text{Magnitude of voltage surge} = \frac{IZ_1Z_2}{Z_1 + Z_2}$$

$$= 5\left[\frac{400 \times 50}{400 + 50}\right] = 222.22\ kV$$

$$\text{Magnitude of current surge in cable} = \frac{222.22}{50}$$

$$= 4.4444 \text{ kA}$$

It is a forward current wave. Both voltage and current surges are positive.

$$\text{Magnitude of current surge in OH line} = \frac{-222.22}{400}$$

$$= -0.55555 \text{ kA}$$

In OH line the voltage surge has positive polarity but current surge has negative polarity.

The surges are shown in Fig. 12.39 (c)

Example 12.15

Fig. 12.40 shows a 300 km long line having a pulse current source with magnitude 500 A and pulse duration 25 μs. The right hand end of line is short circuited and left hand end is terminated in a resistance equal to characteristic impedance. The wave velocity is 300m/μ-sec. Plot the current wave at right end and voltage wave at left end for a time of 1.5 ms. Surge impedance of line is 400 ohm.

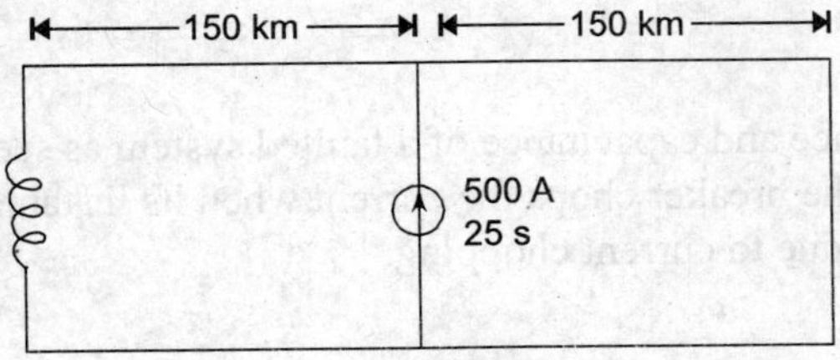

Fig 12.40

Solution

Since the current source is at middle of line, the current is divided into two equal parts 250 A each.

The voltage wave travelling on each section of the line is 250 × 400 = 100000V or 100 kV.

When the forward wave of 100 kV reaches the short circuit at right end a reflected wave of –100 kV is generated so that voltage at the short circuit is zero. The current at shorted end is doubled and thus has a peak value of 500 A. The time of travel in each section of line is 150000/300 = 500 μs or 0.5 ms.

The initial backward wave of 100 kV reaches left hand resistor at 0.5 ms. At $t = 0.5 + 0.5 + 0.5 = 1.5$ ms the –100 kV pulse reaches the resistor. Fig. 12.41 shows the waves. There is no reflection at left hand resistor because $R = Z_0$

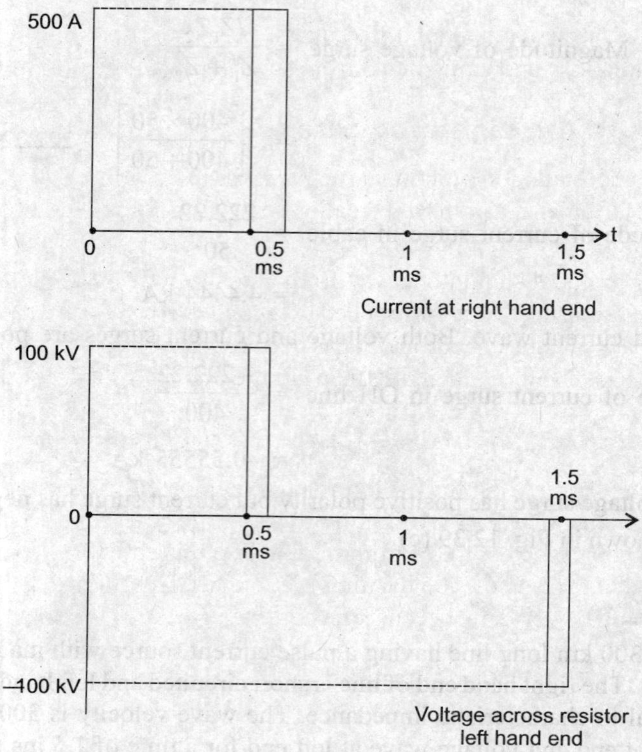

Fig. 12.41

Example 12.16

The effective inductance and capacitance of a faulted system as seen from breaker contacts are 4 mH and 300 μμF. The breaker chops the current when its instantaneous value is 100 A. Find the restriking voltage due to current chopping.

Solution

Using Eq. (12.80)

$$e = 100\left[\frac{4\times10^{-3}}{300\times10^{-12}}\right]^{0.5} \sin \frac{t}{(4\times10^{-3}\times300\times10^{-12})^{0.5}}$$

$$= 365 \times 10^3 \sin (0.913 \times 10^6\, t) \text{ volts}$$

12.31 LIGHTNING PROTECTION OF LOW VOLTAGE INSTALLATIONS

Lightning can cause huge damage to low voltage installations. These installations include high rise buildings, communication lines and equipments and other similar facilities. Lightning surges may travel on power lines and enter these installations through these power lines. A complete lightning protection scheme involves five basic steps.

1. Capture the lightning stroke through an air terminal.
2. Conduct the lightning current to earth through a down conductor
3. Dissipate the energy of lightning stroke in a low impedance earthing system.

4. Avoid earth loops and potential differentials.
5. Protect installations and equipments from surges travelling on power lines.

12.31.1 Capture the Lightning Stroke

The highest point on an installation/building/structure is the most vulnerable point. It is necessary to install a proper air terminal at a preferred location so that the lightning stroke may be attracted to this terminal and not strike other points.

This preferred air terminal should be away from antennae etc. One patented air terminal is called Dynasphere. It consists of a rod surrounded by a conductive sphere. The rod-sphere system ensures minimum space charge and corona. It has been reported that Dynasphere has the ability to capture lightning strokes more efficiently than simple pointed rod air terminal.

12.31.2 Conducting Lightning Current to Earth

The lightning current captured by the air terminal is conducted to the earth through a down conductor. It is necessary that flow of lightning current through mains power supply lines, telephone and other cables is kept to the minimum. It is preferable to use a screened insulated down conductor so that the risk of side flashing is avoided. In particular the inductance of the down conductor should be very low because the high rate of rise of lightning current can cause a very high $L \dfrac{di}{dt}$ voltage.

12.31.3 Dissipate Energy to Earth and Avoid Earth Loops

It is necessary that impedance of earthing system is low so that lightning current can be dissipated efficiently into the earth. A high impedance of earthing system could give rise to a high ground potential rise *i.e.* GPR (GPR is the product of current and impedance of earthing system). To avoid earth loops, there should be a single point connection to the earth network from all equipments in the system. The earthing system may consist of single or multiple driven rods, radial system (consisting of horizontal conductors buried radially in the earth), crowfoot radial system and earth mat. The design of earthing system is discussed in chapter 18.

12.31.4 Protection of Equipments from Surges on Power Lines

High voltage surges may travel on power lines and enter the installations. Moreover high voltage surges may also arise due to capacitive and inductive coupling with nearby lightning strokes. Lightning arresters are necessary for this protection. For this purpose the arresters are classified as under:

Class A	:	arrester for low voltage OH lines
Class B	:	arrester for lightning protection bonding
Class C	:	arrester for overvoltage protection
Class B	:	arrester for mobile use on socket outlets

The arresters for lightning current protection (class B) have high current capabilities. Class C arresters have relatively lower surge current rating.

SUMMARY

12.1 The short circuit current in a 3 phase alternator may be divided into three distinct time periods *i.e.* sub-transient period (lasting for about 2 cycles), transient period (lasting for about 30 cycles) and steady state period.

12.2 When a faulty circuit is opened, a recovery transient occurs. If L or C or both are small, the voltage across circuit breaker contacts rises very quickly and arc may restrike.

12.3 A line is a distributed parameter circuit. It can support travelling waves.

12.4 Wave velocity $= \dfrac{1}{\sqrt{LC}}$

12.5 Surge impedance $= \sqrt{L/C}$

12.6 For OH lines, wave velocity = 3×10^8 m/s and surge impedance is about 400 to 600 ohms.

12.7 For cables, wave velocity $= \dfrac{3\times 10^8}{\sqrt{k}}$ m/s where k is permittivity of insulation. Surge impedance is about 50 ohms.

12.8 A travelling wave is characterised by polarity (positive or negative), crest, front and tail.

12.9 When travelling waves reach a discontinuity, reflection and refraction of waves occurs.

12.10 For an open circuited line, $e_t = 2e_f$, $e_r = e_f$, $i_t = 0$ and $i_r = -i_f$

12.11 For a short circuited line $e_t = 0$, $i_r = 2i_f$, $e_r = -e_f$ and $i_r = i_f$

12.12 When a wave travelling on OH line meets a cable, the surge voltage transmitted on cable is much less than that on OH line.

12.13 Successive reflections can be studied with the help of Bewley lattice diagram.

12.14 Waves suffer attenuation and distortion.

12.15 In an isolated neutral system, arcing grounds occurs and gives rise to intermittent arcing.

12.16 When a capacitance current is switched off, a high frequency transient may occur. This phenomenon is called capacitance switching.

12.17 Interruption of a low current by a circuit breaker gives rise to current chopping transient.

12.18 Lightning strokes cause overvoltages on transmission lines. These strokes can be classified as direct and indirect.

12.19 A ground wire placed above the phase conductors of a transmission line shields the phase conductors from the lightning strokes. A shielding angle of about 30° gives adequate lightning protection.

12.20 A thyrite surge arrester is made of silicon carbide and has a non-linear resistance.

12.21 A surge arrester has the capability to protect the equipments up to suitable distances on both sides.

12.22 Insulation coordination has the aim of correlating insulation of various equipments with the characteristics of protective devices.

12.23 The protective devices must protect the equipments over the whole range of volt-time characteristics.

12.24 The provision of surge protective devices is based on the weakest component in the substation.

SELF TEST QUESTIONS

A. State whether the following are true (T) or false (F).

12.1 Shielding angle for proper shielding of phase conductors of OH lines is kept at about 30°.

12.2 Tower footing resistance should be as high as possible.

12.3 When a short circuit occurs on a 3 phase alternator the subtransient period lasts for about 15 cycles.

12.4 Subtransient reactance of an alternator is lesser than transient reactance.

12.5 When the contacts of a circuit breaker open, an arc is established across the contacts.

12.6 If arc restrikes the circuit interruption time is decreased.

12.7 Shunt conductance is neglected in travelling wave studies on OH lines.

12.8 A function of the form $f(x \pm vt)$ represents a standing wave.

12.9 Surge impedance has the characteristics of resistance.

12.10 The backward voltage and current waves have different polarities.

B. Fill in the blanks with suitable words.

12.11 Wave velocity on cables is ________ than that on OH lines.

12.12 Front time and tail time of surges are generally expressed in ________.

12.13 If a line having surge impedance Z_c is terminated by an impedance Z, the reflection coefficient is ________.

12.14 For an open circuited line, the voltage at open circuit is ________ the forward voltage.

12.15 The unique characteristic of short circuit is that voltage across it is ________.

12.16 When a forward rectangular wave with amplitude E travels on a line with surge impedance Z_c and terminated by inductance L, the voltage across inductance as function of time is ________.

12.17 When a surge travels an OH line towards its junction with a cable, the voltage transmitted on cable is ________ than that of initial forward wave.

12.18 A systematic way of keeping track of all transmitted and reflected waves is through the use of ________.

12.19 For a battery with constant voltage the reflection coefficient is ________.

12.20 The phenomenon of arcing grounds occurs on ________ neutral systems.

ANSWERS

12.1 T	12.2 F	12.3 F	12.4 T	12.5 T	12.6 F
12.7 T	12.8 F	12.9 T	12.10 T	12.11 less	
12.12 μ-secs	12.13 $\dfrac{Z - Z_c}{Z + Z_c}$	12.14 twice	12.15 zero	12.16 $2Ee^{-Z_c t/L}$	
12.17 Less	12.18 Bewley lattice diagram		12.19 –1	12.20 isolated	

JOB INTERVIEW QUESTIONS

12.1 What is the difference between travelling waves and standing waves?

12.2 What do you mean by the terms reflection coefficient and refraction coefficient?

12.3 What is arcing grounds phenomenon?

12.4 What is BIL?

12.5 What is the significance of characteristic impedance?

12.6 Differentiate between direct and indirect lightning strokes.

12.7 Differentiate between subtransient and transient reactance.

12.8 What is current chopping?

12.9 What is the effect of restriking of arc?

12.10 What is insulation coordination?

REVIEW QUESTIONS

12.1 Discuss why:

(a) The function $f(x \pm vt)$ represents a travelling wave.

(b) A travelling wave is usually represented as a step wave in the analysis.

(c) A travelling wave suffers reflection when it reaches a discontinuity.

(d) The indoor transformers are usually connected to the overhead lines through short lengths of cables.

(e) HV transmission lines are provided with ground wires as the topmost conductor.

(f) Present-day surge diverters use non-linear resistance elements.

12.2 Starting from first principles show that surges behave as travelling waves. Find expressions for surge impedance and wave velocity.

12.3 Discuss the phenomenon of wave reflection and refraction. Derive expressions for reflection and refraction coefficients.

12.4 A rectangular surge voltage E travels along a conductor of surge impedance Z_c towards a transition point P. Show that the voltages V_0 and current i_0 at point P satisfy the relation

$$V_0 = 2E - Z_e\, i_0$$

12.5 Develop equivalent circuit for analyzing the behaviour of travelling waves at transition points on transmission lines.

12.6 Discuss the behaviour of a travelling wave when it reaches the end of (a) open circuited (b) short circuited transmission line. Draw diagrams to show voltage and current on the line before and after the wave reaches the end.

12.7 Draw equivalent circuit for finding the transmitted voltage and current surges on a forked line. Derive expressions for the transmitted voltage and currents.

12.8 How can the analysis of a wave travelling on a line terminated by an inductance be carried out? Derive the expression for the voltage across the terminating inductance connected at the end of a line on which a step wave of magnitude E is travelling.

12.9 Develop an expression for the voltage across a capacitance C connected at the end of a line of surge impedance Z_c when a step wave E travels on the line and reaches the capacitance.

12.10 Discuss the use of (a) ground wires and (b) surge diverters for protection of transmission systems against surge voltages.

12.11 Derive an equation for finding the transient current in an *RL* series circuit excited by an ac source. What can be the maximum value of initial current? Under what condition does it occur?

12.12 Draw a typical oscillogram of the current in any phase of a synchronous generator subjected to sudden 3-phase short circuit. Why does the current follow the shape shown in this oscillogram? How can winding reactances be calculated from the above oscillogram?

12.13 Explain the term 'transient recovery voltage'. On what factors does the frequency of the recovery transient depend? What is the significance of rate of rise of recovery voltage?

12.14 What is arcing ground? On which systems does it occur?

12.15 Explain the mechanism of lightning stroke. What are the main characteristics of lightning strokes?

12.16 What types of lightning strokes can occur on transmission systems? How can the overvoltage due to these strokes be calculated?

12.17 How can Bewley lattice be drawn? Discuss its use.

PROBLEMS

12.1 Calculate the velocity of propagation for waves (a) in an overhead line of capacitance 0.147×10^{-10} F/m and inductance 0.75×10^{-6} H/m; (b) in a cable of inductance 0.75×10^{-6} H/m and capacitance 13.3×10^{-10} F/m; (c) estimate the relative permittivity of insulating material in case (b). [(a) 3×10^8 m/sec. (b) 10^8 m/sec. (c) 9]

12.2 A cable has an inner conductor of radius 0.5×10^{-2} m inside a sheath of inner radius 2×10^{-2} m. Find (a) the inductance per metre length, (b) capacitance per unit length, (c) surge impedance, (d) velocity of propagation. The permittivity of insulation is 3.

[(a) 2.77×10^{-7} H/m; (b) 0.12×10^{-9} F/m; (c) 48 ohms; (d) 1.73×10^8 m/sec.]

12.3 A 500 kV surge travels on an overhead line of surge impedance 400 ohms towards its junction with a cable which has a surge impedance of 40 ohms. Find (a) transmitted voltage, (b) transmitted current, (c) reflected voltage and (d) reflected current.

[(a) 90.9 kV; (b) 2.273 kA; (c) – 409.1 kV; (d) 1.023 kA]

12.4 A 300 kV surge travels on a cable with a surge impedance 30 ohms towards its junction with an overhead line which has a surge impedance of 450 ohms. Find (a) e_t, (b) i_t, (c) e_r, (d) i_r.

[(a) 562.5 kV; (b) 1.25 kA; (c) 262.5 kV; (d) – 8.75 kA]

12.5 A surge of 150 kV travels on a line of surge impedance 450 ohms and reaches junction of the line with two branch lines. The surge impedances of branch lines are 400 ohms and 40 ohms. Find the transmitted voltage and currents. [22.43 kV, 0.056 kA, 0.56 kA]

12.6 A surge of 125 kV travels on a line of surge impedance 500 ohms and reaches the junction of the line with two branch lines of surge impedances 450 ohms and 50 ohms. Find the reflected voltage, transmitted voltage, and transmitted currents.[–104.4 kV, 20.6 kA, 0.046 kA, 0.41 kA]

12.7 A step wave of magnitude E travels on a line of surge impedance Z_a and reaches the end of the line where the line is terminated by a resistance R and inductance L in series. Find an expression for the current in the terminating impedance.

$$\left[\frac{2E}{Z_0 + R}\left\{1 - \varepsilon^{-(Z_0+R)t/L}\right\}\right]$$

12.8 A step wave of 200 kV travels on a line having surge impedance of 500 ohms and reaches the end of the line where the line is terminated by an inductance of 2500 μH. Find the voltage across the inductance. [400 $\varepsilon^{-0.2t}$]

12.9 A transmission line has an inductance of 0.933 mH/km and a capacitance of 0.00778, μF/km. This overhead line is connected to an underground cable having an inductance of 0.155 mH/km and a capacitance of 0.187, μF/km. If a surge of crest 50 kV travels on the cable towards its junction with the line, find the surge transmitted along the line. [92.3 kV]

12.10 A step wave of 75 kV travels on the line having a surge impedance of 400 ohms and reaches the line termination where a capacitance of 2500 μF is connected between line and earth. Find the voltage across the capacitance. Derive the formula used. [150 (1 – ε^{-t}) kV]

12.11 In the system of example 12.5 the battery voltage is changed to 110 V and the terminating resistance is changed to 1000 Ω. The surge impedance of line is 400 Ω. Draw reflection lattice and plot the variation of voltage across the terminating resistance.

12.12 Two lengths of cable each 450 m are separated by a 300 m OH line. OH lines are also connected to outer ends of both the cables. The surge impedance of each OH lines is 450 Ω and that of each cable is 50 Ω. Velocity of propagation is 300 m/μ-sec for OH line and 150 m/μ-sec for cable. Plot the voltage at the junction of outer line and cable for 10 μ-sec, if a surge of 100 kV originates on the other outer OH line.

12.13 An overhead line of surge impedance 500 Ω is terminated by a cable of 50 Ω surge impedance. Length of cable is 300 m and it is open circuited at the far end. Velocity of wave on cable is 150 m/μ-sec. A step wave of 200 kV originates in the OH line and travels towards the cable. Plot voltage at mid-point of cable up to 20 μ-sec from the time the wave reaches the junction.

12.14 A transmission line of surge impedance 400 ohms has a capacitance 2500 pF shunted across it in the middle. If a step wave E comes on the line, find the voltage rise at capacitance.

[E $(1 - \varepsilon^{-2t})$]

12.15 In the system of example 12.12 the surge originates on the cable and travels towards resistance *R*. The remaining data is the same. Show the transmitted and reflected voltage and current waves on diagram.

12.16 The effective inductance and capacitance of a faulted system as viewed from the breaker contacts are 5 mH and 400 μμF respectively. The breaker chops the current when it has an instantaneous value of 200 A. Find the amplitude of restriking voltage due to current chopping.

[707.1 kV]

CHAPTER 13

Power System Stability

13.1 INTRODUCTION

The stability of a physical system is referred as its capability to return to the original equilibrium position on the occurrence of a disturbance or to another equilibrium state which is generally in the proximity of the initial equilibrium point. If the state of the system "runs away" so that certain physical variables go on increasing as $t \to \infty$, then the system is referred as unstable. These intuitive ideas are also applicable to a power system.

Suppose at a given time $t = t_0$, the power system is operating in a synchronous equilibrium and frequency equilibrium, *i.e.*, the rotor angles of the different machines with respect to a rotating synchronous reference frame are fixed and the system frequency is constant. The occurrence of a disturbance tends to momentarily alter synchronous equilibrium and the frequency equilibrium.

The disturbance may be small or big and the system may become unstable in either event depending on the operating condition at $t = t_0$. The study of stability in the presence of small disturbances constitutes what is known as "static stability" or "dynamic stability" analysis in the literature. The mathematical model for such a study is a set of linear time-invariant differential equations. When the disturbances are large, the non-linearities inherent in the power system can no longer be ignored and the study of stability under such circumstances constitutes what is known as 'transient stability' analysis. The mathematical model for such a study is a set of non-linear differential equations coupled with a set of non-linear algebraic equations. The dimensionality of the mathematical model both for dynamic and transient stability can be very large for even moderately sized systems.

The terms stability, maintenance of synchronism and power limit are frequently used interchangeably though interpretation is made many times erroneously. A system consisting of a synchronous generator, a line and an induction motor may become unstable but cannot lose synchronism. A system consisting of a synchronous generator, a line and a resistance load has a definite power limit without having a stability limit.

Every power system has a number of synchronous machines operating in parallel. When applied to electric power systems, the term stability denotes a condition in which the various synchronous machines in the system remain in synchronism or in step with each other. Conversely, instability denotes a condition involving loss of synchronism or falling out of step. Thus power system stability is concerned with the conditions necessary for the successful operation of a power system when changing from one stable condition to another due to variations in its operating conditions which can be classified as either small and slow or large and sudden, where slow means a long time action compared with the time constants of the field circuits of the machines, of automatic voltage regulators and of the turbine governors. The stability limit is defined as the maximum power which can flow through a point in the system without causing loss of stability. The stability problem can be subdivided into steady stability, transient stability and dynamic stability.

A power system has steady state stability if after a small slow disturbance it can regain and maintain synchronous speed. The small slow disturbance means normal load variations or changes. The steady state stability limit is the maximum possible flow of power, without loss of stability, when the power is changed very slowly. Similarly, a power system has transient stability if, after a large sudden disturbance, it can regain and maintain synchronous speed. A large sudden disturbance is one which is caused by faults and switching. The transient stability limit refers to the maximum possible flow of power without loss of stability when a large sudden disturbance occurs. Since every system is subjected to transient disturbances and further because the transient stability limit is generally less than the steady state stability limit, transient stability study is of utmost importance and almost the whole of this chapter deals with transient stability.

The transient instability phenomenon is a very fast one and occurs within one second or a fraction of it for a generator close to the location of disturbance. The action of voltage regulators and turbine governors is not included in transient stability studies because these devices are too slow to act during this interval. However, if the rotor swings are large, the regulator and governor may react. Dynamic stability is sometimes referred to this case of transient stability when the regulators and governors are fast acting and are taken into account in the analysis. In dynamic stability studies the system is analyzed for a period of 4–10 seconds following a large disturbance such as a short circuit or loss of generation or loss of load, etc. Dynamic stability study is also a very important area in the power system analysis but is beyond the scope of this text.

The most severe switching action, as far as transient stability is concerned, is the balanced 3-phase fault. This is fortunate from the analytical point of view because in this fault we have to deal with only the positive sequence network. All machines are assumed to be lossless. This assumption is valid for two reasons: (1) The resistance is always very small as compared to the inductive reactance. (2) The results based on this assumption are conservative and the system will always be more stable than our results predict.

The number of synchronous machines in a power system may be quite large. However all the machines in a power exporting area can be reduced to an equivalent generator. Similarly, all the machines in a power importing area can be reduced to an equivalent synchronous motor. The machines which are not connected to the same bus but are not separated by lines of high reactance can also be lumped together and considered as one machine. Thus, a multi-machine problem can be reduced to two-machine problem. The two-machine system can be reduced to one machine connected to an inifinite bus—a constant voltage constant frequency system. This simplifies the analysis to a great extent.

13.2 POWER ANGLE CURVE, TRANSFER REACTANCE

13.2.1 Generator Loaded at its Terminals

Figure 13.1 shows a synchronous machine having a direct axis synchronous reactance X_d.

$E = |E| \angle\delta =$ voltage behind direct axis synchronous reactance of generator

V = terminal voltage of generator

The complex power output of generator is

$$S = VI^*$$

$$= |V| \left[\frac{|E| \angle \delta - |V|}{j X_d} \right]^* \quad (13.1a)$$

$$= \frac{|E||V|}{X_d} \angle (90 - \delta) - \frac{j|V|^2}{X_d} \quad (13.1b)$$

$$= \frac{|E||V|}{X_d} \sin \delta + j \left[\frac{|E||V|}{X_d} \cos \delta - \frac{|V|^2}{X_d} \right] \quad (13.1c)$$

The real power output P_e of the generator is

$$P_e = \text{Re}\,|S| = \frac{|E||V|}{X_d} \sin \delta = P_m \sin \delta \quad (13.2)$$

where $$P_m = \frac{|E||V|}{X_d}$$

Thus, the real power output depends on $|E|$, $|V|$, X_d and power angle δ. The reactive power output Q_e is

$$Q_e = \text{Im}\,|S| = \frac{|E||V|}{X_d} \cos \delta - \frac{|V|^2}{X_d} \quad (13.3)$$

If line to line values of E and V are used in Eqs. (13.2) and (13.3), we directly get 3-phase power. Fig. 13.2 shows the steady state real power variation with power angle for both generator and motor action for constant values of E, V and X_d. This curve is known as power angle curve. The condition of positive values of δ, *i.e.*, E leading V, applies to the generator action and the condition of negative value of δ, *i.e.*, E lagging V, applies to the motor action.

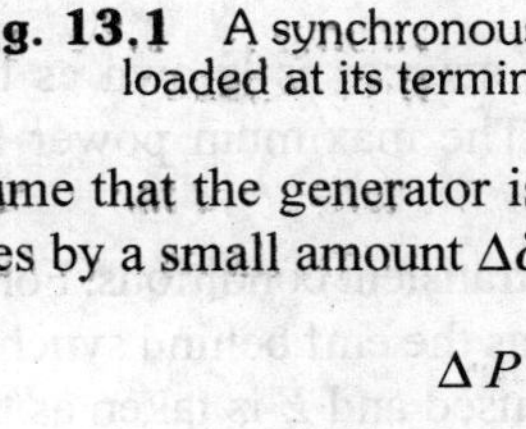

Fig. 13.1 A synchronous machine loaded at its terminal

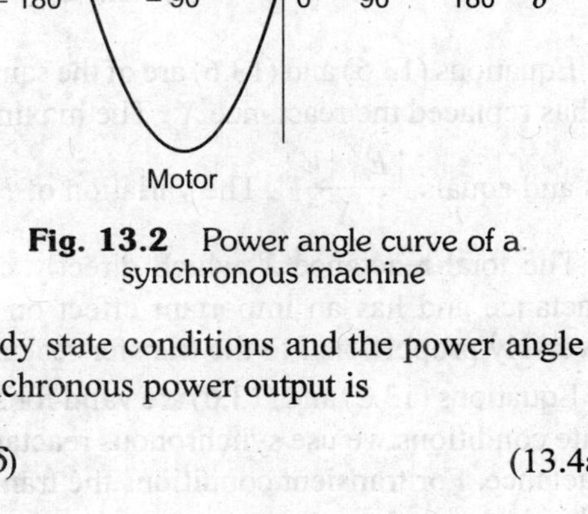

Fig. 13.2 Power angle curve of a synchronous machine

Assume that the generator is working under steady state conditions and the power angle δ increases by a small amount $\Delta\delta$. The increase in synchronous power output is

$$\Delta P = \frac{dP}{d\delta} (\Delta\delta) = P_r (\Delta\delta) \quad (13.4a)$$

where $$P_r = \frac{dP}{d\delta} = \frac{|E||V|}{X_d}\cos\delta \tag{13.4b}$$

The quantity P_r is known as synchronizing power coefficient. So long as $P_r > 0$, *i.e.*, $0 \le \delta \le 90°$, an increase in power angle will result in an increase in power output. This power must be taken kinetically from the rotor (since the generator input has not been increased) and the generator slows down. Thus the generator is stable, *i.e.*, it resists any tendency for δ to change from the value corresponding to the steady state condition. This is not true if $\delta \ge 90°$. The maximum value of δ for successful operation is 90° and the maximum steady state power output equals $\frac{|E||V|}{X_d}$. In actual practice δ is kept around 30°.

13.2.2 Generator Connected to Infinite Bus

Figure 13.3 shows a synchronous machine connected to an infinite bus through a transmission line of reactance X_t. An infinite bus is one whose voltage and frequency are unaffected by a disturbance in the machine under consideration. Thus an infinite bus has a constant voltage and frequency and has zero internal reactance. Let $V = |V| \angle 0$ be the voltage of the infinite bus. The complex power delivered by the generator to the system is

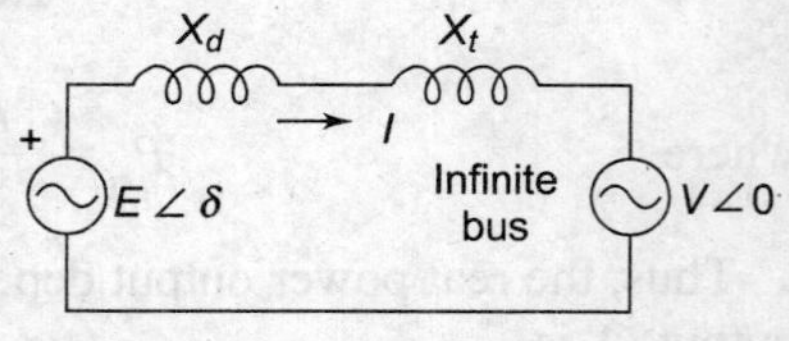

Fig. 13.3 A synchronous machine connected to infinite bus

$$S = VI^* = |V|\left[\frac{|E|\angle\delta - |V|}{j(X_d + X_t)}\right]^* \tag{13.5a}$$

$$= \frac{|E|\,|V|}{X}\angle(90-\delta) - j\frac{|V|^2}{X} \tag{13.5b}$$

$$= \frac{|E||V|}{X}\sin\delta + j\left(\frac{|E|\,|V|}{X}\cos\delta - \frac{|V|^2}{X}\right) \tag{13.5c}$$

where $X = X_d + X_t$

The real power delivered to the system is

$$P_e = \text{Re}\,|S| = \frac{|E|\,|V|}{X}\sin\delta = P_m\sin\delta \tag{13.5d}$$

Equations (13.5) and (13.6) are of the same form as Eqs. (13.1) and (13.2) except that reactance X has replaced the reactance X_d. The maximum steady state power transfer P_m occurs when δ = 90 and equals $\frac{|E|\,|V|}{X}$. The variation of P_e with δ is also similar to Fig. 13.2.

The total reactance X which directly connects the two emf sources is known as transfer reactance and has an important effect on power angle curve. The maximum power limit is inversely proportional to the transfer reactance.

Equations (13.2) and (13.6) are valid for steady state as well as transient conditions. For steady state conditions we use synchronous reactance X_d and E is taken as the emf behind synchronous reactance. For transient conditions the transient reactance X'_d is used and E is taken as the emf behind transient reactance.

13.2.3 Power Transfer through Impedance

In all electrical machines and transmission lines, the resistance is negligible as compared to inductive reactance. However, it is instructive to study the power transfer through an impedance Z

$$I = \frac{|E|\angle\delta - |V|\angle 0}{|Z|\angle\theta} = \frac{|E|}{|Z|}\angle(\delta-\theta) - \frac{|V|}{|Z|}\angle-\theta$$

$$I^* = \frac{|E|}{|Z|}\angle(\theta-\delta) - \frac{|V|}{|Z|}\angle\theta \tag{13.7}$$

$$S = VI^* = \frac{|V|\,|E|}{|Z|}\angle(\theta-\delta) - \frac{|V|^2}{|Z|}\angle\theta \tag{13.8}$$

$$P_e = \text{Re}\,[S] = \frac{|V|\,|E|}{|Z|}\cos(\theta-\delta) - \frac{|V|^2}{|Z|}\cos\theta \tag{13.9}$$

$$Z = R + jX = \sqrt{R^2+X^2}\ \angle\tan^{-1}(X/R)$$

Let $\alpha = (90 - \theta) = \sin^{-1}(R/Z)$, then

$$P_e = \frac{|V||E|}{|Z|}\cos(90-\alpha-\delta) - \frac{|V|^2}{|Z|}\cos(90-\alpha)$$

$$= \frac{|V||E|}{|Z|}\sin(\alpha+\delta) - \frac{|V|^2}{|Z|}\sin\alpha \tag{13.10}$$

P_e, is maximum when $(\alpha+\delta) = \pi/2$, therefore

$$P_{e\max} = \frac{|V||E|}{\sqrt{R^2+X^2}} - \frac{|V|^2}{\sqrt{R^2+X^2}} \times \frac{|R|}{\sqrt{R^2+X^2}} \tag{13.11}$$

If $E = V$

$$P_{e\max} = \frac{|V|^2}{\sqrt{R^2+X^2}} - \frac{|V|^2 R}{R^2+X^2} \tag{13.12}$$

For maximum power transfer

$$\frac{d P_{e,\max}}{d X} = 0 = |V|^2\left[\frac{X}{(R^2+X^2)^{1.5}} - \frac{2XR}{(R^2+X^2)^2}\right]$$

or $$\frac{|V|^2 X}{(R^2+X^2)^2}\,[(R^2+X^2)^{0.5} - 2R] = 0$$

or $$(R^2+X^2)^{0.5} = 2R$$

or $$X = \sqrt{3}\,R \tag{13.13}$$

In actual practice R is very small as compared to X and, therefore, the practical application of Eq. (13.13) is limited.

Equation (13.12) shows that if $X = 0$, power transferred is zero. Thus a finite value of reactance is necessary for transfer of power. Power transfer over a purely resistive line is zero. If $R = 0$, $\alpha = 0$ and the above results become identical with the results derived in sections 13.2.1 and 13.2.2.

Example 13.1

A generator having $X_d = 0.7\ pu$ delivers rated load at a power factor of 0.8 lagging.

(a) Find P_e, Q_e, E and δ. Draw phasor diagram.

(b) The steam valve of the prime mover is opened further so that P_e increases by 20%. Find new values of P_e, Q_e, E and δ. Draw phasor diagram.

(c) The steam valve is resorted to the original position. The exciter is adjusted to raise E by 20%. Find new value of P_e, Q_e, E and δ. Draw phasor diagram.

Solution

(a) Let the subscript $_0$ denote initial condition and subscripts 1 and 2 denote new conditions.

$$V = 1\angle 0,\ I_0 = 1\ (.8 - j.6)\ pu$$

$$E_0 = V + jX_d I_0 = 1 + (j.7)(.8 - j.6) = 1.53\ \angle\ 21.5^\circ$$

Thus $\quad E_0 = 1.53\ pu,\ \delta_0 = 21.5^\circ$

$$P_{eo} = \frac{|E_0|\,|V|}{X_d}\sin\delta_0 = \frac{1.53\times 1}{0.7}\sin 21.5^\circ = 0.8$$

$$Q_{e0} = \frac{|E_0|\,|V|}{X_d}\cos\delta_0 - \frac{|V|^2}{X_d} = \frac{1.53\times 1}{0.7}\cos 21.5^\circ - \frac{1}{0.7} = 0.6\ pu$$

The phasor diagram is shown in Fig. 13.4 (a)

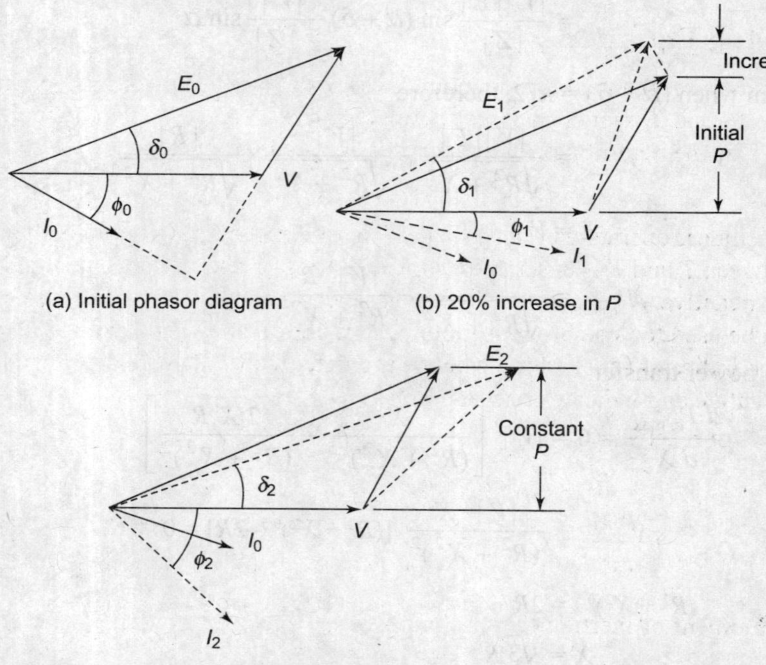

Fig. 13.4 Phasor diagrams for example 13.1

(b) $\quad P_{e1} = 1.2 \times 0.8 = 0.96$ (*i.e.*, 20% increase)

$$E_1 = 1.53 \text{ (no change)}$$

$$0.96 = \frac{1.53 \times 1}{0.7} \sin \delta_1$$

or
$$\delta_1 = \sin^{-1}\left(\frac{0.96 \times 0.7}{1.53}\right) = 26.1^\circ$$

$$Q_{e1} = \frac{1.53 \times 1}{0.7} \cos (26.1) - \frac{1 \times 1}{0.7} = 0.535\ pu$$

The phasor diagram is shown in Fig. 13.4(b)

(c)
$$P_{e2} = 0.8$$
$$E_2 = 1.2\ (1.53) = 1.84\ (20\%\ \text{increase})$$
$$0.8 = \frac{1.84 \times 1}{0.7} \sin \delta_2$$

or
$$\delta_2 = 17.8^\circ$$
$$Q_{e2} = \frac{1.84 \times 1}{0.7} \cos (17.8^\circ) - \frac{1 \times 1}{0.7}$$
$$= 1.07$$

The phasor diagram is shown in Fig. 13.4(c).

13.3 MECHANICS OF ANGULAR MOTION

13.3.1 Swing Equation

A generator receives mechanical torque input T_i via shaft from the prime mover and develops electromagnetic torque T_e which opposes the mechanical torque T_i. The algebraic difference between T_i and T_e is the net torque T_a which causes acceleration of rotor.

$$T_a = T_i - T_e \tag{13.14}$$

When the machine is operating in steady state, T_a is zero and there is no acceleration. Whenever this balance between T_i and T_e is upset, there is acceleration or retardation depending on whether T_a is positive or negative. A synchronous motor has electrical input T_e and mechanical output T_i. Eq. (13.14) can be used for synchronous motor also if both T_i and T_e are taken as negative. Then T_a is positive and indicates acceleration when T_e is greater than T_i. On the other hand, if T_e is less than T_i, T_a is negative and indicates retardation.

From the laws of mechanics

$$I\alpha = T_a \tag{13.15a}$$

or
$$I \frac{d^2\theta}{dt^2} = T_a \tag{13.15b}$$

where I is the moment of inertia in kg-m^2 and α or $\frac{d^2\theta}{dt^2}$ is the angular acceleration in elect. degrees/sec^2.

The angular position θ of the rotor continuously varies with time. It is convenient to measure the angular position with respect to a synchronously rotating reference axis than with respect to a stationary axis. Let

$$\delta = \theta - \omega_s t \tag{13.16}$$

where ω_s = synchronous speed in elect. degree/sec

δ = angular displacement of rotor, from the synchronously rotating reference axis in electric degrees.

Differentiating Eq. (13.16) with respect to t

$$\frac{d\delta}{dt} = \frac{d\theta}{dt} - \omega_s \tag{13.17a}$$

and

$$\frac{d^2\delta}{dt^2} = \frac{d^2\theta}{dt^2} \tag{13.17b}$$

Combining Eqs. (13.14, 13.15b and 13.17b)

$$I\frac{d^2\delta}{dt^2} = T_a = T_i - T_e \tag{13.18}$$

Multiplying Eq. (13.18) by rotor speed ω

$$M\frac{d^2\delta}{dt^2} = P_a = P_i - P_e \tag{13.19}$$

where $M = I\omega$ = angular momentum

$P_i = T_i\,\omega$ is the mechanical power input, corrected for windage and friction losses.

$P_e = T_e\,\omega$ is the electrical power output, corrected for electrical losses.

$P_a = P_i - P_e$ is the accelerating power or difference between input and output, each corrected for losses.

Equation (13.19) is known as swing equation. The electrical power P_e is the real power given by Eq. (13.6). The angle δ for a machine connected to an infinite bus is the same as power angle (also known as torque angle) used in Eq. (13.6) since this angle is the difference between the internal angle of machine and the angle of the synchronously rotating reference frame. For a two-machine system two swing equations are necessary, one for each machine, but as shown in section 13.5.2, a two-machine system can be reduced to a single machine connected to infinite bus.

13.3.2 Inertia Constant

The stored kinetic energy of a rotating body is given by

$$\text{KE} = \frac{1}{2} I\,\omega^2 = \frac{1}{2} M\,\omega \tag{13.20}$$

This equation gives the unit of M as Joule-sec, per radian. The energy is more conveniently expressed in mega-joules and angles are often measured in degrees. Therefore, M can be measured in mega-joule seconds per electrical degree.

Even for machines of similar designs, moment of inertia and momentum vary over a wide range with variation in rating. Therefore, momentum is given in terms of an inertia constant H which is defined as

$$H = \frac{\text{Stored energy in mega - joules}}{\text{Machine rating in MVA, } i.e., \text{G}} \tag{13.21}$$

Thus stored energy in mega-joules equals GH.

From Eqs. (13.20) and (13.21)

$$GH = \frac{1}{2} M\omega = \frac{1}{2} M(360 f)$$

or
$$M = \frac{GH}{180 f} \text{ Mega-joule-sec/elect. degree} \quad (13.22a)$$

$$= \frac{GH}{\pi f} \text{ Mega-joule-sec/elect. rad.} \quad (13.22b)$$

The inertia constant H has a characteristic value for each class of machine. Table 13.1 gives typical values of H for different types of synchronous machines.

Table 13.1 Typical inertia constants

Type of machine	*Inertia constant H (MJ per MVA)*
Turbo-generator	3–9
Water wheel generator	2–4
Synchronous condenser	1–1.25
Synchronous motor	2.0

When a number of synchronous machines are operating in parallel, they can be replaced by a single machine having a rating G equal to the sum of the ratings of all the machines acting together. The inertia constant H of the equivalent machine can be found from the equation:

$$H_e G_e = H_1 G_1 + H_2 G_2 + \dots \quad (13.23)$$

If the inertia constant is to be found on a base of G_b, the required equation is

$$H_{eb} G_b = H_1 G_1 + H_2 G_2 + \dots \quad (13.24a)$$

or
$$H_{eb} = \frac{H_1 G_1}{G_b} + \frac{H_2 G_2}{G_b} \quad (13.24b)$$

When H_1, H_2......... are the inertia constants of individual machines having ratings as G_1 G_2........, H_e is the inertia constant of an equivalent machine having rating G_e (equal to G_1+ G_2 + ...) and H_{eb} is the equivalent inertia constant on a base G_b. It is evident from the above discussion that the value of H for a given rating varies inversely as the base.

Example 13.2

A 200 MVA 11 kV 50 Hz 4 pole turbo generator has an inertia constant of 6 MJ/MVA. (a) Find the stored energy in the rotor at synchronous speed. (b) The machine is operating at a load of 120 MW when the load suddenly increases to 160 MW. Find the rotor retardation. Neglect losses. (c) The retardation calculated above is maintained for 5 cycles. Find the change in power angle and rotor speed in rpm at the end of this period. (d) Another generator 150 MVA, 3000 rpm, having H = 4 MJ/MVA is put in parallel with above generator. Find the inertia constant for the equivalent generator on a base of 100 MVA. *[JNTU 2011]*

Solution

(a) Stored energy = GH = 200 × 6 = 1200 MJ

(b) P_a = 120 – 160 = – 40 MW

$$M = \frac{GH}{180\,f} = \frac{1200}{180 \times 50} = \frac{2}{15} \text{ MJ-sec/elect. deg.}$$

$$M\alpha = P_a$$

or
$$\frac{2}{15}\alpha = -40 \text{ or } \alpha = -300 \text{ elect. deg/sec}^2$$

The retardation is 300 elect. degrees/sec^2

(c) 5 cycles = 0.1 sec.

$$\text{Change in } \delta = \frac{1}{2}\alpha t^2 = \frac{1}{2} = (-300)\,(0.1)^2 = -1.5 \text{ elect. deg.}$$

Since it is a 4-pole machine, 1 revolution corresponds to 4 × 180, *i.e.*, 720 elect. deg.

$$\alpha = -300 \text{ elect. deg/sec}^2 = -(300)\left(\frac{60}{720}\right) = -25 \text{ rpm/sec}$$

Synchronous speed = 1500 rpm

Rotor speed at the end of 5 cycles or 0.1 sec.

$$= 1500 + (-25)\,(0.1) = 1497.5 \text{ rpm}$$

(d)
$$H_{eb} = \frac{H_1\,G_1}{G_b} + \frac{H_2\,G_2}{G_b}$$

$$= \frac{6 \times 200}{100} + \frac{4 \times 150}{100} = 18 \text{ MJ/MVA}$$

13.4 STEADY STATE STABILITY

The steady state stability of a system refers to its ability to maintain synchronism when the system is subjected to slow changes in load or input. The steady state stability limit is the maximum power which can be transferred without loss of synchronism.

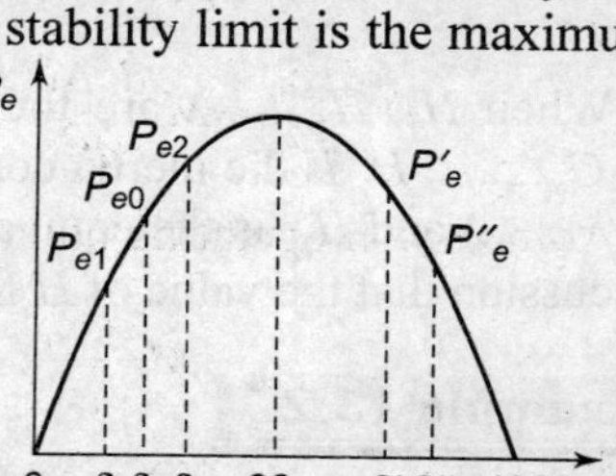

Fig. 13.5 Power angle curve for synchronous generator

The power angle curve of a synchronous generator connected to infinite bus bar is redrawn in Fig. 13.5. Suppose that the generator is operating in steady state at a load angle δ_0 and delivering power P_{eo}. Neglecting losses there is a perfect balance between the mechanical power input and the electrical power output of the generator. If a slight arbitrary change in load takes place reducing the load angle to δ_1, the power output reduces to P_{e1}. The generator input is now greater than output thus causing an acceleration of the rotor and an increase in the value of δ. Since the losses have been neglected, the angle would oscillate about the value δ_0. In actual practice, the losses are always present, the oscillations would be damped and the load angle would return to the original value δ_0. A similar situation takes place if the load angle arbitrarily increases to δ_2. Now the power ouput P_{e2} is more than the input causing a retardation of rotor, the load angle decreases and finally returns to its original value δ_0. Thus, if the system, operating at a load angle δ_0, is subjected to small arbitrary disturbing forces, then regardless of the direction of the disturbing forces, the system returns to its original position.

Now consider the operation of the system at load angle δ'. Let the load angle increase to δ''. The output of the generator decreases to P_e''. Since the input is the same, an acceleration of the rotor takes place causing a further increase in load angle and the generator will finally lose synchronism. The maximum value of load angle for successful operation is 90° and the steady state power limit is $\frac{|E||V|}{X}$. A simple test for the motion to be a damped oscillation is that the synchronizing power coefficient $\frac{|E||V|}{X}$ cos δ must be positive which happens when $0 < \delta <$ 90°.

13.4.1 Steady State Stability Analysis for Change in Load Angle

To analyse steady state stability it is necessary to evaluate dynamic response of a single machine connected to infinite bus. The swing equation is

$$\frac{Md^2\delta}{dt^2} = P_i - \frac{|E||V|}{X}\sin\delta = P_i - P_m \sin\delta \tag{13.25}$$

The swing equation is non-linear but can be assumed to be linear if disturbances are small. Let the initial operating power angle be δ_0 and consider a small deviation $\Delta\delta$ in power angle, *i.e.*

$$\delta = \delta_0 + \Delta\delta \tag{13.26}$$

Substituting Eq. (13.26) into Eq. (13.25) we get

$$M\frac{d^2(\delta_0 + \Delta\delta)}{dt^2} = P_i - P_m \sin(\delta_0 + \Delta\delta)$$

or

$$M\frac{d^2\delta_0}{dt^2} + \frac{Md^2\Delta\delta}{dt^2} = P_i - P_m(\sin\delta_0\cos\Delta\delta + \cos\delta_0\sin\Delta\delta)$$

Since we are considering only small disturbances, $\Delta\delta$ is small. Therefore $\sin\Delta\delta \simeq \Delta\delta$ and $\cos\Delta\delta \simeq 1$. Then we get

$$M\frac{d^2\delta_0}{dt^2} + M\frac{d^2\Delta\delta}{dt^2} = P_i - P_m\sin\delta_0 - P_m\cos\delta_0\Delta\delta \tag{13.27}$$

At the initial operating point δ_0

$$M\frac{d^2\delta_0}{dt^2} = P_i - P_m\sin\delta_0$$

Therefore Eq. (13.27) becomes a linearized equation in terms of changes in power angle *i.e.*

$$M\frac{d^2\Delta\delta}{dt^2} + P_m\cos\delta_0\Delta\delta = 0 \tag{13.28}$$

The quantity P_m cos δ_0 in Eq. (13.28) is the slope of the power angle curve at δ_0. It is called synchronising power coefficient denoted by P_r. Thus

$$P_r = \left.\frac{dP}{d\delta}\right|_{\delta=\delta_0} = P_m\cos\delta_0$$

P_r plays an important role in determining system stability. It is maximum at $\delta_0 = 0$ and is zero at $\delta_0 = 90°$. Substituting P_r in Eq. (13.28) we get

$$M\frac{d^2\Delta\delta}{dt^2} + P_r\Delta\delta = 0 \tag{13.29}$$

The solution is obtained by substituting $p = \frac{d}{dt}$ and $p^2 = \frac{d^2}{dt^2}$. The characteristic equation is

$$Mp^2 + P_r = 0 \tag{13.30a}$$

or

$$p = \pm(-P_r / M)^{0.5} = \pm j\omega_n \tag{13.30b}$$

where $\omega_n = (P_r /_M)^{0.5}$ is the undamped frequency of oscillations in radians/sec. The losses in the system damp the oscillations and the system returns to the steady state after the small disturbance. The system is stable so long as $P_r > 0$ because under this condition both roots of Eq. (13.30a) are imaginary. When $P_r < 0$, the roots of Eq (13.30a) are real and the system becomes unstable. The condition $P_r > 0$ means that $0 < \delta < 90°$. It may be added that the operations of a synchronous motor is stable if $-90° < \delta < 0$.

When there is difference in angular velocity between the motor and the resultant rotating air gap field, an induction motor action takes place and a torque is set up. This torque tends to minimise the difference between the angular velocity of motor and the rotating field. This torque is called damping torque. The damping power P_d is proportional to speed variation. Thus

$$P_d = D\frac{d\delta}{dt} \tag{13.31}$$

where D is the damping coefficient. D can be found either from design data or by test.

Taking damping into account the linearized swing equation becomes

$$M\frac{d^2\Delta\delta}{dt^2} + D\frac{d\Delta\delta}{dt} + P_r\Delta\delta = 0$$

or

$$\frac{d^2\Delta\delta}{dt^2} + \frac{D}{M}\frac{d\Delta\delta}{dt} + \frac{P_r}{M}\Delta\delta = 0 \tag{13.32}$$

Eq (13.32) can be written in standard second order differential equation form. Thus we have

$$\frac{d^2\Delta\delta}{dt^2} + 2\zeta\omega_n\frac{d\Delta\delta}{dt} + \omega_n^2\Delta\delta = 0 \tag{13.33}$$

where natural frequency ω_n is $(P_{r/M})^{0.5}$ and damping ratio is given by

$$\zeta = \frac{D}{2}\left(\frac{1}{M P_r}\right)^{0.5} \tag{13.34}$$

The characteristic equation is

$$s^2 + 2\zeta\omega_n s + \omega_n^2 = 0 \tag{13.35}$$

For normal operating conditions $\zeta < 1$ and the roots of characteristic equation are complex. Thus

$$\begin{aligned} s_1, s_2 &= -\zeta\omega_n \pm j\omega_n\sqrt{1-\zeta^2} \\ &= -\zeta\omega_n \pm j\omega_d \end{aligned} \tag{13.36}$$

ω_d is the damped frequency of oscillations and is given by

$$\omega_d = \omega_n\sqrt{1-\zeta^2} \tag{13.37}$$

When damping is positive and P_r is also positive, the roots of characteristic equation have negligible real part and the system is stable.

Eq. (13.33) can also be solved through state variable method. In this way the analysis can be extended to multimachine systems. Let

$$x_1 = \Delta\delta \text{ and } x_2 = \Delta\omega = \dot{\Delta}\,\delta$$

then $\dot{x}_1 = x_2$ and $\dot{x}_2 = -\omega_n^2 x_1 - 2\zeta\,\omega_n\, x_2$

The above equation can be written in matrix form as

$$\begin{bmatrix} \dot{x}_1 \\ \dot{x}_2 \end{bmatrix} = \begin{bmatrix} 0 & 1 \\ -\omega_n^2 & -2\zeta\,\omega_n \end{bmatrix}\begin{bmatrix} x_1 \\ x_2 \end{bmatrix} \tag{13.38a}$$

or

$$\dot{x}\,(t) = A\,x\,(t) \tag{13.38b}$$

where

$$[A] = \begin{bmatrix} 0 & 1 \\ -\omega_n^2 & -2\zeta\omega_n \end{bmatrix}$$

The desired output is the state variables x_1 and x_2. Thus the output vector $y\,(t)$ is

$$y(t) = \begin{bmatrix} 1 & 0 \\ 0 & 1 \end{bmatrix}\begin{bmatrix} x_1 \\ x_2 \end{bmatrix} \tag{13.39}$$

Taking Laplace transform of Eq. (13.38b) we get

$$s\,X\,(s) - x\,(0) = A\,X\,(s)$$

or

$$X\,(s) = [s\,I - A]^{-1}\,x\,(0) \tag{13.40}$$

where

$$[s\,I - A] = \begin{bmatrix} s & -1 \\ \omega_n^2 & s + 2\zeta\,\omega_n \end{bmatrix}$$

where we substitute for $[s\,I - A]^{-1}$ in Eq. (13.40) we get

$$X(s) = \frac{\begin{bmatrix} s+2\zeta\,\omega_n & 1 \\ -\omega_n^2 & s \end{bmatrix} x(0)}{s^2 + 2\zeta\omega_n\,s + \omega_n^2}$$

where a small disturbance occurs, rotor angle changes by a small amount $\Delta\delta_0$. Then $x_1(0) = \Delta\delta_0$ and $x_2\,(0) = \Delta\omega_0 = 0$. Then we get

$$\Delta\delta(s) = \frac{(s+2\zeta\,\omega_n)\Delta\delta_0}{s^2 + 2\zeta\,\omega_n s + \omega_n^2}$$

$$\Delta\omega(s) = \frac{\omega_n^2\,\Delta\delta_0}{s^2 + 2\zeta\,\omega_n s + \omega_n^2} \tag{13.41}$$

Taking inverse Laplace transform we get zero input response as

$$\Delta\delta = \frac{\Delta\delta_0}{\sqrt{1-\zeta^2}} e^{-\zeta\omega_n t} \sin(\omega_d t + \theta)$$

$$\Delta\omega = -\frac{\omega_n \Delta\delta_0}{\sqrt{1-\zeta^2}} e^{-\zeta\omega_n t} \sin(\omega_d t) \tag{13.42}$$

where ω_d is the damped frequency of oscillations and θ is given by

$$\theta = \cos^{-1}\zeta$$

The motion of motor relative to synchronously revolving field is

$$\delta = \delta_0 + \frac{\Delta\delta_0}{\sqrt{1-\zeta^2}} e^{-\zeta\omega_n t} \sin(\omega_d t + \theta) \tag{13.43a}$$

and rotor angular frequency is

$$\omega = \omega_0 - \frac{\omega_n \Delta\delta_0}{\sqrt{1-\zeta^2}} e^{-\zeta\omega_n t} \sin(\omega_d t) \tag{13.43b}$$

The response time constant is

$$\tau = \frac{1}{\zeta\omega_n} = D/_{2M} \tag{13.44}$$

The setting time is approximately four time constants. Thus

$$t_s \simeq 4\tau$$

Since $\omega_n = (P_r/M)^{0.5}$ and $\zeta = \frac{D}{2}\left(\frac{1}{M P_r}\right)^{0.5}$ and inertia constant H is directly proportional to M, as H increases ω_n and ζ decreases which means longer setting time. Moreover as synchronising power coefficient P_r increases, ω_n increases and ζ decreases.

13.4.2 Steady State Stability Analysis for Increase in Power Input

When the power input changes by a small amount ΔP, the linearised swing equation is

$$M\frac{d^2\Delta\delta}{dt^2} + D\frac{d\Delta\delta}{dt} + P_r\Delta\delta = \Delta P$$

or

$$\frac{d^2\Delta\delta}{dt^2} + \frac{D}{M}\frac{d\Delta\delta}{dt} + \frac{P_r}{M}\Delta\delta = \frac{\Delta P}{M} \tag{13.45}$$

Writing Eq (13. 45) in standard second order differential form we have

$$\frac{d^2\Delta\delta}{dt^2} + 2\zeta\omega_n\frac{d\Delta\delta}{dt} + \omega_n^2 = \Delta u \tag{13.46}$$

The terms ζ and ω_n have already been defined in section 13.4.1 and $\Delta u = \Delta P/_M$

Expressing in state variable form we get

$$x_1 = \Delta\delta \text{ and } x_2 = \Delta\omega = \dot{\Delta}\delta$$

$$\dot{x}_1 = x_2 \text{ and } \dot{x}_2 = -\omega_n^2 x_1 - 2\zeta\omega_n x_2$$

The above equations written in matrix form are

$$\begin{bmatrix} \dot{x}_1 \\ \dot{x}_2 \end{bmatrix} = \begin{bmatrix} 0 & 1 \\ -\omega_n^2 & -2\zeta\,\omega_n \end{bmatrix}\begin{bmatrix} x_1 \\ x_2 \end{bmatrix} + \begin{bmatrix} 0 \\ 1 \end{bmatrix}\Delta u \tag{13.47a}$$

or

$$\dot{x}\,(t) = \mathrm{A}\,x\,(t) + \mathrm{B}\,\Delta u\,(t) \tag{13.47b}$$

The variables x_1 and x_2 are the desired response and output vector, as before, as given by Eq (13.39). Taking Laplace transform of Eq (13.47b) we get

$$s\,\mathrm{X}\,(s) = \mathrm{A}\,\mathrm{X}\,(s) + \mathrm{B}\,\Delta\,\mathrm{U}(s)$$

or

$$\mathrm{X}\,(s) = (s\mathrm{I} - \mathrm{A})^{-1}\,\mathrm{B}\,\Delta\,\mathrm{U}\,(s) \tag{13.48}$$

where $\Delta U(s) = \dfrac{\Delta u}{s}$

Substituting for $(s\mathrm{I} - \mathrm{A})^{-1}$ we get

$$X(s) = \frac{\begin{bmatrix} s+2\zeta\,\omega_n & 1 \\ -\,\omega_n^2 & s \end{bmatrix}\begin{bmatrix} 0 \\ 1 \end{bmatrix}\begin{matrix} \Delta u \\ s \end{matrix}}{s^2 + 2\zeta\,\omega_n\,s + \omega_n^2}$$

The solution in s domain is

$$\Delta\delta(s) = \frac{\Delta u}{s\left(s^2 + 2\zeta\,\omega_n\,s + \omega_n^2\right)}$$

and

$$\Delta\omega(s) = \frac{\Delta u}{s^2 + 2\zeta\;\omega_n s + \omega_n^2}$$

Taking inverse Laplace transform we get

$$\Delta\delta = \frac{\Delta u}{\omega_n^2}\left[1 - \frac{1}{\sqrt{1-\zeta^2}}\,e^{-\zeta\omega_n t}\sin(\omega_d t + \theta)\right] \tag{13.49a}$$

where θ, as before, equals $\cos^{-1}\zeta$ and

$$\Delta\omega = \frac{\Delta u}{\omega_n^2\sqrt{1-\zeta^2}}\,e^{-\zeta\omega_n t}\sin\omega_d t \tag{13.49b}$$

Since $\Delta u = \dfrac{\Delta P}{M}$, the motion of rotor relative to synchronously revolving field (in electrical radians) is

$$\delta = \delta_0 + \frac{\Delta P}{M\omega_n^2}\left[1 - \frac{1}{\sqrt{1-\zeta^2}}\,e^{-\zeta\omega_n t}\sin(\omega_d t + \theta)\right] \tag{13.50a}$$

and rotor angular velocity in radians /sec is

$$\omega = \omega_0 + \frac{\Delta P}{M\omega_n\sqrt{1-\zeta^2}}\,e^{-\zeta\omega_n t}\sin\omega_d t \tag{13.50b}$$

We have written Eq. (13.50) in terms of angular momentum M. It is more convenient to obtain solution in terms of inertia constant H and per unit values. Since, in per unit representation $M = \frac{H}{\pi f}$, Eq. (13.50) takes the form

$$\delta = \delta_0 + \frac{\Pi f \Delta P}{H\omega_n^2}\left[1 - \frac{1}{\sqrt{1-\zeta^2}} e^{-\zeta\omega_n t} \sin(\omega_d t + \theta)\right] \quad (13.51a)$$

and

$$\omega = \omega_0 + \frac{\Pi f \Delta P}{H\omega_n \sqrt{1-\zeta^2}} e^{-\zeta\omega_n t} \sin \omega_d t \quad (13.51b)$$

Example 13.3

A 50 Hz turbo generator having an inertia constant of 9 MJ/MVA is connected to infinite bus. The total transfer reactance is 0.60 $p.u.$ The real power transferred is 0.7 pu at 0.8 pf lagging. Assume voltage of infinite bus as 1 $p.u$ and $D = 0.14$. Suddenly the circuit breakers open and close causing a change of 10° in the rotor angle. Find the values of synchronising power coefficient, undamped angular frequency of oscillations, damping ratio, the equation of motion of rotor, damped angular frequency of oscillations in radians /sec and in Hz. Also find the equation for oscillations in δ and f as functions of time

Solution

$$V = 1\angle 0 \; p.u.$$

$$S = \frac{0.7}{0.8}\angle \cos^{-1} 0.8 = 0.875\angle 36.87° \; p.u.$$

$$I = 0.875 \angle -36.87° \; p.u.$$

$$E = V + jXI$$

$$= 1\angle 0 + j(0.875\angle -36.87°)(0.6)$$

$$= 1 + j0 + 0.525\angle 53.13°$$

$$= 1 + j0 + 0.315 + j0.42$$

$$= 1.315 + j0.42 = 1.38\angle 17.71°$$

Thus $\delta_0 = 17.71° = 0.309$ radians

Synchronising power coefficient $P_r = P_m \cos \delta_0$

$$= \frac{(1.38)(1)}{0.6}\cos(17.71°)$$

$$= 2.191 \text{ p.u.}$$

$$\omega_n = \left[\frac{P_r}{M}\right]^{0.5} = \left[\frac{P_r \Pi f}{H}\right]^{0.5} = \left[\frac{2.191(\Pi)(50)}{9}\right]^{0.5} = 6.18 \text{ rad/sec.}$$

$$\zeta = \frac{D}{2}\left(\frac{1}{MP_r}\right)^{0.5} = \frac{D}{2}\left[\frac{\Pi f}{HP_r}\right]^{0.5} = \frac{0.14}{2}\left[\frac{\Pi \times 50}{9 \times 2.191}\right]^{0.5} = 0.1976$$

Substituting the values of Eq. (13.33) we get

$$\frac{d^2\Delta\delta}{dt^2}+2(0.1976)(6.18)\frac{d\Delta\delta}{dt}+6.18^2\,\Delta\delta=0$$

or

$$\frac{d^2\Delta\delta}{dt^2}+2.442\frac{d\Delta\delta}{dt}+38.19\Delta\delta=0$$

$$\omega_d=\omega_n\left(1-\zeta^2\right)^{0.5}=6.18\left(1-0.1976^2\right)^{0.5}=6.058 \text{ rad/sec.}$$

$$=\frac{6.058}{2\Pi}=0.964 \text{ Hz}$$

Substituting the values in Eqs. (13.43 a and b)

$$\delta=17.71+\frac{10}{\sqrt{1-0.1976^2}}\,e^{-(0.1976)(6.18)\,t}\sin\left(6.058t+\cos^{-1}0.1976\right)$$

or

$$\delta=17.71+10.406e^{-1.221t}\sin\left(6.058t+78.6°\right)$$

$$\omega=\frac{2\pi(50)}{\lambda}-\frac{6.18(10)/360}{\sqrt{1-0.1976^2}}e^{-1.221t}\sin\left(6.058t\right)$$

or

$$f=50-0.1751\,e^{-1.221t}\sin\,(6.058t)$$

Example 13.4

The system of example 13.3 is subjected to a sudden increase in input of 0.1 *p.u*. The remaining data is the same. Derive expressions for oscillations of δ and f as functions of time.

Solution

Substituting all the values in Eqs. (13.50a and b) we get

$$\delta=17.71+\frac{180\times50\times0.1}{9(6.18)^2}\left[1-\frac{1}{\sqrt{1-0.1976^2}}e^{-0.1976\times6.18\,t}\sin\left(6.058t+78.6°\right)\right]$$

or

$$\delta=17.71+2.618\left[1-1.0829e^{-1.221t}\sin\left(6.058t+78.6°\right)\right]$$

$$\omega=50(2\Pi)+\frac{\Pi(50)(0.1)}{(9)(6.18)\sqrt{1-0.1976^2}}e^{-1.221t}\sin 6.058t$$

Dividing by 2Π we get

$$f=50+0.0917\,e^{-1.221t}\sin 6.058t$$

Example 13.5

Figure 13.6 shows a generator transformer unit feeding an infinite bus bar through a transmission line. At the middle of the transmission line a 3-phase inductor of reactance 0.5 *pu* per phase is connected. The generator e.m.f behind synchronous reactance is 1.1 *pu* and the voltage of infinite bus is 1 *pu*. Find (a) transfer reactance and steady state power limit with inductor switch open; (b) transfer reactance and steady state power limit with inductor switch closed; (c) transfer reactance and steady state power limit with inductor replaced by a capacitor of the same *pu* reactance; (d) transfer reactance and steady state power limit if, instead of shunt capacitor, a series capacitor of the same per unit reactance is connected in series with the line at the same location.

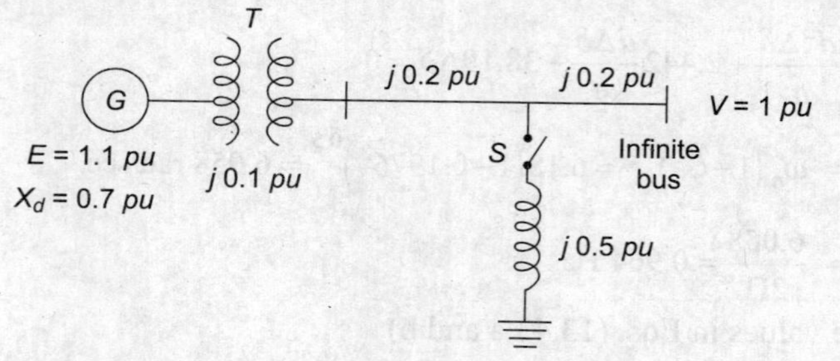

Fig. 13.6 System for example 13.5

Solution

(a) When switch *S* is open

transfer reactance $X = 0.7 + 0.1 + 0.2 + 0.2 = 1.2\ pu$

$$\text{Steady state power limit} = \frac{1.1 \times 1}{1.2} = 0.9167\ pu$$

(b) When switch *S* is closed, the equivalent circuit is shown in Fig. 13.7(a). Converting the star to delta gives network of Fig. 13.7 (b).

The reactance of the branch which connects the two emf sources is transfer reactance and has a value

$$X = \frac{(j1)(j0.2) + (j0.2)(j0.5) + (j1)(j0.5)}{j0.5}$$

$$= j\,1.6\ pu$$

It is not necessary to calculate the reactance of shunt arms of delta because these reactances, being directly across the constant voltage sources, do not contribute to power transfer, although they increase the reactive power output.

$$\text{Steady state power limit} = \frac{1.1 \times 1}{1.6} = 0.6875\ \text{pu}$$

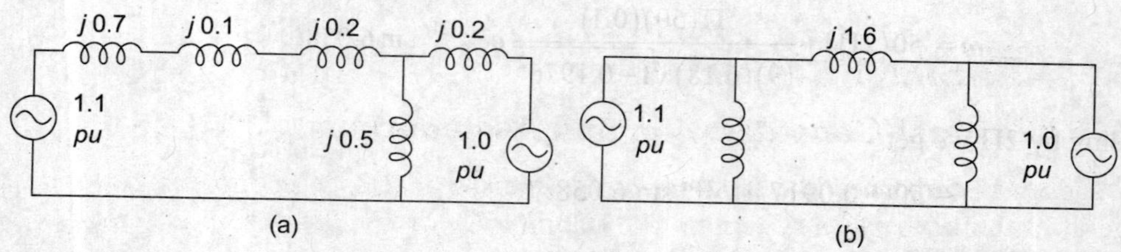

Fig. 13.7

(c) When the inductive reactor is replaced by a capacitor of value 0.5 *pu*, the equivalent circuit is shown in Fig.13.8 (a). Converting the star to delta reduces this network to that of Fig. 13.8 (b). The transfer reactance is $j\,0.8\ pu$.

$$\text{Steady state power limit} = \frac{1.1 \times 1}{0.8} = 1.375\ pu$$

(d) When the shunt capacitor is replaced by a series capacitor, the transfer reactance is

$$X = j0.7 + j0.1 + j0.2 - j\,0.5 + j\,0.2 = j\,0.7\ pu$$

$$\text{Steady state power limit} = \frac{1.1 \times 1}{1.7} = 1.57\ pu$$

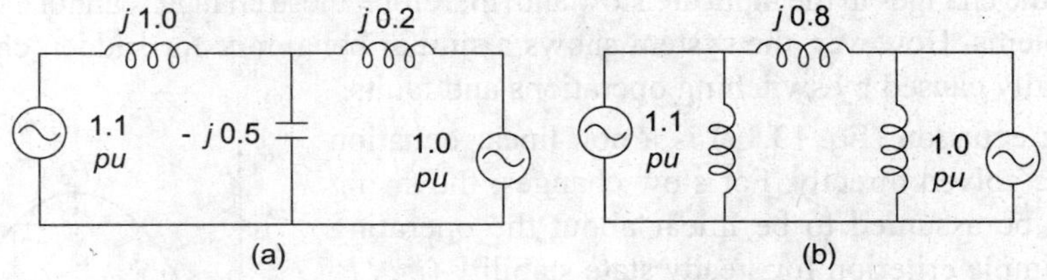

Fig. 13.8

Example 13.6

A 50 Hz generator of reactance 1 *pu* is connected to an infinite bus through a line of reactance 0.5 *pu*. *E* = 1.1 *pu* and *V* = 1 *pu*. The inertia constant is 5 MW-sec/MVA. The generator is loaded to 50% of the maximum power limit. Find the frequency of natural oscillations.

Solution

The operating angle δ_0 is givn by

$$\sin \delta_0 = \frac{P}{P_m} = 0.5 \text{ or } \delta_0 = 30°$$

$$P_r = \frac{|E||V|}{X} \cos \delta_0 = \frac{1.1 \times 1}{1.5} \cos 30° = 0.635 \text{ unit power/elect. rad.}$$

$$M = \frac{H}{\pi \times 50} = \frac{5}{\pi \times 50} \text{ unit power-sec}^2\text{/elect. rad.}$$

From Eq. (13.30 b)

$$\omega_n = (P_r/M)^{0.5} = \left(\frac{0.635 \times \pi \times 50}{5}\right)^{0.5} = 4.466 \text{ rad/sec.} = 0.71 \text{ Hz}$$

13.5 TRANSIENT STABILITY

13.5.1 General Consideration and Assumptions

Many sudden disturbances like fault, change in transfer reactance due to switching, sudden change in load, may cause a system to lose stability even when the system is being operated below its steady state stability limit. The maximum power transfer is limited by the requirement of maintaining transient stability.

Figure 13.9 shows the power angle curve of a generator. Suppose the system is operating in steady state at $P_i = P_{i0}$ and $\delta = \delta_0$. When the mechanical power suddenly increases to P_{i1} at t = 0, δ cannot adjust instantly due to rotor inertia. Thus at t = 0, $\delta = \delta_0$ and input exceeds output. The rotor starts accelerating and δ starts increasing. When $\delta = \delta_1$, the input and output are equal but the rotor cannot stop here due to inertia and δ continues to increase. For $\delta > \delta_1$, input is less than output and rotor retardation takes place. Eventually the rotor may stop at δ_2 and swing back towards δ_1. There will be some oscillations around δ_1. However, these oscillations will damp out and the rotor will stabilize at the new equilibrium value δ_1. The maximum swing of the rotor, beyond δ_1, depends on the amount of sudden increase in input. If during the swing, δ becomes greater than δ_m, input again becomes greater than the output, the rotor will be again accelerated causing δ to increase. Therefore, stability will be lost if δ swings past δ_m.

Generally the changes in the input are slow and, therefore, these changes cannot cause transient stability problems. However, the system shows a similar behaviour for sudden changes in P_e. This is generally caused by switching operations and faults.

The swing equation (Eq. 13.19) is a non-linear equation and cannot be solved directly. For slow changes, the swing equation can be assumed to be linear about the operating point and a simple criterion for steady state stability (*i.e.*, $P_r > 0$) is available. However, for sudden changes, the swing equation can be solved only by numerical methods. The solution of swing equation gives a plot of δ versus t. This graph is known as swing curve. If the swing curve indicates that δ starts decreasing after reaching a maximum value, it is usually assumed that the system is stable and the oscillations of δ about the equilibrium point will become successively smaller and will be eventually damped out.

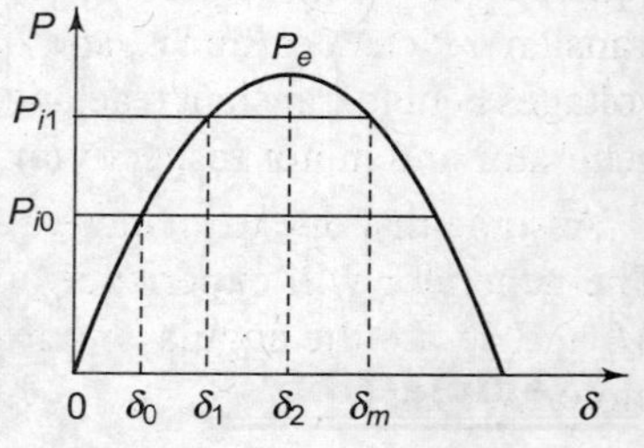

Fig. 13.9 System reaction to sudden load charges

The following assumptions are usually made in solving swing equation:

1. The input remains constant during the entire period of analysis. This assumption is quite valid for a synchronous motor because the speed of the motor does not change significantly unless the motor loses synchronism. The input to a generator is controlled by the governor of the prime mover. The governor acts by sensing a change in speed. Since the change in speed is insignificant (unless stability is lost) and governor action is usually slow, the effect of governor action is ignored in stability studies. This assumption makes the results somewhat pessimistic.
2. The resistances of lines and machines are neglected. The effect of damper windings is also neglected. This assumption also leads to pessimistic results because resistance introduces damping and improves stability. Line shunt capacitance is also neglected.
3. The effect of saliency is neglected. The machine is represented by a constant voltage (*i.e.*, voltage behind direct axis transient reactance) in series with X'_d. This means that the excitation of the machine is assumed to remain constant and the effect of the voltage regulator is ignored.

The above assumptions simplify the calculations to some extent. Even then the calculations are so lengthy that the use of digital computer is preferred. The solution for a multimachine system is possible only through a computer simulation.

For the case of a single machine connected to infinite bus, it is possible to steady transient stability by the application of equal area criterion and the solution of swing equation is not necessary. As mentioned in section 13.1, the machines in a power exporting area can be combined to represent an equivalent generator. Similarly, all the machines in a power importing area can be reduced to an equivalent motor. This two-machine system can be reduced to one machine connected to infinite bus.

13.5.2 Reduction of Two-Machine System to One Machine System

Figure 13.10 shows the equivalent circuit of a two-machine system consisting of a generator feeding a motor through a transmission line having reactance X_t. The mechanical power input to the generator is P_i and its electrical power output is P_e. Since the losses are neglected, the electrical power input to the motor is P_e and its mechanical power output equals P_i, X'_{1d} and X'_{2d} are the direct axis transient reactances, and E_1 and E_2 are the voltages behind transient reactances of the generator and motor respectively.

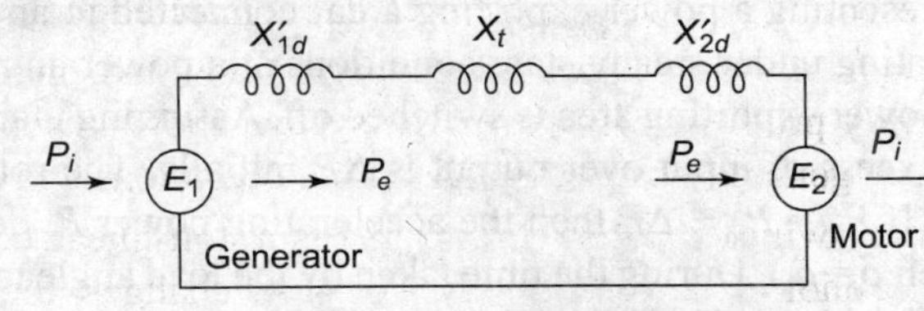

Fig. 13.10 A two-machine system

Assume that due to an increase in the transfer reactance, the power transfer decreases by ΔP. The generator will experience an acceleration and the motor will experience a retardation. If M_1 and M_2 are the angular momentum and δ_1 and δ_2 be the power angles of the generator and motor, we have

$$\frac{d^2\,\delta_1}{d\,t^2} = \frac{\Delta P}{M_1} \tag{13.52a}$$

$$\frac{d^2\,\delta_2}{d\,t^2} = -\frac{\Delta P}{M_2} \tag{13.52b}$$

The relative angles between the two machines is

$$\delta = \delta_1 - \delta_2 \tag{13.53}$$

Expressing the acceleration of the generator relative to the motor as the algebraic difference of their accelerations which must equal the acceleration of the equivalent single generator relative to the infinite bus, the following equation can be obtained:

$$\frac{d^2\,\delta}{dt^2} = \frac{d^2\,\delta_1}{dt^2} - \frac{d^2\,\delta_2}{dt^2} = \Delta P\left(\frac{1}{M_1} + \frac{1}{M_2}\right) = \frac{\Delta P}{M_e} \tag{13.54}$$

Equation (13.54) is the swing equation of the equivalent single generator connected to infinite bus, where M_e is the angular momentum of the equivalent generator and is given by

$$\frac{1}{M_e} = \frac{1}{M_1} + \frac{1}{M_2} \tag{13.55a}$$

or

$$M_e = \frac{M_1\,M_2}{M_1 + M_2} \tag{13.55b}$$

Further, the power transfer is given by

$$P_e = \frac{|E_1||E_2|}{X'_{1d} + X_t + X'_{2d}} \sin\delta = \frac{|E_1|\,|E_2|}{X} \sin\delta \tag{13.56}$$

Thus the two-machine system of Fig. 13.10 is equivalent to a single generator of angular momentum M_e and voltage (behind transient reactance) E_1 connected to an infinite bus of voltage E_2 and the transfer reactance between the generator and infinite bus, *i.e.*, $X = X'_{1d} + X_t + X'_{2d}$.

13.6 EQUAL AREA CRITERION

The stability of single machine swinging with respect to an infinite bus can be studied by the use of equal area criterion. Figure 13.11 (a) shows the power angle curve for an equivalent generator (representing a power exporting area) connected to an infinite bus. Suppose that the system is operating under steady state conditions at a power angle δ_0 when a large local load ΔP within the power exporting area is switched off. Assuming that the mechanical input remains the same, the excess of input over output is ΔP initially. The rotor is accelerated leading to an increase in δ. If $P_{i1} = P_{i0} + \Delta P$ then the accelerating power P_a decreases from ΔP (when $\delta = \delta_0$) to zero (when $\delta = \delta_1$). During the time taken by the load angle to increase from δ_0 to δ_1, the rotor absorbs kinetic energy. This kinetic energy equals the shaded area A_1. At point 2 the accelerating power is zero but the rotor has acquired a speed slightly greater than the synchronous speed and the angle δ continues to increase beyond δ_1. However, as δ becomes greater than δ_1, P_a becomes negative causing the rotor to retard. The rotor swing continues till the load angle is δ_2 and the rotor attains a speed equal to synchronous speed. Neglecting all losses (*e.g.*, resistance, eddy current, damping, etc.) the load angle δ_2 can be obtained from the condition that the kinetic energy gained by rotor during its swing from δ_0 to δ_1 must equal the kinetic energy returned as it swings from δ_1 to δ_2. This leads to the conclusion that area A_1 must be equal to the shaded area A_2. This is referred to as equal area criterion.

The accelerating area A_1 is given by

$$A_1 = \int_{\delta_0}^{\delta_1} (P_{i1} - P_m \sin \delta)\, d\delta$$

$$= P_{i1}(\delta_1 - \delta_0) - P_m(\cos \delta_0 - \cos \delta_1) \tag{13.57a}$$

The decelerating area A_2 is given by

$$A_2 = \int_{\delta_1}^{\delta_2} (P_m \sin \delta - P_{i1})\, d\delta$$

$$= P_m(\cos \delta_1 - \cos \delta_2) - P_{i1}(\delta_2 - \delta_1) \tag{13.57b}$$

For stability $\quad A_1 = A_2$

or $\quad P_m(\cos \delta_0 - \cos \delta_2) = P_{i1}(\delta_2 - \delta_0)$

or $\quad \cos \delta_0 - \cos \delta_2 = (\delta_2 - \delta_0) \sin \delta_1 \qquad (13.58)$

The load angles in Eq. (13.58) must be measured in radians. This equation cannot be solved by ordinary algebraic methods due to its transcendental character.

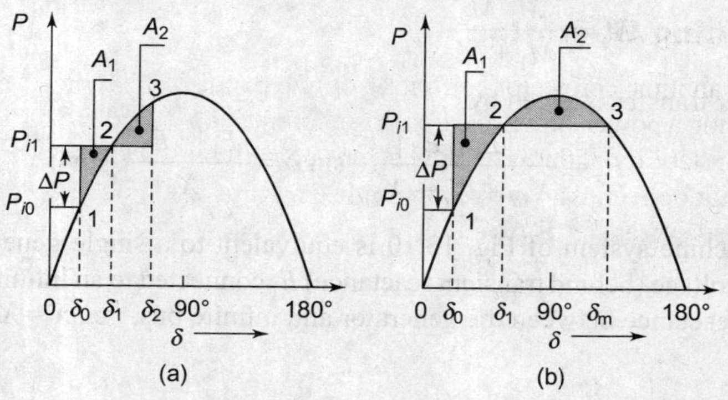

Fig. 13.11 (a) Equal area criterion for a large sudden change in load; (b) limiting case when $\delta_m = 180 - \delta_1$

For a given δ_0 the maximum or limiting value of ΔP is given by the condition that $\delta_m = \pi - \delta_1$ radians (Fig. 13.11b). If δ does not go beyond δ_m, the rotor will swing back and the load angle will stabilize at $\delta = \delta_1$. However, if δ becomes greater than δ_m the power transfer will be lesser than P_{i1} and the rotor will experience further acceleration. This will make a further increase in load angle and the generator will go out of synchronism.

Therefore, the limiting value of ΔP is computed by the condition that $\delta_m = \pi - \delta_1$ radians. So long as δ does not increase beyond δ_m the equal area criterion is satisfied and the system remains within stability limits. Substituting $\delta_2 = \pi - \delta_1$ in Eq. (13.34) gives the condition

$$\cos \delta_0 + \cos \delta_1 = (\pi - \delta_1 - \delta_0) \sin \delta_1 \qquad (13.59)$$

For a given δ_0, solution of δ_1 cannot be obtained in a straightforward manner due to the transcendental nature of equation. An approximate solution can be obtained by counting the squares occupied by the areas A_1 and A_2 and selecting a value of δ_1 (Fig. 13.11b) such that areas A_1 and A_2 are equal.

The equal area criterion can be proved mathematically as follows. From swing equation,

$$\frac{d^2}{dt^2} = \frac{P_a}{M} \qquad (13.60)$$

where P_a is the accelerating power (difference between input and output.)

Multiplying both sides of Eq. (13.36) by $\frac{d}{dt}$

$$\frac{d}{dt} \cdot \frac{d^2}{dt^2} = \frac{1}{2}\frac{d}{dt}\left(\frac{d}{dt}\right)^2 = \frac{P_a}{M} \cdot \frac{d}{dt} \qquad (13.61a)$$

or

$$\frac{d}{dt} = \left[\int \frac{2\,P_a}{M} \cdot \frac{d}{dt}\,dt\right]^{0.5} = \left[\frac{2}{M}\int P_a d\right]^{0.5} \qquad (13.61b)$$

The load angle will have a maximum value when

$$\frac{d}{dt} = 0, \textit{ i.e.}, \text{ when } \int_0^m P_a d = 0 \qquad (13.62)$$

Thus the rotor swings from δ_0 to a maximum value δ_m at which time the relative velocity is zero. The two shaded areas A_1 and A_2 must be equal and of opposite sign (Fig. 13.11b).

13.7 APPLICATIONS OF EQUAL AREA CRITERION

13.7.1 Switching Operation

Figure 13.12 shows an inter-connector consisting of two parallel lines connecting an equivalent generator (representing a power exporting area) to an infinite bus. Equal area criterion can be used to study the transient stability of the system when one of the lines is switched out (simultaneously and instantaneously at both ends). Two power angle diagrams are involved. When both the lines are operating, the power transfer is given by

$$P_e = \frac{|E||V|}{X_d' + \dfrac{X_1 \times X_2}{X_1 + X_2}} \sin \delta \qquad (13.63)$$

This is shown as curve I in Fig. 13.13. When one of the lines (say line 2) is switched out, the transfer reactance increases and the power transfer is given by

$$P'_e = \frac{|E||V|}{X'_d + X_1} \sin \delta \tag{13.64}$$

This is shown as curve II in Fig. 13.13.

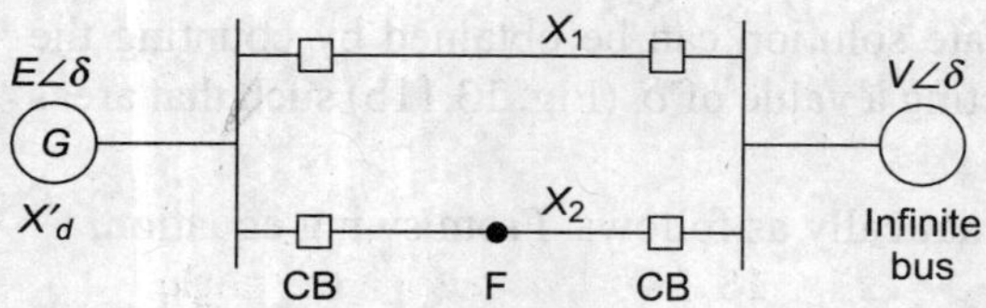

Fig. 13.12 A machine connected to an infinite bus through an inter-connector

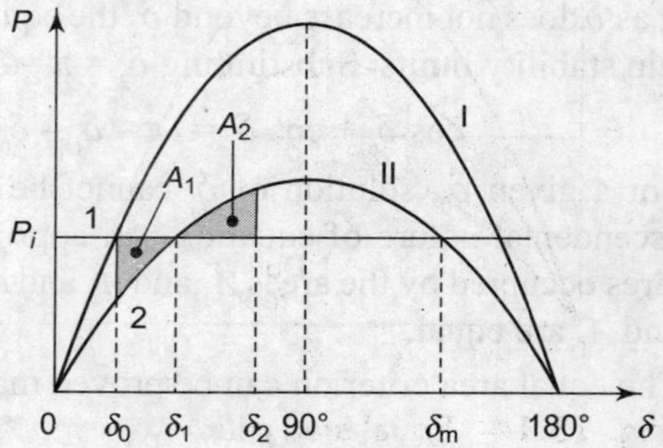

Fig. 13.13 Equal area criterion as applied to the stability study of a switching operation

The input to the generator is P_i and, therefore, the initial conditions are represented by point 1 and the initial load angle is δ_0. When line 2 is switched out the operating point shifts to point 2 on curve II. The power output is less than the power input and the rotor is accelerated. An energy corresponding to area A_1 is put into the rotor. At $\delta = \delta_1$ the input and output are equal but the angle continues to increase because the rotor has acquired a speed slightly greater than the synchronous speed. When $\delta > \delta_1$, deceleration of rotor starts. When $\delta = \delta_2$, the area A_2 equals area A_1, and rotor starts swinging back. After a few oscillations the load angle will stabilize at δ_1. If the initial power transfer is increased (line P_i is shifted upwards in Fig. 13.13) a limit is reached beyond which the decelerating area A_2 cannot be equal to accelerating area A_1. The maximum value which δ can attain without loss of system stability is δ_m and equals $(\pi - \delta_1)$ radians.

13.7.2 Fault and Subsequent Circuit Isolation

Consider the system of Fig. 13.12. Equal area criterion can be used to study the stability of the system when a fault develops at any point F on (say) line 2 and is subsequently cleared by the opening of circuit breakers at both ends of the faulted line. In this case, three power angle curves are involved, first for the pre-fault system, second for the system during fault and third for the system after the faulted line has been switched out (*i.e.*, post-fault condition). These curves are shown in Fig. 13.14. The input is P_i.

The initial load angle δ_0 is determined by the intersection of the input line and the pre-fault output curve (point *a*). When a fault occurs the operating point shifts to *b* on the output curve during fault.

The accelerating power causes δ to increase. At angle δ_c when the operation reaches point *c*, circuit breakers open and clear the fault. The operation shifts to point *e* on the post-fault curve. Now the output power is more than input power and the rotor starts decelerating. The maximum value of the angle δ is δ_2 such that the area A_2 (= area *defg*) equals the area A_1 (= area *abcd*).

A higher input P_i would cause point *f* to move to the right until at stability limit, point *f* would coincide with point *h*. A still higher value of P_i would lead to unstable conditions. Another factor which would cause point f to move to the right is an increase in the time of clearing the fault resulting in a larger clearing angle δ_c. For a given initial load there is a maximum value of clearing angle, known as critical clearing angle δ_{cc}, if stability is to be maintained. If the actual

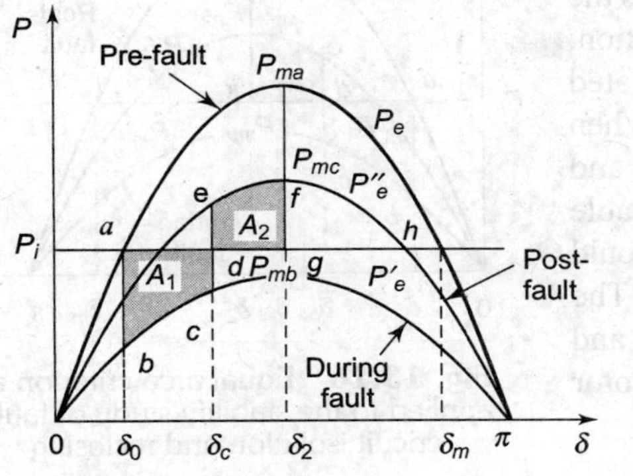

Fig. 13.14 Equal area criterion as applied to the stability study of fault and subsequent circuit isolation

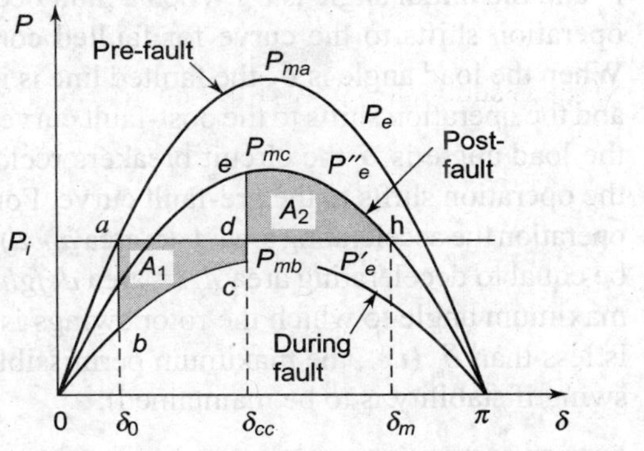

Fig. 13.15 Critical clearing angle

clearing angle δ_c is smaller than δ_{cc} the system is stable, if larger, the system is unstable. When $\delta_c = \delta_{cc}$ the maximum angle up to which rotor swings is δ_m as shown in Fig. 13.15. An expression for δ_{cc} can be derived as under.

Let the amplitudes of pre-fault, during fault and post-fault curves P_{ma}, P_{mb} and P_{mc} respectively. Applying equal area criterion to Fig. 13.15.

$$\int_0^{cc} (P_i - P_{mb} \sin \delta)\, d\delta = \int_{cc}^{m} (P_{mc} \sin \delta - P_i)\, d\,\delta \tag{13.65}$$

Evaluating the integrals we have

$$(P_i\, \delta - P_{mb} \cos \delta) \Big|_0^{cc} = -(P_{mc} \cos \delta + P_i\, \delta) \Big|_{cc}^{m}$$

or

$$P_i\,(\delta_{cc} - \delta_0) + P_{mb}\,(\cos \delta_{cc} - \cos \delta_0) = -P_i\,(\delta_m - \delta_{cc}) - P_{mc}\,(\cos \delta_m - \cos \delta_{cc})$$

whence

$$\cos \delta_{cc} = \frac{P_i\,(\delta_m - \delta_0) - P_{mb} \cos \delta_0 + P_{mc} \cos \delta_m}{P_{mc} - P_{mb}} \tag{13.66}$$

where

$$\delta_m = \pi - \sin^{-1}\left(\frac{P_i}{P_{mc}}\right) \text{ radians}$$

The angles in Eq. (13.66) are in radians. Knowing P_i, P_{mb}, P_{mc} and δ_0, we can find the critical clearing angles δ_{cc}. To find the critical clearing time from δ_{cc} it is necessary to determine the swing curve up to the clearing time. The use of swing curve is not eliminated but is reduced to a minimum by the equal area criterion. A pre-calculated swing curve can also be used (see reference 20).

13.7.3 Fault, Circuit Isolation and Reclosing

Most of the faults are of transient nature. The transmission lines are provided with automatic quick reclosing circuit breakers. When a fault occurs, the faulted line is de-energised to suppress the arc in the fault and reclosed after an interval to improve stability. Figure 13.16 shows the application of equal area criterion for such a case. The input is P_i and the initial angle is δ_0. When a fault occurs the operation shifts to the curve for faulted condition. When the load angle is δ_c, the faulted line is isolated and the operation shifts to the post-fault curve. When the load angle is δ_r the circuit breakers reclose and the operation shifts to the pre-fault curve. For stable operation the accelerating area A_1 (= area *abcd*) should be equal to decelerating area A_2 (= area *defghk*). The maximum angle to which the rotor swings is δ_2 and is less than δ_m (*i.e.*, the maximum permissible rotor swing if stability is to be maintained).

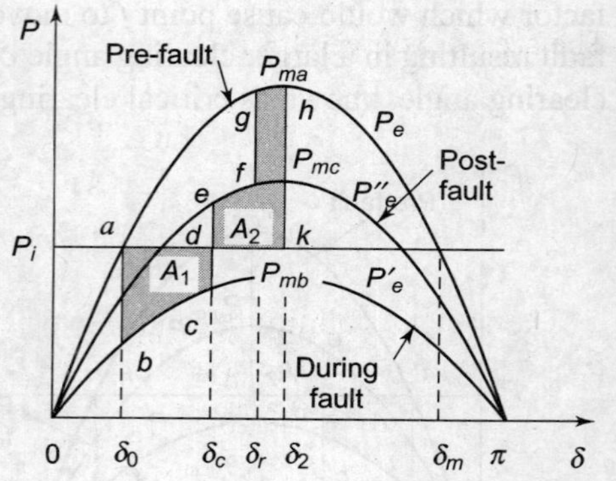

Fig. 13.16 Equal area criterion as applied to the stability study of fault, cricuit isolation and reclosing

Example 13.7

For the system shown in Fig. 13.12, the per unit values of different quantities are

$$E = 1.2,\ V = 1,\ X'_d = 0.2,\ X_1 = X_2 = 0.4$$

The system is operating in equilibrium with $P_i = 1.5\ pu$ when one of the lines is suddenly switched out. Predict whether the system will be stable or not. If the system is stable, find the maximum value which δ attains.

Solution

The power angle curves are shown in Fig. 13.13.

For curve I (*i.e.*, when both the lines are in)

$$\text{Transfer reactance} = 0.2 + \frac{0.4}{2} = 0.4\ pu$$

$$P_e = \frac{|E||V|}{X} \sin\delta = \frac{1.2\times1}{0.4} \sin\delta = 3\sin\delta$$

when $\delta = \delta_0 . P_e = P_i$

or $1.5 = 3\sin\delta_0,\ i.e.,\ \delta_0 = 30° = 0.524$ rad.

For curve II (*i.e.*, when one lines is in)

$$\text{Transfer reactance} = 0.2 + 0.4 = 0.6\ pu$$

$$P'_e = \frac{1.2\times1}{0.6}\sin\delta = 2\sin\delta$$

The value of δ_1 and δ_m are given by the condition $P'_e = P_i$

or $2\sin\delta = 1.5$ or $\sin\delta = 0.75$

$$\delta = 48.6° \quad \text{or} \quad 131.4°$$

$$\delta_1 = 48.6°,\ i.e.,\ 0.848 \text{ rad.}$$

$$\delta_m = 131.4°,\ i.e.,\ 2.293 \text{ rad.}$$

The area A_1 is given by

$$A_1 = \int_{\delta_0}^{\delta_1} (P_i - P_e')\, d\delta = \int_{0.524}^{0.848} (1.5 - 2 \sin \delta)\, d\delta$$

$$= (1.5\,\delta + 2 \cos \delta) \Big|_{0.524}^{0.848} = 0.0773$$

The system stability depends on whether there is enough negative area A_2 in the interval $\delta_1 < \delta < \delta_m$ to match the area A_1:

$$A_{2\,max} = \int_{\delta_1}^{\delta_m} (P_i - P_e')\, d\delta = \int_{0.848}^{2.293} (1.5 - 2 \sin \delta)\, d\delta$$

$$= (1.5\delta + 2 \cos \delta) \Big|_{0.848}^{2.293} = -0.478$$

Since $|A_{2\,max}| > |A_1|$, the system is stable.

The maximum value of rotor swing, *i.e.*, δ_2 is given by the condition that $A_2 = -A_1$

or $$\int_{\delta_1}^{\delta_2} (P_i - P_e')\, d\delta = -0.0773$$

or $$\int_{\delta_1}^{\delta_2} (1.5 - 2 \sin \delta)\, d\delta = -0.0773$$

or $$(1.5\,\delta + 2 \cos \delta) \Big|_{0.848}^{\delta_2} = -0.0773$$

or $$1.5\,\delta_2 + 2\cos \delta_2 = 2.518$$

This equation is non-linear and we use iterative method for solution. The result is $\delta_2 = 1.218$ rad. or 69.8°.

The value of δ_2 can also be found graphically by making the area A_2 numerically equal to A_1 (A simple method is to plot the power angle curves on a graph paper and equate the areas by counting the squares).

Example 13.8

A balanced 3-phase fault occurs at middle point of line 2 when the power transfer is 1.5 *pu* in the system of Fig. 13.12. $E = 1.2$, $V = 1$, $X'_d = 0.2$, $X_1 = X_2 = 0.4$ *pu*.

(a) Determine whether the system is stable for a sustained fault.

(b) The fault is cleared at $\delta = 60°$. Is the system stable ? If so, find the maximum rotor swing.

(c) Find the critical clearing angle.

Solution

Pre-fault Condition

$$\text{Transfer reactance} = 0.2 + 0.2 = 0.4\ pu$$

$$P_e = \frac{1.2 \times 1}{1.4} \sin \delta = 3 \sin \delta$$

During Fault Condition

For fault at F, the equivalent circuit is as shown in Fig. 13.17 (a). Converting the Y network (of reactances 0.2, 0.4 and 0.2 *pu*, into a Δ network, the circuit transforms to that shown in Fig. 13.17 (b). The transfer reactance is 1.0 *pu*. It is not necessary to calculate the reactances of the other two branches of Δ because these are connected across the voltage sources and, therefore, do not affect the power transfer, although they affect the reactive power output of generator.

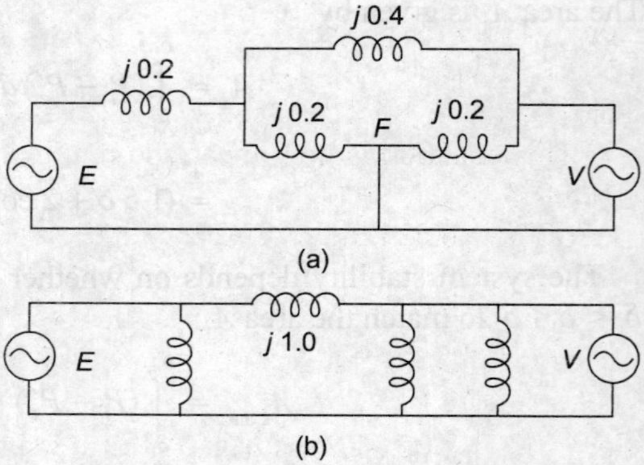

Fig. 13.17 Equivalent circuits for the system in example 13.6

$$P'_e = \frac{1.2 \times 1}{1.0} \sin \delta = 1.2 \sin \delta$$

Post-Fault Condition

After the faulted line has been switched out, the transfer reactance is 0.2 + 0.4, *i.e.*, 0.6 *pu*

$$P''_e = \frac{1.2 \times 1}{0.6} \sin \delta = 2 \sin \delta$$

The three-power angle curves are shown in Fig. 13.14.

(a) The maximum possible value of power transfer during fault conditions is 1.2 *pu* (*i.e.*, peak value of P'_e curve). Since the initial load is 1.5 *pu*, stability is impossible for a sustained fault.

(b) The initial angle δ_0 is given by

$$3 \sin \delta_0 = 1.5 \text{ or } \delta_0 = 30^\circ \quad \text{or} \quad 0.524 \text{ rad.}$$

$$\delta_c = 60^\circ \text{ or } 1.047 \text{ rad.}$$

$$\delta_m = \pi - \sin^{-1}(1.5/2) = 131.4^\circ \text{ or } 2.293 \text{ rad.}$$

$$A_1 = \int_{0.524}^{1.047} (P_i - P'_e)\, d\delta = \int_{0.524}^{1.047} (1.5 - 1.2 \sin \delta)\, d\delta$$

$$= (1.5\,\delta + 1.2 \cos \delta) \Big|_{0.524}^{1.047} = 0.346$$

The maximum available decelerating area $A_{2\max}$ is

$$A_{2\max} = \int_{\delta_c}^{\delta_m} (P_i - P''_e)\, d\delta$$

$$= \int_{1.047}^{2.293} (1.5 - 2 \sin \delta)\, d\delta = (1.5\,\delta + 2 \cos \delta) \Big|_{1.047}^{2.293} = -0.453$$

Since the maximum available decelerating area is numerically greater than the accelerating area, the system is stable. To find the maximum rotor swing δ_2.

$$\int_{\delta_c}^{\delta_z} (P_i - P_e'')\, d\delta = -A_1 = -0.346$$

$$\int_{1.047}^{\delta_2} (1.5 - 2 \sin \delta)\, d\delta = -0.346$$

or $\quad (1.5\,\delta + 2\cos\delta)\Big|_{1.047}^{\delta_2} = -0.346$

or $\quad 1.5\,\delta_2 + 2\cos\delta_2 = 2.225$

This equation is non-linear and requires iterative solution. The result is δ_2 = 1.848 rad. or 105.88°

(c) $\delta_0 = 0.524$ rad., δ_m 2.293 rad., $P_{mb} = 1.2$, $P_{mc} = 2$, $P_i = 1.5$. Using Eq. (13.41)

$$\int_{\delta_0}^{\delta_{cc}} (P_i - P_{mb}\sin\delta)\,d\delta = \int_{\delta_{cc}}^{\delta_m} (P_{mc}\sin\delta - P_i)\,d\delta$$

or
$$\int_{0.524}^{\delta_{cc}} (1.5 - 1.2\sin\delta)\,d\delta = \int_{\delta_{cc}}^{2.293} (2\sin\delta - 1.5)\,d\delta$$

$$(1.5\,\delta + 1.2\cos\delta)\Big|_{0.524}^{\delta_{cc}} = (-2\cos\delta - 1.5\,\delta)\Big|_{\delta_{cc}}^{2.293}$$

whence $\quad \delta_{cc} = 1.196$ rad. or 68.6°

Alternatively we can use Eq. (13.42)

$$\cos\delta_{cc} = \frac{P_i(\delta_m - \delta_0) - P_{mb}\cos\delta_0 + P_{mc}\cos\delta_m}{P_{mc} - P_{mb}}$$

$$= \frac{1.5(2.293 - 0.524) - 1.2\cos(0.524) + 2\cos(2.293)}{2 - 1.2} = 0.3655$$

$$\delta_{cc} = 1.196 \text{ rad. or } 68.6°$$

Example 13.9

In the system of example 13.8, find the critical clearing angle if a 3-phase fault occurs on the line 2 close to the generator bus. All the other parameters remain the same.

Solution

It is evident that no power transfer can take place during the fault condition, *i.e.*, $P_{mb} = 0$

$$\cos\delta_{cc} = \frac{1.5(2.293 - 0.524) - 2\cos(2.293)}{2} = 0.6657$$

$$\delta_{cc} = 0.842 \text{ rad or } 48.24°$$

Example 13.10

A synchronous motor of negligible resistance is receiving 25% of power that it is capable of receiving from an infinite bus. If the load on the motor is suddenly doubled, calculate the maximum value of power angle δ during the swinging of the motor around its new equilibrium position.

Solution

P_e, electrical power input to the motor is given by

$$P_e = P_m \sin\delta$$

P_m is the maximum power which the motor is capable of receiving and is 1 *pu*.

When $\quad \delta = \delta_1,\ P_e = 0.25\ pu$

or $$0.25 = \sin \delta_1$$

or $$\delta_1 = 14.48° \text{ or } 0.253 \text{ rad.}$$

The power angle curve is shown in Fig. 13.18. At the new equilibrium position, *i.e.*, when $\delta = \delta_2$, $P_e = 0.5$

or $$0.5 = \sin \delta_2$$

or $$\delta_2 = 30° \text{ or } 0.524 \text{ rad.}$$

The maximum power angle is δ_3. From $\delta = \delta_1$ to $\delta = \delta_2$, the input to the motor is less than output and the rotor decelerates (*i.e.*, δ increases). From $\delta = \delta_2$ to $\delta = \delta_3$ the input to the motor is more than the output and the rotor accelerates (*i.e.*, δ decreases). The maximum power angle δ_3 can be calculated by equating the area A_1 and A_2

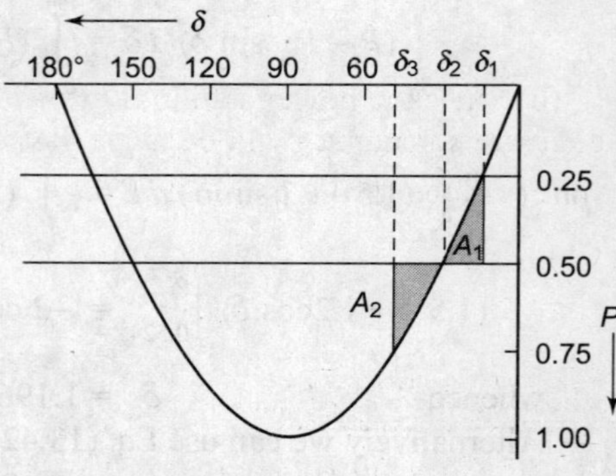

Fig. 13.18

Area $$A_1 = \int_{\delta_1}^{\delta_2} (0.5 - \sin \delta)\, d\delta$$

$$= \Big| 0.5\,\delta + \cos \delta \Big|_{\delta_1}^{\delta_2}$$

$$= \Big| 0.5\,\delta + \cos \delta \Big|_{0.253}^{0.524} = 0.0335$$

Area $$A_2 = \int_{\delta_2}^{\delta_3} (0.5 - \sin \delta)\, d\delta$$

$$= \Big| 0.5\,\delta + \cos \delta \Big|_{\delta_2}^{\delta_3}$$

$$= \Big| 0.5\,\delta + \cos \delta \Big|_{0.524}^{\delta_3}$$

or $$A_2 = 0.5\,\delta_3 + \cos \delta_3 - 0.5 \times 0.524 - \cos (0.524)$$

Using the condition $$A_2 = -A_1$$

$$0.5\,\delta_3 + \cos \delta_3 - 0.262 - 0.866 = -0.0335$$

or $$\cos \delta_3 + 0.5\,\delta_3 = 1.0945$$

This equation is non-linear. We resort to an iterative method for its solution. The result is

$$\delta_3 = 0.81 \text{ rad. or } 46.4°$$

Example 13.11

A power station A consists of two synchronous generators. The generator 1 has a rating of 50 MVA, 50 Hz, 1500 rpm and has an inertia constant of 8 MJ/MVA. The generator 2 has a rating of 100 MVA, 50 Hz, 3000 rpm and has an inertia constant of 4 MJ/MVA. (a) Find the inertia constant for the equivalent generator on a base of 100 MVA: (b) Another power station B has 4 generators two each of the above type. Find the inertia constant for the equivalent generator on a base of 100 MVA. (c) If the two power stations are connected through an inter-connector, find the inertia constant for the equivalent generator connected to infinite bus bar.

Solution

(a) Base MVA 100

$$H_1 = \frac{50 \times 8}{100} = 4 \text{ MJ/MVA}$$

$$H_2 = 4 \text{ MJ/MVA}$$

The inertia constant H_A of the equivalent generator is

$$H_A = H_1 + H_2 = 8 \text{ MJ/MVA}$$

(b) Since the power station B has two generators of each type, the inertia constant H_B of the equivalent generator will be twice that for station A, *i.e.*, 16 MJ/MVA.

(c) H_e, the inertia constant of the equivalent generator connected to infinite bus is

$$H_e = \frac{H_A\, H_B}{H_A + H_B} = \frac{8 \times 16}{24} = 5.333 \text{ MJ/MVA.}$$

Example 13.12

A 100 MVA, 50 Hz alternator is operating at rated speed. The H constant of the machine is 5 kW • sec. per kVA. The load suddenly increases by 50 MW. Due to delay in governor action there is a delay of 0.6 seconds in opening of steam valve. Find the frequency deviation.

Solution

Kinetic energy stored at rated speed

$$= 100 \times 1000 \times 5 = 5 \times 10^5 \text{ kW-sec.}$$

Loss in kinetic energy due to increase in load

$$= 50 \times 1000 \times 0.6 = 30 \times 10^3 \text{ kW-sec.}$$

Kinetic energy ∞ (frequency)^2

$$\text{New frequency} = \left[\frac{5 \times 10^5 - 30 \times 10^3}{5 \times 10^5}\right]^{0.5} \times 50 = 48.477 \text{ Hz.}$$

$$\text{Frequency deviation} = \frac{50 - 48.477}{50} \times 100 = 3.046\%$$

Example 13.13

A 3-phase generator delivers 1.0 *pu* power to an infinite bus through a transmission network when a fault occurs. The maximum power which can be transferred during pre-fault, during fault and post-fault conditions is 1.75 *pu*, 0.4 *pu* and 1.25 *pu*. Find critical clearing angle.

Solution

Using Eq. (13.66)

$$P_i = 1.0\, pu,\ P_{ma} = 1.75\, pu,\ P_{mb} = 0.4\, pu,\ P_{mc} = 1.25\, pu.$$

$$\delta_0 = \sin^{-1}(P_i/P_{ma}) = \sin^{-1}(1/1.75) = 34.85° = .608 \text{ radians}$$

$$\delta_m = \pi - \sin^{-1}(P_i/P_{mc}) = \pi - \sin^{-1}(1/1.25) = 2.214 \text{ radians}$$

$$= 126.87°$$

$$\cos \delta_{cc} = \frac{1(2.214 - 0.608) - 0.4 \cos (34.85^\circ) + 1.25 \cos (126.87^\circ)}{1.25 - 0.4}$$

$$= 0.621$$

$$\delta_{cc} = 51.6^\circ$$

13.8 POINT BY POINT SOLUTION OF SWING EQUATION

The swing equation (derived in section 13.3.1) is

$$\frac{d^2 \delta}{d t^2} = \frac{P_a}{M} = \frac{1}{M} (P_i - P_m \sin \delta) \tag{13.67}$$

Its solution gives a plot of δ versus t. This graph is known as swing curve. If the swing curve indicates that δ starts decreasing after reaching a maximum value, the system can be assumed to be stable. The swing equation is a non-linear equation and a formal solution is not feasible. The point by point solution (also known as step by step solution) is a very simple and common method of solving this equation. In this method the change in δ during a small time interval Δt (say 0.05 sec) is calculated by assuming that the accelerating power P_a calculated at the beginning of an interval is constant from the middle of the preceding interval to the middle of the interval being considered. Let us consider the nth time interval which begins at $t = (n - 1) \Delta t$. The angular position of the rotor at this instant is δ_{n-1} (Fig. 13.19c). The accelerating power $P_{a\,(n-1)}$ and, hence, acceleration α_{n-1} as calculated at this instant is assumed to be constant from $t = \left(n - \frac{3}{2}\right) \Delta t$ to $\left(n - \frac{1}{2}\right) \Delta t$.

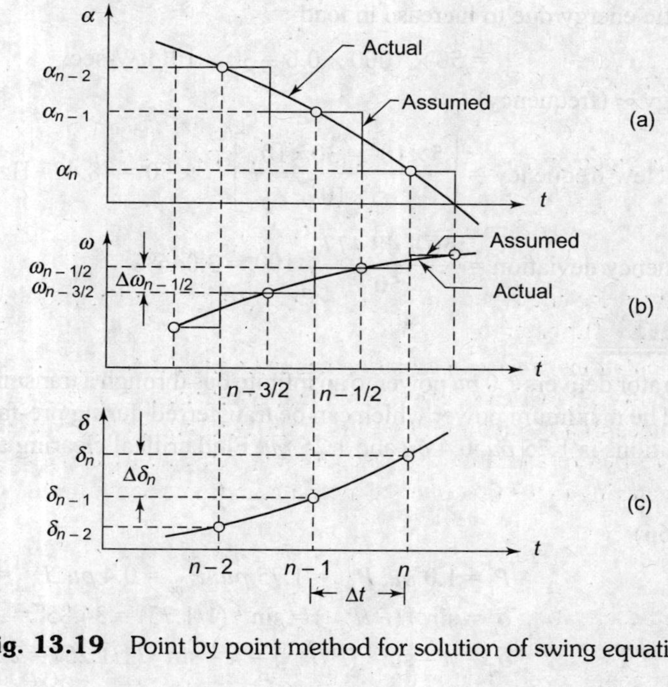

Fig. 13.19 Point by point method for solution of swing equation

During this interval the change in rotor speed (with respect to synchronous speed) can be written as

$$\Delta\omega_{n-1/2} = (\Delta t)\,\alpha_{n-1} = \frac{\Delta t}{M} P_{a(n-1)} \tag{13.68}$$

Thus, the speed at the end of nth interval is

$$\omega_{n-\frac{1}{2}} = \omega_{n-\frac{3}{2}} + \Delta\omega_{n-\frac{1}{2}} \tag{13.69}$$

As a logical outcome of the above assumption, regarding acceleration, the change in speed occurs linearly with time. To simplify the calculations, the change in speed is assumed to occur at the middle of one interval, *i.e.*, $t = (n-1)\,\Delta t$ which is the same instant for which the acceleration was calculated. Then the speed is assumed to remain constant till the middle of the next interval as shown in Fig. 13.19 (b). In other words, the speed is assumed to be constant at the value $\omega_{n-1/2}$ throughout the nth interval from $t = (n-1)\,\Delta t$ to $t = n\,\Delta t$.

The change in angular position of rotor during nth time interval is

$$\Delta\delta_n = (\Delta t)\,\omega_{n-\frac{1}{2}} \tag{13.70}$$

and the value of δ at the end of nth interval is

$$\delta_n = \delta_{n-1} + \Delta\delta_n \tag{13.71}$$

This is shown in Fig. 13.19 (c). Since we are interested in calculating only δ and not ω, a formula for directly calculating $\Delta\delta_n$ can be found. Substituting Eq. (13.68) into Eq. (13.69) and the result in Eq. (13.70) leads to

$$\Delta\delta_n = (\Delta t)\,\omega_{n-\frac{3}{2}} + \frac{(\Delta t)^2}{M} P_{a\,(n-1)} \tag{13 72}$$

By analogy with Eq. (13.70)

$$\Delta\delta_{n-1} = (\Delta t)\omega_{n-3/2} \tag{13.73}$$

Substituting the value of $\omega_{n-3/2}$ from Eq. (13.73) into Eq. (13.72)

$$\Delta\delta_n = \Delta\delta_{n-1} + \frac{(\Delta t)^2}{M} P_{a\,(n-1)} \tag{13.74}$$

Eq. (13.74) gives the increment in angle δ during any interval (say nth) in terms of the increment during $(n-1)$th interval.

During the calculations, a special attention has to be paid to the effects of discontinuities in the accelerating power P_a which occur when a fault is applied or cleared or when a switching operation takes place. If a discontinuity occurs at the beginning of an interval then the average of the values of P_a before and after the discontinuity must be used. Thus, for calculating the increment in δ occurring in the first interval after a fault is applied at $t = 0$, Eq. (13.74) becomes

$$\Delta\delta_1 = \frac{(\Delta t)^2}{M} \cdot \frac{P_{a0+}}{2} \tag{13.75}$$

when P_{a0+} is the accelerating power immediately after occurrence of the fault. (Immediately before the occurrence of fault, the system is in steady state with $P_{a0-} = 0$ and the previous increment in rotor angle is zero). If the fault is cleared at the beginning of mth interval, the value $\frac{1}{2}[P_{a(m-1)-} + P_{a(m-1)+}]$ should be used instead of $P_{a(m-1)}$, where, $P_{a(m-1)-}$ is the accelerating power immediately before clearing and $P_{a(m-1)+}$ is the accelerating power immediately after clearing the fault. If a discontinuity occurs at the middle of an interval, no special procedure is needed. The increment in δ during such an interval is calculated, as usual, from the value of P_a at the beginning of the interval.

The selection of proper value of Δt depends on the degree of accuracy desired in the results. A smaller value of Δt improves the accuracy but makes the calculations more lengthy. Usually $\Delta t = 0.05$ seconds provides good accuracy in results.

Example 13.14

In the system shown in Fig. 13.12 the generator has a rating of 50 MVA and $H = 2.7$ MJ/MVA at rated speed. $E = 1.05$, $V = 1$, $X'_d = 0.2$, $X_1 = X_2 = 0.4$ *pu*. The generator is supplying 50 MW to the infinite bus when a three-phase fault occurs at the middle of line 2. (a) Plot the swing curve for a sustained fault up to a time of 0.5 secs. (b) Plot the swing curve if the fault is cleared in 0.10 seconds by simultaneous opening of the breakers at both the ends of line 2. (c) Find the critical clearing angle and the critical clearing time.

Solution

Taking Base MVA = 50, $G = 1.0$ pu

$$M = \frac{GH}{180\,f} = \frac{1 \times 2.7}{180 \times 50} = 3 \times 10^{-4} \text{ unit power sec}^2\text{/elect. deg.}$$

Pre-fault condition (both lines in)

$$X = 0.4\ pu$$

$$P_e = \frac{1.05 \times 1}{0.4} \sin \delta = 2.625 \sin \delta$$

The initial angle δ_0 is given by

$$2.625 \sin \delta_0 = \frac{50}{50} = 1 \text{ or } \delta_0 = 22.4°$$

During fault condition

$$X = 1.0\ pu \text{ (From Fig. 13.17 b)}$$

$$P'_e = \frac{1.05 \times 1}{1} \sin \delta = 1.05 \sin \delta$$

Post-fault condition (Line 2 switched out)

$$X = 0.6\ pu$$

$$P''_e = \frac{1.05 \times 1}{0.6} \sin \delta = 1.75 \sin \delta$$

Taking $\Delta t = 0.05$ secs

$$P_{a(n-1)} = P_i - P_m \sin \delta_{n-1}$$

$$\frac{(\Delta t)^2}{M} P_{a(n-1)} = \frac{0.05 \times 0.05}{3 \times 10^{-4}} P_{a(n-1)}$$

$$= 8.33\ P_{a(n-1)}$$

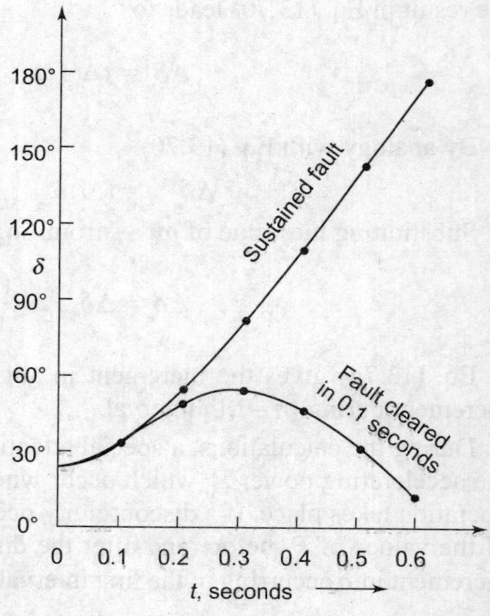

Fig. 13.20 Swing curves for the system of example 13.14

$$\Delta\delta_n = \Delta\delta_{n-1} + 8.33\ P_{a(n-1)}$$

$$\delta_n = \delta_{n-1} + \Delta\delta_n \text{ elect degree}$$

(a) Sustained fault: A discontinuity exists at $t = 0$. Therefore, average value of P_a is used.

At $\quad t = 0_-, P_m = 2.625, P_a = 0$

At $\quad t = 0_+, P_m = 1.05$

$$P_a = P_i - P_m \sin\delta_0 = 1.0 - 1.05 \sin 22.4^\circ = 0.6$$

$$P_{a(\text{average})} = \frac{0 + 0.6}{2} = 0.3$$

The details of calculations are shown in Table 13.2. A representative calculation (at $t = 0.1$ sec.) is shown below.

$$t = 0.1 \text{ sec}, \delta = 32.05^\circ, \sin(32.05^\circ) = 0.53$$

$$P_e = 1.05 \times 0.53 = 0.5565$$

$$P_a = P_t - P_t = 1 - 0.5565 = 0.4435$$

$$\frac{(\Delta t)^2}{M} P_a = 8.33\ P_a = 3.69$$

$$\Delta\delta = 7.15 + 3.69 = 10.84$$

$$\delta\ (\text{at } t = 0.15 \text{ sec}) = 32.05 + 10.84 = 42.89^\circ$$

The swing curve is plotted in Fig. 13.20. It is evident that the system is unstable.

Table 13.2 Point by point solution of swing equation for sustained fault in example 13.12(a)

t *secs*	P_m *pu*	*sin δ*	P_e *pu*	P_a *pu*	*8.33* P_a *elect. deg.*	Δδ *elect. deg.*	δ *elect. deg.*
0_-	2.625	0.381	1.0	0	0		22.4
0_+	1.05	0.381	0.4	0.6			22.4
0_{avg}				0.3	2.5		
						2.5	
0.05	1.05	0.421	0.442	0.558	4.65		24.9
						7.15	
0.10	1.05	0.53	0.5565	0.4435	3.69		32.05
						10.84	
0.15	1.05	0.68	0.714	0.286	2.38		42.89
						13.22	
0.20	1.05	0.83	0.872	0.128	1.07		56.11
						14.29	
0.25	1.05	0.942	0.989	0.011	0.09		70.40
						14.38	
0.30	1.05	0.996	1.046	–0.046	–0.38		84.87
						14.00	
0.35	1.05	0.988	1.0374	–0.0374	–0.31		98.78
						13.69	
0.40	1.05	0.924	0.97	0.03	0.25		112.47
						13.94	
0.45	1.05	0.805	0.845	0.155	1.29		126.41
						15.23	
0.50	1.05	0.62	0.651	0.349	2.91		141.64
						18.14	
0.55	1.05	0.346	0.363	0.637	5.30		159.78
						23.44	
0.60							183.22

(b) Fault cleared in 0.10 secs:

The swing curve till 0.05 seconds is the same as that for a sustained fault. Due to fault clearing, P_m changes from 1.05 at $t = 0.10\_$ to 1.75 *pu* at 0.10_+. Since the discontinuity occurs at the beginning of an interval we calculate average value of P_a at $t = 0.10$ and use this value in computing the change in δ at this time. The rest of the calculations are similar to those for a sustained fault. The results are shown in Table 13.3 and plotted in Fig. 13.20. It is evident that the system is stable.

Table. 13.3 Computations for swing curve for example 13.12 (b)

t secs	P_m *pu*	*sin δ*	P_e *pu*	P_a *pu*	$8.33\ P_a$	Δδ *elect. deg*	δ *elect. deg.*
0_	2.625	0.381	1.0	0	0		22.4
0_+	1.05	0.381	0.4	0.6			22.4
0_{avg}				0.3	2.5		
0.05	1.05	0.421	0.442	0.558	4.65	2.5	24.9
0.10_	1.05	0.53	0.5565	0.4435		7.15	32.05
0.10_+	1.75	0.53	0.9275	0.0725			32.05
0.10_{avg}				0.258	2.15		
0.15	1.75	0.66	1.155	–0.155	–1.29	9.30	41.35
0.20	1.75	0.7588	1.328	–0.328	–2.73	8.01	49.36
0.25	1.75	0.8155	1.427	–0.427	–3.56	5.28	54.64
0.30	1.75	0.8325	1.457	–0.457	–3.81	1.72	56.36
0.35	1.75	0.8118	1.421	–0.421	–3.51	–2.09	54.27
0.40	1.75	0.7509	1.314	–0.314	–2.62	–5.60	48.67
0.45	1.75	0.6488	1.135	–0.135	–1.12	–8.22	40.45
0.50	1.75	0.5167	0.904	0.096	0.80	–9.34	31.11
0.55	1.75	0.3838	0.672	0.328	2.73	–8.54	22.57
0.60						–5.81	16.76

(c) $\delta_0 = 22.4°$ or 0.391 rad., $P_{mb} = 1.05$, $P_{mc} = 1.75$, $P_i = 1$

$$\delta_m = \pi - \sin^{-1}\frac{P_i}{P_{mc}} = 2.53 \text{ rad.}$$

Using Eq. (13.66)

$$\cos\delta_{cc} = \frac{P_i(\delta_m - \delta_0) - P_{mb}\cos\delta_0 + P_{mc}\cos\delta_m}{P_{mc} - P_{mb}}$$

$$= \frac{1(2.53 - 0.391) - 1.05\cos(0.391) + 1.75\cos(2.53)}{1.75 - 1.05} = -0.37785$$

$$\delta_{cc} = 112.2°$$

From the swing curve for sustained fault conditions, critical clearing time is 0.395 seconds.

13.9 DIGITAL SOLUTION OF SWING EQUATION, RUNGE-KUTTA METHOD

A number of numerical techniques are available for the solution of differential equations. These techniques can be employed for digital computations. In using these techniques an nth order differential equation is written as n first order equations. The modified Euler method and Runge-Kutta method are two popular methods used for the solution of swing equation. We discuss only the Runge-Kutta method. The method is self-starting and gives an accuracy of the order of $(\Delta t)^5$. This method uses Taylor's series expansion, truncated after the fourth term (the method is also known as Runge-Kutta fourth order approximation). The swing equation is a second order differential equation.

Consider two first order differential equations in two variables x and y such that

$$\frac{dx}{dt} = f(x, y) \tag{13.76a}$$

$$\frac{dy}{dt} = g(x, y) \tag{13.76b}$$

We start with known initial conditions x^o and y^o and a time interval Δt and calculate the following eight constants:

$$k_1^o = f(x^o, y^o)\,\Delta t \tag{13.77a}$$

$$l_1^o = g(x^o, y^o)\,\Delta t \tag{13.77b}$$

$$k_2^o = f(x^o + 0.5\,k_1^o, y^o + 0.5\,l_1^o)\,\Delta t \tag{13.77c}$$

$$l_2^o = g(x^o + 0.5\,k_1^o, y^o + 0.5\,l_1^o)\,\Delta t \tag{13.77d}$$

$$k_3^o = f(x^o + 0.5\,k_2^o, y^o + 0.5\,l_2^o)\,\Delta t \tag{13.77e}$$

$$l_3^o = g(x^o + 0.5\,k_2^o, y^o + 0.5\,l_2^o)\,\Delta t \tag{13.77f}$$

$$k_4^o = f(x^o + k_3^o,\; y^o + l_3^o)\,\Delta t \tag{13.77g}$$

$$l_4^o = g(x^o + k_3^o, y^o + l_3^o)\,\Delta t \tag{13.77h}$$

We use the above eight constants to compute the changes in x and y as follows:

$$\Delta x^o = \frac{1}{6}\,(k_1^o + 2k_2^o + 2k_3^o + k_4^o) \tag{13.78a}$$

$$\Delta y^o = \frac{1}{6}\,(l_1^o + 2l_2^o + 2l_3^o + l_4^o) \tag{13.78b}$$

Then we update the values of t, x and y

$$t^1 = 0 + \Delta t \tag{13.79a}$$

$$x^1 = x^o + \Delta x^o \tag{13.79b}$$

$$y^1 = y^o + \Delta y^o \tag{13.79c}$$

Then we replace x^o and y^o by x^1 and y^1 and recalculate the k's, l's, Δx and Δy. In general

$$t^{n+1} = (n + 1)\,\Delta t \tag{13.80a}$$

$$x^{n+1} = x^n + \Delta x^n \tag{13.80b}$$

$$y^{n+1} = y^n + \Delta y^n \tag{I3.80c}$$

For using the above method for solving the swing equation of one machine connected to infinite bus we put

$$x = \delta \tag{13.81a}$$

$$y = \omega = \frac{d\delta}{dt} = \text{angular speed.} \tag{13.81b}$$

Then
$$\frac{d\delta}{dt} = f(\delta, \alpha) = \omega \tag{13.82a}$$

and
$$\frac{\delta\omega}{dt} = g(\delta, \omega) = \frac{1}{M}(P_i - P_e) \tag{13.82b}$$

where
$$M = \frac{GH}{\pi f} \text{ Mega-Joule-sec/elect. rad.}$$

The initial value δ_0 is calculated using the pre-fault values

$$P_i = P_e = P_{ma} \sin \delta_0 \tag{13.83a}$$

or
$$\delta_0 = \sin^{-1}(P_i/P_{ma}) \tag{13.83b}$$

The values of P_e used in Eq. (13.82) are

$$\text{For } 0 < t. < t_c \; P_e = P'_e = P_{mb} \sin \delta \tag{13.84}$$

$$\text{For } t > t_c \; P_e = P''_e = P_{mc} \sin \delta \tag{13.85}$$

where t_c is the fault clearing time and P_e, P'_e P''_e P_{ma}, P_{mb}, P_{mc}, denote the quantities as shown in Fig. 13.14.

13.10 LIAPUNOV'S DIRECT METHOD

During the past few years power engineers have shown considerable interest in using Liapunov's direct method for stability studies. This method can be quite helpful in giving a reasonably accurate prediction of stability and critical clearing time.

13.10.1 Liapunov Theorem, Liapunov Function

The power system equations, under the conditions of transient disturbance, are similar to the vector equation

$$\frac{dx}{dt} = f(x, t) \tag{13.86}$$

The aim is to study the stability of the above set of equations and their dependence on initial conditions, x and f are column vectors.

The Liapunov theorem states:

The equilibrium is stable in a domain R of x space (*i.e.*, any trajectory starting within this domain will remain within this domain) if a continuous scalar function $V(x, t)$ can be found such that its total derivative $\frac{dV}{dt}$ for the differential equation (13.86) is negative definite or at least negative semi-definite in R.

The function $V(x, t)$ is called Liapunov function and possesses the following properties:

1. It is positive definite function in R (*i.e.*, it is positive everywhere in the region R and is zero at the origin).
2. It is a function of time t and n variables x_1 x_2 ....x_n of the system represented by Eq. (13.86)
3. $\frac{dV}{dt}$ along the trajectories of the system is negative definite or negative semi-definite in R.

The statement that the derivative $\frac{dV}{dt}$ for the differential [Eq. (13.86)] is negative definite or semi-definite has the following significance: when the total derivative of the function $V(x, t)$ is taken, terms involving $\frac{dx_1}{dt}, \frac{dx_2}{dt} \ldots \frac{dx_n}{dt}$ arise and, thus, Eq. (13.86) is incorporated into $\frac{dV}{dt}$.

The equilibrium is stable if $\frac{dV}{dt}$ is negative definite or at least negative semi-definite.

In a simple system such as a mass and spring and described by the differential equation

$$\ddot{x} + D\dot{x} + x = 0 \tag{13.87}$$

a suitable Liapunov function is

$$V = \frac{\dot{x}^2}{2} + \int_0^x x'\,dx' = 0 \tag{13.88}$$

V represents the total energy in this case. The derivative of V with respect to time is

$$\dot{V} = \dot{x}\,\ddot{x} + x\,\dot{x}$$

$$= \dot{x}\,(\ddot{x} + x) \tag{13.89}$$

Substituting the value of $\ddot{x} + x$ from Eq. (13.87) into Eq. (13.89)

$$\dot{V} = -D\dot{x}^2 \tag{13.90}$$

$\dot{V}$ is observed to be a negative semi-definite function. Note that for the damping factor D to be zero, $\dot{V}$ is zero and V is constant so that the system operates along a path of constant energy and is stable. If D were negative, the system is unstable. However. $\dot{V} = 0$ along the line $\dot{x} = 0$.

13.10.2 Application to Transient Stability Analysis

Figure 13.21 (a) shows a single machine feeding an infinite bus through a double circuit line. Initially both the lines are in and the system is operating at a power angle δ_0. A fault occurs on one of the lines and is cleared by the simultaneous operation of the circuit breakers at the two ends of the faulty line. The pre-fault and post-fault power angle curves are shown in Fig. 13.21(b). After the fault is cleared, the system operates at power angle δ_2. The swing equation for the post-fault condition is

$$M\frac{d^2\delta}{dt^2} = M\frac{d\omega}{dt} = P_i - P_{mc}\sin\delta \tag{13.91}$$

Before the Liapunov function can be written for this system, it is necessary to shift the origin to the final equilibrium position δ_2. This change can be made by the substitution

$$x = \delta - \delta_2 \tag{13.92}$$

Substituting Eq. (13.68) in Eq. (13.67) we have

$$M\frac{d^2\delta}{dt^2} = M\frac{d\omega}{dt} = P_i - P_{mc}\sin(x + \delta_2) \tag{13.93}$$

A suitable Liapunov function for the system is

$$V = \frac{\omega^2}{2} + \frac{1}{M}\int_0^x [-P_i + P_{mc}\sin(x' + \delta_2)]\,dx' \tag{13.94}$$

with $x[-P_i + P_{mc}\sin(x+\delta_2)] = 0, x = 0$ (13.95)

$x[-P_i + P_{mc}\sin(x+\delta_2)] > 0, 0 < x < (\delta_m - \delta_2), 0 > x > -(\delta_2 + \pi)$ (13.96)

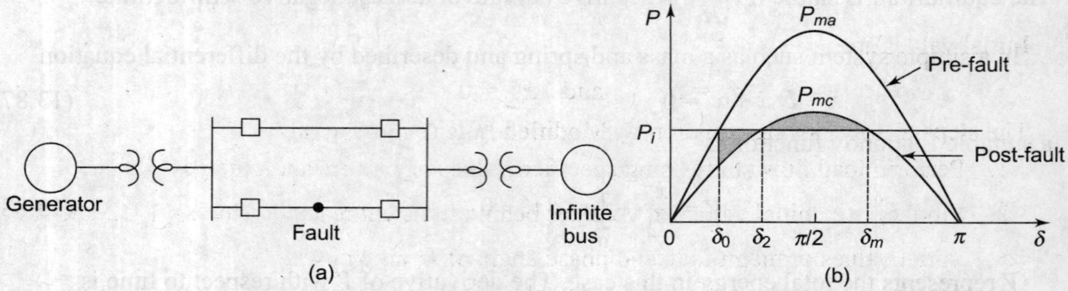

Fig. 13.21 A single machine feeding an infinite bus: (a) Circuit; (b) Power angle curve

The range of validity of the Liapunov function corresponds to the region over which the restoring force is positive. Equation (13.70) is a positive definite function over the specified interval. If the integration indicated in Eq. (13.70) is carried out we have

$$V = \frac{\omega^2}{2} + \frac{1}{M}[-P_i x - P_{mc}\cos(x+\delta_2) + P_{mc}\cos\delta_2] \tag{13.97}$$

The range of values for V are from 0 at $x = 0$, $\omega = 0$ to a maximum at $\omega = 0$, $x = \delta_m - \delta_2$. These two extreme values of V define the stability region. The system can be studied by comparing the actual value of V with the maximum value. The actual value of V is found by substituting the value of ω and x at the last instant of the switching in Eq. (13.97). If the actual value of V is less than the maximum value, the system is stable. The relative degree of stability is indicated whether V is near to its minimum or maximum value. An unstable condition is indicated if $x \geq (\delta_m - \delta_2)$ or V is greater than the maximum stable value for $0 < x < (\delta_m - \delta_2)$.

It is possible to apply the direct method of Liapunov to multi-machine systems also. The critical clearing time can also be found. The effects of saliency, flux decay, governor action and regulator action can be included. For a further discussion on all these aspects, the reader is referred to [44] in the bibliography.

13.11 Modified Euler's Method

Modified Euler's method is a numerical method to solve swing equations for a multi-machine system. This method uses the state variable analysis.

For a system having m generators, the swing equation for kth generator is

$$\frac{d^2\delta_k}{dt^2} = \frac{\pi f}{H_k}\left(P_{Gk}^0 - P_{Gk}\right) \quad \text{for } k = 1, 2........\text{m} \tag{13.98}$$

where P_{Gk}^0 denotes initial power and P_{Gk} denotes power after disturbance.

It is better to rewrite Eq (13.98) using state variable method. For this purpose we write

$$x_{1k} = \delta_k$$

$$x_{2k} = \overset{\circ}{\delta}_k$$

Thus $\overset{\circ}{x}_{1k} = x_{2k}$

Therefore $$\overset{\circ}{x}_{2k} = \frac{\pi f}{H_k}\left(P^0_{Gk} - P_{Gk}\right) \quad \text{for } k = 1, 2\text{..........m} \qquad (13.99)$$

Initial values are

$$x^0_{1k} = \delta^0_k \quad \text{and} \quad x^0_{2k} = 0$$

The steps to solve swing equation by Modified Euler's method are :

1. Perform load flow study using specified values of powers and voltages.
2. Find E^0_k *i.e.* initial values of voltages behind transient reactance for $k = 1, 2....m$. Thus initial values of magnitude and phase angle of E_k are known.
3. Find bus admittance matrix Y_{bus} for all the three conditions of system operation *i.e.* before, during and post-fault.
4. Set time count $r = 0$.
5. Compute generator power outputs using appropriate Y_{BUS} using the following equations:

   $$P_1 = |E_1|^2 G_{11} + |E_1||E_2||Y_{12}|\cos(\partial_1 - \delta_2 - \theta_{12}) + \text{...........} \qquad (13.100)$$

   where $E_1 = |E_1|\angle\delta_1, \quad E_2 = |E_2|\angle\delta_2$

   $Y_{11} = G_{11} + jB_{11}$ and $Y_{12} = |Y_{12}|\angle\theta_{12}$...........

   Thus we get $P^{(r)}_{Gk}$ for $t = t^{(r)}$

   After occurrence of fault we divide time into suitable intervals with Δt as the small time between different discrete time values. A suitable value of Δt is 0.05 seconds.

   Thus time is counted as $t^{(0)}$, $t^{(1)}$ etc. $t^{(1)} - t^{(0)} = \Delta t$

6. Compute $\overset{\circ}{x}^{(r)}_{1k}, \overset{\circ}{x}^{(r)}_{2k}$ for $k = 1, 2$........m using Eq (13.99)
7. Find first state estimates for $t = t^{(r+1)}$ as

   $$x^{(r+1)}_{1k} = x^{(r)}_{1k} + \overset{\circ}{x}^{(r)}_{1k}\,\Delta t$$

   for $k = 1, 2$............m.

   $$x^{(r+1)}_{2k} = x^{(r)}_{2k} + \overset{\circ}{x}^{(r)}_{2k}\,\Delta t$$

8. Find first estimate of $E^{(r+1)}_k$ as

   $$E^{(r+1)}_k = E^0_k\left[\cos x^{r+1}_{1k} + j\sin x^{(r+1)}_{1k}\right]$$

9. Find $P^{(r+1)}_{Gk}$ using appropriate Y_{BUS} and the equation

   $$\delta_n = \delta_{n-1} + \Delta\delta_n$$

10. Compute $\dot{x}_{1k}^{(r+1)}$ and $\dot{x}_{2k}^{(r+1)}$ for $k = 1, 2........m$ using Eq. (13.99)

11. Find average values of state derivatives

$$\dot{x}_{1k}^{(r)}, \text{avg} = \frac{1}{2}\left[\dot{x}_{1k}^{(r)} + \dot{x}_{1k}^{(r+1)}\right] \qquad \text{for } k = 1, 2..........m$$

$$\dot{x}_{2k}^{(r)}, \text{avg} = \frac{1}{2}\left[\dot{x}_{2k}^{(r)} + \dot{x}_{2k}^{(r+1)}\right]$$

12. Find final state estimates for $t = t^{(r+1)}$

$$x_{1k}^{(r+1)} = x_{1k}^{(r)} + \dot{x}_{1k}^{(r)}, \text{avg } (\Delta t)$$

$$x_{2k}^{(r+1)} = x_{2k}^{(r)} + \dot{x}_{2k}^{(r)}, \text{avg } (\Delta t) \qquad \text{for } k = 1, 2,..........m$$

13. Compute final estimate of E_k at $t = t^{(r+1)}$ using the equation

$$E_k^{(r+1)} = |E_k^0|\left[\cos x_{1k}^{(r+1)} + j\sin x_{1k}^{(r+1)}\right]$$

14. Print $x_{1k}^{(r+1)}, x_{2k}^{(r+1)}$ for $k = 1, 2..........m$

15. Test for time for which the swing curve is to be plotted *i.e.* if $r > r_{\text{final}}$. If not $r = r + 1$ and repeat all steps from step 5.

16. If $r > r_{\text{final}}$, print result and stop.

Using the above algorithm the swing curves for all machines $k = 1, 2......m$ can be plotted. If angle δ for any machine continues to increase indefinitely, that machine is unstable and will lose synchronism.

13.12 FACTORS AFFECTING TRANSIENT STABILITY AND TECHNIQUES FOR STABILITY IMPROVEMENT

Transient stability is a very important aspect of modern multimachine power systems. When a multimachine system becomes unstable it is split into two groups of machines which go out of synchronism with each other while the machines within each group maintain synchronism. The grouping may differ with the fault location. Still, for a given fault location, the general behaviour of a multimachine system is similar to that of a two-machine system. As discussed in section 13.5.2, a two-machine system can be considered equivalent to a single machine connected to infinite bus. Therefore, the following conclusions, though based on a single machine connected to infinite bus, apply equally well to a multimachine system.

The problem of maintaining stability is easier if $P_m = \frac{|E||V|}{X}$ is large because for any load, δ_0 is small and for any given increase in load, the increase in load angle $\delta_1 - \delta_0$ is small. The techniques for improving stability can be classified into traditional techniques and new approaches.

13.12.1 Traditional Techniques

(a) *Effect of Generator Design:* A heavy machine has greater inertia and is more stable than a light machine. In the present-day competitive market the trend in generator designing is to achieve more power from smaller machines. However, this trend is undesirable from the stability point of view. In earlier days a large number of small machines were employed to derive large power. This was not a desirable feature from system stability considerations. A salient pole generator operates at lower load angles and is preferred to cylindrical rotor generators from considerations of stability.

(b) *Increase of Voltage:* The amplitude of the power angle curve is directly proportional to the internal voltage of the machine. An increase in voltage increases the stability limit.

(c) *Reduction in Transfer Reactance:* The amplitude of the power angle curve is inversely proportional to the transfer reactance. This reactance can be reduced by connecting more lines in parallel. When two lines are in parallel and a fault occurs on one of the lines, some power is transferred over the healthy line except when the fault is at the receiving end or sending end bus. This power transfer aids stability. Some features of the power system layout and bussing arrangement also help in improving stability. The use of bundled conductor lines helps in reducing reactance and improving stability. The compensation of line reactance by series capacitors is another effective method of improving stability. However, the series compensation should be used only after its effects on system voltage profile, over voltages during faults and protection schemes have been studied.

(d) *Rapid Fault Clearing:* It is seen from the equal area criterion that stability is improved by decreasing the accelerating area A_1 and increasing the decelerating area A_2. This can be achieved by decreasing the fault clearing angle (by using high speed breakers) provided all other factors remain the same.

(e) *Automatic Reclosing:* Most of the faults on the transmission lines are of transient nature and are self-clearing. Modern circuit breakers are mostly of reclosing type. When a fault occurs, the faulted line is de-energised to suppress the arc in the fault and then the circuit breaker recloses, after a suitable time interval. Automatic reclosing increases the decelerating area A_2 (as discussed in section 13.7.3) and, thus, helps in improving stability.

13.12.2 New Approaches

The above mentioned methods of stability improvement have long been in use. With the continual growth of power system, new approaches are needed. Generators of today are smaller in size (due to improved cooling and better design) than generators of the same rating built few years ago. This has made the stability problem all the more difficult. Some new approaches to this problem are as under:

(a) *Quick Valve Opening Action:* One reason for power system instability is the excess energy supplied by the turbine during the disturbance. Obviously if the turbine output is reduced and made equal to that needed by the generator during the disturbance, the stability problem reduces by a considerable degree. However, it is easier said than done. Valves on modern large generators are very heavy and to operate these valves in a time of 0.5 second or less is very difficult. In addition there are side effects to the operation, the valve itself may be damaged due to fast operation. A more recent development in governor design is the electronic governor in which the governor valve is driven electrically and controlled by electronic signals.

(b) *Application of Braking Resistors:* An alternative to quick valve opening action is the application of braking resistors. In this method the input to the turbine remains constant and the generator is artificially loaded by a resistor that substitutes for the loss in load on the generator. When a fault occurs a suitable control scheme comes into action and connects the braking resistor to the generator terminals. A few cycles after the clearance of the fault, the same control scheme disconnects the braking resistors. However, the control schemes available till today are not very reliable. A possible failure or maloperation of the control scheme can make the matter still worse. Control schemes using thyristors have recently been suggested.

(c) *Single Pole Switching:* Most transmission line faults are single phase to ground faults. If the protection scheme and breakers are properly arranged, only the faulted phase of the transmission system needs to be de-energised. Since a large percentage of these faults are transitory, this phase can also be returned to service after it has been de-energised for a sufficient time. The time for which a line should be de-energised is a function of wind velocity, circuit voltage, fault current, etc. It is not feasible to operate the systems over long periods with one phase open. Therefore, some means must be employed for tripping the entire line if one phase remains open for a predetermined time.

(d) *Fast Acting Automatic Voltage Regulator:* The satisfactory operation of synchronous generators of a complex power system at high load angles and during transient condition is very much dependent on the source of excitation for the generators and on the automatic voltage regulator.

The power output of a generator is proportional to internal voltage *E*. Under fault conditions the terminal voltage of the machine decreases. A fast acting voltage regulator increases *E* (so as to keep the terminal voltage constant). A higher value of *E* means a higher generator output.

Figure 13.22 shows the effect of the voltage regulator on the transient stability. Along with the usual pre-fault, during fault and post-fault curves, the post-fault power angle curve when the generator has an automatic voltage regulator (avr) is also shown. Under pre-fault condition the generator is operating at point 1. When a fault occurs the operation shifts to point 2 and the rotor accelerates. With no excitation control the operating point will return to point 3 on the post-fault curve after the fault is cleared. The forward momentum of the generator may take the generator beyond point 4 and if this happens, synchronism will be lost. If the generator has a fast acting avr, the generator will return to point 5. In this case the limiting swing is given by point 6. The decelerating power is much greater and the chances of losing synchronism are reduced considerably. The ability of the avr to change the excitation rapidly so that curve *d* is applicable, depends on the time lag of the regulator circuit. For optimum transient response the time lag should be as small as possible.

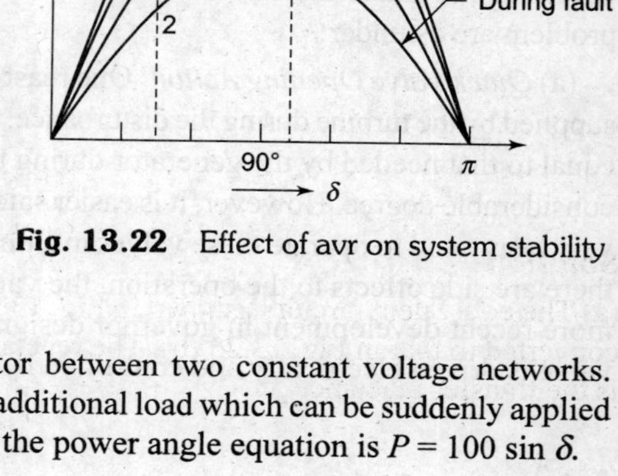

Fig. 13.22 Effect of avr on system stability

Example 13.15

A transmission line is acting as an inter-connector between two constant voltage networks. Determine graphically or otherwise the maximum additional load which can be suddenly applied to this inter-connector already carrying 50 MW if the power angle equation is $P = 100 \sin \delta$.

Solution

Figure 13.23 shows the limiting conditions for stability. P_0 is the initial power and P_1, is the maximum power which can be carried without loss of transient stability

$$50 = 100 \sin \delta_0 \text{ or } \delta_0 = 30^\circ$$

Accelerating area is A_1 and decelerating area is A_2

$$A_1 = \int_{\delta_0}^{\delta_1} (P_1 - 100 \sin \delta)\, d\delta$$

$$= P_1 (\delta_1 - \delta_0) - 100 (\cos \delta_0 - \cos \delta_1)$$

$$A_2 = \int_{\delta_1}^{\delta_2} (100 \sin \delta - P_1)\, d\delta$$

$$= 100 (\cos \delta_1 - \cos \delta_2) - P_1 (\delta_2 - \delta_1)$$

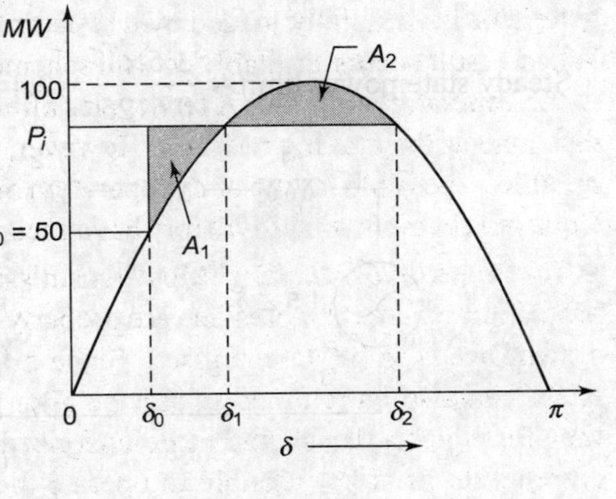

Fig. 13.23

For limiting case $\delta_2 = \pi - \delta_1$. Moreover $P_1 = 100 \sin \delta_1$. Equating areas A_1 and A_2 and substituting these values we get the equation

$$\cos \delta_0 + \cos \delta_1 = (\pi - \delta_1 - \delta_0) \sin \delta_1$$

The angles δ_1 and δ_0 in this equation are in radians. This equation can be solved either graphically (by counting the squares) or by hit and trial. The result is $\delta_1 = 60.4^\circ$

$$P_1 = \sin 60.4^\circ = 86.95 \text{ MW}$$

$$\text{Additional power} = 86.95 - 50 = 36.95 \text{ MW}$$

Example 13.16

Figure 13.24 shows a simple system. Find the maximum steady state power transfer (a) when capacitor is connected (b) when capacitor is replaced by an inductor of the same value. All the reactances are to a common base.

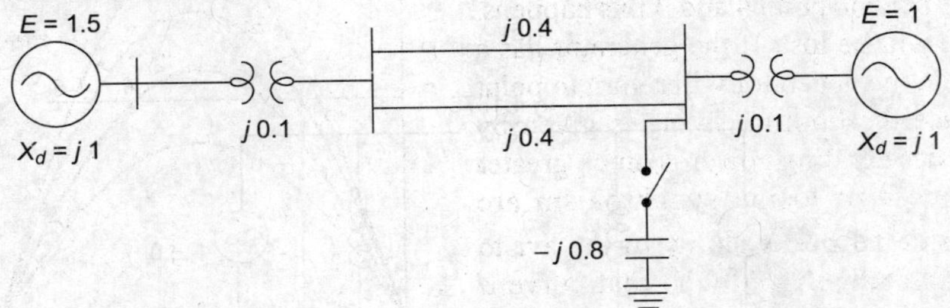

Fig. 13.24

Solution

(a) The equivalent circuit is shown in Fig. 13.25(a). By using star delta transformation it can be converted to that in Fig. 13.25 (b). The reactance of the branch which connects the two sources is the transfer reactance X.

$$X = \frac{(j1.3 \times j1.1) + (j1.3)(-j0.8) + (j1.1)(-j0.8)}{-j0.8} = j\,0.6125\ pu$$

$$\text{Steady state power limit} = \frac{1.5 \times 1}{0.6125} = 2.449\ pu$$

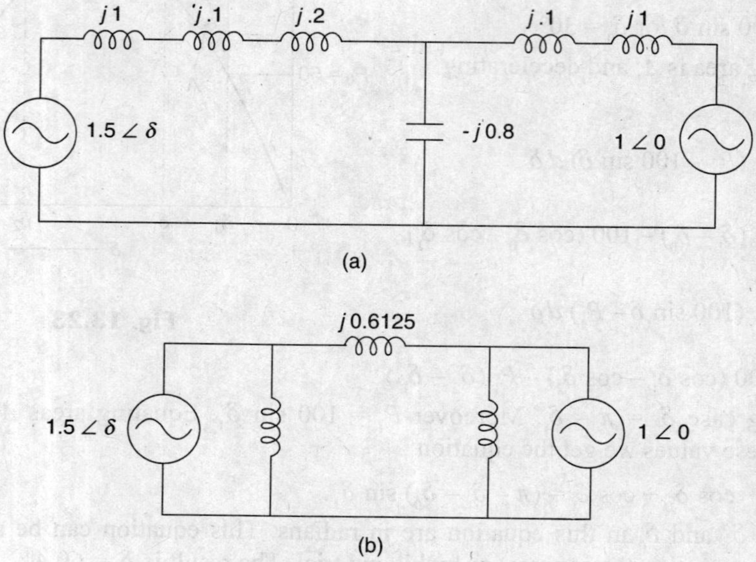

Fig. 13.25

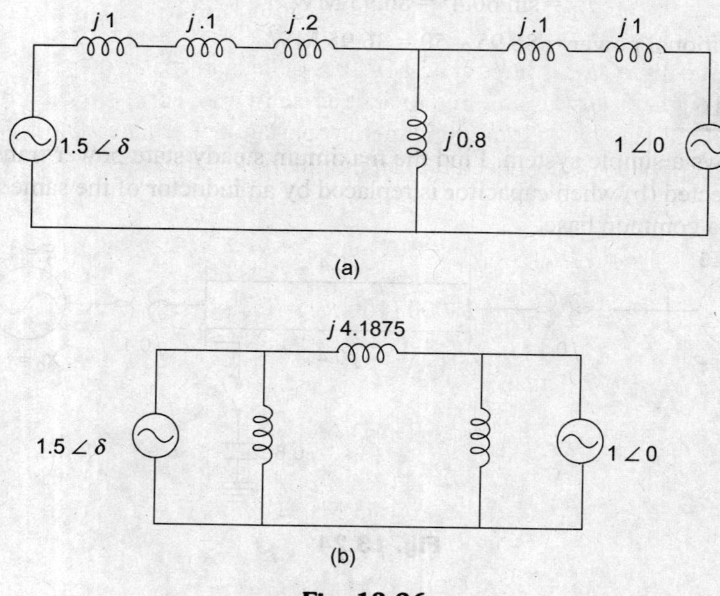

Fig. 13.26

(b) The equivalent circuit is shown in Fig. 13.26(a). Using star delta conversion it can be converted to that in Fig. 13.26(b)

$$X = \frac{(j1.3)\,(j1.1) + (j1.3)\,(j\,0.8) + (j1.1)\,(j\,0.8)}{j\,0.8}$$

or $$X = j\,4.1875\ pu$$

$$\text{Steady state power limit} = \frac{1.5\times 1}{4.1875} = 0.358\ pu.$$

Example 13.17

A 2-pole 50 Hz 11 kV turbo generator has a rating of 60 MW and power factor 0.85 lagging. Its rotor has a moment of inertia of 8800 kg-m^2. Find its inertia constant in MJ/MVA and momentum in MJs/elect. deg.

Solution

$$\omega = 2\,\pi f = 100\,\pi \text{ rad/sec.}$$

$$\text{KE} = \frac{1}{2} I\,\omega^2 = \frac{1}{2}\times 8800\,(100\,\pi)^2$$

$$= 434.265 \times 10^6 \text{ J} = 434.265 \text{ MJ.}$$

$$\text{MVA} = \frac{60}{0.85} = 70.588$$

$$\text{H} = \frac{434.265}{70.588} = 6.152 \text{ MJ/MVA}$$

$$\text{M} = \frac{GH}{180\,f} = \frac{434.265}{180\times 50} = 0.04825 \text{ MJ-s/elect. deg.}$$

Example 13.18

A 4-pole 50 Hz turbo alternator is rated at 45 MW, 0.8 pf lag and has an inertia of 25000 kg-m^2. It is connected through a transmission line to another set whose corresponding data is 2-pole, 60 MW, 50 Hz, 0.75 pf lag, 9000 kg-m^2. Find inertia constant of each machine on its own rating and that of single equivalent set connected to an infinite bus on a base of 100 MVA.

Solution

For the first machine

$$\text{KE} = \frac{1}{2}\times 25000\,(100\,\pi)^2 = 1233.7\times 10^6 \text{ J}$$

$$= 1233.7 \text{ MJ}$$

$$\text{MVA} = \frac{45}{0.8} = 56.25 \text{ MVA}$$

$$\text{H}_1 = \frac{1233.7}{56.25} = 21.93 \text{ MJ/MVA}$$

$$\text{M}_1 = \frac{1233.7}{180\times 50} = 0.137 \text{ MJ-s/elect. deg.}$$

For the second machine

$$\text{KE} = \frac{1}{2}\times 9000\,(100\,\pi)^2 = 444.134\times 10^6 \text{ J}$$

$$= 444.134 \text{ MJ}$$

$$\text{MVA} = \frac{60}{0.75} = 80 \text{ MVA}$$

$$H_2 = \frac{444.134}{80} = 5.55 \text{ MJ/MVA}$$

$$M_2 = \frac{444.134}{180 \times 50} = 0.04935 \text{ MJ-s/elect.deg.}$$

$$M = \frac{M_1 M_2}{M_1 + M_2} = \frac{0.137 \times 0.04935}{0.137 + 0.04935}$$

$$= 0.03628 \text{ MJ-s/elect. deg.}$$

$$GH = M \times 180 \times 50 = 0.03628 \times 180 \times 50$$

$$= 326.52 \text{ MJ}$$

$$\text{Inertia constant of equivalent machine on 100 MVA base} = \frac{326.52}{100} = 3.2652 \text{ MJ/MVA.}$$

Example 13.19

A generator is transferring power to a load through a short line. The power angle equation is $P = P_m \sin \delta$. The initial power is $P_i\, pu$ when a 3-phase fault occurs at the terminals of generator, (a) Use equal area criterion to find equation for critical clearing angle and critical clearing time (b) Find critical clearing angle and critical clearing time if $P_m = 2\, pu$, $P_i = 1\, pu$.

$$H = 6 \text{ MJ/MVA}, f = 50 \text{ Hz.}$$

Solution

(a) Figure 13.27 shows the power angle curve. A_1 and A_2 are the accelerating and decelerating areas, δ_0, δ_{cc} and δ_m denote the initial, critical clearing and maximum values of δ. P_m is the maximum value of power and P_i is initial power.

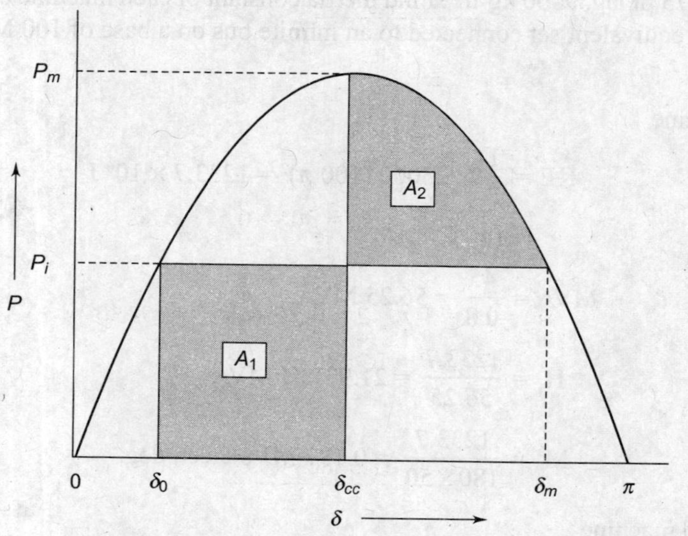

Fig. 13.27

$$A_1 = \int_{\delta_0}^{\delta_{cc}} P_i d\delta = P_i(\delta_{cc} - \delta_0)$$

$$A_2 = \int_{\delta_{cc}}^{\delta_m} (P_m \sin \delta - P_i)\, d\delta$$

$$= P_m(\cos \delta_{cc} - \cos \delta_m) - P_i(\delta_m - \delta_{cc})$$

For stability $A_1 = A_2$

$$P_i(\delta_{cc} - \delta_0) = P_m(\cos \delta_{cc} - \cos \delta_m) - P_i(\delta_m - \delta_{cc}) \quad \text{(i)}$$

It is seen from Fig. 13.27 that $\delta_m = \pi - \delta_0$ and $P_i = P_m \sin \delta_0$. Making these substitutions in Eq. (i) and simplifying we get

$$\delta_{cc} = \cos^{-1}[(\pi - 2\delta_0)\sin \delta_0 - \cos \delta_0] \quad \text{(ii)}$$

During fault, the power transfer is zero. Therefore, swing equation is

$$\frac{d^2\delta}{dt^2} = \frac{P_i}{M}$$

or

$$\frac{d\delta}{dt} = \int_0^t \frac{P_i}{M} dt = \frac{P_i}{M} t$$

and

$$\delta = \frac{P_i}{2M} t^2 + \delta_0$$

At $t = t_{cc}$, $\delta = \delta_{cc}$

or

$$\delta_{cc} = \frac{P_i}{2M}(t_{cc})^2 + \delta_0$$

or

$$t_{cc} = \left[\frac{2M(\delta_{cc} - \delta_0)}{P_i}\right]^{0.5} \quad \text{(iii)}$$

(b)

$$P_i = 1, \quad P_m = 2$$

$$M = \frac{GH}{\pi f} = \frac{1 \times 6}{\pi \times 50}$$

$$1 = 2 \sin \delta_0 \text{ or } \delta_0 = 30° = 0.5236 \text{ rad.}$$

Using Eq. (ii)

$$\delta_{cc} = \cos^{-1}[(\pi - 2 \times 0.5236)\, 0.5 - \cos 30°]$$

$$= 79.56° = 1.3886 \text{ rad.}$$

Using Eq (iii)

$$t_{cc} = \left[\frac{2 \times 6}{\pi \times 50} \; \frac{(1.3886 - 0.5236)}{1}\right]^{0.5}$$

$$= 0.257 \text{ seconds or } 12.85 \text{ cycles.}$$

Example 13.20

A 50 Hz 4-pole turbo generator rated 20 MVA 13.2 kV has an inertia constant of 9 kW- sec/kVA. (a) Find the kinetic energy stored in rotor at synchronous speed, (b) find acceleration if input less rotational losses is 25000 hp and electric power developed is 15000 kW. (c) Find the change in torque angle if this acceleration is maintained for 15 cycles. (d) Find rpm at the end of 15 cycles.

(AMIE 2004)

Solution

(a) $$\text{Stored energy} = 20 \times 10^3 \times 9 \text{ kW-sec} = 180 \times 10^3 \text{ kJ}$$

$$= 180 \text{ mJ}$$

(b) $$\text{Input} = 25000 \text{ hp} = 25000 \times 0.735$$

$$= 18375 \text{ kW}$$

(Assuming that 25000 is metric horse power)

Acceleration power $$P_a = 18375 - 15000$$

$$= 3375 \text{ kW} = 3.375 \text{ MW}$$

$$M = \frac{GH}{180\,f} = \frac{180}{180 \times 50} = \frac{1}{50} \text{ MJ-sec/elect deg.}$$

$$\text{M } \alpha = P_a$$

$$\text{Acceleration } \alpha = \frac{P_a}{M} = \frac{3.375}{1/50} = 168.75 \text{ elect. deg/sec}^2$$

(c) $15 \text{ cycles} = \dfrac{15}{50}$ sec.

$$\text{Change is } \delta = \frac{1}{2}\alpha\, t^2 = \frac{1}{2}(168.75)\left(\frac{15}{50}\right)^2$$

$$= 7.59^\circ$$

(d) Since it is a 4-pole machine, 1 revolution = 4 × 180 = 720°

$$\alpha = 168.75 \text{ elect. deg/ sec}^2 = 168.75\left(\frac{60}{720}\right) = 14.06 \text{ rpm/sec}$$

Synchronous speed = 1500 rpm

Rotor speed at the end of 15 cycles or $\dfrac{15}{50}$ secs

$$= 1500 + 14.06\left(\frac{15}{50}\right) = 1504.218 \text{ rpm}$$

Example 13.21

A generator supplies power to an infinite bus via a line as shown in Fig. 13.28. The machine is delivering 1.0 *p.u.* power and terminal voltages of both generator and infinite bus are 1.0 *p.u.* All reactances are on a common base. Determine power angle equation. *[JNTU 2010]*

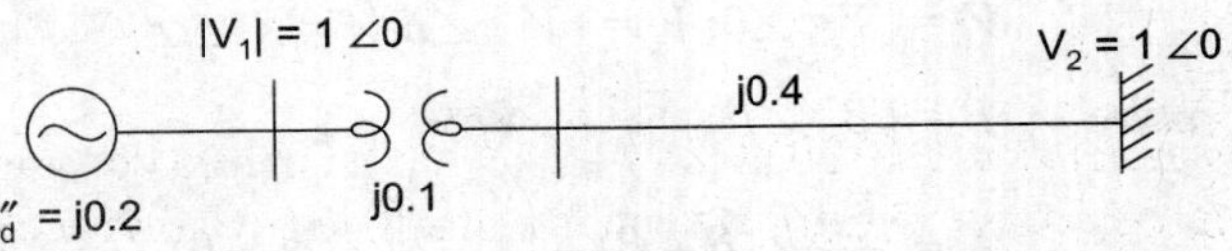

Fig. 3.28

Solution

$$V_1 = 1 \angle 0$$

$$P = 1 + j0$$

Assume that power factor is 0.8 lagging

$$I = 1\ (0.8 - j.6)\ p.u.$$

$$E = 1 \angle 0 + I\,(j X''_d)$$

$$= 1 \angle 0 + (0.8 - j0.6)\ (j0.2)$$

$$= 1 + 0.12 + j0.16 = 1.12 + j0.16$$

$$= 1.13 \angle 8.13°\ p.u.$$

Total transfer reactance between generator and infinite bus $= j0.2 + j0.1 + j0.4 = 0.7\ p.u.$

$$P_m = \frac{|E|\,|V|}{X} = \frac{(1.13)\ (1)}{0.7} = 1.614\ p.u.$$

Power angle equation is

$$P = 1.614 \sin \delta$$

Example 13.22

(a) Discuss the effect of including line resistance and shunt admittance on steady state stability limit.

(b) A 50 Hz 500 km long transmission line with following constants ties up two large areas. $R = 0.11$ ohm/km, $L = 1.45$ mH/km and $C = 0.009$ microfarad per km. Find steady state power limit if $|V_s| = |V_r| = 200$ kV. *[JNTU 2010]*

Solution

(a) Consider the system shown in Fig. 13.29.

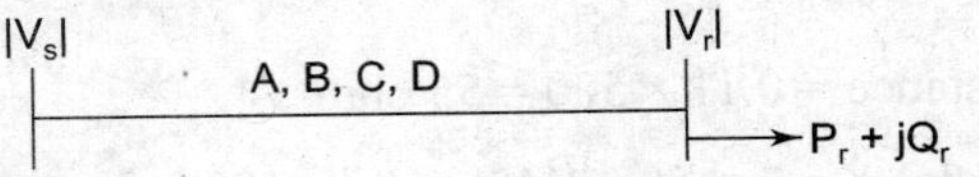

Fig. 13.29

Line equations are

$$V_s = A\,V_r + B\,I_r \tag{i}$$

$$I_s = C\,V_r + D\,I_r \tag{ii}$$

From Eq. (i)

$$I_r = \frac{1}{B} V_s - \frac{A}{B} V_r \tag{iii}$$

Let $V_r = |V_r| \angle 0, V_s = |V_s| \angle \delta, A = A \angle \alpha$

and $B = |B| \angle \beta$

Then
$$I_r = \frac{|V_s|}{|B|} \angle(\delta - \beta) - \frac{|A||V_r|}{|B|} \angle(\alpha - \beta) \quad \text{(iv)}$$

Conjugate of I_r is

$$I_r^* = \frac{|V_s|}{|B|} \angle(\beta - \delta) - \frac{|A||V_r|}{|B|} \angle(\beta - \alpha) \quad \text{(v)}$$

Complex power received is

$$S_r = P_r + j\,Q_r = V_r I_r^*$$

$$= |V_r| \angle 0 \left[\frac{|V_s|}{|B|} \angle(\beta - \delta) - \frac{|A||V_r|}{|B|} \angle(\beta - \alpha) \right] \quad \text{(vi)}$$

Therefore real power received is

$$P_r = \frac{|V_s||V_r|}{|B|} \cos(\beta - \delta) - \frac{|A||V_r|^2}{|B|} \cos(\beta - \alpha) \quad \text{(vii)}$$

If resistance and shunt admittance are neglected $|B| = X$, $\beta = 90°$, $|A| = 1$ and $\alpha = 0$. Then

$$P_r = \frac{|V_s||V_r|}{X} \sin \delta \quad \text{(viii)}$$

From Eq. (vii) maximum received power is

$$P_{r\max} = \frac{|V_s||V_r|}{|B|} - \frac{|A||V_r|^2}{|B|} \cos(\beta - \alpha) \quad \text{(ix)}$$

If line resistance and shunt admittance are neglected, steady state power limit is $\frac{|V_s||V_r|}{X}$. However if line resistance and shunt admittance are included in calculations, the steady state power limit is given by Eq. (ix). It is seen that the limit as given by Eq. (ix) is less than $\frac{|V_s||V_r|}{X}$.

(b) Total resistance = 0.11 × 500 = 55 ohm

$$\text{Total } X = 2\pi (50) (1.45 \times 10^{-3}) (500)$$
$$= 227.77 \text{ ohm}$$
$$Z = R + jX = 55 + j227.77 = 234.32 \angle 76.42$$
$$Y = j (2\pi + 50 \times 0.009 \times 10^{-6} \times 500)$$
$$= 0.001414 \angle 90° \text{ siemens}$$
$$YZ = (0.001414) (234.32) \angle 166.42°$$
$$= 0.331 \angle 166.42°$$

$$A = 1 + \frac{YZ}{2} + \frac{(YZ)^2}{24}$$

$$= 1 + \frac{0.331 \angle 166.42°}{2} + \frac{(0.331 \angle 166.42)^2}{24}$$

$$= 1 + 0.1655 \angle 166.42° + 0.0046 \angle 332.84°$$

$$= 1 - 0.1609 + j\,0.03886 + 0.004 - j\,0.0021$$

$$= 0.8431 + j\,0.03676$$

$$A = 0.8439 \angle 2.5°$$

Thus $|A| = 0.8439$ and $\alpha = 2.5°$

$$B = Z\left[1 + \frac{YZ}{6} + \frac{(YZ)^2}{120}\right]$$

$$B = 234.32 \angle 76.42 \left[1 + \frac{0.331 \angle 166.42}{6} + \frac{(0.331 \angle 166.42)^2}{120}\right]$$

$$= 234.32 \angle 76.42\ [1 + 0.05517 \angle 166.42 + 0.00092 \angle 332.84]$$

$$= 234.32 \angle 76.42\ [1 - 0.0536 + j.013 + .008 - j.004]$$

$$= 234.32 \angle 76.42\ (.9472 + j.0126)$$

$$= (234.32 \angle 76.42)\ (.9473 \angle .76)$$

$$= 221.97 \angle 77.18°$$

$$= 49.25 + j\,216.44$$

Thus $|B| = 221.97$ and $\angle\beta = 77.18°$. Using Eq. (ix) steady state power stability limit is

$$= \frac{(200)(200)}{221.97} - \frac{.8439(200)^2}{221.97}\cos(77.18 - 2.5°)$$

$$= 180.2 - 40.18 = 140.02 \text{ MW}$$

If line resistance and capacitance are neglected steady state power limit $= \dfrac{(200)(200)}{227.77}$

$$= 175.61 \text{ MW}$$

Example 13.23

Find steady state stability limit for a short transmission line having $V_s = 11$ kV/phase, $V_r = 10.5$ kV/phase and $Z = (15 + j25)$ ohm/phase. *[JNTU 2010]*

Solution

$$P_{e,\max} = \frac{|V_s||V_r|}{\left(R^2 + X^2\right)^{0.5}} - \frac{|V_r|^2 |R|}{R^2 + X^2}$$

Substituting the values

$$P_{e,\max} = \frac{(11)(10.5)}{\left(15^2+25^2\right)^{0.5}} - \frac{(10.5)^2(15)}{15^2+25^2}$$

$$= 2.016 \times 10^6 \text{ W or } 2.016 \text{ MW per phase}$$

For 3 phase line, the total power limit is 3 times *i.e.* 6.048 MW

Example 13.24

The steady state power limit of a power system is 100 MW. A generator with constant excitation is suppling 50 MW to system. Find the maximum permissible sudden increase in generator output without causing instability. [*JNTU 2011*]

Solution

$$50 = 100 \sin \delta_0$$

or $$\delta_0 = \sin^{-1} 0.5 = 30°$$

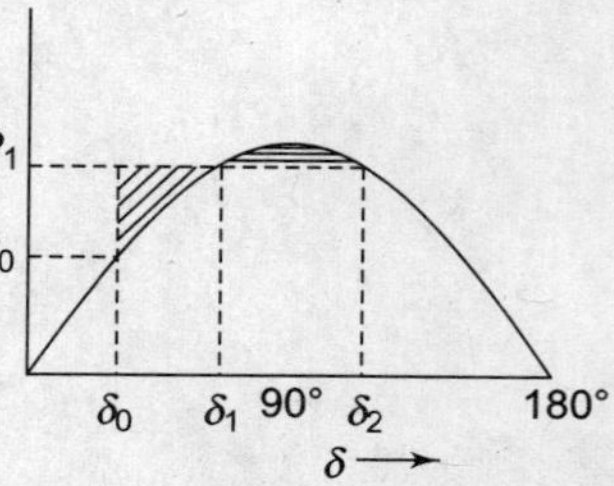

Fig. 13.30

Figure 13.30 shows power angle curve.

Using Eq. (13.59)

$$\cos \delta_0 + \cos \delta_1 = (\pi - \delta_1 - \delta_0) \sin \delta_1 \quad \text{(i)}$$

$$\delta_2 = \pi - \delta_1$$

By hit and trial we get the solution of Eq. (i) as

$$\delta_1 = 60.5°$$

$$P_1 = 100 \sin (60.5°) = 87.04 \text{ MW}$$

Max. sudden increase permissible = 87.04 – 50

= 37.04 MW

This example can also be solved graphically.

Example 13.25

A 2-pole 50 Hz 11 kV turbo generator has a rating of 100 MW p.f. 0.85 lagging. The rotor has a moment of inertia of 10000 kg m². Find *H* and *M*. [*JNTU 2011*]

Solution

$$\omega = 2\pi f = 100\,\pi \text{ rad/sec.}$$

$$\text{MVA} = \frac{100}{0.85} = 117.65 \text{ MVA}$$

$$\text{K.E.} = \frac{1}{2} I\omega^2 = \frac{1}{2}(10000)(100\pi)^2 = 49.35 \times 10^7 \text{ J}$$

$$= 493.5 \text{ MJ}$$

$$H = \frac{493.5}{117.65} = 4.195 \text{ MJ/MVA}$$

$$M = \frac{GH}{180f} = \frac{493.5}{180 \times 50} = 0.0548 \text{ MJ-s/elect. deg.}$$

Example 13.26

An alternator supplies 50 MW to an infinite bus, the steady state power limit is 75 MW. Determine whether system will remain stable if prime mover input is suddenly increased by 25 MW. Neglect losses. [*JNTU 2010*]

Solution

The condition is depicted in Fig. 13.31. Initial angle is δ_0. When input is increased from 50 MW to 75 MW, the rotor starts accelerating and the accelerating area is shown dotted. It is seen that decelerating area is zero. Hence stability cannot be maintained.

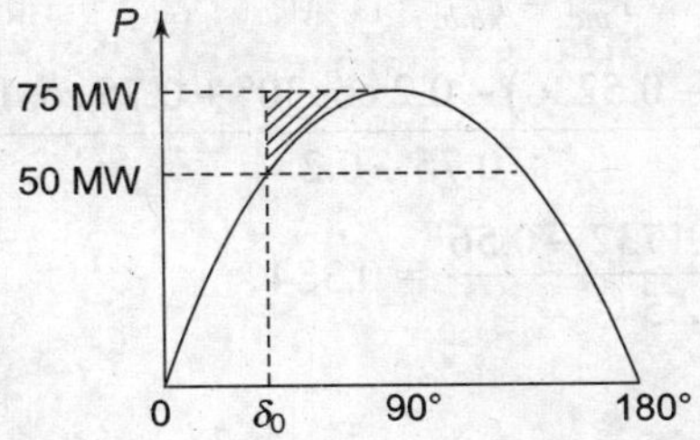

Fig. 13.31

Example 13.27

A 2-pole 50 MVA 11 kV generator is supplying full load at 0.8 p.f. lagging. If the inertia constant is 6 MJ/MVA, find energy stored when the generator is operating at synchronous speed of 3000 rpm. If net input is increased suddenly to 62000 metric h.p., find acceleration. [*RGPV 2009*]

Solution

$$\text{Stored energy} = 6 \times 50 = 300 \text{ MJ}$$

$$\text{Input} = \frac{62000 \times 735.5}{10^6} = 45.6 \text{ MW}$$

$$\text{Output} = 50 \times 0.8 = 40 \text{ MW}$$

$$\text{Accelerating Power} = 45.6 - 40 = 5.6 \text{ MW}$$

$$M = \frac{GH}{180\,f} = \frac{300}{180 \times 50} = \frac{1}{30} \text{ MJ-s/elect deg.}$$

$$M\alpha = 5.6 \text{ or } \alpha = 5.6 \times 30 = 168 \text{ elect. deg/sec}^2$$

Example 13.28

A 50 Hz generator is delivering 50% of the power that it is capable of delivering through a transmission system to an infinite bus. A fault occurs and causes the reactance between generator and infinite bus to become 500% of value before fault. When the fault is isolated the power which can be delivered is 75% of original maximum value. Find critical clearing angle. [*RGPV 2007*]

Solution

Maximum power which can be transmitted is inversely proportional to transfer reactance.

Let $\quad P_{ma} = 1\ p.u. \qquad\qquad P_i = 0.5\ p.u.$

$P_{mb} = \frac{1}{5} = 0.2\ p.u.$ $\qquad P_{mc} = 0.75\ p.u.$

$\delta_0 = \sin^{-1} 0.5 = 30° = 0.5236$ rad.

$\delta_m = \pi - \sin^{-1}(P_i/P_{mc}) = \pi - \sin^{-1}(0.5/0.75)$

$= 2.412$ rad $= 138.2°$

Using Eq. (13.66)

$$\cos \delta_{cc} = \frac{P_i(\delta_m - \delta_0) - P_{mb}\cos\delta_0 + P_{mc}\cos\delta_m}{P_{mc} - P_{mb}}$$

$$= \frac{0.5(2.412 - 0.5236) - 0.2\cos 30° + 0.75\cos 138.2°}{0.75 - 0.2}$$

$$= \frac{0.9442 - 0.1732 - 0.56}{0.55} = 0.384$$

$\delta_{cc} = 67.42°$

Example 13.29

A generator with constant excitation supplies 30 MW to an infinite bus through a transmission system. The steady state stability limit is 60 MW. Find the maximum sudden increase in generator output which can be permitted without loss of stability. *[RGPV 2007]*

Solution

Figure 13.32 shows the power angle diagram. δ_0 is the initial power angle corresponding to 0.5 *p.u.* power transfer. Maximum permissible power transfer is P_1 and the corresponding power angle is δ_1.

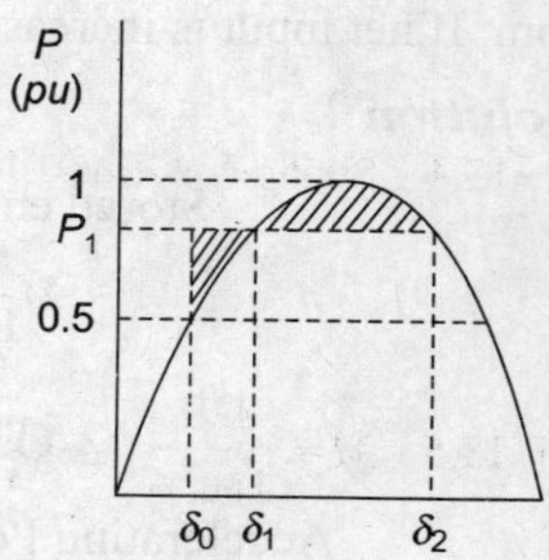

Fig. 13.32

Using Eq. (13.59)

$$\cos\delta_0 + \cos\delta_1 = (\pi - \delta_1 - \delta_0)\sin\delta_1$$

By hit and trial we get the solution as $\delta_1 = 60.5°$

$$P_1 = 60 \sin(60.5°) = 52.22 \text{ MW}$$

Maximum permissible increase = 52.22 – 30

= 22.22 MW

Example 13.30

A 60 Hz generator is delivering 50% of maximum permissible power through a transmission system to an infinite bus. A fault occurs and causes transfer reactance to increase to 400% of the value before fault. When the fault is isolated the maximum power transfer is 75% of original maximum value. Find critical clearing angle using equal area criterion. *[CSVTU 2011]*

Solution

Maximum permissible power transfer is inversely proportional to transfer reactance.

Let $\quad P_{ma} = 1\ p.u.$ $\qquad P_i = 0.5\ p.u.$

$P_{mb} = \frac{1}{4} = 0.25\ p.u.$ $\qquad P_{mc} = 0.75\ p.u.$

$$\delta_0 = \sin^{-1}(0.5) = 30° = 0.5236 \text{ rad.}$$

$$\delta_m = \pi - \sin^{-1}(P_i/P_{mc}) = \pi - \sin^{-1}(0.05/0.75)$$

$$= 138.2° = 2.412 \text{ rad.}$$

Using Eq. (13.66)

$$\cos \delta_{cc} = \frac{P_i(\delta_m - \delta_0) - P_{mb}\cos\delta_0 + P_{mc}\cos\delta_m}{P_{mb} - P_{mc}}$$

$$= \frac{0.5(2.412 - 0.5236) - 0.25\cos 30° + 0.75\cos 138.2°}{0.75 - 0.25}$$

$$\cos \delta_{cc} = \frac{0.9442 - 0.2165 - 0.56}{0.5} = 0.3354$$

$$\delta_{cc} = 70.4°$$

SUMMARY

13.1 The stability of a physical system is its capability to return to an equilibrium position on the occurrence of a disturbance.

13.2 Stability problem can be divided into steady state stability, transient stability and dynamic stability.

13.3 When a generator is connected to an infinite bus,

Real power output $\quad P_e = \dfrac{|E||V|}{X}\sin\delta$

or $\quad P_e = P_m \sin\delta$

13.4 Swing equation

$$M\frac{d^2\delta}{dt^2} = P_a = P_i - P_e$$

13.5 $M = \dfrac{GH}{180 f}$ Mega - Joule - sec/elect deg.

13.6 For a number of machines in parallel

$$H_{eb} = \frac{H_1 G_1}{G_b} + \frac{H_2 G_2}{G_b} + \ldots$$

13.7 Undamped frequency of oscillations

$$\omega_s = \left(\frac{P_r}{M}\right)^{0.5} \text{ rad/sec}$$

13.8 Sudden disturbances may cause a system to become unstable even when it is operating below its steady state stability. This problem is known as transient stability.

13.9 Transient stability can be studied and predicted by equal area criterion.

13.10 Equal area criterion: Kinetic energy gained by rotor during its swing (say from δ_0 to δ_1) must equal the kinetic energy returned as it swings from δ_1 to δ_2.

13.11 Using equal area criterion we can find critical clearing angle. If the fault clearing angle is less than this angle, stability is maintained.

13.12 Swing equation can be solved using point by point method or using a digital method.

13.13 Liapunov's direct method can also be used to study stability problem.

13.14 Traditional techniques for stability improvement include increase of system voltage, reduction of transfer reactance, rapid fault clearing and automatic reclosing.

13.15 New approaches for stability improvement include quick value opening, use of braking resistors, single pole switching and fast acting automatic voltage regulator.

SELF TEST QUESTIONS

A. State whether the following are true (T) or false (F)

13.1 Stored energy in a synchronous machine is equal to GH.

13.2 Swing equation is a non-linear equation.

13.3 For a generator feeding an infinite bus

$$Q_e = \frac{|E||V|}{X}\cos\delta - \frac{|V|^2}{X}$$

13.4 Stability and power limit mean the same.

13.5 Transient instability phenomenon is a very fast one.

13.6 Automatic voltage regulator of a synchronous machine does not have any effect on stability.

13.7 δ is the angular displacement of rotor from synchronously rotating reference axis

13.8 Steady state power limit = $\dfrac{|E||V|}{X}$.

13.9 Equal area criterion cannot be used to study stability problem due to switching action.

13.10 Load angle is maximum when $\dfrac{d\delta}{dt} = 0$.

B. Fill in the blanks with the suitable words

13.11 Proper use of breaking resistors ______ stability.

13.12 Use of high speed breakers ______ stability.

13.13 The swing equation can be solved by ______ method.

13.14 For a single line between a generator and infinite bus the power transferred during a 3 phase fault on line is ______.

13.15 Synchronous machines in a power exporting area can be grouped into an ______.

13.16 The usual value of δ is about ______.

13.17 In terms of M and ω, kinetic energy = ______.

13.18 $I\dfrac{d^2\delta}{dt^2}$ = ______.

13.19 Undamped frequency of oscillations of a synchronous machine = ______.

13.20 Losses cause ______ of oscillations.

ANSWERS

13.1 T	13.2 T	13.3 T	13.4 F
13.5 T	13.6 F	13.7 T	13.8 T
13.9 F	13.10 T	13.11 improves	13.12 improves
13.13 point by point	13.14 zero	13.15 equivalent generator	13.16 30°

13.17 $\frac{1}{2} M\omega$ 13.18 accelerating power 13.19 $(P_{r/M})^{0.5}$ 13.20 damping.

JOB INTERVIEW QUESTIONS

13.1 Differentiate between stability and loss of synchronous.

13.2 Differentiate between steady state and transient stability.

13.3 Name the methods used to solve swing equation.

13.4 What is equal area criterion?

13.5 How can stability of a system be improved?

13.6 What is the effect of automatic voltage regulator on stability?

13.7 Does reactive power have any effect on stability?

13.8 On what factors does maximum power transfer depend?

13.9 What is critical clearing angle?

13.10 What is transfer reactance?

REVIEW QUESTIONS

13.1 Discuss why ?

(a) Transient stability limit is lower than steady state stability limit.

(b) An early fault clearing means better chances of maintaining system stability.

(c) Present-day generators are less stable than the generators made a few years ago.

(d) A system has better chances of maintaining synchronism if the generators are fitted with fast acting automatic voltage regulators.

(e) The use of automatic reclosing circuit breakers improves system stability.

(f) A salient pole machine is more stable than a cylindrical rotor machine.

13.2 Distinguish between steady state, transient and dynamic stability. Derive power angle equation. Define transfer reactance.

13.3 Starting from first principles, derive the swing equation of a synchronous machine. Define inertia constant.

13.4 Explain the concept of equal area criterion. How can it be used to study transient stability? Explain qualitatively.

13.5 Draw diagrams to illustrate the application of equal area criterion to study transient stability for the following cases:

(a) A sudden increase in input of the generator.

(b) A switching operation causing the switching out of one of the circuits, of a double circuit line feeding an infinite bus.

(c) A fault on one of the parallel circuits of a two-circuit line feeding an infinite bus. The fault is very close to the sending end bus and is subsequently cleared by the opening of faulted line.

(d) A fault on one of the parallel circuits of a two-circuit line feeding an infinite bus. The fault is at the middle of one of the lines and is subsequently cleared by opening of the faulted line. The circuit breakers are of reclosing type. Mark the accelerating and decelerating area in each case.

13.6 Discuss the procedure for solving the swing equation using point by point method.

13.7 Using equal area criterion, derive an expression for critical clearing angle for a system having a generator feeding a large system through a double circuit line.

13.8 Give the necessary equations for solving the swing equation by Runge-Kutta method.

13.9 How can the transient stability of a system be improved? Discuss the traditional as well as new approaches to the problem.

13.10 Explain the Liapunov's direct method for stability analysis of a power system. Discuss how critical clearing angle may be calculated using this method.

PROBLEMS

13.1 A large three-phase cylindrical rotor generator is delivering power over a transmission system to an infinite bus bar when a fault occurs on the system. The reactances between the generator and infinite bus before, during and after the fault are 0.4, 1.1 and 0.5 *pu* respectively. Calculate the critical clearing angle if at the time of the fault the power transfer was 1.0 *pu*. Sketch the power angle curves and show the equality of accelerating and decelerating areas. Take $E = V = 1$ *pu*. [109.2°]

13.2 A large 3-phase cylindrical rotor generator is delivering 1.0 *pu* power to an infinite bus through a transmission network. The maximum power which can be transferred for pre-fault, during fault and post-fault conditions is 1.8 *pu*, 0.4 *pu* and 1.3 *pu*. Find the critical clearing angle. [55.2°]

13.3 Fig. 13.33 shows a transmission network. The *pu* reactances of the equipments are as shown. The voltage behind transient reactance of generator is 1.10 *pu* and the voltage of infinite bus is 1.0 *pu*. The system is transmitting 1 *pu* power when fault occurs at the point *F*. Determine (a) transfer reactance for pre-fault, during fault and post-fault condition (b) critical clearing angle. [(a) 0.5. 2.0. 0.75 pu; (b) 67.14°]

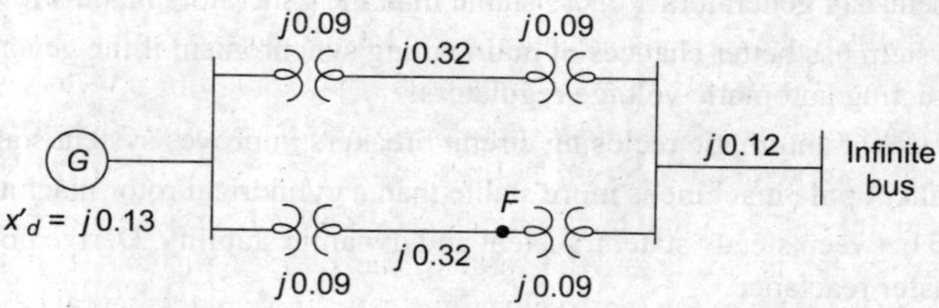

Fig. 13.33

13.4 Plot the swing curve upto a time of 0.5 seconds for the system of problem 13.3 for (a).sustained fault (b) fault cleared by opening of circuit breakers at both ends of faulted line in 0.1 seconds.

13.5 A 2-pole, 50 Hz, 11.5 kV, turbo generator has a rating of 100 MW, power factor 0.85 lagging. Its rotor has moment of inertia of 8800 kg. m². Calculate its inertia constant in MJ/MVA and its momentum in MJ-sec/elect. deg. [3.69, 0.04825]

13.6 A power station has 4 identical sets each rated 50 MVA and each having an inertia constant of 5 MJ/ MVA. A second power station has 3 sets each of 100 MVA and each having an inertia constant of 3 MJ/ MVA. If the two stations are tied together at one end of a long tie line, calculate the inertia constant of a single equivalent set on a base of 100 MVA. [19 MJ/MVA]

13.7 A 100 MVA, 11 kV, 16-pole, 50 Hz water wheel generator has an inertia constant of 3 MJ/MVA. (a) Find the stored energy in the rotor at synchronous speed. (b) The machine is operating at a load of 50 MW when the load suddenly increases to 80 MW. Find the rotor retardation. Neglect losses. (c) The rotor retardation calculated above is maintained for 5-cycle. Find the change in the power angle during this period. [(a) 300 MJ (b) 900 elect, deg./sec². (c) – 4.5°]

13.8 A 50 Hz generator of reactance 0.8 *pu* is connected to an infinite bus through a line of 0.4 *pu* reactance. $E = 1.05$ *pu*, $V = 1.0$ *pu*. The inertia constant is 4 MJ/MVA. The generator is loaded to 70% of the maximum power limit. Find the frequency of natural oscillations. [0.788 Hz]

13.9 The system in problem 13.3 is provided with automatic reclosing circuit breakers. The fault is cleared by the opening of circuit breakers of the faulted line in 0.1 seconds. The circuit breakers reclose after allowing a time of 0.2 seconds for the deionization of the arc. Plot the swing curve up to a time of 0.5 seconds.

13.10 Two large generators *A* and *B* are connected through a network. shown in Fig. 13.34. The generator *A* has an inertia constant of 1.76 MJ/MVA on 100 MVA base and $X'_d = 0.5$ *pu*. The generator *B* has an inertia constant of 14.1 MJ/MVA on 100 MVA base and $X'_d = 0.12$ *pu*. The reactances shown in the figure are also on 100 MVA and rated voltage base. (a) Find the equivalent inertia constant to represent this two-machine system by a single machine system, (b) A 3-phase fault occurs on the middle of one of the parallel circuits. Find the transfer reactance for pre-fault, during fault and post-fault conditions, (c) If $P_i = 0.60$ *pu*, $E_a = 1.1$ *pu*, $E_B = 1.0$ *pu*, find the critical clearing angle for the above fault condition. Draw power angle curves and show the accelerating and decelerating areas. [(a) 1.56 (b) 1.115 ohm, 7.2 ohm, 1.29 ohm, (c) 65.1°]

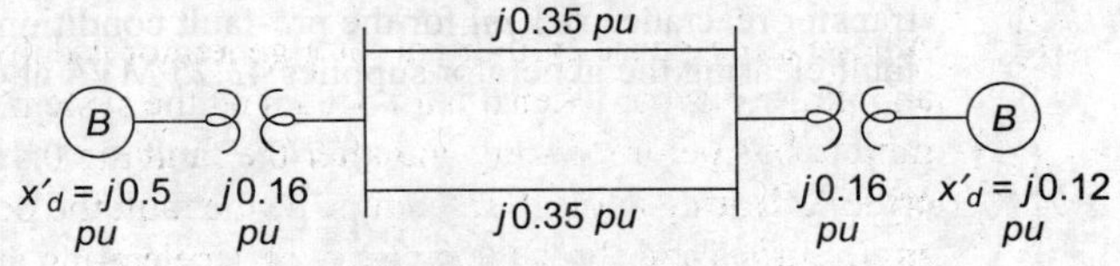

Fig. 13.34

13.11 For the system of problem 13.10 plot the swing curve for a time of 0.5 seconds for (a) sustained fault (b) fault cleared in 0.1 seconds. $P_i = 0.50$ *pu*, $E_A = 1.05$ pu, $E_B = 1.0$ *pu*.

13.12 The kinetic energy of the rotor of a 100 MVA, 4-pole, 50 Hz synchronous machine is 500MJ. At a certain instant the input is 62 MW and the output is 54 MW. Find the accelerating power and acceleration. [8MW, 2.513 rad/s²]

13.13 A 50 Hz generator is delivering power to an infinite bus through a transmission line. The maximum power which can be transferred during pre-fault, during fault and post-fault conditions is 3.6,0.8, and 2.5 *pu*. The initial power transfer is 2 *pu*. Find critical clearing angle. [50.34°]

13.14 A 500 MW generator is operating at a power angle of 8°. By how much can the input power be suddenly increased without loss of stability? [313.42 MW]

13.15 In the system of Fig. 13.35 find the limit of steady state power when (a) capacitor switch is open (b) capacitor switch is closed and the value of capacitor is such that X_c is 1 *pu*. (c) Capacitor is replaced by an inductor of the same value of reactance. Assume internal generator voltage as 1.2 *pu* and motor voltage 1 *pu*. [(a) .49 *pu* (b) 1.243 *pu* (c) .305 *pu*] [*JNTU 2011*]

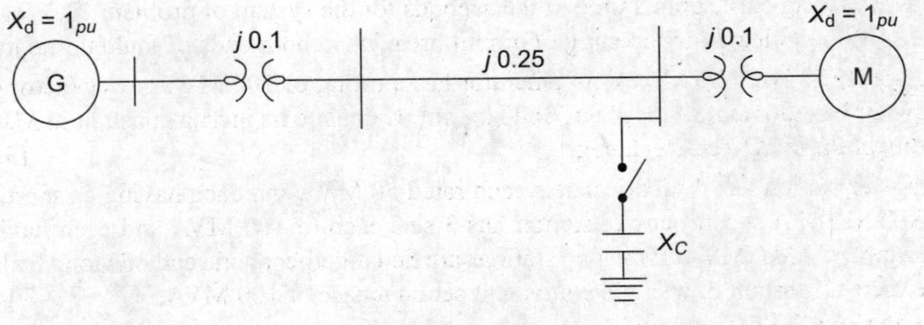

Fig. 13.35

13.16 A 3-phase 50.MVA 50 Hz generator has an angular momentum of 0.05 MJs/elect degree and is loaded 0.8 pu. It is feeding an infinite bus through a transmission system. The maximum power limits before during and after fault are 1.6, 0.2 and 1.2 respectively. Plot swing curve for (a) sustained fault, (b) fault cleared in 0.25 seconds.

13.17 A 3-phase generator is feeding an infinite bus through a transmission system. The maximum power limit before, during and after fault clearing are 5, 2 and 4 *pu* respectively. The initial power transfer is 2.5 pu. Use equal area criterion to find critical clearing angle. [89.9°]

13.18 Find critical clearing angle using the following data: Initial power transfer 1 *pu*, maximum power limits under pre-fault, during fault and post-fault conditions 2, 0.5 and 1.5 *pu* respectively.

[70.29°]

13.19 A generator is transferring power to a load through a short line when a 3-phase fault occurs on the line. The maximum power limit before fault is 2.1 *pu* and initial power transfer is I *pu*. The inertia constant is 5 MJ/ MVA. Find critical clearing angle and critical clearing time. $f = 50$ Hz.

[81.697°, 0.2433 secs]

13.20 A synchronous generator of 100 MVA rating is supplying 62.5 MVA at 0.8 lagging p.f. The transfer reactance is 1 *pu* for the pre-fault conditions and 3 *pu* during the fault condition. After fault clearing the generator supplies 43.75 MVA at 0.8 lagging p.f. Find critical clearing angle.

[68.58°]

CHAPTER 14

Extra High Voltage Transmission

14.1 INTRODUCTION

The generation and consumption of electrical energy has been increasing at a tremendous rate throughout the world during the past many decades. The increasing need of transmitting greater and greater amounts of power has led to a continuous increase in the transmission voltages. As per the current terminology, voltage less than 300 kV are termed as high voltages, those between 300 to 765 kV are termed as extra high voltages (EHV) and those greater than 765 kV are termed as ultra high Voltages (UHV). Some important aspects and characteristics of EHV lines are discussed in this chapter. Almost all these aspects and characteristics are applicable to UHV lines.

14.2 NEED FOR EHV TRANSMISSION

The necessity of increasing the transmission voltages to EHV and UHV range has been primarily due to the following reasons:

(a) *Increase in Size of Generating Units:* In order to reduce the investment cost per kW, the size of generating units has been constantly increased during the past many years. At present 1500 MW turbogenerators are in operation in USA and some other countries. Transmission of large amounts of power over long distances is technically and economically feasible only at voltages in the EHV and UHV range.

(b) *Pithead Steam Plants and Remote Hydro Plants:* The cost of transportation is an appreciable part of the cost of coal, especially for regions situated far away from the coal fields. Sometimes the bottlenecks in transportation tend to starve the steam plants resulting in fall of generation. One method to avoid such situations is to have super thermal plants near coal fields. These steam plants are known as pithead steam plants. In many countries such plants are in operation. In India, too, some such plants are in operation. Similarly, some prospective hydro sites are situated in remote areas. Moreover, the scarcity of sites having favourable characteristics (availability of cooling water, transportation facilities, cheap land and labour, etc.) is forcing many utilities, all over the world, to build large sized steam plants in remote areas safe away from load centres.

EHV transmission systems are needed to transmit large amounts of power over long distances from these plants to load centres.

(c) *Surge Impedance Loading:* The concept of surge impedance loading yields a simple method to compare the approximate load carrying capability of lines of different voltage levels. Surge impedance loading (SIL) is the power that a line carries when each phase is terminated by a load equal to the surge impedance of the line. Surge impedance of a line (symbol Z_c) is equal to $\sqrt{\dfrac{L}{C}}$ where L is the series inductance per unit length and C is the shunt capacitance per unit length. If V is the line to neutral voltage, SIL equals $\dfrac{3V^2}{Z_c}$. At this loading the reactive power consumed by the line inductance (i.e., ωLI^2) is approximately equal to the reactive power generated by the line capacitance (i.e., (ωCV^2).

The surge impedance of a line can be determined from its conductor configuration. The approximate values of surge impedances for lines with single, double, triple and quadruple conductors are 400, 300, 280 and 260 ohms.

A short line can be loaded appreciably above its SIL. However, the load on a long line without series compensation has to be kept at a value even less than the SIL, due to stability considerations. Fig. 14.1 is representative of the practice recommended for the loading of lines with no series compensation. If the load is less than that indicated in this curve, the phase angle across the line is less than 30°. Series compensation should be used to carry loadings heavier than those indicated in this curve, whenever the series compensation is more economical than the use of additional lines or going to next higher voltage. It is evident that as line loads increase, a stage comes when it is necessary to adopt a higher voltage.

(d) *Transmission:* The number of circuits and the land requirement for transmission decreases with the adoption of a higher transmission voltage. For transmitting 600 MW over a distance of 250 km, a single circuit 400 kV line using two conductors per phase and requiring a right of way of about 40 m is sufficient, thus, giving a ratio of $\dfrac{600}{40}$, i.e., 15 MW per metre right of way. If the transmission is at 200 kV, two double circuit lines are needed for transfer of this power. These two lines require a right of way of 60 metres, thus, giving a ratio of 10 MW per metre right of way. It has been reported that a double circuit 400 kV line gives this ratio as about 30 MW per metre right of way and a 1050 kV line gives a ratio of 70 MW per metre right of way. Thus, the advantage of EHV and UHV lines in reducing the right of way requirement can be quite substantial.

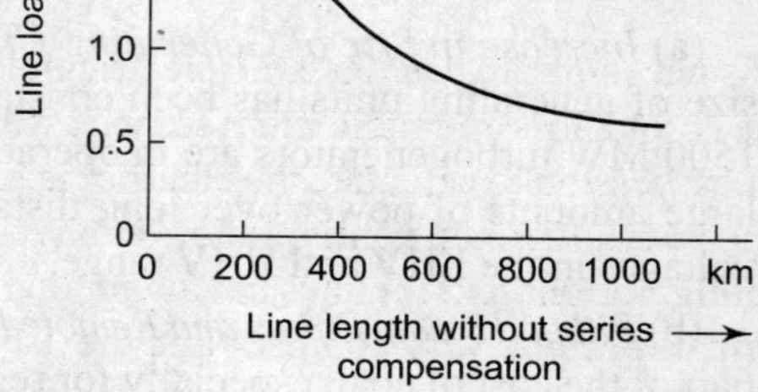

Fig. 14.1 Loading of lines without series compensation

(e) *Line Costs:* The line installation cost per MW per km decreases with increase in voltage level. Moreover, the impact of the cost of losses on the overall transmission costs can be substantially different at different voltage levels. The total line cost (including the cost of losses) per MW per km decreases considerably with the increase in line voltage.

14.3 USE OF BUNDLED CONDUCTORS

For lines upto 220 kV, the size of the conductor is selected on the basis of its continuous and short term current carrying capacity.

However, for EHV lines the corona and radio interference aspects also assume importance. The size of the conductor which gives satisfactory corona and radio interference performance is usually larger than the size corresponding to current carrying capacity.

EHV lines always use bundled conductors to control the voltage gradient at the surface of the conductor so as to reduce corona loss and radio interference. If a single conductor per phase is used and the size of the conductor is selected on the basis of the current carrying capacity alone, the gradient at the surface of the conductor would often be higher than the acceptable value (between 20 —30 kV peak per cm). One solution is to use expanded conductors. These conductors have filler material or air spaces in the interior so that the outside radius is increased. However, due to manufacturing problem and the fact that an expanded conductor must be handled very carefully to avoid distortion (which increase the voltage gradient) these conductors are not used widely. The only other alternative is to use 2 or more sub-conductors per phase so as to give a lower maximum and average voltage gradient than that which a single conductor of the same conductivity would give. The number of sub-conductors per phase can be 2,3,4, or even 8. 400kV lines use only 2 or 4 sub-conductors per phase. The 765 kV lines in USA and other countries use upto 8 sub-conductors per phase. The advantages of bundled conductors have been outlined in section 1.5.

When bundled conductors are used, spacers have to be installed at intervals of 80 —100 m in each span to maintain the spacing between the sub-conductors. This is necessary to prevent the conductors from getting stuck-up due to wind forces or electromagnetic induction introduced by heavy currents. Since the electromagnetic forces are proportional to the square of current and vary in an inverse proportion to spacing, load current is sufficient to hold the conductors in contact once they have come near each other. The increase in surge impedance loading with the increase in the spacing between the sub-conductors is small and, therefore, the intra bundle spacings are not very critical and 45 cms is the most commonly used value throughout the world. It may be added that bundled conductors are subject to greater ice and wind loadings than single conductors of the same total cross-section and, therefore, require somewhat stronger towers.

14.4 CONDUCTOR SURFACE GRADIENTS

It is necessary to calculate the electric field at the surface of the conductors in order to compute the corona loss and radio interference associated with EHV lines. In section 6.8 we have derived an expression to calculate the electric field at the surface of the sub-conductors of a single circuit line with horizontal configuration and having two sub-conductors per phase. For other conductor configurations and for lines having more than two sub-conductors per phase, analytical solutions tend to become very complex and the use of digital computers becomes necessary.

A simple and quite accurate method for calculating surface gradients for bundled conductor lines has been suggested by Edison Electric Institute (USA). To evolve this method, known as base case method, a typical set of EHV line dimensions (conductor diameter, height, phase spacing) for 345, 500 and 735 kV lines having 1,2,3 or 4 sub-conductors per phase was selected. The gradients for each of these lines were calculated using a digital computer. Then the effect of variations in the conductor diameter, conductor height, phase spacing, etc., on the gradient were determined and charts were developed to calculate the multiplying factors for these variations about the base case. The conductor surface gradient for a particular case is found by multiplying the base case gradient by the appropriate multiplying factor. It has been reported that this method gives results which are within 0.6% the actual values. For the details of this method, the reader is referred to [12] of the bibliography.

14.5 CORONA LOSS

Many formulae have been suggested for calculating corona loss on transmission lines. Two of these (Peek's formula and Peterson's formula) are given in Chapter 6. In general the magnitude of fair weather corona loss is generally small for most of the EHV designs. However, some designers consider it to be economically significant because the fair weather loss exists for a very large percentage of total operating time. The corona loss under bad weather conditions may be as high as 100 times the fair weather corona loss. However, the bad weather persists for only a short time during the year.

For estimating the corona losses during bad weather (rainy season) for EHV lines, the following formula has been suggested:

$$\Sigma P = P_{FW} + \left[\frac{V}{1.6} \cdot J r^2 \ \text{In}(1 + 10R)\right] \sum_{1}^{n} (E^5) \qquad (14.1)$$

where ΣP = total 3-phase corona loss, kW per 3-phase km.

P_{FW} = total fair weather corona loss, kW per 3-phase km.

V = line to neutral voltage, kV

$J = 7.04 \times 10^{-10}$ for 400 kV lines

$= 5.35 \times 10^{-10}$ for 500 kV and 765 kV lines

r = sub-conductor radius, cms.

n = total number of sub-conductors (i.e., number of sub-conductors per bundle × 3)

E = gradient at the bottom of each sub-conductor, kV peak/cm.

R = rain rate in mm per hour

The calculation of E at the bottom of each conductor is done as per the method discussed in section 14.4. Through the use of the factor $\sum_{1}^{n} E^5$ the gradient at the bottom of each conductor is calculated individually and raised to the power 5 and then summation is done for all the sub-conductors. Eq. (14.1) can also be used for calculating corona loss under conditions of snow and frost. However, the value of J should be changed to 1.27×10^{-9} and the following equivalent rain rates should be used for these conditions.

For heavy snow $R = 2.54$ mm per hour

For medium snow $R = 0.636$ mm per hour

Light snow $R = 0.0127$ mm per hour

For calculating the total corona loss of a line for one year the calculation of fair weather and bad weather corona losses is not sufficient. A detailed method to include the meteorological conditions at different times during the year to compute the hour to hour corona loss for the total duration has been suggested.

14.6 ENVIRONMENTAL ASPECTS IN EHV AND UHV LINES DESIGN

Some environmental aspects become important in the design of EHV and UHV lines. These aspects need special attention for lines of 765 kV and higher voltages. Apart from visual impact the main environmental impacts are:

1. Route clearance through country side and forests.
2. Biological effects of electrostatic and electromagnetic fields.
3. Audible noise.
4. Ozone emission.
5. Radio and television interference.

14.6.1 Route Clearance

EHV and UHV lines are generally constructed over different terrain including hilly areas, dense forests and farm land and it is important that the minimum right of way is cleared. Generally the height of trees does not exceed 10m. It is necessary to see if trees could be saved by compact designs and efficient stringing methods.

It has been found that a live tree is well grounded through its sap, has small power loss and little chance of burning. However, a dead tree without sap offers a high resistance to flow of induced currents. A conductive surface due to rain and dew may be formed on the dead trees and in the process of current discharge high voltage concentration resulting in arcing and fire may occur. Therefore, dry trees should be cleared from the right of way of transmission lines. Moreover, studies have indicated that the dry tree burning can occur if an induced potential of more than 100 kV occurs. The potential of 100 kV can exist within a 6m radius from 765 kV line conditions. Therefore, atleast 6m tree top clearance from 765 kV line conductors is necessary in forest area.

14.6.2 Biological Effects of Electrostatic and Electromagnetic Fields

EHV and UHV lines generate electrostatic and electromagnetic fields. Voltage gradients at ground level must be kept within safe limits to avoid excessive charging currents, induced voltages and other undesirable effects.

(a) Calculation of electrostatic field

The charges Q on the conductors are determined from the conductor voltages V and Maxwell's potential coefficients P by the equation

$$[Q] = [P]^{-1}\,[V] \tag{14.2}$$

For an n conductor system, P is an $n \times n$ matrix whose diagonal terms P_{nn} and off-diagonal terms P_{nm} are given by

$$P_{nn} = \frac{1}{2\pi\varepsilon}\ln\left(\frac{4\,H_n}{d_n}\right) \tag{14.3}$$

$$P_{nm} = \frac{1}{2\pi\varepsilon}\ln\left(\frac{L'_{nm}}{L_{nm}}\right) \tag{14.4}$$

where H_n is height of conductor n above ground, d_n is diameter of conductor n, L_{nm} is distance between conductors n and m and L_{nm} is the distance between conductor n and image of conductor m.

For a bundled conductor line, it is sufficient to consider an equivalent single conductor whose diameter d_{eq} is given by

$$d_{eq} = D\,(Nd/D)^{1/N} \tag{14.5}$$

where D is the diameter of a circle on which the sub-conductors lie, N is the number of sub-conductors and d is diameter of each sub-conductor.

The voltage gradient at ground level due to conductor n is given by

$$G_n = \frac{q_{rn} + j\,q_{in}}{2\pi\varepsilon}\left(\frac{2H_n}{H_n^2 + L_n^2}\right) \tag{14.6}$$

where q_{rn} and q_{in} are the real and imaginary components of charge on conductor n and L_n is the horizontal distance between the conductor n and the point on ground at which gradient is to be calculated. The total gradient is

$$G = \Sigma\, G_n \tag{14.7}$$

Generalized curves have been developed for computation of maximum voltage gradient at ground level [12].

(b) Effects of electrostatic field

Electrostatic field can induce currents in animals, birds and human beings. The tolerable limits of body currents are discussed in section 18.3.

(c) Electrostatic field limits

The maximum value of electric field, at ground level, should not exceed the following values.

For difficult terrain 20 kV/m

For non-populated areas 15-20 kV/m

For road crossings 10 kV/m

(d) Calculation of electromagnetic field

The magnetic flux density at a point at distance D_n from a conductor n carrying current I_n is given by

$$B_n = \frac{\mu_0}{2\pi D_n} I_n \tag{14.8}$$

where B_n = flux density, T

$\mu_0 = 4\pi \times 10^{-7}$

The field B_n can be resolved into x and y components. The resultant x and y components (considering all the conductors) can be found by summation.

Whereas the electrostatic field increases with increase in line voltage, the magnetic field depends only on the line current. The magnetic field due to a distribution line may be high as that due to an EHV line.

(e) Effects of electromagnetic fields

It has been found that very low frequency magnetic fields even of weak intensities can affect certain cellular processes. When the induced current densities are below 10 mA/m^2 no significant biological effects are produced. Currents 10–100 mA/m^2 cause minor biological effects. Above 100 mA/m^2 (i.e., 60 m T at 50–60 Hz) various stimulation thresholds are exceeded and health hazards exist. Alteration in cell tissues have been reported above 10 mA/m^2.

(f) Limits of electromagnetic field

According to WHO no significant biological effects are observed from induced currents below 10 mA/ m^2 corresponding to about 5 mT flux density. However, International Radiation Protection Association has approved a limit of 0.1 mT.

14.6.3 Audible Noise

Transmission line audible noise has two characteristic components (1) broad band noise (2) pure tone components at frequencies of 100 Hz and multiples. The pure tones are superimposed on the broad band noise.

The broad band noise is caused by a random sequence of pulses produced by partial discharges (corona) in air at the surface of conductor. In dry conditions the conductors usually operate below corona inception level and corona is almost absent. However in wet conditions, water drops on the conductor surface produce corona discharges, each one of them creating a burst of noise. Below 500 kV, audible noise levels are of little consequence. However the lines of 500 kV and above must be designed for a satisfactory audible noise level.

14.7 RADIO NOISE FROM EHV LINES

Radio noise is a very important consideration in the design of EHV and UHV lines. Since radio noise is associated with corona it is mainly dependent on the potential gradients at the conductors. However it is also influenced by all the other factors, which influence corona, such as air density, humidity, wind contaminants, imperfections and precipitation. Extensive studies and radio noise measurements have been carried out in USA and other countries. It has been reported that for lines in the 750 kV class, the fair weather radio noise at 15 m from the line is about 40 – 45 dB, the extreme values being 30 and 70 dB. The higher figures pertain to older lines in the hilly areas. The radio noise in the bad weather is about 10 –25 dB higher than the radio noise during fair weather. The exact calculation of radio noise from a line is a difficult exercise. Computer programs have been developed for this purpose. As an alternative to the computer-based calculations some empirical formulae have also been developed. One of these is

$$RN = 50 + K(E_m - 16.95) + 40 \log \frac{d}{3.93} + F_n + 32 \log \frac{20}{R} + F_{fw} \qquad (14.9)$$

where RN = radio noise in dB above 1 μV/m at 1 MHz

K = 3 for 750 kV class

= 3.5 for others, gradient limits 15 –19 kV/cm

E_m = maximum electric field at conductor, kV-rms/cm

d = sub-conductor diameter, cm

F_n = – 4 for single conductor

$$= 10 \log \frac{n}{4} \text{ for } n > 1,$$

n = number of sub-conductors in a bundle

R = distance from line, m

F_{fw} = 17 for bad weather and zero for fair weather.

The effect of radio noise on the amplitude modulated radio receivers depends on the level of the radio noise and the field strengths of the desired radio signals at the location of the receiver. For good radio reception, the desired signal level should be 20 to 25 dB over the noise level. If radio interference is particularly troublesome for few selective frequencies at some locations, selective filters can be installed on the lines to suppress these frequencies. A quarter wave trap has been used successfully in USA. This is an open ended line section, one quarter wavelength long, connected in shunt with the line. The effect is to place a short circuit on the line for the frequencies to be filtered.

The noise due to corona processes in EHV lines is not a serious problem for frequency modulated radio receivers or TV receivers. However the TV reception in the immediate vicinity of a UHV line is likely to be affected especially during bad weather. Fortunately the higher frequencies in the TV bands are attenuated at a greater rate than those in the standard amplitude modulated bands and lower frequencies.

14.8 INSULATION REQUIREMENTS OF EHV LINES

The insulation strength of transmission lines must be chosen after a careful study of all the types of over voltages which can stress the insulation during the operation of the line. The over voltages can be classified as switching overvoltages, temporary overvoltages and overvoltages due to lightning. For EHV and UHV lines the switching over voltages become the limiting factor in the choice of insulation levels and arrester characteristics.

14.8.1 Switching Overvoltages

The switching surges, vital to the insulation design are initiated due to energization and reclosing, fault interruption, dropping of lines, load rejections and out of phase switching.

The wave shapes of the switching surges are practically infinite. Since the wave shape has a definite influence on the flashover strength, a standard wave shape has been selected for analyzing the line performance. This standard switching surge has a crest time (i.e., time to reach the peak value) of 175, μ-secs. and a tail time (i.e., time to reach half the peak value on the tail) of 3200 μ-secs.

It has been found desirable and possible to reduce the switching surges for EHV systems to about 2.5 *pu* and insulation of EHV lines is usually designed for this strength.

The problem of designing the insulation is complicated due to the great variety of meteorological conditions which the line has to experience and which affect the insulation strength itself. The four major meteorological conditions which affect switching surges are: relative air density, humidity, rain and wind. Appropriate corrections have to be applied for each of these conditions occurring individually and collectively.

Two basic methods are used for the insulation design. These are known as handbook method and reliability method. In the handbook method a certain high value of surge strength (about 2.5 *pu)* and some representative combination of meteorological conditions are selected and the insulation is designed for these conditions with the assumption that if the insulation can withstand these conditions, it will give a satisfactory performance for all conditions. In the reliability method an effort is made to assess the relative frequency of occurrence of all the combinations of electrical and meteorological events. The effect of increase of insulation strength on the line performance is evaluated and the insulation strength is decided on the basis of satisfactory performance and reasonable investment. The reliability method is more accurate but usually requires a digital computer and a comprehensive data for meteorological and transient voltage conditions. The handbook method is considered to be sufficient for most practical purposes and the insulation design by this method can be summarized in the following steps:

(i) Calculation of maximum switching overvoltages to ground. This is generally taken as 2.5 times the peak value of the line to neutral voltage of the system.
(ii) Selection of the number of insulator discs on the basis of experience.
(iii) Calculation of the critical flashover voltage for the above number of insulator discs.
(iv) Multiplication of the critical flashover voltage by suitable factors to account for meteorological conditions and flashover on a number of towers simultaneously.
(v) Computing the withstand probability and revising the design if necessary.

Standard curves have been drawn to aid the design engineer. For the details of the design and these curves, the reader is referred to [10] of the bibliography.

14.8.2 Temporary Power Frequency Overvoltages

The temporary overvoltages at power frequency are due to sudden loss of load, disconnection of inductive loads, connection of capacitive loads, etc. By the use of shunt reactors, etc., it is possible to limit these voltages to 1.5 *pu.*

It has been found that if the insulation design of an EHV line is adequate for switching surges, it is generally adequate for normal power frequency overvoltages. However some flashovers have occurred on some lines without any evidence of switching or lightning. The flashovers generally occur in the early morning hours when dew or fog is present. The cause for these flashovers is an interaction of a thin layer of contaminant on the insulator surface with a deposit of moisture. Therefore, the number of insulator discs selected on the basis of switching surge requirements should be checked for satisfactory performance under different contamination conditions.

14.8.3 Lightning Overvoltages

The problem of predicting the lightning performance of a line is to estimate the most probable number of outages due to lightning per 100 km per year. The lightning outage of a line depends on many probabilities, viz., probability of a thundercloud coming over a line, probability of a lightning stroke occurring from the thundercloud to the line, the probability that the stroke will land at a particular spot on the line (i.e., tower, mid span, etc.), the probability that the magnitude of current due to lightning stroke will be beyond a certain magnitude, the probability that the magnitude and polarity of the power frequency line voltage at the instant of lightning stroke will be such as to aid the initiation of the breakdown streamers across the insulator and the probability of occurrence of a flashover.

On the basis of detailed studies of lightning behaviour of existing lines and research work, graphs have been developed to compute the lightning performance of lines. It has been found that overvoltages due to lightning are not very important for EHV and UHV lines.

14.9 TUNED POWER LINES

Hydroelectric and super thermal power plants are generally located at great distances from the load centres. It is necessary to use extra long distance (ELD) point to point transmission lines for transporting electrical energy from these plants to load centres. The design of ELD lines is becoming increasingly difficult due to the high cost of compensation equipment. To provide sufficiently stable margin of safety, the phase shift over the line should be only about 30°. At 50 Hz, a 500 km line at surge impedance loading has an angle of about 30°. To maintain an angle of 30° for lines longer than 500 km at SIL, 100 per cent compensation is required for the length beyond 500 km. Therefore, as the line length increases the capital cost per unit length of line (including the cost of compensation equipment) increases. In addition to high capital cost, extra long lines produce many operating problems such as surplus vars during light load conditions, ferranti effect and generator self excitation. Tuned power lines provide one solution to this long distance power transmission problem.

We have derived the performance equations of the long lines in Chapter 3. Referring to the nomenclature used in Chapter 3, we have

$$\gamma = \sqrt{zy} = \sqrt{(r + j\omega L)(G + j\omega C)}$$

The shunt conductance G is always negligible. Since r is also small as compared to ωL, r can also be neglected.

$$\text{Then } \gamma = j\,\omega\sqrt{LC} \tag{14.10}$$

The velocity of propagation 'v' of wave on the line is given by

$$v \simeq \frac{1}{\sqrt{LC}} \tag{14.11}$$

The velocity v, frequency f and wavelength λ are related by

$$v = f\lambda \tag{14.12a}$$

Combining Eqs. (14.13, 14.14 & 14.15a)

$$\gamma = j\,(2\,\pi/\lambda) \tag{4.12b}$$

$$\cosh \gamma l = \cosh j\left(\frac{2\pi}{\lambda}\right)l = \cos\left(\frac{2\pi}{\lambda}\right)l \tag{4.13a}$$

$$\sin \gamma l = \sinh j\left(\frac{2\pi}{\lambda}\right)l = j\sin\left(\frac{2\pi}{\lambda}\right)l \tag{4.13b}$$

Then the performance equations of the long line (Eq. 3.42) are

$$\begin{bmatrix} V_s \\ I_s \end{bmatrix} = \begin{bmatrix} \cos\left(\frac{2\pi}{\lambda}\right)l & j\,Z_c \sin\left(\frac{2\pi}{\lambda}\right)l \\ j\,\frac{1}{Z_c}\sin\left(\frac{2\pi}{\lambda}\right)l & \cos\left(\frac{2\pi}{\lambda}\right)l \end{bmatrix} \begin{bmatrix} V_r \\ I_r \end{bmatrix} \tag{14.14}$$

If $\quad l = \dfrac{\lambda}{2}, \lambda, \dfrac{3\lambda}{2}$

$$|V_s| = |V_r| \tag{14.15a}$$

$$|I_s| = |I_s| \tag{14.15b}$$

Under these conditions the receiving end voltage and current are numerically equal to the sending end voltage and current respectively. Such lines are known as tuned lines.

Since $v = 3 \times 10^8$ m/s, the wavelength of a 50 Hz line is

$$\lambda = \frac{3\times10^8}{50} = 6 \times 10^6 \text{ m or } 6000 \text{ km}$$

Therefore, the length of a tuned line can be 3000 km, 6000 km, 9000 km, etc. These lengths are too long for power transmission from the point of view of cost and efficiency. Nevertheless some power engineers have investigated the characteristics of a 3000 km line. This line is also known as half wavelength line. Salient features of half wavelength line are:

1. The general line constants for an exact half wavelength line are $A = D = -1$ and $B = C = 0$. This means that such a line presents zero series impedance and zero shunt admittance.
2. The receiving end voltage and current are equal in magnitude and 180° out of phase with sending end voltage and current, i.e.,
$$V_r = -V_s \text{ and } I_r = -I_s$$
3. The line is neither a sink nor a source of vars (i.e., $V_s \cdot I_s^* \cong V_r I_r^*$)
4. There is no ferranti effect when the line is open circuited at the receiving end.

The electrical length of a half wavelength line is 180°

The electrical length of the line (including transformation equipment) should be about 10° more than one half of a wavelength due to two important reasons. The first reason is to reduce the required sensitivity of the generator var output control and the second is to ensure a half wavelength transmission during temporary reduced frequency operation. The wavelength is inversely proportional to frequency. If frequency drops by 5 per cent, wavelength increases by 5 per cent, thereby electrically shortening the line. If a sufficient margin of safety is not provided, loss of power flow may occur. This means that if the line is electrically lengthened so that its effective electrical length is between 180° and 270°, the system is as stable as the one operating in the first quadrant (i.e., 0 to 90°).

The length of a half wavelength line is 3000 km. This is too long a distance of transmission from the point of view of cost and efficiency. Changing the line parameters such as conductor size, grouping, phase spacing, etc., has practically no effect on wavelength. However, it is possible to increase the electrical length of an actual line by using tuning equipment. Two methods are available for this purpose. In the first method the tuning equipment is placed at the ends of the line. The tuning equipment can be placed in π sections (Fig. 14.2 a) or T sections (Fig. 14.2b). In the second method, shunt capacitors are added along the line as shown in Fig. 14.2 (c).

It has been found that the tuning methods are not very suitable at power frequency. Therefore, tuned power lines are not used. However, tuned lines are used in telephony where higher frequencies are used.

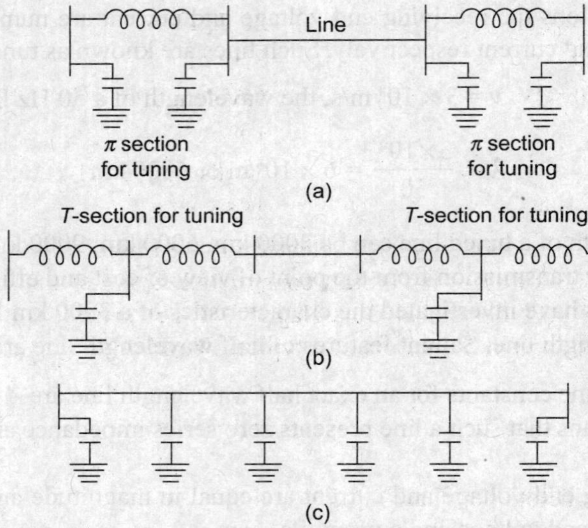

Fig. 14.2 Tuning equipment for power line (a) π section tuning (b) T section tuning (c) Capacitance tuning

14.10 EHV LINES IN INDIA

(a) **400 kV lines :** In India the first 400 kV line was commissioned in 1978. A total of about 62000 circuit kilometers of 400 kV lines have been constructed. Some of these are Dehar Panipat line. Obra-Sultanpur line, Sultanpur-Lucknow line, Obra-Kanpur line, Kanpur-Moradnagar line and Koradi-Kalwa line. Another 28000 circuit kilometres of 400 kV lines are planned for commissioning in the 12th plan. Data of a typical 400 kV line is as under:

Nominal voltage	400 kV
Circuits per phase	1
Number of sub-conductors phase	2
Span	400 m
Conductor name and size	Moose, 54/3.53 mm Aluminium, 7/3.53 mm steel
Sub-conductor diameter	3.177 cm
Bundle configuration	Horizontal
Bundle spacing	450 mm
Interphase spacing	11m
Conductor configuration	Horizontal
Resistance per phase per km at 20 °C	0.0274 ohm
Inductive reactance per phase per km	0.3321 ohm
Shunt admittance per phase per km	3.2983×10^{-6} ohm
Surge impedance	282 ohm
Surge impedance loading	505 MW
Current carrying capacity at 40°C ambient temp.	900 A
Ground wire	2 number, 7/3.66 mm galvanized steel
Ground wire height at tower	30.4 m
Tower height	30.4 m

Conductor height at tower	20.75 m
Conductor sag at 0 °C	8.39 m
Ground wire sag at 0 °C	6.654 m
Shielding angle	10°
Suspension string	23 discs, 255 × 145 mm
Impulse flashover voltage	1550 kV

(b) *765 kV lines* The next higher voltage selected for the Indian grid is 765 kV. About 1200 circuit kilometres of 765 kV lines have been constructed. Some of these are: Anpara-Unnao (UP), Tehri Meerut (UP), Vindhayachal-Bina-Nagda (MP), etc. One 765 kV line from Seoni to Sipat MP was commissioned in 2008 Data of a typical 765 kV line is as under:

Nominal voltage	765 kV
Line length	430 km
Circuits per phase	1
Number of sub-conductors per phase	4
Span	400 m
Conductor name and size mm	Moose, 54/3.53 mm, Aluminium, 7/3.53 Steel
Sub-conductor diameter	3.177 cm
Bundle configuration	Square
Bundle spacing	450 mm
Interphase spacing	14 m
Resistance per phase per km. at 20 °C	.0137 ohm
Inductive reactance per phase per km	0.2815 ohm
Shunt admittance per phase per km	3.95×10^{-6} ohm
Conductor configuration	Horizontal
Surge impedance	267 ohm
Surge impedance loading	2194 MW
Ground wire	2 number, galvanized steel 7/3.66 mm
Tower height	38.74 m
Conductor height at tower	25.5 m
Conductor sag at 0 °C	11.08 m
Ground wire height at tower	38.74 m
Shielding angle	12.68°
Suspension string	32 discs, 280 × 170 mm
Tower footing resistance	10 ohms
Impulse flashover voltage	2400 kV

SUMMARY

14.1 EHV lines are needed to transmit power from pithead generating stations to load centres.

14.2 Lines between 300 and 765 kV are called EHV lines and those with voltage above 765 kV are called UHV lines.

14.3 EHV lines use bundled conductors to reduce corona and ratio interference.

14.4 Many enviormental aspects are important in the designing EHV and UHV lines.

14.5 Electromagnetic and Electrostatic field at ground level must be kept within safe limits: Audible noise should also be low.

14.6 Radio noise from EH lines can be found only by computer insulation or by following empirical formulae.

$$RN = 50 + K\,(E_m - 16.95) + 40 \log \frac{d}{3.93} + F_n + 32 \log \frac{20}{R} + F_{fw}$$

14.7 Insulation level of EHV lines is based on switching overvoltage.

14.8 EHV lines invariably require static var system to maintain proper voltage profile.

14.9 SVS systems are: self saturated reactor. Thyristor controlled reactor — fixed capacitor, Thyristor switched capacitor — Thyristor controlled reactor.

14.10 Series compensation increases transmission capacity and improves stability. However series compensation has a number of problems associated with it.

SELF TEST QUESTIONS

A State whether the following are true (T) or false (F)

14.1 765 kV lines are called UHV lines.

14.2 Pithead plants need long transmission lines to evaluate power.

14.3 Intra bundle spacing is about 80 cm.

14.4 Lines with bundled conductors have low surface voltage gradient.

14.5 EHV lines may cause high electrostatic field.

14.6 Radio noise from EH lines is expressed in dB above 1 μ v/m at 1MHz.

14.7 SVS reduces var flow on lines.

14.8 SVS use Thysistors for switching.

B Fill in the blanks in the suitable words

14.9 Series compensation increases ________

14.10 Series compensation is usually limited to about ________

14.11 For 50 Hz, the wave length is ________

14.12 For a half wave length constants B and C are ________

14.13 The electrical length of a half wave length line is ________

14.14 SIL increases with line ________

14.15 As the line voltage increases the line costs per MW per km ________

ANSWERS

14.1 F	14.2 T	14.3 F	14.4 T	14.5 T	14.6 T
14.7 T	14.8 T	14.9 transmission capacity	14.10 40%	14.11 6000 km	
14.12 zero	14.13 180°	14.14 voltage	14.15 decrease		

JOB INTERVIEW QUESTIONS

14.1 Why do EHV lines need bundled conductors?

14.2 What are switching over voltages?

14.3 What is meant by SVS?

14.4 What is the necessity of EHV lines?

14.5 What are tuned power lines?

14.6 What is series compensation? Why is it required?

14.7 What are the advantages of SVS?

14.8 What is meant by Radio noise from EHV lines?

REVIEW QUESTIONS

14.1 Discuss why?

(a) EHV and UHV lines are necessary to transmit large blocks of power over long distances.

(b) Insulation design of EHV lines is governed by switching surges.

(c) EHV lines invariably need shunt and series compensation.

(d) Series compensation leads to improvement of stability.

(e) An EHV line requires both generation and absorption of vars.

(f) Static compensation is preferred to synchronous compensation in modern power systems.

14.2 What is shunt compensation? Why is it necessary? Discuss the types of static var system used in EHV transmission.

14.3 What are the advantages of series compensation? What are the problems associated with its use?

14.4 On what factors does the design of the insulation systems of a transmission line depend? Which factor is the most important in EHV systems? Briefly discuss the procedure for selecting the number of insulator discs for an EHV line.

14.5 Why is the radio interference aspect important in the design of EHV systems? How can radio interference be estimated?

14.6 Discuss the following:

(a) Use of bundled conductors for transmission lines.

(b) Estimation of conductor surface gradient for EHV lines.

(c) Corona loss under bad weather conditions and its estimation.

14.7 What is a tuned line? What are its features? Discuss the possibility of using tuned lines in power systems.

CHAPTER 15

HVDC Transmission

15.1 INTRODUCTION

The initial application of electricity started with the use of direct current. The first electric power station in the world was commissioned in 1882 in New York (USA). This was a dc station and supplied 110 V dc to an area of 1 mile radius. This station used Edison bipolar dc generators. Within a few years of the commissioning of this power station, many other dc power stations were built and commissioned in many cities of the world.

The invention of transformer and induction motor and the concept of 3-phase ac around 1890 initiated the use of ac. An era of ac power systems started following this development. The transformer being simple, rugged and efficient, made the availability of different voltage levels for generation, transmission, distribution. It made the long distance high voltage power transmission possible. The induction motor, especially the three-phase one, has a very simple and rugged construction and is the most widely used motor. The advantages of 3-phase ac almost eliminated the use of dc systems except for some specific applications in electrolytic processes and adjustable speed motor drives.

The power systems all over the world have been expanded vastly during the past few decades. Now-a-days large blocks of power are transmitted over long distances from the remotely located power stations to urban load centres. The main technical problems of long distance power transmission using ac are: voltage regulation associated with reactive power balance, steady state, transient state and dynamic stability of the system under different load conditions and also under outage conditions. In view of these problems with ac, the dc transmission has staged a comeback, in the form of high voltage dc transmission to supplement the ac transmission system. The first dc link was set up in 1954 between Swedish mainland and the island of Gotland. This was a monopolar, 100 kV, 200 MW, cable system making use of sea return. In 1961 an under water dc link was set up between England and France. This was a bipolar, ± 100 kV, 160 MW, cable system. Since then more and more high voltage dc systems have been set up. The earlier dc schemes used mercury arc converters. Since 1972 all new hvdc systems use thyristors. Table 15.1 shows some HVDC systems in the world. In addition some other HVDC systems are: Chateaugua–Quebec 1000 MW system, North Dakota–Minnesota ± 250 kV, 500 MW system, Vancouver, 682 MW system, Intermountain, Utah - California 1920 MW system, Itapu, Brazil

6300 MW system, (all in American continent), Wyborg, (Russia) 1050 MW system, Kontek (Denmark–Germany) 600 MW system, Baltic cable (Sweden–Germany) 600 MW system, Etzenricht (Germany) 600 MW system, Hokkaido–Honshu (Japan) 600 MW system, Shikoku–Kansai (Japan) 3400 MW system, North and South islands (New Zealand) 1240 MW system and Gezhuba–Shanghai (China) 1200 MW system. At present the world has over 60 HVDC schemes in operation for a total capacity of more than 66000 MW and the capacity is increasing by about 2000 MW every year.

Table 15.1 Some HVDC system in world

No.	*System*	*Rated voltage (kV)*	*Rated power (MW)*	*Length (km)* OH line	*Cable*	*Total*	*Year of Commissioning*	
1.	Gotland (Sweden)	100–150	269	0	96	96	1954/70	Long distance submarine cable
2.	Cross Channel (UK-France)	± 100	2000	0	65	65	1961	Long distance submarine cable
3.	Volgograd Donbass (USSR)	± 400	720	470	0	470	1965	Long distance OH transmission
4.	Konti-Skan (Denmark-Sweden)	250	550	95	85	180	1965	Long distance cable
5.	Sakuma I (Japan)	2 × 125	300	0	0	0	1965	Asynchronous link
6.	Pacific Intertie (USA)	± 400	3100	1362	0	1362	1970	Long distance OH transmission
7.	Nelson River Bipole I (Canada)	± 300 ± 450	1080 2340	890	0	890	1973	Long distance OH transmission
8.	Kingsnorth (UK)	± 266	640	0	82	82	1975	Underground cable
9.	Cabora Bassa (Mazambiquie-South Africa)	± 266	960	1414	0	1414	1975	Long distance OH line
		± 400	1440				1977	
		± 533	1920				1979	
10.	Rihand-Delhi (India)	± 500	1500	810	0	810	1991	Long distance OH line
11.	Vindhyachal (India)	70	2 × 250	0	0	0	1991	Back to back station (Asynchronous link)

The use of an HVDC link in an ac system requires converter stations at each end of the line. The main equipments in a converter station are transformers and thyristor valves. At the sending end the thyristor valves act as rectifiers to convert ac into dc which is transmitted over the line. At the receiving end the thyristor valves act as inverters to convert dc into ac which is utilized at the receiving end. Each converter can function as rectifier or inverter and, thus, power can be transmitted in either direction.

15.2 TYPES OF DC LINKS

The dc links can be classified into the following types:

(a) *Monopolar Link:* This link has only one conductor, usually of negative polarity and uses ground or sea water as the return conductor. The negative polarity is preferred on overhead lines due to lesser radio interference. Figure. 15.1 (a) shows a monopolar link.

(b) *Bipolar Link:* This link has two conductors, one positive and the other negative. At each terminal, two converters of equal rated voltages are connected in series on the dc side. The neutral points (i.e., the junctions between converters) are grounded, at one or both ends (Fig. 15.1b). If both the neutrals are grounded, the two poles operate independently. If the currents in the two conductors are equal, the ground current is zero. If a fault develops on one conductor, the other conductor (along with ground return) can supply half of the rated load. The rated voltage of bipolar link is expressed as (say) ± 500 kV.

(c) *Homopolar Link:* A homopolar link has two (or more) conductors, all having the same polarity (usually negative) and always operates with ground as the return conductor. If a fault develops on one conductor, the converter equipment can be reconnected so that the healthy conductor (having some overload capacity) can supply more than half the rated power. Such a reconnection is very complicated in a bipolar scheme and, therefore, a homopolar scheme is preferable to a bipolar scheme provided continuous ground return does not pose additional problems. Figure. 15.1 (c) shows a homopolar link.

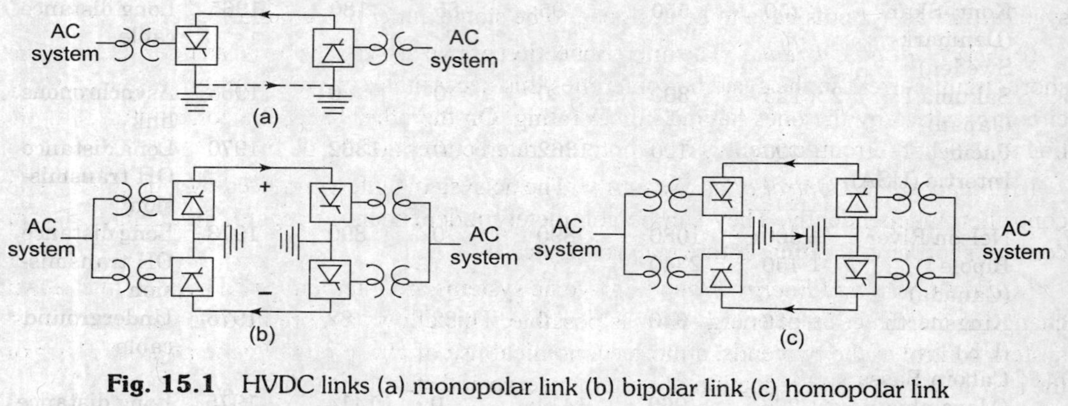

Fig. 15.1 HVDC links (a) monopolar link (b) bipolar link (c) homopolar link

15.3 ADVANTAGES OF DC TRANSMISSION

15.3.1 Technical Advantages

DC transmission possesses many technical advantages over ac transmission. Some of these are:

(a) *Reactive Power Requirement:* The production and consumption of reactive power by the line constitutes a serious problem in long overhead lines. The reactive power generated by the line capacitance equals the reactive power absorbed by the line inductance only when the load impedance is equal to the surge impedance of the line. Since the loads vary with time, the line cannot always be operated at its natural load. If the load is greater than the natural load, net reactive power is consumed by the line and must be supplied from one or both ends. The reactive power required for a given variation of load increases with line length. At times of light load the reactive power absorbed by the line inductance is less than the reactive power generated by line capacitance and net reactive power is generated by the line. Thus, to maintain constant equal voltages at both ends, reactive power must be absorbed during light load periods and must be supplied during high load periods. In dc transmission no reactive power is transmitted over the line. The voltage drop on the line is merely the resistive drop. However, the converters at the two ends of the line draw reactive power from the ac systems. This reactive power varies with the transmitted power but is independent of line length. It has been found that for lines greater than 400 km length, a dc line requires less reactive power than an ac line.

In case of ac transmission by underground cables the situation is different than on overhead lines. Cables are always operated at a load much below the surge impedance load in order to avoid overheating. As a result of this the reactive power generated by the shunt capacitance always exceeds the reactive power absorbed by series inductance. In a cable of about 50 km length, the charging current itself equals the load current leaving no margin for the load current. Therefore, the practical lengths of ac cable is only 40 km or so. DC cables have no such limitation.

(b) *System Stability:* An ac system must remain in synchronism. To maintain stability under transient condition, the length of a 50 Hz uncompensated ac line must be less than about 500 km. If series compensation is used, the length can be somewhat longer than this value.

The dc transmission has no such limitations and no stability problem. Two separate ac systems interconnected by a dc link do not operate in synchronism and may even operate at different frequencies. A breakdown of dc link does constitute a stability problem in ac systems interconnected by the dc link. This stability problem is similar to that caused by a sudden loss of large load for the sending end system and to that caused by a sudden loss of generation for the receiving end system. The ac systems have to be designed to be stable under this condition.

(c) *Short Circuit Current:* The interconnection of two ac systems by an ac line increases the short circuit current in the system. Sometimes this necessitates the replacement of the existing circuit breakers by the ones having higher rating. On the other hand, the contribution of a dc line to the short circuit current is only upto the rated current of the dc line.

(d) *Independent Control of ac Systems:* The ac systems interconnected by a dc line can be controlled independently. They can be completely independent as regards frequency, system control, short circuit rating, future extension, etc.

(e) *Fast Change of Energy Flow:* If two ac systems are interconnected by a dc line, a fast change in the direction of energy flow is possible. This allows a more flexible coordination of system control at the two ends, a more economical use of cheap power generation in either of the two ac systems and reduction of system reserve and standby capacity.

(f) *Lesser Corona Loss and Radio Interference:* The corona loss and radio interference, especially during bad weather, are lower for a dc line as compared to that in an ac line of the same conductor diameter and voltage.

(g) *Greater Reliability:* A two-conductor bipolar dc line is more reliable than a 3-conductor 3-phase ac line because in the event of a fault on one conductor, the other conductor of the dc line can continue to operate with ground return. A monopolar dc line with ground return is simpler than a 3-phase ac line and is equally reliable.

15.3.2 Economic Advantages

DC transmission offers many economic advantages over ac transmission especially for long distance bulk power transmission. Some of these are:

1. DC lines and cables are cheaper than ac lines and cables. However, the dc line terminal equipment is considerably more expensive than ac line terminal equipment. This is because of

the high cost of converters for the dc line. Figure. 15.2 shows a comparison of the total cost for dc and ac transmission. It is seen that for distances above the break-even distance, the total cost of dc transmission is lesser than the total cost of ac transmission. The break-even distance is generally different for different projects. For overhead lines the break-even instance generally lies in the range of 600—900 km. For submarine cables the break-even distance lies in the range of 25 to 50 km and for underground cables it lies in the range of 50 to 100 km. In some cases, it is necessary to adopt dc transmission, even for distances less than the break-even distance, because of technical advantages.

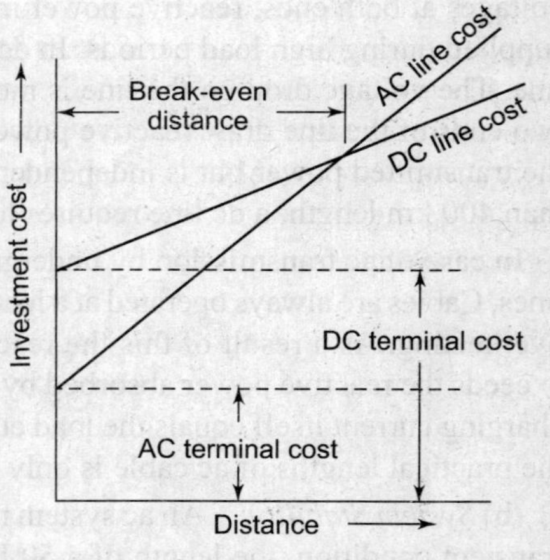

Fig. 15.2 Comparison of the costs of ac and dc transmission

2. A dc line is simpler and cheaper as compared to an ac line. The towers of dc lines are also simpler, cheaper and narrower as compared to the towers of ac lines. The right of way requirement of a dc line is about 20—40 per cent lesser than that of an ac line of the same power transmission capability.

3. The line losses in a dc line are lower than the losses in an ac line of the same power transmission capability.

4. An HVDC transmission line can be built in stages. Initially the line can be built as a monopolar line with ground return. Later on the line can be converted into a bipolar line. This aspect becomes very important when there is a long period between the successive stages because the investment for the future extension is not required to be made initially.

15.4 INCORPORATING HVDC INTO AC SYSTEMS

An HVDC link can be incorporated in an ac system in a number of ways. Each of these schemes is special in its own way and serves a specific purpose which is different from that of the other schemes.

15.4.1 Two Terminal DC Link for Point to Point Transmission

A two-terminal dc link using overhead lines is used to transmit large blocks of relatively cheap power from remote sources (generally hydro electric) to load centres situated far away from the generating stations. Such a scheme enables the development of remote power generation sites. Nelson River Scheme in Manitoba (USA) is a typical example of a two terminal dc link. This 890 km long ± 450 kV, bipolar, 1620 + 1080 MW (stage 1 and 2) link has enabled the development of hydro potential of Nelson river and the transmission of this hydro power to Winnipeg. Other example are: 1415 km, 1920 MW, ± 533 kV, Cabora Bassa (Africa) scheme, 560 MW, ± 500 kV, 1700 km, Inga-Shaba scheme (Zaire).

In some cases, the interconnecting link between ac systems has to use submarine or underground cables. Cable transmission with ac is possible for a distance of not more than 40 km due to the problem of the charging current. For longer distances HVDC is the only solution. The ± 100 kV, 2000 MW, 65 km long English Channel dc link between England and France is operating since 1961. The purpose of this interconnecting link is to take advantage of time zone and generation diversity.

Sometimes the power transmission in congested urban areas has to be done through underground cables. Where distances of more than 40 km are involved, dc transmission is the only feasible solution. The ± 266 kV, 640 MW, 82 km Kingsnorth-London scheme is a typical example of dc link using underground cables.

Some HVDC schemes combine the use of overhead line and submarine cable. The Sardinia-Italian mainland 200 kV, 300 MW, dc link consists of 292 km of overhead line and 121 km of submarine cable.

15.4.2 Back to Back DC Links

A back to back link consists of sending and receiving end converter stations situated close to each other and the dc line is of zero or negligible length. Such schemes are used to couple ac systems of different frequencies (e.g., 50 Hz and 60 Hz) or of different system controls, (i.e., asynchronous systems). An example of 50 Hz to 60 Hz back to back link is the 6300 MW link between Paraguay and Brazil. Another example is the Sakuma 300 MW scheme interconnecting the 50 Hz and 60 Hz systems of Japan. An example of a back to back link between asynchronous ac system (of the same nominal frequency) is the 550 MW Durnrohr scheme in Austria which interlinks the east and west European grids.

15.4.3 DC Link in Parallel with AC Link

Sometimes it is necessary to put a dc link in parallel with an existing ac link. Such a scheme is used to reinforce the ac link and to improve its stability. A dc link has an extremely rapid speed of response (of the order of 50 milli seconds) and, hence, it can change its transmitted power much more rapidly than (say) a turbine-generator can adjust its output. This feature can be used to improve the stability of an existing ac link.

In an ac system several modes of electro-mechanical oscillations may involve single machines (or generating stations) or a number of machines in a system or a number of areas connected by tie lines. The single machine or multi-machine oscillations within one system may have a frequency of 1 Hz or above but inter-area oscillations have low frequencies of 0.1 Hz to 0.5 Hz. Action can be taken to damp the oscillations within a system (or control area) but inter area oscillations pose more severe problems and in such cases a dc link can offer a good solution.

The maximum transmitted power capability of an ac link may be set by its steady state stability limit or transient stability limit. System studies have shown that connecting a dc link in parallel with the ac link raises the steady state and transient stability limits of the ac link. The 3100 MW, 1362 km pacific inter tie (USA) is a typical example of dc link set up in parallel with an ac link to improve the stability of ac link and to increase the power capability of ac link from 210 MW to 2500 MW.

15.4.4 Multi-Terminal DC Links

All large ac power systems are effectively of multi-terminal variety. However, almost all dc links set up so far are two terminal links. The only exception is the Kingsnorth-London scheme in which each half of the sending end power is independently controlled and delivered to two different receiving end stations.

A multi-terminal dc network having a number of rectifier and inverter stations is theoretically possible. Many analogue and digital studies for such networks have been carried out. One of the essential requirements in a multi-terminal dc scheme is the use of large dc circuit breakers.

The two-terminals dc link does not need to be equipped with dc circuit breakers because faults on the link can be cleared by the converter control action. However, a multi-terminal scheme needs dc circuit breakers. So far dc circuit breakers with adequate interrupting capacity have not been built though some prototypes have been developed and tested. A three-terminal HVDC link has been set up in Italy-Corsica-Sardina.

Figure 15.3 shows a three-terminal HVDC link. In this configuration converters 1 and 2 operate as rectifiers while converter 3 operates as inverter. Alternatively, with an automatic reversal of voltage along with firing angle control, converters 1 and 2 can operate as inverters and 3 as rectifier. Reversal of connections of 2 will result in a new configuration in which 1 operates as rectifier and 2 and 3 as inverters. This combination, with automatic reversal of voltage along with firing angle control results in 1 working as inverter while 2 and 3 work as rectifiers. Thus this 3-terminal configuration leads to a greater flexibility of operation.

The 2000 MW Radison Nicolet Sandy Pond three converter link connects the Quebec hydro electric system to New England through 1500 kilometre long transmission system. Power transfer is possible from Quebec to New England as also (during emergencies) from New England to Quebec.

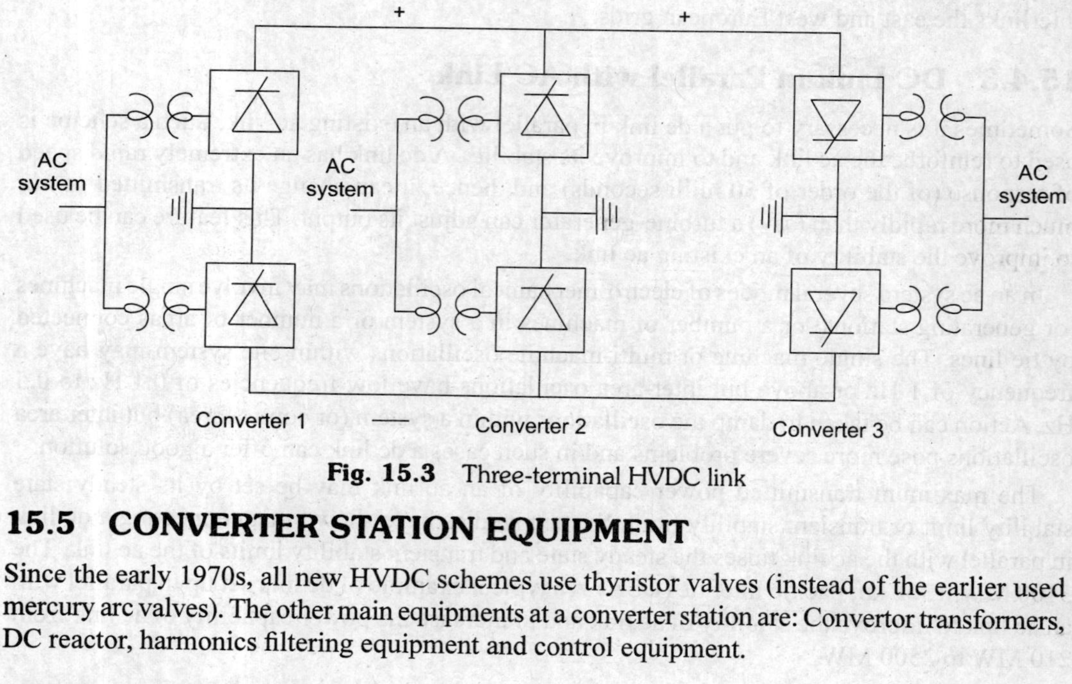

Fig. 15.3 Three-terminal HVDC link

15.5 CONVERTER STATION EQUIPMENT

Since the early 1970s, all new HVDC schemes use thyristor valves (instead of the earlier used mercury arc valves). The other main equipments at a converter station are: Convertor transformers, DC reactor, harmonics filtering equipment and control equipment.

15.5.1 Thyristor Valves

A thyristor has a better operating life and higher consistency of performance than a mercury arc valve but its rating should not be exceeded. The use of thyristors has led to a considerable simplification in the converter station design. The ratings of thyristors have increased remarkably during the last one decade. In the Ecl river project (Canada) set up in 1972 which was the first HVDC scheme to use thyristors, the thyristors had a rating of 70 kW of dc power per thyristor. In the Durnrohr project (Austria) commissioned in 1983, thyristors have a rating of 1000 kW of dc power per thyristor. The current ratings of the present day thyristors are of the magnitude of line current required for transmission. However, the individual thyristor voltage capability is very small as compared to the line voltage. Therefore, a large number of thyristors have to be connected in series to give the proper voltage rating. Thus, a thyristor valve is formed by a large number (typically 50 to 200) thyristors connected in series. The valves are of indoor design, air insulated and air or water cooled. Since the thyristors always fail to an internal short circuit, a thyristor valve is always equipped with a number of extra (redundant) thyristor in series so that even if some of the thyristors fail, the operation of the valve is not affected. In order to achieve a flexibility in design, the valves have a modular construction which makes it possible to use the same properly tested units for several applications. Each valve module usually contains 6 — 12 thyristors with their associated heat sinks, control pulse generating circuits and voltage dividing circuits (resistors and capacitors). The purpose of the voltage dividing circuits is to ensure a uniform voltage distribution along a chain of thyristors when the valve is non-conducting.

All present day HVDC schemes use 12-pulse converter. In a 12-pulse converter there are 12 pulsations (cycles of ripple) of direct voltage per cycle of alternating current Fig. 15.4 shows a converter station with two 12-pulse units forming a bipole. There are two transformers (3-phase) connecting each 12-pulse converter to the ac bus bars. One of the transformers is *Y*/*Y* connected and the other is *Y*/Δ connected in order to produce a 30° phase shift between the lower and upper 6-pulse bridge of the converter.

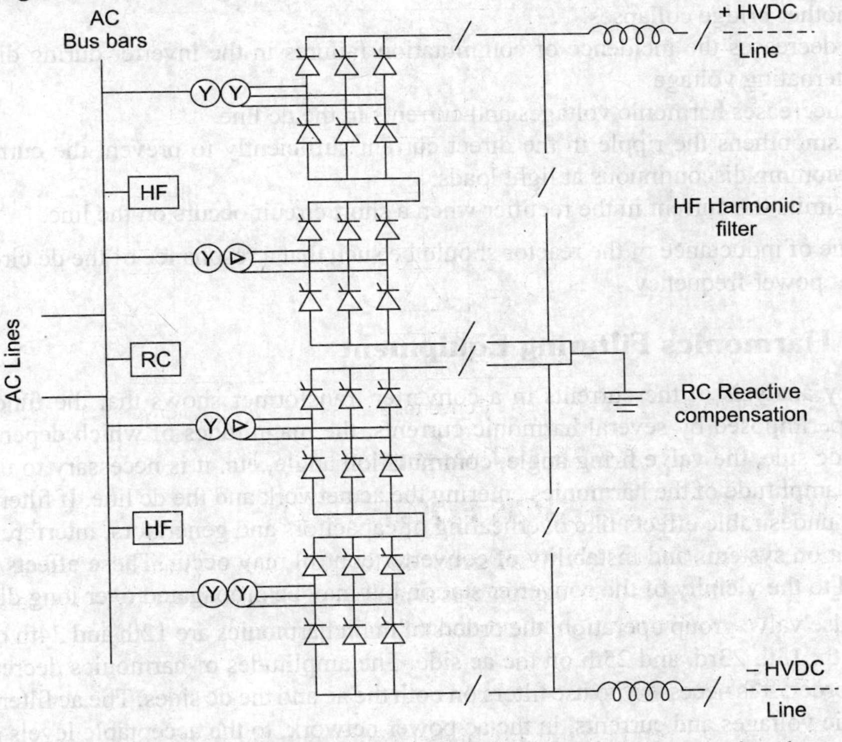

Fig. 15.4 Converter station with two 12 pulse units forming a bipole

15.5.2 Converter Transformer

The converter transformer transforms the ac voltage to a suitable value for feeding the converter. In addition it serves the following functions:

1. Reactive power is supplied to the converter through tap changing.
2. Short circuit currents are controlled by suitable impedance value of these transformers.

3. The reactance of the converter transformer helps in harmonic suppression.
4. By suitable star-star and star-delta connections the required 30° phase shift for 12-pulse operation is achieved.

A converter transformer may be designed as a 3-phase unit or bank of single phase units. For the 12-pulse converter, 3-phase units are used.

A converter transformer has a somewhat different design than that of a normal power transformer because in the converter transformer, the insulation system has to be designed to withstand direct voltage stresses. Moreover, in a converter transformer the currents have a high harmonic content so that special care has to be taken with regard to eddy current losses. Converter transformers used in large HVDC schemes have high ratings, about 300 MVA or even more.

15.5.3 DC Reactor

A dc reactor is connected in series with each pole of a converter station. It inductance lies in the range of 0.4 to 1 H. It serves the following purposes:

1. It prevents consequent commutation failures in the inverter by limiting the rate of increase of direct current during commutation in one bridge when the direct voltage of another bridge collapses.
2. It decreases the incidence of commutation failures in the inverter during dips in the alternating voltage.
3. It decreases harmonic voltages and currents in the dc line.
4. It smoothens the ripple in the direct current sufficiently to prevent the current from becoming discontinuous at light loads.
5. It limits the current in the rectifier when a short circuit occurs on the line.

The value of inductance of the reactor should be such that a resonance of the dc circuit does not occur at power frequency.

15.5.4 Harmonics Filtering Equipment

A frequency analysis of the currents in a converter transformer shows that the fundamental wave is superimposed by several harmonic currents, the magnitudes of which depend on the current on dc side, the valve firing angle, commutation angle, etc. it is necessary to use filters to limit the amplitude of the harmonics entering the ac network and the dc line. If filters are not used, some undesirable effects like overheating of capacitors and generators, interference with communication systems and instability of converter control may occur. These effects may not be confined to the vicinity of the converter station but may be propagated over long distances.

In 12-pulse valve group operation, the orders of main harmonics are 12th and 24th on the dc side and 11th, 13th 23rd, and 25th on the ac side. The amplitudes of harmonics decrease with increasing order. It is necessary to use filters on both the ac and the dc sides. The ac filters reduce the harmonic voltages and currents, in the ac power network, to the acceptable levels and also provide whole or part of the reactive power needed by the converter (the remaining reactive power being supplied by shunt capacitors or by the ac power system). The dc filters serve only to reduce harmonics on the dc side.

The ac filters which have been used in HVDC systems are: tuned filter which is sharply tuned to a harmonic frequency, damped filter which, if shunt connected, offers a low impedance over a broad band of frequencies (say 11th and higher order harmonics).The adoption of 12-pulse valve group operation has resulted in reduction of harmonic current generation and considerable simplification in ac filters. In systems with 6-pulse valve group operation, four tuned filters and one damped filter were used. Now in 12-pulse valve group operation only two damped filters are used. Damped filters are simpler to design and have a lower risk of the resonant over voltages and currents as compared to tuned filters.

The harmonics on the dc side must also be limited before entering the line. The dc reactor is sufficient to limit the magnitude of the harmonics on the dc side.

15.5.5 Control Equipment

The control of firing angle is very important in HVDC systems, Electrical impulses, for firing, have to be sent simultaneously to all thyristors connected in series. Since each thyristor is at a different potential (with respect to ground), the firing impulses must be transmitted to each one by means of an insulating medium. The system, adopted in most of the modern HVDC systems, uses light pulses conducted through fibre optic light guides. At each thyristor level, the light pulse is converted by electronic circuits to an electric pulse which is sent to the thyristor gate.

An important difference between the operations of a dc system and an ac system is that the power transmitted over a dc link is always controlled. The valves of the two converter stations are controlled in such a way that one of the stations (generally the inverter) controls the voltage and the other station (i.e., the rectifier) controls the current. These controls allow the link to maintain constant power. It is also possible to add other input signals to the control system (e.g., a signal derived from ac system frequency) in order to attain the optimum conditions for the connected ac system.

15.5.6 Reactive Power Compensation

No reactive power is transmitted over a dc line. However, the converters at the two ends draw reactive power from the ac systems. The reactive power balance must be maintained at the two ends to ensure that ac voltage is held within specified limits. This requires installation of reactive compensation equipment in the HVDC terminal stations. This equipment may consist of ac filter capacitors (which also reduce ac harmonics), shunt capacitors, synchronous condenser or static var system. A combination of these equipments is also used. The reactive power requirement of a rectifier varies as sin α where α is the delay angle of rectifier. For inverter the reactive power requirement varies as sin γ where γ is the extinction angle of the inverter. For both rectifier and inverter the reactive power is also proportional to active power and the reactive power requirement may be about 0.3 to 0.6 *pu* of active power. Since the active power varies, it is necessary that the equipment should provide a fast control of reactive power compensation.

15.6 GROUND RETURN

A ground return means a return path through ground or sea or both. Most dc transmission lines use ground return for reasons of economy and reliability. The monopolar link and homopolar link use the ground return continuously for carrying the return current. The bipolar link uses ground return for short periods when one of the lines is under repair or maintenance. The ground path has a low resistance and, therefore, low power loss as compared to a metallic conductor path provided the ground electrodes are properly designed. The resistance is low because direct current in the earth, unlike alternating current, does not follow closely the route of the line but spreads over a very large cross-sectional area in both depth and width. The resistance of the ground return path is independent of the length of the lines and is essentially the sum of the grounding resistances of the grounding electrodes at the two ends of the line.

The cost of a monopolar or homopolar link with ground return is much less than the cost of such a link with a return metallic conductor. Therefore, monopolar and homopolar links always use ground return. In a bipolar link the main advantage of the ground return is that the line can be built in stages. Initially the power capability requirement is generally less and the dc line can be set up as a monopolar line with ground return. When the load requirement increases the second line can be set up to complete the bipolar link. Thus, a considerable part of the investment cost is deferred to the second stage. A bipolar link does not use the ground return under normal conditions. However, in the event of one of the lines being under repair or maintenance, the link can be operated as a monopolar link with ground return and about half the rated load of the bipolar link can be supplied even under such a condition.

Ground return has associated problems. The main problems are: design of grounding electrodes for low cost of installation and maintenance, location and screening of electrodes so that ground currents cause negligible electrolytic corrosion of buried and immersed metallic structures, prevention of ground current's interference with the operation of other services and control of surface gradient near the electrode for safety of persons, live stock and fish.

15.7 EARTH ELECTRODE, STATION EARTH

An HVDC system needs a properly designed earth electrode at each terminal. This electrode is situated at a safe distance (about 5 to 30 km) from the terminal station, major pipe lines, substations and populated areas. It is installed in a marshy soil having low resistivity. The earth electrode at one of the stations acts as anode (where dc current from terminal station enters earth) and at the other station it acts as cathode (from which dc current returns from earth to HVDC circuit). The terminal station is connected to the earth electrode through an insulated cable known as earth electrode line. In addition an array of interconnected conductors (known as earth mat or grid) is placed in earth at each terminal for protection of equipment from over voltages and safety of personnel. The neutral of convertor transformers and earth terminal of all equipments are connected to this earth mat. There is no direct metallic connection between earth electrodes and earth mat.

In HVDC schemes return path through sea, the earth electrode is either sea electrode (situated a few kilometres off shore) or shore electrode (situated on the sea shore).

(a) *Material of Earth Electrode:* The electrolytic corrosion of anode is an important consideration in selection of material and design of earth electrode. Since the direction of earth current may reverse (due to change in direction of power transmitted) any one of the electrodes may act as anode. Iron is not used due to its high rate of corrosion. Graphite has a somewhat lower rate of corrosion but its direct burial in earth also causes significant loss of material due to corrosion. Instead graphite electrodes buried in a pit filled with crushed coke are used. The electrode transfers current to coke which distributes it to the earth. This arrangement reduces the loss of material due to corrosion to a negligible value.

(b) *Design of Earth Electrode:* The design aspects are:

1. Current density at electrode surface should not exceed 1.5 A/m.
2. Temperature rise of electrode and surroundings should be limited to 60*C.
3. Earth resistance should be low.
4. Step voltage on the ground surface above the electrode should be within safe limits.

(c) *Shape of Earth Electrodes:* An earth electrode may be a straight electrode, a ring electrode or a radial star electrode buried horizontally in earth at about 1m depth. Fig. 15.5 shows these configurations.

The radial star arrangement uses land area more effectively and is very commonly used. The number of arms is usually 6. An earth electrode may cover an area of about 100 m x 100 m.

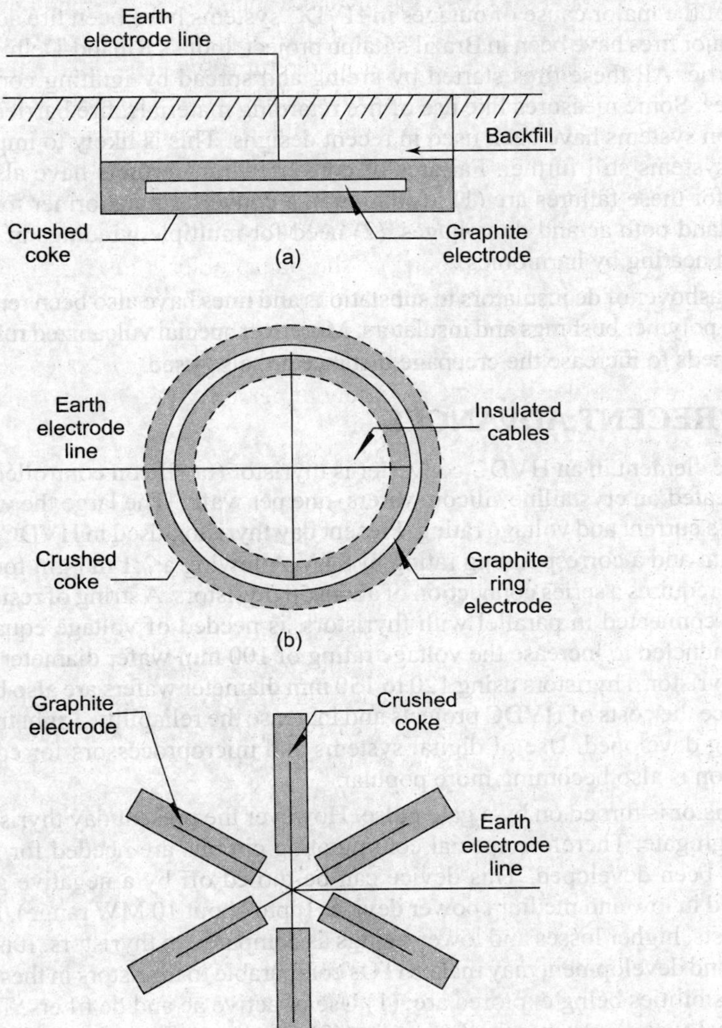

Fig. 15.5 Earth electrodes for HVDC stations (a) straight (b) ring (c) radial star

15.8 RELIABILITY OF HVDC SYSTEMS

Recent studies have indicated that HVDC systems are very reliable. The first HVDC project set up in 1954 used mercury valves but the schemes set up since 1972 use thyristors. All these schemes have pretty low outage rates.

The data collected for 31 utilities, in different countries, indicates that forced unavailability of energy due to converter stations is about 1.62% of time. (Forced unavailability is the percentage of time occupied by forced outages of converter station equipment). The scheduled unavailability of energy (due to planned maintenance outages) is about 5.39% of time. In both these values only about one-fourth are due to faults in thyristor valves and their control equipment. This indicates that HVDC systems have good reliability comparable with that of AC systems.

One of the major cause of outages in HVDC systems have been fire accidents in valve halls. Some major fires have been in Brazil's Itaipu project, India's Rihand-Delhi project and US Pacific DC intertie. All these fires started by arcing and spread by igniting combustible materials in the valves. Some measures like use of fire retarding materials, fire barriers and fire sensing and protection systems have been used in recent designs. This is likely to improve the reliability of HVDC systems still further. Failures of converter transformers have also been reported. The reasons for these failures are (1) insulation in a converter transformer for HVDC systems, has to withstand both ac and dc voltages, (2) need for multiple windings in such transformers (3) localized heating by harmonics.

The flashover of dc insulators in substations and lines have also been reported. Recent designs use solid polymer bushings and insulators. Moreover special vulcanized rubber coating or special rubber sheds to increase the creepage distance are also used.

15.9 RECENT ADVANCES

The basic element in an HVDC converter is thyristor (or silicon controlled rectifier). Thyristors are fabricated on crystalline silicon wafers–one per wafer. The large the wafer, the higher is the thyristor's current and voltage rating. Present day thyristor, used in HVDC, has a wafer diameter of 100 mm and a corresponding rating of 8 kV. Allowing a 2:1 margin for over voltages, a 200 kV valve requires a series connection of about 50 thyristors. A string of resistor-capacitor voltage dividers, connected in parallel with thyristors, is needed of voltage equalization. Research is being conducted to increase the voltage rating of 100 mm wafer diameter thyristors to 10 —12 kV per thyristor. Thyristors using 120 to 150 mm diameter wafers are also being developed. This will reduce the costs of HVDC projects and increase the reliability. Light triggered thyristors are also being developed. Use of digital systems and microprocessors for control, protection and supervision is also becoming more popular.

A thyristor is turned on by a gate pulse. However the present day thyristors cannot be turned off through gate. Therefore special commutating circuits are needed for turning off. Recently GTO has been developed. This device can be turned off by a negative gate pulse. GTOs are being used in low and medium power devices (upto about 10 MW range). Presently GTOs have higher costs, higher losses and lower ratings as compared to thyristors. It is possible that further research and development may make GTOs comparable to thyristors in these three aspects. Some other possibilities being explored are: (1) Use of active ac and dc filters. They can be turned an automatically and suck out the harmonics (2) Outdoor valves (in air tight containers) so that valve hall building is not needed (3) Earth electrodes in 100 —200 m deep holes in order to reduce the earth surface gradients (4) advanced fully digital control systems using optical fibres.

15.10 HVDC SYSTEMS IN INDIA

HVDC has been incorporated in the Indian grid also. About 6500 circuit km of HVDC lines are in operation. The details of three such systems are as under.

(a) *Rihand-Delhi HVDC System:* A ± 500 kV, 1500 MW, 810 km bipolar HVDC line has been set up between Rihand and Delhi to transmit bulk power from Rihand/Singrauli complex to Delhi. It is designed to operate in the following modes: bipolar, monopolar with ground return, monopolar with metallic return.

The system is designed to operate in bipolar mode under normal conditions, both the poles sharing the power equally, thus, resulting in negligible current (less than 10 A) in electrode line/

earth. Under contingencies when one pole goes out of operation, the system automatically switches over to monopolar ground return mode. Thereafter the operator may change over to monopolar metallic return mode where the other conductor is used as return conductor. The modes other than bipolar are not used as primary modes of operation. The earth electrode at Rihand terminal is at Chapki (26.3 km from Rihand) and that at Delhi terminal is at Dankaur (26 km from Delhi).

(b) *Vindhyachal HVDC back to back System:* This back to back link is for exchange of power between Northern and Western regions. Each block of 250 MW is capable of operating independently in either direction and can transfer power in the range of 25 MW to 250 MW depending on system conditions.

This system is provided with indoor air insulated water cooled thyristor valves designed for 12-pulse operation, star-extended delta, 2 winding convertor transformers, ac harmonic filters, capacitors for reactive power requirements, smoothing reactor, zinc oxide arresters, protection system, auxiliary power supply, etc.

(c) *Talcher-Kolar HVDC System:* A 2000 MW, ± 500 kV, bipolar HVDC system has been set up between Talcher and Kolar. It is a East-South interconnector to facilitate exchange of power between east India and south India. It has 1367 kms of ± 500 kV HVDC transmission line, 65 km of earth electrode, 611 kms of 400 kV EHV AC line, 2000 MW HVDC stations at Talcher (Orissa) and Kolar (Karnataka). It will evacuate power from 2000 MW (4 × 500 MW) Talcher II generating station in Orissa to southern states (Tamil Nadu, Karnataka, AP, Kerala and Pondicherry).

Some other HVDC systems in India are :

Chandrapur Padghe ± 500 kV, 1500 MW bipolar line 1504 circuit kilometre

Balia — Bhiwadi ± 500kV, 2500 MW bipolar line 1800 circuit kilometre

Barsur — Lower Sileru 200 kV, 200 MW monopolar line 162 circuit kilometre

Biswanath — Agra ± 600 kV, 4000 MW bipolar line, 3600 circuit kilometre (under construction)

Chandrapur 1000 MW back to back system

Gazuwaha 1000 MW back to back system

Sasavan 500 MW back to back system

Many other HVDC systems are also proposed to be planned and added to the Indian grid.

SUMMARY

15.1 HVDC systems possess many technical and economic advantages over AC systems.

15.2 HVDC lines are cheaper than ac lines. However HVDC terminal equipment is more costly than ac terminal equipment.

15.3 Breakeven distance for overhead lines is in the range of 600—900 km. For submarine cables and underground cables the breakeven distance is 50—100 km.

15.4 Losses in HVDC systems are lesser than the losses in ac line for same transmission capacity.

15.5 HVDC links can be monopolar, bipolar or homopolar.

15.6 HVDC can be incorporated into ac systems through two terminal dc link for point to point transmission, back to back dc links, dc link in parallel with ac link and multi terminal dc link.

15.7 Converter stations equipment includes thyristor valves, converter transformer, dc reactor, harmonic filtering equipment, control equipment and reactive power compensation.

15.8 Many HVDC systems use return path through ground or sea or both.

15.9 HVDC systems need properly designed earth electrode at each terminal.

15.10 Ground electrodes may be either straight or ring or star shaped.

SELF TEST QUESTIONS

A. Indicate whether the following are true (T) or false (F).

15.1 A homopolar dc link has only one conductor.

15.2 Present day HVDC systems use thyristors.

15.3 HVDC systems can be only overhead.

15.4 A converter can work either as rectifier on inverter both not both.

15.5 India does not have HVDC systems.

15.6 A back to back dc link does not have a line.

15.7 A dc line is cheaper than ac line.

15.8 HVDC systems mostly use 12 pulse converters.

B. Fill in the blanks with suitable words

15.9 Break even distance for overhead HVDC system is about ________.

15.10 Two terminal dc link is used for ________ transmission.

15.11 DC terminal stations are ________ costly than ac terminal stations.

15.12 Monopolar HVDC links usually have ________ polarity.

15.13 DC line has ________ corona loss and radio interference than ac line.

15.14 Ground return is needed for ________ and ________ links.

15.15 Temperature rise of earth electrode and surroundings of HVDC systems is limited to about ________.

ANSWERS

15.1	F	15.2	T	15.3	F	15.4	F
15.5	F	15.6	T	15.7	T	15.8	T
15.9	600—900 km	15.10	point to point	15.11	more	15.12	negative
15.13	less	15.14	homopolar, monopolar	15.15	60°C		

JOB INTERVIEW QUESTIONS

15.1 What is meant by HVDC?

15.2 Name some HVDC systems in India?

15.3 What is meant by break even distance?

15.4 What is the function of dc reactor?

15.5 What is the difference between monopolar and homopolar links?

15.6 What is meant by ground return?

15.7 List the various advantages of HVDC systems?

15.8 Why do most HVDC systems use 12 pulse converters?

REVIEW QUESTIONS

15.1 Discuss why ?

(a) DC transmission is superior to ac transmission for long distances.

(b) Thyristor valves have completely replaced mercury are valves in HVDC systems.

(c) Modem HVDC systems use 12 pulse converters.

(d) Most of the present day dc schemes are two terminal links.

(e) DC systems invariably use ground or sea return.

15.2 Discuss the technical and economic advantages of dc systems over ac systems.

15.3 What types of dc links are used in present day systems? Discuss the application of each of these links.

15.4 Briefly describe the following equipment in a converter station

(a) Converter transformer

(b) Thyristor valves

(c) Harmonic filtering equipment

15.5 What is ground return? Why is it used? What are the problems associated with use of ground as return conductor?

CHAPTER 16

Distribution

16.1 COMPARISON OF VARIOUS DISTRIBUTION SYSTEMS

AC three-phase four-wire is the standard distribution system all over the world. Nevertheless it is instructive to compare the volumes of conductor material required in different systems. This comparison is made on the basis that the line length l, the amount of power transferred P and the line losses are the same in each and that the maximum voltage to earth and, therefore, the cost of insulation is also the same in each point. Every system requires one conductor to be grounded and, this is referred to as the neutral (designated n). Since majority of the lines are overhead, the comparison is restricted to only overhead lines.

16.1.1 DC Two-Wire System

Figure 16.1 (a) shows this system. The neutral acts as the return conductor and carries the same current as the positive conductor.

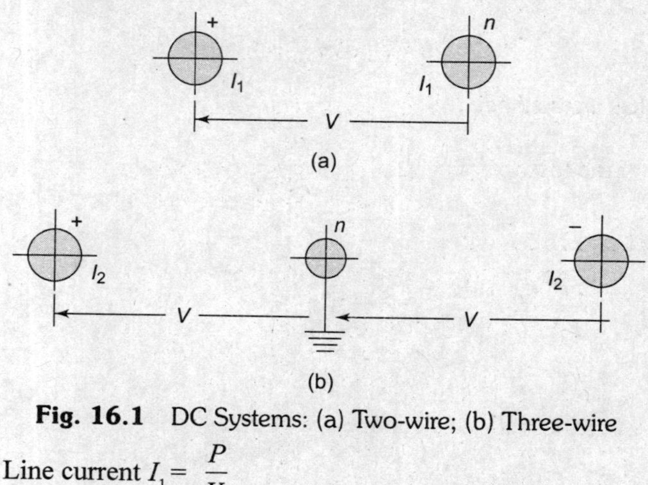

Fig. 16.1 DC Systems: (a) Two-wire; (b) Three-wire

$$\text{Line current } I_1 = \frac{P}{V}$$

If the resistance of each conductor is R_1,

$$\text{Line losses} = 2I_1^2 R_1 = \frac{2P^2}{V^2} \cdot R_1 \tag{16.1}$$

16.1.2 DC Three-Wire System

Figure 16.1 (b) shows this system. Since the voltage from the two outer conductors to the neutral are of opposite polarity, the currents flow in the outer conductors is in opposite directions. For a balanced system the current in the neutral conductor is zero. Then the losses in the neutral are zero and its volume can be neglected.

$$I_2 = \frac{P}{2V}$$

If each conductor has a resistance R_2

$$\text{Line losses} = 2\, I_2^2\, R_2 = \frac{P^2}{2V^2} R_2 \tag{16.2}$$

Equating Eqs. (16.2) and (16.1)

$$2\frac{P^2}{V^2} R_1 = \frac{P^2}{2V^2} R_2$$

or

$$R_2 = 4R_1 \tag{16.3}$$

The resistance is inversely proportional to the area of cross-section. Denoting the area of cross-section of conductors in system 1 (two-wire dc) by A_1 and in this system (three-wire dc) by A_2

$$A_2 = 0.25A_1 \tag{16.4}$$

16.1.3 Single Phase AC

Figure 16.2 (a) shows this system. Since the peak voltage between the conductors is V, the rms voltage is 0.707 V.

If the power factor is cos ϕ

$$I_3 = \frac{P}{0.707\text{ V}\cos\phi}$$

If each conductor has a resistance R_3

$$\text{Line losses} = 2\, I_3^2\, R_3 = \frac{4\,P^2}{V^2\cos^2\phi} R_3 \tag{16.5}$$

Equating Eqs. (16.1 and 16.5)

$$\frac{2\,P^2}{V^2} R_1 = \frac{4\,P^2}{V^2\cos^2\phi} R_3$$

or

$$R_3 = (0.5\cos^2\phi)\, R_1 \tag{16.6}$$

Denoting the area of cross-section of conductors in this system as A_3,

$$A_3 = \frac{2}{\cos^2\phi} \cdot A_1 \tag{16.7}$$

16.1.4 3-Phase 4-Wire AC

Figure 16.2 (b) shows this system. The voltage between each phase conductor and neutral is 0.707 V. If the system is balanced, there would be no current in the neutral. Then the losses in the neutral conductor are zero and its volume can be neglected.

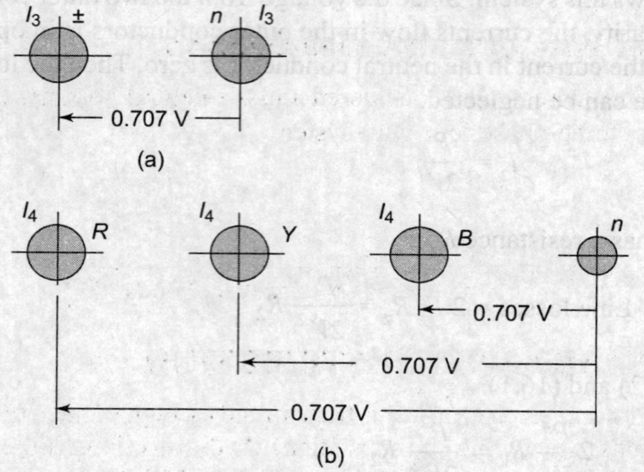

Fig. 16.2 AC systems (a) single-phase (b) three-phase

$$I_4 = \frac{P}{3\,(0.707\ \text{V})\cos\phi}$$

If each conductor has a resistance R_4

$$\text{Line losses} = 3\,I_4^2 R_4 = \frac{2P^2}{3V^2\cos^2\phi}\,R_4 \tag{16.8}$$

Equating Eqs. (16.1 and 16.8)

$$\frac{2\,P^2}{V^2}\,R_1 = \frac{2\,P^2}{3\,V^2\cos^2\phi}\,R_4$$

or

$$R_4 = 3\,(\cos^2\phi)\,R_1 \tag{16.9}$$

Denoting the area of cross-section of each phase conductor as A_4

$$A_4 = \frac{A_1}{3\cos^2\phi} \tag{16.10}$$

In each of the cases (1), (2) and (3) the system has two conductors. In case (4) the system has three conductors. If v_1, v_2, v_3 and v_4 denote the total conductor volume in the four cases, we have $v_1 = 2A_1\,l$, $v_2 = 2\,A_2\,l$, $v_3 = 2A_3\,l$ and $v_4 = 3\,A_4 l$.

Therefore,

$$v_1 : v_2 : v_3 : v_4 = 1 : \frac{1}{4} : \frac{2}{\cos^2\phi} : \frac{1}{2\cos^2\phi} \tag{16.11}$$

Equation (16.11) shows a very important comparison. From the point of view of conductor economy, the three-wire dc system is far superior to other systems. However, because of simpler and cheaper generation and transformation equipment, all modern power systems are ac systems (though some HVDC links are also used.) The three-phase four-wire system requires lesser conductor material than the single phase ac system. An n phase $(n + 1)$ wire ac system $(n > 3)$ requires the same amount of conductor material as a three-phase four-wire system. Since the three-phase system is the simplest of all the n $(n > 3)$ phase systems, all modern power systems are three-phase ac systems.

16.1.5 Neutral Wire of Half Cross-Section

If the volume of neutral wire is also considered and the neutral wire has half the cross-section of phase coductors in three-phase four-wire system

Then, $v_5 = 3.5\, A_4 L$

Therefore,

$$v_1 : v_2 : v_3 : v_4 : v_5 = 1 : \frac{1}{4} : \frac{2}{\cos^2\phi} : \frac{1}{2\cos^2\phi} : \frac{7}{12\cos^2\phi} \tag{16.12}$$

16.2 AC 3-PHASE 4-WIRE DISTRIBUTION

As discussed in section 1.1, electric power is transmitted by high voltage transmission lines from sending end substations to receiving end substations. At the receiving end substations the voltage is stepped down to a lower value (say 66 kV or 33 kV or 11 kV). The secondary transmission system transfers power from these receiving end substations to secondary substations. A secondary substation consists of one or more power transformers together with voltage regulating equipment, buses and switchgear. At the secondary substations voltage is stepped down to 11 kV. The portion of the power network between a secondary substation and consumers' installation is known as distribution system. The distribution system can be classified into primary and secondary distribution system. Some large consumers are given high voltage supply from the receiving end substations or secondary substations.

The area served by a secondary substation can be subdivided into a number of sub-areas. Each sub-area has its primary and secondary distribution system. The primary distribution system consists of main feeders, and laterals. The main feeder runs from the low voltage bus of the secondary substation and acts as the main source of supply to sub-feeders, laterals, or direct connected distribution transformers. The lateral is supplied by the main feeder and extends through the load area with connections to distribution transformers. The distribution transformers are located at convenient places in the load area. They may be located in specially constructed enclosures or may be pole mounted. The distribution transformers for a large multi storyed building may be located within the building itself. At the distribution transformer the voltage is stepped down to 400 V and power is fed into the secondary distribution systems. The secondary distribution system consists of distributors which are laid along the road sides. The service connections to consumers are tapped off from the distributors. The main feeders, laterals and distributors may consist of overhead lines or cables or both. The distributors are three-phase four-wire circuits, the neutral wire being necessary to supply the single phase loads. Most of the residential and commercial consumers are given single phase supply. Some large residential and commercial consumers get three-phase supply. The service connections of consumers are known as service mains.

16.3 TYPES OF PRIMARY DISTRIBUTION SYSTEMS

The primary distribution system may consist of radial feeders, parallel feeders, ring feeders or a network.

16.3.1 Radial Feeder

A radial feeder is the simplest and the most commonly used one. It is used extensively to supply small and medium residential, commercial, industrial and non-critical loads. It derives its name from the fact that the feeder radiates from the secondary substation and branches into sub-feeders and laterals which extend into all parts of the area served. The feeders and sub-feeders are three-phase three-wire (or four-wire) circuits. The laterals may be of three-phase or single phase. The distribution transformers are connected to the primary feeders, sub-feeders and laterals, usually through fused cutouts. The radial feeder is the most economical but the least reliable of the various types since a fault in the feeder means disruption of supply. Figure 16.3 shows a radial primary feeder.

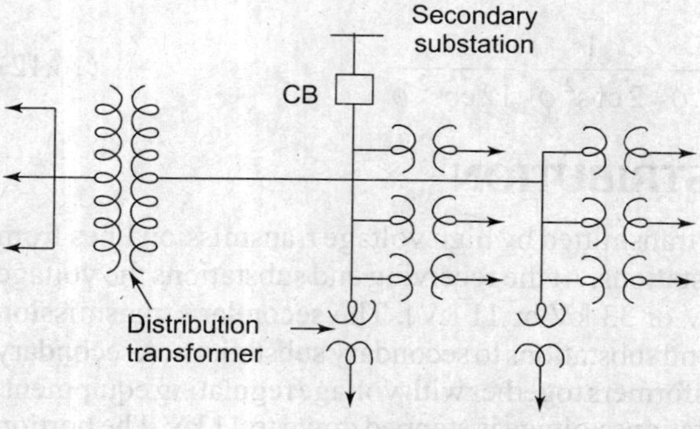

Fig. 16.3 Radial feeder

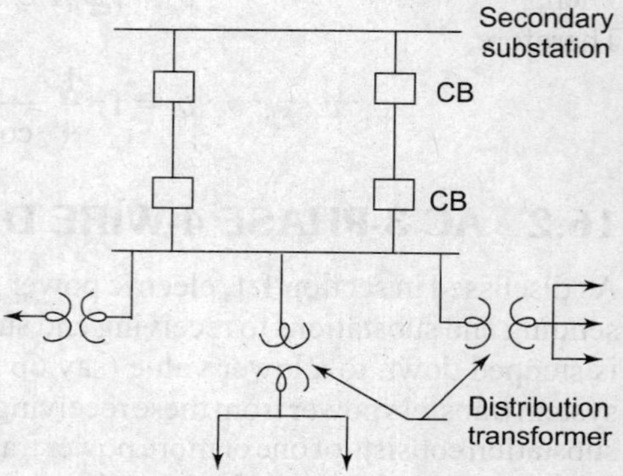

Fig. 16.4 Parallel feeder

16.3.2 Parallel Feeder

A parallel feeder consists of duplicate feed system having two-radial feeders running in parallel. The two feeders may originate from the same or different secondary substations. Each feeder supplies about half of the total load of the area but has a capability to supply the entire load in the event of an outage on the other feeder. Loss of either feeder will result in interruption of service until the load normally supplied by the faulted feeder is transferred to the other feeder by automatic or manually controlled switches. Parallel feeders cost more than radial feeders but substantially reduce the frequency and duration of outages. Figure 16.4 shows a parallel feeder.

16.3.3 Loop Feeder

A system of two or more radial feeders originating from the same or different secondary substations and separately routed through the load areas is known as loop feeder system. If the ends of the two feeders are tied together through normally open switching devices, the resulting arrangement is known as open loop system. If the ends are tied together by means of normally closed switching devices, the result is a ring loop or simply ring feeder. The loop feeder provides a good continuity of service. Feeders and loop components must have sufficient reserve capacity to serve the load that may be transferred under emergency conditions. This system is the most practical for providing reasonable reliability with nearly full utilization of facilities. Figure 16.5 shows a loop feeder.

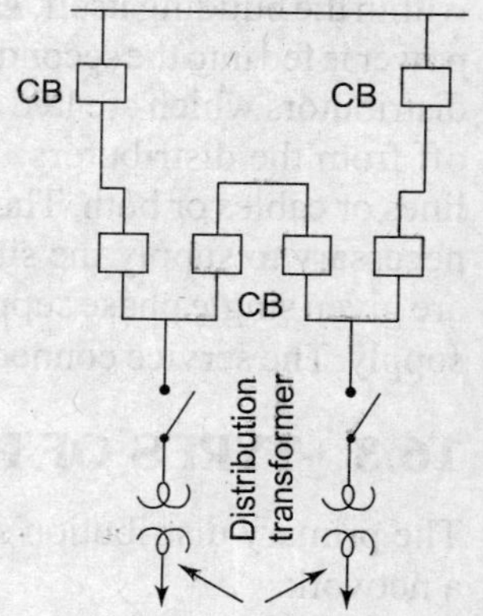

Fig. 16.5 Loop feeder

16.3.4 Primary Network

Figure 16.6 shows a network or grid for primary distribution. The system consists of a number of interconnected feeders. Two or more sub-transmission circuits supply two or more secondary substations from which the feeders take off. Because the feeders are interconnected, power is supplied to all the distribution transformers even though a part of the network may be out of service. Each secondary substation consists of a transformer (to step down the sub-transmission voltage to the primary feeder voltage) and the necessary switchgear to isolate the faulty feeder and to control the feeders. This system provides good flexibility and reliability and is used in large metropolitan areas where continuity of supply is essential. Full advantage of load diversity is also realized.

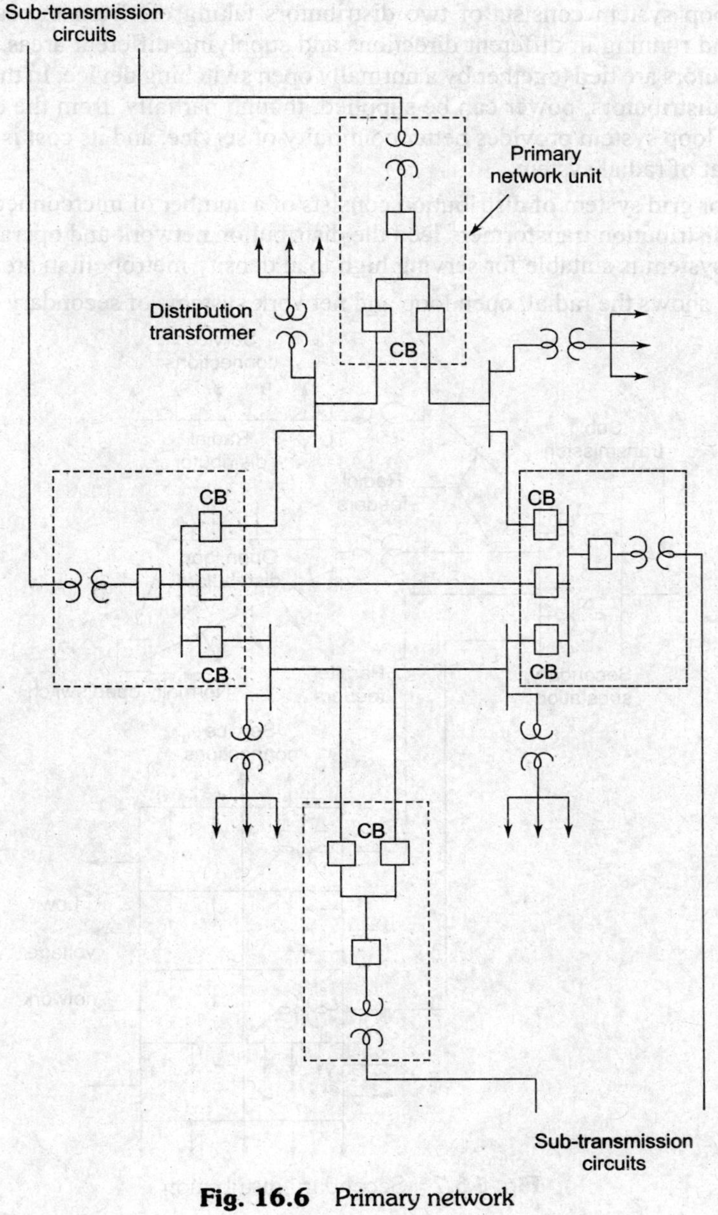

Fig. 16.6 Primary network

16.4 TYPES OF SECONDARY DISTRIBUTION SYSTEMS

The secondary distribution system consists of three-phase four-wire 400 V distributors laid along road sides. The service connections to the consumers are tapped off the distributors at convenient points. The service connections may be single phase two-wire circuits or three-phase four-wire circuits (for large loads). An attempt is made to divide the single phase loads equally (as far as possible) on the three-phases.

The secondary distribution systems generally used are: radial, open loop and network distribution. A radial distributor takes off from the distribution transformer and runs through the area to be served by it. This is the simplest and the least expensive system but has the poorest service reliability.

An open loop system consists of two distributors taking off from the same distribution transformer and running in different directions and supplying different areas. The far ends of the two distributors are tied together by a normally open switching device. In the event of a fault on one of the distributors, power can be supplied, though partially, from the other distributor. Thus, an open loop system provides better continuity of service, and its cost is only marginally higher than that of radial system.

A network or grid system of distribution consists of a number of interconnected distributors. Two or more distribution transformers feed the distribution network and operate continually in parallel. This system is suitable for serving high load density metropolitan areas.

Figure 16.7 shows the radial, open loop and network systems of secondary distribution.

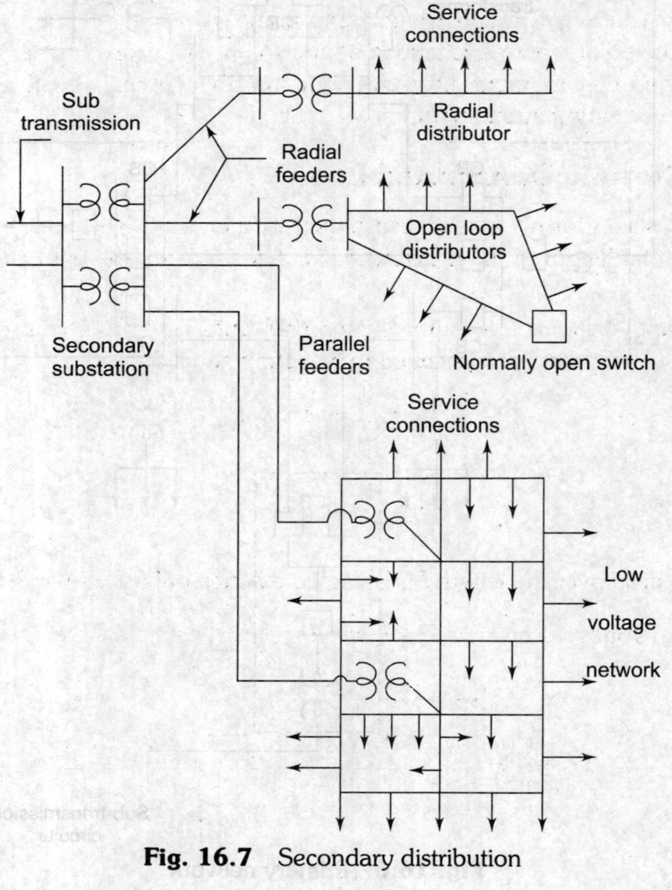

Fig. 16.7 Secondary distribution

16.5 VOLTAGE DROP IN DC DISTRIBUTORS

It is necessary to ensure that the voltage at the consumer premises is within the permissible limits. Therefore, it is necessary to calculate the voltage drop in different parts of the power supply network. The calculations of voltage drops in distributors are pretty lengthy because loads are tapped at many points along the route of the distributor.

16.5.1 Distributor Fed at One End

Figure 16.8 shows a dc distributor having concentrated loads and fed at one end, I_1, I_2, I_3 are the load currents tapped at different points. r is the resistance per unit length of the distributor (go and return).

$$\text{Voltage drop} = (I_1 + I_2 + I_3)(l_1 r) + (I_2 + I_3)(l_2 r) + I_3(l_3 r)$$

$$= I_1 l_1 r + I_2 (l_1 + l_2) r + I_3 (l_1 + l_2 + l_3) r \quad (16.13)$$

Thus, the total voltage drop at the far end of the distributor is the sum of the moments of the various currents tapped off about the feeding point where the moment of a current is defined as the product of current and the total resistance through which it flows upto the point where it is tapped off. It is seen that $l_1 r$ is the resistance of the distributor upto point where I_1 is tapped off, $(l_1 + l_2) r$ is the resistance upto the point where I_2 is tapped of and so on. The drop upto any intermediate point is equal to the sum of the moments upto that point plus the moments of all currents beyond that point assumed to be acting at that point.

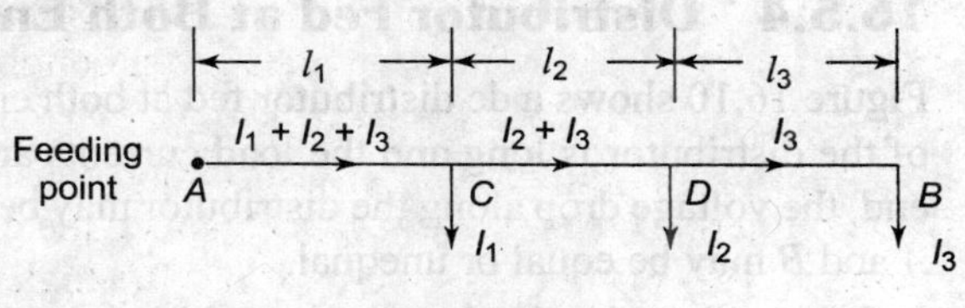

Fig. 16.8

16.5.2 Uniformly Distributed Load

Figure 16.9 shows a uniformly loaded distributor fed at one end. The current tapped off per unit length is i. The total length of the distributor is l. The voltage drop upto a distance u from the feeding point is

Voltage drop upto distance u = (sum of moments upto u) + (moments of loads beyond u assumed to be acting at distance u)

$$= \int_0^u r\, u\, i\, du + i(l - u)\, r\, u$$

$$= i\, l\, r\, u - \frac{1}{2}\, i\, r\, u^2 \quad (16.14)$$

Total voltage drop over the length l is obtained by substituting $u = l$ in Eq. (16.14). Thus

$$\text{Total voltage drop} = i\, r\, l^2 - \frac{1}{2}\, i\, r\, l^2 = \frac{1}{2}\, i\, r\, l^2 = \frac{1}{2}\, I\, r\, l \quad (16.15)$$

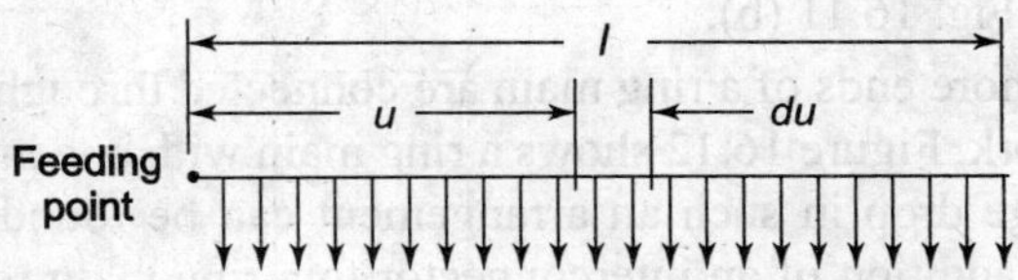

Fig. 16.9

where I is the total current supplied by the distributor and equals $i\,l$. In Eqs (16.14) and (16.15) r is the resistance per unit length of both wires of the distributor (go and return). Eq. (16.15) indicates that a uniformly loaded distributor may be represented by the total load tapped off at the centre of the distributor.

16.5.3 Both Concentrated and Distributed Loads

It is possible that a distributor has concentrated loads at some points and a uniformly distributed load i A/m. Let the resistance per unit length be r and current fed at any point be I. The voltage drop in a length l from the point at which the current I is fed is given by

$$\text{Voltage drop in length } l = \int_0^l r(I - iu)du = rl\left(I - \frac{1}{2}il\right) \tag{16.16}$$

16.5.4 Distributor Fed at Both Ends

Figure 16.10 shows a dc distributor fed at both ends. Such a distributor is used when the length of the distributor is long and the load currents are high. If such a distributor is fed at only one end, the voltage drop along the distributor may be rather high. The voltages at the feeding points A and B may be equal or unequal.

It is evident that load current would be supplied partly from end A and partly from end B. The minimum potential will occur at one of the load points. The current tapped off at the point of minimum potential would come partly from end A and partly from end B.

The currents I_A and I_B can be calculated by the application of Kirchhoff's laws. The total voltage drop along the distributor must be equal to difference of voltages V_A and V_B.

If a uniformly loaded distributor is fed at both the ends the minimum potential may occur at any point. Both concentrated and uniformly distributed loads may also occur simultaneously.

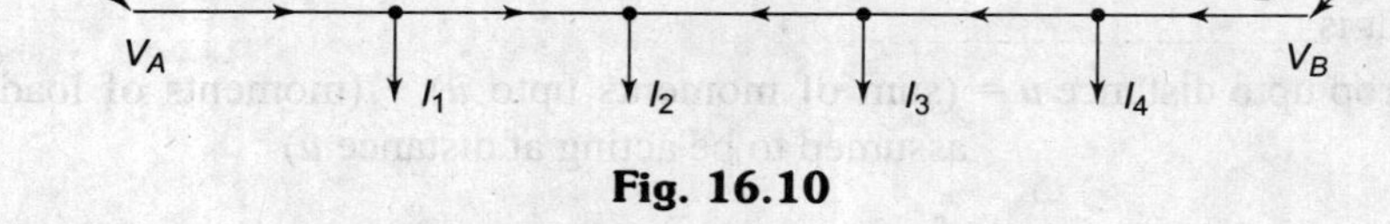

Fig. 16.10

16.5.5 Ring Main Distributor

If a distributor is arranged to form a closed loop, it becomes a ring main. It may be fed at one or more points. A ring main and a secondary network are more or less similar. Such a system has better flexibility and reliability. In case a fault develops in any one section it can be isolated from both sides so that only that section suffers discontinuity of supply. The continuity of supply is maintained for the remaining sections. Fig. 16.11 (a) shows a ring main distributor. The voltage drop in such a distributor can be calculated by considering it as two radial distributors fed at both ends as shown in Fig. 16.11 (b).

Sometimes two or more ends of a ring main are connected through interconnectors, so as to form a complete network. Figure 16.12 shows a ring main with one interconnector. The current distribution and voltage drop in such an arrangement can be found by direct application of Kirchhoff's laws. The addition of an interconnector to a ring main reduce the voltage drop in the distribution system.

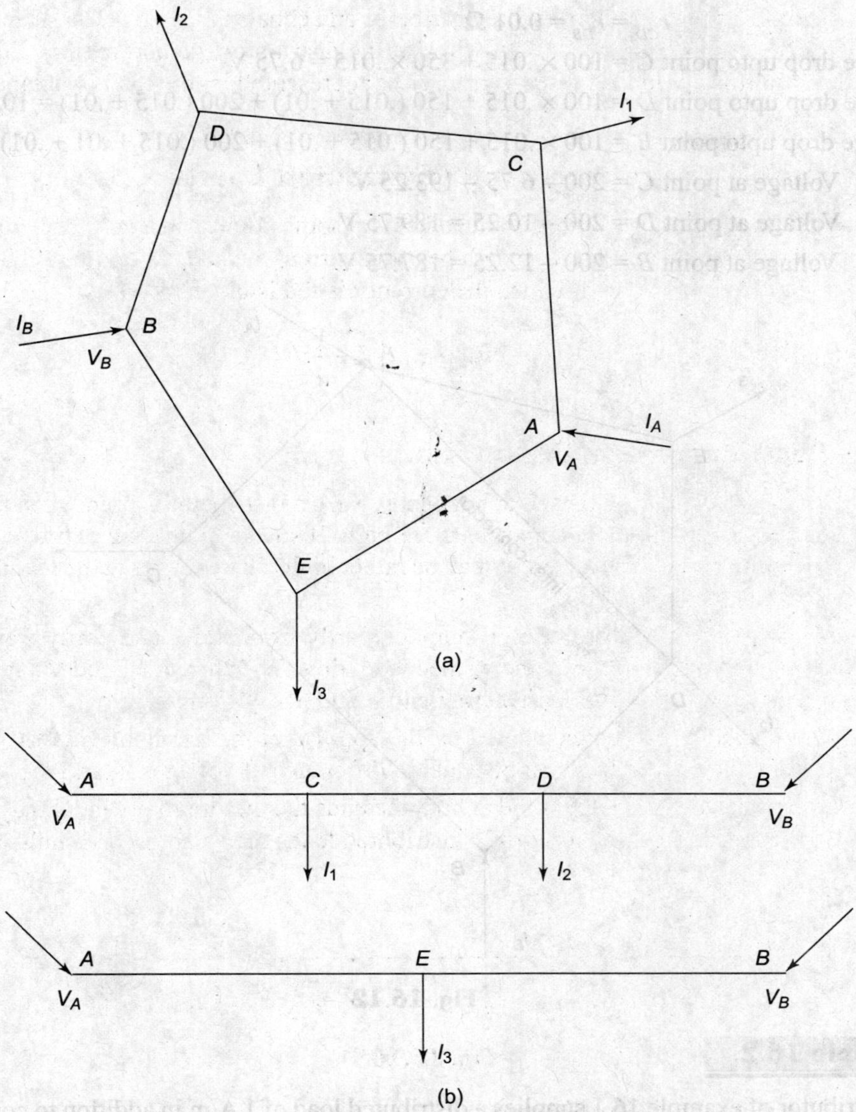

Fig. 16.11

Example 16.1.

The dc distributor in Fig. 16.8 is loaded as under. $I_1 = 100$ A, $I_2 = 150$ A, $I_3 = 200$ A. The lengths are $l_1 = 150$ m, $l_2 = 100$ m, $l_3 = 100$ m. The resistance of both the conductors (go and return) is 0.1 ohm per 1000 m. Find the voltage at points C, D and B if $V_A = 200$ V.

Solution

The resistances of different sections are:

$$r_{AC} = \frac{150 \times 0.1}{1000} = 0.015\ \Omega$$

$$r_{CD} = r_{DB} = 0.01\ \Omega$$

Voltage drop upto point $C = 100 \times .015 + 350 \times .015 = 6.75$ V

Voltage drop upto point $D = 100 \times .015 + 150\,(.015 + .01) + 200\,(.015 + .01) = 10.25$ V

Voltage drop upto point $B = 100 \times .015 + 150\,(.015 + .01) + 200\,(.015 + .01 + .01) = 12.25$ V

Voltage at point $C = 200 - 6.75 = 193.25$ V

Voltage at point $D = 200 - 10.25 = 189.75$ V

Voltage at point $B = 200 - 12.25 = 187.75$ V

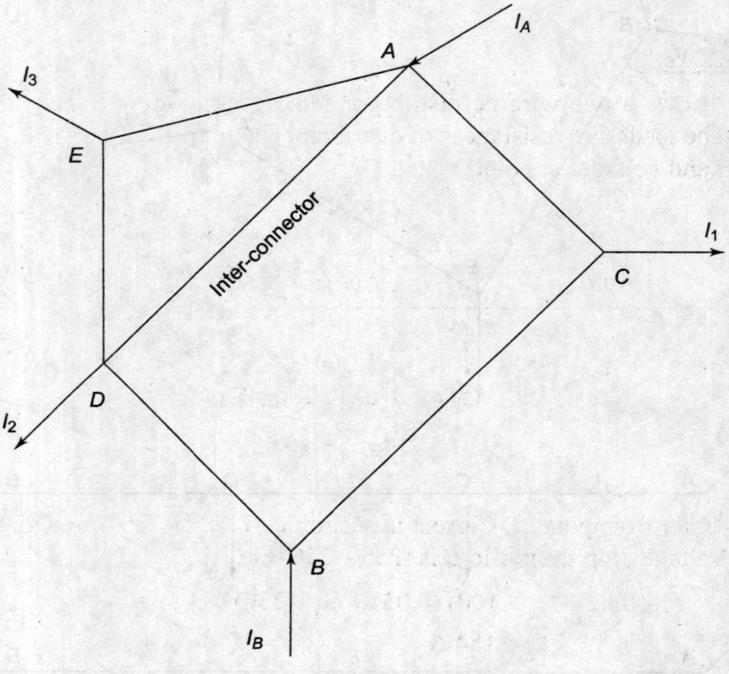

Fig. 16.12

Example 16.2.

The distributor of example 16.1 supplies a distributed load of 1 A/m in addition to concentrated loads. Find the voltages at points C, D and B.

Solution

$$r = 0.1/1000 = 0.0001\ \Omega/\text{m}$$

Current entering section AC

$$= 100 + 150 + 200 + 1\,(150 + 100 + 100) = 800\text{A}$$

Using Eq. (16.16)

$$\text{Voltage drop is section } AC = .0001 \times 150 \left(800 - \frac{1}{2} \times 1 \times 150\right) = 10.875 \text{ V}$$

Voltage at point $C = 200 - 10.875 = 189.125$ V

Current entering section $CD = 800 - 100 - 150 = 550$ A

$$\text{Voltage drop in section } CD = .0001 \times 100 \left(550 - \frac{1}{2} \times 1 \times 100\right) = 5 \text{ V}$$

$$\text{Voltage at point } D = 189.125 - 5 = 184.125 \text{ V}$$

$$\text{Current entering section } DB = 550 - 150 - 100 = 300 \text{ A}$$

$$\text{Voltage drop in section } DB = .0001 \times 100 \left(300 - \frac{1}{2} \times 1 \times 100\right) = 2.5 \text{ V}$$

$$\text{Voltage at point } B = 184.125 - 2.5 = 181.625 \text{ V}$$

Example 16.3.

Figure 16.13 shows a two-wire dc distributor fed at both ends A and B at 255 and 250 V respectively. The loads and resistances of distributor (both wires) are shown. Find the currents fed at A and B and voltages at points C and D.

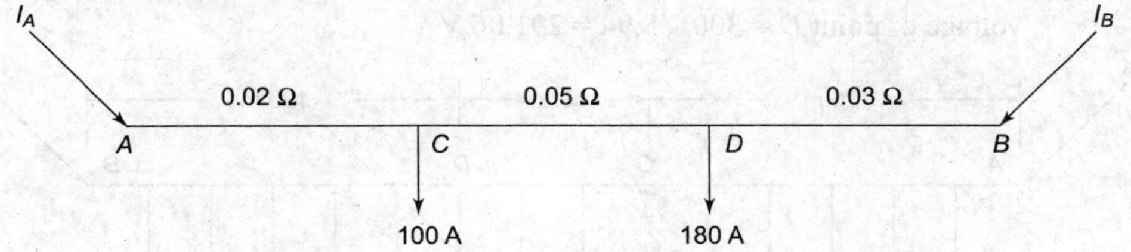

Fig. 16.13

Solution

Let current I_A enter from point A. Current in section C_D is $(I_A - 100)$ A and that in section DB is $(I_A - 280)$ A. Voltage drop from A to B is $(255 - 250)$ or 5 V.

$$I_A \times 0.02 + (I_A - 100)(0.05) + (I_A - 280)(0.03) = 5$$

or $$I_A = 184 \text{ A}$$

$$I_B = -(I_A - 280) = 96 \text{ A}$$

$$V_C = 255 - (184)(0.02) = 251.32 \text{ V}$$

$$V_D = 251.32 - (184 - 100)(0.05) = 247.12 \text{ V}$$

Example 16.4.

A two-wire dc distributor AB 300 m long is fed at both ends A and B. The distributor supplies uniformly distributed load of 0.25 A/m and concentrated loads of 40 A at C and 60 A at D. The distances AC and BD are 120 m each. The loop resistance of distributor is 0.1 ohm/100 m. Both A and B are maintained at 300 V. Find the currents fed at A and B and potentials of points C and D.

Solution

Figure 16.14 shows the distributor. Let current entering section $AC = I_A$

$$r = 0.1/100 = 0.001 \text{ ohm/m}$$

Using Eq. (16.16)

$$\text{Voltage drop in section } AC = 0.001 \times 120\,(I_A - 0.5 \times 0.25 \times 120)$$

Current entering section $CD = I_A - 40 - 0.25 \times 120 = (I_A - 70)$ A

Voltage drop in section $CD = 0.001 \times 60\,(I_A - 70 - 0.5 \times 0.25 \times 60)$

Current entering section $DB = I_A - 70 - 60 - 0.25 \times 60 = (I_A - 145)$ A

Voltage drop in section $DB = 0.001 \times 120\,(I_A - 145 - 0.5 \times 0.25 \times 120)$

Since both A and B are at the same potential, total voltage drop from A to B is zero. Thus

$$0.12\,(I_A - 15) + 0.06\,(I_A - 70 - 7.5) + 0.12\,(I_A - 145 - 15) = 0$$

or
$$I_A = 85.5 \text{ A}$$

Total current supplied by distributor

$$= 60 + 40 + 0.25 \times 300 = 175 \text{ A}$$

$$I_B = 175 - 85.5 = 89.5 \text{ A}$$

Voltage drop in section $AC = 0.12\,(85.5 - 15) = 8.46$ V

Voltage of point $C = 300 - 8.46 = 291.54$ V

Voltage drop in section $BD = 0.12\,(89.5 - 15) = 8.94$ V

Voltage of point $D = 300 - 8.94 = 291.06$ V

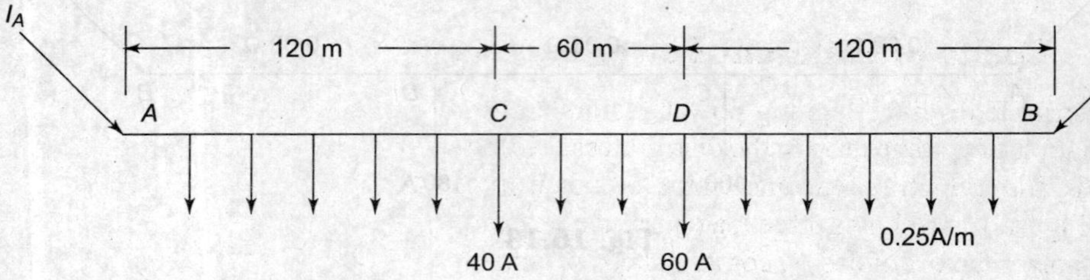

Fig. 16.14

16.6 VOLTAGE DROP IN AC DISTRIBUTORS

The calculation of voltage drop in ac distributors has to take into account the inductive reactance of the distributor. Therefore, the calculations become more difficult than the calculations in dc distributors. If the loads are at unity power factor, the effect of inductive reactance can be neglected and the calculations are exactly similar to those in dc distributors. The distributors feeding residential consumers operate at nearly unity power factor. Therefore, voltage drop calculations for ac distributors feeding residential loads are generally done by neglecting the effect of inductive reactance.

In a balanced three-phase four-wire circuit the current in the neutral wire is zero and only the voltage drop in the phase conductor need be calculated. In actual practice the loads are never balanced. However, the degree of unbalance is quite small and the calculations can be done by neglecting the current in the neutral wire.

16.6.1 Unity Power Factor Loads

Figure 16.15 shows a single phase distributor supplying three unity power factor loads. If l_1, l_2, l_3 are the lengths, r is the resistance and x is the inductive reactance per unit length of the distributor, V_s is the voltage at feeding point V_r is the voltage at far end. Then

$$V_s = V_r + (I_1 + I_2 + I_3)\, l_1\, (r + jx) + (I_2 + I_3)\, l_2\, (r + jx) + I_3\, l_3\, (R + jx)$$
$$= V_r + I_1\, l_1\, (r + jx) + I_2\, (l_1 + l_2)(r + jx) + I_3\, (l_1 + l_2 + l_3)\, (r + jx)$$

$$= V_r + I_1 l_1 r + I_2 (l_1 + l_2) r + I_3 (l_1 + l_2 + l_3) r + j([I_1 l_1 x + I_2 (l_1 + l_2) x + I_3 (l_1 + l_2 + l_3) x]$$

or $$V_s \approx V_r + I_1 l_1 r + I_2 (l_1 + l_2) r + I_3 (l_1 + l_2 + l_3) r$$

$$\text{Voltage drop} = V_s - V_r = I_1 l_1 r + I_2 (l_1 + l_2) r + I_3 (l_1 + l_2 + l_3) r \qquad (16.17)$$

Thus, the total voltage drop at the far end of the distributor is the sum of the moments of the various currents tapped off about the feeding point where the moment of a current is defined as the product of the current and the total resistance through which it flows upto the point where it is tapped off. The drop at any intermediate point is equal to the sum of the moments upto that point plus the moments of all the currents beyond that point assumed to be acting at that point. The quantity $l_1 r$ is the resistance of the distributor upto point where I_1 is tapped off, $(l_1 + l_2) r$ is the resistance upto the point where I_2 is tapped off and so on. It is seen that Eq. (16.17) is the same as Eq. (16.13).

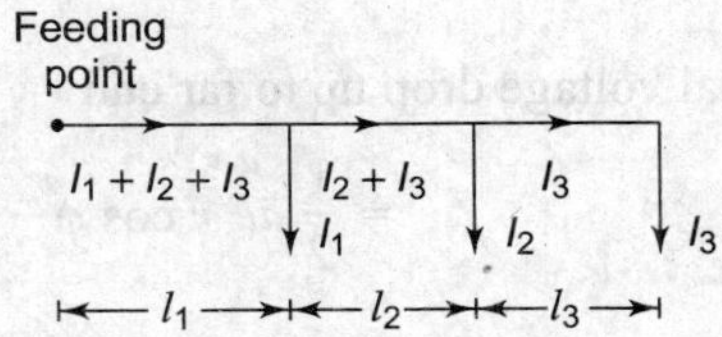

Fig. 16.15 Single phase distributor supplying loads at unity power factor

16.6.2 Loads of Different Power Factors

The industrial loads have lagging power factors and in the calculation of voltage drop in the distributors feeding such loads, the inductive reactance must be taken into account. Figure 16.16 shows three load currents I_1, I_2 and I_3 having power factors of $\cos \phi_1$, $\cos \phi_2$ and $\cos \phi_3$ tapped off a distributor. As in the previous case l_1, l_2 and l_3 are the lengths of the three sections and r and x are the resistance and inductive reactance per unit length. The voltage drop can be calculated by the above mentioned method of moments by replacing r by $(r + jx)$ and the current by I $(\cos \phi - j \sin \phi)$ where $\cos \phi$ is the power factor. The voltage drop upto the far end is given by:

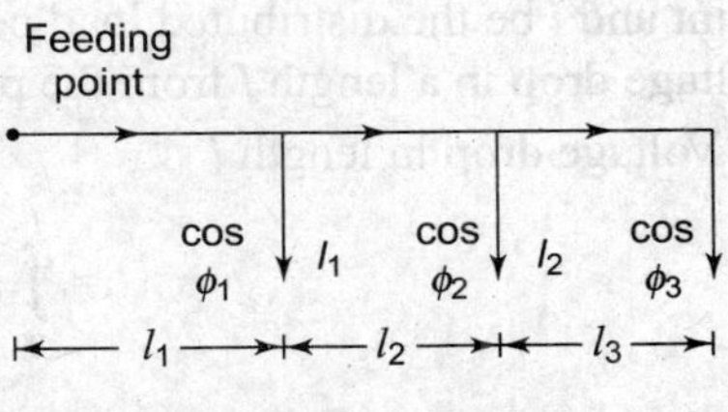

Fig. 16.16 Single phase distributor supplying loads at different power factors

$$\text{Voltage drop} = I_1 (\cos \phi_1 - j \sin \phi_1) l_1 (r + jx) + I_2 (\cos \phi_2 - j \sin \phi_2) \times (l_1 + l_2)(r + jx) + I_3 (\cos \phi_3 - j \sin \phi_3)(l_1 + l_2 + l_3)(r + jx)$$

$$\simeq I_1 l_1 (r \cos\phi_1 + x \sin \phi_1) + I_2 (l_1 + l_2)(r \cos \phi_2 + x \sin \phi_2) + I_3 (l_1 + l_2 + l_3)(r \cos \phi_3 + x \sin \phi_3) \qquad (16.18)$$

A comparison of Eq. (16.17) and Eq. (16.18) shows that the effect of power factor is to replace the term r by $(r \cos \phi + x \sin \phi)$. Thus the voltage drop in ac distributor is also calculated as sum of moments of the various currents about the feeding point but, in this case, the moment is defined as the product of current and the term $(R \cos \phi + X \sin \phi)$ where R and X are the total resistance and reactance through which the current flows.

16.6.3 Uniformly Distributed Load

Let a current i at power factor $\cos \phi$ be tapped per unit length of distributor of total length l.
Voltage drop upto distance u

= (sum of moments upto u) + (moments of load beyond u assumed acting at distance u)

$$= \int_0^u (r \cos \phi \, iu \, du + x \sin \phi \, iu \, du) + i\,(l-u)\, r \cos \phi \, u + i\,(l-u)\, x \sin \phi \, u$$

$$= i\, l\, u\, r \cos \phi - \frac{1}{2}\, ir\, u^2 \cos \phi + il\, u\, x \sin \phi - \frac{1}{2}\, i\, x\, u^2 \sin \phi \qquad (16.19)$$

Total voltage drop up to far end

$$= \frac{1}{2}\, il^2\, r \cos \phi + \frac{1}{2}\, il^2\, x \sin \phi \qquad (16.20)$$

16.6.4 Both Concentrated and Distributed Loads

It is possible that an ac distributor has both concentrated and distributed loads. When these loads are at different power factors, the general expression for voltage drop becomes rather long. One method is to consider the concentrated and distributed loads separately and calculate voltage drops separately using Eqs (16.18) and (16.20). The total voltage drop is obtained by algebraic addition of these separate voltage drops.

If the concentrated and distributed loads are at the same power factor cos ϕ, the expression for voltage drop can be obtained by an extension of Eq. (16.16). Let I be the current fed at any point and i be the distributed load per meter, both these loads being at power factor cos ϕ. The voltage drop in a length l from the point at which the current I is fed is given by

Voltage drop in length l

$$= \int_0^l [(I - iu) \cos \phi \, du + x\,(I - iu) \sin \phi \, du]$$

$$= r \cos \phi \, l \left(I - \frac{1}{2} il \right) + x \sin \phi \, l \left(I - \frac{1}{2} il \right)$$

$$= l \left(I - \frac{1}{2} il \right) (r \cos \phi + x \sin \phi) \qquad (16.21)$$

16.6.5 Uniformly Distributed Load Fed at Both Ends

If an ac distributor of length l having a uniform load i, A/m at power factor cos ϕ is fed at both ends at the same voltage, minimum voltage will occur at mid point.

$$\text{Voltage drop upto mid point} = \frac{1}{8}\,(r\, i\, l^2 \cos \phi + x\, i\, l^2 \sin \phi) \qquad (16.22)$$

16.6.6 Unbalanced Loading

AC distribution supplies a mixture of three-phase and single phase loads. If the degree of unbalance is large, the voltage drop in the neutral wire has also to be taken into account. The voltage drop calculations are done for each of the four-wires separately and then the voltage between each phase conductor and neutral conductor at the far end of the distributor can be calculated.

Example 16.5.

A 400 m long distributor has the following unity power factor loads tapped off *R* phase : 100 A load at 100 m from the feeding point, 120 A load at 250 m from the feeding point and 80 A load at 400 m from the feeding point. The resistance of the conductor is 0.25 ohm per km length. Neglecting the voltage drop in the neutral conductor, find the voltage across the load at the far end if the voltage at the feeding point is 240 V.

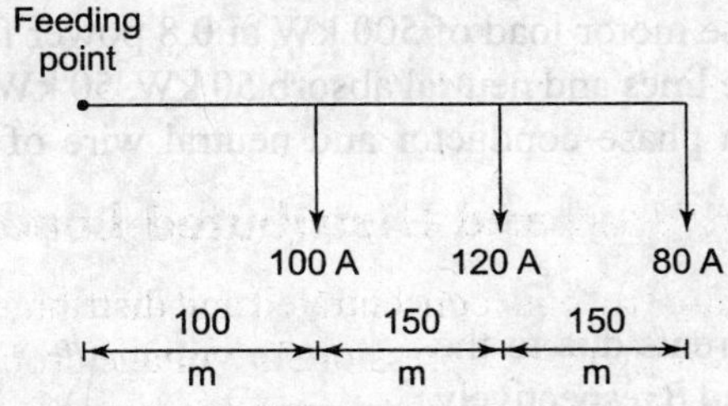

Fig. 16.17

Solution

The distributor is shown in Fig. 16.17. The total voltage drop from the feeding point to the far end is:

$$\text{Total voltage drop} = \frac{0.25}{1000}[100 \times 100 + 120 \times 250 + 80 \times 400] = 18 \text{ V}$$

$$\text{Voltage at the far end} = 240 - 18 = 222 \text{ V}$$

Example 16.6.

The loads on the *R* phase of the distributor of example 16.5 are as under: 100 A, 0.707 lagging p.f. load at 100 m, 120 A, unity p.f. load at 250 m and 80 A, 0.8 lagging p.f. load at 400 m from the feeding point. The resistance and inductive reactance per km length are 0.25 and 0.125 ohm. Neglecting the voltage drop in the neutral wire, find the voltage across the load at the far end if the voltage at the feeding point is 240 V.

Solution

$$\cos\phi_1 = 0.707, \sin\phi_1 = 0.707$$

$$\cos\phi_2 = 1, \sin\phi_2 = 0$$

$$\cos\phi_3 = 0.8, \sin\phi_3 = 0.6$$

$$I_1 = 100 \text{ A}, I_2 = 120 \text{ A}, I_3 = 80 \text{ A}$$

$$r = \frac{0.25}{1000} \text{ ohm per m}$$

$$x = \frac{0.125}{1000} \text{ ohm per m}$$

Substituting the values in Eq. (16.18)

Total voltage drop

$$= 100 \times 100\left(\frac{0.25}{1000} \times 0.707 + \frac{0.125}{1000} \times 0.707\right) + 120 \times 250\left(\frac{0.25}{1000} \times 1 + 0\right)$$

$$+ 80 \times 400 \left(\frac{0.25}{1000} \times 0.8 + \frac{0.125}{1000} \times 0.6 \right) = 18.96 \text{ V}$$

Voltage at the far end = 240 – 18.95 = 221.05 V

Example 16.7.

In a three-phase four-wire distribution system having 240 volts between the line and neutral, there is a balanced three-phase motor load of 500 kW at 0.8 power factor lagging. Lamp loads connected between respective lines and neutral absorb 50 kW, 50 kW and 200 kW respectively. Calculate the current in each phase conductor and neutral wire of the system. Draw phasor diagram.

Solution

Let $I_1 I_2$ and I_3 denote the currents due to the lighting loads on phases *R*, *Y* and *B* respectively and I_m the current in each line due to motor load.

$$I_1 = \frac{50000}{240} = 208.33 \text{ A}$$

$$I_2 = \frac{50000}{240} = 208.33\text{A}$$

$$I_3 = \frac{200000}{240} = 833.33\text{A}$$

$$I_m = \frac{500000}{3 \times 240 \times 0.8} = 868.05\text{A}$$

The currents I_1, I_2 and I_3 are in phase with the respective line to neutral voltages. The currents I_m in the three phases lag behind their respective line to neutral voltages by $\cos^{-1} 0.8$, i.e., 36.87°.

The phasor diagram is shown in Fig. 16.18.

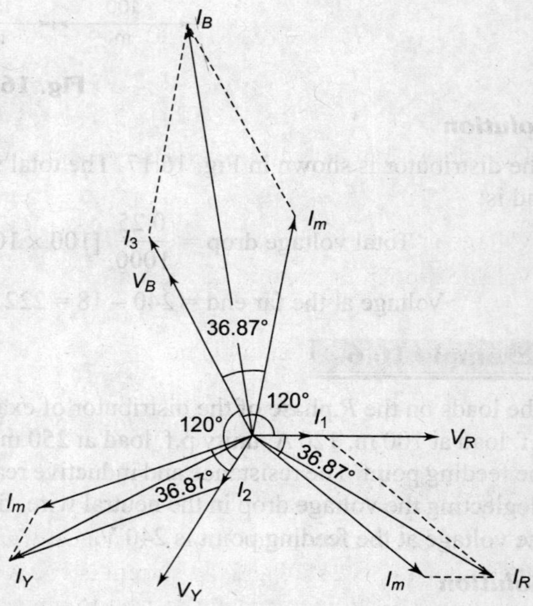

Fig. 16.18 Phasor diagram for example 16.7

The total line currents are

$$I_R = [(208.33)^2 + (868.05)^2 + (2 \times 208.33 \times 868.05 \times 0.8)]^{0.5} = 1042 \text{ A}$$

$$I_Y = [(208.33)^2 + (868.05)^2 + (2 \times 208.33 \times 868.05 \times 0.8)]^{0.5} = 1042 \text{ A}$$

$$I_B = [(833.33)^2 + (868.05)^2 + (2 \times 833.33 \times 868.05 \times 0.8)]^{0.5} = 1614 \text{ A}$$

The resultant of all the currents flows through the neutral wire. The motor current, being balanced, does not contribute to neutral current. To calculate the resultant, the currents, I_1, I_2 and I_3 have to be written as phasors.

$$I_1 = 208.33 \angle 0 = 208.33 + j\,0 \text{ A}$$

$$I_2 = 208.33 \angle -120 = -104.165 - j\,180.42 \text{ A}$$

$$I_3 = 833.33 \angle -240 = -416.66 + j\,721.67 \text{ A}$$

$$I_N = I_1 + I_2 + I_3 = 208.33 - 104.165 - j\,180.42 - 416.66 + j\,721.67$$

$$= -312.495 + j\,541.25 = 624.98 \text{ A.}$$

Example 16.8.

A three-phase four-wire distributor supplies a balanced voltage of 400/230 V to a load consisting of 80 A at power factor 0.8 lagging, 70 A at power factor 0.9 lagging and 50 A at unity power factor on phases *R*, *Y*, *B* respectively. Calculate the voltage at the supply end of phase *R* relative to the load voltage. The resistance of each core is 0.2 ohms.

Solution

Taking the receiving end voltage of *R* phase with respect to neutral as the reference phasor

$$I_R = 80\,(0.8 - j\,0.6) = 64 - j\,48 \text{ A}$$

$$I_Y = 70 \angle (-120^\circ - 25.8^\circ) = -57.89 - j\,39.34 \text{ A}$$

$$I_B = 50 \angle -240^\circ = -25 + j\,43.30 \text{ A}$$

The current in the neutral wire is the phasor sum of line currents.

$$I_N = I_R + I_Y + I_B = -18.89 - j\,44.04 \text{ A}$$

Voltage drop in the line conductor of *R* phase $= 0.2\,(64 - j\,48) = 12.8 - j\,9.6$ V

Voltage drop in the neutral conductor

$$= 0.2\,(-18.89 - j\,44.04) = -3.778 - j\,8.808 \text{ V}$$

Voltage of *R* phase at supply end

$$= 230 + j\,0 + 12.8 - j\,9.6 - 3.778 - j\,8.808$$

$$= 239.002 - j\,18.408 = 239.7 \angle -4.4^\circ \text{ volts}$$

Example 16.9.

A single phase distributor has a resistance of 0.2 ohm and a reactance of 0.3 ohm. At far end the voltage V_b is 240 V and the current is 100 A at p.f. 0.8. lagging. At mid point *a* the current is 100 A at 0.6 lagging p.f. with respect to voltage V_a at *a*. Find supply voltage and phase angle between V_s and V_b.

Solution

The distributor is shown in Fig. 16.19

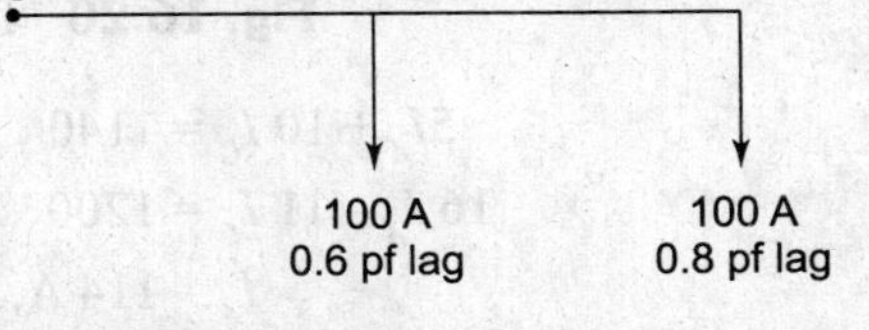

Fig. 16.19 Distributor for example 16.9

$$V_b = 240 \angle 0$$

$$V_a = 240 + 100\,(0.8 - j.6)\,(.1 + j.15)$$

$$= (240 + 17) + j\,(12 - 6)$$

$$= 257 + j\,6 = 257.07 \angle 1.34^\circ$$

The current at *a* is lagging by $\cos^{-1} 0.6$, L.I., 53.13° with respect to V_a. Therefore, phase angle of this current with respect to V_b is (53.13 – 1.34) or 51.79° lagging.

The current in section *sa*

$$= 100\,(\cos 51.79^\circ - \sin 51.79^\circ) + 100\,(.8 - j.6) = 141.85 - j\,138.57 \text{ A}$$

$$V_s = V_a + (141.85 - j\,138.57)\,(.1 + j.15)$$

$$= 257 + j6 + 14.185 + j\,21.28 + 20.78 - j\,13.85$$
$$= 291.965 + j\,13.43 = 292.27 \angle 2.63° \text{ V}$$

Hence, $|V_s| = 292.27$ V and phase angle between V_s and V_b is 2.63°.

Example 16.10.

Find the currents supplied at points A and B of the ring main distributor shown in Fig. 16.20. All loads are at unity p.f.: Assume

(a) $V_A = V_B$ (b) $V_B = V_A + 5$ V

Solution

Assume currents as shown. Total load current = 200 A. If I_A is the current fed at A, then current fed at B would be $200 - I_A$.

(a) $V_A = V_B$. Therefore, voltage drop from point A to point B in either direction is zero. Therefore,

$$.02\,I_A + .01\,I_C + .03\,(I_C - 20) + .04\,(I_C - 70) + .02\,(I_C - 100) - 0.03(200 - I_A) = 0$$
$$.02\,I_A + .05\,(I_A - I_C) + .06\,(I_A - I_C - 100) - 0.03\,(200 - I_A) = 0$$

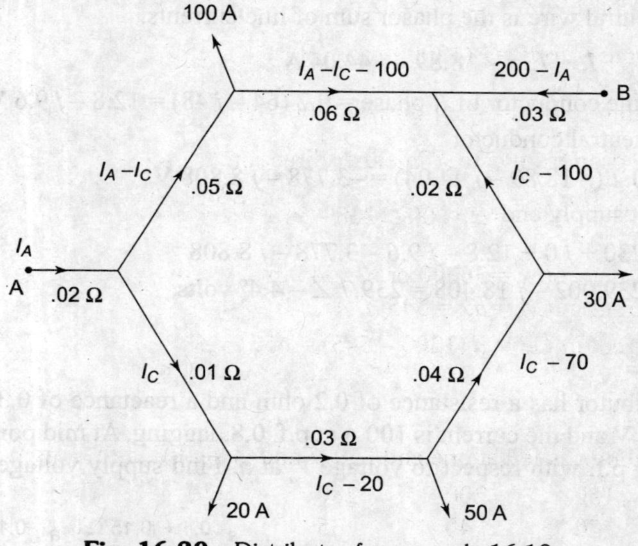

Fig. 16.20 Distributor for example 16.10

or $$5I_A + 10\,I_C = 1140$$
$$16\,I_A - 11\,I_C = 1200$$

or $$I_A = 114 \text{ A}, I_B = 200 - I_A = 86 \text{ A}$$

(b) $V_B = V_A + 5$. Therefore, voltage drop from point A to point B be – 5V.

$$.02\,I_A + .01\,I_C + .03\,(I_C - 20) + .04\,(I_C - 70) + .02\,(I_C - 100) - .03\,(200 - I_A) = -5$$
$$.02\,I_A + .05\,(I_A - I_C) + .06\,(I_A - I_C - 100) - .03\,(200 - I_A) = -5$$

or $$5\,I_A + 10\,I_C = 640$$
$$16\,I_A - 11\,I_C = 700$$

or $$I_A = 65 \text{ A}, I_B = 200 - I_A = 135 \text{ A}$$

Example 16.11.

A single phase distributor AB, 1200 m long is loaded uniformly with unity p.f. current of a A/m. It is supplied from a substation C through two single phase feeders CA and CB which are 900 m and 600 m long respectively. The section of each feeder conductor is 50% more than that of distributor. Find point X of minimum voltage and current in each feeder.

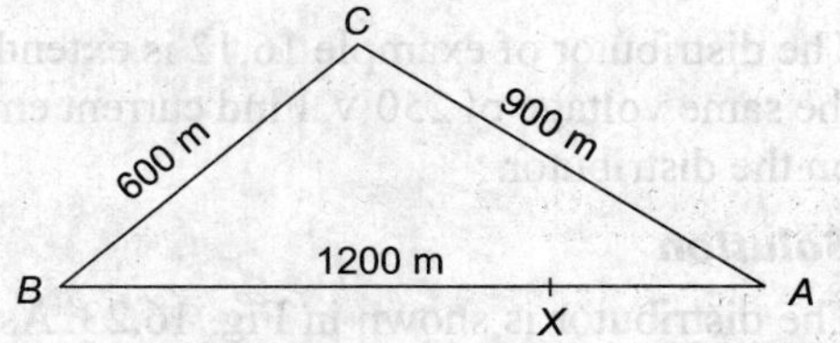

Fig. 16.21 Distributor for example 16.11

Solution

The system is shown in Fig. 16.21.

Let the point X be x m from end A. Then current in feeder $CA = a\,(x)A$ and current in feeder $CB = a\,(1200 - x)A$. Let the resistance of feeder be r ohm/m. Since cross-feeders of feeders is 50% more than that of distributor, resistance of distributor is 1.5 r ohm/m.

Voltage drop from C to $A = r\,(900)\,ax$

Voltage drop from A to $X = \frac{1}{2}\,(1.5r)(ax^2)$

Voltage drop from C to $B = r\,(1200 - x)\,a\,(600)$

Voltage drop from B to $X = \frac{1}{2}\,(1.5\,r\,a)\,(1200 - x)^2$

Equating the voltage drop from C to X in either direction.

$$900\,r\,a\,x + \frac{1.5\,r\,a\,x^2}{2} = a\,r\,600\,(1200 - x) + \frac{1.5ra(1200 - x)^2}{2}$$

or $\quad 66\,x = 36000$ or $x = 545$ m

Current in $CA = ax = 545\,a, A$

Current in $CB = a\,(1200 - 545) = 655\,a, A$

Example 16.12.

A single phase distributor is fed at one end 250 V and is loaded with unity p.f. loads as under:

Distance, m	150	200	280	320	390	450	500
Load, A	20	40	35	25	10	20	30

The resistance of double run is 0.1 ohm. Find voltage at the far end.

Solution

The distributor is shown in Fig. 16.22.

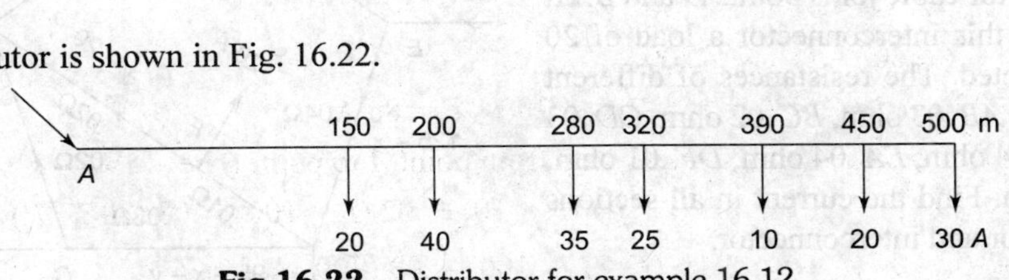

Fig 16.22 Distributor for example 16.12

$$\text{Total voltage drop} = \frac{0.1}{2\times 500}\,(150\times 20 + 200\times 40 + 280\times 35 + 320\times 25 + 390\times 10 + 450\times 20 + 500\times 30)2 = 11.34\text{ V}$$

Voltage at far end = 250 – 11.34 = 238.66 V

Example 16.13.

The distributor of example 16.12 is extended by 150 m by the same sized cable and fed there at the same voltage of 250 V. Find current entering from each feeding point and minimum voltage on the distributor.

Solution

The distributor is shown in Fig. 16.23. Assume that current I is fed at point A. The currents in all the sections are as shown. The current leaving point B is $(I - 180)$ A. Since points A and B are at the same potential, the voltage drop from A to B is zero.

$$\text{voltage drop} = \frac{0.1}{2\times 500}[I \times 150 + (I - 20)\,50 + (I - 60)80 + (I - 95)\,40 + (I - 120)\,70 + (I - 130)\,60 + (I - 150)\,50 + (I - 180)\,150]\,2 = 0$$

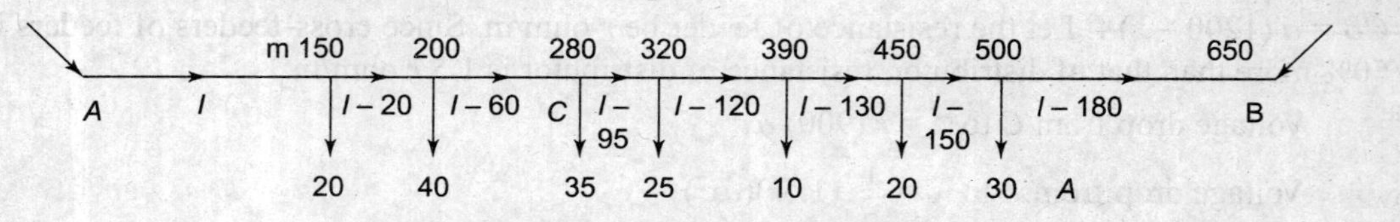

Fig. 16.23 Distributor for example 16.13

or $I = 92.77$A

Current fed at A = 92.77A

Current fed at B = 180 – 92.77 = 87.23 A

Starting from point A, loads of 20 and 40 A are supplied from end A. For the load of 35A a current of (92.77 – 20 – 40), i.e., 32.77 is fed from point A and (35 – 32.77), i.e., 2.23 A from B.

$$\text{Voltage drop} = 2[(92.77 \times 150) + (72.77 \times 50) + (32.77 \times 80)]\ \frac{0.1}{500\times 2} = 4.035 \text{ V}$$

Minimum voltage at point C = 250 – 4.035 = 245.965 V.

Example 16.14.

Four power loads A, B, C, D are connected to a ring distributor and take 20, 30, 25 and 30 A respectively. The ring is supplied from a substation at point E between A and D. An interconnector cable joins points D and B. At point F on this interconnector a load of 20 A is connected. The resistances of different sections are AB .03 ohm, BC .02 ohm, CD .03 ohm, DE .04 ohm, EA .04 ohm, DF .01 ohm, FB .02 ohm. Find the current in all sections of distributor and interconnector.

Solution

Figure 16.24 shows the configuration. Assume current x in section EA and y in section DF. The currents in other sections will be as

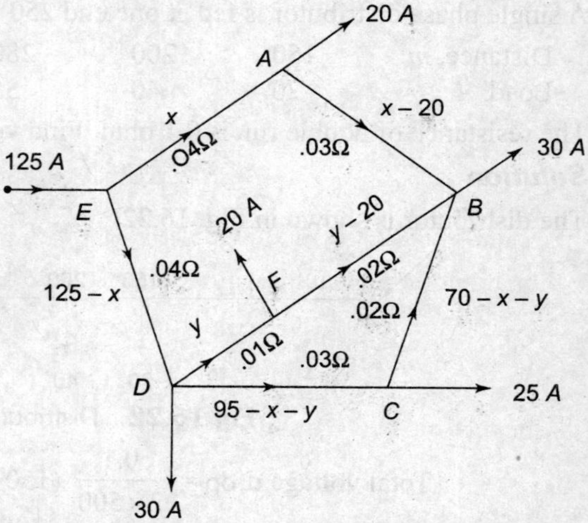

Fig. 16.24 Distributor for example 16.14

shown. The values of x and y can be found by equating voltage drops between points (say) E and B and D and B.

$$.04\,x + .03\,(x - 20) = .04\,(125 - x) + .03\,(95 - x - y) + .02\,(70 - x - y)$$

$$.03\,(95 - x - y) + .02\,(70 - x - y) = .01\,y + .02\,(y - 20)$$

or
$$16\,x + 5\,y = 985$$
$$5x + 8y = 465$$

hence
$$x = 53.93\text{ A},\ y = 24.42\text{ A}$$

The currents in various sections are

EA 53.93A, *AB* 33.93A, *ED* 71.07 A, *DF* 24.42 *DC* 16.65 A, *BC* 8.35 A, *FB* 4.42 A.

Example 16.15.

A single phase ring distributor *ABC* is fed at point *A*. The impedances of sections *AB*, *BC*, *AC* are $2 + j1$, $2 + j3$ and $1 + j2$ ohms respectively. Load at *B* is 40 A at 0.8 p.f. lagging while that at *C* is 60 A at 0.6 p.f. lagging, both with respect to voltage at *A*. Find current in the three sections with respect to voltage at *A*. [*RTM Nagpur Uni. 2010*]

Solution

Figure 16.25 shows the distributor. Assume current $(x - jy)$ in section *AB*. Then the currents in other two sections will be as shown. x and y can be found by equating the total voltage drop around the ring to zero.

Thus,

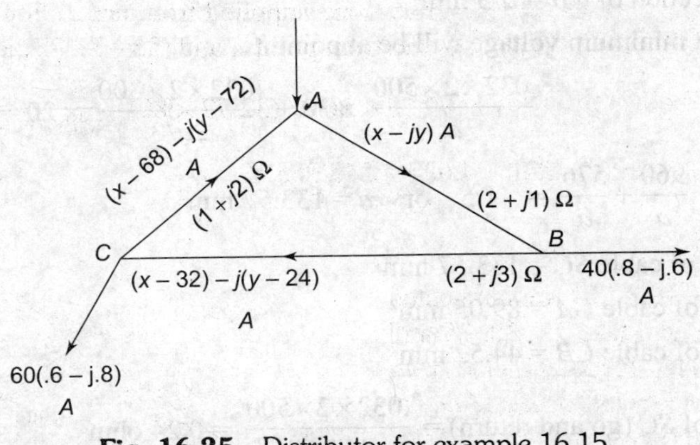

Fig. 16.25 Distributor for example 16.15

$$(x - jy)\,(2 + j1) + [(x - 32) - j\,(y - 24)]\,(2 + j3) + [(x - 68) - j\,(y - 72)]\,(1 + j2) = 0$$

Equating the real and imaginary parts to zero

$$5x + 6y = 348$$
$$6x - 5y = 112$$

whence $x = 39.54$ A and $y = 25.05$ A

Current in $AB = 39.54 - j\,25.05$ A from *A* to *B*

Current in $AC = 28.46 - j\,46.95$ A from *A* to *C*

Current $BC = 7.54 - j\,1.05$ A from *B* to *C*

Example 16.16

Two consumers A and B draw 20 A and 10 A at unity p.f. respectively. The feeding point S is fed at 230 V single phase supply. From S the supply is taken to a point C 500 m away from S. Distances CA and CB are 300 m and 200 m respectively. All the three cables are worked at the same current density. The maximum permissible voltage drop is 5%. (a) Find cross-sectional area of cables (b) percentage power loss.

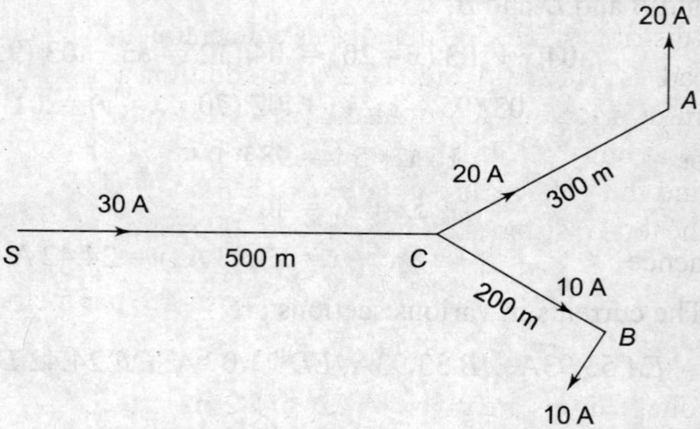

Fig. 16.26 System for example 16.16

Solution

The configuration is shown in Fig. 16.26.

Using aluminium conductor cable, resistance of cable = 0.0286 ohm/m length/mm² area at 20°C. Assuming a temperature rise of 30°C and $\alpha = 0.004$

$$\rho = 0.0286\ (1 + 30 \times 0.004) = 0.32 \text{ ohm/m/mm}^2$$

(a) Let cross-section of cable $SC = a$ mm²

$$\text{Then cross-section of } CA = \frac{a \times 20}{30} = 2a/3 \text{ mm}^2$$

$$\text{Cross-section of } CB = a/3 \text{ mm}^2$$

It is evident that minimum voltage will be at point A.

$$\frac{.032 \times 2 \times 500}{a} \times 30 + \frac{.032 \times 2 \times 300}{2a/3} \times 20 = \frac{5}{100} \times 230$$

$$\frac{960}{a} + \frac{576}{a} = 11.5 \quad \text{or} \quad a = 133.57 \text{ mm}^2$$

Cross-section of cable SC = 133.57 mm²

Cross-section of cable CA = 89.05 mm²

Cross-section of cable CB = 44.52 mm²

$$\text{(b) Resistance of } SC \text{ (go and return)} = \frac{.032 \times 2 \times 500}{133.57} = 0.24 \text{ ohm}$$

$$\text{Resistance of } CA \text{ (go and return)} = \frac{.032 \times 2 \times 300}{98.05} = 0.216 \text{ ohm}$$

$$\text{Resistance of } CB \text{ (go and return)} = \frac{.032 \times 2 \times 200}{44.52} = 0.288 \text{ ohm}$$

$$\text{Total power loss} = 30^2 \times .24 + 20^2 \times .216 + 10^2 \times 0.288 = 331.2 \text{ W}$$

$$\text{Percentage power loss} = \frac{331.2}{230 \times 20 + 230 \times 10} \times 100 = 4.8\%$$

Example 16.17

A distributor *AF*, 500 m long has concentrated load as shown in Fig. 16.27. In addition a uniform load of 1 A/m also exists. All loads are at unity p.f. It is fed at 230 V at point *A*. Find the voltage at points *B*, *C*, *D*, *E* and *F*. The total resistance of distributor is .075 ohm.

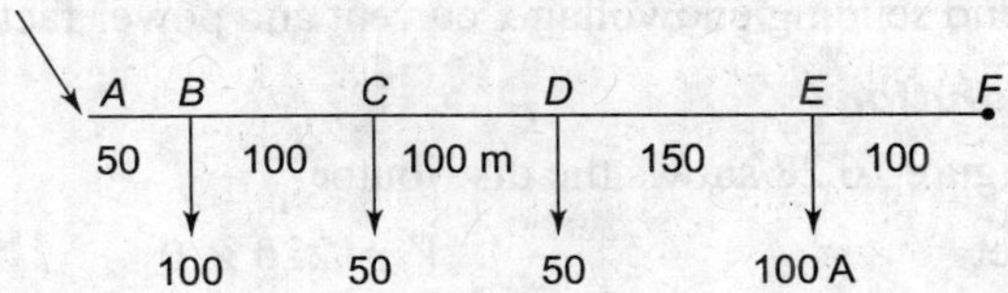

Fig. 16.27 Distributor for example 16.17

$$r = \text{resistance per metre} = \frac{0.075}{500} = 0.15 \times 10^{-3}\ \Omega = \text{m}$$

$$\cos\phi = 1,\ \sin\phi = 0,\ i = 1\ \text{A/m}$$

Voltage drops are given by Eq. (16.21)

Current entering section $AB = 100 + 50 + 50 + 100 + 500 = 800$ A

$$\text{Voltage drop in section } AB = 50\left(800 - \frac{1}{2}\times 1\times 50\right)(0.15\times 10^{-3}) = 5.8125\ \text{V}$$

Current entering section $BC = 800 - 150 = 650$ A

$$\text{Voltage drop in section } BC = 100\left(650 - \frac{1}{2}\times 1\times 100\right)(.15\times 10^{-3}) = 9\ \text{V}$$

Current entering section $CD = 650 - 100 - 50 = 500$ A

$$\text{Voltage drop in section } CD = 100\left(500 - \frac{1}{2}\times 1\times 100\right)(.15\times 10^{-3}) = 6.75\ \text{V}$$

Current entering section $DE = 500 - 100 - 50 = 350$ A

$$\text{Voltage drop in section } DE = 150\left(350 - \frac{1}{2}\times 1\times 150\right)(.15\times 10^{-3}) = 6.1875\ \text{V}$$

Current entering section $EF = 350 - 150 - 100 = 100$ A

$$\text{Voltage drop in section } EF = 100\left(100 - \frac{1}{2}\times 1\times 100\right)(.15\times 10^{-3}) = .75\ \text{V}$$

Voltage at point $B = 230 - 5.8125 = 224.1875$ V
Voltage at point $C = 224.1875 - 9 = 215.1875$ V
Voltage at point $D = 215.1875 - 6.75 = 208.4375$ V
Voltage at point $E = 208.4375 - 6.1875 = 202.25$ V
Voltage at point $F = 202.25 - .75 = 201.5$ V

Example 16.18.

A two-wire distributor 1200 m long supplies a load of 100 A at 0.8 p.f. lagging at its far end and a load of 60 A at 0.9 p.f. lagging at mid point. Both power factors are with respect to voltage at far end which is to be

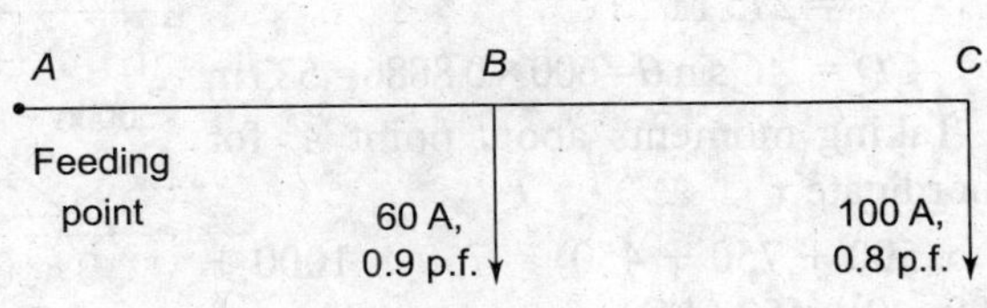

Fig. 16.28

maintained at 220 V. The resistance and reactance of each line are .15 ohm and .2 ohm respectively. Find sending end voltage, current and power factor.

Solution

Figure 16.28 shows the distributor

Let $V_C = 220 \angle 0$

$$Z_{AB} = Z_{BC} = 2 \times \frac{1}{2} \ (.15 + j.2) = (.15 + j.2) \text{ ohm}$$

$$V_B = 220 \angle 0 + 100\ (.8 - j.6)\ (.15 + j.2) = 244 + j\,7 \text{ V}$$

$$I_{AB} = 100\ (.8 - j.6) + 60\ (.9 - j.\ 436) = 134 - j\ 86.16 \text{ A}$$

$$= 159.31 \angle - 32.74^\circ \text{ A}$$

$$V_A = 244 + j\,7 + (134 - j\ 86.16)\ (.15 + j.2) = 281.332 + j\,20.876 \text{ V}$$

$$= 282.1 \angle 4.24^\circ \text{ V}$$

Sending end voltage = 282.1 V

Sending end current = 159.31 A

Sending end power factor = cos (4.24 + 32.74) = 0.799 lagging.

Example 16.19.

In the system shown in Fig. 16.29 the loads at points A, B, C are 500 kW, 750 kW and 450 kW respectively. Find the point that provides the most economical centre of distribution.

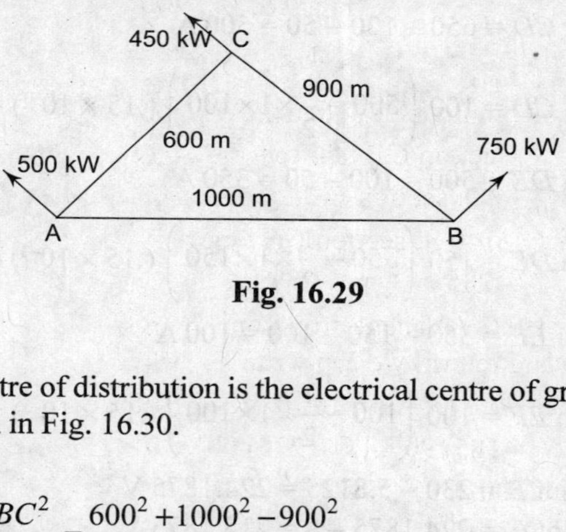

Fig. 16.29

Solution

The most economical centre of distribution is the electrical centre of gravity. Let this point have coordinates *x*, *y* as shown in Fig. 16.30.

For the triangle *ACD*

$$\cos\theta = \frac{AC^2 + AB^2 - BC^2}{(AC)(AB)} = \frac{600^2 + 1000^2 - 900^2}{2(600)(1000)}$$

$$= 0.4583$$

or $\theta = 62.7°$, $\sin\theta = 0.8886$

$$AD = AC\ \cos\ \theta = 600 \times 0.4583$$

$$= 275 \text{ m}$$

$$CD = AC \sin\theta = 600 \times 0.8886 = 533 \text{ m}$$

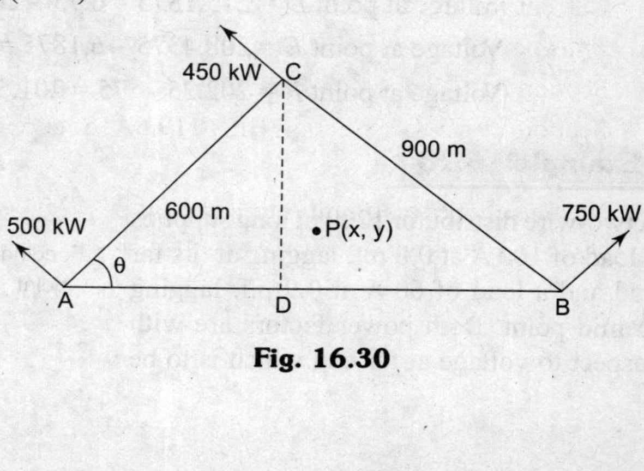

Fig. 16.30

Taking moments about point *A*, for coordinate *x*

$x(500 + 750 + 450) = 750 \times 1000 + 500 \times 0 + 450\ (AD)$

or $\quad 1700\,x = 750000 + 450\,(275)$

or $\quad x = 513.97$

Similarly for co-rdinates y, taking moments about point A

$y \times 1700 = 500(0) + 750\,(0) + 450\,(533)$

or $\quad y = 141$ m

Hence the most economical centre of distribution has coordinates 513.97, 141 m with respect to point A.

Example 16.20.

Fig. 16.31 shows a 2 wire dc distributor fed at both ends at 460 V. Find the minimum cross-sectional area of the cable so that the minimum voltage between the two wires is not less than 440 V. Resistivity of wire is 1/58 ohm per m per sq.mm. Distances are in metres.

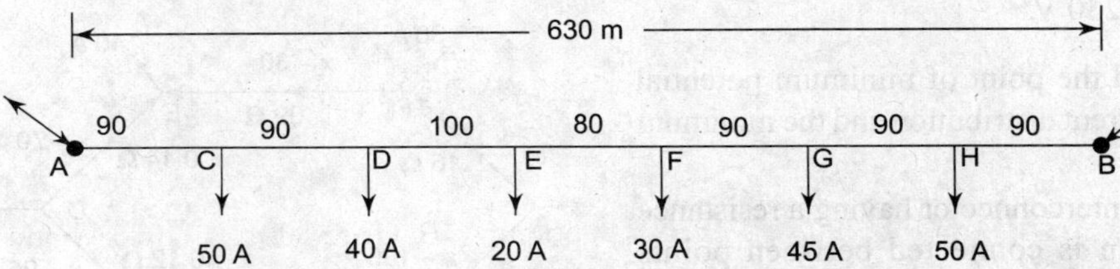

Fig. 16.31

Solution

Let the current supplied from point A be I_A. Then the currents in different sections will be as under

Section AC : Current = I_A, section CD : current = $I_A - 50$

Section DE : Current = $I_A - 90$, section EF : current = $I_A - 110$

Section FG : Current = $I_A - 140\,A$, section GH : current = $I_A - 185$

Section HB : Current = $I_A - 235$

Since both ends are fed at 460 V, Voltage drop from A to B should be zero. Taking r as resistance per m length of cable (go and return) we can write

$$r[I_A \times 90 + (I_A - 50)\,90 + (I_A - 90)\,100 + (I_A - 110)80$$
$$+ (I_A - 140)90 + (I_A - 185)\,90 + (I_A - 235)\,90] = 0$$

or $\quad r[I_A \times 360] = r \times 72700$

or $\quad I_A = 115.4$ A

Current fed from end B = 235 – 115.4 = 119.6A

Thus point F is the point of minimum voltage. Currents in different sections are :

Section AC : 115.4A, Section CD : 65.4A, Section DE : 25.4A

Section EF : 5.4 A, Section BH : 119.6A, Section HG : 69.6A

Section GF : 24.6 A

Voltage chop from point A to F should be (460 – 440) = 20 V

$r[115.4 \times 90 + 65.4 \times 90 + 25.4 \times 100 + 5.4 \times 80] = 20$

or $$r = \frac{20}{19244}$$

For each 1 m (go and return) of cable $r = \frac{\rho \times 2}{a}$

Hence $$\frac{20}{19244} = \frac{\rho \times 2}{a} = \frac{\frac{1}{58} \times 2}{a}$$

$$a = \frac{2}{58} \times \frac{19244}{20} = 33.18 \text{ mm}^2$$

Example 16.21.

A two wire dc distributor ABCDEFA is in the form of a ring main and is loaded as under: 20 A at B, 50 A at C, 25 A at D, 40 A at E, 30 A at F. The resistances are AB = 0.08 ohm, BC = 0.1 ohm, CD = 0.12 ohm, DE = 0.14 ohm, EF = 0.09 ohm and FA = 0.16 ohm. It is fed at point A at 230 V.

(a) Find the point of minimum potential and the current distribution and the minimum voltage.

(b) An interconnector having a resistance of 0.1 ohm is connected between points A and D. Find the point of minimum of potential, voltage at this point and the current distribution.

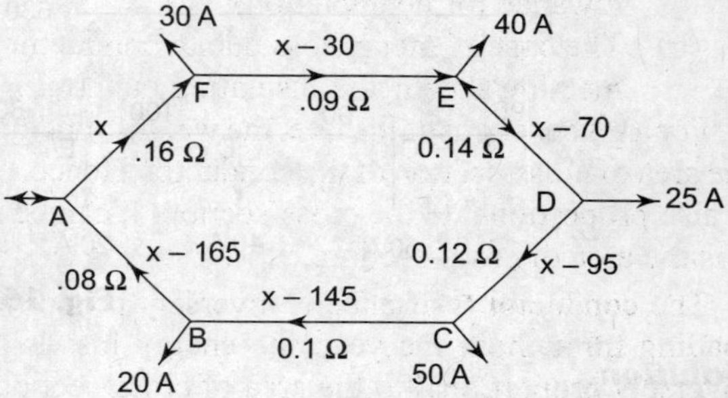

Fig. 16.32

Solution (a)

Let x be the current in section AF. Then currents in other sections are as shown. The total voltage chop along loop AFED CBA should be zero. Then

$0.16\,x + 0.09\,(x - 30) + 0.14\,(x - 70) + 0.12\,(x - 95) + 0.1\,(x - 145) + 0.08\,(x - 165) = 0$

or $x = 74.8$ A

Thus currents in various sections are

AF = 74.8 A, FE = 44.8 A, EB = 4.8 A, AB = 90.2A, BC = 70.2 A, CD = 20.2 A

Point B is the point of minimum potential

Potential at point B = 230 – 74.8 (0.16) – 44.8 (0.09) – 4.8 (0.14) = 213.328 V

(b)

The configuration is shown in Fig. 16.33. Let x be the current in section AF and y in interconnector. Then currents in remaining sections are as shown.

Voltage drop in loop AFEDA = $0.16\,x + 0.09\,(x - 30) + 0.14\,(x - 70) - 0.1y = 0$

or $0.39x - 0.1y - 12.5 = 0$ (*i*)

Voltage drop in loop ABCDA = $0.08\,(165 - x - y) + 0.1\,(145 - x - y) + 0.12\,(95 - x - y) - 0.1y = 0$

or $-0.3x - 0.4y + 39.1 = 0$ (*ii*)

From Eq. (*i*) and (*ii*) we get

$x = 47.9$ A

$y = 61.8$ A

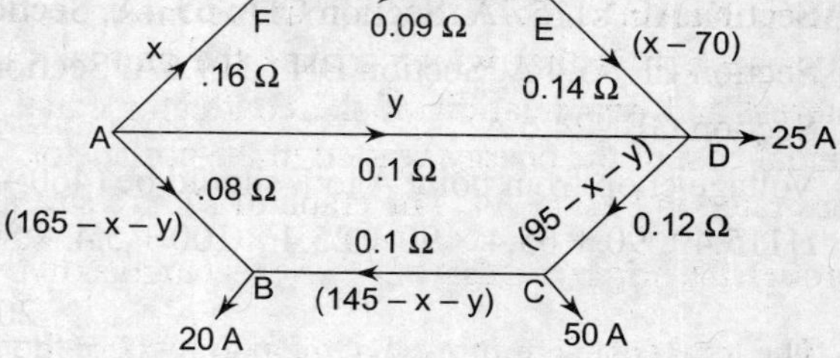

Fig. 16.33

Current distribution is as under

AB = 55.3A, BC = 35.3 A, DC = 14.7 A, AF = 47.9A, FE = 17.9 A, DE = 22.1 A

Potential at point E = 230 – 47.9 × .16 – 17.9 × 0.09 = 220.725 V

Potential at point C = 230 – 55.3 × 0.08 – 35.3 × 0.1 = 222.05 V

Thus minimum potential is 220.725 V at point E.

16.7 KELVIN'S LAW

The design of distributor is governed mainly by the consideration of voltage drop. A distributor cannot be loaded to its current carrying capacity because such a loading would result in an excessive voltage drop. On the other hand, the voltage drop is not such a vital factor in the design of a feeder and, therefore, a feeder can be designed on the basis of current carrying capacity and minimum cost.

The cost of a conductor is made up of two components:

(a) The interest on the capital cost of purchase and installation of conductor (plus an allowance for depreciation).

(b) The cost of energy loss due to conductor resistance and in the case of cables, losses in metallic sheath and insulating material.

For a given length of the line, the weight and, therefore, the cost of conductor is proportional to the area of cross-section of the conductor. Hence, the annual cost due to interest and depreciation is also proportional to the cross-section. It can be written as Rs. $P \times a$ where P is a constant and a is the area of cross-section.

The conductor resistance is inversely proportional to the area of cross-section. For a given loading throughout the year, the energy loss is proportional to the resistance and, therefore, inversely proportional to the area of cross-section. Hence the energy loss can be written as Rs $\frac{Q}{a}$ where Q is a constant.

Total cost $$C = Pa + \frac{Q}{a} \tag{16.23}$$

For the total cost to be minimum $\frac{dC}{da}$ must be zero

$$\frac{dC}{da} = P - \frac{Q}{a^2} = 0$$

or $$a = \left(\frac{Q}{P}\right)^{0.5} \tag{16.24}$$

For this value of cross-section both the components of the total cost become equal, each being equal to $\sqrt{PQ}$. This is Kelvin's law. It states that the most economical cross-section is that which makes the annual value of interest and depreciation of the conductor equal to the annual cost of the energy wasted in the conductor. This is illustrated in Fig. 16.34. The graph of Pa is a straight line through the origin and that of $\frac{Q}{a}$ is a rectangular hyperbola.

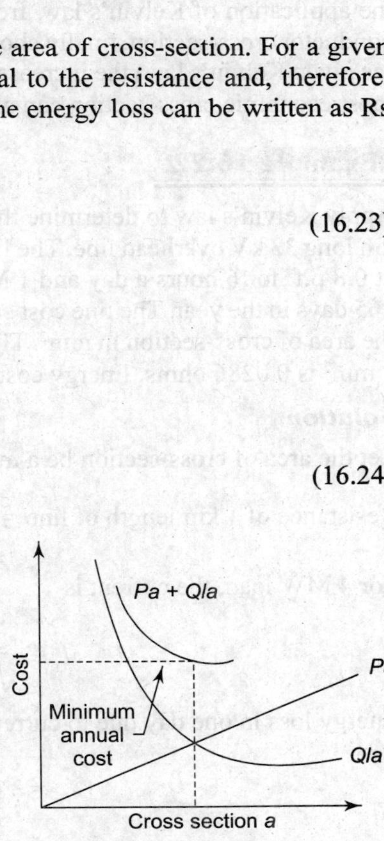

Fig. 16.34 Kelvin's Law

The total cost is minimum at the point where the graphs of Pa and $\frac{Q}{a}$ intersect each other.

16.8 LIMITATIONS OF KELVIN'S LAW

Kelvin's law cannot be precisely applied in practice. As a matter of fact some amount of discretion is necessary in its application. This is due to the fact that the energy costs, interest and depreciation have no relation to the purely physical aspects of the problem such as resistance, temperature rise, voltage drop, etc. If two systems have exactly the same load demand, they should have the same conductor size from the engineering point of view. However, if the energy, interest and depreciation costs are different in the two cases, the application of Kelvin's law will give different conductor sizes for them.

The cost of conductor is not the only cost which varies with the cross-section. The cost of insulation also varies with the conductor cross-section at least to some extent. It is difficult to express the cost of insulation in terms of conductor cross-section.

The current which can be safely carried by cables depends on the permissible temperature rise. Moreover, besides I^2R loss in the conductor, dielectric loss in metallic sheath also takes place all the time irrespective of the current carried by the conductor of the cable. It is very difficult to take all the losses into account while calculating the economical cross-section.

In the case of overhead lines the question of temperature is less important than in the case of cables. Moreover, the only appreciable loss in overhead lines is the $I^2 R$ loss in conductor. Therefore, Kelvin's law is likely to give more acceptable results when applied to overhead lines upto 33 kV. In the case of cables, it is necessary to examine the cross-section, as obtained by the application of Kelvin's law, from the consideration of permissible temperature rise. If the conductor cross-section, finally chosen, differs somewhat from the most economic cross-section (as given Kelvin's law) the increase in cost will not be much. As such small deviations from the most economic cross-section can always be made.

Example 16.22.

Apply Kelvin's law to determine the economic cross-section for the conductor of a 3-phase 10 km long 33 kV overhead line. The line supplies a load of 4 MW at 0.8 p.f. 10 hours a day, 2 MW at 0.8 p.f. for 6 hours a day and 1 MW at 0.8 p.f. for 8 hours a day. The line is used for all the 365 days in the year. The line cost can be taken as Rs (85000 + 2000 a) per km length where a is the area of cross-section in mm^2. The resistance of aluminium conductor of length 1 m and area 1 mm^2 is 0.0286 ohms. Energy cost is Re 0.80 per kWh.

Solution

Let the area of cross-section be a mm^2

$$\text{Resistance of 1 km length of line} = \frac{0.0286}{a} \times 1000 = \frac{28.6}{a} \text{ ohms per phase}$$

For 4 MW load, the current is

$$I_1 = \frac{4 \times 10^6}{\sqrt{(3)} \times 33 \times 10^3 \times 0.8} = 87.48 \text{ A}$$

Energy loss in one day due to current I_1 in 1 km length of line

$$= 3\,(87.48)^2 \left(\frac{28.6}{a}\right)\left(\frac{10}{1000}\right) = \frac{6566.1}{a} \text{ kWh}$$

For 2 MW load, the current is

$$I_2 = \frac{2\times10^6}{\sqrt{(3)}\times33\times10^3\times0.8} = 43.74 \text{ A}$$

Energy loss in one day to current I_2 in 1 km length of line

$$= 3\,(43.74)^2\left(\frac{28.6}{a}\right)\left(\frac{6}{1000}\right) = \frac{984.9}{a} \text{ kWh}$$

For 1 MW load, the current is

$$I_3 = \frac{2\times10^6}{\sqrt{(3)}\times33\times0.8\times10^3} = 21.87 \text{ A}$$

Energy loss due to 1 MW load in one km length of line

$$= 3\,(21.87)^2\left(\frac{28.6}{a}\right)\left(\frac{8}{1000}\right) = \frac{328.2}{a} \text{ kWh}$$

$$\text{Total energy loss in one day} = \frac{1}{a}\,(6566.1 + 984.9 + 328.2) = \frac{7879.2}{a} \text{ kWh}$$

$$\text{Yearly cost of energy losses} = \frac{7879.2}{a}\times365\times0.80 = \text{Rs } \frac{2300726.4}{a}$$

$$\text{Total cost per km of line } 85000 + 2000\,a + \frac{2300726.4}{a}$$

For minimum total cost, the differential of total cost with respect to a should be zero.

Thus $$a^2 = \frac{2300726.4}{2000} = 1150.36$$

or $$a = 33.91 \text{ sq. mm}$$

The standard size nearest to this area (as per IS: 398–1976) is 6/1/2.59 ACSR conductor having an aluminium area of 31.61 mm^2. The conductor size 6/1/2.59 means 6 strands of aluminium and 1 strand of steel each strand having a diameter of 2.59 mm. The current carrying capacity of this conductor for a temperature rise of 40°C is 160 A which is more than the maximum current to be carried by the feeder.

Example 16.23.

A single phase cable has motor loads as shown in Fig. 16.35. The mean motor voltage is 220 V. Find cross-section of each core of cable for a transmission loss of 5%. Feeding point is at *A*.

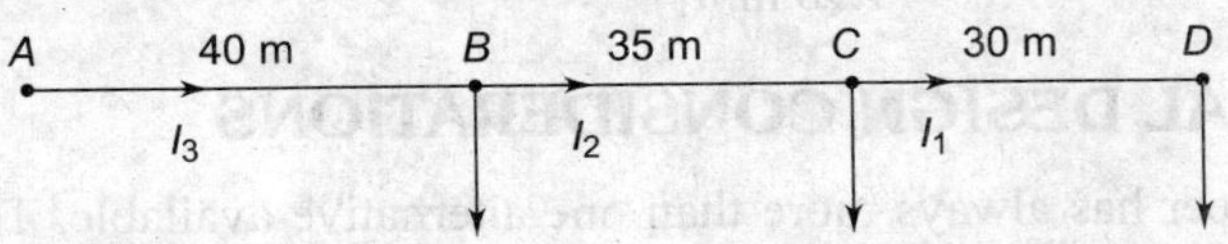

	B	C	D
HP	10	26	5
Efficiency	0.83	0.87	0.81
p.f.	0.82	0.85	0.77

Fig. 16.35 System for example 16.23

Solution

Using Aluminium conductor, resistance of cable = 0.0286 ohm/m length/mm² area, at 20°C. Assuming a temperature rise of 30°C and $\alpha = 0.004$.

$$\rho = .0286\ (1 + 30 \times .004) = 0.032 \text{ ohm/m/mm}^2$$

$$\text{Current of motor at point } D = \frac{5 \times 735.5}{220 \times 0.81 \times 0.77} = 26.8 \text{ A at } 0.77 \text{ p.f.}$$

$$I_1 = \text{current section } CD = 26.8\ (.77 - j.638) \text{ A}$$

$$\text{Current of motor at point } C = \frac{26 \times 735.5}{220 \times .87 \times .85} = 117.54 \text{ at } 0.85 \text{ p.f.}$$

$$= 117.54\ (.85 - j\ .527) \text{ A}$$

$$I_2 = \text{current section } BC$$

$$= 117.54\ (.85 - j.\ 527) + 26.8\ (.77 - j\ .638)$$

$$= 120.54 - j\ 79 = 141.12 \angle -33.24° \text{ A}$$

$$\text{Current of motor at point } B = \frac{10 \times 735.5}{220 \times .83 \times .82}$$

$$= 49.12 \text{ A at } 0.82 \text{ p.f.} = 49.12\ (.82 - j\ 0.57)$$

$$I_3 = \text{current in section } AB = 120.54 - j\ 79 + 49.12\ (.82 - j.\ 57)$$

$$= 160.82 - j\ 107 = 193.16 \angle -33.64° \text{ A}$$

$$\text{Total power} = 220 \times 193.16 \cos (33.64°) \times 10^{-3} = 35.38 \text{ kW}$$

$$\text{Permissible power loss} = \frac{5}{100} \times 35.38 = 1.769 \text{ W}$$

Let cross-section of cable = a mm²

$$\text{Power loss in section } AB = 2\ (193.16)^2 \left(\frac{.032 \times 40}{a}\right) = \frac{9.55 \times 10^4}{a} \text{ W}$$

$$\text{Power loss in section } BC = 2(141.12)^2 \left(\frac{.032 \times 35}{a}\right) = \frac{4.46 \times 10^4}{a} \text{ W}$$

$$\text{Power loss in section } CD = 2(26.8)^2 \left(\frac{.032 \times 30}{a}\right) = \frac{.138 \times 10^4}{a} \text{ W}$$

$$\text{Total power loss} = \frac{14.148 \times 10^4}{a} = 1.769 \times 10^3$$

or $$a = 79.98 \text{ mm}^2$$

16.9 GENERAL DESIGN CONSIDERATIONS

The planning engineer has always more than one alternative available. The choice of proper scheme is the most difficult and complex task because a number of factors have to be taken into account and some of these factors cannot be expressed in economic terms. Economy, reliability, voltage level, regulation and feasibility of expansion are some of the important considerations.

(a) *Economy* The planning engineer must evaluate the economics of different plans. The costs which have to be taken into account include the annual fixed costs (i.e. interest, depreciation, etc.), which can be taken as a certain percentage of total investment, operating expenses (i.e., labour, material and maintenance costs), overhead expenses and the cost of electrical losses. The different steps in economic study are:

1. Determination of total investment for each plan.
2. Determination of annual fixed costs, operating costs and overhead expenses.
3. Determination of cost of electrical losses.
4. Evaluation of the present worth of each plan. The present worth is the amount of money which would have to be invested at the present time to make the necessary funds available as and when required.
5. Selection of the most economic plan.

It is necessary to remember that many a time there are other factors which override economics. Moreover the economic study has its limitations and it should not be given more credibility than its accuracy dictates.

(b) *Reliability* Reliability can be expressed in a number of ways. From the consumers' point of view, the number of interruptions, total interruption time and average interruption time per outage are of main interest.

Many standard terms for expressing reliability of power systems have been framed. Some of these are

ASAI (Average Service Availability Index)

$$= \frac{\text{Customer hours when supply is available}}{\text{Total customer hours}} \tag{16.25}$$

SAIFI (System Average Interruption Frequency Index)

$$= \frac{\text{Total number of customers interruptions in the area}}{\text{Total number of customers in the area}} \tag{16.26}$$

SAIDI (System Average Interruption Duration Index)

$$= \frac{\Sigma \text{ (restoration time for each interruptions)} \times \text{(number of interrupted customers for each interruption)}}{\text{Total number of customers}} \tag{16.27}$$

CAIDI (Customer Average Interruption Duration Index)

$$= \frac{\Sigma \text{ Customer interruption durations}}{\text{Total number of customer interruptions}} \tag{16.28}$$

In addition to above another index is defined for monetary interruptions. IEEE std 1366 defines monetary interruptions as those which can result from each single operation of an interrupting device e.g. recloser.

MAIFI (Monetary average interruption frequency Index)

$$= \frac{\text{Total number of momentary customer interruptions}}{\text{Total number of customers}} \tag{16.29}$$

The above indices give average values. However, the maximum values of these quantities are also important. The service interruption can be due to outage in primary distribution as well as due to outage in secondary distribution. Outages in primary distribution affect a larger number of consumers and, therefore, need special attention. The use of mathematical techniques to express reliability in quantitative terms is progressively increasing. For more information on the application of these techniques for evaluating the reliability of distribution systems, the reader is referred to [53, 68] of the bibliography and Chapter 20.

(c) *Voltage Level and Regulation* Electrical equipments operate satisfactorily only if they are operated at or near the rated voltage. Excessively high or low voltage may lead to equipment failure or unsatisfactory service. The electricity supply regulation requires that the voltage at the consumers premises should remain within ± 6% of the declared voltage. The voltage drop occurs in primary distribution (and the distribution transformer), secondary distribution and service connections. Generally the distribution systems are designed for about 3% voltage drop in the primary distribution and about 6% voltage drop in the secondary distribution. In addition, the following voltage control methods are used singly or in combination.

1. Transformer tap changing under load.
2. Shunt capacitors to improve power factor thereby reducing voltage drop.
3. Series capacitors to compensate the line reactance.

(d) *Feasibility of Expansion* Distribution systems need almost continuous expansion. The expansions are usually carried out in small increments. One of the key concepts in the system design is, therefore, provision for expansion. The design must provide plans and equipment which can accommodate expected and unexpected future developments. The design should be adaptable to changes in different conditions. Some specific changed conditions and improvements which must be taken into account in the design are as under:

1. *Sites and Right of Way* Urbanization of areas tends to limit options of a utility as regards routing new circuit extensions after the initial system has been constructed. System design, therefore, should consider whether sites and rights of way will be available at a reasonable cost in future.

2. *Public Improvement and Area Development* Public improvements such as widened or regraded roads, bridges, sewer and water pipes affect the distribution systems. Many a time the expansion of the distribution systems is limited by these public improvements. It is very desirable if the expansion of the distribution system is deferred or advanced so as to co-ordinate with the public work. The process of removal of old buildings and replacing them by new buildings also affects distribution systems. As areas develop the load densities become higher and reliability requirements become more stringent. Sometimes it necessitates the replacement of overhead systems by underground systems.

3. *Growth of Load* The loads keep on increasing almost continuously. This growth is partly due to increased number of housing, commercial and other units and partly due to the fact that use of electrical gadgets also keeps on increasing. The growth rates may be different in different areas and this fact must be taken into account in the design.

4. *Voltage Conversion* The voltage conversion represents the greatest incremental step in expansion of distribution system. As system expands, a stage comes when it becomes necessary to switchover to a higher voltage. The system designer must look into future and anticipate the time when it would be necessary to switchover to the next higher voltage. The load areas of high growth should be identified so that these may be supplied directly from the higher voltage system.

5. *Capacity Additions* Design, layout and arrangement of electrical equipment like enclosures and supports should be economically and readily expandable at several different times during its service life. The supports and enclosures should be designed not only for initial ratings but also for the subsequent ratings at least in immediate future.

16.10 LOAD ESTIMATION

The estimation of load likely to come on the distribution systems immediately as well as in the future is the first step in the design of the distribution system. Generally the following estimates are necessary.

(a) *Short Term Estimates* The short term estimates refer to the length of time required to plan, construct and commission an addition to the system capacity to cater to an expected increase in load. The shortest period for which estimates are made is one year. The 400 V and 11 kV systems require a time of about two years for planning, construction and commissioning and, therefore, a short term estimate of 2 years is required for such systems. A 33 kV system requires 3 to 4 years and the forecast for a system of this voltage class should be made for the future 4 years.

(b) *Long Term Estimates* If the planning and construction is done only on the basis of the short term estimates, it is often found that the system reinforcement would be necessary within a very short time after the system is commissioned. In view of this long term estimates are also necessary. The long term estimates are generally prepared for about 5 – 7 years for 11 kV systems and for about 8 – 10 years for 33 kV systems.

Many sophisticated mathematical techniques are available for load forecasting. However, the simple extrapolation method, though approximate, is the most widely used method. This method consists in plotting the previously recorded winter maximum demands against time (in years). This produces a rather irregular curve from which a best fit curve can be obtained. The best fit curve can be obtained by trial and error or through digital computer (if a greater accuracy is desired). The best fit curve can be projected forward to give a forecast for the feature years. However, it must be noted that there are several factors like extremely bad weather conditions, discrete load increases, an unusual large growth due to local factors, etc., which may upset the results obtained from the extrapolation method and these factors should be taken into account in load forecasting.

The load estimation for a proposed new locality can be done on the basis of housing, commercial, industrial and miscellaneous load units. An estimate of connected load is the first step. This can be done on the basis of the covered area occupied by each unit. Suitable values of demand factors and diversity factors can be assumed depending on the nature of loads. From the values of connected loads, demand factors and diversity factors the maximum demand can be estimated. For a detailed discussion of this topic the reader is referred to [16, 54] of the bibliography.

A simple, though approximate, estimate of the load can be made by assuming a value of load density for the area. As per load density, an area can be designated as high load density area (having a load density of 2000 kVA per sq. km. or more) medium load density area (having a load density of 200 to 2000 kVA per sq.km) and low load density area (having a load density of less than 200 kVA per sq.km).

16.11 DESIGN OF PRIMARY DISTRIBUTION

16.11.1 Selection of Voltage

Once the load estimate is available, the selection of voltage is the first step in the design of the primary distribution. The voltages most commonly used for this purpose are 11 kV and 33 kV. The simple rule is to select the highest voltage which can be justified economically. A long term plan will always justify a higher voltage than the short term plan. The period of time to be taken for economic comparison is mostly a matter of choice of the planning engineer. Generally a period of 7 – 10 years is considered. It is necessary to prepare a number of different schemes at each of the possible voltages, divide the schemes into the steps to be taken at specified years and reduce the capital costs involved to present values thus obtaining the most economic design. All the alternative schemes should be identical in capacity and reliability. If the costs of two schemes are nearly the same, it may be necessary to capitalize the losses in each and add the cost of losses to the costs. If the total costs are still similar the scheme having higher primary distribution voltage or the scheme having lower initial investment is preferable.

An empirical formula for calculating the optimum voltage for a primary distribution system is

$$V = 5.5\left(\frac{L}{1.6} + \frac{P}{100}\right)^{0.5} \tag{16.30}$$

where V = line voltage in kV

L = distance in km

P = power in kW

The above formula does not have any scientific basis but it gives reasonably good results. The application of this formula would give an odd value of voltage. The standard voltage nearest to this value should be adopted.

It is necessary to ensure that the primary distribution voltage should be such that line voltage regulation is within limits. It may be mentioned that in some cases the voltage regulation may be poor due to high line reactance and low load power factor. In such cases the selection of an uneconomical high voltage can be avoided by using and shunt capacitors.

16.11.2 Choice of Scheme

The selection of primary distribution scheme is governed by the considerations of economy and reliability. The present trend is to extend the sub-transmission circuits to areas as close to the load centres as possible so that full advantage of sub-transmission voltage is achieved. This practice results in short lengths of feeders with consequent saving in voltage drop.

For low and medium load density areas, a radial primary distribution is generally sufficient. Wherever the continuity of supply is of greater importance, parallel feeders or open loop feeders can be used. For high load density areas open loop and ring feeders are generally used. Primary networks are quite popular in advanced countries but are seldom used in India. However, as loads grow, the primary networks are likely to become more popular.

The number of feeders depends on a number of factors. Some of these are: type of scheme (whether radial, parallel, loop or network), the total load to be supplied, the load density, availability of suitable sites for secondary substations, the distance between the secondary substation and the load, permissible voltage drop, etc.

16.11.3 Size of Feeders

The conductor size of a feeder is governed by the current carrying capacity, voltage drop and overall economy.

The current carrying capacity of a conductor depends on the conductor losses and surroundings. The capacity is usually determined for a maximum operating temperature of 75°C. Appendix A gives the current carrying capacity of different ACSR conductors as per IS: 398 – 1976.

The values of conductor resistances at 20°C are also given in Appendix A. From these values the effective resistance at the operating temperature can be calculated. For determining the voltage drop, it is necessary to calculate the inductive reactance of the feeder. The method discussed in chapter 2 can be used. Standard tables which directly give the value of reactance are also available. Then the voltage drop can be calculated. If the voltage drop comes out to be high it is necessary to select another conductor size to reduce the voltage drop. If the voltage drop comes out to be too high, some other method of voltage control (e.g., use of series capacitors) may be necessary.

The value of conductor size obtained above should be checked for overall economy. By the application of Kelvin's law, the most economic conductor size can be calculated. The current in the feeder is at the maximum value only at certain times. At all other times the current is less than the maximum current and it is necessary to use a factor to account for this fact.

Loss factor is defined as

$$\text{Loss factor} = \frac{\text{average power loss}}{\text{power loss at peak load}} \tag{16.31}$$

The actual power loss is

$$\text{Loss} = 3\,(I_{max})^2\,R\,(\text{loss factor}) \tag{16.32}$$

If the load is constant throughout, the loss factor is one. In actual practice the load varies with time of the day. If the actual load curve is known, the loss factor can be calculated. An approximate value of loss factor can be found from the following equation

$$\text{Loss factor} = (0.3 \times \text{Load factor}) + (0.7)\,(\text{Load factor})^2 \tag{16.33}$$

where the load factor is the ratio of average load to peak load.

As discussed in section 16.8, Kelvin's law has a number of limitations. Therefore the size of conductor as determined from this law can be taken only as a rough guide to economy.

Example 16.24.

A distribution station has a peak load of 3000 kW and total annual energy of 10^7 kWh. The peak power loss is 220 kW. Find (a) load factor (b) loss factor (c) actual loss (d) annual cost of energy losses if energy cost is Rs 2.10 per kWh.

Solution

(a) $$\text{Load factor} = \frac{\text{average load}}{\text{peak load}} = \frac{10^7}{8760 \times 3000} = 0.38$$

(b) Using Eq. (16.33)

$$\text{Loss factor} = (0.3 \times 0.38) + 0.7\,(0.38)^2 = 0.215$$

(c) $$\text{Actual loss} = \text{Loss factor} \times \text{Peak power loss}$$

$$= 0.215 \times 220 = 47.3 \text{ kW}$$

(d) Yearly energy loss = 47.3 × 8760 = 414348 kWh

Cost of energy loss = 414348 × 2.10 = Rs. 870130.8

16.12 SUB-STATIONS

16.12.1 Number and Size

The estimation of the number of secondary and distribution substation is a difficult process. The cost of material used in a distribution circuit per kVA of distributed power varies as the square of the linear dimensions of the supply area. If the number of substations is less (i.e., supply area per substation is large) the cost of distribution circuit will be high. If the number of substations is large the cost of distribution network will be less but the substation costs will be high. Table 16.1 shows the size of substations usually employed.

Table 16.1 *Substation sizes*

Type of Consumer	*Size of Substation*	
	Distribution 11000/400 V	*Secondary* 33/11 kV
Domestic, commercial (small and medium) and small industrial	25 – 1000 kVA	–
Commercial (large) and industrial (medium)	300 – 3000 kVA	2 – 10 MVA
Industrial (large) HV supply	–	2 – 30 MVA

Transformers are manufactured in standard sizes. As per IS: 2026 – 1962, the standard sizes of 3-phase transformers are: 25, 40, 63, 100, 160, 250, 400, 630, 1000, 1600, 2500, 4000, 6300, 10000 kVA, etc. The number of transformers for a distribution scheme should neither be too small nor too large.

An important consideration in the selection of number and size of substations is the ever-increasing loads. To accommodate the additional loads an existing system can be reinforced in two ways:

(1) To keep to a fixed number of substations and increase transformer sizes as the load grows.

(2) To keep to the same size of transformer and increase the number of substations.

Neither of the above two approaches can be strictly followed and a mid way compromise of increasing both the number and size in the same ratio is best approach.

16.12.2 Location and Installation

The location of a substation is governed by the voltage level, voltage regulation and cost of primary and secondary distribution systems. The following rules should be followed in deciding the location:

1. Locate the substation as near the load centre as possible so that Σ load $\times$ distance from substation is minimum.
2. Locate the substation so that proper voltage regulation can be achieved.
3. Locate the substation such that proper access to incoming and outgoing lines can be provided.
4. The selected site should have enough space for future expansion.
5. The selected site should be such that the number of customers affected by any outage is minimum.

In addition to the above, the availability of land at reasonable cost and municipal and other restrictions also affect the selection of site.

The substation may be outdoor type or indoor type. The distribution stations are mostly of outdoor type and the secondary substations are generally of indoor type. An outdoor substation may be pole mounted or floor mounted. A pole mounted substation consists of transformer and the isolating switch mounted on an *H* pole or an iron platform formed by four poles. The *H* polemounting is limited to transformers upto 160 kVA capacity. The platform mounting is used upto 400 kVA. The higher size transformers are located in special enclosures built on the ground surface.

16.12.3 Equipment

The main equipments in substations are transformers, circuit breakers, isolating switches, current and potential transformers, bus bars, protective relays, surge arresters and control room equipment, etc.

(a) *Transformers* Transformers, in distribution substations, step down the primary voltage to lower values suitable for distribution purpose. Three-phase oil cooled transformers connected in delta/star are used. Sometimes 3-winding transformers are used so that a different supply voltage may be obtained, from tertiary. The transformers are generally fitted with on load tap changing gear capable of changing the voltage from 110% to 90% of nominal value.

(b) *Circuit breakers* Circuit breakers play an important role in power system protection. Whenever a fault occurs, the relays sense the fault and send a trip signal to the circuit breaker. The circuit breakers disconnect and isolate the faulty section. Oil circuit breakers, air blast circuit breakers and sulphur hexa fluoride circuit breakers are used in power systems. In secondary substations generally oil circuit breakers are used.

(c) *Isolating switches (Isolators)* Whenever maintenance or repair work is to be carried out on an equipment in a substation, it is disconnected from the supply by the isolator. They are normally operated on no load. Isolators are generally interlocked with circuit breakers and earthing switches.

(d) *Current and potential transformers* A current transformer steps down the line current for the purpose of metering and relaying. Generally single turn primary current transformers are used. The line or conductor passes through the core and acts as primary. The secondary, having a large number of turns, is wound on the core. The instrument or relay is connected to the secondary. The secondary rated current is either 1 A or 5 A.

A voltage (or potential) transformer is used to step down the system voltage to a low value for metering and relaying. It may be an iron core type or capacitative voltage transformer (C.V.T.), the CVT being more commonly used. The rated secondary voltages are 110 V or 230 V line to line.

(e) *Bus bars* The incoming and outgoing circuits are connected to bus bars. These may be either rigid Aluminium tubular bus bars supported on post insulators or flexible ACSR stranded conductor bus bars supported from two ends by strain insulators.

(f) *Protective relays* Wherever a fault occurs the protective relay operates and sends a trip signal to circuit breaker. The relays are housed in panels in the control room.

Time graded overcurrent protection is used for feeders. Distance protection and carrier current protection are used for transmission lines. Differential protection is used for transformer protection.

(g) *Surge arresters (Lightning arresters)* Surge arresters protect the substation equipment from lightning and switching surges. Present day power systems invariably use valve type arresters. Their application has been discussed in Chapter 12.

(h) *Earthing switch* The earthing switch is connected between the line conductor and earth and is mounted on the frame of isolator. Normally it is in open position. When the line is disconnected the earthing switch is closed to discharge the trapped charges to earth.

(i) *Station earthing system* The function of station earthing system is to provide a low resistance path for flow of earth fault currents (for proper operation of protection device) and safety of equipment and personnel. It is discussed in Chapter 22.

(j) *Shunt capacitors* They supply the reactive power to the loads and thus help in improving the power factor. They are switched on during peak load conditions and switched off during opp-peak conditions. Shunt capacitor banks are connected in-star or delta across the tertiary winding of the transformer.

(k) *Station battery and charging equipment* A substation needs dc supply for protection and control purposes. This supply is obtained from batteries. 110 V dc supply is used for medium size substations while 240 V supply is used for large substations. The batteries are equipped with charging equipment.

16.12.4 Key Diagram

Figure 16.31 shows a key diagram of a typical primary substation with two 33 kV incoming lines and four 11 kV outgoing feeders. A circuit breaker is connected in each incoming and outgoing line.

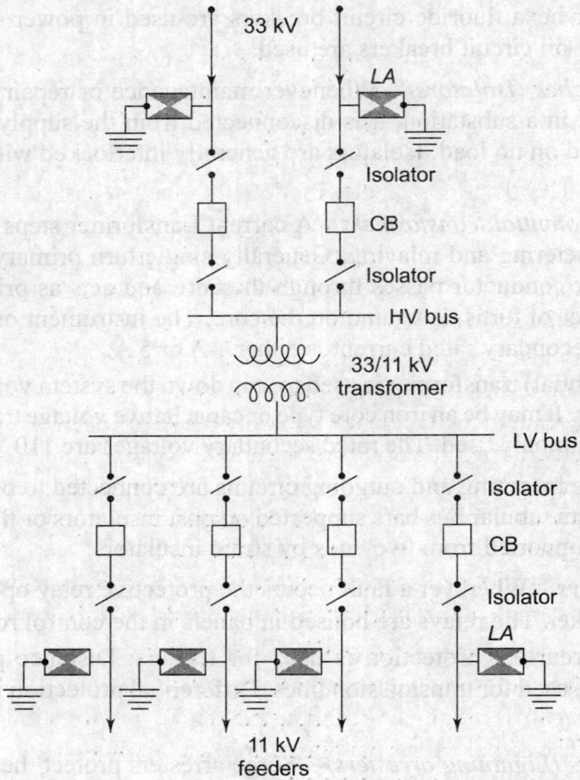

Fig. 16.36 Key diagram of a typical primary substation

On each side of breaker is an isolator. The isolator should never be opened on load, but is generally capable of interrupting the transformer no load current. The feeder protection is by time graded overcurrent relays. The circuit breakers have auto reclosing facility. A lightning arrester is connected on each incoming and outgoing circuit for lightning protection. Each incoming line is connected to HV bus (33 kV) through a circuit breaker and two isolators. Each outgoing feeder is connected 11 kV bus through a circuit breaker and two isolators.

16.13 SECONDARY DISTRIBUTION DESIGN

The distribution transformer steps down the feeder voltage to 400 V. The low voltage distributors take off from the secondary of the distribution transformer. The consumer services are tapped off the distributors. ACSR and AAC (all aluminium conductors) are used for the distributors. The main consideration in the design of distributor is the voltage drop.

The selection of the type of distributor (whether radial, open loop or network) is governed by the available investment and the degree of reliability of supply. In low density urban areas and rural areas radial distributors are used. The medium and high density areas are generally supplied through open loop distributors. In developed countries the areas are generally fed by network distribution schemes. The network distribution schemes are becoming popular in India also.

The secondary distribution system may be overhead or underground or partly overhead and partly underground. An underground system has greater reliability but is more costly than the overhead system. The distribution system in high load density and congested areas is usually underground.

The planning of the secondary distribution system is mostly done by hit and trial. The use of computer methods has been suggested and is likely to become popular in future.

The design of distributors starts with a map of the area. The scale of the map should be such that the map size is sufficiently big. The positions of all roads, streets and buildings should be shown on the map. The positions of distribution stations are first marked on the map. Then distributors are shown along all the roads and streets. The distributors are always laid along the roads and streets so that service connections can be easily given. Moreover, the street lightning fixtures an also be supported on the poles of these distributors. The pole positions should be selected so as to provide good facilities for service connections.

Once a preliminary layout of distributors is decided, the maximum current of each distributor is estimated. This is done from the values of the connected loads of consumers to be fed from that distributor. Suitable demand factor and diversity factor are used. Then the conductor size for the distributor is assumed. An initial guess of the conductor size is made from the considerations of current carrying capacity. Since the current in a distributor decreases, as its distance from the transformer increases, the size of distributor should be continuously decreased. Since the conductors are manufactured only in a few standard sizes, the practical optimum is a stepped distributor made up of sections of successively reducing standard conductors until the smallest size is used in the last section of the distributor. Generally the distributors use 3 or 4 standard sizes.

The next step in the design in the calculation of voltage drop. It is necessary to assume suitable values of power factors before voltage drop can be calculated. For residential and commercial loads, unity power factor is assumed. For industrial and irrigation loads, a power factor of 0.8 lagging is assumed. The actual values of voltage drop are compared with the permissible values. More often than not it will be found that voltage drop on almost all distributors is more than permissible. The conductor sizes of the distributors are then increased to the next higher standard values and voltage drops calculated again. The process is continued till the voltage drop on all distributors is within limits. Sometimes it may be necessary to change the layout and increase the number of distributors so as to satisfy the voltage drop constraint.

16.14 ECONOMICAL DESIGN OF DISTRIBUTORS

It has been found that the total cost of a distribution scheme is almost entirely dependent on the choice of conductor sizes. Therefore, it is necessary to design each distributor with the minimum volume of the conductor material consistent with the voltage regulation requirements. We restrict the analysis for a balanced system (so that only one phase of the distributor need be considered), unity power factor (so that only the resistance of the distributor need be considered in calculating voltage drop) and continuous uniformly distributed current loading along the length of the distributor. The minimum volume of conductor is obtained when the conductor is tapered so that the cross-sectional area at any point is proportional to the square root of the distance from the end of the distributor. This is not possible in practice. The practical optimum is a stepped distributor as shown in Fig. 16.37.

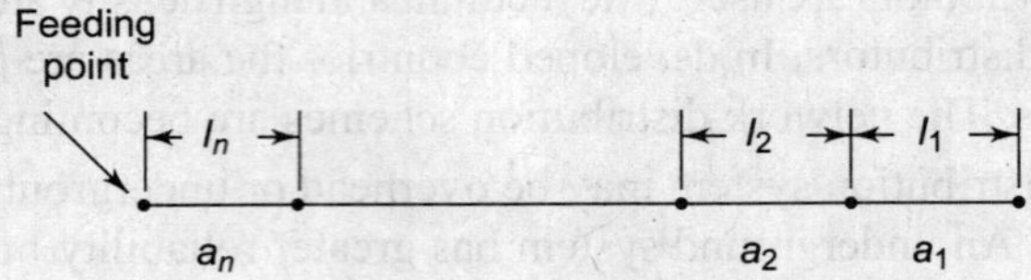

Fig. 16.37 Stepped distributor

Let i = current loading per unit length of distributor

$V.D$ = permissible voltage drop

v = volume of conductor material

ρ = resistivity of conductor material

b = total length of distributor

Referring to Fig 16.37, the conductor volume is

$$v = a_1\, l_1 + a_2\, l_2 + \dots\dots\dots + a_k\, l_k + \dots\dots\dots + a_n\, l_n \tag{16.34}$$

The section lengths must satisfy the condition that:

$$b = l_1 + l_2 + l_3 + \dots\dots\dots + l_k + \dots\dots\dots + l_n \tag{16.35}$$

The voltage drop along the distributor must be equal to the permissible voltage drop.

$$V.D. = \rho i \left[\frac{1}{2}\frac{l_1^2}{a_1} + \frac{l_2}{a_2}(l_1 + 0.5 l_2) + \frac{l_3}{a_3}(l_1 + l_2 + 0.5 l_3) + \dots + \frac{l_n}{a_n}\left(l_1 + l_2 + \dots + l_{n-1} + \frac{l_n}{2} \right) \right] \tag{16.36}$$

The problem becomes that of minimizing the volume v subject to the constraints expressed by Eqs. (16.35 and 16.36). This is a typical problem of calculus of variation and can be solved using the well known mathematical techniques. For details of the procedure and computer method of solution, the reader is referred to (39,63,71) of the bibliography.

16.15 DESIGN OF SECONDARY NETWORK

The method discussed in section 16.13 is not suitable for the planning and design of a network or grid system of secondary distribution. The best method is to use a calculating board because the network characteristics can not be easily determined by analytical methods. It is necessary to have adequate information about loads, available supply, existing facilities, etc., before the design can be prepared.

The first step is to lay down an apparently feasible plan by estimation. In case an existing radial system is to be converted into a network system, it may be sufficient to reconnect the distributors so as to form a network. Then the loads are concentrated, in proper units, at various points in the network so that a reasonable number of circuit elements can be considered. This is shown in Fig. 16.33. Mostly the loads are concentrated at junction points. Where a large individual load is to be served (e.g., load x) it is preferable to consider this load at that point only. If the load near a junction, is very small (point y) no concentrated load should be shown there. While concentrating the loads, it is quite accurate to divide each load between the adjacent sections in inverse proportion to the distances between the load and the junctions. After the loads are concentrated at the junction points, they are converted to ultimate kVA demand, at the transformer (taking diversity factor and power factor into account). The loads are now suitable for representation on a calculating board.

When the loads and secondary network have been determined the location and approximate sizes of distribution transformers is decided. The transformers should be located at the major loads and at the various junctions where the concentrated loads are large. It is generally desirable to select two or three sizes of transformers for the total scheme. Then the size of secondary mains is selected. This size depends on current carrying capacity and voltage regulation. The current carrying capacity is usually fixed at one-half to two-third of the rated current of predominant transformers. The voltage regulation for normal operation can be estimated by calculating the voltage drop to the distributed loads along some sections of main where inspection indicates maximum voltage drop.

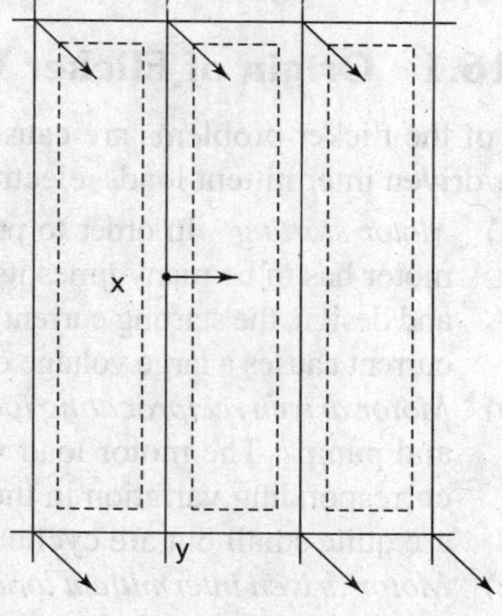

Fig. 16.38 Concentration of loads at suitable points for designing a secondary network using a calculating board

The primary feeder connections to the different distribution transformers should be carefully selected so that, when one feeder is out of service, adjacent transformers are not simultaneously out of service. This is achieved by inter-lacing the primary feeders. Thus a preliminary plan of the location and size of distribution transformers, feeder inter-lacing, secondary grid arrangement and conductor size is established. This plan is simulated on the calculating board and check studies are made to estimate the accuracy of initial estimates. It may be necessary to change some transformer locations or sizes and sizes of some of the conductors to avoid overheating, under-loading or excessive voltage drop. After the necessary changes have been made the final plan is ready. The characteristics of the final plan are determined on the calculating board. All reasonable operating conditions like load division with one primary feeder out of service, etc., are studied. The maximum load on any element of the network, the voltage regulation under various conditions and fault currents are determined. If necessary a revision of the plan is made.

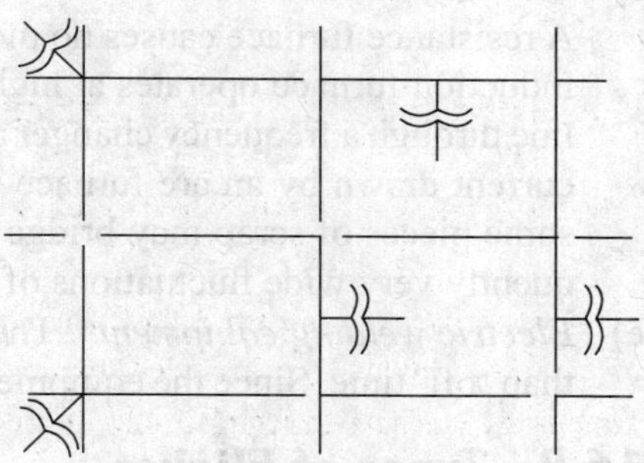

Fig. 16.39 Division of network into radial secondary circuit for design and analysis

Many times the voltage drop calculations on network distribution systems are carried out by assuming the network to be divided into a number of adjacent radial circuits as shown in Fig. 16.39. This method gives reasonable accuracy with minimum and easy calculations.

16.16 LAMP FLICKER

The voltage regulation on a system has two aspects. One is the normal drop in voltage from no load to full load. The other is the momentary sudden change of voltage due to a change of load. This sudden change causes a sudden change in the intensity of illumination of lamps. This is known as lamp flicker. Human eye reacts to even very small variation of light intensity. Therefore lamp flicker is a very common phenomenon in power systems.

16.16.1 Origin of Flicker Voltage

Most of the flicker problems are caused by starting motors, motor driven reciprocating loads, motor driven intermittent loads, electric furnaces and electric welding equipment.

(a) *Motor starting* In order to produce a sufficient starting torque, the starting current of a motor has to be many times its full load current. Depending on the size, number of poles and design, the starting current may be 5 to 10 times the full load current. The high starting current causes a large voltage drop in the system resulting in a momentary voltage dip.

(b) *Motor driven reciprocating loads* These loads consist of air compressors, refrigerators and pumps. The motor load varies cyclically with each power stroke and produces a corresponding variation in the line current. The variations in voltage due to this cause are quite small but are cyclic in nature and, therefore, not acceptable.

(c) *Motor driven intermittent loads* In some motors the nature of the work calls for heavy overloads for short intervals and for cyclic loads of long and irregular period. In saw mills and coal cutters heavy over-loads are common. In punch presses and shears the load undergoes wide variations.

(d) *Electric furnaces* Electric furnaces are of three types: resistances, induction and arc. A resistance furnace causes no more flicker than a resistance load of the same size. An induction furnace operates at high frequency and is therefore, connected to the power line through a frequency changer and consequently presents a steady load. However the current drawn by an arc furnace varies very widely. During the melting down period some pieces of scrap may bridge the electrodes approximating a short circuit. Consequently very wide fluctuations of current may occur during the melting down period.

(e) *Electric welding equipment* The electric welding equipment has a smaller 'on' time than 'off' time. Since the equipment draws a heavy current, a serious flicker may occur.

16.16.2 Types of Flicker

Lamp flickers may be classified as cyclic high frequency flicker, cyclic low frequency flicker, non-cyclic frequent flicker and non-cyclic infrequent flicker.

(a) *Cyclic high frequency flicker* The flicker caused by air compressors, reciprocating pumps and resistance spot welders may have a frequency of 2 to 10 cycles per second. Such a flicker is designated as cyclic high frequency flicker.

(b) *Cyclic low frequency flicker* In this category comes the flicker due to punch presses, shears and arc welding equipment. The frequency may be 10 to 120 per minute.

(c) *Non-cyclic frequent flicker* The loads such as cranes, hoists, X-ray apparatus cause noncyclic frequent flicker. Such a flicker may occur a few times per minutes.

(d) *Non-cyclic infrequent flicker* The loads such as refrigerators, air conditioners, etc., may cause a non-cyclic infrequent flicker occurring a few times per hour.

16.16.3 Remedial Measures

System changes and use of series capacitors are the two important measures used to reduce flicker.

1. *System changes* Flicker voltages are mostly caused by industrial loads. Therefore, separating the loads such that the industrial loads are fed by a separate feeder can reduce the flicker in residential and commercial premises. Voltage fluctuations objectionable

to the residential and commercial consumers are generally acceptable to the industrial plants because there is greater tolerance in industrial work. Moreover the industrial plants have to accept flicker because it is caused by their own operation. Increase in transformer capacity and increase in conductor size of feeders and distributors can go a long way in reducing flicker. However such system changes are justified only if additional capacity is likely to be needed at a later stage.

2. *Use of series capacitors* Capacitors connected in series with lines compensate the inductive reactance voltage drop and help in reducing the flicker caused by all types of fluctuating loads. However, they do not correct the conditions in the system as a whole because their effect is only beyond their point of installation. Series capacitors are economically justified only if the load causing flicker is a large portion of the total load or if the line reactance is more than line resistance and if supply circuits are fairly long.

Example 16.25.

Figure 16.35 (a) shows a distribution system feeding a large induction motor and domestic load. The pu reactances of transformers are on their rated MVA and voltage base. The pu reactance of line is on 10 MVA, 11 kV base. (a) Find the voltage at domestic load point during steady state operating condition (b) Find the voltage at domestic load point during the starting of induction motor if the induction motor takes 5 times full load current at zero p.f. lagging during starting time (c) Suggest a remedy for this dip in voltage during starting of induction motor.

Solution

Let Base MVA = 10

Base voltage 66 kV/11 kV/3.3 kV/0.415 kV in respective sections

P.u. reactance of $T_2 = .08 \times \frac{10}{5} = .16$ p.u.

P.u. reactance of $T_3 = .06 \times \frac{10}{5} = .12$ p.u.

(a) Current drawn by domestic load during steady state conditions $= \frac{1}{10}(0.8 - j\,0.6)$

$$= .08 - j\,.06 \text{ p.u.}$$

Current drawn by induction motor during steady state conditions $= \frac{5}{10}(.8 - j.6) = .4 - j.3$ p.u.

Current through T_1 and line during steady state conditions $= .48 - j.36$ p.u.

Figure 16.40 (b) shows the equivalent circuit. Voltage at domestic load point during steady state conditions

$$= 1 - (.48 - j.36) - (.08 - j.06)\,(j.12)$$

$$= .9208 - j.1056 = .927 \text{ p.u.}$$

(b) Current drawn by induction motor during starting $= -j\,\frac{25}{10} = -j\,2.5$ p.u.

Current through T_1 and line during starting period of induction motor

$$= 0.08 - j\,0.06 - j\,2.5 = .08 - j\,2.56 \text{ p.u.}$$

Voltage at domestic load during starting of induction motor

$$= 1 - (.08 - j\,2.56)\,(j.2) - (.08 - j.06)\,(j.12)$$

$$= 0.4808 - j.0256 \text{ p.u.} = 0.481 \text{ p.u.}$$

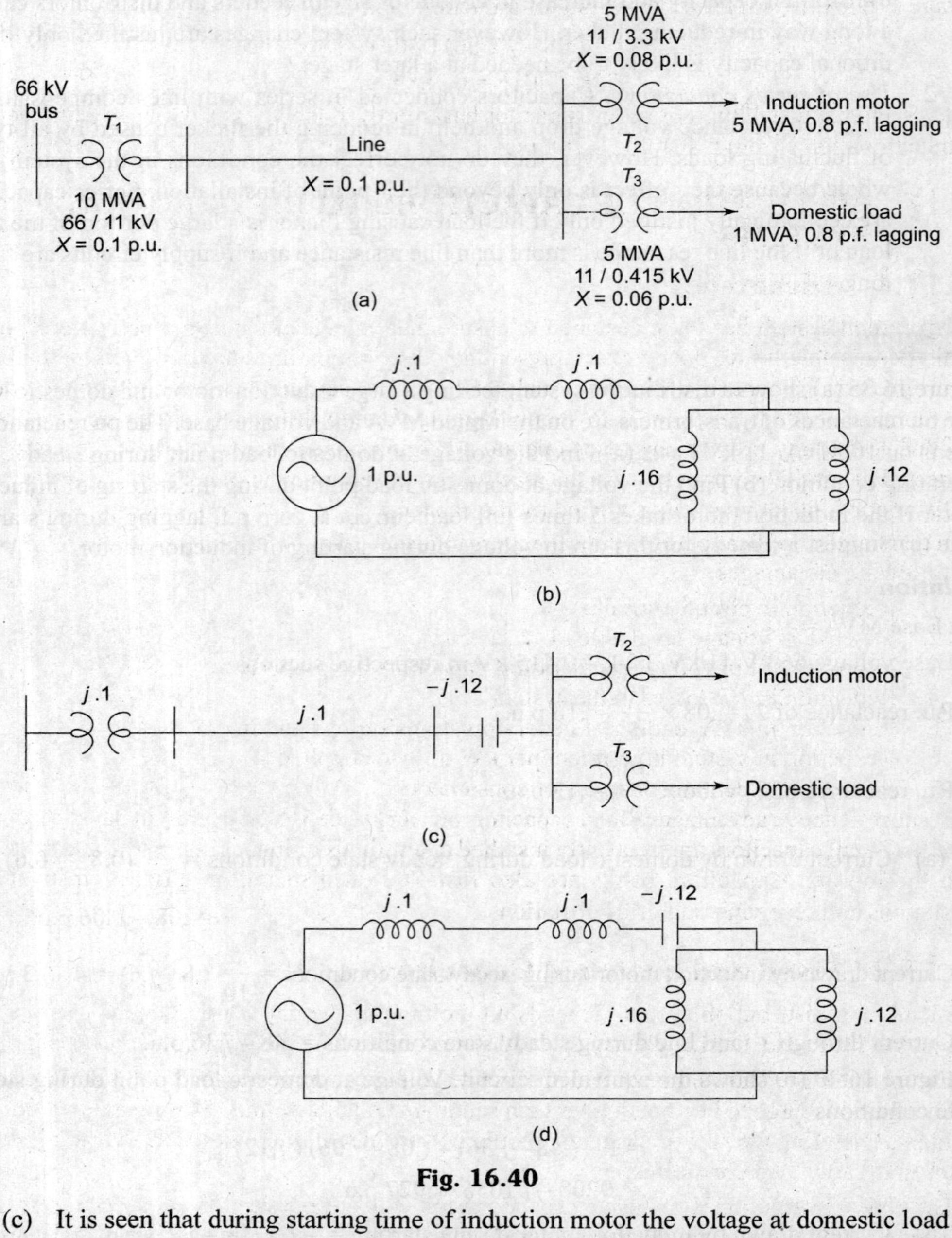

Fig. 16.40

(c) It is seen that during starting time of induction motor the voltage at domestic load has dipped to a very low value. One possible solution, to avoid this dip is to instal a series capacitor in the line as shown in Fig. 16.40 (c). A suitable degree of series compensation is 60%. Since total reactance of T_1 and L_1 is 0.2 *pu*, the series capacitor should have a reactance of 0.12 *pu*. Fig. 16.40 (d) shows the equivalent circuit after installation of series capacitor.

Voltage at domestic load during steady state conditions

$$= 1 - (.48 - j.36)\,(j.08) - (.08 - j.06)\,(j.12)$$

$$= .964 - j.048 = 0.965 \text{ p.u.}$$

Voltage at domestic load during starting of induction motor

$$= 1 - (.08 - j\,2.56)\,(j.08) - (.08 - j.06)\,(j.12)$$

$$= .788 - j.016 = 0.788 \text{ p.u.}$$

It is seen that installation of series capacitor has improved the voltage during starting period of induction motor from 0.481 and 0.788 *pu*.

16.17 APPLICATION OF CAPACITORS TO DISTRIBUTION SYSTEMS

16.17.1 Advantages

A distribution system has to be designed to ensure that voltage at consumer premises remains within permissible limits. Many of the present day distribution lines are designed on the basis of current carrying capacity. Such lines make use of shunt capacitors to limit the voltage drop. The function of shunt capacitor is to supply lagging vars, to the system, at the point where it is connected. Many present day loads operate a lagging power factor. If shunt capacitors are connected in parallel with such loads, the reactive power requirement of the loads can be, almost totally, met by the lagging vars supplied by the capacitors.

In general the application of shunt capacitors to distribution systems leads to one or more of the following advantages:

1. Reduction in circuit current.
2. Increase in voltage level at load.
3. Improvement in voltage regulation if capacitors are properly switched.
4. Reduction in I^2R losses in the system.
5. Reduction in kVA loading of generators, transformers and lines.
6. Reduction in system investment per kW of load supplied.
7. Reduction in demand charges for consumers.

Because of above advantages, shunt capacitors are very widely used in present day distribution systems. Small capacitors are frequently installed at terminals of induction motors and switched with the motors. Capacitors banks are also installed on distribution circuits, distribution substations and receiving end grid substations.

16.17.2 Fixed and Switched Capacitors

The load on a distribution system depends on the time of the day. During evening hours, the system has peak load. During the day the load is moderate while it is generally low during night.

If a capacitor is connected at the end of a line which is not supplying any load, it will cause the receiving end voltage to become higher than sending end voltage thus creating an over voltage problem. Therefore the size of capacitor should be such that it does not result in leading power factor under low load conditions.

The optimum solution is to have a capacitor bank which can adjust its var supply so that the vars supplied match the vars demand of the load exactly. Though it is possible to have such a system, it would be very expensive and complex.

The feasible solution is to have fixed and switched capacitors, the fixed capacitor bank for the low load conditions and the switched capacitor bank for peak load conditions. Such a system would cause leading power factor under zero load condition but this is generally acceptable.

16.17.3 Capacitor Location

The location of capacitor banks on the distribution system is very important. A capacitor in a substation does not reduce the losses on the lines leaving the substation because the vars must still be carried by the lines. The factors which influence the location are: lengths of lines, variation in load, load factor, type of load, etc. The capacitors can be located as under:

1. Group correction at primary of transformer
2. Group correction at secondary of transformer
3. Local correction on motor circuits
4. Local correction on feeders and distributors
5. Local correction on motor terminals.

The group correction is preferred under the following conditions:

(a) When the loads within the plant are widely separated and the loads shift a great deal from one feeder to another. In such a situation the capacitor may be needed in one feeder at one time and in other feeder at an other time. A centrally located group capacitor is a cheaper alternative in this case.

(b) When motor voltages are low. If the installation has low voltage motors, group correction is better. Low voltage capacitors tend to be more costly as compared to high voltage capacitors if kvar rating is the same. Group correction on the primary of transformer is likely to be less costly.

In all other situations localized correction is better. Localized correction leads to optimum utilization of shunt capacitor kvar because losses, in the circuits between transformer and loads, are reduced and voltage drop in these circuits is also reduced. Moreover in this case the capacitors are switched on and off directly along with the loads.

16.17.4 Method of Application

The method of application includes the determination of capacitor kvar and location. The system has to be studied thoroughly before the size and location can be determined so that optimum benefits are achieved. The method is as under:

1. Determine the variation of kW and kVA during a 24-hour period. The minimum kvar requirement (during the minimum load period) determines the fixed capacitor rating. If voltage conditions during light load period permit and leading power factor (during light load) is not objectionable, a little higher fixed capacitor kvar can also be used.
2. Obtain data of circuit parameters and load distribution.
3. Measure voltages at different points on the installation both during peak load and light load conditions. If load distribution cannot be determined accurately, it is assumed that loads are proportional to transformer capacity for peak and light feeder loads. It is generally possible to combine single phase loads into three-phase loads to simplify the calculations. The points of minimum voltage are likely to give the location of capacitors.
4. It is preferable to locate the capacitors near the load centres or near the ends of feeders. Preferably each load point should have an exact capacitor size to supply necessary kvar. This may not be always possible because only standard size capacitors are available.
5. Calculate the reduction in losses in the system because of addition of capacitors.
6. Calculate the release in transformer capacity, feeder capacity and substation capacity.
7. Compare the economic benefits obtained as a result of capacitor application with the cost of capacitor installation.

Generally the above procedure has to be repeated a number of times to achieve the desired maximum benefits.

16.17.5 Control of Switched Capacitors

Since the loads vary during a 24-hour period, most distribution systems require a combination of fixed and switched capacitors. The methods used to control switched capacitors are:

(a) *Voltage control* In systems where voltage varies with load this method is inexpensive and effective. Whenever the voltage tends to decrease below a certain level, the capacitor bank is switched on.

(b) *Current control* The capacitors are switched on when the current tends to increase beyond a certain value. This method requires a current sensing relay which is more expensive than a voltage sensing relay.

(c) *kVAr control* In this method the capacitors are switched on when kvar demand increases. This method requires measurement of kvar and is expensive. However, it is very effective.

(d) *Time* In this method capacitors are switched on and off at fixed time during the day. The method is very simple and inexpensive. However, the abnormal loads cannot be accounted for.

Example 16.26.

A 400 V, 3-phase star connected induction motor draws a current of 40 A at 0.75 p.f. lagging under full load condition. It is desired to improve the power factor to 0.95 lagging. Find kVAr rating, phase current and value of capacitance per phase of a delta connected capacitor bank.

Solution

kW rating of induction motor

$$= \sqrt{3} \times 400 \times 40 \times 0.75 \times 10^{-3} = 20.784 \text{ kW}$$

Power factor angle (initial) $= \cos^{-1}(0.75) = 41.41°$

Initial kVAr $= 20.784 \tan(41.41°) = 18.33$ kVAr

Power factor angle (final) $= \cos^{-1}(0.95) = 18.195°$

kVAr after power factor improvement $= 20.784 \tan(18.195°) = 6.83$ kVAr

kVAr of capacitor bank $= 6.83 - 18.33 = -11.5$ or 11.5 kVAr capacitive

Line current of capacitor bank

$$= \frac{11.5 \times 1000}{\sqrt{3} \times 400} = 16.6 \text{ A}$$

$$\text{Phase current of capacitor bank} = \frac{16.6}{\sqrt{3}} = 9.584 \text{ A}$$

$$\text{Reactance of capacitor} = \frac{400}{9.584} = 41.74 \ \Omega/\text{phase}$$

$$\text{Capacitance} = \frac{1}{2\pi \times 50 \times 41.74} = 76.26 \times 10^{-6} \text{ F/phase or } 76.26 \ \mu\text{F per phase}$$

Example 16.27.

In the induction motor of example 16.26 find the reduction in power loss in the lines feeding the motor due to application of capacitor bank.

Solution

Initial line current = 40 A

Reactive component of initial line current = 40 sin ($\cos^{-1}$ 0.75) = 26.46 A

Line current of capacitor bank = 16.6 A

Final value of line current = $40 \times 0.75 - j(26.46 - 16.6)$

$= 30 - j\ 9.86 = 31.85$ A

$$\frac{\text{New line loss}}{\text{Original line loss}} = \frac{31.58^2}{40^2} = 0.6233$$

Reduction in losses = 1 – 0.6233 = .3767 or 37.67%

Example 16.28.

A distribution system has the following loads

8 AM to 11 PM	1.3 MW at 0.75 lagging p.f.
11 PM to 8 AM	0.4 MW at 0.8 lagging p.f.

Find the kVAr rating of fixed and switched capacitors to improve the power factor to 0.95 lagging during the whole day.

Solution

Fixed capacitor kVAr is determined from requirements under light load condition.

During light load period (11 PM to 8 AM)

Initial kVAr = 0.4×1000 tan ($\cos^{-1}$ 0.8) = 300 kVAr

Final kVAr = 0.4×1000 tan ($\cos^{-1}$ 0.95) = 131.47 kVAr

kVAr of fixed capacitor bank = 300 – 131.47 = 168.53 kVAr

During peak load (8 AM to 11 PM)

Initial kVAr = 1.3×1000 tan ($\cos^{-1}$ 0.75) = 1146.5 kVAr

Final kVAr = 1.3×1000 tan ($\cos^{-1}$ 0.95) = 427.29 kVAr

Total kVAr requirement = 1146.5 – 427.29 = 719.21 kVAr

Switched capacitor kVAr = 719.21 – 168.53 = 550.68 kVAr

Example 16.29.

Figure 16.41 shows a dc ring distributor having interconnector BD Supply is at point A.

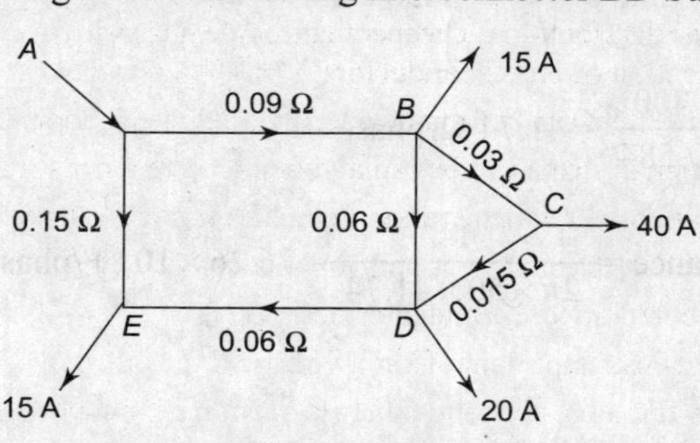

Fig. 16.41

Find (a) current in interconnector (b) voltage drop in interconnector.

[*RTM Napur Univ. 2010*]

Solution

Assume currents as shown in Fig. 16.42 Current fed at A is total current = 90 A

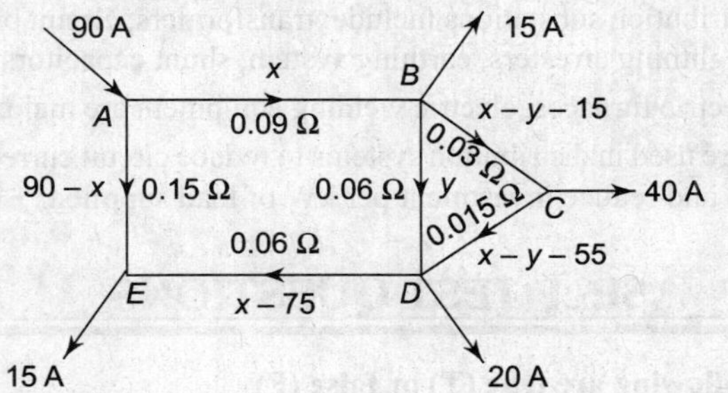

Fig. 16.42

KVL equation around loop ABDEA is

$$0.09x + 0.06y + 0.06\,(x - 75) - 0.15\,(90 - x) = 0$$

or $$0.3x + 0.06y = 18 \quad \text{(i)}$$

Moreover voltage drop along BD = Voltage drop along BCD

or $$0.06y = 0.03(x - y - 15) + 0.015\,(x - y - 55)$$

or $$0.045x - 0.105y = 1.275 \quad \text{(ii)}$$

From Eqs (i) and (ii)

$$x = 57.5\text{ A},\ y = 12.5\text{ A}$$

(a) Current in interconnector = 12.5 A

(b) Voltage drop in interconnector = 0.06 y

= 0.75 V

SUMMARY

16.1. Three phase 4 wire ac distribution is cheaper than single phase distribution. The neutral conductor has half the cross-section of phase conductors.

16.2. Primary distribution system can be radial feeder, parallel feeder, loop feeder and primary network.

16.3. Secondary distribution systems can be radial, open loop and network.

16.4. Voltage drop is an important consideration in the design of distributors.

16.5. Distributor can be fed at one end or at both ends or can be ring main distributor.

16.6. Loads can be distributed or concentrated or both.

16.7. Power factor of loads has important effect a voltage drop.

16.8. As per Kelvin's law the most economical cross-section is one which makes the annual value of interest and depreciation of the conductor equal to annual cost of energy loss in the conductor.

However, this law has many limitations and can not be applied precisely.

16.9. General design considerations of distribution systems are: economy, reliability, voltage level and regulation, feasibility of expansion etc.

16.10. Long term and short term load estimates is an important aspect of design of distribution.

16.11. Number, size and location of distribution substation are important aspects of substation design.

16.12. Equipments at distribution substations include: transformers, circuit breakers, isolators, CT, PT, bus bars, relays, lightning arresters, earthing system, shunt capacitors and battery.

16.13. Motor starting, electric furnaces, electric welding equipment are major causes of voltage flicker.

16.14. Shunt capacitors are used in distribution systems to reduce circuit current, reduce losses, improve voltage regulation and reduce investment per kW of load supplied.

SELF TEST QUESTIONS

A. Indicate whether the following are true (T) or false (F)

16.1 DC 3-wire system is cheaper than AC single phase system.

16.2 In 3-phase 4-wire distribution neutral conductor has half the area of phase conductors.

16.3 In primary network the advantage of load diversity is realised fully.

16.4 Radial feeder is used for critical loads.

16.5 Network secondary distribution is more reliable than other types of secondary distributions.

16.6 Distributors fed at both ends have higher voltage drop than distributors fed at one end.

16.7 If power factor is unity, the inductance of distributor has no effect on voltage drop.

16.8 Loads on ac distributors are generally unbalanced.

16.9 Cost of insulation of cables does not depend on the area of conductor.

16.10 Distribution transformers generally have a voltage rating of 11/0.44 kV.

B. Fill in the blanks with suitable words

16.11 Surge arresters protect the equipment from ________ and ________ over voltages.

16.12 Shunt capacitors lead to ________ in system investment per kW of load supplied.

16.13 Distribution transformer connection is mostly ________.

16.14 At substations dc battery is needed for ________ systems.

16.15 For feeders the relays generally used are ________ relays.

16.16 Lamp flicker can be reduced by the use of ________.

16.17 Two types of shunt capacitors used in distribution systems are ________ and ________.

16.18 Distributors are all ________

16.19 Radial feeder is most ________ but has least ________.

16.20 In a dc distributor having length l, resistance per unit length r and distributed load i ampere per unit length, the total voltage drop is ________.

ANSWERS

16.1 T	16.2 T	16.3 T	16.4 F	16.5 T	16.6 F
16.7 T	16.8 T	16.9 F	16.10 T	16.11 lightning, switching	

16.12 reduction | 16.13 delta-star | 16.14 protection

16.15 over current | 16.16 series capacitors | 16.17 fixed, switched

16.18 3-phase 4-wire | 16.19 unreliable, cost | 16.20 $\frac{1}{2}\ irl^2$

JOB INTERVIEW QUESTIONS

16.1 Why are all distributors 3-phase 4-wire?
16.2 What is lamp flicker?
16.3 What parameters can indicate reliability of distribution systems?
16.4 What are the various types of secondary distribution systems?
16.5 What are the main equipments at a substation?
16.6 What is the function of isolators?
16.7 How does power factor affect the voltage drop?
16.8 What are the advantages of using shunt capacitors on distributors?
16.9 Why is the power factor of distributors mostly lagging?
16.10 What is Kelvin's law?

REVIEW QUESTIONS

16.1 Discuss why?
  (a) AC transmission circuits are 3-phase 3-wire but ac distribution circuits are 3-phase 4-wire.
  (b) Open loop distributors are preferred to radial distributors.
  (c) The design of distributors is done on the basis of permissible voltage drop.
  (d) The distributors generally consist of stepped conductors.
  (e) Motor starting causes lamp flicker in power systems.

16.2 Why are the present day power systems all 3-phase? Why not single phase? Why not 6-phase?

16.3 What are the different types of feeders and distributors? Bring out their relative advantages and disadvantages.

16.4 What general considerations govern the design of distribution systems? Bring out their relative importance.

16.5 Discuss the design of primary distribution system with respect to following: selection of voltage, choice of scheme, size of feeders.

16.6 What is a secondary network? What are its advantages? How can it be designed?

16.7 What is lamp flicker? What are its causes? What types of loads are responsible for it? How can it be reduced?

PROBLEMS

16.1 A 3-phase 4-wire system has 420 V between lines at the sending end. The loads are: phase *R*, 40A at power factor 0.8 lagging, phase *Y*. 30A at unity power factor and phase *B*. 20 A at power factor 0.87 leading. The resistance of each phase conductor is 0.2 ohm and that of neutral is 0.1 ohm. Find the load voltages V_{RN} and V_{BR}. [236.4 V, 408 V]

16.2 A 3-phase 4-wire supply with a line voltage of 400 V is loaded as follows: A 3-phase load of 25 kW at a power factor of 0.8 lagging, single phase lighting loads of 10,15,20 kW on phases *R*,*Y* and *B* respectively. Determine the currents in all the conductors.

[84 A, 105 A, 126 A, 37.8 A]

16.3 A 3-phase 4-wire system is fed with a phase voltage of 230 V at the sending end. Each of the 4 conductors has resistance of 2 ohms and a reactance of 1 ohm. The loads are: A 3-phase balanced load of 30A at a power factor of 0.9 lagging, single phase loads of 20A at power factor of unity, 50 A at power factor of 0.8 lagging and 80 A at power factor of 0.7 lagging on phases

R, Y and B respectively. Calculate the currents in all the 4 conductors and the line to neutral voltages at the receiving end. [$I_R = 48.1A$, $I_Y = 80.6$A, $I_B = 108$ A, $I_N = 56.5$ A, $V_{RN} = 229$ V, $V_{YN} = 170$ V, $V_{BN} = 145.5$ V]

16.4 The R phase of a 3-phase 4-wire distributor has the following lighting loads; 100 A at 100 m, 50 A at 300 m and 50 A at 500 m from the supply end. The resistance of the conductor is 350×10^{-6} ohms per m. The supply voltage is 240 volts between line and neutral wire, calculate the voltages at the different load points. [233 V, 226 V, 222.5 V]

16.5 A dc distributor AB is 500 m long and is fed at point A at 220 V. Concentrated loads are taken from the distributor as under:

Distance from A,	m	150	200	280	320	390	450	500
Current	A	20	40	35	25	10	20	30

The total resistance of double run is 0.2 ohm. Find potential of point B. [197.32 V]

16.6 A uniformly distributed load of 0.2 A/m is added on the distributor of Problem 16.5. The remaining data is the same. Find the potential of point B. [187.32 V]

16.7 A dc two-wire distributor AB is 300 m long. It supplies a uniform load of 0.15 A/m and the following concentrated loads

Distance from A,	m	75	175	225
Load current	A	50	60	40

The potentials of feeding points A and B are 206 V and 200 V respectively. The resistance of each wire is 0.00015 ohm/m. Find the current fed at points A and B. [161 A, 34 A]

16.8 A two-wire dc distributor AB is 500 m long and is fed at 250 V at each end. Its supplies following concentrated loads:

Distance from A,	m	100	250	350	400
Load current,	A	50	60	40	30

The minimum allowable voltage at any point is 245 V. The specific resistance is 1.7×10^{-8} ohm-m. Find the area of cross-section of each conductor. [1 cm^2]

16.9 A 2-wire dc distributor AB is 500 m long and has a resistance of 0.1×10^{-3} Ω/m (go and return). The points C and D are 100 m and 200 m from A and have concentrated loads of 50 A and 100 A respectively. Point E is at distance x from C. The portion CE has a uniformly distributed load of 0.5 A/m. (a) The distributor is fed at A. Find x so that maximum voltage drop to any point may be 3V (b). The distributor is fed both at a and B. The potential at B is 2V higher than that at A. Find potentials of points D and E. The loads are the same as in part (a).

[(a) $x = 73$ m (b) 1V below potential of A]

16.10 A 2-wire dc distributor AB is 500 m long and has a total resistance of 0.075 ohm (both go and return). It is fed at A at 240 V. It has a uniform load of 1 A/m and the following concentrated loads.

Distance from A,	m	100	200	250	500
Load current,	A	60	100	150	50

Find the voltage at the far end. [207.975 V]

16.11 The distributor of Problem 16.10 is converted into an ac distributor. All the concentrated and uniform load are at unity p.f. The line reactance is equal to resistance. Find the voltage at far end.

[207.975 V]

16.12 The distributor of Problem 16.10 is converted to an ac distributor. All the loads are at 0.8 lagging p.f. The line reactance is equal to line resistance. Find the voltage at far end.

[195.165 V]

16.13 A substation S supplies three load points A, B and C through 2-wire dc feeders. Resistances SA, SB and SC are 0.05 ohm, 0.1 ohm and 0.04 ohm respectively. The loads at A, B and C are 50 A,

20 A and 60 A respectively. The load points *A* and *B* and *B* and *C* are interconnected through interconnectors having resistances of 0.1 ohm and 0.15 ohm respectively. If voltage at *S* is 250 V, find voltages at load points *A, B, C*. [V_A = 247.58 V, V_B = 247.75 V, V_C = 247.63 V]

16.14 A ring main distributor *ABCDEFGA* is fed at 500 V at point *A*. The lengths of different sections are *AB* = 150 m, *BC* = 200 m, *CD* = 180 m, *DE* = 250 m, *EF* = 120 m, *FG* = 320 m and *GA* = 100 m. The loads at points *B, C, D, E, F, G* are 50 A, 60 A, 40 A, 50 A, 30 A and 80 A respectively. The resistance of double run is 0.08 ohm per 100 m. Find voltage at all load points.

[V_B = 482.2 V, V_C = 466.46 V, V_D = 461 V, V_E = 461.2 V, V_F = 466.2 V, V_G = 487 V]

16.15 A two-wire ring distributor *ABCDEA* is fed at *A* at 250 V. Loads at points *B, C, D, E* are 12 A, 15 A, 12 A and 15 A. The load points *C* and *E* are interconnected through an interconnector *CFE*. At point *F* on the interconnector a load of 10A is connected. The resistances of both conductors of different sections are: *AB* 0.05 ohm, *BC* 0.04 ohm, *CD* 0.03 ohm, *DE* 0.04 ohm, *EA* 0.05 ohm, *CF* 0.02 ohm, and *FE* 0.01 ohm. Calculate potentials of points *B, C, D, E* and *F*.

[V_B = 248.6 V, V_C = 248 V, V_B = 247.87 V
V_E = 248.18 V, V_F = 248 V]

16.16 A single phase ring main *ABC* is fed at *A*. The impedances of sections *AB, BC, CA* are .2 + *j*. 2, .1 + *j* .2 and .2 + *j* .1 ohm respectively. Load at *B* is 50 A at 0.8 pf lagging and that *C* is 60 A at 0.9 lagging both with respect to voltage at *A*. Find currents in *AB, BC* and *AC*.

[I_{AB} = 37.59 – *j* 31.25 A, I_{AC} = 56.41 – *j* 24.9 A, I_{CA} = 2.41 + *j* 1.25 A]

16.17 A single phase distributor *AB* is fed at point *A*. At far end *B* is a load of 80 A at 0.8 p.f. lagging. At mid point *C* is a load of 100 A at 0.6 p.f. lagging. The total resistance and reactances of distributor are 0.3 ohm and 0.2 ohm respectively. Find the voltages at points *A* and *C*. The phase angles of both the load currents are with respect to voltage at point *B* which is 230 V.

[V_A = 275.9 V, V_C = 244.4 V]

16.18 For the induction motor of example 16.23 find the kVAr rating, line current and value of capacitance per phase if the capacitor bank is star connected. [11.5 kVAr, 16.6 A, 228.83 μF phase]

16.19 A single phase 440 V, 50 Hz motor draws a line current of 40 A at a pf of 0.7 lagging. Find the value of capacitor to be connected in parallel with motor to improve the p.f. to 0.9 lagging.

[108.5 μF]

16.20 A distribution system has the following loads

8 AM to 11PM	0.9 MW at 0.8 lagging p.f.
11 PM to 8 AM	0.3 MW at 0.75 lagging p.f.

Find the kVAr rating of fixed and switched capacitor banks to improve the p.f. to 0.9 lagging during the whole day. [119.28 kVAr, 119.83 kVAr]

CHAPTER 17

Symmetrical Faults

17.1 INTRODUCTION

The steady state operating mode of a power system is balanced 3-phase ac. However, due to sudden external or internal changes in the system, this condition is disrupted. When the insulation of the system fails at one or more points or a conducting object comes in contact with a live point, a short circuit or fault occurs. The causes of faults are numerous, e.g., lightning, heavy winds, trees falling across lines, vehicles colliding with towers or poles, birds, line breaks, etc. A fault involving all the three phases is known as symmetrical (balanced) fault while one involving only one or two phases is known as unsymmetrical fault. Single line to ground, line to line and two-line to ground faults are unsymmetrical (unbalanced) faults. Majority of the faults are unsymmetrical. Fault calculations involve finding the voltage and current distribution throughout the system during the fault. It is important to determine the values of system voltages and currents during fault conditions so that the protective devices may be set to detect the fault and isolate the faulty portion of the system so as to minimize the harmful effects of such contingencies.

When fault occurs at a point in a power system, the corresponding MVA is referred to as the fault level at that point.

17.2 THEVENIN'S EQUIVALENT CIRCUIT

According to Thevenin's theorem any linear network containing any number of voltage sources and impedances can be replaced by a single emf and an impedance. The emf is the open circuit voltage as seen from the terminals (under consideration) and the impedance is the network impedance as seen from these terminals. This circuit consisting of a single emf and impedance is known as Thevenin's equivalent circuit.

The calculation of the fault current can be very easily done by applying this theorem. It is only necessary to find the open circuit emf and network impedance as seen from the fault point. In most of the calculations, the open circuit emf can be assumed to be 1 *pu*.

17.3 SYMMETRICAL FAULT

This type of fault occurs infrequently, as for example, when a line, which has been made safe for maintenance by clamping all the three phases to earth, is accidentally made alive or when, due to slow fault clearance, an earth fault spreads across to the other two phases or when a mechanical excavator cuts quickly through a whole cable. It is an important type of fault in that it results in an easy calculation and generally, a pessimistic answer. The circuit breaker rated MVA breaking capacity is based on 3-phase fault MVA. Since circuit breakers are manufactured in preferred standard sizes, e.g., 250, 500, 750 MVA high precision is not necessary when calculating the 3-phase fault level at a point in a power system. Moreover the system impedances are also never known accurately. In view of this, the following assumptions are made in fault calculations.

1. The emfs of all generators are $1 \angle 0$ per unit. This means that the system voltage is at its nominal value and the system is operating on no load at the time of fault. The effect of this is that all generators can be replaced by a single generator since all emfs are equal and in phase. When desirable the load current can be taken into account, at a later stage, by superposition.
2. Shunt elements in the transformer model that account for magnetizing current and core loss are neglected.
3. Shunt capacitances of the transmission line are neglected.
4. System resistance is neglected and only inductive reactance of the system is taken into account. This assumption is generally made only for hand calculations and educational purposes. For computer solution this approximation is not necessary. The sub-transient reactances of the generators are generally used in calculations. However, if transient current is to be determined, then transient reactance should be used.

   The calculations for a symmetrical fault are easy because the circuit is a completely symmetrical circuit and calculations can be made only for one phase. The steps in the calculations are as under:

1. Draw a single line diagram of the system.
2. Select a common base and find out the per unit reactances of all generators, transformers, lines, etc., as referred to common base.
3. From the single line diagram draw a single line reactance diagram showing one phase and neutral. Indicate all the reactances, etc., on the reactance diagram.
4. Reduce the reactance diagram (by series parallel and star delta transformation) keeping the identity of the fault point intact. Find the reactance of the system as seen from the fault point (Thevenin's reactance).
5. Find the fault current and fault MVA in per unit. Convert these per unit values to actual values.
6. Retrace the steps of calculation to find the current and voltage distribution throughout the network.

Example 17.1.

Figure 17.1 shows a generating station feeding a 132 kV system. Determine the total fault current, fault level and fault current supplied by each alternator for a 3-phase fault at the receiving end bus. The line is 200 km long.

Solution

Take a base of 100 MVA, 11 kV for *LV* side and 132 kV for *HV* side.

per unit reactance of $G_1 = j\,0.15$

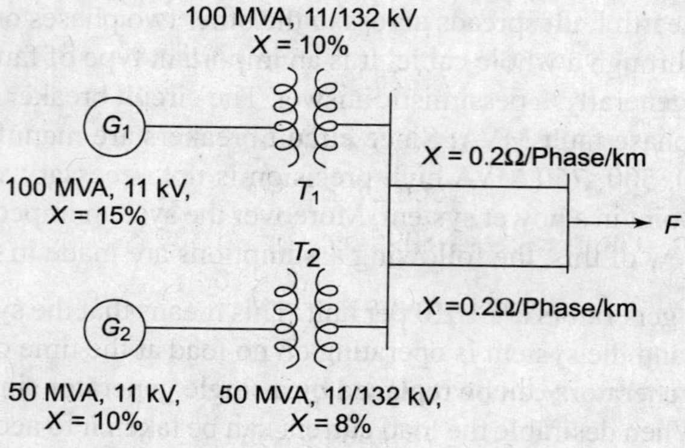

Fig. 17.1 System for example 17.1

$$\text{per unit reactance of } G_2 = j\,0.1 \times \frac{100}{50} = j\,0.2$$

$$\text{per unit reactance of } T_1 = j\,0.1$$

$$\text{per unit reactance of } T_2 = j\,0.08 \times \frac{100}{50} = j\,0.16$$

$$\text{per unit reactance of each line} = \frac{j\,0.2 \times 200 \times 100}{132 \times 132} = j\,0.23$$

The single line reactance diagram is shown in Fig. 17.2 (a). This can be reduced to the diagram shown in Fig. 17.2 (b) and then to Fig. 17.2 (c).

$$\text{Base current for 132 kV side} \quad = \frac{100 \times 100}{\sqrt{3} \times 132} = 437.4 \text{ A}$$

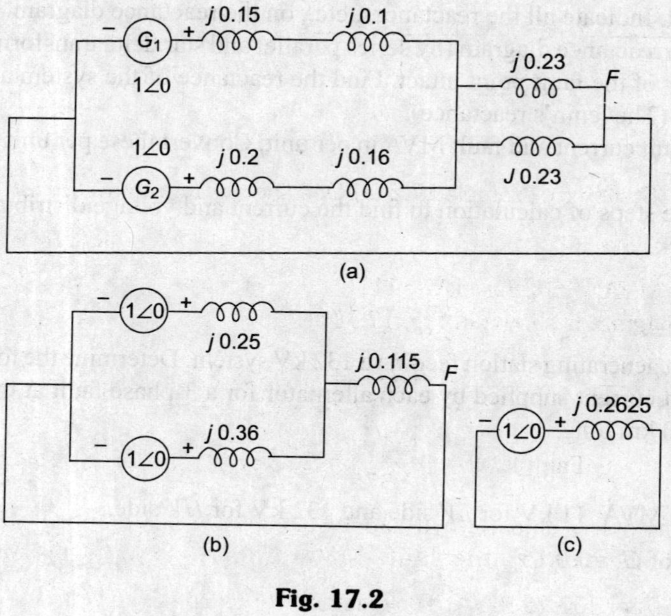

Fig. 17.2

$$\text{Total fault current} = \frac{1\angle 0}{j\,0.2625} = -j\,3.8095\,pu$$

$$= 3.8095 \times 437.4\angle -90°\text{ A}$$

$$= 1666.27\angle -90°\text{ A}$$

$$\text{Fault level} = 3.8095\,pu \quad \text{or} \quad 3.8095 \times 100, \text{ i.e., } 380.95 \text{ MVA}$$

$$\text{Base current for 11 kV side of transformers} = \frac{100 \times 100}{\sqrt{3} \times 11} = 5248.8\text{ A}$$

Total fault current supplied by the two generators

$$= -j\,3.8095 \times 5248.8 = 19995\angle -90°\text{ A}$$

From Fig. 17.2 (b)

$$\text{Fault current supplied by } G_1 = \frac{(19995\angle -90°)(j\,0.36)}{j\,0.25 + j\,0.36} = 11800.3\angle -90°\text{ A}$$

$$\text{Fault current supplied by } G_2 = (19995\angle -90°) - (11800.3\angle -90°)$$

$$= 8194.7\angle -90°\text{ A}$$

Example 17.2.

Figure 17.3 shows a system having 4 alternators each rated at 11 kV, 50 MVA and each having a sub-transient reactance of 15%. Find (a) fault level for a fault on one of the feeders (near the bus) with zero value of reactance *X*, (b) the reactance of the current limiting reactor *X* to limit the fault level to 800 MVA for a fault on one of the feeders (near the bus).

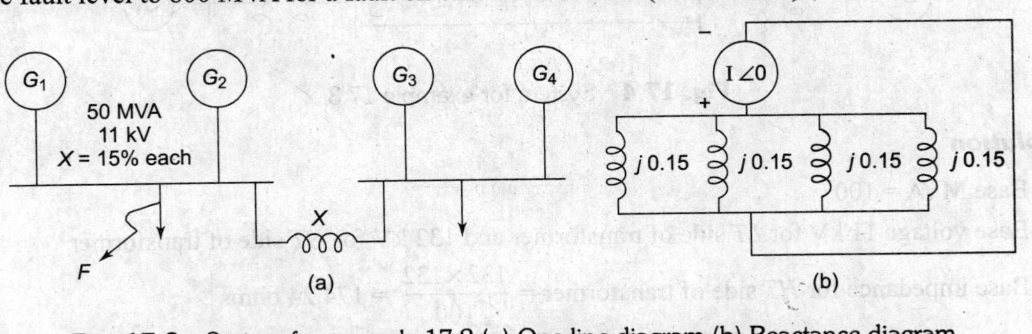

Fig. 17.3 System for example 17.2 (a) One line diagram (b) Reactance diagram

Solution

(a) Take base MVA = 50, Base kV = 11

The reactance diagram is shown in Fig. 17.3 (b)

$$X_{pu} = \frac{j\,0.15}{4} = j\,0.0375$$

$$\text{Fault level} = \frac{1\angle 0}{0.0375} = 26.67\,pu \quad \text{or} \quad 26.67 \times 50, \text{ i.e., } 1333.5 \text{ MVA}$$

(b) The alternators G_1 and G_2 will supply 0.5 × 1333.5, i.e., 666.75 MVA directly to the fault. Therefore, the fault MVA from G_3 and G_4 must be limited to 800 – 666.75 = 133.25 MVA. The reactance of G_3 and G_4 together is $\frac{0.15}{2} = 0.075\,pu$

Thus, $$\frac{1}{X+0.075} = \frac{133.25}{50} = 2.665$$

or $$X + 0.075 = 0.3752 \quad \text{or} \quad X = 0.3002\ pu.$$

Base reactance = (Base Line Voltage in kV)²/Base MVA

$$= \frac{11 \times 11}{50} = 2.42 \text{ ohms}$$

Therefore, 0.3002 *pu* reactance = .3002 × 2.42 = 0.726 ohms

The reactor X should have a reactance of 0.726 ohms.

Example 17.3.

Figure 17.4 shows a power system. Each of the alternators G_1 and G_2 is rated at 100 MVA, 11 kV and has a sub-transient reactance of 20%. Each of the transformers is rated at 100 MVA, 11/132 kV and has a leakage reactance of 5%. The line L_1 has an inductance of 1.2 mH/phase/km and a length of 100 kms. Lines L_2 and L_3 have an inductance of 1 mH/phase/km and a length of 50 kms each. Find the fault MVA and fault current for a fault at bus 5.

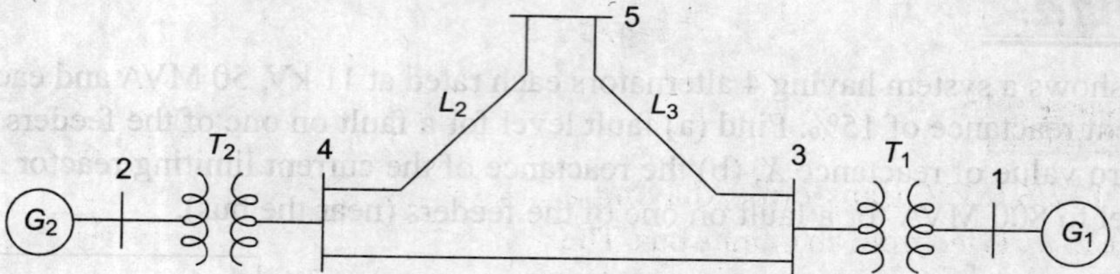

Fig. 17.4 System for example 17.3

Solution

Base MVA = 100

Base voltage 11 kV for *LT* side of transformer and 132 kV for *HT* side of transformer

Base impedance for *HT* side of transformer $= \frac{132 \times 132}{100} = 174.24$ ohms

Reactance of line $L_1 = 2\pi \times 50 \times 1.2 \times 10^{-3} \times 100 = 37.7\ \Omega$

P.u. reactance of line $L_1 = \frac{37.7}{174.24} = 0.216\ pu$

P.u. reactance of line $L_2 = (2\pi \times 50 \times 1 \times 10^{-3} \times 50)/174.24 = 0.09\ pu.$

P.u. reactance of line $L_3 = 0.09\ pu.$

The reactance diagram is shown in Fig. 17.5 (a). Using a delta-star transformation this diagram reduces to Fig. 17.5 (b). Series-parallel combinations simplify the diagram to Fig. 17.5 (c)

$$\text{Fault MVA} = \frac{1}{0.1695} = 5.9\ pu \text{ or } 590 \text{ MVA}$$

$$\text{Fault current} = \frac{1}{0.1695} = 5.9\ pu$$

$$\text{Base current} = \frac{100\times10^6}{\sqrt{3}\times132\times10^3} = 437.4 \text{ A}$$

$$\text{Fault current} = 5.9 \times 437.4 = 2580.66 \text{ A}$$

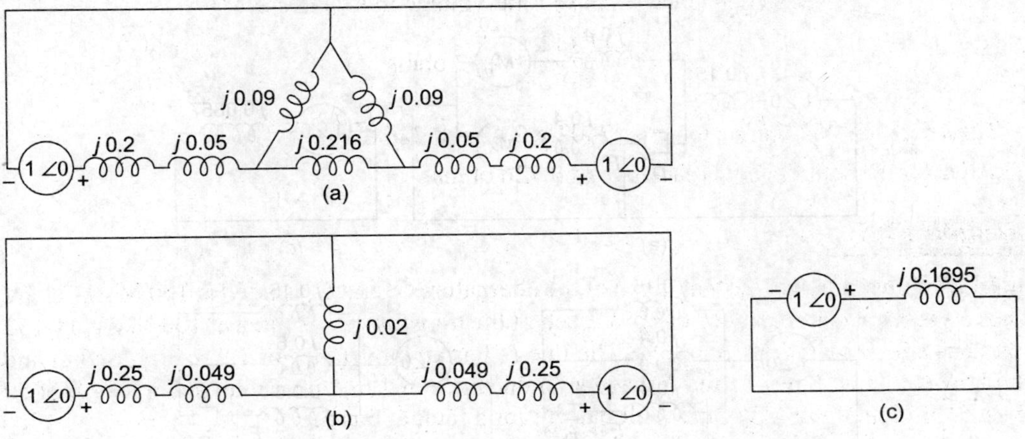

Fig. 17.5

Example 17.4.

A 10 MVA, 132/6.6 kV, transformer having a reactance of 0.15 *pu* is fed from an infinite bus. The transformer feeds 2 motors each 5 MVA, 6.6 kV. Each motor has a transient reactance of 0.3 *pu* and a sub-transient reactance of 0.2 *pu* based on its own rating. A 3-phase fault occurs at the terminals of one motor as shown in Fig. 17.6. Find (a) sub-transient fault current (b) sub-transient current in breaker *D* (c) momentary current rating of breaker *D* (d) current to be interrupted by breaker *D* which has a breaking time of 5 cycles. Assume suitable values for factors to account for dc offset.

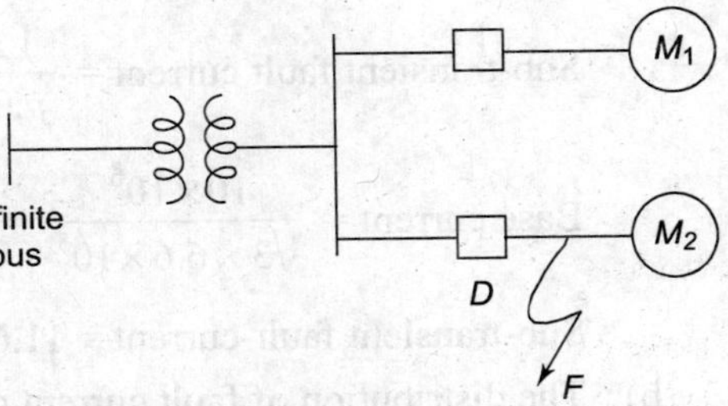

Fig. 17.6 System for example 17.4

Solution

If the external power system is very large as compared to the system of any installation, disturbances within the installation do not affect the voltage and frequency of the external system. In such a situation, the external power source is known as an infinite bus and is represented by an ideal voltage source, i.e., a constant voltage with zero internal impedance. Every power system has a number of such points whose voltage and frequency can be assumed to be constant.

$$\text{Base MVA} = 10 \text{ MVA}$$

$$\text{Base kV for } LT \text{ side} = 6.6 \text{ kV}$$

$$\text{For each motor } X''_d = 0.2 \times \frac{10}{5} = 0.4\ pu$$

$$X'_d = 0.3 \times \frac{10}{5} = 0.6\ pu$$

(a) The reactance diagram for sub-transient conditions is shown in Fig. 17.7 (a). This can be reduced to that of Fig. 17.7 (b) and then to Fig. 17.7(c).

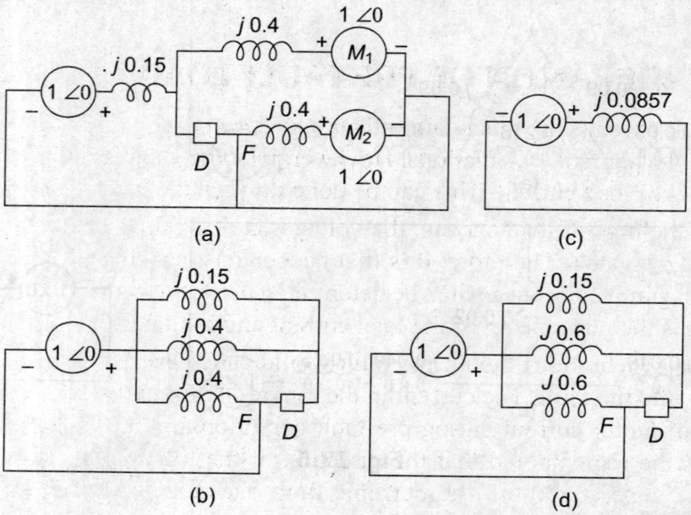

Fig. 17.7

$$\text{Sub-transient fault current} = \frac{1\angle 0}{j\,.0857} = -j\,11.67\ pu$$

$$\text{Base current} = \frac{10\times 10^6}{\sqrt{3}\times 6.6\times 10^3} = 874.8\ \text{A}$$

Sub-transient fault-current = $11.67 \times 874.8 = 10209$ A

(b) The distribution of fault current can be found from Fig. 17.7(a)

$$\text{Fault current from infinite bus} = \frac{1\angle 0}{j\,0.15} = -j\,6.67\ pu$$

$$\text{Fault current from each motor} = \frac{1\angle 0}{0.4} = -j\,2.5\ pu$$

Fault current through breaker D is the current from infinite bus and motor M_1, i.e., $(-j\,6.67) + (-j\,2.5)$ or $-j\,9.17\ pu$ or $9.17 \times 874.8 = 8022$ A.

(c) For finding the momentary current through the breaker, it is necessary to calculate dc offset current. An empirical method to include this effect is to take the momentary current as 1.6 times the symmetrical fault current.

Momentary current rating of breaker = $1.6 \times 8022 = 12835$A

(d) To find the current to be interrupted by the breaker, it is necessary to consider the transient reactance of the motors. The reactance diagram with transient reactance of motors is shown in Fig. 17.7 (d). From this diagram symmetrical current to be interrupted by breaker D

$$= \frac{1\angle 0}{j\,0.15} + \frac{1\angle 0}{j\,0.6} = -j\,8.33\ pu$$

or 8.33 × 874.8, i.e., 7287 A. The effect of dc offset can be included by using a multiplying factor of 1.1. Therefore, current to be interrupted by the breaker = 1.1 × 7287 = 8015.7 A.

17.4 CONSIDERATION OF PRE-FAULT LOAD CURRENT

Generally the fault currents are much larger than load currents and, therefore, the load currents can be neglected during fault calculations. However, in some situations it may be necessary to consider the effect of load current. This can be done through the use of superposition theorem.

As a result of the flow of load current, the voltages at different points in the power network depart from the 1 *pu* value. Therefore, it is first necessary to calculate the pre-fault voltage at the fault point. The fault current can then be determined using this value of the pre-fault voltage. The total current is the phasor sum of the load current and fault current.

If the load is a synchronous motor, the fault would cause the motor to supply power from its stored energy into the fault. The current in the motor circuit, under fault conditions, is again the phasor sum of motor current during pre-fault condition and the fault current contribution of motor. It must be remembered that the direction of the load current and fault current are the same for the generator (both the currents flow out of generator terminals). However, the directions of the two currents are opposite to each other for the motor because the load current flows into the motor terminals while the fault current flows out of the motor terminals. Example 17.5 illustrates this point.

Example 17.5.

An 11 kV 100 MVA alternator having a sub-transient reactance of 0.25 *pu* is supplying a 50 MVA motor having a sub-transient reactance of 0.2 *pu* through a transmission line. The line reactance is 0.05 *pu* on a base of 100 MVA. The motor is drawing 40 MW at 0.8 *p.f.* leading with a terminal voltage of 10.95 kV when a 3-phase fault occurs at the generator terminals. Calculate the total current in generator and motor under fault conditions.

Solution

The circuit is shown in Fig. 17.8

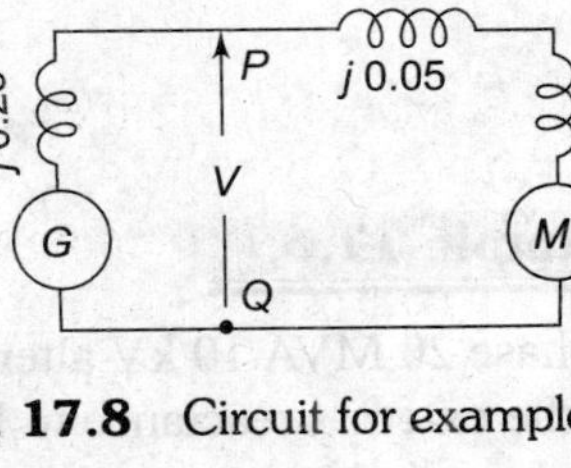

Fig. 17.8 Circuit for example 17.5

Base MVA = 100 MVA

Base voltage = 11 kV

$$\text{Base current} = \frac{100\times10^6}{\sqrt{3}\times11\times10^3} = 5248.88\ \text{A}$$

$$\text{Load current} = \frac{40\times10^6}{\sqrt{3}\times10.95\times10^3\times0.8} = 2636.4\ \text{A}$$

or $\dfrac{2636.4}{5248.8} = 0.502\ pu$

Taking the motor terminal voltage as the reference phasor, the load current is

$$I = 0.502\ (.8 + j\ .6) = .402 + j\ .301\ pu$$

$$\text{Motor terminal voltage} = \frac{10.95}{11} = 0.995\ pu$$

Prefault voltage at alternator terminals (fault location) V is

$$V = .995 + j\ 0 + (.402 + j\ .301)\ (j\ 0.05) = .9799 + \text{j}.0201\ pu$$

Sub-transient reactance of motor on 100 MVA base = 0.4 pu

Thevenin's reactance when seen from fault terminals PQ

$$= \frac{(j0.25)\ (j\ 0.45)}{j0.7} = j\ 0.1607\ pu$$

$$\text{Total fault current } I_f = \frac{0.9799 + j\ 0.0201}{j\ 0.1607} = .125 - j\ 6.09\ pu$$

Fault current supplied by generator, i.e., I_{fg} is

$$I_{fg} = (0.125 - j\ 6.09)\left(\frac{j\ 0.45}{j\ 0.45 + j\ 0.25}\right) = 0.08 - j\ 3.91\ pu$$

Fault current supplied by motor, i.e., I_{fm} is

$$I_{fm} = (0.125 - j\ 6.09)\left(\frac{j\ 0.25}{j\ 0.45 + j\ 0.25}\right) = 0.045 - j\ 2.18\ pu$$

Total post fault generator current, i.e., I_g is

$$I_g = I_{fg} + I = (0.08 - j\ 3.91) + (0.402 + j\ 0.301)$$

$$= 0.482 - j\ 3.609 = 3.64 \angle -82.4°\ pu$$

$$= 5248.8 \times 3.64 = 19105.6 \angle -82.4°\ \text{A}$$

Total post fault motor current, i.e., I_m is

$$I_m = I_{fm} - I = (0.045 - j\ 2.18) - (0.402 + j\ 0.301)$$

$$= -0.357 - j\ 2.48 = 2.505 \angle 261.8°\ pu$$

$$= 5248.8 \times 2.505 = 13148 \angle 261.8°\ \text{A}$$

Example 17.6.

A 3 phase 20 MVA 10 kV alternator has an internal reactance of 5% and negligible resistance. Find the external reactance to be connected in series with each phase so that steady current in each phase does not exceed 8 times full load current. [*JNTU 2011*]

Solution

$$\text{Full load current} = \frac{20 \times 10^6}{\sqrt{3} \times 10 \times 10^3} = 1154.73\ \text{A}$$

$$\text{Fault current} = 8 \times 1154.73 = 9237.84\ \text{A}$$

$$\text{Base impedance} = \frac{10\times10}{20} = 5\ \Omega$$

$$\text{Actual reactance} = \frac{5}{100}\times5 = 0.25\ \Omega$$

$$\text{Total reactance required} = \frac{10\times1000}{\sqrt{3}\times9237.84} = 0.625\ \Omega$$

$$\text{External reactance} = 0.625 - 0.25$$

$$= 0.375 \text{ ohm per phase}$$

Alternatively

$$\text{Base impedance} = \frac{10\times10}{20} = 5\ \Omega$$

$$\text{Total reactance required} = \frac{1}{8} = 12.5\%$$

$$= \frac{12.5}{100}\times5 = 0.625\ \Omega$$

$$\text{External reactance required} = 0.625 - 0.25$$

$$= 0.375 \text{ ohm per phase}$$

Example 17.7.

Consider the system in Fig. 17.9. The percentage reactances are on its own capacity. Find short circuit current for a dead three phase fault at F. *[JNTU 2011]*

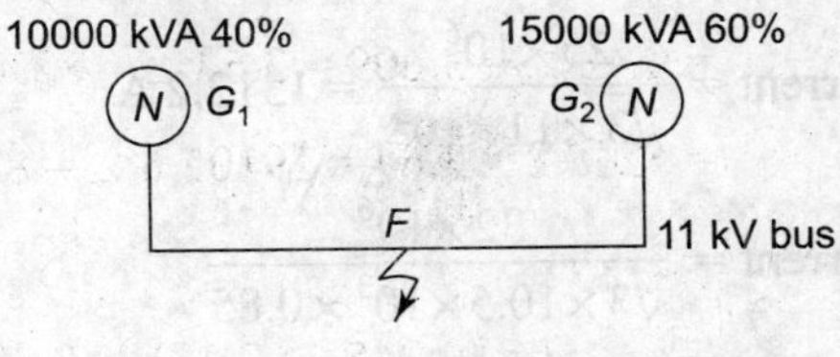

Fig. 17.9

Solution

Let base kVA = 10000 kVA

Percentage reactance of G_2 on new base

$$= \frac{10000}{15000}\times60\% = 40\%$$

Figure 17.10 shows the reactance diagram.

$$\text{Net reactance} = \frac{40\times40}{80} = 20\% = 0.2 \text{ p.u.}$$

$$\text{Short circuit current} = \frac{1}{0.2} = 5 \text{ p.u.}$$

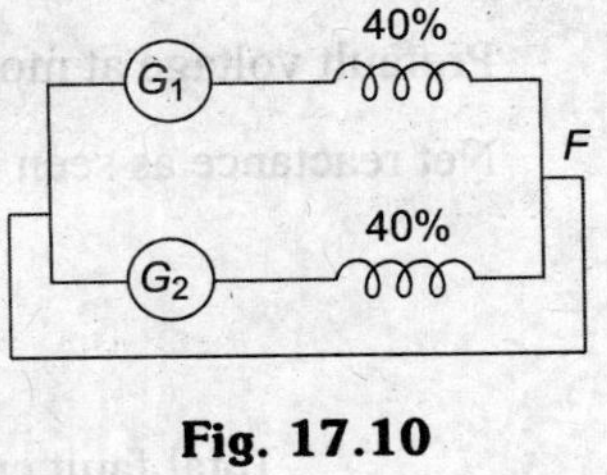

Fig. 17.10

$$\text{Full load current} = \frac{10000}{\sqrt{3}\times 11} = 524.88 \text{ A}$$

$$\text{Fault current} = 524.88 \times 5 = 2624.4 \text{ A}$$

Example 17.8.

A synchronous generator and motor rated at 25 MVA, 11 kV each have subtransient reactance of 15% each. Line reactance is 12% on the machine rating base. The motor is drawing 32 MW at 0.85 leading power factor. When terminal voltage is 10.5 kV 3 phase fault occurs at motor terminals. Find subtransient current in generator, motor and at fault point. [*JNTU 2010*]

Solution

Figure 17.11 shows the circuit

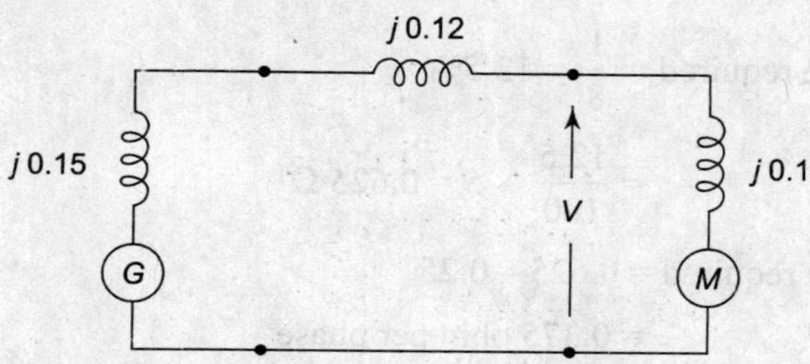

Fig. 17.11

$$\text{Base MVA} = 25 \text{ MVA}$$

$$\text{Base voltage} = 11 \text{ kV}$$

$$\text{Base current} = \frac{25\times 10^6}{\sqrt{3}\times 11\times 10^3} = 1312.2 \text{ A}$$

$$\text{Load current} = \frac{32\times 10^6}{\sqrt{3}\times 10.5\times 10^3\times 0.85}$$

$$= 2070.1 \text{ A} = 1.5776 \text{ p.u.}$$

Taking V as reference phase, load current is

$$\cos\theta = 0.85, \ \sin\theta = 0.527$$

$$I = 1.5776\,(0.85 + j\,0.527)$$

$$= 1.341 + j\,0.831 \text{ p.u.}$$

$$\text{Prefault voltage at motor terminals (fault location)} = \frac{10.5}{11} = 0.9545 \text{ p.u.}$$

Net reactance as seen from fault point

$$= \frac{(j\,0.15)\,(j\,0.27)}{j\,0.15 + j\,0.27} = j\,0.0964 \text{ p.u.}$$

$$\text{Total fault current} = I_f = \frac{0.9545}{j\,0.0964} = -j\,9.90 \text{ p.u.}$$

Fault current supplied by generator

i.e. $$I_{fg} = \frac{(-j\,9.90)\,(j\,0.15)}{j\,0.27 + j\,0.15} = -\,j\,3.54 \text{ pu}$$

Fault current supplied by motor $= \dfrac{(-j9.90)\,(j\,0.27)}{j\,0.42}$

or $$I_{fm} = -j\,6.36 \text{ p.u.}$$

Total post fault generator current $= I_{fg} + I$

$$= -j\,3.54 + 1.341 + j\,0.831$$

$$= 1.341 - j\,2.709 = 3.02\,\angle{-63.66°} \text{ p.u.}$$

$$= 3962.84 \text{ A}$$

Total post fault motor current $= I_{fm} - I$

$$= -j\,6.36 - 1.341 - j\,0.831$$

$$= -\,1.341 - j\,7.191 \text{ p.u.}$$

$$= 7.31\,\angle 259.43° \text{ p.u.}$$

$$= 9592.18 \text{ A}$$

Total fault current $= -j\,9.90$ p.u.

$$= 12990.8 \text{ A}$$

17.5 EFFECT OF SHORT CIRCUIT

Power systems operate in steady state for most of the time. However faults do occur off and on. Fault can be short circuit or open conductor fault. Short circuit can be symmetrical or unsymmetrical.

In general a short circuit causes high current in the system. Sometimes the short circuit current can be as much as 10 times the steady state load current. Such high currents can have many dangerous effects on all components in the system (generators, transformers, lines etc.) Fortunately protective systems consisting of relays and circuit breakers sense the fault and isolate the faulty section within a few cycles of occurrence. But flow of short circuit current for a few cycles before it is interrupted can cause the following effects.

1. *Damage to power system components due to high current.* The high short circuit current may cause severse damage to some components. Generally all power system equipments are designed for about 25% overload capacity. Currents higher than this value cause severse damage. Statistics have indirected some instances of conductors melting under short circuit conditions. High temperature rise associated with these high currents may cause severse damage to insulation in generators, transformers, motors etc. Because of these high currents, high centrifugal forces may occur and cause severe mechanical damage.
2. *Damage due to abnormal voltage conditions.* The voltages during fault conditions are, in general, different from rated values. A fault through zero impedance is called dead short circuit or bolted fault. In this case the voltage at fault point becomes zero.

The voltages at other points in the system will also depart from rated values. On some phase the line to neutral voltage may rise to line to line value. This can cause damage to appliances connected at those points.

17.6 SHORT CIRCUIT CAPACITY

The strength of a bus is expressed in turns of short circuit capacity or short circuit MVA. At any bus the short circuit capacity (SCC) or short circuit MVA is defined as

$$\text{SCC} = \sqrt{3}\ V I_f \times 10^{-6}\ \text{MVA} \tag{17.1}$$

where V is the line to line voltage of bus (in volts), I_f is fault current in amperes.

Generally the rated voltage of bus is taken as V in calculation of SCC. In terms of base MVA, SCC is expressed as

$$\text{SCC} = \frac{\text{Base MVA}}{\text{System reactance as seen from bus}} \tag{17.2}$$

Most of the time resistance is neglected in calculation of SCC because the system reactance is always many times the resistance. Equation (17.2) assumes that voltage is 1 p.u. Since actual voltage may be different from 1 p.u., Eq. (17.2) gives only an approximate value of SCC.

17.7 RATINGS OF CIRCUIT BREAKERS

Whenever a fault occurs, the relay senses the fault and energises the trip circuit of circuit breaker which operates and clears the fault. The interrupting rating of the circuit breaker must atleast be equal to the maximum short circuit capacity at that point. As discussed in chapter 12 the initial short circuit current is the subtransient current (which is calculated using the subtransient reactance of generator). However it must be remembered that this subtransient current is lesser than momentary current at the instant of short circuit. The reason for this is that at the instant of short circuit dc offset is also present. The extent of dc offset depends on the speed of the breakers. In this respect breakers are classified as 8 cycle, 5 cycle, 3 cycle or $1\frac{1}{2}$ cycle. The dc offset can be taken into account by multiplying the symmetrical subtransient current by the following factors.

Breaker interrupting time	Multiplying factor
8 cycles or more	1
5 cycles	1.1
3 cycles	1.2
$1\frac{1}{2}$ cycles	1.5

In calculating subtransient current, the general practice is to use subtransient reactance of synchronous generators and transient reactance of synchronous motors. The contribution of induction motors is generally neglected.

It must be remembered that circuit breaker rating must at least be equal to maximum short circuit current. For most of fault locations, 3 phase symmetrical short circuit current is more than the short circuit current for unsymmetrical faults. However for fault at generator terminals, single line to ground fault current is more than the fault current for 3 phase symmetrical fault.

Other factors included in specifications of circuit breakers are voltage class (e.g. 69 kV) rated continuous current, rated maximum voltage and rated short circuit current at rated maximum voltage and a voltage range factor *k*. This factor *k* determines the range of voltage over which the short circuit current times operating voltage is constant. For 69 kV breaker, the maximum rated voltage is 72.5 kV, the factor k is 1.21. If short circuit current at 72.5 kV is 19 kA, the short circuit current at 60 kV is 19 × 1.21 = 23 kA. At 69 kV, the short circuit current is (72.5/69) (19) = 20 kA. For breakers of voltage class higher than 69 kV, the factor *k* is 1.1.

17.8 BUS BAR ARRANGEMENTS

17.8.1 General

The switching equipment in a generating station can be connected in a variety of ways. The considerations governing the selection of a bus bar arrangement include simplicity, initial cost of equipment, flexibility of operation, possibility of system expansion and ease of maintenance.

All modern generating stations use a unit system of generator-transformer connection. Each generator is connected to its own transformer and the generator-transformer unit is connected through circuit breaker and isolators to the high voltage bus. The paralleling of generators is done on the high voltage side. Since the three phases in a system are similar, the arrangement of connections is depicted through a single line diagram which indicates only one of the three phases.

17.8.2 Single Bus-Bar (Fig. 17.12)

This is the simplest and cheapest arrangement and uses the minimum equipment. However the scheme is not reliable because in the event of a fault on the bus, the station becomes completely dead. If any maintenance work is to be done on the bus, the station has to be shut down. The scheme is used for switch boards of small power plants and substations.

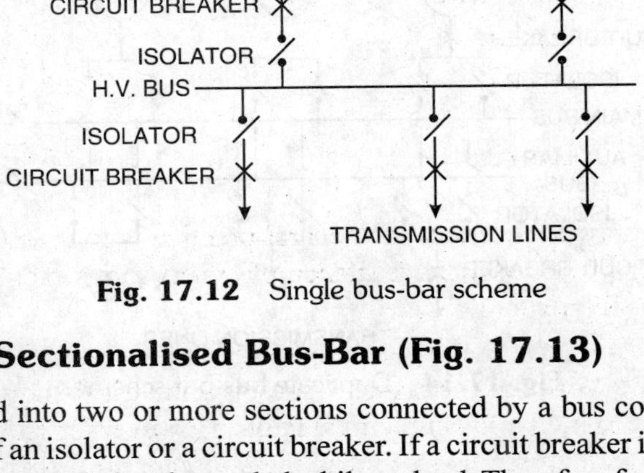

Fig. 17.12 Single bus-bar scheme

17.8.3 Single Sectionalised Bus-Bar (Fig. 17.13)

The bus-bar is divided into two or more sections connected by a bus coupler (B. C.). The bus coupler may consist of an isolator or a circuit breaker. If a circuit breaker is used as a bus coupler the bus sections may be coupled or decoupled while on load. Therefore for full benefits and more reliable operation a circuit breaker as bus coupler is preferable. In the event of a fault, the faulty bus section can be isolated and the continuity of supply maintained at least partly. Obviously this scheme is costlier than a single bus scheme.

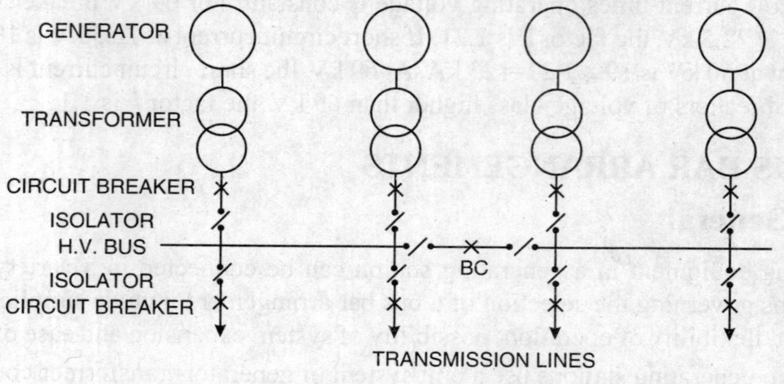

Fig. 17.13 Single sectionalised bus-bar scheme.

17.8.4. Duplicate Bus-Bar (Fig. 17.14)

This arrangement consists of two identical bus bars—one main and the other auxiliary or reserve. Under normal operation all the circuits are connected to the main bus. Whenever a fault develops on the main bus or it is to be taken out for maintenance, the circuits are transferred to the auxiliary bus. Thus a double bus scheme ensures greater reliability and flexibility of operation. However it is costlier than the single sectionalised bus scheme. The transfer of circuits from main bus to the auxiliary bus can be done through the following sequence of steps :

(1) Close the bus coupler.

(2) Close the isolators on the auxiliary bus.

(3) Open isolators on the main bus.

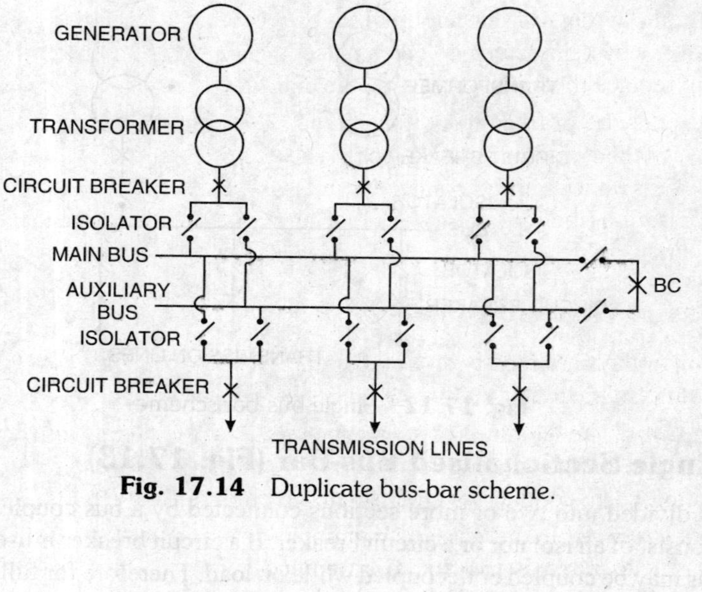

Fig. 17.14 Duplicate bus-bar scheme.

17.8.5 Ring Bus (Fig. 17.15)

This scheme has greater reliability and flexibility of operation. Whenever a fault develops on any section, that section can be isolated. However the extension of the system is difficult when this scheme is adopted.

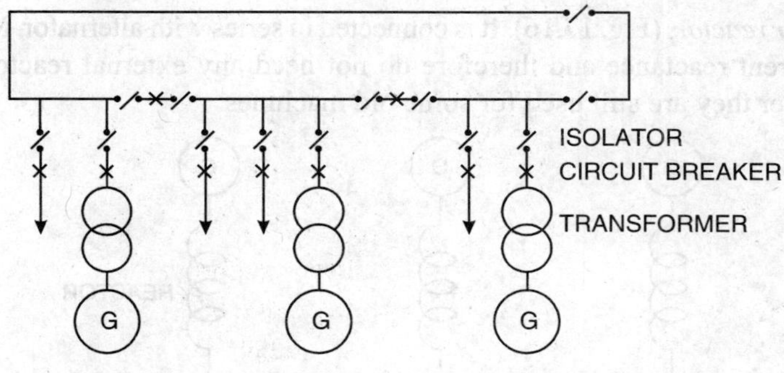

Fig. 17.15 Ring bus scheme

17.9 REACTORS

17.9.1 Application

A series reactor is an inductive coil connected in a circuit or system to limit the short circuit current. The use of shunt reactor is discussed in chapter 25.

The short circuit current in a power system depends on the system generating capacity, voltage at the point of fault and the impedance between the generators and the fault point. Whenever a fault occurs in a system, one or the other circuit breaker interrupts the fault current and isolates the faulty circuit. The circuit breakers in the system are selected so that they can interrupt the maximum short circuit current in the system. However for some systems the short circuit current may be so high that is may be worthwhile to limit the current by the use of reactors connected suitably in the circuit. In that case circuit breakers of lesser breaking capacity would be needed. Moreover whenever a system extension takes place, the short circuit current increases. In case it is desired not to replace the circuit breakers, the current can be limited by the use of reactors.

A series reactor may be of fixed or switched type. A fixed reactor is permanently connected in the circuit. A switched reactor is one whose reactance can be varied so that at any time a desired amount of reactance can be connected in the system. The specifications for a reactor include the type, location (indoor or outdoor), reactance in ohm, kVA rating, rated voltage, rated current and frequency.

17.9.2 Types of Reactors

Reactors used now-a-days are mostly air cored coils. They are built in two types – the dry type reactor and oil immersed reactor.

(*a*) *Dry type reactor*. The windings of a dry type reactor are embedded in vertical supports of specially treated concrete. The winding is bolted to concrete base which is insulated from ground by porcelain-footings. The conductors are insulated with fire proof asbestos braid.

(*b*) *Oil immersed reactor*. In this case the windings are similar to those used for transformers. A non-magnetic shield prevents the flux from entering the steel tank where it would cause heating and losses. The oil helps in cooling and providing insulation. An oil immersed reactor has a smaller size (for the same rating) higher thermal capacity and fewer chances of flashover as compared to dry type. Oil immersed reactors are built for both outdoor and indoor use and for all system voltages.

17.9.3 Reactor Locations

(*a*) *Generator reactor*. (Fig. 17.16). It is connected in series with alternator. Modern alternators have high inherent reactance and therefore do not need any external reactor for limiting the current. However they are still used for some old machines.

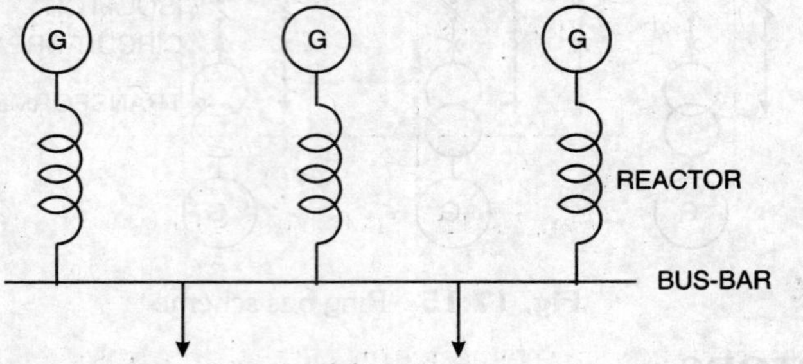

Fig. 17.16 Generator reactors

(*b*) *Feeder reactor*. (Fig. 17.17). A reactor is connected in series with each feeder. Since the number of feeders is generally large, this connection entails the use of a large number of reactors. Moreover since the reactors carry full load current, a constant power loss occurs.

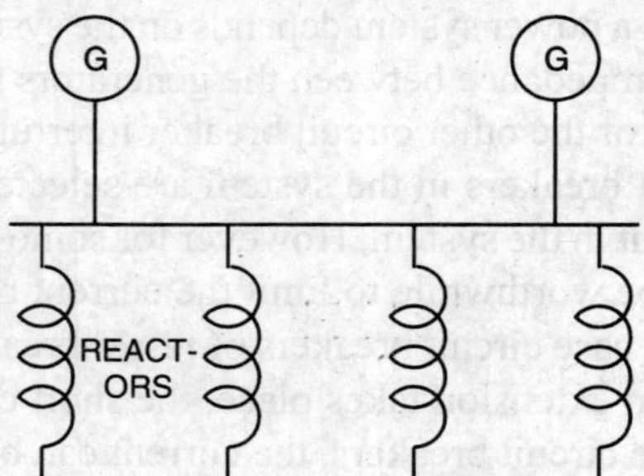

Fig. 17.17 Feeder reactors

(*c*) *Bus bar reactor*. (Fig. 17.18). The reactors are inserted in between bus sections. During normal operation only a small power flows through the reactor and hence very small power loss takes place. During short circuit only that generator feeds the fault directly. The contribution of fault current by other generators is limited by the reactors.

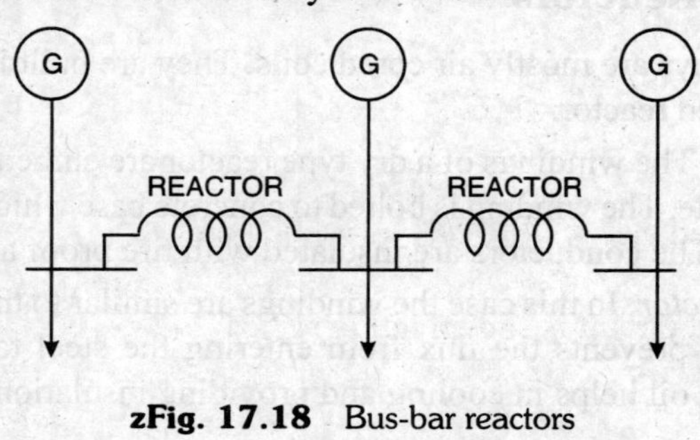

z**Fig. 17.18** Bus-bar reactors

(*d*) *Tie bus-bar reactor*. (Fig. 17.19). The reactors are connected between the buses and a tie bar. This scheme has the advantage that the fault current is independent of the number of generators connected in the system.

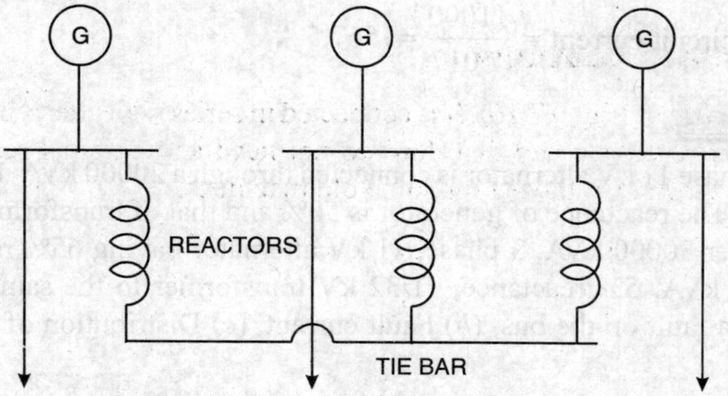

Fig. 17.19 Tie bus-bar reactors

Example 17.9.

Two single phase generators are connected in parallel to a common bus. Generating each 3000 kVA, 10% reactance, 11 kV. Three feeders emanate from the bus. A fault takes place on one of the feeders. The reactance of feeder from bus to the fault point is 5 ohm. Calculate the fault current (*a*) working in p.u. quantities, (*b*) working in actual quantities. Neglect resistances.

Solution (*a*)

Take Base kVA = 3000

Base kV = 11

$$\text{Base current} = \frac{3000}{11} = 272.7 \text{ A}$$

$$\text{Base impedance} = \frac{\text{Base voltage}}{\text{Base current}} = \frac{11000}{272.7} = 40.34 \text{ ohm}$$

$$\text{p.u. feeder reactance} = \frac{5}{40.34} = 0.124$$

The equivalent circuit is shown in Fig. 17.20.

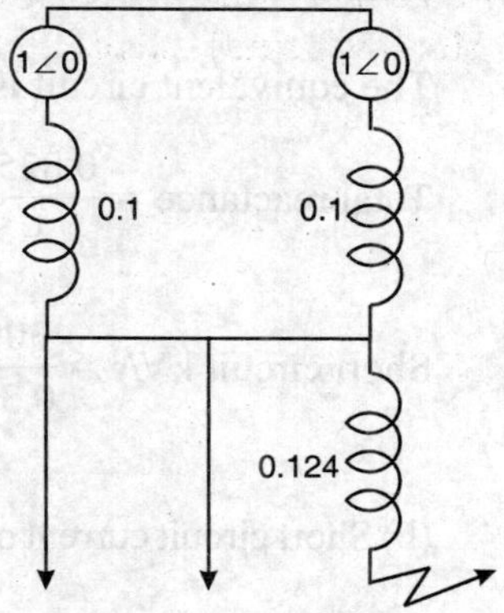

Fig. 17.20 Equivalent circuit

$$\text{Total reactance} = \frac{0.1}{2} + 0.124 = 0.174 \text{ p.u.}$$

$$\text{Short circuit kVA} = \frac{3000}{0.174} \text{ kVA} = 17242 \text{ kVA}$$

$$\text{Short circuit current} = \frac{17242}{11} = 1567.5 \text{ A}$$

(b) Generator impedance $= (40.34)\ \dfrac{10}{100} = 4.034$ ohm

The equivalent circuit is shown in Fig. 17.21.

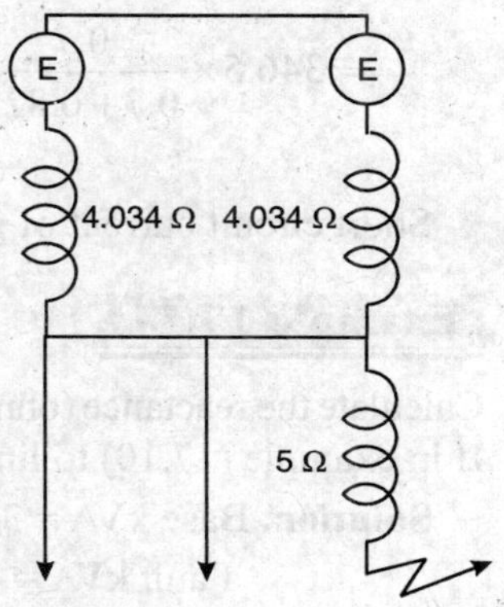

Fig. 17.21 Equivalent circuit

$$\text{Total impedance} = \frac{4.034}{2} + 5 = 7.017 \text{ ohm}$$

$$\text{Short circuit current} = \frac{11000}{7.017} = 1567.5 \text{ A}$$

Example 17.10.

A 20000 kVA 3 phase 11 kV alternator is connected through a 20000 kVA 11/132 kV transformer to a 3 phase bus. The reactance of generator is 50% and that of transformer is 5% on their own kVA base. Another 30000 kVA, 3 phase, 11 kV alternator having 65% reactance is connected through a 30000 kVA, 5% reactance, 11/32 kV transformer to the same bus. Find (*a*) Short circuit MVA for a fault on the bus, (*b*) Fault current, (*c*) Distribution of fault current between the generators. (*P.U. 1986*)

Solution

Taking base kVA = 30000 and base voltage = 11 kV on primary (*i.e.* generator) side and 132 kV on secondary (bus bar) side of transformer

(*a*) Reactance of 1st transformer to new base $= 0.05 \times \frac{1}{3} = 0.075$ p.u.

Reactance of 1st alternator to new base kVA $= 0.5 \times \frac{3}{2} = 0.75$ p.u.

The equivalent circuit is shown in Fig. 17.22.

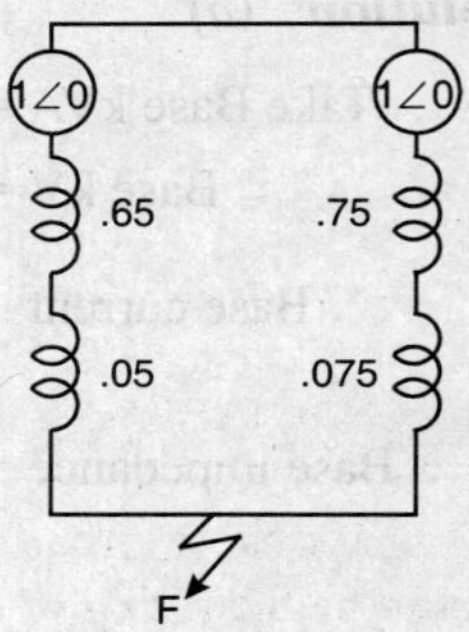

Fig. 17.22 Equivalent circuit

$$\text{Total reactance} = \frac{0.825 \times 0.7}{1.525} \text{ p.u.} = 0.3787 \text{ p.u.}$$

$$\text{Short circuit kVA} = \frac{30000}{0.3787} = 79218 \text{ kVA}$$

(b) Short circuit current on bus bar side $= \frac{79218}{\sqrt{3} \times 132} = 346.5\text{A}$

(c) Short circuit current of generator I

$$= 346.5 \times \frac{0.7}{0.7 + 0.825} \times \frac{132}{11} = 1908.6 \text{ A}$$

Short circuit current of generator II $= 346.5 \times \frac{0.825}{1.525} \times \frac{132}{11} = 2249.4$ A

Example 17.11.

Calculate the reactance (ohmic value) of a reactor to be placed on the secondary side of transformer II in example (17.10) to limit the fault MVA to 50 MVA.

Solution. Base kVA = 30000

Fault kVA = 50000

$$\therefore \text{ total reactance as seen from fault point } = \frac{30000}{50000} = 0.6 \text{ p.u.}$$

Let the reactor added in circuit of generator II have X% reactance on 30000 kVA base (Fig. 17.23)

$$\frac{1}{0.6} = \frac{1}{0.7+X} + \frac{1}{0.825}$$

whence $X = 1.5$ p.u.

$$\text{Base impedance} = \frac{132 \times 132 \times 1000}{30000} = 580.8 \text{ ohms}$$

$$\therefore \quad X = 1.5 \times 580 = 871.2 \text{ ohms.}$$

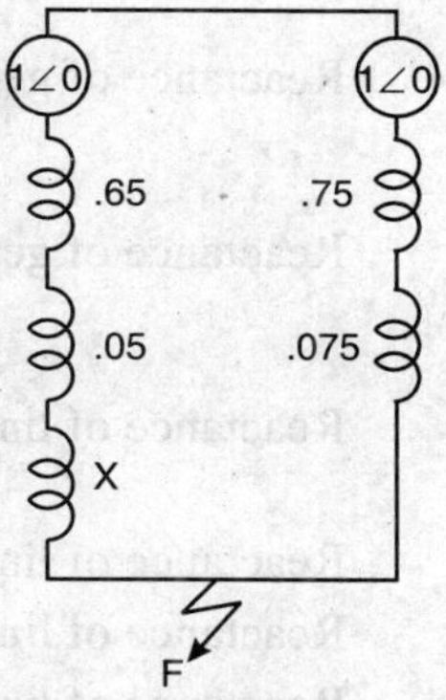

Fig. 17.23

Example 17.12.

Three similar alternators 50 MVA, 50% reactance are connected to separate bus sections. Two reactors each 10% reactance (on 50 MVA base) connect the three bus sections. Calculate the fault MVA for a fault on the middle bus section. (*A.U. 1989*)

Solution

The equivalent circuit is shown in Fig. 17.24.

$$\text{Total reactance} = \frac{1}{\frac{1}{0.6} + \frac{1}{0.6} + \frac{1}{0.5}} = 0.1875 \text{ p.u.}$$

$$\text{Fault MVA} = \frac{50}{0.1875} = 266.7 \text{ MVA}$$

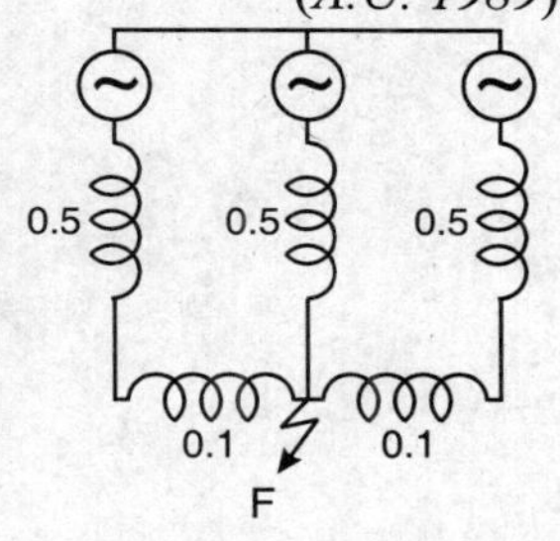

Fig. 17.24

Example 17.13.

A 3 phase short circuit occurs at the position shown on the 33 kV system of Fig. 17.25. What should be the rating of circuit breaker in the faulty feeder. (*B.U. 1991*)

Solution

Taking Base kVA = 20000

Base kV = 33

$$\text{Base impedance} = \frac{33 \times 33 \times 1000}{20000} = 54.45 \text{ ohms}$$

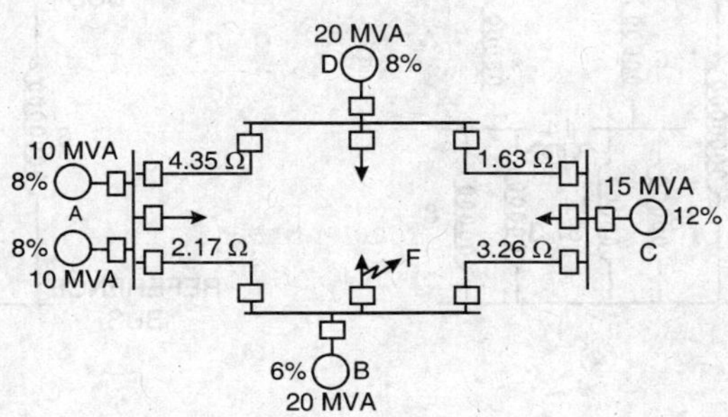

Fig. 17.25 33 kV system.

$$\text{Reactance of generator A on 20 MVA base} = 8 \times \frac{20}{10} = 16\% \text{ for each generator}$$

$$\text{Reactance of generator C to new base} = 12 \times \frac{20}{15} = 16\%$$

$$\text{Reactance of line AB} = \frac{2.17}{54.45} \times 100 = 4\%$$

Reactance of line BC = 6%

Reactance of line CD = 3%

Reactance of line AD = 8%

The equivalent circuit as seen from the fault point is as shown in Fig. 17.26. The values shown in the circuit are in percentage.

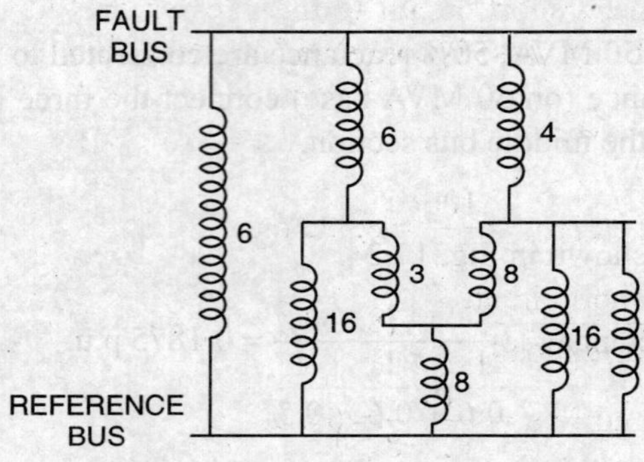

Fig. 17.26

This circuit can be simplified in successive steps as shown in Fig. 17.27 (*a* to *d*).

$$\text{Net reactance} = \frac{6(1.75 + 4.16)}{6 + 1.75 + 4.16} = 2.977\%$$

$$\text{Fault kVA} = \frac{20000}{2.977} \times 100 = 672000 \text{ kVA}$$

Hence, circuit breaker rating = 672 MVA

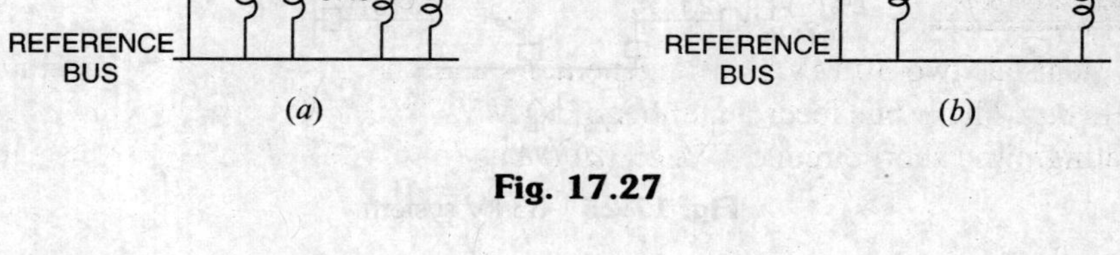

Fig. 17.27

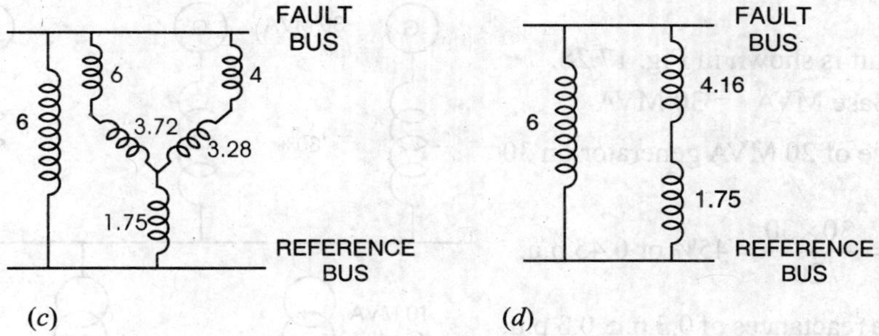

Fig. 17.27

Example 17.14.

A 150 MVA, 11 kV, 3 phase synchronous generator has 12% reactance. (*a*) Find the ratio of mechanical stresses on dead short circuit to the stresses on full load (*b*) If a reactor of 8% reactance on 150 MVA, 11 kV, base is connected in series, find the new ratio of short circuit to full load stresses.

Solution.

$$(a) \text{ Short circuit current } = \frac{100}{12} = 8.333 \text{ p.u.}$$

$$\frac{\text{Mech. stresses on short circuit}}{\text{Mech. stresses on full load}} = (8.333)^2 = 69.44$$

$$(b) \text{ Short circuit current} = \frac{100}{12+8} = 5 \text{ p.u.}$$

$$\frac{\text{Mech. stresses on short circuit}}{\text{Mech. stresses on full load}} = (5)^2 = 25$$

Example 17.15.

A feeder reactor has 4% reactance at rated current. It is carrying rated current at 0.8 lagging power factor. Find the percentage increase in bus bar voltage to compensate for drop in reactor.

Solution.

$$I = 1 \text{ p.u.}$$

$$V = 1 \text{ p.u.}$$

$$X = 4\% = j\,0.04 \text{ p.u.}$$

$$\text{Bus bar voltage} = 1\angle 0 + 1(0.8 - j\,0.6)(j\,0.04)$$

$$= 1 + j0.032 + 0.024 = 1.024 + j0.032 \text{ p.u.} = 1.0245 \angle 1.8°$$

Increase in bus bar voltage = 0.0245 p.u. or 2.45%

Example 17.16.

A power plant has two 30 MVA, 30% generators and one 20 MVA, 30% generator feeding a common *lv* bus. The *lv* bus feeds a number of 10 MVA transformers each with 4% reactance on its own rating. Find short circuit MVA on (*a*) *lv* bus (*b*) *hv* side of each transformer.

(*K.U. 1986*)

Solution

The circuit is shown in Fig. 17.28.

(*a*) Let Base MVA = 30 MVA

Reactance of 20 MVA generator on 30 MVA base

$$= \frac{30\times30}{20} = 45\% \text{ or } 0.45 \text{ p.u.}$$

The three reactances of 0.3 p.u, 0.3 p.u. and 0.45 p.u. are in parallel

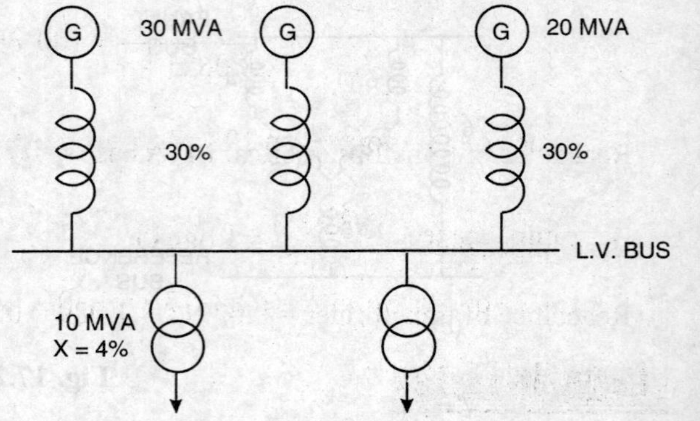

Fig. 17.28

$$\text{Total Reactance} = \frac{1}{\frac{1}{0.3}+\frac{1}{0.3}+\frac{1}{0.45}}$$

$$= 0.1125 \text{ p.u.}$$

Short circuit MVA on L.V. bus

$$= \frac{30}{0.1125} = 266.67 \text{ MVA}$$

(*b*) Reactance of transformer on 30 MVA base = $0.04\times\frac{30}{10} = 0.12$ p.u.

Total reactance as seen from H.V. side of transformer = 0.1125 + 0.12 = 0.2325 p.u.

$$\text{Short circuit MVA} = \frac{30}{0.2325} = 129.03 \text{ MVA}$$

Example 17.17.

A 3 phase power plant has following generators feeding a common bus

2 Generators 30 MVA, 15% reactance each

1 Generator 10 MVA, 12.5% reactance

A 10 MVA transformer steps up the voltage and feeds a 3.3 kV circuit. Find the safe minimum reactance of transformer so that fault level on the secondary bus of the transformer may not exceed 100 MVA. [*P.U. 1990*]

Solution

Let Base MVA = 30

Reactance of 12.5% on 10 MVA base when converted to 30 MVA base

$$= 12.5\times\frac{30}{10} = 37.5\% \text{ or } 0.375 \text{ p.u.}$$

The reactances of the 3 generators are in parallel. The net reactance

$$= \frac{1}{\frac{1}{0.15}+\frac{1}{0.15}+\frac{1}{0.375}} = 0.0625 \text{ p.u.}$$

Let reactance of transformer on 30 MVA base be x p.u. Then

$$\frac{30}{0.0625 + x} = 100 \text{ or } x = 0.2375 \text{ p.u.}$$

Reactance of transformer on 10 MVA base $= 0.2375 \times \dfrac{10}{30} = 0.07917$ p.u.

Base impedance $= \dfrac{3.3 \times 3.3}{10} = 1.089\ \Omega$

Reactance of transformer = 0.07917 × 1.089 = 0.0862 ohm

Example 17.18.

Two 20 MVA, 11 kV, 50 Hz synchronous generators having 20% reactance each are connected to bus A of a station. Two similar generators are connected to bus B. The two busses are connected by a 30 MVA, 20% reactance reactor. A 10 MVA feeder having a reactance of 6% takes off from bus A. (*a*) If a 3 phase fault occurs at far end of feeder find the fault current. (*b*) It is desired to limit the fault current to 50% of the value found above. Find the ohmic value of feeder reactor.

(*J.U. 1991*)

Solution.

(*a*) Let base MVA = 20 MVA,

base kV = 11

Reactance of interconnecting reactor on

20 MVA base $= 20 \times \dfrac{20}{30} = 13.333\%$

= 0.13333 p.u.

Reactance of feeder on 20 MVA base

$= 0.06 \times \dfrac{20}{10} = 0.12$ p.u.

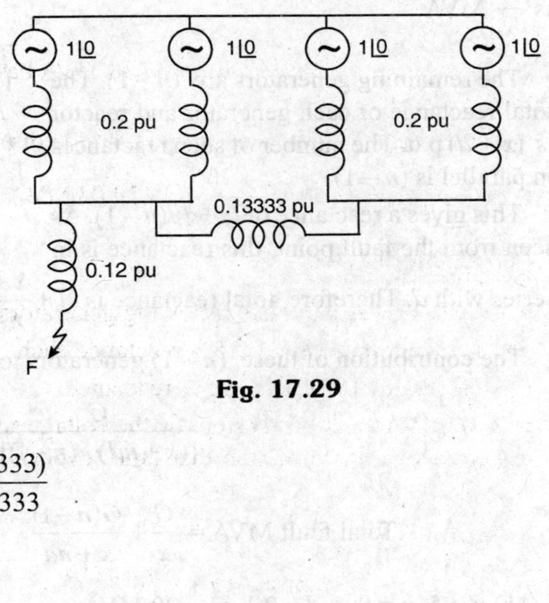

Fig. 17.29

The equivalent circuit is shown in Fig. 17.29.

$$\text{Net reactance} = 0.12 + \frac{(0.1)(0.1 + 0.13333)}{0.1 + 0.1 + 0.13333}$$

= 0.19 p.u.

$$\text{Fault current} = \frac{1\angle 0}{0.19} = 5.263 \text{ p.u.}$$

$$\text{Base current} = \frac{20 \times 1000}{\sqrt{3} \times 11} = 1049.759 \text{ A}$$

Fault current = (1049.759) (5.263) = 5524.88 A

(*b*) p.u. value of required fault current = 0.5 × 5.263 = 2.6315 p.u.

p.u. value of total required reactance $= \dfrac{1}{2.6315} = 0.38$ p.u.

Additional reactance required = 0.38 – 0.19 = 0.19 p.u.

$$\text{Base Impedance} = \frac{11\times 11}{20} = 6.05 \text{ ohm}$$

$$\text{Reactance required} = 0.19 \times 6.05 = 1.1495 \text{ ohm}$$

Example 17.19.

A generating plant has n generating units each rated at G MVA, x p.u. reactance. Each unit has its own bus bar. The bus bars are interconnected by tie bus bar reactors each having a reactance of d p.u. on G MVA base. (*a*) Find an expression for short circuit MVA for a 3 phase fault on one section bus (*b*) Find the value of short circuit MVA for G = 20 MVA, x = 0.4 p.u. and d = 0.1 p.u. if n = 5 and 10 (*c*) Show that for a very large value of n, the maximum value of short circuit MVA will not exceed 250 MVA.

Solution

The configuration is shown in Fig. 17.30.

The contribution towards fault MVA by the generator on whose bus the fault has occurred is $\frac{G}{x}$ MVA.

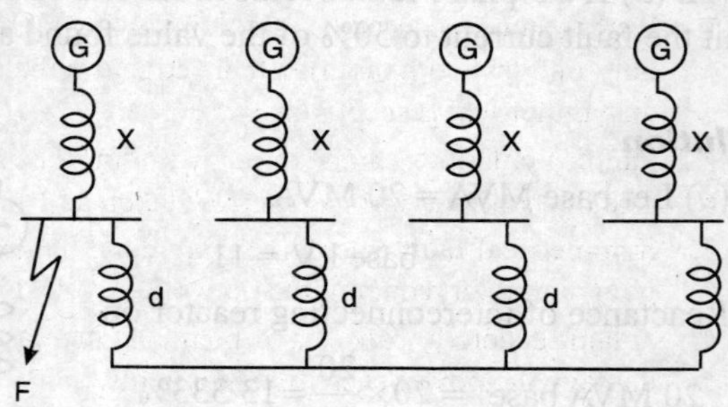

Fig. 17.30

The remaining generators are $(n-1)$. The total reactance of each generator and reactor is $(x+d)$ p.u. The number of such reactances in parallel is $(n-1)$.

This gives a reactance of $(x+d)/(n-1)$. As seen from the fault point, this reactance is in series with d. Therefore, total reactance is $d+\frac{x+d}{n-1} = \frac{x+nd}{n-1}$

The contribution of these $(n-1)$ generators to fault MVA is

$$\frac{G}{[(x+nd)/(n-1)]} = \frac{G(n-1)}{x+nd}$$

$$\text{Total fault MVA} = \frac{G}{x} + \frac{G(n-1)}{x+nd}$$

(*b*) n = 5, x = 0.4, d = 0.1, G = 20 MVA

$$\text{Fault MVA} = \frac{20}{0.4} + \frac{20(5-1)}{0.4+5\times 0.1} = 50 + \frac{20\times 4}{0.9}$$

$$= 50 + 88.89 = 138.89 \text{ MVA}$$

n = 10, x = 0.4, d = 0.1

$$\text{Fault MVA} = \frac{20}{0.4} + \frac{20(10-1)}{0.4+10\times 0.1} = 50 + \frac{20\times 9}{1.4} = 178.57 \text{ MVA}$$

(c) Fault MVA $= \frac{G}{x} + \frac{G(n-1)}{nd + x}$

$$= \frac{G}{x} + \frac{G}{\frac{nd}{n-1} + \frac{x}{n-1}}$$

when n is very large $\frac{nd}{n-1} \simeq d$ and $\frac{x}{n-1} \simeq 0$. Therefore,

$$\text{Fault level} = \frac{G}{x} + \frac{G}{d} = \frac{20}{0.4} + \frac{20}{0.1} = 250 \text{ MVA}$$

SUMMARY

17.1. Faults can be balanced (i.e. those involving all the three phases) or unbalanced (i.e. those involving only one or two phases). Balanced and unbalanced faults are also known as symmetrical and unsymmetrical respectively.

17.2. Fault calculations can be done by finding Thevenin's equivalent circuit as seen from fault point. The open circuit emf is generally taken as 1 p.u.

17.3. Symmetrical fault calculations are done by drawing impedance diagram as seen from the fault point and then reducing the network.

17.4. A fault generally causes a tremendous increase in current in the system. This high current may damage the power system components.

17.5. System MVA under fault conditions is also called short circuit capacity.

17.6. Rating of a circuit breaker should not be less than maximum short circuit capacity.

17.7. Bus bar arrangements can be single bus bar, single sectionalised bus bar, duplicate bus bar and ring bus bar.

17.8. Series reactors are used to reduce short circuit current.

SELF TEST QUESTIONS

A. State whether the following are true (T) or false (F).

17.1 In fault calculations open circuit emf at fault point is generally assumed to be 1 p.u.

17.2 Shunt capacitance are generally neglected in fault calculations.

17.3 In a motor the direction of load current and fault current are the same.

17.4 Fault calculations should be done by using subtransient reactance of generator.

17.5 For circuit breakers having interrupting time of 8 cycles or more, dc offset is neglected.

17.6 S.C.C. = $\sqrt{3}\ V_L I_f \times 10^{-6}$ MVA

17.7 Small power plants generally use single bus bar.

17.8 Series reactors help in reducing short circuit current.

B. Fill in the blanks with suitable words.

17.9 Symmetrical fault involves all the ____________.

17.10 Fault calculations are generally done in ____________.

17.11 S.C.C. = ____________.

17.12 Series reactors ———————— the fault current.

17.13 Circuit breaker rating should be at least equal to ————————.

17.14 The reference bus in short circuit calculations is ————————.

17.15 For synchronous motor ———————— is used in fault calculations.

ANSWERS

17.1 T   17.2 T   17.3 F   17.4 T   17.5 T   17.6 T

17.7 T   17.8 T   17.9 three phases   17.10 p.u.

17.11 $\sqrt{3}\, V_L I_f \times 10^{-6}$ MVA   17.12 reduce   17.13 S.C.C.   17.14 system neutral

17.15 transient reactance

JOB INTERVIEW QUESTIONS

17.1 What is the importance of fault studies?

17.2 How are faults classified?

17.3 What is the difference between symmetrical and unsymmetrical faults?

17.4 What is Thevenin's equivalent circuit?

17.5 What is the functions of series reactor?

17.6 What is meant by short circuit capacity?

REVIEW QUESTIONS

17.1 What is a fault? Why do faults occur? What is the effect of faults?

17.2 Distinguish between symmetrical and unsymmetrical faults. How are symmetrical faults analysed?

17.3 What is short circuit capacity? What is its importance?

17.4 What are series reactors? How are they classified? Describe each type.

17.5 What is a circuit breaker? How is its rating decided?

PROBLEMS

17.1 Three 20 MVA generators each having a reactance of 0.2 p.u. are operating in parallel. A fault occurs at generator bus. Find the fault MVA if fault is symmetrical. [300 MVA]

17.2 Three 10 MVA generators each having a reactance of 0.2 p.u. are operating in parallel. They feed a transmission line through a 30 MVA transformer having p.u. reactance of 0.05. Find fault MVA for a 3 phase fault at sending end of line. [120 MVA]

17.3 A 3 phase, 6 MVA, 6.6 kV generator has a reactance of 0.1 p.u. It is connected to a transformer 6 MVA, 6.6/33 kV with 0.09 p.u. reactance. The transformer feeds a 3 phase line having 0.09 ohm resistance and 0.36 ohm reactance per km. A 3 phase dead short circuit occurs at 10 km distance from the transformer. The generator voltage at the time of fault is 7.2 kV. Find fault current in generator. [2726 A]

17.4 Three 6.6 kV generators of ratings 8 MVA, 2 MVA and 5 MVA have p.u reactances of 0.16, 0.08 and 0.12 respectively. They feed a line having a reactance of 0.125 ohm. A 3 phase fault occurs at far end of line. Find fault MVA. [87.2 MVA]

17.5 A power system has three bus sections with 11 kV generators on a ring system as under:

Section 1: Two generators 30 MVA and 20 MVA with 0.15 and 0.12 reactance p.u. respectively.

Section 2: One generator of 60 MVA and 0.2 p.u. reactance.

Section 3: Two generators of 30 and 60 MVA with 0.18 and 0.12 p.u. reactance respectively. Section 1 and 2, 2 and 3 and 3 and 1 are connected through reactors of 60, 90 and 80 MVA rating with 0.2, 0.3 and 0.24 p.u. reactances respectively. Find fault current for a 3 phase fault at far end of a feeder connected to bus 3. The feeder has 0.01 p.u. resistance and 0.05 p.u. reactance on 30 MVA base. Also find fault current supplied by 60 MVA generator on bus 3. [17800 A, 6760 A]

17.6 A system has 3 similar 3 phase 50 Hz generators each 200 MVA, 11 kV, 0.9 p.u. reactance. Each generator feeds a transformer 11/132 kV, 200 MVA, 0.09 p.u. reactance. The secondaries of transformers feed a common bus. A line having 0.5 ohm resistance and 8 mH inductance per phase takes off from the bus. Find fault MVA for 3 phase fault at end of line. [557.4 MVA]

17.7 Two 22 kV 3 phase 50 Hz alternators feed a common bus. One alternator 60 MVA 0.18 p.u. reactance and the other is 100 MVA, 0.2 p.u. reactance. A step up transformer 22/132 kV, 150 MVA, 0.12 p.u. reactance is connected to station bus. The transformer feeds a 3 phase line having an inductance of 27.8 mH. The receiving end of line feeds a transformer with 132/11 kV, 150 MVA, 0.09 p.u. reactance rating. The 11 kV receiving end bus feeds the following loads:

(i) balanced 3 phase 40 MW 0.8 lagging power factor

(ii) 20 MVA synchronous motor with 0.16 p.u. transient reactance taking 20 MVA at 0.9 p.f. leading. The voltage at receiving end is 10.55 kV. The two generators share the vars equally bus share the active power in proportional to their ratings.

A 3 phase fault with are resistance of 0.0484 ohm between each phase and earth occurs at motor terminals when the system is loaded as above. Find currents in all the three synchronous machines.

$$\begin{bmatrix} \text{For 60 MVA generator } I = 1186 - j\,3398 \text{ A} \\ \text{For 100 MVA generator } I = 1850 - j\,4897 \text{ A} \\ \text{For motor } I = 143 - j\,6453 \text{ A} \end{bmatrix}$$

17.8 A 20 MVA, 11 kV, 3 phase generator having a transient reactance of 0.2 p.u. feeds a 20 MVA 11 kV 3 phase synchronous motor having a transient reactance of 0.5 p.u. through a short line having 0.3025 ohm reactance. A 3 phase fault occurs at motor terminals when it is drawing 15 MW at unity p.f. The voltage at motor terminals is 10.45 kV at the time of fault. Find total current supplied by generator and motor. [787.3 – j 3989 A, – 787.3 – j 1994 · 5 A]

17.9 Figure 17.31 shows a system fed by two generators. A 3 phase fault occurs at receiving end. Find (a) fault level and fault current (b) fault current supplied by G_1 and G_2.

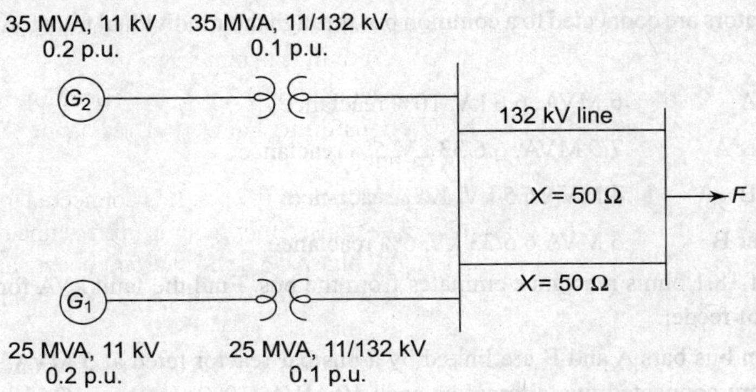

Fig. 17.31

$$\begin{bmatrix} (a)\ 155.55 \text{ MVA } 680.37 \text{ A} \\ (b)\ 3401.875 \text{ A}, 4762.625 \text{ A} \end{bmatrix}$$

17.10 A large generating station has its bus bars divided into four sections. Each section is connected to a tie bar through a reactor having 0.1 p.u. reactance at 25 MVA base. Each section is fed by 25 MVA 0.2 p.u. reactance generator. Find fault level for a fault at bus 1 (a) with reactors connected (b) with reactors short circuited. [(a) 250 MVA (b) 500 MVA]

17.11 A power station has 3 generators having 10 MVA, 20 MVA and 20 MVA rating. Their reactances are 10%, 15% and 15% respectively. They have separate bus bars A, B, C. The bus bars are connected through reactors as under:

A and B 5% reactance on 10 MVA base

B and C 4% reactance on 8 MVA base

Take 100 MVA base and draw reactance diagram. Find fault level and fault current for 3 phase fault at bus A. [203.25 MVA, 10668A]

17.12 A 20 MVA generator having 15% reactance and 10 MVA generator having 10% reactance feed a common bus. The bus feeds 30 MVA, 11/33 kV transformer having 5% reactance. The transformer feeds 33 kV feeder having $(3 + j15)$ ohm impedance. Find fault MVA and fault current for 3 phase fault at end of feeder. [50.207 MVA, 878.42 A]

17.13 Find the reactance per phase connected externally to a generator having 25 MVA, 11 kV, 15% reactance so as to limit fault current to six times full load current for 3 phase fault. Find the value of reactance in ohms and p.u. [0.0807 ohm, 1.667%]

17.14 An alternator has 6% internal reactance. It is rated at 10 MVA, 10 kV. Find the value of a reactor to be put in series with the generator to limit the fault current to 800% of full load current of alternator. [0.65 ohm]

17.15 Two alternators each rated at 3 MVA and 10% reactance are connected to a common bus. The generator bus is connected to an infinite bus bar through a 9000 kVA, 5% reactance transformer. Find the value of a reactor to be connected in between the transformer and generator bus so that the fault MVA on a feeder eminating from generator bus is limited to 150 MVA. [5% on 9000 kVA base]

17.16 Three alternators each 20 MVA\, 15% reactance are connected to their separate bus sections. Three reactors each 7% reactance on 20 MVA base are connected between a tie bar and generator bus sections. (*a*) Calculate fault MVA for a fault on bus section of Generator 1 (*b*) If the reactors are removed and the three bus sections joined together, what would be the fault MVA for the same fault position. [(*a*) 245 MVA (*b*) 400 MVA]

17.17 Two alternators are connected to a common bus through their individual transformers.The ratings are:

Generator A	6 MVA, 6.6 kV, 10% reactance
Transformer A	7.5 MVA, 6.6/33 kV, 5% reactance
Generator B	4 MVA 6.5 kV, 12% reactance
Transformer B	6 MVA 6.6/33 kV, 6% reactance.

A feeder of 18.1 ohms reactance eminates from the bus. Find the fault kVA for a 3 phase fault at the end of feeder. [32000 kVA]

17.18 Two section bus bars A and B are linked by a bus bar reactor rated at 5 MVA, 10% reactance. On bus A are connected two alternators each 10 MVA. 10% reactance. On bus bar B are two alternators each 8 MVA, 12% reactance. Find the fault MVA for a 3 phase fault on bus bar B with (*a*) bus bar reactor in circuit (*b*) bus bar reactor short circuited. [(*a*) 173 MVA (*b*) 333 MVA]

17.19 A generating station has N bus bar sections. To each bus section a generating capacity of G kVA with x% reactance is connected. Each of the bus sections is connected to a tie bar through r% reactance.

(*a*) Show that the fault level for a fault on one bus section will be

$$\left[\frac{G}{x}+\frac{G(N-1)}{(rN+x)}\right]100$$

(*b*) For G = 50000 kVA, x = 20%, r = 10%, find the short circuit kVA for N = 3 and 9.

(*c*) Show that as N becomes large, the maximum fault level would be limited to 750 MVA.

[(*b*) 450000 kVA, 614000 kVA]

17.20 An 11 kV system is shown in Fig. 17.32. The transmission line reactances are to a base of 15000 kVA. For a 3 phase fault at point F, find the fault MVA and fault current. [286.5 MVA, 15 kA]

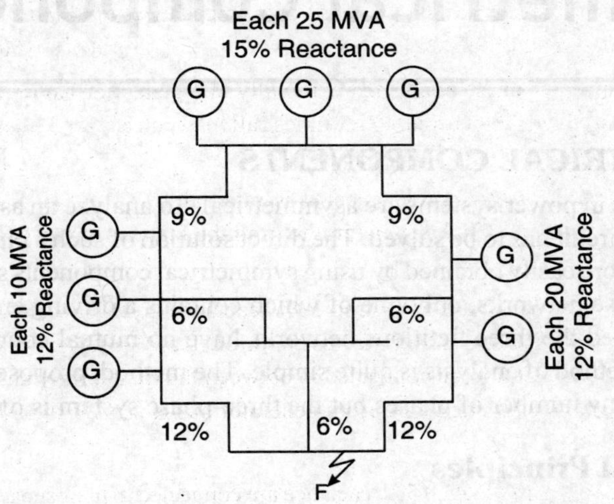

Fig. 17.32

CHAPTER 18

Symmetrical Components

18.1 SYMMETRICAL COMPONENTS

The majority of faults in power systems are asymmetrical. To analyze an asymmetrical fault, an unbalanced 3-phase circuit has to be solved. The direct solution of such a circuit is very difficult. The solution can be more easily obtained by using symmetrical components since this yields three (fictitious) single phase networks, only one of which contains a driving emf. Since the system reactances are balanced the three fictitious networks have no mutual coupling between them. Due to this fact the method of analysis is quite simple. The method, proposed by C.L. Fortescue in 1918, is valid for any number of phases but the three-phase system is of main interest.

18.1.1 General Principles

Any set of unbalanced 3-phase voltages (or current) can be transformed into 3 balanced sets. These three balanced sets are:

1. A positive sequence set of three symmetrical voltages (i.e., all numerically equal and all displaced from each other by 120°) having the same phase sequence *abc* as the original set and denoted by V_{a1}, V_{b1} and V_{c1} as shown in Fig. 18.9 (a).
2. A negative sequence set of three symmetrical voltages having the phase sequence opposite to that of the original set and denoted by V_{a2}, V_{b2} and V_{c2} as shown in Fig. 18.9 (b).
3. A zero sequence set of three voltages, all equal in magnitude and in phase with each other and denoted by V_{a0}, V_{b0} and V_{c0} (Fig. 18.9c).

The positive, negative and zero sequence sets are known as symmetrical components. Fig. 18.9 (d) shows the three-phase voltages obtained by adding the symmetrical components shown in Fig. 18.9 (a) (b) and (c). Thus,

$$V_a = V_{a1} + V_{a2} + V_{a0} \tag{18.1a}$$
$$V_b = V_{b1} + V_{b2} + V_{b0} \tag{18.1b}$$
$$V_c = V_{c1} + V_{c2} + V_{c0} \tag{18.1c}$$

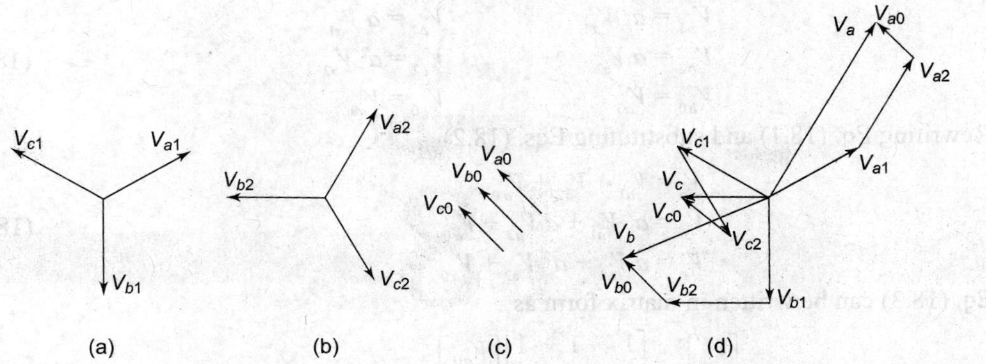

Fig. 18.1 (a) Positive sequence components (b) Negative sequence components (c) Zero sequence components (d) Addition of sequence components to give phase voltages

The application of symmetrical components to power system analysis is of fundamental importance. We can transform an arbitrarily unbalanced condition into symmetrical components, compute the system response by straightforward circuit analysis on simple circuit models and transform the results back into the original phase variables. This approach is superior to the direct (and complicated) solution of the unbalanced three phase circuit.

18.1.2 Operator *a*

We are familiar with the operator j. In polar form $j = 1 \angle 90°$. The multiplication by j has the effect of rotating a phasor forward through 90° in the positive (anticlockwise) direction without changing the magnitude. The operator a is defined as $1 \angle 120°$, an operator which when multiplied with a phasor rotates it through 120° in the positive (anticlockwise) direction without changing the magnitude. Thus,

$$a = 1 \angle 120° = 1\ \varepsilon^{j2\pi/3} = -0.5 + j\,0.866$$

If a phasor is multiplied by a twice in succession, it is rotated through 240°. Three successive multiplications by a rotate the phasor through 360°. Thus,

$$a^2 = 1 \angle 240° = 1 \angle -120° = -0.5 - j\,0.866$$
$$a^3 = 1 \angle 360° = 1$$
$$a^4 = 1 \angle 480° = 1 \angle 120° = a$$

Some important properties of operator a (which can be easily verified) are:

$$1 + a = 1 \angle 60° = 0.5 + j0.866 = -a^2$$
$$1 - a = \sqrt{3} \angle -30° = 1.5 - j0.866$$
$$1 + a^2 = 1 \angle -60° = 0.5 - j0.866 = -a$$
$$1 - a^2 = \sqrt{3} \angle 30° = 1.5 + j0.866$$
$$1 + a + a^2 = 0$$

18.1.3 Matrix Notation

The components of phasors b and c can be written in terms of the components of phasor a by using the operator a. Referring to Fig. 18.1 we see that

$$\begin{aligned} V_{b1} &= a^2 V_{a1} & V_{c1} &= a V_{a1} \\ V_{b2} &= a V_{a2} & V_{c2} &= a^2 V_{a2} \\ V_{b0} &= V_{a0} & V_{c0} &= V_{a0} \end{aligned} \tag{18.2}$$

Rewriting Eq. (18.1) and substituting Eqs. (18.2)

$$\begin{aligned} V_a &= V_{a1} + V_{a2} + V_{a0} \\ V_b &= a^2 V_{a1} + a V_{a2} + V_{a0} \\ V_c &= a V_{a1} + a^2 V_{a2} + V_{a0} \end{aligned} \tag{18.3}$$

Eq. (18.3) can be written in matrix form as

$$\begin{bmatrix} V_a \\ V_b \\ V_c \end{bmatrix} = \begin{bmatrix} 1 & 1 & 1 \\ 1 & a^2 & a \\ 1 & a & a_2 \end{bmatrix} \begin{bmatrix} V_{a0} \\ V_{a1} \\ V_{a2} \end{bmatrix} \tag{18.4}$$

For convenience we write

$$[A] = \begin{bmatrix} 1 & 1 & 1 \\ 1 & a^2 & a \\ 1 & a & a^2 \end{bmatrix} \tag{18.5}$$

Equation (18.4) can be written in compact form as

$$V_{abc} = [A]\, V_{012} \tag{18.6}$$

where $V_{abc} = \begin{bmatrix} V_a \\ V_b \\ V_c \end{bmatrix}$ and $V_{012} = \begin{bmatrix} V_{a0} \\ V_{a1} \\ V_{a2} \end{bmatrix}$

It can be easily verified that

$$[A]^{-1} = \frac{1}{3} \begin{bmatrix} 1 & 1 & 1 \\ 1 & a & a^2 \\ 1 & a^2 & a \end{bmatrix} \tag{18.7}$$

Multiplying both sides of Eq. (18.4) by A^{-1} gives

$$\begin{bmatrix} V_{a0} \\ V_{a1} \\ V_{a2} \end{bmatrix} = \frac{1}{3} \begin{bmatrix} 1 & 1 & 1 \\ 1 & a & a^2 \\ 1 & a^2 & a \end{bmatrix} \begin{bmatrix} V_a \\ V_b \\ V_c \end{bmatrix} \tag{18.8a}$$

or more compactly

$$V_{012} = [A]^{-1}\, V_{abc} \tag{18.8b}$$

Equation (18.8) gives the symmetrical components V_{a0}, V_{a1}, and V_{a2} of the unsymmetrical phasors V_a, V_b, V_c and is a very important relation. All the above equations are valid for currents also. Thus,

$$I_{abc} = [A]\, I_{012} \tag{18.9}$$

$$I_{012} = [A]^{-1}\, I_{abc} \tag{18.10}$$

where $I_{abc} = \begin{bmatrix} I_a \\ I_b \\ I_c \end{bmatrix}$ and $I_{012} = \begin{bmatrix} I_{a0} \\ I_{a1} \\ I_{a2} \end{bmatrix}$

18.1.4 Power Considerations

The total complex power in a 3-phase circuit is

$$S = V_a I_a^* + V_b I_b^* + V_c I_c^* \tag{18.11a}$$

cr

$$S = [V_a\, V_b\, V_c] \begin{bmatrix} I_a \\ I_b \\ I_c \end{bmatrix}^* = \begin{bmatrix} V_a \\ V_b \\ V_c \end{bmatrix}^T \begin{bmatrix} I_a \\ I_b \\ I_c \end{bmatrix}^*$$

$$= [V_{abc}]^T [I_{abc}]^* \tag{18.11b}$$

Using Eqs. (18.6) and (18.9)

$$S = [A\, V_{012}]^T [A\, I_{012}]^* \tag{18.12}$$

We know that the transpose of the product of two matrices is equal to the product of transposes of the matrices in reverse order. Thus,

$$[A\, V_{012}]^T = [V_{012}]^T [A]^T \tag{18.13}$$

Then

$$S = [V_{012}]^T [A]^T [A\, I_{012}]^*$$

$$= [V_{012}]^T [A]^T [A]^* [I_{012}]^* \tag{18.14}$$

Noting that $[A]^T = [A]$ and that a and a^2 are conjugates

$$[A]^T [A]^* = \begin{bmatrix} 1 & 1 & 1 \\ 1 & a^2 & a \\ 1 & a & a^2 \end{bmatrix} \begin{bmatrix} 1 & 1 & 1 \\ 1 & a & a^2 \\ 1 & a^2 & a \end{bmatrix} = 3 \begin{bmatrix} 1 & 0 & 0 \\ 0 & 1 & 0 \\ 0 & 0 & 1 \end{bmatrix} \tag{18.15}$$

$$S = 3\, [V_{012}]^T [I_{012}]^*$$

$$= 3\, (V_{a0} I^*_{a0} + V_{a1} I^*_{a1} + V_{a2} I^*_{a2}) \tag{18.16}$$

Equation (18.16) does not have any cross terms (such as $V_{a0}\, I_{a1}^*$). This is because of the fact that the three sequence networks have no mutual coupling. The factor of 3 arises due to the total no. of 9 components (3 for each phase). It is seen that sum of powers of the three symmetrical components equals the three-phase power or that symmetrical component transformation is power invariant.

18.1.5 Effect on Impedance

We can write the relation between phase voltages and currents as

$$[V_{abc}] = [Z_{abc}]\, [I_{abc}] \tag{18.17}$$

where $[Z_{abc}]$ is 3 × 3 matrix giving the self and mutual impedance in and between the three phases. Substituting Eqs. (18.6) and (18.9) into Eq. (18.17)

$$[A]\, [V_{012}] = [Z_{abc}]\, [A]\, [I_{012}] \tag{18.18}$$

or

$$[V_{012}] = [A]^{-1} [Z_{abc}]\, [A]\, [I_{012}] \tag{18.19}$$

or

$$[V_{012}] = [Z_{012}]\, [I_{012}] \tag{18.20}$$

where

$$[Z_{012}] = [A]^{-1}\, [Z_{abc}]\, [A] \tag{18.21}$$

For almost all power system components the matrix $[Z_{abc}]$ is not diagonal but possesses certain symmetries. These symmetries are such that $[Z_{012}]$ is diagonal either exactly or approximately. This fact is the basis of the simplification achieved by using symmetrical components.

18.1.6 Sequence Impedances of Star Connected Loads

Fig. 18.2 shows a 3 phase balanced load connected in star. The load has self and mutual impedences as shown. The star point is grounded through impedance Z_n.

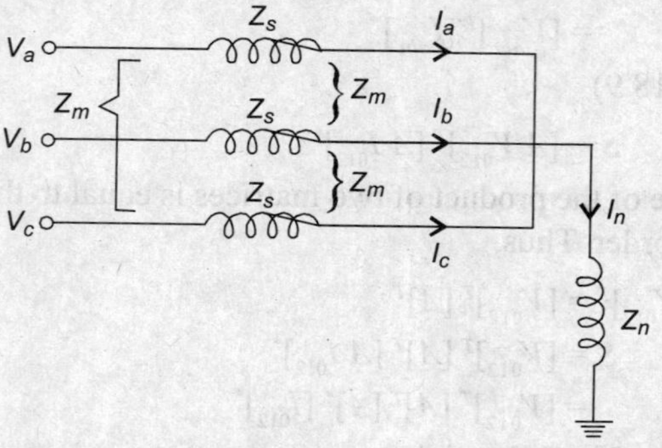

Fig. 18.2 3 Phase Balanced Star Connected Load

V_a, V_b, V_c are line to ground voltages of the three phases. I_a, I_b, I_c are the currents in the three phases, I_n is the natural current.

From kVL, the voltages V_a, V_b, V_c in terms of currents I_a, I_b, I_c, I_n are

$$V_a = Z_s I_a + Z_m I_b + Z_m I_c + Z_n I_n \tag{18.22 a}$$

$$V_b = Z_m I_a + Z_s I_b + Z_m I_c + Z_n I_n \tag{18.22 b}$$

$$V_c = Z_m I_a + Z_m I_b + Z_s I_c + Z_n I_n \tag{18.22 c}$$

Moreover

$$I_n = I_a + I_b + I_c \tag{18.23}$$

Substituting for I_n from Eq (18.23), the set of Eqs (18.22) can be rewritten in matrix form as :

$$\begin{bmatrix} V_a \\ V_b \\ V_c \end{bmatrix} = \begin{bmatrix} Z_s + Z_n & Z_m + Z_n & Z_m + Z_n \\ Z_m + Z_n & Z_s + Z_n & Z_m + Z_n \\ Z_m + Z_n & Z_m + Z_n & Z_s + Z_n \end{bmatrix} \begin{bmatrix} I_a \\ I_b \\ I_c \end{bmatrix} \tag{18.24 a}$$

or

$$V_{abc} = Z_{abc} I_{abc} \tag{18.24 b}$$

$$\text{Thus} \quad Z_{abc} = \begin{bmatrix} Z_s + Z_n & Z_m + Z_n & Z_m + Z_n \\ Z_m + Z_n & Z_s + Z_n & Z_m + Z_n \\ Z_m + Z_n & Z_m + Z_n & Z_s + Z_n \end{bmatrix}$$

From the above value of Z_{abc} we can calculate Z_{012} by using Eq. (18.21). Thus

$$Z_{012} = [A]^{-1}\ [Z_{abc}]\ [A]$$

or

$$Z_{012} = \frac{1}{3} \begin{bmatrix} 1 & 1 & 1 \\ 1 & a & a^2 \\ 1 & a^2 & a \end{bmatrix} \begin{bmatrix} Z_s + Z_n & Z_m + Z_n & Z_m + Z_n \\ Z_m + Z_n & Z_s + Z_n & Z_m + Z_n \\ Z_m + Z_n & Z_m + Z_n & Z_s + Z_n \end{bmatrix} \begin{bmatrix} 1 & 1 & 1 \\ 1 & a^2 & a \\ 1 & a & a^2 \end{bmatrix}$$

After the above multiplication of three matrices, we get

$$Z_{012} = \begin{bmatrix} Z_s + 3Z_n + 2Z_m & 0 & 0 \\ 0 & Z_s - Z_m & 0 \\ 0 & 0 & Z_s - Z_m \end{bmatrix}$$

It is seen that Z_{012} has only diagonal elements. Thus for a balanced load, the three sequences are independent. The current of one sequence produces voltage of only that sequence. Therefore each sequence network can be analysed on per phase basis. If $Z_m = 0$, Z_{012} reduces to

$$Z_{012} = \begin{bmatrix} Z_s + 3Z_n & 0 & 0 \\ 0 & Z_s & 0 \\ 0 & 0 & Z_s \end{bmatrix}$$

Moreover as seen from the above equation for Z_{012} the neutral impedance Z_n appears as $3Z_n$ in the final matrix. This is evidently due to the fact that current flowing through Z_n is $3I_{a0}$. Thus in all zero sequence networks the neutral impedance Z_n appears as $3Z_n$. This aspect is further discussed in section 18.4.3.

Example 18.1

The line currents in a 3-phase system are $I_a = 5 \angle 60°$ A, $I_b = 5 \angle -60°$ A and $I_c = 0$. Find the symmetrical components.

Solution

$$I_a = 5 \angle 60° = 5(0.5 + j\,0.866) = 2.5 + j\,4.33 \text{ A}$$

$$I_b = 5 \angle -60° = 5\,(0.5 - j\,0.866) = 2.5 - j\,4.33 \text{ A}$$

$$\begin{bmatrix} I_{a0} \\ I_{a1} \\ I_{a2} \end{bmatrix} = \frac{1}{3}\begin{bmatrix} 1 & 1 & 1 \\ 1 & a & a^2 \\ 1 & a^2 & a \end{bmatrix}\begin{bmatrix} 2.5 + j\,4.33 \\ 2.5 - j\,4.33 \\ 0 \end{bmatrix} = \begin{bmatrix} 1.667 \angle 0° \\ 3.33 \angle 60° \\ 1.667 \angle 120° \end{bmatrix}$$

Using the above values of I_{a0}, I_{a1}, I_{a2} and Eq. (18.9) we can find I_a, I_b, I_c and check the correctness of results. This is left as an exercise for the reader.

Example 18.2

Figure 18.3 shows the self and mutual inductive reactance in a 3-phase transmission line. Find the matrix $[Z_{012}]$.

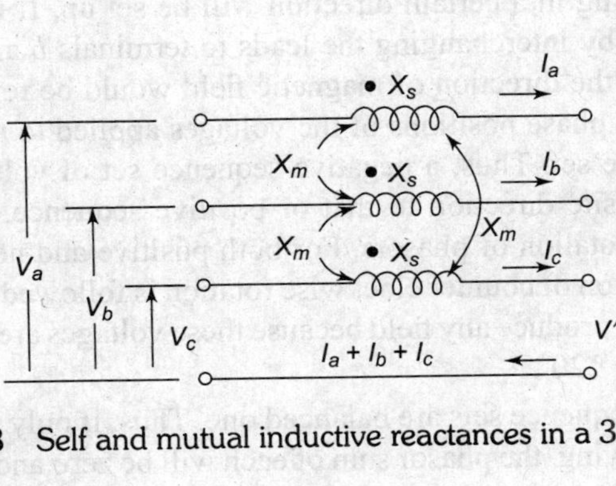

Fig. 18.3 Self and mutual inductive reactances in a 3-phase transmission line

Solution

Applying Kirchhoff's voltage law we have

$$V_a - V_a' = jX_s I_a + jX_m I_b + jX_m I_c$$

$$V_b - V_b' = jX_m I_a + jX_s I_b + jX_m I_c$$

$$V_c - V_c' = jX_m I_a + jX_m I_b + jX_s I_c$$

or

$$\begin{bmatrix} V_a \\ V_b \\ V_c \end{bmatrix} - \begin{bmatrix} V_a' \\ V_b' \\ V_c' \end{bmatrix} = j\begin{bmatrix} X_s & X_m & X_m \\ X_m & X_s & X_m \\ X_m & X_m & X_s \end{bmatrix}\begin{bmatrix} I_a \\ I_b \\ I_c \end{bmatrix}$$

or

$$[V_{abc}] - [V'_{abc}] = [Z_{abc}]\,[I_{abc}]$$

$$[Z_{012}] = [A]^{-1}\,[Z_{abc}]\,[A]$$

$$= \frac{1}{3}\begin{bmatrix} 1 & 1 & 1 \\ 1 & a & a^2 \\ 1 & a^2 & a \end{bmatrix} j\begin{bmatrix} X_s & X_m & X_m \\ X_m & X_s & X_m \\ X_m & X_m & X_s \end{bmatrix}\begin{bmatrix} 1 & 1 & 1 \\ 1 & a^2 & a \\ 1 & a & a^2 \end{bmatrix}$$

$$= \frac{j}{3}\begin{bmatrix} (X_s + 2X_m) & (X_s + 2X_m) & (X_s + 2X_m) \\ (X_s - X_m) & [aX_s + (1+a^2)X_m] & [a^2 X_s + (1+a)X_m] \\ (X_s - X_m) & [a^2 X_s + (1+a)X_m] & [aX_s + (1+a^2)X_m] \end{bmatrix}\begin{bmatrix} 1 & 1 & 1 \\ 1 & a^2 & a \\ 1 & a & a^2 \end{bmatrix}$$

$$= j\begin{bmatrix} X_s + 2X_m & 0 & 0 \\ 0 & X_s - X_m & 0 \\ 0 & 0 & X_s - X_m \end{bmatrix}$$

It is seen that $[Z_{abc}]$ contains mutual elements whereas $[Z_{012}]$ is a diagonal matrix and does not contain any mutual term.

18.2 PHYSICAL SIGNIFICANCE OF SEQUENCE COMPONENTS

A physical significance of sequence voltages may be obtained by considering the fields which results when these sequence voltages are applied to the stator of a 3-phase machine (say an induction motor). If a positive sequence set of voltages is applied to the terminals *a*, *b*, *c* of the machine, a magnetic field revolving in a certain direction will be set up. If now the voltages to terminals *b* and *c* are changed by interchanging the leads to terminals *b* and *c*, it is known from induction motor theory that the direction of magnetic field would be reversed. It is seen that for this condition the relative phase positions of the voltages applied to the motor are the same as for the negative sequence set. Thus, a negative sequence set of voltages produces a rotating field rotating in an opposite direction to that of positive sequence. Phase sequence should not be confused with the rotation of phasors. For both positive and negative sequence components, the standard convention of counter clockwise rotation is followed. The application of zero sequence voltages does not produce any field because these voltages are in phase and the 3-phase windings are displaced by 120°.

The positive and the negative sequence sets are balanced one. Thus, if only positive and (or) negative sequence currents are flowing, the phasor sum of each will be zero and there will be no residual current. However, the zero sequence components of currents in the three phases are in phase and the residual current will be three times the zero sequence current of one phase. In the case of a fault involving ground, the positive and the negative sequence currents balance between themselves while the zero sequence currents flow through the ground and overhead ground wires.

18.3 THREE-SEQUENCE IMPEDANCES AND SEQUENCE NETWORKS

As discussed above the positive sequence currents give rise to only positive sequence voltages, the negative sequence currents give rise to only negative sequence voltages and zero sequence currents give rise to only zero sequence voltages. Therefore, we can regard each sequence current as flowing within its own network through impedances of its own sequence only.

The impedance of a circuit when only positive sequence currents are flowing is called the impedance to positive sequence current or simply the positive sequence impedance. When only negative sequence currents are flowing the impedance is termed as negative sequence impedance. When only zero sequence currents are flowing the impedance is termed as zero sequence impedance.

The analysis of unsymmetrical faults in power systems is carried out by finding t symmetrical components of the unbalanced currents. Since each sequence current causes voltage drop of that sequence only, each sequence current can be considered to flow in independent network composed of impedances to the currents of that sequence only. The sing phase equivalent circuit composed of the impedances to the currents of any one sequence or is called the sequence network for that particular sequence. The sequence networks contain t generated emfs and impedances of like sequence. Thus, for every power system we can fo three-sequence networks. These sequence networks, carrying current I_{a1}, I_{a2} and I_{a0} are th inter-connected to represent the different fault conditions.

18.4 SEQUENCE NETWORKS OF SYNCHRONOUS MACHINES

Figure 18.4 shows an unloaded synchronous machine having its neutral earthed through an impedance Z_n. A fault at its terminals causes currents I_a, I_b and I_c to flow in the lines. If the fault involves earth, a current I_n flows into the neutral from earth. This current flows through the neutral impedance Z_n. Depending on the type of fault one or more of the line currents may be zero.

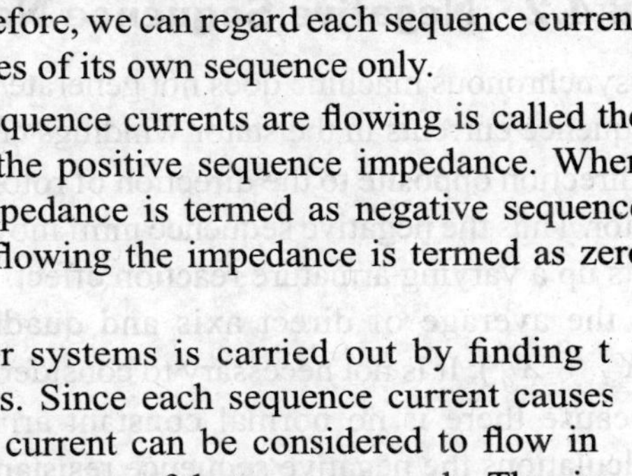

Fig. 18.4 Circuit diagram of an unloadı generator grounded through impedance ź

18.4.1 Positive Sequence Network

The windings of a synchronous machine are symmetrical. Thus, the generated voltages are positive sequence only. The positive sequence network consists of an emf (equal to no lo terminal voltage) in series with the positive sequence impedance Z_1 of the machine. Fig. 18 (a) shows the paths for positive sequence currents in the synchronous machine. The neutr impedance Z_n does not appear in this circuit because the phasor sum of I_{a1}, I_{b1} and I_{c1} is zero ar no positive sequence current can flow through Z_n. Since it is a balanced circuit, the positive sequence network can be drawn on single phase basis as shown in Fig. 18.5 (b). The reference bus for the positive sequence network is the neutral of the generator. The positive sequence impedance Z_1 consists of winding resistance and direct axis reactance. The reactance is the sub-transient reactance X_d'' or transient reactance X_d' or synchronous reactance X_d depending on whether sub-transient, transient or steady state conditions are being studied. As seen in Fig. 18.5 (b) the positive sequence voltage of terminal *a* (with respect to the reference bus) is

$$V_{a1} = E_a - Z_1 I_{a1} \tag{18.25}$$

18.4.2 Negative Sequence Network

A synchronous machine does not generate any negative sequence voltage. The flow of negative sequence currents in the stator windings creates an mmf which rotates at synchronous speed in a direction opposite to the direction of rotor, i.e., at twice the synchronous speed with respect to rotor. Thus the negative sequence mmf moves alternately past the direct and quadrature axis and sets up a varying armature reaction effect. Therefore, the negative sequence reactance is taken as the average of direct axis and quadrature axis sub-transient reactance, i.e., $X_2 = .5\,(X_d'' + X_q'')$. It is not necessary to consider any time variation of X_2 during transient conditions because there is no normal constant armature reaction to be affected. For more accurate calculations the negative sequence resistance, for accounting the power dissipated in the rotor poles or damper winding by double supply frequency induced currents, should also be considered. Fig. 18.5 (*c* and *d*) show the negative sequence current paths and the negative sequence network on single phase basis. The reference bus for the negative sequence network is the neutral of the machine. The negative sequence voltage of terminal a (with respect to the reference bus) is

$$V_{a2} = -Z_2 I_{a2} \tag{18.26}$$

18.4.3 Zero Sequence Network

No zero sequence voltage is induced in a synchronous machine. The flow of zero sequence currents in the stator windings produces three mmfs which are in time phase. If each phase winding produced a sinusoidal space mmf, then with the rotor removed, the flux at a point on the axis of the stator due to zero sequence current would be zero at every instant. When the flux in the air gap or the leakage flux around slots or end connections is considered, no point in these regions is equidistant from all the three-phase windings of the stator. Furthermore the mmf produced by a phase winding departs from a sine wave, by amounts which depend upon the arrangement of the winding.

The zero sequence currents flow through the neutral impedance Z_n and this fact has to be taken into account in drawing the zero sequence network. The current flowing through this impedance Z_n is $3\,I_{a0}$. As seen in Fig. 18.5 (e) the zero sequence voltage drop from point *a* to ground is $-3\,I_{a0}\,Z_n - I_{a0}\,Z_{g0}$ where Z_{g0} is the zero sequence impedance per phase of the generator. Since the current in the zero sequence network (Fig. 18.5f) is I_{a0} this network must have an impedance of $3\,Z_n + Z_{g0}$. Thus,

$$Z_0 = 3\,Z_n + Z_{g0} \tag{18.27}$$

The zero sequence voltage of terminal *a* with respect to the reference bus is

$$V_{a0} = -I_{a0}\,Z_0 \tag{18.28}$$

Typical values of synchronous machine reactance expressed in *pu* of rated MVA are given in Table 18.1

Table 18.1 Typical *pu* values of synchronous machine reactance

	Turbo generator	*Salient pole generator*	*Synchronous condenser*	*Synchronous motor*
X_d	1.0–2.0	0.6–1.5	1.5–2.5	0.8–1.10
X_d'	0.12–0.30	0.2–0.5	0.3–0.5	0.3–0.4
X_d''	0.07–0.20	0.12–0.40	0.2–0.3	0.17–0.2
X_2	0.07–0.20	0.12–0.40	0.17–0.40	0.2–0.35
X_0	0.04–0.15	0.02–0.2	0.02–0.15	0.05–0.07

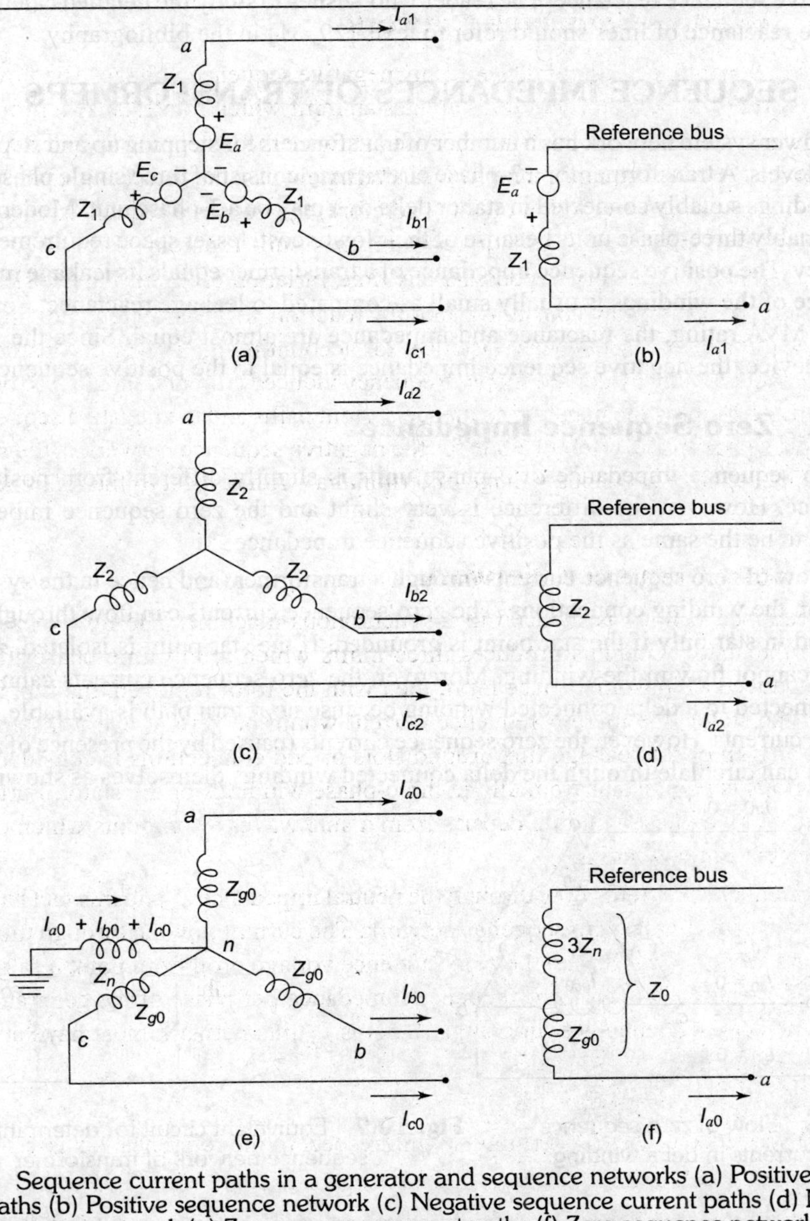

Fig. 18.5 Sequence current paths in a generator and sequence networks (a) Positive sequence current paths (b) Positive sequence network (c) Negative sequence current paths (d) Negative sequence network (e) Zero sequence current paths (f) Zero sequence network

18.5 SEQUENCE IMPEDANCES OF TRANSMISSION LINES

The positive and negative sequence impedances of linear, symmetrical static circuits do not depend on the phase sequence and are, therefore, equal. The formulae for reactance, derived in Chapter 2 are valid for both positive and negative sequence. When only zero sequence currents flow in the lines, the currents in all the phases are identical. These currents return partly through ground and partly through overhead ground wires. The magnetic field due to the flow of zero sequence currents (through lines, ground and round wires) is very different from the magnetic field due to positive sequence currents. The zero sequence reactance of lines is about 2 to 4 times the positive sequence reactance. The reader who wishes to study the detailed calculations of zero sequence reactance of lines should refer to texts [29, 31] in the bibliography.

18.6 SEQUENCE IMPEDANCES OF TRANSFORMERS

Every power system network has a number of transformers for stepping up and stepping down the voltage levels. A transformer for a 3-phase circuit may consist of three single phase transformers with windings suitably connected in star or delta or it may be a 3-phase unit. Modern transformers are invariably three-phase units because of their lower cost, lesser space requirements and higher efficiency. The positive sequence impedance of a transformer equals its leakage impedance. The resistance of the windings is usually small as compared to leakage reactance. For transformers above 1 MVA rating, the reactance and impedance are almost equal. Since the transformer is a static device, the negative sequence impedance is equal to the positive sequence impedance.

18.6.1 Zero Sequence Impedance

The zero sequence impedance of 3-phase units is slightly different from positive sequence impedance. However, the difference is very slight and the zero sequence impedance is also assumed to be the same as the positive sequence impedance.

The flow of zero sequence currents through a transformer (and hence in the system) depends greatly on the winding connections. The zero sequence currents can flow through the winding connected in star only if the star point is grounded. If the star point is isolated, zero sequence currents cannot flow in the winding. Moreover, the zero sequence currents cannot flow in the lines connected to a delta connected winding because no return path is available for these zero sequence currents. However, the zero sequence currents (caused by the presence of zero sequence voltages) can circulate through the delta connected windings themselves as shown in Fig. 18.6.

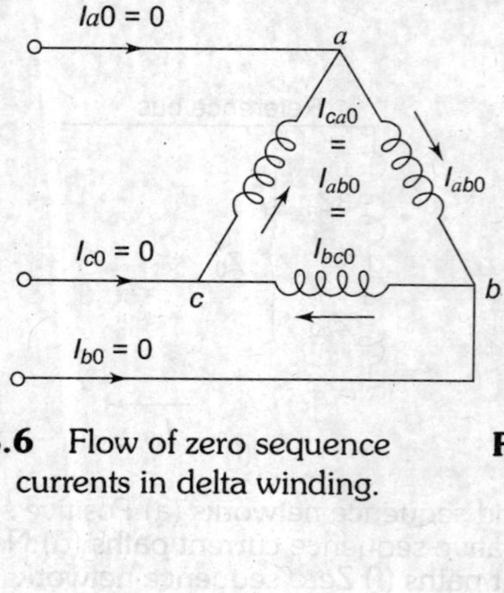

Fig. 18.6 Flow of zero sequence currents in delta winding.

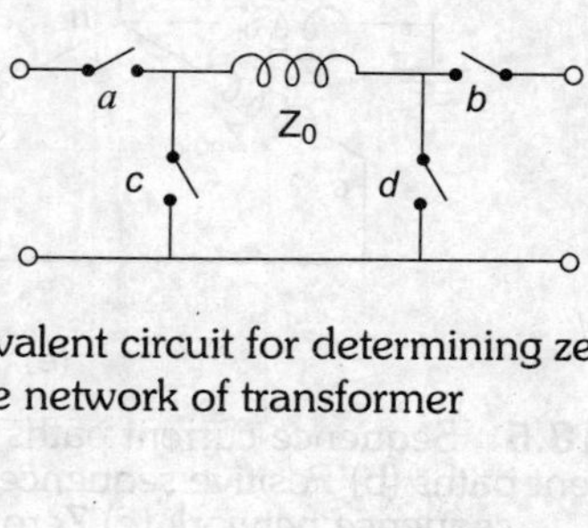

Fig. 10.7 Equivalent circuit for determining zero sequence network of transformer

These various conditions can be taken into account by the use of equivalent circuit shown in Fig. 18.7. Z_0 is the zero sequence impedance of the transformer. The general rules to find the correct connection are:

1. Close switch '*a*' if the primary winding is star connected with neutral grounded.
2. Close switch '*c*' if the primary winding is delta connected. Keep both '*a*' and '*c*' open if the primary winding is star connected with neutral isolated (not grounded).
3. Close switch '*b*' if the secondary winding is star connected with neutral grounded.
4. Close switch '*d*' if the secondary winding is delta connected. Keep both '*b*' and '*d*' open if the secondary winding is star connected with neutral isolated (not grounded).

The correct zero sequence network of the transformer can be very easily determined from the above rules. The switches need not be shown in the final network. Fig. 18.8 shows some typical transformer connections and the zero sequence networks.

(*a*) Y–Y (both neutrals grounded) *connection.* The grounding of both the neutral provides a path for zero sequence current to flow in the primary and secondary. Therefore the zero sequence network shows transformer leakage impedance per phase connected as shown.

(*b*) Y–Δ (neutral grounded) *connection.* The zero sequence currents can flow in the primary because the neutral in grounded. But no zero sequence current can leave delta connected winding. Therefore the primary and secondary sides are isolated as shown.

(*c*) Y–Δ *connection.* Since the neutral is not grounded, zero sequence currents cannot flow in transformer winding. Hence the network has an open circuit as shown.

(*d*) Δ–Δ *connection.* The zero sequence currents can circulate in the delta windings but cannot leave the delta terminals. Therefore the zero sequence network is as shown.

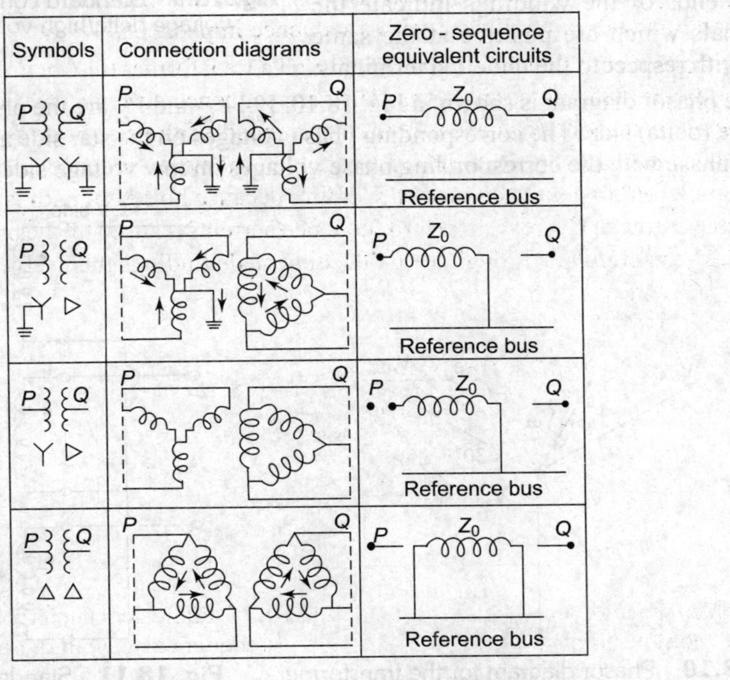

Fig. 18.8 Some transformer connections and their zero sequence networks

18.6.2 Phase Shift in Y–Δ Transformer

A 3-phase transformer has 3 high voltage windings and 3 low voltage windings. As per convention the six terminals of the three high voltage windings are denoted by A_1, A_2, B_1, B_2, C_1 and C_2. The terminals of the three low voltage windings are denoted by a_1, a_2, b_1, b_2, c_1 and c_2. The terminals with subscript 1 are the terminals of the neutral end while those with subscript 2 are of the line end. The polarity markings, are such that emfs are instantanceously in time phase between terminal with corresponding markings, i.e., at the instant when the potential of A_2 with respect to A_1 is at a positive maximum value, that of a_2 with respect to a_1 is also at the positive maximum value.

In a Y – Y or Δ – Δ 3-phase transformer there is no phase shift between the corresponding voltage of any phase to neutral on either side (although a phase shift of 180° can occur if for example A_1, B_1 and C_1 are joined to form the star point on the primary side while a_2, b_2 and c_2 are joined on the secondary side.) However, a delta connected winding is desirable in many power transformers for reasons of harmonic reduction. As such most of the transformers used in power systems are either Y – Δ or Δ – Y.

Fig. 18.9 shows a delta star transformer with delta as the low voltage side. The windings that are drawn opposite to each other are those which are linked magnetically by being wound on the same core. The dots at the ends of the windings indicate the terminals which are positive at the same time with respect to the undotted terminals.

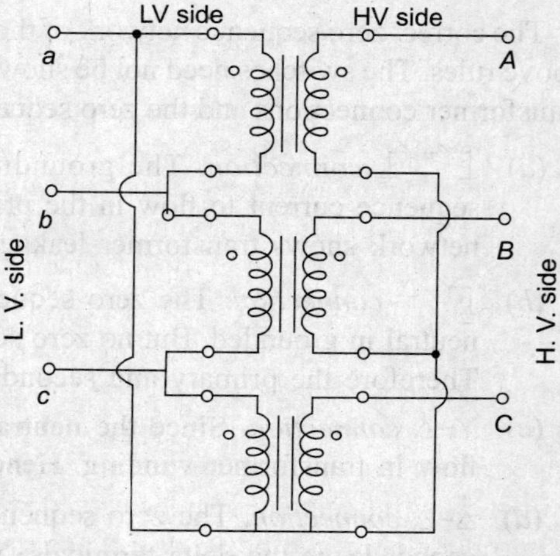

Fig. 18.9 Standard connection diagram for low voltage delta/high voltage star connected transformer

The phasor diagram is shown in Fig. 18.10. V_{ab}, V_{bc} and V_{ca} are the phase voltages on the low voltage (delta) side. The corresponding phase voltages on the star side are V_{AN}, V_{BN} and V_{CN} and are in phase with the corresponding phase voltages on low voltage side.

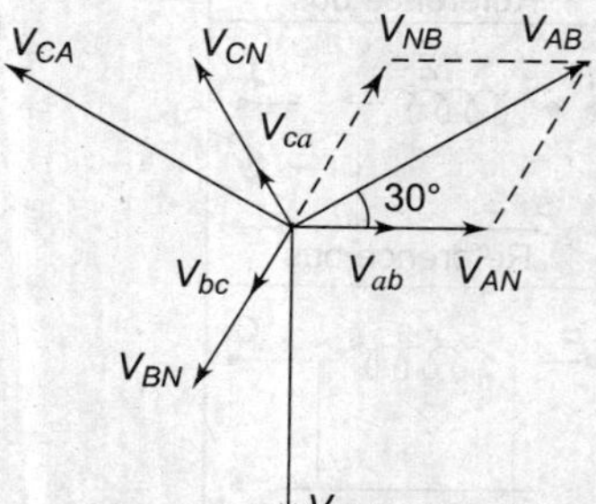

Fig. 18.10 Phasor diagram for the transformer connections of Fig. 18.8

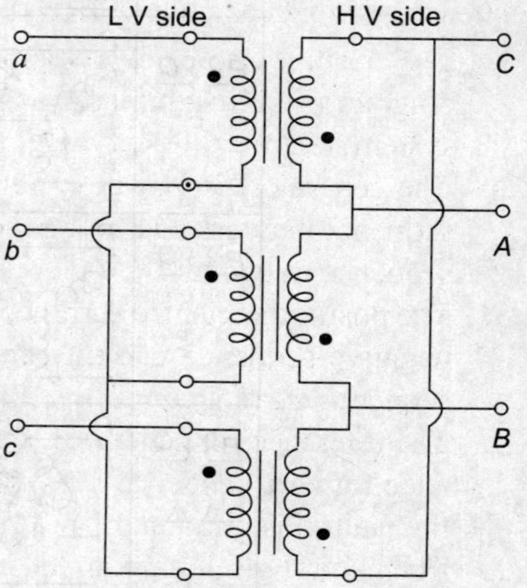

Fig. 18.11 Standard connection diagram for low voltage star/high voltage delta connected transformer

The line voltages on the star side are V_{AB}, V_{BC} and V_{CA}. Note that V_{AB} leads V_{ab} by 30°. The voltages V_{BC} and V_{CA} lead the voltages V_{bc} and V_{ca} by the same amount. Similarly, the positive sequence line currents on the high voltage side lead their corresponding positive sequence line currents on the low voltage side by 30°. The following standard convention is generally adopted.

For either Y–Δ or Δ–Y connections, phases are labelled in such a way that positive sequence quantities, on the high voltage side lead their corresponding positive sequence quantities on the low voltage side by 30°. The effect on the negative sequence is reverse, i.e., the *HV* negative sequence quantities lag the low voltage negative sequence quantities by 30°.

The connection of Fig. 18.10 conform to above convention. The connections for high voltage delta-low voltage star which satisfy the above convention are shown in Fig. 18.11.

18.7 FORMATION OF SEQUENCE NETWORKS

A power system network consists of synchronous machines, transmission lines and transformers. In the preceding sections we have discussed the sequence networks of all these components and by combining them the sequence network of a power system can be drawn. The positive sequence network is, as a matter of fact, the same as the single line reactance diagram as used for the calculation of symmetrical fault current. The reference bus for positive sequence network is the system neutral.

The negative sequence network is similar to the positive sequence network. However, the negative sequence network does not contain any voltage source. The negative sequence impedances for transmission lines and transformers are the same as the positive sequence impedances. But the negative sequence impedance of a synchronous machine may be different from its positive sequence impedance. Any impedance connected between a neutral and ground is not included in the positive and negative sequence networks because the positive and the negative sequence currents cannot flow through such an impedance.

The zero sequence network also does not contain any voltage source. Special care must be exercised in connecting the zero sequence impedance of transformer as discussed in section 18.6.1. Any impedance included between neutral and ground becomes three times its value in a zero sequence network as discussed in section 18.4.3.

In view of discussions in the proceeding sections we can summarise the rules for formation of sequence networks as under:

1. The positive sequence network is the same as single line impedance (or reactance) diagram used in symmetrical fault analysis. The reference bus for this network is the system neutral.
2. The generators in power systems produce balanced voltages. Therefore only positive sequence network has voltage source. These are no voltage sources in negative and zero sequence networks.
3. The positive sequence current can cause only positive sequence voltage drop. Similarly negative sequence current can cause only negative sequence voltage drop and zero sequence current can cause only zero sequence voltage drop.
4. The reference for negative sequence network is the system neutral. However the reference for zero sequence network is the ground. Zero sequence current can flow only if the neutral is grounded.
5. The neutral grounding impedance Z_n appears as $3Z_n$ in the zero sequence network.
6. The three sequence networks are independent and are interconnected suitably depending on the type of fault.

Example 18.3

Fig. 18.12 shows a power system network. Draw zero sequence network for his system.
The system data is as under:

Generator G_1	50 MVA,	11 kV, $X_0 = 0.08$ *pu*
Transformer T_1	50 MVA,	11/220 kV, $X_0 = 0.1$ *pu*
Generator G_2	30 MVA,	11 kV, $X_0 = 0.07$ *pu*
Transformer T_2	30 MVA,	11/220 kV, $X_0 = 0.09$ *pu*

Zero sequence reactance of line is 555.6 ohms.

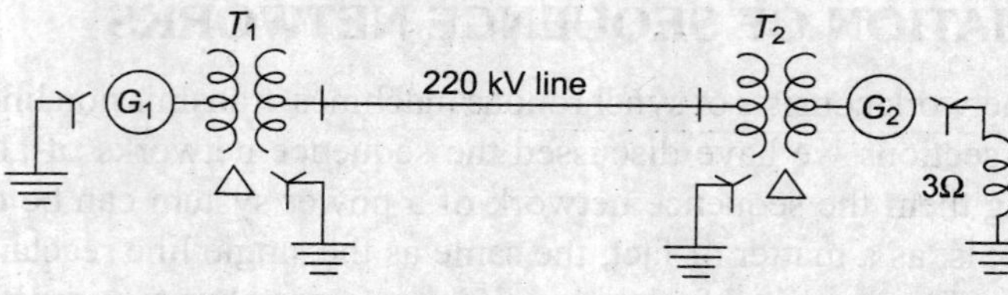

Fig. 18.12 Power system network for example 18.3

Solution

Base MVA = 50 MVA

Base voltage = 11 kV for *LT* side and 220 kV for *HT* side of transformers

$$\text{Base impedance of line} = \frac{220 \times 220}{50} = 968 \text{ ohms}$$

$$pu \text{ impedance for line} = \frac{j\,555.6}{968} = j\,0.574$$

$$pu \text{ impedance of } T_2 = j\,0.09 \times \frac{50}{30} = j\,0.15$$

$$pu \text{ impedance of } G_2 = j\,0.07 \times \frac{50}{30} = j\,0.117$$

$$\text{Base impedance for generator} = \frac{11 \times 11}{50} = 2.42 \text{ ohms}$$

$$pu \text{ impedance of neutral reactor} = \frac{j\,3}{2.42} = j\,1.24$$

The zero sequence network is shown in Fig. 18.13 and is self explanatory. The neutral impedance appears as three times its actual value in the circuit.

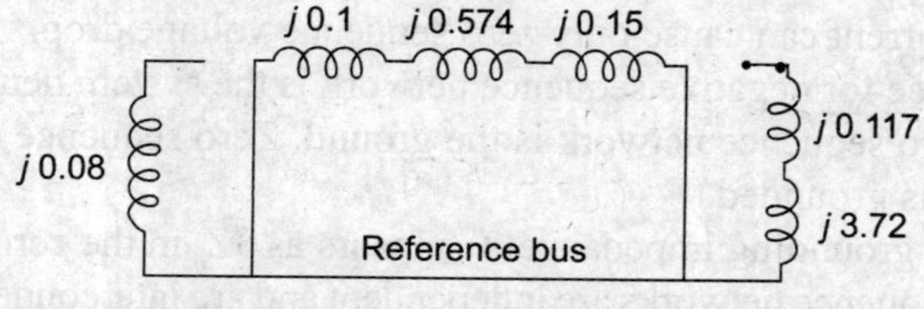

Fig. 18.13 Zero sequence network for the system of Fig. 18.12

Example 18.4

Figure 18.14 shows a power system network. Draw positive sequence network, negative sequence network and zero sequence network. The system data is as under;

Equipment	*MVA rating*	*Voltage rating*	X_1	X_2 p.u.	X_0
Generator G_1	100	11 kV	0.25	0.25	0.05
Generator G_2	100	11 kV	0.2	0.2	0.05
Transformer T_1	100	11/220 kV	0.06	0.06	0.06
Transformer T_2	100	11/220 kV	0.07	0.07	0.07
Line 1	100	220 kV	0.1	0.1	0.3
Line 2	100	220 kV	0.1	0.1	0.3

Solution

Base MVA 100

Base voltage 11 kV for *LT* side of transformers and 220 kV for *HT* side of transformers.

The positive, negative and zero sequence networks are shown in Fig. 18.15. Since positive and negative sequence impedances of all the elements are equal, the positive and negative sequence networks are similar except that the voltage sources are present only in the positive sequence network. The neutral reactance of j 0.03 *pu* appears as j 0.09 in the zero sequence network.

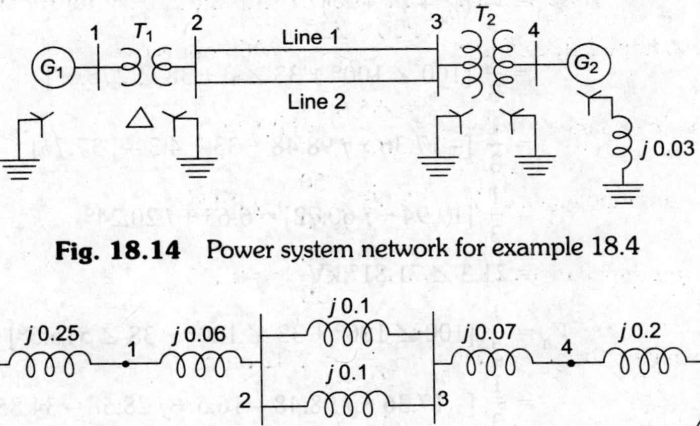

Fig. 18.14 Power system network for example 18.4

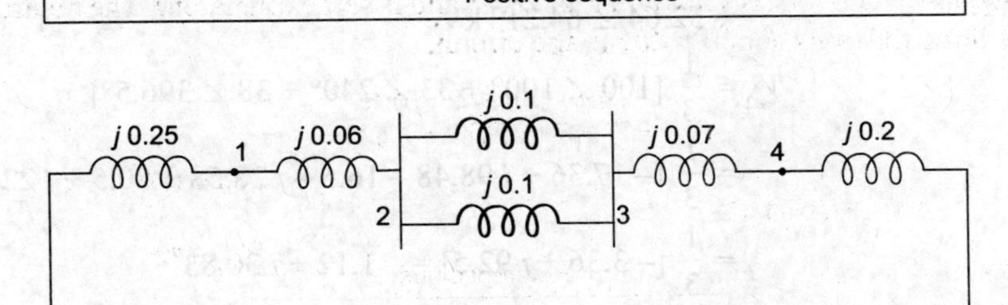

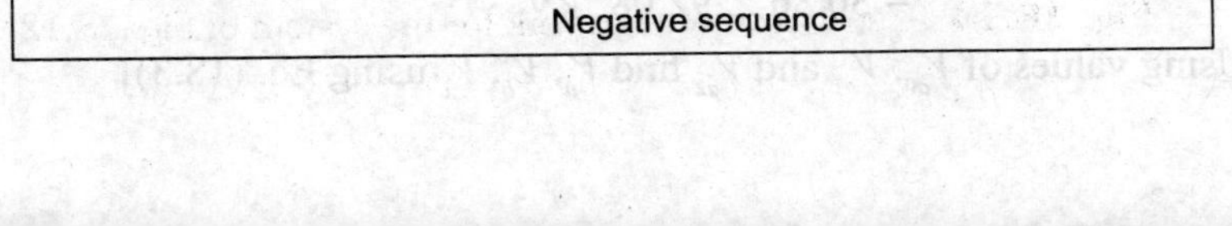

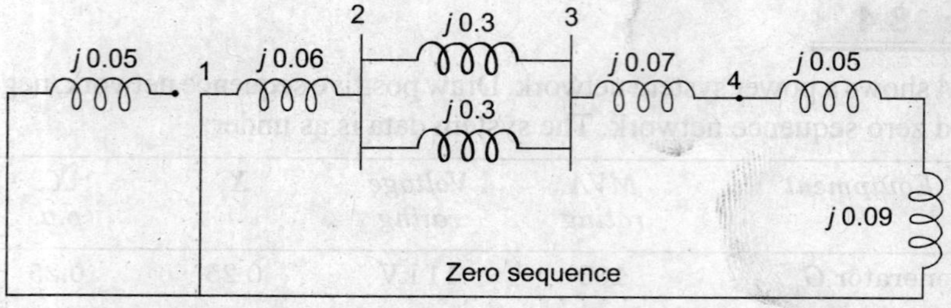

Fig. 18.15 Sequence networks for the system of Fig. 18.14

Example 18.5

The line to line voltages on hV side of a step up transformer are 100 kV, 33 kV and 38 kV for phases a, b, c respectively. Voltage of phase *a* leads that of phase *b* by 100° and lags that of phase *c* by 176.5°. Find symmetrical components of voltage. [JNTU 2011]

Solution

As per given data, V_a, V_b, V_c can be written as

$$V_a = 100 \angle 100° \text{ kV}, \quad V_b = 33 \angle 0 \text{ kV}, \quad V_c = 38 \angle 276.5° \text{kV}$$

Using Eq. (18.8)

$$V_{a0} = \frac{1}{3}(V_a + V_b + V_c)$$

$$= \frac{1}{3}[100 \angle 100° + 33 \angle 0 + 38 \angle 276.5°]$$

$$= \frac{1}{3}[-17.36 + j\,98.48 + 33 + 4.3 - \text{j}\,37.76]$$

$$= \frac{1}{3}[19.94 + j\,60.72] = 6.65 + j\,20.24°$$

$$= 21.3 \angle 71.81° \text{ kV}$$

$$V_{a1} = \frac{1}{3}[100 \angle 100° + 33 \angle 120° + 38 \angle 516.5°]$$

$$= \frac{1}{3}[-17.36 + j\,98.48 - 16.5 + j\,28.58 - 34.85 + j\,15.15]$$

$$= \frac{1}{3}[-68.71 + j\,142.21] = -22.9 + \text{j}\,47.4$$

$$= 52.64 \angle 64.21° \text{ kV}$$

$$V_{a2} = \frac{1}{3}[100 \angle 100° + 33 \angle 240° + 38 \angle 396.5°]$$

$$= \frac{1}{3}[-17.36 + j\,98.48 - 16.5 - j\,28.58 + 30.5 + j\,22.6]$$

$$= \frac{1}{3}[-3.36 + j\,92.5] = -1.12 + j\,30.83$$

$$= 30.86 \angle 92.08° \text{ kV}$$

[Check Using values of V_{a0}, V_{a1} and V_{a2} find V_a, V_b, V_c using Eq. (18.3)]

Example 18.6

For the system shown in Fig. 18.16 draw positive, negative and zero sequence networks. The ratings of components are

Generator G_1: 25 MVA, 11 kV, $X_1 = 0.2$ p.u., $X_2 = 0.15$ p.u. and $X_0 = 0.03$ p.u.

Generaor G_2 : 15 MVA, 11 kV, $X_1 = 0.2$ p.u., $X_2 = 0.15$ p.u., $X_0 = 0.05$ p.u.

Motor M: 25 MVA, 11 kV, $X_1 = X_2 = 0.1$ p.u., $X_0 = 0.1$ p.u.

Transformer T_1: 25 MVA, 11Δ/120 Y kV, X = 10%

T_2: 12.5 MVA, 11Δ/120 Y kV, X = 10%

T_3: 10 MVA, 120Y/11Δ kV, X = 10%

Choose a base of 50 MVA, 11 kV in generator G_1 circuit. [VTU 2011]

Solution

For 50 MVA, 11 kV base in G_1, the reactances of different components are as under

$$G_1:\ X_1 = 0.2 \times \frac{50}{25} = 0.4 \text{ p.u.}$$

$$X_2 = 0.15 \times \frac{50}{25} = 0.3 \text{ p.u.}$$

$$X_0 = 0.03 \times \frac{50}{25} = 0.06 \text{ p.u.}$$

$$G_2:\ X_1 = 0.2 \times \frac{50}{15} = 0.667 \text{ p.u.}$$

$$X_2 = 0.15 \times \frac{50}{15} = 0.5 \text{ p.u.}$$

$$X_0 = 0.05 \times \frac{50}{15} = 0.1667 \text{ p.u.}$$

$$\text{Motor M}\ \ X_1 = X_2 = 0.2 \times \frac{50}{25} = 0.4 \text{ p.u.}$$

$$X_0 = 0.1 \times \frac{50}{25} = 0.2 \text{ p.u.}$$

$$T_1\ \ X = 0.1 \times \frac{50}{25} = 0.2 \text{ p.u.}$$

$$T_2\ \ X = 0.1 \times \frac{50}{12.5} = 0.4 \text{ p.u.}$$

$$T_3\ \ X = 0.1 \times \frac{50}{10} = 0.5 \text{ p.u.}$$

The sequence networks are shown in Fig. 18.17.

$$\text{For}\ \ G_1,\ X_n = 0.05 \times \frac{50}{25} = 0.1 \text{ p.u.}$$

For $M, X_n = 0.05 \times \frac{50}{25} = 0.1$ p.u.

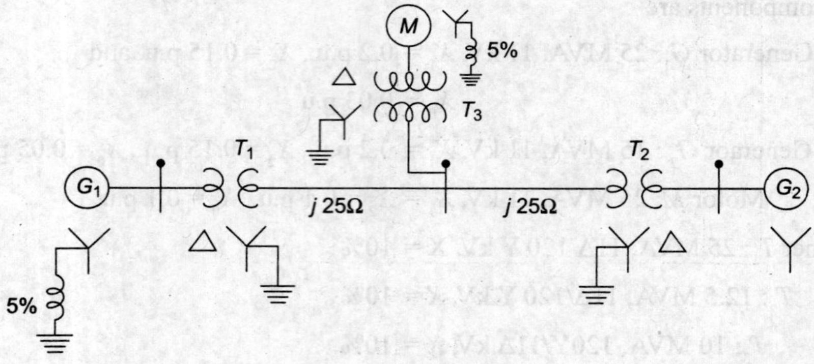

Fig. 18.16

$$\text{Base impedance} = \frac{120^2}{50} = 288 \text{ ohm}$$

$$\text{Line impedance} = j\frac{25}{288} = j\,0.0868 \text{ p.u.}$$

Assume $X_1 = X_2 = X_0$ for each line

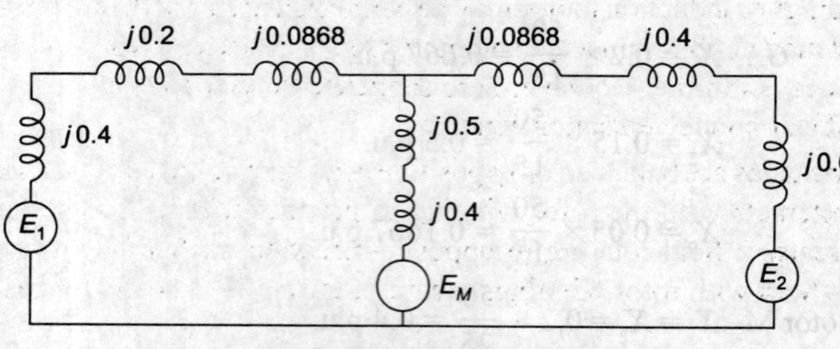

Positive Sequence Network

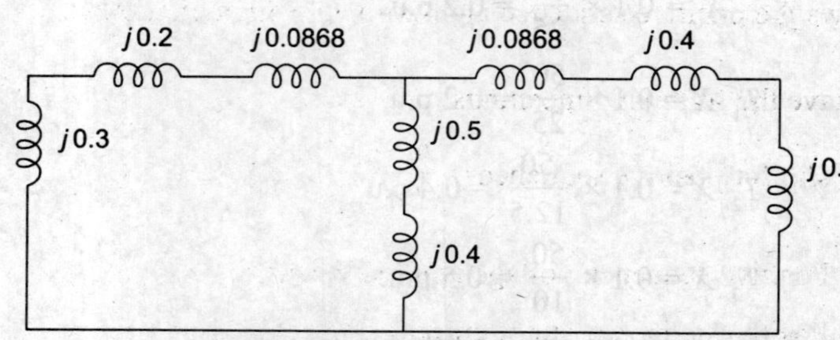

Negative Sequence Network

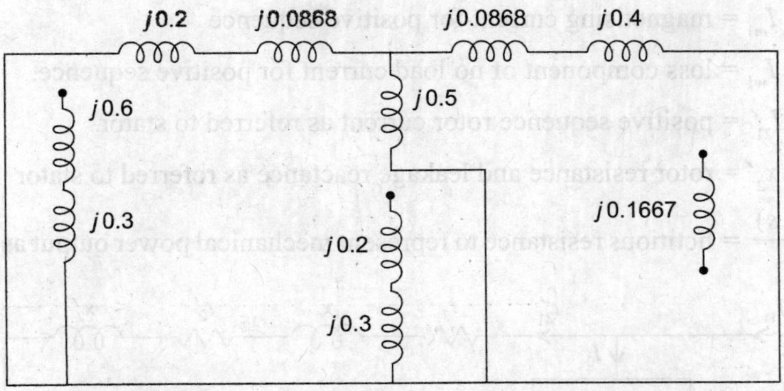

Zero Sequence Network

Fig. 18.17.

18.7 OPERATION OF INDUCTION MOTOR WITH UNBALANCED SUPPLY VOLTAGE

In the analysis of 3-phase induction motor it is assumed that the three stator windings are identical and supply voltages are balanced. The similarity of three windings is a necessary consequence of proper design and manufacture. As such the three stator windings are always similar. However the supply voltages may be sometimes, unbalanced due to disturbances in power supply networks. As an extreme case one of the phase voltages may get disconnected resulting in a situation called single phasing. These conditions can be analysed by the method of symmetrical components.

The stator windings of induction motor may be connected in star or delta. Even when connected in star, the neutral may not be connected and motor fed by 3 phase 3 wire supply. Therefore zero sequence impedance is infinite. However zero sequence voltage may exist at the terminals and at neutral point of star connected stator windings.

When supply voltages are balanced only positive sequence voltages exist and create a rotating field. This field interacts with the rotor winding to produce torque T_1. The negative sequence set of voltages creates a field rotating in opposite direction to that of positive sequence. The interaction of this field with rotor windings creates a torque T_2. The net torque is $T_1 - T_2$.

18.7.1 Positive Sequence Equivalent Circuit

Figure 18.18 shows the positive sequence approximate equivalent circuit of 3 phase induction motor.

The symbols have the following meanings

$$s = \text{slip for positive sequence} = \frac{n_s - n}{n_s}$$

V_{a1} = Positive sequence voltage applied to stator

I_{11} = Positive sequence stator current [First subscript 1 refers to stator and second subscript 1 to positive sequence].

r_1 and x_1 = stator winding resistance and leakage reactance per phase

I_{10} = no load current for positive sequence

I_{m1} = magnetising current for positive sequence

I_{w1} = loss component of no load current for positive sequence.

I_{21}' = positive sequence rotor current as referred to stator.

r_2' and x_2' = rotor resistance and leakage reactance as referred to stator

$\frac{r_2'(1-s)}{s}$ = fictitious resistance to represent mechanical power output as referred to stator.

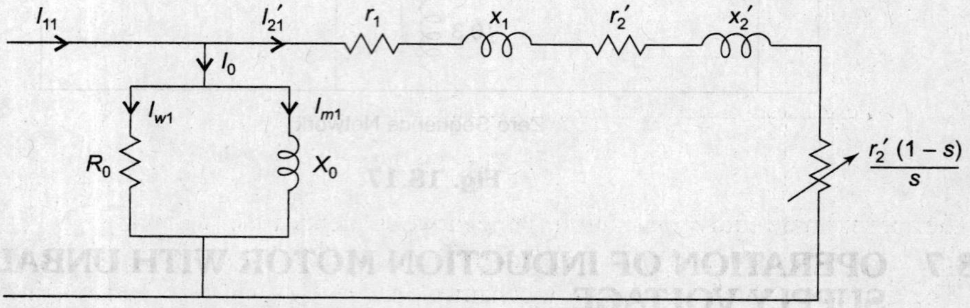

Fig. 18.18 Positive sequence equivalent circuit of 3 phase induction motor

$$\text{Power input} = 3\,V_{a1}\,I_{11}\cos\theta_1 \tag{18.29}$$

where θ_1 is the phase angle between V_{a1} and I_{11}

Positive sequence air gap power

= Positive sequence power input – stator core losses – stator copper losses

Also
$$P_{g1} = \frac{3I_{21}^2\,r_2}{s} = \frac{3I_{21}'^2\,r_2'}{s} \tag{18.30}$$

$$\text{Rotor copper loss} = 3I_{21}^2\,r_2 = 3I_{21}'^2\,r_2' \tag{18.31}$$

$$T_1 = \frac{P_{g1}}{2\pi n_s} \tag{18.32}$$

18.7.2 Negative Sequence Equivalent Circuit

The slip s is defined in terms of the speed of the rotor in the positive direction with respect to the rotating field set up by positive sequence voltages. The rotating field due to negative sequence voltages has a direction opposite to that of positive sequence field. Therefore the slip s_2 for negative sequence equivalent circuit can be found as under

$$s = \frac{n_s - n}{n_s} \tag{18.33}$$

where n and n_s are the actual rotor speed and synchronous speed in revolutions per second.

$$s_2 = \frac{n_s - (-n)}{n_s} = 1 + \frac{n}{n_s} \tag{18.34}$$

Combining Eqs. (18.33) and (18.34) we get

$$s_2 = 2 - s \tag{18.35}$$

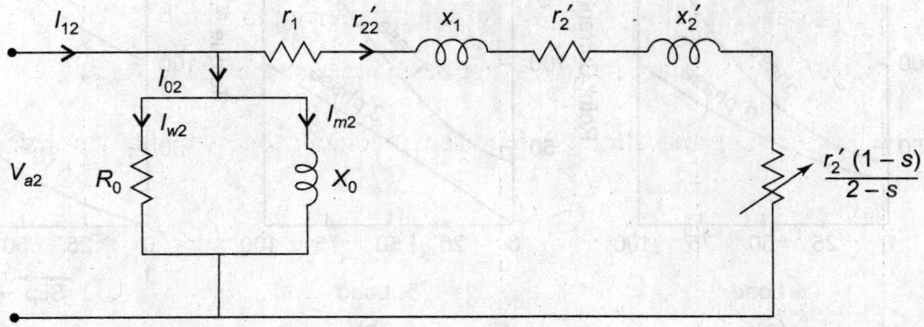

Fig. 18.19.

The approximate equivalent circuit for negative sequence is shown in Fig. 18.19. The applied voltage is V_{a2}. The stator and rotor currents are I_{12} and I_{22}. The rotor current as referred to stator is I_{22}'. The resistance which replaces the mechanical load is $r_2(1-s)/(2-s)$ and this resistance when referred to stator becomes $r_2'(1-s)/(2-s)$. I_{02} denotes no-load current for negative sequence

$$\text{Negative Sequence Power Input} = 3\, V_{a2}\, I_{12} \cos\theta_2 \tag{18.36}$$

where θ_2 is the phase angle between V_{a2} and I_{12}

$$\text{Negative sequence air gap power } P_{g2} = \frac{3I_{22}^2 r_2}{2-s} = \frac{3I_{22}'^2 r_2'}{2-s} \tag{18.37}$$

$$\text{Negative sequence rotor copper-losses} = 3I_{22}^2 r_2 = 3I_{22}'^2 r_2' \tag{18.38}$$

$$T_2 = \frac{P_{g2}}{2\pi n_s} \tag{18.39}$$

$$\text{Net torque} = T_1 - T_2 = \frac{P_{g1} - P_{g2}}{2\pi n_s} \tag{18.40}$$

If Eq. (18.40) is multiplied by $(1 - s)$ we get the shaft output torque.

The zero sequence voltages, even if present, do not produce any resultant field and do not affect the torque or mechanical power output.

18.7.3 Single Phasing

An extreme case of unbalance supply voltage can occur if one phase of supply gets disconnected when the motor is running. It could happen due to blowing of fuse in one phase. If the motor is operating on full load and single phasing occurs, the motor may even stall. If the load is less the motor may continue operating but with poor performance. Both stator and rotor will get over heated. The temperature rise in stator will be very prominent in two phases if it is star connected and in one phase if it is delta connected.

The performance of the motor under the condition of single phasing can also be evaluated using symmetrical components. Figure 18.20 shows some characteristics under this condition. For comparison the characteristics for balanced 3-phase operation are also shown.

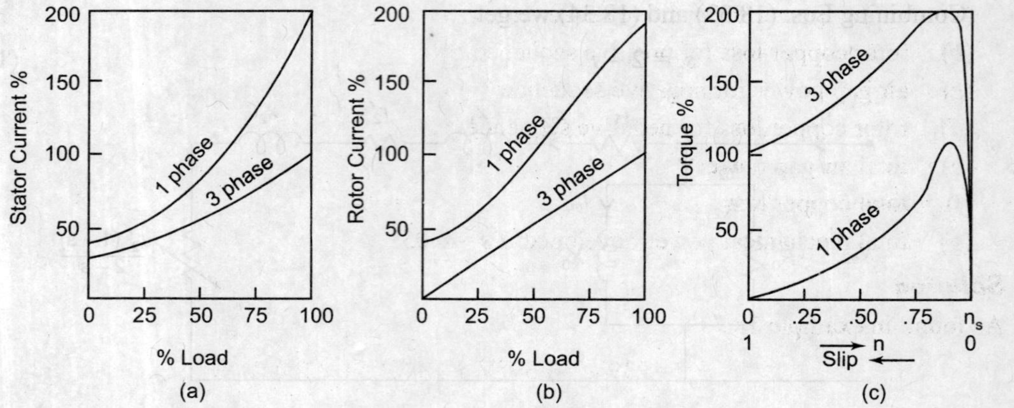

Fig. 18.20 Comparison of performance of 3-phase induction motor for 3-phase and single phase operations

Example 18.7

In an unbalanced 3-phase system the phase voltages are $V_a = 200\ \angle 0$ V, $V_b = 250\ \angle{-100°}$V, $V_c = 300\ \angle 100°$ V. Find the symmetrical components V_{a1}, V_{a2} and V_{a0}.

Solution

Using Eq. (18.8)

$$V_a = 200 + j\,0,\ V_b = -\,43.4 - j\,246.2 \text{ and } V_c = -\,52.1 + j\,295.4$$

$$V_{a0} = \frac{1}{3}\ (V_a + V_b + V_c) = \frac{1}{3}\ (+\,104.5 + j\,49.2)$$

$$= 34.83 + j\,16.4\text{V} = 38.5\ \angle\ 25.2°\text{V}$$

$$V_{a1} = \frac{1}{3}\ (V_a + aV_b + a^2V_c)$$

$$V_{a1} = \frac{1}{3}\ [200\ \angle 0 + (1\ \angle 120°)\ (250\ \angle{-100°}) + (1\ \angle 240°)\ (300\ \angle 100°)]$$

$$= \frac{1}{3}\ [200\ \angle 0 + 250\ \angle 20° + 300\ \angle 340°]$$

$$= 238.94 - j\,5.7 = 239\ \angle -\,1.37°\text{V}$$

$$V_{a2} = \frac{1}{3}\ (V_a + a^2V_b + aV_c)$$

$$= \frac{1}{3}\ [200\ \angle 0 + 250\ \angle 140° + 300\ \angle 220°]$$

$$= -\,73.77 - j\,10.71 = 74.54\ \angle 188.26°\text{V}$$

Example 18.8

The parameters of a 3-phase induction motor are $r_1 = 0.7\ \Omega$, $x_1 = 1.3\ \Omega$, $r_2' = 0.4\ \Omega$, $x_2' = 0.6\ \Omega$, $X_0 = 35\ \Omega$, $R_0 = 150\ \Omega$. If the supply voltages are $V_a = 200\ \angle 0$ V, $V_b = 250\ \angle -100°$V and $V_c = 300\ \angle 100°$V find

(*a*) air gap power for positive sequence
(*b*) rotor copper loss for positive sequence
(*c*) air gap power for negative sequence
(*d*) rotor copper loss for negative sequence
(*e*) total air gap power
(*f*) total copper loss
(*g*) total mechanical power developed if $s = 0.05$.

Solution

As found in example 18.7

$$V_{a1} = 239 \angle -1.37°\text{V},\ V_{a2} = 74.54 \angle 188.26°\text{V}$$

(*a*) Positive sequence rotor current (as referred to stator)

$$I_{21}' = \frac{V_{a1}}{\left(r_1 + \frac{r_2'}{s}\right) + j(x_1 + x_2')}$$

$$= \frac{239 \angle -1.37°}{\left(.7 + \frac{.4}{.05}\right) + j(1.3 + .6)} = \frac{239 \angle -1.37°}{8.74 \angle 13.36°}$$

or

$$I_{21}' = 26.73 \angle -14.73° \text{ A}$$

$$P_{g1} = 3I_{21}'^2 \frac{r_2'}{s} = 3(26.73)^2 \left(\frac{.4}{.05}\right) = 17147.83 \text{ W}$$

(*b*) Positive sequence rotor copper losses $= 3I_{21}'^2 r_2'$

$$= 3\,(26.73)^2\,(.4) = 857.39 \text{ W}$$

(*c*)

$$I_{22}' = \frac{V_{a2}}{\left(r_1 + \frac{r_2'}{2-s}\right) + j(x_1 + x_2')}$$

$$= \frac{74.54 \angle 188.26}{0.7 + \frac{0.4}{2 - 0.5} + j\,(1.3 + .6)} = \frac{74.54 \angle 188.26°}{.905 + j1.9}$$

$$= \frac{74.54 \angle 188.26°}{2.1 \angle 64.53°} = 35.49 \angle 123.73° \text{ A}$$

$$P_{g2} = 3I_{22}'^2 \frac{r_2'}{2-s} = 3(35.49)^2 \left(\frac{.4}{1.95}\right) = 775.1 \text{ W}$$

(*d*) Negative sequence rotor copper-losses

$$= 3I_{22}'^2 r_2' = 3(35.49)^2\,(0.4)$$

$$= 1511.45 \text{ W}$$

(*e*) Total air gap power $= 17147.83 + 775.1$

$$= 17922.93 \text{ W}$$

(*f*) Total copper-losses = 857.39 + 1511.45

$$= 2368.84 \text{ W}$$

(*g*) Total mechanical power developed

$$= 17922.93 - 2368.84 = 15554.09 \text{ W}$$

SUMMARY

18.1 Symmetrical components are of great help in analysing 3 phase unbalanced circuits.

18.2 Any unbalanced set of 3 phase voltages can be resolved into 3 balanced sets called positive sequence, negative sequence and zero sequence sets.

18.3 The three sequence networks do not have any mutual coupling.

18.4 $S = 3\,(V_{a0} J_{a0}^* + V_{a1} I_{a1}^* + V_{a2} I_{a2}^*)$

18.5 Positive and negative sequence impedances of static equipments are equal.

18.6 Zero sequence network is affected by the method of neutral grounding.

18.7 Phase shift between sequence voltages (and currents) occurs as we go from one side to other of Δ–Y or Y–Δ transformer.

18.8 Symmetrical components can also be used to study operation of 3 phase induction motor from unbalanced supply.

18.9 When a 3 phase induction motor is rotating and one of the phases gets switched off the motor will continue to rotate.

18.10 Single phase results in poor performance.

SELF TEST QUESTIONS

A. State whether the following are true (T) or false (F)

18.1 $a = 1 \angle 60°$

18.2 $1 - a^2 = 1.5 \div j\,0.866$

18.3 $V_c = a\,V_{a1} + a^2\,V_{a2} + V_{a0}$

18.4 $S = V_{a1} I_{a1}^* + V_{a2} I_{a2}^* + V_{a0} I_{a0}^*$

18.5 Only zero sequence currents flow in ground.

18.6 $V_{a1} = E_{a1} - I_{a1} Z_1$

18.7 Negative sequence voltage generated by a synchronous machine is zero.

18.8 The current flowing through neutral grounding impedance Z_n is I_{a0}.

B. Fill in he blanks with suitable words

18.9 At fault point V_{a2} = __________.

18.10 At fault point Z_0 = __________.

18.11 Zero sequence current can flow in a star connected winding only if star point is __________.

18.12 In a transmission line zero sequence reactance is __________ than positive sequence reactance.

18.13 In a star-delta transformer the positive sequence line current on high voltage side __________ the positive sequence current on low voltage side by 30°.

18.14 If T_1 and T_2 are torque due to positive and negative sequence voltages in a 3 phase induction motor, net torque T = __________.

18.15 Single phase operation of 3 phase induction motor results in __________ performance.

ANSWERS

18.1 F 18.2 T 18.3 T 18.4 F 18.5 T 18.6 T

18.7 T 18.8 F 18.9 $-I_{a2}Z_2$ 18.10 $Z_{g0} + 3Z_n$

18.11 grounded 18.12 more 18.13 lead 18.4 $T_1 - T_2$ 18.15 poor

JOB INTERVIEW QUESTIONS

18.1 What is meant by sequence voltages?

18.2 What are symmetrical components?

18.3 What is the need of symmetrical components?

18.4 Which sequence current flows through ground and why?

18.5 How does neutral grounding affect zero sequence impedance?

18.6 What is meant by single phasing of 3 phase induction motor?

18.7 What is meant by the term-power invariant.

REVIEW QUESTIONS

18.1 What is the necessity of using symmetrical components in analysis of unbalanced circuits? Write equations for phase voltage interms of symmetrical components and vice versa.

18.2 Derive the equations between phase currents and their symmetrical components.

18.3 Show that symmetrical component transformation is power invariant.

18.4 Discuss the significance of zero sequence circuit.

18.5 Why does a generator produce only positive sequence voltage? Discuss

18.6 Why should Z_n appear as $3Z_n$ in zero sequence equivalent circuit.

18.7 How can performance of 3 phase induction motor be evaluated using symmetrical components?

PROBLEMS

18.1 The line currents in a 3 phase system are $I_a = 60 + j\,40$, $I_b = -80 + j\,20$, $I_c = 20 - j\,60$. Find symmetrical components of line current.

[$I_{a1} = 6.9 - j\,8.87$ A, $I_{a2} = 53 + j\,48.8$ A, $I_{a0} = 0$]

18.2 The line to neutral voltages in a 3 phase sysem are $V_{an} = 200 \angle 0$, $V_{bn} = 600 \angle 100°$V, $V_{cn} = 400 \angle 270°$V. Find V_{a1}, V_{a2}, V_{a0}.

[$V_{a1} = 211.3 \angle 197°$V, $V_{a2} = 370 \angle 0$ V, $V_{a0} = 71.2 \angle 63.3°$V]

18.3 Using the values of I_{a1}, I_{a2} and I_{a0} in problem 18.1 find I_a, I_b and I_c.

18.4 Using the values of V_{a1}, V_{a2} and V_{a0} in Problem 18.2 find V_{an}, V_{bn} and V_{cn} in Problem 18.2.

18.5 In the network of example 18.3 both transformers are converted into Δ-Δ transformers. Redraw zero sequence network. The remaining data is the same.

18.6 In the system of Fig. 18.14 both transformers are converted into Δ-Δ connections. Redraw all the three sequence networks.

18.7 In the network of Fig. 18.16, transformer T_1 is changed to Y–Δ, T_2 to Δ–Y and T_3 to Y–Y. In all three transformers star neutral is solidly grounded. Redraw all the three sequence networks.

18.8 A 3 phase induction motor has $r_1 = 0.8$ ohm, $x_1 = 1.4$ ohm, $r_2' = 0.4$ ohm $x_2' = 0.7$ ohm, $X_0 = 40$ ohm, $R_0 = 160$ ohm, the supply voltages are $V_a = 150 \angle 0$V, $V_b = 250 \angle -90°$V, $V_c = 350 \angle 110°$V. If slip = 0.04 find air gap power for positive and negative sequence and total mechanical power developed. [14043.5 W, 1324.12 W, 12210.63 W]

CHAPTER 19

Unsymmetrical Faults

19.1 UNSYMMETRICAL FAULTS

Our basic approach to the analysis of unsymmetrical faults is to consider the general situation shown in Fig. 19.1 which shows the three lines of the three-phase power system at the point of fault. The general terminals brought out are for purposes of external connections which simulate the fault. Appropriate connections of the three stubs represent the different faults, e.g., connecting stub '*a*' to ground produces a single line to ground fault, through zero impedance, on phase '*a*'. The currents in stubs *b* and *c* are then zero and I_a is the fault current. Similarly, the connection of stubs *b* and *c* produces a line to line fault, through zero impedance, between phases *b* and *c*, The current in stub *a* is then zero and I_b is equal to $-I_c$. The positive assignment of phase quantities is important. It is seen that the currents flow out of the system. The three general sequence circuits are shown in Fig. 19.2 (a). The ports indicated correspond to the general 3-phase entry port of Fig. 19.2. A suitable inter-connection of the three-sequence networks (depending on the type of the fault) yields the solution to the problem.

The sequence networks of Fig. 19.2 (a) can be replaced by equivalent sequence networks of Fig. 19.2 (b). Z_0, Z_1 and Z_2 indicate the sequence impedances of the network looking into the fault.

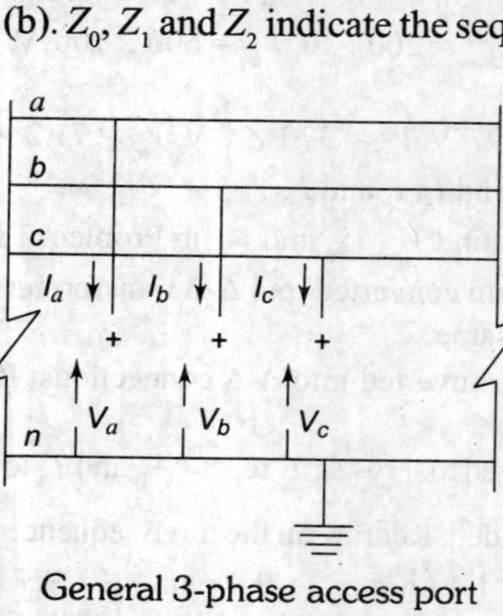

Fig. 19.1 General 3-phase access port

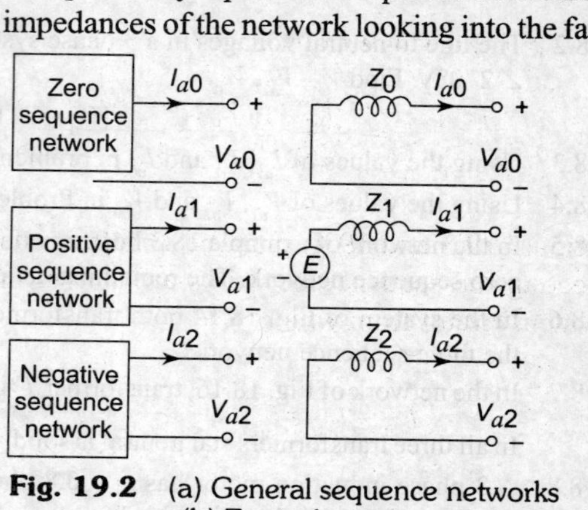

Fig. 19.2 (a) General sequence networks
(b) Equivalent sequence networks

19.2 SINGLE LINE TO GROUND FAULT

The termination of the three-phase access port as shown in Fig. 19.3 brings about a condition of single line to ground fault through a fault impedance Z_f. Typically Z_f is set to zero in all fault studies. We include Z_f in the analysis for the sake of generality. The terminal conditions at the fault point give the following equations:

$$I_b = 0 \tag{19.1a}$$

$$I_c = 0 \tag{19.1b}$$

$$V_a = I_a Z_f \tag{19.1c}$$

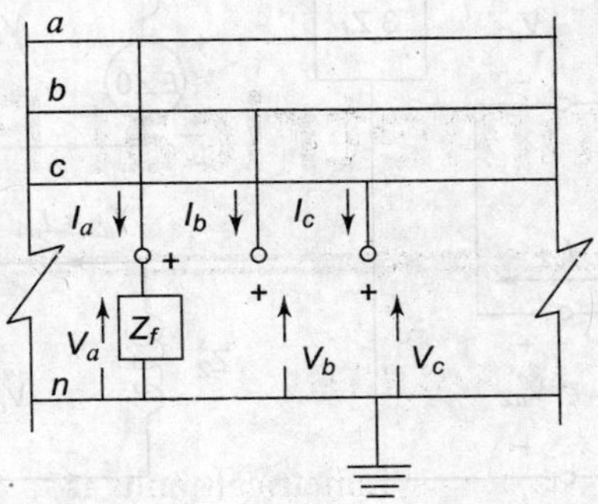

Fig. 19.3 Single line to ground fault

The sequence currents at the fault point can be calculated by Eq. (18.10)

$$\begin{bmatrix} I_{a0} \\ I_{a1} \\ I_{a2} \end{bmatrix} = \frac{1}{3}\begin{bmatrix} 1 & 1 & 1 \\ 1 & a & a^2 \\ 1 & a^2 & a \end{bmatrix}\begin{bmatrix} I_a \\ 0 \\ 0 \end{bmatrix} \tag{19.2}$$

Thus,
$$I_{a0} = I_{a1} = I_{a2} = \frac{I_a}{3} \tag{19.3}$$

From Eqs. (19.1 c and 19.3)

$$I_{a1} = \frac{V_a}{3Z_f} = \frac{V_{a0} + V_{a1} + V_{a2}}{3\,Z_f} \tag{19.4}$$

Equations (19.3 and 19.4) must be simultaneously satisfied. These conditions can be met by connecting the sequence networks are shown in Fig. 19.4 (a). Fig. 19.4 (b) shows the final equivalent circuit from which

$$I_{a0} = I_{a1} = I_{a2} = \frac{E}{Z_1 + Z_2 + Z_0 + 3\,Z_f} \tag{19.5a}$$

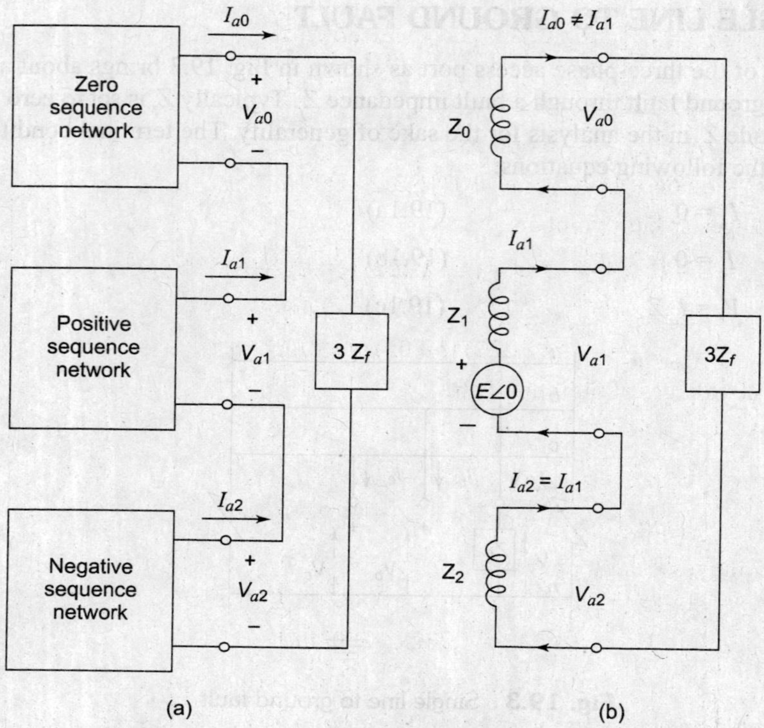

Fig. 19.4 Connections of sequence networks for a single line to ground fault

$$I_a = I_{a0} + I_{a1} + I_{a2} = \frac{3\,E}{Z_1 + Z_2 + Z_0 + 3\,Z_f} \tag{19.5b}$$

Using the above value of sequence currents, we can find V_{a1}, V_{a2} and V_{a0} from Eqs. (18.25, 18.26 and 18.28). V_b and V_c can then be found.

Example 19.1.

A 30 MVA, 11 kV generator has $Z_1 = Z_2 = j\,0.2\,pu$, $Z_0 = j\,0.05\,pu$. (a) A line to ground fault occurs on the generator terminals. Find the fault current and line to line voltages during fault conditions. Assume that the generator neutral is solidly grounded and that the generator is operating at no load and at rated voltage at the occurrence of fault. (b) Find the line current for a 3-phase fault.

Solution

Base MVA = 30 MVA

Base voltage = 11 kV

$$\text{Base current} = \frac{30 \times 10^6}{\sqrt{3} \times 11 \times 10^3} = 1574.6\text{ A}$$

$$Z_f = 0$$

(a) Since the fault is at generator terminals, no network reduction is necessary. The values of Z_1, Z_2 and Z_0 of the generator can be directly used to get the results,

$$I_{a1} = I_{a2} = I_{a0} = \frac{E}{Z_1 + Z_2 + Z_0} = \frac{1 \angle 0}{j0.2 + j0.2 + j0.05}$$

$$= -j\,2.222\ pu$$

Fault current $= I_a = 3\,I_{a1} = -j\,6.666\ pu$

$= 6.666 \times 1574.6$, i.e., $10496.3 \angle -90°$ A

Sequence voltages at fault point are

$$V_{a1} = E - I_{a1} Z_1 = 1 \angle 0° - (-j\,2.222)\,(j\,0.2) = 0.5556\ pu$$
$$V_{a2} = -I_{a2} Z_2 = -(-j\,2.222)\,(j\,0.2) = -0.4444\ pu$$
$$V_{a0} = -I_{a0} Z_0 = -(-j\,2.222)\,(j\,0.05) = -0.1111\ pu$$

Line to neutral voltages at fault point are

$$\begin{bmatrix} V_a \\ V_b \\ V_c \end{bmatrix} = \begin{bmatrix} 1 & 1 & 1 \\ 1 & a^2 & a \\ 1 & a & a^2 \end{bmatrix} \begin{bmatrix} -0.1111 \\ +0.5556 \\ -0.4444 \end{bmatrix} \begin{bmatrix} 0 \\ -0.1667 - j0.866 \\ -0.1667 + j0.866 \end{bmatrix}$$

Line to line voltages at fault point are

$$V_{ab} = V_a - V_b = 0.1667 + \mathrm{j}\,0.866 = 0.882 \angle 79.1°\ pu$$
$$V_{bc} = V_b - V_c = -j\,1.732 = 1.732 \angle 270°\ pu$$
$$V_{ca} = V_c - V_a = -0.1667 + \mathrm{j}\,0.866 = 0.882 \angle 100.9°\ pu$$

Since the generated line to neutral voltage has been taken as 1 *pu* in the calculation of sequence currents, the above voltages are in *pu* of line to neutral voltage. Therefore, the actual values of line to line post fault voltages are

$$V_{ab} = \frac{11}{\sqrt{3}} \times 0.882 \angle 79.1° = 5.6 \angle 79.1°\ \text{kV}$$

$$V_{bc} = \frac{11}{\sqrt{3}} \times 1.732 \angle 270° = 11 \angle 270°\ \text{kV}$$

$$V_{ca} = \frac{11}{\sqrt{3}} \times 0.882 \angle 100.9° = 5.6 \angle 100.9°\ \text{kV}$$

Before the fault each line voltage was equal to 11 kV. It is seen that a fault causes the post fault line voltages to be different from pre-fault voltages.

(b) For a 3-phase fault, line current $= \dfrac{1 \angle 0}{j\,0.2} = 5 \angle -90\ pu$

Actual value of line current $= 5 \times 1574.6 \angle -90°$ or $7873 \angle -90°$ A

Example 19.2.

A fault occurs at the bus 2 of the network of example 18.4. Reduce the sequence networks to give the Thevenin's equivalents looking into bus 2.

Solution

The positive sequence network (Fig. 18.15) can be reduced in steps as shown in Fig. 19.5 (a, b, c)

The negative sequence impedances are the same as positive sequence impedances. The zero sequence network (Fig. 18.15) can be reduced as shown in Fig. 19.6 (a, b, c).

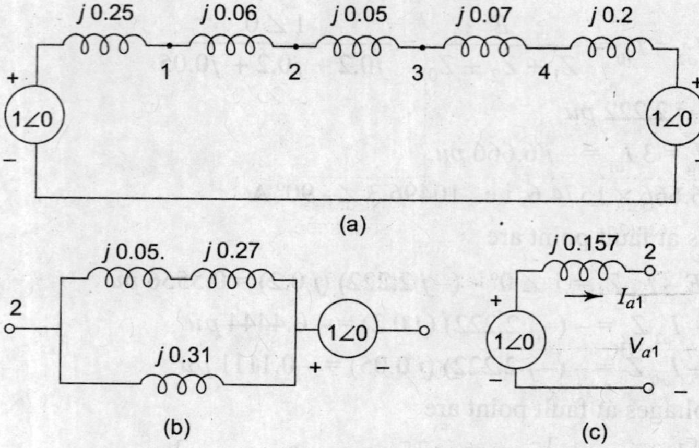

Fig. 19.5

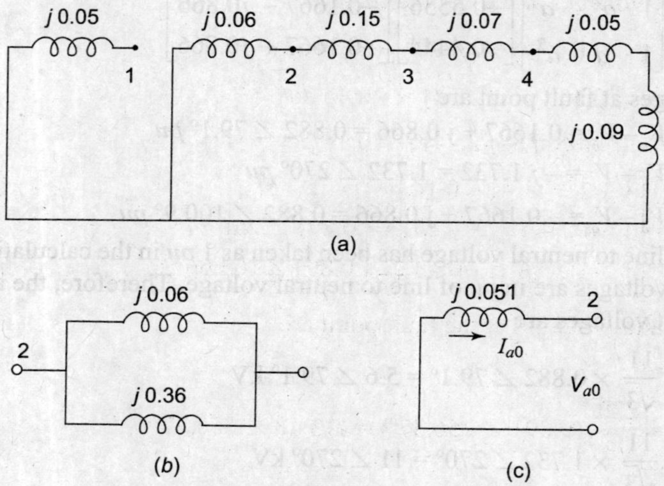

Fig. 19.6

The Thevenin equivalents looking into bus 2 are shown in Fig. 19.7

Example 19.3.

A single line to ground fault occurs at bus 2 of the network of examples 18.4 and 19.2. Find (a) fault current (b) line to neutral voltages at fault point (c) currents and voltages at generator terminals.

Solution

The sequence impedances as seen from the fault point are shown in Fig. 19.7

$$Z_1 = j\,0.157, Z_2 = j\,0.157, Z_0 = j\,0.051, Z_f = 0$$

(a) $$I_{a1} = I_{a2} = I_{a0} = \frac{1\angle 0}{j\,0.157 + j0.157 + j0.051} = -j\,2.74\,pu$$

$$\begin{bmatrix} I_a \\ I_b \\ I_c \end{bmatrix} = \begin{bmatrix} 1 & 1 & 1 \\ 1 & a^2 & a \\ 1 & a & a^2 \end{bmatrix} \begin{bmatrix} -j2.74 \\ -j2.74 \\ -j2.74 \end{bmatrix} = \begin{bmatrix} -j8.22 \\ 0 \\ 0 \end{bmatrix}$$

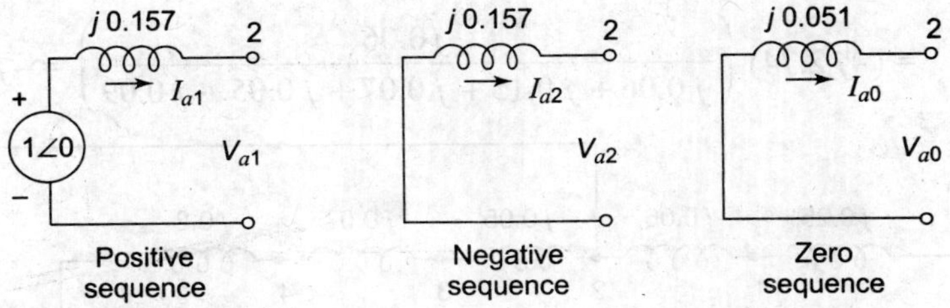

Fig. 19.7 Thevenin equivalent circuits of sequence networks of example 19.2

$$\text{Base current} = \frac{100 \times 10^6}{\sqrt{3} \times 220 \times 10^3} = 262.4 \text{ A}$$

$$\text{Fault current} = 8.22 \times 262.4 = 2156.93 \text{ A}$$

(b) The sequence voltages at the fault point are

$$V_{a1} = 1 - (j\,0.157)\,(-j\,2.74) = 0.57\ pu$$

$$V_{a2} = -\,(j\,0.157)\,(-j\,2.74) = -\,0.43\ pu$$

$$V_{a0} = -\,(j\,0.051)\,(-j\,2.74) = -\,0.14\ pu$$

$$\begin{bmatrix} V_a \\ V_b \\ V_c \end{bmatrix} = \begin{bmatrix} 1 & 1 & 1 \\ 1 & a^2 & a \\ 1 & a & a^2 \end{bmatrix} \begin{bmatrix} -0.14 \\ +0.57 \\ -0.43 \end{bmatrix} = \begin{bmatrix} 0 \\ 0.891 \angle 256.37^\circ \\ 0.891 \angle 103.63^\circ \end{bmatrix}$$

Thus, line to neutral voltages at the fault point are

$$V_a = 0$$

$$V_b = \frac{220}{\sqrt{3}}\,(0.891 \angle 256.37^\circ) = 113.18 \angle 256.37^\circ \text{ kV}$$

$$V_c = \frac{220}{\sqrt{3}}\,(0.891 \angle 103.63^\circ) = 113.18 \angle 103.63^\circ \text{ kV}$$

(c) For calculation of currents and voltages at generator terminals the networks of Fig. 19.6 are inadequate. In the networks of Fig. 19.5 (a) and 19.6 (a) the identity of generator terminals has not been lost. Therefore, we interconnect these networks (alongwith negative sequence network) for the single line to ground fault as shown in Fig. 19.8. The negative sequence portion of the network in Fig. 19.8 is the same as the positive sequence portion except that the voltage sources are not present.

Generator G_2 (Bus 4).

The component of I_{a1} flowing towards bus 2 from generator G_2

$$= (-j\,2.74)\left(\frac{j\,0.31}{j\,0.31 + j\,0.32}\right) = -j\,1.35\ pu$$

The component of I_{a2} flowing towards bus 2 from generator G_2 is also equal to $(-j\,1.35)$ *pu*.

The component of I_{a0} flowing towards bus 2 from generator G_2 is

$$= (-j\,2.74)\left(\frac{j\,0.06}{j\,0.06 + j\,0.15 + j\,0.07 + j\,0.05 + j\,0.09}\right) = -j\,0.39\ pu$$

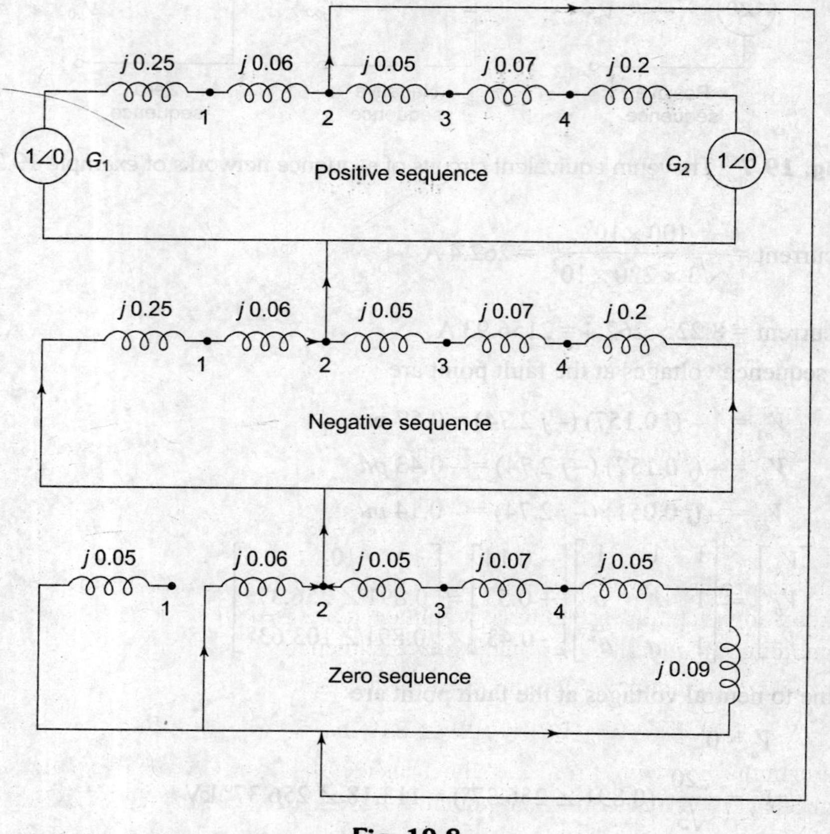

Fig. 19.8

The line currents at terminals of generator G_2 are

$$\begin{bmatrix} I_a \\ I_b \\ I_c \end{bmatrix} = \begin{bmatrix} 1 & 1 & 1 \\ 1 & a^2 & a \\ 1 & a & a^2 \end{bmatrix}\begin{bmatrix} -j0.39 \\ -j1.35 \\ -j1.35 \end{bmatrix} = \begin{bmatrix} 3.09\angle 270^\circ \\ 0.96\angle 90^\circ \\ 0.96\angle 90^\circ \end{bmatrix}$$

$$\text{Base current for generators} = \frac{100\times 10^6}{\sqrt{3}\times 11000} = 5248.8\ \text{A}$$

Therefore, $I_a = 3.09 \times 5248.8 = 16218.8$ A

$I_b = 0.96 \times 5248.8 = 5038.8$ A

$I_c = 0.96 \times 5248.8 = 5038.8$ A

The sequence voltages at the terminals of G_2 are

$$V_{a0} = -(-j\,0.39)\,(j\,0.05 + j\,0.09) = -\,0.0546$$

$$V_{a1} = 1 - (-j\,1.35)\,(j\,0.2) = -\,0.73$$

$$V_{a2} = -(-j\,1.35)\,(j\,0.2) = -\,0.27$$

The line to neutral voltages at terminals of G_2 are

$$\begin{bmatrix} V_a \\ V_b \\ V_c \end{bmatrix} = \begin{bmatrix} 1 & 1 & 1 \\ 1 & a^2 & a \\ 1 & a & a^2 \end{bmatrix} \begin{bmatrix} -0.0546 \\ 0.73 \\ -0.27 \end{bmatrix} = \begin{bmatrix} 0.405 \angle 0^\circ \\ 0.91 \angle 251.8^\circ \\ 0.91 \angle 108.2^\circ \end{bmatrix}$$

Thus, actual values of line to neutral voltages at the terminals of G_2 are

$$V_a = \frac{11}{\sqrt{3}} \times 0.405 = 2.57 \text{ kV}$$

$$V_b = \frac{11}{\sqrt{3}} \times 0.91 = 5.78 \text{ kV}$$

$$V_c = \frac{11}{\sqrt{3}} \times 0.91 = 5.78 \text{ kV}$$

Generator G_1 (Bus 1)

The component of I_{a1} flowing from G_1 to the bus 2

$$-(-j\,2.74)\left(\frac{j0.32}{j(0.31)+(0.32)}\right) = -j\,1.39\,pu = 1.39 \angle 270^\circ\,pu$$

Because of the presence of Δ–Y transformer, a phase shift in the positive sequence current occurs. As discussed in section 18.6.2 the low voltage quantities are to be shifted 30º backward of the corresponding *HV* quantities for the positive sequence.

Therefore, the component of I_{a1} flowing from G_1 to bus 2

$$= 1.39 \angle (270^\circ - 30^\circ) = 1.39 \angle 240^\circ\,pu$$

The component of I_{a2} flowing from G_1 to bus 2 also equals $(-j\,1.39\,pu)$. However, the phase shift in this case is + 30º. Hence, this component is $1.39 \angle (270^\circ + 30^\circ) = 1.39 \angle 300^\circ\,pu$.

The zero sequence network is open at bus 1. Hence, no zero sequence current flows through generator G_1.

The line currents at the terminals of G_1 are

$$\begin{bmatrix} I_a \\ I_b \\ I_c \end{bmatrix} = \begin{bmatrix} 1 & 1 & 1 \\ 1 & a^2 & a \\ 1 & a & a^2 \end{bmatrix} \begin{bmatrix} 0 \\ 1.39 \angle 240^\circ \\ 1.39 \angle 300^\circ \end{bmatrix} = \begin{bmatrix} 2.40 \angle 270^\circ \\ 2.40 \angle 90^\circ \\ 0 \end{bmatrix}$$

Base current = 5248.8 A

Therefore, $I_a = 2.4 \times 5248.8 = 12597$ A

$I_b = 2.4 \times 5248.8 = 12597$ A

$I_c = 0$

The sequence voltages at the terminals of G_1 are

$$V_{a0} = 0$$

$$V_{a1} = 1 - (1.39 \angle 270)\,(j\,0.25) = 0.6525\,pu$$

$$V_{a2} = -(1.39 \angle 270)\,(j\,0.25) = -0.3475\,pu = 0.3475 \angle 180^\circ$$

Taking into account the phase shift

$$V_{a1} = 0.6525 \angle (0 - 30^\circ) = 0.6525 \angle -30^\circ\ pu$$

$$V_{a2} = 0.3475 \angle (180 + 30^\circ) = 0.3475 \angle 210^\circ\ pu$$

The line to neutral voltages at the terminals of G_1 are

$$\begin{bmatrix} V_a \\ V_b \\ V_c \end{bmatrix} = \begin{bmatrix} 1 & 1 & 1 \\ 1 & a^2 & a \\ 1 & a & a^2 \end{bmatrix} \begin{bmatrix} 0 \\ 0.6525 \angle -30^\circ \\ 0.3475 \angle 210^\circ \end{bmatrix} = \begin{bmatrix} 0.566 \angle 298^\circ \\ 0.566 \angle 242^\circ \\ 1 \angle 90^\circ \end{bmatrix}$$

Therefore, actual values of lines to neutral voltages at the terminals of G_1 are

$$V_a = \frac{11}{\sqrt{3}} \times 0.566 = 3.59 \text{ kV}$$

$$V_b = \frac{11}{\sqrt{3}} \times 0.566 = 3.59 \text{ kV}$$

$$V_c = \frac{11}{\sqrt{3}} \times 1 = 6.35 \text{ kV}$$

Example 19.4.

A 3-phase, 37.5 MVA, 33 kV alternator having $X_1 = 0.18\ pu$, $X_2 = 0.12\ pu$ and $X_0 = 0.10\ pu$, based on its rating, is connected to a 33 kV overhead line having $X_1 = 6.3$ ohms, $X_2 = 6.3$ ohms and $X_0 = 12.6$ ohms per phase. A single line to ground fault occurs at the remote end of the line. The alternator neutral is solidly grounded. Calculate fault current.

Solutions

$$\text{Base MVA} = 37.5 \text{ MVA}$$

$$\text{Base voltage} = 33 \text{ kV}$$

$$\text{Base impedance} = \frac{33 \times 33 \times 1000}{37.5 \times 1000} = 29.04 \text{ ohm}$$

$$\text{Total } X_1 = j\,0.18 + \frac{j6.3}{29.04} = j\,0.397 \text{ ohm}$$

$$\text{Total } X_2 = j\,0.12 + \frac{j6.3}{29.04} = j\,0.337 \text{ ohm}$$

$$\text{Total } X_0 = j\,0.10 + \frac{j12.6}{29.04} = j\,0.534 \text{ ohm}$$

$$\text{Fault current} = \frac{3 \times 1 \angle 0^\circ}{j0.397 + j0.337 + j0.534} = -j\,2.366\ pu$$

$$\text{Base current} = \frac{37.5 \times 10^6}{\sqrt{3} \times 33000} = 656.08 \text{ A}$$

$$\text{Fault current} = 2.366 \times 656.08 = 1552.3 \text{ A}$$

19.3 COMPARISON OF SINGLE LINE TO GROUND AND 3-PHASE FAULT CURRENT

Assuming that the impedance Z_1, Z_2 and Z_0 are pure reactances (X_1, X_2 and X_0) and $Z_f = 0$, the fault currents are

$$\text{Three-phase fault } I_a = \frac{E}{j\,X_1} \tag{19.6}$$

$$\text{Line to ground fault } I_a = \frac{3\,E}{j\,X_1 + j\,X_2 + j\,X_0} \tag{19.7}$$

Three cases of practical importance are as under:

(a) Fault at generator terminals with generator neutral solidly grounded. For a generator $X_0 << X_1$. Moreover we can assume $X_1 = X_2$ (for sub-transient conditions X_1 and X_2 are exactly equal). If these values are substituted in Eqs. (19.6) and (19.7), it is seen that the current for a single line to ground fault is more than the current for a 3-phase fault. Therefore, we conclude that a single line to ground fault at the terminals of a generator with solidly grounded neutral is more severe than a 3-phase fault at the same location.

(b) Fault at generator terminals with neutral grounded through reactance X_n. The method of neutral grounding does not affect three phase fault current. However, the fault current for a single line to ground fault is:

$$I_a = \frac{3\,E}{j\,(X_1 + X_2 + X_0 + 3\,X_n)} \tag{19.8}$$

The relative severity of 3-phase fault and the line to ground fault depends on the value of X_n. If X_n is large, the current for a line to ground fault is lesser than the current for a 3-phase fault. If X_n is very small, the current for a line to ground fault may be more than that for a 3-phase fault. Since a term $3X_n$ appears in the denominator of Eq. (19.8) even a small value of X_n is sufficient to make the value of the single line to ground fault current lesser than 3-phase fault current. If $X_n = \frac{1}{3}(X_1 - X_0)$, the line currents for the two cases are equal. If $X_n < \frac{1}{3}(X_1 - X_0)$, a single line to ground fault is more severe than a 3-phase fault. If $X_n > \frac{1}{3}(X_1 - X_0)$, a 3-phase fault is more severe than a single line to ground fault.

(c) Fault on transmission line. For a transmission line $X_0 >> X_1$ so that for a fault on a line sufficiently away from the generator terminals, 3 phase fault current is more than the current for a single line to ground fault.

19.4 LINE TO LINE FAULT

The termination of the three-phase access port as shown in Fig. 19.9 simulates a line to line fault through a fault impedance Z_f.

The terminal conditions are

$$I_a = 0 \tag{19.9a}$$

$$I_b = -I_c \tag{19.9b}$$

$$V_b = V_c + Z_f\, I_b \tag{19.9c}$$

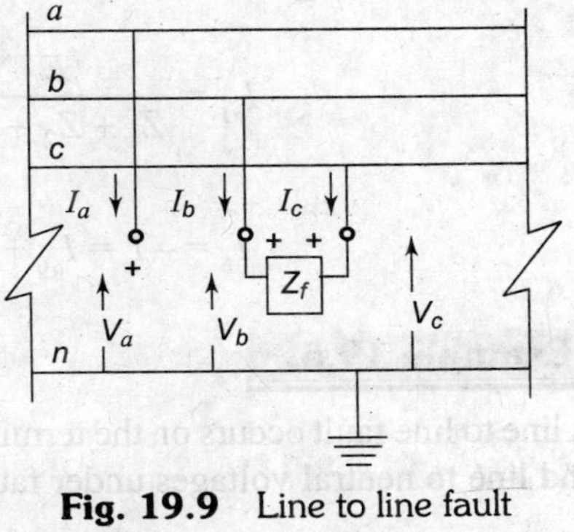

Fig. 19.9 Line to line fault

The sequence currents at the fault point are

$$\begin{bmatrix} I_{a0} \\ I_{a1} \\ I_{a2} \end{bmatrix} = \frac{1}{3}\begin{bmatrix} 1 & 1 & 1 \\ 1 & a & a^2 \\ 1 & a^2 & a \end{bmatrix}\begin{bmatrix} 0 \\ I_b \\ -I_b \end{bmatrix}$$

Whence

$$I_{a0} = 0 \tag{19.10}$$

$$I_{a2} = -I_{a1} \tag{19.11}$$

Replacing the actual quantities by their symmetrical components, we can write Eq. (19.9c) as

$$V_{a0} + a^2 V_{a1} + a V_{a2} = V_{a0} + a V_{a1} + a^2 V_{a2} + Z_f(I_{a0} + a^2 I_{a1} + aI_{a2}) \tag{19.12}$$

Since $I_{a0} = 0$ (and $V_{a0} = 0$) and $I_{a2} = -I_{a1}$, Eq. (19.12) can be simplified as

$$(a^2 - a) V_{a1} = (a^2 - a) V_{a2} + (a^2 - a) I_{a1} Z_f$$

or

$$V_{a1} = V_{a2} + I_{a1} Z_f \tag{19.13}$$

Equations (19.11 and 19.13) are simultaneously satisfied by connecting the sequence networks as shown in Fig. (19.10a). The simplified equivalent circuit is shown in Fig. 19.10 (b).

From Fig. 19.10 (b)

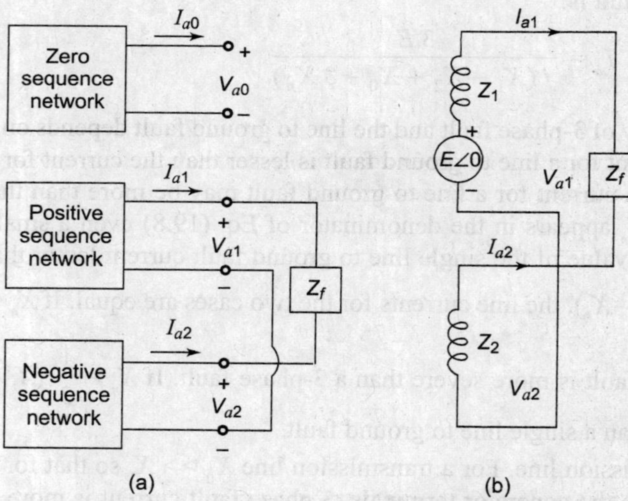

Fig. 19.10 Connection of sequence networks for line to line fault

$$I_{a1} = \frac{E}{Z_1 + Z_2 + Z_f} \tag{19.14}$$

$$I_b = -I_c = I_{a0} + a^2 I_{a1} + a I_{a2} = (a^2 - a)I_{a1} = \frac{-j\sqrt{3}\,E}{Z_1 + Z_2 + Z_f} \tag{19.15}$$

Example 19.5.

A line to line fault occurs on the terminals of the generator of example 19.1. Find the line currents and line to neutral voltages under fault conditions.

Solution

$$E = 1\angle 0^\circ\ pu,\ Z_1 = Z_2 = j\,0.2\ pu,\ Z_f = 0$$

$$I_{a1} = \frac{1\angle 0}{j\,(0.2+0.2)} = -j\,2.5\ pu$$

$$I_{a2} = -I_{a1} = j\,2.5\ pu$$

The line currents are

$$I_a = 0$$

$$I_b = -j\,\sqrt{3}\ I_{a1} = 4.33\angle 180^\circ\ pu$$

$$I_c = -I_b = 4.33\angle 0^\circ\ pu$$

Base current = 1574.6 A

$$I_a = 0$$

$$I_b = 1574.6 \times 4.33\angle 180^\circ = 6818\angle 180^\circ\ \text{A}$$

$$I_c = 1574.6 \times 4.33\angle 0^\circ = 6818\angle 0^\circ$$

$$V_{a1} = V_{a2} = 1 - (-j\,2.5)\,(j\,0.2) = 0.5\ pu$$

$$V_{a0} = 0$$

The line to neutral voltages at fault point are

$$\begin{bmatrix} V_a \\ V_b \\ V_c \end{bmatrix} = \begin{bmatrix} 1 & 1 & 1 \\ 1 & a^2 & a \\ 1 & a & a^2 \end{bmatrix} \begin{bmatrix} 0 \\ 0.5 \\ 0.5 \end{bmatrix} = \begin{bmatrix} 1 \\ -0.5 \\ -0.5 \end{bmatrix}$$

Since the line to neutral voltage has been taken as 1 *pu* in the calculation of sequence currents, the actual values of line to neutral voltages are

$$V_a = \frac{11}{\sqrt{3}} = 6.35\ \text{kV}$$

$$V_b = (-.5)\ \frac{11}{\sqrt{3}} = -\,3.175\ \text{kV}$$

$$V_c = (-.5)\ \frac{11}{\sqrt{3}} = -\,3.175\ \text{kV}$$

19.5 DOUBLE LINE TO GROUND FAULT

The termination of the three-phase access port as shown in Fig. 19.11 simulates a double line to ground fault through fault impedance Z_f.

The terminal conditions are:

$$I_a = 0 \qquad (19.16a)$$

$$V_b = V_c = (I_b + I_c)\,Z_f \qquad (19.16b)$$

It follows from Eq. (19.16a) that

$$I_{a0} + I_{a1} + I_{a2} = 0 \qquad (19.17)$$

From Eq. (19.16b)

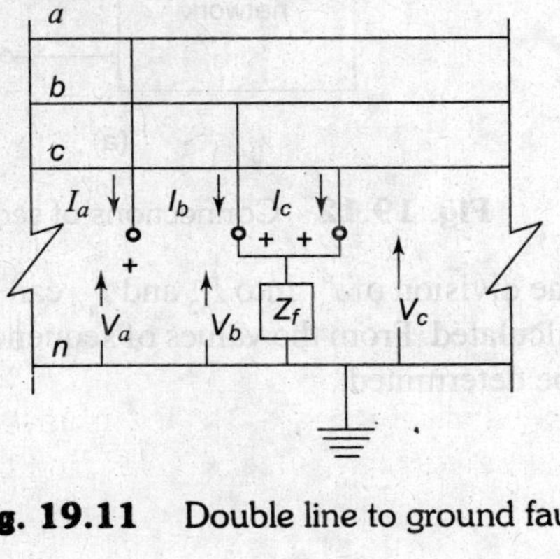

Fig. 19.11 Double line to ground fault

$$V_{a0} + a^2 V_{a1} + a\,V_{a2} = V_{a0} + a\,V_{a1} + a^2 V_{a2} \tag{19.18a}$$

or
$$V_{a1} = V_{a2} \tag{19.18b}$$

Replacing V_b, I_b and I_c by their sequence components we can write Eq. (19.16b) as

$$V_{a0} + a^2 V_{a1} + a\,V_{a2} = (I_{a0} + a^2 I_{a1} + a\,I_{a2} + I_{a0} + a\,I_{a1} + a^2 I_{a2})\,Z_f \tag{19.19a}$$

Combining Eqs (19.18b), (19.19a) and (19.17) and simplifying, we have

$$V_{a0} - V_{a1} = [2I_{a0} + (a^2 + a)\,(I_{a1} + I_{a2})]\,Z_f \tag{19.19b}$$

or
$$V_{a0} - V_{a1} = 3\,I_{a0}\,Z_f \tag{19.19c}$$

Equations (19.17), (19.18b) and (19.19c) are simultaneously satisfied by connecting the sequence networks as shown in Fig. (19.12a). The simplified equivalent circuit is shown in Fig. (19.12b). From this circuit we can write

$$I_{a1} = \frac{E}{Z_1 + \dfrac{Z_2\,(Z_0 + 3\,Z_f)}{Z_2 + Z_0 + 3\,Z_f}} \tag{19.20}$$

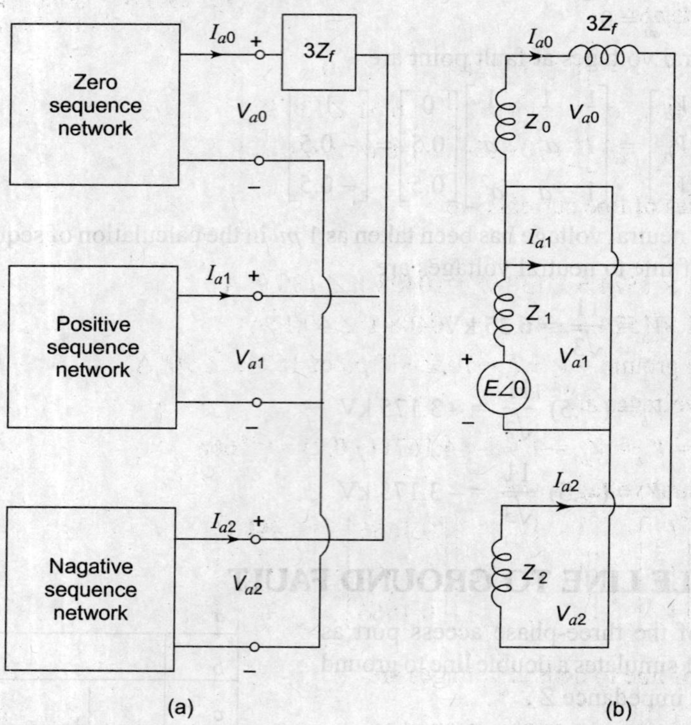

Fig. 19.12 Connections of sequence networks for double line to ground fault

The division of I_{a1} into I_{a2} and I_{a0} can be found from Fig. 19.12 (b). Then V_{a1}, V_{a2} and V_{a0} can be calculated. From the values of sequence currents and voltages, the line currents and voltages, can be determined.

Example 19.6.

A double line to ground fault occurs on the terminals of generator of example 19.1. Determine the line currents, fault current and line to neutral voltages under fault conditions.

Solutions

Figure 19.13 shows the sequence networks connected for a double line to ground fault when $Z_f = 0$

$$I_{a1} = \frac{1\angle 0}{j\,0.2 + \dfrac{(j0.2)(j\,0.05)}{j\,0.25}} = -j\,4.167\ pu$$

$$I_{a2} = -(-j\,4.167)\,\frac{j\,0.05}{j\,0.05 + j\,0.2} = j\,0.8334\ pu$$

$$I_{a0} = -(-j\,4.167)\,\frac{j\,0.2}{j\,0.05 + j\,0.2} = j\,3.3336\ pu$$

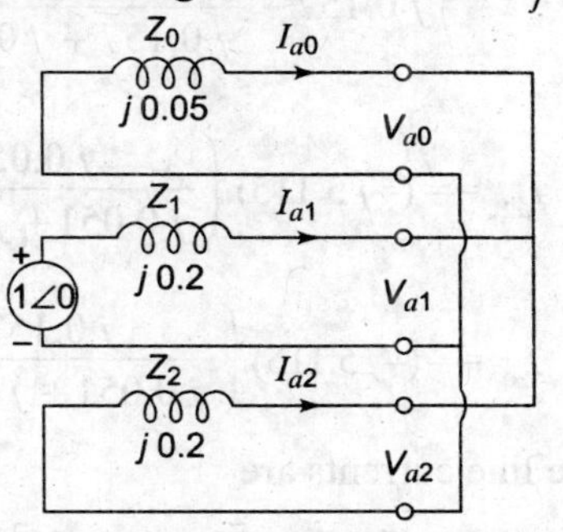

Fig. 19.13 Connection of sequence networks for Example 19.6

The line currents are

$$\begin{bmatrix} I_a \\ I_b \\ I_c \end{bmatrix} = \begin{bmatrix} 1 & 1 & 1 \\ 1 & a^2 & a \\ 1 & a & a^2 \end{bmatrix} \begin{bmatrix} +j3.3336 \\ -j4.167 \\ +j0.8334 \end{bmatrix} = \begin{bmatrix} 0 \\ 6.61\angle 130.9^\circ \\ 6.61\angle 49.1^\circ \end{bmatrix}$$

The actual values of line currents are

$$I_a = 0$$

$$I_b = 6.61 \times 1574.6\ \angle 130.9^\circ = 10408.1\ \angle\ 130.9^\circ\ \text{A}$$

$$I_c = 6.61 \times 1574.6\ \angle 49.1^\circ = 10408.1\ \angle\ 49.1^\circ\ \text{A}$$

Fault current to ground $= I_b + I_c = 10\ \angle\ 90^\circ\ pu$ or $15746\ \angle\ 90^\circ$ A

The sequence voltages are

$$V_{a1} = V_{a2} = V_{a0} = 1 - (-j\,4.167)(j\,0.2) = 0.1666$$

The line to neutral voltages are

$$\begin{bmatrix} V_a \\ V_b \\ V_c \end{bmatrix} = \begin{bmatrix} 1 & 1 & 1 \\ 1 & a^2 & a \\ 1 & a & a^2 \end{bmatrix} \begin{bmatrix} 0.1666 \\ 0.1666 \\ 0.1666 \end{bmatrix} = \begin{bmatrix} 0.4998 \\ 0 \\ 0 \end{bmatrix}$$

Actual values of line to neutral voltages are

$$V_a = \frac{11}{\sqrt{3}} \times 0.4998 = 3.174\ \text{kV}$$

$$V_b = V_c = 0$$

Example 19.7.

A double line to ground fault occurs at bus 2 of the system in examples 18.4 and 19.2. Find the line currents under fault conditions. Assume that the system is on no load at the time of fault.

Solution

$$Z_1 = j\,0.157\,pu,\ Z_2 = j\,0.157\,pu,\ Z_0 = j\,0.051\,pu,\ Z_f = 0$$

The connection of sequence networks to represent a double line to ground fault is shown in Fig. 19.14.

$$I_{a1} = \frac{1\angle 0}{j\,0.157 + \dfrac{(j0.157)\,(j\,0.051)}{j\,0.157 + j\,0.051}} = -j\,5.115\,pu$$

$$I_{a2} = -(-j\,5.115)\left(\frac{j\,0.051}{j\,0.051 + j\,0.157}\right) = j\,1.254\,pu$$

$$I_{a0} = -(-j\,5.115)\left(\frac{j\,0.157}{j\,0.051 + j\,0.157}\right) = +j\,3.861\,pu$$

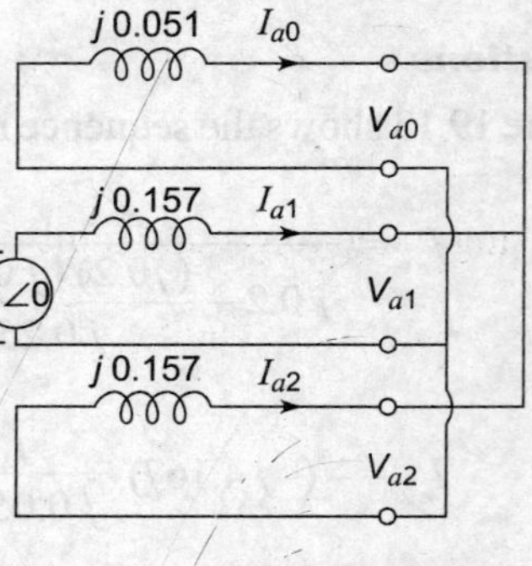

Fig. 19.14 Connection of sequence networks for example 19.7

The line currents are

$$\begin{bmatrix} I_a \\ I_b \\ I_c \end{bmatrix} = \begin{bmatrix} 1 & 1 & 1 \\ 1 & a^2 & a \\ 1 & a & a^2 \end{bmatrix} \begin{bmatrix} +j3.861 \\ -j5.115 \\ +j1.254 \end{bmatrix} = \begin{bmatrix} 0 \\ 8\angle 133.6° \\ 8\angle 46.4° \end{bmatrix}$$

Base current for 220 kV system = 262.4 A

The actual values of line currents are

$$I_a = 0$$
$$I_b = 8 \times 262.4 \angle 133.6° = 2099.2 \angle 133.6°$$
$$I_c = 8 \times 262.4 \angle 46.4° = 2099.2 \angle 46.4°\ A$$

Example 19.8.

A 50 MVA, 11 kV, 3-phase alternator was subjected to different types of faults. The fault currents were: 3-phase fault 1870 A, line to line fault 2590 A, single line to ground fault 4130 A

The alternator neutral is solidly grounded. Find the *pu* values of the three-sequence reactances of the alternator.

Solution

For 3-phase fault

$$\text{Fault current} = \frac{\text{Line to neutral voltage, i.e., } E}{X_1}$$

$$1870 = \frac{11000/\sqrt{3}}{X_1}$$

or $$X_1 = 3.396 \text{ ohms}$$

For line to line fault

$$\text{Fault current} = \frac{\sqrt{3}\,E}{X_1 + X_2}$$

$$2590 = \frac{\sqrt{3} \times 11000/\sqrt{3}}{X_1 + X_2}$$

or
$$X_1 + X_2 = \frac{11000}{2590} = 4.247 \text{ ohms}$$

$$X_2 = 4.247 - 3.396 = 0.851 \text{ ohms}$$

For single line to ground fault

$$\text{Fault current} = \frac{3\,E}{X_1 + X_2 + X_0} = \frac{\sqrt{3} \times 11000}{X_1 + X_2 + X_0}$$

$$4130 = \frac{\sqrt{3} \times 11000}{X_1 + X_2 + X_0}$$

or
$$X_1 + X_2 + X_0 = \frac{\sqrt{3} \times 11000}{4130} = 4.613 \text{ ohms}$$

$$X_0 = 4.613 - 4.247 = 0.366 \text{ ohms}$$

$$\text{Base impedance} = \frac{11 \times 11}{50} = 2.42 \text{ ohms}$$

$$X_1 = \frac{3.396}{2.42} = 1.4\ pu$$

$$X_2 = \frac{0.851}{2.42} = 0.35\ pu$$

$$X_0 = \frac{0.366}{2.42} = 0.15\ pu$$

19.6 CONSIDERATION OF PRE-FAULT LOAD CURRENT IN UNSYMMETRICAL FAULT ANALYSIS

As in the case of symmetrical faults, the principle of superposition can be used to include the effect of load current in unsymmetrical fault analysis. It is first necessary to calculate the pre-fault voltage at the fault point. Then the sequence networks are drawn and properly inter-connected depending on the type of fault. The sequence components of fault current and their distribution in different parts of the network are then calculated. The sequence components of fault current are then superimposed on the sequence components of load current. It must be remembered that the load current is completely of positive sequence character. Then the currents in different phases can be calculated. The phase shift introduced by *Y*–Δ transformer must be taken into account. Example 19.9 illustrates the procedure.

Example 19.9.

An 11 kV, 100 MVA, alternator with solidly grounded neutral has positive sequence and negative sequence reactances of 0.2 *pu* each and zero sequence reactance of 0.05 *pu*. It is supplying an 11 kV, 50 MVA motor with positive and negative sequence reactances of 0.2 *pu* and zero sequence reactance of 0.05 *pu* through a short line. The neutral of the motor is also solidly grounded. The transmission line has positive and negative sequence reactances of 0.05 *pu* and zero sequence reactance of 0.15 *pu* on 100 MVA base. The motor is drawing 40 MW, at 0.8 *pf* leading with a terminal voltage of 10.95 kV when a single line to ground fault occurs at the generator terminals of phase *a*. Calculate the total current in the phase *a* of generator and motor under fault conditions.

Solution

$$\text{Base MVA} = 100 \text{ MVA, Base voltage} = 11 \text{ kV}$$

$$\text{Base current} = \frac{100 \times 10^6}{11 \times \sqrt{3} \times 10^3} = 5284.8 \text{ A}$$

Taking the motor terminal voltage as the reference phasor

$$\text{motor terminal voltage} = \frac{10.95}{11} = 0.995 \angle 0 \; pu$$

$$\text{Load current} = \frac{40 \times 10^6}{\sqrt{3} \times 10.95 \times 0.8 \times 10^3} = 2636.4 \text{ A}$$

$$= \frac{2636.4}{5248.8} \angle \cos^{-1} 0.8 = 0.402 + j\,0.301 \; pu$$

Pre-fault voltage at alternator terminals (fault location) is

$$V = 0.995 \angle 0 + (0.402 + j\,0.301)\,(j\,0.05) = 0.9799 + j\,0.0201 \; pu$$

Positive (and negative) sequence reactance of motor on 100 MVA base = 0.4 pu

Zero sequence reactance of motor on 100 MVA base = 0.1 pu. Fig. 19.15 (a) shows the circuit. Fig. 19.15 (b) shows the interconnection of sequence networks.

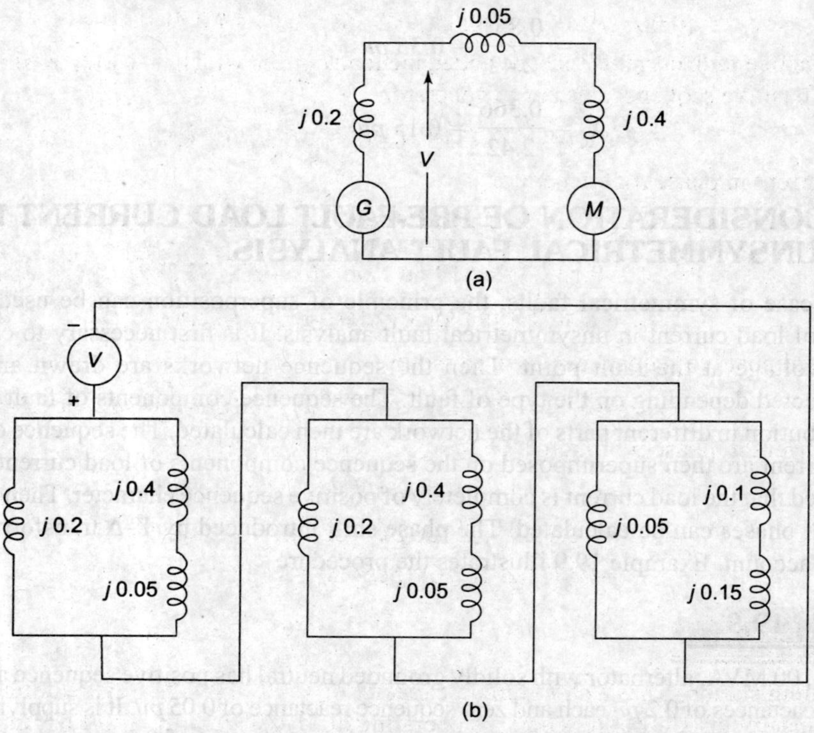

Fig. 19.15 (a) Circuit (b) Interconnection of sequence networks for the system of Example 19.9

The sequence components of fault current are

$$I_{a1} = I_{a2} = I_{a0} = \frac{0.9799 + j0.0201}{\dfrac{(j\,0.2)\,(j\,0.45)}{j\,0.2 + j\,0.45} + \dfrac{(j\,0.2)\,(j\,0.45)}{j\,0.2 + j\,0.45} + \dfrac{(j\,0.05)\,(j\,0.25)}{j\,0.05 + j\,0.25}}$$

$$= \frac{0.9799 + j0.0201}{j(0.1385 + 0.1385 + 0.0417)} = 0.06 - j\,3.075\ pu$$

Component of positive sequence fault current through generator

$$= (0.06 - j\,3.075)\left(\frac{j\,0.45}{j\,0.65}\right) = 0.04 - j\,2.13\ pu$$

Component of positive sequence fault current through motor

$$= (0.06 - j\,3.075) - (0.04 - j\,2.13) = 0.02 - j\,0.945\ pu$$

The negative sequence fault current components through the generator and motor are the same as the respective positive sequence components.

Component of zero sequence fault current through generator

$$= (0.06 - j\,3.075)\left(\frac{j\,0.25}{j\,0.3}\right) = 0.05 - j\,2.56\ pu$$

Component of zero sequence fault current through motor

$$= (0.06 - j\,3.075) - (0.05 - j\,2.56) = 0.01 - j\,0.515\ \text{pu}$$

To the above fault currents must be added the load current which is wholly positive sequence. The total positive sequence current in generator

$$= (0.04 - j\,2.13) + (0.402 + j\,0.301) = 0.442 - j\,1.829$$

The current in phase a of generator

$$= 0.442 - j\,1.829 + 0.04 - j\,2.13 + 0.05 - j\,2.56$$

$$= 0.532 - j\,6.519\ \text{pu} \quad \text{or} \quad 6.54\angle -85.3^\circ\ pu$$

or $$6.54 \times 5284.8 \text{ i.e., } 34562.6\angle -85.3^\circ\ \text{A}$$

The total positive sequence current in motor

$$= (0.02 - j\,0.945) - (0.402 + j\,0.301) = -\,0.382 - j\,1.246\ pu$$

The current in phase a of motor

$$= (-\,0.382 - j\,1.246) + (0.02 - j\,0.945) + (0.01 - j\,0.515)$$

$$= -\,0.352 - j\,2.706\ pu \quad \text{or} \quad 2.73\angle 262.6^\circ\ pu$$

or $$(2.73)\angle 262.6)\,(5284.8) = 14427.5\angle 262.6\ \text{A}$$

Example 19.10.

A generating station A has a short circuit capacity of 1000 MVA. Another station B has a short circuit capacity of 650 MVA. They are operating at 11 kV. Find the short circuit MVA if they are interconnected by a cable of 0.5 ohm reactance per phase.

Solution

$$\text{Let Base MVA} = 100 \text{ MVA}$$

$$\text{Base kV} = 11 \text{ kV}$$

$$pu \text{ reactance of } A = \frac{100}{1000} = 0.1\ pu$$

$$pu \text{ reactance of } B = \frac{100}{650} = 0.1538\ pu$$

$$\text{Base impedance} = \frac{11 \times 11}{100} = 1.21 \text{ ohm}$$

$$pu \text{ reactance of cable} = \frac{0.5}{1.21} = 0.413\ pu$$

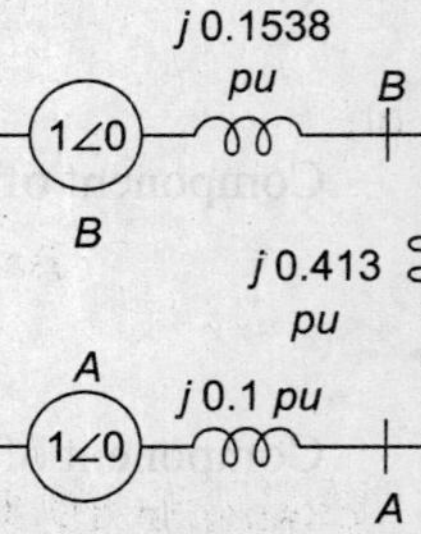

Fig. 19.16

Fig. 19.16 shows the reactance diagram. It is evident that a 3-phase fault at bus A will give maximum short circuit MVA

$$\text{Reactance as seen from bus } A = \frac{(0.1)(0.1538 + 0.413)}{0.1 + 0.1538 + 0.413} = 0.085\ pu$$

$$\text{Short circuit MVA} = \frac{100}{0.085} = 1176.47 \text{ MVA}$$

Example 19.11.

A station operating at 33 kV is divided into two sections A and B interconnected through a reactor X. Section A has three generators each 15 MVA having 15% reactance. Section B is fed from a grid through a 75 MVA transformer with 8% reactance. The circuit breakers have a rupturing capacity of 750 MVA. Find the reactance of reactor X to prevent the breakers being overloaded if a 3-phase fault occurs on outgoing feeder connected to A.

Solution

$$\text{Let Base MVA} = 75 \text{ MVA}$$

Reactance of each generator on 75 MVA base

$$= \frac{75}{15} \times 0.15 = .75\ pu$$

The circuit (on a common base) is shown in Fig. 19.17. The internal impedance of the grid is zero. The reactance diagram is shown in Fig. 19.18. Since fault level is 750 MVA, reactance as seen from fault point

$$= \frac{75}{750} = 0.1\ pu$$

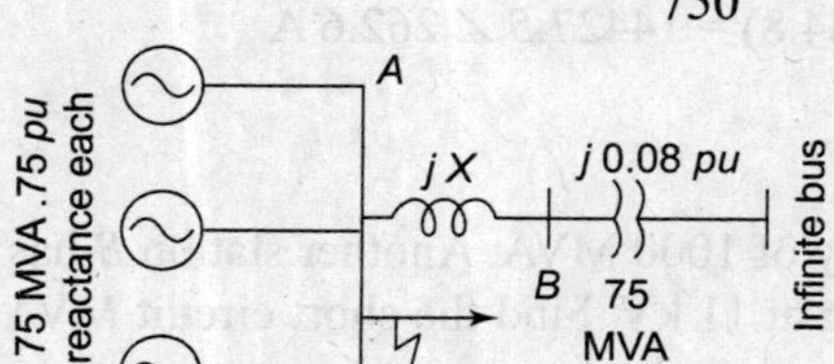

Fig. 19.17

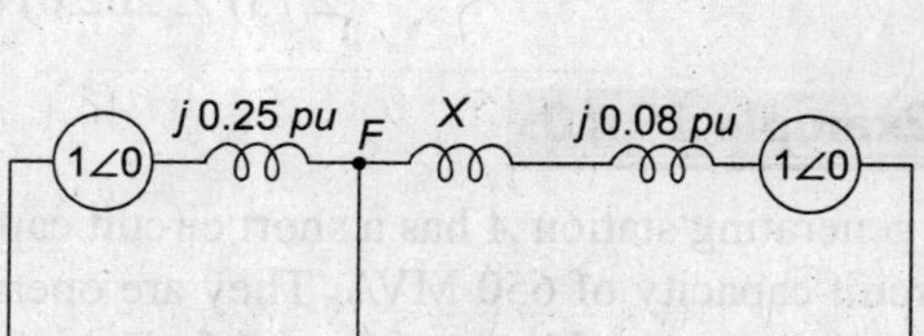

Fig. 19.18

From Fig. 19.18, reactance

$$= \frac{(0.25)\,(0.08 + X)}{0.25 + 0.08 + X}$$

Therefore $\dfrac{(0.25)\,(0.08 + X)}{0.25 + 0.08 + X} = 0.1$

or, $X = .0867\ pu$

$$\text{Base impedance} = \frac{(\text{kV}^2)}{\text{MVA}} = \frac{33 \times 33}{75} = 14.52 \text{ ohm}$$

$$X = 14.52 \times 0.0867 = 1.259 \text{ ohm}$$

Example 19.12.

Three 6.6 kV 3-phase 10 MVA alternators are connected to a common bus. Each alternator has a positive sequence reactance of 0.15 *pu*. The negative and zero sequence reactances are 75% and 30% of positive sequence reactance. A single line to ground fault occurs on the bus. Find the fault current if (a) all the alternator neutrals are solidly grounded (b) one alternator neutral is solidly grounded and the other two neutrals are isolated (c) one alternator neutral is grounded through 0.3 ohm resistance and the other two neutrals are isolated.

Solution

Base MVA = 10 MVA

Base voltage = 6.6 kV

$$\text{Base Current} = \frac{10 \times 10^6}{\sqrt{3} \times 6.6 \times 10^3} = 874.8 \text{ A}$$

$$\text{Base impedance} = \frac{(6.6)^2}{10} = 4.356 \text{ ohm}$$

$$X_1 = j\,0.15\ pu$$
$$X_2 = (j\,0.15)\,(0.75) = j\,.1125\ pu$$
$$X_0 = (j\,0.15)\,(0.3) = j\,.045\ pu$$

(a) The connections of sequence networks are as shown in Fig. 19.19.

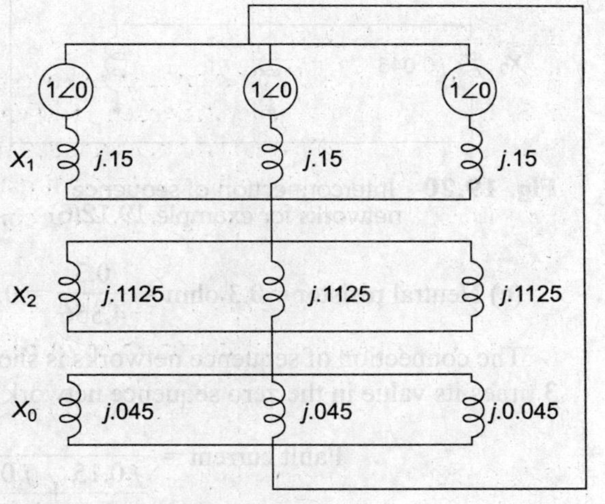

Fig. 19.19 Interconnection of sequence networks for example 19.12(a)

$$\text{Fault current} = \frac{3E}{X_1 + X_2 + X_0}$$

$$= \frac{3 \times 1\angle 0}{j\left(\dfrac{0.15}{3} + \dfrac{0.1125}{3} + \dfrac{0.045}{3}\right)}$$

$$= \frac{3\angle 0}{j\,0.1025} = 29.268\ pu$$

$$= 29.268 \times 874.8 = 25603.6\text{A}$$

(b) The connections of sequence networks are shown in Fig. 19.20.

$$\text{Fault current} = \frac{3E}{X_1 + X_2 + X_0} = \frac{3\angle 0}{j\left(\frac{0.15}{3} + \frac{0.1125}{3} + 0.045\right)}$$

$$= \frac{3\angle 0}{j\,0.1325} = 22.64\ pu$$

$$= 22.64 \times 874.8 = 19805.5\ \text{A}$$

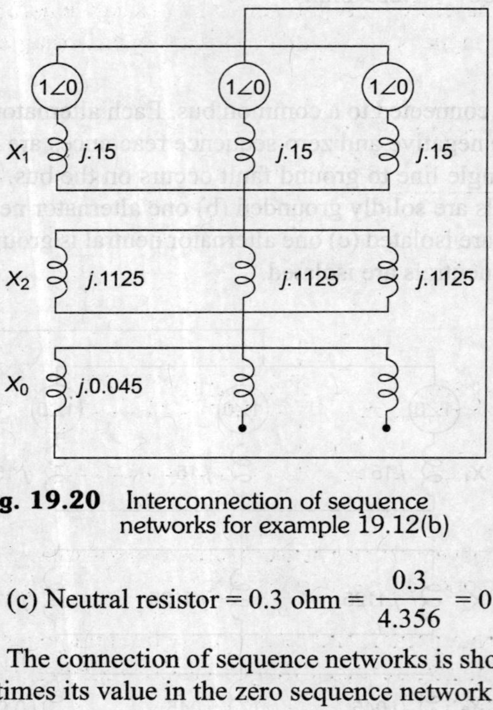

Fig. 19.20 Interconnection of sequence networks for example 19.12(b)

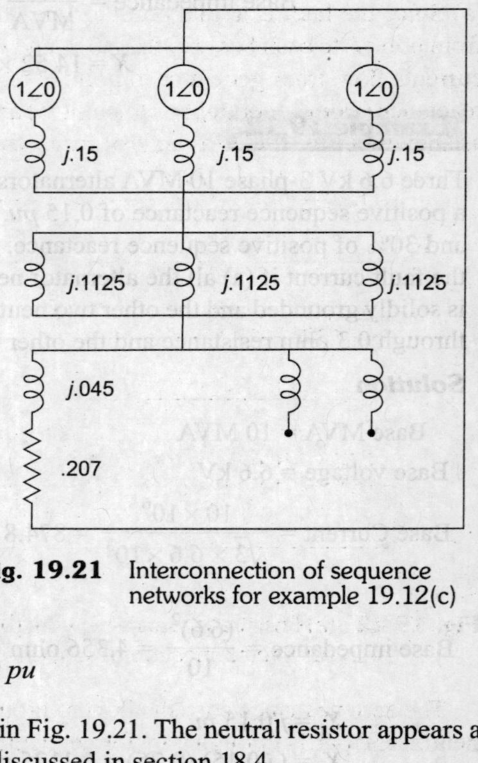

Fig. 19.21 Interconnection of sequence networks for example 19.12(c)

(c) Neutral resistor = 0.3 ohm $= \frac{0.3}{4.356} = 0.069\ pu$

The connection of sequence networks is shown in Fig. 19.21. The neutral resistor appears as 3 times its value in the zero sequence network as discussed in section 18.4.

$$\text{Fault current} = \frac{3\angle 0}{\frac{j\,0.15}{3} + \frac{j\,0.1125}{3} + 0.207 + j\,0.045}$$

$$= \frac{3\angle 0}{0.207 + j\,0.1325} = 12.206\ pu$$

$$= 12.206 \times 874.8 = 10677.8\ A$$

Example 19.13.

A 3-phase generator is connected to a star/delta transformer as shown in Fig. 19.22. The positive, negative and zero sequence impedances of alternator are $j\,0.1$, $j\,0.1$ and $j\,0.05\ pu$ respectively and those for the transformer are $j\,0.05\ pu$ each. The positive, negative and zero sequence reactances of the line are $j\,0.4$, $j\,0.4$ and $j\,0.8\ pu$. repectively. Determine the fault current and voltages at fault point when a single line to ground fault occurs at point P at mid point of line and the alternator neutral is (a) grounded (b) isolated.

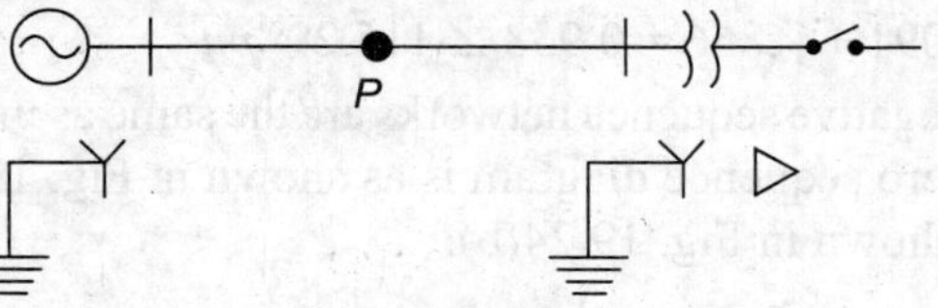

Fig. 19.22

Solution

(a) Since the fault is at mid point of line, the positive, negative and zero sequence impedances from point P to each bus are $j\,0.2$, $j\,0.2$, and $j\,0.4$ respectively. The positive and negative sequence currents flow from generator to point P only. Therefore, only the generator reactance and line reactance (from generator bus to point P) appear in the positive and negative sequence diagrams as shown in Fig. 19.23 (a) and (b) respectively.

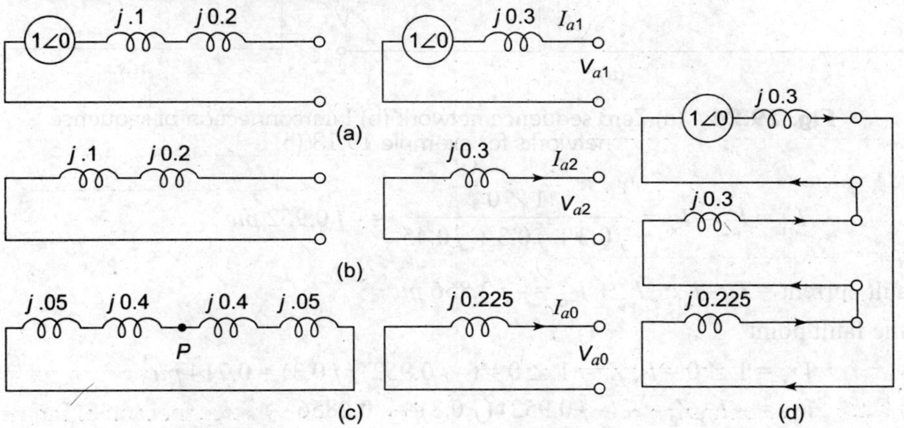

Fig. 19.23 (a) Positive sequence network (b) Negative sequence network (c) Zero sequence network (d) Interconnection of sequence networks for example 19.13 (a)

The zero sequence currents flow to ground from point P and return partly through generator neutral and partly through transformer neutral. The zero sequence diagram is shown in Fig. 19.23 (c). The connection of sequence networks is shown in Fig. 19.23 (d).

$$I_{a1} = I_{a2} = I_{a0} = \frac{1 \angle 0}{j0.3 + j0.3 + j0.225} = -j\,1.212\;pu$$

$$\text{Fault current} = I_a = I_{a1} + I_{a2} + I_{a0} = -j\,3.636\;pu$$

At the fault point

$$V_{a1} = 1 \angle 0 - I_{a1} Z_1 = 1 \angle 0 - (-j\,1.212)\,(j\,0.3) = 0.6363\;pu$$

$$V_{a2} = -I_{a2} Z_2 = (-j\,1.212)\,(j\,0.3) = -0.3636\;pu$$

$$V_{a0} = -I_{a0} Z_0 = (-j\,1.212)\,(j\,0.225) = -0.2727\;pu$$

$$\begin{bmatrix} V_a \\ V_b \\ V_c \end{bmatrix} = \begin{bmatrix} 1 & 1 & 1 \\ 1 & a^2 & a \\ 1 & a & a^2 \end{bmatrix} \begin{bmatrix} -0.2727 \\ 0.6363 \\ -0.3636 \end{bmatrix} = \begin{bmatrix} 0 \\ -0.4091 - j0.866 \\ -0.4091 + j0.866 \end{bmatrix}$$

$$V_a = 0$$
$$V_b = -.4091 - j\,.866 = 0.958 \angle 244.71^\circ \; pu$$
$$V_c = -.4091 + j\,.866 = 0.958 \angle 115.29^\circ \; pu$$

(b) The positive and negative sequence networks are the same as in part (a). Since the generator neutral is isolated, the zero sequence diagram is as shown in Fig. 19.24(a). The connections of sequence networks are shown in Fig. 19.24(b).

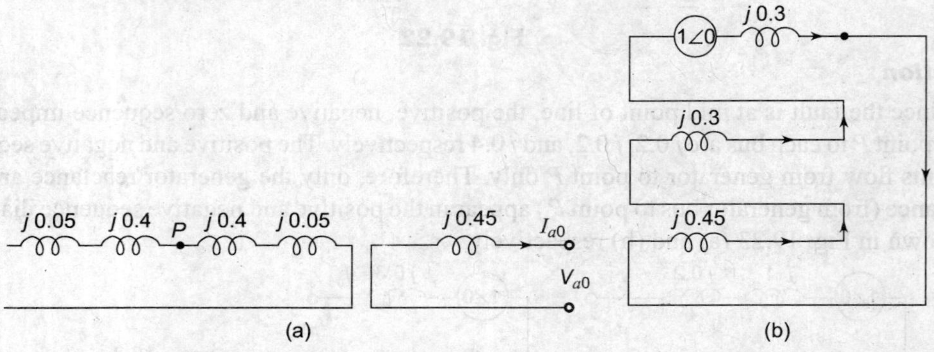

Fig. 19.24 (a) Zero sequence network (b) Interconnection of sequence networks for example 19.13 (b)

$$I_{a1} = I_{a2} = I_{a0} = \frac{1 \angle 0}{j0.3 + j0.3 + j0.45} = -j\,0.952 \; pu$$

Fault current $= I_a = I_{a1} + I_{a2} + I_{a0} = -j\,2.856 \; pu$

At the fault point

$$V_{a1} = 1 \angle 0 - I_{a1} Z_1 = 1 \angle 0 - (-j\,0.952)\,(j\,0.3) = 0.714 \; pu$$
$$V_{a2} = -I_{a2} Z_2 = -(-j\,0.952)\,(j\,0.3) = -0.2856$$
$$V_{a0} = -I_{a0} Z_0 = -(-j\,0.952)\,(j\,0.45) = -0.4284$$

$$\begin{bmatrix} V_a \\ V_b \\ V_c \end{bmatrix} = \begin{bmatrix} 1 & 1 & 1 \\ 1 & a^2 & a \\ 1 & a & a^2 \end{bmatrix} \begin{bmatrix} -0.4284 \\ 0.714 \\ -0.2856 \end{bmatrix} = \begin{bmatrix} 0 \\ -0.6426 - j0.8656 \\ -0.6426 + j0.8656 \end{bmatrix}$$

$$V_a = 0$$
$$V_b = -0.6426 - j\,0.8656 = 1.078 \angle 233.41^\circ \; pu$$
$$V_c = -0.6426 + j\,0.8656 = 1.078 \angle 126.59^\circ \; pu$$

Example 19.14.

A generator with grounded neutral has sequence impedances of Z_1, Z_2 and Z_0 and generated emf E. If a single line to ground fault occurs on terminals of phase a, find V_b and V_c. Assume $Z_f = 0$.

Solution

As derived in section 19.2

$$I_{a0} = I_{a1} = I_{a2} = \frac{E}{Z_1 + Z_2 + Z_0}$$

$$V_{a1} = E - I_{a1} Z_1$$

$$V_{a2} = -I_{a2} Z_2$$

$$V_{a0} = -I_{a0} Z_0$$

$$V_b = a^2 V_{a1} + a V_{a2} + V_{a0}$$

$$= a^2 (E - I_{a1} Z_1) + a(-Ia_2 Z_2) + (-I_{a0} Z_0)$$

$$= a^2 E - I_{a1} (a^2 Z_1 + a Z_2 + Z_0)$$

$$= a^2 E - \frac{E}{Z_1 + Z_2 + Z_0} (a^2 Z_1 + a Z_2 + Z_0)$$

$$V_b = E\left[\frac{(a^2 - a) Z_2 + (a^2 - 1) Z_0}{Z_1 + Z_2 + Z_0}\right]$$

$$V_c = a V_{a1} + a^2 V_{a2} + V_{a0}$$

$$= a (E - I_{a1} Z_1) + a^2 (-I_{a2} Z_2) + (-I_{a0} Z_0)$$

$$= a E - I_{a1} (a Z_1 + a^2 Z_2 + Z_0)$$

$$= a E - \frac{E}{Z_1 + Z_2 + Z_0} (a Z_1 + a^2 Z_2 + Z_0)$$

or
$$V_c = E\left[\frac{(a - a^2) Z_2 + (a - 1) Z_0}{Z_1 + Z_2 + Z_0}\right]$$

Example 19.15.

A 3-phase generator rated 15 MVA, 13.2 kV has a solidly grounded neutral. Its positive, negative and zero sequence reactances are 40%, 30% and 5% respectively (a) Find the value of reactance to be connected in the neutral circuit so that fault current for a single line to ground fault (of zero fault impedance) does not exceed rated line current (b) Find the value of resistance to be connected in the neutral circuit to serve the same purpose.

Solution

(a)
$$E = 1 \angle 0, I_f = 1$$

$$X_1 = 0.4, X_2 = 0.3, X_0 = 0.05$$

$$1 = I_f = \frac{3E}{j(X_1 + X_2 + X_0 + 3X_n)} = \frac{3}{j(0.4 + 0.3 + 0.05 + 3X_n)}$$

or
$$X_n = 0.75\ pu$$

$$\text{Base impedance} = \frac{13.2^2}{15} = 11.615\ \Omega$$

$$X_n = 0.75 \times 11.616 = 8.712\ \Omega$$

(b)
$$1 = \frac{3}{j(0.4 + 0.3 + 0.05) + 3R_n}$$

or $3R_n + j\,0.75 = 3$ or $R_n = 0.968\ pu$

or $R_n = 0.968 \times 11.616 = 11.24\ \Omega$

Example 19.16.

Draw zero sequence diagram of the power system shown in Fig. 19.25(a). All reactances are in *pu* to the same base.

Solution

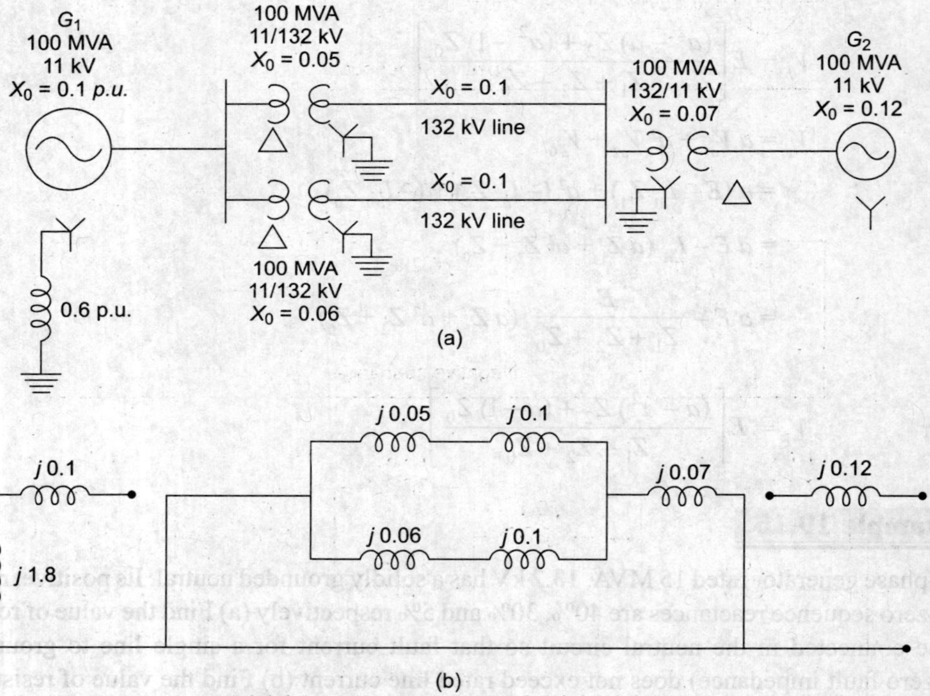

Fig. 19.25

The zero sequence diagram is shown in Fig. 19.25 (b). The neutral reactance of 0.6 *pu* in the neutral connection of G_1 appears as $j\,3 \times 0.6 = j\,1.8$ *pu*. The zero sequence reactance of G_2 is isolated because the neutral of G_2 is not grounded. The remaining diagram is self explanatory.

Example 19.17.

Draw positive, negative and zero sequence reactance diagrams for the system shown in Fig. 19.26 (a). The system data is as under

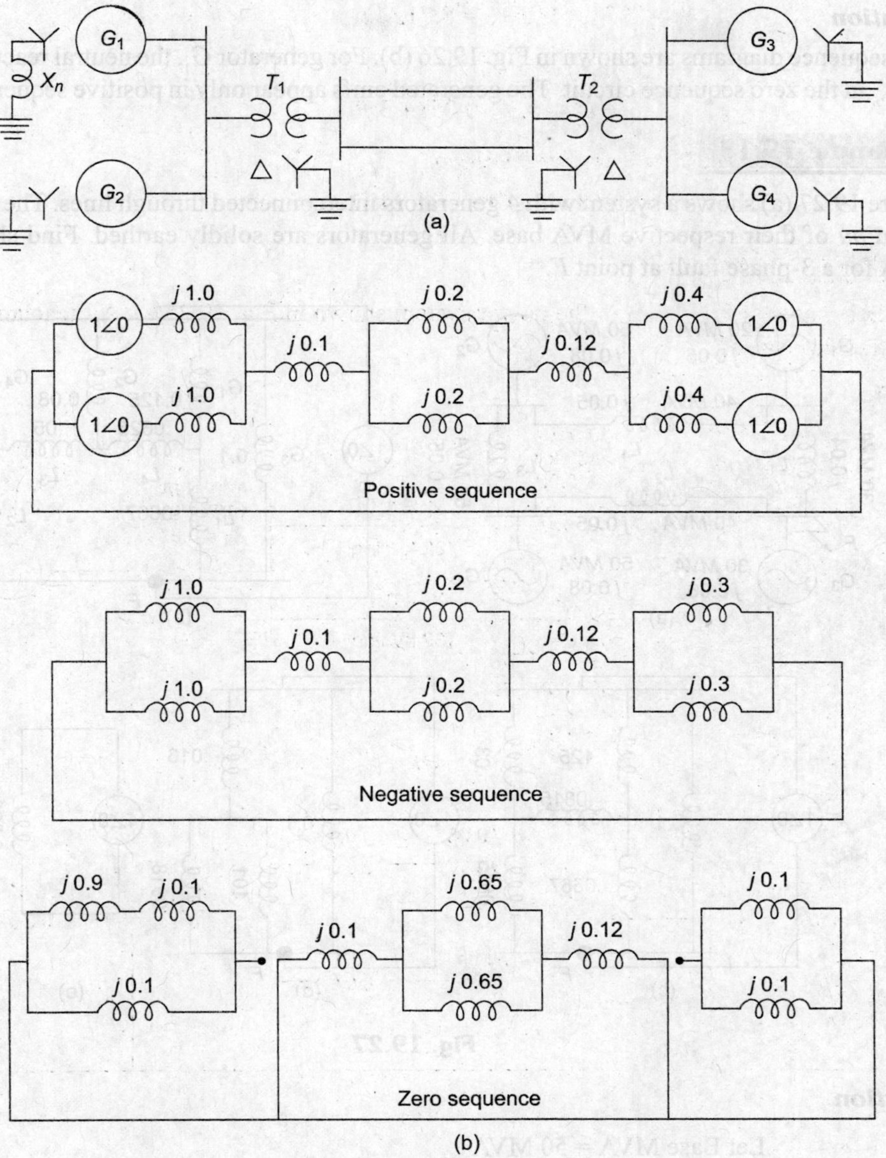

Fig. 19.26

Generator G_1: $X_1 = X_2 = 1.0$, $X_0 = 0.1$, $X_n = 0.3$ pu.

Transformer T_1: $X_1 = X_2 = X_0 = 0.1$

Lines (each): $X_1 = X_2 = 0.2$, $X_0 = 0.65$

Transformer T_2: $X_1 = X_2 = X_0 = 0.12$

Generator G_2; $X_1 = X_2 = 1.0$, $X_0 = 0.1$

Generator G_3, G_4; $X_1 = 0.4$, $X_2 = 0.3$, $X_0 = 0.1$

Solution

The sequence diagrams are shown in Fig. 19.26 (b). For generator G_1, the neutral reactor appears as $3X_n$ in the zero sequence circuit. The generated emfs appear only in positive sequence circuit.

Example 19.18.

Figure 19.27 (a) shows a system with 4 generators interconnected through lines. The reactances are in *pu* of their respective MVA base. All generators are solidly earthed. Find short-circuit MVA for a 3-phase fault at point *F*.

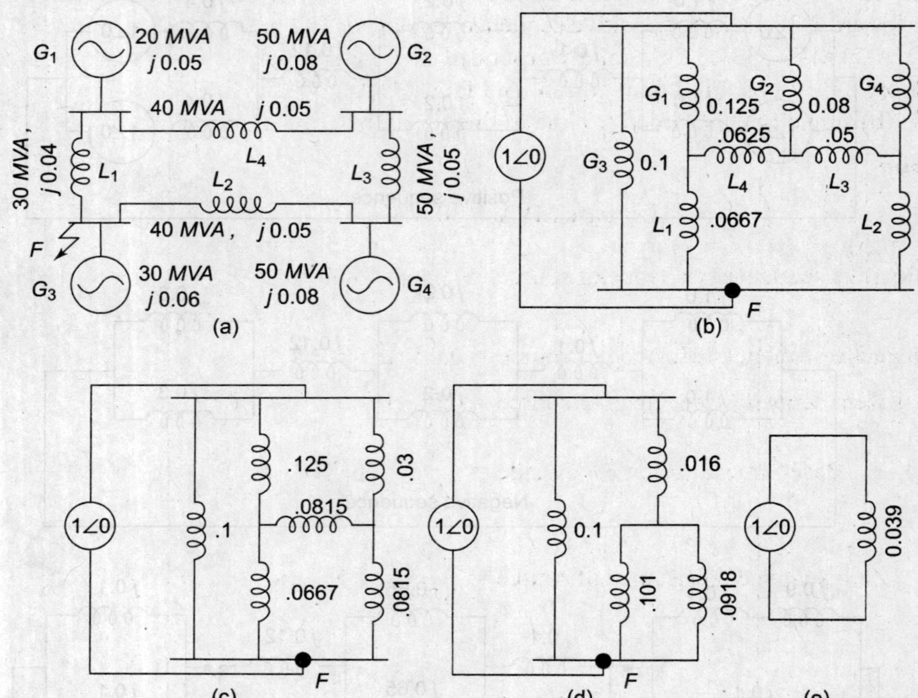

Fig. 19.27

Solution

Let Base MVA = 50 MVA

$$\text{Reactance of } G_1 = j\,0.05 \times \frac{50}{20} = j\,0.125\ pu$$

$$\text{Reactance of } G_3 = j\,0.06 \times \frac{50}{30} = j\,0.1\ pu$$

$$\text{Reactance of } L_1 = j\,0.04 \times \frac{50}{30} = j\,0.0667\ pu$$

$$\text{Reactance of } L_2, L_4 = j\,0.05 \times \frac{50}{40} = j\,0.0625\ pu$$

The reactance diagram is shown in Fig. 19.27 (b). The pu reactances are to a common base of 50 MVA (The prefix *j* for reactances has been omitted). It can be simplified to Fig. 19.27 (c) by using star-delta transformation. It can be further simplified to Fig. 19.27 (d) by using another star delta transformation and finally to Fig. 19.27 (e)

$$\text{Fault MVA} = \frac{50}{0.039} = 1282.05 \text{ MVA}$$

Example 19.19.

A 33 kV bus bar has a 3-phase fault level of 1000 MVA. The negative and zero sequence source reactances are 2/3 and 1/3 of positive sequence reactance. The zero sequence source resistance is 60 Ω. A 30 MVA, 33/132 kV solidly grounded Δ/Y transformer having a reactance of 0.1 *pu* is fed from 33 kV bus. Find fault current and fault MVA at 132 kV bus for following faults (a) 3-phase (b) single line to ground (c) line to line (d) double line to ground.

Solution

Let base MVA = 30 MVA

$$\text{Positive sequence reactance of source} = \frac{30}{1000} = 0.03\ pu$$

Negative sequence reactance of source = 0.02 *pu*

Zero sequence reactance of source = 0.01 *pu*

$$\text{Base impedance on 33 kV side} = \frac{33^2}{30} = 36.3\ \Omega$$

$$\text{Zero sequence resistance of source} = \frac{60}{36.3} = 1.653\ pu$$

$$\text{Base current for 132 kV side} = \frac{30 \times 10^6}{\sqrt{3} \times 132 \times 10^3}$$

$$= 131.22 \text{ A}$$

System is shown in Fig. 19.28(a). Fig. 19.28 (b) shows the sequence diagrams

$$X_1 = j\,0.13,\ X_2 = j\,0.12,\ X_0 = j\,0.1$$

(a) 3-Phase fault.

$$X_1 = j\,0.13\ pu$$

$$\text{Fault MVA} = \frac{30}{j\,0.13} = 230.77 \text{ MVA}$$

$$\text{Fault current} = \frac{1}{j\,0.13} = 7.692\ pu$$

$$= 7.692 \times 131.22 = 1009.34 \text{ A}$$

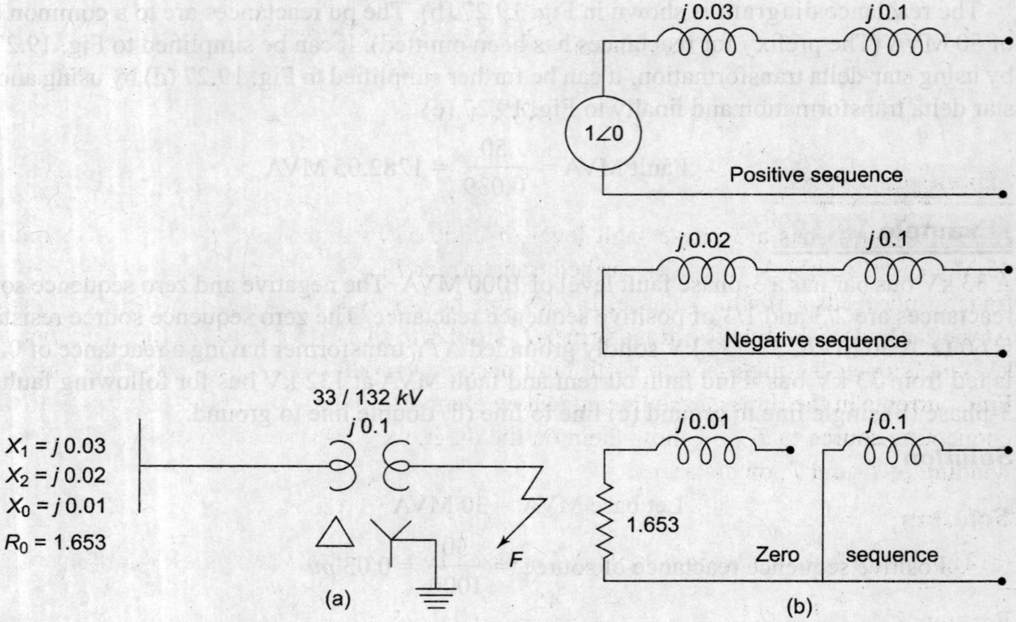

Fig. 19.28

(b) Single line to ground fault

$$I_f = \frac{3 \times 1 \angle 0}{j\,(0.13 + 0.12 + 0.1)} = 8.571\ pu$$

$$= 8.571 \times 131.22 = 1124.69 \text{ A}$$

$$\text{Fault level} = \frac{\sqrt{3} \times 132 \times 1124.69 \times 10^3}{10^6} = 257.13 \text{ MVA}$$

$$= 30 \times 8.571 = 257.13 \text{ MVA}$$

(c) *L-L* fault

$$I_f = \frac{\sqrt{3} \times 1 \angle 0}{j\,(0.13 + 0.12)} = 6.928\ pu$$

$$= 6.928 \times 131.22 = 909.1 \text{ A}$$

$$\text{Fault level} = \frac{\sqrt{3} \times 132 \times 10^3 \times 909.1}{10^6} = 207.84 \text{ MVA}$$

$$= 30 \times 6.928 = 207.84 \text{ MVA}$$

(d) Double line to ground fault

$$I_f = \frac{1 \angle 0}{j\left[0.13 + \dfrac{0.12 \times 0.1}{0.12 + 0.1}\right]} = 5.419\ pu$$

$$= 5.419 \times 131.22 = 711.08 \text{ A}$$

$$\text{Fault level} = \frac{\sqrt{3} \times 132 \times 10^3 \times 711.08}{10^6} = 162.57 \text{ MVA}$$

$$= 30 \times 5.419 = 162.57 \text{ MVA}$$

Example 19.20.

A 33 kV source has a 3-phase fault level of 2000 MVA and has $x_1 = x_2 = x_0$. It supplies a 45 MVA 33/132 kV, Δ/Y, solidly earthed transformer T_1 with a leakage reactance of 10%. The transformer feeds a 60 km, 3-phase, 132 kV line having $x_1 = x_2 = 0.4$ Ω/km and $x_0 = 1.0$ Ω/km. At the far end of the line is a transformer T_2 similar to that at the sending end. The remote 33 kV end is open. A single line to earth fault occurs at middle of line (a) Find fault current (b) Find currents in the three-phase line and show them on a diagram (c) Find currents in the lines connecting source to T_1 and show them on the diagram (d) Show the currents in the *lV* are *hV* winding of T_1 and T_2 on diagrams.

Solution

(a) The system is shown in Fig. 19.29 (a). Let base MVA = 45 MVA and base kV for line = 132 kV

For source $$x_1 = x_2 = x_0 = \frac{45}{2000} = j\,0.0225\ pu.$$

$$\text{Base impedance for line} = \frac{132^2}{45} = 387.2\ \Omega$$

For line

$$x_1 = x_2 = \frac{j\,0.4 \times 60}{387.2} = j\,0.062\ pu$$

$$x_0 = \frac{j\,60}{387.2} = j\,0.155\ pu$$

Since the fault is at the middle of line, the positive, negative and zero sequence reactances from fault point *P* to each transformer are 50% of total values. The positive and negative sequence currents flow from source to point *P*. Therefore the reactances of source, T_1 and line from T_1 to *P* appear in the positive and negative sequence circuits. The zero sequence currents flow from point *P* to ground and return through earthed star points of both transformers. In view of this the positive, negative and zero sequence circuits are shown in Fig. 19.29 (b). The interconnections of sequence networks is shown in Fig. 19.29 (c). From fault point *P* to earth

$$I_{a1} = I_{a2} = I_{a0} = \frac{1\angle 0}{j0.1535 + j0.1535 + j0.08875} = -j\,2.527\ pu$$

$$\text{Fault current} = I_a = I_{a1} + I_{a2} + I_{a0}$$

$$= -j\,7.581\ pu$$

$$\text{Base current} = \frac{45 \times 10^6}{\sqrt{3} \times 132 \times 10^3} = 196.83 \text{ A}$$

$$\text{Fault current} = 7.581 \times 196.83 = 1492.\ 17 \text{ A}$$

(b) From T_1 to P

$$I_{A1} = -j\,2.527 = 2.527 \angle -90^\circ\ pu$$

$$I_{A2} = 2.527 \angle -90^\circ\ pu$$

$$I_{A0} = (2.527 \angle -90) \left[\frac{j0.1775}{j0.1775 + j0.1775}\right]$$

$$= 1.2635 \angle -90\ pu$$

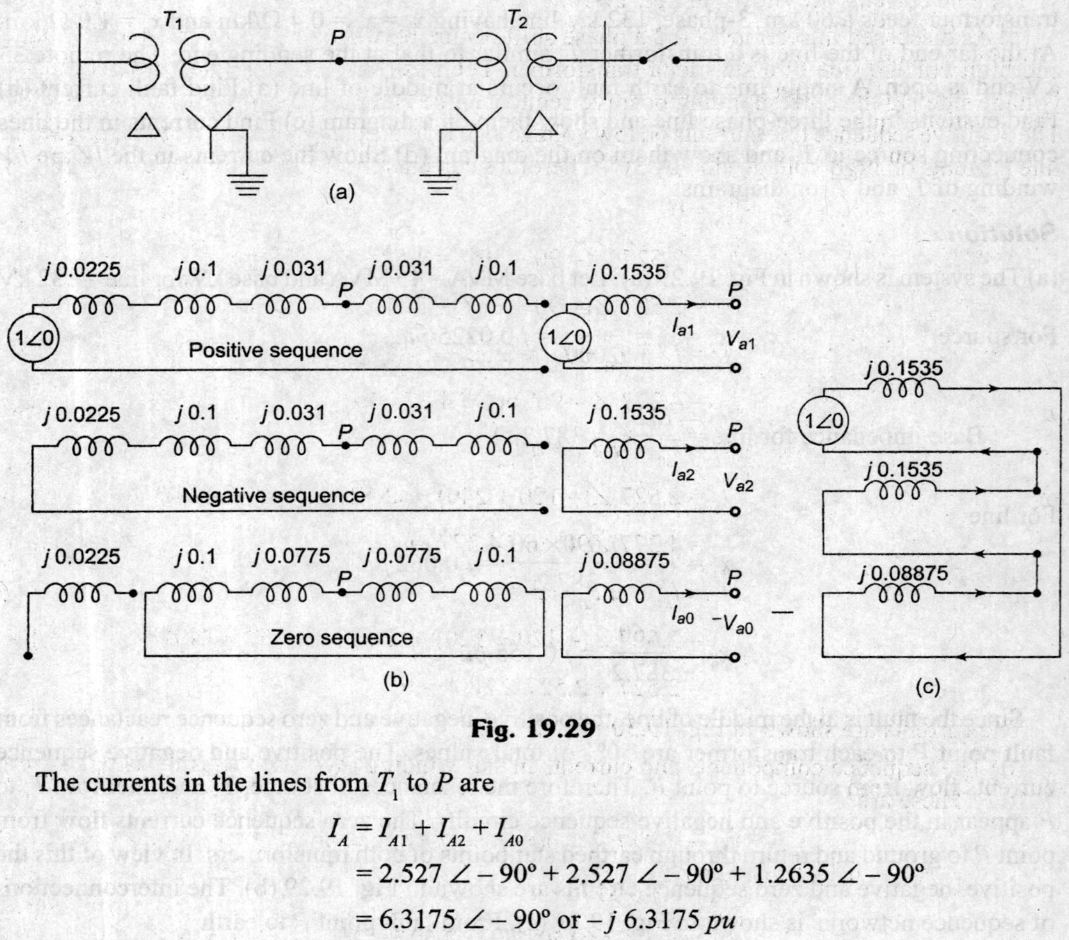

Fig. 19.29

The currents in the lines from T_1 to P are

$$I_A = I_{A1} + I_{A2} + I_{A0}$$

$$= 2.527 \angle -90^\circ + 2.527 \angle -90^\circ + 1.2635 \angle -90^\circ$$

$$= 6.3175 \angle -90^\circ \text{ or } -j\,6.3175\ pu$$

$$I_B = a^2 I_{A1} + a\,I_{A2} + I_{A0}$$

$$= 2.527 \angle(-90+240)^\circ + 2.527 \angle(-90+120)^\circ + 1.2635 \angle -90^\circ$$

$$= 2.527 \angle 150^\circ + 2.527 \angle 30^\circ + 1.2635 \angle -90^\circ$$

$$= 1.2635 \angle 90^\circ \text{ or } j\,1.2635\ pu$$

$$I_C = a\,I_{A1} + a^2 I_{A2} + I_{A0}$$

$$= 2.527 \angle(-90+120)^\circ + 2.527 \angle(-90+240)^\circ + 1.2635 \angle -90^\circ$$

$$= 1.2635 \angle 90^\circ \text{ or } j\,1.2635\ pu$$

From T_2 to P

$$I_{A1} = I_{A2} = 0$$

$$I_{A0} = 2.527 \angle -90^\circ \left[\frac{0.1775}{j0.1775 + j0.1775}\right]$$

$$= 1.2635 \angle -90^\circ \ pu$$

$$I_A = I_B = I_C = 1.2635 \angle -90^\circ \text{ or } -j\,1.2635 \ pu$$

The line currents are shown in Fig. 19.30.

(c) As discussed in section 18.6.2 a phase shift occurs between line currents on low voltage and high voltage side of a star-delta transformer. The positive sequence line currents on low voltage side lag the corresponding positive sequence line currents on high voltage side by 30°. The negative sequence line currents on low voltage side lead the corresponding negative sequence line currents on high voltage side by 30°. Therefore currents from source to T_1 are

$$I_{a0} = 0$$

$$I_{a1} = 2.527 \angle(-90 - 30)^\circ = 2.527 \angle -120^\circ \ pu$$

$$I_{a2} = 2.527 \angle(-90 + 30)^\circ = 2.527 \angle -60^\circ \ pu$$

$$I_a = I_{a0} + I_{a1} + I_{a2} = 2.527 \angle -120^\circ + 2.527 \angle -60^\circ$$

$$= 4.377 \angle -90^\circ \text{ or } -j\,4.377 \ pu$$

$$I_b = I_{a0} + a^2 I_{a1} + a\,I_{a2}$$

or

$$I_b = 2.527 \angle(-120 + 240)^\circ + 2.527 \angle(-60 + 120)^\circ$$

$$= 4.377 \angle 90^\circ \text{ or } j\,4.377 \ pu$$

$$I_c = I_{a0} + a\,I_{a1} + a^2 I_{a2}$$

$$= 2.527 \angle(-120 + 120)^\circ + 2.527 \angle(-60 + 240)^\circ$$

$$= 2.527 + 2.527 \angle 180^\circ = 0$$

The currents are shown in Fig. 19.30

(d) The sequence components and currents in star winding of T_1 are the same as those from T_1 to P. These are

$$I_{A1} = 2.527 \angle -90^\circ, I_{A2} = 2.527 \angle -90^\circ, I_{A0} = 1.2635 \angle -90^\circ$$

$$I_A = -j\,6.3175 \ pu$$

$$I_{B1} = 2.527 \angle 150^\circ, I_{B2} = 2.527 \angle 30^\circ, I_{B0} = 1.2635 \angle -90^\circ$$

$$I_B = j\,1.2635 \ pu$$

$$I_{C1} = 2.527 \angle 30^\circ, I_{C2} = 2.527 \angle 150^\circ, I_{C0} = 1.2635 \angle -90^\circ$$

$$I_c = j\,1.2635 \ pu$$

The sequence components and phase currents in delta windings of T_1 are $1/\sqrt{3}$ times their corresponding values in star winding. Thus,

$$I_{ab_1} = \frac{2.527}{\sqrt{3}} \angle -90^\circ = 1.459 \angle -90^\circ \text{ pu} = I_{ab_2}$$

$$I_{ab_0} = \frac{1.2635}{\sqrt{3}} \angle -90^\circ = 0.7295 \angle -90^\circ \, pu$$

$$I_{bc_1} = 1.459 \angle 150^\circ, I_{bc_2} = 1.459 \angle 30^\circ, I_{bc_0} = 0.7295 \angle -90^\circ \, pu$$

$$I_{ca_1} = 1.459 \angle 30^\circ, I_{ca_2} = 1.459 \angle 150^\circ, I_{ca_0} = 0.7295 \angle -90^\circ \, pu$$

$$I_{ab} = I_{ab_1} + I_{ab_2} + I_{ab_0} = -j\,3.6475 \, pu$$

$$I_{bc} = I_{bc_1} + I_{bc_2} + I_{bc_0} = j\,0.7295 \, pu$$

$$I_{ca} = I_{ca_1} + I_{ca_2} + I_{ca_0} = j0.7295 \, pu$$

These currents are shown in Fig. 19.30.

The sequence components and phase currents in star winding of T_2 are the same as those from T_2 to P. These are

$$I_{A_1} = I_{A_2} = I_{B_1} = I_{B_2} = I_{C_1} = I_{C_2} = 0$$

$$I_{A_0} = I_{B_0} = I_{C_0} = 1.2635 \angle -90^\circ \, pu$$

$$I_A = I_B = I_C = 1.2635 \angle -90^\circ \text{ or } -j\,1.2635 \, pu$$

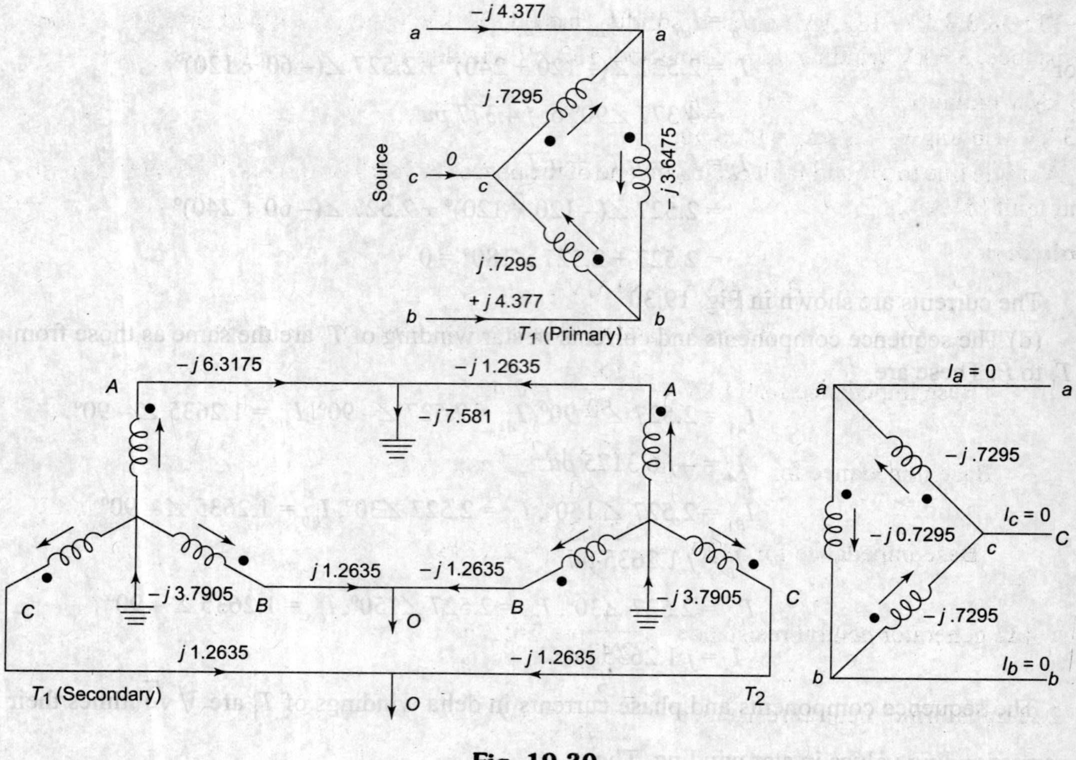

Fig. 19.30

The sequence components and phase currents in delta windings of T_2 are $1/\sqrt{3}$ times the corresponding values in star windings.

$$I_{ab_1} = I_{ab_2} = I_{bc_1} = I_{bc_2} = I_{ca_1} = I_{ca_2} = 0$$

$$I_{ab_0} = I_{bc_0} = I_{ca_0} = \frac{1.2635}{\sqrt{3}} \angle -90^\circ = 0.7295 \angle -90^\circ \; pu$$

$$I_{ab} = I_{bc} = I_{ca} = 0.7295 \angle -90^\circ \; pu$$

These currents (Fig. 19.30) circulate in the delta windings of T_2. The line currents I_a, I_b, I_c are zero. This is, evidently, due to the fact that the switch on 33 kV side of T_2 is open.

Example 19.21.

A 50 MVA generator, 11 kV, 3-phase, is feeding an 11/132 kV, 50 MVA transformer. The transformer feeds a 132 kV line at the far end of which is a 132/33/3.3 kV 3 winding transformer. The data is as under:

Generator: 50 MVA, 11 kV, $x_1 = 0.4 \; pu$, $x_2 = 0.3 \; pu$

$x_0 = 0.1 \; pu$. star connected, star point earthed through 4 Ω resistance.

Two winding transformer: 50 MVA, 11/132 kV,

Δ/Y, star point solidly earthed, $x_1 = x_2 = x_0 = 0.08 \; pu$.

Line : 132 kV, $x_1 = x_2 = 20 \; \Omega$, $x_0 = 50 \; \Omega$

Three winding transformer: 50 MVA,

132/33/3.3 kV, 132 kV winding solidly earthed, 33 kV winding earthed through 2 ohm resistance, 3.3 kV winding delta connected. 132 kV winding $x_1 = x_2 = x_0 = 0.05 \; pu$,

33 kV winding $x_1 = x_2 = x_0 = 0.04 \; pu$,

3.3 kV winding $x_1 = x_2 = x_0 = 0.06 \; pu$.

A single line to ground fault occurs on one of the phases at 33 kV terminals. Find fault current and fault MVA

Solution

$$\text{Base MVA} = 50 \text{ MVA}$$

$$\text{Base kV} = \text{rated kV for all the four voltage levels}$$

$$\text{Base impedance for 11 kV} = \frac{11 \times 11}{50} = 2.42 \; \Omega$$

$$\text{Base impedance for 132 kV} = \frac{132 \times 132}{50} = 348.48 \; \Omega$$

$$\text{Base impedance for 33 kV} = \frac{33 \times 33}{50} = 21.78 \; \Omega$$

$$4 \; \Omega \text{ generator neutral resistance} = \frac{4}{2.42} = 1.653 \; pu$$

$$2 \; \Omega \text{ transformer neutral resistance} = \frac{2}{21.78} = .0918 \; pu$$

The system is shown in Fig. 19.31 (a). The sequence networks are shown in Fig. 19.31 (b),

The zero sequence network is based on equivalent circuit of Fig. 18.7

Impedance of line

$$x_1 = x_2 = j\,20\ \Omega = \frac{j\,20}{348.48} = j\,0.0574\ pu$$

$$x_0 = j\,50 = \frac{j\,50}{348.48} = j.\,1435\ pu$$

The neutral resistances appear as three times their values in zero sequence circuit. All the three sequence circuits can be reduced to give

$$X_1 = j\,0.6274,\ X_2 = j\,.5274$$

$$Z_0 = .2754 + j.\,0892\ pu.$$

$$\text{Fault current} = \frac{3 \times 1\angle 0}{j0.6274 + j0.5274 + 0.2754 + j0.0892}$$

$$= \frac{3}{0.2754 + j1.244} = 2.355\ pu.$$

$$\text{Base current at 33 kV} = \frac{50 \times 10^6}{\sqrt{3} \times 33 \times 10^3} = 874.8\ \text{A}$$

$$\text{Fault current} = 2.355 \times 874.8 = 2060.15\ \text{A}$$

$$\text{Fault MVA} = 2.355\ pu = 2.355 \times 50\ \text{MVA} = 117.75\ \text{MVA}$$

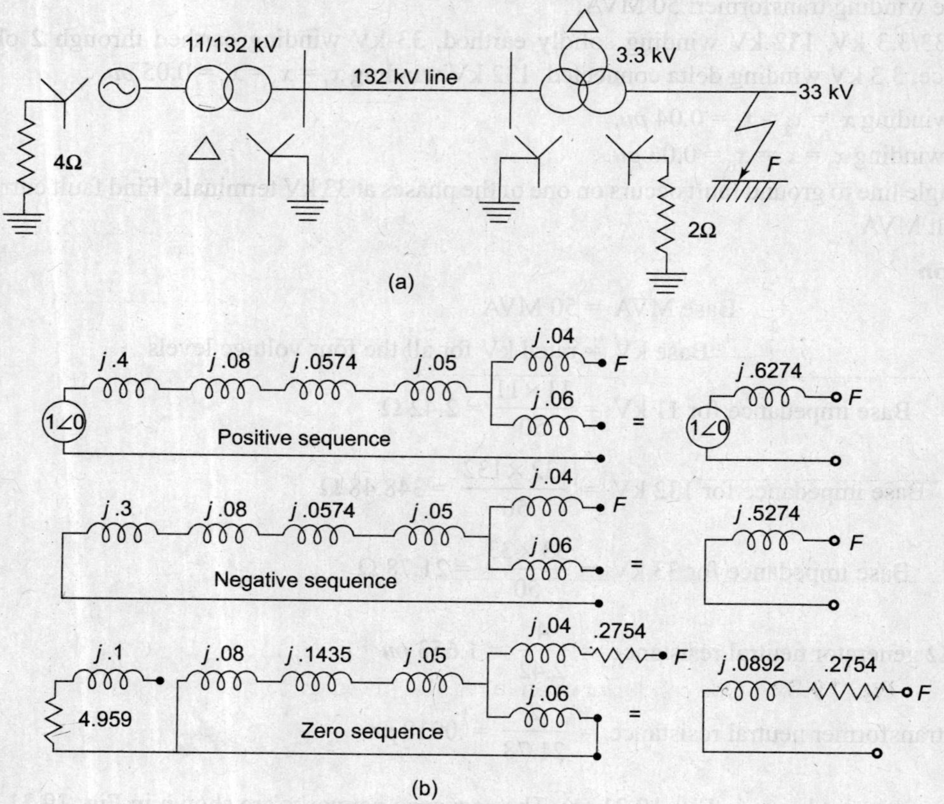

Fig. 19.31

19.7 OPEN CONDUCTOR FAULTS

A break in one or two conductors in a 3-phase circuit leads to an open conductor fault. When fuses are used in 3-phase circuits, the fuses in one or two phases may blow off due to over load thus leading to an open conductor fault. These unbalanced conditions can also be analyzed by the method of symmetrical components.

19.7.1 One Conductor Open

Figure 19.32 (a) shows a 3-phase line with conductor a broken leading to one conductor open conditions. The terminal conditions at the fault point give the following equations

$$I_a = 0 \tag{19.21}$$

$$v_a = V_a - V'_a \tag{19.22a}$$

$$v_b = V_b - V'_b = 0 \tag{19.22b}$$

$$v_c = V_c - V'_c = 0 \tag{19.22c}$$

The voltage drops v_a, v_b and v_c can be resolved into symmetrical components v_{a1}, v_{a2} and v_{a0}.

$$v_{a1} = \frac{1}{3}(v_a + v_b + v_c) = \frac{v_a}{3} \tag{19.23a}$$

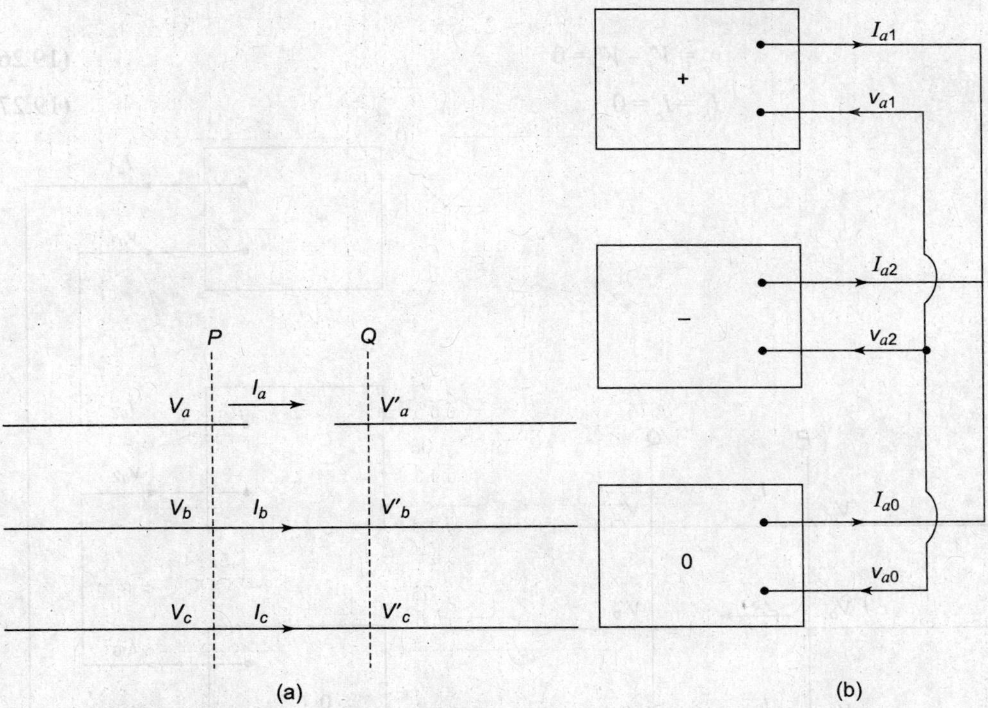

Fig. 19.32 One conductor open (a) circuit (b) sequence equivalent circuit

$$v_{a2} = \frac{1}{3}(v_a + v_b + v_c) = \frac{v_a}{3} \tag{19.23b}$$

$$v_{a0} = \frac{1}{3}(v_a + v_b + v_c) = \frac{v_a}{3} \tag{19.23c}$$

Thus

$$v_{a1} = v_{a2} = v_{a0} \tag{19.24}$$

From Eq. (19.21)

$$I_{a1} + I_{a2} + I_{a0} = 0$$

or

$$I_{a1} = -(I_{a2} + I_{a0}) \tag{19.25}$$

Equation (19.24) indicates that this open conductor condition leads to equal series voltage drops in all the three sequence networks in the direction of current flow from *P* to *Q*. In other words one open conductor condition leads to equal series voltage rises, in the three sequence networks, from *Q* to *P* tending to send currents from *Q* to *P*. These sequence current flow, due to these voltage rises, are opposed by series impedances of the equipments.

Equations (19.24) and (19.25) lead to the connection of sequence networks as shown in Fig. 19.32 (b). The positive, negative and zero sequence circuits shown in Fig. 19.32 (b) are as seen from points *PQ* of Fig. 19.32 (a).

19.7.2 Two Conductors Open

Figure 19.33(a) shows the 3-phase line with break in conductors *b* and *c*. The terminal conditions are

$$v_a = V_a - V'_a = 0 \tag{19.26}$$

$$I_b = I_c = 0 \tag{19.27}$$

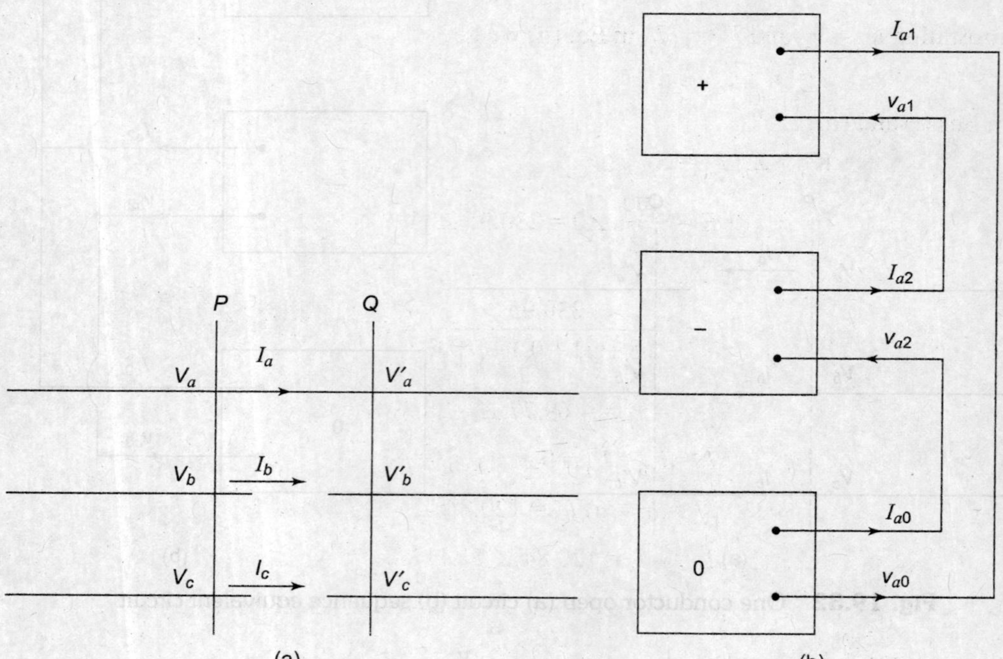

Fig. 19.33 Two conductors open (a) circuit (b) sequence equivalent circuit

From Eq. (19.26)

$$v_{a1} + v_{a2} + v_{a0} = 0 \quad (19.28)$$

From Eq. (19.27)

$$I_{a1} = I_{a2} = I_{a0} = \frac{I_a}{3} \quad (19.29)$$

Equations (19.28) and (19.29) lead to the connection of sequence networks shown in Fig. 19.33 (b)

Example 19.22.

A 3-phase induction motor has $Z_1 = 2.8 + j\,1\ \Omega$ and $Z_2 = 0.1 + j\,0.6\ \Omega$. It is fed by a 400 V 3-phase balanced supply. When the motor is running the fuse in phase A flows causing an open circuit in phase A. Assuming that motor continues to run find the line currents and voltages $V_{a'a}$, V_{an}, V_{bn}, V_{cn} and V_{nn} (Fig. 19.34a). Neglect source impedance.

Solution

The unsymmetrical line section can be replaced by voltages $V_{a'a}$, $V_{b'b}$, $V_{c'c}$, which can be resolved into symmetrical components v_{a1}, v_{a2}, v_{a0}, Fig. 19.34(b) shows the connection of sequence networks. The zero sequence network is not connected because the neutral wire is absent. As derived in section 19.7.1, $v_{a1} = v_{a2} = v_{a0}$ and $I_{a2} = -I_{a1}$ KVL equations for positive and negative sequence networks are

$$Z_1 I_{a_1} + v_{a_1} = E \quad \text{(i)}$$

$$Z_2 I_{a_2} + v_{a_2} = 0 \quad \text{(ii)}$$

Substituting $v_{a_2} = v_{a_1}$ and $I_{a_2} = -I_{a_1}$ in Eq. (ii) we get

$$-Z_2 I_{a_1} + v_{a_1} = 0 \quad \text{(iii)}$$

From Eqs (i) and (iii)

$$(Z_1 + Z_2)\, I_{a1} = E$$

$$E = \frac{400}{\sqrt{3}} \angle 0 = 230.95 \angle^{\circ}\ \text{V}$$

$$I_{a1} = \frac{230.95}{2.8 + j1 + 0.1 + j0.6} = 69.77 \angle -28.89^{\circ}\ \text{A}$$

$$I_{a2} = -I_{a1} = -69.77 \angle -28.89^{\circ}\ \text{A}$$

$$I_a = I_{a1} + I_{a2} = 0$$

$$I_b = (a^2 - a)\, I_{a1} = 120.84 \angle -118.89^{\circ}\ \text{A}$$

$$I_c = -I_b = 120.84 \angle 61.11^{\circ}\ \text{A}$$

$$v_{a2} = -Z_2 I_{a2} = -(.1 + j.6)(-69.77 \angle -28.89^{\circ})$$

$$= 42.42 \angle 51.65^\circ \text{ V}$$

$$V_{a'a} = v_{a1} + v_{a2} + v_{a0} = 3\, v_{a2}$$

$$= 127.26 \angle 51.65^\circ \text{ V}$$

$$V_{an} = Z_1 I_{a1} + Z_2 I_{a2} = (Z_1 - Z_2)\, I_{a1}$$

$$= (2.7 + j.4)\,(69.77 \angle{-28.89^\circ})$$

$$= 190.44 \angle{-28.46^\circ} \text{ V}$$

$$V_{bn} = a^2 Z_1 I_{a1} + a\, Z_2 I_{a2} = (a^2 Z_1 - a\, Z_2)\, I_{a1}$$

$$= 189.1 \angle -118.13^\circ \text{ V}$$

$$V_{cn} = a\, Z_1 I_{a1} + a^2 Z_2 I_{a2} = (a\, Z_1 - a^2 Z_2)\, I_{a1}$$

$$= 218.38 \angle 121.84^\circ \text{ V}$$

$$V_{Nn} = v_{a0} = v_{a2} = 42.42 \angle 51.65^\circ \text{ V}$$

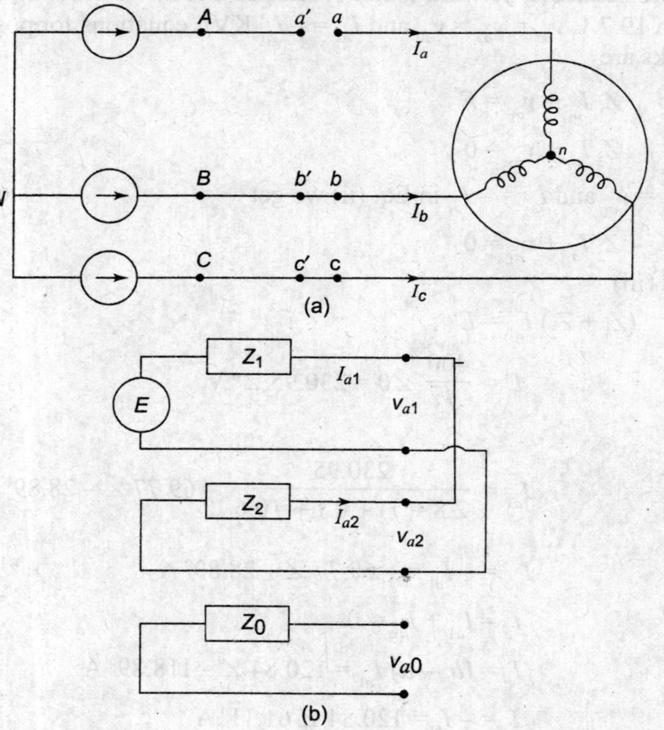

Fig. 19.34

Example 19.23.

A generating station has three generators each of 10 MVA capacity connected through reactors of 8% to a common bus bar as shown in Fig. 19.35. If a fault develops on one generator find fault MVA and compare with the case when no reactors are used. (AMIE 2001)

Solution

Fig. 19.35 shows the circuit.

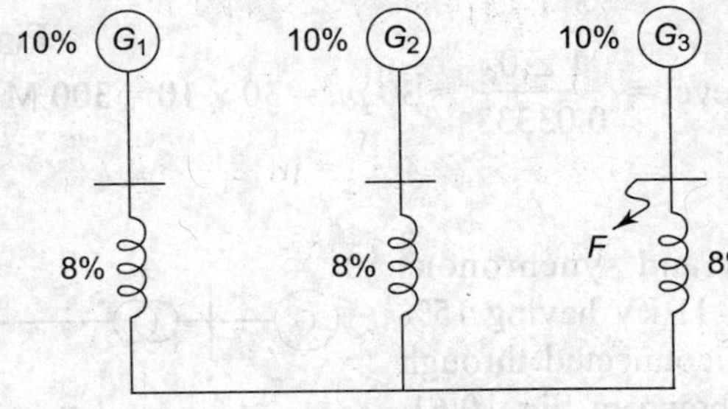

Fig. 19.35

Fig. 19.36 (a) shows the equivalent circuit.

The equivalent circuit can be reduced to that shown in Fig. 19.36 (b) and then to that shown in Fig. 19.36 (c).

$$X = \frac{(0.09 + 0.08)\,(0.10)}{0.09 + 0.08 + 0.10} = \frac{(0.17)\,(0.1)}{0.27} = 0.06296$$

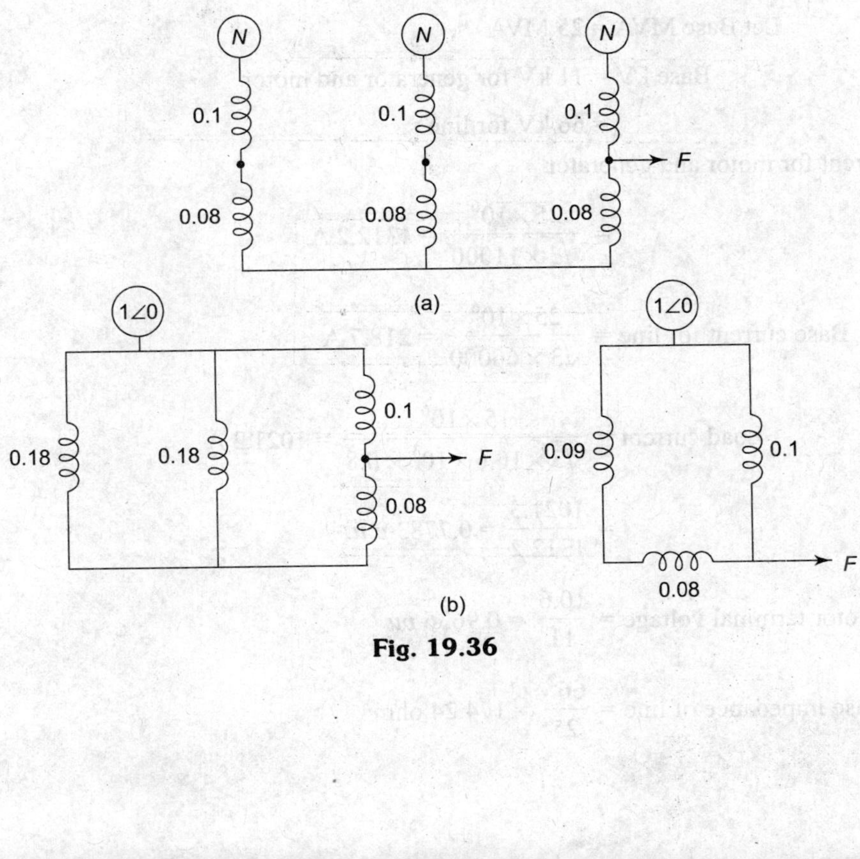

Fig. 19.36

$$\text{Fault level} = \frac{1\angle 0}{0.06296} = 15.883\ pu$$

$$= 10 \times 15.883 = 158.83 \text{ MVA}$$

When reactors are not used, the circuit is shown in Fig. 19.37.

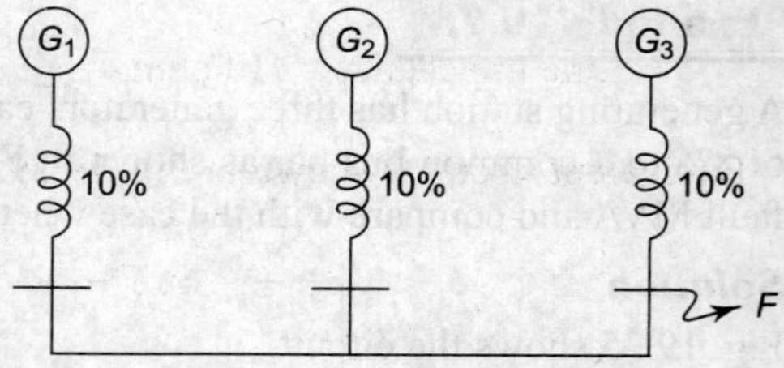

Fig. 19.37

$$X = \frac{0.1}{3} = 0.03333$$

$$\text{Fault level} = \frac{1\angle 0}{0.03333} = 30\ pu = 30 \times 10 = 300 \text{ MVA}$$

Example 19.24.

A synchronous generator and synchronous motor each rated 25 MVA, 11 kV having 15% subtransient reactance are connected through transformers and line as shown in Fig. 10.61. Transformers are rated at 25 MVA, 66 kV. The motor is drawing 15 MW at 0.8 power factor leading and at a terminal voltage of 10.6 kV when a symmetrical fault occurs at motor terminals. Find subtransient current in generator, motor and fault. Assume line impedance equal to $0 + j10$ ohm and transformer reactance equal to $j0.10\ pu$.

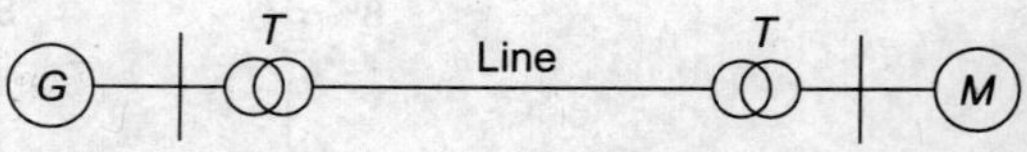

Fig. 19.38

(AMIE 2000)

Solution

Let Base MVA = 25 MVA

Base kV = 11 kV for generator and motor

= 66 kV for line

Base current for motor and generator

$$= \frac{25 \times 10^6}{\sqrt{3} \times 11000} = 1312.2 \text{ A}$$

$$\text{Base current for line} = \frac{25 \times 10^6}{\sqrt{3} \times 66000} = 218.7 \text{ A}$$

$$\text{Load current} = \frac{15 \times 10^6}{\sqrt{3} \times 10.6 \times 10^3 \times 0.8} = 1021.3 \text{ A}$$

$$= \frac{1021.3}{1312.2} = 0.7783\ p.u.$$

$$\text{Motor terminal voltage} = \frac{10.6}{11} = 0.9636\ pu$$

$$\text{Base impedance of line} = \frac{66^2}{25} = 174.24 \text{ ohm}$$

$$\text{Line impedance} = j10 \text{ ohm} = \frac{j10}{174.24} = 0.0574\, pu$$

Equivalent circuit is shown in Fig. 19.39.

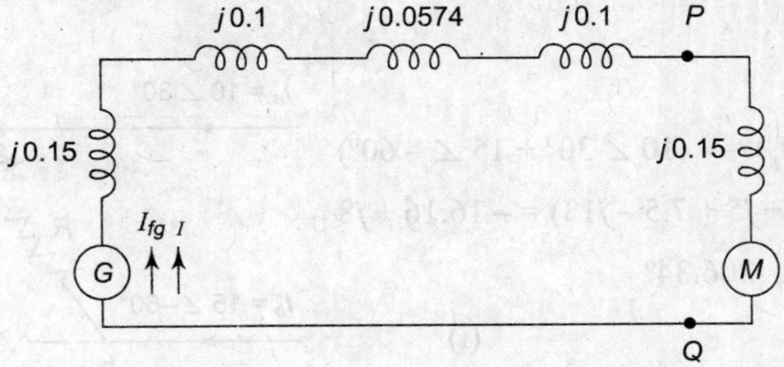

Fig. 19.39

Thevenin's reactance as seen from fault terminals PQ

$$= \frac{(j0.15)\,(j0.15 + j0.1 + j0.0574 + j0.1)}{j0.15 + j0.15 + j0.1 + j0.0574 + j0.1}$$

$$= \frac{(j0.15)\,(j0.4074)}{j0.5574} = j0.1096 \text{ pu}$$

$$\text{Total fault current} = \frac{0.9636}{j0.1096}$$

$$= -j8.792 \text{ pu}$$

$$= (8.792)\,(218.7) = 1922.8 \text{ A}$$

Fault current supplied by generator

$$I_{fg} = (-j8.792)\left(\frac{j0.15}{j0.5574}\right) = -j2.366\, pu$$

$$I_{fm} = \text{Fault current supplied by motor} = (-j8.792) - (-j2.366)$$

$$= -j6.426\, pu$$

$$I = \text{Load current supplied by generator}$$

$$= 0.7783\,(0.8 + j0.6) = 0.623 + j0.467\, pu$$

Total post fault generator current

$$= I_{fq} + I = -j2.366 + 0.623 + j0.467$$

$$= 0.623 - j1.899\, pu = 1.999\, \angle 71.84^\circ\, pu$$

$$= (1312.2)\,(1.999) = 2623 \text{ A}$$

Total post fault motor current

$$= I_{fm} - I = -j6.426 - (0.623 + j.467)$$

$$= -0.623 - j5.96 = 5.99\, \angle -264^\circ$$

$$= (5.99)\,(1312.2) = 7860 \text{ A}$$

Example 19.25.

A balanced delta connected load is fed from unbalanced 3 phase supply as shown in Fig. 19.40. Find (a) symmetrical components of line currents (b) symmetrical components of delta currents.

Solution

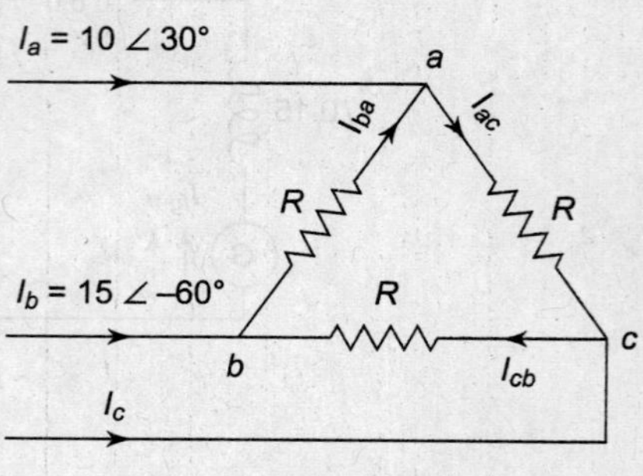

Fig. 19.40

$$I_a + I_b + I_c = 0$$

or $$I_c = -(I_a + I_b) = -(10 \angle 30° + 15 \angle -60°)$$

or $$I_c = -(8.66 + j5 + 7.5 - j13) = -16.16 + j8$$

$$= 18.03 \angle -206.34°$$

$$I_a = I_{ac} - I_{ba} \quad \text{(i)}$$

$$I_b = I_{ba} - I_{cb} \quad \text{(ii)}$$

$$I_c = I_{cb} - I_{ac} \quad \text{(iii)}$$

Applying KVL to closed delta

$$I_{ac} R + I_{cb} R + I_{ba} R = 0$$

or $$I_{ac} + I_{cb} + I_{ba} = 0$$

or $$I_{cb} = -I_{ac} - I_{ba} \quad \text{(iv)}$$

From Eqs (i) and (ii)

$$I_a + I_b = I_{ac} - I_{cb} \quad \text{(v)}$$

From Eqs (iv) and (v)

$$I_a + I_b = I_{ac} + I_{ac} + I_{ba} \quad \text{(vi)}$$

Adding Eqs (i) and (vi)

$$2I_a + I_b = 3I_{ac}$$

or $$I_{ac} = \frac{1}{3}(2I_a + I_b) = \frac{1}{3}(17.32 + j10 + 7.5 - j13)$$

or $$I_{ac} = \frac{1}{3}(25.82 - j3) = 8.27 - j1 = 8.33 \angle -6.89° - 7.9 + j7$$

$$I_{cb} = I_c + I_{ac} = -16.16 + j8 + 8.27 - j1 - 7.9 + j7$$

$$= 10.555 \angle 138.5°$$

$$I_{ba} = I_{ac} - I_a = 8.27 - j1 - 8.66 - j5$$

$$= -0.39 - j6 = 6.01 \angle -93.72°$$

(a) $$I_{a_0} = \frac{1}{3}(I_a + I_b + I_c) = 0$$

$$I_{a_1} = \frac{1}{3}(I_a + aI_b + a^2 I_c)$$

$$= \frac{1}{3}\ [10\ \angle 30^\circ + (1\ \angle 120^\circ)\ (15\ \angle -60^\circ) + (1\ \angle -120^\circ)\ (18.03\ \angle -206.34^\circ)]$$

$$= \frac{1}{3}\ [10\ \angle 30^\circ + 15\ \angle 60^\circ + 18.03\ \angle 33.66^\circ]$$

$$= \frac{1}{3}\ [8.66 + j5 + 7.5 + j13 + 15 + j10) = \frac{1}{3}\ (31.16 + j28)$$

$$= 10.39 + j9.33$$

$$I_{a_2} = \frac{1}{3}\ (I_a + a^2 I_b + aI_c)$$

$$= \frac{1}{3}\ [10\ \angle 30^\circ + (1\ \angle -120^\circ)\ (15\ \angle -60^\circ) + (1\ \angle 120^\circ)\ (18.03\ \angle -206.34^\circ)]$$

$$= \frac{1}{3}\ [10\ (30^\circ + 15\ \angle -180^\circ + 18.03\ \angle -86.34^\circ]$$

$$= \frac{1}{3}\ (8.66 + j5 - 15 + 1.15 - j18) = \frac{1}{3}\ (-5.19 - j13)$$

$$= -1.73 - j4.33$$

[Check $\quad I_a = I_{a_0} + I_{a_1} + I_{a_2}$]

(b) $$I_{ac_0} = \frac{1}{3}\ (I_{ac} + I_{cb} + I_{ba})$$

$$= \frac{1}{3}\ (8.27 - j1 - 7.9 + j7 - 0.39 - j6) = 0$$

$$I_{ac_1} = \frac{1}{3}\ [I_{ac} + a\, I_{cb} + a^2\, I_{ba}]$$

$$= \frac{1}{3}\ [8.33\ \angle -6.89^\circ + (1\ \angle 120^\circ)\ (10.555\ \angle 138.5) + (1\ \angle -120^\circ)\ (6.01\ \angle -93.72^\circ)$$

$$= \frac{1}{3}\ [8.33\ \angle -6.89^\circ + 10.555\ \angle 258.5^\circ + 6.01\ \angle -213.72^\circ]$$

$$= \frac{1}{3}\ [8.27 - j1 - 2.1 - j10.34 - 5 + j3.336]$$

$$= \frac{1}{3}\ (1.17 - j8) = 0.39 - j2.67$$

$$I_{ac_2} = \frac{1}{3}\ [I_{ac} + a^2 I_{cb} + a\, I_{ba}]$$

$$= \frac{1}{3}\ [8.33\ \angle -6.89^\circ + (1\ \angle -120^\circ)\ (10.555\ \angle 138.5^\circ)$$

$$+ (1\ \angle 120^\circ)\ (6.01\ \angle -93.72^\circ)]$$

$$= \frac{1}{3}\ [8.33\ \angle -6.89^\circ + 10.555\ \angle 18.5^\circ + 6.01\ \angle 26.28^\circ]$$

$$= \frac{1}{3}[8.27 - j1 + 10 + j3.35 + 5.39 + j2.66]$$

$$= \frac{1}{3}(23.66 + j5.01) = 7.89 + j1.67$$

[Check $I_{ac} = I_{ac_0} + I_{ac_1} + I_{ac_2}$]

SUMMARY

19.1 Faults can be balanced (i.e. those involving all the three phases) or unbalanced (i.e. those involving only one or two phases)

19.2 The three sequence networks do not have any mutual coupling.

19.3 Positive and negative sequence impedances of static equipments (i.e. transformers, lines etc.) are equal.

19.4 Zero sequence network is affected by the method of neutral grounding.

19.5 Phase shift between sequence voltages (and currents) occurs as we go from one side to the other side of Δ-Y or Y-Δ transformer.

19.6 Single line to ground fault is the most common fault.

19.7 The effect of prefault load current can be taken into account by the principle of superposition.

19.8 Open conductor faults can also be analysed by the method of symmetrical components.

SELF TEST QUESTIONS

A. State whether the following are true (T) or false (F).

19.1 In fault calculations open circuit emf at fault point is generally assumed to be 1 pu.

19.2 Shunt capacitances of lines are neglected in fault calculations.

19.3 In a motor the direction of fault current and load current are the same.

19.4 $1 - a^2 = 1.5 + j0.866$

19.5 $V_c = a V_{a1} + a^2 V_{a2} + V_{a0}$

19.6 $S = V_{a0} I^*_{a0} + V_{a1} I^*_{a1} + V_{a2} I^*_{a2}$

19.7 Only zero sequence current flows through ground.

19.8 $V_{a1} = E_a - I_{a1} Z_1$

19.9 Negative sequence voltage generated by a synchronous machine is zero.

19.10 The current flowing through neutral grounding impedance Z_n is I_{a0}.

B. Fill in the blanks with suitable words.

19.11 At fault point V_{a2} = ______

19.12 At fault point Z_0 =

19.13 Zero sequence current can flow through a star connected winding only if the star point is______.

19.14 For an overhead line the zero sequence reactance is______ than positive sequence reactance.

19.15 In a Y-Δ transformer the positive sequence line currents on high voltage side ______ the positive sequence currents a low voltage side by 30º.

19.16 In a generator with solidly grounded neutral, single line to ground fault current is ______ than 3 phase fault current.

19.17 In a ______ fault the zero sequence network is not connected.

19.18 Only the ______ sequence network contains a voltage source.

19.19 The reference bus for positive sequence network is system ______.

19.20 The total number of sequence voltage components in a 3 phase system is ______.

ANSWERS

19.1 T	19.2 T	19.3 F	19.4 T	19.5 T
19.6 F	19.7 T	19.8 T	19.9 T	19.10 F
19.11 $-Z_2 I_{a2}$	19.12 $3Z_n + Z_{g0}$	19.13 grounded	19.14 more	19.15 lead
19.16 more	19.17 line to line	19.18 positive	19.19 neutral	19.20 nine

JOB INTERVIEW QUESTIONS

19.1 Why are fault studies important?

19.2 What is the difference between symmetrical and unsymmetrical faults?

19.3 What are symmetrical components?

19.4 How does neutral grounding affect fault calculations?

19.5 Which sequence current flows through ground and ground wires and why?

19.6 Why is resistance generally neglected in fault calculations?

19.7 Which fault is more severe if it occurs at generator terminals and why?

19.8 How can the effect of load current be taken into account in fault calculations?

REVIEW QUESTIONS

19.1 Give reasons:

(a) For a fault at alternator terminals, a single line to ground fault is generally more severe than a 3-phase fault.

(b) For faults on transmission lines, a 3-phase fault is more severe than other faults.

(c) The neutral grounding impedance Z_n appears as 3 Z_n, in the zero sequence equivalent circuit.

(d) The positive and negative sequence equivalent circuits are not affected by the method of neutral grounding.

(e) The analysis of unsymmetrical faults can be more easily done with the help of symmetrical components than by a direct solution of the unbalanced circuit.

(f) The phase shift in the positive sequence and negative sequence quantities through a star delta transformer are opposite to each other.

19.2 Distinguish between symmetrical and unsymmetrical faults. List the steps in the symmetrical fault calculations. How can the load current be taken into account in fault calculations?

19.3 Discuss the principle of symmetrical components. Derive the necessary equations to convert (a) phase quantities into symmetrical components (b) symmetrical components into phase quantities.

19.4 Draw a general circuit which can be used to determine the zero sequence network of a two-winding transformer. Using this circuit, draw the zero sequence networks of (a) star-star transformer with star points grounded (b) delta-delta transformer.

19.5 Derive the necessary equation to determine the fault current for a single line to ground fault. Draw a diagram showing the inter-connection of sequence networks.

19.6 Show that positive and negative sequence currents are equal in magnitude but out of phase by 180° in a line to line fault. Draw a diagram showing inter-connection of sequence networks for this type of fault.

19.7 Draw a diagram showing interconnection of sequence networks for a double line to ground fault. Derive equations for sequence currents.

19.8 In what respects are the fault calculations, for a fault on the alternator terminals, different from the fault calculations for a fault in a power system network?

PROBLEMS

19.1 Three 10 MVA generators each having a reactance of 0.2 pu are operating in parallel. They feed a transmission line through a 30 MVA transformer having a per unit reactance of 0.05. Find the fault MVA for a fault at the sending end of line. [120 MVA]

19.2 An alternator has the following sequence impedances:
$Z_1 = 0 + j\,1.0$, $Z_2 = 0.1 + j.20$, $Z_0 = 0 + j\,1.0$ ohm. The line to neutral voltage at the generator terminals is 1000 V. A fault between yellow and blue phases occurs. Find the fault current and line to neutral voltage of healthy phase. [1440 A, 371 V]

19.3 A 3-phase 33 kV, 37.5 MVA alternator is connected to a 33 kV overhead line which develops an earth fault on one phase at the remote end. The positive, negative and zero sequence reactances of the alternator are 0.18, 0.12 and 0.1 per unit of its rating while those of the line are 6.3, 6.3 and 12.6 ohm per conductor. Find (a) fault current (b) line to neutral voltages at alternator terminals.
(a) $I_a = 1.555$ kA (b) $V_{an} = 13 \angle 0$ kV
[$V_{bn} = 18 \angle -119.16°$ kV, $V_{cn} = 18 \angle 119.16°$ kV]

19.4 A line to line fault occurs between two phases at the remote end of line of the system in problem 19.3. Find (a) fault current (b) line to neutral voltage of the healthy phase at fault point.
[(a) 1.545 kA (b) 17.5 kV]

19.5 A 20 MVA, 11 kV, 3-phase alternator having a transient reactance of 0.2 *pu* feeds a 20 MVA, 11 kV, 3-phase synchronous motor having a transient reactance of 0.5 *pu* through a short line. The line has a reactance of 0.3025 ohm. A 3-phase fault occurs at the motor terminals when the motor is taking an input of 15 MW at unity *p.f.* The line voltage across motor terminals at the time of fault is 10.45 kV. Find the total current supplied by the generator and motor to the fault.
[$787.3 - j\,3989$ A, $-787.3 - j\,1994.5$ A]

19.6 The line currents in a 3-phase supply are $I_a = 12 + j24$ A, $I_b = 16 - j2$ A and $I_c = -4 - j\,6$ A. The phase sequence is *abc*. Calculate the sequence components of currents.
[$I_{a1} = 15.1$ A, $I_{a2} = 4.75$ A, $I_{a0} = 9.61$ A]

19.7 A 10 MVA 50 Hz 3-phase generator has 0.3, 0.1 and 0.05 *pu* positive, negative and zero sequence reactances. It feeds a 3-phase line having 1 cm diameter conductors in equilateral triangle configuration. The interphase spacing is 5 m. The generator has an open circuit emf of 30 kV when a line to line fault occurs at 20 km distance from generator bus. Find currents in the lines.
[555.55 A, 555.55 A, 0]

19.8 In the system of example 19.13 find the current circulating in the delta windings of the transformer at the receiving end. [0.35 pu]

19.9 In the system of example 19.21, the three-winding transformer at the receiving end is replaced by a two-winding transformer connected in a star-star with solidly earthed neutrals on both sides. This transformer is 50 MVA, 132/33 kV, 5% reactance. Find the fault current for a single line to ground fault at secondary terminals of this transformer. [1946.45 A]

19.10 In the system of example 19.13 find fault current for a line to line fault. [2.8867 *pu*]

19.11 In the system of example 19.13, find positive sequence current for a double line to ground fault. [2.333 *pu*]

19.12 A 3-phase 10 MVA, 6600 V alternator with earthed neutral has $Z_1 = 0.5 + j\,4.7$, $Z_2 = 0.2 + j\,0.6$ and $Z_0 = j\,0.48$ ohm. It supplies a 3-phase 3-wire cable having $Z_1 = 0.36 + j.25$, $Z_2 = 0.36 + j.25$ and $Z_0 = 2.9 + j.95$ ohm. Find (a) fault current (b) voltages of healthy phases at fault point for a single line to ground fault at the far end of cable. [1364.2 A, 2633.1 V, 3252.8 V]

19.13 A 3-phase generator A having positive, negative and zero sequence reactances of 0.3, 0.2 and 0.05 *pu* has an earthed neutral. It feeds a 3-phase line through a transformer T_1. The transformer has $x_1 = x_2 = x_0 = 0.12$ *pu* and is connected star-star with both neutrals earthed. For the line $x_1 = x_2 = 0.15$ and $x_0 = 0.35$ *pu*. The other end of line is connected to a transformer T_2 having $x_1 = x_2 = x_0 = 0.1$ *pu*. A generator *B* feeds T_2. The sequence reactances of generator *B* are 0.25, 0.15 and 0.03 *pu*. The transformer T_2 is also connected star-star with both neutrals earthed. A line to line fault occurs on secondary side of T_1. Find fault current in *pu*. All the reactances are on the same base. [4.224 pu]

19.14 Find currents flowing into fault from the three lines for a double line to ground fault at the same location in problem 19.13. [0, 4.78 pu, 4.78 pu]

19.15 A 20 MVA generator having 15% reactance and a 10 MVA generator having 10% reactance are connected to a common bus. The bus feeds a 30 MVA, 11/33 kV transformer with 5% reactance. A 33 kV feeder takes off from transformer and has an impedance of (3 + *j* 15) Ω. If a 3-phase fault occurs at far end of feeder, find fault MVA and fault current. The generators are rated at 11 kV. [50.207 MVA, 878.42 A]

19.16 A single line to earth fault occurs on a feeder. The sequence impedances upto fault point are $0.3 + j\,0.6$, $0.3 + j\,0.55$ and $1 + j\,0.78$ *pu*. The fault resistance is 0.66 pu. If voltage is $1 \angle 0$, find fault current and voltage of faulty phase at fault point. [0.738 pu, $0.429 - j\,0.23$ *pu*]

19.17 A 3-phase 13.8 kV, 30 MVA star connected generator has a subtransient reactance of 0.15 *pu*. It feeds a step up transformer consisting of three single phase units each 10 MVA, delta-star, 13.2 – 66.4/115 kV having leakage reactance of 0.077 and resistance 0.005 *pu*. The transformer supplies a transmission line 40 km long having a resistance of 0.2 Ω/km and positive sequence reactance 0.8 Ω/km. The step down transformer, at the end of line, consists of 3 single phase units each 8.333 MVA, connected delta-star 110–3.98/6.9 kV, resistance 0.008 *pu* and leakage reactance 0.08 *pu*. The load consists of a 3-phase synchronous motor 25 MVA, 6.6 kV having a sub-transient reactance of 0.25 *pu* and a 3-phase 10 MW, 0.7 lagging p.f. shunt impedance load. (a) Find current in line and voltages at its two ends when voltage at load is 6.6 kV, the motor is drawing 10 MW at unity pf and the shunt impedance load is connected (b) If a 3-phase fault occurs at motor terminals, find initial symmetrical line current in the generator. [(a) 123 A, 110 kV, 115 kV (b) 3660 A]

CHAPTER 20

Digital Techniques in Faults Calculations

20.1 INTRODUCTION

The symmetrical and unsymmetrical fault calculations were discussed in Chapter 19. The network reduction method discussed there can be used only for very small systems. The actual power systems are very large in general. Therefore, a systematic procedure suitable for digital computer calculations is necessary. As discussed in Chapter 10, system studies can be carried out using Y_{bus} or Z_{bus}. Y_{bus} is used for load flow computations. For fault calculations (also known as short circuit studies), the use of Z_{bus} is preferable. As compared to load flow studies, the short circuit studies are simpler because a direct solution is possible and no iterations are required. However, all the three sequence components and sequence networks are involved in the short circuit studies. Y_{bus} can be formed by the methods discussed in Chapter 19 and Z_{bus} can be formed by obtaining inverse of Y_{bus}. However, this procedure is unnecessary and rather wasteful. A step by step procedure for obtaining Z_{bus} directly is preferable.

20.2 BUS IMPEDANCE MATRIX

Figure 20.1 shows an n bus passive linear network. The voltage of ith bus (with respect to reference) is V_i and the current entering the ith bus is I_i. Our knowledge of network theory tells us that this network can be described by the following system of equations:

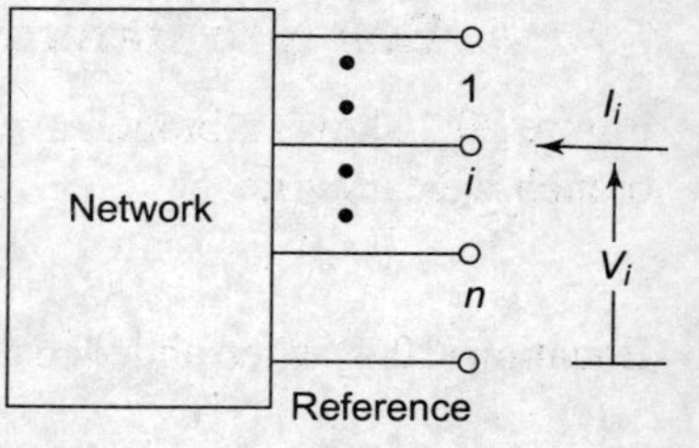

Fig. 20.1 General n port network

$$\begin{bmatrix} V_1 \\ V_2 \\ \vdots \\ V_i \\ \vdots \\ V_n \end{bmatrix} = \begin{bmatrix} Z_{11} & Z_{12} & \ldots Z_{1n} \\ Z_{21} & Z_{22} & \ldots Z_{2n} \\ \ldots & \ldots & \ldots \\ Z_{i1} & Z_{i2} & \ldots Z_{in} \\ \ldots & \ldots & \ldots \\ Z_{n1} & Z_{n2} & \ldots Z_{nn} \end{bmatrix} \begin{bmatrix} I_1 \\ I_2 \\ \vdots \\ I_i \\ \vdots \\ I_n \end{bmatrix} \tag{20.1}$$

or simply $V_{bus} = [Z_{bus}] I_{bus}$ (20.2)

where V_{bus} is the bus voltage $(n \times 1)$ matrix, I_{bus} is the bus current $(n \times 1)$ matrix and Z_{bus} is the bus impedance $(n \times n)$ matrix. The general entry Z_{ik} of Z_{bus} can be obtained from the equation.

$$Z_{ik} = \left. \frac{V_i}{I_k} \right| \quad (20.3)$$

$$I_1 = I_2 .... = I_n = 0$$

$$I_k \neq 0$$

Eq. (20.3) suggests a method for calculating Z_{ik}. Terminate all the buses in open circuits except the kth bus which is terminated in a current source of strength I_k. Solve for V_i and calculate Z_{ik}. However, this method is not suitable because calculation of V_i is generally very difficult. The following step by step method is very simple and suitable for computer work.

20.3 ALGORITHM FOR FORMULATING $[Z_{bus}]$

An algorithm for formulating $[Z_{bus}]$ is described in terms of modifying an existing bus impedance matrix designated as $[Z_{bus}]_{old}$. The new modified matrix is designated as $[Z_{bus}]_{new}$. The network consists of a reference bus and a number of other buses. When a new element having self impedance z_s is added, a new bus may be created (if the new element is a tree branch) or a new bus may not be created (if the new element is a link). Each of these two cases can be subdivided into two cases so that z_s may be added in the following ways:

1. Adding z_s from a new bus to reference.
2. Adding z_s from a new bus to an old bus.
3. Adding z_s from an old bus to reference.
4. Adding z_s between two old buses.

The case of addition of z_s to create two new buses simultaneously has been carefully avoided. Such a case can always be bypassed by properly selecting the order in which the elements are added. In the following discussion i and k denote old buses and q denotes new bus. The bus impedance matrix is always diagonally symmetric, i.e., $Z_{ik} = Z_{ki}$. The sign convention for the voltages and currents is as shown in Fig. 20.1. We assume that the elements do not have mutual coupling.

20.3.1 Type 1 Modification (Addition of Tree Branch z_s from a New Bus q to Reference)

Figure 20.2 shows a branch z_s connected between reference bus and a new bus q. For this new branch we can write

$$V_q = z_s I_q \quad (20.4)$$

Equation (20.4) when added to the set of Eq. (20.1) gives the following complete set of equations.

$$\begin{bmatrix} V_1 \\ \vdots \\ V_i \\ \vdots \\ V_n \\ V_q \end{bmatrix} = \begin{bmatrix} Z_{11} & Z_{12} & .....Z_{1n} & 0 \\ \multicolumn{4}{c}{\dotfill} \\ Z_{i1} & Z_{i2} & .....Z_{in} & 0 \\ \multicolumn{4}{c}{\dotfill} \\ Z_{n1} & Z_{n2} & .....Z_{nn} & 0 \\ 0 & 0 & ....0 & z_s \end{bmatrix} \begin{bmatrix} I_1 \\ \vdots \\ I_i \\ \vdots \\ I_n \\ I_q \end{bmatrix} \quad (20.5)$$

$[Z_{bus}]_{old}$ is an $n \times n$ matrix. The addition of the new branch means the addition of an $(n+1)$th row and $(n+1)$th column. As seen from Eq. (20.5) $Z_{iq} = Z_{qi} = 0$ for $i = 1, 2, ...i, ...n$

$$Z_{qq} = z_s$$

$[Z_{bus}]_{new}$ can, therefore, be written as

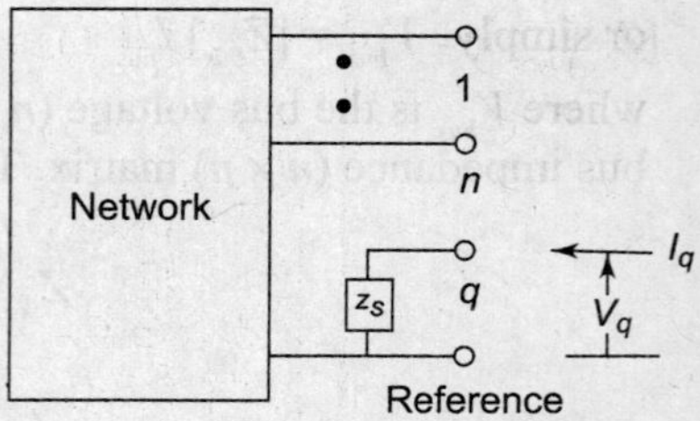

Fig. 20.2 Type 1 modification

$$[Z_{bus}]_{new} = \left[\begin{array}{c|c} [Z_{bus}]_{old} & \begin{matrix} 0 \\ : \\ 0 \end{matrix} \\ \hline 0.........0 & z_s \end{array}\right] \tag{20.6}$$

20.3.2 Type 2 Modification (Addition of Tree Branch z_s from a New Bus q to Old Bus k)

Figure 20.3 shows this modification. The equation for the new branch is

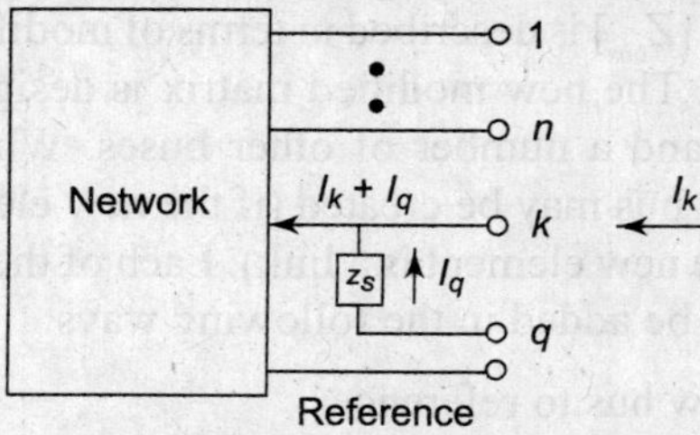

Fig. 20.3 Type 2 modification

$$\begin{aligned} V_q &= z_s I_q + V_k \\ &= z_s I_q + Z_{k1} I_1 + Z_{k2} I_2 + ... + Z_{kk}(I_k + I_q) + ... + Z_{kn} I_n \\ &= Z_{k1} I_1 + Z_{k2} I_2 + ... + Z_{kk} I_k + ... + Z_{kn} I_n + (Z_{kk} + Z_s) I_q \end{aligned} \tag{20.7}$$

The equation for V_1 (i.e., first equation of the set of Eq. 20.1) can be written as under keeping in mind that new current at kth bus is $I_k + I_q$.

$$\begin{aligned} V_1 &= Z_{11} I_1 + Z_{12} I_2 + ... + Z_{1k}(I_k + I_q) + ... + Z_{1n} I_n \\ &= Z_{11} I_1 + Z_{12} I_2 + ... + Z_{1k} I_k + ... + Z_{1n} I_n + Z_{1k} I_q \end{aligned} \tag{20.8}$$

The equations for $V_2, V_3 ... V_n$ also get modified in a similar manner. When Eq. (20.7) is added to the modified equations for $V_1, V_2 ... V_n$ the complete new set of equations is given by

$$\begin{bmatrix} V_1 \\ V_2 \\ : \\ V_k \\ : \\ V_n \\ V_q \end{bmatrix} = \begin{bmatrix} Z_{11}\ Z_{12} ...... Z_{1k} ... Z_{1n}\ Z_{1k} \\ Z_{21}\ Z_{22} ...... Z_{2k} ... Z_{2n}\ Z_{2k} \\ .... \\ Z_{k1}\ Z_{k2} ...... Z_{kk} ... Z_{kn}\ Z_{kk} \\ .... \\ Z_{n1}\ Z_{n2} ...... Z_{nk} ... Z_{nn}\ Z_{nk} \\ Z_{k1}\ Z_{k2} ...... Z_{kk} ... Z_{kn}\ (Z_{kk} + z_s) \end{bmatrix} \begin{bmatrix} I_1 \\ I_2 \\ : \\ I_k \\ : \\ I_n \\ I_q \end{bmatrix} \tag{20.9}$$

$[Z_{bus}]_{new}$ is an $(n+1) \times (n+1)$ matrix. An observation of Eq. (20.9) shows that

$$[Z_{bus}]_{new} = \left[\begin{array}{c|c} [Z_{bus}]_{old} & \begin{matrix} Z_{1k} \\ Z_{2k} \\ \vdots \\ Z_{kk} \\ Z_{nk} \end{matrix} \\ \hline Z_{k1}\ Z_{k2} \ldots Z_{kk} \ldots Z_{kn} & Z_{kk} + z_s \end{array}\right] \tag{20.10}$$

20.3.3 Type 3 Modification (Addition of a Link z_s between an old Bus *k* and Reference)

Figure 20.4 shows this modification. This case can be treated as an extension of type 2 modification. Initially let z_s be connected between a new bus q and an old bus k and then let the new bus q be connected to the reference bus so that $V_q = 0$. The new set of equation is

$$\begin{bmatrix} V_1 \\ V_2 \\ \vdots \\ \vdots \\ V_n \\ \hline 0 \end{bmatrix} = \left[\begin{array}{c|c} [Z_{bus}]_{old} & \begin{matrix} Z_{1k} \\ Z_{2k} \\ \vdots \\ Z_{nk} \end{matrix} \\ \hline Z_{k1}\ Z_{k2} \ldots Z_{kn} & Z_{kk} + z_s \end{array}\right] \begin{bmatrix} I_1 \\ I_2 \\ \vdots \\ \vdots \\ I_n \\ \hline I_q \end{bmatrix} \tag{20.11}$$

If I_q is eliminated from Eq. (20.11), i.e., the last row and last column of the above impedance matrix are eliminated (see appendix B for node elimination by matrix algebra), we have

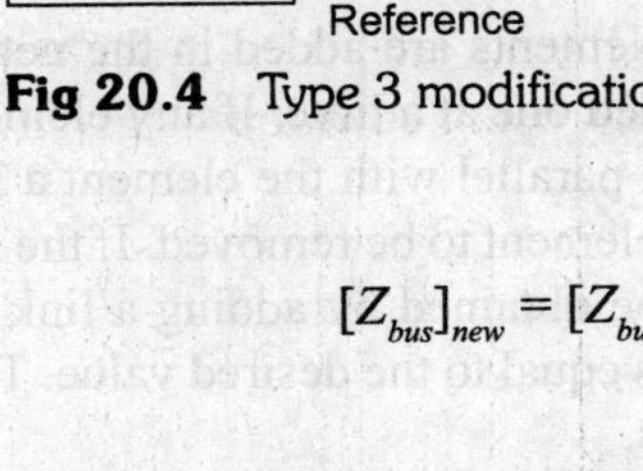

Fig 20.4 Type 3 modification

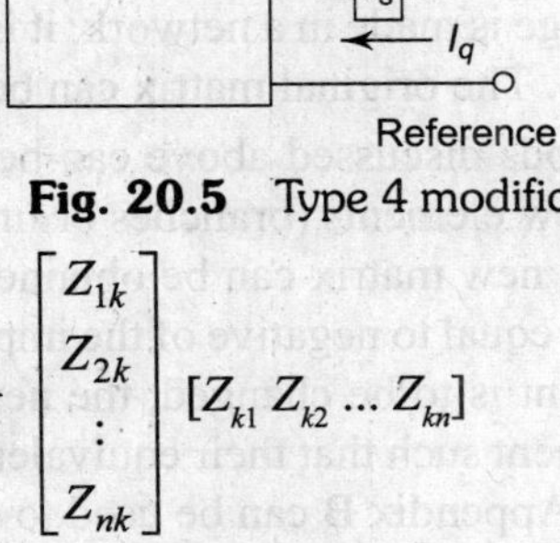

Fig. 20.5 Type 4 modification

$$[Z_{bus}]_{new} = [Z_{bus}]_{old} - \frac{1}{Z_{kk} + z_s} \begin{bmatrix} Z_{1k} \\ Z_{2k} \\ \vdots \\ Z_{nk} \end{bmatrix} [Z_{k1}\ Z_{k2} \ldots Z_{kn}] \tag{20.12}$$

20.3.4 Type 4 Modification (Addition of Link z_s between two Old Buses *i* and *k*)

Referring to Fig. 20.5 the Equation for V_1 can be written as

$$V_1 = Z_{11} I_1 + Z_{12} I_2 + \ldots Z_{1i} (I_i + I_q) + Z_{1j} I_j + Z_{1k} (I_k - I_q) + \ldots + Z_{1n} I_n \tag{20.13a}$$

Rearranging

$$V_1 = Z_{11} I_1 + Z_{12} I_2 + ... + Z_{1n} I_n + (Z_{1i} - Z_{1k}) I_q \tag{20.13b}$$

Equations similar to Eq. (20.13b) can be written for all buses, It is to be noted further that

$$V_k = z_s I_q + V_i \tag{20.14a}$$

or

$$Z_{k1} I_1 + Z_{k2} I_2 + ... + Z_{ki} (I_i + I_q) + Z_{kj} I_j + Z_{kk} (I_k - I_q) + ...$$
$$= z_s I_q + Z_{i1} I_1 + Z_{i2} I_2 + ... Z_{ii} (I_i + I_q) + Z_{ij} I_j + Z_{ik} (I_k - I_q) + ... \tag{20.14b}$$

Rearranging

$$0 = (Z_{i1} - Z_{k1}) I_1 + ... + (Z_{ii} - Z_{ki}) I_i + (Z_{ij} - Z_{kj}) I_j$$
$$+ (Z_{ik} - Z_{kk}) I_k + ... + (z_s + Z_{ii} - Z_{ik} - Z_{ki} + Z_{kk}) I_q \tag{20.14c}$$

Equation (20.13b) along with similar equation for V_2, V_3 ... V_n and Eq. (20.14c) can be written in the following form:

$$\begin{bmatrix} V_1 \\ V_2 \\ \vdots \\ \vdots \\ V_n \\ \hline 0 \end{bmatrix} = \left[\begin{array}{c|c} [Z_{bus}]_{old} & \begin{matrix} Z_{1i} - Z_{1k} \\ Z_{2i} - Z_{2k} \\ \vdots \\ \end{matrix} \\ \hline Z_{i1} - Z_{k1} & z_s + Z_{ii} + Z_{kk} - 2Z_{ik} \end{array}\right] \begin{bmatrix} I_1 \\ I_2 \\ \vdots \\ \vdots \\ I_n \\ \hline I_q \end{bmatrix} \tag{20.15}$$

Eliminating I_q from Eq. (20.15) as per the method explained in Appendix *B*.

$$[Z_{bus}]_{new} = [Z_{bus}]_{old} - \frac{1}{z_s + Z_{ii} + Z_{kk} - 2Z_{ik}} \begin{bmatrix} Z_{1i} - Z_{1k} \\ Z_{2i} - Z_{2k} \\ \vdots \\ \vdots \\ Z_{ni} - Z_{nk} \end{bmatrix} [(Z_{i1} - Z_{k1}) ... (Z_{in} - Z_{kn})] \tag{20.16}$$

20.3.5 Modification of [Z_{bus}] for Changes in Network

When a change is made in a network, it is not necessary to frame the new bus impedance matrix from abinitio. The original matrix can be modified to give the new matrix.

The methods discussed above can be used if new elements are added in the network. Any number of new elements (branches or links) can be added one at a time. If any element is to be removed, the new matrix can be obtained by adding in parallel with the element a link whose impedance is equal to negative of the impedance of the element to be removed. If the impedance of any element is to be changed, the new matrix can be obtained by adding a link in parallel with the element such that their equivalent impedance is equal to the desired value. The method discussed in Appendix B can be used to eliminate a bus.

Example 20.1.

Figure 20.6 shows a 4-bus system. Treating bus 4 as the reference bus obtain [Z_{bus}].

Solution

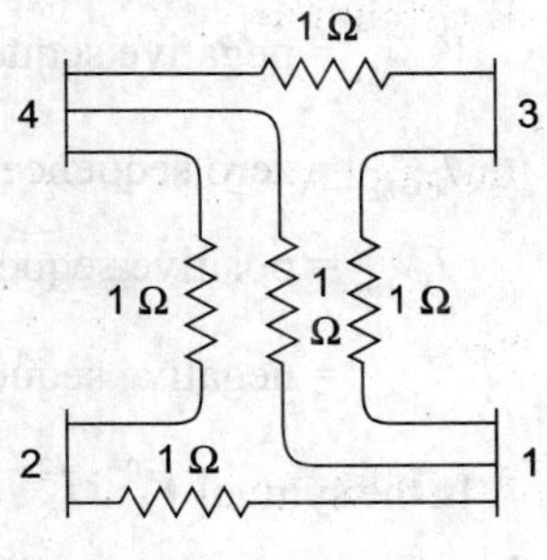

Fig.20.6

1. We start with the 1 ohm element between buses 1 and 4. Type 1 modification gives

$$[Z_{bus}] = [1]$$

2. Add the 1 ohm element between buses 2 and 4. Type 1 modification gives

$$[Z_{bus}] = \begin{bmatrix} 1 & 0 \\ 0 & 1 \end{bmatrix}$$

3. Now add the 1 ohm element between buses 3 and 4. Type 1 modification gives

$$[Z_{bus}] = \begin{bmatrix} 1 & 0 & 0 \\ 0 & 1 & 0 \\ 0 & 0 & 1 \end{bmatrix}$$

4. Next add the 1 ohm element between the old buses 1 and 2. Type 4 modification gives

$$[Z_{bus}] = \begin{bmatrix} 1 & 0 & 0 \\ 0 & 1 & 0 \\ 0 & 0 & 1 \end{bmatrix} - \frac{1}{1+1+1}\begin{bmatrix} 1 \\ -1 \\ 0 \end{bmatrix} [1 - 1\ 0]$$

$$= \begin{bmatrix} 1 & 0 & 0 \\ 0 & 1 & 0 \\ 0 & 0 & 1 \end{bmatrix} - \frac{1}{3}\begin{bmatrix} 1 & -1 & 0 \\ -1 & 1 & 0 \\ 0 & 0 & 0 \end{bmatrix} = \begin{bmatrix} \frac{2}{3} & \frac{1}{3} & 0 \\ \frac{1}{3} & \frac{2}{3} & 0 \\ 0 & 0 & 1 \end{bmatrix}$$

5. Finally add the 1 ohm element between the old buses 1 and 3. Type 4 modification gives

$$[Z_{bus}] = \begin{bmatrix} \frac{2}{3} & \frac{1}{3} & 0 \\ \frac{1}{3} & \frac{2}{3} & 0 \\ 0 & 0 & 1 \end{bmatrix} - \frac{1}{1+\frac{2}{3}+1}\begin{bmatrix} \frac{2}{3} \\ \frac{1}{3} \\ -1 \end{bmatrix}\begin{bmatrix} \frac{2}{3} & \frac{1}{3} & -1 \end{bmatrix}$$

$$= \begin{bmatrix} \frac{2}{3} & \frac{1}{3} & 0 \\ \frac{1}{3} & \frac{2}{3} & 0 \\ 0 & 0 & 1 \end{bmatrix} - \frac{3}{8}\begin{bmatrix} \frac{4}{9} & \frac{2}{9} & \frac{-2}{3} \\ \frac{2}{9} & \frac{1}{9} & \frac{-1}{3} \\ \frac{-2}{3} & \frac{-1}{3} & 1 \end{bmatrix} = \begin{bmatrix} \frac{1}{2} & \frac{1}{4} & \frac{1}{4} \\ \frac{1}{4} & \frac{5}{8} & \frac{1}{8} \\ \frac{1}{4} & \frac{1}{8} & \frac{5}{8} \end{bmatrix}$$

20.4 SEQUENCE MATRICES

To start the discussion of short circuit studies, we define the following notations:

V_{0-bus} = zero sequence bus voltage vector ($n \times 1$), general entry V_k^0

V_{1-bus} = positive sequence bus voltage vector ($n \times 1$), general entry V_k^1

V_{2-bus} = negative sequence bus voltage vector ($n \times 1$), general entry V_k^2,

I_{0-bus} = zero sequence bus current vector ($n \times 1$), general entry I_k^0

I_{1-bus} = positive sequence bus current vector ($n \times 1$), general entry I_k^1

I_{2-bus} = negative sequence bus current vector ($n \times 1$), general entry I_k^2

In the symbol V_k^0, I_k^2, etc., the subscript refers to the bus number and the super script indicates the sequence as shown in Fig. 20.7. All these sequence components are for phase *a* of the network. The subscript *a* has been dropped for convenience in writing. Thus, $V_{0\text{-}bus}$ means $V_{a0\text{-}bus}$ and so on. It is seen that the direction of current, in Fig. 20.7 is opposite to that shown in Fig. 20.1 However, the direction shown in Fig. 20.7 is more natural (because fault currents, in general, flow out of the buses) and agrees with that used in Chapter 19.

The sequence impedance matrices required for the short circuit studies are:

$[Z_{0-bus}]$ = zero sequence bus impedance matrix ($n \times n$), general entry Z_{ik}^0

$[Z_{1-bus}]$ = positive sequence bus impedance matrix ($n \times n$), general entry Z_{ik}^1

$[Z_{2-bus}]$ = negative sequence bus impedance matrix ($n \times n$), general entry Z_{ik}^2

The sequence impedance matrices can be formulated by the method discussed in section 20.3. All these matrices refer to the passive sequence network. As discussed in Chapter 19, the positive sequence impedance network contains active sources. While forming the positive sequence bus impedance matrix the emfs of sources are assumed to be short circuited.

A somewhat simplified, although approximate, short circuit study can be made by neglecting the prefault currents. This means that all the bus voltages are 1 *pu* immediately before the fault. The equations relating the sequence quantities are:

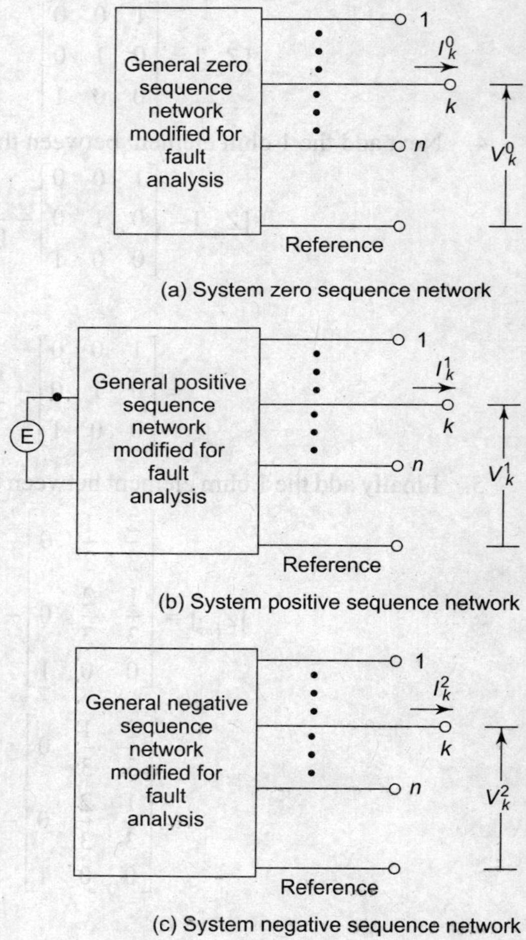

(a) System zero sequence network

(b) System positive sequence network

(c) System negative sequence network

Fig. 20.7 Sequence negative sequence network for fault analysis

$$V_{0-bus} = -[Z_{0-bus}]\, I_{0-bus} \quad (20.17a)$$

$$V_{1-bus} = E_{bus} - [Z_{1-bus}]\, I_{1-bus} \quad (20.17b)$$

$$V_{2-bus} = -[Z_{2-bus}]\, I_{2-bus} \quad (20.17c)$$

Note the negative sign in the above equations (Whereas Eq. 20.1 does not have a negative sign). This is due to the fact that the direction of current in Fig. 20.7 is opposite to that in Fig. 20.1. Since the prefault currents are neglected, vector E contains $1\angle 0$ in all entries. The currents are all zero till the network is terminated externally. At a time only one bus (i.e., the faulted bus k) is terminated. Thus, only I_k^0, I_k^1, I_k^2 have non-zero entry. Very frequently $[Z_{1-bus}]$ and $[Z_{2-bus}]$ are assumed to be identical to reduce computer memory requirement.

20.5 EQUATIONS FOR SHORT CIRCUIT STUDIES

The equations for short circuit studies are developed by using Eqs. (20.17) and terminating the network at the faulted bus (bus k).

20.5.1 Symmetrical Fault

For a symmetrical fault the negative sequence and zero sequence are absent,

i.e., V_{0-bus}, V_{2-bus}, I_{0-bus} and I_{2-bus} are zero.

$$V_k^1 = E - (Z_{k_2}^1 I_1^1 + Z_{k_1}^1 I_2^1 + \ldots.. + Z_{kk}^1 I_k^1 + \ldots\ldots + Z_{kn}^1 I_n^1) \tag{20.18}$$

But all currents except the current at the faulted bus, i.e., I_k^1 are zero. Therefore,

$$V_k^1 = E - Z_{kk}^1 I_k^1 \tag{20.19}$$

If Z_f is the fault impedance

$$V_k^1 = I_k^1 Z_f \tag{20.20}$$

From Eqs. (20.19 and 20.20)

$$I_k^1 = \frac{E}{Z_{kk}^1 + Z_f} \tag{20.21}$$

The voltage at ith bus is

$$V_i^1 = E - Z_{ik}^1 I_k^1 = E\left(1 - \frac{Z_{ik}^1}{Z_{kk}^1 + Z_f}\right)$$

$$\text{for } i = 1, 2, \ldots\ldots.. n \tag{20.22}$$

20.5.2 Single Line to Ground Fault (Fig. 19.3)

We know that

$$I_i^0 = I_i^1 = I_i^2 = 0 \quad \text{for } i = 1, 2 \ldots n \tag{20.23a}$$

$$i \neq k$$

and

$$I_k^0 = I_k^1 = I_k^2 \tag{20.23b}$$

$$V_k^0 + V_k^1 + V_k^2 = 3Z_f I_k^1 \tag{20.24}$$

From Eq. (20.17)

$$V_k^0 = -Z_{kk}^0 I_k^0$$

$$V_k^1 = E - Z_{kk}^1 I_k^1 \tag{20.25}$$

$$V_k^2 = -Z_{kk}^2 I_k^2$$

Combining Eqs. (20.23b, 20.24 and 20.25)

$$I_k^1 = \frac{E}{Z_{kk}^0 + Z_{kk}^1 + Z_{kk}^2 + 3Z_f} \tag{20.26}$$

The sequence components of bus voltages at *i*th bus (*i* = 1, 2, ... *n*) are

$$V_i^0 = -Z_{ik}^0 I_k^0 = -Z_{ik}^0 I_k^1 \tag{20.27a}$$

$$= \frac{-Z_{ik}^0 E}{Z_{kk}^0 + Z_{kk}^1 + Z_{kk}^2 + 3Z_f} \tag{20.27b}$$

$$V_i^1 = E - Z_{ik}^1 I_k^1 \tag{20.28a}$$

$$= E\left[1 - \frac{Z_{ik}^1}{Z_{kk}^0 + Z_{kk}^1 + Z_{kk}^2 + 3Z_f}\right] \tag{20.28b}$$

$$= E\left[\frac{Z_{kk}^0 + Z_{kk}^1 + Z_{kk}^2 + 3Z_f - Z_{ik}^1}{Z_{kk}^0 + Z_{kk}^1 + Z_{kk}^2 + 3Z_f}\right] \tag{20.28c}$$

$$V_i^2 = -Z_{ik}^2 I_k^2 = -Z_{ik}^2 I_k^1 \tag{20.29a}$$

$$= \frac{-Z_{ik}^2 E}{Z_{kk}^0 + Z_{kk}^1 + Z_{kk}^2 + 3Z_f} \tag{20.29b}$$

20.5.3 Line to Line Fault (Fig. 19.9)

For a line to line fault

$$V_{0-bus} = I_{0-bus} = 0 \tag{20.30}$$

Equations (20.23a and 20.25) are still valid

$$V_k^1 = V_k^2 + I_k^1 Z_f \tag{20.31}$$

Substituting the values for V_k^1 and V_k^2 from Eq. (20.25) into Eq. (20.31)

$$E - Z_{kk}^1 I_k^1 = -Z_{kk}^2 I_k^2 + I_k^1 Z_f \tag{20.32}$$

We know that for a line to line fault $I_k^2 = -I_k^1$. Making this substitution in Eq. (20.32)

$$E - Z_{kk}^1 I_k^1 = Z_{kk}^2 I_k^1 + I_k^1 Z_f$$

or

$$I_k^1 = \frac{E}{Z_{kk}^1 + Z_{kk}^2 + Z_f} \tag{20.33}$$

The positive and negative sequence voltages at ith bus ($i = 1, 2 \dots n$) are

$$V_i^1 = E - Z_{ik}^1 I_k^1 \tag{20.34a}$$

$$= E - Z_{ik}^1 \frac{E}{Z_{kk}^1 + Z_{kk}^2 + Z_f} \tag{20.34b}$$

$$= E\left[\frac{Z_{kk}^1 + Z_{kk}^2 + Z_f - Z_{ik}^1}{Z_{kk}^1 + Z_{kk}^2 + Z_f}\right] \tag{20.34c}$$

$$V_i^2 = -Z_{ik}^2 I_k^2 \tag{20.35a}$$

$$= + Z_{ik}^2 I_k^1 \tag{20.35b}$$

$$= \frac{Z_{ik}^2 E}{Z_{kk}^1 + Z_{kk}^2 + Z_f} \tag{20.35c}$$

20.5.4 Double Line to Ground Fault (Fig. 19.11)

From the equivalent circuit shown in Fig. 19.12, it can be inferred that

$$V_k^1 = V_k^2 \tag{20.36a}$$

$$V_k^0 - V_k^1 = 3\, I_k^0 Z_f \tag{20.36b}$$

and
$$I_k^0 + I_k^1 + I_k^2 = 0 \tag{20.37}$$

Equations (20.23a and 20.25) are still valid. From Eqs. (20.25, 20.36 and 20.37)

$$I_k^1 = \frac{E - V_k^1}{Z_{kk}^1} \tag{20.38a}$$

$$I_k^2 = -\frac{V_k^2}{Z_{kk}^2} = -\frac{V_k^1}{Z_{kk}^2} \tag{20.38b}$$

$$I_k^0 = -\frac{V_k^0}{Z_{kk}^0} = \frac{-V_k^1}{Z_{kk}^0 + 3Z_f} \tag{20.38c}$$

Substituting Eqs. (20.38) into Eq. (20.37) we have

$$-\frac{V_k^1}{Z_{kk}^0 + 3Z_f} + \frac{E - V_k^1}{Z_{kk}^1} - \frac{V_k^1}{Z_{kk}^2} = 0$$

or
$$V_k^1 = \frac{E\left(Z_{kk}^0 + 3Z_f\right) Z_{kk}^2}{Z_{kk}^1\left(Z_{kk}^2 + Z_{kk}^0 + 3Z_f\right) + Z_{kk}^2\left(Z_{kk}^0 + 3Z_f\right)} \tag{20.39}$$

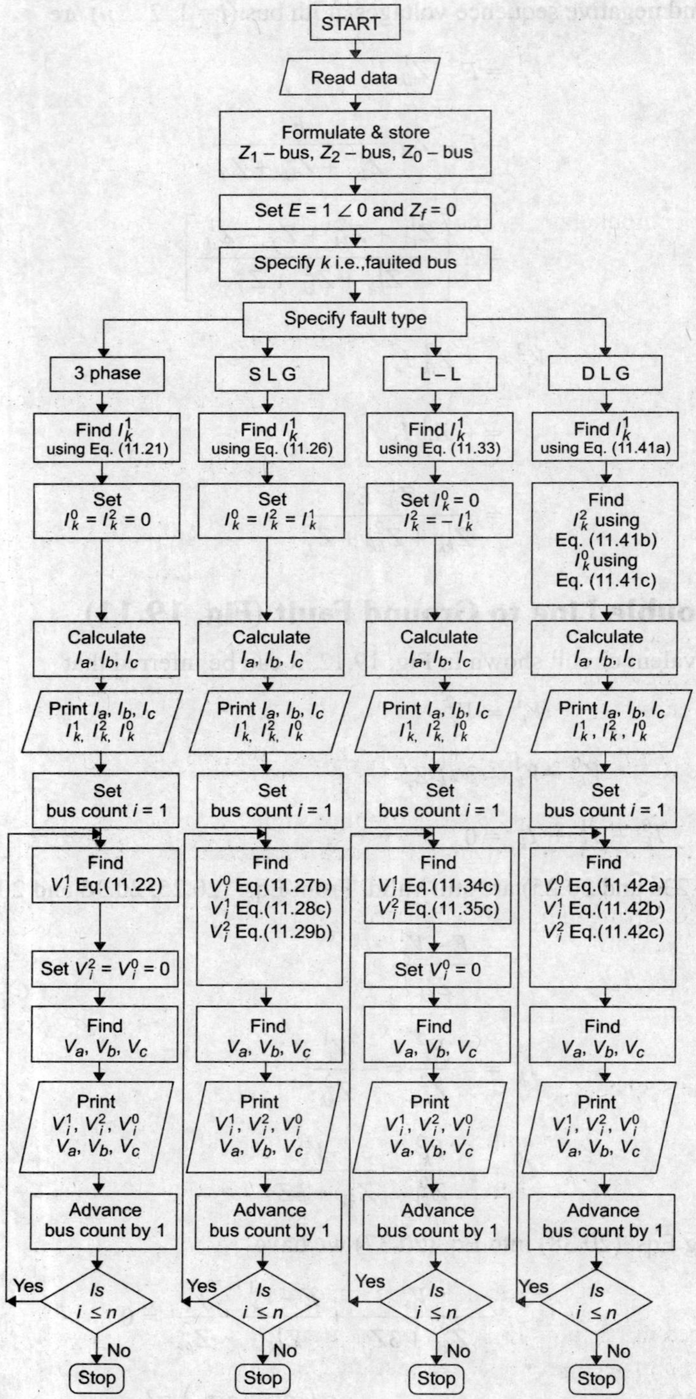

Fig. 20.8 Flowchart for short circuit studies

Let
$$Z_{kk}^{1}\left(Z_{kk}^{2}+Z_{kk}^{0}+3Z_{f}\right)+Z_{kk}^{2}\left(Z_{kk}^{0}+3Z_{f}\right)=\Delta \qquad (20.40)$$

From Eq. (20.38)

$$I_k^1 = \left(Z_{kk}^2 + Z_{kk}^0 + 3Z_f\right) E/\Delta \tag{20.41a}$$

$$I_k^2 = -E\left(Z_{kk}^0 + 3Z_f\right)/\Delta \tag{20.41b}$$

$$I_k^0 = -E\, Z_{kk}^2/\Delta \tag{20.41c}$$

The sequence components of voltages at *i*th bus ($i = 1, 2 \ldots n$) are

$$V_i^0 = -\,Z_{ik}^0 I_k^0 = E\, Z_{ik}^0 Z_{kk}^2/\Delta \tag{20.42a}$$

$$V_i^1 = E - Z_{ik}^1 I_k^1$$

$$= E\,[\Delta - Z_{ik}^1\,(Z_{kk}^2 Z_{kk}^0 + 3Z_f)]/\Delta \tag{20.42b}$$

$$V_i^2 = -Z_{ik}^2 I_k^2$$

$$= Z_{ik}^2\,[Z_{kk}^0 + 3Z_f)]E/\Delta \tag{20.42c}$$

From the values of sequence quantities, the values of phase quantities can be calculated using Eqs. (18.6 and 18.9).

Figure 20.8 shows a simplified computer flowchart for calculating fault current and bus voltages. This flowchart summarizes the application of the equations developed above. For detailed flow chart, the reader is referred to [11, 24, 37] of the bibliography.

20.6 CALCULATIONS OF LINE CURRENTS

The aim of a short circuit study is to determine fault current, bus voltages and line currents under fault conditions. The line currents can be calculated from the values of bus voltages determined in section 20.5. Referring to Fig. 20.9 the current in the line joining *i*th and *k*th buses is

$$I_{ik}^m = \frac{V_i^m - V_k^m}{z_s^m} \tag{20.43}$$

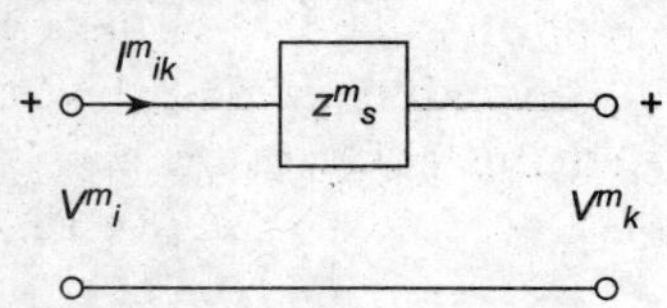

Fig. 20.9 mth sequence component current in the line joining ith and kth buses

where *m* is the sequence, $m = 0, 1, 2$

I_{ik}^m = *m*th sequence current from *i*th bus to *k*th bus

V_i^m = *m*th sequence voltage at *i*th bus

V_k^m = *m*th sequence voltage at *k*th bus

z_s^m = *m*th sequence impedance of branch connecting *i*th and *k*th buses.

As discussed in Chapter 19, a *Y*/Δ or Δ/*Y* transformer causes a phase shift. This phase shift must be taken into account in the calculations.

When the $[Z_{bus}]$ method is used for fault calculation, it may appear that the calculations are more laborious than the calculations using the method discussed in Chapter 19. When the fault calculations are to be done only at one bus of a simple system, the method of Chapter 19 may yield quicker result. However, the actual power system networks are very large (having 100 buses or even more). For studies on such actual networks the Z_{bus} method is far superior. Moreover, once Z_{bus} has been computed, short circuit studies can be very conveniently done for all the buses. The effects of making changes in network can also be conveniently found out.

Example 20.2.

Formulate positive and negative sequence impedance matrices for the network shown in Fig. 18.14.

Solution

The positive sequence equivalent circuit is shown in Fig. 18.15 (a) Z_{1-bus} can be formed as follows by following the step by step method outlined in section 20.3.

1. We start with $j\,0.25$ element between reference bus and bus 1. Type 1 modification gives

$$[Z] = [j\,0.25]$$

2. Add the $j\,0.06$ element between old bus 1 and new bus 2. Type 2 modification gives

$$[Z] = \begin{bmatrix} j0.25 & j0.25 \\ j0.25 & j0.31 \end{bmatrix}$$

3. Add $j\,0.05$ element between old bus 2 and new bus 3. Type 2 modification gives

$$[Z] = \begin{bmatrix} j0.25 & j0.25 & j0.25 \\ j0.25 & j0.31 & j0.31 \\ j0.25 & j0.31 & j0.36 \end{bmatrix}$$

4. Add $j\,0.07$ element between old bus 3 and new bus 4. Type 2 modification gives

$$[Z] = \begin{bmatrix} j0.25 & j0.25 & j0.25 & j0.25 \\ j0.25 & j0.31 & j0.31 & j0.31 \\ j0.25 & j0.31 & j0.36 & j0.36 \\ j0.25 & j0.31 & j0.36 & j0.43 \end{bmatrix}$$

5. Add $j\,0.2$ element between old bus 4 and reference. We have taken into account all the elements. The matrix obtained as a result of this type 3 modification is $[Z_{1-bus}]$.

$$[Z_{1-bus}] = \begin{bmatrix} j0.25 & j0.25 & j0.25 & j0.25 \\ j0.25 & j0.31 & j0.31 & j0.31 \\ j0.25 & j0.31 & j0.36 & j0.36 \\ j0.25 & j0.31 & j0.36 & j0.43 \end{bmatrix}$$

$$- \frac{1}{j0.43 + j0.2} \begin{bmatrix} j0.25 \\ j0.31 \\ j0.36 \\ j0.43 \end{bmatrix} [j0.25 \quad j0.31 \quad j0.36 \quad j0.43]$$

$$= \begin{bmatrix} j0.25 & j0.25 & j0.25 & j0.25 \\ j0.25 & j0.31 & j0.31 & j0.31 \\ j0.25 & j0.31 & j0.36 & j0.36 \\ j0.25 & j0.31 & j0.36 & j0.43 \end{bmatrix}$$

$$-\frac{1}{j0.63}\begin{bmatrix} -0.0625 & -0.0775 & -0.09 & -0.1075 \\ -0.0775 & -0.0961 & -0.1116 & -0.1333 \\ -0.09 & -0.1116 & -0.1296 & -0.1548 \\ -0.1075 & -0.1333 & -0.1548 & -0.1849 \end{bmatrix}$$

$$[Z_{1-bus}] = \begin{matrix} \\ 1 \\ 2 \\ 3 \\ 4 \end{matrix}\begin{bmatrix} 1 & 2 & 3 & 4 \\ j0.1507937 & j0.1269841 & j0.1071429 & j0.0793651 \\ j0.1269841 & j0.1574603 & j0.1328571 & j0.0984127 \\ j0.1071429 & j0.1328571 & j0.1542857 & j0.1142857 \\ j0.0793651 & j0.0984127 & j0.1142857 & j0.1365079 \end{bmatrix}$$

Since all the negative sequence reactances are the same as the positive sequence reactances, $[Z_{1-bus}] = [Z_{2-bus}]$.

Example 20.3.

Formulate $[Z_{0-bus}]$ for the network shown in Fig. 18.14

Solution

The zero sequence equivalent circuit is shown in Fig. 10.28(a)

$[Z_{0-bus}]$ is obtained as under:

1. We start with $j\,0.05$ element between new bus 1 and reference. Type 1 modification gives

$$[Z] = [j\,0.05]$$

2. Add $j\,0.06$ element between new bus 2 and reference. Type 1 modification gives

$$[Z] = \begin{bmatrix} j0.05 & 0 \\ 0 & j0.06 \end{bmatrix}$$

3. Add $j\,0.15$ element between old bus 2 and new bus 3. Type 2 modification gives

$$[Z] = \begin{bmatrix} j0.05 & 0 & 0 \\ 0 & j0.06 & j0.06 \\ 0 & j0.06 & j0.21 \end{bmatrix}$$

4. Add $j\,0.07$ element between old bus 3 and new bus 4. Type 2 modification gives

$$[Z] = \begin{bmatrix} j0.05 & 0 & 0 & 0 \\ 0 & j0.06 & j0.06 & j0.06 \\ 0 & j0.06 & j0.21 & j0.21 \\ 0 & j0.06 & j0.21 & j0.28 \end{bmatrix}$$

5. Add $j\,0.14$ element between old bus 4 and reference. We have now taken into account all the elements and the matrix obtained as a result of this type 3 modification is $[Z_{0-bus}]$

$$[Z_{0-bus}] = \begin{bmatrix} j0.05 & 0 & 0 & 0 \\ 0 & j0.06 & j0.06 & j0.06 \\ 0 & j0.06 & j0.21 & j0.21 \\ 0 & j0.06 & j0.21 & j0.28 \end{bmatrix}$$

$$-\frac{1}{j.42}\begin{bmatrix} 0 \\ j0.06 \\ j0.21 \\ j0.28 \end{bmatrix}\begin{bmatrix} 0 & j0.06 & j0.21 & j0.28 \end{bmatrix}$$

$$= \begin{bmatrix} j0.05 & 0 & 0 & 0 \\ 0 & j0.06 & j0.06 & j0.06 \\ 0 & j0.06 & j0.21 & j0.21 \\ 0 & j0.06 & j0.21 & j0.28 \end{bmatrix} - \frac{1}{j0.42}\begin{bmatrix} 0 & 0 & 0 & 0 \\ 0 & -0.0036 & -0.0126 & -0.0168 \\ 0 & -0.0126 & -0.0441 & -0.0588 \\ 0 & -0.0168 & -0.0588 & -0.0784 \end{bmatrix}$$

$$[Z_{0-bus}] = \begin{array}{c} \\ 1 \\ 2 \\ 3 \\ 4 \end{array}\begin{bmatrix} 1 & 2 & 3 & 4 \\ j0.05 & 0 & 0 & 0 \\ 0 & j0.0514286 & j0.03 & j0.02 \\ 0 & j0.03 & j0.105 & j0.07 \\ 0 & j0.02 & j0.07 & j0.0933333 \end{bmatrix}$$

Example 20.4.

A 3-phase dead short circuit occurs at bus 2 of the system shown in Fig. 18.17. Find (a) fault current (b) line to neutral voltages at bus 3. Assume that prefault voltages at all buses are 1 *pu*.

Solution

$[Z_{1-bus}]$ has been formulated in example 20.2. $Z_f = 0$

(a) Using Eq. (20.21)

$$I_2^1 = \frac{1\angle 0}{Z_{22}^1} = \frac{1\angle 0}{j0.1574603} = 6.350807\angle -90°\ pu$$

Base current = 262.4 A (see example 19.3)

$$I_2^1 = 262.4 \times 6.350807\angle -90° = 1666.45\angle -90°\text{ A}$$

The negative and zero sequence currents are absent. (Why?). The line currents due to 3-phase fault at bus 2 are

$$I_a = 1666.45\angle -90°\text{ A}$$
$$I_b = 1666.45\angle -210°\text{ A}$$
$$I_c = 1666.45\angle 30°\text{ A}$$

(b) The voltage at bus 3 can be found from Eq. (20.22)

$$V_3^1 = E\left(1 - \frac{Z_{32}^1}{Z_{22}^1}\right) = (1\angle 0)\left[1 - \frac{j0.1328571}{j0.1574603}\right]$$

$$= 0.15625\angle 0\ pu$$

Line to neutral base voltage for bus 3 $= \frac{220}{\sqrt{3}}$ kV

$$V_3^1 = \frac{220}{\sqrt{3}} \times 0.15625 = 19.847 \angle 0 \text{ kV}$$

The negative and zero sequence voltages are absent. The line to neutral voltages of different phases at bus 3 are

$$V_a = 19.847 \angle 0° \text{ kV}$$
$$V_b = 19.847 \angle -120° \text{ kV}$$
$$V_c = 19.847 \angle 120° \text{ kV}$$

Example 20.5.

A single line to ground fault occurs at bus 3 of the network in Fig. 18.14. Find (a) fault current (b) line to neutral voltages at bus 2. Assume that the prefault voltage at all buses is 1 *pu*.

Solution

$[Z_{1-bus}]$ and $[Z_{0-bus}]$ have been formulated in examples 20.2 and 20.3

(a) $Z_f = 0$. Using Eq. (20.26)

$$I_3^1 = I_3^2 = I_3^0 = \frac{E}{Z_{33}^0 + Z_{33}^1 + Z_{33}^2} = \frac{1 \angle 0}{j0.105 + j(2 \times 0.1542857)}$$
$$= 2.41796 \angle -90° \; pu$$

Base current = 262.4 A

$$I_3^1 = I_3^2 = I_3^0 = 262.4\,(2.41796 \angle -90°) = 634.47 \angle -90° \text{ A}$$

The line currents at bus 3 are

$$\begin{bmatrix} I_a \\ I_b \\ I_c \end{bmatrix} = \begin{bmatrix} 1 & 1 & 1 \\ 1 & a^2 & a \\ 1 & a & a^2 \end{bmatrix} \begin{bmatrix} 634.47 \angle -90° \\ 634.47 \angle -90° \\ 634.47 \angle -90° \end{bmatrix} = \begin{bmatrix} 1903.41 \angle -90° \\ 0 \\ 0 \end{bmatrix}$$

(b) Using Eq. (20.27a), the zero sequence voltage at bus 2 is

$$V_2^0 = -Z_{23}^0 I_3^0 = -(j\,0.03)\,(2.41796 \angle -90°) = -0.0725388 \; pu$$

Using Eq. (20.28a) the positive sequence voltage at bus 2 is

$$V_2^1 = E - Z_{23}^1 I_3^1 = 1 \angle 0 - (j\,0.1328571)\,(2.417976 \angle -90°)$$
$$= 0.6787568 \; pu$$

Using Eq. (20.29a), the negative sequence voltage at bus 2 is

$$V_2^2 = -Z_{23}^2 I_3^2 = -(j\,0.1328571)\,(2.41796 \angle -90°)$$
$$= -0.321243 \; pu$$

The *pu* values of line to neutral voltages at bus 2 are

$$\begin{bmatrix} V_a \\ V_b \\ V_c \end{bmatrix} = \begin{bmatrix} 1 & 1 & 1 \\ 1 & a^2 & a \\ 1 & a & a^2 \end{bmatrix} \begin{bmatrix} -0.0725388 \\ 0.6787568 \\ -0.321243 \end{bmatrix} = \begin{bmatrix} 0.284975 \angle 0 \\ 0.90175 \angle 253.82° \\ 0.90175 \angle 106.18° \end{bmatrix}$$

The actual values of line to neutral voltages at bus 2 are

$$V_a = \frac{220}{\sqrt{3}}(0.284975) = 36.2 \angle 0 \text{ kV}$$

$$V_b = \frac{220}{\sqrt{3}}(0.90175 \angle 253.82) = 114.54 \angle 253.82° \text{ kV}$$

$$V_c = 114.54 \angle 106.18° \text{ kV}$$

20.7 CONSIDERATION OF PREFAULT CURRENTS

It has been assumed in the preceding analysis that the prefault currents in the system are zero. This assumption is not strictly correct. However, the error caused by this assumption, in the fault computations, is small. This is due to the fact that prefault load currents are usually very small as compared to fault currents. Moreover, the load currents and fault currents are nearly in quadrature and their phasor sum is nearly equal to the larger component, i.e., the fault current. An additional reason for making this assumption is that a fault can occur at any time and there is no way of predicting the loading condition of the system at the instant of fault.

However, it is necessary to consider the prefault current when the magnitude of fault current is small. The magnitudes of prefault currents and bus voltages are known from the results of the load flow studies. One method to include the effect of prefault current is to superimpose these currents to the fault currents and use the actual values of bus voltages for the matrix E_{bus} in Eq. 20.17 (b). For further information on this topic the reader is referred to [14] of the Bibliography.

20.8 MUTUALLY COUPLED BRANCHES IN Z_{BUS}

Sometimes transmission lines have mutual coupling with other lines in the system. Thus it is necessary to incorporate mutually coupled branches in Z_{bus}. Moreover, it may also be necessary to eliminate a mutually coupled branch from an existing Z_{bus}. The mutually coupled branches can be treated as under.

20.8.1 Addition of Mutually Coupled z_b between a New Bus q and an Old Bus p

Figure 20.10 shows a branch z_a already existing between buses m and n. A new branch z_b between old bus p and a new bus q is added. This new branch z_b has mutual coupling z_m with z_a. KVL equations for the voltage drops across the two branches are:

$$v_a = z_a I_a + z_m I_b \qquad (20.44a)$$

$$v_b = z_m I_a + z_b I_b \qquad (20.44b)$$

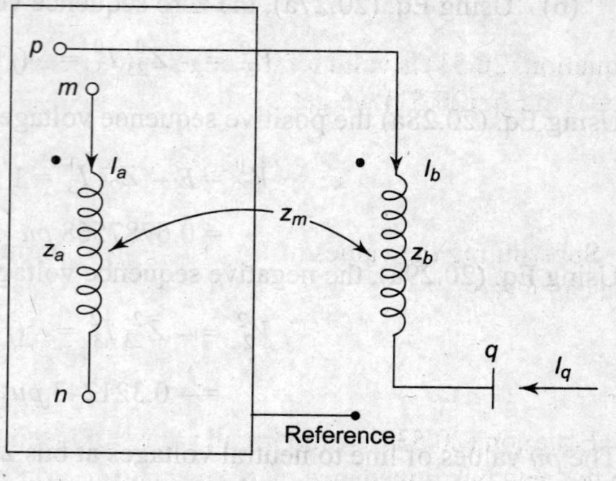

Fig. 20. 10

The current I_a is the branch current through z_a, i.e., from bus m to n. The current I_b from bus p to q equals $-I_q$. Eliminating I_a in Eqs. (20.44) and substituting $-I_q$ for I_b we get

$$v_b = \frac{z_m}{z_a} v_a + \left(\frac{z_m^2}{z_a} - z_b\right) I_q \qquad (20.45)$$

Evidently $v_a = (V_m - V_n)$ and $v_b = (V_p - V_q)$. Making these substitutions in Eq. (20.45) we get

$$V_q = V_p - \frac{z_m}{z_a}(V_m - V_n) - \left(\frac{z_m^2}{z_a} - z_b\right) I_q \tag{20.46}$$

Eq. (20.46) gives the voltage at new bus q after inclusion of the effect of mutual coupling.

The complete bus impedance matrix for the augmented network (after addition of new bus q and the mutually coupled branches) is

$$\begin{bmatrix} V_1 \\ V_2 \\ \vdots \\ V_i \\ \vdots \\ V_n \\ V_q \end{bmatrix} = \left[\begin{array}{ccccc|c} Z_{11} & Z_{12} & \cdots & Z_{1n} & & Z_{1q} \\ Z_{21} & Z_{22} & \cdots & Z_{2n} & & Z_{2q} \\ \vdots & \vdots & & \vdots & & \vdots \\ Z_{i1} & Z_{i2} & \vdots & Z_{in} & & Z_{iq} \\ \vdots & \vdots & & \vdots & & \vdots \\ Z_{n1} & Z_{n2} & \vdots & Z_{nn} & & Z_{nq} \\ \hline Z_{q1} & Z_{q2} & \vdots & Z_{qn} & & Z_{qq} \end{array}\right] \begin{bmatrix} I_1 \\ I_2 \\ \vdots \\ I_i \\ \vdots \\ I_n \\ I_q \end{bmatrix} \tag{20.47}$$

If we can find elements $Z_{1q} \ldots Z_{qq}$, the new Z_{bus} is determined. The ith row in Eq. (20.47) is

$$V_i = Z_{i1} I_1 + Z_{i2} I_2 + \ldots + Z_{ii} I_i + \ldots + Z_{in} I_n + Z_{iq} I_q \tag{20.48}$$

Equation (20.48) can be written as

$$V_i = V_{i0} + Z_{iq} I_q \tag{20.49}$$

where
$$V_{i0} = Z_{i1} I_1 + Z_{i2} I_2 + \ldots + Z_{ii} I_i + \ldots + Z_{in} I_n = \sum_{j=1}^{n} Z_{ij} I_j \tag{20.50}$$

If we set $i = m, n, p$ and q in Eq. (20.49) we get Eqs. for V_m, V_n, V_p and V_q. Substituting these values of V_m, V_n, V_p and V_q in Eq. (20.46) we get

$$V_{q0} + Z_{qq} I_q = [V_{p0} + Z_{pq} I_q] - \frac{z_m}{z_a}[(V_{m0} - V_{n0}) + (Z_{mq} - Z_{nq}) I_q] - \left[\frac{z_m^2}{z_a} - z_b\right] I_q \tag{20.51}$$

Equation (20.51) is valid for the augmented network even if the currents are not injected. Setting $I_q = 0$ in Eq. (20.51) we get

$$V_{q0} = V_{p0} - \frac{z_m}{z_a}(V_{m0} - V_{n0}) \tag{20.52}$$

Substituting the values of V_{m0}, V_{n0}, V_{p0}, V_{q0} from Eq. (20.50) into Eq. (20.52) and simplifying we can write

$$Z_{qj} = Z_{pj} - \frac{z_m}{z_a}(Z_{mj} - Z_{nj}) \tag{20.53}$$

Equation (20.53) is valid for all values of j from 1 to n except $j = q$. Thus all the new elements in the new bus impedance matrix except Z_{qq}, can be found from Eq. (20.53).

Thus to find Z_{qj}, we subtract each element of row n from the corresponding element of row m, multiply the difference by z_m/z_a and subtract the result from the corresponding element of row p. This gives the element Z_{qj}. The element $Z_{jq} = Z_{qj}$.

To find Z_{qq}, we set all currents except I_q equal to zero in Eq. (20.51) and then equate coefficients of I_q on both sides in Eq. (20.51). The result is

$$Z_{qq} = Z_{pq} - \frac{z_m}{z_a}\left[Z_{mq} - Z_{nq}\right] - \left[\frac{z_m^2}{z_a} - z_b\right] \tag{20.54}$$

20.8.2 Addition of Mutually Coupled z_b between Existing Bus p and Reference

This can be treated as an extension of the method discussed in section 20.8.1. Initially we add element z_b (having mutual impedance z_m with element z_a), between a new bus q and an old bus p,. Then we set $V_q = 0$, i.e., eliminate qth row and qth column of the Z_{bus} using the method of node elimination discussed in appendix B.

20.8.3 Addition of Mutually Coupled z_b between Two Old Buses p and k

This can be treated as a combination of methods discussed in sections 20.8.1 and 20.3.4. First we add the element z_b between old bus p and a new bus q (using the method discussed in section 20.4.1). Then we connect bus q to old bus k, i.e., add a branch of zero impedance between buses q and k. Thus $V_q - V_k$ is set to zero. Then we eliminate qth row and qth column by the method of node elimination discussed in appendix B.

20.8.4 Removal of Mutually Coupled z_b between Old Buses p and k

For this modification we add, between old buses p and k, a branch with impedance $-z_b$ and having the same mutual coupling with z_a as that between z_a and z_b. This is shown in Fig. 20.11. The branch $-z_b$ carrying current I'_b has mutual coupling z_m with element z_a. The mutual coupling between z_a and z_b is also z_m. First we add $-z_b$ between old bus p and a new bus q. Then we connect new bus q to old bus k thus eliminating bus q. It is seen that

$$v_a = z_a I_a + z_m I_b + z_m I'_b \tag{20.55a}$$

$$v_b = z_m I_a + z_b I_b + 0 \tag{20.55b}$$

$$v_b = z_m I_a + 0 + (-z_b)\, I'_b \tag{20.55c}$$

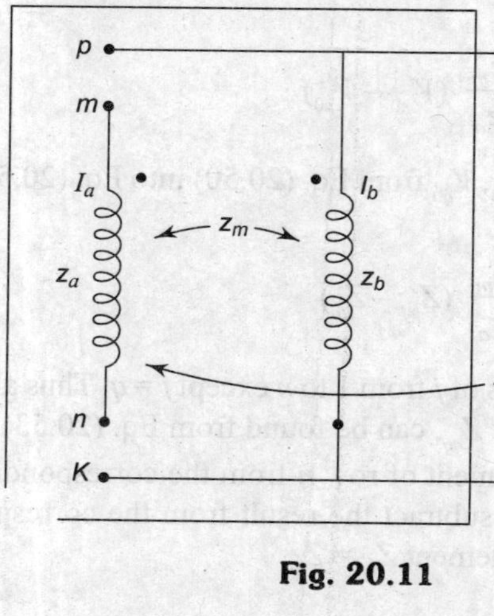

Fig. 20.11

From Eq. (20.55b and c)

$$0 = z_b\,(I_b + I'_b) \qquad (20.56)$$

Which means that $I_b + I'_b$ is zero

Substituting $I_s + I'_b = 0$ in Eq. (20.55a)

$$v_a = z_a I_a \qquad (20.57)$$

Eq. (20.57) shows that the mutual coupling of z_a with z_b has been removed. Thus we have removed mutually coupled z_b between old buses p and k.

Example 20.6.

Figure 20.12 shows a 4-bus system. Treat bus 4 as reference. Find $[Z_{bus}]$

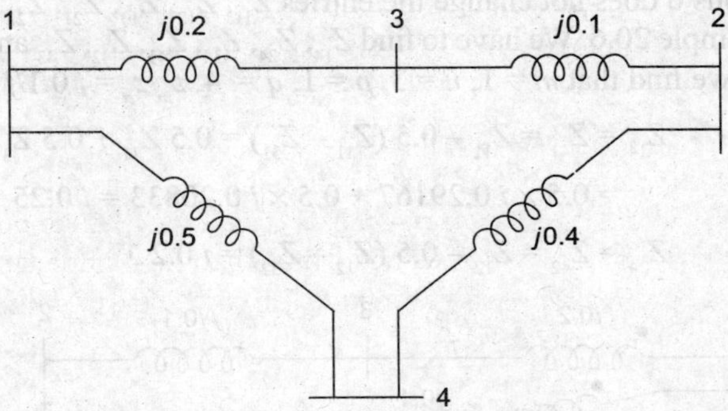

Fig. 20.12

Solution

We start with j 0.5 Ω element between buses 1 and 4. Type 1 modification gives

$$[Z_{bus}] = [j\,0.5]$$

Add j 0.4 Ω element between buses 2 and 4. Type 1 modification gives,

$$[Z_{bus}] = \begin{bmatrix} j0.5 & 0 \\ 0 & j0.4 \end{bmatrix}$$

Add j 0.2 Ω element between buses 1 and 3. Type 2 modification gives

$$[Z_{bus}] = \begin{bmatrix} j0.5 & 0 & j0.5 \\ 0 & j0.4 & 0 \\ j0.5 & 0 & j0.7 \end{bmatrix}$$

Add j 0.1 Ω element between old buses 2 and 3. Type 4 modification gives

$$[Z_{bus}] = \begin{bmatrix} j0.5 & 0 & j0.5 \\ 0 & j0.4 & 0 \\ j0.5 & 0 & j0.7 \end{bmatrix} - \frac{1}{j.1 + j.4 + j.7}\begin{bmatrix} -j.5 \\ j.4 \\ -j.7 \end{bmatrix}[-j.5 \quad j.4 \quad -j.7]$$

$$= \begin{bmatrix} j0.5 & 0 & j0.5 \\ 0 & j0.4 & 0 \\ j0.5 & 0 & j0.7 \end{bmatrix} - (-j.83333)\begin{bmatrix} -.25 & .2 & -.35 \\ .2 & -.16 & .28 \\ -.35 & .28 & -.49 \end{bmatrix}$$

$$[Z_{bus}] = \begin{bmatrix} j.29167 & j.16667 & j.20833 \\ j.16667 & j.26667 & j.23333 \\ j.20833 & j.23333 & j.29167 \end{bmatrix}$$

Example 20.7.

The system of Fig. 20.12 is modified by designating bus 4 as bus 0. A new element having impedance $j\,0.5$ is added between old bus 1 and a new bus 4. This branch has a mutual coupling of $j\,0.1$ with the branch between buses 1 and 3. Find the Z_{bus} for this augmented system.

solution

The augmented system is shown in Fig. 20.13. Since bus 4 was taken as reference in example 20.2, designating it as bus 0 does not change the entries $Z_{11}, Z_{12}, Z_{13}, Z_{21}, Z_{22}, Z_{23}, Z_{31}, Z_{32}, Z_{33}$ of the Z_{bus} evaluated in example 20.6. We have to find $Z_{14}, Z_{24}, Z_{34}, Z_{41}, Z_{42}, Z_{43}$ and Z_{44} to define new Z_{bus}. Using Eq. (20.53) we find that $m = 1, n = 3, p = 1, q = 4, z_m/z_a = j\,0.1/j_{0.2} = 0.5$. Therefore

$$Z_{14} = Z_{41} = Z_{11} - 0.5\,(Z_{11} - Z_{31}) = 0.5\,Z_{11} + 0.5\,Z_{31}$$
$$= 0.5 \times j\,0.29167 + 0.5 \times j\,0.20833 = j\,0.25$$

$$Z_{24} = Z_{42} = Z_{12} - 0.5\,(Z_{12} - Z_{32}) = j\,0.2$$

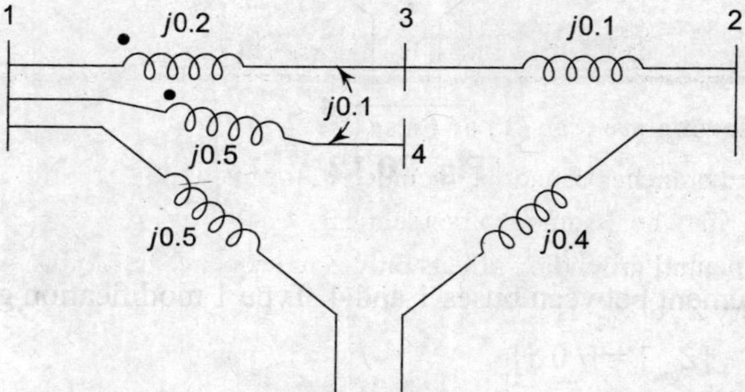

Fig. 20.13

$$Z_{34} = Z_{43} = Z_{13} - 0.5\,(Z_{13} - Z_{33}) = j\,0.25$$

To find Z_{44} we use Eq. (20.54)

$$Z_{44} = Z_{14} - 0.5\,(Z_{14} - Z_{34}) - \left[\frac{z_m^2}{z_a} - z_b\right]$$

$$= j\,0.25 - 0.5\,(j\,0.25 - j\,0.25) - \left[\frac{j\,0.1 \times j\,0.1}{j\,0.2} - j\,0.5\right]$$

$$= j\,0.25 - j\,0.05 + j\,0.5 = j\,0.7$$

Therefore $$[Z_{bus}] = \begin{bmatrix} j0.29167 & j0.16667 & j0.20833 & j0.25 \\ j0.16667 & j0.26667 & j0.23333 & j0.2 \\ j0.20833 & j0.23333 & j0.29167 & j0.25 \\ j0.25 & j0.2 & j0.25 & j0.7 \end{bmatrix}$$

SUMMARY

20.1 Digital fault studies are carried out using Z_{bus}.

20.2 Algorithm for formulating $[Z_{bus}]$ is described in terms of modifying an existing bus impedance matrix.

20.3 The modifications of an existing bus impedance matrix can take any of the following four forms:
- (a) Adding z_s from a new bus to reference
- (b) Adding z_s from a new bus to an old bus
- (c) Adding z_s from an old bus to reference
- (d) Adding z_s between two old buses.

20.4 If a power system network is changed, the original Z_{bus} matrix can be modified to give the new matrix.

20.5 Mutually coupled lines can also be incorporated in formulating Z_{bus}.

20.6 In all fault calculations fault impedance Z_f is generally neglected.

20.7 If Z_{bus} is formulated, fault calculations can be done for fault on any of the buses.

20.8 For unsymmetrical fault analysis it is necessary to formulate Z_{1-bus}, Z_{2-bus} and Z_{0-bus} matrices.

20.9 The method of neutral grounding affects Z_{0-bus} only.

20.10 If magnitude of fault current is small, the pre-fault current assumes greater importance.

SELF TEST QUESTIONS

A. State whether the following are true (T) or False (F)

20.1 Mutually coupled branches cannot be included in formulating Z_{bus}.

20.2 Z_{1-bus} and Z_{2-bus} may be assumed to be identical.

20.3 The method of neutral grounding affects only zero sequence bus impedance matrix.

20.4 The general entry Z_{ik} of Z_{bus} is

$$Z_{ik} = \left.\frac{V_i}{I_k}\right|_{I_1 = I_2 \ldots\ldots = I_n = 0 \text{ and } I_k \neq 0}$$

20.5 Addition of a tree branch changes an $n \times n$ matrix $[Z_{bus}]_{old}$ to $(n+1) \times (n+1)$ matrix.

20.6 Addition of a link between an old bus and reference increases the size of $[Z_{bus}]$.

20.7 V_{0-bus} is a $(n \times 1)$ matrix.

20.8 Fault currents, in general, flow into the buses.

B. Fill in the blanks with suitable words.

20.9 In formulating Z_{bus} the number of elements added at one time is ________.

20.10 For a single line to ground fault at kth bus, the relation between sequence currents is ________.

20.11 The line current I_{ik}^{m} = ________

20.12 In terms of $[Z_{0-bus}]$ and I_{0-bus}, V_{0-bus} = ________

20.13 If Z_f is the fault impedance and I_k^1 is 3 phase fault current at kth bus, V_k^1 = ________.

20.14 If a system has 15 buses, than for fault at 8th bus, I_7^1 = ________.

20.15 Only the ________ sequence network contains active sources.

ANSWERS

20.1 F   20.2 T   20.3 T   20.4 T   20.5 T   20.6 F

20.7 T   20.8 F   20.9 one   20.10 $I_k^0 = I_k^1 = I_k^2$ 20.11 $\dfrac{V_i^m - V_k^m}{z_s^m}$

20.12 $-[Z_{0-bus}]I_{0-bus}$ 20.13 $I_k^1 Z_f$ 20.14 zero   20.15 positive

JOB INTERVIEW QUESTIONS

20.1 What is the advantage of digital fault analysis?

20.2 What is meant by Z_{bus}?

20.3 For unsymmetrical faults how many bus impedance matrices have to be formulated?

20.4 Which bus impedance matrix is affected by the method of neutral grounding?

20.5 When should prefault currents be included in fault analysis?

REVIEW QUESTIONS

20.1 The Z_{bus} method is very suitable for fault studies on large systems. Why?

20.2 Z_{bus} can be obtained by inverting Y_{bus}. Is this Y_{bus} the same as used for load flow studies? Discuss.

20.3 In the step by step formulation of Z_{bus} it is not necessary to include the case of an element connected between two new buses. Why?

20.4 Why are prefault currents usually neglected in fault computations?

20.5 In all fault calculations Z_f is usually neglected. Why?

20.6 The passive positive and negative sequence bus impedance matrices are sometimes assumed to be identical. Why?

20.7 Give a systematic procedure for formulating Z_{bus} deriving all necessary equations.

20.8 Derive the necessary equations for calculating the fault current and bus voltages for (a) 3-phase fault (b) single line to ground fault (c) line to line fault (d) double line to ground fault.

20.9 How can mutually coupled branches be included in formulating Z_{bus}? Derive all necessary equations for various cases.

PROBLEMS

20.1 In the network of Fig. 20.6, bus 4 is redesignated as bus 1. Treat the new bus 1 as reference. Formulate Z_{bus}.

$$\begin{bmatrix} 5/8 & 1/8 & 1/4 \\ 1/8 & 5/8 & 1/4 \\ 1/4 & 1/4 & 1/2 \end{bmatrix}$$

20.2 Formulate Z_{bus} for the system of Fig. 20.14. Treat bus 1 as reference.

$$\begin{bmatrix} 2 & 1 \\ 1 & 2 \end{bmatrix}$$

20.3 Formulate $[Z_{bus}]$ for the system shown in Fig. 20.15.

$$\begin{bmatrix} 1 & 1 & 1 \\ 1 & 2 & 2 \\ 1 & 2 & 3 \end{bmatrix}$$

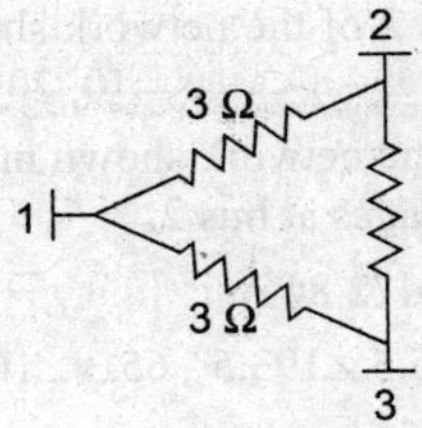

Fig. 20.14

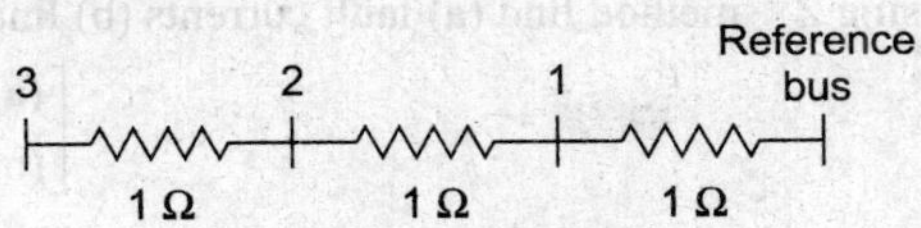

Fig. 20.15

20.4 The system of Fig. 20.15 is modified by adding a new bus number 4 which is connected to the bus 2 through a 7-ohm resistance. Find the bus impedance matrix for the modified network.

$$\begin{bmatrix} 1 & 1 & 1 & 1 \\ 1 & 2 & 2 & 2 \\ 1 & 2 & 3 & 2 \\ 1 & 2 & 2 & 9 \end{bmatrix}$$

20.5 The network of Fig. 20.15 is modified by adding a 6-ohm resistance between reference bus and bus 3. The new element has no mutual coupling with any of the existing elements. Find $[Z_{bus}]$ for the modified network.

$$\begin{bmatrix} \frac{8}{9} & \frac{7}{9} & \frac{2}{3} \\ \frac{7}{9} & \frac{14}{9} & \frac{4}{3} \\ \frac{2}{3} & \frac{4}{3} & 2 \end{bmatrix}$$

20.6 The network of Fig. 20.15 is modified by adding a resistance of 5 ohm between buses 1 and 3. Neglect mutual coupling. Find $[Z_{bus}]$ for the modified network.

$$\begin{bmatrix} 1 & 1 & 1 \\ 1 & \frac{13}{7} & \frac{12}{7} \\ 1 & \frac{12}{7} & \frac{17}{7} \end{bmatrix}$$

20.7 Modify the result obtained in problem 20.6 by adding a resistance of – 5 ohm between buses 1 and 3. This change modifies the network back to that shown in Fig. 20.15. Check that by this modification, the bus impedance matrix also returns to the original value obtained in problem 20.3.

20.8 A 3-phase dead short circuit occurs at bus 3 of system shown in Fig. 18.14. Find (a) fault currents (b) line to neutral voltage at bus 2. Neglect prefault load currents.

$$\begin{bmatrix} \text{(a)}\, 1700.7 \angle -90°, 1700.7 \angle -210°, 1700.7 \angle -30° \text{ A} \\ \text{(b)}\, 17.64 \angle 0°, 17.64 \angle -120°, 17.64 \angle 120° \text{kV} \end{bmatrix}$$

20.9 A single line to ground fault occurs at bus 2 of the network shown in Fig. 18.14. Using Z_{bus} method find (a) fault current (b) line to neutral voltages at the fault point.

[(a) 2148.77 A (b) 0, 113.18 $\angle 256.37°$, 113.18 $\angle 103.63°$ kV]

20.10 A double line to ground fault occurs on phases *b* and *c* at bus 2 of the network shown in Fig. 18.14. Using $[Z_{bus}]$ method find fault currents. [0, 2090, 2090 A]

20.11 A line to line fault occurs between phases *b* and *c* at bus 3 of the network shown in Fig. 18.14. Using Z_{bus} method find (a) fault currents (b) line to neutral voltages at bus 2.

$$\begin{bmatrix} \text{(a) } 0, -1472.84, 1472.84 \text{ A} \\ \text{(b) } 127.02 \angle 0°, 65.3 \angle 193.5°, 65.3 \angle 166.50° \text{ kV} \end{bmatrix}$$

CHAPTER

21

Design of Transmission Lines

21.1 INTRODUCTION

The design of a transmission line involves a number of technical and economic aspects. The power capacity and distance of transmission are specified. The voltage regulation and efficiency are also specified. The design details include line voltage, size of phase conductors, span, spacing and configuration of conductors, number and size of earth wires, number of insulators, clearances, sag under operating and erection conditions, etc. Once these design features are available, the voltage regulation and efficiency can be calculated. In case these quantities are not within the prescribed limits, a revision of the design is necessary.

21.2 CHOICE OF VOLTAGE

The cost and performance of the line depend, to a great extent, on the line voltage. An empirical formula for the optimum voltage is

$$V = 5.5\left(\frac{L}{1.6}+\frac{P}{100}\right)^{0.5} \tag{21.1}$$

where V = line voltage in kV

L = distance in km

P = power in kW

A standard voltage nearest to this value should be adopted.

The above formula gives only a preliminary estimate. As discussed in section. 1.6, the choice of the most economical voltage requires a detailed study of many technical and economic aspects. Once a preliminary estimate is available a detailed analysis is necessary. This becomes all the more necessary when the final choice is likely to fall in the EHV/UHV range.

21.3 SELECTION OF CONDUCTOR SIZE

The cost of conductor is about 30 to 45 per cent of the total cost of the line. Moreover the cost of towers, foundations and line losses also depend on the conductor size. A proper selection of the size of phase conductors is, therefore, very important.

Overhead transmission lines invariably use ACSR conductors. These conductors are manufactured in a variety of sizes (Appendix A).

The size of the conductor should be such that it can carry the rated current continuously without excessive rise in temperature. The temperature affects the sag and the loss of the tensile strength (due to annealing) of the conductor. For copper and aluminium, annealing starts at about 100°C and the operating temperature should be the well below this value. The standard practice is to design the line for a conductor temperature of 75°C.

The temperature rise of the conductor depends on the conductor heating due to I^2R loss and heat dissipation. In overhead lines heat is dissipated by convection and radiation. A steady temperature will be reached when

$$I^2R = (W_c + W_r)A_s \tag{21.2}$$

where I = rms value of conductor current, amperes

R = ac resistance of conductor, ohms/meter length

W_c = heat loss due to convection, watts per m² surface area

W_r = heat loss due to radiation, watts per m² surface area

A_s = conductor surface area per metre length

The heat loss due to convection is given by the equation

$$W_c = \frac{5.727\sqrt{pv}}{(T_a)^{0.123}\sqrt{d}}\,\Delta t \text{ watts/m}^2 \tag{21.3}$$

where p = pressure in atmospheres, T_a is the temperature of air in °K, v is velocity of air in m/sec, d is the diameter of the conductor in m and Δt is the difference between the temperatures of conductor and air. The above formula is valid if $v > 0.15$ m/sec and $d > 0.75 \times 10^{-2}$ m.

The heat loss due to radiation is proportional to the difference of the fourth power of the temperature of the conductor and the surroundings. This loss can be found from the equation

$$W_r = 57039.9.e.\left[\left(\frac{T_1}{1000}\right)^4 - \left(\frac{T_2}{1000}\right)^4\right] \text{ w/m}^2 \tag{21.4}$$

where T_1 is the conductor temperature in °K, T_2 is the temperature of surroundings °K and e is the relative emissivity of the surface (e = 1.0 for black body and about 0.5 for oxidized copper).

To use Eqs (21.2), (21.3) and (21.4) it is necessary to assume suitable values of maximum safe conductor temperature, air temperature and velocity and temperature of surroundings. Once these assumptions are made, the ampacity (ampere carrying capacity) of a specific conductor can be determined. The ampacity values have been determined for the different conductor sizes and are available in conductor tables given in reference books. The ampacities of some commonly used ACSR conductors are given in Appendix A.

The above criteria for the selection of conductor size is applicable for lines upto 220 kV. For EHV and UHV lines the considerations of corona and radio interference have also to be taken into account. Experience has shown that a conductor selected on the basis of corona and radio interference considerations would normally be thicker than the one selected from the point of view of current carrying capacity.

An economical method of reducing corona and radio interference is to use bundled conductors. EHV and UHV line always use bundled conductors. 400 kV lines, in India, use two sub-conductors per phase. In other countries bundles of upto eight sub-conductors have been used in actual practice. An experimental UHV line has 18 sub-conductors. However, bundles of two and four sub-conductors are the most common.

In chapter 14, the calculation of corona loss and radio interference for EHV an UHV lines is described briefly. For a detailed description the reader is referred to (12) in the bibliography.

21.4 CHOICE OF SPAN, NUMBER OF CIRCUITS, CONDUCTOR CONFIGURATION

The considerations governing the choice of span, conductor configuration and clearances have been discussed in section 6.5. For a high voltage line the economical value of span lies in the range of 200 to 300 m. For 400 kV lines spans ranging from 350 to 400 m are used. For river and ravine crossings exceptionally long spans upto 800 m or so have been satisfactorily employed.

A transmission line may be a single circuit line or a double circuit line. A double circuit line has a higher power transfer capability and greater reliability than a single circuit line. Each circuit of a double circuit line is usually designed for 75% of the line capacity. In India, both single circuit and double circuit lines exist in the EHV and high voltage class (66 kV, 132 kV, 220 kV and 400 kV). In foreign countries also both double and single circuit lines exist. The number of circuits for a proposed line can be determined from the surge impedance loading.

Experience has shown that a vertical configuration is the most economical for double circuit lines. For single circuit lines, horizontal or *L* type configurations are used.

21.5 INSULATION DESIGN

The insulation design affects the performance of the line to a great extent. Line insulation should be sufficient to take care of switching overvoltages, temporary overvoltages and atmospheric overvoltages.

The insulation level of the transmission lines is based on the switching surge expectancy on the system. The maximum switching surge overvoltage to the ground is taken as 2.5 *pu* and the insulation is designed for this voltage. In addition adequate protection against atmospheric overvoltages (direct lighting strokes) is provided. In EHV and UHV lines overvoltages due to switching surges assume a greater importance than atmospheric overvoltages. The basic procedure for insulation design has been outlined in section 14.8. For a detailed design procedure and calculation of lightning trip out rate of the transmission lines the reader is referred to [10, 31] of the Bibliography.

21.6 SELECTION OF GROUND WIRE

The primary function of ground wires is to shield the phase conductors from the lightning strokes. They are placed above the phase conductors and are grounded at every/alternate tower. Thus they help in dissipating the lightning currents to the ground.

The selection of the number and configuration of the ground wires is of great importance in the protection of transmission lines against direct strokes. The number of ground wires may be one or two. A shielding angle (Fig. 12.29) of about 30° is considered to be adequate for high voltage lines. However, for high voltage lines in areas with low lightning hazards, shielding angles upto 45° have been used. EHV lines are usually provided with two ground wires and the shielding angle for such lines is kept at about 20°. To prevent back flashover from the earthed metal to the phase conductors, the tower footing resistance should not exceed 10 ohms. The vertical separation between the ground wires and phase conductors should be greater at mid span than at the supports, i.e., the ground wire should have lesser sag as compared to the phase conductors. The material most commonly used for ground wires is galvanized steel.

A ground wire should be able to carry the maximum expected lightning current, without undue heating. It should also have sufficient mechanical strength. Experience has shown that if a ground wire is mechanically strong, it can carry the maximum lightning current without excessive heating. Therefore, the size of the ground wire is generally decided on the basis of mechanical strength.

21.7 EVALUATION OF LINE PERFORMANCE

The parameters of the line can be calculated as discussed in Chapter 2. These parameters are then used to evaluate the efficiency and regulation. It is sufficiently accurate to represent the line by a nominal T or π circuit for the efficiency and regulation calculations. However, if the line is very long, the calculations should be based on *ABCD* constants. If the efficiency and regulation are not within the prescribed values, it may be necessary to revise the design by selecting a thick conductor cross-section and changing the conductor configuration. In some cases it may be necessary to use a higher transmission voltage in the revised design. The shunt compensation requirements for peak load and off peak conditions can be calculated as discussed in chapter 3. The corona loss and radio interference can be calculated as discussed in chapters 6 and 14. The radio interference calculations are usually necessary only for EHV and UHV lines. The sag and tension calculations can be done as discussed in chapter 5.

Example 21.1.

It is proposed to transmit 100 MW at 0.9 power factor lagging over a distance of 200 km. The line efficiency and regulation at full load should be better than 95% and 15% respectively. Work out the following details of the transmission line. Make suitable assumptions.

(a) Select line voltage and number of circuits.

(b) Choose proper conductor and span for this line.

(c) Select a suitable value of inter-phase spacing and a suitable configuration of conductors.

(d) Calculate line parameters. Estimate the line efficiency and regulation for full load condition.

(e) Estimate corona loss.

(f) Find the capacity of shunt compensation equipment to improve the receiving end power factor to 0.95 lagging.

(g) Estimate line efficiency and regulation for full load at 0.95 power factor lagging.

(h) The line will be erected at a temperature of 30°C in still air condition. It is desired that a factor of safety of 2.5 should be maintained under bad weather condition when the temperature is 5°C and wind load is 378 N/m² of projected area. Find the sag and tension under erection condition. Also find the sag under the bad weather condition.

(i) Select a suitable number and size of ground wires for this line.

Solution

(a) Using Eq (21.1) the optimum line voltage is

$$V = 5.5\left(\frac{200}{1.6} + \frac{100 \times 1000}{100}\right)^{0.5} = 184.5 \text{ kV}$$

The nearest standard line voltage is 220 kV. Therefore, it should be a 220 kV line.

The surge impedance of a single circuit 220 kV line is about 400 ohms.

$$\text{Surge impedance loading} = \frac{220 \times 220}{400} = 121 \text{ MW}$$

Since the required power transfer is less than SIL, a single circuit is sufficient.

(b) $$I_r = \frac{100 \times 10^6}{\sqrt{3} \times 220000 \times 0.9} = 291.6 \text{ A}$$

Let the ambient temperature be 35°C. Therefore, temperature rise of 40°C can be allowed. Referring to Appendix A, a suitable conductor for this current is ACSR 6/1/4.5 mm conductor.

It is necessary to calculate the line losses and the line efficiency to check the suitability of this conductor. Line losses are approximately equal to $3\,(I_r)^2 R$ where R is the total line resistance per phase at 75°C. For the ACSR 6/1/4.50 mm conductor, the resistance at 20°C is 0.302 ohms/km. To calculate the resistance at 75°C we use Eq. (2.3)

$$\frac{r_{75}}{r_{20}} = \frac{1/\alpha_0 + 75}{1/\alpha_0 + 20} = \frac{228 + 75}{228 + 20}$$

$$r_{75} = 0.302 \times \frac{308}{228} = 0.37 \text{ ohms/km}$$

$$R = 0.37 \times 200 = 74 \text{ ohms}$$

$$\text{Line efficiency} = \frac{100 \times 10^6}{100 \times 10^6 + 3 \times 291.6 \times 291.6 \times 74} = 0.84$$

The efficiency is very poor. Hence this conductor size is not suitable.

Referring again to Appendix *A* choose the ACSR conductor 30/7/3.71 mm. The resistance of this conductor at 20°C is 0.0898 ohms/km.

$$r_{75} = 0.0898 \times \frac{303}{248} = 0.11 \text{ ohms/km}$$

$$R = 0.11 \times 200 = 22 \text{ ohms}$$

$$\text{Line efficiency} = \frac{100 \times 10^6}{100 \times 10^6 + 3 \times 291.6 \times 291.6 \times 22} = 0.947$$

The ACSR conductor 30/7/3.71 mm has much a higher current rating than the rated current of the proposed line. Therefore, the maximum temperature will be less than 75°C and the line efficiency will be higher than 95%. The characteristics of this conductor are:

Number of aluminium strands 30, each having a diameter of 3.71 mm. Number of steel strands 7, each having a diameter of 3.71 mm, over all diameter 25.97 mm.

Weight 1492.4 kg/km. Ultimate strength 135.5 kN.

The considerations governing the selection of span have been discussed in sections 6.5 and 21.4.

Experience has shown that a span of 300 m is suitable for a 220 kV line. Hence we keep the span as 300 m.

(c) As per the values given in section 6.5, an interphase spacing of 6 meters is suitable for a 220 kV line. The conductor configuration can be horizontal or *L* type. Choose horizontal configuration and 6 metre spacing between adjacent phases (12 metre between the two outer phases).

(d) $$D_{eq} = \sqrt[3]{6 \times 6 \times 12} = 7.5595 \text{ m}$$

$$r = 0.5 \times 25.97 \text{ mm} = 12.985 \times 10^{-3} \text{ m}$$

$$\text{Geometric mean radius } r' = 0.7788 \times 12.985 \times 10^{-3} = 10.11 \times 10^{-3} \text{ m}$$

To calculate line inductance we use Eq. (2.56 C)

$$L = 0.4605 \log \frac{7.5595}{10.11\times10^{-3}} = 1.32 \text{ m H/km}$$

$$Z = 22 + j\,2\,\pi \times 50 \times 1.32 \times 10^{-3} \times 200$$
$$= 22 + j\,82.9 = 85.77 \angle 75.13° \text{ ohms/phase}$$

To calculate line capacitance we use Eq. (2.89b)

$$C = \frac{0.02412}{\log \dfrac{7.5595}{12.985\times10^{-3}}} = 8.72 \times 10^{-3}\ \mu\text{F/km}$$

$$Y = j\,\omega\,Cl = 2\,\pi \times 50 \times 8.72 \times 10^{-3} \times 10^{-6} \times 200 \angle 90°$$
$$= 547.9 \times 10^{-6} \angle 90° \text{ siemens/phase}$$

Since the length of the line is 200 km, sufficiently accurate results can be obtained by the nominal π or T representation. Let we use nominal π representation.

$$1 + \frac{ZY}{2} = 1 + \frac{1}{2}\,(85.77 \angle 75.13°)\,(547.9 \times 10^{-6} \angle 90°)$$
$$= 0.977 \angle 0.35°$$

$$Y\left(1 + \frac{ZY}{4}\right) = (547.9 \times 10^{-6} \angle 90°)\,[1 + \frac{1}{4}\,(85.77 \angle 75.13°)\,(547.9 \times 10^{-6} \angle 90°)]$$
$$= 541.7 \times 10^{-6} \angle 90.2°$$

$$V_r = \frac{1}{\sqrt{3}}\,(220000) = 127017 \angle 0° \text{ V}$$

$$I_r = 291.6 \angle -\cos^{-1} 0.9 = 291.6 \angle -25.84 \text{ °A}$$

Using Eq. (4.22)

$$V_s = V_r\left(1 + \frac{ZY}{2}\right) + I_r\,Z$$
$$= (127017 \angle 0°)\,(0.977 \angle 0.35°) + (291.6 \angle -25.84°)\,(85.77 \angle 75.13°)$$
$$= 124096 \angle 0.35° + 25010 \angle 49.29°$$
$$= 141783.5 \angle 8.0° \text{ V}$$

$$I_s = V_r Y\left(1 + \frac{ZY}{4}\right) + I_r\left(1 + \frac{ZY}{2}\right)$$
$$= (127017 \angle 0°)\,(514.7 \times 10^{-6} \angle 90.2°) + (291.6 \angle -25.84°)\,(0.977 \angle 0.35°)$$
$$= 68.8 \angle 90.2° + 284.9 \angle -25.49° = 262.5 \angle -11.8°\text{A}$$

Sending end power factor = cos (8 + 11.8) = 0.941 lagging

Sending end power = 3 × 141783.5 × 262.5 × 0.941 × 10^{-6} = 105.07 MW

$$\text{Line efficiency} = \frac{100}{105.07} \times 100 = 95.17\%$$

$$V_r \text{ at no load} = \frac{|V_s|}{\left|1 + \dfrac{ZY}{2}\right|} = \frac{141783.5}{0.977} = 145121 \text{ V}$$

$$\text{Regulation} = \frac{145121-127017}{127017} \times 100 = 14.25\%$$

(e) Taking $\theta = 50°C$ and p, i.e., pressure = 74 cm of mercury and using Eq. (7.1b).

$$\delta = \frac{3.86\,p}{273+\theta} = \frac{3.86\times 74}{273\times 50} = 0.884$$

Let $m_0 = 0.84$. Using Eq. (7.9)

$$V = \frac{3\times 10^6}{\sqrt{2}} \times 12.985 \times 10^{-3} \times 0.884 \times 0.84 \ln \frac{7.5595}{12.985\times 10^{-3}} = 130.25 \text{ kV}$$

$$\frac{V}{V_d} = \frac{127017\times 10^3}{130.25\times 10^3} = 0.975$$

From the table in section 6.4, F = 0.05

Using Eq. (7.14)

$$\text{Corona loss} = \frac{21\times 10^{-6}\, f V^2}{\left(\log_{10} \dfrac{D_{eq}}{r}\right)^2} \times F$$

$$= \frac{21\times 10^{-6} \times 50 (127.017)^2}{\left(\log \dfrac{7.5595}{12.985\times 10^{-3}}\right)^2} \times 0.05 = 0.11 \text{ kW/phase/km}$$

A corona of less than 0.2 kW/phase/km is considered to be tolerable. Hence the corona loss for this line is within limits.

Total corona loss = 0.11 × 3 × 200 = 66 kW

(f) When cos ϕ = 0.9 lagging, tan ϕ_r = 0.4843

Receiving end reactive power = 100 × 0.4843 = 48.53 Mvar lagging.

When cos ϕ_r = 0.95 lagging, tan ϕ_r = 0.3287

Receiving end reactive power = 100 × 0.3287 = 32.87 Mvar lagging.

Capacity of shunt capacitors to improve the receiving end power factor from 0.9 lagging to 0.95 lagging = 48.43 – 32.87 = 15.56 Mvar leading.

(g) When the receiving end power factor has been improved to 0.95 lagging

$$V_r = 127017 \angle 0°$$

$$I_r = \frac{100\times 10^6}{\sqrt{3}\times 220000\times 0.95} = 276.24 \angle (-\cos^{-1} 0.95) = 276.24 \angle -18.20° \text{ A}$$

$$V_s = V_r\left(1+\frac{ZY}{2}\right) + I_r Z$$

$$= (127017 \angle 0°)\,(0.977 \angle 0.35°) + (276.24 \angle -18.20°)\,(85.77 \angle 75.13°)$$

$$= 124096.0 \angle 0.35° + 23693.1 \angle 56.93 = 138563.8 \angle 8.55° \text{ V}$$

$$I_s = V_r Y\left(1+\frac{ZY}{4}\right) + I_r\left(1+\frac{ZY}{2}\right)$$

$$= (127017\ \angle 0°)\ (541.7 \times 10^{-6}\ \angle\ 90.2°) + (276.24\ \angle -18.20°)\ (0.977\ \angle\ 0.35°)$$

$$= 68.8\ \angle\ 90.2° + 270 - 17.85° = 257.1\ \angle -3.1° \text{A}$$

Sending end power factor $= \cos(8.55 + 3.1) = 0.98$ lagging

Sending end power $= 3 \times 138563.8 \times 257.1 \times 0.98 \times 10 = 104.74$ MW

$$\text{Line efficiency} = \frac{100}{104.74} \times 100 = 95.5\%$$

$$V_r \text{ at no load} = \frac{|V_s|}{\left|1+\frac{ZY}{2}\right|} = \frac{138563.8}{0.977} = 141825.8 \text{ V}$$

$$\text{Regulation} = \frac{141825.8 - 127017}{127017} = 11.66\%$$

(h) Referring to section 6.7

$$d = 25.97 \text{ mm} = 2.597 \times 10^{-4} \text{m}$$

$$A = 37 \times \frac{\pi}{4}\left(\frac{3.71}{1000}\right)^2 = 4 \times 10^{-4} \text{ sq m}$$

Let modulus of elasticity $E = 91.4 \times 10^9$ N/m^2

and $\alpha = 18.44 \times 10^{-6}$ per °C

$$w = 1492.4 \text{ kg/km} = 14.64 \text{ N/m}$$

For bad weather condition (subscript 1)

$$F_w = 378 \times 2.597 \times 10^{-2} = 9.82 \text{ N/m}$$

$$F_{t1} = (14.64^2 + 9.82^2)^{0.5} = 17.63 \text{ N/m}$$

$$T_1 = \frac{135.5 \times 1000}{2.5} = 54200 \text{ N},\ \theta_1 = 5°\text{C}$$

For erection condition (subscript 2)

$$F_{t2} = 14.64 \text{ N/m}$$

$$T_2 = ?\ \theta_2 = 30\ °\text{C}$$

Using Eq. (6.20)

$$T_2^2\left[T_2 - \left\{T_1 - \alpha\, AE\,(\theta_2 - \theta_1) - \frac{AEF_{t1}^2\, l^2}{24 T_1^2}\right\}\right] = \frac{AEF_{t2}^2\, l^2}{24}$$

$$\alpha AE\,(\theta_2 - \theta_1) = 18.44 \times 10^{-6} \times 4 \times 10^{-4} \times 91.4 \times 10^9 \times (30 - 5) = 16854.2$$

$$\frac{AEF_{t1}^2\, l^2}{24\, T_1^2} = \frac{4\times10^{-4} \times 91.4\times10^9\ (17.63)^2\ (300)^2}{24(54200)^2} = 14505.7$$

$$\frac{AEF_{t2}^2 l^2}{24} = \frac{4\times10^{-4}\times91.4\times10^9 (14.64)^2 (300)^2}{24} = 29384.6\times10^9$$

$$T_2^2 [T_2 - 54200 + 16854.2 + 14505.7] = 29384.6\times10^9$$

or $$T_2^2 (T_2 - 22840.1) = 29384.6\times10^9$$

or $$T_2^3 - 22840.1 T_2^2 - 29384.6\times10^9 = 0$$

Whence $T_2 = 40650$ N

Sag under erection condition $= \dfrac{14.64\times300\times300}{8\times40650} = 4.05$ m

Sag under bad weather condition $= \dfrac{17.63\times300\times300}{8\times54200} = 3.66$ m

Vertical sag under bad weather condition

$$= 3.66 \cos\left[\tan^{-1}\frac{9.82}{14.64}\right] = 3.04$$

(i) The number of ground wires depends on the degree of lightning protection for which the line is to be designed. For high voltage lines with *L* type configuration one ground wire provides satisfactory lightning protection. However, the use of two ground wires results in a lower value of shielding angle and better lightning protection. For lines with horizontal configuration of conductors, two ground wires are needed for proper lightning protection. Therefore, the proposed line should be provided with two ground wires.

Experience has shown that stranded galvanized steel wire of 7/3.66 mm size has good mechanical strength. Hence use 2 ground wires of 7/3.66 mm galvanized steel wires.

SUMMARY

21.1 Line voltage depends on power to be transmitted and line length.

21.2 Conductor size of transmission lines should be such that it can carry the rated current continuously without excessive temperature rise.

21.3 Generally a temperature rise of 30ºC is used to calculate conductor size.

21.4 Spans of transmission lines are about 300–500 m

21.5 The maximum switching surge is taken as 2.5 *pu*.

21.6 Size of ground wire should be such that it is mechanically strong.

21.7 Line performance calculations include evaluation of losses, voltage regulation, corona loss, sag and tension, radio interference and shunt compensation requirement.

SELF TEST QUESTIONS

A. State whether the following are true (T) or False (F)

21.1 Selection of line voltage is governed by line length and MW capacity.

21.2 The conductor cross-section of transmission line is governed by temperature size considerations.

21.3 For double circuit lines horizontal configuration is most common.

21.4 Insulation design of EHV line is based on lightning surges.

21.5 Ground wires shield the phase wires from lightning strokes.

21.6 Only one ground size is generally used.

21.7 Line parameters affect the line efficiency and voltage regulation.

21.8 Radio interference calculation are necessary only for EHV and UHV lines

B. Fill in the blanks with suitable words.

21.9 Temperature size of ____________ is assumed for calculating the size of conductor of transmission lines.

21.10 Magnitude of maximum switching surges is is assumed to be ____________.

21.11 For long lines performance calculations are done by using ____________.

21.12 In transmission lines, ground wire is placed ____________ the phase conductors.

21.13 Ground were is generally made of ____________.

21.14 Shielding angle of EHV line is about ____________.

21.15 EHV line use ____________ conductors.

ANSWERS

21.1 T	21.2 T	21.3 F	21.4 F
21.5 T	21.6 F	21.7 T	21.8 T
21.9 30°C	21.10 2.5 *pu*	21.11 ABCD Constants	21.12 above
21.13 galvanised steel		21.14 20°	21.15 Bundled

JOB INTERVIEW QUESTIONS

21.1 How is the voltage of a new line decided?

21.2 What is meant by shielding angle?

21.3 What is the importance of corona loss in line design

21.4 What is the use of ground wire in transmission lines?

21.5 What is meant by insulation level of transmission line

REVIEW QUESTIONS

21.1 Discuss the procedure to design a proposed line.

21.2 What consideration govern the size of conductor of a transmission line?

21.3 How is line performance calculated ?

21.4 What is the effect of span and conductor configuration on line performance.

21.5 What are switching overvoltages ? What is their role in transmission line design?

PROBLEMS

21.1 It is proposed to transmit 50 MW at 0.95 power factor over a distance of 100 km. The line efficiency and regulation should be better than 95% and 10% respectively. Select line voltages, number of circuits, conductor size, span, interphase spacing and configuration of conductors. Calculate line parameters, efficiency and regulation of the line.

21.2 Work out the following details of a double circuit line to supply a 250 MW, 0.9 lagging power factor load situated at a distance of 150 km from a generating station.

(a) Select line voltage, conductor size, span, interphase spacing and configuration of conductors.

(b) Calculate line parameters, efficiency, regulation and corona loss for this line.

21.3 Design a double circuit transmission line to transmit 150 MW at 0.9 lagging power factor over a distance of 150 km. Work out the following details:

(a) Select line voltage, conductor size, interphase spacing and configuration of conductors.

(b) Calculate line parameters, efficiency, regulation and corona loss for this line.

21.4 The double circuit line of problem 17.3 is connected through a step up transformer to a 150 MW generator at the sending end. At the receiving end it is connected to an infinite bus. The voltage of the infinite bus is to be 1 pu and the sending end voltage can be taken as 1.05 pu. Assume suitable values of transformer reactance, generator reactance, and inertia constant of generator and an initial load of 0.75 pu. Carry out the following stability studies:

(a) Plot swing curves using point by point method for a 3-phase fault at the middle of one of the lines for sustained fault and for fault clearing times of 0.2 and 0.4 seconds.

(b) Plot curves for transient power limit as a function of clearing angle and as a function of clearing time for the above fault location.

(c) Find stable transient power limit for instantanous clearance and for fault clearning time of 0.2 and 0.5 seconds.

CHAPTER 22

Power System Earthing

22.1 OBJECTIVE

A number of points from generators to consumer's installations are earthed (or grounded). The objects of earthing are:

1. To ensure that no part of equipments, other than live parts, should assume a potential which is dangerously different from that of surroundings.
2. To allow sufficient current to flow safely for proper operation of protective devices.
3. To limit over voltages between neutral and ground and between line and ground.
4. To suppress dangerous potential gradients. The earth potential gradients, if not suppressed, may cause incorrect operation of control and protective devices and may also cause shock or injury to personnel.

Earthing can be divided into neutral earthing and equipment earthing. Neutral earthing deals with the earthing of system neutral to ensure system security and protection. Equipment earthing deals with earthing of non-current carrying parts of the equipment to ensure safety of personnel and protection against lightning.

22.2 DEFINITIONS

(a) *Earth electrode:* A rod, pipe, plate or an array of conductors, embedded in earth horizontally or vertically. In distribution system the earth electrode may consist of a rod, about 1 m long, driven vertically into ground. For substations an elaborate earthing system known as earthing grid or earth mat is used.

(b) *Earth current:* The current dissipated by earth electrode into the ground.

(c) *Resistance of earth electrode:* The resistance offered by the earth electrode to the flow of current into the ground. This resistance is not the ohmic resistance of the electrode but represents the resistance of the mass of earth surrounding the earth electrode. Numerically it is equal to the ratio of the potential of earth electrode with respect to a remote point, to the current dissipated by it.

(d) *Step potential:* The potential difference shunted by a human body between two accessible points on the ground separated by the distance of one pace assumed to be equal to one meter.

(e) *Touch potential:* The potential difference between a point on the ground and a point on an object likely to carry fault current (e.g., frame of equipment) and which can be touched by a person.

(f) *Mesh potential:* The maximum touch potential within a mesh of the grid.

(g) *Transferred potential:* A special case of touch potential where a potential is transferred into or out of the substation.

22.3 TOLERABLE LIMITS OF BODY CURRENTS

The effect of electric current passing through vital organs of the body depend on magnitude, duration and frequency of current. The most dangerous consequence is a heart condition known as ventricular fibrillation which results in stoppage of blood circulation.

(a) *Effect of magnitude of current* The threshold of perception is a current of 1 mA. Currents in the range of 1–6 mA are known as 'let go current' because these currents, though unpleasant, do not impair the ability of a person, holding an energised object, to release it. Currents in the 9–25 mA range may be painful and impair the ability to release energised object. Still higher currents make breathing difficult. However, if the current is less than about 60 mA, the effects are not permanent and disappear when current is interrupted. Currents higher than 60 mA may lead to ventricular filbrillation, injury and death.

(b) *Effect of duration of current* The magnitude of 50 Hz tolerable current is related to duration. According to tests reported by Dalziel, 99.5% of persons of 50 kg weight can safely withstand the current given by equation.

$$I_B = 0.116/\sqrt{t} \tag{22.1}$$

where I_B is the rms value of body current in amperes and t is the time in seconds. If the weight of body is 70 kg, the equation for tolerable current is

$$I_B = 0.157/\sqrt{t} \tag{22.2}$$

Eqs. (22.1) and (22.2) are valid for $0.03 < t < 3$ seconds.

(c) *Effect of frequency* The tolerable currents mentioned above for 50–60 Hz. It has been found that human body can tolerate about 5 times higher direct current. At high frequencies (3000 – 10000 Hz.) still higher currents can be tolerated.

22.4 SOIL RESISTIVITY

The resistivity of soil depends on the following factors:

(a) *Soil type:* The type of soil governs the resistivity to a large extent. Some typical values are: Sea water 2.5 ohm-m, Tap water 20 ohm-m, clay 50 ohm-m, sand clay mixture 100 ohm-m, sand 2000 ohm-m, wet concrete 100 ohm-m, dry concrete 10000 ohm-m, rock 10000 ohm-m.

(b) *Moisture content:* Electrical conduction in soils is electrolytic. Soil resistivity decreases with increase in moisture content as shown in Fig. 22.1 (a). However, when moisture is more than 20%, the change in resistivity is negligible. The amount of water which a soil can absorb depends on soil type, grain size and compactness.

(c) *Temperature:* When temperature is more than 0°C, its effect on soil resistivity is negligible. At 0°C, water in the soil starts freezing and resistivity increases.

(d) *Salt content:* The composition and amount of soluble salts also affect resistivity considerably as shown in Fig. 22.1 (b).

(e) *Magnitude of current:* If the value of current being dissipated by the soil is high, it may cause significant drying of soil and increase in its resistivity.

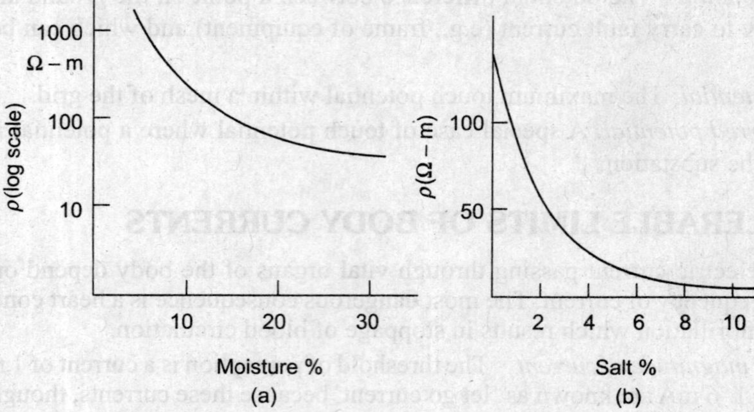

Fig. 22.1 Variation of soil resistivity with (a) moisture (b) salt content

The soil resistivity at a particular location also changes with depth. Generally, the lower layers of soil have greater moisture content and lower resistivity. However, if the lower layer contains hard and rocky soil, resistivity may increase with depth.

Since soil resistivity depends on a number of factors, it is always advisable to make proper resistivity measurements at the proposed site of earthing system.

22.5 EARTH RESISTANCE

22.5.1 Hemispherical Electrode

Fig. 22.2 (a) shows a buried hemisphere. The total resistance can be divided into three parts (i) resistance of conductor (ii) contact resistance between the surface of electrode and main body of earth (iii) resistance of body of earth surrounding the electrode.

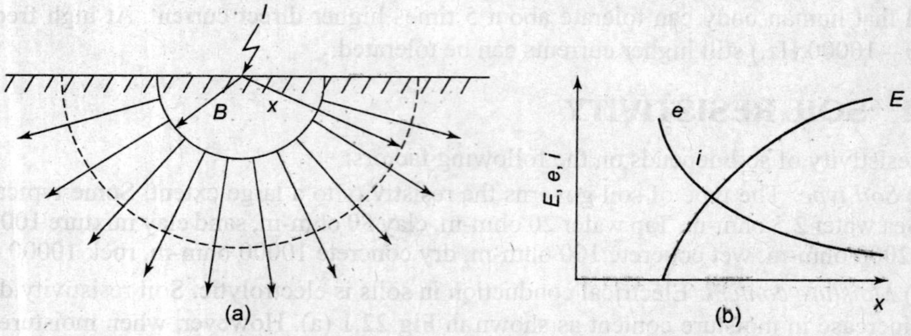

Fig. 22.2 (a) Hemispherical electrode (b) Variation of e, i, E, with x

The resistance of conductor and contact resistance are negligible. Therefore, the main part of resistance is that of body of earth surrounding the electrode.

I, the current dissipated by electrode spreads out radically in the earth. The current density *i* at a distance *x* from the centre of hemisphere is

$$i = \frac{I}{2\pi x^2} \tag{22.3}$$

As per ohm's law electric field strength e due to current density i is:

$$e = \rho\, i = \frac{\rho I}{2\pi x^2} \tag{22.4}$$

The voltage is the line integral of field strength e from the surface of sphere of radius B to the distance x. Therefore,

$$E = \int_B^x e\,dx = \frac{\rho I}{2\pi}\int_B^x \frac{dx}{x^2} = \frac{\rho I}{2\pi}\left[\frac{1}{B} - \frac{1}{x}\right] \tag{22.5}$$

The variation of e, i and E with distance x is shown in Fig. 22.2(b). The voltage between the hemispherical electrode and a point at infinity (i.e., $x = \infty$) is

$$E = \frac{\rho I}{2\pi B} \tag{22.6}$$

The earth resistance is

$$R = \frac{E}{I} = \frac{\rho I}{2\pi B} \tag{22.7}$$

22.5.2 General Equation

Consider a system of two electrodes. Let V_1 and V_2 be the potentials of these electrodes and V be the potential at any point in the medium of resistivity ρ. Let ψ be the electrostatic potential in the analogous problem in electrostatics with air as medium.

The current flow normal to surface at any point of the electrode surface is $\frac{1}{\rho}\frac{\delta V}{\delta n}$. Total flow outwards from the electrode is

$$-\frac{1}{\rho}\iint \frac{\partial V}{\partial n}\cdot ds = -\frac{1}{\rho}\iint \frac{\partial \Psi}{\partial n}\cdot ds$$

where ds is an element of the electrode surface. If Q is the charge on this electrode in the analogous electrostatic case, then by Gauss theorem

$$-\iint \frac{\partial \Psi}{\partial n}\cdot ds = 4\pi Q$$

The total flow of current is

$$I = \frac{4\pi Q}{\rho} \tag{22.8}$$

If capacity between the electrodes in air in the analogous electrostatics case is C, then

$$\psi_1 - \psi_2 = V_1 - V_2 = \frac{Q}{C} \tag{22.9}$$

The resistance between electrodes is R and equals

$$R = \frac{V_1 - V_2}{I} = \frac{Q}{C}\left(\frac{\rho}{4\pi Q}\right) = \frac{\rho}{4\pi C} \tag{22.10}$$

If C is the capacity of single electrode, the return electrode being at infinity, then R is the earthing resistance of single electrode.

Fig. 22.3 shows a sphere of radius B. The capacity of sphere in air is equal to radius, i.e., B. Therefore,

$$R = \frac{\rho}{4\pi B} \tag{22.11}$$

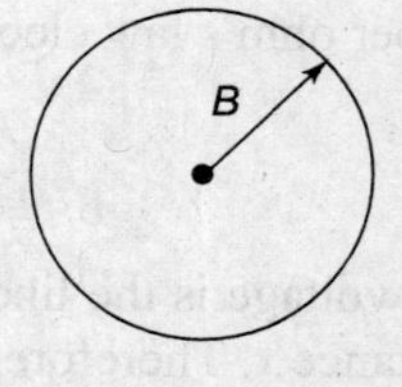

Fig. 22.3 Spherical electrode

If the electrode is a hemisphere buried with lower half in earth, the resistance would be doubled. Thus,

$$R = \frac{\rho}{2\pi B} \tag{22.12}$$

The general equation for any electrode is

$$R = \frac{\rho}{2\pi C} \tag{22.13}$$

where C is the electrostatic capacity of the electrode along with its image above the surface of earth, the combined electrode being considered as in air.

22.5.3 Driven Rod

Fig. 22.4 shows a driven rod of length l and diameter d along with its image. The electrode can be considered equivalent to an ellipsoid of revolution in which major axis is very large as compared to minor axis. On the basis of this equivalence, the capacity of rod and its image is

$$C = \frac{2l}{2\ln\frac{4l}{d}} = \frac{l}{\ln\frac{4l}{d}} \tag{22.14}$$

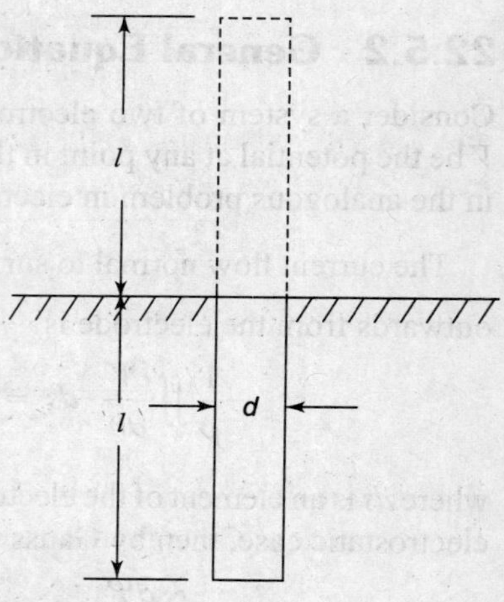

Fig. 22.4 Driven rod

Using Eq. (22.13), the earthing resistance is

$$R = \frac{\rho}{2\pi l}\ln\frac{4l}{d} \tag{22.15}$$

An alternate expression for earthing resistance of a driven rod, found by uniform current dissipation method, is

$$R = \frac{\rho}{2\pi l}\left(\ln\frac{8l}{d} - 1\right) \tag{22.15a}$$

It is seen from Eq. (22.15) that the resistance is nearly inversely proportionate to length l. The diameter d is of minor significance since l/d is the argument of logarithm.

For complex configurations it is useful to replace the rod electrode by an equivalent hemisphere having the same resistance R. From Eqs. (22.7 and 22.15), the radius B of equivalent hemisphere is

$$B = \frac{l}{\ln\frac{4l}{d}} \tag{22.16}$$

22.5.4 Multiple Rod Electrodes

Figure 22.5 shows two rods connected in parallel and at a distance D from each other. Each rod can be replaced by its equivalent hemisphere of radius B carrying a charge Q. The potential of either rod is

$$V = \frac{Q}{B} + \frac{Q}{D} = \frac{Q}{B}\left(1 + \frac{B}{D}\right) = \frac{Q}{B}(1+\alpha) \quad (22.17)$$

where $\alpha = B/D$

Total charge is $2Q$. Therefore, capacity C is

$$C = \frac{2Q}{V} = \frac{2B}{1+\alpha} \quad (22.18)$$

The earthing resistance of two parallel rods is

$$R = \frac{\rho}{2\pi C} = \frac{\rho}{4\pi B}(1+\alpha) \quad (22.19)$$

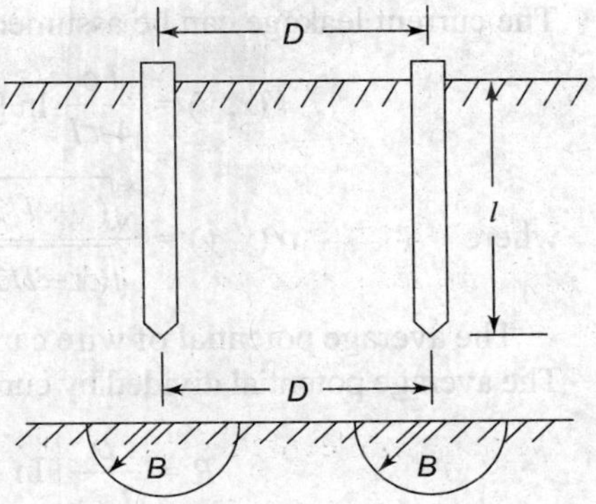

Fig. 22.5 Two driven rods

Since the resistance of one rod is $\rho/2\,\pi\,B$ we get

$$\frac{\text{Resistance of two rods in parallel}}{\text{Resistance of one rod}} = \frac{1+\alpha}{2} \quad (22.20)$$

Similarly, it can be shown that for three rods in a straight line

$$\frac{\text{Resistance of three rods in parallel}}{\text{Resistance of one rod}} = \frac{2+\alpha-4\alpha^2}{6-7\alpha} \quad (22.21)$$

and for four rods in a straight line

$$\frac{\text{Resistance of four rods in parallel}}{\text{Resistance of one rod}} = \frac{12+6\alpha-21\alpha^2}{48-40\alpha} \quad (22.22)$$

and for three rods arranged at corners of an equilateral triangle

$$\frac{\text{Resistance of three parallel rods in triangle}}{\text{Resistance of one rod}} = \frac{1+2\alpha}{3} \quad (22.23)$$

In Eqs. (22.20–22.23), α is the ratio B/D.

For four rods at the corners of a square of side D,

$$\frac{\text{Resistance of four rods at corners of square}}{\text{Resistance of one rod}} = \frac{1+2.707\alpha}{4} \quad (22.24)$$

where $\alpha = B/D$

22.5.5 Buried Horizontal Wire

Let a horizontal wire of length l and radius a extend along x axis from $x = -l/2$ to $x = l/2$ and be buried at infinite depth in homogeneous soil of resistivity ρ ohm-m. The current fed into the wire is I. Let the current in the wire at $x = u$ be $I(u)$. The potential at a point (x, y) in the surrounding medium due to current leaving the wire element du at $x = u$ is

$$dV(x, y) = \frac{d\,I(u)}{du}\,\frac{\rho}{4\pi}\,[(x-u)^2 + y^2]^{-0.5}\,du \quad (22.25)$$

The potential due to current leakage along the entire wire is

$$V(x, y) = \frac{\rho}{4\pi}\int_{-l/2}^{l/2} [(x-u)^2 + y^2]^{-0.5}\,\frac{dI(u)}{du}\,du \quad (22.26)$$

The current leakage can be assumed to be constant so that $dI(u)/du = I/l$. Then

$$V(x, y) = \frac{I\rho}{4\pi l} \ln [\alpha(x, y)] \tag{22.27}$$

where
$$\alpha(x, y) = \frac{\sqrt{(x + l/2)^2 + y^2} + (x + l/2)}{\sqrt{(x - l/2)^2 + y^2} + (x - l/2)}$$

The average potential of wire can be found by integrating Eq. (22.27) from $x = 0$ and $x = l/2$. The average potential divided by current I gives the resistance R of the buried wire. The result is

$$R = \frac{\rho}{2\pi l}\left[\ln\left\{\frac{l}{a}(1 + \sqrt{1 + (a/l)^2}\right\} + \frac{a}{l} - \sqrt{1 + (a/l)^2}\right] \tag{22.28}$$

when $l >> a$

$$R = \frac{\rho}{2\pi l}\left(\ln\frac{2l}{a} - 1\right) \tag{22.29}$$

Eq. (22.29) is valid if the wire is buried at infinite depth. To find the resistance of wire buried at surface of earth, imagine a plane through the axis of above wire. The medium, the wire and current to one side of plane may be removed without disturbing the potential. Since the total current is halved while potential is the same, the resistance of a wire at surface of ground is twice that given by Eq. (22.29) or

$$R = \frac{\rho}{\pi l}\left(\ln\frac{2l}{a} - 1\right) \tag{22.30}$$

The resistance of wire buried at depth h can be found by assuming two wires separated by distance 2 h in infinite medium. The result is

$$R = \frac{\rho}{\pi l}\left(\ln\frac{2l}{\sqrt{2\,ah}} - 1\right) \tag{22.31}$$

where $h << l$

Thus, the resistance of a wire of radius a and burried at depth h is the same as that of a wire of radius $(2ah)^{0.5}$ buried at surface.

For n radial wires each of length l buried symmetrically at the surface of earth such that angle between any two adjacent wires is the same, the resistance is

$$R = \frac{\rho}{n\pi l}\left[\ln\frac{2l}{a} - 1 + N(n)\right] \tag{22.32}$$

where
$$N(n) = \sum_{m=1}^{n-1} \ln\frac{1 + \sin\pi m/n}{\sin\pi m/n}$$

If the above radial wires are buried at depth h, the radius a is replaced by $(2ah)^{0.5}$

If a strip is used instead of wire, its resistance can be found by replacing the strip with a wire having the same circumference.

Example 22.1.

(a) Find the earthing resistance of a driven rod length 3 m and diameter 2 cm if soil resistivity is 60 ohm-m.

(b) Find the radius of a hemispherical electrode which has the same resistance as the above driven rod.

Solution

(a) $$R = \frac{\rho}{2\pi l} \ln \frac{4l}{d} = \frac{60}{2\pi(3)} \ln \left(\frac{4 \times 3}{2 \times 10^{-2}} \right) = 20.36 \text{ ohm}$$

(b) $$\frac{\rho}{2\pi B} = 20.36$$

or $$B = \frac{\rho}{2\pi(20.36)} = \frac{60}{2\pi(20.36)} = 0.469 \text{ m}$$

Example 22.2.

(a) Find the resistance of two driven rods each of length 3 m, diameter 2 cm and buried in a soil of resistivity 60 ohm-m. The spacing between rods is 3 m.
(b) Find the resistance of three such rods arranged in a straight line with 3 m as spacing between adjacent rods.
(c) Find the resistance of three such rods arranged in an equilateral triangle with 3 m side
(d) Find the resistance of four rods arranged at the corners of a square of side 3 m.

Solution

Using the results of example 22.1

(a) $$\alpha = B/D = 0.469/3 = 0.156$$

Resistance of two rods in parallel $= \left(\frac{1+\alpha}{2} \right) (20.36) = 11.768$ ohm

(b) Resistance of three rods in parallel $= \frac{2 + \alpha - 4\alpha^2}{6 - 7\alpha} (20.36)$

$$= \frac{2 + 0.156 - 4(0.156)^2}{6 - 7(0.156)} (20.36) = 8.54 \text{ ohm}$$

(c) Resistance of 3 rods at corners of a triangle

$$= \frac{1 + 2\alpha}{3} (20.36) = \frac{1 + 2(0.156)}{3} (20.36) = 8.9 \text{ ohm}$$

(d) Resistance of 4 rods at the corners of square

$$= \frac{1 + 2.707\alpha}{4} (20.36) = \frac{1 + 2.707(0.156)}{4} (20.36) = 7.24 \text{ ohm}$$

Example 22.3.

A wire of length 3 m and radius 0.25 cm is buried in a soil of resistivity 100 ohm-m. Find the earthing resistance if (a) wire is buried at surface of earth (b) wire is buried at 0.5 m depth (c) wire is buried at infinite depth.

Solution

(a) Using Eq. (22.30)

$$R = \frac{100}{\pi \times 3} \left(\ln \frac{2 \times 3}{0.25 \times 10^{-2}} - 1 \right) = 71.97 \text{ ohm}$$

(b) Using Eq. (22.31)

$$R = \frac{100}{\pi \times 3}\left(\ln \frac{2\times 3}{(2\times 0.25\times 10^{-2}\times 0.5)^{0.5}} - 1\right) = 40.19 \text{ ohm}$$

(c) When depth is infinite, the resistance is half of the value at surface. Hence

$$R = 35.98 \text{ ohm}$$

22.5.6 Earthing Grid

The earthing system is high voltage and EHV substations consists of a number of interconnected bare conductors buried horizontally at a depth of about 0.5 m. Such a system, known as earthing grid, provides common earth for all devices and metallic structures in the substation. This arrangement is often supplemented by a number of vertical rods, each about 3 m long, at some joints.

When the grid depth h is less than 0.25 m, the following Laurant's formula gives a fairly accurate value of earthing resistance of the grid (without vertical rods)

$$R = \frac{\rho}{4}\sqrt{\frac{\pi}{A}} + \frac{\rho}{L} \tag{22.33}$$

where R is the grid resistance in ohms, A is the ground area occupied by the grid, in m^2 and L is the total length of buried conductors, in m.

For $h > 0.25$ m, a more accurate formula is

$$R = \rho\left[\frac{1}{L} + \frac{1}{\sqrt{20A}}\left(1 + \frac{1}{1 + h\sqrt{20/A}}\right)\right] \tag{22.34}$$

The addition of ground rods decreases the resistance slightly. For details of exact calculation of resistance of grid with rods, the reader should refer to [18] of the bibliography.

22.6 TOLERABLE STEP AND TOUCH VOLTAGE

When a fault occurs, the flow of current to earth results in voltage gradient on the surface of the earth in the vicinity of earthing system. This voltage gradient may affect a person in two ways, viz., step or foot to foot contact and hand to both feet or touch contact.

22.6.1 Step Voltage

Figure. 22.6 shows the circuit for step contact. The potential difference shunted by the body is limited to the maximum value between two accessible points, on the ground surface, separated

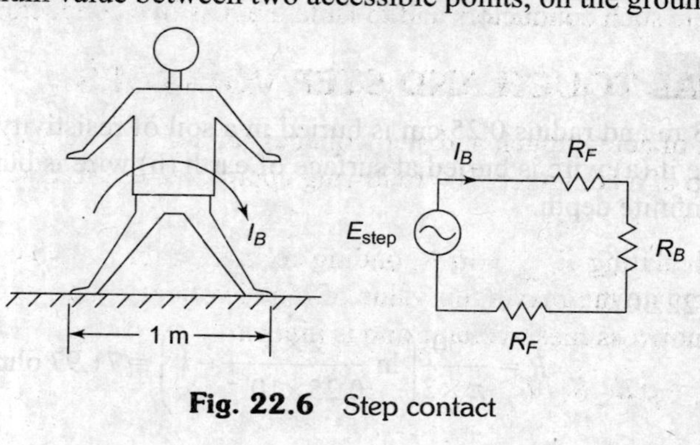

Fig. 22.6 Step contact

by one pace (assumed by equal to 1 m). R_f is the earthing resistance of one foot (on ohms), and R_B is the resistance of body (in volts)

A human foot can be taken as equivalent to a circular plate electrode with a radius of about 0.08 m and its earth resistance may be assumed to be 3 ρ_s where ρ_s is the resistivity of soil near the surface of earth. R_B is assumed to be 1000 ohms. Therefore, tolerable value of step voltage is

$$E_{step} = (R_B + 2\,R_F)\,I_B \text{ volts} \tag{22.35}$$

Substituting the values of R_B, R_F and I_B (from Eq. 22.1), we get

$$E_{step} = (1000 + 6\rho_s)\,0.116/\sqrt{t} \text{ volts} \tag{22.36}$$

22.6.2 Touch Voltage

Figure 22.7 shows the circuit for touch contact. The maximum earth potential difference intercepted would be that which would occur over a distance along the ground equal to maximum possible horizontal reach. Tolerable value of touch potential is

$$E_{touch} = (R_B + 0.5\,R_F)\,I_B = (1000 + 1.5\,\rho_s)\,0.116/\sqrt{t} \text{ volts} \tag{22.37}$$

Eqs. (22.36) and (22.37) are for 50 kg weight. For 70 kg weight, the factor 0.157 replaces the factor 0.116 in these equations. The actual step and touch voltages should be less than the values given by Eqs. (22.36) and (22.37).

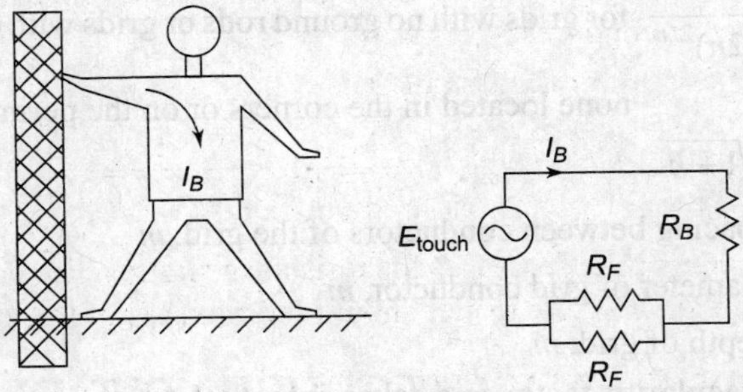

Fig. 22.7 Touch contact

22.6.3 Transferred Potential

A special situation occurs when a person standing within the station area touches a conductor grounded at a remote point or a person standing at a remote point touches a conductor connected to station grid. This special case of touch potential is known as transferred potential and can have value. It is impossible to design an earthing grid for safe transferred potential. The only remedy is to isolate such conductors and to lable them as live conductors.

22.7 ACTUAL TOUCH AND STEP VOLTAGES

Exact calculation of actual touch and step voltages for a grid is very difficult and can be done only with the help of a computer. The following approximate formulae give reasonable degree of accuracy.

Instead of calculating E_{touch} corresponding to maximum possible horizontal reach, it is preferable to calculate the maximum value of touch voltage found within a mesh of the grid. This voltage is known as mesh voltage and is given by

$$E_m = \rho\,K_m\,K_i\,I/L \tag{22.38}$$

where E_m = mesh voltage, volts

ρ = soil resistivity, ohm-m

I = Total current dissipated by grid

$L = L_c + 1.1.5\, L_r$ for grids with rods along perimeter, m

$= L_c + L_r$ for grids with no rods or rods evenly spread all over, m

L_c = total length of horizontal grid conductors, m

L_r = total length of ground rods, m

$K_i = 0.656 + 0.172\, n$

n = number of parallel conductors in one direction (excluding cross connections)

$$K_m = \frac{1}{2\pi}\left[\ln\left(\frac{D^2}{16hd} + \frac{(D+2h)^2}{8Dd} - \frac{h}{4d}\right) + \frac{K_{ii}}{K_h}\ln\frac{8}{\pi(2n-1)}\right] \qquad (22.39)$$

K_{ii} = 1 for grids with rods along perimeter or for grids with rods in grid corners as well as both along perimeter and throughout grid area

$= \dfrac{1}{(2n)^{2/n}}$ for grids with no ground rods or grids with only a few ground rods, none located in the corners or on the perimeter

$K_h = \sqrt{1+h}$

D = spacing between conductors of the grid, m

d = diameter of grid conductor, m

h = depth of grid, m

K_i is known as irregularity factor and takes into account the non-uniformity of current dissipation by the grid.

The step voltage E_s is given by

$$E_s = \rho\, K_s\, K_i\, I/L \qquad (22.40)$$

where
$$K_s = \frac{1}{\pi}\left[\frac{1}{2h} + \frac{1}{D+h} + \frac{1}{D}(1 - 0.5^{n-2})\right] \qquad (22.41)$$

It has been found that step voltage is in general less dangerous than mesh voltage. Therefore, the mesh voltage is used as the basis for design. The total conductor length L is calculated from Eq. (22.38) and then checked for safe step voltage.

22.8 DESIGN OF EARTHING GRID

The following steps should be followed for the design of an earthing grid:

(a) Data needed for design

1. Substation ground area
2. Soil resistivity at the site
3. Fault clearing time
4. Maximum grid current
5. Resistivity of soil at surface

(b) Design of earthing system
1. Selection of electrode material
2. Determination of conductor size
3. Preliminary design
4. Determination of conductor length required for control of voltage gradient
5. Calculation of resistance and ground potential rise
6. Checking of step potential and revision of design, if necessary.

22.8.1 Substation Ground Area and Soil Resistivity

Generally the substation plan is prepared before the grid design. The earthing grid should cover as much ground area as possible. Soil resistivity can be determined using the four probe method (section 22.12). A thin surface layer of crushed rock helps in limiting the body current. A value of 3000 ohm-m is usually assumed for resistivity of crush rock. However, it is preferable to measure the resistivity of the material to be put on the surface.

22.8.2 Fault Clearing Time

The fault clearing time is governed by system stability considerations and depends on protection and switchgear equipment. Generally a value of 0.5 seconds is assumed. The size of the conductor is based on a time of 1 second.

22.8.3 Determination of Maximum Grid Current

A single line to ground fault is more common and causes more fault current as compared to a double line to ground fault. Therefore, the design is based on single line to ground fault current. For calculation of this current grid resistance and fault resistance are assumed to be zero. Then

$$I_g = S_f I_f \tag{22.42}$$

where I_g = symmetrical grid current, A

I_f = rms value of symmetrical ground fault current, A

S_f = current division factor relating the magnitude of fault current to that of its portion flowing in the earthing grid.

Only a part of the total fault current flows through the grid. The remaining current flows through the overhead ground wires and other ground return paths. The factors S_f gives this current division and can be calculated, if the system data is known. From the value of I_g the design value of grid current is found as

$$I = G_f D_f I_g \tag{22.43}$$

where I = maximum grid current

G_f = A factor to account for increase in fault current due to system growth during life span of grid.

D_f = Decrement factor to account for asymmetry of the fault current wave.

Typical values of G_f, assumed in design, lie in the range of 1.2 to 1.5 depending on the rate of growth of the system.

The decrement factor D_f can be calculated if system data is known. For fault duration of two cycles or less, the ac component of asymmetrical wave depends on subtransient reactance. For fault duration of more than 2 cycles, transient reactance is used for calculations. It is assumed that maximum possible dc offset is present. If data is not available the following values may be assumed

Fault duration (seconds)	0.008	0.1	0.25	0.5 or more
D_f	1.65	1.25	1.1	1.0

22.8.4 Selection of Electrode Material

The material for grounding grid should have good conductivity, be mechanically rugged and resist fusing and deterioration of joints.

Copper was very commonly used in the past. It has high conductivity and is resistant to underground corrosion. However, a grid of copper forms a galvanic cell with other buried structures and pipes and is likely to hasten the corrosion of the latter. Aluminium is not used because of corrosion problem.

The material most commonly used presently is galvanized steel. The use of steel avoids galvanic action between grid and other underground structures and pipes. However, the grid itself may suffer corrosion. The zinc coating gets destroyed over a period of time. It is desirable that a suitable allowance for corrosion be made while determining the size of conductor. Generally strips are used because they can be bent, bolted and welded more easily than round bars. However, when the soil is highly corrosive it is preferable to use circular rods and not strips because a circular rod has minimum surface area for a given cross-sectional area.

22.8.5 Determination of Conductor Size

The size of conductor is given by the equation

$$A_c = I\left[\frac{0.00104\,\rho_m \alpha_m t}{\delta\, s \log_{10}\dfrac{(1+\alpha_m\,\theta_m)}{1+\alpha_m\,\theta_a}}\right]^{0.5} \tag{22.44}$$

where A_c = Cross-sectional area, mm^2

I = maximum grid current, A

ρ_m = resistivity of material, μ ohm-cm

α_m = temperature coefficient of material, per °C

t = fault clearing time, seconds

δ = density of material, gm/cm^3

s = specific heat of material, cal/gm °C

θ_m = maximum allowable temperature, °C

θ_a = ambient temperature, °C

For copper, δ = 8.89 gm/cm^3, s = 0.092 cal/gm °C

α_m = 0.00427 per °C, ρ_m = 1.72 μ ohm-cm, θ_m = 1084 °C

For Steel, δ = 7.86 gm/cm^3, s = 0.114 cal/gm/°C

α_m = 0.00423 per °C, ρ_m = 15 μ ohm-cm

θ_m = 620 °C for welded joints

θ_m = 310 °C for bolted joints.

When the above values for steel are substituted in Eq. (22.44) and $\theta_a = 40$ °C, we get

$$A_c = 12.25 \times 10^{-3}\, I\sqrt{t} \text{ for welded joints}$$

$$= 15.8 \times 10^{-3}\, I\sqrt{t} \text{ for bolted joints.}$$

Eq (22.44) is based on short time current carrying capacity of conductors.

For adequate mechanical strength the minimum size recommended is:

Copper 54.48 mm², Steel 24 mm²,

Steel strips minimum thickness 3 mm.

The following corrosion allowance is recommended:

(i) In soils having resistivity more than 100 ohm-cm, no allowance.

(ii) In soils having resistivity from 25 to 100 ohm-cm, 15% allowance.

(iii) In soils having resistivity less than 25 ohm-cm, 30% allowance.

22.8.6 Preliminary Design

In the preliminary design a continuous grounding conductor should surround the entire substation are. Additional grounding conductors should be placed in parallel lines at spacings of about 5 m. Cross connections should be provided for multiple paths at about 5 m spacing. A number of ground rods, each about 3 m long, are put. The fault current divided by 500 gives an approximate number of ground rods. These should be distributed uniformly over the grid. A ground rod at each junction of the grid, where the lead from surge arrester is connected, is desirable.

22.8.7 Conductor Length for Gradient Control

The conductor length required to keep mesh voltage within safe limits is obtained by equating actual E_{mesh} (Eq. 22.38) to tolerable E_{touch} (Eq. 22.37). The result is

$$L = \frac{\rho K_m K_i I\sqrt{t}}{(1000 + 1.5\rho_s)(0.116)} \tag{22.45}$$

If the length of conductor in the preliminary design is less than that given by Eq. (22.45) a revision in design is necessary. This revision may include change in layout, use of more ground rods and use of closer spacing of conductors. In case the length given by Eq. (22.45) is too small to allow for connections to all equipments, more conductor may be required even though it is not necessary for safety. In case the length given by Eq. (22.45) is too large to be economically justified, it may be necessary to use some special methods for limiting the fault current and diverting a greater position of fault current to other paths through decreased tower footing resistance of towers in the vicinity of station, etc.

22.8.8 Calculation of Resistance and GPR

The resistance of the grid can be calculated by using Eq. (22.34). The product of grid resistance and maximum grid current, *I*, is the ground potential rise (GPR). If GPR is less than tolerable mesh voltage, it is a very safe design. However, it has been found that generally GPR is higher than tolerable mesh voltage. In such a case, there is a need of guarding personnel and communication equipment against transferred potential.

22.8.9 Checking of Step Potential

The tolerable step potential and actual step potential can be calculated from Eqs. (22.36) and (22.40). If actual step potential is more than tolerable, a revision in design is necessary.

The final step in design includes provision of closer meshes in the area frequently visited by operating personnel, additional ground rods for surge arrester leads, etc., and provision of risers for connections to equipments.

Example 22.4.

Design an earthing grid for a 220 kV substation. Use following data:

Soil resistivity = 55 ohm-m, substation area = 47.50 m × 31.5 m

Maximum grid current = 5000 A, Fault clearing time = 0.5 s

Resistivity of soil at surface = 3000 ohm-m.

Use steel conductor and welded joints.

Solution

(a) *Conductor size:* Let ambient temp. = 40 °C.

$$A_c = 12.25 \times 10^{-3} \times 5000\sqrt{0.5} = 43.3 \text{ mm}^2$$

corrosion allowance = 15%

conductor size = 43.3 × 1.15 = 49.8 mm²

$$d = 7.96 \text{ mm} = 7.96 \times 10^{-3} \text{ m}$$

(b) *Preliminary design:* Let the grid cover the entire station area. Assume spacing between conductors = 5.25 m and $h = 0.5$ m

$$A = 47.5 \times 31.5 = 1496.25 \text{ m}^2$$

$$L = 10 \times 31.5 + 7 \times 47.5 = 647.5 \text{ m}$$

Let there be no ground rods. Using Eq. (22.34)

$$R = 55\left[\frac{1}{647.5} + \frac{1}{\sqrt{20 \times 1496.25}}\left(1 + \frac{1}{1 + 0.5\sqrt{20/1496.25}}\right)\right] = 0.703 \text{ ohm}$$

(c) Conductor length for gradient control

Using Eq. (22.45)

$$n = 10$$

$$K_i = 0.656 + 0.172 \times 10 = 2.376$$

$$K_{ii} = \frac{1}{(2 \times 10)^{2/10}} = 0.55$$

$$K_h = \sqrt{1 + 0.5} = 1.225$$

$$K_m = \frac{1}{2\pi}\left[\ln\left(\frac{5.25 \times 5.25}{16 \times 0.5 \times 7.96 \times 10^{-3}} + \frac{(5.25 + 2 \times 0.5)^2}{8 \times 5.25 \times 7.96 \times 10^{-3}} - \frac{0.5}{4 \times 7.96 10^{-3}}\right) + \frac{0.55}{1.225}\ln\frac{8}{\pi(2 \times 10 - 1)}\right] = 0.856$$

$$\text{Length required} = \frac{55 \times 0.856 \times 2.376 \times 5000 \times \sqrt{0.5}}{(1000 + 1.5 \times 3000)\,(0.116)} = 619.9 \text{ m}$$

Since the length in preliminary design is more than conductor length required for control of gradient, the design is safe from the consideration of mesh potential.

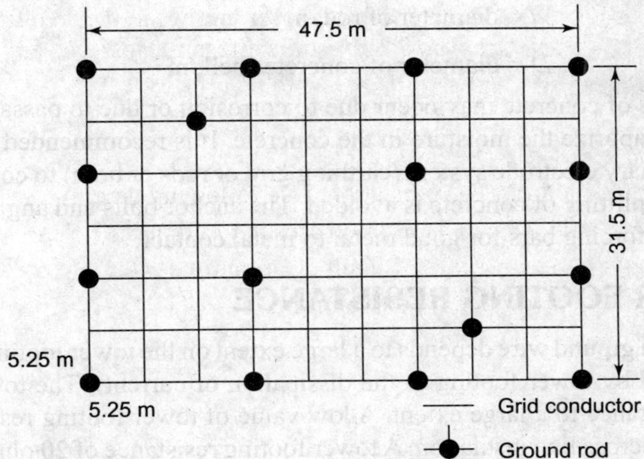

Fig. 22.8 Earthing grid

(d) *Step potential*

$$\text{Tolerable } E_{step} = (1000 + 6 \times 3000)\ 0.116/\sqrt{0.5} = 3117 \text{ V}$$

$$K_s = \frac{1}{\pi}\left[\frac{1}{2 \times 0.5} + \frac{1}{5.25 + 0.5} + \frac{1}{5.25}(1 - 0.5^8)\right] = 0.434$$

$$\text{Actual } E_{step} = 55 \times 0.434 \times 2.376 \times 5000/647.5 = 437.95 \text{ V}$$

The grid is safe from the consideration of step potential.

(e) $$\text{Tolerable } E_{touch} = (1000 + 1.5 \times 3000)\ 0.116/\sqrt{0.5} = 902 \text{ V}$$

$$\text{Ground potential rise} = 5000 \times 0.703 = 3515 \text{ V}$$

It is necessary to guard personnel and communication equipment against transferred potential.

(f) *Final design:* In the final design, add some ground rods distributed over the grid area. This will increase the safety margin. Fig. 22.8 shows the final layout.

22.9 CONCRETE ENCASED ELECTRODES

Concrete, being hygroscopic, attracts moisture. When concrete is buried in earth it behaves as a semiconducting medium having a resistivity of 30 to 90 ohm-m. In medium and high resistivity soils a rod encased in concrete has a lower resistance than that of a similar electrode directly buried in earth. The earthing resistance of vertical rod encased in concrete is given by

$$R_{cE',\text{ rod}} = \frac{1}{2\pi l}\left[\rho_c \ln\frac{D}{d} + \rho\left(\ln\frac{8l}{D} - 1\right)\right] \quad (22.46)$$

where $R_{cE\text{ rod}}$ = earthing resistance of concrete encased rod, ohm

ρ_c = resistivity of concrete, ohm-m

ρ = resistivity of soil, ohm-m

l = length of rod, m

d = diameter of rod, m

D = diameter of concrete shell, m

However, splitting of concrete may occur due to corrosion or due to passage of a very high current which may vaporize the moisture in the concrete. It is recommended that there should be a satisfactory primary electrode system (earthing grid or rods or both) to conduct most of the fault current so that splitting of concrete is avoided. The anchor bolts and angle stubs should be connected to the reinforcing bars for good metal to metal contact.

22.10 TOWER FOOTING RESISTANCE

The effectiveness of a ground wire depends to a large extent on the tower footing resistance (i.e., the resistance offered by tower footing to the dissipation of current). The tower top potential depends on this resistance to a large extent. A low value of tower footing resistance results in less voltage stresses across line insulation. A tower footing resistance of 20 ohms for EHV lines and 10 ohms for HV lines provides sufficient lightning protection.

The foundation of tower is generally of grillage type and the foundation itself may provide a sufficiently good earthing system. However, it is usual practice to provide one or two driven rods also at each tower footing to achieve a low resistance. In soils of high resistivity (sandy soil and rocks) the use of driven rods may not give adequate low resistance grounds. In such cases, the use of counterpoise is recommended: A counterpoise is a conductor, buried horizontally in ground, tied to the tower base and run parallel or at some angle to the line. The aim is to reduce the resistance by increasing the area of earth in contact with the grounding system. Fig. 22.9 shows some commonly used arrangements of counterpoises.

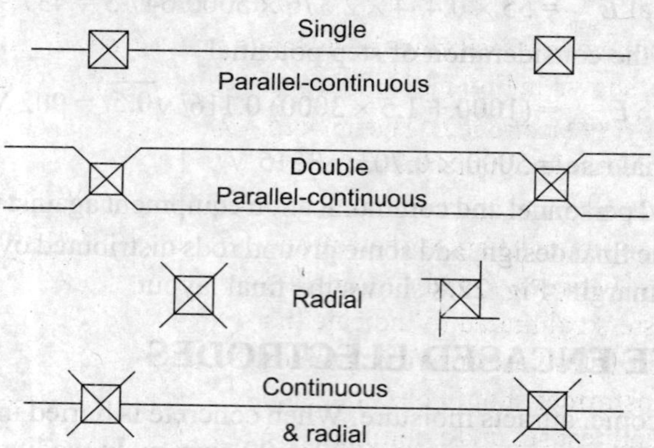

Fig. 22.9 Arrangement of counterpoises

A counterpoise, at right angles to the line reduces only the tower footing resistance while a counterpoise parallel to the line also reduces, though by a small amount, by mutual coupling, the potential difference between conductor and tower top. The depth of burial has only a very small effect on resistance. It is usually buried at a depth of about 0.5 metre. The earthing resistance can be calculated by using formulae given in section 22.5.5.

The degree of lightning protection depends on the impulse impedance and not the power frequency grounding resistance. Investigations have revealed that impulse resistance of a driven rod is less than its power frequency resistance (due to the effect of soil ionization). However, the inductance of counterpoise causes its impulse impedance to be more than its power frequency resistance. The ratio of impulse impedance to power frequency resistance may be about 1.5 for counterpoise of short length but may be about 10 for a long counterpoise. Therefore, it is preferable to use several wires, each of small length, than a single long wire.

22.11 MEASUREMENT OF EARTH RESISTANCE R

The measurement of earth resistance of any electrode is done by the fall of potential method. Fig. 22.10 shows the test set up. E is the earth electrode under test. Two auxiliary electrodes P and C are placed at suitable distances from E. A current I is passed between electrode E and C and the voltage V between E and P is measured. If C is at adequate distance from E, it would have no appreciable effect on the resistance of E. Since current flowing into electrode P is very small, this electrode too has negligible effect on resistance.

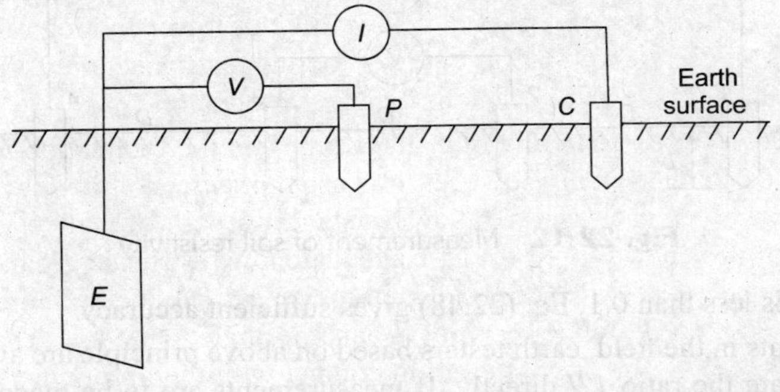

Fig. 22.10 Measurement of earth resistance of electrode E

The measurements are made by varying the distance of electrode P from E. The variation of ratio V/I with distance is shown in Fig. 22.11. The portion of the curve which is nearly horizontal gives the earth resistance R of electrode E. The upward slope in the curve near electrode C indicates the effect of resistance of electrode C.

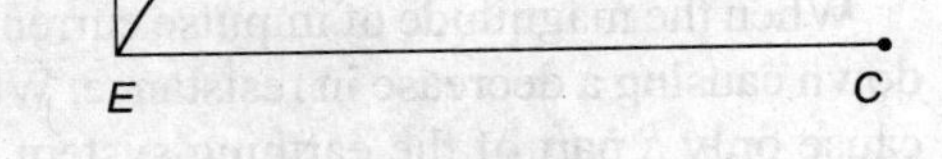

Fig. 22.11

It can be shown that optimum location of electrode P is 0.62 of the distance between E and C. For field measurements earth testers calibrated to indicate the ratio V/I directly are used. If tests are done using mains supply, an isolating transformer should be connected between main supply and test set up to eliminate the effect of residual earth currents.

22.12 MEASUREMENT OF SOIL RESISTIVITY

Soil resistivity is measured by four electrode method suggested by F. Wenner. Fig. 22.12 shows the test set up. Four metal rods whose diameter is small as compared to length ae driven into earth at equal distances. A known current I is passed between electrodes 1 and 4 and potential drop V is measured between electrodes 2 and 3.

The current I produces an electric field proportional to current density and soil resistivity. The voltage V is proportional to this field. Therefore, the soil resistivity is proportional to the ratio V/I and is given by

$$\rho = \frac{4\pi DV/I}{1 + \dfrac{2D}{\sqrt{D^2 + 4b^2}} - \dfrac{2D}{\sqrt{4D^2 + 4b^2}}} \tag{22.47}$$

where ρ = soil resistivity, ohm-m

D = horizontal distance between 2 successive electrodes, m

b = depth of burial, m

If $b << D$ Eq. (22.47) simplified to

$$\rho = 2\pi D V/I \tag{22.48}$$

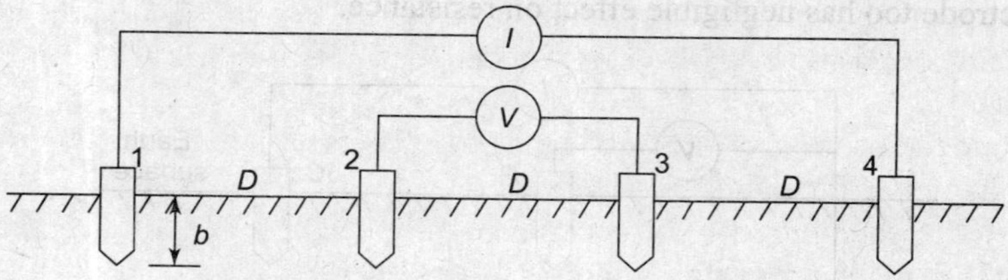

Fig. 22.12 Measurement of soil resistivity

If the ratio b/D is less than 0.1, Eq. (22.48) gives sufficient accuracy.

For measurements in the field, earth testers based on above principle are available. They are calibrated to indicate the ratio V/I directly. If measurements are to be made using the mains supply an isolating transformer should be connected between the mains supply and test set up so that residual earth currents do not affect the results. A number of readings with different values of spacing and the test set up in different directions should be taken. The average value of ρ obtained from these measurements should be used in design.

22.13 IMPULSE BEHAVIOUR OF EARTHING SYSTEMS

The degree of protection provided by an earthing system under lightning discharge conditions depends on its impulse impedance (i.e., impedance under impulse conditions) and not the power frequency resistance. The impulse impedance is different from the power frequency resistance due to effect of (1) soil ionization (2) electrode inductance.

When the magnitude of impulse current is high, the soil in the vicinity of electrode may break-down causing a decrease in resistance. When the length of electrode is large, its inductance may cause only a part of the earthing system to be effective in dissipating current and, thus, result in an increase in impedance.

The relative contributions of these two factors depend on the type of earthing system. The effect of soil ionization is very predominant in driven rods so that the impulse resistance of a driven rod is always less than its power frequency resistance. The effect of electrode inductance is present in counterpoise and grids so that their impulse impedance is always more than power frequency resistance. If impulse impedance of a grid is very high it may not give satisfactory protection against direct lightning strokes.

For a detailed analysis of impulse behaviour the reader is referred to [49, 50] of the bibliography.

22.14 NEUTRAL EARTHING

Neutral earthing does not have any influence on the operation of a 3-phase system under balanced steady state conditions. However, the voltages and currents during ground fault conditions are greatly affected by the method used to connect neutral to earth.

22.14.1 Isolated Neutral or Ungrounded System

The main feature of an unrounded neutral system is its ability, in some cases, to clear ground faults without interruption. This self clearing feature disappears when the length of line becomes appreciable. The main features of an isolated neutral system are:

(a) In the event of a ground fault, the voltages across healthy phases become equal to line to line values, and arcing grounds may occur. This phenomenon has been discussed in section 12.16.

(b) The ground fault current is limited to capacitive ground fault current. The use of ground fault relaying is difficult and unreliable.

(c) The danger to equipment on the occurrence of line to ground fault is appreciable and danger to life in the proximity of fault is often prolonged.

(d) Radio interference may be appreciable during faults or when neutral is displaced.

22.14.2 Effectively Earthed System

The term 'effectively grounded' has replaced the older term 'solidly grounded'. A transformer neutral may be connected to earth directly without any impedance between neutral and earth. However, the transformer capacity may be so small as compared to the system size that the neutral grounding of transformer may not be effective in stabilizing the voltages from phases to ground in the event of a ground fault.

As per IEEE definition, a system (or a portion of it) is said to be effectively grounded when the ratio of zero sequence reactance to positive sequence reactance is not greater than 3 and the ratio of zero sequence resistance to positive sequence reactance is not greater than 1 for all conditions of operation. The main features of effectively grounded neutral system are:

1. The ground fault current is high. Its maximum value may sometimes exceed even the 3-phase short circuit current as discussed in section 10.14.
2. In the event of single line to ground fault, the maximum voltage of healthy phases does not exceed 80% of line to line voltage.
3. Arcing grounds cannot occur because the short circuit current eliminates the influence of capacitive current.
4. Since the ground fault current is high, danger to personnel in the vicinity of fault is high.
5. Ground fault relaying is simple and satisfactory.
6. Even transient ground faults may get converted into short circuit.
7. The heavy ground fault current may cause considerable damage.
8. Because of high ground fault current, the interference due to electromagnetic induction with neighbouring communication circuits may be high.

22.14.3 Resistance Earthing

When neutral is grounded through a resistance, arcing grounds are eliminated. The value of resistance should be such that ground fault current is less than 3-phase short circuit current. This is necessary so that the power loss in the grounding resistance is not excessive. However, this results in neutral displacement and the maximum voltage across healthy phases may become equal to line to line value. The other main characteristics of resistance grounded systems are:

1. The ground fault current is low but higher than capacitive ground fault current.
2. The ground fault relaying is simple and satisfactory.
3. The power dissipation in the grounding resistance may improve system stability because this power dissipation reduces the accelerating power.
4. Since the ground fault current is lower than that is effectively grounded systems, the inductive interference with parallel communication circuits is also lesser than that in effectively grounded systems.
5. Transient ground faults are converted into controlled current faults.

The resistor used for neutral earthing is usually a metallic resistance mounted on insulators in a metallic frame. For voltages below 6.6 kV liquid resistors are also used. The neutral earthing resistor is designed to carry its rated current for a short period, usually 30 seconds.

22.14.4 Reactance Earthing

As per IEEE definition reactance grounding means grounding through impedance the principle element of which is reactance. This grounding is used when the zero sequence reactance of system is so low as to cause excessive ground fault current.

Let X_1, X_2, X_0 be the positive, negative and zero sequence reactances of the system and V the voltage per phase.

If X_0 is negligible as compared to $X_1\ (= X_2)$

$$\text{Ground fault current} = \frac{3V}{X_1 + X_2 + X_0} = \frac{3V}{2X_1}$$

$= 1.5$ times 3-phase short circuit current

If $X_0 = X_1\ (= X_2)$

$$\text{Ground fault current} = \frac{3V}{X_1 + X_2 + X_0} = \frac{V}{X_1}$$

$=$ 3-phase short circuit current

If $X_0 = X_1\ (= X_2)$ and a reactance X_e is connected between neutral and earth

$$\text{Ground fault current} = \frac{3V}{X_1 + X_2 + X_0 + 3X_e} = \frac{V}{X_1 + X_e}$$

As the value of X_e is increased, the ground fault current decreases and neutral displacement increases. If X_e is very small, the system behaves as an effectively grounded system. If X_e is very high, the system behaves as an isolated neutral system. The main characteristics of a system with reactance grounding are:

1. Ground fault current is reduced but is much larger than capacitive ground fault current.
2. The voltages across healthy phases are between 0.8 to 1 per unit of line to line voltage.
3. Arcing grounds are avoided.
4. Transient ground faults are converted into controlled current faults.
5. Ground fault relaying is simple and satisfactory.
6. Since ground fault current is reduced, inductive interference with parallel communication circuits is also reduced.

22.14.5 Arc Suppression Coil Earthing (Resonant Earthing)

It was invented by W. Peterson in 1916 and is, therefore, also known as Peterson coil grounding. The neutral is grounded through a tapped reactor whose value is tuned to system capacitance.

When a line to ground fault occurs, a lagging reactive current flows from the faulted conductor to ground and returns to system through the neutral reactor. At the same time, capacitance currents also flow from healthy phases to ground. The lagging fault current and leading capacitance currents are almost 180° out oı phase. By a proper selection of the value of neutral reactor the two currents can be made almost equal so that the current in the fault is almost zero and arc cannot be maintained and is extinguished. The combination of neutral reactance and line capacitance acts as a parallel resonant circuit. Fig. 22.13 (a) shows the circuit. If V is the line to neutral voltage.

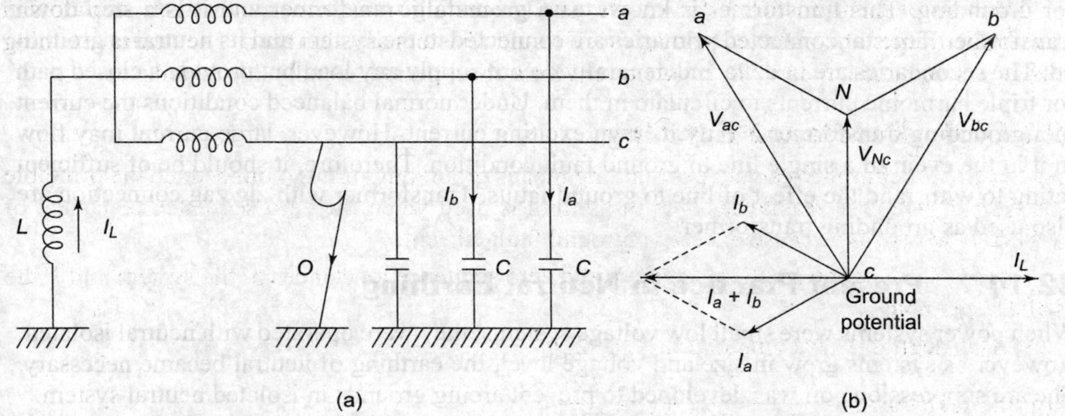

Fig. 22.13 Arc suppression coil earthing (a) circuit (b) phasor diagram

$$I_a = I_b = \sqrt{3}\ V\omega C \tag{22.49}$$

$$I_a + I_b = \sqrt{3}\,(\sqrt{3}\ V\omega C) = 3\ \omega CV \tag{22.50}$$

$$I_L = V/\omega L \tag{22.51}$$

For compensation of arc current

$$\frac{V}{\omega L} = 3\ \omega CV \text{ or } L = \frac{1}{3\omega^2 L}$$

and

$$X_L = \frac{1}{3\omega C} \tag{22.52}$$

Fig. 22.13 (b) shows the phasor diagram. Some characteristics of this earthing are:

1. During normal conditions, the suppression coil has little effect on balance of system voltages to ground and insignificant continuous losses.
2. Transient ground faults are suppressed. More than 70% of these faults are cleared without necessitating circuit breaker operation.
3. During sustained ground fault, the current is a small percentage of capacitive ground fault current. Burning and heating effects are minimized and damage to equipment is restricted.
4. Service can be continued for long periods with one phase grounded till the faulty section is disconnected.

5. Voltage gradient on ground surface in the vicinity of fault is greatly reduced. Therefore, danger to life in the proximity of fault is negligible.
6. Inductive interference with parallel communication circuits is low in magnitude but duration may be long.
7. Relaying requires special provision and is generally restricted to indication.
8. During a single line to ground fault, the voltages across healthy phases are equal to line to line value.

22.14.6 Grounding Transformer

Sometimes the neutral of a power transformer is not available for grounding (e.g., when a $\Delta - \Delta$ transformer is used). In such situations a special small size $Y - \Delta$ transformer is solely used for grounding. This transformer is known as a grounding transformer and it is a step down transformer. The star connected primaries are connected to the system and its neutral is grounded. The secondaries are in delta and generally do not supply any load but provide a closed path for triple harmonic currents to circulate in them. Under normal balanced conditions the current in a grounding transformer is only its own exciting current. However, large current may flow in it in the event of a single line to ground fault condition. Therefore, it should be of sufficient rating to withstand the effect of line to ground faults. Transformer with zig zag connection are also used as grounding transformer.

22.14.7 Present Practice in Neutral Earthing

When power systems were small low voltage systems, they were operated with neutral isolated. However as systems grew in size and voltage level, the earthing of neutral became necessary. The arc suppression coil was developed to prevent arcing grounds in isolated neutral system.

In most present day high voltage and extra high voltage systems, the transformer neutrals are effectively grounded. This grounding limits the voltages of healthy phases to line to neutral values in the event of ground faults and also eliminates arcing grounds.

The effective neutral earthing results in high magnitudes of earth fault current. However the modern protection systems are very sensitive and fast operating so that the faults are cleared in very short times.

The neutrals of generators are grounded through a resistance to limit stator fault current. The value of this resistor determines the percentage of generator winding left unprotected.

22.15 CORROSION

22.15.1 Phenomenon

Corrosion is the deterioration (and destruction) of metals and alloys by electrochemical reactions with their environments. The essential phenomenon is the same in all metals and alloys but the degree is different in each. Ferrous products corrode very easily.

Iron occurs in nature as Fe_2O_3 ore and this is a stable form. When iron is manufactured from the ore by a series of complex processes, it enters an unstable condition and corrosion starts so that it may pass into stable state i.e. oxide.

All commonly used metals and alloys are classified in an order called standard electromotive series given in table 22.1. Hydrogen is considered as neutral element. Magnesium is most reactive the index (i.e. standard electrode potential) being – 2.34. Gold never suffers any corrosion (index + 1.42). The reactivity decreases until it is minimum for lead (index – 0.13). Lead is highly resistant towards corrosion and is, therefore, used extensively in storage batteries and as lining in acid tanks. The electromotive series reveals that any metal is capable of displacing one below it in the series e.g. iron can displace copper through the reaction

$$Fe + CuSO_4 = FeSO_4 + Cu \tag{22.52}$$

Table 22.1 Galvanic Series

Element	*Standard electrode*	*Potential*
Magnesium (Mg)	– 2.34	Anodic or least noble or corroded end
Aluminium (Al)	– 1.66	
Manganese (Mn)	– 1.05	
Zinc (Zn)	– 0.76	
Chromium (Cr)	– 0.70	
Iron (Fe)	– 0.44	
Cadmium (Cd)	– 0.40	
Cobalt (Co)	– 0.28	
Nickel (Ni)	– 0.25	
Tin (Sn)	– 0.14	
Lead (Pb)	– 0.13	
Hydrogen (H_2)	0.00	
Copper (Cu)	+ 0.35	
Mercury (Hg)	+ 0.80	
Silver (Ag)	+ 0.80	
Platinum (Pt)	+ 1.20	
Gold (Au)	+ 1.42	Cathodic or most noble or protected end

When two metals are in contact with each other (say through soil) the metal towards the top of the series functions as anode and undergoes corrosion while the metal towards the bottom of the series functions as cathode and is protected. Consider the case of galvanised steel sheet. Galvanising process adds a coating of zinc which is towards the top of the series as compared to iron. Hence zinc gets corroded in preference to iron. Therefore, until the zinc coating is completely destroyed by corrosion, iron is protected.

22.15.2 Theory of Corrosion

The conditions necessary for corrosion to occur are: presence of electrolytic and presence of anode and cathode. An electrolyte is any liquid having ions e.g. water. One drop of rain water contains billions of positively charged hydrogen ions and an equal number of negatively charged hydroxyl (OH) ions. Due to the difference in potential between anode and cathode, an electric current is set up between the electrodes via the electrolyte. The current leaves anode and flows through electrolyte to cathode. Anode suffers corrosion and cathode is protected. The process of corrosion of iron can be explained as under:

Before corrosion starts all atoms of iron are electrically neutral. An atom has a positively charged nucleus surrounded by electrons revolving in definite orbits. When corrosion starts two of the electrons in one atom of iron are strongly pulled by two positively charged ions of electrolyte (i.e. water). The electromotive force is sufficiently strong to detach the electrons. The negatively charged electrons neutralise the positively charged hydrogen ions producing hydrogen gas. The chemical reactions are

$$Fe - 2e = Fe^{++} \text{ (anode reaction)} \quad (22.53)$$

$$2H^{+} + 2e = H_2 \text{ (reduction at cathode)} \quad (22.54)$$

The overall reaction is

$$Fe + 2H^{+} = Fe^{++} + H_2 \quad (22.55)$$

The iron atom which has lost two electrons becomes an electropositive iron ion. In this way countless number of iron atoms are disintegrated. The iron ions react with negatively charged hydroxyl (OH) ions producing

$$Fe^{++} + 2\,(OH) = Fe(OH)_2 \quad (22.56)$$

Iron is very corrosive. Since iron constitutes about 10 times the quantity of all non-ferrous metals and alloys, the study of corrosion of iron is very important. Hundreds of crores of rupees worth of iron is rendered useless by corrosion every year.

22.15.3 Factors Governing Corrosion of Metals Buried in Soil

The corrosion of metal buried in soils depends on the intrinsic properties of the metal (or alloy), type of dissimilar metals which are in the soil where the earthing conductor is buried, areas of anode and cathode, and characteristics of soil.

(a) Intrinsic properties of metals

All materials used as grounding electrodes suffer corrosion. Copper is intrinsically more noble and stable than aluminium and iron. Iron is unstable in most environments, readily reacts with oxygen, moisture and other constituents of soil. The average rate of corrosion of iron is six times that of copper.

(b) Type of dissimilar metals

Corrosion is due to galvanic current in the soil. This current depends on the open circuit emf between the two metals. As is obvious from the values of electrode potential (table 22.1) corrosion of iron will be more when coupled to copper than when coupled with lead.

(c) Area of anode and cathode

Corrosion occurs due to the galvanic current. This current is a function of the areas of anode and cathode.

(d) Characteristics of soil

The galvanic current depends on the open circuit emf between the two metals and the resistance of path through the soil. If soil resistivity is low, the resistance of path will be low and magnitude of galvanic current would be high. High galvanic current means high rate of corrosion.

Electrical resistivity of soil is a measure of the different physiochemical characteristics of soil. An increase in moisture and salt content decreases the resistivity. From the point of view of corrosion soils are classified in four categories given in table 22.2.

Table 22.2 Classification of soils as regards corrosion

Soil resistivity	*Soil category*
Less than 25 ohm-m	highly corrosive
25 ohm-m to 50 ohm-m	moderately corrosive
50 ohm-m to 100 ohm-m	corrosive
More than 100 ohm-m	mildly corrosive

22.15.4 Galvanic Current

The rate of dissolution of metal due to corrosion depends on galvanic current which can be approximated by the equation

$$I(t) = \frac{E_{c0} - E_{a0}}{R + \dfrac{AI(t)}{A_a} + \dfrac{CI(t)}{A_c}} \tag{22.57}$$

Where $I(t)$ = galvanic current

E_{c0} = open circuit potential at cathode

E_{a0} = open circuit potential at anode

R = ohmic resistance of current path without anodic and cathodic polarisation

$AI(t)$ = shift in anode potential in positive direction due to concentration of reaction products at the anode

$CI(t)$ = shift in cathode potential in negative direction due to concentration of reaction products at cathode.

A_c = area of cathode

A_a = area of anode

The initial rate of galvanic current is equal to $(E_{c0} - E_{a0})/R$. As time passes the reaction products tend to get concentrated at anode and cathode and the current decreases. Decrease of current means lesser rate of corrosion.

22.15.5 Protection Against Corrosion

Corrosion causes continuous dissolution of metal into the soil. Therefore, it is necessary to protect the earthing electrodes. The methods available for protection are:

1. Corrosion resistant metallic coating.
2. Provision of extra allowance in conductor size
3. Use of circular rods instead of strips
4. Use of sacrificial anodes.
5. Cathodic protection

(a) Corrosion resistant metallic coating

Steel is given a zinc coating to protect it from corrosion. However this coating tends to get destroyed with passage of time.

(b) Provision of extra allowance is conductor size

The size of earth electrode is given by Eq. (22.44). It is advisable to provide an extra allowance in the conductor size to take care of corrosion. This extra allowance is decided as under:

(i). Mildly corrosive soils (resistivity more than 100 ohm-m). In these soils the rate of corrosion is less than 0.04% per year. No extra allowance is needed in this case.

(ii) Corrosive and moderately corrosive soils (resistivity 25 ohm-m to 100 ohm-m). The rate of corrosion in these soils is taken as 1% per year for first 10 years, 0.5% per year for the next 10 years and negligible after that. Allowance of 15% is recommended in these soils.

(iii) Highly corrosive soils (resistivity less than 25 ohm-m). The rate of corrosion is these soils is taken as 2% per year for first 10 years, 1% per year for next 10 years and negligible after that. An allowance of 30% is recommended in these soils.

(c) Use of spherical rods instead of strips

Generally galvanised steel strips are used for earthing because the strips can be bent, bolted and welded more easily than round bars. However it is important to remember that in highly corrosive soils it is preferable to use circular rods and not strips. Corrosion rate depends on surface area and a circular rod has minimum surface area for a given cross-sectional area.

(d) Use of sacrificial Anodes

Sacrificial anode are made of metal which is more electronegative than steel. The metals mostly used are magnesium, zinc, aluminium and magnesium alloys. A sacrificial anode is attached to the grid. It helps in maintaining the grid at negative potential. Therefore positive ions do not leave the grid surface and the material of grid is protected.

(e) Cathodic protection

A dc voltage is applied between the steel grid (to be protected) and an auxiliary anode. The positive terminal of dc voltage is connected to the auxiliary anode and negative terminal to the steel grid. The dc current is adjusted so that a potential difference of about – 0.85 V is maintained between grid and soil. This potential difference is sufficient to stop the tendency of metal ions to go to the soil. The dc source is generally a rectifier whose input is connected to ac mains. The auxiliary anode is generally made of graphite or scrap iron.

Cathodic protection is a costly method. It is rarely used for protection of earthing grids. But in highly corrosive soils (resistivity less than 5 ohm-m) it may become necessary. However cathodic protection is very commonly used for corrosion protection of pipes carrying petroleum products and buried in soils of low resistivity.

SUMMARY

22.1 A properly designed earthing system is essential for operation of protective devices, lightning protection and safety of personnel.

22.2 For a person of 50 kg weight the safe magnitude of 50 Hz current is given by

$$I_B = 0.116/\sqrt{t}$$

22.3 Soil resistivity depends on soil type, moisture content, salt content and temperature.

22.4 Earthing resistance of a hemisphere of diameter B is

$$R = \frac{\rho}{2\pi B}$$

22.5 For a driven rod of diameter d and length l,

$$R = \frac{\rho}{2\pi l}\left(\text{In}\frac{4l}{d}\right)$$

22.6 Multiple driven rods and buried horizontal wires are also used for earthing.

22.7 At substations the earthing system consists of an array of interconnected bare conductors buried horizontally at about 0.5 in depth. This is called an earthing grid. It is supplemented by a number of vertical rods, each about 3 m long, at some joints.

22.8 Resistance of an earthing grid (without rod)

$$R = \frac{\rho}{4}\sqrt{\frac{\pi}{A}} + \frac{\rho}{L}$$

22.9 Tolerable step voltage

$$E_{\text{step}} = (1000 + 6\,\rho_s)\,\frac{0.116}{\sqrt{t}} \text{ volts}$$

22.10 Tolerable touch voltage

$$E_{\text{touch}} = (1000 + 1.5\,\rho_s)\,\frac{0.116}{\sqrt{t}} \text{ volts}$$

22.11 The touch voltage is maximum within a mesh of the grid. This is called mesh voltage.

22.12 Actual step and mesh voltages must be less than tolerable step and mesh voltages for proper safety.

22.13 Galvanized steel is the most commonly used material for earthing grids and rods.

22.14 The conductor size of earthing grids is determined from maximum allowable temperature consideration.

22.15 Maximum Ground Potential Rise (GPR) should be 5 kV.

22.16 Concrete behaves as a semiconducting medium having a resistivity of 30 to 90 ohm-m.

22.17 Tower footing resistance governs the effectiveness of ground wire.

22.18 Soil resistivity can be measured by Wenner method.

22.19 Neutral earthing systems can be: Isolated neutral, solidly earthed neutral, resistance earthing, reactance earthing and arc suppression coil earthing.

22.20 The impulse behaviour of earthing system can be very different from power frequency behaviour.

SELF TEST QUESTIONS

A. State whether the following are true (T) or false (F)

22.1 Isolated neutral system may cause arcing grounds.

22.2 If neutral point of a system is not available, a grounding transformer can be used.

22.3 Magnitude of 50 Hz tolerable current is related to duration.

22.4 Presence of moisture increases soil resistivity.

22.5 Presence of salt has no effect on resistivity.

22.6 Earth resistance of a hemispherical electrode is directly proportional to radius.

22.7 For a vertical driven rod, the earthing resistance is mainly dependent on length of rod.

22.8 Transferred potential is always low.

22.9 Actual mesh voltage must always be less than tolerable touch potential.

22.10 Concrete encased earthing electrodes do not provide any earthing.

B. Fill in the blanks with suitable words

22.11 The range of let go currents is ____________.

22.12 For tap water resistivity is about ____________.

22.13 Safe value of GPR is about ____________.

22.14 Resistivity of rocky soil is ____________.

22.15 Earthing resistance of a hemisphere buried deep in soil is ____________ than that of the hemisphere on surface of earth.

22.16 In design of earthing system the conductor length for gradient control is calculated from ____________ consideration.

22.17 The conductor size for bolted joints is ____________ than that for welded joints.

22.18 Material for earthing grid is generally ____________.

22.19 A suitable value of TFR is ____________ ohms.

22.20 A horizontally buried earth wire is called ____________.

ANSWERS

22.1 T	22.2 T	22.3 T	22.4 F	22.5 F
22.6 F	22.7 T	22.8 F	22.9 T	22.10 F
22.11 1-6 mA	22.12 20 Ω-m		22.13 5 kV	22.14 very high
22.15 less	22.16 touch voltage		22.17 more	22.18 galvanized steel
22.19 ten	22.20 counterpoise			

JOB INTERVIEW QUESTIONS

22.1 What is the need of an earthing system?

22.2 What are touch and step voltages?

22.3 What is an earthing grid?

22.4 What is tower footing resistance?

22.5 What is meant by GPR?

22.6 What factors govern soil resistivity?

22.7 What is grounding transformer?

22.8 What is neutral earthing?

22.9 What is the difference between solidly grounded and isolated neutral systems?

22.10 What is meant by 'let go' currents.

REVIEW QUESTIONS

22.1 Explain the terms: earth electrode, resistance of earthing electrode, step potential, touch potential.

22.2 What factors govern soil resistivity? Draw graphs to explain your answer.

22.3 Derive expressions for earthing resistance of (a) hemispherical electrode (b) driven rod.

22.4 What is a counterpoise? Where is it used?

22.5 Explain the procedure to design an earthing grid.

22.6 What is an isolated neutral system? What are its features?

22.7 Why is it necessary to earth neutral? Explain the difference between resistance and reactance grounding.

PROBLEMS

22.1 Find the earthing resistance of a hemispherical electrode buried at the surface of earth having a soil resistivity of 60 ohm-m. [12.73 ohm]

22.2 Find the earthing resistance of a driven rod of length 2.5 m and diameter 1.5 cm if soil resistivity is 50 ohm-m. [20.7 ohm]

22.3 Find the resistance of

(a) two driven rods in a straight line with a spacing of 4 m.

(b) three driven rods at corners of equilateral triangle with 4 m side

length of each rod 2.5 m, diameter 1.5 cm, soil resistivity 50 ohm-m. [(a) 11.345 ohm (b) 8.23 ohm]

22.4 Design an earthing grid for a 132 kV substation. Soil resistivity 70 ohm-m, Fault current 7000 A, substation area 55 m × 45 m. Assume suitable data.

CHAPTER 23

Voltage Stability

23.1 INTRODUCTION

It is desirable that voltage at different buses in a power system is equal to the nominal value (i.e., 1 *pu*) at all times. Many phenomena tend to make the actual voltage different from the nominal value. An unacceptable voltage level means voltage instability. If the voltage departs too much from the nominal value, the phenomenon is known as voltage collapse. Large system disturbances may lead to voltage instability or even voltage collapse. The system voltage may decline progressively leading to unacceptable voltage level. Sometimes voltage instability may be oscillatory in nature. Power systems must be planned, designed, installed and operated in such a manner so as to avoid voltage instability and voltage collapse.

23.2 DEFINITIONS

Voltage stability can be classified into three categories and definitions are analogous to power system stability definitions.

A power system, in a given operating state, is small disturbance voltage stable if, after a small disturbance, the voltages at load points are identical or close to the pre-disturbance values.

A power system, in a given operating state and subject to a given disturbance, is voltage stable if voltages at load points approach post disturbance equilibrium values.

A power system, in a given operating state and subject to a given disturbance, faces voltage collapse if post disturbance equilibrium values are below acceptable level.

Lack of voltage stability means voltage instability. Generally voltage instability occurs during and after large disturbances.

Sometimes the term 'voltage security' is also used. The ability of a system to maintain the voltage, at load points, at acceptable level is known as voltage security.

It is generally necessary to study the voltage stability problem for different durations from a fraction of a second to several minutes. The study for a short duration, upto a few seconds, is known as transient voltage stability. When the study spans a longer interval, it is referred to as long term voltage stability.

23.3 REACTIVE POWER TRANSMISSION

We have discussed the loose physical interaction between MW and Mvar flows in a system in chapters 4 and 10. The reactive power transferred over a line is directly proportional to $(|V_s| - |V_r|)$, i.e., voltage drop along the line and is independent of power angle (on the other hand the real power transfer is directly proportional to sin δ). This means that the major cause of voltage drop on the line is the transfer of reactive power over the line. To maintain a good voltage profile and to maintain voltage stability, the control of reactive power is necessary.

We have derived the power formulae for transmission lines in chapter 4 (Eqs 4.74 and 4.75). To simplify the formulae we can neglect the line resistance (i.e., assume the line to be loss-less). Then the line generalized circuit constants for the nominal π model become

$$A = |A| \angle 0, B = jX = X \angle 90°$$

Substituting these values in Eqs (4.74 and 4.75) we have

$$P_r = \frac{|V_s||V_r|}{|X|} \cos(90 - \delta) \tag{23.1a}$$

$$Q_r = \frac{|V_s||V_r|}{|X|} \sin(90 - \delta) - \frac{|A||V_r|^2}{|X|} \tag{23.1b}$$

$$P_s = -\frac{|V_r||V_s|}{|X|} \cos(90 + \delta) \tag{23.2a}$$

$$Q_s = \frac{|A||V_r|^2}{|X|} - \frac{|V_r||V_s|}{|X|} \sin(90 + \delta) \tag{23.2b}$$

The above equations indicate that when line is long (i.e., X is high) the power angle δ (which is the phase angle between V_s and V_r) would be high. However the transmission of reactive power over the line causes V_s and V_r to be unequal. Thus it is not possible to transmit reactive power over the line without excessive voltage gradient. In some cases it may be necessary to feed reactive power into the line at both the sending and receiving ends, i.e., the line may become a sink of lagging reactive power.

23.4 SHORT CIRCUIT CAPACITY

An important aspect of power systems is its short circuit capacity which is the product of system voltage and fault current. Since the system voltage is assumed to be 1 *pu* and fault current is the ratio of system voltage to impedance (or reactance) as seen from fault point, the short circuit capacity is inversely proportional to system impedance (or reactance) as seen from fault point.

A high short circuit capacity indicates that the system is strong or stiff. A low short circuit capacity means that the system is weak. Switching on or off a load, a shunt capacitor or a shunt reactor does not have much effect on a strong system. However any such switching action may affect a weak system appreciably, especially if the load, shunt capacitor or reactor switched on or off has high capacity. The short circuit capacity indicates system voltage strength. A system having high short circuit capacity has a higher degree of voltage stability than the one with a low short circuit capacity.

The ratio of system short circuit capacity to the rating of a device is known as short circuit ratio of the device. A 500 MW device connected to a system with a short circuit strength of 2000 MVA has a short circuit ratio of 4.

Another term used in power systems, especially HVDC systems, is effective short circuit ratio (ESCR). In calculating short circuit ratio, only the inherent system strength is considered. However in calculating *ESCR,* the effect of shunt compensation at the device location is also included. A synchronous condenser increase the short circuit capacity. On the other hand a shunt capacitor reduces the effective short circuit ratio.

Example 23.1.

A 400 kV line has $X = 96\ \Omega$ and $Y = 0.001 \angle 90°$ siemens. Base MVA is 500 MVA. If real power transfer is 2 *pu* find Q_s and Q_r if (a) $V_s = 1.0\ pu$ and $V_r = 0.95\ pu$ (b) $V_s = 1.0\ pu$ and $V_r = 0.9\ pu$ (c) $V_s = 1.0\ pu$ and $V_r = 1.0\ pu$. The line can be assumed to be lossless. Use nominal π representation.

Solution

$$Z = 96 \angle 90°\ \Omega.\ Y = 0.001 \angle 90°\ \Omega$$

Take 400 kV and 500 MVA base

$$\text{Base impedance} = \frac{400 \times 400}{500} = 320\ \Omega$$

$$A = 1 + \frac{ZY}{2} = 1 + \frac{96 \times 0.001}{2} \angle 180° = 0.952 \angle 0$$

$$B = Z = \frac{96 \times \angle 90°}{320} = 0.3 \angle 90°\ pu$$

(a) $P_r = 2\ pu$, $V_s = 1.0$ and $V_r = 0.95$

Using Eqs (23.1) and (23.2)

$$2 = \frac{1 \times 0.95}{0.3} \cos (90 - \delta)$$

or

$$\delta = 39.17°$$

$$Q_r = \frac{1 \times 0.95}{0.3} \sin (90 - 39.17) - \frac{0.952 \times 0.95^2}{0.3}$$

$$= -0.409\ pu$$

$$Q_s = \frac{0.952\ (1)^2}{0.3} - \frac{1 \times 0.95}{0.3} \sin (90 + 39.17°)$$

$$= 0.718\ pu$$

(b) $P_r = 2\ pu$, $V_s = 1.0$ and $V_r = 0.9\ pu$. Using Eqs (23.1) and (23.2)

$$2 = \frac{1 \times 0.9}{0.3} \cos (90 - \delta)$$

or

$$\delta = 41.81°$$

$$Q_r = \frac{1 \times 0.9}{0.3} \sin (90 - 41.81°) - \frac{0.952 \times 0.9^2}{0.3} = -0.334\ pu$$

$$Q_s = \frac{0.952 \times (1)^2}{0.3} - \frac{1 \times 0.9}{0.3} \sin (90 + 41.81°) = 0.937\ pu$$

(c) $P_r = 2\,pu,\ V_s = 1.0\,pu = V_r$

$$2 = \frac{1\times1}{0.3}\cos(90-\delta)$$

or $$\delta = 36.87°$$

$$Q_r = \frac{1\times1}{0.3}\sin(90-36.87°) - \frac{0.952\times1}{0.3} = -0.506\,pu$$

$$Q_s = \frac{0.952\times1}{0.3} - \frac{1}{0.3}\sin(90-36.87°) = +0.506\,pu$$

Example 23.2.

Fig 23.1 shows a simple representation of a system with both sending end and receiving end voltage of 10 kV (rms). The power transmitted is 12.5 MW. The line inductive reactance is 4 ohms and frequency is 50 Hz.

(*a*) Draw phasor diagram showing all voltages and currents. Final power angle

(*b*) Find the value of C so that power flow is at unity power factor.

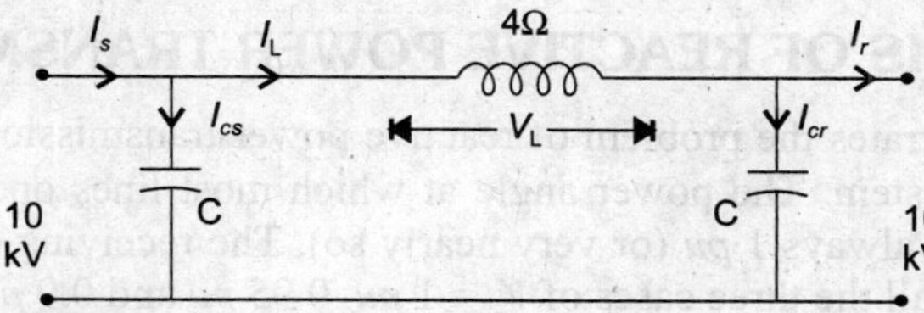

Fig. 23.1

Solution

$$|V_s| = |V_r| = |V| = 10000\ V$$

$$P = \frac{|V_s||V_r|}{X_L}\sin\delta$$

or $$12.5\times10^6 = \frac{(10000)^2}{4}\sin\delta$$

or $$\sin\delta = 0.5 \quad \text{or} \quad \delta = 30°$$

Phasor diagram is shown in Fig. 23.2

V_L = Voltage drop across line inductance = $2|V|\sin\delta/2$ = 5176.42

Reactive power absorbed by line inductance

$$= Q_L = \frac{(5176.4)^2}{4} = 6698779.2\ \text{VAr}$$

For unity power factor power flow, half of Q_L should be absorbed by each capacitor

$$Q_C = \frac{Q_L}{2} = \frac{6698779.2}{2} = \frac{|V|^2}{X_c} = |V|^2\,\omega C$$

or $$C = \frac{6698779.2}{2(2\pi)(50)(10000)^2} = 0.0001066F$$

$$= 106.6\ \mu F$$

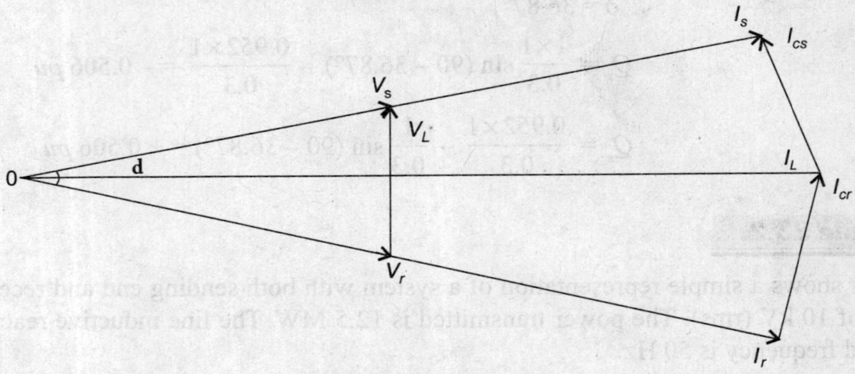

Fig. 19.2. Phasor diagram

23.5 PROBLEMS OF REACTIVE POWER TRANSMISSION

Example 23.1 demonstrates the problem of reactive power transmission. The data and results pertain to an actual system. The power angle at which most lines operate is about 40°. The sending end voltage is always 1 *pu* (or very nearly so). The receiving end voltage should not be less than 0.9 *pu*. In all the three cases of $V_r = 1$ *pu*, 0.95 *pu* and 0.9 *pu*, Q_s is positive and Q_r is negative. This means that the generators at the sending end are feeding lagging vars into the line but no lagging vars are being receiving at the load end. Negative Q_r means that lagging vars are required to be pumped into the line at the receiving end also. Thus the line has become a sink of lagging vars. Thus real power can be transmitted over a long line. But reactive power can not be transmitted as the results of example 23.1 demonstrate. It is possible that the real power transfer, power angle and frequency in a system are satisfactory but the reactive power conditions are unsatisfactory leading to voltage instability.

The series reactance of the line consumes lagging vars while the shunt capacitance consumes leading vars. The shunt capacitance leading vars are proportional to V^2. A decrease in voltage causes a reduction in these leading vars thus causing a greater difference between the lagging and leading vars. This action causes a further decrease in voltage and thus the effect is cumulative. It may finally lead to voltage collapse.

It is necessary that real and reactive line losses in the line are kept to the minimum. Real line losses represent a waste of energy while reactive losses cause increased investment for shunt compensation equipment, etc. These losses can be written as

$$\text{Real line losses} = 3\, I^2R = 3\, \frac{P^2 + Q^2}{V^2}\,(R) \tag{23.3}$$

$$\text{Reactive line losses} = 3\, I^2X = 3\, \frac{P^2 + Q^2}{V^2}\,(X) \tag{23.4}$$

In Eqs (23.3 and 23.4), *P* and *Q* denote the per phase real and reactive power transfer over the line, *V* is the per phase voltage, *R* is line resistance per phase and *X* is line series reactance per phase. Eq (23.3) indicates that real line losses also increase with increase in Q. To keep real and reactive line losses to the minimum, the reactive power transfer over the line should be kept to the minimum.

It is desirable that receiving end voltage, under peak load conditions, is maintained close to 1.0 *pu.* If reactive power is being transferred over the line, the sending end voltage has to be higher than 1.0 *pu.* (the reactive power transfer $\propto |V_s - V_r|$). When the line breakers open, the receiving end voltage becomes V_s/A. Since the generalized circuit constant A is less than 1, the receiving end voltage becomes more than the sending end voltage thus leading to voltage instability conditions. The temporary overvoltage due to load rejection, is mainly dependent on the reactive power transfer. To keep the temporary overvoltage to a minimum value, the reactive power transfer should be minimized.

Transfer of reactive power over the lines should be minimized for the following reasons:

1. Transfer of reactive power leads to voltage instability.
2. Transfer of reactive power results in greater real and reactive line losses.
3. The reactive power required to be transferred has to be supplied by the generators and transformers. This increases the size of generators and transformers.
4. If reactive power transfer is decreased, the generators, transformers and lines can carry greater real power for the same maximum temperature rise leading to higher utilization of equipments and greater economy.
5. Reactive power transfer can lead to temporary overvoltage in the event of load rejection.

Example 23.3

A short transmission line is being fed by a source having a source reactance of 0.22 *pu* on 1000 MW base. The line reactance is 0.15 *pu* on 1000 MW and rated voltage base. The line is supplying 1 *pu* MW, at 0.8 lagging p.f., at rated voltage. What will be the voltage at the receiving end if the receiving end line breaker opens suddenly.

Solution

Neglecting the line resistance

$$V_s = V_r \angle 0 + jXI_r$$
$$= V_r \angle 0 + jX \frac{P_r - jQ_r}{V_r}$$

or
$$V_s = V_r + \frac{XQ_r}{V_r} + j\frac{XP_r}{V_r}$$

When the receiving end breaker opens with sending end energised, the receiving end voltage becomes equal to V_s.

$$V_r = 1 \angle 0, P_r = 1\ pu.$$
$$\cos\phi_r = 0.8,$$
$$Q_r = P_r \tan\phi_r = 0.75\ pu$$
$$X = 0.22 + 0.15 = 0.37\ pu.$$
$$V_s = 1 \angle 0 + \frac{0.37 \times 0.75}{1} + \frac{j(0.37 \times 1)}{1}$$
$$= 1 \angle 0 + 0.2775 + j\,0.37$$
$$= 1.33 \angle 16.1^\circ\ pu$$

The temporary overvoltage due to load rejection will become equal to 1.33 *pu*. It is seen that this overvoltage is mainly due to reactive power transfer. If the load is at unity p.f. (so that $Q_r = 0$), the temporary overvoltage will be

$$V_s = 1\angle 0 + j\,(0.37 \times 1) = 1.066\ \angle 20.37°\ pu$$

23.6 ROTOR ANGLE STABILITY AND VOLTAGE STABILITY

We have discussed the problem of power system stability in chapter 13. To make the distinction clear, the term rotor angle stability can be used in place of power system stability.

The rotor angle stability and voltage stability problems are, to some extent, interlinked but in many aspects quite different also. Rotor angle stability is basically generator stability while voltage stability means load stability. In a large interconnected power system voltage instability may occur while the rotor angle stability is still maintained and the generators continue to remain in synchronism.

Transient voltage stability is generally interlinked to transient rotor angle stability while small disturbance voltage stability is linked to steady state (small or slow disturbance) rotor angle stability.

Rotor angle stability is mainly interlinked to real power transfer whereas voltage stability is mainly related to reactive power transfer. However, reactive power control affects rotor angle stability also to some extent.

Voltage stability problems arise mainly in the event of faults. Rotor angle stability problems also arise during and after faults.

Power transfer over transmission lines is limited due to the rotor angle stability. Applications of some modern techniques is useful in removing the power transfer limits imposed by transient rotor angle stability. Some of these techniques are fast acting automatic voltage regulator, rapid fault clearing, quick valve opening action, high performance excitation systems, etc. If the limits imposed by transient rotor angle stability are removed, the limits due to thermal capacity or voltage stability may dictate the power transfer over the lines.

23.7 SURGE IMPEDANCE LOADING (NATURAL LOAD)

We have discussed the concept of surge impedance loading in chapter 4.

The characteristic impedance of a line, denoted by Z_c is given, by

$$Z_c = \sqrt{z/y} \tag{23.5}$$

where Z_c = characteristic impedance, Ω

$z = r + j\,\omega L$ = line series impedance per phase per unit length

$y = j\,\omega C$ = line shunt admittance per unit length per phase to neutral.

If the line resistance is neglected,

$$Z_c = \sqrt{\frac{L}{C}} \tag{23.6}$$

When the line resistance is neglected, the characteristic impedance is known as surge impedance.

A line is said to be operating at natural load (or surge impedance load) when it is terminated by a resistance equal to characteristic impedance. If the line to line voltage at receiving end is V_r the receiving end current I_r under the condition of surge impedance loading is

$$I_r = \frac{V_r}{\sqrt{3}\,Z_c} \tag{23.7}$$

Surge impedance loading is also the ideal loading. The voltage and current are equal at all points along the line (i.e., line voltage and current profiles are flat). The voltage and current are in phase at all points along line. Moreover no reactive power is generated or absorbed at the line ends. The total reactive power absorbed by line inductance is equal to total reactive power generated by line capacitance, i.e.,

$$I^2 \omega L = \frac{V^2}{\omega C}$$

or
$$\frac{V}{I} = \sqrt{\frac{L}{C}} = Z_c \tag{23.8}$$

A relation between actual power received and surge impedance loading can be derived as under. If all quantities are expressed in *pu*.

Receiving end power $\quad S_r = V_r\, I_r^*$ (23.9)

Using Equivalent π circuit (Fig. 4.14)

$$I_r = \frac{V_s - V_r}{Z'} - \frac{V_r Y'}{2} = \frac{2V_s - 2V_r - Z'Y'V_r}{2Z'}$$

$$I_r^* = \left[\frac{2V_s - 2V_r - Z'Y'V_r}{2Z'}\right]^*$$

$$S_r = V_r I_r^* = |V_r|^2 \left[\frac{2\frac{V_s}{V_r} - 2 - Z'Y'}{2Z'}\right]^*$$

$$SIL = 1/Z_c$$

$$\frac{S_r}{SIL} = \frac{|V_r|^2}{1/Z_c}\left[\frac{2\frac{V_s}{V_r} - 2 - Z_c \sinh(\gamma l)\frac{2}{Z_c}\tanh\left(\frac{\gamma l}{2}\right)}{2Z_c \sinh(\gamma l)}\right]^*$$

$$= \frac{Z_c}{Z_c^*}|V_r|^2\left[\frac{\frac{V_s}{V_r} - 1 - \sinh(\gamma l)\frac{(\cosh \gamma l - 1)}{\sinh \gamma l}}{\sinh(\gamma l)}\right]^*$$

$$\frac{S_r}{SIL} = \frac{Z_c}{Z_c^*}|V_r|^2\left[\frac{\frac{V_s}{V_r} - \cosh \gamma l}{\sinh \gamma l}\right]^*$$

If line resistance is neglected, the real part of γ, i.e., attenuation constant α becomes zero and $\gamma = j\beta$. Then $Z_c = Z_c^*$ and

$$\frac{S_r}{SIL} = |V_r|^2 \left|\frac{\left(\frac{V_s}{V_r}\right)^* - \cos \beta l}{\sin \beta l}\right| \tag{23.10}$$

The ratio S_r/SIL is a function of line length, receiving end voltage and ratio V_s/V_r. Fig. 23.3 shows this ratio as a function of line length for lines without any var compensation. This ratio depends on the short circuit capacity of the system. A system with high short circuit capacity has a higher S_r/SIL ratio as compared to one with low short circuit capacity.

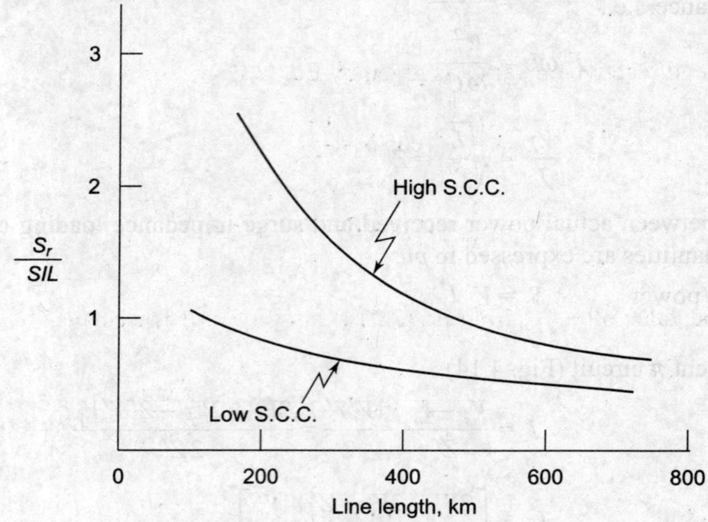

Fig. 23.3 Variation of Sr/SIL with line length

23.8 VOLTAGE STABILITY LIMIT

Consider the case of power transmission over a loss less line. Let V_s and V_r be the sending end and receiving end voltages and X be the transfer reactance. The real power P_r at the receiving end is (Eq. 4.80)

$$P_r = \frac{V_s V_r}{X} \sin \delta \tag{23.11}$$

$$\frac{d P_r}{d V_s} = \frac{V_r}{X} \sin \delta + \frac{V_s V_r}{X} \cos \delta \frac{d \delta}{d V_s}$$

Since the line is lossless, $P_s = P_r$. If real power generation is constant, $\frac{d P_r}{d V} = 0$ so that we get

$$\frac{V_r}{X} \sin \delta + \frac{V_s V_r}{X} \cos \delta \frac{d \delta}{d V_s} = 0$$

or

$$\frac{d \delta}{d V_s} = -\frac{\tan \delta}{V_s} \tag{23.12}$$

At the point of maximum power transfer $\delta = 90°$ so that as $\delta \to 90°$

$$\frac{d \delta}{d V_s} \to -\infty \tag{23.13}$$

Eq. (23.13) indicates the critical point on the δ versus V_s curve. The above analysis assumes constant receiving end voltage V_r. If the sending end voltage V_s is assumed constant, the resulting equation is

$$\frac{d\delta}{dV_r} = \frac{\tan\delta}{V_r} \tag{23.14}$$

The receiving end reactive power Q_r is given by Eq. (3.86)

$$Q_r = \frac{V_s V_r}{X}\cos\delta - \frac{V_r^2}{X} \tag{23.15}$$

$$\frac{dQ_r}{dV_r} = \frac{1}{X}\left[V_s \cos\delta - V_s V_r \sin\delta \frac{d\delta}{dV_r} - 2V_r\right] \tag{23.16}$$

Substituting the value of $\frac{d\delta}{dV_r}$ from Eq (23.14) into Eq (23.16) we have

$$\frac{dQ_r}{dV_r} = \frac{1}{X}\left[V_s \cos\delta + V_s V_r \sin\delta\left(\frac{\tan\delta}{V_r}\right) - 2V_r\right]$$

$$= \frac{1}{X}\left[V_s \cos\delta + V_s \frac{\sin^2\delta}{\cos\delta} - 2V_r\right]$$

$$= \frac{1}{X}\left[\frac{V_s(\cos^2\delta + \sin^2\delta)}{\cos\delta} - 2V_r\right]$$

or

$$\frac{dQ_r}{dV_r} = \frac{1}{X}\left(\frac{V_s}{\cos\delta} - 2V_r\right) \tag{23.17}$$

At steady state power angle stability, $\delta = 90°$ so that as $\delta \to 90°$

$$\frac{dQ_r}{dV_r} \to \infty \tag{23.18}$$

Equation (23.18) represents steady state voltage stability limit. This limit is reached when dQ_r/dV_r becomes infinite or dV_r/dQ_r becomes zero. Thus the rotor angle stability limit under steady state conditions is coincident with steady state voltage stability limit.

The voltage stability at the receiving end is also affected by the load. In the above analysis the contribution of loads has been neglected. For an exact analysis the effect of load must be taken into account.

Example 23.4.

A transmission line with a transfer reactance X has V_s and V_r as the sending and receiving end voltages. Find the maximum power and the critical voltage (i.e., voltage when power is maximum) if load power factor is (a) 1 (b) 0. Also find the expression for dQ_s/dQ_r when, load power factor is zero.

Solution

Let $\dfrac{P_r X}{V_s^2} = p,\ \dfrac{Q_r X}{V_s^2} = q$ and $\dfrac{V_r}{V_s} = v$

Substituting the above in Eqs. (23.11) and (23.15) we get

$$p = v \sin \delta \tag{23.19}$$

$$q = v \cos \delta - v^2 \tag{23.20}$$

since $v^2 (\cos^2 \delta + \sin^2 \delta) = v^2$, Eqs. (23.19) and (23.20) give

$$p = [v^2 - v^2 \cos^2 \delta]^{0.5} = \left[v^2 - (q + v^2)^2\right]^{0.5}$$

(a) when p.f is l, $Q_r = 0$ and $q = 0$

$$p = (v^2 - v^4)^{0.5} \tag{23.21}$$

$$\frac{dp}{dv} = 0.5\,(v^2 - v^4)^{-0.5}\,(2v - 4v^3)$$

Setting $dp/dv = 0$ we get the equation for critical voltage,

$$2v^2 = 1$$

or $$v_{\text{crit}} = 1/(\sqrt{2}) = 0.707$$

and $$p_{\text{max}} = 0.5 \text{ and } \delta = 45°$$

(b) $$P = 0 \text{ and } \delta = 0$$

$$q = v \cos \delta - v^2 = v - v^2$$

$$\frac{dq}{dv} = 1 - 2v$$

Setting dq/dv equal to zero gives the critical condition

$$v_{\text{crit}} = 0.5 \text{ or } \frac{V_r}{V_s} = 0.5$$

and $$q_{\text{max}} = 0.25$$

Q_s, i.e., sending end reactive power is

$$Q_s = Q_r + I^2 X = Q_r + \frac{Q_s^2}{V_s^2}.X$$

or $$Q_s^2 - \frac{V_s^2}{X} Q_s + \frac{V_s^2}{X} Q_r = 0$$

or $$Q_s = 0.5 \left[\frac{V_s^2}{X} \pm \sqrt{\frac{V_s^4}{X^2} - \frac{4V_s^2}{X} Q_r}\right]$$

or $$Q_s = 0.5 \left[\frac{V_s^2}{X} \pm V_s \left(\frac{V_s^2}{X^2} - \frac{4Q_r}{X}\right)^{0.5}\right]$$

$$\frac{dQ_s}{dQ_r} = \frac{1}{\left[1 - \dfrac{4XQ_r}{V_s^2}\right]^{0.5}} = \frac{1}{\left(1 - \dfrac{Q_r}{Q_{r,\max}}\right)^{0.5}}$$

The ratio dQ_s/dQ_r is generally known as 'voltage collapse proximity indicator'. This factor indicates the nearness of actual voltage conditions to impending voltage collapse. At no load this factor is 1 and at the condition of $Q_r = Q_{r,\max}$ this factor is infinity. As this factor tends to become very high, voltage collapse may occur.

23.9 GRAPHICAL METHODS

Two sets of graphs used for voltage stability study are *P-V* graph and *V-Q* graph.

Figure 23.3 shows a transmission line having voltages V_s and V_r at the two ends and feeding a load $P_r + j\,Q_r$ at the receiving end bus. The line is lossless and has a total series inductive reactance X. [If necessary the line resistance can also be used to prepare the graph].

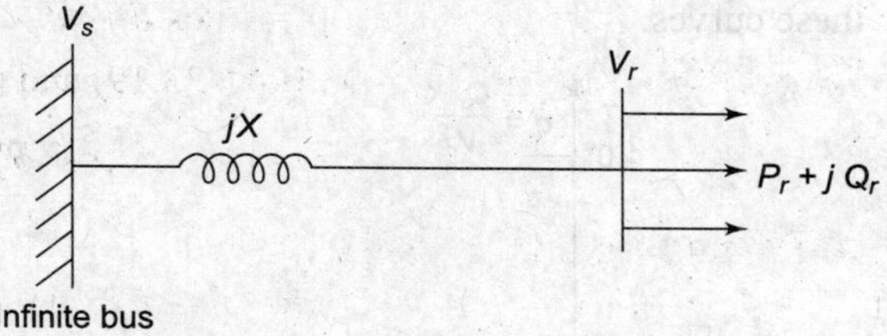

Fig. 23.4

23.9.1 PV Curve

It is customary to use normalized values in preparation of these curves. Thus the receiving end power P_r is represented in terms of p (where $p = P_r X/V_s^2$) and receiving end voltage V_r is represented by v (= V_r/V_s). Using transmission line equations we can find v for different values of p and power factor.

For a large system, the values can be found from load flow study. The *P-V* curve shows a plot of bus voltage v as a function of p for a specified power factor. The curves are plotted for a number of power factors, lagging as well as leading. Figure 19.5 shows a family of such curves. It is seen that critical voltage is high for high values of p. For leading power factors the maximum power and critical voltage are higher.

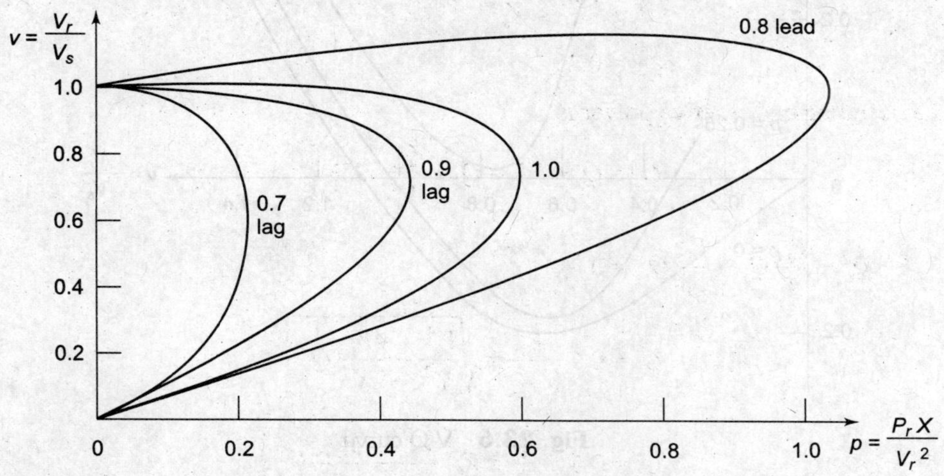

Fig. 23.5 PV curve

23.9.2 V-Q Curve

Another set of curves shows the relation between reactive power Q_r and receiving end voltage V_r for different values of P_r. It is more convenient to use the normalized values, i.e., q, v and p. These curves can be obtained from the pv curves. A particular value of p and power factor gives the corresponding value of q and two values of v. Thus the vq curves can be plotted and are shown in Fig. 23.6. For a large system the results of load flow study can be used for preparing these curves.

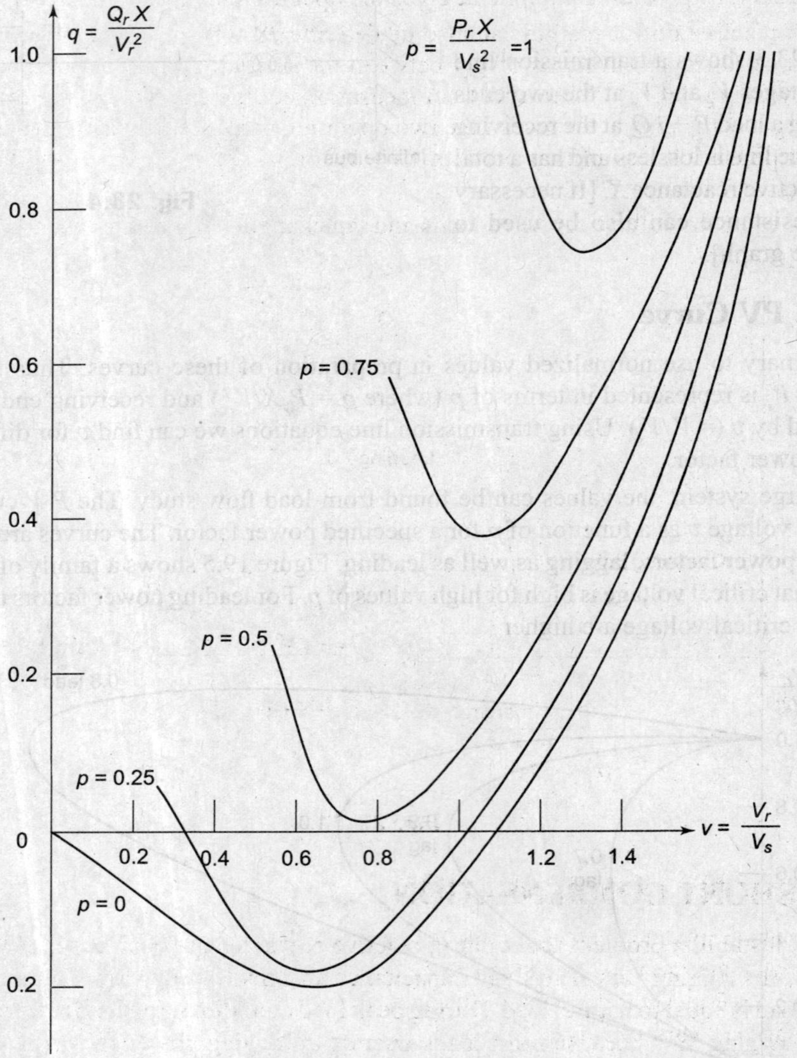

Fig. 23.6 V-Q curve

For voltage stability study it is more convenient and preferable to use *V-Q* curves due to the following reasons.

1. The slope of *V-Q* curve gives the stiffness of the bus. (Stiffness is the factor $\Delta V/\Delta Q$).

2. Since bus voltage is directly related to reactive power, a *V-Q* curve gives better insight into voltage stability phenomenon.
3. If necessary the reactive power of generators can also be plotted on the same graph. The point where the slope of *V-Q* curve becomes less steep and the bottom of the curve is approached indicates that the reactive power limit of the generator has been reached.
4. The solution to voltage instability problem lies in use of shunt compensation equipment (static capacitors, synchronous condensors, shunt reactors). The characteristics of these devices can also be plotted on the *V-Q* curves. The intersection of *V-Q* curves of the bus and the compensation equipment gives the operating point.(Fig. 23.7 b).
5. The reactive power margin can also be determined. The reactive power margin is the vertical distance (in terms of Q_r) between the operating point and the bottom of the curve. (Fig. 23.7a). When shunt compensation equipment is used, the reactive power margin is the distance between the two operating points (either without and with shunt compensation equipment or with two different shunt compensation equipments (Fig. 23.7 b).
6. The effect of voltage sensitive loads and tap changing can also be incorporated on the *V-Q* curves.

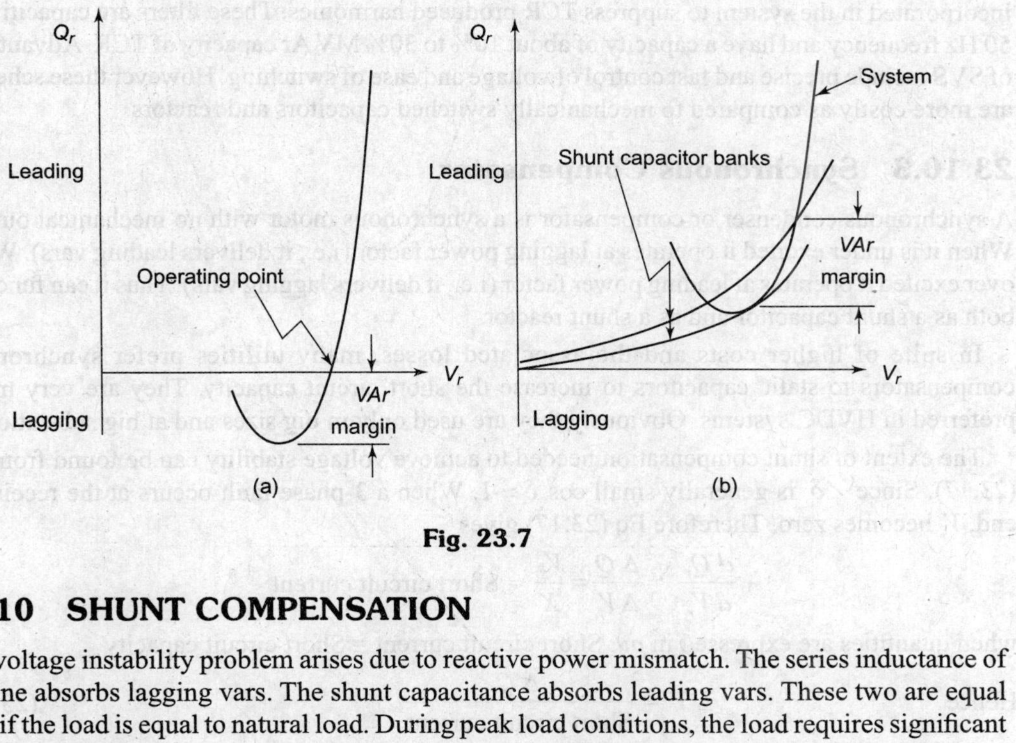

Fig. 23.7

23.10 SHUNT COMPENSATION

The voltage instability problem arises due to reactive power mismatch. The series inductance of the line absorbs lagging vars. The shunt capacitance absorbs leading vars. These two are equal only if the load is equal to natural load. During peak load conditions, the load requires significant amount of lagging vars because most loads operate at lagging power factor. If these lagging vars are supplied by the generators at the sending end, the generators and other equipments may get overloaded and voltage at the receiving end may be too low. Therefore it is only logical that these lagging vars are generated by using shunt capacitors at the receiving end. During off peak conditions net vars are generated by the line. To obtain voltage stability these vars must be absorbed by using shunt reactors. Shunt compensation involves the use of shunt capacitors and shunt reactors to avoid voltage instability.

19.10.1 Mechanically Switched Shunt Capacitors and Reactors

Mechanically switched shunt capacitors and reactors are used in substations to provide voltage stability. The switching is done by circuit breakers which are actuated by voltage relays. When voltage tends to be too low, shunt capacitors are switched in. When it tends to be too high, shunt reactors are switched in. Current limiting series reactors are also used to minimize switching transients. This system has the main advantage of low cost. However the switching of.capacitors and reactors may not be fast enough to avoid voltage instability and voltage collapse.

Nowadays microprocessors are being increasingly used for switching logic.

23.10.2 Static Var System, Static Var Compensator

The terms static var system and static var compensator generally mean the same system. Both the terms are used to describe the same equipment. However, according to CIGRE, a static var system is one in which the compensator also controls the mechanical switching of shunt capacitors and reactors.

SVS include self saturated reactor, thyristor controlled reactor—fixed capacitor (TCR-FC) and thyristor switched capacitor—thyristor controlled reactor (TSC—TSR). Harmonic filters are incorporated in the system to suppress TCR produced harmonics. These filters are capacitive at 50 Hz frequency and have a capacity of about 10% to 30% MV Ar capacity of TCR. Advantages of SVS include precise and fast control of voltage and ease of switching. However these schemes are more costly as compared to mechanically switched capacitors and reactors.

23.10.3 Synchronous Compensator

A synchronous condenser or compensator is a synchronous motor with no mechanical output. When it is under excited it operates at lagging power factor (i.e., it delivers leading vars). When over excited it operates at leading power factor (i.e., it delivers lagging vars). Thus it can function both as a shunt capacitor and as a shunt reactor.

In spite of higher costs and the associated losses, many utilities prefer synchronous compensators to static capacitors to increase the short circuit capacity. They are very much preferred in HVDC systems. Obviously they are used only in big sizes and at big substations.

The extent of shunt compensation needed to achieve voltage stability can be found from Eq (23.17). Since $\angle\delta$ is generally small cos $\delta \simeq 1$. When a 3-phase fault occurs at the receiving end, V_r becomes zero. Therefore Eq (23.17) gives

$$\frac{dQ_r}{dV_r} = \frac{\Delta Q}{\Delta V} = \frac{V_s}{X} = \text{Short circuit current}$$

when quantities are expressed in *pu*, Short circuit current = Short circuit capacity

Hence
$$\Delta V = \frac{\Delta Q}{\text{Short circuit current}} \tag{23.22}$$

where ΔV = Voltage fluctuation

ΔQ = MV Ar capacity of SVS

23.11 SERIES COMPENSATION

Generally series compensation is used in long lines to increase transmission capacity, improvement of system stability and to obtain correct load division between parallel circuits. However, series compensation is being increasingly recommended on short lines also to improve voltage stability.

Series capacitors are generally installed on special platforms at one or both ends of the line. The total equipment consists of the series capacitors, spark gap protection, metal oxide resistor (for protection), current limiting reactor, bypass switch, control and protection equipment. Figure 23.6 shows these equipments.

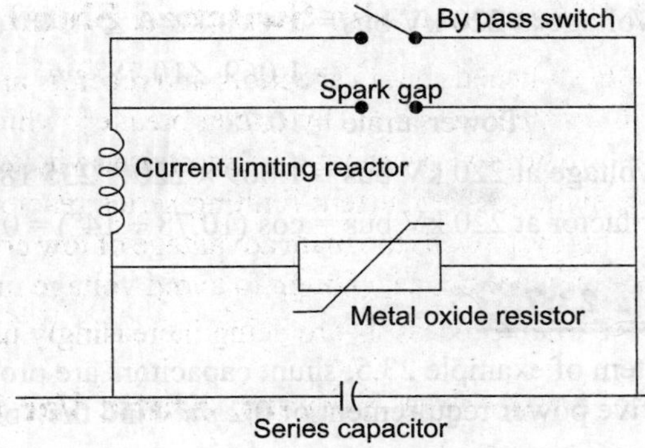

Fig. 23.8

The reactive power contribution of a series capacitor is proportional to I^2. Hence they are most effective at peak loads. If total series compensation is 50%, it is preferable to have about 25% compensation at each end.

Series compensation increases the effective natural load (shunt compensation lowers it). The changed natural load is given by

$$\frac{SIL'}{SIL} = \left[\frac{1-K_{sh}}{1-K_{se}}\right]^{0.5} \tag{23.23}$$

where SIL = surge impedance loading without compensation

SIL' = surge impedance loading with compensation

K_{sh} = degree of shunt compensation (positive for shunt reactors)

K_{se} = degree of series compensation

Example 23.5.

The short circuit capacity at a bus is 8000 MVA. Find the capacity of SVS to achieve voltage stability if voltage fluctuation is ± 3%.

Solution

Using Eq (23.22)

$$\Delta Q = \pm \frac{3}{100} \times 8000 = \pm 240 \text{ MVAr}$$

Example 23.6.

A 220 kV bus supplies a load of 0.8 + *j* 0.2 *pu* through a short line and a transformer. The *pu* reactance of line and transformer are 0.2 and 0.05 *pu*. If the voltage at load terminals is 1 *pu*, find voltage and power factor at 220 kV bus. Base MVA = 100, Base kV = 220 kV.

Solution

$$V = 1\angle 0, \; X = 0.25$$

$$|I| = \frac{P^2 + Q^2}{V^2} = \frac{0.8^2 + 0.2^2}{1} = 0.8246 \; pu$$

$$I = 0.8246 \angle -14^\circ \; pu$$

$$\text{Voltage at 220 kV bus} = 1 + (0.8246 \angle -14°)(0.25 \angle 90°)$$
$$= 1.069 \angle 10.78° \; pu$$
$$\text{Power angle} = 10.78°$$
$$\text{Voltage at 220 kV bus} = 1.069 \times 220 = 235.18 \text{ kV}$$
$$\text{Power factor at 220 kV bus} = \cos(10.78 + 14°) = 0.908 \text{ lagging}$$

Example 23.7.

In the system of example 23.5, shunt capacitors are provided at the receiving end to supply the total reactive power requirement of 0.2 *pu*. Find the voltage and power factor at 220 kV bus.

Solution

$$V = I \angle 0, \qquad X = 0.25$$
$$I = 0.8 \angle 0 \; pu$$
$$\text{Voltage at 220 kV bus} = 1 + (0.8 \angle 0)(0.25 \angle 90°)$$
$$= 1.02 \angle 11.31° \; pu$$
$$= 224.4 \text{ kV}$$
$$\text{Power factor at 220 kV bus} = \cos(11.31°) = 0.98 \text{ lagging.}$$

Example 23.8.

A 350 km long line has a total series impedance of 180 ∠75° Ω per phase and a total shunt admittance to neutral per phase of 1×10^{-3} ∠90° Siemens (a) Find *ABCD* parameter of the line (b) The line is provided with 60% series compensation. Find *ABCD* parameters of the compensated line.

Solution

(a) Using the results of example 4.15

$$Z_c = 424 \angle -7.5° \; \Omega$$
$$\sinh \gamma l = 0.0505 + j\,0.4083 = 0.4114 \angle 82.95°$$
$$\cosh \gamma l = 0.914 + j\,0.0226 = 0.9143 \angle 1.42°$$
$$A = D = \cosh \gamma l = 0.9143 \angle 1.42°$$
$$B = Z' = Z_c \sinh \gamma l = 174.44 \angle 75.45° \; \Omega$$
$$C = \frac{\sinh \gamma l}{Z_c} = \frac{0.4114 \angle 82.95°}{424 \angle -7.5°} = 0.00097 \angle 90.45°$$

(b) Series Compensation affects only the series impedance of the line. The shunt admittance is not affected. The new value of series impedance is the constant *B* for the compensated line.

$$\text{Inductive reactance of line} = 180 \sin 75° = 173.867 \; \Omega$$
$$\text{Series compensation} = 60\%$$
$$B = 174.44 \angle 75.45 - j\,0.6 \times 173.867$$
$$= 43.82 + j\,64.53 = 78 \angle 55.82°$$

For the equivalent π curcuit

$$\frac{Y'}{2} = \frac{1}{Z_c}\left[\frac{\cosh \gamma l - 1}{\sinh \gamma l}\right] = \frac{0.9143 \angle 1.42° - 1}{174.44 \angle 75.45°}$$

$$= \frac{0.089 \angle 165.3°}{174.44 \angle 75.45°} = 0.00051 \angle 89.85°$$

Using (4.22)

$$A = D = 1 + (78 \angle 55.92)\,(0.00051 \angle 89.85)$$
$$= 0.9671 + j\,0.02238 = 0.9674 \angle 1.33°$$
$$C = 2 \times 0.00051 \angle 89.85° + [78 \angle 55.82°]\,[0.00051 \angle 89.85]^2$$
$$= 0.00102 \angle 89.85° + 0.00002 \angle 235.52°$$
$$= 0.001 \angle 90.5°\ \text{S}$$

Example 23.9.

Find the constants of nominal π circuit for the uncompensated and compensated line of example 23.8.

Solution

For the uncompensated line

$$B = Z = 180 \angle 75°\ \Omega$$

$$A = D = 1 + \frac{ZY}{Z} = 1 + \frac{(180 \angle 75°)\,(1 \times 10^{-3} \angle 90°)}{2}$$

$$= 1 + 0.09 \angle 165° = 0.913 \angle 1.46°$$

$$C = Y\left(1 + \frac{ZY}{4}\right) = 1 \times 10^{-3} \angle 90° \left[1 + \frac{(180 \angle 75°)\,(1 \times 10^{-3} \angle 90°)}{4}\right]$$

$$= (1 \times 10^{-3} \angle 90)\,(1 + 0.045 \angle 165°) = 0.957 \times 10^{-3} \angle 90.7°\ \text{S}$$

For the compensated line

$$B = Z = 180 \angle 75° - j\,0.6 \times 173.867$$
$$= 46.59 + j\,69.55 = 83.71 \angle 56.18°\ \Omega$$

$$A = D = 1 + \frac{ZY}{Z} = 1 + [83.71 \angle 56.18°]\left[\frac{(1 \times 10^{-3} \angle 90°)}{2}\right]$$

$$= 0.965 \angle 1.38°$$

$$C = 1 \times 10^{-3} \angle 90° \left[1 + \frac{(83.71 \angle 56.18°)\,(1 \times 10^{-3} \angle 90°)}{4}\right]$$

$$= 0.983 \times 10^{-3} \angle 90.7°\ \text{S}$$

It is seen that both for the uncompensated and compensated lines, the constants of equivalent and nominal π circuit are nearly equal.

Example 23.10.

The transmission line of example 4.11 is provided with a shunt inductor under no load conditions at the receiving end.

The shunt inductor compensates for 60% of total shunt admittance of the line (a) Find the voltage regulation of the compensated line. (b) Compare the above value of voltage regulation with the corresponding value for the uncompensated line

Solution

(a) *ABCD* constants of the line as found in example 4.11 are

$$\begin{bmatrix} 0.8705\ \angle 2.33° & 187.03\ \angle 75.15° \\ .0013494\ \angle 90.722° & 0.8705\ \angle 2.33° \end{bmatrix}$$

The admittance of the shunt inductor

$$= -j\,0.6 \times 0.001411 = 0.0008466\ \angle{-90°}\text{S}$$

Using the results of section 3.12, the *ABCD* constants of shunt inductor are

$$\begin{bmatrix} 1 & 0 \\ 0.0008466\ \angle -90° & 1 \end{bmatrix}$$

Constant *A* of the combination of line and shunt inductor is

$$A = 0.8705\ \angle 2.33° + (187.03\ \angle 75.15)\ (.0008466\ \angle{-90°}) = 1.02\ \angle{-0.3°}$$

$$|V_r| \text{ at no load} = \frac{|V_s|}{|A|} = \frac{124795.23}{1.02} = 122348.3\text{ V}$$

$$\text{Voltage regulation} = \frac{122348.3 - 127021}{127021} \times 100 = -3.68\%$$

(b) When the line is not compensated

$$|A| = 0.8705$$

$$|V_r| \text{ at no load} = \frac{124795.23}{.8705} = 143360.4\text{ V}$$

$$\text{Voltage regulation} = \frac{143360.4 - 127021}{127021} = 12.86\%$$

It is seen that provision of shunt inductor has made a very considerable effect on voltage regulation.

Example 23.11.

Figure 23.9 shows a busbar *D* connected to three infinite bus bars *A*, *B* and *C*. Bus bars *A* and *B* operate at 220 kV while *C* and *D* operate at 132 kV. The transformer reactances are on 200 MVA base. System resistance can be neglected. Assume 200 MVA and 220 kV at bus *A* as base. Draw reactance diagram. If the voltage at bus *D* falls by 4 kV, find the var injection at bus *D* to bring back the bus voltage to its original value.

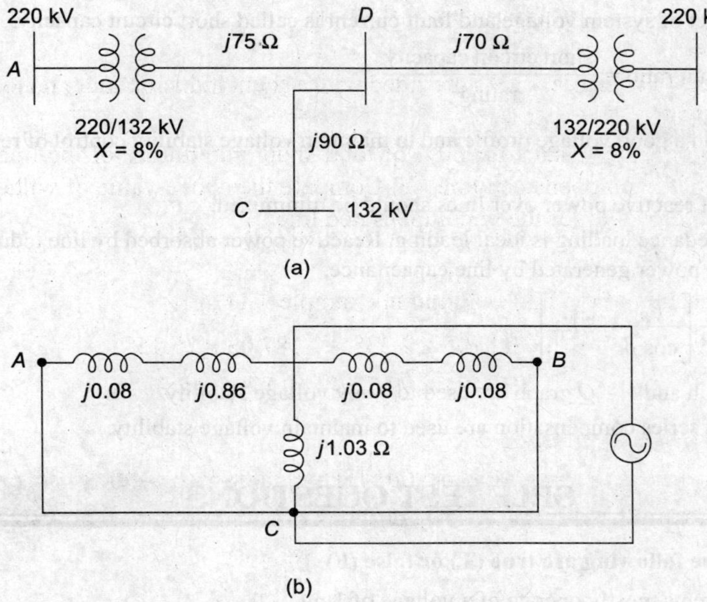

Fig 19.9

Solution

$$\text{Base MVA} = 200\text{ MVA}$$

$$\text{Base impedance in 132 kV circuit} = \frac{132 \times 132}{200} = 87.12\ \Omega$$

$$j\,75\ \Omega = j\,75/87.12 = j\,0.86\ pu$$

$$j\,70\ \Omega = j\,70/87.12 = j\,0.8\ pu$$

$$j\,90\ \Omega = j\,90/87.12 = j\,1.03\ pu$$

$$\text{Reactance from } A \text{ to } D = j\,.08 + j\,.86 = j\,.94\ pu$$

$$\text{Reactance from } B \text{ to } D = j\,.08 + j\,.8 = j\,.88\ pu$$

The parallel combination of $j\,0.94$, $j\,0.88$ and $j\,1.03$ gives a reactance of $j\,0.315\ pu$. The reactance diagram is shown in Fig. 19.9(b)

$$\text{Short circuit MVA at } D = \frac{200}{0.315} = 634.92\text{ MVA}$$

$$\frac{\Delta Q}{\Delta V} = \frac{634.92}{132} = 4.81\text{ MVAr/kV}$$

$$\Delta Q = 4.81 \times 4 = 19.24\text{ MVAr}$$

SUMMARY

23.1 When the voltage of a bus departs too much from 1 *pu*, the phenomenon is called voltage collapse.

23.2 An unacceptable voltage level means voltage instability.

23.3 The product of system voltage and fault current is called short circuit capacity.

23.4 Short circuit ratio = $\dfrac{\text{short circuit capacity}}{\text{rating}}$.

23.5 To maintain a good voltage profile and to maintain voltage stability control of reactive power is necessary.

23.6 Transfer of reactive power over lines should be minimized.

23.7 Surge impedance loading is ideal loading. Reactive power absorbed by line inductance is equal to reactive power generated by line capacitance.

23.8 $\dfrac{dQ_r}{dV_r} = \dfrac{1}{X}\left[\dfrac{V_s}{\cos\delta} - 2V_r\right]$.

23.9 $P-V$ graph and $V-Q$ graph are used to study voltage stability.

23.10 Shunt and series compensation are used to maintain voltage stability.

SELF TEST QUESTIONS

A. State whether the following are true (T) or false (F)

23.1 Power systems mostly operate of a voltage of 1 *pu*.

23.2 Voltage instability may be oscillatory or of long duration.

23.3 Major cause of voltage drop on a line is transfer of reactive power.

23.4 S. C. Ratio = $\dfrac{\text{Rating}}{\text{S. C. capacity}}$.

23.5 In the event of load rejection, reactive power transfer can cause voltage collapse.

23.6 When load = SIL, voltage and current are in phase of all points on the line.

23.7 As line length increases the ratio S_r/SIL increases.

23.8 SVS leads to power losses on line.

B. Fill in the blanks with suitable words.

23.9 In terms of z and y, Z_c = ________.

23.10 When reactive power transferred over a line is more, the difference between V_s and V_r is ________.

23.11 Synchronous condenser ________ short circuit capacity.

23.12 If reactive losses are more, the cost of SVS is ________.

23.13 In terms of L and C, Z_c = ________.

23.14 As $\delta \to \infty$, $\dfrac{dQ_r}{dV_r} \to$ ________.

23.15 The graphs used to study voltage stability are ________ and ________ graphs.

ANSWERS

23.1 T	23.2 T	23.3 T	23.4 F	23.5 F	23.6 T
23.7 F	23.8 T	23.9 $\sqrt{z/y}$	23.10 more	23.11 increases	
23.12 more	23.13 $\sqrt{L/C}$	23.14 ∞	23.15 $P-V$, $V-Q$		

JOB INTERVIEW QUESTIONS

23.1 What is meant by voltage instability?

23.2 What is voltage collapse?

23.3 What are the advantages of using SVS?

23.4 Compare series and shunt compensation?

23.5 What is meant by short circuit ratio?

REVIEW QUESTIONS

23.1 Explain the term 'Voltage stability'. Why does voltage instability occur in power systems? How is voltage stability classified?

23.2 What are the problems of reactive power transfer over transmission lines? Is it possible that reactive power injection is necessary at both ends of the lines?

23.3 Differentiate between the terms 'Short circuit capacity' and 'Effective short circuit capacity' as applied to power systems. What is short circuit ratio? What is the significance of these terms?

23.4 Why should reactive power transfer ove the lines be minimized?

23.5 Differentiate between voltage stability and rotor angle stability.

23.6 What is surge impedance loading? What is its significance? Can every line be loaded to this level? Can a line be loaded to a level higher than this level? Explain.

23.7 What is voltage stability limit? When does it occur? Derive an expression for dQ_r/dV_r for a line.

23.8 Draw *p-v* and *v-q* graphs for a typical line. What is the use of these graphs? Which of them is more useful for voltage stability studies?

23.9 What is shunt compensation? How is it different from series compensation? Discuss the advantages and disadvantages of these.

PROBLEMS

23.1 An EHV line has constants $A = 0.93 \angle 0$ and $B = 0.4 \angle 90°$ *pu* to a base of 500 MVA and 400 kV. If $P_r = 1.5$ *pu*, find Q_s and Q_r if (a) $V_s = V_r = 1$ *pu* (b) $V_s = 1.05$ *pu* and $V_r = 1$ *pu*.

[(a) $Q_s = +.325$ *pu*, $Q_r = -0.325$ *pu*, (b) $Q_s = +.4091$ *pu*, $Q_r = -.1708$ *pu*]

23.2 For the transmission line of Problem 23.1 find the maximum value of P_r. What are the value of Q_s and Q_r under this condition. Assume $V_s = V_r = 1$ *pu*. [2.5 *pu*, + 2.325 *pu*, – 2.325 *pu*]

23.3 A short transmission line is fed by a source having a reactance of 0.25 *pu* on 200 MW and 132 kV base. The line reactance is 0.18 *pu* on the same base. The line is supplying rated current at rated voltage and at 0.8 p.f lagging (a) What will be the voltage at the receiving end if the receiving breaker opens suddenly. (b) What will be the receiving end voltage under similar situation if the load is unity pf. [(a) 1.304 *pu* (b) 1.088 *pu*]

23.4 The short circuit capacity at a 400 kV bus is 12000 MVA. Find the capacity of SVS to achieve voltage stability if voltage fluctuation is ± 4%.

[± 480 MVA]

23.5 A 132 kV bus supplies a load of $0.9 - j\,0.3$ *pu* through a short line and a transformer having *pu* reactance of 0.25 and 0.06 *pu* respectively. Find the voltage and p.f. at sending end bus if the voltage at receiving is 1 *pu*. Base MVA = 200 MVA, Base kV = 132 kV.

[148.9 kV, .841 lagging]

23.6 In the system of Problem 23.5, capacitors are provided at the receiving end to supply the total reactive power of 0.3 *pu*. Find voltage and p.f at sending end. [137.04 kV, 0.963 lagging]

23.7 In the system of example 23.10, the bus bar *A* and *B* operate at 400 kV while those at *C* and *D* operate at 220 kV. The reactance are as under. Transformer T_1 5% Transformer T_2 6%, Line *AD* 0.7 *pu*, Line *BD* 0.8 *pu* and Line *CD* 1.05 *pu*. Base voltages are equal to rated values Base MVA 800. Find the var injection at bus *D* if the voltage at this bus falls by 8 kV. [100.35 MV Ar]

23.8 A 220 kV 3-phase overhead line has a total series impedance of 38.0 + *j* 150 ohm and a total shunt admittance of *j* 1.5 × 10^{-3} siemens. (a) Find *ABCD* constants (b) Line is provided with 60% series compensation. Find *ABCD* constants of the compensated line. Use nominal π representation.

$$\left[\begin{array}{l}(a)\ A = D = 0.89\ \angle 1.84^\circ \\ \quad B = 154.74\ \angle 75.78^\circ \text{ ohm} \\ \quad C = 1.416\times10^{-3}\ \angle 90.9^\circ \text{ siemens} \\ (b)\ A = D = .955\ \angle 1.7^\circ,\ B = 71.02\ \angle 57.7^\circ\Omega \\ \quad C = 1.466\times10^{-3}\ \angle 90.84^\circ \text{ siemens}\end{array}\right]$$

CHAPTER

24

Reliability of Transmission and Distribution Systems

24.1 DEFINITION OF RELIABILITY

Reliability is the probability of a device performing its function adequately for the intended period of time under specified operating conditions.

This definition consists of 4 basic parts: probability, adequate performance, time and operating conditions.

Thus we can quantitatively define reliability as the probability that the device will meet the qualitative definition. If T is time to failure (T is itself variable), the reliability that the device will not fail before time t, i.e.,

$$R(t) = P(T > t) \tag{24.1}$$

It is assumed that the device is working satisfactorily at time $t = 0$. However no device can work indefinitely without failure, i.e.,

$$R(0) = 1 \quad \text{and} \quad R(\infty) = 0 \tag{24.2}$$

$R(t)$ is gerneraly a decreasing function between the limits of 1 and 0

The probability value is the first index of relibility and in most cases it is considered to be most significant and sufficient index. However many other indices are also used. Some of these are:

1. Expected number of failures in a specified period
2. Average time between failures
3. Expected down time
4. Expected loss in revenue due to failure
5. Expected loss in plant output due to failure.

24.2 OUTAGE

An outage refers to that state of device when it is not available to preform its intended function due to some event directly associated with that device. An outage of component may or may not cause an outage of the system or interruption of service depending on system configuration and the degree of redundancy in the system. An outage may be a forced outage or a scheduled outage.

A forced outage is an outage which results from emergency conditions directly associated with a component requiring that the component be taken out of service immediately either automatically or as soon as the switching operation can be performed or an outage caused by improper operation of equipment or human error.

A scheduled outage (also known as maintenance outage) is one which results when a component is deliberately taken out service at a selected time, usually for the purpose of preventive maintenance or overhaul or repair. A simple test of determine whether an outage should be classified as forced or scheduled is as under: If it is possible to defer an outage, the outage is a scheduled one, otherwise it is a forced on. Deferring an outage may be desirable, for example, to prevent overloading of facilities or interruption of service to consumers.

Still another category of outages is momentary outage. This is the outage caused by the reclosing breaker to clear a temporary fault. This outage lasts for a very short time, only a few seconds. More than one temporary outage may also occur in quick succession. Till a few years back these outages were not considered important. However with more automation such outages may cause havoc in the form of computer shut downs and damage to industrial processes.

The forced and momentary outages in transmission and distribution systems can be due to a number of reasons. Some of these are lighning, tree contact, wind, animals, vehicle accidents, construction activity, etc.

24.3 BATH TUB CURVE

Figure 20.1 shows a hazard function typical of many devices. This curve depicts the three different regions in the total life of a component. In the initial period, known as de-bugging or the period of infant mortality, the failures may be high due to errors in design or manufacture. It is expected that these errors are detected or removed in an initial operating period. This region shows a decreasing failure rate. Then comes the normal operating life or useful life. It is expected that the errors in design and manufacture have been corrected in the debugging period. Therefore the failure rate in the useful life period is only due to chance failures which are comparatively low. The failure rate in this period is more or less constant. The last region is the wear out period or old age when the failures come at an ever increasing rate. Usually the repairs in this period are very costly and it is preferable to replace the component by a new one. It is possible to select a probability distribution for each of these three regions from the Weibull distribution which can be expressed as

$$\lambda(t) = K\, t^{\beta-1} \text{ for } t > 0 \tag{24.3}$$

Where $\lambda(t)$ is the hazard or failure rate and K and β are constants. If $\beta > 1$, the failure rate increases with time (wear out stage). If $\beta = 1$, the failure rate is constant (useful life period). If $0 < \beta < 1$, the failure rate decreases (debugging region).

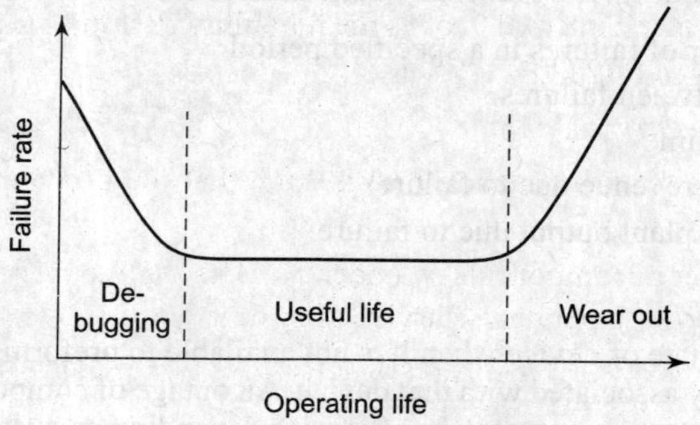

Fig. 24.1 Bath tub curve

Figure 24.1 shows the above three stages in the life of a component. It is known as bath tub curve because its shape resembles a bath tub. Power system components can be kept within useful life period for a pretty long time by careful and proper periodic maintenance.

24.4 TWO STATE MODEL

Figure 24.2 shows a two state model representation of the life history of repairable component (during the useful life period). After some operation a fault develops and the component goes to the down state. It is repaired and brought back to the up state. After being in up state for a certain period a fault develops again and the component goes to the down state. The cycle of up and down states is repeated, the component being available during the up state and being under repair the during the down state.

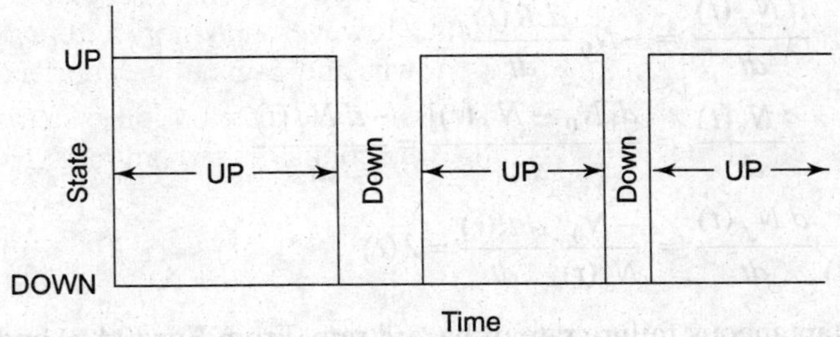

Fig. 24.2 Two state model

24.5 FAILURE AND REPAIR RATE

The number of failures per unit time is known as failure rate (symbol λ) of the component. Similarly the number of repairs per unit time is known as repair rate (symbol μ). For most components the failure and repair rates, during useful life period, can be assumed to be constant.

The life time distribution of a component with constant failure rate is exponential. The converse of this is also true, i.e., failure rate associated with exponential distribution is constant. A component with exponential life time distribution has a constant probability of failure in any time interval dt, i.e., the probability of failure in any interval t to $t + dt$ is constant if the component is operating at time t (irrespective of the value of t). Only exponential distribution has this property.

24.6 PROBABILITY DENSITY FUNCTION

Probability of non-failure in the interval 0 to t is the reliability $R(t)$ and the probability of failure in this interval is the unreliability $Q(t)$.

$$R(t) + Q(t) = 1 \tag{24.4}$$

The failure probability density function $f(t)$ is such that the probability of failure in the interval t to $t + dt$ is $(dt)\,[f(t)]$.

Consider a fixed number of components N_0 under test. During the interval 0 to t, components $N_f(t)$ fail and there fore $N_s(t)$ components survive after time t.

$$\text{Probability of survival} = R(t) = \frac{N_s(t)}{N_0} \tag{24.5}$$

As the test proceeds, the number $N_s(t)$ becomes smaller and smaller and $R(t)$ decreases.

$$R(t) = \frac{N_s(t)}{N_0} = \frac{N_0 - N_f(t)}{N_0} = 1 - \frac{N_f(t)}{N_0} \tag{24.6}$$

$$\frac{d\,R(t)}{dt} = -\frac{1}{N_0}\frac{d\,N_f(t)}{dt} \tag{24.7}$$

As dt approaches zero we get instantaneous probability, i.e., probability density function $F(t)$.

$$\frac{d\,R(t)}{dt} = -f(t) \tag{24.8}$$

From Eq. (24.7)

$$\frac{d\,N_f(t)}{dt} = -N_0\frac{d\,R(t)}{dt} \tag{24.9}$$

$$\frac{d\,N_s(t)}{dt} = \frac{d[N_0 - N_f(t)]}{dt} = \frac{-d\,N_f(t)}{dt} \tag{24.10}$$

$$\frac{1}{N_s(t)}\frac{d\,N_f(t)}{dt} = \frac{-N_0}{N_s(t)}\frac{d\,R(t)}{dt} = \lambda(t) \tag{24.11}$$

where $\lambda(t)$ is the instantaneous failure rate or hazard rate. From Eqs (24.5) and (24.11) we get

$$\lambda(t) = -\frac{1}{R(t)}\frac{d\,R(t)}{dt} \tag{24.12}$$

From Eqs (24.8) and (24.12)

$$\lambda(t) = \frac{f(t)}{R(t)} \tag{24.13}$$

If failure rate is independent of time, $\lambda(t)$ is written as λ.

24.7 PROBABILITIES OF SURVIVAL AND FAILURE

The probability of a component surviving a time t in a constant failure environment is denoted by $R(t)$ and is given by

$$R(t) = e^{-\lambda t} \tag{24.14}$$

The failure density function $f(t)$ is

$$f(t) = \frac{-d\,R(t)}{dt} = \lambda\, e^{-\lambda t} \tag{24.15}$$

$Q(t)$, i.e., the probability of failure in time is t given by

$$Q(t) = \int_0^t f(t)\,dt = \int_0^t \lambda e^{-\lambda t}\,dt = 1 - e^{-\lambda t} \tag{24.16}$$

The probability of survival in time t is $R(t)$

$$R(t) = \int_t^\infty f(t)\,dt = \int_t^\infty \lambda\, e^{-\lambda t}\,dt = e^{-\lambda t} \tag{24.17}$$

It $\lambda t << 1$, then $Q(t) = \lambda t$ and $R(t) = 1 - \lambda t$

Figure 24.3 shows a variation of $f(t)$ with t. This is known as exponential failure density function $f(t)$.

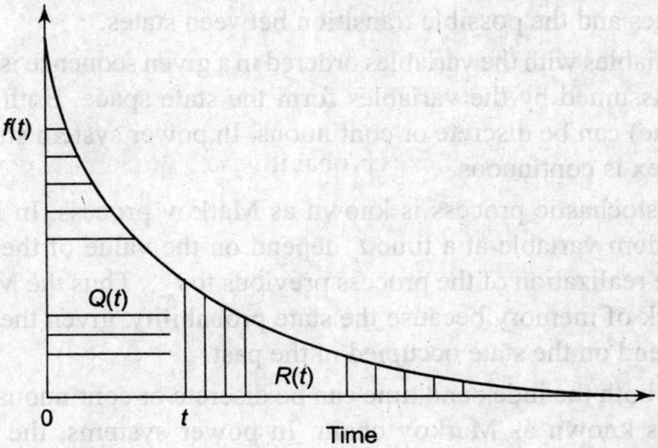

Fig. 20.3 Exponential failure density function

24.8 MEAN TIME TO FAILURE, MEAN DOWN TIME

If x is a continuously random variable with a probability density function $f(x)$, the expected value $E(x)$ is given by

$$E(x) = \int_0^\infty x\, f(x)\, dx \tag{24.18}$$

For the failure density function $f(t)$, the expected value $E(t)$ is given by

$$E(t) = \int_0^\infty t.\, f(t)\, dt = \int_0^\infty t \cdot \lambda e^{-\lambda t}\, dt \tag{24.19}$$

Integrating by parts, we get

$$E(t) = \left| -t\, e^{-\lambda t} \right|_0^\infty + \int_0^\infty e^{-\lambda t}\, dt$$

$$= 0 + \left| \frac{e^{-\lambda t}}{-\lambda} \right|_0^\infty = \frac{1}{\lambda} \tag{24.20}$$

This expected value is known as mean time to failure (MTTF) or mean up time. Thus MTTF is reciprocal of failure rate.

Similarly we can define a repair density function and calculate the expected value. This expected value is known as mean down time or mean repair time and is given by

$$\text{Mean down time} = \int_0^\infty t\, \mu\, e^{-\mu t}\, dt = \frac{1}{\mu} \tag{24.21}$$

Thus the mean down time is reciprocal of repair rate μ. Another term used in reliability studies is mean time between failure or MTBF. This is the sum of MTTF and mean down time. MTBF is also known as mean cycle time.

24.9 CONTINUOUS MARKOV PROCESS

One method of reliability evaluation is state space method. In this method the system is represented by its states and the possible transition between states.

A set of random variables with the variables ordered in a given sequence is known as stochastic process. The values assumed by the variables form the state space. Both the state space and their indices (e.g. time) can be discrete or continuous. In power system studies the state space is discrete but the index is continuous.

A special class of stochastic process is known as Markov process. In Markov process the probabilities of a random variable at a time t_n depend on the value of the random variable at time t_{n-1} but not on the realization of the process previous to t_{n-1}. Thus the Markovian process is characterized by a lack of memory because the state probability, given the present state of the process, does not depend on the state occupied in the past.

In Markov process both the index and time can be discrete or continuous. A Markov process with discrete index is known as Markov chain. In power systems, the index, i.e., time is continuous. Therefore, the process is a continuous Markov process.

Figure 24.4 shows the state space diagram of a single rewardable component whose failure and repair rates are characterized by exponential distributions. Let

$P_0(t)$ = Probability that the component is in state 0 at time t.

$P_1(t)$ = Probability that the component is in state 1 at time t.

We consider the transitions for a time interval Δt. During this small interval the probability of occurrence of more than 1 event can be assumed to be zero. The probabilities of failure and repair in this interval Δt are $\lambda \cdot \Delta t$ and $\mu \cdot \Delta t$.

Probability that the component is in state 0 at time $t + \Delta t$

= [Probability that the component is in state 0 as time t] [Probability that the state does not change from 0 to 1 in time Δt)

+ [Probability that the component is in state 1 at time t] [Probability that the component is repaired in time Δt].

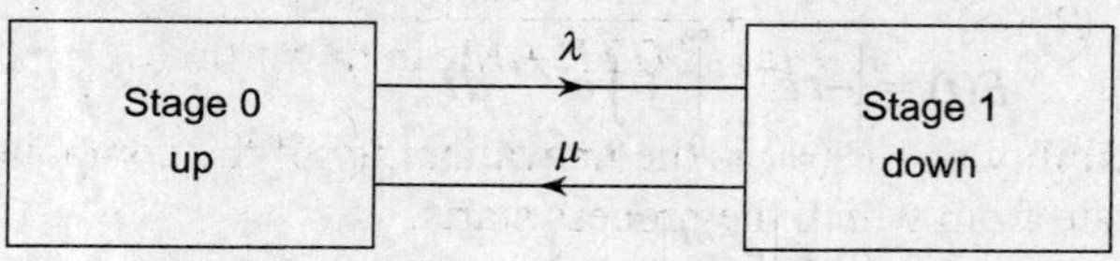

Fig. 24.4 State space diagram

Thus

$$P_0(t+\Delta t) = P_0(t)[1-\lambda\Delta t] + P_1(t)[\mu\Delta t] \qquad (24.22)$$

Similarly the probability $P_1(t+\Delta t)$, i.e., the probability that the component is in state 1 at time $(t+\Delta t)$ can be written as

$$P_1(t+\Delta t) = P_1(t)[1-\mu\Delta t] + P_0(t)[\lambda\Delta t] \qquad (24.23)$$

Eqs (24.22) and (24.23) can be written as

$$\frac{P_0(t+\Delta t) - P_0(t)}{\Delta t} = -\lambda P_0(t) + \mu P_1(t) \qquad (24.24)$$

$$\frac{P_1(t+\Delta t) - P_1(t)}{\Delta t} = \lambda P_0(t) - \mu P_1(t) \tag{24.25}$$

As $\Delta t \to 0$, LHS of Eq. (24.24) is $\frac{d P_0(t)}{dt}$ or $P_0'(t)$ and LHS of Eq. (24.25) is $P_0'(t)$. Thus

$$\begin{bmatrix} P_0'(t) \\ P_0'(t) \end{bmatrix} = [P_0(t) \quad P_1(t)] \begin{bmatrix} -\lambda & \lambda \\ \mu & -\mu \end{bmatrix} \tag{24.26}$$

Eq. (24.26) can be solved by classical method or Laplace Transform method. The solution is

$$P_0(t) = \frac{\mu}{\lambda+\mu}[P_0(0) + P_1(0)] + \frac{e^{-(\lambda+\mu)t}}{\lambda+\mu}[\lambda P_0(0) - \mu P_1(0)] \tag{24.27}$$

$$P_1(t) = \frac{\lambda}{\lambda+\mu}[P_0(0) + P_1(0)] + \frac{e^{-(\lambda+\mu)t}}{\lambda+\mu}[\mu P_0(0) - \lambda P_0(0)] \tag{24.28}$$

where $P_0(0)$ and $P_1(0)$ are the initial conditions and $P_0(0) + P_1(0) = 1.0$

If the process starts form state 0, i.e., the system is in state 0 at time 0, $P_0(0) = 1$ and $P_1(0) = 0$. Then Eqs. (24.27) and (24.28) simplify to

$$P_0(t) = \frac{\mu}{\lambda+\mu} + \frac{\lambda e^{-(\lambda+\mu)t}}{\lambda+\mu} \tag{24.29}$$

$$P_1(t) = \frac{\lambda}{\lambda+\mu} + \frac{\lambda e^{-(\lambda+\mu)t}}{\lambda+\mu} \tag{24.30}$$

When $t \to \infty$, the probabilities are known as limiting state probabilities. These are

$$P_0(\infty) = \frac{\mu}{\lambda+\mu} = \frac{MTTF}{MTTF + \text{Mean down time}} \tag{24.31}$$

$$P_1(\infty) = \frac{\mu}{\lambda+\mu} + \frac{\text{Mean down time}}{MTTF + \text{Mean down time}} \tag{24.32}$$

$P_0(\infty)$ is the availability and $P_1(\infty)$ is the unavailability of the component. These values are independent of the state from which the process starts.

24.10 RELIABILITY OF SERIES AND PARALLEL SYSTEMS

24.10.1 Series System

When n components having reliability R_1, R_2, R_3 etc., are connected in series, then all the components must be available for system success. The system reliability is given by

$$R_s = R_1 . R_2 . R_3 \ldots R_n \tag{24.33}$$

or

$$R_s = e^{-\lambda 1^t} \cdot e^{-\lambda 2^t} \cdot e^{-\lambda 3^t} \cdots e^{-\lambda n^t}$$

or

$$R_s = e^{-\sum_{i=l}^{n} \lambda_i t} \tag{24.34}$$

where R_s is the system reliability and λ_1, λ_2 ... are the failure rates of the components.

24.10.2 Parallel System

If a system has n components in parallel, the chances of all the components failing simultaneously is given by

$$Q_s = Q_1\, Q_2\, Q_3 \cdots Q_n$$

and

$$R_s = 1 - Q_s \tag{24.35}$$

Eq. (24.35) assumes that the system is fully redundant. Genrally a system is not fully redundant. Then the reliability of the system depends on the number of parallel components required for system success. If a system has 4 identical components, each having reliability (or availability) R and unavailability Q ($= 1 - R$), the probabilities associated with finding 0, 1, 2, 3, 4 components available at the same time are given by the terms in the binomial expansion of $(R + Q)^4$.

$$(R + Q)^4 = R^4 + 4R^3Q + 6R^2Q^2 + 4RQ^3 + Q^4 \tag{24.36}$$

The probabilities of system success are given as under:

Number of components required for system success	*Probability of system success*
4	R^4
3	$R^4 + 4R^3Q$
2	$R^4 + 4R^3Q + 6R^2Q^2$
1	$R^4 + 4R^3Q + 6R^2Q^2 + 4RQ^3$

24.11 TWO STATE FLUCTUATING ENVIRONMENT

The transmission and distribution systems are mostly outdoor and operate in an environment which keeps on changing. The failure and repair rates during bad weather conditions can be considerably higher than the values prevailing during fair weather conditions. For reliability calculations of such systems a two state fluctuating environment is assumed. This is a simple representation of weather conditions as the complete weather data is generally not available. This representation divides the weather into normal and stormy weather. A wind velocity more than a specified value is assumed to constitute a storm. Heavy clouds and thunder storms also constitute stormy weather. The weather data is scanned to determine the total time for which the normal and stormy weather prevail during the study period and the distributions of normal and stormy weather durations.

24.12 MARKOV APPLICATION TO SINGLE REPAIRABLE COMPONENT OPERATING IN TWO STATE FLUCTUATING ENVIRONMENT

Figure 24.5 shows the state space diagram of a repairable component operating in two state fluctuating environment. Let

λ, μ = normal weather failure and repair rates of the component

λ', μ' = stormy weather failure and repair states of the component

$m = \dfrac{1}{S}$ where S is the expected duration of stormy weather period

$n = \dfrac{1}{N}$ where N is the expected duration of normal weather period.

The differential equations, in matrix form, are as under:

$$\begin{bmatrix} P_0'(t) \\ P_1'(t) \\ P_2'(t) \\ P_3'(t) \end{bmatrix} = [P_0(t) \quad P_1(t) \quad P_2(t) \quad P_3(t)] \begin{bmatrix} -(\lambda+n) & n & \lambda & 0 \\ m & -(m+\lambda') & 0 & \lambda' \\ \mu & 0 & -(\mu+n) & n \\ 0 & \mu' & m & -(\mu'+m) \end{bmatrix} \quad (24.37)$$

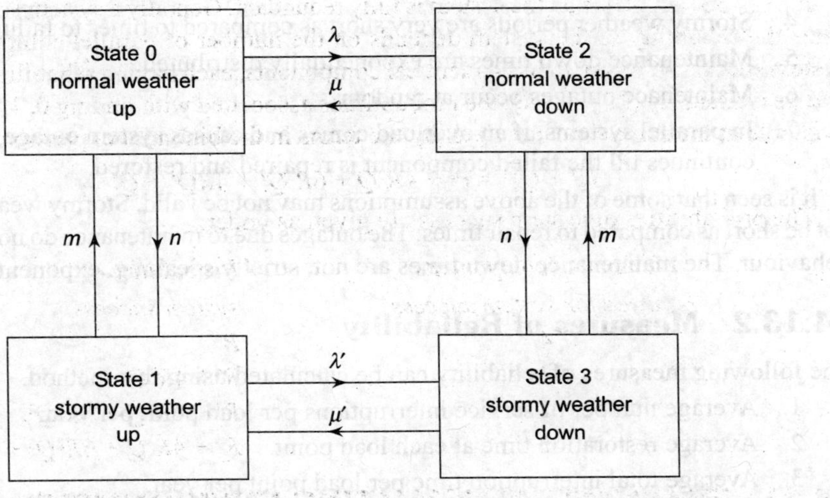

Fig. 24.5 State space diagram of a single repairable component operating in two state fluctuating environment

The steady state limiting probabilities can be found from the following equations.

$$-(\lambda + n) P_0 + m P_1 + \mu P_2 = 0$$
$$n P_0 - (m + \lambda') P_1 + \mu' P_3 = 0$$
$$\lambda P_0 - (\mu + n) P_2 + m P_3 = 0$$
$$\lambda' P_1 + n P_2 - (\mu' + m) P_3 = 0$$
$$P_0 + P_1 + P_2 + P_3 = 1$$

For this system

$$P \text{ (system operating)} = P_0 + P_1$$
$$P \text{ (system failed)} = P_2 + P_3$$

These probabilities, mean time to failure and mean down time can be evaluated using the Markow approach. The calculations tend to be very lengthy even for this simple one component system. As the number of components increases, the calculations become all the more lengthy. Moreover the data of failure and repair rates and the weather data is never available to a precise degree. Therefore it is preferable to use an approximate for reliability calculations of transmission and distribution systems.

24.13 APPROXIMATE METHOD

The transmission and distribution systems consist of a large number of components in series and parallel. The use of Markov approach leads to very complex equation. The use of approximate method leads to reasonably accurate results with minimal effort.

24.13.1 Assumptions

1. Times to failure (i.e., period between failures) and repair times are exponentially distributed both during normal weather and stormy weather.
2. The durations of normal weather and stormy weather periods are exponentially distributed.
3. Repair durations are very short as compared to times to failure and periods between storms.
4. Stormy weather periods are very short as compared to times to failure and repair times.
5. Maintenance down times are exponentially distributed.
6. Maintenace outages occur at random.
7. In parallel systems, if an overload comes and causes system outage, the system outage continues till the failed component is repaired and restored.

It is seen that some of the above assumptions may not be valid. Stormy weather durations may not be short as compared to repair times. The outages due to maintenance do not observe a random behaviour. The maintenance down times are not, strictly speaking, exponentially distributed.

24.13.2 Measures of Reliability

The following measures of reliability can be calculated using this method.

1. Average number of service interruptions per load point per year.
2. Average restoration time at each load point.
3. Average total interruption time per load point per year.
4. Maximum expected number of interruptions experienced by any load point.
5. Maximum expected restoration time experienced by any load point.
6. Probability that any load point will be out of service at any time longer than a specified time.

24.13.3 Component Data Required

This method requires the following data:

1. Normal weather failure rate data for all components. This is expressed as number of failures per year of normal weather per unit of apparatus or per kilometer of line. It is calculated as

$$\lambda_k = C/Y \tag{24.39}$$

where λ_k = normal weather failure rate of kth component.

C = number of non-storm associated failures.

Y = total normal weather exposure time of component

2. Stormy weather forced outage rate of components whose failure is affected by weather. It is calculated as

$$\lambda'_k = C'/Y' \tag{24.40}$$

where λ_k = stormy weather failure rate of kth component.

C' = number of storm-associated failures.

Y' = Summation of stormy weather exposure times of component.

3. Temporary forced outage of all components if temporary outages are to be included in the analysis. Since the duration of these outages is short; it is not necessary to separate them into normal weather and stormy weather.
4. Maintenance outage rates of all components. It is calculated as

$$\lambda''_k = C''/Y'' \tag{24.41}$$

where λ''_k = maintenance outage rate of kth component

C'' = number of maintenance outages

Y'' = Summation of observation period

5. Distribution of component repair times for forced and maintenance outages denoted by r_k and r''_k respectively. It is normally not possible to determine separate repair times for normal and stormy weather. Therefor, r_k denotes repair time for all forced outages.
6. Distribution of normal weather and stormy weather durations.

24.13.4 Analysis for Series System

Let

$\lambda_1, \lambda_2 ..... \lambda_n$ = normal weather component failure rate per year

$\lambda'_1, \lambda'_2 .... \lambda'_n$ = stormy weather component failure rate per year

$\lambda''_1, \lambda''_2 .....\lambda''_n$ = component maintenance outage rate per year

$r_1, r_2 ....r_n$ = expected value of repair time for forced outages, years

$r''_1, r''_2 ...... r''_n$ = expected value of maintenace time, years

N = expected value of normal weather duration, years

S = expected value of stormy weather duration, years

The approximate over all failure rate λ_{fk} for any individual kth component is

$$\lambda_{fk} = \frac{N}{N+S}\lambda_k + \frac{S}{N+S}\lambda'_k \text{ outages/year} \tag{24.42}$$

The series system forced outage rate is

$$\lambda_{fe} = \sum_{k=1}^{n} \lambda_{fk} \text{ outages/year} \tag{24.43}$$

The series system maintenance outage rate is

$$\lambda''_e = \sum_{k=1}^{n} \lambda''_k \text{ outages/year} \tag{24.44}$$

If the series systems is in parallel with other components, it is necessary to determine the equivalent stormy and normal weather failure rates for the series system. The equivalent component is referred to by subscript, e. Thus

$$\lambda_e = \sum_{k=1}^{n} \lambda_k = \text{failures/year of normal weather} \tag{24.45}$$

$$\lambda'_e = \sum_{k=1}^{n} \lambda'_k = \text{failures/year of stormy weather} \tag{24.46}$$

Expected values of down times, for the series system, in respect of forced and maintenance outages are

$$r_{fe} = \frac{\sum_{k=1}^{n} \lambda_{fk}\, r_k}{\lambda_{fe}} \text{ years} \tag{24.47}$$

$$r''_e = \sum_{k=1}^{n} \frac{\lambda''_k\, r''_k}{\lambda''_e} \text{ years} \tag{24.48}$$

The reliability measures for the series system are

1. Annual outage rate = $\lambda_{fe} + \lambda''_e$ outage/year (24.49)
2. Expected down time per outage

$$= \frac{\lambda_{fe}\, r_{fe} + \lambda''_e\, r''_e}{\lambda_{fe} + \lambda''_e} \text{ years} \tag{24.50}$$

3. Average total outage time per year

$$= \lambda_{fe}\, r_{fe} + \lambda''_e\, r''_e \tag{24.51}$$

4. Probability P that an outage will last longer than t hours is

$$P = \sum_{k=1}^{n} \frac{\lambda_{fk} e^{-t/(8760 r_k)} + \lambda''_k e^{-t/(8760 r''_k)}}{\lambda_{fe} + \lambda''_e} \tag{24.52}$$

Using Eqs (24.49) to (24.52) the reliability measures for each load point can be calculated. From these results the average values of these indices for the whole system can be determined. The maximum values of above indices indicate the poorest service reliability at any load point.

24.13.5 Analysis for Parallel System

The data needed for reliability of a parallel system is the same as that for the series system. Each of the parallel elements may consist of a number of components in series. These series components are reduced to an equivalent element for further combination. The following analysis pertains to two such equivalent elements denoted as 1 and 2. Generally the transmission and distribution systems have only two parallel circuits. Moreover it is assumed that each element of the parallel system is capable of carrying the full load individually so that an outage will occur only if both the elements are not available simultaneously. However methods are available for calculating outages due to overloads also if each element is not designed for full load.

The derivation for the two elements in parallel is done in 4steps.

1. Initial failure occurs during normal weather and the second failure also occurs in normal weather.

$$\text{Failure rate} = \frac{N}{N+S}\left[\lambda_1\left(1-\frac{r_1}{N}\right)(\lambda_2\, r_1) + \lambda_2\left(1-\frac{r_2}{N}\right)(\lambda_1\, r_2)\right] \tag{24.53}$$

where $\dfrac{N}{N+S}$ = long run fraction of time that the weather is normal

λ_1, λ_2 = normal weather failure rates of elements 1 and 2 respectively, failures/year

$1-\frac{r_1}{N} \simeq e^{-r_1/N}$ = probability that a storm does not occur when element 1 is being repaired

r_1, r_2 = expected repair times of elements 1 and 2 respectively; years

$\lambda_2 r_1 \simeq 1-e^{-\lambda 2r1}$ = probability that element 2 fails during the repair of element 1.

The repair times r_1 and r_2 are generally very small as compared to N. There Eq. (24.53) simplifies to

$$\text{Failure rate} = \frac{N}{N+S}\lambda_1 \lambda_2 (r_1 + r_2) \tag{24.54}$$

2. Initial failure occurs during normal weather and second failure occurs during stormy weather.

$$\text{Failure rate} = \frac{N}{N+S}\left[\lambda_1\left(\frac{r_1}{N}\right)(\lambda_2' S) + \lambda_2\left(\frac{r_2}{N}\right)(\lambda_1' S)\right] \tag{24.55}$$

where λ_1', λ_2' = stormy weather failure rates of elements 1 and 2 respectively.

$\frac{r_1}{N} \simeq 1 - e^{r1/N}$ = probability that a strom occurs during the repair of element 1

$\lambda_2' S \simeq 1 - e^{\lambda 2'S}$ = probability that component 2 fails during storm of duration S.

3. Initial failure occurs during stormy weather and second failure is also during stormy weather.

$$\text{Failure rate} = \frac{S}{N+S}[(\lambda_1')(S\lambda_2') + \lambda_2'(S\lambda_1')] \tag{24.56}$$

where $\frac{S}{N+S}$ = long run fraction of time that the weather is stormy

4. Initial failure occurs during stormy weather and second failure occurs during normal weather

$$\text{Failure rate} = \frac{S}{N+S}[\lambda_1'(1-S\lambda_2')(\lambda_2 r_1) + \lambda_2'(1-S\lambda_1')(\lambda_1 r_2)] \tag{24.57}$$

where

$1 - S\lambda_2' = e^{-s\lambda' 2}$ = probability that component 2 does not fail during storm of duration S.

The product $S\lambda_2'$, << 1. Therefore $1 - S\lambda_2' \simeq 1$. Similarly $1 - S\lambda_1' \simeq 1$. Therefore, Eq. (24.57) simplifies to

$$\text{Failure rate} = \frac{S}{N+S}[\lambda_1' \lambda_2 r_1 + \lambda_2' \lambda_1 r_2] \tag{24.58}$$

The complete expression for failure rate, due to forced outages, of a 2-element system is the summation of Eqs (24.54, 24.55, 24.56 and 24.58). Thus

$$\text{Failure rate } \lambda_{fe} = \frac{N}{N+S}\left[\lambda_1 \lambda_2 (r_1 + r_2) + \frac{S}{N}(\lambda_1 \lambda_2' r_1 + \lambda_2 \lambda_1' r_2) + \frac{S}{N}(\lambda_1' \lambda_2 r_1 + \lambda_2' \lambda_1 r_2) + \frac{2S^2}{N}\lambda_1' \lambda_2'\right] \tag{24.59}$$

Similarly the failure rate due to maintenance outages can be calculated. It is assumed that maintenance work is undertaken only during normal weather. Evidently such an outage occurs if one of the elements has been taken out of service for maintenance and a forced outage occurs on the second element. The failure rate due to maintenance outages is

$$\text{Failure rate } \lambda_e'' = \lambda_1''\lambda_2 r_1'' + \lambda_2''\lambda_1 r_2'' \tag{24.60}$$

where λ_1'', λ_2'' = maintenance outages rates of elements 1 and 2 respectively, failures/year

r_1'', r_2'' = maintenance down times of elements 1 and 2 respectively, years

The total failure rate due to both forced and maintenance outages is summation of Eqs. (24.59) and (24.60).

The expected value of parallel system down time due to overlapping forced outages is given by

$$r_{fe} = \frac{r_1 r_2}{r_1 + r_2} \text{ years} \tag{24.61}$$

The expected value of down time when an element forced outage overlaps the maintenance outage of the second parallel element is given by

$$r_e'' = \left[\frac{\lambda_1''\lambda_2 r_1''}{\lambda_1''\lambda_2 r_1'' + r_2''\lambda_1 r_2''}\right]\left(\frac{r_2 r_1''}{r_2 + r_1''}\right) + \left[\frac{\lambda_2''\lambda_1 r_2''}{\lambda_1''\lambda_2 r_1'' + \lambda_2''\lambda_1 r_2''}\right]\left(\frac{r_1 r_2''}{r_1 + r_2''}\right) \text{ years} \tag{24.62}$$

In the Eq. (24.62), the first factor of the first term is the fraction of system outages involving element maintenance outages is which element 2 failed when element 1 was under maintenance and the second factor of first term gives the expected down time if element 2 fails when element 1 is under maintenance. The second term in Eq. (24.62) accounts for the situation when a forced outage of element 1 overlaps a maintenance of element 2.

The expected down time per outage and average total outage time per year can be calculated from Eqs. (24.50 and 24.51) using the above values of $\lambda_{fe}, \lambda_e'', r_{fe}$ and r_e''

For the 2-element parallel system the probability that a single outage lasts longer than t hours is

$$P = \frac{1}{\lambda_{fe} + \lambda_e''}\left[\lambda_{fe}\, e^{-t/(8760 r_{fe})} + \lambda_1''\lambda_2 r_1'' e^{-t\Big/\left(8760\frac{r_2 r_1''}{r_2 + r_1''}\right)} + \lambda_2''\lambda_1 r_2'' e^{-t\Big/\left(8760\frac{r_1 r_2''}{r_1 + r_2''}\right)}\right] \tag{24.63}$$

24.14 RELIABILITY PLANNING

An electrical power system consists of generation, transmission and distribution systems. The ultimate goal in reliability analysis is the evaluation of the entire system so that overall reliability measures can be obtained. However this is rather too ambitious and difficult task because of a large number of components and complex interconnections. Therefore, it is more prudent to calculate reliability measures separately for the generation, transmissions distribution systems. An added advantage of this approach is that the weakest links and components can be spotted more easily and remedial measures taken for reliability improvement.

Reliability improvement is possible through the use of better components and provision of redundancy. Redundancy in generation means additional generation capacity. Redundancy in transmission means that ties between stations and load centres should be stronger. This extra transmission capacity can be used to avoid overloading during usual operating conditions. Redundancy in distribution systems means duplication of certain components and use of better bus schemes.

Distribution systems have a large number of components in series and a number of switching devices. A system with perfect switching devices is most desirable. However most of switching devices are rather imperfect and are responsible for most of the unreliability.

24.15 PREPARATION OF RELIABILITY MODELS

The preparation of a proper analytical model for reliability studies requires the consideration of many factors some of which are as under.

(a) *Composition of system:* The model depends on the number of and types of components and their interconnection.

(b) *System failure criteria:* The failure criteria depends on the part of the system being analyzed. It is usually necessary to carry out a failure effect analysis to define the system failure criteria. In one system the continuity of supply, irrespective of voltage, available, may be sufficient. In another study a lower than a specified voltage may mean unreliability.

(c) *Assumptions in modelling:* The assumptions largely determine the method of analysis. The assumptions should be realistic but should not be so liberal as to give very wrong results. The assumptions are needed in many diverse areas, e.ge., weather model, system states to be neglected in the study, load conditions, common mode failure, maintenance, etc.

(d) *Selection and preparation of model:* The analytical model should be realistic as well as simple. It must take into account all the important generation and load conditions as well as weather effects and possible interconnections.

The various steps, in reliability study are:

1. Define the system
2. Define the failure criteria
3. Make suitable assumptions
4. Prepare the system model
5. Carry out failure effect analysis
6. Evaluate reliability indices
7. Analyze the results
8. Prepare reliability improvement plans

A reliability study can be undertaken only when a proper and reliable data bank of failure rates and repair times for various components is available.

Example 24.1.

A radial system consists of a transformer and a distributor 0.4 km long. The failure rates for transformer and distributor are 0.8 failures per year and 8.2 failures per km per year respectively. The repair times for transformer and distributor are 6 hours and 5 hours respectively. Find (a) failure rate of system (b) down time per outage (c) total outage time per year.

Solution

(a) Failure rate $= 0.8 + 8.2 \times 0.4 = 4.08$ outages/year

(b) Down time $= \dfrac{0.8 \times 6 + 8.2 \times 0.4 \times 5}{4.08} = 5.196$ hours per outage

(c) Total outage time $= 4.08 \times 5.196 = 21.2$ hours/year

Example 24.2.

A system consists of 4 components having reliabilities of 0.95, 0.92, 0.98 and 0.88. All the components are in series. Find the reliability of the system.

Solution

Using Eq. (24.33)

$$\text{System reliability} = 0.95 \times 0.92 \times 0.98 \times 0.88 = 0.7537$$

Example 24.3.

Find the reliability of the system of example 20.2 when another system, of 4 similar components in series, is connected in parallel with the system.

Solution

Using Eq. (24.35)

$$Q_s = (1 - 0.7537)\,(1 - 0.7537) = 0.06066$$

$$R_s = 1 - Q_s = 1 - 0.06066 = 0.93934$$

Example 24.4.

A circuit is formed by 3 components *A*, 2 components *B* and 1 component *C* in series. The failure rates of components are

$$\lambda_A = 3 \times 10^{-3} \text{ failures/hour}$$

$$\lambda_B = 2 \times 10^{-3} \text{ failures/hour}$$

$$\lambda_C = 4 \times 10^{-3} \text{ failures/hour}$$

The components follow exponential distribution. Find the reliability of the circuit for an operating time of 20 hours. Also find MTTF.

Solution

For the whole circuit

$$\lambda = 3\,\lambda_A + 2\,\lambda_B + \lambda_C$$

$$= 3 \times 3 \times 10^{-3} + 2 \times 2 \times 10^{-3} + 4 \times 10^{-3}$$

$$= 17 \times 10^{-3} \text{ failures/hour}$$

$$\text{Reliability} = e^{-17 \times 10^{-3} \times 20} = 0.71177$$

$$MTTF = \frac{1}{17 \times 10^{-3}} = 58.823 \text{ hours}$$

Example 24.5.

System is formed by 3 components *A*, *B*, *C* in series. Their reliabilities are 0.8, 0.86 and 0.92 respectively. Find the reliability of the system.

Solution

$$\text{System reliability} = 0.8 \times 0.86 \times 0.92 = 0.633$$

Example 24.6.

It is desired to improve the reliability of the system of example 20.5 by using system redundancy. A similar system consisting of 3 components A, B, C in series is added in parallel with existing system. Find the reliability of the new system.

Solution

$$Q_1 = Q_2 = (1 - 0.633) = 0.367$$

$$\text{System reliability} = 1 - (0.367)\,(0.367) = 0.865$$

Example 24.7.

It is desired to improve the reliability of the system of example 24.5 by using component redundancy. Element 1 has 2 components A in parallel, elements 2 has 2 components B in parallel and element 3 has 2 components C in parallel. Elements 1, 2, 3 are in series. Find the system reliability.

Solution

$$Q_1 = (1 - .8)\,(1 - .8) = 0.04$$

$$R_1 = 1 - Q_1 = 0.96$$

$$Q_2 = (1 - .86)\,(1 - .86) = .0196$$

$$R_2 = (1 - .0196) = 0.9804$$

$$Q_3 = (1 - .92)\,(1 - .92) = .0064$$

$$R_3 = 1 - Q_3 = 0.9936$$

$$R_s = R_1\,R_2\,R_3 = 0.96 \times 09804 \times 0.9936 = 0.935$$

Example 24.8.

The reliability of a component is 0.3. How many such components be connected in parallel to achieve an overall reliability of at least 0.85.

Solution

When n components are connected in parallel and each has a reliability R, the system reliability (ie., at least one component is available) is given by

$$R_s = 1 - (1 - R)^n$$

or

$$n = \frac{\log_e (1 - R_s)}{\log_e (1 - R)}$$

$$R_s = 0.85,\ R = 0.3$$

$$n = \frac{\log_e (1 - 0.85)}{\log_e (1 - 0.3)} = \frac{-1.897}{-0.35667} = 5.3$$

Let the system have 6 components in parallel.

$$R_s = 1 - (1 - .3)^6 = 0.882$$

Example 24.9.

A distribution system consists of a 132/ 11 kV transformer, 11 kV breaker, 11 kV feeder and 11/0.44 kV distribution transformer. It is operating in a two state fluctuating environment in which the expected values of normal weather period duration and stormy weather period duration are 120 hours and 3 hours respectively. The outage rates and repair times are as under

Equipment	λ	λ'	λ''	r	r''
		Outages/year		hr	hr
132/11 kV transformer	0.1	0.1	0.5	6	8
Circuit breaker	0.15	0.15	0.4	4	5
Feeder	1.2	5.2	2	5	3
11/.44 kV transformer	0.3	0.3	0.6	5	4

Find the annual outage rate, down time per outage and total outage time per year of the system.

Solution

For the feeder

$$\lambda_f = \frac{120}{123} \times 1.2 + \frac{3}{123} \times 5.2 = 1.298 \text{ outages/year}$$

Using Eq. (24.43)

$$\lambda_{fe} = 0.1 + 0.15 + 1.298 + 0.3 = 3.5 \text{ outages/year}$$

Using Eq. (24.44)

$$\lambda''_e = 0.5 + 0.4 + 2 + 0.6 = 3.5 \text{ outages/year}$$

Using Eq. (24.47)

$$r''_e = \frac{0.1 \times 6 + 0.15 \times 4 + 1.298 \times 5 + 0.3 \times 5}{1.848} = 4.973 \text{ hours}$$

Using Eq. (24.48)

$$r''_e = \frac{0.5 \times 8 + 0.4 \times 5 + 2 \times 3 + 0.6 \times 4}{3.5} = 4.11 \text{ hours}$$

$$\text{Annual outage rate} = 1.848 + 3.5 = 5.348 \text{ outage/year}$$

$$\text{Down time per outage} = \frac{1.848 \times 4.973 + 3.5 \times 4.11}{5.348} = 4.41 \text{ hours}$$

$$\text{Total outage time per year} = 5.348 \times 4.41 = 23.58 \text{ hours/year}$$

Example 24.10.

A double circuit 220 kV line operates in a 2 state fluctuating environment having expected normal weather and stormy weather durations of 110 hours and 4 hours respectively. Both circuits are similar. The outage data for each circuit is

$$\lambda = 3 \text{ outages/year of normal weather}$$

$$\lambda' = 9 \text{ outages/year of stormy weather}$$

$$\lambda'' = 1 \text{ outage/year}, r = 6 \text{ hours}, r'' = 8 \text{ hours}$$

Find annual outage rate, down time per outage and total outage time per year.

Solution

Using Eqs. (24.59 to 24.62) and substituting $\lambda_1 = \lambda_2 = 3,\ \lambda_1^1 = \lambda_2' = 9$

$$r_1 = r_2 = \frac{6}{8760}, r''_1\ r''_2 = \frac{8}{8760}$$

$$S = \frac{4}{8760}, N = \frac{110}{8760}, \lambda''_1 = \lambda''_2 = 1$$

$$\lambda_{fe} = \frac{110}{114}\left[\frac{3\times3\times12}{8760} + \frac{4}{110}\,\frac{(2\times3\times9\times6)}{8760}\right.$$

$$\left. + \frac{4}{110}\,\frac{(2\times3\times9\times6)}{8760} + \frac{2\times4\times4}{110\times8760}\times9\times9\right]$$

$$= 0.01707 \text{ outages/year}$$

$$\lambda''_e = \frac{1\times3\times8}{8760} + \frac{1\times3\times8}{8760} = .00548 \text{ outages/year}$$

$$r_{fe} = \frac{\frac{6}{8760}\times\frac{6}{8760}}{\frac{12}{8760}} = \frac{3}{8760} \text{ years} = 3 \text{ hours}$$

$$r''_e = \frac{\frac{6}{8760}\times\frac{8}{8760}}{\frac{6}{8760}+\frac{8}{8760}} = \frac{3.429}{8760} \text{ years} = 3.429 \text{ hours.}$$

Annual outage rate = .01707 + 0.00548 = 0.02255 outages/year

Total outage time per year

$$= 0.01707 \times 3 + .00548 \times 3.429 = 0.07 \text{ hours/year}$$

$$\text{Down time per outage} = \frac{0.07}{0.02255} = 3.1 \text{ hours.}$$

SUMMARY

24.1 If T is time to failure

$$R(t) = P\,(T > t)$$

24.2 $R(0) = 1$ and $R(\infty) = 0$

24.3 Outages in power system can be forced, scheduled and temporary.

24.4 Bath tub curve depicts three different regions in the life of a component.

24.5 Every component/system undergoes through alternate up and down states.

24.6 Failure rate (λ) is number of failures per unit time. Repair rate (μ) is number of repairs per unit time.

24.7 $R(t) = e^{-\lambda t}$

24.8 $E(t) = \dfrac{1}{\lambda}$

24.9 Reliability can be studied by continuous Markov process.

24.10 $P_0(\infty) = \dfrac{\text{MTTF}}{\text{MTTF + Mean Down time}}$

$$P_1(\infty) = \dfrac{\text{Mean Down time}}{\text{MTTF + Mean Down time}}$$

24.11 Many power system components operate in two state fluctuating environment.

24.12 For power system components times to failure and repair times are assumed to be exponentially distributed.

24.13 Approximate method is a simple method to evaluate reliability of transmission and distribution systems .

SELF TEST QUESTIONS

A. state whether the following are true (T) or false (F)

24.1 Failure rate is the number of failures per unit time.

24.2 Reliability studies are always done for useful life period.

24.3 Momentary outages last for very short time.

24.4 If $\lambda t << 1$, $R(t) = \lambda t$

24.5 As dt approaches zero

$$\frac{d\,R(t)}{dt} = -f(t)$$

24.6 $E(t) = \lambda$

24.7 $P_0(t + \Delta t) = P_0(t)\,[1 - \lambda\,\Delta t] + P_1(t)\,[\mu\,\Delta t]$

24.8 MTTF = μ

B. Fill in the blanks with suitable words

24.9 Mean down time = ______

24.10 MTTF is reciprocal of ______

24.11 A Markov process with discrete index is called______

24.12 In continuous Markov process, ______ is continuous.

24.13 In terms of μ and λ, $P_0(\infty)$ = ______

24.14 If n components having reliabilities R_1, R_2, R_3 ... are in series, system reliability = ______

24.15 If S is expected duration of stormy weather period, the transition rate from stormy to fair weather is ______

ANSWERS

24.1 T	24.2 T	24.3 T	24.4 F	24.5 T
24.6 F	24.7 T	24.8 F	24.9 $\dfrac{1}{\mu}$	24.10 failure rate
24.11 Markov chain	24.12 Time	24.13 $\dfrac{\mu}{\lambda + \mu}$	24.14 $R_1 R_2 R_3 \ldots R_n$	24.15 $\dfrac{1}{S}$

JOB INTERVIEW QUESTIONS

24.1 What are reliability indices?
24.2 What is meant by useful life period?
24.3 What is MTTF?
24.4 What is Markov process?
24.5 What is the effect of weather on failure rate?
24.6 What is meant by two state fluctuating environment?
24.7 What is two state model?
24.8 Define probability density function?

REVIEW QUESTIONS

24.1 Explain the terms 'Reliability', 'Outage', 'MTTF', 'MTBF'.
24.2 Distinguish between forced and scheduled outages.
24.3 Draw typical bath tub curve. Explain the three regions of the curve. For which region are reliability calculations valid? Why?
24.4 Draw a two state model of equipment. Define failure rate and repair rate.
24.5 Explain the term ' Probability density function'. What is exponential failure density function?
24.6 What is continuous Markov process?
24.7 Use continuous Markov process to find the probabilities of being available and not being available for a single repairable component having failure rate λ and repair rate μ.
24.8 What is two state fluctuating environment?
24.9 Use approximate method to find expressions for various reliability indices for a system having *n* components in series. Assume two state fluctuating environment.

PROBLEMS

24.1 A radial system consists of a trnaformer and a feeder. The feeder is 3.2 km long. The failure rates for transformer and feeder are 1.2 failures/year and 0.6 failures/km/year respectively. The repair times for transformer and feeder are 7 hours and 6 hours per outage. Fing (a) failure rate of system (b) total outage time per year (c) down time per outage.
[(a) 3.12 outages/year (b) 19.92 hours/year (c) 6.385 hours/outage]

24.2 A system consists of 3 components in series. The individual reliabilities of components are 0.96, 0.82 and 0.87 respectively. Find the reliability of the system. [0.68486]

24.3 A system has 2 elements in parallel. Each element consists of 2 components in series. The reliabilities of components are 0.84 and 0.9 respectively. Find the reliability of the system. [0.94046]

24.4 A circuit is formed by 2 components *A*, 1 component *B* and 3 components *C* all connected in series. The failure rates of component *A*, *B*, *C*, are 2×10^{-3}, 5×10^{-3} and 0.6×10^{-3} failures/hour. Find (a) MTTF (b) system reliability for an operating time of 10 hours. Assume exponential distribution. [(a) 92.59 hours (b) 0.8976]

24.5 A system has 2 components *A* and *B* in series. Their reliabilites are 0.87 and 0.91 respectively. Fing the system reliability. [0.7917]

24.6 It is desired to improve the reliability of system of Problem 24.5 by using system redundancy. Another system having components *A* and *B* in series is added in parallel with existing system. Find the raliability of the new system. [0.9566]

24.7 It is desired to improve the reliability of system of Problem 24.5 by using component redundancy. Element 1 has 2 components A in parallel and element 2 has 2 components B in parallel. Elements 1 and 2 are in series. Find system reliability. [0.975]

24.8 The reliability of component is 0.2. How many such components be connected in parallel so that at least one is avilable with a reliability of 0.8 or more. [8]

24.9 A distribution system consists of a 66/11 kV power transformer T_1, 11 kV breaker, 11 kV feeder, 3.2 km long and 11/0.44 kV distribution transformer T_2. It is operating in a 2 state fluctuating environment in which the expected values of normal weather period and stromy weather period are 100 hours and 5 hours respectively. The outage rates and repair times and as under:

Equipment	λ	λ'	λ''	r	r''
		outages/year		hour	hour
Transformer T_1	0.2	0.2	0.8	8	10
Feeder	1.6	7	2	6	8
Breaker	0.1	0.1	0.2	3	2
Transformer T_2	0.4	0.4	1	5	6

Find annual outage rate, total outage time per year and expected down time per outage

[6.557 outages/years, 45.44 hours/year, 6.93 hours.]

24.10 A double circuit 132 kV line operates in a 2 state fluctuating enviroment having expected normal weather and storing weather durations of 120 hours and 5 hours respectively, Both the circuits are similar. The outage data for each circuit is; $\lambda = 4$ outages per year of normal weatehr, $\lambda' = 10$ outages per year of stormy weather, $\lambda'' = 1$ outage/year, $r = 5$ hours, $r'' = 10$ hours. Find annual outage rate, down time per outage and total outage time per year.

[0.03488 outages/year, 2.72 hours, 0.0948 hours/year]

CHAPTER

25

Load Frequency and MVAR Voltage Control

25.1 TYPES OF INTERCONNECTION

Majority of today's power systems have interconnection facilities with one another. Both small and large utilities find interconnection of tremendous benefit.

System interconnection may be of two types: integrated and unintegrated. The two types differ in the degree of coordination between the individual systems interconnected with each other.

An integrated interconnected system operates as if the total system is one. The identity of individual systems is almost lost. All operations like planning increased generation facilities, installation of new equipment, increase in reserve capacity, incremental loading of generators, coordination of building and maintenance programmes, increase in transmission facilities etc. are done in a coordinated way. Such a system has a central office to deal with accounting, dispatching, preparation of loading and maintenance schedule, laying down standard practices etc.

In an unintegrated interconnected system the identity of individual utilities is not lost. Each utility is primarily responsible for meeting the load requirements of the area served by it. There is no central control office. However the different utilities forming the pool have some contracts with each other. These contracts include supply of definite quantity of energy at definite times, mutual help at difficult times, use of some special facilities created by one utility by the other utilities etc.

An integrated interconnection results in maximum overall economy. However such an organisation tends to become unwieldy especially when more than two utilities are members of the group. In view of this most of the system interconnections are unintegrated. The contracts between different utilities forming a pool include one or more items like emergency assistance, economy interchange, joint maintenance, planning, inadvertent interchange etc.

25.2 ADVANTAGES OF INTERCONNECTION

System interconnection leads to several advantages both to the members of the pool and the users of electricity. It is necessary that advantages and savings be examined thoroughly and weighed against the cost of system interconnections. This analysis is complicated by the fact that some benefits cannot be evaluated in terms of money.

25.2.1 Increase in Service Reliability

Every equipment is prone to outage during its operation. System interconnection leads to increase in reliability of electric supply. In the event of one utility falling short of generation, due to forced outage of generators or unexpected increase in demand, another utility of the pool can come to its rescue by making its surplus power available. An interconnection between the local transmission system of two utilities can lead to an increase in service reliability at considerably less expense than that which both would have to incur in the construction of additional lines. When the lines are fed from more than one generation source, the reliability is still further improved.

25.2.2 Addition and Replacement of Transmission Facilities

Load growth necessitates additional transmission facilities. The questions like cost of right of way and installation of line and the territory to be served can be more economically and easily decided when more than one utility is involved. The utilities can share the cost and use of the lines. This would evidently lead to optimum use of facilities.

25.2.3 Installed Capacity Savings

An interconnection is likely to decrease the installed capacity needed to meet the load requirements. In majority of cases it happens that the maximum demand of one area does not coincide with the maximum demand of another area. This could be due to difference in sun times and weather conditions of the two areas. It could also be due to the fact that in one area the major load is industrial while in the other it is residential and commercial. If two such areas are interconnected, the diversity of loads would cause the maximum combined demand to be less than the sum of the individual maximum demands.

25.2.4 Decrease in Spinning Reserve Requirements

Every utility has to keep some generation capacity as reserve to take into account the capacity reduction due to forced outages and scheduled maintenance. As a thumb rule a capacity equal to the capacity of the largest unit is kept as reserve. When two systems operate separately each has to maintain a separate reserve capacity. If they are interconnected it may be sufficient to have a smaller total reserve for the combined system for the same degree of reliability.

25.2.5 Operating Savings

When two different utilities are operating separately, each would run its units most economically. However when the two are interconnected and the optimum scheduling is decided for the combined system, the operating costs are likely to decrease. Thus energy interchange between the two would result in savings for both.

25.3 ROLE OF SYSTEM FREQUENCY IN REAL POWER CONTROL

25.3.1 System Frequency

System frequency is one quantity which is common to all components in an interconnected system. If steady state frequency is constant, it means that the system generation is meeting the load requirements. If a power system has to operate satisfactorily its frequency must be maintained at constant value. Maintaining the frequency constant ensures proper operation of induction and synchronous motors. The motors driving the auxiliary system of generating plants will work satisfactorily only if frequency is constant. Frequency is a basic quantity that can be measured and applied to the control of active power in the system. Since all generators are synchronous machines, they are locked together at synchronous speed. The different generators in the system may be running at different speeds depending on the number of poles. The hydroelectric generators operate at much lower speed than steam turbine driven generators. But all generators have same electrical speed. When system load increases the frequency will decrease and cause the speed governing system to increase the mechanical power input so as to restore the balance between real power generated and load. When system load decreases the frequency increases and causes a decrease in mechanical power input so as to again restore the balance between generation and load. Thus frequency is the basic quantity applied to active power control in the system. Earlier speed governing system used rotating flyballs to activate a hydraulic system for increasing or decreasing the throttle valves of prime movers. Nowadays electronic governors are used to accompolish the same task.

When machines have to operate in parallel, the governors must have drooping characteristics, *i.e.* as load increases, the speed decreases and as load decreases, the speed increases. Earlier generating units used governor speed control to maintain system frequency and real power balance. Nowadays automatic generation control (AGC) is universally used. AGC system can maintain system frequency within very close limits. AGC system are provided with computer capability to maintain real power balance, ensure economic dispatch, hydro thermal coordination, maintain tie line loadings at proper values so that interconnected system operate satisfactorily and economically.

25.3.2 Control of Interconnected System

We have discussed in chapter 2 the problem of economic load allocation between different plants in a system. The system will be in economic dispatch when the incremental cost of the power delivered to the load is the same for each generator. Since the incremental transmission losses are a function of both the active and reactive loading of transmission lines, the incremental cost is also a function of both the active and reactive power.

The operation of an interconnected system has therefore to be co-ordinated from three angles:

1. At any time the total system generation must equal the system load plus transmission losses.
2. At any time the balance between production and absorption of lagging vars must be maintained.
3. The generation at different plants must be scheduled so that the incremental cost of power delivered to the load is the same for each generator.

Having identified the problem we proceed to discuss the procedure to satisfy the above constraints.

25.4 PHILOSOPHY OF REAL AND REACTIVE POWER CONTROL

Power system load flow studies bring out the following properties of power system networks:

1. The changes in real bus power affect mainly the bus voltage phase angles (and therefore real line flows) and have negligible effect on bus voltage magnitudes and reactive line flows.

2. The changes in reactive bus powers affect mainly the bus voltage magnitudes (and reactive line flows) and have negligible effect on bus voltage phase angles and real line flows.
3. The changes in reactive bus powers at a bus have a very strong effect on the voltage magnitude at that bus but have a mild effect on voltage magnitudes at distant buses.

The above system properties lead us to the following two methods of real and reactive power control in power systems.

(1) *Load Frequency or Real Power Control*. This is also referred to as Megawatt frequency or p.f. control. The aim of this control is to maintain real power balance in the system through control of system frequency. Whenever the real power demand changes, a frequency change occurs. This frequency error is amplified, mixed and changed to a command signal which is sent to the turbine governor. The governor operates to restore the balance between the input and output by changing the turbine input.

(2) *Reactive Power Control or Automatic Voltage Control*. This is also referred to as Mega var voltage or QV control. The aim of this control is to maintain the system voltage between limits by adjusting the excitation of machines. The automatic voltage regulator senses the difference between a rectified voltage derived from stator voltage and a reference voltage. This error signal is amplified and fed to the excitation circuit. The change of excitation maintains the vars balance in the network.

The above two control channels operate more or less independent of each other. Moreover the p.f. loop is rather slow in action due to inertia of mechanical parts whereas the QV loop is very fast. Fig. 25.1 shows the two control channels for maintaining the real and reactive power balance in the system.

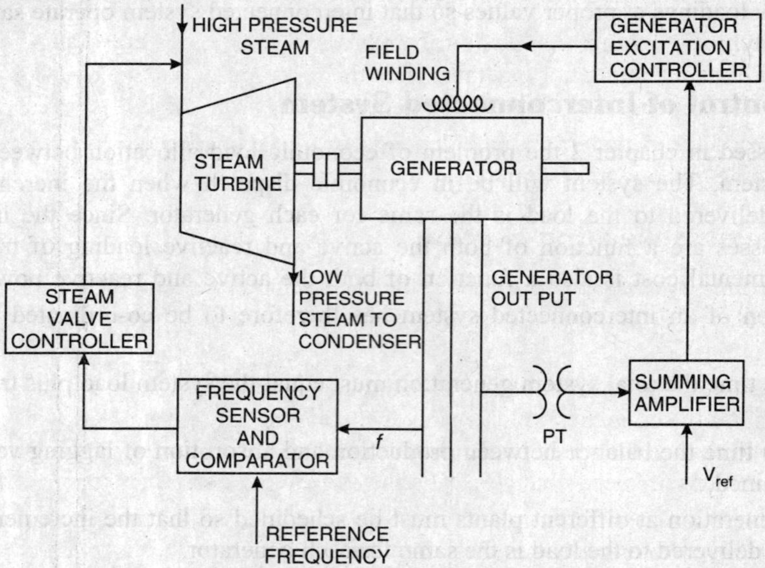

Fig. 25.1 Load frequency and automatic voltage control channels of a turbo-alternator.

REAL POWER CONTROL (SINGLE AREA)

25.5 EFFECT OF REAL POWER ON SYSTEM FREQUENCY

The a.c. motors run at speeds which are related to the system frequency. It is necessary to maintain the system frequency within ±2 Hz of the 50 Hz value to ensure satisfactory performance of motors. However the main reason for keeping the system frequency within strict bounds is that if the frequency error is kept within strict limits, the balance between generation and absorption of real power can be maintained in a satisfactory way. Thus a control on system frequency enables the control of real power in the system.

Every alternator in the system runs synchronously and generates real power. The total real power output of all alternators equals the load plus transmission loss. Within each alternator there is a perfect balance between the prime mover torque and the load torque. If a change in system load occurs, the frequency tends to fluctuate. The kinetic energy tries to keep a momentary energy balance till the change in frequency (or speed) is detected by the governor which changes the input to the prime mover and the balance between the input and output is again restored. Modern power systems maintain the system frequency with ±0.25 Hz of the 50 Hz value.

25.6 CONTROL AREA CONCEPT

A vast power system can be decomposed into a number of load frequency control areas tied together through tie lines. The problem of real power control can be studied by first studying the problem for a single area and then extending the treatment to a two or multi area system. Within each control area the alternators in the different power plants speed up or slow down together thus forming a coherent group. The electricity networks of each state can be thought of as separate control areas for load frequency control purposes.

25.7 BLOCK DIAGRAM OF SINGLE AREA LOAD FREQUENCY CONTROL (AUTOMATIC GENERATION CONTROL)

The different machines within a single control area form a coherent group. Whenever a change in load occurs, all the machines act together. Thus the response of a single control area can be studied through the response of a single turbo generator (along with its governor).

The response of turbo generator unit depends on the response of the speed governing mechanism, turbine and power system.

25.7.1 Speed Governing System

The speed governing system acts to translate the speed changes resulting from load variation into steam control. Fig. 25.2 shows the speed governing system of a turbo-alternator. It consists of the following :

(*i*) *Speed changer*. It is a sort of bias applied to the governing system and usually takes the form of an adjustment of spring tension. It enables the adjustment of the speed governing system to change the speed load characteristics of the machine. As shown in Fig. 25.2, an adjustment of speed changer shifts the speed load characteristics parallel to itself, up or down.

(*ii*) *Speed responsive element-generally a fly ball mechanism*. This is mounted on the main turbine shaft. Centrifugal force due to the rotation of the shaft causes the flyballs to move outward. A reduction in load causes an increase in speed which means a greater outward motion of the balls.

The Governor has two inputs (*a*) $\Delta P'$ *i.e.,* change in speed changer position or reference setting (*b*) Δf *i.e.,* change in frequency. An increase in $\Delta P'$ causes an increase in output whereas a decrease in Δf results in more output. Therefore the governor output command ΔP_0 can be written as

$$\Delta P_0 = \Delta P' - \frac{1}{R}\Delta f \qquad \text{... (25.1)}$$

The constant R is known as speed regulation of Governor and has the units of Hz per MW or Hz/pu MW. Laplace transform of Eq. (25.1) gives

$$\Delta P_0(s) = \Delta P'(s) - \frac{1}{R}\Delta F(s) \qquad \text{... (25.2)}$$

where $\Delta P_0(s)$ = Laplace transform of ΔP_0

$\Delta P'(s)$ = Laplace transform of $\Delta P'$

and $\Delta F(s)$ = Laplace transform of Δf.

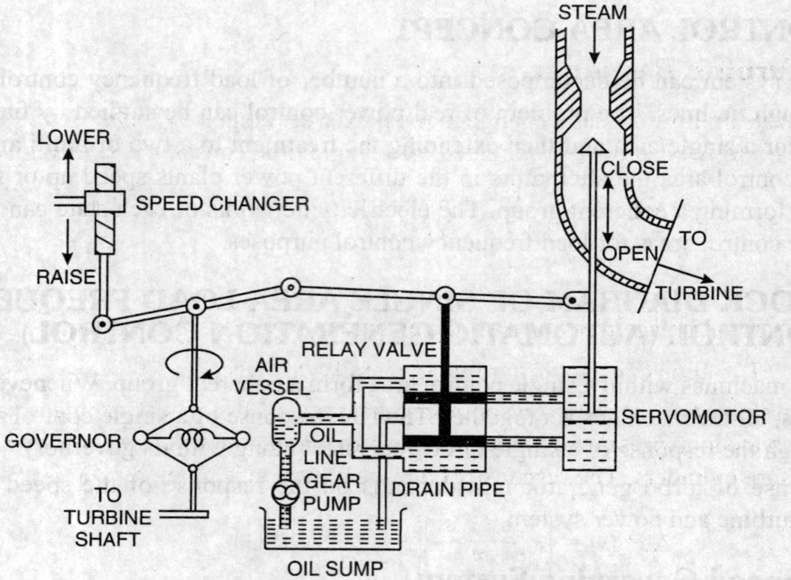

Fig. 25.2 Speed governing system of a turbine.

(*iii*) *Relay valve and servomotor*. The relay valve allows oil under pressure to flow to either side of the piston of servomotor whose motion changes the steam valve opening. The input position Δx_i of the relay valve increases due to an increased command ΔP_0 but decreases due to increased valve output ΔP_v. Equal increases in ΔP_0 and ΔP_v cause no net change in Δx_i. Therefore,

$$\Delta x_i = \Delta P_0 - \Delta P_v \qquad \text{... (25.3)}$$

For small changes in Δx_i, the oil flow in servomotor is proportional to Δx_i. Therefore the equation giving the position of the piston of servomotor is

$$\Delta P_v = k_g \int \Delta x_i dt \qquad \text{... (25.4)}$$

The constant k_g depends on the oil pressure and geometry of relay valve and servomotor. Taking Laplace transform of Eqs. (25.3) and (25.4) and eliminating Δx_i we get

$$\Delta P_v(s) = \frac{\Delta P_0(s)}{1+T_g s} \quad \text{... (25.5)}$$

where $\Delta P_v(s)$ = Laplace transform of ΔP_v.

$T_g = \frac{1}{k_g}$ is the time constant of the relay valve and servomotor.

Typical value of T_g is 100 milli-seconds.

Substituting Eq. (25.2) into Eq. (25.5) we get

$$\Delta P_v(s) = \left[\Delta P'(s) - \frac{1}{R}\Delta F(s)\right]\left[\frac{1}{1+T_g s}\right] \quad \text{... (25.6)}$$

25.7.2 Turbine Response

The change in steam flow causes a change in the power developed by the turbine. The response of a non-reheat turbine can be approximated by a single time constant T_t and is given by

$$\Delta P_v(s) = \frac{1}{1+T_t s}\Delta P_v(s) \quad \text{... (25.7)}$$

where

$\Delta P_m(s)$ = Laplace transform of the change in power developed by the turbine.

T_t = time constant of turbine (generally 0.25 sec. or so).

All modern steam turbines have one or more reheat stages. In a reheat turbine, a part of power is developed in the high pressure cylinder and the remaining power is developed in intermediate and low pressure cylinders. The response of a re-heat turbine is given by

$$\Delta P_m(s) = \left(\frac{1+AT_r s}{1+T_r s}\right)\left(\frac{1}{1+T_t s}\right)\Delta P_v(s) \quad \text{... (25.8)}$$

where A = proportion of torque developed in high pressure cylinder.

T_r = time lag associated with reheater.

Generally A varies from 0.25 to 0.5 and T_r from 3 to 5 secs.

25.7.3 Power System Response

The change in power developed by the turbine causes a change in the alternator output ΔP_g. The difference between the change in alternator output and change in load (*i.e.* ΔP_L) tends to change the system frequency. The difference $\Delta P_g - \Delta P_L$ is absorbed by the power system in two ways :

(*i*) Increase in stored kinetic energy of alternator rotor.

(*ii*) Increase in consumption in the power system. The motor load in the system increases with increase in speed (or frequency).

The power system response is given by

$$\Delta F(s) = \left[\Delta P_g(s) - \Delta P_L(s)\right]\frac{K_p}{1+T_p s} \quad \text{... (25.9)}$$

where $\Delta P_g(s)$ = Laplace transform of ΔP_g, change in generator output.

$\Delta P_L(s)$ = Laplace transform of ΔP_L, change in system load.

K_p = power system gain

T_p = power system time constant.

Typical values of K_p and T_p are 100 Hz/pu MW and 20 seconds.

25.7.4 Complete Block Diagram

The response of governor, turbine and power system as given by Eqs. (25.6, 25.8 and 25.9) can be combined to give the block diagram of single area load frequency control. This block diagram with illustrative transfer functions, is shown in Fig. 25.3.

25.7.5 Static Response of Automatic Load Frequency Control Loop

The automatic load frequency control loop shown in Fig. 25.3 has two inputs $\Delta P'(s)$ and $P_L(s)$ and one output $\Delta F(s)$. Let

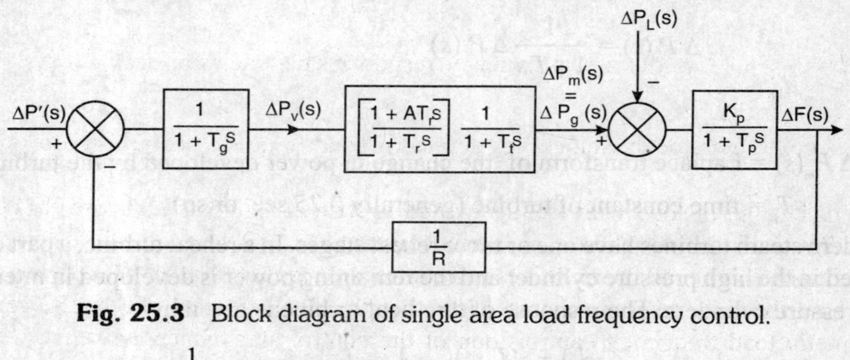

Fig. 25.3 Block diagram of single area load frequency control.

$$\frac{1}{1+T_g s} = \text{Transfer function of relay valve and servomotor} = G_r$$

$$\left[\frac{1+AT_r s}{1+T_r s}\right]\left(\frac{1}{1+T_t s}\right) = \text{Transfer function of turbine} = G_t$$

and
$$\frac{K_p}{1+T_p s} = \text{Transfer function of power system} = G_p$$

Then $$\left\{\left[\Delta P'(s) \frac{1}{R} \Delta F(s)\right] G_r G_t - \Delta P_L(s)\right\} G_p = \Delta F(s) \quad \text{...(25.10)}$$

For a fixed speed changer position, $\Delta P'(s) = 0$. Then

$$\left[-\frac{1}{R} \Delta F(s)\, G_r G_t - \Delta P_L(s)\right] G_p = \Delta F(s)$$

or
$$\Delta F(s) = \frac{-G_p}{1+\dfrac{G_r\, G_t\, G_p}{R}} \Delta P_L(s) \quad \text{... (25.11)}$$

For a step load change $\Delta P_L = M$, the Laplace transform of the load change *i.e.*, $\Delta P_L(s)$ is

$$\Delta P_L(s) = \frac{M}{s}$$

Therefore
$$\Delta F(s) = \frac{-G_p}{1+\dfrac{G_r\, G_t\, G_p}{R}}\left(\frac{M}{s}\right) \qquad \text{... (25.12)}$$

The static frequency drop Δf_0 can be found by using final value theorem.

$$\Delta f_0 = \underset{s\to 0}{Lt}\; s\,\Delta F(s) = -\frac{K_p\, M}{1+K_p/R}$$

or
$$\Delta f_0 = \frac{-M}{\dfrac{1}{K_p}+\dfrac{1}{R}} = \frac{-M}{\beta} \qquad \text{... (25.13)}$$

where $\beta = \dfrac{1}{K_p}+\dfrac{1}{R}$ is the area frequency response characteristic and has the units of pu MW/Hz. Another parameter, sometimes defined in load frequency control studies is the $\dfrac{\partial P_L}{\partial f}$ *i.e.*, rate of change of load with frequency and is written as D. Its units are MW/Hz. Time constant T_p equals $2H/fD$ where H is inertia constant.

As mentioned earlier all the machines in a control area act together. If the individual generator turbines have the same response characteristic then the block diagram of Fig. 25.3 can represent the automatic load frequency control loop of the control area. Since consumer demands are changing from instant to instant, time lags are present in matching generation with demand. An increase in demand is met by three inherent properties in the power system.

1. As frequency falls, the kinetic energy of the rotating mass of generators decreases. This released energy meets the increased demand.
2. A drop in frequency means a reduction in frequency sensitive (*i.e.*, motors) load. Generally a one per cent decrease in frequency causes about one per cent decrease in load.
3. The speed governing system acts to increase the turbine input.

Initially the second and third components are zero but as speed drops they account for more and more contribution. Eventually the steady state condition will be reached when the increased output from generators matches the load and the speed will level off at a slightly lower new value.

25.7.6 Dynamic Response of Automatic Load Frequency Control Loop

The static response gives information about the static frequency drop. The dynamic response gives the change in frequency as a function of time for a step change in load.

The dynamic response can be found by taking Laplace inverse of Eq. (25.12). However each of the terms G_r, G_t and G_p involves one time constant. As such the denominator in Eq. (25.12) is of third order and the mathematical analysis becomes tedious.

Eq. (25.12) can be simplified by making reasonable assumptions. The time constants T_r and T_t are about 1 second each whereas T_p has a value of about 20 seconds. Therefore T_r and T_t can be assumed zero. Then Eq. (25.12) simplifies to

$$\Delta F(s) = \frac{-G_p}{1+\dfrac{G_p}{R}}\left(\frac{M}{s}\right)$$

or

$$\Delta F(s) = \frac{-\dfrac{K_p}{1+T_p s}}{1+\dfrac{K_p}{R(1+T_p s)}}\left(\frac{M}{s}\right)$$

or

$$\Delta F(s) = \frac{-K_p\, RM}{RT_p s\left(s+\dfrac{R+K_p}{RT_p}\right)}$$

or

$$\Delta F(s) = \frac{-K_p RM}{R+K_p}\left[\frac{1}{s}-\frac{1}{s+\dfrac{R+K_p}{RT_p}}\right]$$

Taking inverse Laplace transform we get

$$\Delta f(t) = \frac{-K_p RM}{R+K_p}\left[1-\varepsilon^{-\left(\frac{R+K_p}{RT_p}\right)t}\right] \qquad \text{... (25.14)}$$

Fig. 25.4 shows the dynamic response of single area ALFC to a step increase in load. Δf_0 is the static frequency drop. The approximate response is given by Eq. (25.14) and is obtained if time constants T_r and T_t are neglected. The exact response (as obtained if T_r and T_t are not neglected) is also shown. It is seen that time constant of the closed loop system is pretty small. Let $R = 3$ Hz/pu MW, $K_p = 100$ Hz/pu MW and $T_p = 20$ seconds. Then time constant $(= RT_p/(R+K_p))$ equals 0.583 seconds which is very small as compared to 20 seconds. A reduction in the value of R makes the response faster. If R is reduced, the static frequency error is also reduced.

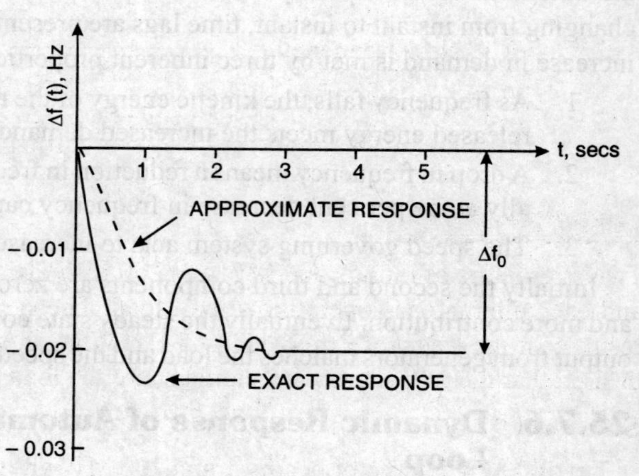

Fig. 25.4 Dynamic response of ALFC to a step increase in load.

Example 25.1.

The speed of a 100 MW alternator drops by 4% from no load to full load. (*a*) Find the speed regulation R of the alternator. Express the results in Hz per unit MW and Hz per MW (*b*) If frequency drops by 0.1 Hz, find change in power output.

Solution

(*a*) $$R = \frac{4\times 50}{100\times 1} = 2 \text{ Hz/p.u. MW}$$

$$R = \frac{4\times 50}{100\times 100} = 0.02 \text{ Hz/MW}$$

(*b*) $$\Delta P = -\frac{1}{R}\Delta f = -\frac{1}{0.02}(-0.1) = 5\,\text{MW}$$

Thus a decrease in frequency by 0.1 Hz means an increase in output by 5 MW.

Example 25.2.

A 100 MVA 50 Hz alternator is operating at rated speed. The H constant of the machine is 5 kW-sec per kVA. The load suddenly increases by 50 MW. Due to delay in governor action there is a delay of 0.5 seconds in opening of steam valve. Find the frequency deviation.

Solution

The kinetic energy stored at rated speed

$$= 100\times 1000\times 5 = 5\times 10^5 \text{ kW–sec.}$$

The loss in kinetic energy due to increase in load = $50 \times 1000 \times 0.5 = 25 \times 10^3$ kW–sec.

Kinetic energy $\propto$ (frequency)^2

$$\text{New frequency} = \left(\frac{5\times 10^5 - 25\times 10^3}{5\times 10^5}\right)^{0.5} \times 50 = 48.734 \text{ Hz}$$

$$\text{Frequency deviation} = \frac{50 - 48.734}{50}\times 100 = 2.532\%$$

Example 25.3.

Two alternators of ratings 500 MW and 200 MW are feeding a system. The frequency is 50 Hz and each alternator is operating at half load. The system load increases by 140 MW and as a result of it the frequency drops to 49.5 Hz. What should be the value of R for each unit if it is desired that the two alternators should share the load in proportion to their ratings ?

Solution

Let 500 MW alternator be unit 1 and 200 MW alternator be unit 2. Then $\Delta P_1 = 100$ MW and $\Delta P_2 = 40$ MW. Let R_1 and R_2 be the regulations for the 500 MW and 200 MW unit respectively. Since there is no change in reference setting $\Delta P' = 0$. Using Eq. (25.2)

$$100 = -\frac{1}{R_1}(-0.5) \text{ or } R_1 = 0.005 \text{ Hz/MW}$$

$$40 = -\frac{1}{R_2}(-0.5) \text{ or } R_2 = 0.0125 \text{ Hz/MW}$$

Example 25.4.

A control area has a total rated capacity of 10000 MW. The regulation R for all the units in the area is 2 Hz/pu MW. A one per cent change in frequency causes a one per cent change in load. If the system is operating at half of rated capacity and the load increases by 2% (*a*) find the static frequency drop. (*b*) If the speed governor loop were open, what would be the frequency drop ?

Solution

(*a*)

$$R = 2 \text{ Hz/p.u. MW}$$

$$D = \frac{\partial P_L}{\partial f} = \frac{0.01 \times 5000}{0.01 \times 50} = 100 \text{ MW/Hz}$$

$$= \frac{100}{10000} = 0.01 \text{ p.u. MW/Hz}$$

$$\beta = D + \frac{1}{R} = 0.01 + \frac{1}{2} = 0.51 \text{ p.u. MW/Hz}$$

$$M = \frac{2}{100} \times 5000 = 100 \text{ MW} = 0.01 \text{ p.u. MW}$$

Using Eq. (25.13)

$$\Delta f_0 = \frac{0.01}{0.51} = -0.01961 \text{ Hz.}$$

(*b*) Since speed governor loop is open, R is infinite.

$$\beta = D = 0.01 \text{ p.u. MW/Hz}$$

$$\Delta f_0 = -\frac{0.01}{0.01} = -1 \text{ Hz}$$

It is seen that in the absence of automatic load frequency control loop the drop in frequency would be much higher than in the case when ALFC is present.

Example 25.5.

Find the primary ALFC loop parameters for a control area having the following data :

Control area capacity = 10000 MW

Normal operating load = 5000 MW

Inertia constant = H = 5 seconds

Regulation R = 3 Hz/pu MW

Assume that load frequency dependency is linear and 1% change in frequency corresponds to 1% change in load.

Solution

$$D = \frac{0.01 \times 5000}{0.01 \times 50} = 100 \text{ MW/Hz}$$

$$= \frac{100}{10000} = 0.01 \text{ p.u. MW/Hz}$$

$$K_p = \frac{1}{D} = 100 \text{ Hz/p.u. MW}$$

$$T_p = \frac{2H}{f\,D} = \frac{2 \times 5}{50 \times 0.01} = 20 \text{ seconds}$$

Example 25.6.

In the system of example 25.4 find the contributions made by frequency drop and increased generation to meet the increase in load.

Solution

As found in example 25.4, the static frequency drop is 0.01961 Hz. Since D is 100 MW/Hz, the contribution of frequency drop to meet increase in load is

$$0.01961 \times 100 = 1.961 \text{ MW}$$

Since R is 2Hz/pu MW, the increase in generation is

$$\frac{0.01961}{2} = 0.009805 \text{ p.u. MW} = 0.009805 \times 10000 = 98.05 \text{ MW}$$

Both the above components add upto 100 MW which is the total increase in load.

25.8 EQUIPMENT FOR LOAD FREQUENCY CONTROL

The controller is an impedance bridge which is at balance when the frequency is 50 Hz. When the frequency varies (by a value as small as 0.005 Hz) an unbalance current flows, the direction of current depending on whether the change in the frequency is positive or negative. The galvanometer of the frequency controller deflects to close a certain set of contacts and this causes the motion of the governor motor. The duration of impulse to the governor motor depends on the frequency deviation. The greater the frequency deviation, the greater is the duration of impulse, greater is the governor action and greater is the change in prime mover input. In addition to the above feature, a manual knob is also provided.

25.9 INTEGRAL CONTROL (SECONDARY ALFC LOOP)

The single area control technique discussed above can result in considerable steady state frequency drop from no load to full load. The steady state frequency can be adjusted to the desired value by adjusting the speed changer setting of the governor. A better control technique has been developed to improve the dynamic response of the system. In this control (known as integral control) the speed changer setting is adjusted automatically by a signal obtained by amplifying and integrating the frequency error. In load frequency control application, the frequency change in an area is known as Area Control Error (ACE). Thus, in integral control, the speed changer setting is adjusted automatically by a signal of the type $\int$ACE dt. The integral control gives zero steady state error. A block diagram of this control is shown in Fig. 25.5.

If the frequency decreases, the integrator calls for an increase in power. The gain K_i controls the response. The polarity of the integral controller must be chosen so that a positive frequency error gives a negative or decrease generation command and a negative frequency error gives a positive or increase generation command. Generally electronic integrators, of the type used in analog computers are used.

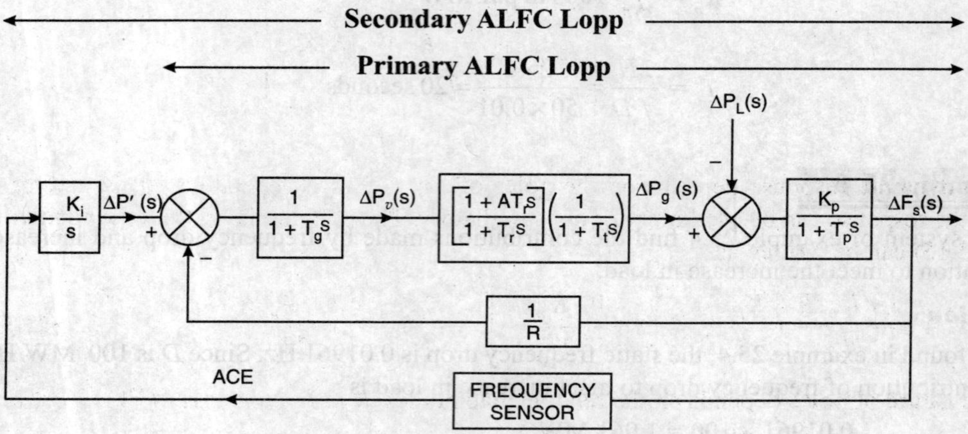

Fig. 25.5 Integral load frequency control.

When load increases the frequency decreases. During the first few instants the integral controller has not had time to act and the system response is the same as that of primary ALFC loop (Fig. 25.5).

After some time the integral controller comes into action and brings the frequency back to normal value.

25.9.1 Dynamic Response of Integral Controller

The evaluation of exact response (by including the effect of time constants T_r and T_t) is very complex. An approximate analysis can be done by assuming T_r and T_t equal to zero. Moreover we assume that speed changer acts instan-taneously. These assumptions affect the dynamic response to some extent but the static response is not affected.

The speed changer setting is controlled by a signal obtained by first amplifying and then integrating frequency error. A positive frequency error causes a negative or decrease generation command. Therefore we can write.

$$\Delta P'(s) = \frac{-K_i}{s} \Delta F(s)$$

For a step change in load, $\Delta P_L = M$

and $$\Delta P_L(s) = \frac{M}{s}$$

Using control system theory (block diagram algebra) we can write the following expression for the block diagram of Fig. 25.5.

$$\Delta F(s) = \frac{-K_p}{T_p}\left[\frac{M}{s^2 + s\left\{\dfrac{\left(1+\dfrac{K_p}{R}\right)}{T_p}\right\} + \dfrac{K_i K_p}{T_p}}\right] \quad ...(25.15)$$

The dynamic response depends on the poles of Eq. (25.15). Depending on the nature of poles, the response can be oscillatory or non-oscillatory. The polynomial in the denominator of Eq. (25.15) can be written as

$$s^2 + s\frac{1+K_p/R}{T_p} + \frac{K_i K_p}{T_p} = \left[s + \frac{1+K_p/R}{2T_p}\right]^2 + \frac{K_i K_p}{T_p} - \left[\frac{1+K_p/R}{2T_p}\right]^2 \quad ...(25.16)$$

The nature of poles depends on the relative magnitudes of different terms in Eq. (25.16).

If $\dfrac{K_i K_p}{T_p} > \left[\dfrac{1+K_p/R}{2T_p}\right]^2$ Eq. (25.16) can be expressed in the form $(s+\alpha)^2 + \omega^2$ where both α are ω positive and real. $\Delta F(s)$ has a conjugate complex pole pair in the stable s plane. The response $\Delta f(t)$ contains damped oscillatory terms of the type $\varepsilon^{-\alpha t}\sin \omega t$ and $\varepsilon^{-\alpha t}\cos \omega t$.

If $\dfrac{K_i\, K_p}{T_p} < \left[\dfrac{1+K_p/R}{2T_p}\right]^2$, Eq. (25.16) can be written in the form $(s+\beta_1)(s+\beta_2)$ where both β_1 and β_2 are real and positive. $\Delta F(s)$ has a real pole pair in the stable s plane. The response $\Delta f(t)$ is non-oscillatory and contains terms like $\varepsilon^{-\beta_1 t}$ and $\varepsilon^{-\beta_2 t}$.

The critical value of K_i is given by the equation

$$\frac{K_i\, K_p}{T_p} = \left[\frac{1+K_p/R}{2\,T_p}\right]^2$$

or

$$K_{i,crit} = \frac{(1+K_p/R)^2}{4\,T_p\,K_p} \quad ...(25.17)$$

Fig. 25.6 shows the dynamic response of integral controller for two values of K_i. One value of K_i gives oscillatory response. A lower value of K_i gives non-oscillatory (damped) response. Both these responses are shown in Fig. 25.6. The steady state frequency error (in the absence of integral controller) is also shown.

If the value of K_i is less than critical, we get non-oscillatory but sluggish response. The time taken by the system to bring frequency to the initial value will be quite large. However it is better to use a sub critical (*i.e.* less than critical) value of K_i so that the generator may not chase rapid load changes causing equipment wear.

It is also seen from Fig. 25.4 and 25.6 that the response during the first few instants is the same in the two cases. This is because of the time taken by the integral controller to come into action. Once the integral controller comes into action the frequency comes back to the initial value and there is no static frequency drop.

The automatic load frequency control of Fig. 25.3 is also known as primary ALFC loop. The integral controller (Fig. 25.6) is also known as secondary ALFC loop.

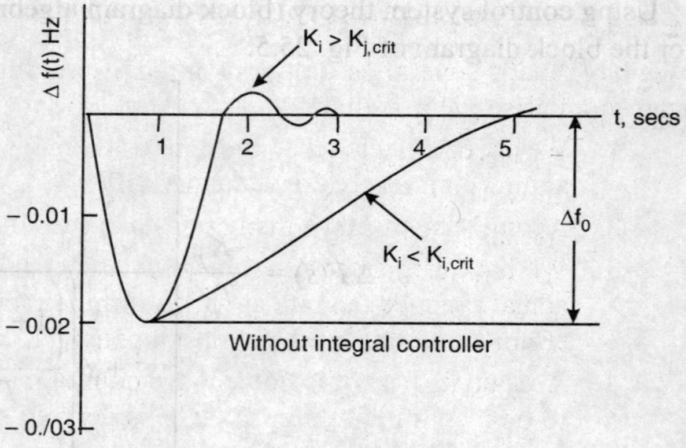

Fig. 25.6 Dynamic response of integral controller.

25.10 CONTROL SYSTEM SPECIFICATIONS

The specifications for single area ALFC can be summarised as under:

1. The system should have sufficient stability. In a feedback system stability is linked to error specifications. If we keep the error specifications tight. The system will tend to be unstable. As such the error specifications should be such that the system has good stability.
2. As and when the load changes, the frequency error should finally become zero. During the transient period, the frequency deviation should be minimal. This is called isochronous control. Use of integral control ensures that static frequency error will be zero. But transient frequency error will always be there. The time error of synchronous clocks is proportional to integral of frequency error. Thus

$$\Delta t = \frac{1}{f^\circ}\int_0^t \Delta f \, dt \text{ seconds} \qquad \text{... (25.18)}$$

The integral as given by Eq. (25.18) will not become zero automatically. When an increase in load occurs frequency tends to be below 50 Hz. If load decreases, frequency tends to be above 50 Hz.

Let frequency drop to 49.99 Hz due to a step increase in load. The time error is

$$\Delta t = \frac{1}{50}\int_0^{3600} (50 - 49.99)\, dt = \frac{-36}{50} = -0.72 \text{ seconds per hour}$$

In actual practice the time error is found by comparing the system time with that of Bureau of standards. When a time error of ±3 seconds has accumulated the speed changer is adjusted by about 0.02 Hz for a long time so that time error may be reduced to zero.

3. The control system should minimize the time error.
4. The final requirement of the control system is to ensure economic dispatch. The different generators should operate on equal incremental costs. This is done by economic dispatch control scheme. The response of this control scheme is pretty slow and the response time is about one minute or so.

25.11 ASSUMPTIONS IN ANALYSIS OF INTEGRAL CONTROL

We have made several assumptions in analysing the action of integral controller. Briefly we can summarise them as under.

1. We have neglacted the time constants T_g, T_r and T_t. Thus we have assumed that governor and turbine respond instantaneously.
2. Actual systems have many non-linearites like dead zone etc. We have neglected them.
3. We have neglected the dynamics associated with hydraulic amplifier and turbine. In actual practice the rate at which turbine input can increase is finite. Similarly the generator cannot pick up load instaneously.
4. Area control error is not a continuous signal in actual practice but we have assumed it to be a continuous signal. ACE is derived by measuring frequency in a sampled data fashion. The sampling frequency must be high so that the ACE can be assumed to be a continuous signal.

25.12 ECONOMIC DISPATCH AND AUTOMATIC GENERATION CONTROL

The AGC shown in Fig. 25.5 meets the first three specifications discussed in section 25.10. To satisfy the fourth constraint (*i.e.* economic dispatch) all generators must operate at equal incremental cost.

Primary ALFC loops makes caoarse adjustment of frequency whenever load demand changes. The generators in the control area react to the change in load by sharing the change in proportion to their capacity, The response time of primary ALFC loop is about 2 to 25 seconds.

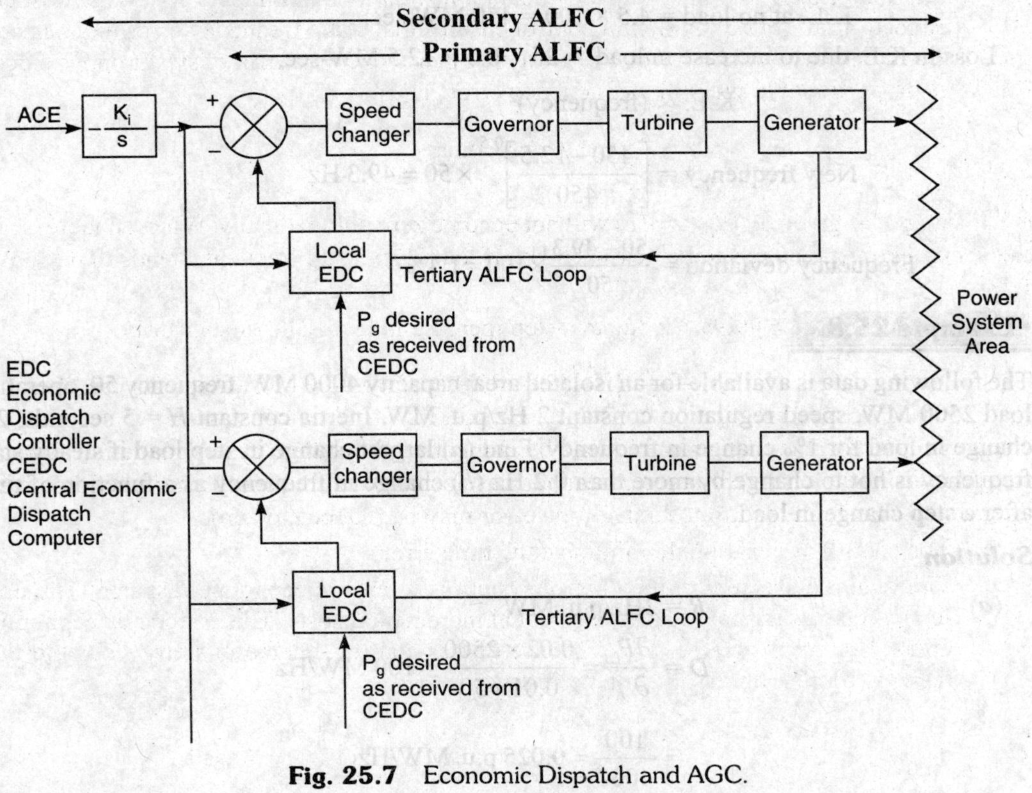

Fig. 25.7 Economic Dispatch and AGC.

Secondary ALFC (integral control) makes fine adjustment of frequency to reduce static frequency drop to zero. This resetting of integral controller occurs after primary ALFC action is over. The response time of integral controller is about 0.5 minute.

Economic dispatch is taken care by a tertiary control loop. This loop requires the solution of optimum dispatch equation i.e.

$$(IC)_1 = (IC)_2 = ... = \lambda$$

Where $(IC)_1, (IC)_2$ ... are the incremental costs of generator 1, 2 ... and λ is Lagrange multiplier. This solution is done by the digital computer in energy control centre. The energy control centre is connected to different power plants through communication channels. The mega watt settings of different generators are fed to the computer every 5 minute or so. There settings are compared with optimal settings as decided by solution of optimum dispatch equations. Based on this comparison the MW outputs of different generators are adjusted to ensure economic operation. This adjustment is also through speed changer.

Fig. 25.7 shows a simple block diagram of Economic Dispatch and AGC.

Example 25.7.

A 100 MVA, 50 Hz generator is operating at no load at 3000 rpm. A load of 25 MW is suddenly applied to the machine. Due to inertia the valve does not open immediately but after 0.5 seconds. Inertia constant of generator is 4.5 MW-sec per MVA. Find frequency deviation before the valve opens to meet the load demand. Assume no change in load due to change in frequency.

Solution

$$\text{K.E. at no load} = 4.5 \times 100 = 450 \text{ MW-sec.}$$

$$\text{Loss in K.E. due to increase in load} = 25 \times 0.5 = 12.5 \text{ MW-sec.}$$

$$\text{K.E.} \propto (\text{frequency})^2$$

$$\text{New frequency} = \left[\frac{450 - 12.5}{450}\right]^{0.5} \times 50 = 49.3 \text{ Hz}$$

$$\text{Frequency deviation} = \frac{50 - 49.3}{50} \times 100 = 1.4\%$$

Example 25.8.

The following data is available for an isolated area: capacity 4000 MW, frequency 50, operating load 2500 MW, speed regulation constant 2 Hz/p.u. MW, Inertia constant $H = 5$ seconds, 2% change in load for 1% change in frequency. Find (*a*) largest change in step load if steady state frequency is not to change by more than 0.2 Hz (*b*) change in frequency as a function of time after a step change in load.

Solution

(*a*)

$$R = 2 \text{Hz/p.u. MW}$$

$$D = \frac{\partial P_L}{\partial f} = \frac{0.02 \times 2500}{0.01 \times 50} = 100 \text{ MW/Hz}$$

$$= \frac{100}{4000} = 0.025 \text{ p.u. MW/Hz}$$

$$\beta = D + \frac{1}{R} = 0.025 + 0.5 = 0.525 \text{ p.u. MW/Hz}$$

Using Eq. (25.13)

$$\Delta f_0 = -\frac{M}{\beta}$$

or $$M = -(-0.2)\,(0.525) = 0.105 \text{ p.u. MW}$$

or $$0.105 \times 4000 = 420 \text{ MW}$$

(*b*) $$K_p = \frac{1}{D} = 40 \text{ Hz/p.u. MW}$$

$$T_p = \frac{2H}{f\,D} = \frac{2\times 5}{50\times 0.025} = 8 \text{ seconds}$$

$$\frac{R+K_p}{RT_p} = \frac{2+40}{2\times 8} = 2.625 \text{ seconds}$$

Using Eq. (25.14)

$$\Delta f(t) = -0.2\,(1-\varepsilon^{-2.625t})$$

Example 25.9.

A single area system has the following data: area capacity = 4000 MW, operating load = rated area capacity, f = 50 Hz, H = 5 secs., R = 2.5%. 1% change in load for 1% change in frequency. Find (*a*) frequency response and static frequency error in the absence of secondary loop (integral control) if a step increase of 80 MW in load occurs (*b*) critical value of K_i if integral controller is used.

Solution

$$R = 2.5\% = \frac{2.5\times 50}{100}$$

$$= 1.25 \text{ Hz/pu MW}$$

$$D = \frac{\partial P_L}{\partial f} = \frac{0.01\times 4000}{0.01\times 50} = 80 \text{ MW/Hz}$$

$$= \frac{80}{4000} = 0.02 \text{ p.u. MW/Hz}$$

$$M = \frac{80}{4000} = 0.02$$

$$K_p = \frac{1}{D} = 50 \text{ Hz/p.u. MW}$$

$$T_p = \frac{2H}{f\,D} = \frac{2\times 5}{50\times 0.02} = 10 \text{ seconds}$$

$$\frac{R+K_p}{RT_p} = \frac{1.25+50}{1.25\times 10} = 4.1$$

$$\frac{K_p\ RM}{R+K_p} = \frac{50\times1.25\times0.02}{1.25+50} = 0.0244$$

Using Eq (25.14)

$$\Delta f(t) = -\,0.0244(1-\varepsilon^{-4.1t})$$

The static frequency error is $-\,0.0244$ Hz

(*b*) Using Eq. (25.17)

$$K_{i,\ Crit.} = \frac{(1+K_p/R)^2}{4\,T_pK_p}$$

$$= \frac{(1+50/1.25)^2}{4\times10\times50} = 0.8405 \text{ p.u. MW/Hz}$$

Example 25.10.

The time constant of the governing system of a steam turbine is 0.2 seconds while that of turbine is 2 seconds. $H = 5$ seconds, $R = 4\%$ (Hz/p.u. MW), Load changes by 1% for 1% change in frequency. The rated capacity is 1500 MW and $f = 50$ Hz. (*a*) Find an equation expressing $\Delta F(s)$ in terms of step change M in load and various system parameters. (*b*) Find largest step change in load if frequency is not to change by more than 1% in steady state if integral controller is not used. (*c*) static frequency error for this step change in load if integral controller is used.

Solution

$$\text{Transfer function of governor} = G_r = \frac{1}{1+T_g s} = \frac{1}{1+0.2\,s}$$

$$\text{Transfer function of turbine} = G_t = \frac{1}{1+T_t s} = \frac{1}{1+2s}$$

$$R = 4\% = \frac{4\times50}{100} = 2 \text{ Hz/p.u. MW}$$

$$D = \frac{\partial P_L}{\partial f} = \frac{0.01\times1500}{0.01\times50} = 30 \text{ MW/Hz}$$

$$= \frac{30}{1500} = 0.02 \text{ p.u. MW/Hz}$$

$$K_p = \frac{1}{D} = 50 \text{ Hz/p.u. MW}$$

$$T_p = \frac{2H}{f\,D} = \frac{2\times5}{50\times0.02} = 10 \text{ seconds}$$

Transfer function of power system

$$G_p = \frac{K_p}{1+T_p s} = \frac{50}{1+10\,s}$$

(*a*) Using Eq. (25.12)

$$\Delta F(s) = \frac{-\dfrac{50}{1+10s}}{1+\dfrac{1}{2}\left(\dfrac{1}{1+0.2s}\right)\left(\dfrac{1}{1+2s}\right)\left(\dfrac{50}{1+10s}\right)}\left(\frac{M}{s}\right)$$

(*b*) Using final value theorem, the static frequency drop Δf_0 is

$$\Delta f_0 = \underset{s\to 0}{\text{Lt}}\; s\Delta F(s) = -\frac{K_p M}{1+K_p/R}$$

or

$$\Delta f_0 = \frac{-50\,M}{1+50/2} = -1.923\,M$$

$$\Delta f_0 = 1\% = \frac{1\times 50}{100} = 0.5\,\text{Hz}$$

Therefore

$$M = \frac{-0.5}{1.923} = -\,0.26 \text{ p.u. MW}$$

$$= -\,(0.26)\,(1500)\text{MW}$$

or

$$M = -\,390 \text{ MW}$$

The minus sign is because of the fact that if frequency is to increase by 1%, the step change in load must be negative.

(*c*) When integral controller is used, static frequency error is zero.

MULTI-AREA LOAD FREQUENCY CONTROL

25.13 TWO SYSTEMS CONNECTED BY TIE LINE

Fig. 25.8 shows two large power systems A and B interconnected through a tie line. The tie line transfers power from system *A* to *B* and *vice-versa*. The aim of the control is to regulate the frequency in both the areas as also to regulate the power flow over the tie line as per the agreement between the utilities of systems *A* and *B*.

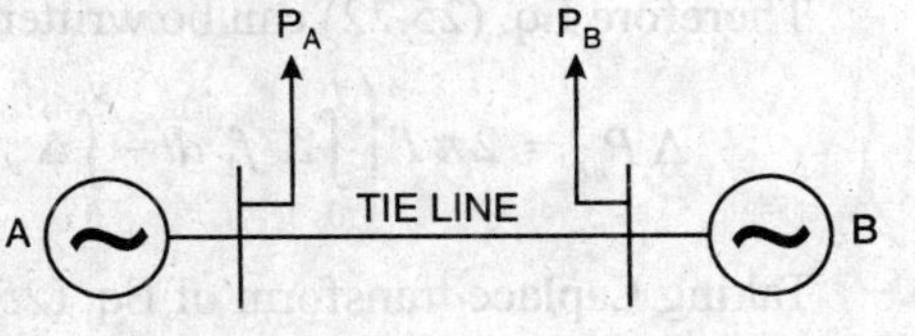

Fig. 25.8 Interconnection of two systems.

The control areas intertied by transmission lines can take advantage of the spinning mass of large number of generators and differences in incremental costs between units. Power flow over the tie lines can be on short term basis (if a control area suffers a decrease in generation due to forced outage of equipment) or on long term basis if there is a mutually beneficial sale of energy between the utilities of the two areas.

A small system can suffer extensive transient power swings due to load changes. If the system capacity is 1000 MW and a 200 MW load change occurs, this 20% change in load may make the system unstable and a complete blackout may result. However if this system were a part of a bigger system, say 50000 MW, a 200 MW change in load means only a 0.4% change. The drop in frequency would be very small and the flow of power over the tie line would save the system from collapse.

25.14 BLOCK DIAGRAM REPRESENTATION OF TIE LINE

The power flow over the line of Fig. 25.8 is

$$P_{AB} = \frac{|V_A||V_B|}{X}\sin(\delta_A - \delta_B) \qquad \text{... (25.19)}$$

where P_{AB} = Power transfer from system A to system B

$|V_A|$ and $|V_B|$ = Voltage magnitudes at ends A and B

δ_A and δ_B = phase angles of voltages V_A and V_B

X = reactance of tie line

For small deviations in angles δ_A and δ_B, the change in power transfer ΔP_{AB} is

$$\Delta P_{AB} = \frac{|V_A||V_B|}{X}\cos(\delta_A - \delta_B)(\Delta\delta_A - \Delta\delta_B) \qquad \text{... (25.20)}$$

The synchronising coefficient T of tie line is defined as

$$T = \frac{|V_A||V_B|}{X}\cos(\delta_A - \delta_B) \qquad \text{... (25.21)}$$

Then
$$\Delta P_{AB} = T(\Delta\delta_A - \Delta\delta_B) \qquad \text{... (25.22)}$$

The frequency deviation Δf can be expressed in terms of reference angle $\Delta\delta$ by the equation

$$\Delta f = \frac{1}{2\pi}\frac{d}{dt}(\delta + \Delta\delta) = \frac{1}{2\pi}\frac{d}{dt}(\Delta\delta) \qquad \text{... (25.23)}$$

or
$$\Delta\delta = 2\pi\int_0^t \Delta f\, dt \qquad \text{... (25.24)}$$

Therefore Eq. (25.22) can be written as

$$\Delta P_{AB} = 2\pi T\left(\int_0^t \Delta f_1\, dt - \int_0^t \Delta f_2\, dt\right) \qquad \text{... (25.25)}$$

Taking Laplace transform of Eq. (25.25)

$$\Delta P_{AB}(s) = \frac{2\pi T}{s}[\Delta F_1(s) - \Delta F_2(s)] \qquad \text{... (25.26)}$$

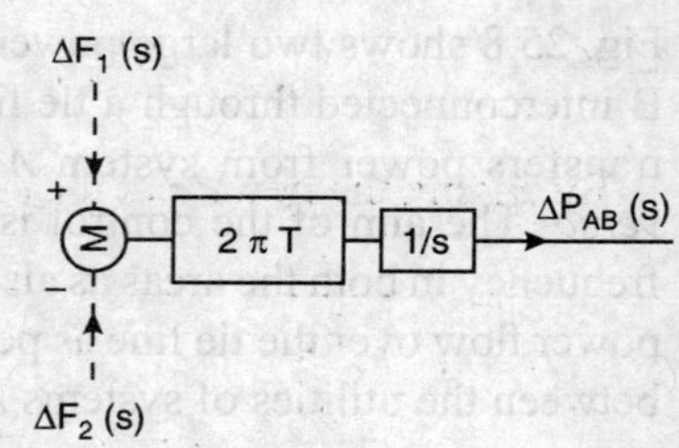

Eq. (25.26) gives the block diagram of Fig. 25.9.

Fig. 25.9. Tie line representation.

25.15 ALFC FOR TWO AREA SYSTEM

Fig. 25.10 shows the block diagram of automatic load frequency control system for a two area system. The dashed portion represents the integral controllers to reduce the static frequency error to zero. The control error for each area consists of a linear combination of frequency and tie line error. α_A and α_B are constants which specify the frequency bias for the purpose of the line load bias control. G_{rA} and G_{rB} are the transfer functions (of the relay valves and servomotors) of areas A and B. Similarly G_{tA}, G_{tB}, G_{pA}, G_{pB}, R_A and R_B refer to transfer functions of turbines, power systems and regulations of the two areas. The blocks of transfer functions $1/s$ and $2\pi T$ represent the tie line. The block of transfer function -1 accounts for the fact that ΔP_{AB} is imported by area B from area A.

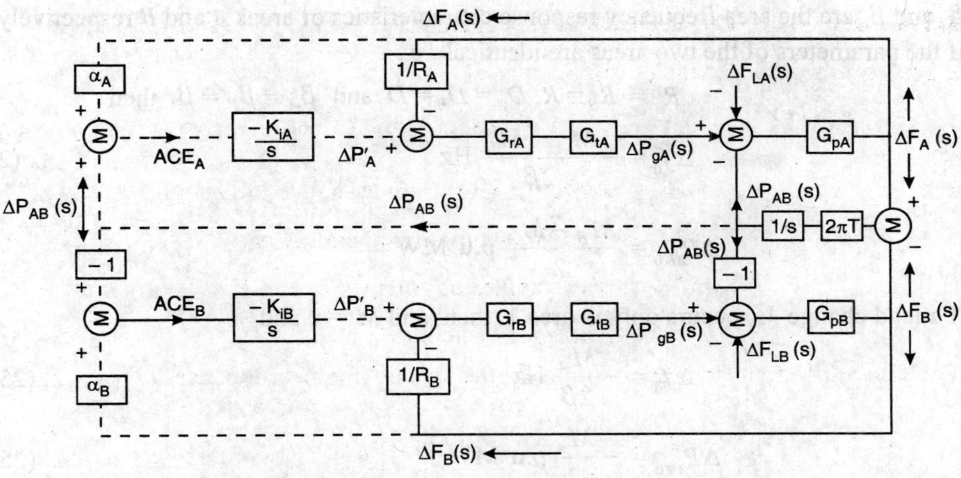

Fig. 25.10 ALFC for Two Area System.

25.16 RESPONSE OF TWO AREA SYSTEM

25.16.1 Static Response

The static response of the two area system is determined for fixed speed changer position. The dashed portions of Fig. 25.10 need not be considered in determining static frequency change Δf_0 and tie line power change $\Delta P_{AB,0}$.

Let the loads in the two areas increase by $\Delta P_{LA} = M_A$ and $\Delta P_{LB} = M_B$ respectively. Then

$$\Delta P_{gA,0} = \frac{-1}{R_A} \Delta f_0 \qquad \text{... (25.27)}$$

and

$$\Delta P_{gB,0} = \frac{-1}{R_B} \Delta f_0$$

By adding the powers at the summing junctions we get

$$-\frac{1}{R_A} \Delta f_0 - M_A = D_A \Delta f_0 + \Delta P_{AB,0} \qquad \text{... (25.28a)}$$

$$-\frac{1}{R_B} \Delta f_0 - M_B = D_B \Delta f_0 + \Delta P_{AB,0} \qquad \text{...(25.28b)}$$

Solving Eq. (25.28)

$$\Delta f_0 = -\frac{M_A + M_B}{\beta_A + \beta_B} \text{ Hz} \qquad \text{... (25.29)}$$

and

$$\Delta P_{AB,0} = \frac{\beta_A M_B - \beta_B M_A}{\beta_A + \beta_B} \text{ p.u. MW}$$

where

$$\beta_A = D_A + \frac{1}{R_A} \qquad \text{... (25.30a)}$$

and

$$\beta_B = D_B + \frac{1}{R_B} \qquad \text{... (25.30b)}$$

β_A and β_B are the area frequency response characteristics of areas A and B respectively.

If the parameters of the two areas are identical *i.e.,*

$$R_A = R_B = R, D_A = D_B = D \text{ and } \beta_A = \beta_B = \beta, \text{ then}$$

$$\Delta f_0 = -\frac{M_A + M_B}{2\beta} \text{ Hz} \quad \text{... (25.31)}$$

and

$$\Delta P_{AB,0} = \frac{M_B - M_A}{2} \text{ p.u. MW}$$

If a load change M_A occurs only in area 1, then and $M_B = 0$ and

$$\Delta f_0 = -\frac{M_A}{2\beta} \text{ Hz} \quad \text{... (25.32}a\text{)}$$

$$\Delta P_{AB,0} = -\frac{M_A}{2} \text{ p.u. MW} \quad \text{... (25.32}b\text{)}$$

Eq. (25.32 *a*) indicates that static frequency drop would be only 50% of that which would occur if the areas were operating alone. Moreover 50% of the increase in load in area A would be supplied by area B through the tie line. The negative sign in Eq. (25.32 *b*) indicates that power flow is from area B to area A.

Example 25.11.

A 5000 MW area A is interconnected with a 10000 MW area B. The parameters of each based on its own capacity base are

$$R = 2 \text{ Hz/p.u. MW}$$

$$D = 0.01 \text{ p.u. MW/Hz}$$

Area B experiences a load increase of 100 MW. Find the static frequency drop and the change in the tie line power

Solution

Taking 10000 MW as base

$$R_A = 2\times\frac{10000}{5000} = 4 \text{ Hz/p.u. MW}$$

$$D_A = 0.01\times\frac{5000}{10000} = 0.005 \text{ p.u. MW/Hz}$$

$$R_B = 2 \text{ Hz/p.u. MW}$$

$$D_B = 0.01 \text{ p.u. MW/Hz.}$$

$$\beta_A = D_A + \frac{1}{R_A} = 0.005 + \frac{1}{4} = 0.255 \text{ p.u. MW/Hz}$$

$$\beta_B = 0.01 + \frac{1}{2} = 0.51 \text{ p.u. MW/Hz}$$

$$M_A = 0$$

$$M_B = \frac{100}{10000} = 0.01 \text{ p.u. MW}$$

Using Eqs. (25.29)

$$\Delta f_0 = \frac{-0.01}{0.255+0.51} = -0.01307 \text{ Hz}$$

$$\Delta P_{AB,\,0} = \frac{0.255\times 0.01}{0.255+0.51} = 0.003333 \text{ p.u. MW}$$

$$= 33.33 \text{ MW}$$

25.16.2 Dynamic Response

We have seen that the evaluation of dynamic response of a single area system is not easy and only an approximate solution (with some simplifying assumptions) was discussed in section 25.7.6. The evaluation of dynamic response of a two area system requires the solution of a seventh order polynomial and is thus very complex. However an approximate solution is possible if following assumptions are made:

1. The two areas are similar and of equal capacity.
2. Time constants T_r and T_t are assumed to be zero.
3. System damping is neglected. This means that load does not vary with frequency *i.e.* $D_A = D_B = 0$
4. The dashed portion of Fig. 25.10 (representing the action of integral controller) is not considered.

Then the change in tie line power can be expressed as

$$\Delta P_{AB}(s) = \frac{\pi f T}{H}\left[\frac{\Delta P_{LA}(s) - \Delta P_{LB}(s)}{s^2 + \dfrac{fs}{2RH} + \dfrac{2\pi f T}{H}}\right] \qquad \text{... (25.33)}$$

The following inferences can be drawn from Eq. (25.33)

1. The polynomial in the denominator is of the form
$$s^2 + 2\alpha s + \omega^2 = (s+\alpha)^2 + \omega^2 - \alpha^2$$
where both α and ω are positive. Therefore the system is stable and response is damped.
2. When a disturbance (*i.e.* change in load in either of the areas) occurs the system will oscillate at the damped angular frequency ω_0 given by
$$\omega_0 = (\omega^2 - \alpha^2)^{0.5} = \left[\frac{2\pi f T}{H} - \left(\frac{f}{4RH}\right)^2\right]^{0.5} \qquad \text{... (25.34)}$$
3. The value of α governs the degree of damping. α depends on f, H and R. Out of these quantities f and H are constant. Thus damping depends on the value of regulation R. A high value of R results in low value of α and weak damping. If R is low, damping is strong.
4. If R is infinite *i.e.* speed governor loops are open the system will have undamped oscillations of angular frequency ω_0.

When dashed portion of Fig. 25.10 (*i.e.* integral controller) is absent, the system will have static frequency error. The use of integral controller helps in elimination of static frequency error.

Example 25.12.

A two area system has the following data:

Area A, rated capacity 500 MW,
speed regulation 2.5 Hz/p.u. MW

Area B, rated capacity 2000 MW,
speed regulation 2 Hz/p.u. MW,

In each area 1% change in load occurs for 1% change in frequency. Find steady state change in frequency and change in tie line power if 20 MW change in load occurs in (*a*) area A (*b*) and B

Solution

Choose 2000 MW base

$$D_A = \frac{0.01\times 500}{0.01\times 50} = 10 \text{ MW/Hz}$$

$$= \frac{10}{2000} = 0.005 \text{ p.u. MW/Hz}$$

$$D_B = \frac{0.01\times 2000}{0.01\times 50} = 40 \text{ MW/Hz}$$

$$= \frac{40}{2000} = 0.02 \text{ p.u. MW/Hz}$$

$$R_A = 2.5 \text{ Hz/p.u. MW on 500 MW base}$$

Therefore for 2000 MW base

$$R_A = 2.5\times\frac{2000}{500} = 10 \text{ Hz/p.u. MW}$$

$$R_B = 2 \text{ Hz/p.u. MW}$$

$$\beta_A = D_A + \frac{1}{R_A} = 0.005 + \frac{1}{10} = 0.105 \text{ p.u. MW/Hz}$$

$$\beta_B = D_B + \frac{1}{R_B} = 0.02 + \frac{1}{2} = 0.52 \text{ p.u. MW/Hz}$$

(*a*)

$$M_B = \frac{20}{2000} = 0.01$$

$$M_B = 0$$

Using Eq. (25.29)

$$\Delta f_0 = -\frac{0.01}{0.105 + 0.52} = -0.016 \text{ Hz}$$

$$\Delta P_{AB,0} = \frac{-0.52\times 0.01}{0.105 + 0.52} = -0.00832 \text{ p.u. MW}$$

$$= (-0.00832)(2000) = -16.64 \text{ MW}$$

The minus sign indicates that power transfer is from area B to area A

(*b*) $M_A = 0$

$$M_B = \frac{20}{2000} = 0.01$$

Using Eq. (25.29)

$$\Delta f_0 = -0.016 \text{ Hz}$$

$$\Delta P_{AB,0} = \frac{0.105 \times 0.01}{0.105 + 0.52} = 0.00168 \text{ p.u. MW}$$

$$= 0.00168 \times 2000 = 3.36 \text{ MW}$$

The power transfer is from area *A* to area *B*.

Example 25.13.

A system has two similar areas of equal capacity 4000 MW. $R = 2$ Hz/pu MW and $H = 5$ seconds. The tie line has a capacity of 600 MW and is operating at a power angle of 40°. Find (*a*) the frequency of oscillation of the system (*b*) steady state change in tie line power if a step change of 100 MW load occurs in area *B* (*c*) frequency of oscillation of the system if speed governor loop is open. The system frequency is 50 Hz.

Solution

(*a*) Using Eq. (25.21)

$$T = \frac{600}{4000} \cos 40° = 0.115$$

Using Eq. (25.34)

$$\omega_0 = \left[\frac{2\pi \times 50 \times 0.115}{5} - \left(\frac{50}{4 \times 2 \times 5}\right)^2\right]^{0.5}$$

$$= 2.38 \text{ radians/sec.}$$

(*b*) Since the two areas are similar, each area will supply half of increase in load. This is also evident from Eq. (25.29) because $\beta_A = \beta_B$.

Therefore,

$$\Delta P_{AB,0} = 50 \text{ MW from area } A \text{ to area } B$$

(*c*) $R = \infty$

Using Eq. (25.34)

$$\omega_0 = \left(\frac{2\pi \times 50 \times 0.115}{5}\right)^{0.5}$$

$$= 2.688 \text{ rad/sec.}$$

25.17 TYPES OF AUTOMATIC LOAD FREQUENCY CONTROLS FOR INTERCONNECTED POWER SYSTEMS

Depending on whether one or all the systems have been assigned the job of maintaining the frequency constant and extent of tie line load required, three methods are available for load frequency control of interconnected systems. The equipment used is the same as in the case of single area systems. In addition a tie line load recorder controller is also necessary. The load on tie line is measured by a thermal converter which operates the tie line load recorder-controller. A knob on this controller sets the scheduled amount of power transfer. Whenever the interchange is above or below the set value, a set of contacts in the load controller come into service. These contacts are in series with load frequency controller circuit. These contacts open and block all those impulses which would make the tie line load to further deviate from the set value. However all impulses which aid in keeping scheduled interchange are allowed to go to the governor motor. Generally the same equipment can be used for all the following types of control by putting the selector switch in the proper position.

25.17.1 Flat Frequency Control

The job of maintaining the frequency constant is assigned to one machine or one station. This machine or station absorbs all changes in the system load while all other machines (and stations) carry constant loads. There is no control on tie line power (the tie line load recorder controller is out of circuit) and frequency controllers are used on only the master machines. On master machines the frequency controller can usually detect as small as 0.005 Hz change in frequency and initiate control action. Flat frequency control has the advantage that new and more efficient machines can be made to carry the base load and the less efficient machines can act as the master machines for taking up load changes. The main disadvantage of this type of control is that it results in random variations in tie line powers.

25.17.2 Flat Tie Line Load Control

The aim of this control is to keep the tie line power constant irrespective of load demands. The increase in demands of an area is met by increase in generation in that area. This control is used when a small system and a large system are interconnected through a tie line. The large system maintains the system frequency and the small system is controlled to keep the tie line power constant. It is not suitable when two or more large systems are interconnected because in such cases (and with this control) the tie line power and frequency deviation have a tendency to swing back and forth (in addition to the swings at tie line natural frequency) following a load change. Needless to add that in this type of control, the control equipment consists of the frequency controller only (similar to the one used for single area systems) at the larger system and tie line power controller-recorder at the smaller system.

25.17.3 Tie Line Load Bias Control

This is the most widely used method on large inter connections. All power systems assist in regulating frequency and tie line power flow irrespective of where the disturbance originates. Each power system has control units of almost identical operating characteristics.

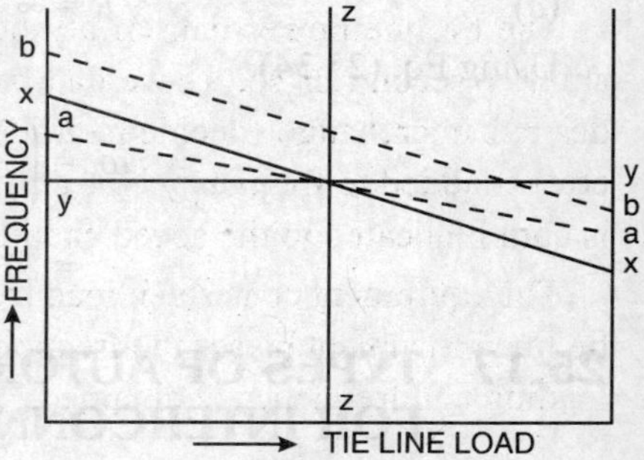

Fig. 25.11 Frequency tie-line load characteristics.

Frequency tie line load relationship is shown in Fig. 25.11. Solid line *yy* represents flat frequency characteristic. If the system were operating on this characteristic, the tie line load would change over very wide limits. Line *zz* shows flat tie line load characteristic. In this case the tie line load is held constant irrespective of frequency. In the tie line load control with frequency bias the actual characteristic is adjustable between the above two extreme limits and occupies positions *xx*. The slope of line *xx* can be adjusted. It can be made less drooping (*i.e.,* line *aa*) or be moved parallel to itself (*i.e.,* line *bb*) so that any line load can be scheduled and maintained thereafter. Thus this control provides more flexibility for a wide range of operating conditions.

The control signals (for each area) are proportional to the change in frequency as well as change in tie line power. As in the case of single area systems, integral control is preferable because it gives zero steady state error. The area control errors for a two area system are given by

$$\text{ACE}_A = \Delta P_{AB} + \alpha_A \Delta f_A \quad \text{... (25.35)}$$

$$\text{ACE}_B = \Delta P_{BA} + \alpha_B \Delta f_B$$

where

ACE_A = Area control error of system A

ACE_B = Area control error of system B

ΔP_{AB} = Change in power transfer from A to B (positive or negative)

$\Delta P_{BA} = -\Delta P_{AB}$

α_A and α_B are constants which specify the frequency bias and are determined from a knowledge of the size of the system.

The commands to the speed changers are of the form

$$\Delta P_A' = -K_{iA} \int (\Delta P_{AB} + \alpha_A f_A)\ dt$$

$$\Delta P_B' = -K_{iB} \int (\Delta P_{BA} + \alpha_B f_B)\, dt \quad \text{... (25.36)}$$

The constants K_{iA} and K_{iB} are integrator gains. The minus sign in Eqs. (25.36) is essential because the generation in each area must increase if either its frequency error or tie line power increment is negative.

Most of the time a control area is interconnected with many other areas through several tie lines. Let their be a total of *m* tie lines. Then for the *i*th control area, the net interchange is the sum of power transfer over all the *m* tie lines. The area control error ACE_i of the *i*th area should be proportional to total exchange of power and can be expressed as

$$\text{ACE}_i = \sum_{j=1}^{m} \Delta P_{ij} + \alpha_i \Delta f_i \quad \text{... (25.37)}$$

The tie line power data of all the lines are sampled continuously at sampling intervals of about 1 second or so. These data are added in an energy control centre and compared with desired interchange (decided earlier by mutual agreement). The total tie line power transfer error is added to frequency bias error $\alpha_i \Delta f_i$ to give the area control error. This ACE command is communicated to the speed changers of all the generators in the area.

The equipment consists of load frequency controller and tie line load recorder controller. The tie line instrument biases the frequency controller till a desired relationship between the tie line loading and frequency is obtained.

25.18 AUTOMATIC LOAD DISPATCHING

The operation of a modern power system has become very complex. It is necessary to maintain frequency and voltage within limits in addition to ensuring reliability of electric supply. For maintaining the frequency and voltage within limits it is necessary to match the generation of active and reactive power with the demand. To ensure reliability of supply it is necessary to put additional generation capacity into the system in the event of an outage of generating equipment at some station. Over and above this, it is necessary to ensure that the cost of electric supply is kept to the minimum.

In a multi-area integrated system it is necessary to control generation and transmission from a national control centre. This control centre co-ordinates the working of area control centres. The functions of an area control centre include load prediction, data collection, load frequency control within the area, economic dispatch within the area, system switching, maintenance scheduling etc. An area control centre has access to telemetered values of all major tie line power flows, outputs of all big generating units and voltage profile at major load points. The national control centre acts as a co-ordinating centre responsible for inter-area energy transfer and operating plans for weeks and months ahead.

In a manual load dispatch system, the control engineer at the area control centre obtains the telemetered information mentioned above, calculates the station outputs for different plants and advises the different stations to increase or decrease the generation.

With systems becoming more and more complex, it is very difficult to satisfy all the system constraints with manual control. It has become necessary to aid the operating engineers by means of digital computers. The use of on-line computers for *load* dispatching is increasing day by day. The on-line computer assisted control system can be open loop systems or closed loop systems.

The open loop systems supply upto date information to the control engineer to enable him to take corrective action. The computer has all the information such as expected loads, restrictions on plants, power flows at different load points, system voltage profile, the area generation and demand, incremental costs, automatic annunciation of switchgear operation, overload indication etc. The on-line data is stored at minute intervals, for upto 30 minutes or so, so that animated pictures of system conditions leading upto any incident can be studied and recorded if necessary. The computer calculates the generation requirements at different plants and this information is transmitted to the plants to take corrective action. Thus in an open loop application the computer serves as a tool for high speed security checking and load prediction.

The closed loop on line computer systems perform all the above functions of data storage and security checking and in addition take corrective action. Examples of closed loop applications include automatic switching at sub-stations, automatic starting, loading and shut down of generators.

The close loop operation allows a great variety of emergency conditions to be catered successfully and more efficiently. Automatic control involves the provision of a machine controller connected to the governor of each turbo generator unit. These machine controllers are controlled by central process control computer at the area control centre. The computer calculates the expected demand and, through the different machine controllers, schedules the loading on each unit to meet the demand with maximum economy subject to system security. The optimum economy for a given combination of generators is achieved when the incremental cost of received power is the same for all generators as discussed in chapter 11.

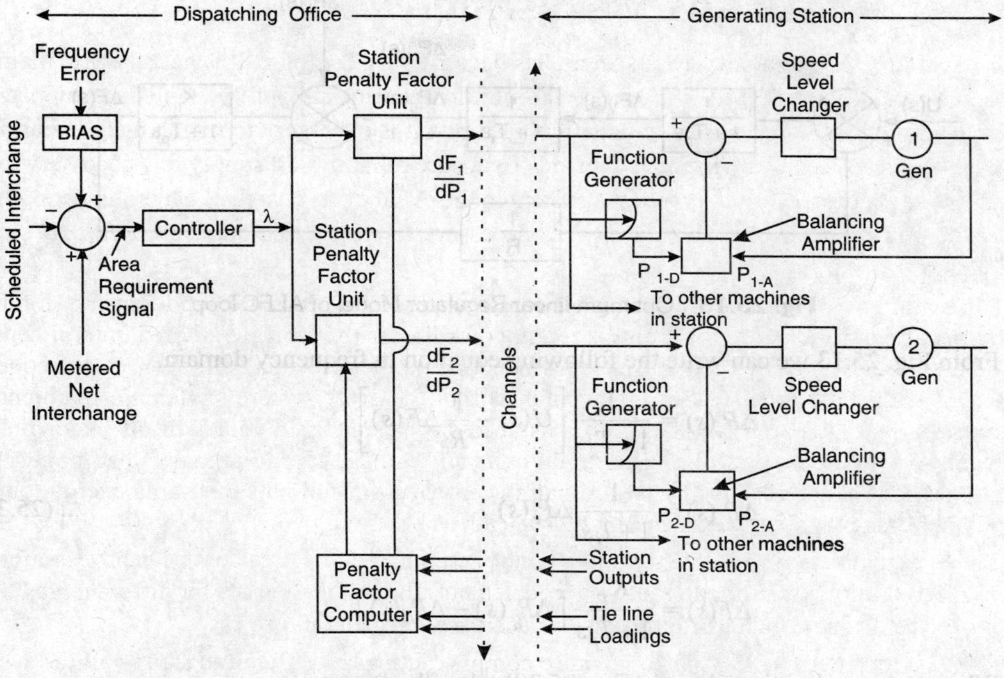

Fig. 25.12 Automatic load dispatch system.

Many developed countries in the world have switched over to automatic load dispatching with closed loop on line computers for power system operation and control. Fig. 25.12 shows a simple diagram of automatic load dispatch system. For each generator the desired output P_D and actual output P_A are compared by balancing amplifier and a correction signal is sent which brings about a change in output of the unit in agreement with desired output. The area requirement signal is developed from the frequency and net interchange errors. The controller develops a correcting signal which is sent to the speed governing system of all the units under control. If a change in load occurs, an area requirement signal is immediately generated and this signal calls for a change in generation. As incremental cost of received power changes, the station penalty factors get adjusted so that all generators operate at the same incremental cost. By modifying the incremental cost signal by the penalty factor the signal sent to each unit is the proper incremental cost for the station. The penalty factor computer requires for its input a knowledge of station loads, tie line flows etc. The function generator is an electronic storage device which indicates the economic schedule corresponding to operation of all units at the same incremental cost.

25.19 STATE VARIABLE MODEL OF ALFC LOOP

The design of ALFC has to fulfill many objectives like fast response, stability etc. The optimum linear regulator (OLR) design meets these objectives satisfactorily. The design of OLR requires the ALFC equations to be expressed in state variable form. Fig 25.13 shows block diagram of ALFC loop. To simplify the analysis we have neglected the effect of reheat stage of turbine.

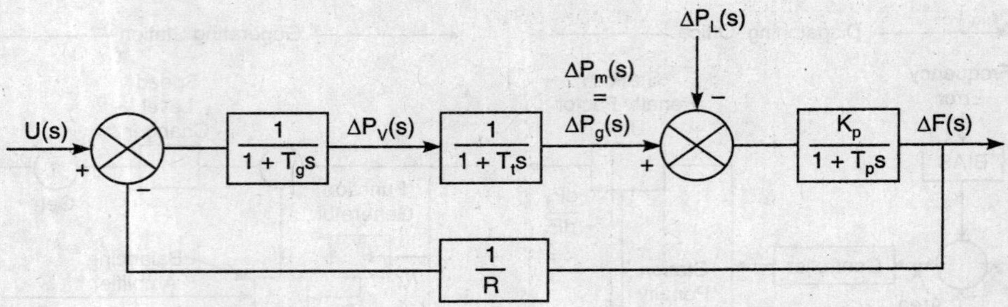

Fig. 25.13 Optimum linear Regulator Model of ALFC loop.

From Fig. 25.13 we can write the following equation in frequency domain.

$$\Delta P_v(s) = \frac{1}{1+T_g s}\left[U(s) - \frac{1}{R}\,\Delta F(s)\right]$$

$$\Delta P_g(s) = \frac{1}{1+T_t s}\,\Delta P_v(s) \qquad \text{... (25.38)}$$

$$\Delta F(s) = \frac{K_p}{1+T_p s}\left[\Delta P_g(s) - \Delta P_L(s)\right]$$

When converted to time domain Eq. (25.38) takes the form

$$\Delta P_v + T_g\,\frac{d}{dt}(\Delta P_v) = u - \frac{\Delta f}{R}$$

$$\Delta P_g + T_t\,\frac{d}{dt}(\Delta P_g) = \Delta P_v \qquad \text{... (5.39)}$$

$$\Delta f + T_p\,\frac{d}{dt}(\Delta f) = K_p(\Delta P_g - \Delta P_L)$$

The state vector x consisting of state variables x_1, x_2 and x_3 can be written as

$$X = \begin{bmatrix} x_1 \\ x_2 \\ x_3 \end{bmatrix} = \begin{bmatrix} \Delta P_v \\ \Delta P_g \\ \Delta f \end{bmatrix} \qquad \text{... (25.40)}$$

Eq. (25.39) when re-arranged and put in matrix form is

$$\begin{bmatrix} \mathring{x}_1 \\ \mathring{x}_2 \\ \mathring{x}_3 \end{bmatrix} = \begin{bmatrix} -\frac{1}{T_g} & 0 & -\frac{1}{R\,T_g} \\ \frac{1}{T_t} & \frac{-1}{T_t} & 0 \\ 0 & \frac{K_p}{T_p} & \frac{-1}{T_p} \end{bmatrix} \begin{bmatrix} x_1 \\ x_2 \\ x_3 \end{bmatrix} + \begin{bmatrix} \frac{1}{T_g} \\ 0 \\ 0 \end{bmatrix} u + \begin{bmatrix} 0 \\ 0 \\ -\frac{K_p}{T_p} \end{bmatrix} p \qquad \text{... (25.41)}$$

Eq. (25.41) is the standard state variable form

$$\overset{\circ}{x} = Ax + Bu + Cp \qquad \text{... (25.42)}$$

where $$\overset{\circ}{x} = \begin{bmatrix} \overset{\circ}{x}_1 \\ \overset{\circ}{x}_2 \\ \overset{\circ}{x}_3 \end{bmatrix}, \; A = \begin{bmatrix} -\dfrac{1}{T_g} & 0 & \dfrac{-1}{RT_g} \\ \dfrac{1}{T_t} & \dfrac{-1}{T_t} & 0 \\ 0 & \dfrac{K_p}{T_p} & \dfrac{-1}{T_p} \end{bmatrix}$$

$$B = \begin{bmatrix} \dfrac{1}{T_g} \\ 0 \\ 0 \end{bmatrix}, \; C = \begin{bmatrix} 0 \\ 0 \\ \dfrac{-K_p}{T_p} \end{bmatrix}, \; p = \Delta P_L$$

Fig. 25.14 shows the block diagram representation of Eq. (25.42).

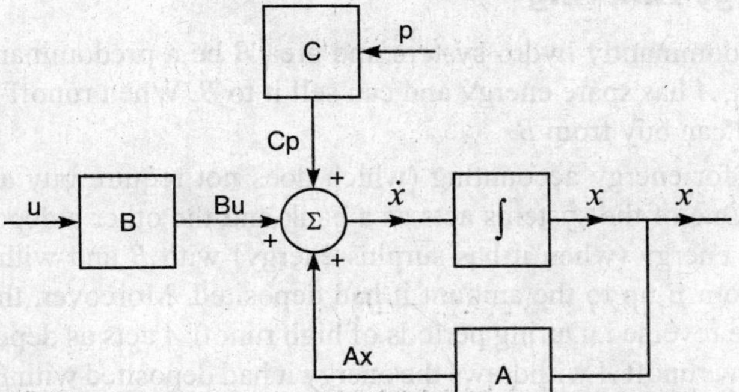

Fig. 25.14 Block diagram of linear state variable model.

For a two area system (areas A and B) we will have two control and disturbing forces. The control forces are u_1 and u_2 and disturbing forces are $\Delta P_L A$ and $\Delta P_L B$. Then

$$u = \begin{bmatrix} u_1 \\ u_2 \end{bmatrix}, \; p = \begin{bmatrix} p_1 \\ p_2 \end{bmatrix} = \begin{bmatrix} \Delta P_{LA} \\ \Delta P_{LB} \end{bmatrix}$$

and the state variable form is

$$\overset{\circ}{x} = Ax + Bu + Cp \qquad \text{... (25.43)}$$

25.20 TYPES OF POWER/ENERGY INTERCHANGE

Operation in an interconnected manner leads to reduction in costs and other benefits. However, interchange of power and energy may be necessary for a number of other reasons also.

25.20.1 Capacity Interchange

Every utility keeps on adding new generation capacity to meet the ever increasing demand of consumers. Nevertheless this may not be possible at all times due to variety of reasons *e.g.,* financial constraint, delay in procurement of equipment, delay in construction etc. Under such circumstances the utility may enter into a capacity interchange agreement with a neighbouring utility (which has surplus capacity). A capacity interchange agreement may also be necessary to cover the reserve requirement of a system.

25.20.2 Diversity Interchange

When two areas have peak power requirements at different times (due to characteristics and types of loads in the two areas as also due to the fact that the two areas span different time zones), they may enter into diversity interchange agreement. The interchange of power in such a case would be in both directions. When system A is facing peak load, it would buy power from system B (Which does not have its peak at this time). When system B is facing peak load, it would buy power from A. Such an interchange agreement may also be possible if the seasons for peak power in areas A and B are different. System A may have its peak occuring in summer while B may have its peak in winter. During summer B will sell power to A while A, would sell power to B during winter. Such diversity interchanges can bring immense benefits to both the systems.

25.20.3 Energy Banking

Let area A be a predominantly hydro system and area B be a predominantly thermal system. When runoff is high, A has spare energy and can sell it to B. When runoff is low A would face energy shortage and can buy from B.

Another method for energy accounting (which does not require buy and sell agreements) is energy banking. One of the systems acts as a bank and the other a depositer. The depositer (say A) can deposit energy (when it has surplus energy) with B and withdraw energy (when it faces shortage) from B up to the amount it had deposited. Moreover, the roles of bank and depositer can also be reversed. During periods of high runoff A acts as depositer and B as bank. During periods of low runoff A withdraws the energy it had deposited with B. Sometimes B may deposit energy with A and withdraw it later. Now B acts as depositer and A as bank.

25.20.4 Emergency Interchange

Sometimes a utility may face unprecedented forced outages of its generating units or an unforeseen high demand. To meet the energy requirement under such circumstances it may have an emergency interchange agreement with a neighbouring utility. Generally such agreements, are only for energy supply for short periods and the price of such an energy supply is also high.

25.20.5 Inadvertent Exchange

The automatic generation control (AGC) systems are not perfect. The error in AGC may lead to an inadvertent exchange of power. In most interconnected systems, such inadvertent receipt of energy is paid back at a similar time next day or next week.

It is possible that the inadvertent exchange goes beyond the normal practice (It may be done deliberately to meet energy shortages due to fuel or water constraints). This condition is often called 'leaning on the ties'. This energy exchange is also paid back so that the account is cleared. However, if the leaning is for longer duration or involves large amounts of energy, some compensations may also be done.

25.21 POWER POOL

Interchange of power between two systems can be beneficial to both systems. It is also possible that a system is interconnected with a number of systems. Signing up interchange agreements between each pair of systems can be a difficult task. It may be more advantageous to form a power pool. A power pool, formed by some utilities, has a central dispatch office from where the pool is administered. All tasks regarding interchange and settlements of disputes etc., are assigned to the pool office. The formation of a pool and finalisation of terms and conditions binding on the pool members is generally a complex issue. However, the formation of a pool leads to the following advantages:

1. Decrease in operating costs.
2. Savings in reserve capacity requirements.
3. Help from pool in unit commitment.
4. Minimization of costs of maintenance scheduling.
5. More reliable operation.

Formation of power pool is beset with a number of problems and constraints. Some of these are:

1. Pool agreement may be very complex.
2. Costs associated with establishing central dispatch office.
3. Reluctance of pool members to relinquish their rights as regards individual transactions.
4. Complexity as regards dealing with regulatory authorities, if pool operates in more than one state.
5. Fear of some members that their freedom will be restricted, if pool is formed.
6. Effort by each member of the pool to maximize its savings.

It is important to note that formation of pool does not merge different utilities forming the pool into one company. This is merely an arrangement to maximize the savings and attain more efficient utilization of facilities. Pool agreements always contain clauses for reimbursement to members for many of their contributions beyond the agreed ones.

Sometimes power pools are set up without a central dispatch office. The central office acts only as a brokerage house for transactions between the members. The decisions regarding payments and disbursements are, generally, the most important ones in power pools.

25.21.1 Energy Broker

Energy broker is like any other broker (share broker, property broker etc.). Many utilities find it advantageous to deal through an energy broker. The energy broker has all the necessary information regarding surplus energy available with different utilities and can arrange several transaction almost simultaneously. The energy broker has all the quotations for selling and purchasing and has the position of supply, demand and prices available with him.

One form of energy broker is bulletin board. The different utilities set up a data exchange network and send offers to buy or sell energy. The pool members access the bulletin board as and when required and contact the members whose offers are attractive. In another energy broker scheme, the members send hourly sell and buy offers to the broker who matches them and arranges the transactions. In all these transactions the seller is compensated for the incremental generation cost and the financial advantages of the transactions are shared by the buyer and seller.

Let $\frac{dC_1}{dP_1}$ = incremenntal cost of seller (Rs/MWh)

$\frac{dC_2}{dP_2}$ = decremental cost of buyer (Rs/MWh)

$$\text{Transaction rate} = 0.5\left(\frac{dC_1}{dP_1} + \frac{dC_2}{dP_2}\right) \text{Rs/MWh}$$

In the above equation the decremental cost dC_2/dP_2 is the decrease in operating cost when the generation is reduced by a small amount.

25.21.2 Allocation of Pool Savings

When energy exchange occurs between two utilities the rate can be easily decided on the basis of incremental and decremental costs as discussed above. When a power pool has been formed, all excess available energy is sold to the pool and all the requirements are also purchased from the pool. For calculations of allocations two methods are in use in USA.

1. Allocations are made in real time mode. The costs and savings are allocated periodically on the basis of incremental and decremental costs.
2. Allocations are made, after the energy transfers, on the basis of total production costs. The total costs are then added and total pool savings are calculated as the difference between this cost and the actual cost determined by central dispatch office. The pool savings are divided among members as per predetermined rules.

Example 25.14.

Two synchronous generators are operating in parallel. Their capacities are 300 MW and 400 MW. The droop characteristics of their governors are 4% and 5% from no load to full load. Assuming that they are operating at 50 Hz at no load, how would they share a load of 600 MW and what will be the frequency at this load. Assume free governor action. [*Mumbai Uni 2010*]

Solution

Let load on first generator be L_1. Then load on second is $(600 - L_1)$. The frequency when they are supplying load should be the same.

$$50 - \frac{4 \times 50}{100}\left(\frac{L_1}{300}\right) = 50 - \frac{5 \times 50}{100}\left(\frac{600 - L_1}{400}\right)$$

which gives $L_1 = 290.32$ MW = Load on first generator

$600 - L_1 = 309.68$ MW = Load on second generator.

$$\text{Frequency at load} = 50\left(1 - \frac{290.32}{300} \times \frac{4}{100}\right) = 48.06 \text{ Hz}.$$

Example. 25.15.

A 100 MVA generator is operating at full load at 50 Hz. Load suddenly reduces to 50 MW. Due to time lag in governor the steam value begins to close after 0.4 seconds. Find change in frequency. $H = 5$ kWs/kVA of generator capacity.

Solution

Kinetic energy stored at rated speed = 100 × 1000 × 5 = 5 ×10^5 kWs

Excess kinetic energy due to decrease in load = 50 × 1000 × 0.4 = 20000 kWs.

$$\text{kinetic energy} \propto (\text{frequency}^2)$$

$$\text{New frequency} = \left[\frac{5\times10^5 + 20000}{5\times10^5}\right]^{0.5} \times 50$$

$$= 50.99 \text{ Hz.}$$

$$\text{Frequency deviation} = \frac{50.99-50}{50}\times100 = 1.98\% .$$

Example. 25.16.

A 100 MVA generator is operating at full load. $H = 5$ kWsecs/KVA. The load reduces to 60 MW suddenly. Due to delay in governor action the steam value begins to close 0.35 seconds later. Find percentage frequency deviation.

Solution

Initial kinetic energy = 100 × 1000 × 5 = 5 × 10^5 kW secs.

Excess kinetic energy due to decrease in load

$$= (100 - 60) \times 1000 \times 0.35$$

$$= 14000 \text{ kWs}$$

$$\text{kinetic energy} \propto (\text{frequency}^2)$$

$$\text{New frequency} = \left[\frac{5\times10^5 + 14000}{5\times10^5}\right]^{0.5} \times 50 = 50.695 \text{ Hz}$$

$$\text{Frequency deviations} = \frac{50.695-50}{50}\times100 = 1.39\%$$

Example. 25.17.

All the generators and motors in a power system represent a total kinetic energy of 1500 MJ as measured at 50 Hz frequency. At what rate will frequency increase if the system experiences a momentary power surplus of 5 MW.

Solution

Initial kinetic energy = 1500 MJ

Excess kinetic energy in one second due to power surplus = 5 × 1 = 5 MJ

$$\text{kinetic energy} \propto (\text{frequency})^2$$

$$\text{New frequency} = \left(\frac{1500+5}{1500}\right)^{0.5} \times 50 = 50.083 \text{ Hz}$$

The change in frequency = 0.083 Hz/sec.

Hence frequency increases by 0.083 Hz/sec.

Example 25.18.

Determine primary ALFC loop parameters for a control area having following parameters.

Total rated capacity = 2000 MW

Normal operating load = 1000 MW

Inertia constant $H = 5$

Regulation $R = 2.4$ Hz/Pu MW

Assume that load increases 1% for 1% increase in frequency.

Solution

$$D = \frac{0.01 \times 1000}{0.01 \times 50} = 20 \text{ MW/Hz}$$

$$= \frac{20}{2000} = 0.01 \text{ pu MW/Hz}$$

$$K_p = \frac{1}{D} = 100 \text{ Hz/pu MW}$$

$$T_p = \frac{2H}{fD} = \frac{2 \times 5}{50 \times 0.01} = 20 \text{ seconds}$$

Example. 25.19.

A single area control system has the following data: $T_p = 10$ sec, $T_g = T_t = 0$, $K_p = 100$ Hz/puMW. $R = 3$Hz/CuMW $\Delta P_D = 0.1$ pu., $K_i = 0.1$ Compute the time error caused by the above step disturbance Prove that error is reduced by increasing K_i. Express the result in seconds and cycles if the system frequency is 50 Hz.

Solution. Using Eq. (25.15) and putting $\Delta P_D = M$

$$\Delta F(s) = \frac{-100}{10}\left[\frac{0.1}{s^2 + s\left\{\dfrac{1+\dfrac{100}{3}}{10}\right\} + \dfrac{0.1 \times 100}{10}}\right]$$

$$= -10\left[\frac{0.1}{s^2 + 3.433s + 1}\right] = \frac{-1}{s^2 + 3.433s + 1}$$

$$= \frac{-1}{(s+3.111)(s+0.3215)}$$

$$= \frac{0.358}{s+3.111} - \frac{0.358}{s+3215}$$

Taking inverse Laplace transform

$$\Delta f(t) = 0.358\,(e^{-3.111t} - e^{-.3215t})$$

Time error is obtained by integrating $\Delta f(t)$ from o to t. For total time error we integrate $\Delta f(t)$ from o to $\propto$.

$$\underset{t\to\infty}{Lt}\int_0^t \Delta f(t)dt = \underset{s\to 0}{Lt}\,[\Delta f(s)] = \underset{s\to 0}{Lt}\,\frac{-1}{s^2+3.433s+1}$$

$$= -1 \text{ cycle}$$

For 50 Hz frequency.

$$\text{Total time error} = 1 \text{ cycle} = \frac{1}{50} = 0.02 \text{ secs.}$$

It is seen from Eq. (25.15) that time error is inversely proportional to K_i. Hence increasing K_i decreases time error.

Example 25.20.

Two 50 Hz power stations are connected by an interconnecting cable of negligible impedence so that the stations operate in parallel. Station *A* has a full load capacity of 15 MW and speed regulation of 3%. Station *B* has a capacity of 4 MW and speed regulation of 4%. The load on bus bar of station *B* is 4 MW and that on *A* is 10 MW. Find generated output of each station, power transmitted by cable and operating frequency. Assume free governor action and no load frequency of 50 Hz at both stations.

Solution

Total load = 14 MW

Let load on station $A = L_1$ MW

Load on station $B = (14 - L_1)$ MW

$$\text{Operating frequency of } A = 50 - \frac{0.03}{15} L_1$$

$$\text{Operating frequency of } B = 50 - (14 - L_1)\left(\frac{0.04}{4}\right)$$

Since they are operating in parallel, operating frequency is the same. Thus,

$$50 - 50 \times \frac{0.03}{15} L_1 = 50 - (14 - L_1)\left(\frac{0.04}{4}\right)(50)$$

Solution is $L_1 = 11.67$ MW

Thus generation at station $A = 11.67$ MW

Generation at station $B = (14 - 11.67) = 2.33$ MW

Power transmitted by cable $= 11.67 - 10 = 1.67$ MW from station A to B.

$$\text{Operating frequency} = 50 - \frac{0.03}{15}(11.67)(50)$$

$$= 48.833 \text{ Hz.}$$

Example 25.21.

Two generators rated 300 MW and 400 MW are operating in parallel. The droop characteristics of their governors are 4% and 6% respectively from no load to full load. The speed changes are set so that a load of 400 MW is shared among the generators at 50 Hz in the ratio of their ratings. What are the no load frequencies of the generators.

Solution

$$\text{Load on 300 MW generator} = \frac{400 \times 300}{300 + 400} = 171.43 \text{ MW}$$

$$\text{Load on 400 M generator} = \frac{400 \times 400}{700} = 228.57 \text{ MW}$$

No load frequency of 300 MW generator is given by $f_0 \left(\dfrac{1 - 0.04 \times 171.43}{300}\right) = 50$

or $\quad f_{01} = 51.17$ Hz

No load frequency of 400 MW generator is given by

$$f_{02} \frac{(1 - 0.06 \times 228.57)}{400} = 50$$

or $\quad f_{02} = 51.74$ Hz.

25.22 AUTOMATIC LOAD SHEDDING AND RESTORATION

A power system should be so designed and operated that uncontrolled widespread power interruptions do not affect its operation. The goal is to design the system to meet a specified security level and avoid widespread system cascading disturbances. Regardless of this goal, conditions can arise in the system which result is large disproportionality between available generation and load. Generators are frequency sensitive and fail when frequency drops below a critical value. In the event of the load being more than generation (which may be due to an outage of a generating unit or a sudden unexpected increase in load) a situation of declining frequency will develop. To arrest this frequency decline-for situations when the rate of change is rapid and the magnitude is significant-means must be provided to rapidly balance load and available generation. This can be accomplished by shedding specific amounts of load utilizing frequency sensitive relays. In the usual application of load shedding programme a predetermined amount of load is shed over a specified frequency range. To implement this programme solid state frequency relays are used. These relays have a high degree of accuracy and can be calibrated at frequency settings within a narrow band. A typical load shedding programme is

Relay Trip Setting	Frequency	Percentage of Load Shed
1	49.5 Hz	$3\frac{1}{3}\%$
2	49.4 Hz	$3\frac{1}{3}\%$
3	49.3 Hz	$3\frac{1}{3}\%$
4	49.1 Hz	5%
5	49 Hz	5%
6	48.9 Hz	5%

When the frequency drops to 49.5 Hz, $3\frac{1}{3}\%$ of the load is shed. A further fall in frequency to 49.4 Hz results in a further $3\frac{1}{3}\%$ load shedding and so on.

An important aspect of load shedding is the selection of circuits to be disconnected. For this purpose a service priority rating is established for different circuits. Medical facilities, radio and TV stations, communication facilities, defence installations etc. are allotted a high priority rating. The circuit having the lowest priority rating is the first one to be disconnected. In the second step the circuit having the next higher rating is disconnected and so on.

Along with an automatic load shedding programme an efficient means must be provided to restore the load quickly when the system state is such that the load can be accepted. Restoration can be done manually or through frequency sensitive relays. The automatic restoration has the advantage of being self-protecting since its operation depends on the system capability to accept load. When capability is exceeded, load restoration stops. The restoration is also done in stages. Each restoration stage should occur when the frequency has come back to 50 Hz and there should be isolation between stages through time delays. The restored loads are generally reconnected in the sequence that they were shed. The stages which are designed to shed first are restored first in order to minimize the transient in the event of subsequent disturbance. The amount of restoration per stage should be less than one third of the amount of shedding per stage. This is necessary to ensure that relaxation oscillations do not occur due to the reaction between load shedding and restoration transients.

25.23 POWER SYSTEM INTERCONNECTION IN INDIA

India has a vast power system network consisting of a large number of hydro, thermal and other power plants and 765 kV, 400 kV, 220 kV and 132 kV transmission lines and sub-stations. It is proposed to integrate the whole system to ensure the benefits of interconnection.

The load despatch system operates at four levels. At level 1 is the national load despatch centre (NLDC) at Delhi. At level 2 are the five regional load despatch centres: northern (NRLDC) at Delhi, western (WRLDC) at Bombay, southern (SRLDC) at Bangalore, eastern (ERLDC) at Calcutta and north eastern (NERLDC) at Shillong. Level 3 consists of state load despatch centres (SLDC) of all states. The load despatch centres at central projects are also included in this level. At level 4 are area or sub-load dispatch centres (sub-LDC).

The national load despatch centre is being set up. However load despatch centres at the other 3 levels have been set up. Each regional load despatch centre co-ordinates the activities of the state load despatch centres of that region. Tie lines exist between different regions for transfer of power.

The functions being performed by each regional electricity board are:

1. To plan and review the progress of power development programs in the region.
2. To ensure integrated operation of the power system in the region so that the generation and load requirements match at every instant and maximum possible benefits accrue to the region.
3. To prepare and co-ordinate maintenance scheduling of the power plants in the region.
4. To determine optimal generation scheduling at different power plants in the region.
5. To determine power transfer between different states in the region under different contingency conditions.
6. To evolve suitable tariffs for exchange of power between different states in the region.
7. To plan and monitor the installation of reactive power compensation in the region.
8. To carry out system studies for operational strategies and system losses.

The regional load despatch centres are being equipped with speech, telemetry, teleprinter and back up communication facilities, on line real time computer system and peripheral devices and under frequency relays for automatic load shedding.

Some problems in satisfactory working of system interconnection in India are as under.

1. *Lack of operational discipline among constituents*. Each constituent in a region is expected to maintain a balance between its own generation (plus share from central projects) and demand. However due to overall shortage in installed capacity, the constituents tend to overdraw share from central stations. This leads to a disturbance in grid parameters and system stability problems.

2. *Frequency regulation*. It is known that for proper system operation frequency should be maintained at least between 49.8 Hz and 50.2 Hz. However due to acute shortage during peak load conditions frequency tends to be as low as 48.5 Hz. During off-peak conditions frequency goes up to about 51 Hz. Proper load management, use of under frequency relays for automatic load shedding and accelerated setting up of new generation units is needed.

3. *Voltage control*. It is necessary that a proper voltage profile is maintained in the system. However due to reactive power unbalance, very low and high voltage conditions are common. The problem of low voltage is especially severe in summer season due to inductive loads of air conditioners, coolers, fans etc. and in agricultural period due to inductive load of pumps. It is known that these problems can be tackled by meeting reactive power requirement at the load centres. The programme of installation of shunt capacitors needs a big thrust.

4. *Load dispatch and communication*. The present load dispatch and communication facilities are inadequate. The present power line carries communication (PLCC) system cannot be expanded due to congestion in the available frequency spectrum, difficulty in cascade tandem operation etc. It is proposed to add multi channel digital microwave system and fibre optic communication system to provide better load dispatch and communication facilities.

5. *Lack of Spinning reserve*. At the time of peak loads, the demand is more than the generation and there is no spinning reserve in the system. Better load management and installation of additional generation facilities is needed.

6. *Metering and Instrumentation*. Adequate metering and instrumentation is needed at different levels. It is necessary to install sequential recorders, disturbance recorders, time of day meters etc.

7. *Inadequate transmission facilities*. The transmission system has not been strengthened to proper level. This has resulted in problems of power evacuation from large pithead super thermal stations. Additional investment in transmission system is needed. There should be adequate redundancy in transmission system for satisfactory operation of grid.

Example 25.22.

Two power plants A and B are interconnected by a short line. Capacity of A is 200 MW and that of B is 100 MW. Their speed regulations (from no load to full load) are 1.5% and 3% respectively. The load on bus of each station is 100 MW. Find the generation at each plant and transfer of power through line.

Solution

Speed regulation of plant A is

$$R_A = \frac{1.5 \times 50}{100 \times 200} = 0.00375 \text{ Hz/MW}$$

Speed regulation of plant B is

$$R_B = \frac{3 \times 50}{100 \times 100} = 0.015 \text{ Hz/MW}$$

Let generation at plant A be P_A. Then generation at plant B is $2 \times 100 - P_A$

$$0.00375\, P_A = 0.015\,(200 - P_A)$$

or
$$P_A = 160 \text{ MW}, \qquad P_B = 40 \text{ MW}$$

Transfer of power from plant A to B over the interconnector

$$= 160 - 100 = 60 \text{ MW}$$

Example 25.23.

A short line having an impedance of $1.5 + j\,2.5$ ohm interconnects two plants A and B. Each plant operates at 11 kV and the two voltages are equal in magnitude and phase. It is proposed to transfer 20 MW power at 0.8 p.f. from plant A to B. Find the voltage boost needed at plant A to achieve this power transfer.

Solution

The current transfer is

$$I = \frac{20 \times 10^3}{\sqrt{3} \times 11 \times 0.8} = 1312.2 \angle -36.87°$$

$$= 1312.2\,(0.8 - j\,0.6) = 1049.76 - j\,787.32 \text{ A}$$

Voltage drop in the interconnector

$$= (1049.76 - j\,787.32)\,(1.5 + j\,2.5)$$

$$= 3542.94 + j1443.42 \text{ V}$$

The voltage boost needed at station A is $3542.94 + j1443.42$ V.

Example 25.24.

A 3 phase interconnector connecting two stations has an impedance of $(3 + j\,9)$ per cent. Find the phase angle between two station voltages and pu real and reactive power received when the voltages at two stations are (*a*) 1 pu each (*b*) 1.05 and 1 pu. The current in interconnector is rated full load current.

Solution

$$Z = 3 + j\,9\% = 0.03 + j\,0.09 \text{ p.u.}$$

$$I = 1 \text{ p.u.}$$

$$IZ = 0.03 + j\,0.09 = 0.095\ \angle 71.60°$$

(*a*) The phasor diagram is shown in Fig. 25.15

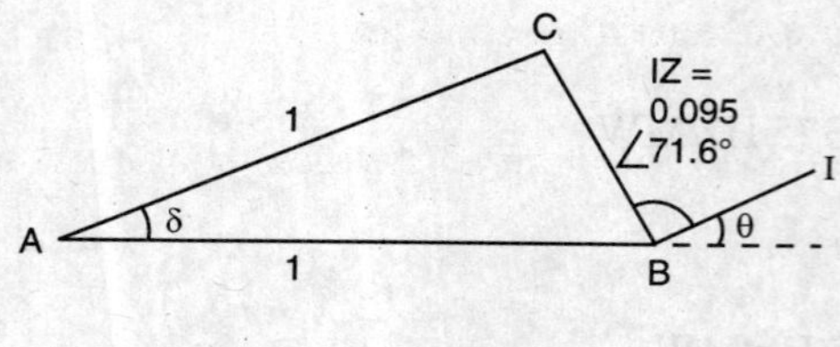

Fig. 25.15

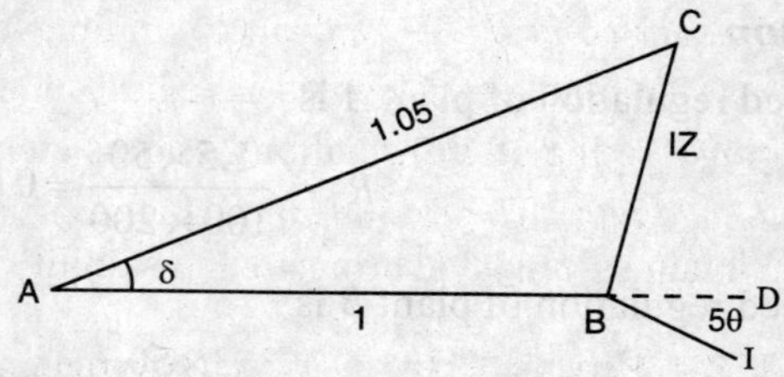

Fig. 25.16

$$1^2 + 1^2 - 2 \times 1 \times 1 \times \cos\delta = (0.095)^2$$

The solution is $\quad \cos\delta = 5.445°$

Moreover

$$1^2 = 1^2 + (0.095)^2 - 2 \times 1 \times 0.095\ (\cos \angle ABC)$$

The solution is $\quad \angle ABC = 87.277°$

$$\angle\theta = 180 - 87.227° - 71.6 = 21.1°$$

$$\text{Real power transfer} = 1 \times 1 \times \cos\theta = 0.933 \text{ p.u.}$$

$$\text{Reactive power transfer} = 1 \times 1 \times \sin\theta = 0.36 \text{ p.u. leading}$$

(*b*) The phasor diagram is shown in Fig. 25.16

$$1.05^2 + 1^2 - 2 \times 1.05 \times 1 \times \cos\delta = (0.095)^2$$

The solution is $\delta = 45°$

Moreover

$$1.05^2 = 1^2 + (0.095)^2 - 2 \times 1 \times 0.095\ (\cos \angle DBC)$$

The solution is $\angle DBC = 60.53°$

$$\angle\theta = 71.6 - 60.53 = 11.07°$$

$$\text{Real power} = 1 \times 1 \times \cos 11.07° = 0.9814 \text{ p.u.}$$

$$\text{Reactive power} = 1 \times 1 \times \sin 11.07° = 0.192 \text{ p.u. lagging}$$

25.24 ROLE OF INFORMATION TECHNOLOGY IN ENERGY MANAGEMENT

Proper and efficient energy management requires lot of data and information about operating conditions . Computers are extensively used for this purpose. The major areas in which computers can help are as under:

25.24.1 Supervisory Control and Data Acquisition System (SCADA)

This system has a master station which communicates with remote terminal units (RTU). These units send the status and measurements to master station which sends the control commands. Tasks of SCADA are:

Data acquisition : To provide data, measurements and status information to operator.

Plotting : To plot different measurements in real time regime.

Supervisory control : To control circuit breakers remotely.

Alarms : To send alarm signals as regards undesirable operating conditions.

Logging : To log all informations, alarms etc.

Load Shedding : To provide automatic and manual load shedding in emergent conditions so as to maintain system synchronism and stability.

25.24.2 Power Plant Maintenance

All the required information as regarding maintenance schedules, maintenance costs, availability of spares and material and equipment performance can be easily stored and called back immediately when required. This data base is essential for proper plant maintenance.

25.24.3 Automatic Generation Control (AGC)

AGC helps in proper system operation. The generation is matched against demand and units operated as per economic dispatch. Load frequency control is used for matching real power with demand. Power allocation between different generating units is done as per equal incremental costs criterion. Purchase and sale of power to neighbouring systems is decided in real time so that the system operates optimally.

AGC also looks after automatic dispatch including the effect of transmission losses. The generation from different units is optimised taking into account their incremental costs and location (the location affects the transmission loss).

25.24.4 Fuel Management

Cost of fuel is a major part of generation cost of thermal and gas turbine plants. Proper fuel management requires information about load forecast, generation scheduling, fuel costs, transportation cost, fuel contracts, fuel usage, etc. All this information is stored in the computer and recalled when required.

25.24.5 Load Management

Load management is essential for proper and economic operation of a system. All utilities employ one or more load management techniques to flatten the load curve. SCADA provides on line real time information about loads on different feeders and positions of sectionalising switches, circuit configurations, voltage levels, TOD tariff rates and uses this information to control consumer loads.

25.24.6 Security Control

Security control system has the following functions :

1. To determine system state.
2. To prepare list of contingencies along with action to be taken in each contingency.
3. To determine preventive actions so as to avoid contingencies.

25.24.7 Distribution Automation System

With computerised control, the operator has all the information as regards voltage profile, currents power factors etc. This information is obtained through telemetry from all the RTUS. The operator can issue instructions based on this information, for optimising power flows, switch in/out shunt capacitors, change transformer tappings etc. All these actions result in better service to consumers, improve voltage at consumer premesis, improve reliability, reduce losses. Distribution automation is being used in all advanced countries and has been made possible only because of information technology.

25.24.8 Metering and Billing

Intelligent metering to detect theft and tampering is possible only because of information technology. Computerised billing when combined with intelligent metering leads to neat and transparent accounts and elimination of human error. Spot billing is also possible. Information like average electricity use, history of payments, change in customer's electricity use with season are readily available. Faster credit control and better financial management is possible.

25.25 DEFINITION OF REACTIVE POWER

Reactive power is an important quantity in *ac* networks. To define reactive power let us consider a series $R - L$ circuit excited by a single phase ac source. Let i be the instantaneous current and v_L be the voltage drop across inductance. The product of i and v_L is q_x and is found as under.

$$v = V_m \sin \omega t \qquad \text{... (25.44)}$$

$$i = I_m \sin (\omega t - \theta) \qquad \text{... (25.45)}$$

where θ is the power factor angle and equals $\tan^{-1} (X/R)$

$$q = i\, v_L = i\left(L \frac{di}{dt} \right)$$

$$= [I_m \sin (\omega t - \theta)][\omega L\, I_m \cos (\omega t - \theta)]$$

$$= \frac{I_m^2\, \omega L}{2} \sin 2 (\omega t - \theta)$$

or

$$q = I^2\, \omega L \sin 2 (\omega t - \theta) \qquad \text{... (25.46)}$$

The average of θ over one cycle is zero. During certain part of cycle energy is supplied by the source to the inductance (and stored in the magnetic field). During the remaining part of the cycle this energy is returned to the source. The amplitude of this double frequency sine wave of Eq. (25.46) is $I^2\omega L$ or $I^2 X_L$. It is known as reactive power (symbol Q) and has the units of vars (volt – ampere reactive). Bigger units for Q are k var and M var. Thus

$$Q = I^2 X_L = I^2 Z \sin \theta = VI \sin \theta \qquad \text{... (25.47)}$$

The reactive power does not contribute anything as far as work done or energy transferred from source to the device is concerned. Yet it contributes to the loading of the equipment. The loading of equipment is proportional to I and hence proportional to apparent power IV. The reactive power is a component of apparent power.

In a single phase circuit, the complex power S is

$$S = VI^* = (V\angle O)\ (I\angle -\theta)^* = (V\angle O)\ (I\angle \theta)$$

or
$$S = P + jQ = VI\angle\theta$$

or
$$S = VI\cos\theta + jVI\sin\theta \quad \text{... (25.48)}$$

$$P = Re[S] = VI\cos\theta \quad \text{... (25.49)}$$

$$Q = Im[S] = VI\sin\theta \quad \text{... (25.50)}$$

In a 3 phase circuit

$$P = 3V_{\text{phase}}\ I_{\text{phase}}\ \cos\theta = \sqrt{3}\,V_L\,I_L\cos\theta \quad \text{... (25.51)}$$

$$Q = 3V_{\text{phase}}\ I_{\text{phase}}\ \sin\theta = \sqrt{3}\,V_L\,I_L\sin\theta \quad \text{... (25.52)}$$

25.26 POSITIVE AND NEGATIVE REACTIVE POWER

We have seen the reactive power associated with an inductor in section 6.1. An *RC* circuit also behaves in a similar way. But one fundamental difference exists between an $R-L$ circuit and an $R-C$ circuit. The current in an $R-L$ circuit lags the voltage but the current in an $R-C$ circuit leads the voltage. The reactive power associated with an inductor in I^2X_L and is assigned positive sign. The reactive power associated with a capacitor is I^2X_C and is assigned negative sign. Thus the complex power associated with an $R-L$ circuit is $P+jQ$ while the complex power associated with an *RC* circuit is $P-jQ$. For both the positive and negative reactive power, the units are the same vars, kvars and Mvars.

Fig. 25.17 shows a 3 phase *ac* system feeding a network. The network may also have active sources in addition to passive elements.

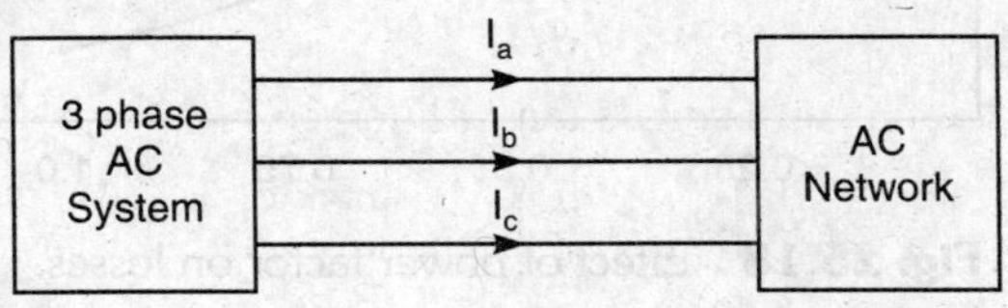

Fig. 25.17 3 Phase AC System feeding an ac network.

The significance of positive and negative P and Q is as under.

P positive	*AC* system supplies real (active power) to network.
P negative	*AC* network supplies real (active) power to 3 phase *AC* system.
Q positive	*AC* system supplies lagging reactive power to *AC* network. The network consists of elements which are predominantly inductive.
Q negative	*AC* system supplies leading reactive power to *AC* network. The network consists of elements which are predominantly capacitive. In other words the ac network is generating positive reactive power and supplying it to the 3 phase ac system.

25.27 SYSTEM LOSSES DUE TO REACTIVE POWER FLOW

As seen in section 25.25 the real power in a 3 phase ac system is $\sqrt{3}\ V_L I_L \cos\theta$. Therefore, current I_L (for a certain value of P) is

$$I_L = \frac{P}{\sqrt{3}\ V_L \cos\theta} \qquad \text{... (25.53)}$$

As seen in Eq. (25.53) the circuit current I_L is minimum if cos θ is unity. As power factor cos θ becomes less than 1, the circuit current I_L increases. The system losses are $3I_L^2R$ where R is the resistance. At unity power factor the losses are minimum. As power factor cos θ becomes less than 1, the losses increase. If cos θ is 0.5 the current increases to twice the value of the current at unity power factor. Since losses are proportional to square of current, the losses at 0.5 power factor are four times the losses at unity power factor. Fig. 25.18 shows the increase in losses due to decrease in power factor. When power factor is unity, the losses are minimum.

When p.f. is 0.5, the current (for the same active power P) becomes double and losses become four times. When *p.f.* is 0.25, the current is four times and losses are 16 times. In actual practice the p.f. is around 0.8 and losses are about 56% higher than the losses at unity power factor.

The system has to transmit reactive power due to the fact that most loads operate at lagging power factor.

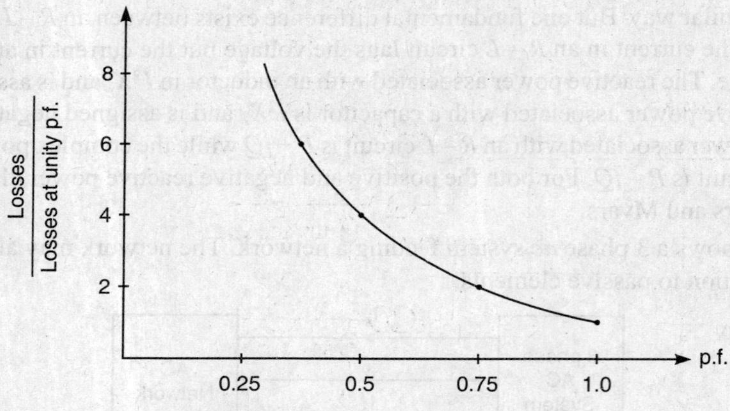

Fig. 25.18 Effect of power factor on losses.

25.28 REACTIVE POWER REQUIREMENTS DURING PEAK AND OFF-PEAK HOURS

We have discussed the variable load problem in earlier chapters. Only the real power (MW) demand was considered there. In power system operation the variable load problem becomes all the more complex because the reactive power requirements are likely to be different at different times of the day and in different seasons. The reactive power requirements have to be met to keep the system voltages at proper levels.

A typical power system is shown in Fig. 25.19. The power system operates at different voltage levels viz. 400 kV, 220 kV, 132 kV, 66 kV, 33 kV and 11 kV. Large, modern and efficient generators are connected to 400 and 220 kV buses. Smaller, olders and less efficient generators are connected to 132 and 66 kV buses. The 33 and 11 kV voltage levels represent distribution networks. Some times some local generators are also connected to 33 and 11 kV buses. The different voltage levels are interconnected through transformers and distribution networks.

During peak load hours it may be necessary to run almost all the available generating units to meet the system requirements. During off peak hours the less efficient plants would be shut down so that generation cost is kept to the minimum. Each generator is equipped with automatic voltage regulator so that the voltage at different buses is kept within limits at all system loading conditions.

Most loads need lagging reactive power to supply the magnetising current of transformers, induction motors etc. As the active power requirements increase the reactive power requirements also increase. Moreover the reactive power requirements are larger in summer than in winter. At any moment of time the var demand on the network is closely related in magnitude and location to the voltage at consumer terminals.

At times of peak load the vars absorbed by the series inductance of lines are larger than vars generated by shunt capacitance of lines. In an extreme case it may be necessary for the line to import vars at both ends (*i.e.,* the delivered load is at leading power factor) in order to reduce the voltage drop. In addition the consumers demands for lagging vars must be met. It is therefore necessary that generators deliver enough vars to the system to meet the requirements. Since the generator rated power factor is around 0.8 lagging, there is a limit to the maximum vars which the generators can deliver to the system. This necessitates the use of shunt capacitors and synchronous condensers for generating lagging vars.

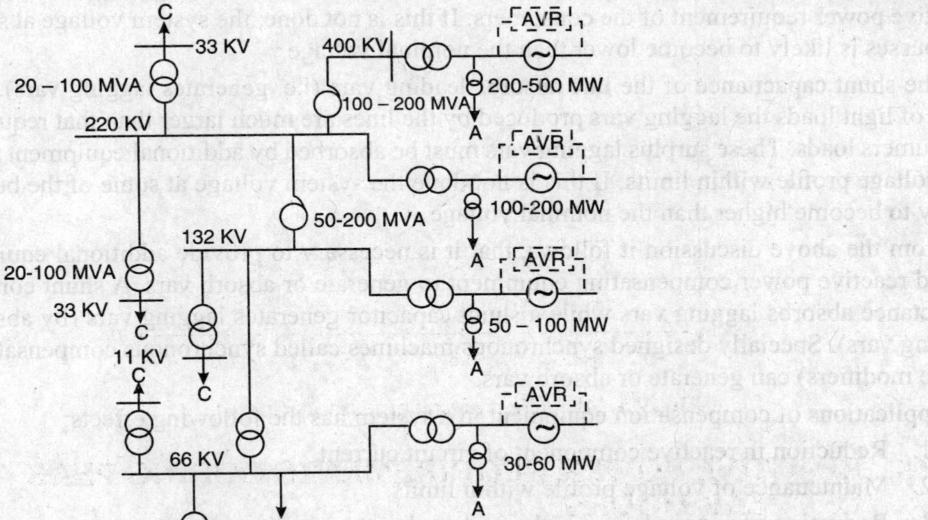

Fig. 25.19 A typical power system.

During off peak period the line generates lagging vars (*i.e.,* absorbs leading vars). Generally the lagging vars produced by the lines are many times that absorbed by the consumers under the off peak conditions. Therefore the remaining lagging vars must be absorbed by running the generators under-excited. Sometimes even this may not be sufficient and it may be necessary to adopt additional means like switching out unnecessary lines (especially cables), staggering taps on paralleled transformers and installation of shunt reactors. Generally it has been found that balance between generation and absorption of reactive power during off peak hours is more difficult than during peak load hours.

We have discussed the problem of maintaining balance between total system generation and system load plus transmission losses. The other problem in power system is to maintain balance between production and absorption of logging vars. Having identified the problem of vars balance we proceed to discuss the sources of vars and their use in power systems.

25.29 LINE VOLTAGE REGULATION AND COMPENSATION

The voltage, relative to nominal voltage, at the different busses of the power system is known as voltage profile. Most of the equipments, especially at 400 kV level and above are operated close to the present limits of design. In view of this the modern power systems are operated with a voltage profile of ±5%.

If follows that the transmission lines have to transfer power with a voltage drop of less than 10%. The major cause of voltage drop in the line is the flow of reactive power. Moreover reactive currents cause I^2R losses in the system but produce no revenue.

Most loads absorb positive (*i.e.*, lagging) vars to supply the magnetizing current of transformers, induction motors, etc. At any moment the maximum vars which can be transferred over the lines are fixed by the voltage profile. If follows that at times of peak loads the vars demanded by the loads greatly exceeds the vars which can be transmitted over the lines from the consideration of voltage drop. Therefore, additional equipment is necessary to generate lagging vars to meet the reactive power requirement of the consumers. If this is not done, the system voltage at some of the busses is likely to become lower than the nominal voltage.

The shunt capacitance of the line absorbs leading vars (i.e. generates lagging vars). At the time of light loads the lagging vars produced by the lines are much larger than that required by consumers loads. These surplus lagging vars must be absorbed by additional equipment to keep the voltage profile within limits. If this is not done the system voltage at some of the busses is likely to become higher than the nominal voltage.

From the above discussion it follows, that it is necessary to provide additional equipment, called reactive power compensating equipment to generate or absorb vars. A shunt connected inductance absorbs lagging vars while a shunt capacitor generates lagging vars (by absorbing leading vars). Specially designed synchronous machines called synchronous compensators (or phase modifiers) can generate or absorb vars.

Applications of compensation equipment in a system has the following effects:

1. Reduction in reactive component of circuit current.
2. Maintenance of voltage profile within limits.
3. Reduction of copper losses in the system due to reduction of current.
4. Reduction in investment in the system per kW of load supplied.
5. Decrease in kVA loading of generators and circuits. This decrease in kVA loading may relieve an overload condition or release capacity for additional load growth.
6. Improvement in power factor of generators.
7. Reduction in kVA demand charges for large consumers.

25.30 SOURCES OF REACTIVE POWER (VARS)

The different sources of reactive power (vars) are

1. Synchronous machines

2. Shunt static capacitors/shunt compensation
3. Series capacitors/Series compensation
4. Synchronous condensers
5. Shunt Reactors
6. Facts Controllers

25.31 VAR GENERATION BY SYNCHRONOUS MACHINES

(*a*) Synchronous Generators

Fig. 25.20 shows the equivalent circuit of cylindrical rotor synchronous generator under steady state conditions. E is the induced emf, R is armature winding resistance, X_s is synchronous reactance (per phase) and I is current per phase.

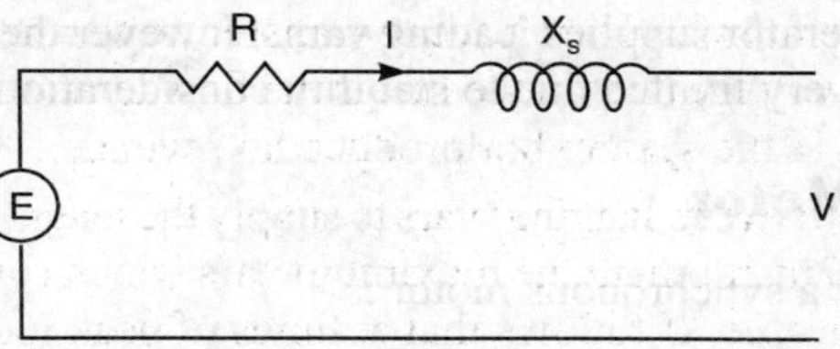

Fig. 25.20 Equivalent circuit of cylindrical rotor alternator.

$$E = V + I\,(R + jX_s) \qquad \text{... (25.54)}$$

Fig. 25.21. shows the phasor diagram at lagging and leading power factor conditions.

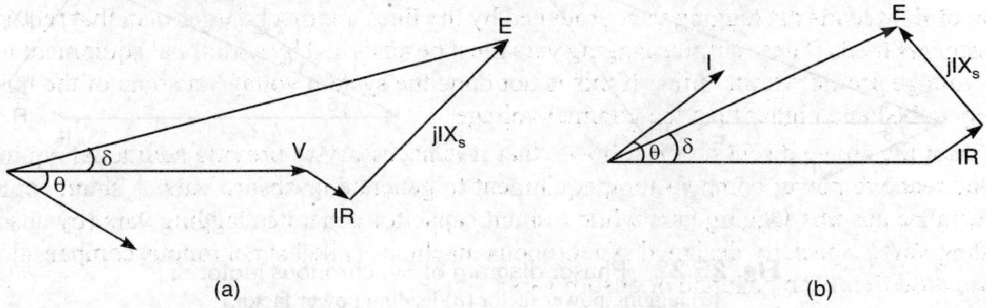

Fig. 25.21 Phasor diagram of cylindrical rotor alternator at (*a*) lagging power factor (*b*) leading power factor.

It is seen that the excitation E under lagging power factor condition is more than V where as E, under leading power factor condition is less than V. This phasor diagram shows that excitation of the alternator has a strong influence on the power factor of the machine.

The conclusions are

1. At unity power factor the generator is delivering only active power and zero reactive power. However at power factor other than unity, the alternator delivers active power as well as reactive power.
2. As the power factor departs more and more from unity, the amount of active power decreases and the amount of reactive power increases. The following table shows the MW and Mvar available from a 100 MVA alternator at different power factors.

Power factor	1	0.9	0.8	0.7	0.5
MW	100	90	80	70	50
Mvar	0	43	60	70	86.5

It is seen from the above table that an alternator can be a big source of vars supply. However it must be remembered that as the vars output of an alternator increases, the MW output decreases. A decrease of active power output means that more generators would be needed to meet the real power demand of the consumers. As such it is not advisable to use the generator as sources of vars. Generally the generator is rated at 0.8 pf. and the power factor should be kept as near to 0.8 as possible. The adjustment of power factor can be done by controlling the excitation of the machine.

3. At lagging power factor i.e. when the generator is overexcited ($E > V$) the generator supplies lagging vars. At leading power factor i.e. when the generator is under excited ($E < V$), the generator supplies leading vars. However the ability of generator to supply leading vars is very limited due to stability considerations.

(*b*) Synchronous Motor

The voltage equation for a synchronous motor is

$$V = E + IR + jIX_s \qquad \text{... (25.55)}$$

Fig. 25.22 shows the phasor diagram under lagging and leading power factor conditions.

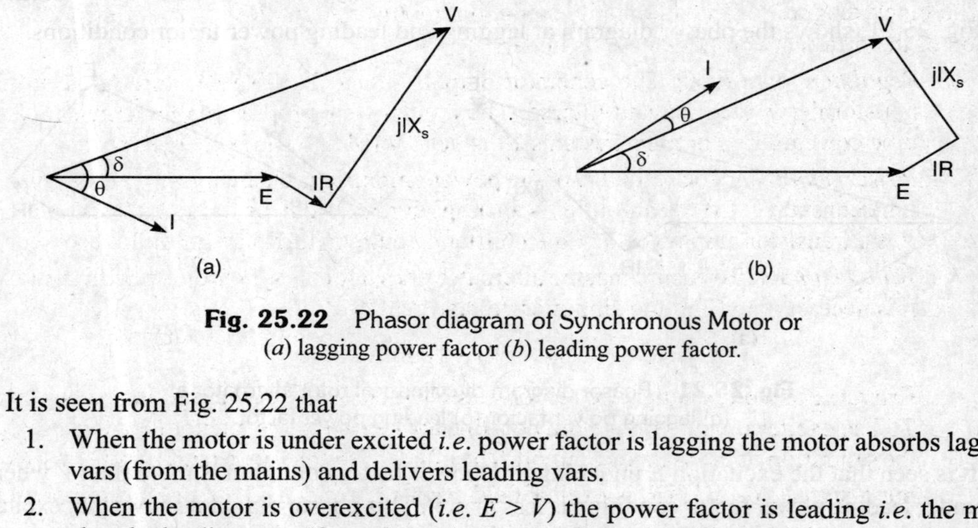

Fig. 25.22 Phasor diagram of Synchronous Motor or (*a*) lagging power factor (*b*) leading power factor.

It is seen from Fig. 25.22 that

1. When the motor is under excited *i.e.* power factor is lagging the motor absorbs lagging vars (from the mains) and delivers leading vars.
2. When the motor is overexcited (*i.e.* $E > V$) the power factor is leading *i.e.* the motor absorbs leading vars (from the mains) and delivers lagging vars.
3. Synchronous motor can be made to deliver lagging or leading vars to the system as per the system requirement. This can be easily done by controlling the excitation.
4. Synchronous motors which do not supply any mechanical load and are used for supply of vars are known as synchronous condensers.

25.32 AUTOMATIC EXCITATION CONTROL (IEEE Type I)

The reactive power balance is maintained through an automatic control of excitation of the alternators. As discussed in sections 6.10 the reactive power balance affects the bus voltage magnitude. The automatic voltage regulators sense the change in the alternator terminal voltage and bring it back to the rated value by adjusting the alternator excitation.

The output voltage of alternator obtained through a potential transformer is rectified, filtered and compared with a reference voltage V_i. The voltage error signal is fed to an amplifier (which may be an amplidyne, a magnetic amplifier or a transistor amplifier). The amplifier output is fed to the exciter field circuit and controls the exciter field current through a buck or boost action. A control of exciter field current means a control of excitation current of alternator.

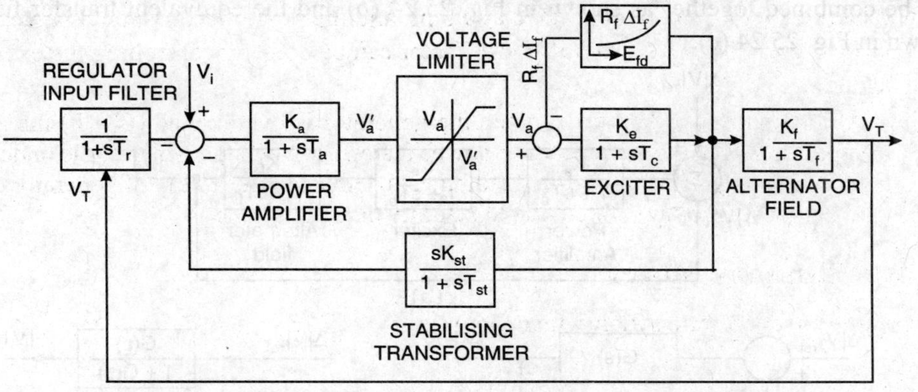

Fig. 25.23 Block diagram of automatic excitation control for alternators.

The block diagram of the automatic excitation control system is shown in Fig. 25.23. The action of the different components is as under:

1. *Regulator input filter*. The generator output voltage V_T, obtained through a potential transformer is rectified and filtered. The regulator input filter can be represented by a time constant T_r. For most systems T_r is very small and can be neglected.
2. *Power amplifier*. The response of the power amplifier is characterised by a gain K_a and time constant T_a. If the amplifier is an amplidyne, T_a may be around 0.2 seconds but if it is a transistor amplifier, the time constant would be less than 20 milli-seconds.
3. *Voltage limiter*. To ensure that the alternator does not lose synchronism with the system, it is necessary to limit the amplifier voltage such that
$$V_{min} \leq V_a \leq V_{max}.$$
4. *Exciter*. For a d.c. exciter, the response is characterised by a gain K_e and time constant T_e. The exciter output voltage is a non-linear function of the exciter field current due to the saturation of the magnetic circuit. The effect of saturation is represented by subtracting a signal $R_f \,\Delta I_f$ (R_f is exciter field resistance and ΔI_f represents the incremental change in exciter field current) from the voltage V_a.
5. *Stabilizing transformer*. To improve the overall response, a stabilizing loop (which consists of a transformer) is added to the control circuit. The stabilizing transformer feeds back a fraction of the derivative (hence *s* in the numerator of the transfer function of stabilizing transformer) of exciter output to the amplifier.
6. To complete the control circuit, the transfer function of the alternator field has to be added to the block diagram. For open circuit conditions the alternator field response can be represented by a gain K_f and time constant T_f.

25.33 RESPONSE OF AUTOMATIC EXCITATION CONTROL SYSTEM

25.33.1 Static Response

To analysis the performance of the automatic excitation system the block diagram of Fig. 25.23 can be simplified by neglecting the time constant of the regulator input filter and the effect of voltage limiter and stabilising transformer. This is shown in Fig. 25.24 (*a*). The different blocks can be combined together as shown in Fig. 25.24 (*b*) and the equivalent transfer function is shown in Fig. 25.24 (*c*).

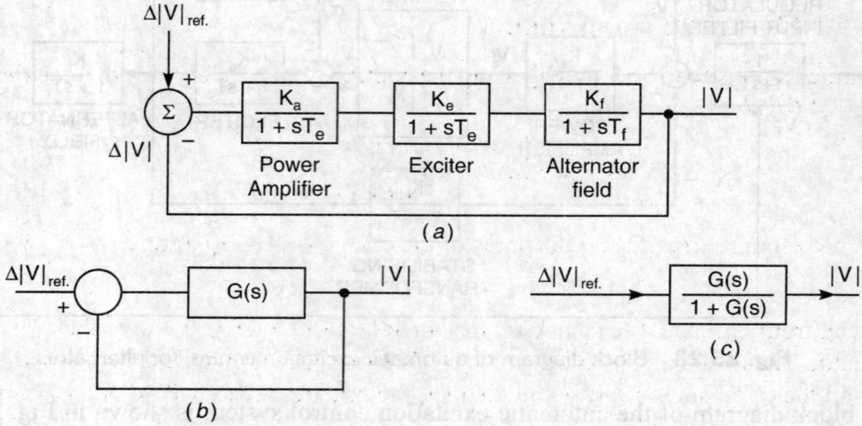

Fig. 25.24 Automatic Excitation System of Alternator (*a*) Block diagram (*b*) After combining blocks (*c*) Equivalent transfer function

It is seen from Fig. 25.24 (*a* and *b*) that

$$G(s) = \frac{(K_a)\,(K_e)\,(k_f)}{(1+sT_a)\,(1+sT_e)\,(1+sT_f)}$$

or

$$G(s) = \frac{K}{(1+sT_a)\,(1+sT_e)\,(1+sT_f)} \quad \text{... (25.56)}$$

Where K is the open loop gain and is equal to $K_a.K_e.K_f$.

The automatic excitation system must

1. Keep the terminal voltage $|V|$ constant (*i.e.* within specified limits)
2. Should respond quickly whenever $|V|$ tends to change.
3. Should be stable.

Since the terminal voltage $|V|$ cannot be held absolutely constant the only possibility is that the error Δe should should be less than a certain percentage x of $\Delta|V|_{ref}$. Writing the constant values with subscript 0 we get

$$\Delta e_0 = \Delta|V|_{ref,0} - \Delta|V|_0 < \frac{x}{100}\cdot\Delta|V|_{ref\ 0} \quad \text{... (25.57)}$$

For steady state conditions, the transfer function is obtained by setting $s = 0$. Then

$$\Delta e_0 = \Delta|V|_{ref,0} - \Delta|V|_0 = \Delta|V|_{ref,\,0} - \frac{G(0)}{1+G(0)}\Delta|V|_{ref,\,0}$$

or
$$\Delta e_0 = \Delta |V|_{ref,\,0}\left[\frac{1}{1+G(0)}\right] = \Delta |V|_{ref,\,0}\left[\frac{1}{1+K}\right] \quad \text{... (25.58)}$$

Eq. (25.58) shows that if loop gain K is increased, the static error Δe_0 decreases. For a specified accuracy of x%, the loop gain is obtained by substitution Eq. (25.58) in Eq. (25.57). Thus

$$K > \frac{100}{x} - 1 \quad \text{... (25.59)}$$

If it is desired that x should be less than 0.5%, K should at least be equal to 199.

25.33.2 Dynamic Response

The dynamic response can be found by using the control system theory. For the transfer function of Fig. 25.24 (*c*), the time response is

$$\Delta |V|(t) = \alpha^{-1}\left[\Delta V_{\text{ref}}(s)\frac{G(s)}{1+G(s)}\right] \quad \text{... (25.60)}$$

The time response depends on the poles of the characteristic equation *i.e.*

$$1 + G(s) = 0 \quad \text{... (25.61)}$$

It is seen from Fig. 25.24 (*a*) that $G(s)$ has three terms $(1 + sT_a)$, $(1 + sT_e)$ and $(1 + sT_f)$ in the denominator. Thus we will get 3 poles s_1, s_2, s_3. The response will depend on the nature of these poles. If all the three poles are real and distinct, the time response will be of the form

$$A_i e^{s_1 t},\ A_2 e^{s_2 t},\ A_3 e^{s_3 t}$$

If two of the poles are complex conjugate, the transient response will have an oscillatory term of the form.

$$Ae^{st} \sin(\omega t + \alpha)$$

If the system is stable the transient terms should decay speedily with time. This will be ensured if the poles lie in the left half plane. The speed at which the transient terms decay depends on the real part of the roots of characteristic equation *i.e.* higher the magnitude of real parts, more rapid is the decay of the transient term. Most of the time the response is determined mostly by the dominant term which means the pole located nearest to the $j\omega$ axis in the s plane.

25.33.3 Stability Compensation

To achieve high accuracy it is necessary to keep the gain K high. But a high gain causes poor dynamic response. The standard technique, in control systems, to improve dynamic response and system stability is to add series and/or feedback compensation. Stabilising transformer is one form of feed back compensation.

Stability can be improved by adding a series phase lead compensator having a transfer function of the form

$$G_C = 1 + sT_C$$

When the transfer function $(1 + sT_C)$ is added to the circuit of Fig. 25.23 the overall open loop transfer function becomes

$$G(s) = \frac{K(1+sT_C)}{(1+sT_a)(1+sT_e)(1+sT_f)} \quad \text{... (25.62)}$$

It is seen from Eq. (25. 62) that static loop gain is still equal to K. However T_C can be selected to improve system stability (say) by making T_C equal to T_e. Then the open loop transfer function becomes

$$G(s) = \frac{K}{(1+sT_a)(1+sT_f)} \quad \text{... (25.63)}$$

The dynamic response corresponding to transfer function of Eq. (25.63) is much better than that corresponding to transfer function of Eq. (25.56).

25.34 SHUNT COMPENSATION, STATIC VAR SYSTEMS

Shunt Compensation is used in all high voltage, EHV systems to supply reactive power and improve voltage profile. Shunt compensation may consist of statics compensation (using capacitors and reactors) or synchronous compensation (using synchronous motor). As compared to synchronous compensation, the static compensation has the advantages of higher speed of response absence of fault infeed to the system, greater reliability and lower maintenance (because there are no moving parts), lower cost and simpler erection. Because of these reasons static compensation is being increasingly preferred to the synchronous compensation. The static compensation is usually referred to as static var system (SVS) or static var compensator (SVC) in modern terminology.

25.34.1 Functions of SVS

The most typical functions of SVS in EHV transmission can be summarized as under:

(*a*) *Voltage Support along the Line*: The vars generated by a line (i.e., ωCV^2) equal the vars consumed by the line (i.e. ωLI^2) and net vars are zero only when the load is equal to natural load. If the load is greater than natural load, net vars are consumed by the line. For maintaining equal voltages at the two ends of the line, equal amount of vars must be supplied at each end. If the load on the line is less than the natural load (and this usually happens during off-peak conditions), net vars are generated by the line. For maintaining equal voltages at the two ends, these vars must be absorbed by using shunt reactor at the two ends of the line.

(*b*) *Improved Stability:* Interconnected systems consist of two or more areas connected by tie lines. System faults and loss of generation cause system stability problems. Studies have indicated that by varying the vars supply at a stable point in the network, system stability can be improved. The thyristor switched shunt capacitors are especially useful for this purpose.

(*c*) *Control of Overvoltages:* Shunt compensation can control overvoltages under transient and dynamic conditions thereby allowing a reduction in design insulation levels. This function requires SVS to swing rapidly far into absorption mode and thyristor controlled reactor is best suited for this purpose.

(*d*) *Reduction of Vars Flow:* The provision of shunt compensation leads to a reduction in the unnecessary circulation of vars in the network leading to reduction of system losses and increased availabity of equipment rating.

25.34.2 Types of Static Var Systems

An EHV line requires generation as well as absorption of vars. An SVS consists of two elements in parallel (a reactor and a bank of capacitors) one or both of which are variable. Three designs have been suggested.

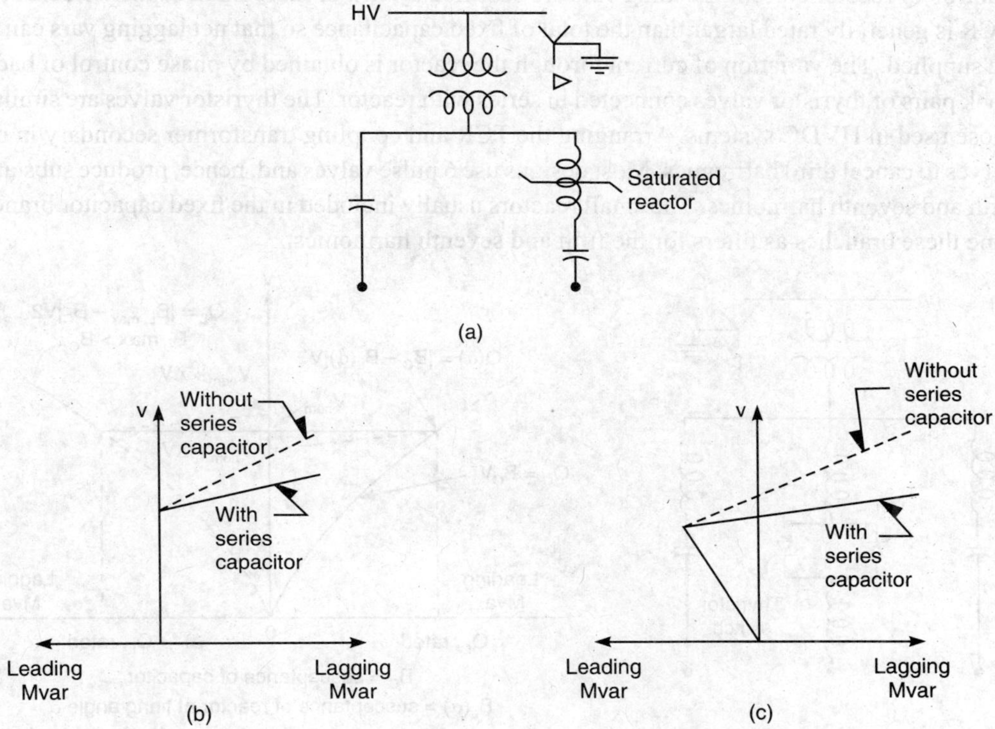

Fig. 25.25 Self saturated shunt reactor (*a*) connection diagram (*b*) characteristics without shunt capacitor (*c*) characteristics with shunt capacitor.

(*a*) *Self Saturated Reactor* (*SR*) (Fig. 25.25*a*) The variable element is an iron-cored reactor working in the saturated region and a capacitor bank is in parallel. It does not require any control system and automatically adjusts its loading to that required by the system. To match its characteristics to the system requirement, it may be necessary to reduce the slope reactance of the saturated reactor by connecting a slope correction capacitor in series. The saturated reactor, being of transformer type construction, has a very high short time overload capability and is, therefore, very suitable for holding down system dynamic over voltage such as those resulting from load rejection.

The saturated reactor is a self regulating devoice that responds only to changes in its terminal voltage. The voltage regulation characteristics is dependent on the natural saturation characteristic of the iron core reactor, the existence and rating of a series capacitor for saturation slope compensation and the existence and rating of shunt capacitor. Fig. 25.25(*b*) shows the characteristics without the shunt capacitor and Fig. 25.25(*c*) shows the characteristics with the shunt capacitor.

The saturated reactors for ratings upto 132 kV can be directly connected to the bus bar. However an intermediate transformer is needed for higher voltages.

(*b*) *Thyristor Controlled Reactor* (*TCR*)*-Fixed Capacitor* (*FC*) The TCR-FC system, shown in Fig 25.26(*a*), provides continuously controllable lagging to leading vars through thyristor control of reactor current. Leading vars are supplied by two or more fixed capacitor banks. The TCR is generally rated larger than the total of fixed capacitance so that net lagging vars can also be supplied. The variation of current through the reactor is obtained by phase control of back to back pairs of thyristor valves connected in series with reactor. The thyristor valves are similar to those used in HVDC systems. Arranging the TCR and coupling transformer secondary in delta serves to cancel third harmonics. Most designs use 6 pulse valves and, hence, produce substantial fifth and seventh harmonics. The small reactors usually included in the fixed capacitor branches tune these branches as filters for the fifth and seventh harmonics.

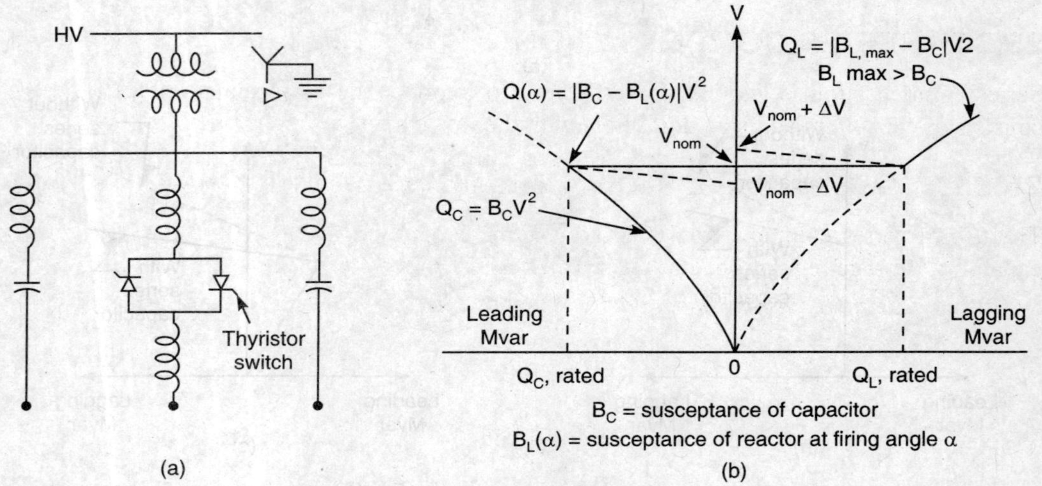

Fig. 25.26 Thyristor controlled reactor (TCR) and fixed capacitor (FC) for shunt compensation (*a*) connection diagram (*b*) characteristics

The reactive power output versus terminal voltage steady state characteristics is shown in Fig. 25.26(*b*). The steady state relationship between the output vars and the applied voltage can be linear over the desired voltage band of regulation. The slope of the characteristics in the regulation interval is determined by the control of the firing angle. Outside the regulation interval, the characteristic is the same as that of a capacitor (low voltage) or an inductor (high voltage).

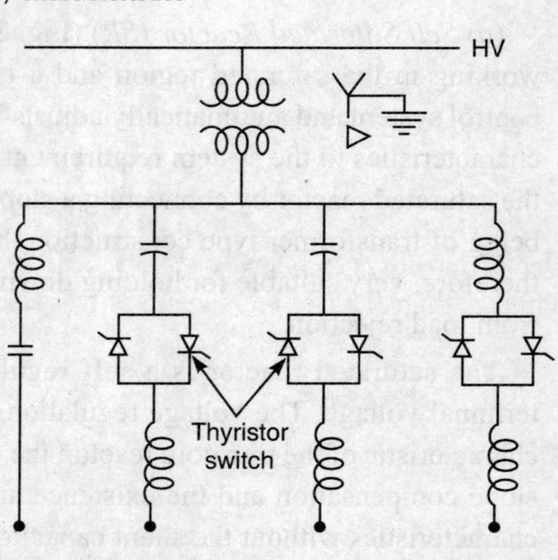

Fig. 25.27 Thyristor switched capacitor (TCS) and thyristor controlled reactor (TCR).

(*c*) *Thyristor Switched Capacitor (TSC)*: *Thyristor Controlled Reactor* (*TCR*) The TSCTCR compensator is shown in Fig. 25.27. It consists of a thyristor controlled reactor bark (or banks) and *n* capacitor banks each in series with a solid state (thyristor) switch. The number *n* of capacitor banks is determined from practical considerations which include operating voltage level, maximum compensation requirement, current rating of solid state switches, etc. The solid state switch in series with the reactor bank is composed of reverse parallel connected thyristor pair (in actual practice many such pairs are connected in series to provide enough voltage rating) and is capable of continuously controlling the current in the reactor from zero to maximum rated value. Similarly, the switch in series with each capacitor bank consists of reverse parallel connected thyristor pairs. The switches in capacitor banks allow the switching in of the required number of capacitor banks in discrete steps. Thus, TSC-TCR scheme has the capability to provide the leading and lagging vars both independently and rapidly. The characteristic is similar to that of TCR-FC scheme Fig. 25.26(*b*). The advantages of TSC-TCR scheme over the TCR-FC scheme are: improved performance during large system disturbances and lower losses. However, this scheme is about 15 – 20%, more costly as compared to TCR-FC scheme.

In addition to the switched capacitors, a small number of fixed capacitors, serving as tuned filters, are also included in this scheme, as shown in Fig. 25.27.

25.35 SERIES COMPENSATION

Series compensation is another powerful tool to improve the performance of EHV lines. It consists of capacitors connected in series with the line at suitable locations.

25.35.1 Advantage of Series Compensation

The use of series compensation leads to a number of advantages as regards power transfer capability, voltage regulation, etc. These can be summarized as under:

(*a*) *Increase in Transmission Capacity:* The power transfer P_1 over a line is given by

$$P_1 = \frac{|V_s||V_r|}{X_L} \sin \delta \qquad \text{... (25.64)}$$

Where $V_s \angle \delta$ = sending end voltage

$V_r \angle 0$ = receiving end voltage

X_L = inductive reactance of the line

δ = phase angle between V_s and V_r

If a series capacitor having capacitance X_c is inserted, the net series reactance becomes $X_L - X_c$ and the power transfer P_2 is given by

$$P_2 = \frac{|V_s||V_r|}{X_L - X_C} \sin \delta \qquad \text{... (25.65)}$$

From Eqs. (25.64) and (25.65) it is seen that for the same magnitudes of V_s, V_r and δ, P_2 is higher than P_1. The increase in power transfer is given by

$$\frac{P_2}{P_1} = \frac{X_L}{X_L - X_C} = \frac{X_L}{X_L\left[1 - \dfrac{X_C}{X_L}\right]} = \frac{1}{1-K} \qquad \text{... (25.66)}$$

The factor K equals $\dfrac{X_C}{X_L}$ and is known as degree of compensation. Studies on transmission systems have revealed that in view of line cost, the effect of lines operating in parallel and problems of system protection, the economic degree of compensation lies in the range of 40% to 70%.

(*b*) *Improvement of System Stability*: From Eqs. (25.64) and (25.65) it is seen that for the same amount of power transfer and the same values of V_s and V_r, the phase angle δ in the case of series compensated lines is less than that for the uncompensated line. It is known that a lower angle δ means better system stability. Some other methods for improvement of system stability are also used. These include reducing the reactance of generators and transformers, use of bundled conductors and increasing the number of circuits in parallel. Series compensation generally offers the most economic solution with respect to both the steady state and the transient stability.

(*c*) *Load Division between Parallel Circuits*: The reduction of series reactance by series compensation makes it a useful tool to balance the loading in parallel circuits. When a system is to be strengthened by the addition of a new line or when one of the existing circuits is to be adjusted for parallel operation in order to achieve maximum power transfer and minimize losses, series compensation can be used to a great advantage. It has been reported that in Sweden, the cost of the first series compensation in the 420 kV system was entirely recovered due to the decreased losses in the 220 kV system operating in parallel with the 420 kV system.

25.35.2 Location of Series Capacitors

The choice of the location of series capacitors depends on many technical and economic considerations. In each case, a special system study concerning load flow, stability, transient overvoltages, protection requirements, system voltage profile, etc., is necessary before the optimum location can be chosen. Three commonly used locations are:

(*a*) *Location along the Line:* In this method the capacitor bank is located at the middle of the line (if the installation consists of one bank) or at one third distance along the line (if the installation consists of two banks). This location has the advantages of better voltage profile along the line, lesser short circuit current through the capacitor in the event of a fault and simpler protection of the capacitor. The capacitor stations are generally unattanded. This location is popular in Sweden, Canada, South Africa, etc.

(*b*) *Location at One or Both Ends of Line section on the Line Sides in the Switching Station:* The main advantage of this location is that the capacitor installation is near the manned substations. However, this location requires more advanced line protection. For the same degree of compensation, more installed M-var capacity is needed than for the location along the line. This location is widely used in USA.

(*c*) *Location between Bus Bars within the Switching Station:* The advantage of this location is that the capacitor installation is concentrated at a few accessible places within the switching station. However this location requires a rather complicated and expensive switching arrangement within the substation. This location has been used in the erstwhile Soviet Union and Argentina.

25.35.3 Problems Associated with Series Compensation

The use of series compensation introduces a few problems too. Some of these are:

(*a*) *Sub-synchronous Resonance:* The series capacitor introduces a sub-synchronous frequency (proportional to the square root of the compensation) in the system. In some cases this frequency may interact with weak steam turbine generator shafts and give rise to high torsional stresses. In hydro-turbine generators the risk of sub-synchronous resonance is small because the torsional frequencies are about 10 Hz or even less.

(*b*) *Ferroresonance:* When an unloaded or a lightly loaded transformer is energized through a series compensation line, ferroresonance may occur. The frequency of oscillation is an integral multiple of the system frequency. This can be suppressed by using shunt resistors across the capacitors or by short circuiting the capacitor temporarily through an isolator or a by pass breaker.

(*c*) *Line Protection:* Series compensation can lead to mal-operation of the distance relay of the line protection if the degree of compensation and capacitor location are not proper. To ensure correct operation of the distance protection the series compensation is limited to about 40% (when the installation is located at the middle of the line) and 30% per bank (when the installations are located at one-third and two-third distances along the line).

(*d*) *High Recovery Voltage:* Series capacitors produce high recovery voltages across the circuit breaker contacts.

In spite of the above problems series compensation is very widely used for EHV systems in the 400 – 500 kV range in many countries of the world.

25.36 SHUNT REACTOR

During light load or no load conditions the voltage at the receiving end of the line may even exceed the sending end voltage. This is due to the charging current drawn by the shunt capacitance of the line. To limit this voltage rise shunt reactors have to be used. It is necessary that the shunt reactor is switched on only during light or no load conditions. Shunt reactor is also a part of SVS. The specification for a shunt reactor include type (indoor or outdoor), reactance in ohms, kvar rating, rated voltage, rated current and frequency.

Reactors are mostly air cored. They are built in two types: dry type and oil immersed

(*a*) *dry type reactor*: The windings of a dry type reactor are embedded in vertical supports of specially treated concrete. The winding is bolted to concrete base which is insulated from ground by porcelain footings. The conductors are insulated with fire proof asbestos braid.

(*b*) *oil immersed reactor*: The windings are similar to those used for transformers. A non magnetic shield prevents the flux from entering the steel tank where it can cause losses and heating. The oil helps in cooling and insulation. An oil immersed reactor has a smaller size (for the same rating) higher thermal capacity and fewer chances of flashover as compared to dry type. They are built for both outdoor and indoor use.

25.37 SYNCHRONOUS CONDENSERS

When the kVAR requirement is small, it can be met through static capacitors. However when requirement exceeds 10,000 kVAR it is generally more economical to use the synchronous condensers.

A synchronous condenser is essentially an over excited synchronous motor. Generally it does not supply any active mechanical power. The excitation of the machine is varied to provide the necessary amount of the leading kVAR. The advantages and disadvantages of using synchronous condensers as compared to static capacitors are as under:

1. A synchronous condenser can supply kVAR equal to its rating and absorb kVAR upto 50 per cent of its capacity. Thus a synchronous condenser of certain kVAR is equal to a static capacitor of that kVAR and a shunt reactor of 50 per cent kVAR.

2. By the use of synchronous condenser a finer control is possible than by the use of static capacitors.
3. A synchronous condenser can be overloaded for short periods but a static capacitor cannot be overloaded.
4. A momentary drop in voltage causes the synchronous condenser to supply greater kVAR to the system whereas in the case of static capacitor, the kVAR supplied is reduced.
5. The inertia of the synchronous condenser improves the system stability and reduces the effect of sudden changes in load.
6. The power loss in a synchronous condenser is much greater than that in a capacitor.
7. For small kVAR requirements, static capacitors are preferable and economical. For requirements above 10,000 kVAR or so synchronous condensers are more economical.
8. Static capacitor installations can be distributed in the system. Thus capacitors can be located near the loads and are more effective. However small size synchronous condensers are very uneconomical. As such the synchronous condensers have to be installed at one point only.
9. The rating of a static capacitor bank can be changed very easily as per requirements. Capacitor units can be added to the bank or taken away from it. This is not possible with synchronous condensers.
10. Installation of a static capacitor bank is easy.
11. A failure of one unit of capacitor bank affects that unit only. The remaining units continue to do their job. However failure of a synchronous condenser means loss of total condenser capacity. On the other hand the failure rate of a synchronoous condenser is very small as compared to the failure rate of a capacitor bank.
12. Synchronous condenser adds to the short circuit currents in the system and increases the circuit breaker ratings.

Synchronous condensers are mostly used by utilities at large sub-stations to improve the power factor and voltage regulation. Machines upto 100 MVAR ratings or even higher have been used. The field current is regulated automatically to give a desired voltage level. A typical instance is of 150 MW to be transmitted over a distance of 240 km. If the receiving end power factor is 0.85, the sending end power factor is 0.68 and sending end voltage 1.5 times receiving end voltage. Addition of 75 MVAR synchronous condenser at receiving end improves the sending end power factor to 0.88 and reduces the voltage drop in transmission line by 50 per cent. In addition the synchronous condenser reduces the switching surges due to the sudden connection or disconnection of the line to the system.

25.38 COMPARISON OF SERIES AND SHUNT COMPENSATION (SVS)

25.38.1 Advantags of Series Compensation

1. Series compensation is some what cheaper than SVS
2. Losses in series compensation are lower than in SVS
3. Series compensation lowers the critical or collapse voltage. Thus chances of voltage collapse are reduced.
4. Series compansation can help in load division in parallel circuits.
5. Series capacitors are generally self regulatory and require little or no control equipment.
6. Series compensation improves system stability (both steady state and transient state)

25.38.2 Disadvantages of Series Compensation

1. Series compensation is effective only during heavy loads. During light load conditions shunt compensation has to be provided.
2. Whenever an outage occurs on a line with series compensation, the series compensation has to be removed. This may cause overloading of other parallel lines.
3. If series compensation is to be added to an existing system it is generally necessary to have it on all lines in parallel.
4. System outages may cause high voltage on one side of series capacitors.
5. Additional expenses are needed to solve the problems of sub-synchronous resonance, line protection etc.

25.38.3 Advantages of SVS

1. SVS provides a better control of voltage profile.
2. It is effective even in off peak conditions.
3. SVS provides fast control over temporary overvoltages.

25.38.4 Disadvantages of SVS

1. It is more expensive than series compensation scheme.
2. Its overload capability is limited.
3. To reduce the overall cost, it is preferable to use mechanically switched capacitor banks in addition to SVS.

It is prudent to have a combination of both the series and shunt compensation for best overall advantages.

Example 25.25.

A 3 phase long line has constant $A = 0.98\angle 3°$ and $B = 110\angle 75°$ ohm per phase. (*a*) If the load is 50 MVA, 0.8 power factor lagging, find the capacity of shunt compensation equipment if voltages at the two ends of the line are 132 kV each (*b*) Find the capacity of shunt compensation equipment if the voltages at the two ends are to be maintained at 132 kV under no load condition.

Solution

(*a*) Using Eq. (4.74*a*)

$$50 \times 0.8 = \frac{132\times 132}{110}\cos(\beta-\delta) - \frac{0.98(132)^2}{110}\cos(75°-3°)$$

or $$40 = 158.4\cos(\beta-\delta) - 47.97$$

or $$(\beta-\delta) = 56.26°$$

Using Eq, (4.74*b*)

$$Q_r = \frac{132\times 132}{110}\sin(56.26°) - \frac{0.98(132)^2}{110}\sin(75°-3°)$$

$$= -15.91 \text{ MVar}$$

Thus for the given operating conditions, a leading Mvar of 15.91 must be drawn from the line along with real power of 40 MW. Since the load requires 50 × 0.6 *i.e.,* 30 MVar lagging, the static capacitors must deliver (30 + 15.91) *i.e.*, 45.91 MVar lagging (or absorb 45.91 MVar leading). The capacity of shunt compensation equipment is, therefore, 45.91 MVar.

(*b*) Using Eq. (4.74*a*)

$$0 = \frac{132 \times 132}{110} \cos(\beta - \delta) - \frac{0.98\,(132)^2}{110} \cos(75° - 3°)$$

or $$0 = 158.4 \cos(\beta - \delta) - 47.97$$

or $$(\beta - \delta) = 72.37°$$

Using Eq, (4.74*b*)

$$Q_r = \frac{132 \times 132}{110} \sin(72.37°) - \frac{0.98\,(132)^2}{110} \sin(72°)$$

$$= 3.33 \text{ MVar}$$

Thus under no load condition the line delivers 3.33 MVar at the receiving end. This reactive power must be absorbed by shunt reactor at the receiving end. Thus the capacity of shunt reactor for no load condition, is 3.33 MVar.

25.39 VOLTAGE CONTROL USING TAP CHANGING TRANSFORMERS

Another method of voltage control is through the use of tap changing transformers. Almost all transformers are provided with taps on the winding for adjusting the ratio of transformation. Taps are usually provided on the high voltage winding to enable a fine control of voltage. Generally the tap changing can be done only when the transformer is in de-energised state. However in some cases tap changing is also possible when the transformer is energised and such transformers are known as tap changing under load (TCUL) transformers. These transformers make it possible to maintain a constant voltage level on important busses in the system.

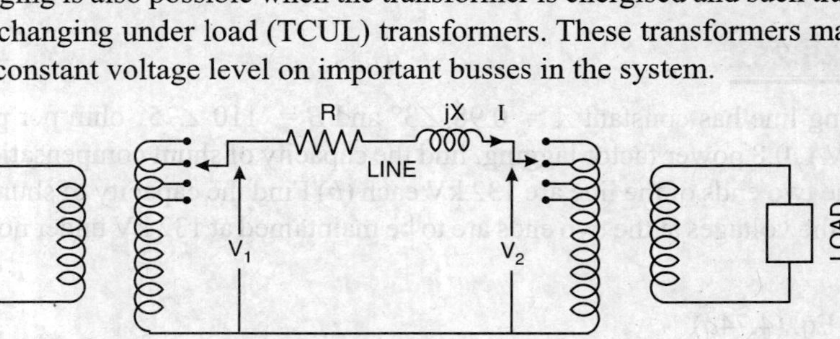

Fig. 25.28 Voltage control of a line using tap changing transformer.

Fig. 25.28 shows a line, with impedance $R + jX$ having tap changing transformers at its two ends. Let t_s and t_r be the fractions of the nominal transformation *i.e.*, tap ratio/nominal ratio *e.g.* a transformer with nominal ratio 11/33 kV when tapped to convert 11 kV to 36 kV has t_s = 36/33 = 1.09. V_1 and V_2 are the nominal voltages at the sending end and receiving end respectively. When taps are used, the actual sending end and receiving end voltages would be $t_s V_1$ and $t_r V_2$. The product $t_s t_r$ is generally made unity so that overall voltage level remains in the same order and the minimum range of taps is used. The tap ratios t_s and t_r can be determined as under

$$t_s V_1 = t_r V_2 + IZ \simeq t_r V_2 + IR\cos\theta + IX\sin\theta \qquad \text{... (25.67)}$$

Where θ is the power factor angle of load and equals $\tan^{-1}(Q/P)$

Since $I\cos\theta = \dfrac{P}{t_r V_2}$ and $I\sin\theta = \dfrac{Q}{t_r\,V_2}$, we get

$$t_s V_1 = t_r V_2 + \frac{RP}{t_r V_2} + \frac{XQ}{t_r V_2} \qquad \text{... (25.68)}$$

Substituting $t_r = 1/t_s$ in Eq. (25.68)

$$t_s = \frac{1}{V_1}\left[\frac{V_2}{t_s} + \frac{t_s(RP + XQ)}{V_2}\right]$$

or

$$t_s^2 = \frac{V_2}{V_1} + \frac{t_s^2\,(RP + XQ)}{V_1 V_2}$$

or

$$t_s^2\left[1 - \frac{RP + XQ}{V_1 V_2}\right] = \frac{V_2}{V_1} \qquad \text{... (25.69)}$$

For complete compensation $V_1 = V_2$ and t_s can be found from Eq. (25.69). $t_r = 1/r_s$. In Eq. (25.69) all quantities are in per phase.

Example 25.26.

A 220 kV line has tap changing transformers at both ends. The transformer at the sending end has a nominal ratio 11/220 kV and that at the receiving end 220/11 kV. The line impedance is 20 + *j* 60 ohm and the load at the receiving end is 100 MVA, 0.8 power factor lagging. If the product of the two off-nominal tap settings is 1, find the tap settings to give 11 kV at load bus.

Solution

$$P = \frac{100 \times 0.8 \times 10^6}{3} = 26.67 \times 10^6 \text{ W}$$

$$Q = \frac{100 \times 0.6 \times 10^6}{3} = 20 \times 10^6 \text{ Vars}$$

$$V_1 = V_2 = \frac{220}{\sqrt{3}} \times 10^3 = 127.02 \times 10^3 \text{ V}$$

Using Eq. (25.69)

$$t_s^2\left[1 - \frac{20 \times 26.67 \times 10^6 + 60 \times 20 \times 10^6}{(127.02 \times 10^3)^2}\right] = 1$$

or

$$t_s = 1.058 \text{ and } t_r = \frac{1}{t_s} = 0.95$$

25.40 COMBINED USE OF TAP CHANGING TRANSFORMERS AND REACTIVE POWER INJECTION

Tap changing transformers can also be used in conjunction with var injection using static capacitors or synchronous condensers. Fig. 25.29(*a*) shows the usual configuration where the static var system is connected to the tertiary of the transformer. The equivalent circuit is shown in Fig. 25.29(b). The three winding transformer can be represented by three reactances

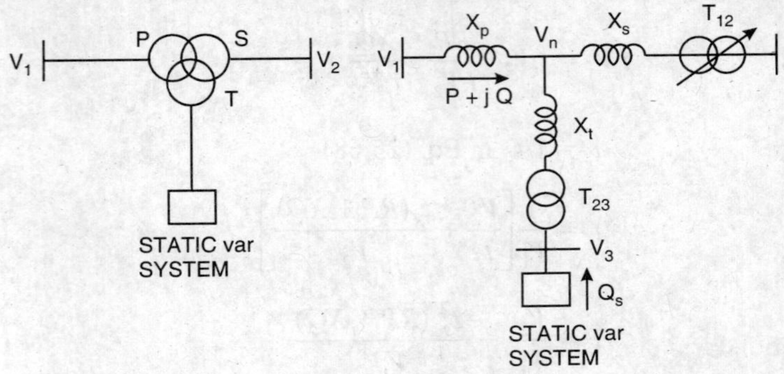

Fig. 25.29

connected in star (resistances have been neglected). The reactance of the line between V_1 and transformer or V_2 and transformer can be combined with reactances X_p and X_s respectively. If the transformer is at the receiving end of the line, the reactance X_s would be small as compared to X_p and can, therefore, be neglected. V_n is the line to neutral voltage of the star point. The power flow from primary to secondary is $P + jQ$. The active power of static var equipment is assumed to be zero and its reactive power is Q_s per phase.

$$V_1 = V_n + IX_p \sin\theta = V_n + \frac{X_p Q}{V_n} \quad \text{... (25.70)}$$

For given values of V_1, X_p and Q, the value of V_n can be determined. Generally the value of V_n is determined for more than one condition of system operation and transformation ratio and tap settings are chosen to give satisfactory voltage level at full load as well as at no load.

Example 25.27.

A three winding transformer is rated at 132/33/11 kV. The MVA ratings of the three windings are 75, 50 and 25 MVA respectively. The reactance of primary is 0.12 p.u. on 132 kV, 75 MVA base. A static var compensating equipment is connected to the tertiary winding. Neglecting losses find the tap settings to give satisfactory operation under the following two conditions:

(a) Load of 60 MW, 50 Mvar with primary and secondary voltage limits of 125 and 33 kV. The static var system is disconnected.

(b) No load condition, SVS absorbing 25 Mvar and primary and secondary voltage limits 140 kV and 33 kV.

Solution

Since X_p is given in pu, all calculations will be done in pu. Base kV = 132, Base MVA = 75

(a) $$V_1 = 125/132 = 0.947 \text{ p.u.}$$

$$Q = \frac{50}{75} = 0.667 \text{ p.u.}$$

Using Eq. (25.70)

$$0.947 = V_n + \frac{0.12 \times 0.667}{V_n}$$

or $\quad V_n^2 - 0.947\,V_n + 0.08 = 0$

or $\quad V_n = 0.853 \text{ pu} = 0.853 \times 132 = 112.6 \text{ kV}$

(*b*) $\quad V_1 = 140/132 = 1.06 \text{ p.u.}$

$$Q = \frac{25}{75} = 0.333 \text{ p.u.}$$

Therefore $\quad 1.06 = V_n + \dfrac{0.12 \times 0.333}{V_n}$

or $\quad V_n = 1.021 \text{ p.u.} = 134.77 \text{ kV}$

Transformation ratio under load condition = 112.6/33 = 3.412

Transformation ratio under no load condition = 134.77/33 = 4.084

The actual ratio can be taken as mean of the above two values *i.e.*, 3.748 varying by ± 0.336 or $3.748 \pm 8.96\%$. Hence tap setting required is $\pm 8.96\%$.

25.41 REGULATING TRANSFORMER

A special type of transformer designed for small adjustments of voltage is known as regulating transformer. The MVA rating of regulating transformer is quite small and it can be used to change the voltage at a point by a small amount. It can be brought into the circuit by the closure of a relay whose operation can be controlled by system voltage or system current. A regulating transformer can be used for voltage magnitude control or phase angle control.

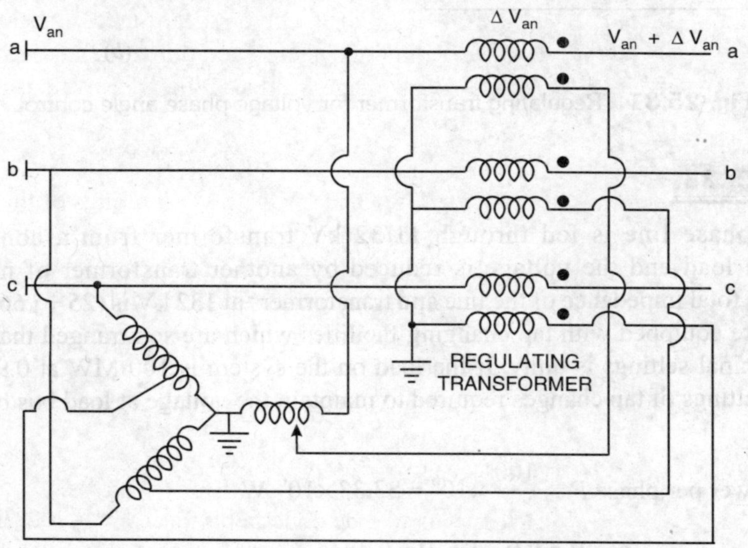

Fig. 25.30 Regulating transformer for voltage magnitude control.

25.41.1 Regulating Transformer for Voltage Magnitude Control

Fig. 25.30 shows a typical arrangement to use a regulating transformer for voltage magnitude control in a 3 phase circuit. A 3 phase transformer provides an adjustable voltage to the primary windings of the regulating transformer. The secondary windings of the regulating transformer are connected in series with the lines. Thus a voltage magnitude $|\Delta V|$ is added to the voltage of each phase. $|\Delta V|$ for each phase is in phase with the voltage of that phase. As mentioned earlier, the changes in voltage magnitudes mainly affect the reactive power flow. Therefore this method can be used to control the flow of reactive power in the system.

25.41.2 Regulating Transformer for Voltage Phase Angle Control

Fig. 25.31 (*a*) shows the use of a regulating transformer for phase angle control. The voltage added to each phase voltage is displaced by ± 90° from it. As seen from Fig. 25.31 (*b*) this arrangement does not cause any change in the magnitude of phase voltages. However the phase angles are changed and the change in phase angle is approximately proportional to the magnitude of added voltage $|\Delta V|$. As mentioned earlier the phase angles control the flow of real power in the system. Therefore this method can be used to control real power flow in the system.

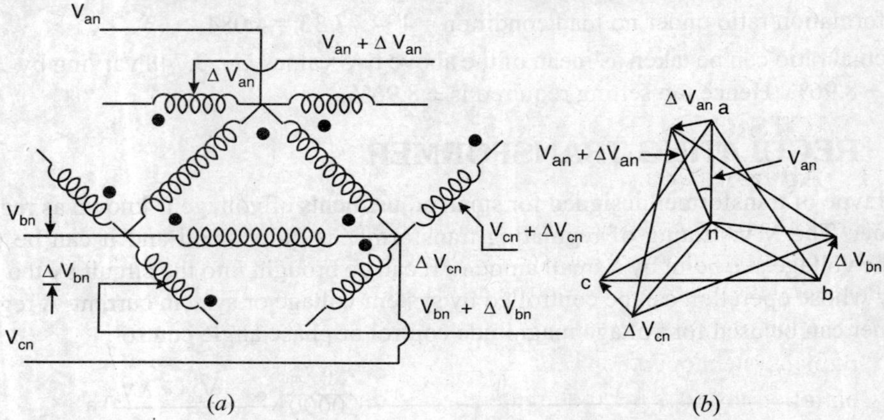

Fig. 25.31 Regulating transformer for voltage phase angle control.

Example. 25.28.

A 132 kV 3 phase line is fed through 11/32 kV transformer from a constant 11 kV supply. At the load end the voltage is reduced by another transformer of nominal ratio 132/11 kV. The total impedance of the line and transformers at 132 kV is (25 + j 66) ohms. Both transformers are equipped with tap changing facilities which are so arranged that the product of two off-nominal settings is unity. If the load on the system is 100 MW at 0.9 p.f. lagging calculate the settings of tap changes required to maintain the voltage at load bus bar at 11 kV.

Solution

$$\text{Active Power per phase } P = \frac{100}{3} \times 10^6 = 33.33 \times 10^6 \text{ W}$$

$$\cos\theta = 0.9 \quad \tan\theta = 0.484$$

$$\text{Reactive Power per phase } Q = \frac{100 \times 0.484}{3} \times 10^6 = 16.14 \times 10^6 \text{ vars}$$

Line reactance $R = 25$ ohms per phase

Line reactance $X = 66$ ohms per phase

Using Eq. (25.69).

$$V_1 = V_2 = \frac{132}{\sqrt{3}} \times 10^3 = 76.21 \times 10^3 V$$

$$t_s^2 \left[1 - \frac{RP + XQ}{V_1 V_2}\right] = \frac{V_2}{V_1}$$

or $$t_s^2 \left[1 - \frac{25 \times 33.33 \times 10^6 + 66 \times 16.14 \times 10^6}{76.21 \times 76.21}\right] = 1$$

or $$t_s^2 (1 - 0.3268) = 1 \text{ or } t_s = 1.2188$$

$$t_r = \frac{1}{t_s} = 0.82$$

25.42 FACTS CONTROLLERS

25.42.1 Advantages

The invention of fast switching devices i.e. thyristors and other similar components prompted power system engineers to device Facts (Flexible AC Transmission) controllers for power system control. In general the use of Facts Controllers leads to one or more of the following advantages.

1. Reduction of system losses
2. Optimal system operation
3. Control of power as per requirement
4. Increase in line capacity
5. Increase in transient stability limit
6. Flexibility in operation
7. Reduction of vars flow in lines thus allowing more active power flow over the lines
8. Improved voltage profile of the lines
9. Elimination of problems like voltage fluctuations, sub synchronous resonance etc.

25.42.2 Types

Facts controllers can be divided into following types.

1. Series controllers
2. Shunt controllers
3. Combined series – series controllers
4. Combined series – shunt controllers

Static var system or SVS (also called static var compensator or SVC) is a facts controller and has been discussed in section 25.34. Some other important facts controllers are discussed in the following section.

25.42.3 Static Phase Shifting Transformer (SPST)

Fig. 25.32 illustrate the principle of SPST. For the purpose of clarity it has been shown only for one phase. The connections for the other two phases are similar.

The primary of excitation transformer is fed from the system. The secondary of ET feeds a converter. The converter feeds the primary of boost transformer. The secondary of boost transformer is connected in series with the line.

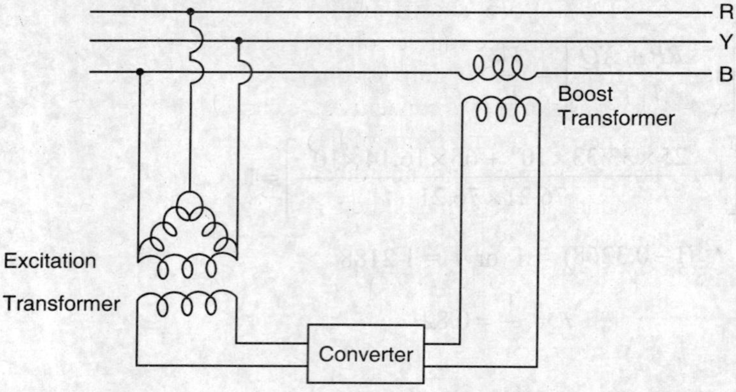

Fig. 25.32 Static Phase Shifting Transformer.

It is seen that by addition of a voltage is series with the line (through the boost transformer). the magnitude of line voltage as well as phase angle can be changed. The change of phase angle is possible because the voltage V_{BY} (*i.e.* $V_B - V_Y$) in the three phase system leads the phase voltage V_R by 90°. Similarly V_{RB} leads V_Y and V_{YR} leads V_B by 90°. The converter can control the magnitude and polarity of the quadrature voltage injected into the line by the boost transformer.

Fig. 25.33 shows a single line diagram of the converter used in *SPST*. This converter may consists of two 3 phase voltage source converters as shown. As in any converter configuration the function of capacitor is to eliminate the harmonics.

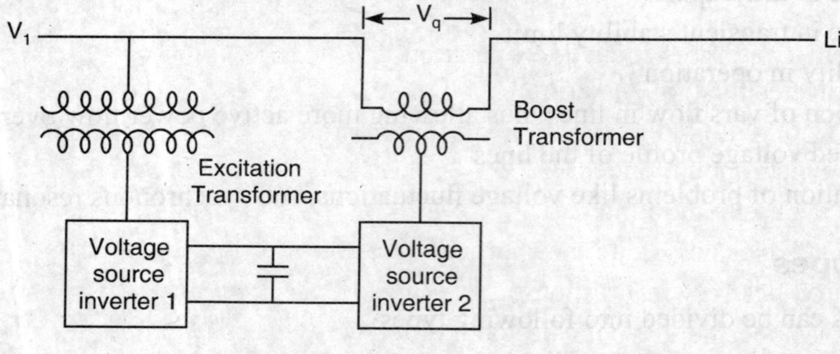

Fig. 25.33 Voltage Source Convertor for SPST.

V_1 is the initial line to neutral voltage of the line and V_q is the voltage injected into the line through the boost transformer. The voltage V_q is, in general, having a phase difference $\angle\theta$ with respect to V_1 *i.e.*

$$V_q = V_1 \angle\theta$$

25.42.4 Static Synchronous Compensator (STATCOM)

The function of STATCOM is similar to that of SVS (or SVC). It can inject lagging or leading vars into the power system.

A STATCOM is similar to a synchronous condenser. In a synchronous condenser the reactive power is generated by controlling the field current. However in a STATCOM the same result is obtained by using voltage source converter.

A simple diagram of STATCOM is shown in Fig. 25.33. The dc voltage of the battery is converted into ac voltage by the voltage source converter using GTO. This ac voltage is injected into the line through the transformer. If output voltage of E_S of converter is more than line voltage, the converter draws negative (*i.e.* capacitive) reactive power and thus supplies lagging vars to the system. This is needed during times of heavy loads. However if E_S is less than line voltage, the converter absorbs lagging vars and thus generates leading vars. This is needed at times of low load. If E_S is equal to line voltage, the STATCOM is floating i.e. it is not supplying any reactive power to the system.

Fig. 25.35 shows the $V-I$ characteristics of STATCOM

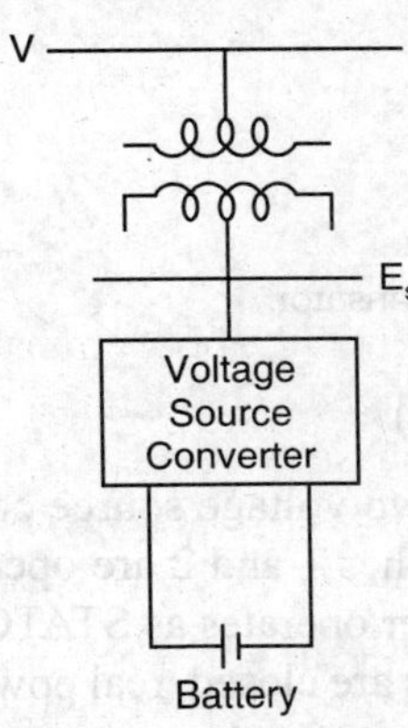

Fig. 25.34 STATCOM.

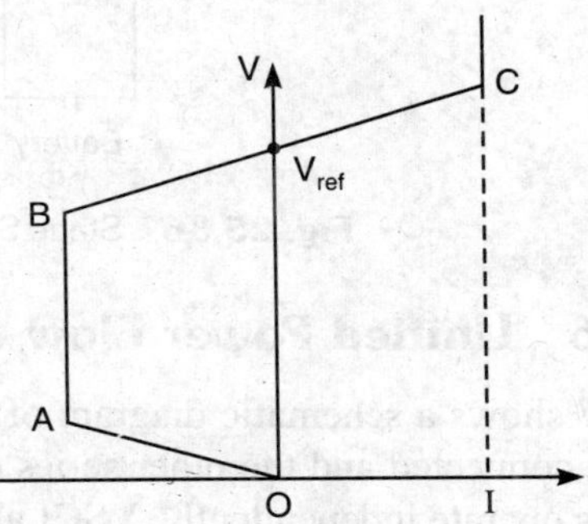

Fig. 25.35 *V–i* Characteristics of STATCOM.

Positive current I indicates inductive operation and negative current indicates capacitive operation. V_{ref} indicates floating condition.

Though STATCOM does the function of synchronous condenser it has the following advantages over synchronous condenser.

1. STATCOM has no rotating parts.
2. STATCOM does not add to the short circuit current of the system.
3. STATCOM has very fast response.
4. STATCOM does not have the problem of loss of synchronism.
5. STATCOM requires less maintenance.

25.42.5 Static Synchronous Series Compensator. (SSSC or S^3C)

The effect of SSSC in electric power transmission is similar to that of series compensation. In series compensation the series capacitor compensates the inductive reactance of the line. If line current is I, the voltage across series capacitor is $-jX_C I$ where X_C is the capacitive reactance of the series capacitor. Since the voltage drop across line induclance X_L is $+j\,IX_L$, the two voltage drops cancel each other thus reducing the effect of line inductance. Due to this the power transfer capability of the line is increased.

Fig. 25.36 shows a simple diagram to illustrate the working of SSSC. The supply from a dc source is converted to ac by the converter. The ac is fed to the line through the coupling transformer. Thus a quadrative voltage is injected into the line. The voltage injected into the line lags behind the line current by 90° thus providing series compensation. The voltage and current for controlling the operation of converter are derived from the system through *PT* and *CT* as shown in Fig. 25.36.

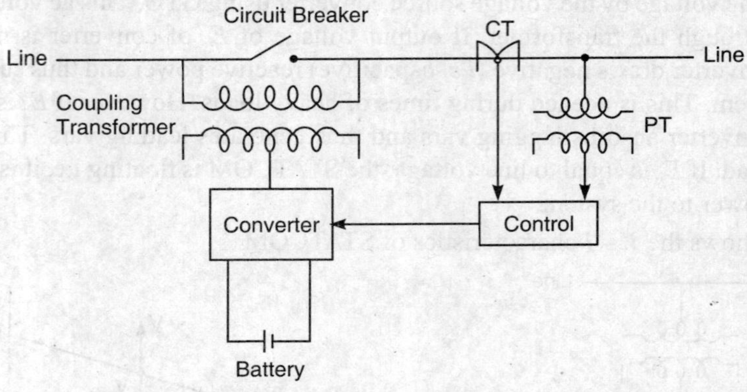

Fig. 25.36 Static Synchronous Series Compensator.

25.42.6 Unified Power Flow Converter (UPFC)

Fig. 25.37 shows a schematic diagram of UPFC. It consists of two voltage source converters, one shunt connected and the other series connected. When switches 1 and 2 are open the two converters operate independently. VSCI along with its transformer operates as STATCOM and VSC2 along with its transformer operates as SSSC. It the switches are closed, real power can be exchanged between the converters. UPFC enables control of both real and reactive power flow in the line. The word unified signifies that all parameters which affect power flow in the line (i.e. voltage, phase angle, impedance) can be controlled. The dc source for the two converters may be separate or common. The shunt converter provides the necessary power and also injects reactive current at the converter bus.

The concept of combining convertors can be extended to provide more flexibility e.g. GUPFC (generalised UPFC) has three converters, one shunt connected and two series connected. The total number of configurations, even while using two converters, can be upto 11 depending on the requirement. UPFC can be connected either at the sending end or receiving end of line.

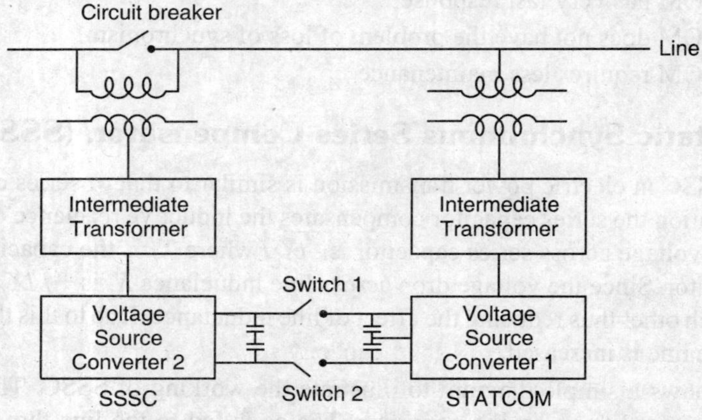

Fig. 25.37 Schematic diagram of UPFC.

Example 25.29.

Find primary ALFC loop parameters for a control area from the following data

Total rated area capacity = 2000 MW

Normal operating load = 1000 MW

Inertia constant $H = 5$

Regulation = 2.4 Hz/p.u. MW

Assume that load frequency dependency is linear and 1% change in frequency causes 1% change in load. *[RGPV Bhopal 2009]*

Solution

$$D = \frac{0.01 \times 1000}{0.01 \times 50} = 20 \text{ MW/ Hz}$$

$$= \frac{20}{2000} = 0.01 \text{ p.u. MW/Hz}$$

$$K_p = \frac{1}{D} = 100 \text{ Hz/pu MW}$$

$$T_p = \frac{2H}{fD} = \frac{2 \times 5}{50 \times 0.01} = 20 \text{ seconds}$$

Example 25.30.

Two synchronous generators operate in parallel and supply a total load of 400 MW. Their capacities are 200 MW and 500 MW and both have governor droop characteristics of 4% from no load to full load. Find load taken by each machine and system frequency.

[RGPV Bhopal 2007]

Solution

Let load supplied by machine 1 be L_1. Then load supplied by machine 2 is $(400 - L_1)$ MW.

$$50 - \frac{L_1}{200} \times 0.04 = 50 - \frac{(400 - L_1)}{500} \times 0.04$$

or $$L_1 = 114.3 \text{ MW}$$

Load supplied by 200 MW generator = 114.3 MW

Load supplied by 500 MW generator = 400 – 114.3

= 285.7 MW

$$\text{frequency} = 50 - \frac{.04 \times 114.3}{200} = 49.977 \text{ Hz}$$

Example 25.31.

Two generating stations have full load capacity of 200 MW and 75 MW respectively. The interconnector connecting the two stations has a generator of 25 MW full capacity. The percentage changes of speeds of A, B, C are 5, 4 and 3 respectively. Loads at bus bars A and B are 75 MW and 30 MW. Find load on set C and direction of flow of power. *[RGPV Bhopal 2007]*

Solution

System is shown in Fig. 25.38.

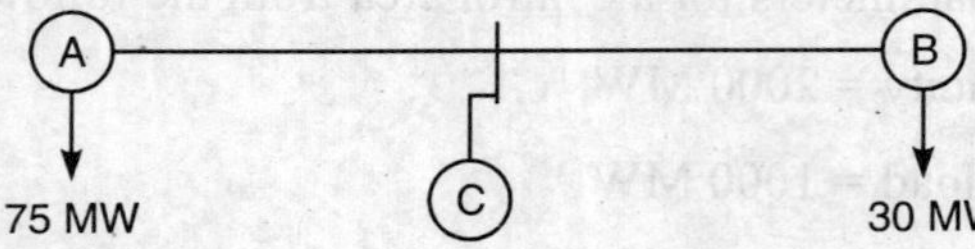

Fig. 25.38

All the three generators must operate at same speed so that frequency is same.

$$\text{Let load on } A = L_1$$

$$\text{load on } B = L_2$$

$$\text{Total load} = 75 + 30 = 105 \text{ MW}$$

$$\text{Load on } C = 105 - (L_1 + L_2)$$

Then $$50 - \frac{0.05}{200} L_1 = 50 - \frac{0.04}{75} L_2$$

or $$L_1 = 2.133\, L_2 \qquad (i)$$

Also $$\frac{0.03\left[105 - (L_1 + L_2)\right]}{25} = \frac{0.04\, L_2}{75} \qquad (ii)$$

From Eqs. (*i*) and (*ii*)

$$L_1 = 62.60 \text{ MW}, L_2 = 29.35 \text{ MW}$$

$$105 - (L_1 + L_2) = 13.05 \text{ MW}$$

Thus station *C* is generating 13.05 MW. Out of this power, station *C* is supplying (75 – 62.60) *i.e.* 12.40 MW to station bus at *A* and (13.05 – 12.40) *i.e.* 0.65 MW to station bus *B*.

Example 25.32.

A 230 kV line is fed by 33/230 kV transformer from 33 kV system. Fig. 25.39 shows the configuration. The line impedance is 30 + *j* 80 ohms. Both transformers have tap changing facilities and tappings are arranged so that two off nominal tap settings have a product of 1. System load is 150 MW at 0.9 p.f. find settings of tap changers to maintain the voltage of load bus at 33 kV. [*RGPV Bhopal 2007*]

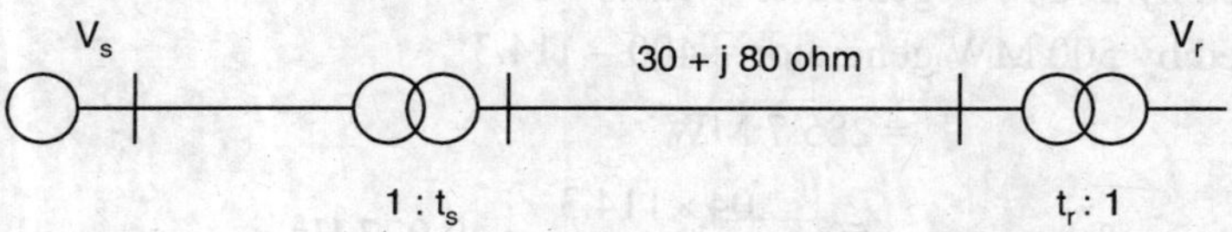

Fig. 25.39

Solution

$$P = \frac{150 \times 10^6}{3} = 50 \times 10^6 \text{ W}$$

$$\tan \phi = \tan (\cos^{-1} 0.9) = 0.484$$

$$Q = \frac{150 \times 10^6 \tan \phi}{3} = 24.2 \times 10^6 \text{ Var}$$

$$V_s = V_r = \frac{230}{\sqrt{3}} \times 10^3 = 132.79 \times 10^3 \text{ V}$$

$$t_s^2 \left[1 - \frac{RP + XQ}{V_s V_r} \right] = \frac{V_r}{V_s} = 1$$

$$t_s^2 \left[1 - \frac{30 \times 50 \times 10^6 + 80 \times 24.2 \times 10^6}{\left(132.79 \times 10^3\right)^2} \right] = 1$$

$$t_s = 1.114, \quad t_r = \frac{1}{t_s} = 0.89767$$

Example 25.33.

For a certain turbine generator set $R = 0.04$ p.u. based on generator rating of 100 MVA and 60 Hz. The generator frequency decreases by 0.02 Hz and the system adjusts to steady state operation. By how much does the turbine power output increase. *[RGPV Bhopal 2006]*

Solution

$$R = 0.04 \text{ p.u.} = 0.04 \times 60 = 2.4 \text{ Hz/p.u. MW}$$

$$R = \frac{2.4}{100} = 0.024 \text{ Hz/MW}$$

$$\Delta P = -\frac{1}{R} \Delta f = -\frac{1}{0.024}(-0.02) = 0.8333 \text{ MW}$$

Example 25.34.

For the LFC block diagram shown in Fig. 25.40 find steady state frequency error if $R = 3$ Hz/p.u. MW and $D = 0.01$ p.u. MW/Hz for unit step disturbance. *[RGPV Bhopal 2009]*

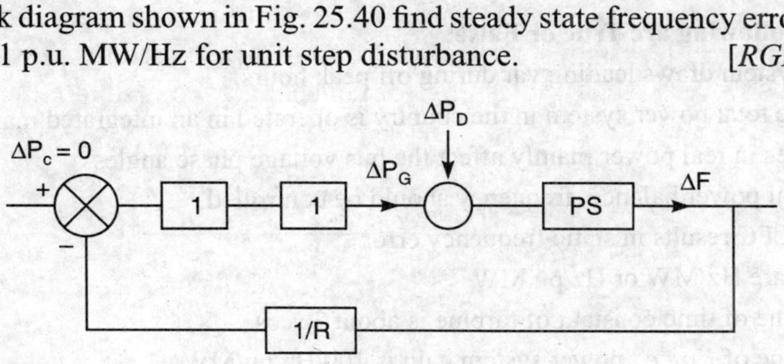

Fig. 25.40

Solution

$$\beta = D + \frac{1}{R} = 0.01 + \frac{1}{3} = 0.3433 \text{ p.u. MW/Hz}$$

For unit step disturbance $M = 1$

$$\Delta f_0 = \frac{M}{\beta} = \frac{1}{0.3433} = 2.913 \text{ Hz}$$

SUMMARY

25.1 Power system interconnection leads to service reliability, saving in installed capacity, decrease in spinning reserve requirements and savings in operating costs.

25.2 A power system draws lagging vars during peak load hours and leading vars during off-peak hours.

25.3 Real power balance is achieved through control of frequency (*i.e.*, load frequency control) and reactive power is controlled through control of voltage (*i.e.*, automatic voltage control).

25.4 Single area load frequency control has speed governing system. This system acts to translate the speed changes resulting from load variations into steam control.

25.5 Primary ALFC leads to static frequency error.

25.6 Secondary ALFC or integral control eliminates static frequency drop.

25.7 Two area load frequency control system uses primary or secondary ALFC for each area and tie line for power transfer.

25.8 Tie line power control may be flat frequency control, flat tie line load control or tie line load bias control.

25.9 Automatic excitation control for an alternator consists of CT, PT, stabilising transformer, regulator input filter, amplifier and voltage limiter.

25.10 Static var system (SVS or static var compensator SVC) consists of Thyristor switched capacitor-Thyristor controlled reactor (TSC-TCR).

25.11 Voltage control can also be achieved through tap changing transformer. Tap changing under load (TCUL) is also possible.

25.12 Regulating transformers for voltage magnitude control and voltage phase angle control are also used.

25.13 Facts controllers help in active and reactive power control.

TEST POINT QUESTIONS

State whether the following are True or False:

25.1 A power system draws leading var during off peak hours.

25.2 In India the total power system in the country is operated in an integrated manner.

25.3 The changes in real power mainly affect the bus voltage phase angles.

25.4 To keep real power balance, frequency should be controlled.

25.5 Primary ALFC results in static frequency error.

25.6 Units of R are Hz/MW or Hz/pu MW.

25.7 Typical value of time constant of turbine is about 3 secs.

25.8 Typical value of Kp *i.e.*, power system gain is 100 Hz/pu MW.

25.9 Time constant of ALFC is about 5 seconds.

25.10 Integral control eliminate static frequency drop.

25.11 In tie line load bias control, the control signal for each area is proportional to change in frequency as well as change in line power.

25.12 MVA rating of regulating transformer is very small.

25.13 TCUL means tap changing under load.

25.14 TSC-TCR can supply only lagging vars.

ANSWERS

25.1 T	25.2 F	25.3 T	25.4 T	25.5 T	25.6 T
25.7 F	25.8 T	25.9 F	25.10 T	25.11 T	25.12 T
25.13 T	25.14 F				

JOB INTERVIEW QUESTIONS

25.1 What is reactive power? What is its importance in a.c. systems?

25.2 What is meant by *p-f* and *QV* control strategies?

25.3 Name the mains parts of single area ALFC system.

25.4 What is meant by power system gain and power system time constant?

25.5 What is meant by speed regulation constant of governor?

25.6 What is single area frequency response characteristic?

25.7 What are primary and secondary ALFC?

25.8 What is static frequency drop of primary ALFC? How can it be eliminated?

25.9 How is an increase in real power demand met in the power system?

25.10 What is the advantage of integral control?

25.11 What is meant by the line load bias control?

25.12 Explain 'automatic load dispatching'.

25.13 What is TCUL?

25.14 How does regulating transformer work.

25.15 What is meant by SVS?

REVIEW QUESTIONS

25.1 Discuss why

(*a*) A power system needs shunt compensation both at times of peak load and off-peak load.

(*b*) The nature of shunt compensation during peak load is different from that during off-peak load.

(*c*) The integral control eliminates static frequency drop.

(*d*) Interconnection of systems results in lesser static frequency drop.

(*e*) Modern systems are usually operated as interconnected systems.

(*f*) When two systems are interconnected the total installed capacity required to meet load requirement is lesser than when they are operated independently.

(*g*) Every power system needs an efficient and quick acting load shedding and restoration programme.

25.2 Discuss the advantages and need of system interconnection. Differentiate between integrated and unintegrated interconnected systems.

25.3 Discuss the requirements of reactive power during peak and off-peak hours.

25.4 Draw a block diagram showing the load frequency and voltage control channels. How is it that the *p.f.* and *QV* loops operate independent of each other?

25.5 Differentiate between single area and multi area load frequency control.

25.6 Draw a block diagram, with illustrative transfer functions, of a single area load frequency control system. Explain the functions of different components of such a control system.

25.7 What do you understand by the term 'Area Control Error'? How does this error differ in a multi area system as compared to that in a single area system?

25.8 Write a short note on 'Automatic Load Dispatching'.

25.9 Write a short note on 'Computer applications in power system operation'.

25.10 How is automatic voltage control different from load frequency control?

25.11 Draw a block diagram of automatic voltage control system of an alternator. Discuss the functions of the different components of the control system.

25.12 Write a short note on load frequency control equipment.

25.13 Differentiate between flat frequency, flat tie line load and tie line load bias control methods for load frequency control of two area systems.

25.14 Discuss the concept of voltage control by reactive power injection.

25.15 Draw a simple connection diagram of a static var system and explain its working. Draw is characteristic curve.

25.16 What are tap changing transformers? How do they help in controlling bus voltages in a power system?

25.17 What is regulating transformer? For what purposes can it be used in power systems. Draw suitable diagrams to illustrate your answer.

25.18 Derive an expression for the static frequency change of a single area ALFC. How can the static frequency change be compensated?

PROBLEMS

25.1 The speed of a 500 MW generator drops by 5% from no load to full load. (*a*) Find speed regulation *R*. Express the results in Hz per unit MW and Hz per MW (*b*). If frequency drops by 0.1 Hz, find the change in power output.

$$\left[\begin{array}{l}(a)\ 2.5\ \text{Hz/pu MW}\ (b)\ 0.5\times10^{-2}\ \text{Hz/ MW}\\(b)\ 20\ \text{MW}\end{array}\right]$$

25.2 Two alternators of ratings 100 MW and 200 MW are feeding a system. The frequency is 50 Hz and each alternator is running at half load. The system load increase by 75 MW. Find the value of *R* for each unit so that they may share the load in proportion to their ratings. Assume that as a result of load increase the frequency drops to 49.75 Hz.

[$R_1 = 0.01$ Hz/MW, $R_2 = 0.005$ Hz/MW]

25.3 A control area has a rated capacity of 20 GW. The regulation *R* for all units in the area is 2.5 Hz/pu MW. A one per cent change in load causes a frequency change of one per cent. If the system is operating at full load and the load decreases by 100 MW find the static frequency change (*a*) if speed governor loop is closed (*b*) if speed governor loop is open.

[(*a*) 0.0119 Hz (*b*) 0.25 Hz]

25.4 Two 60 MW generators are operating in parallel. The machine A experiences a speed drop of 4% from no load to full load. The speed drop for machine *B* is 3%. Find (*a*) load on each machine if total load is 100 MW (*b*) the percentage adjustment of no load speed so that machines may share the load equally.

$$\begin{bmatrix} (a)\ 42.86\text{ MW and } 57.14\text{ MW} \\ (b)\ 0.83\%\text{ increase in no load speed of machine } A \end{bmatrix}$$

25.5 A 100 MVA 50 Hz alternator is operating on full load at rated frequency. The H constant of the machine is 4 kW-sec. per kVA. The load suddenly reduces by 40 MW. A delay of 0.6 seconds occurs in closing the steam valve. Determine the new frequency.

[Frequency = 51.48 Hz]

Appendix-A

Aluminium Conductors Steel Reinforced (ACSR) IS: 398–1976

Stranding and wire diameter mm				Cross-sectional area of aluminium Sq. mm	Overall conductor diameter mm	Resistrance at 20 °C ohm/km	App. ultimate strength kN	Current carrying capacity (A) for a temp. rise of		Weight of conductor kg/km	Code name
aluminium		steel									
No.	dia.	No.	dia.					30°C	40°C		
1	2	3	4	5	6	7	8	9	10	11	12
6	2.11	1	2.11	20.98	6.33	1.403	7.61	112	125	84.7	Squirrel
6	2.36	1	2.36	25.91	7.08	1.098	9.37	130	148	106.0	Gopher
6	2.59	1	2.59	31.61	7.77	0.935	11.12	144	160	127.7	Weasel
6	3.0	1	3.0	41.87	9.0	0.6795	14.95	170	190	171.3	Ferret
6	3.35	1	3.35	52.88	10.05	0.556	18.25	192	218	213.6	Rabbit
6	3.66	1	3.66	62.32	10.98	0.4565	21.70	210	236	255.0	Mink
6	3.99	1	3.99	74.07	11.97	0.3841	25.80	230	260	303.0	Beaver
6	4.09	1	4.09	78.83	12.27	0.3736	26.91	240	270	318.4	Racoon
6	4.22	1	4.22	83.92	12.66	0.3430	28.67	250	280	339	Otter
6	4.50	1	4.50	94.21	13.55	0.302	32.60	265	300	385.4	Cat
6	4.72	7	1.57	105	14.15	0.281	32.41	280	320	394	Dog
6	5.28	7	1.76	131.37	15.84	0.219	40.58	360	410	493	Leopard

(Contd...)

(Table Contd...)

1	2	3	4	5	6	7	8	9	10	11	12
30	2.36	7	2.36	128.1	16.52	0.222	56.50	350	400	603.8	Tiger
30	2.59	7	2.59	158.1	18.13	0.188	67.34	400	455	727.4	Wolf
30	2.79	7	2.79	183.4	19.53	0.159	77.98	440	500	844	Lynx
30	3.0	7	3.0	212.1	21.0	0.140	89.67	470	540	974	Panther
30	3.18	7	3.18	238.26	22.26	0.122	100.1	525	605	1097	Lion
30	3.35	7	3.35	264.40	23.45	0.11	110.9	570	640	1219	Bear
30	3.71	7	3.71	316.5	25.97	0.0898	135.5	640	720	1492.4	Goat
30	3.99	7	3.99	366.1	27.93	0.0777	157.0	700	800	1726.3	Sheep
30	4.27	7	4.27	429.6	29.89	0.068	178.8	810	920	1977	Deer
42	3.50	7	1.96	404.1	26.88	0.0736	88.79	770	880	1282	Kundah
42	4.13	7	2.30	562.7	31.68	0.05265	120.16	890	1020	1790	Markulla
54	3.18	7	3.18	428.9	28.62	0.0691	130.32	805	905	1621	Zebra
54	3.35	7	3.35	475.96	30.17	0.0623	144.7	830	930	1802	Camel
54	3.53	7	3.53	528.5	31.77	0.0563	159.6	860	970	1998	Moose

Appendix-B

AN INTRODUCTION TO MATRICES

Definitions

Matrix. A matrix is a rectangular array of numbers called elements arranged in a systematic manner. The elements may be real or complex. An element in ith row and jth column is written as a_{ij}. A matrix having m rows and n column is known as $m \times n$ matrix. Thus,

$$A = \begin{bmatrix} a_{11} & a_{12} & . & . & a_{1n} \\ a_{21} & a_{22} & . & . & a_{2n} \\ . & . & . & . & . \\ . & . & . & . & . \\ a_{m1} & a_{m2} & . & . & a_{mn} \end{bmatrix}$$

When $m = n$, i.e., the number of rows and columns are equal, the matrix is known as a square matrix. A matrix having a single column and m rows (*i.e.*, $m \times 1$ matrix) is a column matrix. A matrix having a single row and n columns (*i.e.*, $1 \times n$) is known as row matrix.

Diagonal Matrix. A square matrix (*i.e.*, $n \times n$ matrix) in which all off diagonal elements are zero, e.g.,

$$A = \begin{bmatrix} a_{11} & 0 & 0 & 0 \\ 0 & a_{22} & 0 & 0 \\ 0 & 0 & a_{33} & 0 \\ 0 & 0 & 0 & a_{44} \end{bmatrix}$$

Identity or Unit Matrix. If each of the diagonal elements of a diagonal matrix equals 1 (the non-diagonal elements are zero), it is called an identity or unit matrix and is written as $[I]$

$$[I] = \begin{bmatrix} 1 & 0 & 0 \\ 0 & 1 & 0 \\ 0 & 0 & 1 \end{bmatrix}$$

Sparse Matrix. If many elements of a matrix are zeros it is called a sparse matrix.

Null Matrix. If each element of a matrix is zero, it is a null matrix.

Transpose of a Matrix. If the rows and columns of an $m \times n$ matrix A are interchanged the resultant $n \times m$ matrix is called the transpose of matrix A and is written as A^t. Thus if

$$A = \begin{bmatrix} a_{11} & a_{12} & a_{13} \\ a_{21} & a_{22} & a_{23} \end{bmatrix}$$

Then
$$A^t = \begin{bmatrix} a_{11} & a_{21} \\ a_{12} & a_{22} \\ a_{13} & a_{23} \end{bmatrix}$$

It is obvious that $(A^t)^t = A$

If a matrix is equal to its transpose, *i.e.*, $= A^t, = A$ the matrix is known as symmetric matrix.

Conjugate Matrix. If each element of a matrix A is replaced by its conjugate (*i.e.*, $a + jb$ is replaced by $a - jb$) the resultant matrix is the conjugate matrix and is designated as A^*.

If all the elements of a matrix are real, $A = A^*$. If all the elements of a matrix are purely imaginary, $A = -A^*$.

Determinant of a Matrix. For each square matrix there exists a determinant which is formed by taking the determinant of the elements of the matrix. It is a scalar quantity, *i.e.*, if

$$A = \begin{bmatrix} 1 & -3 & 2 \\ 2 & -1 & 5 \\ 4 & 0 & 2 \end{bmatrix}$$

$$\det(A) = (1)\begin{vmatrix} -1 & 5 \\ 0 & 2 \end{vmatrix} - (-3)\begin{vmatrix} 2 & 5 \\ 4 & 2 \end{vmatrix} + (2)\begin{vmatrix} 2 & -1 \\ 4 & 0 \end{vmatrix}$$

$$= 1(-2) + 3(4 - 20) + 2(4) = -42$$

Minor. The minor M_{ij} of an $n \times n$ matrix is the determinant of $(n - 1) \times (n - 1)$ matrix formed by deleting ith row and jth column of the $n \times n$ matrix.

Cofactor. The cofactor C_{ij} of element a_{ij} of a matrix A is defined as

$$C_{ij} = (-1)^{i+j} M_{ij}$$

Adjoint Matrix. The adjoint matrix of a square A is formed by replacing each element a_{ij} of the matrix A by its cofactor C_{ij} and then transposing. It is denoted by *Adj. A*.

Singular and Non-singular matirx. A matirx is singular if its determinant is zero and non-singular if its determinant is non-zero.

Matrix Operations

Equality of Matrices. Two matrices A and B are equal if each element a_{ij} of matrix A is equal to the corresponding element b_{ij} of matrix B. (Evidently matrices A and B must have the same dimensions).

Then $A = B$.

Addition and Subtraction of Matrices. The addition and subtraction of matrices is defined when they have the same dimensions. The sum or the differences of two matrices A and B is matirx C of the same dimensions, *i.e.*,

$$A \pm B = C$$

where each element of C is

$$c_{ij} = a_{ij} \pm b_{ij} \text{ for all values of } i \text{ and } j$$

Multiplication of Matrix by a Scalar. When a matrix is multiplied by a scalar, the elements of the resultant matirx are equal to the product of the original elements and the scalar, e.g.,

$$kA = B$$

where $b_{ij} = k\, a_{ij}$ for all values of i and j.

Multiplication of Matrices. The multiplication of two matrices A and B is defined only if the number of columns in the first matrix A is equal to the number of rows in the second matrix B. Thus, when a matrix A of dimension $m \times p$ is multiplied with a matrix B of dimension $p \times n$, the resultant matrix C has the dimension $m \times n$. Any element c_{ij} of the resultant matrix C is found as

$$c_{ij} = a_{ij}\, b_{ij} + a_{i2}\, b_{2j} + \ldots\ldots + a_{ip}\, b_{pj}$$

or

$$c_{ij} = \sum_{k=1}^{P} a_{ik}\, b_{kj}$$

where $i = 1, 2 \ldots m$ and $j = 1, 2 \ldots n$. For example

$$AB = \begin{bmatrix} a_{11} & a_{12} \\ a_{21} & a_{22} \\ a_{31} & a_{32} \end{bmatrix} \begin{bmatrix} b_{11} & b_{12} \\ b_{21} & b_{22} \end{bmatrix} \begin{bmatrix} a_{11}\, b_{11} + a_{12}\, b_{21} & a_{11}\, b_{12} + a_{12}\, b_{22} \\ a_{21}\, b_{11} + a_{22}\, b_{21} & a_{21}\, b_{12} + a_{22}\, b_{22} \\ a_{31}\, b_{11} + a_{32}\, b_{21} & a_{31}\, b_{12} + a_{32}\, b_{22} \end{bmatrix}$$

In the above product A premultiplies B or B post-multiplies A. The product BA is not defined because the number of columns of B is not equal to the number of row of A.

Inverse of a Matrix. Division does not exist in matrix algebra but the object of division is achieved by obtaining the inverse of a matrix. If A is a non-singular square matrix, its inverse A^{-1} is defined by the equation

$$A^{-1} A = A\, A^{-1} = I$$

where I is the identity or unit matrix.

A^{-1} can be computed from the equation

$$A^{-1} = \frac{[Adj\ A]}{|A|}$$

Where $[Adj.\ A]$ is the adjoint of A and $|A|$ is the determinant of A. It is evident that if a matix is singular its inverse is not defined because the determinant of a singular matrix is zero.

Example

Find $\quad A^{-1}$ if $A = \begin{bmatrix} 1 & 2 & 4 \\ -1 & 0 & 3 \\ 3 & 1 & -2 \end{bmatrix}$

$$\det(A) = |A| = (1)\begin{bmatrix} 0 & 3 \\ 1 & -2 \end{bmatrix} - (2)\begin{bmatrix} -1 & 3 \\ 3 & -2 \end{bmatrix} + (4)\begin{bmatrix} -1 & 0 \\ 3 & 1 \end{bmatrix} = 7$$

$$Adj.\,A = \begin{bmatrix} \begin{vmatrix} 0 & 3 \\ 1 & -2 \end{vmatrix} & -\begin{vmatrix} -1 & 3 \\ 3 & -2 \end{vmatrix} & \begin{vmatrix} -1 & 0 \\ 3 & 1 \end{vmatrix} \\ -\begin{vmatrix} 2 & 4 \\ 1 & -2 \end{vmatrix} & \begin{vmatrix} 1 & 4 \\ 3 & -2 \end{vmatrix} & -\begin{vmatrix} 1 & 2 \\ 3 & 1 \end{vmatrix} \\ \begin{vmatrix} 2 & 4 \\ 0 & 3 \end{vmatrix} & -\begin{vmatrix} 1 & 4 \\ -1 & 3 \end{vmatrix} & \begin{vmatrix} 1 & 2 \\ -1 & 0 \end{vmatrix} \end{bmatrix}^t$$

$$Adj.\,A = \begin{bmatrix} -3 & 7 & -1 \\ 8 & -14 & 5 \\ 6 & -7 & 2 \end{bmatrix}^t = \begin{bmatrix} -3 & 8 & 6 \\ 7 & -14 & -7 \\ -1 & 5 & 2 \end{bmatrix}$$

$$A^{-1} = \frac{1}{7}\begin{bmatrix} -3 & 8 & 6 \\ 7 & -14 & -7 \\ -1 & 5 & 2 \end{bmatrix} = \begin{bmatrix} -\frac{3}{7} & \frac{8}{7} & \frac{6}{7} \\ 1 & -2 & -1 \\ -\frac{1}{7} & \frac{5}{7} & \frac{2}{7} \end{bmatrix}$$

Matrix Partitioning. The elementary rules of matrix algebra lend themselves to useful extension. Instead of handling individual matrix elements, when any matrix operation is to be performed, it is possible to manipulate with groups of elements. The matrices are first partitioned into sub-matrices. The resulting sub-matrices obey the same rules as individual elements. However, the dimension of the sub-matrices must be conformable for the matrix operation to be performed. Let two matrices A and B be partitioned as

$$A = \begin{bmatrix} a_{11} & a_{12} & a_{13} \\ a_{21} & a_{22} & a_{23} \\ \hdashline a_{31} & a_{32} & a_{33} \end{bmatrix} = \begin{bmatrix} A_1 \\ A_2 \end{bmatrix}$$

$$B = \left[\begin{array}{cc:c} b_{11} & b_{12} & b_{13} \\ b_{21} & b_{22} & b_{23} \\ b_{31} & b_{32} & b_{33} \end{array}\right] = [B_1 \vdots B_2]$$

$$AB = \begin{bmatrix} A_1 \\ \hdashline A_2 \end{bmatrix} [B_1 \vdots B_2] = \begin{bmatrix} A_1 B_1 & A_1 B_2 \\ A_2 B_1 & A_2 B_2 \end{bmatrix}$$

$$= \begin{bmatrix} a_{11} b_{11} + a_{12} b_{21} + a_{13} b_{31} & a_{11} b_{12} + a_{12} b_{22} + a_{13} b_{32} & a_{11} b_{13} + a_{12} b_{23} + a_{13} b_{33} \\ a_{21} b_{11} + a_{22} b_{21} + a_{23} b_{31} & a_{21} b_{12} + a_{22} b_{22} + a_{23} b_{32} & a_{21} b_{13} + a_{22} b_{23} + a_{23} b_{33} \\ \hdashline a_{31} b_{11} + a_{32} b_{21} + a_{33} b_{31} & a_{31} b_{12} + a_{32} b_{22} + a_{33} b_{32} & a_{31} b_{13} + a_{32} b_{23} + a_{33} b_{33} \end{bmatrix}$$

The dotted lines indicated the extent of sub-products. It is evident that if A is partitioned as:

$$\left[\begin{array}{c:cc} a_{11} & a_{12} & a_{13} \\ a_{21} & a_{22} & a_{23} \\ a_{31} & a_{32} & a_{33} \end{array}\right] \quad \text{and } B \text{ as} \quad \left[\begin{array}{c:cc} b_{11} & b_{12} & b_{13} \\ b_{21} & b_{22} & b_{23} \\ b_{31} & b_{32} & b_{33} \end{array}\right]$$

the product AB cannot be obtained. Thus, the partitioning has to be done properly.

Node Elimination by Matrix Algebra. To understand the method of node elimination by matrix algebra, let us write the equations for a 3-node network in impedance form.

$$V_1 = z_{11} I_1 + z_{12} I_2 + z_{13} I_3$$
$$V_2 = z_{21} I_1 + z_{32} I_2 + z_{33} I_3$$
$$V_3 = z_{31} I_1 + z_{32} I_2 + z_{33} I_3$$

It is desired to eliminate node 3. Connect this to reference node so that V_3 becomes zero. Then

$$0 = z_{31} I_1 + z_{32} I_2 + z_{33} I_3$$

or

$$I_3 = -\frac{z_{31}}{z_{33}} I_1 - \frac{z_{32}}{z_{33}} I_2$$

Substituting this value of I_3 in the equations for I_1 and I_2 we have

$$V_1 = z_{11} I_1 + z_{12} I_2 + z_{13}\left(-\frac{z_{31}}{z_{33}} I_1 - \frac{z_{32}}{z_{33}} I_2\right)$$

or

$$V_1 = \left(z_{11} - \frac{z_{13}\, z_{31}}{z_{33}}\right) I_1 + \left(z_{12} - \frac{z_{13}\, z_{32}}{z_{33}}\right) I_2$$

$$V_2 = z_{21} I_1 + z_{22} I_2 + z_{23}\left(-\frac{z_{31}}{z_{33}} I_1 - \frac{z_{32}}{z_{33}} I_2\right)$$

$$= \left(z_{21} - \frac{z_{23}\, z_{31}}{z_{33}}\right) I_1 + \left(z_{22} - \frac{z_{23}\, z_{32}}{z_{33}}\right) I_2$$

Thus, the new impedance matrix is

$$\begin{bmatrix} z_{11} - \dfrac{z_{13}\, z_{31}}{z_{33}} & z_{12} - \dfrac{z_{13}\, z_{32}}{z_{33}} \\ z_{21} - \dfrac{z_{23}\, z_{31}}{z_{33}} & z_{22} - \dfrac{z_{23}\, z_{32}}{z_{33}} \end{bmatrix}$$

$$= \begin{bmatrix} z_{11} & z_{12} \\ z_{21} & z_{22} \end{bmatrix} - \frac{1}{z_{33}} \begin{bmatrix} z_{13} \\ z_{23} \end{bmatrix} [z_{31} \quad z_{32}]$$

This result can be generalized as under. Let

$$[Z] = \begin{bmatrix} z_{11} & z_{12} & : & : & : & : & z_{1n} \\ z_{21} & z_{22} & : & : & : & : & z_{2n} \\ .. & .. & : & : & : & : & .. \\ z_{n1} & z_{n2} & : & : & : & : & z_{nn} \end{bmatrix}$$

If it is desired to eliminate *n*th row and *n*th column of the above matrix, the new matrix $[Z]_{new}$ is given by

$$[Z]_{new} = \begin{bmatrix} z_{11} & z_{12} & : & : & z_{1,\,n-1} \\ z_{21} & z_{22} & : & : & z_{2,\,n-1} \\ : & : & : & : & : \\ : & : & : & : & : \\ z_{n-1,\,1} & z_{n-1,\,2} & : & : & z_{n-1,\,n-1} \end{bmatrix} - \frac{1}{z_{nn}} \begin{bmatrix} z_{1n} \\ z_{2n} \\ z_{n-1,\,n} \end{bmatrix} \begin{bmatrix} z_{n-1,\,1} & z_{n-1,\,2} & \cdot\cdot & z_{n-1,\,n-1} \end{bmatrix}$$

In general if *l*th row and *l*th column are to be eliminated (*i.e. l*th node is to be eliminated) in an $n \times n$ matrix, then

$$Z_{ij,\ new} = Z_{ij,\ old} - \frac{Z_{il}\ Z_{lj}}{z_{ll}} \quad \text{for } i, j = 1, 2 \ldots n$$

where $Z_{ij,\ new}$ = element in *i*th row and *j*th column of final matrix (*i.e.*, after node elimination)

$Z_{ij,\ old}$ = element in *i*th column row and *j*th column of initial matrix (*i.e.*, before node elimination)

Z_{il} = element in *i*th row and *l*th column of initial matrix

Z_{lj} = element in *l*th row and *j*th column of initial matrix

Z_{ll} = element in *l*th row and *l*th column of initial matrix

Appendix-C

OBJECTIVE QUESTIONS

1. The highest transmission voltage used in India is
   (a) 220 kV (b) 400 kV
   (c) 765 kV (d) 500 kV
2. The highest voltage used in AC Transmission in USA is
   (a) 1000 kV (b) 765 kV
   (c) 525 kV (d) 500 kV
3. The number of subconductors in each phase of EHV lines in India is
   (a) 4 (b) 3
   (c) 2 (d) 8
4. The maximum number of subconductors in each phase of EHV lines in USA is
   (a) 4 (b) 2
   (c) 8 (d) 3
5. The spacing between subconductors of a bundle is approximately
   (a) 20 cm (b) 40 cm
   (c) 80 cm (d) 10 cm
6. As per Indian Electricity rules the maximum difference between declared voltage and actual voltage at consumer premises should be
   (a) ± 6% (b) ± 5%
   (c) ± 8% (d) ± 10%
7. The approximate cost ratio of a 220 kV underground cable transmission and 220 kV overhead transmission is
   (a) 25 (b) 13
   (c) 5 (d) 50
8. The transmission lines generally use
   (a) copper conductor
   (b) all aluminium conductor
   (c) ACSR conductor
   (d) none of the above
9. The ratio of the diameters of an ACSR conductor and a copper conductor for the same resistance per unit length is
   (a) equal to 1 (b) greater than 1
   (c) less than 1 (d) abount 0.1
10. In USA the next higher transmission voltage is likely to be
    (a) 1100 kV (b) 765 kV
    (c) 1500 kV (d) 2000 kV
11. For a stranded conductor, the ratio of GMR to actual radius is
    (a) equal to 1
    (b) equal to 0.7788
    (c) less than 0.7788
    (d) equal to 1.1
12. As frequency increases, the magnetic flux inside a conductor
    (a) increases
    (b) decreases
    (c) remains constant
    (d) any of the above
13. The inductance of a line is minimum when
    (a) GMD is high
    (b) GMR is high
    (c) Both GMD and GMR are high
    (d) GMD is low and GMR is high
14. As the line inductance increases, the transmission capacity
    (a) increases

(b) decreases
(c) remains the same
(d) any of the above

15. The transmission capacity of a line at 50 Hz frequency as compared to that at 60 Hz frequency is
(a) higher (b) lower
(c) the same (d) any of the above

16. As the conductor diameter is increased an appreciable decrease occurs in
(a) line resistance
(b) line inductance
(c) line capacitance
(d) all the above

17. As the length of a line increases, the charging current
(a) increases
(b) decreases
(c) remain the same
(d) any of the above

18. As the frequency is increased, the charging Mvar
(a) decreases
(b) increases
(c) remains the same
(d) none of the above

19. As the frequency is increased, the skin effect
(a) decreases
(b) increases
(c) remains the same
(d) any of the above

20. As the spacing between the phase conductors increases, the line resistance
(a) increases
(b) decreases
(c) remains the same
(d) any of the above

21. As the spacing between the phase conductors increases, the line inductance
(a) increases
(b) decreases
(c) remains the same
(d) any of the above

22. As the spacing between the phase conductors increases, the line capacitance
(a) increases
(b) decreases
(c) remains the same
(d) any of the above

23. The presence of ground causes the line capacitance to
(a) increase by about 10%
(b) decrease by about 10%
(c) increase by about 0.2%
(d) none of the above

24. As compared to the sending end voltage, the receiving end voltage of a short line under no load condition is
(a) higher (b) lower
(c) the same (d) any of the above

25. As compared to regulation at unity power factor, the regulation at 0.8 leading power factor is
(a) higher (b) lower
(c) the same (d) any of the above

26. The regulation of a line at full load 0.8 power factor lagging is 11%. The regulation at full load 0.8 power factor leading can be
(a) about 20% (b) about 15%
(c) about 3% (d) about 30%

27. For a shortline with R/X ratio 1.0, the regulation will be zero when the load power factor is
(a) unity (b) 0.707 leading
(c) 0 leading (d) 0.5

28. For a long line, the receiving end voltage under no load condition is
(a) equal to sending end voltage
(b) less than sending end voltage
(c) more than sending end voltage
(d) any of the above

29. For a transmission line
$$V_s = A\,V_r + B\,I_r$$
$$I_s = C\,V_r + D\,I_r$$
Then I_r equals
(a) $-C\,V_s + A\,I_s$ (b) $D\,V_s + A\,I_s$
(c) $D\,V_s - B\,I_s$ (d) $A\,V_s + D\,I_s$

30. For the line of questions 29, V_r equals
(a) $D\,V_s - B\,I_s$ (b) $C\,V_s + A\,I_s$
(c) $D\,V_s + C\,I_s$ (d) $D\,V_s + B\,I_s$

31. The shunt admittance of a line is y siemens per unit length. The shunt

admittance for a length l will be
(a) yl (b) yl^2
(c) y/l (d) yl^3

32. The dc resistance of a conductor is R. The resistance at 50 Hz frequency is likely to be
(a) R (b) $1.04\ R$
(c) $1.2\ R$ (d) $2\ R$

33. The constant $1/a_o$ for aluminium is
(a) 228 (b) 240
(c) 210 (d) 110

34. The increase in resistance due to nonuniform distribution of current in the conductor is known as
(a) proximity effect
(b) skin effect
(c) ferranti effect
(d) none of the above

35. For a line with conductor spacings of 6 m, 6 m and 12 m, D_{eq} is
(a) 7.56 m (b) 10.23 m
(c) 8.02 m (d) 11 m

36. For a single circuit 3-phase line the inductance per phase per km is of the order of
(a) 1 mH (b) 10 mH
(c) 0.2 mH (d) 50 mH

37. For a single circuit 3-phase line the capacitance per phase per km is of the order of
(a) 0.01 μF (b) 0.05 μF
(c) 0.001 μF (d) 0.1 μF

38. The per unit impedance of a line, to a 50 MVA 132 kV base, is 0.4. The per unit impedance to a 100 MVA, 132 kV base will be
(a) 0.2 (b) 0.4
(c) 0.8 (d) 1.0

39. An equipment has a per unit impedance of 0.9 *pu* to a base of 20 MVA, 33 kV. The per unit impedance to the base of 20 MVA, 11 kV will be
(a) 0.9 (b) 2.7
(c) 0.3 (d) 8.1

40. For a good voltage profile under load conditions, a long line needs
(a) shunt capacitors at receiving end
(b) shunt reactors at receiving end
(c) shunt reactors at sending end
(d) all the above

41. For a good voltage profile under no load condition a long line needs
(a) shunt capacitors at receiving end
(b) shunt reactors at receiving end
(c) shunt resistance at receiving end
(d) all the above

42. The SIL for a single circuit 220 kV line is around
(a) 120 MW (b) 90 MW
(c) 220 MW (d) 500 MW

43. The surge impedance of a single circuit high voltage line is of the order of
(a) 400 ohms (b) 300 ohms
(c) 500 ohms (d) 1000 ohms

44. The SIL of a line is proportional to
(a) V^2 (b) V
(c) $\sqrt{V}$ (d) V^4

45. Ferranti effect in transmission line is due to the presence of
(a) line inductance
(b) line capacitance
(c) line capacitance and inductance
(d) line resistance and capacitance

46. The real power transfer over a line mainly depends on
(a) power angle (b) $|V_s| - |V_r|$
(c) V_s (d) V_r

47. The reactive power transfer over a line mainly depends on
(a) power angle (b) V_r
(c) $|V_s| - |V_r|$ (d) V

48. The 11 kV lines use
(a) pin insulators
(b) suspension insulators
(c) both pin and suspension insulators
(d) none of the above

49. For a 220 kV line, the number of discs in an insulator string is around
(a) 5 (b) 15
(c) 10 (d) 20

50. The string efficiency of a high voltage line is around
(a) 1.0 (b) 0.8
(c) 0.3 (d) 0.1

51. If the cross arms of a high voltage line are made longer, the string efficiency
(a) increases
(b) decreases
(c) remains the same
(d) any of the above

52. In actual practice the string efficiency is improved by using
(a) capacitance grading
(b) grading ring
(c) discs of different sizes
(d) any of the above

53. For a 400 kV line the number of discs in an insulator string is around
(a) 31 (b) 22
(c) 16 (d) 10

54. As compared to power frequency dry flashover voltage, the power frequency wet flashover voltage is
(a) higher (b) lower
(c) the same (d) any of the above

55. The span of high voltage lines is around
(a) 100 m (b) 300 m
(c) 500 m (d) 1000 m

56. For a 400 kV line the spacing between phase conductors is around
(a) 11 m (b) 8 m
(c) 14 m (d) 20 m

57. An increase in temperature of line conductors results in
(a) increase in length
(b) increase in tension
(c) decrease in sag
(d) increase in length and decrease in tension

58. Modern high voltage lines are designed with a safety factor of about
(a) 1.5 (b) 2.0
(c) 3.0 (d) 4.0

59. Frequency of resonant vibrations lies in the range of
(a) 50 – 100 Hz (b) 10 – 20 Hz
(c) 100 – 200 Hz (d) 500 Hz

60. At NTP, the breakdown strength of air is
(a) 30 kV – peak/cm
(b) 30 kV – rms/cm
(c) 40 kV – peak/cm
(d) 50 kV/cm

61. As pressure decreases the disruptive critical voltage
(a) decrease
(b) increases
(c) remains the same
(d) any of the above

62. As temperature increases the disruptive critical voltage
(a) decreases
(b) increases
(c) remains the same
(d) any of the above

63. As the spacing between conductors decreases, the disruptive critical voltage
(a) decreases
(b) increases
(c) remains the same
(d) any of the above

64. As the conductor radius increases the disruptive critical voltage
(a) decreases
(b) increases
(c) remains the same
(d) any of the above

65. As pressure decreases the corona loss
(a) decreases
(b) increases
(c) remains the same
(d) any of the above

66. The corona loss in a 50 Hz system is 0.2 kW per phase per km. At a frequency of 60 Hz the corona loss would be
(a) 0.17 kW/phase/km
(b) 0.24 kW/phase/km
(c) 0.22 kW/phase/km
(d) 0.3 kW/phase/km

67. As compared to plains the corona loss in hilly areas is
(a) more (b) less
(c) the same (d) any of the above

68. As compared to the corona loss in a copper conductor line, the corona loss in a line with ACSR conductor and the same current carrying capacity, is
(a) less (b) more
(c) the same

69. Which of the following faults would cause a greater electromagnetic interference between the power line and a nearby communication line
(a) 3-phase fault
(b) single line to ground
(c) line to line fault

70. The insulating material most commonly used for power cables is
(a) PVC (b) paper
(c) rubber

71. The insulation resistance of a 2 km long cable is 150 MΩ. For a length of 20 km, the insulation resistance will be
(a) 1500 MΩ (b) 15 MΩ
(c) 300 MΩ (d) 150 MΩ

72. In a 3-core cable, the capacitance between 2 conductors (with sheath earthed) is 3 μF. The capacitance per phase is
(a) 1.5 μF (b) 3 μF
(c) 6 μF

73. The loss angle of a cable is δ. The power factor is
(a) cos δ (b) sin δ
(c) tan δ (d) sec δ

74. To obtain the minimum value of stress in cable the ratio R/r should be
(a) 2.13 (b) 2.718
(c) 1.96 (d) 1.1

75. In a cable the sheath radius is R and conductor radius is r. As r changes from 0.5 R, to 0.25 R, the maximum voltage gradient in the dielectric
(a) decreases by about 6%
(b) increases by about 6%
(c) increase by about 15%
(d) increases by about 25%

76. The thermal resistivity of soil is around
(a) 0.2 °C-m/Watt
(b) 1.5 °C-m/Watt
(c) 5 °C-m/Watt
(d) 10° C-m/Watt

77. The thermal resistivity of cable dielectric is around
(a) 0.2 °C-m/Watt(b) 1.5 °C-m/Watt
(c) 5 °C-m/Watt (d) 10° C-m/Watt

78. In a power system the maximum number of buses are
(a) PQ buses (b) PV buses
(c) slack buses (d) any of the above

79. At a load bus in power system, the quantities specified are
(a) P and V (b) P and $|V|$
(c) P and δ (d) P and Q

80. At a generator bus in power systems, the quantities specified are
(a) P and V (b) P and $|V|$
(c) P and δ (d) P, V and δ

81. The slack bus is
(a) generator bus
(b) load bus
(c) none of the above
(d) any of the above

82. For accurate load flow calculations on large power systems, the best method is
(a) G-S method
(b) NR method
(c) De-coupled method
(d) any of the above

83. For fast load flow calculations, the best method is
(a) G-S method (b) NR method
(c) De-coupled method
(d) any of the above

83. For fast load flow calculations, the best method is
(a) G-S method
(b) NR method in rectangular coordinates
(c) NR method in polar coordinates
(d) De-coupled method

84. An acceleration factor is used in load flow studies by
(a) G-S method
(b) NR method
(c) De-coupled method
(d) all the above

85. A suitable value of acceleration factor for load flow studies is
(a) 2.2 (b) 1.6
(c) 1.1 (d) 3.1

86. For a fault at generator terminals the fault current is maximum for

(a) 3-phase fault (b) L-L fault
(c) SLG fault (d) DLG fault

87. A single line to ground fault means
(a) A fault between phase *R* and ground
(b) A fault between phase *Y* and ground
(c) A fault between phase *B* and ground
(d) A fault between any one phase and ground

88. The zero sequence fault currents are absent when the fault is
(a) SLG (b) L-L
(c) DLG (d) all the above

89. The method of neutral grounding affects the
(a) positive sequence network
(b) negative sequence network
(c) zero sequence network
(d) all the above

90. The symmetrical components are used in the fault analysis because
(a) the number of equations becomes smaller
(b) the sequence networks do not have mutual coupling
(c) the results are required in terms of symmetrical components
(d) all the above

91. Fault level means
(a) fault current
(b) voltage at the point of fault
(c) fault MVA
(d) fault power factor

92. In load flow studies a PV bus is treated as a PQ bus when
(a) voltage at the bus becomes high
(b) reactive power goes beyond limit
(c) phase angle becomes high
(d) any of the above

93. Fault calculations using computer are usually done by
(a) Y_{bus} method
(b) Z_{bus} method
(c) none of the above
(d) any of the above

94. Y_{bus} as used in the load flow studies and Z_{bus} as used for short circuit studies
(a) are the same
(b) are inverse of each other
(c) are not related
(d) any of the above

95. The passive positive and negative sequence bus impedance matrices are
(a) exactly similar
(b) usually assumed to be similar
(c) completely different
(d) a or c

96. The velocity of propagation of electromagnetic waves on overhead lines is
(a) 3×10^8 m/sec
(b) 3×10^8 km/sec
(c) 3×10^8 m/hour
(d) none of the above

97. In a cable with the insulation permittivity 4, the velocity of propagation of electromagnetic waves is
(a) 3×10^8 m/sec
(b) 0.75×10^8 m/sec
(c) 1.5×10^8 m/sec
(d) 10^8 m/sec

98. The surge impedance of a cable is about
(a) 300 ohms (b) 50 ohms
(c) 5 ohms (d) 2 ohms

99. A 25 km long overhead line has a surge impedance of 400 ohms. For 100 km length the surge impedance will be
(a) 100 ohms (b) 400 ohms
(c) 1600 ohms (d) 25 ohms

100. A voltage wave + 200/1/50 means
(a) a positive wave having a crest of 200 kV, front 1 sec. and tail 50 secs.
(b) a positive wave having a crest of 200 kV, front of 1 μ-sec and tail of 50 μ-sec.
(c) a positive wave having a crest of 200 V, front of 1 μ-sec and tail of 50 μ-sec.
(d) none of the above

101. The tail time of a wave means
(a) time from beginning of wave to half the crest value on the tail
(b) time from the crest of the wave to half the crest value on the tail
(c) time from the crest of the wave to zero value on the tail.

(d) none of the above

102. If a wave is represented by a step wave the analytical results are
(a) optimistic (b) pessimistic
(c) none of the above
(d) any of the above

103. When a wave reaches an open circuit
(a) the voltage at the termination is double the incident voltage
(b) the current at the termination is double the incident current
(c) both the voltage and current at the termination are double their incident values
(d) any of the above

104. When a wave reaches a short circuit
(a) the voltage at the termination is zero
(b) the current at the termination is zero
(c) both the voltage and current at the termination are zero.
(d) any of the above

105. A line of surge impedance 400 ohms is terminated by a resistance of 400 ohms. The reflected quantities are
(a) 0
(b) equal to incident quantities
(c) half the incident quantities
(d) double the incident quantities

106. A line of surge impedance Z branches into two lines each of surge impedance Z. The transmitted voltage is
(a) 0
(b) equal to incident voltage
(c) two third incident voltage
(d) thrice the incidence voltage

107. A suitable value of shielding angle for HV lines is
(a) 60° (b) 30°
(c) 20° (d) 10°

108. A suitable value of shielding angle for EHV lines is
(a) 60° (b) 20°
(c) 45° (d) 10°

109. The lines with horizontal configuration usually have
(a) one ground wire
(b) two ground wires
(c) four ground wires
(d) no ground wire

110. A backward wave means
(a) a wave travelling in the negative direction
(b) a negative voltage wave
(c) a negative current waves
(d) either (a) or (b)

111. In transmission lines the ground wires are placed
(a) above the phase conductors
(b) below the phase conductors
(c) in between the phase conductors
(d) (a) or (b)

112. The power systems are operated with power angle around
(a) 10° (b) 30°
(c) 70° (d) 80°

113. The units of inertia constant H are
(a) J. per MVA
(b) kV per MVA
(c) MJ per MVA
(d) kA per MVA

114. The inertia constant H of a turbo generator of 200 MVA is 6.
The value of H corresponding to a base of 300 MVA will be
(a) 9 (b) 4
(c) 6 (d) 13.5

115. Two synchronous machines having inertia constants M_1 and M_2 are swinging together. The inertia constant of the combination is
(a) $M_1 + M_2$
(b) $M_1 - M_2$
(c) $M_1M_2/(M_1 + M_2)$
(d) any of the above

116. A 90 MVA, 11 kV water wheel generator has an inertia constant $H = 3$. The stored energy in the rotor at synchronous speed is
(a) 30 kJ (b) 270 kJ
(c) 270 MJ (d) 810 MJ

117. The addition of a synchronous compensator in the system
(a) improves systems stability
(b) has no effect on stability

(c) decreases system stability
(d) (a) or (b)

118. A synchronous machine having $E = 1.1$ *pu* is feeding an infinite bus with voltage of 1 *pu*. If the transfer reactance is 0.5 *pu*., the steady state power limit is
(a) 1.1 *pu* (b) 0.55 *pu*
(c) 2.2 *pu* (d) 2.6 *pu*

119. The series compensation
(a) increases stability limit
(b) decreases stability limit
(c) has no effect on stability limit
(d) (a) or (b)

120. A synchronous generator is a feeding an infinite bus through a line. At the middle of the line a shunt reactor is added. The stability limit will
(a) increase
(b) decrease
(c) remain the same
(d) any of the above

121. A synchronous generator is feeding an infinite bus through a line. At the middle of the line a shunt capacitor is added. The stability limit will
(a) increase
(b) decrease
(c) remain the same
(d) any of the above

122. A synchronous generator is feeding an infinite bus. A shunt reactor is added at the infinite bus. The stability limit will
(a) increase
(b) decrease
(c) remain the same
(d) any of the above

123. A synchronous generator is feeding an infinite bus. At the infinite bus a shunt capacitor is added. The stability limit will
(a) increase
(b) decrease
(c) remain the same
(d) any of the above

124. A synchronous generator is feeding an infinite bus through a line. At the generator bus a shunt capacitor is added. The stability limit will
(a) increase
(b) decrease
(c) remain the same
(d) any of the above

125. The use of high speed breakers
(a) improves system stability
(b) decreases system stability
(c) has no effect on system stability
(d) (a) or (b)

126. A compared to a cylindrical rotor generator, a salient pole generator is
(a) more stable
(b) less stable
(c) equally stable
(d) none of the above

127. The addition of SVS in a system
(a) increases fault current
(b) decreases fault current
(c) has no effect on fault current
(d) (a) or (c)

128. The addition of synchronous compensator in a system
(a) increases fault current
(b) decreases fault current
(c) has no effect on fault current
(d) (a) or (c)

129. The insulator strength of EHV lines is mainly governed by
(a) switching over voltages
(b) lightning over voltages
(c) power frequency over voltages
(d) both (a) and (b)

130. A suitable degree of series compensation is
(a) 20% (b) 50%
(c) 80% (d) 10%

131. For 50 Hz, the minimum length of a tuned line is
(a) 1000 km (b) 2000 km
(c) 3000 km (d) 500 km

132. The electrical length of a half tuned wave length line is
(a) 120º (b) 180º
(c) 90º (d) 45º

133. HVDC transmission lines are more economical for

(a) long distance transmission
(b) short distance transmission
(c) inter-connected systems
(d) all the above

134. As compared to an ac line the corona and *RI* on a dc line are
(a) lower (b) more
(c) the same (d) (a) or (b)

135. In HVDC converter station equipment using thyristors, it is necessary to use a large number of thyristors in series because
(a) current ratings of thyristors are low
(b) voltage ratings of thyristors are low
(c) thyristors always fail to an internal open circuit
(d) all the above

136. In 12 pulse valve group operation, the most troublesome harmonics on ac side are
(a) 11th and 13th (b) 23rd and 25th
(c) 5th and 7th (d) 3rd

137. The conductor of an EHV line is selected on the basis of
(a) current carrying capacity
(b) corona and *RI* performance
(c) line voltage
(d) all the above

138. In ACSR conductors, the diameters of aluminium and steel strands are generally equal because
(a) such conductors have lesser tendency for vibrations
(b) such conductors are cheaper
(c) such conductors can be manufactured more easily
(d) none of the above

139. The maximum permissible value of fair weather corona loss for a HV line is
(a) 0.6 kW/3-phase km
(b) 1.2 kW/3-phase km
(c) 0.3 kW/phase km
(d) 5 kW/3-phase km

140. The size of the ground wire is mainly governed by the consideration of
(a) mechanical strength
(b) maximum lightning current
(c) system rated current
(d) (a) or (c)

141. The inductance of a single phase two conductor transmission line per km becomes double when
(a) radius of conductor is doubled
(b) distance between conductors is doubled
(c) distance between conductors is increased four times
(d) distance between conductors is increased as square of original distance.

142. Transmission lines are transposed to
(a) reduce transmission loss
(b) reduce skin effect
(c) reduce conductor vibrations
(d) reduce interference with neighbouring communication lines.

143. For a transmission line with resistance R ohms, reactance X ohms and negligible capacitance, parameter A is
(a) 0 (b) 1
(c) $R + jX$ (d) $(R + jX)^{0.5}$

144. Bundled conductors in EHV lines provide
(a) reduced capacitance
(b) increased capacitance
(c) reduced inductance
(d) increased inductance

145. The charging current in transmission line increases due to corona effect because corona increases
(a) line current
(b) power loss in the line
(c) effective line voltage
(d) effective conductor diameter

146. The non-uniform distribution of voltage gradient across the units in a suspension insulator is due to
(a) unequal distance between cross arm and units
(b) unequal self capacitance of units
(c) swinging of insulator in air
(d) existence of stray capacitance between link pins of insulator units and tower body.

147. Whenever the conductors are dead ended or there is a change in direction of transmission line, the insulator are
(a) pin type
(b) suspension type
(c) strain type
(d) shackle type

148. The insulation resistance of a single core cable is 150 MΩ/km. The insulation resistance for 3 km length is
(a) 16.67 MΩ (b) 50 MΩ
(c) 150 MΩ (d) 450 MΩ

149. The voltage regulation of a long EHV line at no load is
(a) negative (b) positive
(c) zero (d) unity

150. Corona loss increases with
(a) increase in supply frequency and conductor size
(b) increase in supply frequency and reduction in conductor size
(c) decrease in supply frequency and reduction in conductor size
(d) decrease in supply frequency and increase in conductor size

151. The electrostatic stress in a single core belted cable is
(a) minimum at conductor surface and maximum at sheath
(b) maximum at conductor surface and minimum at sheath
(c) same throughout insulation layer
(d) maximum both at conductor surface and at sheath and minimum at armouring layer

152. The alternate layers in ACSR conductor are spiralled is opposite directions to
(a) present unwinding
(b) reduce inductance
(c) increase capacitance
(d) make it mechanically stronger

153. The efficiency of transmission increases as
(a) system voltage and load power factor both decrease
(b) system voltage and load power factor both increase
(c) system voltage increases and load power factor decreases
(d) system voltage decreases and load power factor increases

154. A stringing chart is useful
(a) in the location of transmission towers
(b) in determination of number of units in insulator string
(c) in calculating sag at different temperatures
(d) for checking uplift force on towers

155. The tower footing resistance is greatly influenced by driven rods having
(a) large cross-section
(b) large length
(c) good material
(d) none of the above

156. The abnormal voltage due to arcing grounds is reduced by
(a) ground wire (b) capacitor
(c) arcing horns (d) peterson coil

157. The most commonly used insulation in high voltage cables is
(a) impregnated paper
(b) insulating oil
(c) impregnated cloth
(d) mica strips

158. Suspension type insulators are used for voltages beyond
(a) 11 kV (b) 33 kV
(c) 230 kV (d) 400 kV

159. If sag on a transmission is increased, the tension is
(a) decreased
(b) increased
(c) not affected
(d) may decrease or increase

160. In substations the most commonly used arrester is
(a) thyrite (b) horn gap
(c) rod gap (d) any of the above

161. The real power transfer over a line mainly depends on
(a) phase angle between the voltages of two terminal buses
(b) difference in voltage magnitudes of two terminal buses

(c) line capacitance
(d) none of the above

162. The steady state stability limit of a synchronous generator can be increased by
(a) increasing its reactance
(b) increasing its moment of inertia
(c) increasing its excitation
(d) none of the above

163. A line to line fault is restricted by
(a) positive and negative sequence impedances
(b) positive and zero sequence impedances
(c) negative and zero sequence impedances
(d) positive, negative and zero sequence impedances

164. A cable of surge impedance 50 ohms is terminated by a resistance of 50 ohms. The reflected quantities are
(a) zero
(b) equal to incident quantities
(c) half the incident quantities
(d) double the incident quantities

165. Insulation coordination depends on
(a) volt current current
(b) current-time curve
(c) volt time curve
(d) none of the above

166. Time taken by a surge to travel a 600 km long overhead line is
(a) 0.6 s (b) 0.002 s
(c) 0.02 s (d) 0.01 s

167. An 11 kV alternator with 8% reactance will have 2% reactance at
(a) 22 kV (b) 44 kV
(c) 5.5 kV (d) 1.1 kV

168. The fact that conductor carries more current on the surface as compared to core is known as
(a) skin effect (b) corona
(c) permeability (d) none of the above

169. Which distribution system is more reliable
(a) ring main
(b) tree
(c) radial
(d) all are equally reliable

170. A bus bar is rated by
(a) current only
(b) current and voltage only
(c) current, voltage and frequency
(d) none of the above

171. Bundled conductors are used to
(a) reduce line capacitance to ground
(b) reduce weight of line
(c) reduce regulation
(d) reduce corona loss

172. The effect of electric shock on human body depends on
(a) voltage
(b) current
(c) duration of contact
(d) all the above

173. Which of the following equipment is not installed in a substation
(a) exciters
(b) shunt reactors
(c) voltage transformers
(d) series capacitors

174. Chances of corona are maximum during
(a) summer heat
(b) winter
(c) dry weather
(d) humid weather

175. Corona is associated with
(a) hissing noise
(b) ozone formation
(c) bluish luminiscence
(d) all the above

176. Impulse voltage is mathematically a double exponential. Hence practical impulse generators are
(a) single stage
(b) multi stage RC circuit
(c) single stage RL circuit
(d) RLC circuit

177. If V_s and V_r are sending and receiving end voltages and d is power angle, reactive power transfer mainly depends on
(a) $|V_s| - |V_r|$ (b) δ
(c) $|V_s|$ (d) $|V_r|$

178. The lightning arrester is usually connected between
(a) phases
(b) phase and earth
(c) in series with line
(d) none of the above

179. The phenomenon of arcing ground occurs for a line to ground fault in a power system when generator neutral is
(a) solidly grounded
(b) isolated and line is long
(c) earthed through resistance and line is short
(d) both (b) and (c)

180. If two synchronous generators are synchronised and operating in parallel, loss of synchronism results in
(a) stalling of generators
(b) coiled fluctuations in current
(c) coiled fluctuations in voltage
(d) none of the above

181. Which domestic utility item has the highest power rating
(a) refrigerator
(b) ceiling fan
(c) mixi
(d) electric iron

182. Current rating is not necessary in case of
(a) isolators
(b) circuit breakers
(c) load break switches
(d) both (b) and (c)

183. The accepted value of dielectric strength of transformer oil is
(a) 30 kV
(b) 30 kV/cm
(c) 30 kV/cm for one minute
(d) none of the above

184. Zero sequence currents flow in a transmission line when there is
(a) double line to ground fault
(b) an over voltage on the line due to charged cloud
(c) line of line fault
(d) 3 phase fault

185. A small increment in prime mover power of a synchronous generator connected to an infinite bus
(a) increases reactive power generation
(b) decreases reactive power generation
(c) decreases real power generation
(d) increases the bus voltage phase angle

186. Transmission of power between India and Sri Lanka may be done through submarine
(a) ac or dc link
(b) dc link
(c) ac link
(d) not possible

187. The function of lead sheath in a cable is to
(a) provide mechanical protection
(b) control field distribution
(c) prevent entry of moisture
(d) all the above

188. For the transformer connection shown in Fig. C-1,

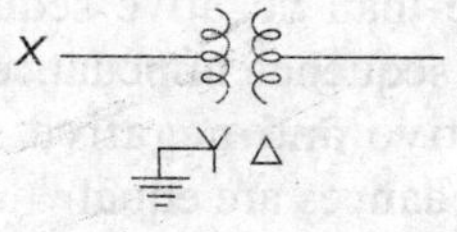

Fig. C-1

the correct zero sequence network is

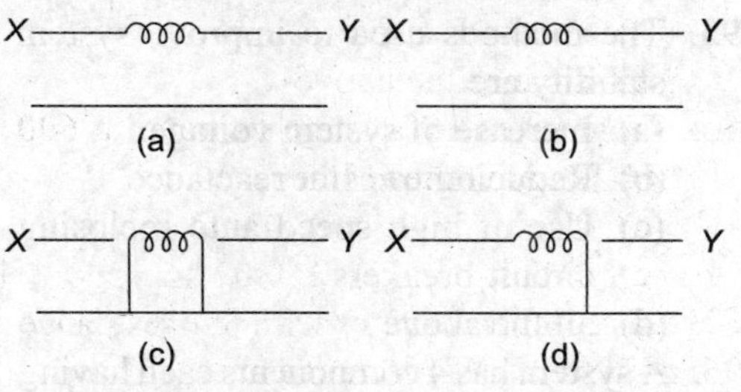

189. The earthing transformer is used to
(a) avoid harmonics in transformers
(b) to provide artificial neutral where neutral point is not accessible
(c) to improve system stability
(d) to measure voltage

190. A synchronous condenser is usually
(a) dc generator
(b) overexcited synchronous motor
(c) under excited synchronous motor
(d) induction motor

191. Back to back connection is used in HVDC to

(a) provide asynchronous tie
(b) point to point long distance power transmission
(c) reduce voltage drop
(d) increase system voltage

192. Which line support has least life
(a) wooden poles
(b) RCC poles
(c) fabricated steel poles
(d) towers

193. As compared to ac transmission, dc transmission is free from
(a) inductance
(b) capacitance
(c) phase displacement
(d) all the above

194. For a power transformer
(a) positive, negative and zero sequence impedances are all equal
(b) positive sequence impedance is more than negative sequence and zero sequence impedances
(c) positive and negative sequence impedances are equal
(d) positive sequence impedance is less than negative and zero sequence impedances.

195. The methods used to improve system stability are
(a) Increase of system voltage
(b) Reducing transfer reactance
(c) Use of high speed auto reclosing circuit-breakers
(d) all the above

196. A system has 4 components each having a reliability of 0.96. All the components are in series. The system reliability is
(a) 0.98
(b) 0.96
(c) 0.921
(d) 0.849

197. A system has a failure rate of 0.01 failures/hour. MTTF is
(a) 0.01 hours
(b) 1 hour
(c) 10 hours
(d) 100 hours

198. A component has a reliability of 0.8. If two such components are connected in parallel, the system reliability is
(a) 0.64
(b) 0.8
(c) 0.2
(d) 0.96

199. A system has a failure rate of 4 outages 1 year. If down time is 5 hours/outage, outage time per year is
(a) 5 hours
(b) 10 hours
(c) 20 hours
(d) 1025 hours

200. A component has a reliability of 0.3. If system reliability must be 0.85 or more, the minimum number of components in parallel must be
(a) one
(b) four
(c) six
(d) eight

201. In substations instrument transformers are used for
(a) protection
(b) measurement
(c) both protection and measurement
(d) either (a) or (b)

202. The use of electric welding equipment in a domestic installation is likely to cause
(a) voltage rise (b) voltage drop
(c) lamp flicker (d) none of the above

203. Which of the following advantages accrue due to use of shunt capacitors in distribution systems
(a) improvement of voltage regulation
(b) reduction of I^2R losses
(c) reduction in kVA loading of equipment
(d) all the above

204. The earthing resistance of a 3 m long driven rod in a sail resistivity of 50 ohm-m is about
(a) 50 Ω (b) 16 Ω
(c) 10 Ω (d) 2 Ω

205. One 3 m driven rod in a soil gave an earthing resistance of 16 ohms. If two

such rods spaced at 3 m are used the earthing resistance is likely to be about
(a) 8 ohms (b) 10 ohms
(c) 16 ohms (d) 32 ohms

206. The addition of ground rods in an earthing grid
(a) decreases the earthing resistance appreciably
(b) has no effect on earthing resistance
(c) decreases the earthing resistance slightly
(d) increases the earthing resistance slightly

207. In earthing grid design the total conductor length required for proper gradient control is calculated from
(a) the tolerable value of touch voltage
(b) the tolerable value of step voltage
(c) the tolerable value of earthing resistance
(d) any of the above

208. For adequate mechanical strengths the minimum thickness of steel strips for earthing grid should be
(a) 10 mm (b) 5 mm
(c) 3 mm (d) 1 mm

209. Which configuration has minimum surface area for a given cross-sectional area
(a) square (b) rectangular
(c) circular (d) either (a) or (b)

210. Transferred potential is a special case of
(a) touch potential
(b) step potential
(c) either (a) or (b)
(d) none of the above

211. For measurement of earthing resistance of a given earth electrode
(a) the collecting electrode should be very near the electrode under test
(b) the collecting electrode should touch the electrode under test
(c) the collecting electrode should be far away from electrode under test
(d) any of the above

212. In a soil resistivity measurement test the distance between successive electrodes was D, current I and voltage V. Then soil resistivity =
(a) $\frac{\pi DV}{I}$ (b) $\frac{2\pi DV}{I}$
(c) $\frac{4\pi DV}{I}$ (d) $\frac{\pi D^2 V}{I}$

213. Which of these leads to voltage instability
(a) transfer of active power
(b) transfer of reactive power
(c) use of shunt capacitors
(d) any of the above

214. A long unloaded line
(a) generates inductive reactive power
(b) generates capacitive reactive power
(c) either (a) or (b)
(d) none of the above

215. A synchronous condenser
(a) increases short circuit capacity
(b) decreases short circuit capacity
(c) has no effect or short circuit capacity
(d) either (a) or (b)

ANSWERS TO OBJECTIVE QUESTIONS

1. (c)	2 (b)	3. (a)	4. (c)	5. (b)	6. (a)
7. (b)	8. (c)	9. (b)	10. (a)	11. (c)	12. (b)
13. (d)	14. (b)	15. (a)	16. (a)	17. (a)	18. (b)
19. (b)	20. (c)	21. (a)	22. (b)	23. (c)	24. (c)
25. (b)	26. (c)	27. (b)	28. (c)	29. (a)	30. (a)
31. (a)	32. (b)	33. (a)	34. (b)	35. (a)	36. (a)
37. (a)	38. (c)	39. (d)	40. (a)	41. (b)	42. (a)
43. (a)	44. (a)	45. (c)	46. (a)	47. (c)	48. (a)
49. (b)	50. (b)	51. (a)	52. (b)	53. (b)	54. (b)
55. (b)	56. (a)	57. (d)	58. (b)	59. (a)	60. (a)
61. (a)	62. (a)	63. (a)	64. (b)	65. (b)	66. (b)
67. (a)	68. (a)	69. (b)	70. (b)	71. (b)	72. (c)
73. (b)	74. (b)	75. (b)	76. (b)	77. (c)	78. (a)
79. (d)	80. (b)	81. (a)	82. (b)	83. (d)	84. (a)
85. (b)	86. (c)	87. (d)	88. (b)	89. (c)	90. (b)
91. (c)	92. (b)	93. (b)	94. (c)	95. (b)	96. (a)
97. (c)	98. (b)	99. (b)	100. (b)	101. (a)	102. (b)
103. (a)	104. (a)	105. (a)	106. (c)	107. (b)	108. (b)
109. (b)	110. (a)	111. (a)	112. (b)	113. (c)	114. (b)
115. (c)	116. (c)	117. (a)	118. (c)	119. (a)	120. (b)
121. (a)	122. (c)	123. (c)	124. (c)	125. (a)	126. (a)
127. (c)	128. (a)	129. (a)	130. (b)	131. (c)	132. (b)
133. (a)	134. (a)	135. (b)	136. (a)	137. (b)	138. (a)
139. (a)	140. (a)	141. (d)	142. (d)	143. (b)	144. (c)
145. (d)	146. (d)	147. (c)	148. (b)	149. (a)	150. (b)
151. (b)	152. (d)	153. (b)	154. (c)	155. (b)	156. (d)
157. (a)	158. (b)	159. (a)	160. (a)	161. (a)	162. (c)
163. (a)	164. (a)	165. (c)	166. (b)	167. (b)	168. (a)
169. (a)	170. (b)	171. (d)	172. (d)	173. (a)	174. (d)
175. (d)	176. (b)	177. (a)	178. (d)	179. (b)	180. (a)
181. (d)	182. (a)	183. (c)	184. (a)	185. (d)	186. (b)
187. (c)	188. (d)	189. (b)	190. (b)	191. (a)	192. (a)
193. (d)	194. (a)	195. (d)	196. (d)	197. (d)	198. (d)
199. (c)	200. (c)	201. (c)	202. (c)	203. (d)	204. (b)
205. (b)	206. (c)	207. (a)	208. (c)	209. (c)	210. (a)
211. (c)	212. (b)	213. (b)	214. (a)	215. (a)	

Bibliography

BOOKS

1. Anderson, P.M., *Analysis of Faulted Power Systems*, Iowa State Press, Ames, Iowa, 1973.
2. Anderson, P.M. and Foud, A.A., *Power Systems Control and Stability*, Iowa State Press, Ames, Iowa, 1977.
3. Barnes, C.C., *Power Cables: Their Design and Installation*, Chapman and Hall, London, 1966.
4. Bewley, L.V., *Travelling Waves on Transmission Systems*, Dover, New York, 1963.
5. Brown, H.E., *Solution of Large Networks by Matrix Methods*, Wiley, New York, 1975.
6. Bryerly, R.T. and Kimbark, E.W., *Stability of Large Electric Power Systems*, IEEE Press, New York, 1974.
7. Clarke, E., *Circuit Analysis of AC Power Systems*, Vol. I, Wiley, New York, 1943.
8. Clarke, E., *Circuit Analysis of AC Power Systems*, Vol. II, G.E.C., New York, 1950.
9. Cotton H. and Barber H., *Transmission and Distribution of Electric Energy*, B.I. Publishers, New Delhi, 1970.
10. Diesendorf, W., *Insulation Coordination in High Voltage Electric Power Systems*, Butterworth, London, 1974.
11. Elgerd, O.I., *Electric Energy Systems—An Introduction*, Tata McGraw Hill, New Delhi, 1973.
12. EPRI, *Trans. Line Ref. Book*, 345 kV and above, Second Ed., 1982.
13. Greenwood, A., *Electric Transients in Power Systems*, Wiley, New York, 1971.
14. Gross, C.A., *Power System Analysis*, Wiley, New York, 1979.
15. Guile, A.E. and Paterson, W., *Electric Power Systems*, Vol. I and II, Pergamon Press, New York, 1977.
16. Gupta, B.R., *Generation of Electrical Energy*, S. Chand and Co., New Delhi, 2005.
17. International Conference on Large High Voltage Electric Systems (CIGRE), *Interference Produced by Corona Effects of Electric Systems*, Paris, 1974.
18. IEEE, *Guide for Safety in AC Substation Grounding No. 80, 1986*.
19. Kimbark, E.W., *Direct Current HVDC Transmission*, Vol. I, Wiley, New York, 1971.
20. Kimbark, E.W., *Power System Stability*, Vol. I, II and III, Wiley, New York, 1948.
21. McCombe, J. and Haigh, F.R., *Overhead Line Practice*, Macdonald, London, 1966.
22. Neuens Wander, J.R., *Modern Power Systems*, International Text Book Co., Scranton Pa., 1971.

23. Say, M.G., *Electrical Engineer's Reference Book*, Newnes-Butterworths, London, 1973.
24. Stagg, G.W. and El-Abiad, A.H., *Computer Methods in Power System Analysis*, McGraw-Hill, New York, 1968.
25. Stevenson, W.D., *Elements of Power System Analysis*, McGraw-Hill, New York, 1975.
26. Taylor, E.O. and Boal, G.A., *Electric Power Distribution* 415 V–33 kV, Edward Arnold, London, 1966.
27. Thapar, B., Gupta, B.R. and Khera, L.K., *Power System Transients and High Voltage Principles*, Mohindra Capital, Chandigarh, 1985.
28. Waddicor, H., *Principles of Electric Power Transmission*, Chapman and Hall, London, 1966.
29. Wagner, C.F. and Evans, R.D., *Symmetrical Components*, McGraw-Hill, New York, 1933.
30. Weedy, B.M., *Electric Power Systems*, Wiley, London, 1978.
31. Westinghouse Electric Corporation, *Electrical Transmission and Distribution-Reference Book*, East Pittsburgh, Pa., 1964.
32. Zaborszky, J. and Rittenhouse, J.W., *Electric Power Transmission: The Power System in Steady State*, Ronald, New York, 1954.

PAPERS

33. Augwood, D.R., *Performance of Extruded Dielectric (XLPE) Cable*, IEEE Trans., May-June 1978, No. 3, Vol. PAS—97, pp. 659–668.
34. Braess, D. and Grebe E., *A Numerical Analysis of Load Flow Calculation Methods*, IEEE Trans., 1981 July, No. 7, Vol. PAS—100, pp. 3642–3647.
35. Brown, G.W., *Designing EHV Line to a Given Outage Rate–Simplified Techniques*, IEEE Trans., March-April 1971, No. 2, Vol. PAS—97, pp. 379–383.
36. Brown, H.E. *et al.*, *Digital Calculation of Three Phase Short Circuits by Matrix Method*, AIEE Trans., 1960, Vol. 79, p. 1277.
37. Brown, H.E. and C.E. Person, *Short Circuit Studies of Large Systems by the Impedance Matrix Method*, Proc. PICA, 1967, p. 335.
38. Clade, J., *et al.*, *Factors on which to Base Comparison in Choosing the Highest Voltage Level for a Transmission Project*, Cigre 1978, Report 31–35.
39. Davies, M., *Design of lv. Distributors from Standard Cable Sizes*, Proc. IEE. 1965, No. 5, Vol. 112, pp. 949–965.
40. Dommel, H.W. and W.F. Tinney, *Optimal Power Flow Solution*, IEEE Trans. October 1968, Vol. PAS—87, p. 1866.
41. Dunlop R.D., Gutman R. and Marchenko, P.P., *Analytical Development of Loadability Characteristics for EHV and UHV Transmission Lines*, IEEE Trans. March-April 1979, No. 2, Vol. PAS–98, pp. 606–617.
42. Ehmcke, B., *HVAC Transmission*, Siemens Power Engg. III (1981) Special Issue "High Voltage Technology", pp. 4–10.
43. Ekstrom, A., Juhlin, L.E. and Liss G., *Parallel Connections of Convertors for HVDC Transmission*, IEEE Trans., May-June 1978, No. 3, Vol. PAS—97, pp. 714–724.
44. El-Abiad, A.H. and Nagappan, K., *Transient Stability Regions of Multimachine Power Systems*, IEEE Trans., Feb. 1966. No. 2, Vol. PAS—85, pp. 169–179.
45. Fouad A.A. and Stranton S.E., *Transient Stability of a Multi Machine Power System*, Part I and II, IEEE Trans., 1981 July, No. 7, Vol. PAS—100, pp. 3408–3424.

46. Gonen T. and Foote B.L., *Distribution System Planning using Mixed Integer Programming*, IEEE Proc., Part C, 1981, Vol. 128, p. 370.
47. Goodrich, R.D., *A Universal Power Circle Diagram*, AIEE Trans. 1951, Vol. 70, p. 2042.
48. Gupta, B.R., *et al.*, *Application of Capacitors in Distribution Circuits*, All India Seminar on Distribution and Utilization of Electrical Energy, The Institution of Engineers (India) Bombay, 1975.
49. Gupta, B.R., *Impulse impedance of grounding grids* IEEE Trans. PAS—99, Dec. 1980 pp. 2357–2362.
50. Gupta, B.R., *Impulse impedance of rectangular grids,* IEEE Trans. on Power Delivery Jan. 1992, pp. 214–217.
51. Gupta, B.R., *Reliability of Electric Power Systems*, Workshop Organized by Central Board of Irrigation and Power (India), Nov. 1995.
52. Gupta, B.R. and Singhal V., A New Approach to Control Potential Gradients in Indoor H V Substations, Journal of IE (India), Vol. 75, Feb. 1995.
53. IEEE Committee Report, *Bibliography on Distribution System Reliability*, IEEE Trans, March-April 1978, No. 2, Vol. PAS—97, pp. 545–548.
54. IEEE Committee Report, *Load Forecasting Bibliography*, *Phase II*, IEEE Trans. 1981 July, No. 7, Vol. PAS—100, pp. 3217–3220.
55. Iliceto, F. and Cinieri, E., *Comparative Analysis of Series and Shunt Compensation Schemes for AC Transmission Systems*, IEEE Trans. Nov-Dec. 1977, No. 6, PAS–96.
56. Jancke, G., Fahlen, N. and Nerf, O., *Series Capacitors in Power Systems*, IEEE Trans, May-June 1965, No. 3, Vol. PAS-94.
57. Kimbark, E.W., *Improvement of Power System Stability*, IEEE Trans., 1969, Vol. PAS-88, p. 773.
58. Kirn, W.G. and Adler R.B., *A Distribution System Cost Model and its Application to Optimal Conductor Sizing*, IEEE Trans. Feb. 1982 No. 2, Vol. PAS-101, pp. 271–275.
59. Larsen, E.V. and Baker, D.H., *Series Compensation Operating Limits—A New Concept for Subsynchronous Resonance Stability Analysis*, IEEE Trans. Sept-Oct. 1980, No. 5, Vol. PAS-99, pp. 1855–1863.
60. Laughton, M.A., *Decomposition Techniques in Power System Network Load Flow Analysis Using the Nodal Impedance Matrix*, Proc. IEE 1968, Vol. 115, p. 539.
61. Povh, D., *HVDC Transmission*, Siemens Power Eng. III (1981) Special Issue *High Voltage Technology*, pp. 11–18.
62. Reichman, J., *Bundled Conductor Voltage Gradient Calculations*, AIEE Trans. 1959 pt. III, Vol. 78, p. 598.
63. Snelson, J.K. and Carson, M.J., *Logical Design of Branched l. v. Distributors*. Proc. IEE, 1970, No. 2, Vol. 117, pp. 415–20.
64. Stott, B., *Decoupled Newton Load Flow*, IEEE Trans., 1972, Vol. PAS—91, p. 1955.
65. Stott, B., *Review of Load-Flow Calculation Methods*, Proc. IEEE July 1974, p. 916.
66. Stott, B. and Alsac, O., *Fast Decoupled Load Flow*, IEEE Trans., 1974, Vol. PAS—93, p. 859.
67. Thanawala, H.L., Kelham, W.O. and Williams P., *The Application of Static Shunt Reactive Compensators in Conjunction with line Series Capacitors to increase the Transmission Capabilities of Long AC lines*, CIGRE, 1976, Paper 31-09.
68. Thaper, B. and Gupta, B.R., *Reliability of Distribution Systems, Proceedings of 40th Research Session* C.B.I.P., 1970, Vol. IV (C).

69. Tinney, W.F. and Walker, J.W., *Direct Solutions of Sparse Network Equations by Optimally Ordered Triangular Factorizations*, Proc. IEEE, November 1967, Vol. 55, p. 1801.
70. Tinney, W.F. and Hart, C.E., *Power Flow Solution by Newton's Method*, IEEE Trans. November 1967, No. 11, Vol. PAS-86, p. 1449.
71. Walkden, F.W., *Design of Low Voltage Distributors*, Proc., IEE 1982, Vol. 129, Part C, pp. 101–103.
72. Ward, J.B. and Hale, H.W., *Digital Computer Solution of Power Flow Problems*, AIEE Trans. June 1956, pt. 3, Vol. 75, p. 398.

Index

A

B

C

D